Lecture Notes in Computer Science 2439

Edited by G. Goos, J. Hartmanis, and J. van Leeuwen

Springer
Berlin
Heidelberg
New York
Hong Kong
London
Milan
Paris
Tokyo

Juan Julián Merelo Guervós
Panagiotis Adamidis
Hans-Georg Beyer
José-Luis Fernández-Villacañas
Hans-Paul Schwefel (Eds.)

Parallel Problem Solving from Nature – PPSN VII

7th International Conference
Granada, Spain, September 7-11, 2002
Proceedings

Springer

Volume Editors

Juan Julián Merelo Guervós
Escuela Técnica Superior de Ingeniería Informática
18071 Granada, Spain
E-mail: jmerelo@geneura.ugr.es

Panagiotis Adamidis
Technological Educational Institute of Thessaloniki
Department of Informatics, 54101 Thessaloniki, Greece
E-mail: adamidis@it.teithe.gr

Hans-Georg Beyer
Hans-Paul Schwefel
University of Dortmund, Department of Informatics XI
44221 Dortmund, Germany
E-mail: beyer@ls11.cs.uni-dortmund.de
hps@udo.edu

José-Luis Fernández-Villacañas
Universidad Carlos III, Department of Signal Theory and Communications
Madrid, Spain
E-mail: pepe@tsc.uc3m.es

Cataloging-in-Publication Data applied for

Die Deutsche Bibliothek - CIP-Einheitsaufnahme

Parallel problem solving from nature : 7th international conference ;
proceedings / PPSN VII, Granada, Spain, September 7 - 11, 2002. Juan Julián
Merelo Guervós ... (ed.). - Berlin ; Heidelberg ; New York ; Barcelona ;
Hong Kong ; London ; Milan ; Paris ; Tokyo : Springer, 2002
(Lecture notes in computer science ; Vol. 2439)
ISBN 3-540-44139-5

CR Subject Classification (1998): F.1-2, C.1.2, D.1.3, I.2.8, I.2.6, I.2.11, J.3

ISSN 0302-9743
ISBN 3-540-44139-5 Springer-Verlag Berlin Heidelberg New York

Springer-Verlag Berlin Heidelberg New York
a member of BertelsmannSpringer Science+Business Media GmbH

http://www.springer.de

Printed in Germany

Typesetting: Camera-ready by author, data conversion by Olgun Computergraphik
Printed on acid-free paper SPIN 10870237 06/3142 5 4 3 2 1 0

Preface

We are proud to introduce the proceedings of the Seventh *International Conference on Parallel Problem Solving from Nature, PPSN VII*, held in Granada, Spain, on 7–11 September 2002. PPSN VII was organized back-to-back with the Foundations of Genetic Algorithms (FOGA) conference, which took place in Torremolinos, Malaga, Spain, in the preceding week.

The PPSN series of conferences started in Dortmund, Germany [1]. From that pioneering meeting, the event has been held biennially, in Brussels, Belgium [2], Jerusalem, Israel [3], Berlin, Germany [4], Amsterdam, The Netherlands [5], and Paris, France [6]. During the Paris conference, several bids to host PPSN 2002 were put forward; it was decided that the conference would be held in Granada with Juan J. Merelo Guervós as General Chairman.

The scientific content of the PPSN conference focuses on problem-solving paradigms gleaned from natural models, with an obvious emphasis on those that display an innate parallelism, such as evolutionary algorithms and ant-colony optimization algorithms. The majority of the papers, however, concentrate on evolutionary and hybrid algorithms, as is shown in the contents of this book and its predecessors. This edition of the conference proceedings has a large section on applications, be they to classical problems or to real-world engineering problems, which shows how bioinspired algorithms are extending their use in the realms of business and enterprise.

In total, this volume contains 90 papers, which were selected from the 181 papers submitted to the conference organizers. This means that slightly fewer than half the papers were accepted for inclusion in the conference proceedings, which, at the same time, means that some papers of good quality could not be selected for publication. Each paper was reviewed by at least three persons, and, in some cases, up to five reviews were necessary to reach a decision. Most papers had four reviews. Thus, we are very grateful to the volunteer reviewers who offered their scientific expertise and time in order to come up with a decision that was as fair as possible. We want also to thank all authors of submitted papers for their participation.

The submission procedure was developed and maintained by Pedro Castillo Valdivieso and Juan J. Merelo Guervós, inspired by the one developed by them for the previous PPSN conferences. The submission, reviewing and Chairperson's information systems were written in Perl and used a PostgreSQL relational database system as the back-end. The submission process was mostly smooth, and the system could take the heaviest loads without a glitch. All in all, the system dealt with around 50 000 requests, with several requests per minute at the moments of heaviest load.

Paper assignment was made using a combined greedy/evolutionary algorithm that took into account keyword matches among reviewers and papers. The assignment could be considered successful, since few reviewers declined to review their assigned papers because of a lack of relevant expertise; at the same time, the average confidence in decisions was 1.7±0.7 (that is, between "Somewhat High" and "Very High", but closer to the former).

As usual, PPSN VII was a poster-only conference; that is, all papers were presented as posters to facilitate personal discussion and the exchange of ideas between the presenter and the audience. Although this might imply a smaller audience than in the case of an oral presentation, the presenter of a paper has a better chance of getting in touch with the people most interested in her/his topic. Consequently, posters are not "second-class" papers, as they are usually considered in some other conferences – they are just the means of presenting. The 90 papers presented in the conference were grouped into five sessions of about 18 papers each. To simplify the orientation within a poster session and to allow the audience to get a global overview of all sessions, each poster session was introduced by a person belonging to the organizing committee who gave a brief overview of all papers presented within a session.

Only the three invited speakers presented a one-hour oral presentation of their research results, geared towards providing inspiration to the conference attendees. Alexander Nareyek (CMU), Roderic Guigó (IMIM, Barcelona, Spain), and William Hart (Sandia Labs, NM, USA), gave keynote speeches on topics that impinge on natural computation: the human genome project, applications of artificial intelligence to computer games, and the relationship between the fields of evolutionary computation and optimization.

Before the technical sessions began on September 9th, two one-day events took place. Five workshops were organized at the Palacio de Congresos on Sept. 7th, and eleven tutorials took place in the same place the next day. We would like to thank the corresponding chairs, David W. Corne (University of Reading, UK) and Marc Schoenauer (INRIA, France).

Finally, we would like to thank the Departamento de Arquitectura y Tecnología de Computadores of the University of Granada, and the regional and national governments who provided monetary support to the conference. Julio Ortega took the job of filling all forms, and submitting and following-up on them, and we are deeply indebted for this. EvoNet, the network of excellence in evolutionary computation sponsored by the European Union, provided support in the form of travel grants for five students.

Pedro Castillo took the tedious job of helping to maintain the website. The other members of the GeNeura team (Maribel García, José Carpio, Víctor Rivas, Javi García, Brad Dolin, Linda Testa, Gustavo Romero, Jaime Anguiano, Louis Foucart) also had to put up with some tasks related to the fact that their boss was busy at something else (although that very fact, of course, also relieved them of some burden). The other members of the Departamento de Arquitectura y Tecnología de Computadores and the Departamento de Ciencias de Computación e Inteligencia Artificial also provided invaluable support. We are also very grateful to the users of the submission system who were able to spot bugs and communicate them to us in good time.

We expect that these proceedings will help to take the natural computation field a bit further, to make it more aware of the problems out there, and, finally, to make it conscious of its existence as a whole field, not as a fractured set of fieldlets.

Granada, June 2002

Juan Julián Merelo Guervós
Panagiotis Adamidis
Hans-Georg Beyer
José Luis Fernández Villacañas
Hans-Paul Schwefel

References

1. Hans-Paul Schwefel, Reinhart Männer, editors. *Proc. 1st Conf. on Parallel Problem Solving from Nature*, Vol. 496 of *Lecture Notes in Computer Science*, Dortmund, Germany, October 1–3, 1991, Springer-Verlag.
2. Reinhart Männer, Bernhard Manderick, editors. *Proc. 2nd Conf. on Parallel Problem Solving from Nature, Brussels, Belgium*, September 28–30, 1992, Elsevier.
3. Reinhart Männer, Yuval Davidor, Hans-Paul Schwefel, editors. *Proc. 3rd Conf. on Parallel Problem Solving from Nature, PPSN III*, Vol. 866 of *Lecture Notes in Computer Science*, Springer-Verlag, 1994.
4. Hans-Michael Voigt, Werner Ebeling, Ingo Rechenberg, Hans-Paul Schwefel, editors. *Proc. 4th Conf. on Parallel Problem Solving from Nature*, Vol. 1141 of *Lecture Notes in Computer Science*, Dortmund, Germany, September 1996, Springer-Verlag.
5. Thomas Back, Agoston E. Eiben, Marc Schoenauer, editors. *Proc. 5th Conf. on Parallel Problem Solving from Nature*, Vol. 1498 of *Lecture Notes in Computer Science*, Dortmund, Germany, September 1998, Springer-Verlag.
6. Marc Schoenauer, Kalyanmoy Deb, Günter Rudolph, Xin Yao, Evelyne Lutton, Juan Julián Merelo, Hans-Paul Schwefel, editors. *Parallel Problem Solving from Nature, PPSN VI*, Vol. 1917 of *Lecture Notes in Computer Science*, Springer-Verlag, 2000.

PPSN VII Conference Committee

Conference Chair
Juan J. Merelo Guervós
University of Granada, Spain

Program Co-Chairs
Panagiotis Adamidis
Technological Educational Institute of Thessaloniki, Greece

Hans-Georg Beyer
University of Dortmund, Germany

and
José Luis Fernández Villacañas
Universidad Carlos III, Madrid, Spain

Electronic Program Chair
Pedro Castillo
University of Granada, Spain

Proceedings Chair
Hans-Paul Schwefel
University of Dortmund, Germany

Finance Chair
Julio Ortega
University of Granada, Spain

Tutorial Chair
Marc Schoenauer
INRIA, France

Workshop Chair
David Corne
University of Reading, UK

PPSN VII Steering Committee

Kenneth De Jong
George Mason University, USA

Agoston E. Eiben
Free University, Amsterdam, The Netherlands

Juan J. Merelo Guervós
University of Granada, Spain

Marc Schoenauer
INRIA, France

Hans-Paul Schwefel
University of Dortmund, Germany

Hans-Michael Voigt
Gesellschaft zur Förderung angewandter Informatik e.V. (GFaI), Germany

Xin Yao
University of Birmingham, UK

PPSN VII Program Committee

Agapie, Alexandru
Alander, Jarmo
Alba, Enrique
Altenberg, Lee
Arnold, Dirk
Bäck, Thomas
Banzhaf, Wolfgang
Barbosa, Helio
Ben Hamida, Sana
Berny, Arnaud
Bersini, Hugues
Bidaud, Philippe
Bonarini, Andrea
Booker, Lashon B.
Braunschweig, Bertrand
Bull, Larry
Burke, Edmund
Cagnoni, Stefano
Cantu-Paz, Erick
Coello, Carlos
Collet, Pierre
Cordon, Oscar
Cottam, Ron
Cotta-Porras, Carlos
Darwen, Paul James
Dasgupta, Dipankar
Davis, Dave
Deb, Kalyanmoy
De Jong, Ken
Delahaye, Daniel
Dorigo, Marco
Dozier, Gerry
Durand, Nicolas
Eiben, Gusz
Elorza-Tenreiro, Javier
Fernandez, Francisco
Fillipic, Bogdan
Fleming, Peter
Floreano, Dario
Fogarty, Terence
Fogel, Gary
Fonlupt, Cyril
Fonseca, Carlos
Freisleben, Bernd
Fukuda, Toshio
Furuhashi, Takeshi
Gallard, Raul
Gero, John
Ghosh, Ashish
Giannakoglou, Kyriakos
Goodman, Erik
Gottlieb, Jens
Graña, Manuel
Greenwood, Garry
Hao, Jin-Kao
Hart, Emma
Harvey, Inman
Hervás, C.
Herdy, Michael
Herrera, Francisco
Hidalgo, Ignacio
Horn, Jeff
Husbands, Phil
Iba, Hitoshi
Isasi, Pedro
Jansen, Thomas
Juille, Hughes
Julstrom, Bryan
Kang, Lishan
Kazarlis, Spyros
Keane, Andy J.
Keijzer, Maarten
Keller, Robert
Kennedy, Jim
Kita, Hajime
Knowles, Joshua
Kok, Joost
Krink, Thiemo
Kushchu, Ibrahim
Lamont, Gary
Lanzi, Pier Luca
Larrañaga, Pedro
Lattaud, Claude
Leriche, Rodolphe
Leung, Kwon Sak
Li, Yun
Liardet, Pierre
Louchet, Jean
Lozano, Manuel
Luchian, Henri
Lutton, Evelyne
Marchiori, Elena
Martin, Worthy
Merkle, Larry
Merz, Peter
Meyer, Jean-Arcady
Miller, Julian
Moraga, Claudio
Muehlenbein, Heinz
Oates, Martin
O'Reilly, Una May
Oudeyer, Pierre-Yves
Paechter, Ben
Paredis, Jan
Parmee, Ian
Paton, Ray C.
Pelikan, Martin
Petridis, Vasilios
Petrowski, Alain
Poli, Riccardo
Porto, Bill
Raidl, Guenther
Ratle, Alain
Reeves, Colin

Reynolds, Robert
Robilliard, Denis
Rojas, Ignacio
Rosca, Justinian
Ross, Peter
Rowe, Jonathan
Rowland, Jem
Roy, Rajkumar
Rudolph, Guenter
Ryan, Conor
Salomon, Ralf
Sánchez, Luciano
Sareni, Bruno
Sarma, Jayshree
Schaffer, David
Schwefel, Hans-Paul
Sebag, Michele
Sen, Sandip
Sendhoff, Bernhard
Sinclair, Mark C.
Sipper, Moshe
Smith, Alice E.
Smith, Jim
Spears, Bill
Talbi, El Ghazali
Tateson, Richard
Tettamanzi, Andrea
Thierens, Dirk
Tomassini, Marco
Tsahalis, Demosthenes
Tuson, Andrew
Valls-Ferrán, José-M.
Venturini, Gilles
Whitley, Darrell
Wu, Annie
Zalzala, Ali
Zitzler, Eckart

PPSN VII Tutorials

Carlos A. Coello Coello, *CINVESTAV-IPN, Mexico*
Evolutionary Multiobjective Optimization: Past, Present and Future

David Wolfe Corne, *University of Reading, UK*
Natural Computation in Bioinformatics

Natalio Krasnogor, *University of Reading, UK*
Memetic Algorithms

Jose A. Lozano and Pedro Larrañaga, *University of the Basque Country, Spain*
Optimization by learning and Simulation of Probabilistic Graphical Models

Evelyne Lutton and Jacques Lévy Véhel, *INRIA - Rocquencourt, France*
Fractals and Evolutionary Algorithms

Daniel Merkle, *University of Karlsruhe, Germany*
and Martin Middendorf, *Catholic University of Eichstätt-Ingolstadt, Germany*
Ant Colony Optimization

Franz Rothlauf, *University of Bayreuth, Germany*
Representations for Genetic and Evolutionary Algorithms

Moshe Sipper, *Ben-Gurion University, Israel*
Go Forth and Replicate

Wolfgang Stolzmann, *DaimlerChrysler AG, Germany*
and Pier Luca Lanzi, *Politecnico di Milano, Italy*
An Introduction to Learning Classifier Systems

Darrell Whitley, *Colorado State University, USA*
Evaluating Evolutionary Algorithms

PPSN VII Workshops

International Workshop on Learning Classifier Systems 2002 – IWLCS 2002
Wolfgang Stolzmann, Pier Luca Lanzi, and Stewart Wilson

International Workshop on Memetic Algorithms III – WOMA-III
Natalio Krasnogor, William Hart, and Jim Smith

Multiobjective Problem Solving from Nature II – MPSN-II
Joshua Knowles

Neural and Evolutionary Computation in the Biosciences – ENB
Gary Fogel and David Corne

Real World Applications II – RWOEC
Rajkumar Roy and Ashutosh Tiwari

Table of Contents

Evolutionary Algorithms Theory

Representation/Codification Issues

Variation Operators: Analysis, New Techniques

Evolutionary Techniques: Coevolution

Multiobjective Optimization

Evolutionary Algorithms: New Techniques

Hybrid Algorithms: Neurogenetic Algorithms, Evolutionary Techniques Applied to Neural Nets

Hybrid Algorithms: Memetic, Other

Learning Classifier Systems

Comparison of Different Techniques

Evolutionary Algorithm Implementations

Applications

Other Bioinspired Algorithms: Cellular Automata, Ant Colony Optimization

Evolutionary Algorithms Theory

Random Dynamics Optimum Tracking with Evolution Strategies

Dirk V. Arnold and Hans-Georg Beyer

Department of Computer Science XI, University of Dortmund
44221 Dortmund, Germany
{arnold,beyer}@ls11.cs.uni-dortmund.de

Abstract. Dynamic optimization is frequently cited as a prime application area for evolutionary algorithms. In contrast to static optimization, the objective in dynamic optimization is to continuously adapt the solution to a changing environment – a task that evolutionary algorithms are believed to be good at. At the time being, however, almost all knowledge with regard to the performance of evolutionary algorithms in dynamic environments is of an empirical nature. In this paper, tools devised originally for the analysis in static environments are applied to study the performance of a popular type of recombinative evolution strategy with cumulative mutation strength adaptation on a dynamic problem. With relatively little effort, scaling laws that quite accurately describe the behavior of the strategy and that greatly contribute to its understanding are derived and their implications are discussed.

1 Introduction

Dynamic optimization is distinguished from static optimization in that in the former, the objective function is not constant but varies with time. Dynamic optimization problems arise in many areas of engineering and computer science, as for example in the determination of optimal control policies or in connection with online job scheduling where new jobs arrive in the course of the optimization. While the goal in static optimization is to locate an optimal solution rapidly and accurately, the objective in dynamic optimization is to track a moving target as closely as possible. Strategies for dynamic optimization thus need to continuously adapt to changes in the environment.

In enumerations of potential domains of application of evolutionary algorithms, dynamic optimization often takes one of the top spots. At the time being, almost all knowledge with regard to the capabilities of evolutionary algorithms in dynamic environments is based on empirical observations. An extensive survey of the literature of the field along with a collection of benchmark functions and a discussion of methods that have been proposed to improve the performance of evolutionary algorithms in dynamic environments has been compiled by Branke [7]. Angeline [1] compares empirically the tracking performance of an evolutionary algorithm employing a form of mutative self-adaptation with that of a strategy using a simple heuristic for mutation strength adaptation. The

J.J. Merelo Guervós et al. (Eds.): PPSN VII, LNCS 2439, pp. 3–12, 2002.

fitness environment considered is a three-dimensional, spherically symmetric objective function that is shifted periodically either in a random fashion or on a linear or a spherical path. Angeline observes that the self-adaptation mechanism is not without problems in the dynamic case. In that same fitness environment, Bäck [4] compares different variants of mutative self-adaptation and presents evidence that seems to indicate that the lognormal self-adaptation used in evolution strategies performs better than the variant of self-adaptation commonly used in evolutionary programming. Salomon and Eggenberger [13] compare the performance of evolution strategies with that of a breeder genetic algorithm on the sphere, an ellipsoid, and Rastrigin's function, where the coordinates are shifted by a constant increment in every time step. The search space dimensionalities they consider for the sphere are $N = 10$ and $N = 30$. Without quantifying the term, they find that the sensitivity to the particular implementation of the strategy and to its parameter values is much lower for the tracking task than it is in a static environment. Without providing details, they also report to have observed that recombination is not beneficial for tracking a moving target. Weicker and Weicker [14] contrast self-adaptation of a single mutation strength with that of N mutation strengths and adaptation of the full mutation covariance matrix and find that in more rapidly changing environments, the adaptation of more than a single mutation strength becomes unreliable. Droste [9] presents a first rigorous analysis of the performance of a $(1 + 1)$-strategy on a discrete, dynamic objective function. However, focus in that paper is not on the tracking behavior of the strategy but rather on the expected time required to first reach the optimum. Finally, Bürger [8] reviews references from the field of population genetics that are concerned with moving optima. Both the objective and the genetic operators considered in those references differ substantially from the focus of the present paper.

While useful for providing the reader with a first idea of the capabilities and the limitations of evolutionary algorithms in dynamic environments, the results of empirical studies are not always easy to interpret. It is not clear how difficult the task of tracking is. Many of the experiments reported in the references listed above have been conducted with large populations in low-dimensional search spaces – a case that should arguably be comparatively easy to handle. It would therefore be desirable to have scaling laws that describe the influence of the parameters of the strategies and of the fitness environment. Not only would such scaling laws yield an improved understanding of the behavior of the strategies and their operators and parameters, but they would also allow for the analytical calculation of optimal strategy parameter values and for the comparison of the performance of different strategy variants. In the realm of evolution strategies, much work has been done towards deriving such scaling laws in simple static environments. Many of the tools developed and the main results can be found in the monographs by Beyer [6] and by Rechenberg [12]. In the present paper, we will see that the tools developed for static environments can be applied to the analysis of evolution strategies for dynamic optimization in real-valued search spaces with relatively little effort. In particular, the tracking performance of the

$(\mu/\mu, \lambda)$-ES with cumulative mutation strength adaptation is studied analytically for a spherically symmetric objective function the center of which is shifted randomly in every time step. For that purpose, in Sect. 2 we briefly introduce and motivate the choice of strategy and fitness environment. In Sect. 3, we analyze the behavior of the strategy in the environment thus introduced for fixed mutation strength. In Sect. 4, the performance of the mutation strength adaptation scheme is investigated. Finally, Sect. 5 concludes with a summary and a discussion of directions for future research.

2 Preliminaries

In all of what follows, we assume real-valued objective functions $\mathbb{R}^N \to \mathbb{R}$. The $(\mu/\mu, \lambda)$-ES with isotropic normal mutations in every time step generates λ offspring candidate solutions from a population of μ parents, where $\lambda > \mu$, and subsequently replaces the parental population by the μ best of the offspring. Generation of an offspring candidate solution consists in adding a vector $\sigma\mathbf{z}$, where $\mathbf{z}$ consists of independent, standard normally distributed components, to the centroid of the parental population. The standard deviation σ of the components of vector $\sigma\mathbf{z}$ is referred to as the *mutation strength*, vector $\mathbf{z}$ as the *mutation vector*. The average of those mutation vectors that correspond to offspring candidate solutions that are selected to form the population of the next time step is the *progress vector* $\langle\mathbf{z}\rangle$. Note that due to the definition of global intermediate recombination, $\sigma\langle\mathbf{z}\rangle$ connects consecutive centroids of the population. The choice of strategy is motivated both by the fact that it is relatively amenable to mathematical analysis and by its proven good performance in static settings.

No evolution strategy in real-valued search spaces is complete without a mutation strength adaptation mechanism. It is necessary for the mutation strength to be adapted continuously to fit the local characteristics of the objective function. Two mechanisms that are commonly used for the adaptation of the mutation strength are *mutative self-adaptation* and *cumulative mutation strength adaptation*. While the former is the more popular approach counting the number of publications, we choose to analyze cumulative mutation strength adaptation as it is known that mutative self-adaptation is unable to make full use of the genetic repair effect in combination with global intermediate recombination. Also note that while the original algorithm by Hansen and Ostermeier [11, 10] adapts the entire mutation covariance matrix, the variant considered here uses isotropic mutations and therefore only a single mutation strength.

The cumulative mutation strength adaptation mechanism relies on the conjecture that if the mutation strength is below its optimal value consecutive steps of the strategy tend to be parallel, and if the mutation strength is too high consecutive steps tend to be antiparallel. For optimally adapted mutation strength, the steps taken by the evolution strategy are uncorrelated. So as to be able to reliably detect parallel or antiparallel correlations between successive steps, information from a number of time steps needs to be accumulated. For the

$(\mu/\mu, \lambda)$-ES, the *accumulated progress vector* $\mathbf{s}$ is defined by $\mathbf{s}^{(0)} = \mathbf{0}$ and the recursive relationship

$$\mathbf{s}^{(t+1)} = (1-c)\mathbf{s}^{(t)} + \sqrt{c(2-c)}\sqrt{\mu}\langle\mathbf{z}\rangle^{(t)} \, , \tag{1}$$

where c is a constant determining how far back the "memory" of the accumulation process reaches. The mutation strength is updated according to

$$\sigma^{(t+1)} = \sigma^{(t)} \exp\left(\frac{\|\mathbf{s}^{(t+1)}\|^2 - N}{2DN}\right) \, , \tag{2}$$

where D denotes a damping constant. The term N in the numerator of the argument to the exponential function is the mean squared length of the accumulated progress vector if consecutive progress vectors are stochastically independent. The constants c and D are set to $1/\sqrt{N}$ and $\sqrt{N}$, respectively, according to recommendations made by Hansen [11].

The *sphere model* is the set of all functions $f : \mathbb{R}^N \to \mathbb{R}$ with

$$f(\mathbf{x}) = g(\|\hat{\mathbf{x}} - \mathbf{x}\|) \, ,$$

where $g : \mathbb{R} \to \mathbb{R}$ is a strictly monotonic function of the distance $R = \|\hat{\mathbf{x}} - \mathbf{x}\|$ of a candidate solution $\mathbf{x}$ from the target $\hat{\mathbf{x}}$. The sphere model usually serves as a model for fitness landscapes at a stage where the population of candidate solutions is in relatively close proximity to the target and is most often studied in the limit of very high search space dimensionality. So as to study the tracking behavior of evolutionary algorithms, several authors ([1, 4, 13]) have added a dynamic component to the sphere model by stipulating that the target $\hat{\mathbf{x}}$ varies with time. Several modes of motion of the target are conceivable and have been explored empirically. Examples include random motion, linear motion, and circular motion in search space. For the present paper, we restrict ourselves to considering random motion and assume that the target at time step $t+1$ is given by $\hat{\mathbf{x}}^{(t+1)} = \hat{\mathbf{x}}^{(t)} + \delta\hat{\mathbf{z}}$, where vector $\hat{\mathbf{z}}$ consists of N independent, standard normally distributed components. The standard deviation δ is a measure for the speed of the target. The same model has been considered by Angeline [1], Bäck [4], and Branke [7].

3 Dynamic Sphere

The analysis of the behavior of evolution strategies on the sphere model is greatly simplified by the symmetries inherent in both the strategies and the environment. Following an idea introduced by both Beyer [5] and Rechenberg [12], a vector $\mathbf{z}$ originating at search space location $\mathbf{x}$ can be written as the sum of two vectors $\mathbf{z}_A$ and $\mathbf{z}_B$, where $\mathbf{z}_A$ is parallel to $\hat{\mathbf{x}} - \mathbf{x}$ and $\mathbf{z}_B$ is in the hyperplane normal to that. We will refer to $\mathbf{z}_A$ and $\mathbf{z}_B$ as the *central* and the *lateral components* of vector $\mathbf{z}$, respectively. We define the *signed length* z_A of the central component of vector $\mathbf{z}$ to equal $\|\mathbf{z}_A\|$ if $\mathbf{z}_A$ points towards the target and to equal $-\|\mathbf{z}_A\|$ if $\mathbf{z}_A$ points away from it. Figure 1 illustrates the decomposition.

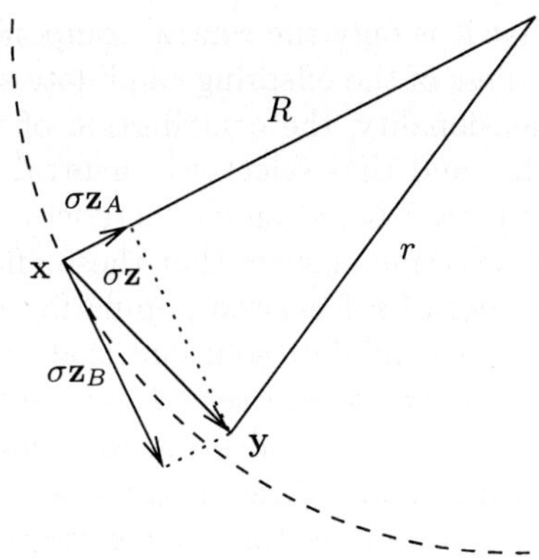

Fig. 1. Decomposition of a vector $\mathbf{z}$ into central component $\mathbf{z}_A$ and lateral component $\mathbf{z}_B$ for the sphere model. Vector $\mathbf{z}_A$ is parallel to $\hat{\mathbf{x}} - \mathbf{x}$, vector $\mathbf{z}_B$ is in the hyperplane perpendicular to that. The starting and end points, $\mathbf{x}$ and $\mathbf{y} = \mathbf{x} + \sigma\mathbf{z}$, of vector $\sigma\mathbf{z}$ are at distances R and r from the target $\hat{\mathbf{x}}$, respectively

Using elementary geometry and denoting the respective distances of $\mathbf{x}$ and $\mathbf{y} = \mathbf{x} + \sigma\mathbf{z}$ from the target by R and r, it is easily seen that

$$\begin{aligned} r^2 &= (R - \sigma z_A)^2 + \sigma^2 \|\mathbf{z}_B\|^2 \\ &= R^2 \left(1 - \frac{2\sigma}{R} z_A + \frac{\sigma^2}{R^2} \|\mathbf{z}\|^2\right) . \end{aligned}$$

Moreover, we will see that in high-dimensional search spaces, progress of the strategies requires $\sigma \ll R$, making it possible to expand the term in parentheses around unity and to cut off after the linear term, yielding

$$r = R - \sigma z_A + \frac{\sigma^2}{2R} \|\mathbf{z}\|^2 + \ldots \tag{3}$$

for the distance of $\mathbf{y}$ from the target.

Let us first consider the case that $\mathbf{z}$ is a mutation vector. Then, as mutations are isotropic, we can without loss of generality assume that $z_A = z_1$ and $\mathbf{z}_B = (0, z_2, \ldots, z_N)^{\mathrm{T}}$, where the z_i, $i = 1, \ldots, N$, are independently drawn from a standardized normal distribution. The squared length $\|\mathbf{z}_B\|^2$ of the lateral component is the sum of squares of $N-1$ terms and as such χ^2_{N-1}-distributed. Recall that the χ^2_{N-1}-distribution has mean $N-1$ and standard deviation $\sqrt{2(N-1)}$. With increasing search space dimensionality, the influence of the central component $\mathbf{z}_A$ on the squared length $\|\mathbf{z}\|^2$ of a mutation vector decreases more and more. In the limit of infinite search space dimensionality, the quotient $(\|\mathbf{z}\|^2 - N)/N$ tends to zero, and thus $\|\mathbf{z}\|^2$ is well approximated by N. Therefore, according to (3), the difference between the distance from the target of the centroid of the parental population and the distance from the target of an offspring candidate solution has normal distribution with mean $-\sigma^2 N/2R$ and with variance σ^2. Note that it is only the central component of a mutation vector that determines the fitness of the offspring candidate solution; in the limit of infinite search space dimensionality, the contribution of the lateral component is the same for all offspring and thus selectively neutral. This has been shown more formally using an approach based on characteristic functions in [2].

For the sphere model, selection ensures that those offspring candidate solutions with the smallest values of r form the population of the next time step. Recombination averages those candidate solutions that are selected. The lateral components of the mutation vectors are selectively neutral in that they contribute a constant to the fitness of the offspring they generate. Thus, they are independent. The lateral component of the progress vector is the average of the lateral components of the mutation vectors that correspond to those offspring candidate solutions that are selected. It is easy to see that its squared length is N/μ, where the division by μ is due to the independence of the vectors being averaged. According to (3), the lateral component of the progress vector increases the distance to the target by $\sigma^2 N/2\mu R$. The reduction of that term by the factor μ in the denominator has been termed the *genetic repair effect*.

The central component of the progress vector is the average of those μ of the λ mutation vectors that have the largest signed lengths of the central components. The expected signed length of the central component of the progress vector is thus $c_{\mu/\mu,\lambda}$, where $c_{\mu/\mu,\lambda} \geq 0$ is the expected average of the first μ order statistics of a sample of λ standard normally distributed, independent random variables. An integral expression for $c_{\mu/\mu,\lambda}$ has been derived in [6]. According to (3), the central component of the progress vector reduces the distance to the target by $\sigma c_{\mu/\mu,\lambda}$. Note that while the *gain term* resulting from the central components is linear in σ, the *loss term* resulting from the lateral components grows quadratically. The range of useful mutation strengths is thus limited, requiring $\sigma \ll R$ as stipulated above.

So far, we have not considered the motion of the target. For the dynamic sphere, in every time step, the target takes a random step $\delta\hat{\mathbf{z}}$, where $\hat{\mathbf{z}}$ consists of N independent, standard normally distributed components. In effect, that is the same as the centroid of the population taking a random step $-\delta\hat{\mathbf{z}}$. The consequences of such a step have been investigated above: it leads to a increase in distance from the target by $\delta^2 N/2R$. Thus, the expected distance between the target and the centroid of the population at the next time step is

$$\mathrm{E}\left[R^{(t+1)}\right] = R^{(t)} - \sigma c_{\mu/\mu,\lambda} + \frac{\sigma^2 N}{2\mu R} + \frac{\delta^2 N}{2R}\,, \tag{4}$$

where the second and third terms on the right hand side result from the central and lateral components of the progress vector and the fourth summand is a consequence of the motion of the target.

After initialization effects have faded, the distance from the target of the centroid of the population has a time-invariant distribution. A good approximation to the mean of that distribution is obtained from (4) by ignoring fluctuations and simply demanding that $\mathrm{E}[R^{(t+1)}] = R^{(t)} = R$. Solving for R yields

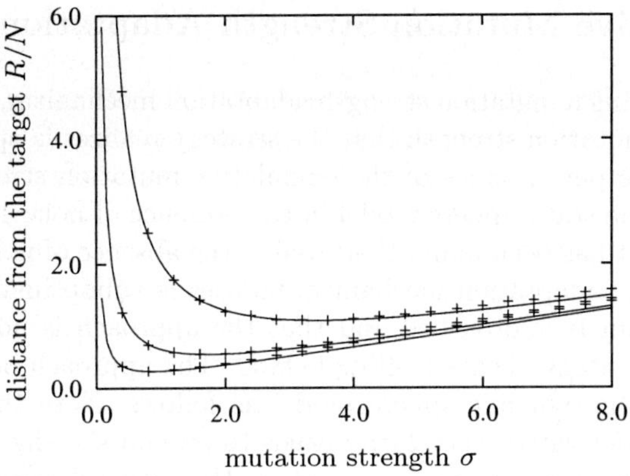

Fig. 2. Distance from the target R/N for a (3/3, 10)-ES with, from bottom to top, $\delta = 0.5$, 1.0, and 2.0. Search space dimensionality is $N = 40$. The crosses are measured values, the lines reflect predictions from (5)

$$R = \frac{N}{2c_{\mu/\mu,\lambda}} \left(\frac{\sigma}{\mu} + \frac{\delta^2}{\sigma} \right) . \tag{5}$$

Figure 2 shows that the resulting predictions are quite accurate. The accuracy increases further with increasing search space dimensionality. It can be seen that it somewhat decreases with increasing mutation strength as some of the assumptions made in the derivation of (3) are violated, but is very good in the range of mutation strengths where R is small.

Using (5), the optimal mutation strength can be found by computing the derivative with respect to σ and finding a root thereof. From

$$\frac{\partial R}{\partial \sigma} = \frac{N}{2c_{\mu/\mu,\lambda}} \left(\frac{1}{\mu} - \frac{\delta^2}{\sigma^2} \right)$$

it follows that $\sigma = \sqrt{\mu}\delta$ is optimal, and (5) shows that the corresponding distance from the target is

$$R = \frac{N\delta}{\sqrt{\mu} c_{\mu/\mu,\lambda}} . \tag{6}$$

That is, the distance from the target increases linearly with the speed of the target and can be decreased by increasing the population size. This decrease is due to the use of global intermediate recombination. To the reader acquainted with the theory of evolution strategies on the sphere model, it will not come as a surprise that from $\sigma = \sqrt{\mu}\delta$ with (6) it follows that the optimal mutation strength is $\sigma = \mu c_{\mu/\mu,\lambda} R/N$. This relationship has long been known to hold for the static sphere model. We now know that it holds as well when tracking a moving target with random dynamics.

4 Cumulative Mutation Strength Adaptation

Clearly, when using a mutation strength adaptation mechanism, there is no guarantee that the mutation strength that the strategy realizes is optimal; indeed, it often is not. The performance of the cumulative mutation strength adaptation mechanism on the static sphere model in the presence of noise has recently been analyzed in [3]. It has been found that even in the absence of noise, the mutation strength that the adaptation mechanism realizes is suboptimal due to the fact that the optimum is approached and that the approach is not instantaneous, leading to the strategy always trailing behind. The approach pursued in [3] can be adapted to the dynamic sphere model as follows. Note that due to space limitations, the derivation necessarily needs to remain sketchy at times.

As mutation vectors and progress vectors, the accumulated progress vector $\mathbf{s}$ can be written as the sum of its central and lateral components, $\mathbf{s}_A$ and $\mathbf{s}_B$. We write s_A for the signed length of the central component and as in Sect. 3 assume without loss of generality that at time t, the direction to the target is such that $s_A = s_1$. Using (1) and for notational simplicity writing $\mathbf{z}$ instead of $\langle\mathbf{z}\rangle$ for the progress vector, we have

$$\begin{aligned}\|\mathbf{s}^{(t+1)}\|^2 &= \sum_{i=1}^{N}\left((1-c)s_i^{(t)} + \sqrt{c(2-c)}\sqrt{\mu}z_i^{(t)}\right)^2 \\ &= (1-c)^2\|\mathbf{s}^{(t)}\|^2 + 2(1-c)\sqrt{c(2-c)\mu}\sum_{i=1}^{N} s_i^{(t)} z_i^{(t)} + c(2-c)\mu\|\mathbf{z}^{(t)}\|^2 \ .\end{aligned}$$

The signed length of the central component of a vector equals the inner product of that vector with a vector of length unity from the centroid of the population to the target. Therefore, writing $\mathbf{x}$ for the centroid of the population and using (1), we have

$$\begin{aligned}s_A^{(t+1)} &= \left((1-c)\mathbf{s}^{(t)} + \sqrt{c(2-c)}\sqrt{\mu}\mathbf{z}^{(t)}\right)^{\mathrm{T}} \frac{\hat{\mathbf{x}}^{(t)} + \delta\hat{\mathbf{z}}^{(t)} - \mathbf{x}^{(t)} - \sigma^{(t)}\mathbf{z}^{(t)}}{R^{(t+1)}} \\ &= \frac{R^{(t)}}{R^{(t+1)}}\Bigg[(1-c)\left(s_A^{(t)} + \frac{\delta}{R^{(t)}}\sum_{i=1}^{N} s_i^{(t)}\hat{z}_i^{(t)} - \frac{\sigma^{(t)}}{R^{(t)}}\sum_{i=1}^{N} s_i^{(t)} z_i^{(t)}\right) \\ &\qquad + \sqrt{c(2-c)}\sqrt{\mu}\left(z_A^{(t)} + \frac{\delta}{R^{(t)}}\sum_{i=1}^{N}\hat{z}_i^{(t)} z_i^{(t)} - \frac{\sigma^{(t)}}{R^{(t)}}\|\mathbf{z}^{(t)}\|^2\right)\Bigg] \ .\end{aligned}$$

These two equations together with (2) describe the development from one time step to the next of the squared length of the accumulated progress vector, the signed length of its central component, and the mutation strength. In analogy to the way we proceeded in Sect. 3, we ignore fluctuations and assume that all quantities can be replaced by their expected values. Omitting time superscripts, for the stationary case it follows that

$$\|\mathbf{s}\|^2 = (1-c)^2\|\mathbf{s}\|^2 + 2(1-c)\sqrt{c(2-c)}\sqrt{\mu}s_A c_{\mu/\mu,\lambda} + c(2-c)N \ , \qquad (7)$$

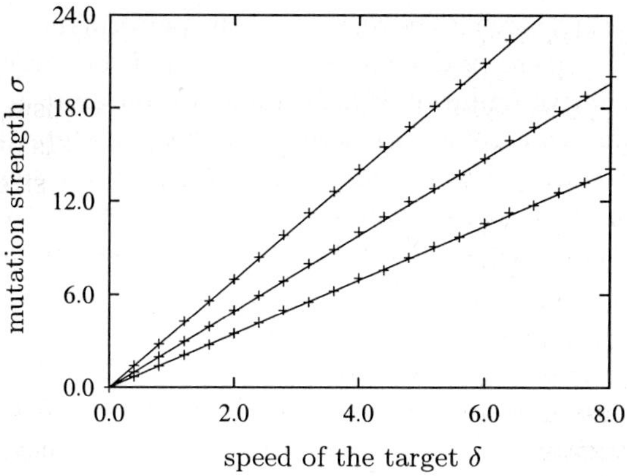

Fig. 3. Mutation strength σ as a function of the speed of the target δ for, from bottom to top, a $(3/3, 10)$-ES, a $(6/6, 20)$-ES, and a $(12/12, 40)$-ES. Search space dimensionality is $N = 40$. The crosses are measured values realized by cumulative mutation strength adaptation, the lines reflect optimal values $\sigma = \sqrt{\mu}\delta$

where we have used the fact that $\mathrm{E}[z_i] = 0$ for $i \neq 1$ due to symmetry reasons, that

$$s_A = (1-c)s_A + \sqrt{c(2-c)}\sqrt{\mu}\left(c_{\mu/\mu,\lambda} - \frac{\sigma N}{\mu R}\right) , \tag{8}$$

where $\sigma s_A c_{\mu/\mu,\lambda}/R$ has been neglected as compared to s_A as $\sigma \ll R$, and that

$$\sigma = \sigma \exp\left(\frac{\|\mathbf{s}\|^2 - N}{2DN}\right) . \tag{9}$$

Solving (7), (8), and (9) for the mutation strength yields $\sigma = \mu c_{\mu/\mu,\lambda} R/N$. Figure 3 demonstrates the accuracy of the result. The resulting distance R to the target is given by (6). Comparison with the optimal mutation strength $\sigma = \sqrt{\mu}\delta$ derived in Sect. 3 shows that cumulative mutation strength adaptation achieves optimal performance on the sphere model with random dynamics of the target.

5 Conclusions

In this paper, we have studied the tracking performance of the $(\mu/\mu, \lambda)$-ES on a variant of the sphere model with random dynamics of the target. A scaling law that describes the dependence of the distance from the target on the mutation strength has been found and solved analytically for the optimal mutation strength. It has then been shown that cumulative mutation strength adaptation works perfectly in that the optimal mutation strength is realized by the algorithm. Thus, our results do not support the observation made by Branke [7] that in dynamic optimization, evolutionary algorithms lose the diversity necessary for exploring the search space.

A goal of future research is the analysis of the performance of the $(\mu/\mu, \lambda)$-ES on a variant of the sphere model where the motion of the target is deterministic and linear. Cumulative mutation strength adaptation aims at achieving a step length that eliminates correlations between successive steps. Linear motion of the target can be expected to introduce correlations in the sequence of steps to be taken and thus to lead to qualitatively new insights in the performance of cumulative mutation strength adaptation in dynamic environments.

Acknowledgments

The first author is supported by the Deutsche Forschungsgemeinschaft (DFG) under grant Be1578/6-3. The second author is supported by the DFG as a Heisenberg fellow under grant Be1578/4-2.

References

[1] P. J. Angeline. Tracking extrema in dynamic environments. In *Proc. of the Sixth International Conference on Evolutionary Programming*, pages 335–345. Springer Verlag, Heidelberg, 1997.
[2] D. V. Arnold and H.-G. Beyer. Local performance of the $(\mu/\mu_I, \lambda)$-ES in a noisy environment. In W. N. Martin and W. M. Spears, editors, *Foundations of Genetic Algorithms 6*, pages 127–141. Morgan-Kaufmann, San Francisco, 2000.
[3] D. V. Arnold and H.-G. Beyer. An analysis of cumulative mutation stength adaptation. In preparation, 2002.
[4] T. Bäck. On the behavior of evolutionary algorithms in dynamic environments. In *Proc. of the 1998 International Conference on Evolutionary Computation*, pages 446–451. IEEE Press, Piscataway, NJ, 1998.
[5] H.-G. Beyer. Toward a theory of evolution strategies: Some asymptotical results from the $(1 \overset{+}{,} \lambda)$-theory. *Evolutionary Computation*, 1(2):165–188, 1993.
[6] H.-G. Beyer. *The Theory of Evolution Strategies.* Natural Computing Series. Springer Verlag, Heidelberg, 2001.
[7] J. Branke. *Evolutionary Optimization in Dynamic Environments.* Kluwer Academic Publishers, Dordrecht, 2001.
[8] R. Bürger. *The Mathematical Theory of Selection, Recombination, and Mutation.* John Wiley & Sons, Chichester, 2000.
[9] S. Droste. Analysis of the $(1+1)$ EA for a dynamically changing objective function. In *Proc. of the 2002 Congress on Evolutionary Computation.* IEEE Press, Piscataway, NJ, 2002.
[10] N. Hansen and A. Ostermeier. Completely derandomized self-adaptation in evolution strategies. *Evolutionary Computation*, 9(2):159–195, 2001.
[11] N. Hansen. *Verallgemeinerte individuelle Schrittweitenregelung in der Evolutionsstrategie.* Mensch & Buch Verlag, Berlin, 1998.
[12] I. Rechenberg. *Evolutionsstrategie '94.* Frommann-Holzboog, Stuttgart, 1994.
[13] R. Salomon and P. Eggenberger. Adaptation on the evolutionary time scale: A working hypothesis and basic experiments. In *Proc. of the Third Conference on Artificial Evolution*, pages 251–262. Springer Verlag, Heidelberg, 1997.
[14] K. Weicker and N. Weicker. On evolution strategy optimization in dynamic environments. In *Proc. of the 1999 Congress on Evolutionary Computation*, pages 2039–2046. IEEE Press, Piscataway, NJ, 1999.

On the Behavior of Evolutionary Global-Local Hybrids with Dynamic Fitness Functions

Roger Eriksson and Björn Olsson

Dept. of Computer Science, University of Skövde
Box 408, 54128 Skövde, Sweden
{roger.eriksson,bjorn.olsson}@ida.his.se

Abstract. This paper investigates the ability of evolutionary global-local hybrid algorithms to handle dynamic fitness functions. Using a model where fitness functions vary in ruggedness as well as in whether changes occur gradually or abruptly, we evaluate the performance of Baldwinian and Lamarckian hybrid strategies and find them capable of locating and tracking a moving global optimum.

1 Introduction

Evolutionary global-local hybrids (hereafter referred to as hybrids) are Evolutionary Algorithms (EAs) that include local search methods to increase their efficiency, either in reducing the time to reach an adequate solution quality or increasing the maximum quality in specified time. The role of evolution in such hybrids is to perform global search, which seeks promising regions of the search space, whereas the role of local search is to fine-tune promising solutions.

Hybrids have primarily been used in domains where special local search methods have been developed. When such local search methods are used, increases in efficiency can often be attributed to their exploitation of problem specific knowledge, (see e.g. [1] and [2] for successful applications in continuous and combinatorial domains respectively). The inclusion of special local search methods therefore seem to be an efficient means of incorporating problem specific knowledge. Another advantage of hybrids described in the literature is the acceleration of evolution from a smoothing of the fitness landscape [3] [4]. A less described advantage is the improvement of the fine-tuning capabilities.

Previous work on hybrids deals with static fitness functions, and their ability to deal with dynamics is largely unknown. This paper aims to remedy this situation by investigating the ability of hybrids to deal with dynamic fitness functions. If hybrids have this capacity, special measures such as the ones described in [5] may not have to be used and we would therefore not have to compromise the static efficiency. According to [6] there is a tension between designing and tuning EAs for static fitness vs. dynamic functions - improved efficiency on one invariably decreases the efficiency on the other. Our hypothesis is that hybrids are able to handle dynamic fitness functions, without using any special measures.

J.J. Merelo Guervós et al. (Eds.): PPSN VII, LNCS 2439, pp. 13–22, 2002.

The need to deal with dynamic fitness functions arises because many real-world problems include time-varying changes to the fitness function. Unfortunately, traditional EAs are generally poor at dealing with such dynamics since they operate by converging the population on the most promising region of the search space. A number of different approaches addressing this problem have been suggested (see e.g. [5] for a survey).

As outlined by De Jong in [6], real-world dynamic fitness functions do not appear to change in arbitrarily random ways but exhibit patterns of change that can be discovered and exploited. De Jong offers a categorisation of the corresponding fitness landscapes into:

- drifting landscapes, exhibiting gradual movement of the optima;
- landscapes undergoing significant morphological changes, where high fitness peaks shrink and new peaks emerge from previously uninteresting regions;
- cyclic landscapes, which repeatedly visit a limited number of states; and
- landscapes which change abruptly and discontinuously.

This paper focuses on gradually and abruptly changing landscapes, and we investigate if two types of hybrids can handle these dynamic landscapes. Following [3], we term these two types of hybrids Baldwinian and Lamarckian. A more thorough description of the landscapes and algorithms tested appears in Chapter 2, followed by experimental results in Chapter 3 and conclusions in Chapter 4.

2 Experimental Setup

The purpose of the experiments reported here is to investigate how evolutionary global-local hybrids deal with dynamic fitness landscapes. For practical reasons the scope of this investigation is limited in various ways defined below, e.g. in that it only deals with a particular global search method and a particular local search method. We feel however that the design decisions taken are relatively commonplace and general. We also decided to evaluate the methods on randomly generated fitness functions with particular characteristics. This differs from the traditional use of test functions for evaluating EAs in that our results can be generalised to all fitness functions with such characteristics and not just to a particular test function. This approach has recently been used in several other studies of EAs with dynamic fitness functions, e.g. [7], [8], and [9]. The remainder of this section describes the fitness functions and how they were generated, followed by a description of the methods evaluated, the experimental conditions, and the metrics used for analysing the methods.

2.1 Fitness Functions

Fitness functions are constructed from component functions, each defining the shape of a peak in a multidimensional landscape. Dynamics are then achieved by altering the coordinates of the peaks over time. In this study a generation is considered as the smallest unit of time, which means that all individuals are evaluated on the same landscape. The alterations are performed according to a fixed policy that defines the type of dynamics.

Here, a fitness function is a mapping from $[x_{min}, x_{max}]^n \subseteq \mathbb{R}^n$ to $\mathbb{R}$, where $x_{min} = 0$ and $x_{max} = 100$, defined as:

$$f(\boldsymbol{x}, t) = \max_{i=1\ldots k} g_i(\boldsymbol{x}, a_i, \boldsymbol{c}_i(t), w_i), \tag{1}$$

where g_i is a component function defining the shape of a peak with amplitude a_i, coordinates $\boldsymbol{c}_i(t)$, and width w_i at time step t, according to:

$$g_i(\boldsymbol{x}, a_i, \boldsymbol{c}_i(t), w_i) = a_i \exp(-d(\boldsymbol{x}, \boldsymbol{c}_i(t))^2/(2w_i^2)), \tag{2}$$

where d is the Euclidean distance function.

The goal is to acquire the fittest solution and track its progression through the search space as closely as possible. Here, g_1 represents the single global optimum, with $w_1 = 20$, and $a_1 = 100$. Additional functions represent local optima with $w_i = 4$, and $a_i = \min(90, g_1(\boldsymbol{c}_i(t), a_1, \boldsymbol{c}_1(t), w_1) + h_i)$, where h_i is uniformly selected from $[0, \max(0, 20(90 - g_1(\boldsymbol{c}_i(t), a_1, \boldsymbol{c}_1(t), w_1))/90)]$. This has the effect of preventing local optima in coordinates where $g_1 \geq 90$. The upper bound of h_i is set in proportion to g_1 to assure that the amplitudes of local optima confer information about the location of g_1. The landscapes used in the experiments either have local optima (denoted rugged) or not (denoted smooth). The dimensionality and the maximum number of component functions were chosen to be as challenging as possible and yet requiring reasonable computational resources. In rugged landscapes the number of component functions was set to 10^n to yield a density of optima equivalent to that of a one-dimensional fitness function with 10 component functions. Landscapes with $n = 4$ then turned out to be challenging enough to reveal limitations in all of the algorithms and yet requiring reasonable computational resources.

Dynamics are achieved as follows. First all peaks are set at uniformly distributed coordinates. Then every Δg generations the coordinates of all peaks are changed by adding $\delta_j = s \cdot N_j(0, 1)$ to each $c_{ij}(t)$, where s is the severity of the change, and $N_j(0, 1)$ are normally distributed random numbers not causing $c_{ij}(t+1)$ to go out of bound in any dimension $j = 1 \ldots n$. An assumption of this type of dynamics is that there is a similarity between the evaluated worth of the new coordinates and the previous ones. Three different dynamics are used in the experiments: static, gradual, and abrupt. Landscapes with static dynamics do not change at all. Landscapes with gradual dynamics have $\Delta g = 1$, and $s = 1$, whereas those with abrupt dynamics have $\Delta g = 20$, and $s = 20$. These dynamics are similar to those used in [9].

2.2 Evolutionary Algorithms

Three different evolutionary algorithms were used in the experiments. These differ in the way solutions are adapted, which is done either by Plain, Baldwinian, or Lamarckian evolution. Algorithms using Plain evolution do not incorporate local search, whereas the other algorithms are evolutionary global-local hybrids.

Each candidate solution is directly represented by a real-valued vector $\boldsymbol{x} \in \mathbb{R}^n$, also operated on by the search operators. Variation is performed by uniform crossover with probability 1, followed by normally distributed mutation of each gene x_i with probability 0.05 according to:

$$x'_i = max(x_{min}, min(x_{max}, x_i + 9.0 \cdot N_i(0,1))) \ . \tag{3}$$

The population consists of 100 individuals whose initial gene values are sampled uniformly from $[x_{min}, x_{max}] \subseteq \mathbb{R}$. Selection is performed by binary tournaments with replacement, as described in [10], and all individuals are replaced in each generation. Each run is terminated after 200 generations.

The hybrid algorithms apply local search to every individual in the population prior to their evaluation. After evaluation the corresponding locally optimal solution is then either discarded (denoted Baldwinian evolution) or replaces the individual (denoted Lamarckian evolution). The probability of replacement is 0 in Baldwinian evolution, and 1 in Lamarckian evolution. Other probabilities are not considered in this paper for practical reasons. The local search method used by the hybrid algorithms is the Polak-Ribiere variant of the conjugate gradient method described in [11], using Brent's method for line optimisations. Local search terminates when the last line optimisation fails to decrease the function value by more than $\pm 10^{-6}$, and a line optimisation terminates when accurate to $\pm 10^{-10}$. This choice of termination criteria was based on the observation that lower accuracies caused local search to terminate before reaching the local optimum if the individual was too distant.

3 Experiments and Results

A total of 18 experiments were conducted, one for each combination of evolutionary algorithm, landscape dynamics, and ruggedness, as described above. Each experiment consists of 10 runs with different random seeds on each of 10 landscapes generated with different random seeds, i.e. 100 runs in total for each of the 18 experiments. In addition to the best and average fitness of the population, we also report the average innate fitness, defined as the average fitness value of the population before local search is applied. We also report the diversity of the population in each generation. The diversity of a population with p individuals, of which u are unique, is here defined as:

$$\text{Diversity} = \frac{2}{p(u-1)\sqrt{n(x_{max} - x_{min})^2}} \sum_{i=1}^{u} \sum_{j=i+1}^{u} d(\boldsymbol{x}_i, \boldsymbol{x}_j), \tag{4}$$

where $d(\boldsymbol{x}_i, \boldsymbol{x}_j)$ is the Euclidean distance between unique individuals $\boldsymbol{x}_i$ and $\boldsymbol{x}_j$. An assumption of this definition is that duplicate solutions do not contribute to the diversity in the population.

An algorithm is said to be capable of handling a landscape if it is able to acquire an optimal or near-optimal solution and then track its progression through the search space until the run terminates.

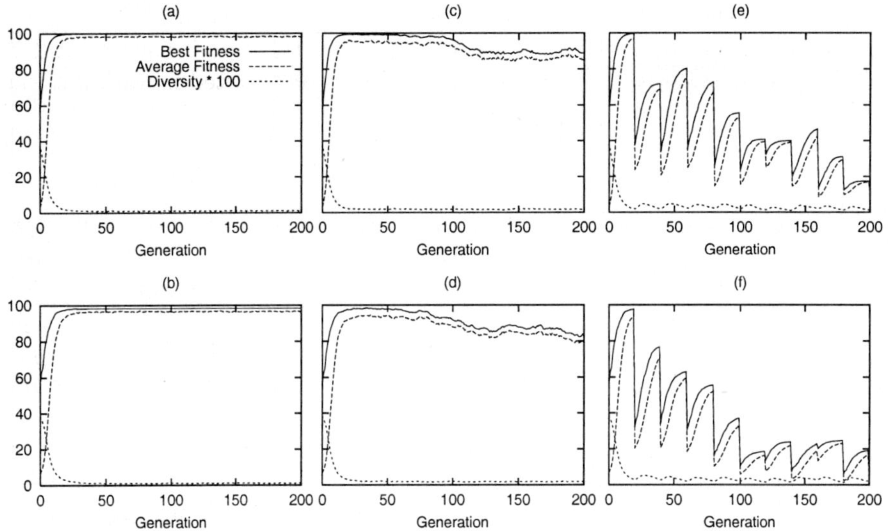

Fig. 1. Plain evolution in landscapes of type: (a) smooth static, (b) rugged static, (c) smooth gradual, (d) rugged gradual, (e) smooth abrupt, and (f) rugged abrupt.

3.1 Plain Evolution

Fig. 1(a) shows the behaviour of Plain Evolution (PE) in smooth static landscapes. PE quickly converges on the optimum in all runs. In rugged static landscapes some runs converge on local optima, as can be seen from the best fitness not quite reaching the maximum value in fig. 1(b).

In smooth gradual landscapes (fig. 1(c)), PE initially locates and tracks the optimum, but after converging close to the optimum it gradually loses its tracking ability. The behaviour in rugged gradual landscapes (fig. 1(d)) is similar, with peak values of best and average fitness being somewhat lower and decaying more quickly, showing that PE is unable to handle the gradual landscapes.

In smooth abrupt landscapes PE locates the initial position of the optimum, but after converging close to the optimum it fails to adapt to landscape changes. In fig. 1(e) the initial increase in best and average fitness up to near optimal values before the first change is followed by a failure to recover from this and subsequent changes. Including local optima yields similar behaviour (fig. 1(f)), but where the peak values of best and average fitness are somewhat lower before the first landscape change. Clearly, PE cannot handle the abrupt landscapes.

3.2 Baldwinian Evolution

Fig. 2(a) shows the behaviour of Baldwinian Evolution (BE) in smooth static landscapes. The best fitness is constantly at its maximum, since there is always a solution that can be fully adapted by local search. The average fitness quickly

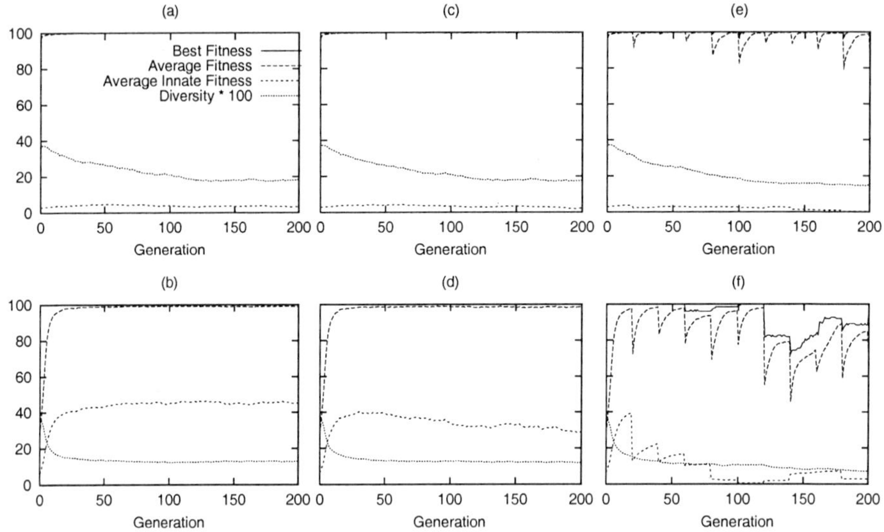

Fig. 2. Baldwinian evolution in landscapes of type: (a) smooth static, (b) rugged static, (c) smooth gradual, (d) rugged gradual, (e) smooth abrupt, and (f) rugged abrupt.

reaching the maximum shows that the population is quickly filled with such solutions in all runs. The only selective pressure is for solutions that can be adapted by local search. This can be seen from the average innate fitness increasing in the first generations when there are solutions that cannot be adapted by local search, and stagnating when the number of such solutions decreases. When all solutions can be adapted, and therefore are equally fit, diversity continues reducing somewhat, probably due to selection noise.

Including local optima has no effect on the best fitness, which is constantly at its maximum value (fig. 2(b)). This means that there are solutions in the global optimum's basin of attraction in the initial population in all runs. However, the number of such solutions is lower than in smooth landscapes, which gives an initially lower average fitness. The number of solutions outside the global optimum's basin of attraction quickly decreases in the following generations and the average fitness reaches its maximum. On smooth landscapes the only selective pressure was for solutions that could be adapted by local search, but in this case, due to the existence of local optima, there is also pressure for solutions in the global optimum's basin of attraction. This can be seen from the average innate fitness increasing in initial generations when there are many solutions outside the global optimum's basin of attraction, and stagnating as the number decreases. After this stagnation, when solutions are equally fit, diversity still decreases somewhat, which is again probably due to selection noise.

In smooth gradual landscapes BE locates and tracks the optimum perfectly, as can be seen in fig. 2(c) from the best fitness being at the maximum through all runs. Note that the behaviour is almost identical to that in static landscapes.

The fact that average innate fitness and diversity have similar values to those in the static landscapes indicates that adaptation to changes is mainly performed by local search. Even when local optima are included (fig. 2(d)), BE locates and tracks the optimum perfectly. The behaviour is again almost identical to that in static landscapes. A difference between fig. 2(c) and 2(d) is that in the latter the average innate fitness decreases after reaching its peak value, showing that adaptation is increasingly performed by local search over the generations.

In smooth abrupt landscapes BE locates and tracks the optimum reasonably well in all runs. This can be seen in fig. 2(e) from the best fitness always being at or close to its maximum value through the run. After each change, the number of solutions that cannot be fully adapted by local search increases, which is seen as dips in the average fitness every 20th generation. However, the average fitness always recovers before the next change. Average innate fitness and diversity have similar values to those observed in static and gradual landscapes, indicating that adaptation to changes is mainly performed by local search.

In rugged landscapes BE initially manages to locate and track the optimum perfectly. This can be seen in fig. 2(f) from the best fitness always being at its maximum in the first 60 generations. However, after the change in generation 60 it fails to find the optimum solution before the next change. From this point there is also little or no progress in average innate fitness after each change, showing that adaptation is mainly performed by local search in subsequent generations.

3.3 Lamarckian Evolution

Fig. 3(a) shows the behaviour of Lamarckian Evolution (LE) in static landscapes without local optima. The behaviour is similar to that of BE, with the best fitness being constantly at its maximum value, meaning that there is always a solution available that can be adapted by local search. For LE, the number of such solutions in the population increases at an even higher rate. This can be seen from the average fitness reaching its maximum already in generation 2. Note that when LE is used, each solution is replaced by the optimum of its basin of attraction, when evaluated. In this case all solutions except those that cannot be adapted by local search are located in the global optimum's basin of attraction, and therefore nearly all solutions are replaced by the globally optimal solution already in the initial population. This affects the diversity that falls to its minimum value in generation 2. The average innate fitness quickly reaches a near maximum value and then stays approximately at this value in subsequent generations. Average innate fitness never reaches its maximum value since mutations keep pushing individuals off the global optimum.

As for BE, the inclusion of local optima has no effect on the best fitness measure since it is constantly at its maximum (fig. 3(b)). This means that there are solutions in the global optimum's basin of attraction in the initial population in all runs. Also, the number of solutions in the global optimum's basin of attraction in the initial population is lower in this case than in the previous case where there where no local optima, which can be seen from an initially lower average fitness in fig. 3(b). The number of solutions outside the global

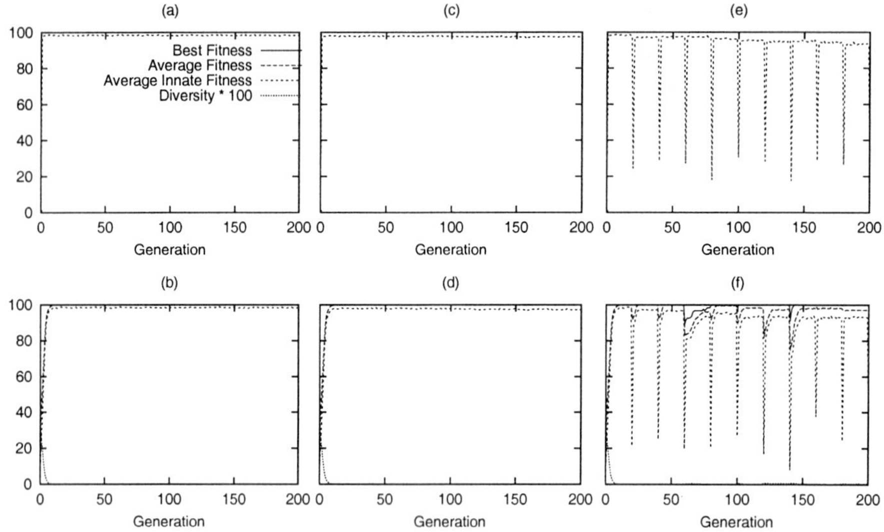

Fig. 3. Lamarckian evolution in landscapes of type: (a) smooth static, (b) rugged static, (c) smooth gradual, (d) rugged gradual, (e) smooth abrupt, and (f) rugged abrupt.

optimum's basin of attraction quickly decreases in the following generations as can be seen from the average fitness reaching its maximum value in generation 12, at which the diversity falls to its minimum. As can be seen in fig. 3(c) and 3(d) the behaviour of LE in the gradual landscapes is almost identical to that in the static landscapes, i.e. it handles the landscapes perfectly in all runs.

LE also manages to locate and track the moving optimum perfectly in the abrupt landscapes without local optima (fig. 3(e)). The only visible difference in behaviour between abrupt and gradual landscapes is that the average innate fitness in the abrupt landscapes drops to a low value immediately after a change has occurred and then recovers to a near optimal value in the following generation. The reason for this drop is that solutions are more likely to be displaced far from the global optimum after a change has occurred in the abrupt landscapes. The instant recovery of the average innate fitness in the following generation is because all solutions can be adapted by local search, as can be seen from the constant maximum value of the average fitness. Note that the average innate fitness drops to a lower level when LE is used than when BE is used. The reason is that in LE the population is completely converged when a change occurs, which is not the case in BE. When local optima are included, LE manages to locate and track the moving optimum reasonably well in all runs. This can be seen in fig. 3(f) from the best fitness always being at or near its maximum value through the run. Although the best fitness drops below the maximum value in generations 60 and 140, it always recovers completely before the next change.

4 Discussion and Conclusions

The results show that PE is incapable of handling the dynamic landscapes, which seems to be due to changes occuring when the population is too converged. BE, on the other hand, handled the gradual landscapes perfectly. It was also capable of handling smooth abrupt landscapes, but it could only handle the first two changes in the rugged abrupt landscapes. The failure of BE to adapt after the first two changes seems to be caused by a converged population, despite the fact that BE maintains diversity in the global optimum's basin of attraction. We hypothesise that this residual diversity is not enough for crossover to adapt the individuals and that mutation is too slow to compensate. That would also explain why the average innate fitness decreases after reaching its peak. LE was found to be capable of handling both gradual and abrupt landscapes. This supports our hypothesis that hybrids are able to handle dynamic landscapes without any special measures. It is hard to say why LE is capable of handling all landscapes as opposed to BE, since LE quickly causes the population to converge. We hypothesise that LE, by back-substituting the result of local search to the individuals, to some degree compensates for the reduced capability of evolution to adapt the individuals. This would be a result of it being able to achieve adaptation of the individuals within a basin of attraction without adaptation by evolution, and in that it can speed up adaptation by evolution, e.g. the relatively slow adaptation by mutation in a converged population. We also hypothesise that it is advantageous to have individuals located at the centre of the global optimum's basin of attraction when changes occur.

Although comparing the algorithms is not an objective of this paper it is tempting to draw the conclusion that local search improves the tracking behaviour as well as the best fitness value in each generation. However, the metrics used in this study are not directly comparable since the algorithms with local search require at least two function evaluations per individual, while plain evolution only requires one. The number of function evaluations per individual for BE and LE was found to be fairly independent of the distance to the local optimum, and stayed close to (but did not exceed) 50 in all of the runs.

For a fair comparison of the algorithms we need to adjust the number of generations between the changes so that they get comparable numbers of function evaluations to adapt. We therefore conducted two additional experiments with PE on abrupt landscapes, where the number of generations between changes was extended so that the number of function evaluations was at least as many as those of BE and LE. However, fig. 4 shows that PE fails to improve the best fitness values from fig. 1(e) and 1(f). We can therefore conclude that local search improves both the tracking behaviour and the best fitness values of the evolutionary algorithms in the abrupt landscapes.

Since the hybrids are able to handle the dynamic landscapes without special measures, and PE is not, it seems that adding local search is a viable alternative to methods designed explicitly to deal with dynamics. In hybrids used for other reasons – assuming a static landscape – we have seen that both BE and LE are able to deal with dynamics, which is obviously useful if the landscape actually is

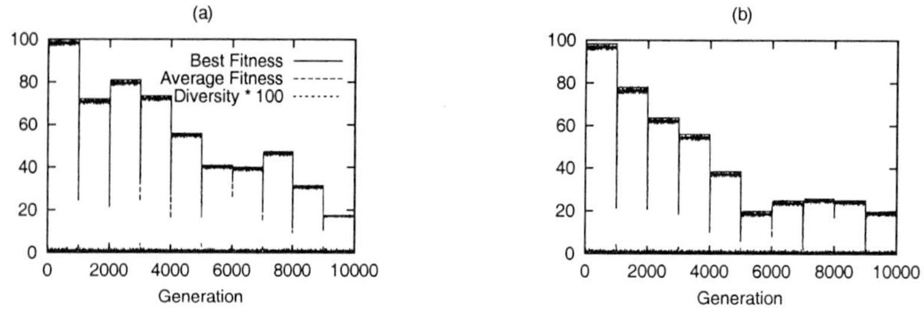

Fig. 4. PE in abrupt landscapes at equal number of fitness evaluations as for BE and LE: (a) smooth landscapes, (b) rugged landscapes.

dynamic. We are, however, just beginning to explore the behaviour of hybrids in dynamic landscapes, and a number of issues need further investigation, including the sensitivity of results to the length of runs and the dimensionality and complexity of the landscapes. In the fairly simple landscapes used in this study limitations were revealed suggesting the possibility of efficiency enhancements.

References

1. Hart, W.E.: Adaptive global optimization with local search. PhD thesis, University of California, San Diego (1994)
2. Land, M.: Evolutionary algorithms with local search for combinatorial optimization. PhD thesis, University of California, San Diego (1998)
3. Hinton, G.E., Nowlan, S.J.: How learning can guide evolution. Complex Systems **1** (1987) 495–502
4. Whitley, D., Gordon, V.S., Mathias, K.: Lamarckian evolution, the Baldwin effect and function optimization. In Davidor, Y., Schwefel, H.P., eds.: Proc. of the PPSN III, Springer (1994) 6–15
5. Branke, J.: Evolutionary approaches to dynamic optimization problems - updated survey. In: Proc. of the GECCO-2001 Workshop on Evolutionary Algorithms for Dynamic Optimization Problems. (2001) 27–30
6. De Jong, K.A.: Evolving in a changing world. In Raś, Z.W., Skowron, A., eds.: 11th Int. Symp. on Foundations of Intelligent Systems, Springer (1999) 512–519
7. Branke, J.: Memory enhanced evolutionary algorithms for changing optimization problems. In Angeline, P.J., Michalewicz, Z., Schoenauer, M., Yao, X., Zalzala, A., eds.: Proc. of the CEC. Volume 3., IEEE Press (1999) 1875–1882
8. Morrison, R.W., De Jong, K.A.: A test problem generator for non-stationary environments. In Angeline, P.J., Michalewicz, Z., Schoenauer, M., Yao, X., Zalzala, A., eds.: Proc. of the CEC. Volume 3., IEEE Press (1999) 2047–2053
9. Grefenstette, J.J.: Evolvability in dynamic fitness landscapes: A genetic algorithm approach. In Angeline, P.J., Michalewicz, Z., Schoenauer, M., Yao, X., Zalzala, A., eds.: Proc. of the CEC. Volume 3., IEEE Press (1999) 2031–2038
10. Blickle, T., Thiele, L.: A comparison of selection schemes used in evolutionary algorithms. Evolutionary Computation **4** (1997) 361–394
11. Press, W.H., Teukolsky, S.A., Vetterling, W.T., Flannery, B.P.: Numerical recipes in C: the art of scientific computing. 2nd edn. Cambridge University Press (1992)

Measuring the Searched Space to Guide Efficiency: The Principle and Evidence on Constraint Satisfaction

Jano I. van Hemert[1] and Thomas Bäck[1,2]

[1] Leiden Institute of Advanced Computer Science – Leiden University,
Niels Bohrweg 1, 2333 CA, Leiden
{jvhemert,baeck}@cs.leidenuniv.nl
[2] Nutech Solutions GmbH
Martin-Schmeisser-Weg 15, D-44227 Dortmund
baeck@nutechsolutions.de

Abstract. In this paper we present a new tool to measure the efficiency of evolutionary algorithms by storing the whole searched space of a run, a process whereby we gain insight into the number of distinct points in the state space an algorithm has visited as opposed to the number of function evaluations done within the run. This investigation demonstrates a certain inefficiency of the classical mutation operator with mutation-rate $1/l$, where l is the dimension of the state space. Furthermore we present a model for predicting this inefficiency and verify it empirically using the new tool on binary constraint satisfaction problems.

1 Introduction

Genetic operators are often taken for granted: they are used without knowledge of the statistical properties behind the process. For almost every operator it is very difficult to analytically determine its properties. In most cases doing some quick runs with a choice of favourite operators seems sufficient to produce a satisfactory result. Unfortunately this lack of knowledge may result in lower performance in the final algorithm.

As an evolutionary algorithm is developed it grows, and at the same time its complexity increases. When its development has finished, its creator starts a quest for the most efficient parameters. This optimisation problem is in principle unsolvable, but it is also necessary as many (even most) parameter settings are unreasonable or lead to an inefficient algorithm.

In the optimisation process we need to take measurements that somehow reflect the efficiency of the algorithm. Furthermore, due to stochastic differences we need to do multiple runs with each setting of parameters. Measurements on these experiments are often restricted to the accuracy, i.e., the percentage of times a solution is found within a time limit, and to the speed, i.e., the average number of evaluations it takes to find a solution. Although these properties are probably seen as most important by end users, they do not tell the developer

J.J. Merelo Guervós et al. (Eds.): PPSN VII, LNCS 2439, pp. 23–32, 2002.

much about what is going on inside. In general, we would like to gain insight into what is actually going wrong—especially when a system does not live up to expectations.

Here we propose a tool that provides a developer or a researcher with some insight into the process of an algorithm that is searching for a solution. This tool also aids in the verification of models of simplified evolutionary algorithms, of which we show an example.

In the next section we shall define the *searched* space (as opposed to the *search* space) of an evolutionary algorithm. Then, in Section 3, we define a mutation operator. In Section 4 we model the chance that this operator will produce unchanged copies and verify this model empirically in Section 5. Section 6 elaborates on the use of the tool in experiments on binary constraint satisfaction. We finish with conclusions in Section 7.

2 The Searched Space

When an evolutionary algorithm is performing its search, it generates candidate solutions. In essence, these are points from the whole possible state space[1]. When working with discrete domains we can define this state space in terms of the alphabet $D = \{d_1, \ldots, d_m\}$ and the number of variables l, also known as the chromosome length.

Definition 1. *State space: The total* state space *an evolutionary algorithm is able to search in is defined as* $M = D_1 \times \ldots \times D_l$. *We assume all* D_i *are equal, thus we know that* $M = D^l$ *and* $|M| = |D|^l = m^l$.

Evolutionary algorithms are stochastic optimisation tools, which means that a solution might not be found. The only proof of global convergence to a solution is made under the assumption that we have unlimited time [4]. Thus for practical reasons a limit is always set on the execution time of an evolutionary algorithm, most often in the form of a maximum number of evaluations. This maximum directly provides an upper bound to the number of points of the state space we are able to visit. In reality, however, evolutionary algorithms in discrete domains tend to visit less points than the maximum they are allowed to.

Definition 2. *Searched space of an algorithm: The* searched space $S \subset M$ *is the set of points taken from the state space that is visited by a particular evolutionary algorithm during a run.*

Insight into the searched space of an algorithm might help us in evaluating its performance. For instance, if the size of the searched space is one then the algorithm has only visited one point of the state space. Except if it found the solution in the first guess this is generally not what we want. At the other extreme, if an algorithm's searched space is just as large as the number of evaluated points in the state space, we could argue that it has fully used its potential.

[1] We use the term *state space* as opposed to *search space* to prevent confusion.

Nevertheless, in the case of an evolutionary algorithm, this is most probably not what we would get, as such population-based algorithms tend to lose population diversity: that is, the individuals in a population will become more alike as the number of evaluations increases. To measure this we introduce *resampling ratio*:

Definition 3. *Resampling ratio: First we define a* revisit *as a point in the state space that we have seen before, i.e., it is already present in the searched space. The* resampling ratio *is defined as the total number of revisits in a run divided by the total number of evaluations in the same run: resampling ratio = revisits/evaluations.*

Of course due to its stochastic nature, an evolutionary algorithm has many searched spaces. With a probability extremely close to one, for every run with a different setting of the random number generator, another searched space is produced. It is not the precise content of the searched space we examine here, but its size related to the total number of evaluations performed.

3 Mutation k/l

A well known way of mutating chromosomes over a discrete alphabet is by changing every gene of the chromosome with some fixed probability $p_{mutation} = k/l$. This probability is often fixed by $k = 1$ because of both theoretical and empirical evidence that show this is a near optimal setting for the mutation rate for many problems with $m = 2$ [2, 7].

The mutation operator is shown in Algorithm 1. For every gene in the chromosome a dice is thrown with l sides. If it shows a value equal or lower than k we generate a new value for the gene by drawing a value uniform randomly from the alphabet minus the current value. This way we make sure that a different value will be chosen.

Algorithm 1 Mutation operator with probability k/l

```
for gene = 1 to l do
  if (uniform_random(l) <= k) then
    c[gene] = uniform_random(D − c[gene])
  end if
end for
```

To test and analyse the mutation operator we need to embed it in an evolutionary algorithm. Eventually we want to analyse the behaviour of the whole algorithm. As this behaviour is influenced by probabilities we will need to keep things as simple as possible. For this reason we will retain from solving a problem. Consequently no selection is performed as we have no fitness to optimise. As we select nothing we can do without a population, or in other words we use a population size of one and replace its only member in every generation with its offspring generated by the mutation operator. The whole process is written down in Algorithm 2. The lack of a selection procedure makes this algorithm not qualified to be categorised as an evolutionary algorithm. However, we would like to point out here that the resampling ratio will become only higher if a selection process is added.

Algorithm 2 Main algorithm that simulates a (1,1) strategy without selection

```
for i = 1 to l do
  c[i] = uniform_random(D_i)
end for
C = {c}
evaluations = 0
revisits = 0
while evaluations < 10,000 do
  c = mutate(c)                                   // See Algorithm 1
  evaluations = evaluations+1
  if c ∈ C then
    revisits = revisits+1
  else
    C = C ∪ {c}
  end if
end while
```

During the run we take two important measurements; the number of evaluations and the number of revisits. The set C symbolises the searched space so far created by the evolutionary algorithm, it contains every unique point that we have generated so far. Using this we can calculate the resampling ratio as revisits/evaluations or, alternatively, as $(|C| - \text{evaluations})/\text{evaluations}$.

4 A Simple Model

We show how the resampling ratio can be predicted for the algorithm in the previous section by using the conjecture that all of the revisited points are caused by the mutation operator producing unchanged individuals. That is

$$\text{resampling ratio} = P(\text{mutate}(c) = c).$$

Examining Algorithm 1 we can calculate the probability that chromosome c is not changed. Every gene $c[i]$ we look at in turn has an independent chance of being changed.

Thus we can multiply every probability of a gene not being changed to get the probability P(chromosome unchanged). We are left with calculating that one gene is unchanged. Whenever the mutation operator decides to change a gene it makes sure the gene gets a different value. Therefore, we need only consider the chance the gene is left unchanged.

$$P(\text{chromosome unchanged}) = (1 - p_{mutation})^l = \left(\frac{l-k}{l}\right)^l \tag{1}$$

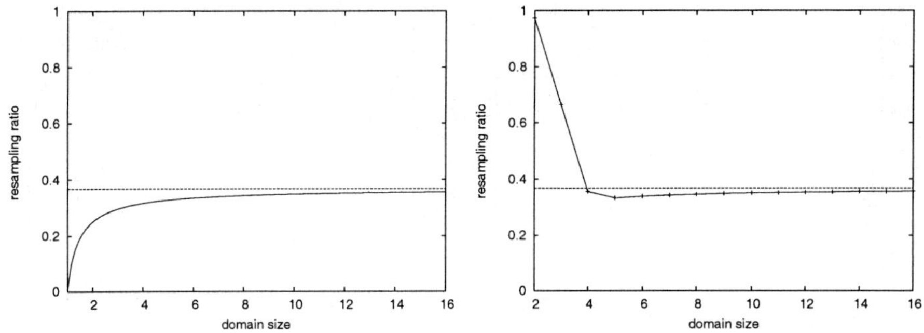

Fig. 1. Modelling the chance of not changing a chromosome (left). Simulation of the mutation operator with $p_{mutation} = 1/l$, thus $k = 1$. Every point is averaged over 25 independent runs (right)

If we plot (1) against the length of the chromosome and setting $k = 1$ and $m = 16$ we get Figure 1. It shows that the chance of not changing converges to approximately 0.37 when l increases. We can further show this convergence by looking at the limit when l approaches infinity:

$$\lim_{l \to +\infty} \left(\frac{l-k}{l} \right)^l = e^{-k} \tag{2}$$

For a mutation rate of $1/l$ we have $k = 1$ and that leads to:

$$P(\text{chromosome unchanged}) \approx 0.37,$$

which is very high. We propose to use a rate that allows less than 1% of unchanged chromosomes which we can achieve with $k = 5$ when l is large or a bit lower for smaller values of l.

5 Verifying the Simple Model

We want to verify our model from Section 4 by simulating the algorithm presented in Section 3. We vary the chromosome length over $2, \ldots, 16$ and let the algorithm run for 10,000 evaluations, setting $k = 1$ and $m = 16$. For every setting we do 25 independent runs with different random seeds.

If we plot the measured resampling ratio we get Figure 1 in which we see that indeed the number of revisited points in terms of ratio to the searched space approaches 0.37 when l increases. The error between the model and the measured values is quite low for a chromosome length higher than four.

The general outline of the figure corresponds with our model presented earlier except for small chromosome lengths. This behaviour can easily be explained if we examine the size of the state space in relation to the size of the searched space, as depicted in Figure 2. When the values for chromosome length become

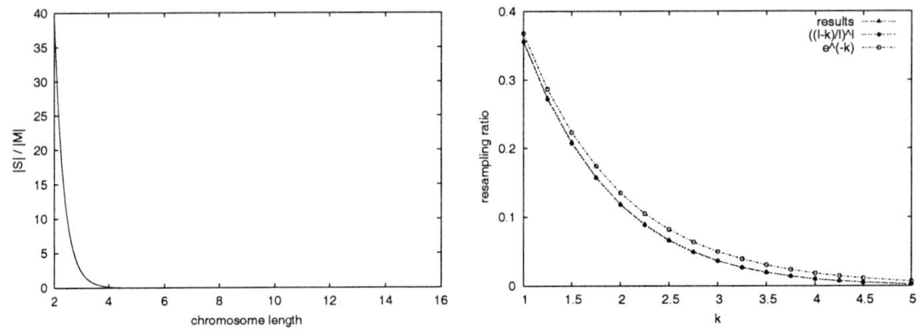

Fig. 2. Size of the searched space divided by the size of the state space (left). Varying k with $m = l = 16$ and 25 independent runs per reported result (right)

too small the size of the state space will become smaller than the number of generated points, which inevitably leads to more revisited points.

As a second test we fix m and l at 16 and vary the mutation rate $p_{mutation} = k/l$. We perform tests with $k = 1, \ldots, 5$ where for every setting of k we do 25 independent runs. Figure 2 shows the resampling ratio together with two estimations using our model (almost a perfect fit) and using the limit our model converges to.

6 Practical Evidence with CSPs

To test our hypothesis of a negative influence of the mutation rate when it is too small, we do a number of experiments on binary constraint satisfaction problems. We shall look at the behaviour of the performance and accuracy of an evolutionary algorithm when trying to solve increasingly difficult problems. We do eight experiments where we test all combinations of crossover on/off, parent selection on/off and survivor selection on/off. Each experiment will be repeated with five different mutation rates to examine the impact of k.

Constraint satisfaction problems (CSPs) form a class of models representing problems that have as common properties, a set of variables and a set of constraints. The variables should be instantiated from a discrete domain while making sure the constraints that restrict certain combinations of variable instantiations to exist, are satisfied. Examples of CSPs are k-graph colouring, 3-SAT and n-Queens.

In general, a CSP has a set of constraints where each constraint can be over any subset of the variables. Here, we focus on binary CSPs: a model that only allows constraints over a maximum of two variables. At first, this seems a restriction, but Tsang has shown [8] this is not the case proving that any CSP can be rewritten to a binary CSP. Solving the general CSP corresponds then to finding a solution for the binary form.

The last ten years many different approaches have been tried to solve constraint satisfaction problems using evolutionary algorithms. These include

Table 1. Features of the evolutionary algorithm that is used to solve binary CSPs. The features in bold font are varied in the experiments

feature	*value*
representation	integer
chromosome length (l)	number of variables in CSP
fitness	number of conflicts
population size	100
evolutionary model	steady-state
parent selection	linear ranked bias (2.0) [10] *or* randomly selected
survivor selection	replace worst *or* randomly selected
mutation operator	See Algorithm 1
mutation rate	varied k in k/l
crossover operator	uniform crossover *or* none
stop criterion	solution found or 100,000 evaluations

heuristics [3], adaptive schemes [5] and local search operators [6]. Here we fall back on the simplest approach, that is, an integer representation where each gene corresponds to one variable in the CSP. The gene can take any value from the variable's domain.

Our experiments consist of running our evolutionary algorithm with the features in Table 1 on a set of randomly created instances of binary CSPs. The table shows two selection steps, parent selection and survivor selection. The first is used to determine who is to be subjected to the genetic operators, while the second determines who is allowed to go to the next generation. These problem instances are created such that some are more difficult to solve than others. We create the test suite using the RandomCsp library [9] and choose Model E [1] as the process whereby the instances are constructed. The difficulty of the problem instances created by this process is controlled using a parameter p.

To verify the influence of different features of our evolutionary algorithm we perform eight experiments, where in each experiment we test five different mutation rates. Each test consists of many runs on many problem instances. Such a test is performed in a manner that gives insight into the behaviour of the algorithm when we use it to solve increasingly more difficult constraint satisfaction problems.

We use the same number of variables ($l = 15$) and the same domain size ($m = 15$) for each problem instance, which provides a state space with a size of 15^{15}. The parameter p determines the difficulty of an instance, which gives an overbounded estimate of the ratio of conflicts in an instance. It is varied from 0.10 to 0.38 [2] in steps of 0.02, thereby increasing the difficulty of finding solutions. At every step we create 25 instances and we let the evolutionary algorithm do 10 independent runs on each instance. This totals the number of runs for each experiment to 5 mutation-rates $*$ 15 steps $*$ 25 instances $*$ 10 runs $= 18,750$ runs.

[2] The value 0.38 is close to the point beyond which we can no longer produce solvable problem instances.

no crossover, no parent selection, survivor selection

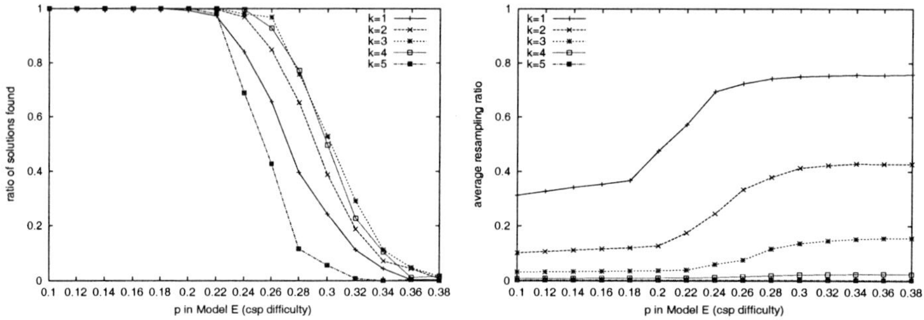

no crossover, parent selection, no survivor selection

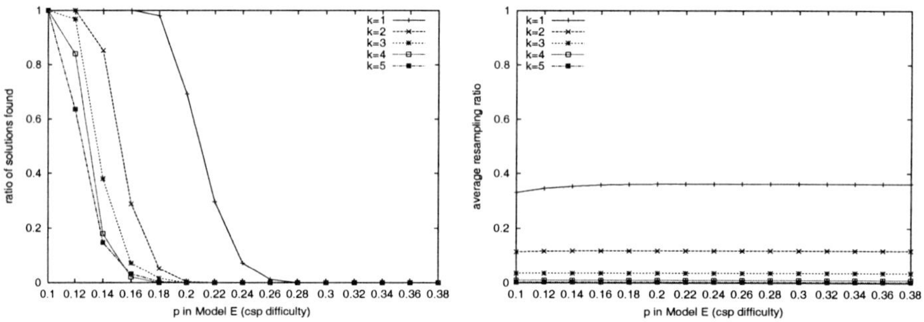

no crossover, parent selection, survivor selection

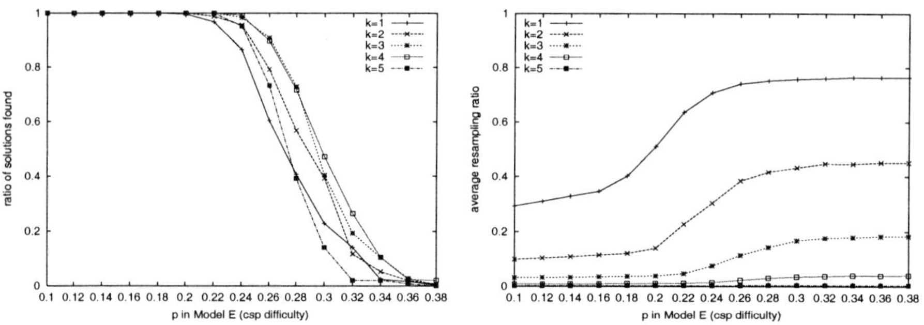

crossover, no parent selection, survivor selection

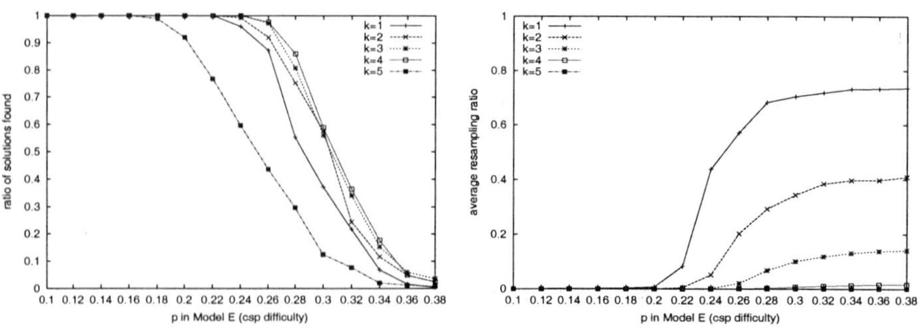

crossover, parent selection, no survivor selection

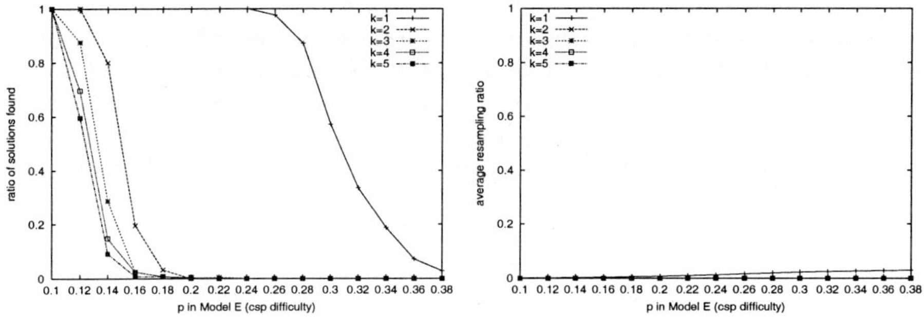

crossover, parent selection, survivor selection

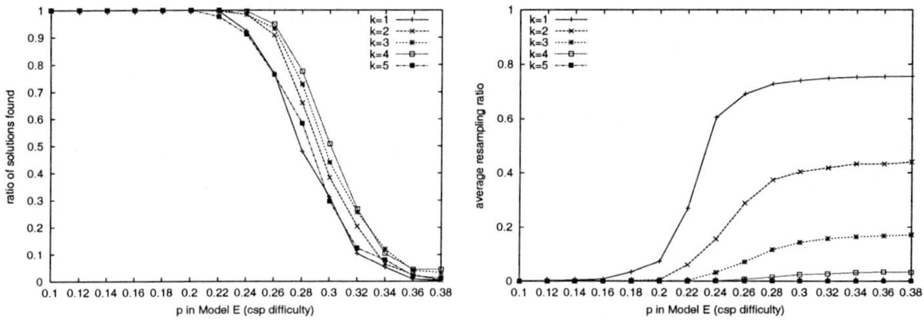

During each run we measure the success rate and the resampling ratio. We present averages of these measures per experiment per setting of the mutation rate. The preceding figures are presented as follows: every row is one experiment, with two graphs, the average success rates and the average resampling ratio. Every figure contains five plots, one for each of the mutation rate settings $p_{mutation} = k/l$.

In two experiments the success rate drops to zero at $p = 0.18$, both of these have no selection methods. We leave out the results of these experiments, but we mention a difference between the resampling ratio in both experiments. With crossover the resampling ratio drops to zero for all settings of k, while without crossover we get the values predicted by the simple model.

The four experiments that have survivor selection show good performance in success rate. In all of these the best mutation rates are either $k = 3$ or $k = 4$. Using (1) and the fact that $l = 15$ we know that P(chromosome unchanged) is 0.0352 and 0.0095 for these two settings of k, very close to our hypothesis.

When using parent selection and not using survivor selection, the best mutation rate is $k = 1$. Furthermore, performance drops significantly when k is increased. The effect of crossover is mainly visible where we only have parent selection. There it considerably boosts performance on success rate. At the same time we witness a very low resampling ratio.

Whenever we see a good performance in success rate, we measure a low resampling ratio. However, when we look at the experiment with crossover and without any selection, we observe a low resampling ratio, but a very poor performance in success rate.

7 Conclusions

We have presented a simple measuring tool that measures independently of the evolutionary algorithm or any other algorithm that works by a repeated process of generating points in the state space. Using this tool and a theoretical model we have analysed a well known standard mutation strategy, moreover we have found that this strategy might lead to inefficient behaviour in a simplified evolutionary algorithm without selection.

To test our tool's usefulness in practical environments, we have performed a study on solving binary constraint satisfaction problems. Here the outcome is that selection and mutation rate, when wrongly balanced, can have a devastating effect on the performance. If the selection is strong and the mutation rate too weak, we observe a high resampling ratio, which means that only few points in the state space have been generated. This is a good indication of what is causing the low performance.

Although the resampling ratio might be a good way of explaining what goes wrong, it is not suitable for detecting when we may expect good performance. Thus maintaining a healthy resampling ratio is not a guarantee for a successful evolutionary algorithm in itself.

References

[1] D. Achlioptas, L.M. Kirousis, E. Kranakis, D. Krizanc, M.S.O. Molloy, and Y.C. Stamatiou. Random constraint satisfaction a more accurate picture. In G. Smolka, editor, *Principles and Practice of Constraint Programming — CP97*, pages 107–120. Springer-Verlag, 1997.

[2] T. Bäck. The interaction of mutation rate, selection, and self-adaptation within a genetic algorithm. In R. Manner and B. Manderick, editors, *Parallel Problem Solving from Nature II*, pages 85–94. Springer, 1992.

[3] S.H. Clearwater and T. Hogg. Problem structure heuristics and scaling behavior for genetic algorithms. *Journal of Artificial Intelligence*, 81:327–347, 1996.

[4] A.E. Eiben, E.H.L. Aarts, , and K.M. Van Hee. Global convergence of genetic algorithms: an infinite markov chain analysis. In *Proceedings of the First International Conference on Parallel Problem Solving from Nature*, pages 4–12. Springer, Berlin, 1991.

[5] A.E. Eiben, J.K. van der Hauw, and J.I. van Hemert. Graph coloring with adaptive evolutionary algorithms. *Journal of Heuristics*, 4(1):25–46, 1998.

[6] E. Marchiori and A. Steenbeek. A genetic local search algorithm for random binary constraint satisfaction problems. In *ACM Symposium on Applied Computing*, pages 458–462, 2000.

[7] H. Mühlenbein. How genetic algorithms really work: Mutation and hill-climbing. In R. Manner and B. Manderick, editors, *Parallel Problem Solving from Nature II*, pages 15–25. Springer, 1992.

[8] E. Tsang. *Foundations of Constraint Satisfaction*. Academic Press, 1993.

[9] J.I. van Hemert. *Documentation of the RandomCsp library*. Leiden University, randomcsp version 1.5 edition, 2001. Available from `http://www.liacs.nl/~jvhemert/randomcsp`.

[10] D. Whitley. The GENITOR algorithm and selection pressure: Why rank-based allocation of reproductive trials is best. In J. David Schaffer, editor, *Proceedings of the Third International Conference on Genetic Algorithms (ICGA'89)*, pages 116–123, San Mateo, California, 1989. Morgan Kaufmann Publishers, Inc.

On the Analysis of Dynamic Restart Strategies for Evolutionary Algorithms

Thomas Jansen

George Mason University, Fairfax, VA 22030, USA
tjansen@gmu.edu

Abstract. Since evolutionary algorithms make heavy use of randomness it is typically the case that they succeed only with some probability. In cases of failure often the algorithm is restarted. Of course, it is desirable that the point of time when the current run is considered to be a failure and therefore the algorithm is stopped and restarted is determined by the algorithm itself rather than by the user. Here, very simple restart strategies that are non-adaptive are compared on a number of examples with different properties. Circumstances under which specific types of dynamic restart strategies should be applied are described and the potential loss by choosing an inadequate restart strategy is estimated.

1 Introduction

Evolutionary algorithms (EAs) are randomized search heuristics applicable to a wide range of problems. We concentrate on optimization here though we do not think that the considerations presented are limited to that case. Due to the random elements employed in EAs one typically observes behavior that varies from run to run. In order to increase the probability of finding a good or even optimal solution of the considered optimization problem, it is common practice to do restarts. Then the choice of the point of time when the EA is stopped and restarted is crucial. In principle, one can distinguish three different classes of ways to choose a restart strategy just like for other parameters in EAs, e. g., mutation probability or population size [1]. The first class is a static setting where a certain number of steps is fixed in advance. Then one can discuss whether it is better to spend a lot of time in one long run or in multiple short runs [10]. Unlike other parameters such a static decision is unusual for restarting. The second class comprises dynamic restart strategies. Here, the point of time for restarts is determined by some fixed schedule that depends on the number of steps counted over all runs or, equivalently, on the number of steps in the current run and the sum of number of steps in all previous runs. In spite of the increased flexibility compared to static choices dynamic restart strategies are seldom used in EAs. A practical example proving the usefulness of dynamic, yet non-adaptive, strategies are cooling schedules used for simulated annealing [9]. What is common practice for determining the point of time for restarts are strategies from the third class, namely adaptive restart strategies. Here, the restart strategy may depend on

J.J. Merelo Guervós et al. (Eds.): PPSN VII, LNCS 2439, pp. 33–43, 2002.

the complete history of the current and all previous runs. There are various different implementations ranging from quite simple ones to quite sophisticated statistical approaches [7]. We note that for EAs it is not necessary to restart the algorithm completely. Partial restarts in the sense of re-initializing parts of the population are possible [11]. Since we concentrate on restart strategies and not on the underlying search algorithm, we will not discuss such mixed strategies.

We consider dynamic restart strategies, although they are not that commonly used. We do this for several good reasons. Obviously, the three classes of strategies form a hierarchy: static choices can be seen as degenerated dynamic ones, dynamic strategies can be described as degenerated adaptive ones. Note that self-adaptive strategies are a special case of adaptive strategies. Considering this hierarchy we believe that it is sensible to prove results on the simpler cases before analyzing the more complex classes. Second, due to their simpler structure such strategies are analyzable and allow for rigorous proofs of non-trivial statements. Third, it will turn out that already simple dynamic restart strategies can increase the performance of an EA enormously. Finally, typical adaptive restart strategies explicitly express some assumptions about the objective function and are therefore relatively easy to fool by constructing functions that do not meet these assumptions. In fact, we will present one example where dynamic restart strategies can be far superior to adaptive strategies that wait until the EA "gets stuck" before restarting it.

Next we discuss briefly the baseline of theoretical analysis of EAs and build our model in accordance to this. In Sect. 3 we present the EA we use to exemplify our restart strategies and the different dynamic restart strategies we analyze. In Sect. 4 we describe our example functions and motivate our choices. In Sect. 5 we derive results on one single run of the EA considered. We use these results for the comparison of the dynamic restart strategies in Sect. 6.

2 Performance Analysis of EAs

We consider exact optimization of static pseudo-Boolean functions $f\colon \{0,1\}^n \to \mathbb{R}$. We assume that f is defined for all $n \in \mathbb{N}$ (or at least infinitely many) and are interested in the asymptotic behavior of an EA on f for n growing to infinity. We neglect the choice of a stopping criterion and assume that the EA eventually finds some global optimum with probability 1 within some number of steps as in fact many EAs do [15]. Formally, let $p_{f,t}(n)$ denote the probability that some fixed EA optimizes an objective function $f\colon \{0,1\}^n \to \mathbb{R}$ in at most t steps. For all choices $p_{f,t}(n)$ is monotonic increasing with t. We consider only EAs where $p_{f,t}(n)$ is strongly monotonic increasing and $\lim\limits_{t\to\infty} p_{f,t}(n) = 1$ holds for all f.

One typical measure is the expected optimization time, i. e., the expected number of steps until some global optimum is found. Sometimes a function f is typically efficiently optimized by some EA whereas in very rare cases it needs extremely long to optimize f. The expected optimization time may be exponential in this case, misleadingly indicating that f is difficult for this EA. However, when applying an appropriate restart strategy, this cannot happen: Assume that after some number of steps T_s the EA has success probability $p_{f,T_s}(n)$ where T_s is not too large and $p_{f,T_s}(n)$ is not too small. Then a restart strategy stopping and restarting the EA after T_s steps leads to an acceptable expected optimization time of $T_s/p_{f,T_s}(n)$.

Note, that in our approach we do not pose some limit on the computational cost that optimization may take. This is fundamentally different from studies where the computational cost is fixed and the goal is to find a restart strategy that maximizes the expected gain while respecting this limit [2, 5, 10]. Our approach is more similar to the analysis of algorithms, where one is interested in the run time of an algorithm needed to perform a well-defined task without imposing a predefined limit of the time the algorithm may spend.

3 Algorithmic Framework

In principle, any EA can serve as basis for the comparison. In order to allow for an easy comparison it is desirable that the EA itself is well understood and easy to analyze. However, it should exhibit a behavior that can be assumed to be typical for EAs. This rules out extremely simple constructs like pure random search that show no similarity to common EAs. We consider the (1+1) EA to be a good compromise. It uses a population of size 1, bit-wise mutation with mutation probability $1/n$, and a deterministic selection, known as plus-selection from evolution strategies. This algorithm is often studied and well-understood [3, 6, 14, 15]. It is known that it often performs at least comparable to much more complicated and sophisticated EAs [8, 12]. We do not claim that the (1+1) EA behaves like more sophisticated EAs that may use crossover. But there are objective functions where more sophisticated EAs behave like the (1+1) EA on the classes of functions we introduce in the next section. Then, the restart strategies we investigate will have the same effects on the EAs as they have on the (1+1) EA. Thus, simplifying the analysis by choosing a simple EA does not restrict the applicability of the obtained results.

Algorithm 1 ((1+1) EA with dynamic restarts)

1. $r := 0$
2. $r := r+1$; $t_r := 0$ Choose $x \in \{0,1\}^n$ uniformly at random.
3. Create y from x by independently for each bit flipping it with probability $1/n$.
4. If $f(y) \geq f(x)$, then $x := y$.
5. $t_r := t_r + 1$
6. If restart$(r) = t_r$, then continue at 2 else continue at 3.

Let T denote the minimal value of $\sum t_i$ when x is some global maximum. Then, we analyze the expected optimization time $\mathrm{E}(T)$. We concentrate on restart strategies where the length of intervals between two consecutive points of time when a restart is done is increasing, thus each run not being shorter than the preceeding run. This seems reasonable since long runs are a potential waste of time when already short runs have a good success probability. We consider two types of increasing dynamic restart strategies to be of special interest.

Definition 1. *An* additive restart strategy *restart*$_a$: $\mathbb{N} \rightarrow \mathbb{N}$ *is defined by* $s_0, s \in \mathbb{N}$ *and is given by* *restart*$_a(r) := s_0 + (r-1) \cdot s$. *A* multiplicative restart strategy *restart*$_m$: $\mathbb{N} \rightarrow \mathbb{N}$ *is defined by* $s_0 \in \mathbb{N}$, $s \in \mathbb{R}$ *with* $s > 1$ *and is given by* *restart*$_m(r) := \lceil s_0 \cdot s^{r-1} \rceil$.

In order to further simplify our considerations we investigate one special additive restart strategy r_a and one special multiplicative restart strategy r_m, only. We believe that important properties of additive and multiplicative restart strategies in general are captured by these two examples.

Definition 2. *The additive restart strategy* r_a *is defined by* $s_0 = n \log n$, $s = n \log n$. *The multiplicative restart strategy* r_m *is defined by* $s_0 = n \log n$, $s = 2$.

The length of the first run of the (1+1) EA is $s_0 = n \log n$ for both strategies. This choice is motivated by the fact that optimization of any function that has a unique global optimum takes $\Omega(n \log n)$ steps on average [3]. Therefore, we consider it to be counterproductive to stop the (1+1) EA earlier.

4 Example Functions

We want to compare dynamic restart strategies in different situations. Thus we are interested in functions exemplifying different situations as clearly as possible. This can be done best with artificial, constructed objective functions. We are only interested in functions where at least a perfect restart strategy can achieve polynomial expected optimization time. Note, though, that the choice of an inappropriate restart strategy can cause an overall inefficient optimization.

Definition 3. *For* $n, k \in \mathbb{N}$ *with* $k = O(1)$ *let the functions* $\mathrm{SP}_k\colon \{0,1\}^n \rightarrow \mathbb{R}$, $\mathrm{SP2}_k\colon \{0,1\}^n \rightarrow \mathbb{R}$, *and* $\mathrm{HSP}\colon \{0,1\}^n \rightarrow \mathbb{R}$ *be defined by*

$$\mathrm{SP}_k(x) := \begin{cases} n \cdot (i+1) \text{ if } x = 1^i 0^{n-i} \text{ with} \\ \qquad\qquad i \in \{0, k, 2k, \ldots, \lceil n/(3k) \rceil k\}, \\ n - \sum\limits_{i=1}^{n} x_i \text{ otherwise,} \end{cases}$$

$$\mathrm{SP2}_k(x) := \begin{cases} n \cdot (i+1) \text{ if } \left(x = 1^i 0^{n-i}\right) \vee \left(x = 0^{n-i} 1^i\right) \text{ with} \\ \qquad\qquad i \in \{0, k, 2k, \ldots, (\lceil n/(3k) \rceil - 1) k\}, \\ n \cdot (n+1) \text{ if } x = 1^i 0^{n-i} \text{ with } i = \lceil n/(3k) \rceil k, \\ n - \sum\limits_{i=1}^{n} x_i \text{ otherwise.} \end{cases}$$

$$\mathrm{HSP}_k(x) := \begin{cases} n \cdot (i+1) \text{ if } x = 1^i 0^{n-i} \\ \qquad\qquad \text{with } i \in \{0, k, 2k, \ldots, \lceil n/(3k) \rceil k\}, \\ n + \sum\limits_{i=1}^{n} x_i \text{ if } x_1 = x_2 = \cdots = x_{\lceil 2n/3 \rceil} = 0, \\ n - \sum\limits_{i=1}^{n} x_i \text{ otherwise.} \end{cases}$$

For SP_k there is a kind of short path starting in 0^n. All points on the path are of the form 1^i0^{n-i} and are formed of $\lceil n/(3k) \rceil$ consecutive blocks of length k. If a block consists only of 1-bits and this holds for all blocks to its left, too, this is rewarded by n. For all points not on the path, each 1-bit in the string reduces the function value by 1 making the first point of the path easy to find. Obviously, the expected optimization time $\mathrm{E}(T)$ will increase with k.

$\mathrm{SP2}_k$ is very similar to SP_k but with two short paths, both starting in 0^n. One of the form 1^i0^{n-i} leading to the unique global optimum, the other of the form 0^i1^{n-i} leading to a local optimum. For symmetry reasons, the (1+1) EA will reach the local optimum in about half of the runs. Then, a direct mutation of at least $(2/3)n - k$ bits is necessary to reach the optimum. Thus, the (1+1) EA is very efficient in about half of the runs and very inefficient in about the other half of the runs and thus very inefficient on average.

HSP_k has a short path similar to the function SP_k which is easy to find. However, since once being at the beginning of the path it is much more likely to go away from the path then to follow it, we expect that the success probability $p_{\mathrm{HSP}_k,t}(n)$ is rather small for each $t = n^{O(1)}$.

5 Analysis Without Restarts

First, we investigate the behavior of the (1+1) EA without restarts. Results on the success probability $p_{f,t}(n)$ are useful when analyzing restart strategies. The more precise the upper and lower bounds on the success probability are, the more precise statements on the expected optimization time of the (1+1) EA in combination with different restart strategies can be concluded.

Theorem 1. *There exist two constants $c_1, c_2 \in \mathbb{R}^+$, such that given $n, k \in \mathbb{N}$ with $k = O(1)$ for the (1+1) EA on $\mathrm{SP}_k\colon \{0,1\}^n \to \mathbb{R}$ the following holds:*

$$\forall t \leq c_1 \cdot n^{k+1}\colon p_{\mathrm{SP}_k,t}(n) = e^{-\Omega(n)}$$
$$\forall t \geq c_2 \cdot n^{k+1}\colon p_{\mathrm{SP}_k,t}(n) = 1 - e^{-\Omega(n)}$$

Proof. We consider a run and divide it into two disjoint phases. The first phase starts with the random initialization and ends when the current string x for the first time is of the form 1^i0^{n-i} with $i \in \{0, k, 2k, \ldots, \lceil n/(3k) \rceil k\}$. The second phase starts when the first phase ends and ends when the current string x is equal to the unique global optimum. We claim that the probability that the first phase does not end within the first $\lceil 2en^2 \rceil$ steps is bounded above by $e^{-n/4}$. We ignore steps ending the first phase by leading to some 1^i0^{n-i} with $i > 0$. Obviously, this can only increase the probability of not ending the first phase. The current string x has Hamming distance $d(x)$ to the all zero string 0^n with $d(x) > 0$. The probability that the child y has Hamming distance at most $d(x)-1$ to 0^n is bounded below by $(1/n)(1-1/n)^{n-1} \geq 1/(en)$, since it is sufficient to mutate exactly one 0-bit in x. Such a child y has larger function value and will replace x. After at most n such replacements the Hamming distance is reduced to 0 and the first phase ends. Thus we are in the situation that we make random experiments where each experiment is a success with probability at least $1/(en)$. The expected number of success in $\lceil 2en^2 \rceil$ trials is lower bounded by $2n$. By Chernoff bounds [13] the probability not to have at least n successes in $\lceil 2ne^2 \rceil$ trials is bounded above by $e^{-n/4}$.

We claim that the probability that the second phase does not end within the first $\lceil 2en^k \lceil n/(3k) \rceil k \rceil$ steps of this phase is bounded above by $e^{-n/12}$. In the second phase mutating exactly at most k 0-bits in the first block in x that contains 0-bits yields a child y that replaces x. The probability of such a mutation is bounded below by $(1/n)^k(1-1/n)^{n-k} \geq 1/\left(en^k\right)$. After at most $\lceil n/(3k) \rceil k$ such steps the global optimum is found. We argue as above and see that the probability not to have at least $\lceil n/(3k) \rceil k$ such steps within $2en^k \lceil n/(3k) \rceil k$ steps is bounded above by $e^{-\lceil n/(3k) \rceil k/4}$. Together this proves the second statement.

We now know that the number of different points in the search space encountered before x becomes some point $1^i 0^{n-i}$ with $i \in \{0, k, 2k, \ldots, \lceil n/(3k) \rceil k\}$ is bounded above by $\lceil 2en^2 \rceil$ with probability $1 - e^{-n/4}$. Thus, the probability that for the first x of such form $i \geq n/12$ holds is bounded above by $2en^2 \cdot (n/4)/\binom{n}{n/12} < e^{-n/12}$. Thus, with probability at least $1 - e^{-n/12}$ at least $n/4$ blocks that are all zero have to become all one. In a mutation that flips one all zero block thereby generating an all one block, other blocks might flip as well. However, the probability that in such a step l additional blocks become all one is bounded above by $(1/n)^{kl}$. Thus, we expect in $n/9$ such steps in total less than $(n/9) + (1/n)^{kl-1}$ blocks to become all one. Due to Chernoff bounds, the probability that in total more then $(2n/9) + 2(1/n)^{kl-1} < n/4$ blocks become all one is bounded above by $e^{-n/25}$. One such step has probability at most $1/n^k$. Thus, we expect to have at most $n/20$ such mutations in $n^{k+1}/20$ steps. Again by Chernoff bounds, the probability to have at least $n/10$ such mutations in $n^{k+1}/20$ steps is bounded above by $e^{-19n/1000}$. This implies that with probability $1 - e^{-9n/500}$ after $n^{k+1}/20$ steps the global optimum is not reached. □

Similarities between SP_k and $\mathrm{SP2}_k$ yield a similar result and allow us to reuse parts of the proofs of Theorem 1 in the proof of the next theorem.

Theorem 2. *There exist two constants* $c_1, c_2 \in \mathbb{R}^+$, *such that given any* $n, k \in \mathbb{N}$ *with* $k = O(1)$ *for the (1+1) EA on* $\mathrm{SP2}_k \colon \{0,1\}^n \to \mathbb{R}$ *the following holds:*

$$\forall t \leq c_1 \cdot n^{k+1} \colon p_{\mathrm{SP2}_k,t}(n) = e^{-\Omega(n)}$$
$$\forall t \geq c_2 \cdot n^{k+1} \colon p_{\mathrm{SP2}_k,t}(n) = 1/2 - O(1/n)$$
$$\forall t \leq n^{n/4} \colon p_{\mathrm{SP2}_k,t}(n) = 1/2 - \Omega(1/n)$$

Proof. The first statement can be proved similarly to the proof of the first statement in Theorem 1. The proof of the second statement is similar to the proof of the second statement of Theorem 1. We extend the first phase until the first point of time when the (1+1) EA reaches some x with $\mathrm{SP2}_k(x) > \mathrm{SP2}_k(0^n)$. Since the path $1^i 0^{n-i}$ contains one more element then the path $0^i 1^{n-i}$, the first x on the path belongs to the $1^i 0^{n-i}$ with probability at least $1/2$. We consider one step where x changes and want to estimate the probability that the path is left in favor of the other path. Assume that $x = 1^i0^{n-i}$ is the current string. Then at least $2i$ bits have to mutate simultaneously in order to change the path. The probability for such a mutation is bounded above by $1/n^{2i}$. In order to reach the next point on the path it is sufficient to mutate at most k bits in the next block that contains some 0-bits. The probability for such a mutation is bounded below by $(1/n)^k(1-1/n)^{n-k} \geq 1/\left(en^k\right)$. The value of i depends on the current position of the path. It is at least k in the beginning and is increased by at least k after each such step. Thus, the probability to leave the path before reaching the optimum is bounded above by $2e/n$.

The proof of the third statement is a combination of the proof of the second statement in this proof and of the second statement in the proof of Theorem 1. The path 0^i1^{n-i} is entered and not left with probability $(1/2) - O(1/n)$. Once the local optimum $0^{(\lceil n/(3k)\rceil-1)k}1^{n-(\lceil n/(3k)\rceil-1)k}$ is reached, a mutation of at least $(2/3)n - k$ bits is necessary. Such a mutation has probability at most $n^{-(2/3)n-k}$. Thus, the probability that such a mutations happens within $n^{n/4}$ steps is bounded above by $n^{(n/4)-(2/3)n+k} \leq n^{-n/3}$. □

HSP_k and SP_k are similar, too. But in HSP_k the Hamming distance to the path can be increased by having 1-bits in the right third of x. Thus, it is possible not to reach the path at all. This implies important differences.

Theorem 3. *There exist two constants $c_1, c_2 \in \mathbb{R}^+$, such that given any $n, k \in \mathbb{N}$ with $k = O(1)$ for the (1+1) EA on $\mathrm{HSP}_k\colon \{0,1\}^n \to \mathbb{R}$ the following holds:*

$$\forall t \leq c_1 \cdot n^{k+1}\colon p_{\mathrm{HSP}_k,t}(n) = e^{-\Omega(n)}$$
$$\forall t \geq c_2 \cdot n^{k+1}\colon p_{\mathrm{HSP}_k,t}(n) = \Omega\left(1/n^k\right)$$
$$\forall t \leq n^{n/4}\colon p_{\mathrm{HSP}_k,t}(n) = 1 - O\left(1/n^k\right)$$

Proof. The first statement can be proved similarly to the proof of the first statement in Theorem 1. For the second and third statement it is crucial to give estimations of the probability for reaching the path 1^i0^{n-i}. We know from the proof of Theorem 1 that it is unlikely to find the path far from its beginning in 0^n. From the ballot theorem [4] it follows that in fact with probability $\lceil 2n/3\rceil/n$ the all zero string 0^n is found. Now, let as assume that the current string x is 0^n. Then, the next child that replaces its parent can either be a point 1^i0^{n-i} on the path or some point y different from 0^n with $y_1 = y_2 = \cdots = y_{\lceil 2n/3\rceil} = 0$. Note, that once a path point is reached all subsequent current strings can only be path points. Obviously, the probability to reach a path point is bounded above by $n^{-k} + n^{-2k} + n^{-3k} + \cdots = O\left(1/n^k\right)$ while we reach some other point with probability at least $(1/3)(1-1/n)^{n-1} = \Omega(1)$. On the other hand, the probability to reach the path is bounded below by $(1/n)^k(1-1/n)^{n-k} = \Omega\left(1/n^k\right)$. Thus, the first point reached is on the path with probability $\Theta\left(1/n^k\right)$. Once on the path the path is never left and we conclude from the proof of Theorem 1 that with probability $1 - e^{-\Omega(n)}$ within $\Theta\left(n^{k+1}\right)$ steps we reach the global optimum.

For the third statement it is sufficient to note that the Hamming distance to points on the path can only increase. In particular, as long as the Hamming distance to the path is bounded above by $n/6$, the Hamming distance is increased with probability $\Omega(1)$ in each step. Thus with probability $1 - O\left(1/n^k\right)$ the point $0^{\lceil 2n/3 \rceil}1^{\lfloor n/3 \rfloor}$ is reached before reaching any point on the path. Then, a mutation of at least $\lfloor n/3 \rfloor$ bits simultaneously is necessary to reach the path and the third statement follows similar to the proof of Theorem 2. □

6 Comparison of Dynamic Restart Strategies

Now we apply the results from the previous section to obtain results for the (1+1) EA with restarts. These results represent the behavior of such dynamic restart strategies for any EA with a success probability converging to 1, converging to some positive constant, and converging to 0 polynomially fast, like the (1+1) EA on SP_k, $\mathrm{SP2}_k$, and HSP_k, respectively.

Theorem 4. *For $n, k \in \mathbb{N}$ with $k = O(1)$ the (1+1) EA with dynamic restart strategy r_m has expected optimization time $E(T) = \Theta\left(n^{k+1}\right)$ on SP_k.*

Proof. The lower bound follows from Theorem 1. For the upper bound we consider the first uninterrupted run of length at least $c_2 n^{k+1}$ where c_2 is the constant from Theorem 1. After $\lceil (k \log n) - (\log\log n) + \log c_2 \rceil$ restarts we have one run of at least this length. The expected number of such runs before the global optimum is found is bounded above by 2. Thus, $\mathrm{E}(T)$ is bounded above by $\sum_{i=0}^{\lceil (k \log n) - (\log\log n) + \log c_2 \rceil + 1} 2^i \cdot n \log n \leq 8c_2 \cdot n^{k+1}$. □

The order of growth of $\mathrm{E}(T)$ with r_m is optimal. Thus, for an easy to optimize function it is good to have a restart strategy quickly increase the length of runs.

Theorem 5. *For $n, k \in \mathbb{N}$ with $k = O(1)$ the (1+1) EA with dynamic restart strategy r_a has expected optimization time $E(T) = \Theta\left(n^{2k+1}/\log n\right)$ on SP_k.*

Proof. We can proof the upper bound similar to the proof of Theorem 4. After $\lceil c_2 n^k/(\log n) \rceil$ restarts, we have a run of length at least $c_2 n^{k+1}$. Thus, $\mathrm{E}(T)$ is bounded above by $\sum_{i=1}^{\lceil c_2 n^k/(\log n) \rceil + 1} i \cdot n \log n < 2c_2^2 \cdot n^{2k+1} \log n$.

For the lower bound we consider the first $\lfloor c_1 n^k/(\log n) \rfloor$ runs of the (1+1) EA. All runs have length at most $\lfloor c_1 n^k/(\log n) \rfloor \cdot n \log n \leq c_1 n^{k+1}$. Thus, all runs have success probability $e^{-\Omega(n)}$. The total length of theses runs is at least $\sum_{i=1}^{\lfloor c_1 n^k/(\log n) \rfloor} i \cdot n \log n = \Omega\left(\frac{n^{2k+1}}{\log n}\right)$. □

In comparison with r_m the restart strategy r_a performs poorly. This is due to the long time that is needed to have runs of sufficient length.

Theorem 6. *For $n, k \in \mathbb{N}$ with $k = O(1)$ the (1+1) EA with dynamic restart strategy r_m has expected optimization time $E(T) = \Theta\left(n^{k+1}\right)$ on $\mathrm{SP2}_k$.*

Proof. The proof can be done in the same way as the proof of Theorem 4. □

Theorem 7. *For $n, k \in \mathbb{N}$ with $k = O(1)$ the (1+1) EA with dynamic restart strategy r_a has expected optimization time $E(T) = \Theta\left(n^{2k+1}/\log n\right)$ on* SP2_k.

Proof. The proof can be done in the same way as the proof of Theorem 6. □

Using additive or multiplicate restart strategies, it makes no difference for the order of growth of $\mathrm{E}(T)$ whether the success probability converges to 1 or to some positive constant.

Theorem 8. *For $n, k \in \mathbb{N}$ with $k = O(1)$ the (1+1) EA with dynamic restart strategy r_m has expected optimization time $E(T) = 2^{\Omega(n/\log n)}$ on* HSP_k.

Proof. Due to Theorem 3, on average $\Theta\left(n^k\right)$ runs of length $> c_1 n^{k+1}$ are needed to optimize HSP_k. After $n/\log n$ restarts the last run has length $2^{n/\log n} n \log n$. Since this is $o\left(n^{n/4}\right)$, the success probability for each run is still $o\left(1/n^k\right)$. So, the optimum is not found after these restarts with probability $1 - o(1)$. □

Theorem 9. *For $n, k \in \mathbb{N}$ with $k = O(1)$ the (1+1) EA with dynamic restart strategy r_a has expected optimization time $E(T) = \Theta\left(n^{2k}\log n\right)$ on* HSP_k.

Proof. The proof of the upper bound follows roughly the same lines as the proof of the upper bound of Theorem 5. After $\lceil c_2 n^k/(\log n)\rceil$ restarts the length of the run (and all following runs) is bounded below by $c^2 n^{k+1}$. Thus, this run and all following runs have success probability $\Theta\left(1/n^k\right)$ according to Theorem 3. Thus, the expected number of runs that is needed on average is bounded above by $O\left(n^k + n^k/(\log n)\right) = O\left(n^k\right)$. Then the total length of all runs is $O\left(n^{2k}\log n\right)$.

The lower bound can be proved in the same way. What is needed additionally is the third statement from Theorem 3. As in the proof of Theorem 8 we see from there that the steps taken after the $(c_2 n^{k+1})$-step in all runs do not add something significant to the success probability since we are considering only a polynomial number of runs and a polynomial number of steps. □

Compared to r_a the strategy r_m performs very poorly. This is due to the fact that the length of the runs increases so quickly. Thus, for difficult functions where the probability of finding an optimum at all is very small even for quite long runs additive restart strategies are preferred.

7 Conclusions

We presented a framework for the evaluation of restart strategies different from other settings: We compared the expected optimization time instead of fixing the computational cost in advance. We restricted ourselves to dynamic restart strategies, which are more complex than fixing some point of time for a restart but less complex than adaptive restart strategies. We know that in applications adaptive restart strategies are "state of the art." But we believe that it makes sense to begin theoretical investigations with a simpler and still interesting class.

We described two classes of dynamic restart strategies and investigated one additive strategy and one multiplicative strategy in detail. We employed the simple (1+1) EA as the underlying search algorithm and presented three classes of example functions with very different success probabilities. The time that is typically spent optimizing these examples can be controlled via a parameter k.

We saw that it made almost no difference whether the success probability within a polynomial number of steps is near to 1 or some other positive constant. In both cases r_m, which quickly increases the length of each run, is superior. However, if the success probability stays close to 0 very long, r_a is by far superior. For r_a the expected optimization is polynomial, for r_m it is exponential. It is easy to see that all functions that can be optimized with multiplicative restart strategies in expected polynomial time can be optimized with additive restart strategies in expected polynomial time, but the degree of the polynomial can be larger. Here the quotient was of order $\Theta\left(n^k/\log n\right)$, with k any positive integer.

We saw that considering expected optimization time is reasonable when comparing restart strategies. The investigation of adaptive restart strategies within this framework is subject to future research. Also open is the empirical comparison of r_a and r_m not only on our examples but also in practical settings.

Acknowledgments

The author thanks Paul Wiegand for helpful discussions. The author was supported by a fellowship within the post-doctoral program of the German Academic Exchange Service (DAAD).

References

1. Th. Bäck. An overview of parameter control methods by self-adaptation in evolutionary algorithms. *Fundamenta Informaticae*, 35:51–66, 1998.
2. E. Cantú-Paz. Single vs. multiple runs under constant computation cost. In *Proc. of he Genetic and Evolutionary Computation Conf. (GECCO 2001)*, page 754. Morgan Kaufmann, 2001.
3. S. Droste, Th. Jansen, and I. Wegener. On the analysis of the (1+1) evolutionary algorithm. CI 21/98, SFB 531, Univ. Dortmund, 1998. To appear in: TCS.
4. W. Feller. *An Introduction to Probability Theory and Its Applications.* Wiley, 1968.
5. A. S. Fukunaga. Restart scheduling for genetic algorithms. In *Parallel Problem Solving from Nature (PPSN V)*, LNCS 1498, pages 357–366. Springer, 1998.
6. J. Garnier, L. Kallel, and M. Schoenauer. Rigorous hitting times for binary mutations. *Evolutionary Computation*, 7(2):173–203, 1999.
7. M. Hulin. An optimal stop criterion for genetic algorithms: A Bayesian approach. In *Proc. of the Seventh International Conf. on Genetic Algorithms (ICGA '97)*, pages 135–143. Morgan Kaufmann, 1997.
8. A. Juels and M. Wattenberg. Hillclimbing as a baseline method for the evaluation of stochastic optimization algorithms. In *Advances in Neural Information Processing Systems 8*, pages 430–436. MIT Press, 1995.
9. S. Kirkpatrick, C. D. Gelatt, and M. P. Vecchi. Optimization by simulated annealing. *Science*, 220:671–680, 1983.
10. S. Luke. When short runs beat long runs. In *Proc. of he Genetic and Evolutionary Computation Conf. (GECCO 2001)*, pages 74–80. Morgan Kaufmann, 2001.

11. J. Maresky, Y. Davidor, D. Gitler, Gad A., and A. Barak. Selectively destructive restart. In *Proc. of the Sixth International Conf. on Genetic Algorithms* (*ICGA '95*), pages 144–150. Morgan Kaufmann, 1995.
12. M. Mitchell, J. H. Holland, and S. Forrest. When will a genetic algorithm outperform hill climbing? In *Advances in Neural Information Processing Systems.* Morgan Kaufmann, 1994.
13. R. Motwani and P. Raghavan. *Randomized Algorithms.* Cambridge University Press, 1995.
14. H. Mühlenbein. How genetic algorithms really work. Mutation and hillclimbing. In *Proc. of the 2nd Parallel Problem Solving from Nature* (*PPSN II*), pages 15–25. North-Holland, 1992.
15. G. Rudolph. *Convergence Properties of Evolutionary Algorithms.* Dr. Kovač, 1997.

Running Time Analysis of Multi-objective Evolutionary Algorithms on a Simple Discrete Optimization Problem

Marco Laumanns[1], Lothar Thiele[1], Eckart Zitzler[1],
Emo Welzl[2], and Kalyanmoy Deb[3]

[1] ETH Zürich, Computer Engineering and Networks Laboratory, CH–8092 Zürich
{laumanns,thiele,zitzler}@tik.ee.ethz.ch
http://www.tik.ee.ethz.ch/aroma

[2] ETH Zürich, Institute of Theoretical Computer Science, CH–8092 Zürich
welzl@inf.ethz.ch

[3] Department of Mechanical Engineering, Indian Institute of Technology Kanpur
Kanpur, PIN 208 016, India
deb@iitk.ac.in

Abstract. For the first time, a running time analysis of population-based multi-objective evolutionary algorithms for a discrete optimization problem is given. To this end, we define a simple pseudo-Boolean bi-objective problem (LOTZ: leading ones - trailing zeroes) and investigate time required to find the entire set of Pareto-optimal solutions. It is shown that different multi-objective generalizations of a (1+1) evolutionary algorithm (EA) as well as a simple population-based evolutionary multi-objective optimizer (SEMO) need on average at least $\Theta(n^3)$ steps to optimize this function. We propose the fair evolutionary multi-objective optimizer (FEMO) and prove that this algorithm performs a black box optimization in $\Theta(n^2 \log n)$ function evaluations where n is the number of binary decision variables.

1 Introduction

Evolutionary Algorithms (EAs) are probabilistic search heuristics that mimic principles of natural evolution. They are often used to solve optimization problems, in particular those with multiple objectives, for an overview see e.g. [1]. In multi-objective optimization, the aim is to find or to approximate the set of Pareto-optimal (or non-dominated) solutions.

Existing theoretic work on convergence in evolutionary multi-objective optimization has so far mainly dealt with the limit behavior [9, 10, 12, 11, 4, 5, 14]. Under appropriate conditions for the variation and the selection operators, global convergence to the Pareto set can be guaranteed in the limit.

In addition to that, we are often interested in a quantitative analysis, specifically the expected running time for a given class of problems and the success probability for a given optimization time. For single-objective evolutionary algorithms many such results are contained in [8]. For the optimization of pseudo-Boolean functions an extensive theory has been built up by Wegener et al., see

J.J. Merelo Guervós et al. (Eds.): PPSN VII, LNCS 2439, pp. 44–53, 2002.

e.g. [16], Droste, Jansen, and Wegener [2, 3], or for a methodological overview [15].

Results on the running time of evolutionary algorithms in the multi-objective case are rare. Scharnow et al. [13] analyze a (1+1)-EA under multiple, non-conflicting objectives. The purpose of this paper is to present a first analysis of different population-based multi-objective evolutionary algorithms (MOEAs) on a two-objective model problem. In particular, the following results are described in the paper:

- The well known "Leading Ones" problem is generalized to two dimensions. The new problem class is called LOTZ (Leading Ones - Trailing Zeros).
- A simple evolutionary multi-objective optimization algorithm is defined (SEMO - Simple Evolutionary Multi-objective Optimizer). Its expected running time on the above problem is shown to be $\Theta(n^3)$.
- The algorithm is improved by a fair sampling strategy of the population (FEMO - Fair Evolutionary Multi-objective Optimizer). Its running time on the above problem is $\Theta(n^2 \log n)$ with a high probability of $1 - O(1/n)$ and its expected running time is $O(n^2 \log n)$.

The model problem and its characteristics are introduced in section 2. It is a multi-objective extension of the "Leading Ones" problem which has been thoroughly analyzed for example in [8] and [3]. The algorithms are described and analyzed in sections 3 and 4. They are instances of a steady state $(\mu + 1)$-EA with variable population size and differ in the manner how the parents are sampled from the population.

2 The Model Problem

As the example problem for this analysis, we consider the maximization of a 2-dimensional vector valued function, LOTZ, which maps n binary decision variables to 2 objective functions.

Definition 1. *The pseudo-Boolean function* LOTZ : $\{0,1\}^n \rightarrow \mathbb{N}^2$ *is defined as*

$$\text{LOTZ}(x_1, \ldots, x_n) = \left(\sum_{i=1}^{n} \prod_{j=1}^{i} x_j, \sum_{i=1}^{n} \prod_{j=i}^{n} (1 - x_j) \right)$$

The abbreviation LOTZ stands for "Leading Ones, Trailing Zeroes" and means that we want to simultaneously maximize the number of leading ones and trailing zeroes in a bit-string. The first component, the LEADINGONES function, has been analyzed in detail in [8] and [3].

As there is no single search point that maximizes both components simultaneously, we want to find the whole set of non-dominated points based on the concept of Pareto optimality, here defined for an m-objective maximization problem with binary decision variables.

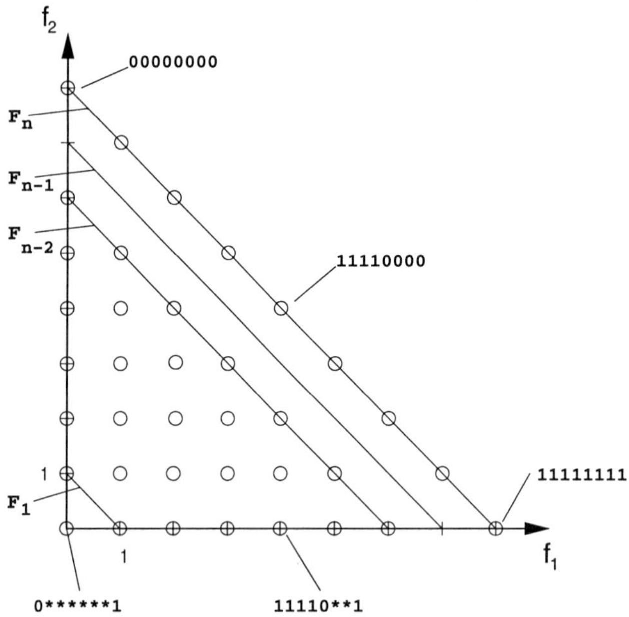

Fig. 1. Objective space of the LOTZ function with $n = 8$

Definition 2 (Pareto optimality, Pareto set). *Let $f : X \to F$ where $X \subseteq \{0,1\}^n$ and $F \subseteq \mathbb{R}^m$. A decision vector $x^* \in X$ is* Pareto optimal *if there is no other $x \in X$ that* dominates *x^*. x dominates x^*, denoted as $x \succ x^*$ if $f_i(x) \geq f_i(x^*)$ for all $i = 1, \ldots, m$ and $f_i(x) > f_i(x^*)$ for at least one index i. The set of all Pareto optimal decision vectors X^* is called* Pareto set. *$F^* = f(X^*)$ is the set of all Pareto optimal objective vectors and denoted as the* Pareto front.

The objective space of this problem can be partitioned into $n+1$ sets $F_i, i = 0, \ldots, n$ (see Fig. 1). The index i corresponds to the sum of both objective values, i.e. $(f_1, f_2) \in F_i$ if $i = f_1 + f_2$. Obviously, F_n represents the Pareto front F^*. The sub-domains X_i are defined as the sets containing all decision vectors which are mapped to elements of F_i. They are of the form $1^a 0 *^{(n-i-2)} 1 0^b$ with $a + b = i$ for $i < n$, and $1^a 0^b$ with $a + b = n$ for X_n.

The cardinality of the Pareto set $X^* = X_n$ is $|X_n| = n + 1$ and we also have $n + 1$ Pareto optimal objective vectors as $|F_n| = n + 1$. The next set F_{n-1} is empty. For the remaining sets with $i = 0, \ldots, n-2$ we have $|F_i| = i + 1$ and $|X_i| = |F_i| \cdot 2^{n-2-i}$. As a consequence, the decision space X contains 2^n different elements, which are mapped to $|F_n| + \sum_{i=0}^{n-2} |F_i| = 1/2 \cdot n^2 + 1/2 \cdot n + 1 = O(n^2)$ different objective vectors.

2.1 How Difficult Is It to Find the Whole Pareto Set?

How long does it take to optimize the LOTZ function? Droste et al. [3] have proved that the expected running time of a (1+1)-EA on LEADINGONES is

$\Theta(n^2)$. Using the same algorithm with an appropriate generalization of the acceptance criterion (either accepting only dominating offspring or by using a weighted sum as a scalar surrogate objective) will certainly lead to finding *one* element of the Pareto set in the same amount of time.

To find all different Pareto optimal points with such a (1+1) EA we can consider the multi-start option, i.e. to run to the EA several times, and collect all non-dominated solutions in an archive. For the acceptance criterion based on the dominance relation, the random variable describing the number of ones in the final solution of each single run follows a binomial distribution with $p = 0.5$. Hence the probability of finding the "outer" points of the Pareto set decreases exponentially. This would mean that the running time of this strategy until all Pareto optimal points are found is exponentially large in n.

Another possibility would be to use the multi-start option together with a weighted sum of the objective values. However, an appropriate choice of the weights is very difficult. In our case, equal weights would lead to the same situation as before, with a very low probability to reach the outer points. Any other selection of weights will let the sequence of search points converge to one of the outer points of the Pareto set. The remaining points must be found "on the way", but the probability of such events is not easy to calculate. Even if we could supply $n+1$ different weights corresponding to each of the $n+1$ Pareto optimal points, this strategy would still need $(n+1) \cdot \Theta(n^2) = \Theta(n^3)$ steps.

Contrary to the multi-start option, one could relax the acceptance criterion such that only dominated offspring are rejected. This allows to accept other Pareto-optimal solutions once an element of the Pareto set has been found and one could continue this process until all Pareto-optimal solutions have been visited at least once. This concept has been implemented in the PAES algorithm [6]. It finally leads to a random walk on or around the Pareto set, and the running time of the algorithm could be calculated based on the cover time of the random walk. A random walk on the path given by the $n+1$ Pareto-optimal solutions has a cover time of $\Theta(n^2)$. As each move of this random walk needs a successful mutation, whose probability is bounded above by $2/n$, the whole process again needs $\Theta(n^3)$ steps.

A last possibility would be to use a simple strategy known from classical multi-objective function optimization. In this case, we optimize only one objective, e.g. the number of leading ones, and constrain the other objective to be strictly larger than its value obtained in the previous optimization run. Therefore, we find all $n+1$ Pareto vectors in $n+1$ runs of a single-objective EA with an additional constraint. At the best, this strategy again needs $\Theta(n^3)$ steps.

The above discussion indicates that a (1+1) strategy may not be the best approach for solving multi-objective optimization problems. Moreover, most of the current multi-objective optimization algorithms use the concept of an archive that maintains a set of Pareto optimal vectors of all decision vectors visited so far. This indicates that the concept of a population is vital in multi-objective evolutionary optimization. In the next sections, we propose and analyze two simple *population-based* steady state EAs.

Algorithm 1 Simple Evolutionary Multi-objective Optimizer (SEMO)

1: Choose an initial individual x uniformly from $X = \{0,1\}^n$
2: $P \leftarrow \{x\}$
3: **loop**
4: Select one element x out of P uniformly.
5: Create offspring x' by flipping a randomly chosen bit.
6: $P \leftarrow P \setminus \{z \in P | x' \succ z\}$
7: **if** $\nexists z \in P$ such that $(z \succ x' \vee f(z) = f(x'))$ **then**
8: $P \leftarrow P \cup \{x'\}$
9: **end if**
10: **end loop**

3 A Simple Evolutionary Multi-objective Optimizer (SEMO)

At first, we analyze a simple population-based multi-objective EA. This algorithm contains a population of variable size that stores all non-dominated individuals. From this population a parent is drawn according to some probability distribution and mutated by flipping a randomly chosen bit. For Algorithm 1 we consider a uniform distribution for selecting the parent.

An appropriate archiving strategy [7] is assumed to prevent the population from growing exponentially. For this study it suffices to ensure that a solution is only accepted if it has different objective values (line 7).

3.1 Running Time Analysis of SEMO Applied to LOTZ

The running time of an algorithm equals the number of necessary evaluations of the objective function. For the analysis we divide the run of the SEMO into two distinct phases: the first phase lasts until the first Pareto-optimal individual has entered the population, and the second phase ends when the whole Pareto set has been found.

Lemma 1 (Expected running time for phase 1). *The expected running time of Alg. 1 until the first Pareto-optimal point is found is $O(n^2)$.*

Proof. Note that during this first phase the population will consist of one individual only, as a mutation changing the objective values yields either a dominating or a dominated individual. Hence, if an offspring is accepted, it will replace the parent from which it was produced. We consider the partition of the search space into distinct subsets X_i as defined in section 2 and note that from any subset X_i only points in $X_j, j > i$ are accepted. As there is always a one-bit mutation leading to the next subset, the probability of improvement is at least $1/n$. As there are at most $n-1$ such steps necessary (X_{n-1} is empty) the expected time is at most n^2. □

Lemma 2 (Expected running time for phase 2). *After the first Pareto-optimal point is found, the expected running time of Alg. 1 until all Pareto-optimal points are found is* $\Theta(n^3)$ *and the probability that the running time is less than* $n^3/c(n)$ *is less than* $(8e/c(n))^{n/2}$.

Proof. We partition this phase into $n-1$ different sub-phases. Sub-phase i lasts from the time when $i-1$ Pareto-optimal solutions have been found to the time when the next solution is found. T_i is a random variable denoting the duration of sub-phase i and the random variable T is the sum of these times. As we always have a contiguous subset of the Pareto set, only the individuals corresponding to the outer points of this subset can create a new Pareto-optimal point. The probability $p_s(i)$ to sample such a candidate in phase i is at least $1/i$ and at most $2/i$. A subsequent mutation has a success probability of at least $1/n$ and at most $2/n$. Hence, $ni/4 \leq E(T_i) \leq ni$. As $T = \sum_{i=1}^{n-1} T_i$, $1/8n^3 - 1/8n^2 \leq E(T) \leq 1/2n^3 - 1/2n^2$.
To derive a lower bound of the running time which holds with a high probability we consider the run after $n/2$ Pareto-optimal solutions have already been found. In this case the probability to find a new Pareto-optimal solution is at most $4/n^2$. If we allow $n^3/c(n)$ trials, the expected number of successes S is at most $4n/c(n)$. With Chernoff's inequality, the probability that we reach the required $n/2+1$ successes to find the remaining solutions can be bounded as

$$P(S > n/2) \leq \left(\frac{e^{\frac{1}{8}c(n)-1}}{(\frac{1}{8}c(n))^{\frac{1}{8}c(n)}}\right)^{4n/c(n)} \leq \left(\frac{8e}{c(n)}\right)^{\frac{1}{2}n}$$

□

From the concatenation of the two phases the following Corollary can be derived.

Corollary 1 (Expected running time Alg. 1). *The expected running time of Alg. 1 until all Pareto-optimal points are found is* $\Theta(n^3)$.

For this problem, the simple population-based SEMO is at least as good as any potential multi-objective adaptation of the (1+1)-EA discussed in the previous section. But is there room for further improvement? As it takes about n^2 steps to find *one* Pareto-optimal point, there is no hope to find the whole Pareto set in less time. But the time to generate all Pareto-optimal points can be reduced substantially.

4 The Fair Evolutionary Multi-objective Optimizer (FEMO)

The main weakness of the SEMO for the optimization problem under consideration lies in the fact that a large number of mutations are allocated to parents whose neighborhood has already been explored sufficiently. On the other hand, an optimal sampling algorithm would use always the most promising parent at

Algorithm 2 Fair Evolutionary Multi-objective Optimizer (FEMO)

```
1: Choose an initial individual x uniformly from X = {0,1}^n
2: w(x) ← 0                                          {Initialize offspring count}
3: P ← {x}
4: loop
5:    Select one element x out of {y ∈ P|w(y) ≤ w(z) ∀z ∈ P} uniformly.
6:    w(x) ← w(x) + 1                                 {Increment offspring count}
7:    Create offspring x' by flipping a randomly chosen bit.
8:    P ← P \ {z ∈ P|x' ≻ z}
9:    if ∄z ∈ P such that (z ≻ x' ∨ f(z) = f(x')) then
10:      P ← P ∪ {x'}
11:      w(x) ← 0                                     {Initialize offspring count}
12:   end if
13: end loop
```

the border of the current population. Of course, this information is not available in a black box optimization scenario.

The uniform sampling leads to a situation, where the Pareto-optimal individuals have been sampled unevenly depending on when each individual entered the population. The following *fair* sampling strategy guarantees that the end all individuals receive about the *same* number of samples.

Algorithm 2 implements this strategy by counting the number of offspring each individual produces (line 6). The sampling procedure deterministically chooses the individual which has produced the least number of offspring so far, ties are broken randomly (line 5).

4.1 Running Time Analysis of FEMO Applied to LOTZ

For the analysis of Algorithm 2 we focus only on the second phase as the first phase is identical to the simple Algorithm 1 described before.

Once the first two Pareto-optimal points are found, there is exactly one possible parent for each of the remaining $n-1$ points. We are interested in the number of mutations that must be allocated to *each* of these $n-1$ parents in order to have at least one successful mutation each that leads to the desired child. Lemma 3 and 4 provide upper and lower bounds on the probability that a certain number of mutations per parent are sufficient. In Theorem 1 these probabilities are used to bound the running time of the FEMO algorithm.

Lemma 3 (Minimal success probability). *Let p be the success probability for each single mutation and $c > 0$ an arbitrary constant. With probability at least $1 - n^{1-c}$ all $n-1$ remaining offspring have been constructed in at most $c \cdot 1/p \cdot \log n$ mutation trials for each corresponding parent.*

Proof. For each individual, the probability of having at least $c \cdot 1/p \cdot \log n$ non-successful mutation is bounded above by

$$(1-p)^{c\cdot 1/p\cdot \log n} = (1-\frac{1}{1/p})^{c/p\log n} = (1-\frac{1}{1/p})^{1/p^{c\log n}} \le \left(\frac{1}{e}\right)^{c\log n} = \frac{1}{n^c}$$

There are $n-1$ individuals that must be produced with the given number of trials. These events are independent, so the probability that at least one individual needs more than $c/p \cdot \log n$ trials is bounded above by $\frac{n-1}{n^c} \le n^{1-c}$. □

Lemma 4 (Maximal success probability). *Let $k \in \{1,\dots,n\}$, $a = k/n$, and $c > 0$ be an arbitrary constant. The probability that $k = a \cdot n$ individuals are produced in $c \cdot 1/p \cdot \log n$ mutation steps each is not larger than $(e^a)^{-n^{(1-c-c/n)}}$.*

Proof. The probability that a parent has created a certain offspring within the first $t = c \cdot 1/p \cdot \log n$ trials is $1-(1-p)^t$. The probability that this happens independently for a selection of k such pairs can thus be bounded as

$$(1-(1-p)^t)^k \le (1-\frac{1}{n^{c\frac{n+1}{n}}})^{an} \le e^{-\frac{an}{n^{c(n+1)/n}}} = (e^a)^{-n^{(1-c-c/n)}}$$

□

Now we can translate the number of mutations that are needed into the running time of Algorithm 2.

Theorem 1 (Running time bounds). *With probability at least $1-O(1/n)$ the number of objective function evaluations T Algorithm 2 needs from the discovery of the first two Pareto-optimal points until the whole Pareto set has been found lies in the interval $[1/4 \cdot 1/p \cdot n\log n, 2 \cdot 1/p \cdot n \log n]$. Hence, $Prob\{T = \Theta(1/p \cdot n\log n)\} = 1 - O(1/n)$. Furthermore, $E(T) = O(1/p \cdot n \log n)$.*

Proof. Let the Pareto-optimal points be indexed according to the order in which they have entered the set P. Let $k \in \{0,\dots,n\}$ be the index of the individual that required the largest number of mutations to be produced. We apply Lemma 3 with $c = 2$ and notice that this individual k did not need more than $2/p \cdot \log n$ trials with probability $1 - O(1/n)$.
What remains to be shown for the upper bound is that no node will be *sampled* more than t times during the algorithm. This can be guaranteed since there is always a candidate $x \in P$ with $w(x) \le t$ (the element that has most recently been added to P). Hence, any element whose weight has reached t will never be sampled again. As there are n such elements, each of which is sampled at most t times, the total number of samples (steps) the algorithm takes does not exceed $T = n \cdot t = 2 \cdot 1/p \cdot n \log n$.

For the lower bound we apply Lemma 4 with $c = 1/2$ and $k = n/2$. With a probability of $1 - \sqrt{e}^{-n^{(0.5-0.5/n)}}$ there is an individual in the second half which needs at least $1/2 \cdot 1/p \cdot \log n$ trials. Hence, all individuals in the first half have been sampled at least $1/2 \cdot 1/p \cdot \log n - 1$ times each. Of course, all individuals in the second half must be sampled at least once. The summation over all nodes gives a total number of samples of at least $1/4 \cdot 1/p \cdot n \log n$ with probability $1 - O(1/n)$.

Using the probability bound from Lemma 3 the expected running time can be bounded as

$$\begin{aligned}(1/p \cdot n \log n)^{-1} E(T) &\leq 1 \cdot P\{0 \leq T < 1\} + 2 \cdot P\{1 \leq T < 2\} + \dots \\ &\leq 2 + \sum_{c=3}^{\infty} c \cdot P\{T \geq k-1\} \\ &\leq 2 + \sum_{c=1}^{\infty} (c+2) n^{-c} \\ &\leq 2 + \frac{n}{(n-1)^2} + \frac{2}{n-1} \quad .\end{aligned}$$

Hence, $E(T) = O(1/p \cdot n \log n)$. □

As before, the time to find the first one (or two) elements of the Pareto set can be neglected and the total running time is mainly determined by Theorem 1. For our case the mutation success probability is $p = 1/n$, which leads with a high probability to a running time of $\Theta(n^2 \log n)$. This is a considerable improvement in comparison to any multi-start strategy of a single objective EA and to the SEMO algorithm.

5 Concluding Remarks

In this paper we have given first analytical results on the running time of evolutionary algorithms for multi-objective optimization problems. We have defined a bi-objective model problem and discussed different concepts of how to find the whole set of Pareto-optimal solutions for this problem.

For a simple steady-state evolutionary multi-objective optimizer (SEMO) a running time of $\Theta(n^3)$ was proven, which is at least as good as any strategy based on using a scalar surrogate objective function or multi-objective adaptations of the (1+1)-EA.

We proposed FEMO, a new evolutionary multi-objective optimizer involving an archive or population and a fair sampling strategy. This algorithm improves the running time substantially as it is able to find the whole Pareto set in $\Theta(n^2 \log n)$ steps.

The FEMO algorithm uses information collected during the run and stored in a population. For the first time it could be shown analytically that this concept of a population-based EA leads to a provable advantage on a multi-objective optimization problem compared to standard approaches based on scalarizing functions.

In addition to its good running time behavior, the algorithm is simple and problem- and representation-independent so that it might be easily and successfully applied to practical problems as well.

Acknowledgments

The research has been funded by the Swiss National Science Foundation (SNF) under the ArOMA project 2100-057156.99/1.

References

1. K. Deb. *Multi-objective optimization using evolutionary algorithms.* Wiley, Chichester, UK, 2001.
2. S. Droste, T. Jansen, and I. Wegener. A rigorous complexity analysis of the (1+1) evolutionary algorithm for separable functions with Boolean inputs. *Evolutionary Computation*, 6(2):185–196, 1998.
3. S. Droste, T. Jansen, and I. Wegener. On the analysis of the (1+1) evolutionary algorithm. *Theoretical Computer Science*, 276(1–2):51–81, 2002.
4. T. Hanne. On the convergence of multiobjective evolutionary algorithms. *European Journal Of Operational Research*, 117(3):553–564, 1999.
5. T. Hanne. Global multiobjective optimization with evolutionary algorithms: Selection mechanisms and mutation control. In *Evolutionary Multi-Criterion Optimization (EMO 2001), Proc.*, LNCS 1993, pages 197–212, Berlin, 2001. Springer.
6. J. D. Knowles and D. W. Corne. Approximating the non-dominated front using the Pareto Archived Evolution Strategy. *Evolutionary Computation*, 8(2):149–172, 2000.
7. M. Laumanns, L. Thiele, K. Deb, and E. Zitzler. Combining convergence and diversity in evolutionary multi-objective optimization. *Evolutionary Computation*, 10(3), 2002.
8. G. Rudolph. *Convergence Properties of Evolutionary Algorithms.* Verlag Dr. Kovač, Hamburg, 1997.
9. G. Rudolph. Evolutionary search for minimal elements in partially ordered sets. In *Evolutionary Programming VII – Proc. Seventh Annual Conf. on Evolutionary Programming (EP-98)*, San Diego CA, 1998. The MIT Press, Cambridge MA.
10. G. Rudolph. On a multi-objective evolutionary algorithm and its convergence to the pareto set. In *IEEE Int'l Conf. on Evolutionary Computation (ICEC'98)*, pages 511–516, Piscataway, 1998. IEEE Press.
11. G. Rudolph. Evolutionary Search under Partially Ordered Fitness Sets. In *Proceedings of the International NAISO Congress on Information Science Innovations (ISI 2001)*, pages 818–822. ICSC Academic Press: Millet/Sliedrecht, 2001.
12. G. Rudolph and A. Agapie. Convergence properties of some multi-objective evolutionary algorithms. In *Congress on Evolutionary Computation (CEC 2000)*, volume 2, pages 1010–1016, Piscataway, NJ, 2000. IEEE Press.
13. J. Scharnow, K. Tinnefeld, and I. Wegener. Fitness landscapes based on sorting and shortest paths problems. This volume.
14. D. A. Van Veldhuizen. *Multiobjective Evolutionary Algorithms: Classifications, Analyses, and New Innovations.* PhD thesis, Graduate School of Engineering of the Air Force Institute of Technology, Air University, June 1999.
15. I. Wegener. Methods for the analysis of evolutionary algorithms on pseudo-boolean functions. Technical Report CI-99/00, SFB 531, Universität Dortmund, 2000.
16. I. Wegener. Theoretical aspects of evolutionary algorithms. In *ICALP 2001*, volume 2076 of *LNCS*, pages 64–78. Springer-Verlag, 2001.

Fitness Landscapes Based on Sorting and Shortest Paths Problems

Jens Scharnow, Karsten Tinnefeld*, and Ingo Wegener*

FB Informatik, LS2, Univ. Dortmund, 44221 Dortmund, Germany
wegener@ls2.cs.uni-dortmund.de

Abstract. The analysis of evolutionary algorithms is up to now limited to special classes of functions and fitness landscapes. It is not possible to describe those subproblems of NP-hard optimization problems where certain evolutionary algorithms work in polynomial time. Therefore, fitness landscapes based on important computer science problems as sorting and shortest paths problems are investigated here. Although it cannot be expected that evolutionary algorithms outperform the well-known problem specific algorithms on these simple problems, it is interesting to analyze how evolutionary algorithms work on these fitness landscapes which are based on practical problems. The following results are obtained:

- Sorting is the maximization of "sortedness" which is measured by one of several well-known measures of presortedness. The different measures of presortedness lead to fitness landscapes of quite different difficulty for EAs.
- Shortest paths problems are hard for all types of EA, if they are considered as single-objective optimization problems, while they are easy as multi-objective optimization problems.

1 Introduction

Our aim is to contribute to a theory of evolutionary algorithms (EAs) which analyzes the expected optimization time of EAs on important and interesting problems and classes of fitness functions. Nowadays, it is a vision to explain the success of EAs on hard problems by identifying those instances of the problem where the considered EA finds the optimum in expected polynomial time. In order to develop tools for such results EAs have to be analyzed in various situations.

Fitness landscapes considered as typical ones have been described and analyzed in many papers (for an overview see Bäck, Fogel, and Michalewicz (1997)). Moreover, interesting classes of fitness functions have been investigated, e.g., separable functions (Droste, Jansen, and Wegener (1998a)), monotone polynomials of small degree (Wegener (2001)), long-path functions (Horn, Deb, and Goldberg (1994), Rudolph (1997), Droste, Jansen, and Wegener (1998b)), and

* This work was supported by the Deutsche Forschungsgemeinschaft (DFG) as part of the Collaborative Research Center "Computational Intelligence" (SFB 531).

J.J. Merelo Guervós et al. (Eds.): PPSN VII, LNCS 2439, pp. 54–63, 2002.

royal road functions (Mitchell, Holland, and Forrest (1994), Jansen and Wegener (2001a)). However, these are artificial functions and problems.

Here we choose the approach to investigate EAs on the most basic and important computer science problems, namely sorting (the maximization of the sortedness) and shortest path problems. We do not and cannot expect EAs to outperform Quicksort or Dijkstra's algorithm. However, universal problem solvers like EAs should be efficient on these simple problems.

Sorting can be considered as the problem of maximizing the sortedness. Hence, the fitness of a permutation can be measured by one of the well-known measures of presortedness. In Section 2, the corresponding fitness landscapes are introduced and appropriate mutation operators are discussed. The analysis in Section 3 shows that most measures of presortedness contain enough information to direct the optimization by EAs. However, there is a well-known measure of presortedness leading to a fitness landscape with large plateaus of search points having equal fitness such that the optimization process gets stuck on such a plateau.

Shortest path problems are multimodal optimization problems and EAs get stuck in local but not global optima (for certain instances). In Section 4, we describe this fitness landscape and an alternative as multi-objective optimization problem. Moreover, we prove that only the multi-objective optimization problem description directs the search of EAs efficiently. This is the first result of this type for EAs.

2 Fitness Landscapes Based on Sorting Problems

Given a sequence of n distinct elements from a totally ordered set, sorting is the problem of maximizing the sortedness. Hence, the search space is the set of all permutations π on $\{1, \ldots, n\}$. For our investigations we can identify π with the sequence $(\pi(1), \ldots, \pi(n))$ and the identity permutation is the optimal one. If we try to solve sorting as such an optimization problem, we need a fitness function f such that $f(\pi)$ describes the sortedness of π. Such measures have been introduced in the discussion of so-called adaptive sorting algorithms (see, e.g., Moffat and Petersson (1995)). We will investigate the five best-known measures of presortedness and the corresponding fitness landscapes.

- INV(π) measures the number of pairs (i, j), $1 \leq i < j \leq n$, such that $\pi(i) < \pi(j)$ (pairs in correct order),
- HAM(π) measures the number of indices i such that $\pi(i) = i$ (elements at the correct position),
- RUN(π) measures the number of indices i such that $\pi(i+1) < \pi(i)$ (number of maximal sorted blocks minus 1) leading to a minimization problem,
- REM(π) equals the largest k such that $\pi(i_1) < \cdots < \pi(i_k)$ for some $i_1 < \cdots < i_k$ (length of the longest sorted subsequence),
- EXC(π) equals the minimal number of exchanges (of pairs $\pi(i)$ and $\pi(j)$) to sort the sequence again leading to a minimization problem.

Since we cannot see any advantage of crossover for sorting problems, we have investigated only mutation-based EAs. Our mutation operator is based on the following two simple operations:

- exchange(i, j) which exchanges the elements at the positions i and j,
- jump(i, j) where the element at position i jumps to position j while the other elements between position i and position j are shifted in the appropriate direction, e.g., jump(5,2) applied to (6,4,3,1,7,2,5) produces (6,7,4,3,1,2,5).

Mutation should allow any π' to be produced from π with positive probability. The usual mutation operator on the search space $\{0,1\}^n$ flips each bit with probability $1/n$. This implies that the number of flipping bits is asymptotically Poisson distributed with parameter $\lambda = 1$. Therefore, we have chosen the following mutation operator where at least one local change is performed:

- Choose s according to a Poisson distribution with parameter $\lambda = 1$ and perform sequentially $s+1$ exchange or jump steps where for each step (i, j) is chosen randomly among all pairs (k, l), $1 \leq k, l \leq n$, $k \neq l$, and it is decided randomly whether exchange(k, l) or jump(k, l) is performed.

We also may consider only exchange or only jump steps. Finally, we have decided to analyze the following evolutionary algorithm shortly called $(1+1)$EA which resembles the well-known (1+1) evolution strategy:

- Choose the first search point π randomly.
- Repeat: Produce π' by mutation from π and replace π by π' if π' is not worse than π ($f(\pi') \geq f(\pi)$ in the case of a maximization problem and $f(\pi') \leq f(\pi)$ otherwise).

In applications, one needs a stopping criterion. Here we consider the infinite stochastic process $(\pi_1, \pi_2, \pi_3, \ldots)$ where π_t equals the permutation π after the t-th step of the above algorithm and investigate the random variable called optimization time which equals the first point of time t when π_t is optimal.

3 The Analysis of the (1+1)EA on the Fitness Landscapes Based on Sorting Problems

Here we analyze the performance of the (1+1)EA on the different landscapes described by the fitness functions introduced in Section 2.

Theorem 1. *The expected optimization time of the (1+1)EA on the fitness landscape described by INV is bounded above by $O(n^2 \log n)$ and bounded below by $\Omega(n^2)$.*

Proof. Let π be the current search point, $1 \leq i < j \leq n$, and $\pi(i) > \pi(j)$, i.e., (i, j) is an incorrect pair. Then exchange(i, j) improves the fitness. Each specific exchange operation has a probability of $1/(2en(n-1))$ to be the only operation ($1/e$ is the probability of performing a single operation, $1/2$ the probability of choosing an exchange operation and $2/(n(n-1))$ the probability of choosing the pair (i, j) or the pair (j, i)). Hence, the probability of increasing the fitness in the presence of m incorrect pairs is at least $m/(en(n-1))$ leading to an expected waiting time (for an improvement) of $O(n^2/m)$. Since $1 \le m \le n(n-1)/2$ and since the fitness is never decreased, the expected optimization time can be estimated by

$$c \sum_{1 \le m \le n(n-1)/2} n^2/m = cn^2 H(n(n-1)/2)$$

for $H(N) := 1 + 1/2 + \cdots + 1/N \le \ln N + 1$ leading to the proposed bound $O(n^2 \log n)$.

For the lower bound we only consider the final step leading to the optimum. Independent of the number of exchange or jump operations chosen in this step, it is necessary that the last operation changes a not sorted sequence into a sorted one. This is possible by at most one exchange and at most two jump operations (jump$(i, i+1)$ and jump$(i+1, i)$ have the same effect). Hence, the success probability of the final step is bounded above by $3/(n(n-1))$ and, therefore, the expected waiting time is bounded below by $\Omega(n^2)$, i.e., by cn^2 for some $c > 0$ and large n. □

In the beginning the fitness typically is increased in successful steps by more than one. This may change in the final stage of the optimization process. The sequence $(2, 3, \ldots, n, 1)$ has $n-1$ incorrect pairs and, in successful steps, the number of incorrect pairs is halved on the average. This leads to an expected optimization time of $O(n^2)$. The sequence $(2, 1, 4, 3, \ldots, n, n-1)$ has $n/2$ incorrect pairs and, in successful steps, the number of incorrect pairs is decreased only by one with large probability. It can be shown that the expected optimization time on this string equals $\Theta(n^2 \log n)$. It is more typical to have in the final stage more small disordered subblocks than to have only few elements which belong to many incorrect pairs. Hence, we believe that the expected optimization time equals $\Theta(n^2 \log n)$.

Corollary 1. *The expected optimization time of the* $(1+1)EA$ *is* $\Omega(n^2)$ *for each fitness landscape based on the sorting problem, i.e., with a unique global optimum.*

Proof. The proof is contained in the proof of Theorem 1, since there we only used the fact that there is a unique global optimum. □

Theorem 2. *The expected optimization time of the* $(1+1)EA$ *on the fitness landscape described by REM equals* $\Theta(n^2 \log n)$.

Proof. If $\text{REM}(\pi) = k$ and j is a position not belonging to the longest sorted subsequence, then there is a jump operator where j jumps to a position where it increases the length of the sorted subsequence from k to $k+1$. The number of these positions j equals $n-k$ and, therefore, the success probability (increasing the fitness), if $\text{REM}(\pi) = k$, is at least $(n-k)/2en(n-1)$ leading to an expected waiting time of $O(n^2/(n-k))$. Summing up these values for $k \in \{1,\ldots,n-1\}$ we obtain the proposed bound.

For the lower bound it is essential that a single jump can change the REM value by at most 1. The jumping element is first taken from the sequence (which can decrease the length of any sorted subsequence by 1) and then inserted somewhere (which can increase the length of sorted subsequences by 1). Since an exchange can be simulated by two jumps, exchange steps can increase the REM value by at most 2. Let us consider the situation where $\text{REM}(\pi) \geq n - n^{1/2}$ for the first time. Then $\text{REM}(\pi) \leq n - n^{1/2}/2$ with overwhelming probability, since at least $n^{1/2}/4$ jumps and exchanges in one step are very unlikely. If an element of the longest sorted subsequence which has two neighbors from this sorted subsequence jumps, this decreases the REM value by 1 while the probability that an element outside the sorted subsequence jumps to a position increasing the REM value equals $1/(n-1)$ for each of these values. We can conclude that the probability of increasing the REM value by a step with at least 10 exchanges and/or jumps is so small that this does not happen within $O(n^2 \log n)$ steps with overwhelmig probability. This implies that we may consider only steps with a single jump operation (increasing the expected optimization time at most by a constant factor). If $\text{REM}(\pi) = k$, the success probability is bounded above by $(n-k)/(n(n-1))$ leading to a waiting time of $\Omega(n^2/(n-k))$. Summing up for $k = n - n^{1/2}/2$ to $k = n-1$, we obtain the proposed lower bound. □

The upper bound of Theorem 2 holds for an arbitrary initial search point π. However, jump operations are essential while the bounds of Theorem 1 for INV even hold if we perform only jumps or only exchanges. Consider the situation of $\pi^* = (2,\ldots,n,1)$ with $\text{REM}(\pi^*) = n-1$. The plateau of search points π with $\text{REM}(\pi) = n-1$ can be characterized by the "wrong element" i which can be at one of the $n-1$ "wrong positions" $p \neq i$ while all other elements are in sorted order. Which exchange operations are accepted? Only those which exchange i with one of the two neighbored elements and those which exchange i with one of the elements $i-1$ or $i+1$. Hence, the probability of an accepted exchange step is bounded above by $4/(n(n-1))$. The probability of accepting a step with more than one exchange step is negligible. However, the sum of all $|j - \pi(j)|$ is decreased at most by 2 by a successful exchange step and we need $(n-1)/2$ of these steps for π^* leading already to a waiting time of $\Omega(n^3)$. However, using the methods introduced by Jansen and Wegener (2001b) we can prove that the expected optimization time equals $\Theta(n^4)$ showing that the choice of the mutation operator is essential.

Theorem 3. *The expected optimization time of the* $(1+1)EA$ *on the fitness landscape described by HAM is bounded above by* $O(n^2 \log n)$.

Proof. The upper bound follows in the same way as in the proof of Theorem 2 by proving that for $\text{HAM}(\pi) = k$ there are at least $n-k$ single exchange operations increasing the HAM value. If element i is at position $p \neq i$, then exchange(i,p) increases the HAM value. Element i reaches its correct position and the element leaving position i was not at its correct position. □

Considering only exchanges we can also prove a lower bound of $\Omega(n^2 \log n)$ similarly to the lower bound proof of Theorem 2. Here the HAM value can be increased by a single exchange step at most by 2. If HAM(π) is large, it is unlikely that a step with more than 10 exchanges increases the HAM value. Moreover, only exchanges of pairs (i, j) where $\pi(j) = i$ and/or $\pi(i) = j$ can increase the HAM value.

Considering only jumps we can prove an upper bound of $O(n^4 \log n)$, since two special jumps can simulate an exchange step. However, the probability of choosing these two jumps is bounded by $O(1/n^4)$. A lower bound for jump operations has to take into account that a single jump $((2, 3, \ldots, n, 1) \rightarrow (1, 2, 3, \ldots, n))$ can increase the HAM value from 0 to n. However, we do not believe that jumps help for this fitness landscape. Hence, for HAM exchanges play the role which is played by jumps for REM.

Theorem 4. *The expected optimization time of the* $(1+1)EA$ *on the fitness landscape described by EXC is bounded above by* $O(n^3)$.

Proof. Obviously, EXC(π) $\leq n - 1$ for all π. Since EXC(π) $= 0$ only for the optimal π, it is sufficient to decrease the EXC value for at most $n - 1$ times. The expected waiting time until the EXC value is decreased is bounded by $O(n^2)$, since, by definition, there is at least one pair (i, j) such that its exchange decreases the EXC value. This proves the theorem. □

Since two special jumps can simulate a given exchange step, we obtain for jumps an $O(n^5)$ bound. It is not surprising that exchanges seem to be better than jumps for this fitness function which is defined via the minimal number of exchange operations.

All the fitness functions based on INV, REM, HAM, or EXC lead to fitness landscapes which are easy for simple EAs. The expected optimization time can depend essentially on the chosen mutation operator. Hence, it seems to be useful to allow jumps and exchanges. However, we still have to investigate the fitness landscape based on the number of runs (RUN). This is the perhaps best-known measure of presortedness, since Mergesort is based on the idea of reducing the number of runs efficiently. We prove that the expected optimization time is under some reasonable assumptions exponential.

Theorem 5. *Let* $n^{1/2} \leq k \leq n/2$ *and let* π *be chosen randomly among all sequences with two runs of length* k *and* $n - k$ *resp. The expected optimization time of the (1+1)EA on the fitness landscape described by RUN and the random initial string* π *is bounded below by* $2^{\Omega(n^{1/2})}$ *and the success probability within* $2^{n^{1/2}}$ *steps is exponentially small.*

Proof. First, we investigate another algorithm which applies in each step one randomly chosen jump operation and accepts the new string if it has two runs or a single run. In the second case the algorithm stops with success. Let the string be described by the two runs $a_1 < \cdots < a_k$ and $b_{k+1} < \cdots < b_n$ where $\{a_1, \ldots, a_k\}$ is a random subset of $\{1, \ldots, n\}$ of size k. Only if $a_k = k$, we have only one run and can stop. This happens for the initial string with the exponentially small probabaility of $1/\binom{n}{k}$. Otherwise, we choose a random element for a jump. Each a_i has a unique correct position in the b-run. All other jumps (except one special case) lead to three runs and are not accepted. The special case is $a_1 = 1, \ldots, a_{k-1} = k-1, b_{k+1} = k$ and a_k jumps to position k (or the dual case). The crucial observation is the following. If we exclude the special situations and count only accepted steps (this may be done for lower bounds) we have for each element the same chance to jump to the correct position in the other run. Hence, we obtain with probability k/n runs of lengths $k+1$ and $n-k-1$ and with probability $1-k/n$ runs of lengths $k-1$ and $n-k+1$. Moreover, the elements of the first run are chosen randomly among all objects. As long as the critical parameter k, the length of the shorter run, is at least $m := n^{1/2}/2$ the probability of finding the optimum or a special case as described above is bounded by $n/\binom{n}{m} = 2^{-\Omega(n^{1/2}\log n)}$. Hence, this probability is exponentially small for $2^{n^{1/2}}$ steps.

In order to reach a value of $k = m$ we have to start from $k = 2m$ and to reach $k = m$ before reaching $k = 3m$. A necessary condition is to have at least m steps decreasing k among the next $3m$ steps. However, the probability of a decreasing step is bounded by $3m/n = O(n^{-1/2})$. By Chernoff bounds, the probability for the necessary condition is bounded by $2^{-\Omega(n^{1/2}\log n)}$ and, therefore, the probability that this happens within $2^{n^{1/2}}$ steps is exponentially small.

Single exchange steps are almost useless. The exchange of the last element of the first run and the first element of the second run is also a jump and may change the lengths of the runs. All other accepted exchange steps exchange two elements from both runs which does not change the lengths of the runs.

What is the effect of many jumps within an operation? With overwhelming probability we do not perform more than $n^{\varepsilon}/4$ jumps in one step. This holds for the Poisson distribution for each $\varepsilon > 0$. Therefore, an element jumping to a position with a distance of more than $n^{\varepsilon}/4$ from each of the two correct positions in the runs has to jump again such that we do not obtain more than two runs. If we choose r different elements which have to jump, the probability that the resulting sequence has at most two runs is bounded above $(n^{\varepsilon}/n)^r = n^{-(1-\varepsilon)r}$ while this probability is at least $1/(n-1)$ for $r = 1$. Since the number of jumps equals $X+1$ where X has a Poisson distribution with $\lambda = 1$, the probability of letting at least r elements jump is bounded by $(2/(r-1)!)$. Hence, we may assume that all successful steps where $r \geq 2$ elements jump shorten the shorter run by r. Since the expected decrease of the shorter run during $3m$ steps caused by this assumption is only $O(mn^{-(1-\varepsilon)\cdot 2})$, we can generalize our estimates above (with minor changes) to obtain the same results for the (1+1)EA. □

The last question is whether it is likely that we reach a situation with two runs where the shorter one has a length of at least $n^{1/2}$ and where the probability that this run contains almost only very small or almost only very large elements is tiny. In each situation it is more likely to increase the length of a short run than to increase the length of a long run. However, for short runs there is a non-vanishing probability of merging runs. Altogether, it seems to be very likely that we reach a situation as described above. However, this has not been proved rigorously here.

Our experiments have confirmed our statements discussed above. For $20 \leq n \leq 40$, in almost all cases the length of the shorter run was at least $0.4n$ for the first string with two runs. For $n = 40$, we have obtained the first string with two runs on the average only after approximately $9 \cdot 10^7$ steps. For $n = 60$ this happens only after $18 \cdot 10^7$ steps and the optimum is not reached after $5.7 \cdot 10^9$ steps. The fitness landscape based on RUN is difficult already for quite small n.

4 Fitness Landscapes Based on Shortest Paths Problems

Given n nodes and a distance matrix with positive entries d_{ij} the single-source-shortest-paths problem (SSSP) is the problem to compute for each node $v \neq s$ a shortest path from node s to node v. Using Dijkstra's algorithm SSSP can be solved in time $O(n^2)$. It seems to be a good idea to consider the search space of all $v = (v_1, \ldots, v_{n-1}) \in \{1, \ldots, n\}^{n-1}$ where $v_i \neq i$ with the following interpretation. We set $s = n$ and obtain a graph where i has the single predecessor v_i. If this graph has a cycle, the individual is invalid with fitness ∞. Otherwise, we get a tree rooted at s describing for each $i \neq s$ an s-i-path. As single-objective problem the fitness equals the sum of the lengths of all s-i-paths. This may lead to a needle-in-the-haystack landscape. If $d_{i,i-1} = 1$ and $d_{ij} = \infty$, if $j \neq i - 1$, only the optimal tree where $v_i = i + 1$ has a fitness smaller than ∞.

Now we consider SSSP as a multi-objective optimization problem where we have $n-1$ objectives namely the lengths of the s-i-paths. We use the same search space with the fitness function $f(v) = (f_1(v), \ldots, f_{n-1}(v))$ where $f_i(v)$ is the length of the s-i-path in the graph described by v, if this graph contains such a path, and $f_i(v) = \infty$, otherwise. The aim is minimization. The theory on shortest paths problems implies that this multi-objective minimization problem has a unique Pareto optimal fitness vector $f^* = (l_1^*, \ldots, l_{n-1}^*)$ containing the lengths of shortest s-i-paths and all Pareto optimal vectors describe graphs representing a system of shortest paths. (A vector is called Pareto optimal if it is minimal w.r.t. the partial order $\leq$ on $(\mathbb{R} \cup \{\infty\})^{n-1}$ where $(a_1, \ldots, a_{n-1}) \leq (b_1, \ldots, b_{n-1})$ iff $a_i \leq b_i$ for all i.)

We may design variants of evolutionary algorithms for multi-objective minimization. However, a very simple variant again called (1+1)EA is an efficient SSSP solver. We use the following mutation operator:

- Choose s according to a Poisson distribution with parameter $\lambda = 1$ and perform sequentially $s + 1$ flips. A flip chooses randomly a position $p \in \{1, \ldots, n-1\}$ and replaces v_p by a random node $w \neq p$.

The (1+1)EA chooses the first search point v randomly, creates v' by the mutation operator and replaces v by v' iff $f(v') \leq f(v)$.

Theorem 6. *The expected optimization time of the multi-objective (1+1)EA on SSSP is bounded above by $O(n^3)$.*

We prove a more sophisticated bound. Let t_i be the smallest number of edges on a shortest s-i-path, $m_j := \#\{i \mid t_i = j\}$, and $T = \max\{j \mid m_j > 0\}$. Then we prove the upper bound

$$en^2 \sum_{1 \le j \le T} (\ln m_j + 1).$$

This bound has its maximal value $\Theta(n^3)$ for $m_1 = \cdots = m_{n-1} = 1$. We also obtain the bound $O(n^2 T \log n)$ which in the typical case where T is small is much better than $O(n^3)$.

Proof. The proof is based on the following simple observation. Whenever $f_i(v) = l_i^*$, we only accept search points v' where $f_i(v') = l_i^*$. Hence, we do not forget the length of shortest paths which we have found (although we may switch to another shortest path). Now we assume that we have a search point v where $f_i(v) = l_i^*$ for all i where $t_i < t$. Then we wait until this property holds for all i where $t_i \le t$. For each node i where $t_i = t$ and $f_i(v) > l_i^*$ there exists a node j such that $t_j = t - 1$, j is the predecessor of i on a shortest s-i-path of length t, and $f_j(v) = l_j^*$. Then a mutation where only v_i is flipped and obtains the value j is accepted and leads to a search point v' where $f_i(v') = l_i^*$. The probability of such a mutation equals $1/(e(n-1)^2)$ ($1/e$ the probability of flipping exactly one position, $1/(n-1)$ the probability of flipping the correct position, and $1/(n-1)$ the probability of flipping it to the right value). If we have r such nodes, the success probability is at least $r/(e(n-1)^2)$ and the expected waiting time is bounded above by en^2/r. The largest value for r is m_t and we have to consider each of the values $m_t, \ldots, 1$ at most once. Hence, the total expected time of this phase is bounded above by $en^2(1 + \frac{1}{2} + \cdots + \frac{1}{m_t}) \le en^2(\ln m_t + 1)$. Since t can take the values $1, \ldots, T$ we have proved the proposed bound. □

It is easy to show that the expected optimization time equals $\Theta(n^3)$ in the extreme case which is a needle-in-the-haystack landscape for single-objective optimization.

5 Conclusion

Robust problem solvers should also solve the well-known simple optimization problems efficiently. This has been investigated for the sorting problem (maximizing the sortedness based on some measure of presortedness) and shortest-paths problems. For four out of five fitness landscapes described by the best-known measures of presortedness simple EAs work very efficiently, although different types of local changes are essential. However, the last measure of presortedness leads to a fitness landscape which is difficult for EAs. There are instances of shortest-paths problems which are difficult for black-box single-objective optimization. The modeling of the SSSP as multi-objective optimization problem reflects the structure of the problem and the fitness vector reveals enough information to direct the search of a simple EA. Usually, multi-objective optimization is only used if no single-objective optimization problem contains the whole structure of the problem. Here it has been shown that a multi-objective problem model may lead to a simpler problem.

References

1. Bäck, T., Fogel, D.B., and Michalewicz, Z. (Eds.) (1997). *Handbook of Evolutionary Computation.* Oxford University Press.
2. Droste, S., Jansen, T., and Wegener, I. (1998a). A rigorous complexity analysis of the (1+1) evolutionary algorithm for separable functions with Boolean inputs. Evolutionary Computation 6, 185–196.
3. Droste, S., Jansen, T., and Wegener, I. (1998b). On the optimization of unimodal functions with the (1+1) evolutionary algorithm. Parallel Problem Solving from Nature – PPSN V, LNCS 1498, 13–22.
4. Horn, J., Goldberg, D.E., and Deb, K. (1994). Long path problems. Parallel Problem Solving from Nature – PPSN III, LNCS 866, 149–158.
5. Jansen, T., and Wegener, I. (2001a). Real royal road functions - where crossover provably is essential. Genetic and Evolutionary Computation Conf. – GECCO, 375–382.
6. Jansen, T., and Wegener, I. (2001b). Evolutionary algorithms - how to cope with plateaus of constant fitness and when to reject strings of the same fitness. IEEE Trans. on Evolutionary Computation 5, 589–599.
7. Mitchell, M., Holland, J.H., and Forrest, S. (1994). When will a genetic algorithm outperform hill climbing. In J. Cowan, G. Tesauro, and J. Alspector (Eds.): *Advances in Neural Information Processing Systems.* Morgan Kaufman.
8. Petersson, O., and Moffat, A. (1995). A framework for adaptive sorting. Discrete Applied Mathematics 59, 153–179.
9. Rudolph, G. (1997). How mutations and selection solve long path problems in polynomial expected time. Evolutionary Computation 4, 195–205.
10. Wegener, I. (2001). Theoretical aspects of evolutionary algorithms. Int. Colloq. on Automata, Languages, and Programming – ICALP, LNCS 2076, 64–78.

Performance Measures for Dynamic Environments

Karsten Weicker

University of Stuttgart, Institute of Computer Science
Breitwiesenstr. 20–22, 70565 Stuttgart, Germany,
Karsten.Weicker@informatik.uni-stuttgart.de

Abstract. This article investigates systematically the utility of performance measures in non-stationary environments. Three characteristics for describing the goals of a dynamic adaptation process are proposed: accuracy, stability, and recovery. This examination underpins the usage of the best fitness value as a basis for measuring the three characteristics in scenarios with moderate changes of the best fitness value. However, for dynamic problems without coordinate transformations all considered fitness based measures exhibit severe problems. In case of the recovery, a newly proposed window based performance measure is shown to be best as long as the accuracy level of the optimization is rather high.

1 Introduction

Research on evolutionary algorithms applied to non-stationary problems dates back to the work by Goldberg and Smith (1987) and is still a topic of increasing popularity. In the past 15 years a vast variety of problems was considered. However, most of the research is rather problem or technique oriented and, therefore, deals with a specific niche in the space of the dynamic problems. There are only few classifications of dynamic problems available trying to establish a foundation for systematic research of dynamic optimization. Moreover, the even more important question concerning the comparability of several evolutionary algorithms acting in dynamic environments is still unanswered. In static optimization, there exist persistent reference quantities like the global optima and their fitness values. Not only these quantities might change over time in dynamic optimization, there are also certain variants of the focus concerning the actual goals of such an optimization. Within a formal framework this article specifies different goals of dynamic optimization, summarizes and develops a number of possible performance measures, and examines empirically how well these measures reflect the different goals for problems from various problem classes.

2 A Classification of Dynamic Problems

There are only few very coarse grained classifications distinguishing alternating (or cyclic) problems, problems with changing morphology, drifting landscapes, and abrupt and discontinuous problems (cf. Collard, Escazut, & Gaspar,

J.J. Merelo Guervós et al. (Eds.): PPSN VII, LNCS 2439, pp. 64–73, 2002.

1997; De Jong, 2000). Other classifications are based on the parameters of problem generators (e.g. Branke, 1999). A third type of classification by Weicker (2000) decomposes the dynamic problem, that is defined on the search space Ω, into several components and the change of each component i is determined by a sequence of coordinate transformations $\left(c_i^{(t)}\right)_{t\in\mathbb{N}}$ with $c_i^{(t)} : \Omega \to \Omega$ and $d(c_i^{(t)}(\omega_1), c_i^{(t)}(\omega_2)) = d(\omega_1, \omega_2)$ for all $\omega_1, \omega_2 \in \Omega$ and fitness rescalings $\left(r_i^{(t)}\right)_{t\in\mathbb{N}}$ with $r_i^{(t)} \in \mathbb{R}_0^+$.

An exemplary non-stationary fitness function is

$$f^{(t)}(A) = \max_{1\leq j\leq hills} \begin{cases} 0, & \text{if } d(A, opt_j) > 150 \\ maxfit_j \frac{150-d(A,opt_j)}{150}, & \text{otherwise} \end{cases}$$

with $A \in \Omega = [-500, 500] \times [-500, 500]$, euclidean distance d, and *hills* randomly chosen local optima $opt_j \in \Omega$ – each local optima corresponds to one component. The coordinate transformation for each component j is a linear translation of length *coordsev* into a direction dir_j which is randomly determined at the beginning and as soon as a point outside of Ω would be created. The fitness rescaling is a factor $fitchange_j$ which is added to $maxfit_j$. Again, $fitchange_j \in [-fitsev, fitsev]$ is chosen randomly at the beginning and when $maxfit_j$ would leave the range $[0, 1]$. In non-alternating problems the maximum hill with $j = 1$ must have maximal fitness in $[0.5, 1]$ and all other hills in $[0, maxfit_1]$.

Weicker (2000) defines several properties within the framework. In this work, it is relevant whether the coordinate transformation is the identity mapping, i.e. the coordinates are static. If this is not the case the coordinate dynamics are constant over a period of time. Moreover, the fitness rescalings may be an identity mapping too. Here it is of interest whether the overall best fitness value may alter between various components.

Based on these properties the following problem classes are considered:

Class 1: coordinate translation, no fitness rescaling, no alternation
Problem instance: $hills = 5$, $coordsev = 7.5$, $fitsev = 0$
Various hills are moving while their height remains constant and the best hill remains best.

Class 2: coordinate translation, fitness rescaling, no alternation
Problem instance: $hills = 5$, $coordsev = 7.5$, $fitsev = 0.01$
Various hills are moving while their height is changing, but the best hill remains best.

Class 3: no coordinate translation, fitness rescaling, alternation
Problem instance: $hills \in \{2, 5\}$, $coordsev = 0$, $fitsev = 0.01$
The hills are not moving but changing their height leading to alternating best hills.

Class 4: coordinate translation, fitness rescaling, alternation
Problem instance: $hills \in \{2, 5\}$, $coordsev = 7.5$, $fitsev = 0.01$
The hills are moving while changing their height and different hills take the role of the best hill at different generations.

The problem instances with 2 hills are chosen such that there is at least one alternation while both hills are changing their height into the same direction. This additional characteristic is supposed to be problematic when measuring the performance. Note, that the fitness severity is chosen moderately in all classes.

3 Goals of Dynamic Optimization

The goal of an evolutionary search process in a dynamic environment is not only to find an optimum within a given number of generations but rather a perpetual adjustment to changing environmental conditions. Besides the accuracy of an approximation at time t, the stability of an approximation is of interest as well as the recovery time to reach again a certain approximation level.

The *optimization accuracy* at time t for a fitness function F and optimization algorithm EA is defined as

$$accuracy_{F,EA}^{(t)} = \frac{F(best_{EA}^{(t)}) - Min_F^{(t)}}{Max_F^{(t)} - Min_F^{(t)}} \qquad (1)$$

where $best_{EA}^{(t)}$ is the best candidate solution in the population at time t, $Max_F^{(t)} \in \mathbb{R}$ the best fitness value in the search space, and $Min_F^{(t)} \in \mathbb{R}$ the worst fitness value in the search space. Note, that the accuracy is only well defined if the fitness function is non-constant at each time step, i.e. the fitness landscape is not a plateau covering the complete search space at any generation. The optimization accuracy ranges between 0 and 1, where accuracy 1 is the best possible value. It is also independent of fitness rescalings. This formula was introduced by Feng et al. (1997) as a performance measure in stationary environments.

As a second goal, *stability* is an important issue in optimization. In the context of dynamic optimization, an adaptive algorithm is called stable if changes in the environment do not affect the optimization accuracy severely. Even in the case of drastic changes an algorithm should be able to limit the respective fitness drop. The stability at time t is defined as

$$stab_{F,EA}^{(t)} = \max\{0, accuracy_{F,EA}^{(t)} - accuracy_{F,EA}^{(t-1)}\} \qquad (2)$$

and ranges between 0 and 1. A value close to 0 implies a high stability. In the case of mere drifting landscapes, e.g. classes 1 and 2, it is a means to gain insight into the ability to track the moving optimum by observing the stability over a period of time. However, the stability must not serve as the sole criteria since it makes no statement on the accuracy level. In classes 3 and 4, the stability at the generations where changes occur are of interest.

An additional aspect is the ability of an adaptive algorithm to react quickly to changes. An algorithm has ε-reactivity at time t

$$react_{F,A,\varepsilon}^{(t)} = \min\left\{t' - t \,\middle|\, t < t' \leq maxgen,\ t' \in \mathbb{N},\ \frac{accuracy_{F,A}^{(t')}}{accuracy_{F,A}^{(t)}} \geq (1-\varepsilon)\right\}$$
$$\cup \{maxgen - t\}$$

where *maxgen* is the number of generations. A smaller value implies a higher reactivity. This aspect of adaptation is especially of interest if the problem has short phases of big severity alternating with extensive phases of no severity with regard to the coordinate transitions or if the problem is alternating concerning the fitness rescalings (with rather low severity for the coordinates).

4 Performance Measures

In the previous section, the goals of optimization in dynamic environments are carefully defined using the accuracy. The accuracy is always a useful value, even in cases where the fitness of best approximation by the algorithm is decreasing, a situation where the overall best fitness could decrease or increase. But if the overall best fitness is not known the accuracy cannot be computed. Since this value might change, there is no common basis for assessment of the quality of a solution. A good fitness value at one time can be a bad fitness value at another time – but this is not transparent. Then, other means to assess the performance of an algorithm are necessary.

Those performance measures can be classified by the knowledge they need.

- Knowledge on the position of the optimum is available (which is usually only the case in academic benchmarks).
- Knowledge on the best fitness value is available.
- No global knowledge is available.

In order to enable a fair comparison of different algorithm with respect to the question how they reach the goal of dynamic optimization, a single value for the algorithm's performance is gained by averaging over all generations (see De Jong, 1975). In the following sections only the measures for one generation are given.

4.1 Measures for Optimization Accuracy

The definition of the accuracy was used by Mori, Kita, and Nishikawa (1996) as a performance measure averaged over a number of generations T. An almost similar performance measure was used by Trojanowski and Michalewicz (1999) where the normalization using the worst fitness value was omitted. They considered also only those time steps before a change in the environment occurs. If more emphasis is to be put on the detection of the optimum, Mori, Kita, and Nishikawa (1998) proposed a different weighting of those successful generations. A more gradual approach seems to be the usage of the square error to the best fitness value proposed by Hadad and Eick (1997).

In the case that no global knowledge is available, the following performance measures are discussed and examined in the remainder of the paper.

$$currentBest^{(t)}_{F,EA} = \max\{F(\omega) \mid \omega \in P^{(t)}_{EA}\}$$
$$currentBestOffl^{(t)}_{F,EA} = \max_{1\leq t'\leq t} currentBest^{(t')}_{F,EA}$$

$$currentAverage^{(t)}_{F,EA} = \frac{1}{|P^{(t)}_{EA}|} \sum_{\omega \in P^{(t)}_{EA}} F(\omega)$$

$$windowAcc^{(t)}_{F,EA,W} = \max\left\{ \frac{F(\omega) - windowWorst}{windowBest - windowWorst} \;\middle|\; \omega \in P^{(t)}_{EA} \right\} \text{ with}$$

$$windowBest = \max\{F(\omega) \mid \omega \in P^{(t')}_{EA},\ t - W \leq t' \leq t\}$$

$$windowWorst = \min\{F(\omega) \mid \omega \in P^{(t')}_{EA},\ t - W \leq t' \leq t\}$$

The majority of publications uses the best fitness value $currentBest^{(t)}_{F,EA}$ to assess the quality of the algorithm (e.g. Angeline, 1997; Cobb, 1990; Dasgupta & McGregor, 1992; Goldberg & Smith, 1987; Grefenstette, 1992; Hadad & Eick, 1997; Lewis, Hart, & Ritchie, 1998; Mori et al., 1996; Vavak, Fogarty, & Jukes, 1996). This measure is better suited to dynamic problems than $currentBestOffl^{(t)}_{F,EA}$, the usual basis for offline performance (De Jong, 1975), that compares incommensurable values from different generations (cf. Grefenstette, 1999). Branke (1999) uses a mixed approach where those values from generations without environmental change are compared which requires global knowledge on any possible change in the environment.

The average fitness value $currentAverage^{(t)}_{F,EA}$ is used as a performance measure e.g. by Dasgupta and McGregor (1992), Goldberg and Smith (1987), Mori et al. (1996). Averaged over generations this leads to the online performance measure of De Jong (1975), which was originally defined as the average over all function evaluations since the start of the algorithm. Presumed that the population size is constant and the algorithm is generational, the online performance may be defined as mean $currentAverage^{(t)}_{F,EA}$. Online performance reflects the focusing of the search on optimal regions (see Grefenstette, 1992, 1999). In the online performance actually each new created individual is supposed to contribute a high fitness value. However, Cobb (1990) argued that this conception might not be suited for many dynamic problems because focusing too much on good fitness values might have negative effects on the adaptability.

Another approach to measure the accuracy without knowing the actual best possible fitness is based on the assumption that the best fitness value will not change much within a small number of generations. As a consequence a local window of interest $W \in \mathbb{N}$ is introduced and the accuracy $windowAcc^{(t)}_{F,EA,W}$ is measured within this window. This window based measure has not been used in the experiments reported in the literature.

Alternatively to the fitness based measures, genotype or phenotype based measures may be used to approximate the optimization accuracy. Though independent of fitness rescalings, they require full global knowledge of the position of the current optimum. There are two variants: Weicker and Weicker (1999) used the minimal distance of the individuals in the population to the current optimum $\omega^* \in \Omega$ and Salomon and Eggenberger (1997) used the distance of the mass center ω_{center} of the population to ω^*.

Table 1. Average accuracy and standard deviation for the genetic algorithm with and without hypermutation

	w/out hypermut.		w/ hypermut.	
problem	avg	sdv	avg	sdv
class 1	0.45	0.023	0.87	0.0049
class 2	0.45	0.018	0.87	0.0035
class 3	0.82	0.035	0.96	0.0054
class 3 (2 hills)	0.97	0.0029	0.99	0.00086
class 4	0.46	0.025	0.87	0.0031
class 4 (2 hills)	0.41	0.023	0.86	0.0019

$$bestDist^{(t)}_{F,EA} = \max\left\{ \frac{maxdist - d(\omega^*, \omega)}{maxdist} \;\middle|\; \omega \in P^{(t)}_{EA} \right\}$$

$$centerDist^{(t)}_{F,EA} = \frac{maxdist - d(\omega^*, \omega_{center})}{maxdist}$$

Where the first approach seems to be straightforward to assess the approximation quality, the second performance measure is more difficult to interpret. It requires that the population as a whole describes very closely the region of the optimum.

4.2 Measures for Stability and Reactivity

In this paper, the measures for stability and reactivity are defined by replacing the accuracy in the definition of stability or reactivity by an accuracy measure from the previous section. However, the genotype/phenotype based measures are omitted in this examination since they also require global knowledge.

5 Empirical Results

5.1 Experimental Setup

To optimize the dynamic problems two genetic algorithms are used. Both algorithms are based on a standard genetic algorithm where each search space dimension is encoded using 16 bits, the crossover rate is 0.6, the bit flipping mutation is executed with probability $\frac{1}{32}$, a tournament selection with tournament size 2 is used, and the algorithm runs for 200 generations. In addition to this standard algorithm, a version using hypermutation with a fixed rate of 0.2 is used (cf. Grefenstette, 1999). Table 1 shows the accuracy averaged over 10 problem instances and 50 independent experiments for each instance as well as the respective standard deviation. The GA with hypermutation performs superior – however the performance of both algorithms should be expressed by a performance measure equally well.

5.2 Statistical Examination of the Measures

The goal of this investigation is to find some empirical evidence for the question how good the various measures approximate the exact adaptation characteristics.

Table 2. Ranking based on pairwise hypothesis tests concerning the MSE of the curves

	standard GA						GA with hypermutation					
	Class 1	Class 2	Class 3	Class 3 (2 hills)	Class 4	Class 4 (2 hills)	Class 1	Class 2	Class 3	Class 3 (2 hills)	Class 4	Class 4 (2 hills)
Accuracy:												
best fitness	1	1	1	4	1	1	1	2	1	4	1	1
average fitness	2	2	2	3	2	2	4	3	2	3	3	3
window based	3	3	5	5	3	5	2	1	5	5	2	2
shortest distance	4	4	3	1	4	3	3	3	4	1	4	4
distance of center	5	4	3	2	5	4	5	5	2	2	5	5
Stability:												
best fitness	1	1	1	1	1	1	1	1	1	1	1	1
average fitness	2	2	2	2	2	2	2	2	2	2	2	2
window based	3	3	3	3	3	3	3	3	3	3	3	3
0.05-Recovery:												
best fitness	1	1	1	1	1	1	1	1	1	1	2	1
average fitness	3	3	3	3	3	3	3	3	3	3	3	3
window based	2	2	2	2	2	2	2	1	1	1	1	1

A first approach is based on the assumption that the curves of the performance measures should match the curves of the respective exact values to guarantee a meaningful statement of the performance measure. The second approach considers the averaged performance values only and tests how well they correlate to the averaged exact values.

In the first approach, the measurements are normalized $(g(t) - E_g)/\sqrt{V_g}$ where E_g is the expectancy value and V_g the variance. This makes the values of different performance measures comparable since the values are independent of the range of the values. To assess the similarity of the curves of the exact values h' and the normalized performance measure g', the mean square error $MSE_{g',h'} = \sum_{t=1}^{maxgen}(g'(t) - h'(t))^2$ is computed. In order to get a statistical confidence of one measure over another, a hypothesis test is carried out using the 500 independent mean square errors of each performance measure. Those pairwise hypothesis tests are used to establish a ranking concerning the suitability of the performance measures. Student's t-test is used as a hypothesis test with a significant error probability of 0.05. Table 2 shows the results of this analysis.

In the second approach, the averaged measures at the end of a optimization run are used to determine how well the algorithm performed on the problem. Therefore, a statistical test is used to compute the correlation of the approximated measures to the exact measures. The input data for the correlation analysis are the averaged performance values of the 50 different runs of an algorithm on a problem instance. (Since the recovery measures depend highly on the successive generations, the values of generation 150 are used for those measures

Table 3. Percentage of problem instances with a high correlation to the exact averaged value

	standard GA						GA with hypermutation					
	Class 1	Class 2	Class 3	Class 3 (2 hills)	Class 4	Class 4 (2 hills)	Class 1	Class 2	Class 3	Class 3 (2 hills)	Class 4	Class 4 (2 hills)
Accuracy:												
best fitness	1.0	1.0	1.0	1.0	1.0	1.0	1.0	1.0	1.0	1.0	1.0	1.0
average fitness	1.0	1.0	1.0	1.0	1.0	1.0	0.8	1.0	1.0	1,0	1.0	1.0
window based	0.9	0.9	0.6	0.0	0.8	0.4	0.9	0.9	0.5	0.0	1.0	1.0
shortest distance	0.7	0.7	0.9	1.0	0.7	0.8	0.9	0.7	0.9	1.0	0.8	0.6
distance of center	0.7	0.7	0.9	1.0	0.7	0.9	0.7	0.4	0.9	0.5	0.6	0.3
Stability:												
best fitness	1.0	1.0	1.0	1.0	1.0	1.0	1.0	1.0	1.0	1.0	1.0	1.0
average fitness	1.0	1.0	0.4	0.0	1.0	1.0	0.0	0.2	0.0	0.0	0.0	0.0
window based	0.4	0.4	0.2	0.2	1.0	0.5	0.1	0.5	0.4	0.2	0.5	0.3
0.05-Recovery:												
best fitness	1.0	1.0	1.0	1.0	1.0	1.0	1.0	0.6	1.0	1.0	1.0	0.8
average fitness	1.0	1.0	0.4	0.2	1.0	1.0	0.3	0.3	0.0	0.0	0.1	0.1
window based	0.9	0.8	0.8	1.0	0.6	0.4	1.0	1.0	1.0	1.0	1.0	1.0

instead of the final performance values in generation 200). As statistical method Spearman's rank correlation is used. A series of data is considered to be highly correlated if the Spearman's rank correlation is positive and the two-sided significance level of its deviation from zero is less than 0.001. The correlation is computed for each of the ten instances of a problem class. Table 3 shows the percentage of instances where a high correlation between exact value and performance measure could be identified.

5.3 Discussion of the Results

The results concerning the accuracy show, that the averaged best fitness values are a good indicator for all classes and both high and low quality algorithms. However, the examination of the MSE shows that all fitness based measures have severe problems with class 3 (2 hills) where only fitness rescalings are occuring, in a misleading way. Especially the windows based measure has severe problems with all class 3 instances. The MSE of the GA with hypermutation on class 2 indicates that the window based measures can be a better indicator than best fitness although this is not approved by the averaged performance values.

Also, the stability is measured best using best fitness values. The average fitness shows very poor results with the averaged performance values. The windows based measure is insufficient regarding the MSE.

Concerning the recovery, the window based measure proofs to be equivalent or superior to the best fitness in case of the high quality experiments (GA with hypermutation).

The good results of the averaged best fitness for the accuracy contradicts partly the examination of the MSE. This indicates that especially in dynamic environments averaged performance measures should be used and interpreted carefully to rule out statistical effects.

6 Conclusions

This paper presents the first systematic approach to examine the usefulness of performance measures in time-dependent non-stationary environments. The goals of an adaptation process are discussed in detail and accuracy, stability, and recovery are proposed as key characteristics. Existing performance measures from literature are reviewed and a new window based performance measure is proposed.

On a wide set of dynamic problems the measures are examined for an algorithm with high accuracy and an algorithm generating low accuracy. Altogether the best fitness value proves to be the best performance measure for problems with moderate fitness severity – deficiencies exist for problems without coordinate transitions and as a basis for recovery measures. In the latter case, the window based measure exhibits a superior performance.

Future work has to examine problems with more severe fitness rescalings or additional problem characteristics. Also an investigation concerning the validity and strength of averaging performance values in dynamic domains is necessary.

References

Angeline, P. J. (1997). Tracking extrema in dynamic environments. In P. J. Angeline, R. G. Reynolds, J. R. McDonnell, & R. Eberhart (Eds.), *Evolutionary Programming VI* (pp. 335–345). Berlin: Springer. (Lecture Notes in Computer Science 1213)

Branke, J. (1999). *Evolutionary algorithms for dynamic optimization problems: A survey* (Tech. Rep. No. 387). Karlsruhe, Germany: Institute AIFB, University of Karlsruhe.

Cobb, H. G. (1990). *An investigation into the use of hypermutation as an adaptive operator in genetic algorithms having continuous, time-dependent nonstationary environments* (Tech. Rep. No. 6760 (NLR Memorandum)). Washington, D.C.: Navy Center for Applied Research in Artificial Intelligence.

Collard, P., Escazut, C., & Gaspar, A. (1997). An evolutionary approach for time dependant optimization. *International Journal on Artificial Intelligence Tools, 6*(4), 665–695.

Dasgupta, D., & McGregor, D. R. (1992). Nonstationary function optimization using the structured genetic algorithm. In R. Männer & B. Manderick (Eds.), *Parallel Problem Solving from Nature 2 (Proc. 2nd Int. Conf. on Parallel Problem Solving from Nature, Brussels 1992)* (pp. 145–154). Amsterdam: Elsevier.

De Jong, K. (2000). Evolving in a changing world. In Z. Ras & A. Skowron (Eds.), *Foundation of intelligent systems* (pp. 513–519). Berlin: Springer.

De Jong, K. A. (1975). *An analysis of the behavior of a class of genetic adaptive systems.* Unpublished doctoral dissertation, University of Michigan.

Feng, W., Brune, T., Chan, L., Chowdhury, M., Kuek, C. K., & Li, Y. (1997). *Benchmarks for testing evolutionary algorithms* (Tech. Rep. No. CSC-97006). Glasgow, UK: Center for System and Control, University of Glasgow.

Goldberg, D. E., & Smith, R. E. (1987). Nonstationary function optimization using genetic algorithms with dominance and diploidy. In J. J. Grefenstette (Ed.), *Proc. of the Second Int. Conf. on Genetic Algorithms* (pp. 59–68). Hillsdale, NJ: Lawrence Erlbaum Associates.

Grefenstette, J. J. (1992). Genetic algorithms for changing environments. In R. Männer & B. Manderick (Eds.), *Parallel Problem Solving from Nature 2 (Proc. 2nd Int. Conf. on Parallel Problem Solving from Nature, Brussels 1992)* (pp. 137–144). Amsterdam: Elsevier.

Grefenstette, J. J. (1999). Evolvability in dynamic fitness landscapes: a genetic algorithm approach. In *1999 Congress on Evolutionary Computation* (pp. 2031–2038). Piscataway, NJ: IEEE Service Center.

Hadad, B. S., & Eick, C. F. (1997). Supporting polyploidy in genetic algorithms using dominance vectors. In P. J. Angeline, R. G. Reynolds, J. R. McDonnell, & R. Eberhart (Eds.), *Evolutionary Programming VI* (pp. 223–234). Berlin: Springer. (Lecture Notes in Computer Science 1213)

Lewis, J., Hart, E., & Ritchie, G. (1998). A comparison of dominance mechanisms and simple mutation on non-stationary problems. In A. E. Eiben, T. Bäck, M. Schoenauer, & H.-P. Schwefel (Eds.), *Parallel Problem Solving from Nature – PPSN V* (pp. 139–148). Berlin: Springer. (Lecture Notes in Computer Science 1498)

Mori, N., Kita, H., & Nishikawa, Y. (1996). Adaptation to a changing environment by means of the thermodynamical genetic algorithm. In H. Voigt, W. Ebeling, & I. Rechenberg (Eds.), *Parallel Problem Solving from Nature – PPSN IV (Berlin, 1996) (Lecture Notes in Computer Science 1141)* (pp. 513–522). Berlin: Springer.

Mori, N., Kita, H., & Nishikawa, Y. (1998). Adaptation to a changing environment by means of the feedback thormodynamical geentic algorithm. In A. E. Eiben, T. Bäck, M. Schoenauer, & H.-P. Schwefel (Eds.), *Parallel Problem Solving from Nature – PPSN V* (pp. 149–158). Berlin: Springer. (Lecture Notes in Computer Science 1498)

Salomon, R., & Eggenberger, P. (1997). Adaptation on the evolutionary time scale: A working hypothesis and basic experiments. In J.-K. Hao, E. Lutton, E. Ronald, M. Schoenauer, & D. Snyders (Eds.), *Artificial Evolution: Third European Conf., AE'97* (pp. 251–262). Berlin: Springer.

Trojanowski, K., & Michalewicz, Z. (1999). Searching for optima in non-stationary environments. In *1999 Congress on Evolutionary Computation* (pp. 1843–1850). Piscataway, NJ: IEEE Service Center.

Vavak, F., Fogarty, T. C., & Jukes, K. (1996). A genetic algorithm with variable range of local search for adaptive control of the dynamic systems. In *Proc. of the 2nd Int. Mendelian Conf. on Genetic Algorithms.* ?

Weicker, K. (2000). An analysis of dynamic severity and population size. In M. Schoenauer, K. Deb, G. Rudolph, X. Yao, E. Lutton, J. J. Merelo, & H.-P. Schwefel (Eds.), *Parallel problem solving from nature – PPSN VI* (pp. 159–168). Berlin: Springer.

Weicker, K., & Weicker, N. (1999). On evolution strategy optimization in dynamic environments. In *1999 Congress on Evolutionary Computation* (pp. 2039–2046). Piscataway, NJ: IEEE Service Center.

Representation/Codification Issues

Direct Representation and Variation Operators for the Fixed Charge Transportation Problem

Christoph Eckert and Jens Gottlieb

SAP AG, Neurottstr. 16, 69190 Walldorf, Germany
{c.eckert,jens.gottlieb}@sap.com

Abstract. The fixed charge transportation problem (FCTP) has been tackled by evolutionary algorithms (EAs) using representations like permutations, Prüfer numbers, or matrices. We present a new direct representation that restricts search to basic solutions and allows using problem-specific variation operators. This representation is compared w.r.t. locality and performance to permutations and Prüfer numbers. It clearly outperforms all other EAs and even reaches the solution quality of tabu search, the most successful heuristic for the FCTP we are aware of.

1 Introduction

A transportation problem (TP) involves shipping certain amounts of a commodity from a set of sources, $S = \{1, \ldots, m\}$, to a set of destinations, $D = \{1, \ldots, n\}$. Each source $i \in S$ has a capacity a_i and each destination $j \in D$ has a demand b_j. The goal is to determine the amounts x_{ij} shipped from i to j, such that total transportation costs are minimized and capacity and demand constraints are met. While the linear TP is solvable in polynomial time due to its linear cost function [8], the *fixed charge transportation problem* (FCTP) is NP-complete [6] since it considers fixed costs in addition to linear costs. The FCTP is stated as

$$\text{minimize} \quad \sum_{i \in S} \sum_{j \in D} c_{ij} x_{ij} + f_{ij} g(x_{ij}) \tag{1}$$

$$\text{with} \quad g(x_{ij}) = \begin{cases} 1 & \text{if } x_{ij} > 0 \\ 0 & \text{else} \end{cases} \qquad \text{for } i \in S \text{ and } j \in D \tag{2}$$

$$\text{subject to} \quad \sum_{j \in D} x_{ij} = a_i \qquad \text{for } i \in S \tag{3}$$

$$\sum_{i \in S} x_{ij} = b_j \qquad \text{for } j \in D \tag{4}$$

$$x_{ij} \geq 0 \qquad \text{for } i \in S \text{ and } j \in D, \tag{5}$$

where variable and fixed costs are given by c_{ij} and f_{ij}, respectively. The FCTP is a linear TP if $f_{ij} = 0$ for all i, j, and it is trivial if $m = 1$ or $n = 1$ because in this case only one solution exists, which is the global optimum. Therefore, we assume $f_{ij} > 0$ for some i and j, $m \geq 2$, and $n \geq 2$.

J.J. Merelo Guervós et al. (Eds.): PPSN VII, LNCS 2439, pp. 77–87, 2002.

An essential property of the FCTP is derived from linear programming theory [8]: the global optimum is a basic solution of the underlying linear TP. Thus, it is promising to restrict search to these candidates, as done in the most successful heuristic we are aware of, the tabu search procedure from Sun et al. [11].

Evolutionary algorithms (EAs) were proposed for different TPs – including the FCTP – but only the decoder-based approaches using permutations [12] or Prüfer numbers [2, 7] guarantee producing basic solutions. The matrix-based EA [4] directly modifies transportation plans, but only guarantees at most $m+n-1$ positive entries in the transportation plan (x_{ij}), which may be also a non-basic solution. Prüfer numbers exhibit poor performance compared to permutations, which are slightly inferior to the matrix representation.

The goal of this paper is to develop a compact encoding that represents only basic solutions of the linear TP and that enables direct manipulation of transportation plans. This would allow working *directly* – i.e. without use of a decoder – in the same search space that is *indirectly* explored by the decoders based on permutations or Prüfer numbers. We base our study on publically available benchmarks[1] that have been discussed in [4, 11].

Section 2 reviews representations that have been suggested for the FCTP and introduces the new direct representation. The representations are examined concerning their locality in section 3. Section 4 presents empirical results, also including a comparison with tabu search. Conclusions are given in section 5.

2 Representations for the FCTP

2.1 The Permutation Representation

Permutations were applied to the linear TP [12] and the FCTP [2, 4]. Transportation plans (x_{ij}) consist of $m \cdot n$ entries, which can be indexed by $\{1, \ldots, m \cdot n\}$. A permutation of these indices is decoded to (x_{ij}) by traversing the variables x_{ij} in the order determined by the permutation, and assigning to each x_{ij} the maximum value with respect to the constraints (3) and (4). This decoding procedure generates only basic solutions, and all basic solutions are represented by at least one permutation. Uniform order based crossover and swap mutation are suitable operators for this representation [2].

2.2 The Prüfer Number Representation

The Prüfer number representation relies on a bijective mapping between spanning trees in a complete graph with r nodes and strings of $r-2$ node labels, which are called Prüfer numbers [9]. Since basic solutions for the FCTP correspond to trees in the transportation graph, Prüfer numbers were proposed for the FCTP by Li et al. [7]. As their decoding procedure may produce infeasible solutions, additional repair mechanisms that allow decoding arbitrary Prüfer numbers into feasible transportation plans were suggested [2]. This representation was used with one-point crossover and inversion as mutation operator [2,

[1] `http://www.in.tu-clausthal.de/~gottlieb/benchmarks/fctp`

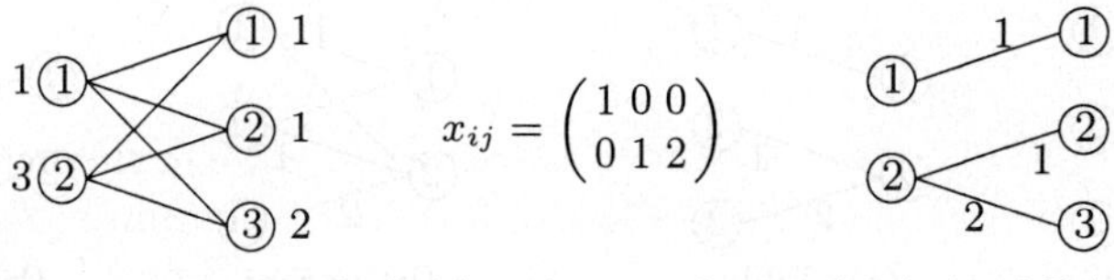

$$x_{ij} = \begin{pmatrix} 1\,0\,0 \\ 0\,1\,2 \end{pmatrix}$$

Fig. 1. A transportation graph (left), a basic solution (x_{ij}) (middle), and its forest consisting of the edges $E(x) = \{(1,1), (2,2), (2,3)\}$ (right).

7]. Prüfer numbers are clearly outperformed by the permutation representation, which is explained by their poor locality.

2.3 The Matrix Representation

Matrices offer a natural way to encode FCTP solution candidates, because they directly represent transportation plans. They have been used by Vignaux and Michalewicz for the linear transportation problem [12]. However, applying their operators to the FCTP causes high fixed costs, since they tend to produce solutions with many non-zero entries. Significantly better results for the FCTP were reported by Gottlieb and Paulmann, who used operators generating solutions with at most $m + n - 1$ positive entries [4]. Their approach is the most successful EA for the FCTP until now. Nevertheless, it does not produce only basic solutions and hence we are confident that there is room for further improvement.

2.4 A New Direct Representation

A basic solution has at most $m+n-1$ positive entries in (x_{ij}), and the positive entries form a tree in the transportation graph, respectively a forest if there are less than $m + n - 1$ non-zero entries. Thus, a basic solution x is represented by the set of its edges $E(x) = \{(i, j) \mid x_{ij} > 0, i \in S, j \in D\}$ and the corresponding amounts. This encoding is very compact and allows efficient evaluation of the solution. An example is shown in figure 1 for a FCTP instance with $m = 2$, $n = 3$, $a_1 = 1$, $a_2 = 3$, $b_1 = 1$, $b_2 = 1$ and $b_3 = 2$. In general, random basic solutions, which are needed to initialize a population, can be generated by decoding random permutations as decribed in section 2.1. The forest in figure 1 is constructed e.g. by traversing the variables in the order x_{11}, x_{12}, x_{21}, x_{22}, x_{23}.

Edge-Insert Cycle-Redistribution Mutation (EICR). Given the basic solution x, a new edge $(i, j) \notin E(x)$ is selected randomly from $S \times D$ and inserted into $E(x)$. We distinguish two cases, which are depicted exemplarily in figure 2. In case (a) there is a cycle in $E(x) \cup \{(i, j)\}$, e.g. because $|E(x)| = m + n - 1$. Then we can redistribute the amounts along that cycle by alternately decreasing and increasing by the same value. This causes $x_{ij} > 0$ and $x_{rs} = 0$ for some edge $(r, s) \in E(x)$, which is therefore deleted.

In case (b) there is no cycle, but we can easily produce one as follows. Since $(i, j) \notin E(x)$ and each node must have an incident edge in $E(x)$, we can randomly

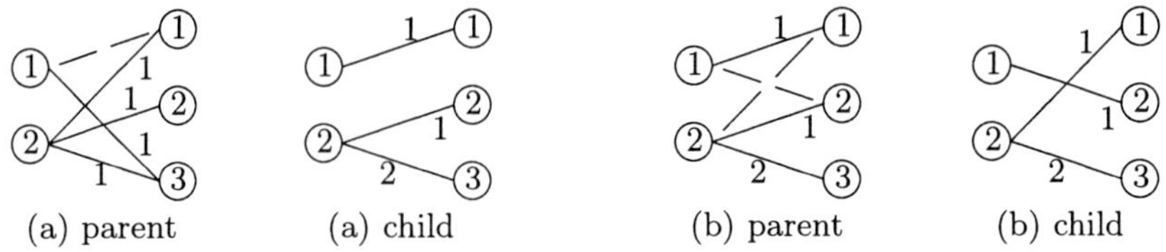

Fig. 2. Examples for EICR: Inserting $(1,1)$ causes $(1,3)$ and $(2,1)$ being removed (case (a)); insertion of $(1,2)$ causes inserting $(2,1)$ and removal of $(1,1)$ and $(2,2)$ (case (b)).

select two edges (i,s) and (r,j) with positive amounts. Inserting (i,j) and (r,s) into $E(x)$ yields a cycle of length 4, along which redistribution can take place as in case (a). Here, at least one of the edges (i,s) and (r,j) is removed. This operator produces a feasible new solution, which is always a basic solution.

Node Local Optimize Mutation (NLO). A node is selected randomly from $S \cup D$, and then all edges are checked that are incident to the selected node but not contained in the current basic solution. For each edge, its insertion is evaluated as done in case (a) of the EICR mutation; an edge is skipped if it does not immediately yield a cycle (unlike case (b) of EICR). Then, the edge with the best resulting objective value is selected. This resembles local optimization according to the $(1,\lambda)$-strategy, where λ offspring of the current solution are evaluated and the best offspring is selected. Note that only one node is selected at a time and that NLO is completed after processing this node. We also consider the $(1+\lambda)$-strategy, where the offspring compete with the current solution, and refer to the $(1,\lambda)$ variant as NLO1, and to the $(1+\lambda)$ version as NLO2.

Distance-Cut Crossover (DCX). The motivation of this crossover is that the child of two parents should inherit characteristics common in both parents, and that remaining properties should stem from at least one of the parents. In order to prevent too much similarity with a parent, the child should also have roughly the same distance from both parents. Note that a similar idea has been proposed for the traveling salesman problem, too [1].

Given the parents x and y, we insert half of the edges from $E(y) \setminus E(x)$ into $E(x)$. The edges are inserted sequentially, and each is processed identically as in the EICR mutation. Although this procedure does not guarantee to preserve all edges from $E(x) \cap E(y)$, the child is mainly formed by edges from $E(x) \cup E(y)$. Figure 3 shows an example for DCX and an instance with $m = n = 3$.

3 Locality Analysis

3.1 Phenotypic and Genotypic Distance Metrics

An individual is viewed as duality of its *genotype* and *phenotype*, where the phenotype is the solution itself – i.e. the transportation plan (x_{ij}) – and the

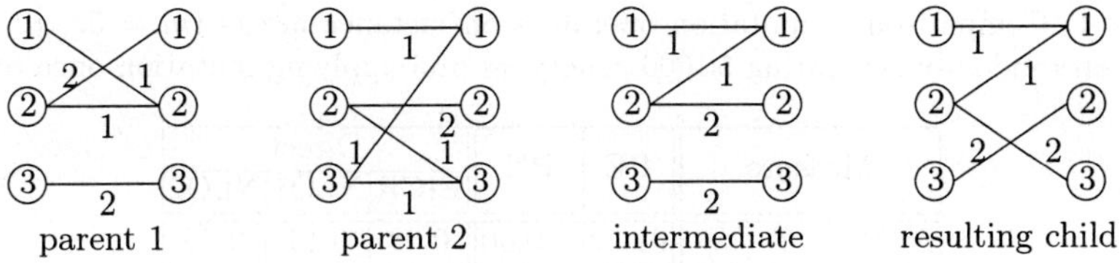

Fig. 3. Example for DCX: The edges (1, 1) and (2, 3) are inserted from the second parent into the first one, yielding the intermediate and the resulting child, respectively.

genotype is its representation, like e.g. a permutation, a Prüfer number, or a set of edges and the corresponding amounts. Locality means that small changes in genotype cause small changes in phenotype. Previous studies [2, 3, 5] showed that locality is crucial for good performance of evolutionary search, and that it can be quantified by similarity measures for both, genotypes and phenotypes.

The *phenotypic distance* of two transportation plans x, y is defined as

$$d_P(x,y) := |E(x) \setminus E(y)| + |E(y) \setminus E(x)| \ , \tag{6}$$

which counts the number of different edges used by x and y. We compared alternative phenotypic distances for the FCTP, including the one used in [2]. Since they are strongly correlated, we decided to use the most simple one.

In order to compare different representations, the representation-independent *genotypic distance* is defined implicitly as follows:

1. Two identical genotypes have distance 0, i.e. $d_G(X, X) = 0$ for genotype X.
2. Two distinct genotypes X, Y have distance $d_G(X, Y) = k$, $k \geq 1$, if X can be transformed to Y by k mutations.

Note that both d_P and d_G satisfy the metric conditions, namely identity, symmetry, and the triangular inequality.

3.2 Mutation Innovation

Given two genotypes X, Y with $d_G(X, Y) = 1$ and the corresponding phenotypes x and y, the *mutation innovation*

$$MI(X,Y) := d_P(x,y) \tag{7}$$

characterizes the phenotypic effects caused by a single application of the mutation operator. Table 1 analyzes the distribution of *MI* for a representative instance and the permutation representation (PE), the Prüfer number representation (PN) and the direct representation using different mutation operators.

The probability that a mutation operator does not affect the phenotype at all is measured by $P(MI = 0)$. While the mutation operator used by PE mostly affects only the genotype, which implies high redundancy of that encoding, the other representations are really innovative since the offspring's phenotype mostly

Table 1. Comparison of mutation operators on instance n370e ($m = 50$, $n = 100$), based on randomly generating 50 000 genotypes and applying mutation once to each.

Measure	PE	PN	Direct		
			EICR	NLO1	NLO2
$P(MI = 0)$ [%]	91.52	0.00	0.08	0.14	0.69
$E(MI \mid MI > 0)$	8.73	205.31	2.01	2.06	2.06
$\sigma(MI \mid MI > 0)$	7.50	39.47	0.11	0.24	0.25

differs from its parent. The operator NLO2 yields higher $P(MI = 0)$ than NLO1, because it prefers the parent over its offspring if the latter has worse fitness.

The main measure for characterizing locality is $E(MI|MI > 0)$, which represents the expected mutation innovation in the case that some phenotypic property has actually been affected. PN has poor locality, because a single application of the mutation operator produces an offpspring that is expected to differ in more than 200 edges from its parent in the instance n370e. The permutation representation achieves much higher locality since only about 8 edges are changed during mutation. As distinct phenotypes differ in at least two edges, the locality of the direct representation can be perceived as nearly perfect for all operators.

The standard deviation $\sigma(MI|MI > 0)$ is another indicator of locality. While we observe a high value for Prüfer numbers, the effects of mutation are more predictable for the other representations. The direct representation yields small values, showing the operators to have strongly limited impact on the phenotype.

3.3 Crossover Innovation and Crossover Loss

When analyzing crossover, we have to take into account that the offspring strongly depends on its parents' similarity. Therefore, we examine the effects of crossover for parents of different genotypic distance. Given parent X^1, we produce X^2 by k subsequent mutations, implying $d_G(X^1, X^2) \leq k$. Then, we apply crossover and analyze the offspring Y concerning its parents. Let x^1, x^2 and y be the associated phenotypes, then the *crossover innovation* is defined as

$$CI(X^1, X^2, Y) := \min(d_P(y, x^1), d_P(y, x^2)) , \tag{8}$$

the phenotypic distance between the offspring and the closer parent. This measures the ability of crossover to introduce new phenotypic material. Another important measure is the *crossover loss*,

$$CL(X^1, X^2, Y) := \left|E(y) \setminus (E(x^1) \cup E(x^2))\right| + \left|(E(x^1) \cup E(x^2)) \setminus E(y)\right| , \tag{9}$$

which quantifies both parents' phenotypic properties that the offspring did not inherit during crossover. Figure 4 shows interesting measures based on CI and CL for a representative instance, different parental genotypic distances k, and the permutation representation (PE), the Prüfer number representation (PN) and the direct representation using DCX and EICR as mutation.

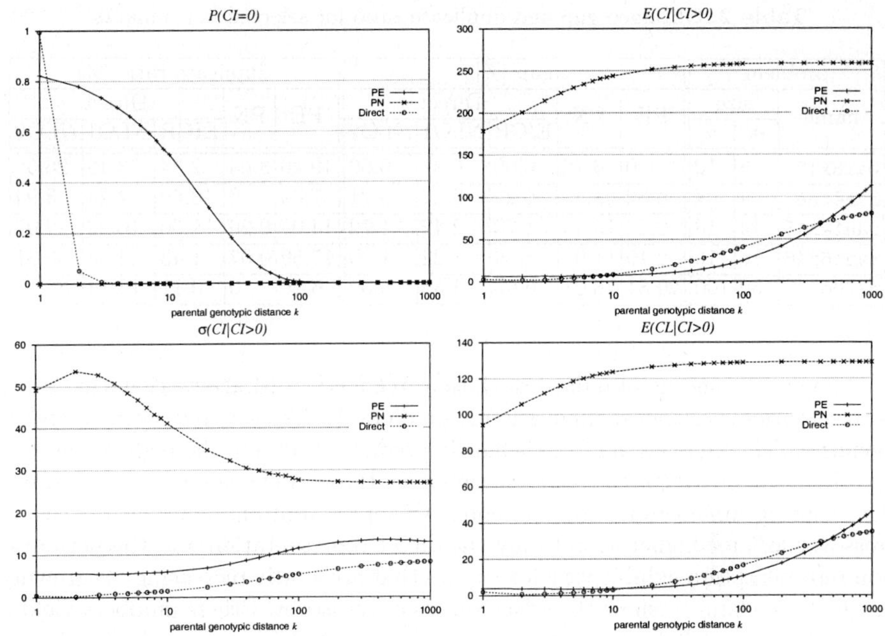

Fig. 4. Empirical comparison of crossover operators on instance n370e ($m = 50$, $n = 100$), based on randomly generating 50 000 genotypes X_1, transforming each X_1 to X_2 by k mutations, and producing offspring Y by applying crossover to X_1, X_2.

The measure $P(CI = 0)$ represents the probability for crossover producing a child that is phenotypically identical to one of its parents. In the case of Prüfer numbers, even for highest parental genotypic similarity $k = 1$ – i.e. both parents differ only by a single mutation – crossover never produces offspring identical to a parent. The permutation representation's redundancy causes relatively high values for $k > 1$. Only the direct representation always produces an offspring identical to one of its parents for $k = 1$, which we perceive as ideal. For higher k crossover is mostly innovative, which shows that the direct representation is more compact and less redundant than the permutation representation. We remark that $P(CI = 0) = 1$ for $k = 0$ (not shown in the figure) since all crossovers preserve the parental phenotype if the parents are genotypically identical. Both, the direct and the permutation representation are smooth regarding the transition from $k = 0$ to $k = 1$, whereas Prüfer numbers behave extremely as crossover and mutation cannot preserve phenotypic properties.

The degree of innovation in case of actual innovation, $E(CI|CI > 0)$, mainly characterizes locality. Increasing values are expected (and confirmed by the figure) for increasing parental genotypic distance k. While PE and the direct representation exhibit a comparable, smooth increase that indicates locality, PN yields a dramatically high value even for $k = 1$, underlining a lack of locality.

Table 2. Obtained gap and duplicate ratio for selected benchmarks.

problem			gap [%]					duplicate ratio [%]				
name	size		PE	PN	Direct			PE	PN	Direct		
	m	n			EICR	NLO1	NLO2			EICR	NLO1	NLO2
bal8x12	8	12	0	0.12	0.00	0.01	0.00	19.40	3.04	2.84	23.19	26.56
ran4x64	4	64	0.82	21.55	0.29	0.33	0.21	7.69	0	0.50	4.24	3.94
ran14x18	14	18	2.12	13.44	1.73	2.42	1.72	14.07	0.20	1.50	16.22	22.07
ran16x16	16	16	2.19	11.01	1.39	1.22	1.90	17.59	0.07	1.43	13.34	36.91
n370e	50	100	20.81	78.01	6.66	4.35	5.09	46.26	0	0.11	13.54	83.05

The corresponding standard deviation $\sigma(CI|CI > 0)$ shows that the direct representation's crossover innovation is predictable, like PE that yields a higher deviation. The innovation of PN is hardly predictable due to the high deviations.

The measure $E(CL|CI > 0)$ characterizes the loss of phenotypic material if the offspring differs from both parents. All representations yield an increasing crossover loss for higher k, but only the direct representation and the permutation representation exhibit very low values that are slowly increasing for growing k, which we claim to show the effectiveness of crossover. Prüfer numbers yield a very high loss of phenotypic material, even for such small values like $k = 10$.

4 Results

4.1 Comparison of the Representations

The representations are evaluated in a performance comparison on a set of benchmark problems. We are using the same general EA setup as in our previous study [2]: population size 100 (200 for n370e), parent selection by tournaments of size 2, crossover probability 1, mutation probability 1, steady-state replacement with phenotypic duplicate elimination[2], and a limit of 1 000 000 non-duplicate solutions.

Solution quality is measured by the gap, defined as $min^{EA}/opt - 1$ where min^{EA} and opt are the best solution found by the EA and the best known solution[3], respectively. Search efficiency is characterized by the duplicate ratio, the ratio of rejected duplicates among all generated solutions. Table 2 presents the average results obtained in 15 runs for each instance and representation.

Prüfer numbers (PN) yield worse gaps – even for the smaller problems – than permutations (PE), as discussed and explained earlier in [2]. For most small problems, the new direct representation is superior to PE. On the most challenging instance n370e, which has a size that is larger by several orders

[2] The worst individual in the population is replaced by the offspring, if its phenotype is not already represented by some genotype in the population. This requires only moderate additional computational efforts and significantly improves quality [10].

[3] The global optimum is known except for the largest instance n370e. Thus, gaps presented for n370e are lower bounds on the real gaps w.r.t. the global optimum.

of magnitude, all variants of the direct representation clearly outperform the permutation representation. The comparison among the direct representation's variants is inconclusive on the smaller instances; on the most difficult instance, the best gap is obtained by NLO-1, followed by NLO-2 and EICR.

While Prüfer numbers produce only very few duplicates, due to their poor locality, some of the other representations produce a significant amount of duplicates. The permutation representation produces many duplicates, corresponding to the high values reported for $P(MI = 0)$ in section 3.2. The direct representation is very efficient when EICR is used, but many duplicates are produced by the variants of NLO. The strategy of NLO-2, which allows the parent being preferred to its offspring, causes a higher duplicate ratio than NLO-1. In particular for `n370e`, the duplicate ratio of NLO-2 is not acceptable since most offspring are rejected, making the search process inefficient.

Among the variants of the direct representation, NLO-1 is more promising than NLO-2, because the former yields a better gap and a much smaller duplicate ratio on `n370e`. There is a trade-off for EICR and NLO-1 concerning solution quality and duplicate ratio. We discuss this trade-off by considering the total CPU time needed for an EA run. The CPU times – not presented in detail here – are summarized as follows. The fastest EAs are PN and the direct representation with EICR, which are computationally equivalent. NLO-1 and NLO-2 need twice the CPU time of EICR, and PE is three times slower than PN. In order to compare EICR with NLO-1, we made EA runs with EICR and a time limit identical to the time needed by an NLO run. Given the same CPU time, EICR was not able to reach the solution quality of NLO-1 or NLO-2, and therefore NLO-1 appears to be the best choice.

We also made experiments with the matrix representation that was the best EA until now [4]: Its solution quality for `n370e` was clearly inferior to the new representation, independent of which mutation operator is used. Thus, we perceive the direct representation using NLO-1 as the most effective representation.

4.2 Comparison with Tabu Search

The solution quality obtained by the tabu search approach of Sun et al. [11] could not be reached by any EA until now. In order to check whether the new direct representation is competitive to tabu search, we analyzed the EA with NLO-1 in more detail. We found further improvements using a crossover probability 0.6 and a special parent-child replacement scheme: The offspring is rejected if its phenotype is already represented in the current population, and otherwise it replaces the worst parent. That means that the parent is always replaced in case of mutation, and the worst parent is replaced if the offspring is produced by both, crossover and mutation. Increasing the limit to 10 000 000 non-duplicate solutions yields an additional improvement, but another increase of this limit does not help, due to convergence of the population.

After all these fine-tuning steps, we obtained the results presented in table 3. The gap for tabu search (TS) is based on a single run, personally communicated to us by M. Sun, and the results for the EA are based on 15 runs per instance.

Table 3. Comparison of the best evolutionary algorithm with tabu search on instances `n370*` of size $m = 50$, $n = 100$; instance `n370d` is omitted as it is identical to `n370e`.

Instance	0	1	2	3	4	5	6	7	8	9	a	b	c	e
gap EA/best [%]	1.48	1.26	1.41	1.41	1.22	1.71	1.23	2.17	1.62	1.38	1.11	1.96	1.03	1.95
gap TS/best [%]	3.76	3.35	2.91	2.30	1.91	2.28	3.35	2.64	1.99	1.62	3.11	2.91	1.46	2.83
gap TS/EA [%]	2.25	2.07	1.48	0.87	0.68	0.56	2.10	0.47	0.36	0.24	1.98	0.92	0.43	0.86
EA wins	15	15	15	12	12	13	15	11	13	11	14	13	11	15

We observe a high potential of the EA since most runs beat tabu search and the EA obtains a better average gap on all instances. Note that the EA has significantly improved all best known solutions.

However, we do not claim that the EA outperforms tabu search, because it is quite difficult to compare the computational costs since different machines were used and the tabu search code was not available to us. Since we allowed much CPU time on a machine that we estimate as being much faster than the machine Sun et al. were using, our results must be interpreted carefully. Our goal was to demonstrate the ability of the EA to produce high-quality solutions.

5 Conclusions

We have developed a new direct representation and several variation operators for the FCTP, which enable direct evolutionary search in the space of basic solutions of the underlying TP. Our locality analysis shows the new representation providing much more locality than decoder-based approaches. As its locality is nearly perfect, the new representation allows highly effective exploration of phenotypic neighbourhoods of promising solutions. This results in high solution quality, superior to all other EAs applied to the FCTP so far. Fine-tuning even allows beating the solution quality of tabu search, and new best solutions have been found on all instances that could not be solved by exact methods until now.

As the overall results are excellent, it is challenging to investigate how this EA could be further improved. We mention two ideas that may be worth studying, (i) the incorporation of more powerful local search into the EA, and (ii) using an adaptive mechanism to drive the population away from local optima that have dominated the search for many subsequent generations.

References

1. B. Freisleben and P. Merz. New genetic local search operators for the traveling salesman problem. In *Proc. of PPSN IV*, 890 – 899, 1996
2. J. Gottlieb and C. Eckert. A comparison of two representations for the fixed charge transportation problem. In *Proc. of PPSN VI*, 345 – 354, 2000
3. J. Gottlieb, B. A. Julstrom, G. R. Raidl and F. Rothlauf. Prüfer numbers: a poor representation of spanning trees for evolutionary search. In *Proc. of Genetic and Evolutionary Computation Conference*, 343 – 350, 2001

4. J. Gottlieb and L. Paulmann. Genetic algorithms for the fixed charge transportation problem. In *Proc. of 5th IEEE Int. Conf. on Evol. Comp.*, 330 – 335, 1998
5. J. Gottlieb and G. R. Raidl. Characterizing locality in decoder-based EAs for the multidimensional knapsack problem. In *Proc. of Artificial Evolution*, 38 – 52, 1999
6. G. M. Guisewite and P. M. Pardalos. Minimum concave-cost network flow problems: Applications, complexity, and algorithms. *Annals of Operations Research*, Vol. 25, 75 – 100, 1990
7. Y. Li, M. Gen and K. Ida. Fixed charge transportation problem by spanning tree-based genetic algorithm. *Beijing Mathematics*, Vol. 4, No. 2, 239 – 249, 1998
8. G. L. Nemhauser and L. A. Wolsey. *Integer and combinatorial optimization.* Wiley, 1998
9. C. C. Palmer and A. Kershenbaum. Representing trees in genetic algorithms. In *Proc. of 1st IEEE Conf. on Evolutionary Computation*, 379 – 384, 1994
10. G. R. Raidl and J. Gottlieb. On the importance of phenotypic duplicate elimination in decoder-based evolutionary algorithms. In *Late Breaking Papers at the Genetic and Evolutionary Computation Conference*, 204 – 211, 1999
11. M. Sun, J. E. Aronson, P. G. McKeown and D. Drinka. A tabu search heuristic procedure for the fixed charge transportation problem. *European Journal of Operational Research*, Vol. 106, 441 – 456, 1998
12. G. A. Vignaux and Z. Michalewicz. A genetic algorithm for the linear transportation problem. *IEEE Transactions on Systems, Man, and Cybernetics*, Vol. 21, No. 2, 445 – 452, 1991

On the Utility of Redundant Encodings in Mutation-Based Evolutionary Search

Joshua D. Knowles[1] and Richard A. Watson[2]

[1] IRIDIA - CP 194/6, Université Libre de Bruxelles
Ave. F. D. Roosevelt, 50, 1050 Brussels, Belgium
jknowles@ulb.ac.be
http://iridia.ulb.ac.be/~jknowles/
[2] DEMO, Volen Center for Complex Systems
MS018, Brandeis University, Waltham, MA 02454, USA
richardw@cs.brandeis.edu

Abstract. A number of recent works in the evolutionary computation field have suggested that introducing large amounts of genetic redundancy may increase the evolvability of a population in an evolutionary algorithm. These works have variously claimed that the reliability of the search, the final fitness achieved, the ability to cope with changing environments, and the robustness to high mutation rates, may all be improved by employing this strategy. In this paper we dispute some of these claims, arguing that adding *random* redundancy cannot be generally useful for optimization purposes. By way of example we report on experiments where a proposed neutral encoding scheme (based on random Boolean networks) is compared to a direct encoding in two mutation-only EAs, at various mutation rates. Our findings show that with the appropriate choice of per-bit mutation rate, the evolvability of populations using the direct encoding is no less than with the redundant one.

1 Introduction

The neutral theory of evolution [7] states that a large fraction of mutations are fitness neutral. At the molecular level, many genotypes (RNA) effectively map to functionally equivalent phenotypes (proteins) and can form *neutral networks* [10] of genotypes separated by only one (or a few) point mutations. At a higher level also, genetic regulatory networks [2] may exhibit neutrality between the organization of the networks and their characteristic expression patterns. The dynamics of populations evolving on such neutral networks are different to those on adaptive landscapes [6]: they are characterized by long periods of 'fitness stasis' (where variation acts to increase the genetic diversity of the population, causing it to diffuse over the neutral network) punctuated by shorter periods of rapid fitness increase (and loss of diversity) due to the discovery of a 'portal' [14] to a network of higher fitness. These dynamics, it is suggested, mean that populations may be better able to evolve to higher fitness configurations and less likely to fall prey to the usual problems of entrapment at local optima.

J.J. Merelo Guervós et al. (Eds.): PPSN VII, LNCS 2439, pp. 88–98, 2002.

Some researchers in the evolutionary computation (EC) community ([1], [4], [11], [12], [13], [15], [18]) have begun to address neutral theory in their work. Some (e.g. [1],[15]) suggest that certain real-world problems, when represented naturally, will exhibit a high degree of neutrality, and thus standard models of the dynamics of populations evolving on (non-neutral) adaptive landscapes may not be relevant. Others ([4], [11], [12], [13]) suggest that introducing artificial redundancy into the representations of optimization problems may improve the evolvability of solutions in an evolutionary algorithm. In particular, the latter have variously claimed that the reliability of the search [12], the final fitness level achieved, the ability to cope with changing environments, and the robustness to high mutation rates [4], of EAs may all be improved by genetic redundancy.

The idea that a search process, that would otherwise be trapped at a local optimum, may be able to traverse, or diffuse over, a neutral network and escape to higher-fitness regions of the search space is intuitively appealing. However, from an engineering point of view, we must be careful in embracing neutrality as a generally advantageous property of a search space. There are, after all, other ways to escape from local optima or decrease the likelihood of becoming stuck at one—for example, using different variation operators, by employing diversity maintenance techniques, and not least, by choosing another encoding scheme. At present, it is poorly understood exactly what kind of neutrality is advantageous, and under what circumstances such an advantage will be shown—it is clear that not all types of genetic redundancy will be useful [5].

In [4], Ebner *et al.* describe several valuable conceptual features of redundant encodings and the neutral networks that such encodings might afford. For example, they suggest that the frequency distribution of phenotypes, the 'reachability' of phenotypes, the 'innovation rate' (the number of new phenotypes encountered as a function of neutral walk length), and the extent of the neutral networks are important characteristics of a redundant encoding and they provide corresponding statistical measures for these features. Their discussion and examinations using various artificially redundant encodings are valuable in understanding the possible nature of redundancy and neutrality in different genotype to phenotype mappings, and in assessing the possible utility of intrinsically redundant mappings in natural and engineering domains. However, we suggest that it is still not clear when an encoding incorporating neutrality should be preferred over a non-redundant encoding in cases where there is a choice. In this paper we wish to point out that some simple suggestions of when neutrality will be advantageous are inadequate, and incidentally, to contrast this with some other possibilities that may be genuinely advantageous but have not yet been demonstrated.

The remainder of this paper is organized as follows. In the next section we discuss why, in general terms, we would not expect the performance of EAs to be improved by the introduction of arbitrary neutral networks. We also highlight a number of results reported in the literature, which in our view do not represent convincing demonstrations of the advantages of the employed redundant encodings. In section 3 we describe a number of experiments that compare a redundant versus a direct encoding, regarding their performance in otherwise identical mutation-only EAs. Three distinct types of problem, H-IFF, MAX-SAT, and *NK* landscapes are used for the comparison. The results of these experiments are reported in section 4, and we review our findings in section 5.

2 Against Arbitrary Neutral Networks

Why should we believe that neutrality in a problem space is advantageous from a search or optimization point of view? One simple way to understand the effect of neutrality is that it may afford better exploration for a search process because, for any method that follows fitness increases, a redundant encoding may have better connectivity to other (perhaps fitter) phenotypes and thus fewer local optima. All else being equal, even an arbitrary increase in connectivity—neutral networks connecting arbitrary phenotypes—certainly provides a potential advantage for optimization—[4], for example, provides illustrations of this advantage. However, in engineering domains we are not necessarily required to keep 'all else equal'. If we have a choice about whether to include redundancy in an encoding scheme, the advantages of doing so must be weighed against the potential advantages of other alternatives. For example, we could provide 'connectivity' from any local optimum to any point in the search space by using a simple hill-climber that, in addition to considering neighbouring points in the search space, also occasionally considers random points in the search space. Alternatively, we might suppose that it would be preferable for neutral networks to increase connectivity to nearby (but not immediately adjacent) points in the search space rather than arbitrary points—but similarly, this could be provided by a simple mutation hill-climber using a larger mutation rate.

A useful question for focusing our reasoning is the following. Why, if at all, should a search method in a problem space with neutrality be able to find points of higher fitness faster or more reliably than it would with additional random exploration on a non-neutral space? For example, the expected time for a random walk to change M variables to a specific configuration is exponential in M, as is the expected time to jump from one point to another specific point directly under M-point mutation. In some scenarios, the exploration of a population on a neutral network might not be described accurately by a random walk, but favourable analyses are thus far absent. Of course, conceivably, a particular encoding, 'natural' or artificial, might provide some non-random bias in the connectivity of points that has an *a priori* bias appropriate for a particular problem domain. However we do not yet have any suggestions for what kind of neutrality is useful for what class of problems. Another possibility is that, in certain scenarios, the structure of neutral pathways might be somehow reorganized adaptively as evolution progresses. These possibilities, a useful *a priori* bias and/or an adaptive bias, in the connectivity of neutral networks could provide a genuine adaptive advantage in certain classes of problems—but they have not yet been demonstrated. In the meantime, it is difficult to see why any kind of arbitrary increase in connectivity provided by random neutral networks, whether inherent in the natural encoding of a problem space or added artificially, would provide an optimization advantage over much simpler search methods on non-redundant encodings.

To provide a preliminary illustration of this reasoning we examine, by way of example, a redundant encoding proposed by [4] and [13]. One particular encoding reported in this work, based on the operation of a random Boolean network (RBN), appears to be favourable with respect to the various statistical measures mentioned above. Furthermore, some experiments report results that seem to show that this mapping aids evolutionary search in multi-modal landscapes [4], [11]. However, it seems likely that the increased connectivity provided by the RBN encoding is arbitrarily structured, and that accordingly, the optimization advantage seen in this experimental work could be provided by a simpler method of random exploration. In general, an increased mutation rate is a fairly unbiased means to increase random exploration (though not completely, see [3]). Therefore, it seems unlikely that an arbitrary artificial redundant encoding could, in general, be better than increased mutation. One possible disadvantage of a higher mutation rate is an inability to retain good solutions should they be discovered—however, the use of some elitism (retaining the best solution found so far) easily remedies this problem. Accordingly, the use of higher mutation rates, in hill-climbers and GAs with elitism (steady-state variety), is the alternative that we examine in some detail below.

3 Experimental Method

3.1 Encodings. In the experiments that follow, we will compare a direct encoding and a redundant encoding of binary strings. In this context, we use the term "phenotype" to refer to the binary string of M bits that the fitness function evaluates. The term "genotype" is reserved for the binary string that is under the direct influence of the variation operators of the evolutionary algorithm. In the direct encoding, the genotype is identical to the phenotype, and no 'mapping' step is necessary. In contrast, the genotype of the redundant encoding is of length $L > M$ bits, and represents a particular random Boolean network (RBN), and its initial state (described below). The RBN is then iterated for a fixed number of steps, and its final state represents the phenotype for which the genotype encodes.

3.2 Aims. The experiments are designed to evaluate the following aspects of performance of EAs using the random Boolean network encoding: (i) the distribution of fitness levels achieved on relatively long runs, for the size of problem; (ii) the rate of fitness increase; (iii) the robustness to the choice of mutation rate; and (iv) changes in (i) and (ii) with respect to any controllable problem features and problem size. We assess these aspects of performance by comparing identical EAs operating on a direct encoding and an RBN encoding of the binary phenotypes.

3.3 The RBN Encoding. An RBN is an iterated dynamical system described over a set of V Boolean variables. The state of each variable at time t is dependent on the state of W other variables at time $t-1$. The new state is the result of a Boolean function over these variables. In our experiments (following the setup described in greater detail in [4]), the initial state of the variables, the wiring determining which variables are dependent on which others, and the Boolean state-update functions for each variable are all specified in the genotype and are subject to evolutionary adaptation. The value of W is fixed at 3. The number of variables V equals the phenotype length M, and the phenotype is determined by running the Boolean network for 20 iterations and reading off the final state of the variables. The problems we address are of length 32 and 64 bits. With the RBN encoding, the genotypes associated with these phenotypes are of length 768 and 1728 bits respectively.

3.4 Mutation Rates. In our experiments we aim to evaluate the performance of the RBN encoding in comparison to the use of a direct encoding, *given an appropriate choice of mutation rate.* The question whether the RBN encoding facilitates greater exploration (and thus evolvability) when mutation is restricted to a single point per-genotype mutation is regarded here as irrelevant, since this kind of mutation would never be used in practical EAs designed for use on multi-modal fitness landscapes. Thus, we choose to compare the performance of the two encodings in EAs using *per-bit mutation*, and we attempt to find an appropriate rate of mutation for both encodings. All of our experiments are carried out using per-bit (flip) mutation rates of $1/L$, $2/L$, $4/L$ and $8/L$.

3.5 Problems. The problems that we use are all unconstrained binary string problems, allowing straightforward use of the RBN encoding. None of the problems intrinsically possesses substantial neutral networks. Our interest here is in evaluating if the *introduction* of neutrality through a redundant genotype-phenotype mapping helps perform search in general, and not with the question of how to perform search on landscapes that have intrinsic neutrality, such as Royal stairways or *NKp* landscapes. Hence, for our problems we use *NK* landscapes [6], H-IFF, initially proposed in [17] and described more fully in [16], and MAX-SAT [8]. The *NK* landscapes allow us to observe if relative performance is affected by varying degrees of ruggedness or correlation. We use an *NK* landscape generator, [9], available as freeware, to generate landscapes of size $N = 32$. We generate one problem each at levels of $K = 2, 4, 8, 16$, and these are labelled NK2, NK4, NK8, and NK16, respectively, in the results section. The H-IFF problem is hill-climber-hard, in that the two global optima (the all 0s string and the all 1s string) are difficult to find by mutation, starting from a random string, because contiguous blocks of both 0s and 1s of size two or greater are rewarded by the hierarchical fitness function. Thus, strings tend to evolve toward local optima (strings comprising sizable blocks of both 0s and 1s) with a very large Hamming distance from the two global optima. This problem is very rugged and it has been shown that mutation cannot be guaranteed to succeed in time less than exponential in the problem size [16]. We use H-IFF instances of length 32 and 64 bits, labelled H-IFF32 and H-IFF64. MAX-SAT is a well-known NP-hard combinatorial optimization problem, and its real-world applications are many. Our two instances are un-weighted 3-SAT problems generated using an algorithm described in [8], for which freeware is available. The first has 32 variables and 200 clauses (SAT32-200) and the second has 64 variables and 280 clauses (SAT64-280), and both are satisfiable. We treat these 3-SAT instances as MAX-SAT optimization problems, with fitness calculated as the proportion of satisfied clauses.

3.6 The EAs: Hill Climber and Genetic Algorithm. We evaluate the performance of the RBN in (i) a random mutation hill-climber (HC) and (ii) a steady-state mutation-only genetic algorithm (GA). HC uses a (1+1) population update: at each iteration, one random mutant of the current solution is generated (independently of previous steps) and the mutant replaces the current solution iff it is at least as fit. GA uses a population size of 100 and uses a steady-state population update. At each iteration, a 2-tournament is used for selection and replacement. In this, two members of the population are chosen at random without replacement. The fitter of the two becomes the parent. The parent is then replaced in the population, copied and a mutant of the copy is generated and evaluated. The mutant then replaces the less fit of the two chromosomes from the tournament, iff it is at least as fit. The two EAs using the random Boolean network encoding are identical to HC and GA except for the encoding, and are labelled HC-NN and GA-NN, respectively. We further use the shorthand GA-NNm1 to denote GA-NN used with a mutation rate of $1/L$, and similarly for the other EAs and mutation rates. With all the algorithms we run until the global optimum has been found (if known) or 4 million iterations have been performed.

4 Simulation Results

Figure 1 presents four plots which give an insight into the difference between algorithm runs with and without the RBN encoding. On NK2, all algorithms reached the global optimum within the allotted number of iterations so that there is no statistical difference in the final fitness values achieved. However, as can be seen from the plot top left, maximum fitness is achieved much faster using the direct encoding than with redundancy, in both GA and HC algorithms (the best mutation rates for each from those rates tested were selected). The plot at the top right shows the evolution of genetic diversity, defined as the sample average probability that two randomly selected individuals differ at a randomly chosen locus, as in [4]. In GA-NN's population, diversity grows over time until it stabilises near the maximum value of 0.5, indicating diffusion through the neutral network of the global optimum. In contrast, GA has lost all genetic diversity. However, the greater diversity in GA-NN's population did not increase its ability to find 'shortcuts' towards the optimum; it got there *more slowly.* In the plot bottom left, the number of phenotypically neutral mutations are plotted against iterations. The massive redundancy in the RBN encoding means the number of neutral mutations is higher than in the direct encoding, at each

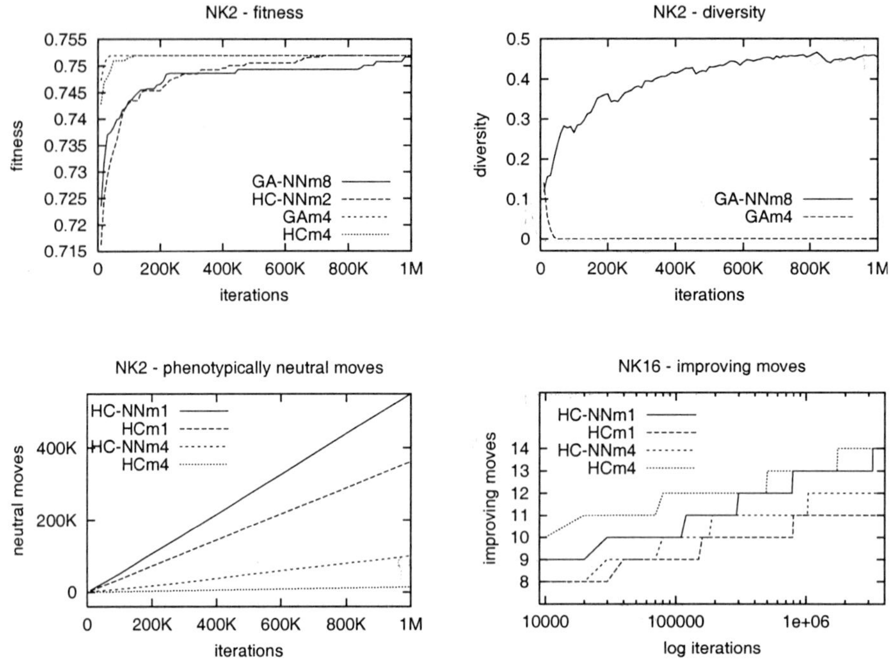

Fig. 1. Plots showing the evolution of (top left) fitness and (top right) diversity on NK2, and the cumulative number of (bottom left) phenotypically neutral moves. For NK16 the cumulative number of fitness-improving moves is shown (bottom right) against log iterations. All values are the mean of 30 independent runs. See text for discussion of these plots.

mutation rate. However, in the plot bottom right (for the NK16 problem), where we can see the cumulative number of mutations that lead to increases in fitness, the characteristics of the HCs employing the RBN encoding do not appear to benefit from the extra neutrality. In fact, we observe that the dynamics look very similar in the four algorithms: a similar number of fitness improvements is seen and the waiting times between them increase exponentially. The plots on these problems are typical for all of the problems considered here.

The final fitness values that were recorded seem to indicate that the random Boolean network encoding does not significantly benefit either GA or HC on any problem, when an appropriate mutation rate is selected. Figures 2 and 3 present these results in the form of boxplots[1]. On NK16 (Figure 2), it appears that GAm4 achieves the highest median value and, together with GAm2, it has the best distribution of final values. With the hill-climbers, there appears to be a marginal advantage to the redundant encoding. We also observe that HC-

[1] Box plots graphically represent the distribution of a random variable. The centre line represents the median, the length of the shaded box indicates the interquartile range (IQR) and the whiskers extend out to the largest and smallest observations within 1.5 times the IQR. Outliers are shown as small circles.

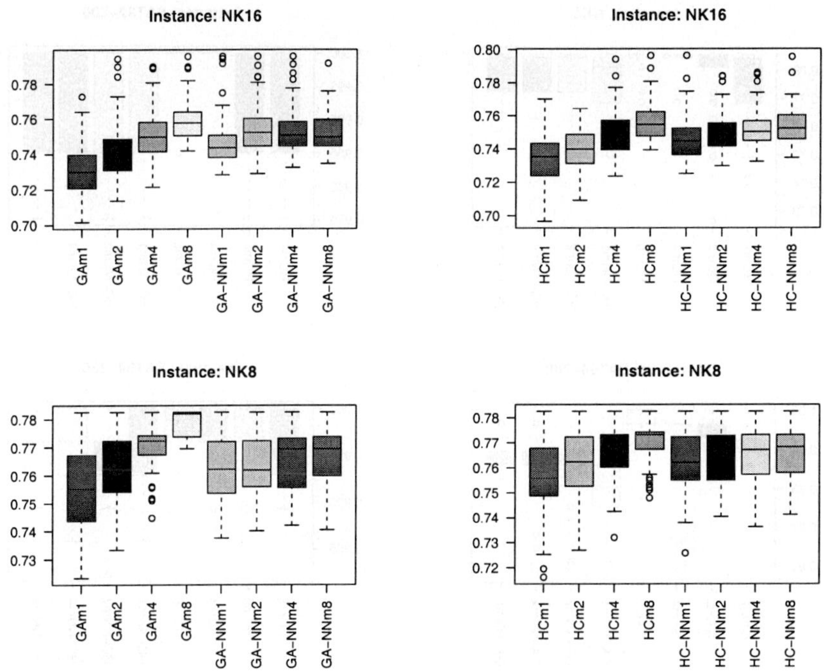

Fig. 2. Results from 100 independent runs on the two more rugged *NK* landscapes.

NN is more robust with respect to different mutation rates than is HC on this problem. On NK8, GAm8 performs best. Once again there is more robustness to mutation-rate choice in GA-NN, but the spread in final values is also greater. There is little to choose between the performance of the best HC and HC-NN on this problem although one might argue that the standard hill-climber with the highest mutation rate (HCm8) seems to be best. On NK4, (Figure 3, top left), HC appears to perform better than HC-NN. On the MAX-SAT problems (also Figure 3) there is a clear difference in favour of the non-redundant encoding on all plots shown[2]. The results on the H-IFF problem shown in Figure 3 are for experiments with 64 bit H-IFF as opposed to the 16 bits used in [4]. Our results demonstrate that on this size of problem the redundant encoding performs worse than the non-redundant encoding, possibly indicating that performance does not scale with the use of redundancy. Our results on H-IFF32 (not shown[2]), are also consistent with this. Generalizing from all the plots, there does not seem to be a difference in the spread of final fitness values achieved, using the standard and redundant encodings, indicating that the RBN does not appear to increase the reliability of the search.

[2] GA results for SAT32-200 and NK4 are not shown because the global optimum is found reliably by both GA and GA-NN at some of the mutation rates. These plots, and others, can be obtained from the first author on request.

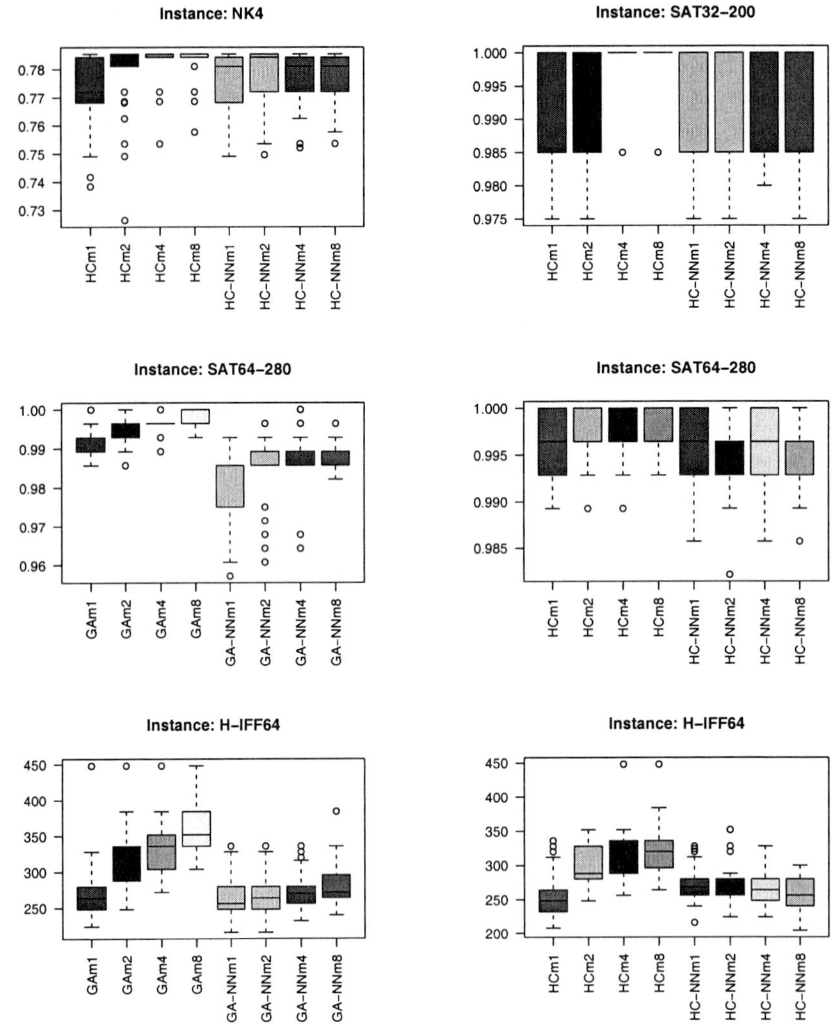

Fig. 3. Results from 100 independent runs each on selected problems.

5 Conclusion

In this paper we have argued against the theory that adding *random* genetic redundancy to the encoding of a problem will generally increase the search performance of a correctly configured EA. The evidence in the literature given to support the use of redundant encodings, we argued, only shows the increased connectivity of the redundant space, and is not a sound demonstration of the utility of the technique. In our experiments we found that when HCs and GAs are provided with massively redundant genotypes based on a random Boolean network, their performance does not improve, in terms of the final fitness achieved on relatively long runs. In addition, the RBN encoding does not appear to affect the number of improving moves to reach a particular fitness value, or the distribution in waiting times between them. Neither was the spread of final fitness values achieved tighter for the RBN encoding (i.e. increased reliability was not observed). On the positive side, the performance of the RBN-encoded algorithms did seem to be more robust to the choice of per-bit mutation rate on most problems. The findings reported here are of course very limited, and other experiments may give different results. However, the consistent finding across the problems tackled was that with an appropriate per-bit mutation rate, escape from local optima and the improvement of fitness occur just as well in a direct encoding as with the redundancy provided by the RBN encoding.

Acknowledgments

JK gratefully acknowledges the support of a European Commission Marie Curie Fellowship, contract number: HPMF-CT-2000-00992.

References

1. L. Barnett. Netcrawling—optimal evolutionary search with neutral networks. In *Proceedings of the 2001 Congress on Evolutionary Computation (CEC2001).* IEEE Service Centre, Piscataway NJ, 2001.
2. S. Bornholdt and K. Sneppen. Robustness as an evolutionary principle. *Proceedings of the Royal Society of London Series B-Biological Sciences*, 267:2281–2286, 2000.
3. S. Bullock. Smooth operator? Understanding and visualising mutation bias. In J. Kelemen and P. Sosik, editors, *Sixth European Conference on Artificial Life*, pages 602–612, 2001.
4. M. Ebner, M. Shackleton, and R. Shipman. How neutral networks influence evolvability. *Complexity*, 2002. (In press).
5. I. Harvey and A. Thompson. Through the labyrinth evolution finds a way: A silicon ridge. In *Proceedings of ICES96*, pages 406–422. Springer-Verlag, 1996.
6. S. A. Kauffman. Adaptation on rugged fitness landscapes. In D. Stein, editor, *Lectures in the Sciences of Complexity*, pages 527–618. Redwood City: Addison-Wesley, 1989. SFI Studies in the Sciences of Complexity, Lecture Volume I.
7. M. Kimura. *The Neutral Theory of Molecular Evolution.* Cambridge University Press, 1983.
8. M. Motoki and R. Uehara. Unique solution instance generation for the 3-satisfiability (3-SAT) problem. Technical Report C-129, Dept. of Math and Comp. Sciences, Tokyo Institute of Technology, 1999.
9. M. A. Potter. An *NK*-landscape generator, 1997. Presented at the Workshop on Test Problem Generators for Evolutionary Algorithms, at the 7th ICGA.
10. C. M. Reidys and P. F. Stadler. Neutrality in fitness landscapes. *Appl. Math. Comput.*, 117:321–350, 1998.
11. M. Shackleton, R. Shipman, and M. Ebner. An investigation of redundant genotype-phenotype mappings and their role in evolutionary search. In *Proceedings of the 2000 Congress on Evolutionary Computation*, pages 493–500, Piscataway, NJ, 2000. IEEE Service Center.

12. R. Shipman. Genetic redundancy: Desirable or problematic for evolutionary adaptation? In *Proceedings of the 4th International Conference on Artificial Neural Networks and Genetic Algorithms (ICANNGA)*. Springer-Verlag, 1999.
13. R. Shipman, M. Shackleton, M. Ebner, and R. Watson. Neutral search spaces for artificial evolution: a lesson from life. In *Artificial Life VII: Proceedings of the Seventh International Conference*. MIT Press, 2000.
14. E. van Nimwegen and J. P. Crutchfield. Metastable evolutionary dynamics: Crossing fitness barriers or escaping via neutral paths? *Bulletin of Mathematical Biology*, 62(5):799–848, Sep 2000.
15. V. K. Vassilev and J. F. Miller. The advantages of landscape neutrality in digital circuit evolution. In *ICES*, pages 252–263, 2000.
16. R. A. Watson. Analysis of recombinative algorithms on a non-separable building-block problem. In *Foundations of Genetic Algorithms VI*, pages 69–89. Morgan Kaufmann, 2001.
17. R. A. Watson, G. S. Hornby, and J. B. Pollack. Modeling building-block interdependency. In *Parallel Problem Solving from Nature - PPSN V*, pages 97–106. Springer-Verlag, 1998.
18. T. Yu and J. Miller. Neutrality and the evolvability of Boolean function landscape. In *Proceedings of the 4th European Conference on Genetic Programming (EuroGP)*, volume LNCS 2038, page 204 ff. Springer-Verlag, 2001.

Binary Representations of Integers and the Performance of Selectorecombinative Genetic Algorithms

Franz Rothlauf*

Department of Information Systems
University of Bayreuth, Germany
franz.rothlauf@uni-bayreuth.de

Abstract. When using representations for genetic algorithms (GAs) every optimization problem can be separated into a genotype-phenotype and a phenotype-fitness mapping. The genotype-phenotype mapping is the used representation and the phenotype-fitness mapping is the problem that should be solved.
This paper investigates how the use of different binary representations of integers influences the performance of selectorecombinative GAs using only crossover and no mutation. It is illustrated that the used representation strongly influences the performance of GAs. The binary and Gray encoding are two examples for assigning bitstring genotypes to integer phenotypes. Focusing the investigation on these two encodings reveals that for the easy integer one-max problem selectorecombinative GAs perform better using binary encoding than using Gray encoding. This is surprising as binary encoding is affected with problems due to the Hamming cliff and because there are proofs that show the superiority of Gray encoding. However, the performance of selectorecombinative GAs using binary representations of integers is determined by the resulting building blocks and not by the structure of the search space resulting from the Hamming distances between the individuals. Therefore, the performance difference between the encodings can be explained by analyzing the fitness of the resulting schemata.

1 Introduction

Integer optimization problems are important in many real-world applications. We know from previous work (Liepins & Vose, 1990; Rothlauf, 2001) that the choice of a proper representation (genotype-phenotype mapping) is crucial for the performance of genetic and evolutionary algorithms (GEAs). When solving integer problems, binary representations of integers like Gray or binary encoding are often used.

In this work we want to investigate how binary representations of integers influence the performance of selectorecombinative genetic algorithms (GAs) which

* Also with Illinois Laboratory of Genetic Algorithms, University of Illinois at Urbana-Champaign, USA.

J.J. Merelo Guervós et al. (Eds.): PPSN VII, LNCS 2439, pp. 99–108, 2002.

only use crossover and selection and no mutation. The results show large differences in GA performance using different binary representations. Furthermore, we see that selectorecombinative GAs perform better using binary encoding than using Gray encoding. This behavior of selectorecombinative GAs can be explained by analyzing the fitness of the resulting schemata.

The paper is structured as follows. In the following section we provide the basis and requisites for our investigations. Section 3 examines how different types of binary representations of integers influence the performance of selectorecombinative GAs. We calculate the number of possible representations and present empirical results. In section 4 we focus on the influence of binary and Gray encoding on GA performance. To explain the experimental results presented in subsection 4.1 we analyze in subsection 4.2 the fitness of the resulting schemata for one specific problem. The paper ends with concluding remarks.

2 Binary Representations for Integer Optimization Problems

We present in subsection 2.2 the integer optimization problem we want to solve, and in subsection 2.3 binary representations of integers.

2.1 Separating Representations from Optimization Problems

The following subsection provides some basic definitions for our discussion of representations. When using some kind of representation, every optimization problem can be decomposed into a genotype-phenotype mapping f_g, and a phenotype-fitness mapping f_p (Liepins & Vose, 1990).

We define Φ_g as the genotypic search space where the genetic operators such as recombination or mutation are applied to. An optimization problem on Φ_g could be formulated as follows: The search space Φ_g is either discrete or continuous, and the function $f(\boldsymbol{x}) : \Phi_g \to \mathbb{R}$ assigns an element in $\mathbb{R}$ to every element in the genotype space Φ_g. The optimization problem is defined by finding the optimal solution $\hat{\boldsymbol{x}} = \max_{\boldsymbol{x} \in \Phi_g} f(\boldsymbol{x})$, where $\boldsymbol{x}$ is a vector of decision variables (or alleles), and $f(\boldsymbol{x})$ is the fitness function. The vector $\hat{\boldsymbol{x}}$ is the global maximum.

When using a representation we have to introduce – in analogy to nature – phenotypes and genotypes. Thus, the fitness function f can be decomposed into two parts. The first maps the genotypic space Φ_g to the phenotypic space Φ_p, and the second maps Φ_p to the fitness space $\mathbb{R}$. Using the phenotypic space Φ_p we get:

$$
\begin{aligned}
f_g(\boldsymbol{x}_g) &: \Phi_g \to \Phi_p, \\
f_p(\boldsymbol{x}_p) &: \Phi_p \to \mathbb{R},
\end{aligned}
$$

where $f = f_p \circ f_g = f_p(f_g(\boldsymbol{x}_g))$. The genotype-phenotype mapping f_g is the used representation. f_p represents the fitness function and assigns a fitness value $f_p(\boldsymbol{x}_p)$ to every individual $\boldsymbol{x}_p \in \Phi_p$. The genetic operators are applied to the individuals in Φ_g that means on the level of genotypes.

2.2 An Integer Optimization Problem

This subsection defines the integer optimization problem we want to use for our investigations. We want to define an easy problem which is defined on the phenotypes independently of the used representation.

We assume that the fitness function f_p assigns a real number to every individual $x_p \in \mathbb{N}$. Therefore, we get for f_p:

$$f_p(x_p) : \mathbb{N} \rightarrow \mathbb{R}.$$

For our investigation we want to defined an integer-specific variation of the one-max problem as

$$f_p(x_p) = x_p. \tag{1}$$

2.3 Binary Representations of Integers

In this subsection we briefly discuss the binary and Gray encoding as examples for binary representations of integers.

If we encode integer phenotypes using binary genotypes we have to ask why we do not use integer genotypes for encoding integer phenotypes. In general, instead of using binary strings with cardinality $\chi = 2$, higher χ-ary alphabets could be used for the genotypes. When using a χ-ary alphabet for the genotypic alleles, we are able to encode with one allele χ different phenotypes instead of only 2 different phenotypes when using a binary alphabet.

Binary Encoding. When using the *binary encoding*, each integer value $x_p \in \Phi_p = \{0, 1, 2, \ldots, x_{p,max}\}$ is represented by a binary string $\boldsymbol{x}_g$ of length $l = \log_2(x_{p,max})$. The genotype-phenotype mapping f_g is defined as $x_p = f_g(\boldsymbol{x}_g) = \sum_{i=0}^{l-1} 2^i x_{g,i}$, with $x_{g,i}$ denoting the ith bit of $\boldsymbol{x}_g$.

The binary encoding has problems associated with the *Hamming cliff* (Schaffer, Caruana, Eshelman, & Das, 1989). The Hamming cliff describes the effect that some neighboring phenotypes (the phenotypes have a distance of one) are represented by completely different genotypes (the distance between the genotypes is much larger than one). The distance d between two genotypes $\boldsymbol{x}_g$ and $\boldsymbol{y}_g$ is defined by using the Hamming distance as $d_{\boldsymbol{x}_g,\boldsymbol{y}_g} = \sum_{i=0}^{l-1} |x_{g,i} - y_{g,i}|$ and denotes the number of different alleles in the two genotypes. The distance between two phenotypes x_p and y_p is defined as $d_{x_p y_p} = |x_p - y_p|$.

Gray Encoding. To overcome problems with the Hamming cliff and the different contribution of the alleles to the fitness of an individual when using the binary encoding, the *Gray encoding* was developed (Gray, 1953; Caruana & Schaffer, 1988; Schaffer, Caruana, Eshelman, & Das, 1989). When using Gray encoding the average contribution of an allele to the represented integer is the same for all alleles in the bitstring.

The Gray encoded bitstring itself can be constructed in two steps. At first, the phenotype is encoded using the binary encoding, and subsequently the binary encoded string can be converted into the corresponding Gray encoded string. The binary string $\boldsymbol{x} \in \{0,1\}^l = \{x_0, x_1, \ldots, x_{l-1}\}$ is converted to the corresponding Gray code $\boldsymbol{y} \in \{0,1\}^l = \{y_0, y_1, \ldots, y_{l-1}\}$ by the mapping $\gamma : \mathbb{B}^l \rightarrow \mathbb{B}^l$:

$$y_i = \begin{cases} x_i & \text{if } i = 0, \\ x_{i-1} \oplus x_i & \text{otherwise,} \end{cases}$$

where $\oplus$ denotes addition modulo 2. A Gray encoded string has the same length l as a binary encoded string and the encoding is redundancy-free. Furthermore, the representation overcomes the problems with the Hamming cliff. Every two neighboring phenotypes ($d_{\boldsymbol{x}_g, \boldsymbol{y}_g} = 1$) are encoded by neighboring genotypes ($d_{x_p, y_p} = 1$). This property gives Gray encoding an advantage over the binary encoding when using mutation-based search operators (compare also subsection 4.1).

3 Performance of Crossover-Based GAs Using Binary Representations

In the following we show for the integer one-max problem how the performance of selectorecombinative GAs depends on the used representation.

3.1 Counting the Number of Binary Representations of Integers

We want to calculate the number of different genotype-phenotype mappings f_g. If we use a redundancy-free encoding the number of genotypes is the same as the number of phenotypes. When using a binary representation of length l we are able to represent 2^l different phenotypes using a bitstring of length l. Therefore, the number of possible genotype-phenotype mappings is $2^l!$. The number of different representations is increasing exponentially with increasing l. One example for a possible genotype-phenotype mapping is given in Figure 1.

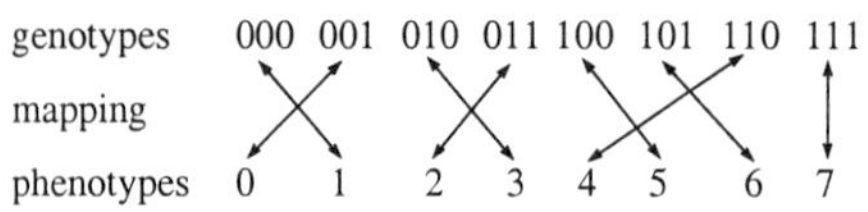

Fig. 1. A random genotype-phenotype mapping.

If we use a binary representation and we encode eight different phenotypes with a genotype of length $l = 3$, there are $2^3! = 40\,320$ different representations. Encoding 16 different phenotypes with a bitstring of $l = 4$ already results in more than 10^{13} different genotype-phenotype mappings. Therefore, to be able to systematically investigate how GA's performance depends on the used encoding we must limit ourselves to a genotypic string length of $l = 3$ and assume without loss of generality that the phenotype $x_p = 0$ is always assigned to the individual $\boldsymbol{x}_g = 000$. Then, the number of different genotype-phenotype mappings is reduced to $(2^l - 1)! = 7! = 5040$. Every genotype-phenotype mapping represents a different representation.

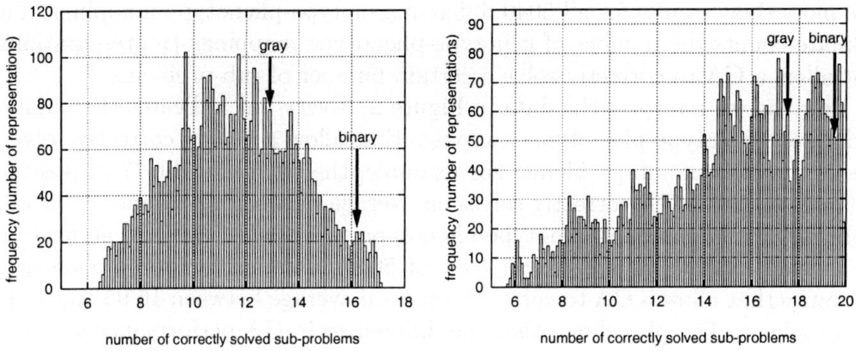

(a) population size $n = 20$

(b) population size $n = 40$

Fig. 2. Experimental results of the frequency of the number of correctly solved sub-problems at the end of a run for all possible genotype-phenotype mappings. We present results for 20 concatenated 3-bit problems. The genotype $\boldsymbol{x}_g = 000$ is always assigned to the phenotype $x_p = 0$ so there are $(2^3 - 1)! = 5040$ different possible genotype-phenotype mappings. We use a GA with tournament selection without replacement of size 2, uniform crossover, and no mutation. We perform 250 runs for each of the 5040 possible encodings.

3.2 Experimental Results

We present empirical results concerning the performance of selectorecombinative GAs using different types of representations for the integer one-max problem defined in subsection 2.2.

For our investigation we concatenate 20 integer one-max problems of size $l = 3$. Each of the 20 phenotypic integers $x_p \in \{0, \ldots, 7\}$ corresponds to 3 bits in the genotype. Therefore, the length of a genotype is $l_{\boldsymbol{x}_g} = 60$. The fitness of an individual is calculated as the sum over the fitness of the 20 sub-problems. The fitness of one sub-problem is calculated according to equation 1.

For our investigation we use a selectorecombinative GA using only uniform crossover and no mutation. For selection we use tournament selection without replacement of size 2. The population size is set either to $n = 20$ (Figure 2(a)) or $n = 40$ (Figure 2(b)). We performed 250 runs for each of the 5040 different genotype-phenotype mappings, and each run was stopped after the population was fully converged. A sub-problem is correctly solved if the GA is able to find the best solution $x_p = 7$. The average number of correctly solved sub-problems at the end of the run gives us a measurement of the GA's performance using one specific representation. The more sub-problems can be solved correctly, the higher the GA performance. As we limit ourselves to genotypes of length $l = 3$ and assign $x_p = 0$ always to $\boldsymbol{x}_g = 000$ there are 5040 different genotype-phenotype mappings (representations).

Figure 2 presents the results of our experiments for the integer one-max problem. We show the distribution of the number of correctly solved sub-problems at the end of a GA run when using different types of genotype-phenotype mappings.

The plots show results for all 5040 different genotype-phenotype mappings. The ordinate counts the number of genotype-phenotype mappings (representations) that allow a GA to correctly solve a certain number of sub-problems.

How can we interpret the data in Figure 2? Every bar indicates the number of different genotype-phenotype mappings that allow a GA to correctly solve a specific number of sub-problems. For example, the bar of height 77 at position 12.9 means that a GA correctly solves on average between 12.85 and 12.95 sub-problems for 77 different genotype-phenotype mappings. The bar at position 17.0 means that there are only 7 (out of 5040) different genotype-phenotype mappings that allow a GA to correctly solve on average between 16.95 and 17.05 sub-problems. The plot shows that the differences in GA performance are large and that a GA with 20 individuals solves dependent on the used representation between 6.5 and 17.1 sub-problems out of 20.

If we compare Figure 2(a) with Figure 2(b) we see that with increasing population size there are still large differences in GA performance. The shapes of the distributions are similar and are shifted with increasing population size n towards a higher number of correctly solved sub-problems. To be able to illustrate how the performance of GAs depends on the different representations the population size n must be chosen relatively small. Using larger n would allow GAs to solve all 20 sub-problems (due to the easiness of the problem) and we would be not able to illustrate the performance differences using different types of representations.

We see that different representations, that means assigning the genotypes $\boldsymbol{x}_g \in \{0,1\}^3$ in a different way to the phenotypes $x_p \in \{0,\dots 7\}$, change the performance of GAs. For some representations GA performance is high, whereas for some representations GA performance is low.

4 Performance of Binary and Gray Encoding

After we have seen that GA's performance strongly depends on the used representation, we focus in the following on the performance of two specific representations, namely Gray and binary encoding.

When using binary representations of integers there are $2^l!$ different genotype-phenotype mappings. Gray and binary encoding are two specific representations. The arrows in Figure 2 indicate the performance of selectorecombinative GAs using these two types of encodings. A GA with $n = 20$ using Gray encoding correctly solves on average only 12.9 of the 20 sub-problems whereas when using binary encoding on average 16.2 out of the 20 sub-problems are solved. It can be seen that GAs using binary encoding perform much better than GAs using Gray encoding. With increasing population size n both encodings perform better but there is still a large performance difference between both encodings.

4.1 Experimental Results

To investigate the performance differences more closely we compare in Figure 3 GA performance when using binary encoding and Gray encoding. We show the

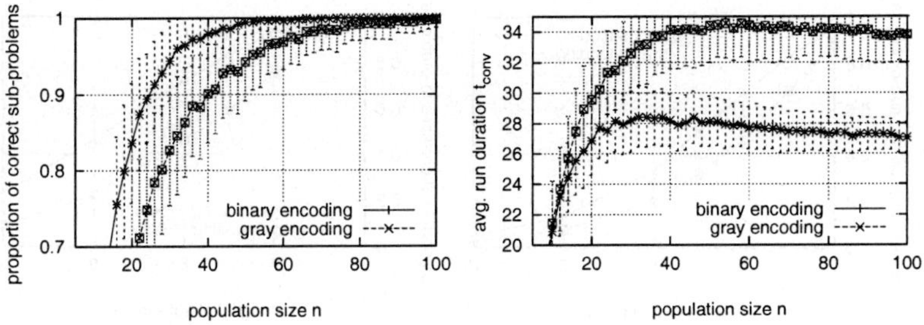

(a) Average proportion of correct building blocks at the end of a GA run

(b) Number of generations

Fig. 3. GA performance for the integer one-max problem. Each sub-problem has length 3 and we concatenated $m = 20$ sub-problems. We show the average proportion of correctly solve sub-problems at the end of a GA run (Figure 3(a)) and the average length of a run (Figure 3(b)). GAs using the binary encoding are able to solve more sub-problems and converge after a shorter number of generations. The error bars indicate the standard deviation of the results.

average proportion of correctly solved sub-problems at the end of the run (Figure 3(a)) and the number of generations (Figure 3(b)) over the population size. As before, we concatenated $m = 20$ sub-problems of length $l = 3$. Furthermore, we use the same parameters for the GA as described in subsection 3.2 but only change the population size n. The results confirm our previous observations and show that selectorecombinative GAs using the binary encoding not only solve a higher proportion of correct sub-problems but also solve the sub-problems in shorter time.

To investigate how GA performance varies over the length of the sub-problems we show in Figure 4 results for the proportion of correctly solved sub-problems over the length l. The length l of the sub-problems varies from $l = 2$ to $l = 10$. Therefore, the number of different integers that can be represented varies from $2^2 = 4$ to $2^{10} = 1024$. As before we concatenate 20 sub-problems of length l and use the same GA parameters as in subsection 3.2. The length of a genotype is $l_{x_g} = 20 * l$.

GA performance declines with increasing length l of the sub-problems. For small problems ($l = 2$) GAs are able to correctly solve most of the 20 sub-problems, whereas for large problems ($l = 10$) only a small fraction of sub-problems can be solved. Comparing Gray and binary encoding shows that independently of l GAs using binary encoding outperform GAs using Gray encoding. With larger population size $n = 40$ (Figure 4(b)) GA performance increases. However, using binary encoding still results in better GA performance.

The performance differences between Gray and binary encoding are surprising because we already noted in subsection 2.3 that the Gray encoding has no problems with the Hamming cliff and the contribution of the alleles is uniformly. Furthermore, other work has shown that the Gray encoding shows some advan-

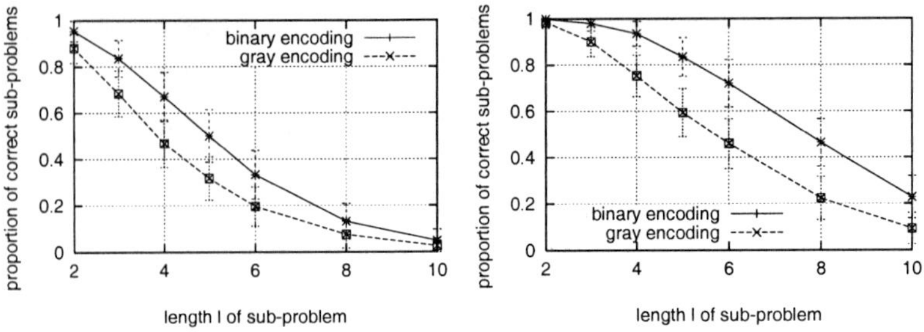

(a) population size $n = 20$ (b) population size $n = 40$

Fig. 4. Proportion of correctly solved sub-problems at the end of a GA run versus the length l of the sub-problems for the binary and Gray encoding. The plots are for the integer one-max problem and a population size of $n = 20$ (Figure 4(a)) and $n = 40$ (Figure 4(b)). With increasing length l the performance of GAs is reduced. When using Gray encoding the performance of a GA declines much stronger than when using binary encoding. The error bars indicate the standard deviation of the results.

tage in comparison to binary encoding. (Rana & Whitley, 1997; Whitley & Rana, 1997; Whitley, 1999). This work formulated a Free Lunch theorem for the use of Gray encoding and mutation-based search approaches. GEAs using mutation as the main search operator perform better when using the Gray encoding than when using the binary encoding. The proof actually shows that the number of local optima introduced by using the Gray encoding is smaller than when using the binary encoding. However, this proof is not in contrast to the results herein which are obtained for selectorecombinative GAs and not for mutation-based search algorithms. When using the Gray encoding all phenotypes with distance $d^p = 1$ are also neighboring genotypes ($d^g = 1$). Therefore, when using mutation-based search approaches and Gray encoding, a small mutation of a genotype always results in the corresponding phenotype and the performance of mutation-based search approaches on easy problems must be higher when using Gray encoding.

However, in this work we focus not on mutation-based search methods but use crossover as main search operator. Therefore, the correct method to measure problem difficulty is to use schema analysis (Holland, 1975; Goldberg, 1989). The performance of selectorecombinative GAs is determined by the building blocks resulting from the used representation.

4.2 Schemata Analysis for Binary and Gray Encoding

This subsection analyzes the fitness of the schemata resulting from the use of Gray versus binary encoding. We perform a static schema analysis and do not consider the actual schemata a GA sees during the run. The analysis of the fitness of the schemata reveals that using binary encoding makes the integer one-max problem easier than using Gray encoding. Therefore, the performance of selectorecombinative GAs is higher when using binary encoding.

Table 1. Schemata fitness for the integer one-max problem using binary versus Gray encoding. The integer one-max problem is completely easy for the binary encoding. Using Gray encoding results in a more difficult problem, because some of the high quality schemata have the same fitness as misleading schemata.

	order	3	2			1			0
binary	schema	111	11*	1*1	*11	**1	*1*	1**	***
	fitness	**7**	**6.5**	**6**	**5**	**4**	**4.5**	**5.5**	**3.5**
	schema		01*	0*1	*01	**0	*0*	0**	
	fitness		2.5	2	3	3	2.5	1.5	
	schema		10*	1*0	*10				
	fitness		4.5	5	4				
	schema		00*	0*0	*00				
	fitness		0.5	1	2				
Gray	schema	100	10*	1*0	*00	1**	*0*	**0	***
	fitness	**7**	**6.5**	**5.5**	**3.5**	**5.5**	**3.5**	**3.5**	**3.5**
	schema		11*	1*1	*11	0**	*1*	**1	
	fitness		4.5	**5.5**	**3.5**	1.5	**3.5**	**3.5**	
	schema		01*	0*1	*01				
	fitness		2.5	1.5	**3.5**				
	schema		00*	0*0	*00				
	fitness		0.5	1.5	**3.5**				

In Table 1, we present the average fitness of the schemata for the integer one-max problem using binary and Gray encoding for $l = 3$. The numbers reveal that for the integer one-max problem with binary encoding all schemata containing the global optimum $\boldsymbol{x}_g = 111$ are superior to their competitors. Therefore, the integer one-max problem is easy and selectorecombinative GAs show a high performance. The schema analysis for Gray encoding reveals that the schemata containing the global optimum $\boldsymbol{x}_g = 100$ are not always superior to their competitors. Therefore, the problem is not completely easy any more, and GAs perform worse in comparison to using binary encoding.

The results show, that for selectorecombinative GAs some easy problems like the presented integer one-max problem, are easier to solve when using the binary encoding as when using the Gray encoding. When using selectorecombinative GAs, neither the Hamming distances between the individuals nor problems with Hamming cliffs are relevant for GA performance, but the schema analysis answers the question if a problem is easy or difficult.

5 Conclusions

This paper investigates how binary representations of integers influence the performance of selectorecombinative GAs.

It is well known, that when using representations every optimization problem can be separated into a genotype-phenotype mapping (the used representation) and a phenotype-fitness mapping (the optimization problem that should be solved). This paper illustrates for binary representations of integers, that the choice of a proper genotype-phenotype mapping is crucial for GA's success. The use of different representations results in large differences in GA performance.

The binary and Gray encoding are two well known possible binary representations of integers. Focusing on these two representations reveals for the easy integer one-max problem, that for selectorecombinative GAs not Hamming distances between individuals, but schemata are important. We illustrate that the analysis of the fitness of the resulting schemata for the easy integer one-max problem can be used for explaining the differences in performance. It reveals, that the use of the binary encoding results in building blocks of lower order than the use of the Gray encoding. Therefore, when using Gray encoding the integer one-max problem is more difficult and the performance of selectorecombinative GAs is lower.

Our empirical analysis of GA performance has shown that the binary encoding results in higher performance than the Gray encoding. However, Figure 2(a) reveals that there are representations that even outperform the binary encoding. If we can theoretically describe the properties of these encodings and systematically construct such representations, we would be able to increase the performance of GAs and solve integer problems more efficiently.

References

Caruana, R. A., & Schaffer, J. D. (1988). Representation and hidden bias: Gray vs. binary coding for genetic algorithms. In Laird, L. (Ed.), *Proceedings of the Fifth International Workshop on Machine Learning* (pp. 153–161). San Mateo, CA: Morgan Kaufmann.

Goldberg, D. E. (1989). *Genetic algorithms in search, optimization, and machine learning.* Reading, MA: Addison-Wesley.

Gray, F. (1953, March). Pulse code communications. U.S. Patent 2632058.

Holland, J. H. (1975). *Adaptation in natural and artificial systems.* Ann Arbor, MI: University of Michigan Press.

Liepins, G. E., & Vose, M. D. (1990). Representational issues in genetic optimization. *Journal of Experimental and Theoretical Artificial Intelligence*, *2*, 101–115.

Rana, S. B., & Whitley, L. D. (1997). Bit representations with a twist. In Bäck, T. (Ed.), *Proceedings of the Seventh International Conference on Genetic Algorithms* (pp. 188–195). San Francisco: Morgan Kaufmann.

Rothlauf, F. (2001). *Towards a theory of representations for genetic and evolutionary algorithms: Development of basic concepts and their application to binary and tree representations.* Doctoral dissertation, University of Bayreuth/Germany.

Schaffer, J. D., Caruana, R. A., Eshelman, L. J., & Das, R. (1989). A study of control parameters affecting online performance of genetic algorithms for function optimization. In Schaffer, J. D. (Ed.), *Proceedings of the Third International Conference on Genetic Algorithms* (pp. 51–60). San Mateo, CA: Morgan Kaufmann.

Whitley, D. (1999). A free lunch proof for gray versus binary encodings. In Banzhaf, W., Daida, J., Eiben, A. E., Garzon, M. H., Honavar, V., Jakiela, M., & Smith, R. E. (Eds.), *Proceedings of the Genetic and Evolutionary Computation Conference: Volume 1* (pp. 726–733). San Francisco, CA: Morgan Kaufmann Publishers.

Whitley, D., & Rana, S. (1997). Representation, search, and genetic algorithms. In *Proceedings of the 14th National Conference on Artificial Intelligence (AAAI-97)* (pp. 497–502). AAAI Press/MIT Press.

Variation Operators: Analysis, New Techniques

Parallel Varying Mutation in Deterministic and Self-adaptive GAs

Hernán E. Aguirre and Kiyoshi Tanaka

Faculty of Engineering, Shinshu University
4-17-1 Wakasato, Nagano, 380-8553 JAPAN
{ahernan,ktanaka}@gipwc.shinshu-u.ac.jp

Abstract. In this work we study varying mutations applied either serial or parallel to crossover and discuss its effect on the performance of deterministic and self-adaptive varying mutation GAs. After comparative experiments, we found that varying mutation parallel to crossover can be a more effective framework in both deterministic and self-adaptive GAs to achieve faster convergence velocity and higher convergence reliability. Best performance is achieved by a parallel varying mutation self-adaptive GA.

1 Introduction

Parameter control methods modify the values of the strategy parameters during the run of the algorithm by taking into account the actual search process. These methods are an alternative form to the common practice of tuning parameters "by hand" and are considered as one of the most important and promising areas of research in evolutionary algorithms[1]. One of the approaches for parameter control in genetic algorithms (GAs) seeks to combine crossover with (higher) varying mutation rates during the course of a run. It has been shown that deterministically varying mutation rates over the generations and/or across the representation can improve the performance of GAs[2],[3],[4]. Self-adaptive mutation rate schedules inspired from Evolution Strategies[5] and Evolutionary Programming[6] have also been proposed to control the mutation rate of generational and steady state GAs[4],[7],[8]. The principle of self-adaptation incorporates strategy parameters into the representation of individuals evolving simultaneously strategy parameters and object variables. It is regarded as the method having the advantage of reducing the number of exogenous parameters[8] and is thought to be the most promising way of combining forms of control (parameters co-adaptation)[1].

From the application of operators standpoint, deterministic, adaptive, and self-adaptive varying mutation GAs have been mostly designed similar to a canonical GA. That is, crossover is applied with probability p_c and then follows mutation. Under these standard varying mutation approaches, higher mutations are mostly applied serial to crossover. This rises questions regarding the interference that one operator could cause to the other and its possible impact on the

J.J. Merelo Guervós et al. (Eds.): PPSN VII, LNCS 2439, pp. 111–121, 2002.

performance and robustness of standard varying mutation algorithms in general and self-adaptive mutation GAs in particular.

An alternative to standard varying mutation methods is to design approaches that apply background mutation after crossover (or none at all) and higher mutations only parallel to crossover. Such an approach could give an efficient framework to achieve better balances for varying mutation and crossover in which the strengths of the operators can be kept without interfering one with the other.

From this point of view, we explore a model of generational GA that applies varying mutations parallel to crossover & background mutation putting the operators in a cooperative-competitive stand with each other by subjecting their offspring to extinctive selection[9],[10]. In previous reports we have discussed the relevance of extinctive selection, adaptation, and mutation strategy[11],[12] to the performance of parallel varying mutation GAs.

In this work, we built on [11] and especially focus on deterministic and self-adaptive mutation GAs. After comparative experiments with several difficult, large, and highly constrained 0/1 multiple knapsacks problems[13], we found that varying mutation parallel to crossover can be a more effective framework in both deterministic and self-adaptive GAs to achieve faster convergence velocity and higher convergence reliability. Best performance is achieved by a parallel varying mutation self-adaptive GA.

2 A GA with Serial Varying Mutation

A standard varying mutation GA, similar to canonical GAs, applies crossover with probability p_c followed by mutation with probability p_m per bit. In the absence of crossover $(1-p_c)$, mutation is applied alone. From the application of operators standpoint, it can be said that the probability of crossover p_c enables an implicit parallel application of two operators. One of the operators is crossover followed by mutation (CM) and the other one is mutation alone (M). It should be noted that mutation in both CM and M is governed by the same mutation probability p_m and applies the same "bit by bit" mutation strategy. **Fig. 1 (a)** illustrates a standard varying mutation GA.

Since p_c is usually set to 0.6, and higher values are often used[1], it turns out that mutation is mostly applied serial to crossover. In canonical GAs p_m is small, therefore the amount of diversity introduced by mutation either through CM or M is modest. For the same reason, the disruption that mutation causes to crossover in CM is also expected to be small. In varying mutation GAs, however, mutations are higher and the combined effect of crossover and mutation in CM and the effect of mutation alone in M should be carefully reconsidered.

In the case of CM, besides those cases in which crossover and mutation aggregate in a positive manner or are neutral, those cases in which one of the operators is working well but is being hampered by the other should also be taken into account. For example, if mutation rates were high, although crossover could be doing a good job it is likely that some of the just created favorable recombinations would be immediately lost, before they become fix in the off-

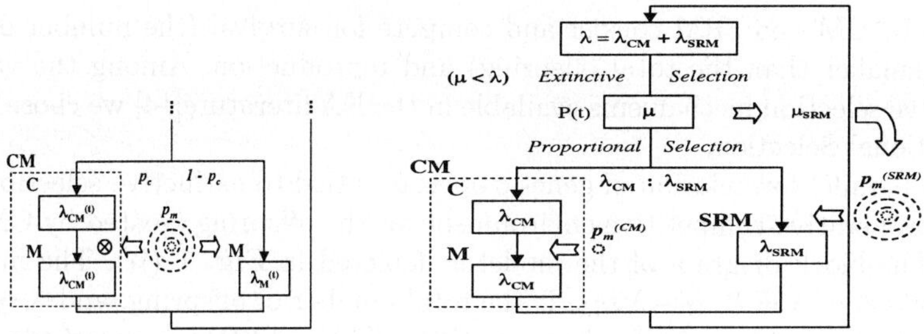

(a) A Standard Varying Mutation GA (b) A GA with Parallel Varying Mutation

Fig. 1. Standard and Parallel Varying Mutation GAs

spring, due to the high disruption introduced by mutation. We can think of this case as a mutation interference with crossover in the *creation* of beneficial recombinations. On the other hand, mutation could be working well but crossover may produce poor performing individuals affecting the survivability of beneficial mutations that can contribute to the search. We can think of this case as a crossover interference with mutation in the introduction of beneficial mutations.

In the case of mutation alone M, its instantaneous effectiveness depends only upon itself and does not diminish the effectiveness of other operator. High mutations in M, when are harmful, will have a negative impact on the *propagation* of beneficial recombinations already present in the parent population, but will not affect their *creation* by crossover as high mutation can do it in CM. In the following we refer to standard varying mutation GAs as *varying mutation serial to crossover.*

3 A GA with Parallel Varying Mutation

3.1 Parallel Genetic Operators

An alternative to standard varying mutation GAs is to explicitly differentiate the mutation operator applied parallel to crossover from the mutation operator applied after crossover. We explore a model of GA that in addition to crossover followed by background mutation (CM) it also explicitly applies parallel varying mutation [9],[10]. To clearly distinguish between mutation operators the parallel varying mutation operator is called *Self-Reproduction with Mutation* (SRM). SRM parallel to CM implicitly increases the levels of cooperation to introduce beneficial mutations and create beneficial recombinations. It also sets the stage for competition between operators' offspring. In the following we refer to GAs that explicitly apply varying mutations only parallel to crossover as *varying mutation parallel to crossover.*

3.2 Extinctive Selection

The model also incorporates the concept of extinctive selection that has been widely used in Evolution Strategies. Through extinctive selection the offspring created by CM and SRM coexist and compete for survival (the number of parents is smaller than the total offspring) and reproduction. Among the various extinctive selection mechanisms available in the EA literature[14] we chose (μ, λ) Proportional Selection.

The parallel formulation of genetic operators tied to extinctive selection creates a cooperative-competitive environment for the offspring created by CM and SRM. The block diagram of the model is depicted in **Fig. 1 (b)**. The number of parents is μ, $\lambda = \lambda_{CM} + \lambda_{SRM}$ is the total number of offspring, and λ_{CM} and λ_{SRM} are the number of offspring created by CM and SRM, respectively. Both λ_{CM} and λ_{SRM} are deterministically decided at the beginning of the run.

3.3 Mutation Rate Control in SRM

In this work we use deterministic and self-adaptive mutation rate controls in SRM. The deterministic approach implements a time-dependent mutation schedule that reduces mutation rate in a hyperbolic shape, originally proposed in [4] and expressed by

$$p_m^{(t)} = \left(r_o + \frac{n - r_o}{T - 1} t \right)^{-1} \tag{1}$$

where T is the maximum number of generations, $t \in \{0, 1, \cdots, T-1\}$ is the current generation, and n is the bit string length. The mutation rate $p_m^{(t)}$ varies in the range $[1/r_o, 1/n]$. In the original formulation $r_o = 2$. Here we included r_o as a parameter in order to study different ranges for mutation. In the deterministic approach the mutation rate calculated at time t is applied to all individuals created by SRM.

To include self-adaptation, each individual incorporates its own mutation probability within the representation. SRM to produce offspring first mutates the mutation probability of the selected individual and then mutates the object variable using the individual's mutated probability. In this work we use the self-adaptive approach originally proposed in [4],[8], which uses a continuous representation for the mutation rate and mutates the mutation probability of each individual by

$$p_m^{(t)}(i) = \left(1 + \frac{1 - p_m^{(t-1)}(i)}{p_m^{(t-1)}(i)} exp(-\gamma N(0,1)) \right)^{-1} \tag{2}$$

where i indicates the i-th individual, γ is a learning rate that control the speed of self-adaptation, and $N(0,1)$ is a normally distributed random number with expectation zero and standard deviation one. Note that individuals selected to reproduce with SRM at generation t could have been created either by SRM or CM at generation $t-1$. Since the mutation rate of each individual is mutated only by SRM, individuals created by CM do not carry an updated mutation rate. Thus, the mutation rate of individuals that were created by CM at generation $t-1$ is first updated by

$$p_m^{(t-1)}(j) = \frac{1}{\mu_{SRM}} \sum_{k=1}^{\mu_{SRM}} p_m^{(t-1)}(k) \tag{3}$$

where j indicates an individual created by CM at $(t-1)$, k indicates the individuals created by SRM at $(t-1)$ that survived extinctive selection, and μ_{SRM} is the number of offspring created by SRM that survived extinctive selection. In the case that no offspring created by SRM survived extinctive selection, $p_m^{(t-1)}(j)$ is set to the mutation value of the best SRM's offspring. SRM will mutate this updated mutation in order to mutate the object variable.

4 Experimental Setup

The following GAs are used in our simulations. A simple canonical GA that applies crossover followed by background mutation, denoted as cGA. A GA with deterministic varying mutation serial (parallel) to crossover, denoted as hGA (GA-hM). A GA with self-adaptive varying mutation serial (parallel) to crossover, denoted as sGA (GA-sM). The GAs use either Proportional Selection or (μ,λ) Proportional Selection. This is indicated by appending to the name of the GA (μ) or (μ,λ), respectively[1]. All algorithms use fitness linear scaling and mating is restricted to $(\boldsymbol{x}_i, \boldsymbol{x}_j)$, $i \neq j$, so a solution will not cross with itself. For cGA, hGA, and sGA $p_c = 0.6$ and for GA-hM and GA-sM the ratio for offspring creation is set to $\lambda_{CM} : \lambda_{SRM} = 1 : 1$. Background mutation is set to $p_m^{(CM)} = 1/n$. The learning rate for self-adaptation is set to $\gamma = 0.2$

In our study we use difficult, large and highly constrained, 0/1 multiple knapsack problems[2] [13]. A 0/1 multiple knapsack problem consists of m knapsacks (constraints) and n objects (size of the search space: 2^n). Each knapsack is of different capacity and each object has associated a profit. Also, there is a set of weights for each object, one per knapsack. The objective of the problem is to find the combination of objects such that profit is maximized but no knapsack is overfilled with objects' weights. Besides m and n, other parameter of the problem is the tightness ratio ϕ between knapsack capacities and object weights (which implies a ratio between the feasible region and the whole search space). By varying m, n, and ϕ, 0/1 multiple knapsack problems allows us to carefully observe the behavior and scalability of the algorithms in these three important aspects that are correlated to the difficulty of a problem.

The initial population is randomly initialized with a 0.25 probability for 1s. The fitness function and penalty term is the same used in [15]. Results are averaged over 50 runs and the number of generations is set to $T = 5000$.

5 Deterministic Varying Mutation

Deterministic mutation varies mutation rates with exactly the same schedule whether it is applied serial (hGA) or parallel to crossover (GA-hM) and there-

[1] a simple GA with (μ,λ) Proportional Selection is denoted GA(μ,λ)

[2] http://mscmga.ms.ic.ac.uk/jeb/orlib/info.html

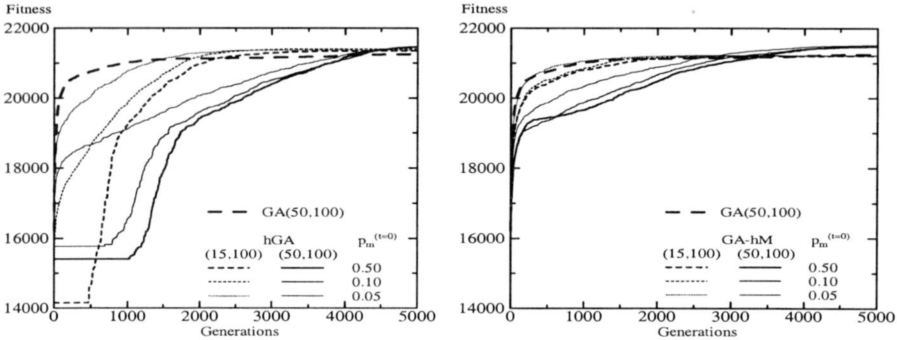

(a) Serial to Crossover (b) Parallel to Crossover

Fig. 2. Deterministic Varying Mutation ($m = 30$, $n = 100$, $\phi = 0.25$)

fore is an ideal candidate to isolate and observe the impact of higher mutations in both models of GAs. Experiments are conducted using various populations $(\mu, \lambda) = \{(15, 100), (50, 100), (100, 100)\}$ and initial mutation probabilities $p_m^{(t=0)} = \{0.50, 0.10, 0.05\}$. **Fig. 2 (a)** and **(b)** plot the average fitness of the best-so-far individual over the generations illustrating the convergence behavior by hGA and GA-hM, respectively. Results by GA (50,100) are also included for comparison.

From **Fig. 2 (a)** we can observe that hGA's convergence becomes faster increasing extinctive pressure (reduce μ while keeping constant λ). However, better final results were given by (μ,λ)=(50,100) rather than by (μ,λ)=(15,100). Setting initial mutation probability to lower values also helps to speed up convergence. These, however, do not help to increase the quality of the final results. Note the initial *flat* periods in which the fitness of the best-so-far individual did not improve. This is a clear indication of the disruption caused by high mutation after crossover.

From **Fig. 2 (b)** we can see that increasing extinctive selection and reducing initial mutation probability in GA-hM produce similar effects to those remarked for hGA. Looking at both **Fig. 2 (a)** and **Fig. 2 (b)** becomes apparent that varying mutation parallel to crossover is less disruptive than varying mutation serial to crossover. Contrary to hGA, in the case of GA-hM there are no initial *flat* periods and in all cases GA-hM converges faster than hGA for similar values of (μ,λ) and $p_m^{(t=0)}$. Also, as a consequence of this less disruptiveness, the initial value set for varying mutation in GA-hM has a smaller impact on convergence speed than it does in hGA. See for example GA-hM(50,100) for $p_m^{(t=0)} = 0.5$ and $p_m^{(t=0)} = 0.05$ and compare it with hGA for similar settings. Thus, GA-hM is more robust than hGA to initial setting of mutation rate.

In GA-hM, similar to hGA, a (μ, λ)=(50,100) extinctive ratio gives better final results than (μ, λ)=(15,100). In fact, note that GA-hM(15,100)'s final quality is not better than GA(50,100)'s that does not apply varying mutations. A (15,100) extinctive selection turns out to be too strong for GA-hM. A less

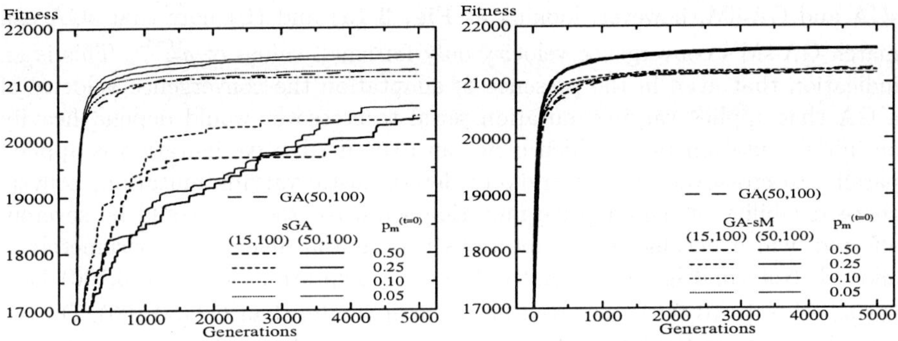

(a) Serial to Crossover (b) Parallel to Crossover

Fig. 3. Self-Adaptive Varying Mutation $p_m^{(t=0)}(i) = p_m^{max}$ ($m = 30$, $n = 100$, $\phi = 0.25$)

strong selection pressure, such (50,100), gives a better chance to hM's offspring to compete with CM's offspring, which in turn helps to improve the search process.

6 Self-adaptive Varying Mutation

A self-adaptive scheme uses one mutation rate per individual, which are usually set at $t = 0$ to random values in the range allowed for mutation. Two important ingredients of self-adaptation are the diversity of parameter settings and the capability of the method to adapt the parameters. It has been indicated that some of the implementations of self-adaptation exploit more the diversity of parameter settings rather than adapting them. However, it has also been argued that the key to the success of self-adaptation seems to consist in using at the same time both a reasonably fast adaptation and reasonably large diversity to achieve a good convergence velocity and a good convergence reliability, respectively[8].

To observe the influence that the serial/parallel application of varying mutations could have on the self-adaptive capability itself we avoid initial diversity of parameters. Experiments are conducted using populations $(\mu, \lambda) = \{(15, 100), (50, 100)\}$ and mutation ranges of $p_m = [p_m^{min}, p_m^{max}] = [1/n, \{0.50, 0.25, 0.10, 0.05\}]$. In all cases initial mutation for each individual is set to the maximum value allowed for the range, $p_m^{(t=0)} = p_m^{max}$. **Fig. 3 (a)** and **(b)** plot the average fitness of the best-so-far individual over the generations illustrating the convergence behavior by sGA and GA-sM, respectively. Results by GA(50,100) are also included for comparison.

From **Fig. 2** and **Fig. 3** it is worth noting the following. (i) Self-adaptive mutation increases convergence speed compared to deterministic mutation either serial or parallel to crossover. Looking at **Fig. 3 (a)** and **Fig. 2 (a)**, note that in sGA the initial *flat* periods observed in hGA have disappeared completely. Also, looking at **Fig. 3 (b)** and **Fig. 2 (b)** we can see that GA-sM(50,100)'s fitness picks up much earlier than GA-hM(50,100)'s for similar values of $p_m^{(t=0)}$. Between sGA and GA-sM, however, looking at **Fig. 3 (a)** and **(b)** note that sGA can match GA-sM's convergence velocity only for small values of $p_m^{(t=0)}$. This is an indication that even in the presence of adaptation the convergence velocity of a GA that applies varying mutation serial to crossover would depend heavily on initial mutation rates, which is not an issue if adaptive mutation is applied parallel to crossover. (ii) Contrary to deterministic varying mutation, convergence reliability of self-adaptive mutation serial to crossover could be severely affected, which becomes quite notorious if no initial diversity of parameters is allowed. Note in **Fig. 3 (a)** that only the configurations of sGA(50,100) having $p_m^{(t=0)} = \{0.10, 0.05\}$ achieved better final results than GA(50,100). On the other hand, the initial lack of diversity of parameters does not affect convergence reliability of GA-sM. Note in **Fig. 3 (b)** that for the same selection pressure convergence reliability of GA-sM is similar for all values of $p_m^{(t=0)}$. (iii) Similar to deterministic varying mutation, better results are achieved by $(\mu, \lambda) = (50, 100)$ rather than by $(\mu, \lambda) = (15, 100)$.

Next, we allow for initial diversity of parameters setting $p_m^{(t=0)}$ to a random value between the minimum and maximum value allowed for mutation. In this case, the disruption that higher serial mutation causes to crossover becomes less apparent due to the initial diversity of parameters and convergence speed is similar for both sGA and GA-sM. Convergence reliability of sGA also improves. However, the negative impact on reliability remains quite significant for sGA (see **7**). **Fig. 4 (a)** and **(b)** illustrates the fitness transition and the average flipped bits (Log scale) by sGA and GA-sM both with random initial mutation rates between [1/n,0.50]. Results for hGA and GA-hM are also included in **Fig. 4 (a)** for comparison. From these figures note that sGA converges to lower fitness and reduces mutation rates faster than GA-sM.

The self-adaptation principle tries to exploit the indirect link between favorable strategy parameters and objective function values. That is, appropriate parameters would lead to fitter individuals, which in turn are more likely to survive and hence propagate the parameter they carry with them to their offspring. A GA that applies varying mutation parallel to crossover as GA-sM can interpret better the self-adaptation principle and achieve higher performance because (i) inappropriate mutation parameters do not disrupt crossover, and (ii) it preserves mutation rates (see **Eq. (3)**) that are being useful to the search. A GA that applies varying mutation serial to crossover as sGA, however, can mislead the mutation rate control because (i) appropriate parameters can be eliminated due to ineffective crossover operations, and (ii) in sGA an appropriate parameter implies parameters that would not affect greatly crossover. Thus, in sGA there is a selective bias towards smaller mutation rates.

7 Convergence Reliability

To obtain a broader perspective on the performance of the GAs we apply them to several knapsacks problems varying ϕ, m, and n. Each combination of ϕ, m, and n defines a subclass of problem. Here we use totally 7 combinations of problem

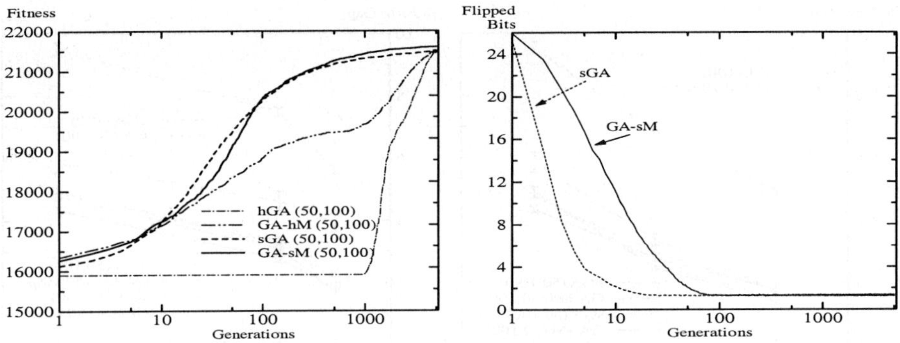

(a) Convergence Velocity (b) Average Number of Flipped Bits

Fig. 4. Convergence Velocity and Average Number of Flipped Bits ($m = 30$, $n = 100$, $\phi = 0.25$). $p_m^{(t=0)} = 0.5$ for hGA and GA-hM. $p_m^{(t=0)}(i) = rand[1/n, 0.5]$ for sGA and GA-SM

parameters ($m = 30$,$n = 100$, and $\phi = \{0.75, 0.50, 0.25\}$; $n = 100$, $\phi = 0.25$, and $m = \{5, 10\}$; $m = 30$, $\phi = 0.25$, and $n = \{250, 500\}$) and one random problem for each combination. **Fig. 5 (a)**, **(b)**, and **(c)** plot the percentage error gap between the best solutions' average in 50 runs and the optimal value given by the linear programming relaxation (the optimal integer solutions are unknown)[13]. The vertical bars overlaying the mean curves represent 95% confidence intervals.

Some important observations are as follows. (i) The performance of a simple GA (50,100) is by far superior to the canonical cGA(100), indicating that extinctive selection is an important factor to increase the performance of GAs in these constrained problems. (ii) GAs that combine extinctive selection with varying mutations give better results than a simple GA with extinctive selection. The difference in performance is more apparent for problems with higher difficulty. (iii) The serial application of varying mutation seems not to affect much the convergence reliability of GAs with deterministic schedules for mutation (see hGA and GA-hM), but it seems to be detrimental to the convergence reliability of the self-adaptive GA (see sGA and GA-sM).

Finally, we test the statistical significance of the results achieved by hGA, GA-hM and sGA, GA-sM presented in **Fig. 5 (a)**, **(b)**, and **(c)**. **Fig. 5 (d)** shows results of the 6 corresponding two-factor factorial ANOVA, where *Source* indicates the source of variation, *df* the degrees of freedom, F is the ratio between the mean square treatment and the mean square error, and *Pval* is the *p value* (the smallest significant level α that would allow rejection of the null hypothesis). Inspection of the p values under hGA,GA-hM reveals that there is some indication of an effect due to the serial/parallel application of deterministic varying mutation, since $Pval = 0.066$ and $Pval = 0.061$ for ϕ and m are not much greater than $\alpha = 0.05$. However, looking under sGA,GA-sM, there is indication of an strong main effect of applying self-adaptive varying mutation serial/parallel to crossover.

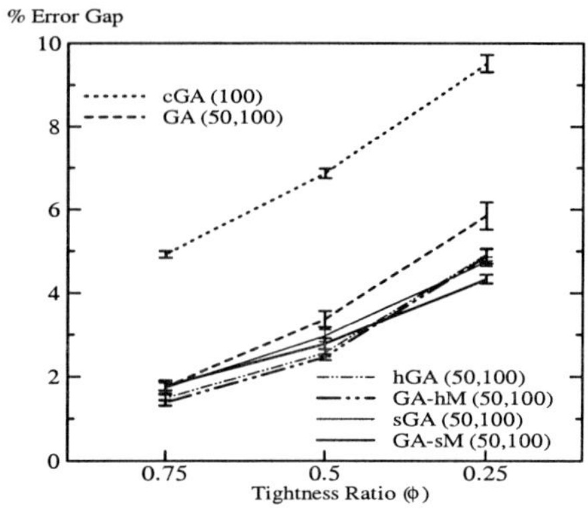

(a) Ratio ϕ ($m = 30$, $n = 100$)

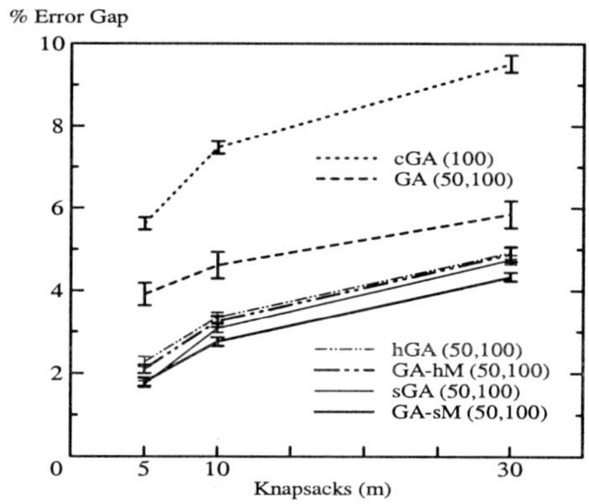

(b) Knapsacks m ($n = 100$, $\phi = 0.25$)

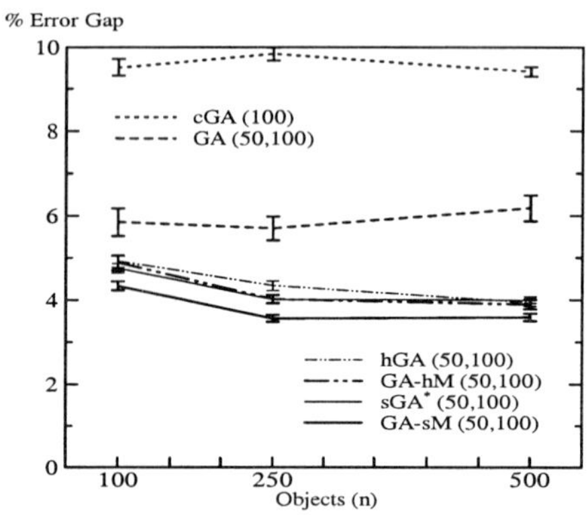

(c) Objects n ($m = 30$, $\phi = 0.25$)

		hGA, GA-hM		sGA, GA-sM	
Source	*df*	*F*	*Pval*	*F*	*Pval*
GA	1	3.40	0.066	13.33	0.000
ϕ	2	1957.01	0.000	1001.62	0.000
GA-ϕ	2	0.18	0.832	7.13	0.001
GA	1	3.53	0.061	29.08	0.000
m	2	763.67	0.000	1464.82	0.000
GA- m	2	0.41	0.663	12.36	0.000
GA	1	6.76	0.010	121.05	0.000
n	2	182.71	0.000	168.08	0.000
GA-n	2	6.66	0.002	0.09	0.910
MSE	294				

(d) Factorial ANOVA

Fig. 5. Convergence Reliability

8 Conclusions

We have studied the application of varying mutation either serial or parallel to crossover and discussed its effect on the performance of deterministic and self-adaptive varying mutation GAs. Experiments were conducted with several 0/1 multiple knapsacks problems. We found that mutation parallel to crossover is more effective than mutation serial to crossover. In the case of deterministic mutation GAs, a GA with varying mutation parallel to crossover showed faster convergence and higher robustness to initial settings of mutation rate than a GA with varying mutation serial to crossover. Also, an ANOVA gave some indication of higher convergence reliability by the parallel application of deterministic varying mutation. In the case of self-adaptive GAs, the convergence velocity of a parallel self-adaptive mutation GA was matched by a serial self-adaptive mutation GA only when initial diversity of parameters was allowed. Convergence reliability was higher for the parallel varying self-adaptive mutation GA with or without initial diversity of parameters. An ANOVA gave a strong indication in this direction. Among deterministic and self-adaptive varying mutation GAs, best performance was achieved by a parallel varying mutation self-adaptive GA.

References

1. A. E. Eiben, R. Hinterding, and Z. Michalewicz, "Parameter Control in Evolutionary Algorithms", *IEEE Transactions on Evolutionary Algorithms*, vol.3, no.2, pp.124-141, 1996.
2. T. Fogarty, "Varying the Probability of Mutation in the Genetic Algorithm", *Proc. 3rd Int'l Conf. on Genetic Algorithms*, Morgan Kaufmann, pp.104-109, 1989.
3. Z. Michalewicz, *Genetic Algorithms + Data Structures = Evolution Programs*, Springer-Verlag, third revised and extended edition, 1996.
4. T. Bäck and M. Schutz, "Intelligent Mutation Rate Control in Canonical Genetic Algorithms", *Proc. 9th Int. Symp. ISMIS'96*, Lecture Notes on Artificial Intelligence, vol. 1079, Springer, pp.158-167, 1996.
5. I. Rechenberg, "Cybernetic Solution Path of an Experimental Problem", Royal Aircraft Establishment Library,1965.
6. L.J. Fogel, A.J. Owens and M.J Walsh, *Artificial Intelligence through Simulated Evolution*, Wiley, 1966.
7. J. Smith and T. C. Fogarty, "Self Adaptation of Mutation Rates in a Steady State Genetic Algorithm", *Proc. IEEE Int'l. Conf. on Evolutionary Computation*, pp.318-326, 1996.
8. T. Bäck, "Self-adaptation", *Handbook of Evolutionary Computation*, IOP and Oxford University Press, pp. C7.1:1-13, 1997.
9. H. Aguirre, K. Tanaka and T. Sugimura, "Cooperative Model for Genetic Operators to Improve GAs", *Proc. IEEE Int'l Conf. on Information, Intelligence and Systems*, pp.98-106, 1999.
10. H. Aguirre, K. Tanaka, T. Sugimura and S. Oshita, "Cooperative-Competitive Model for Genetic Operators: Contributions of Extinctive Selection and Parallel Genetic Operators", *Proc. Late Breaking Papers 2000 Genetic and Evolutionary Computation Conference*, pp.6-14, 2000.
11. H. Aguirre, and K. Tanaka, "Parallel Varying Mutation Genetic Algorithms", *Proc. IEEE Int'l. Conf. on Evolutionary Computation*, pp.795-800, 2002.
12. M. Shinkai, H. Aguirre, and K. Tanaka, "Mutation Strategy Improves GA's Performance on Epistatic Problems", *Proc. IEEE Int'l. Conf. on Evolutionary Computation*, pp.968-973, 2002.
13. P.C. Chu and J. E. Beasley, "A Genetic Algorithm for the Multidimensional Knapsack Problem", *Journal of Heuristics*, vol.4, pp.63-86, 1998.
14. T. Bäck, *Evolutionary Algorithms in Theory and Practice*, Oxford Univ. Press, 1996.
15. H. Aguirre, K. Tanaka and S. Oshita, "Increasing the Robustness of Distributed Genetic Algorithms by Parallel Cooperative-Competitive Genetic Operators", *Proc. 2001 Genetic and Evolutionary Computation Conference*, Morgan Kaufmann, pp. 195-202, 2001.

Self-organizing Maps for Pareto Optimization of Airfoils

Dirk Büche[1], Gianfranco Guidati[2], Peter Stoll[2], and Petros Koumoutsakos[1,3]

[1] Institute of Computational Science, Swiss Federal Institute of Technology (ETH)
CH-8092 Zürich, Switzerland
{bueche,petros}@inf.ethz.ch
http://www.icos.ethz.ch

[2] Alstom (Switzerland) AG, Segelhof
CH-5405 Dättwil, Switzerland
{gianfranco.guidati,peter.stoll}@power.alstom.com
http://www.alstom.ch

[3] NASA Ames Research Center, Moffett Field, CA, USA
http://www.arc.nasa.gov

Abstract. This work introduces a new recombination and a new mutation operator for an accelerated evolutionary algorithm in the context of Pareto optimization. Both operators are based on a self-organizing map, which is actively learning from the evolution in order to adapt the mutation step size and improve convergence speed. Standard selection operators can be used in conjunction with these operators.
The new operators are applied to the Pareto optimization of an airfoil for minimizing the aerodynamic profile losses at the design operating point and maximizing the operating range. The profile performance is analyzed with a quasi 3D computational fluid dynamics (Q3D CFD) solver for the design condition and two off-design conditions (one positive and one negative incidence).
The new concept is to define a free scaling factor, which is multiplied to the off-design incidences. The scaling factor is considered as an additional design variable and at the same time as objective function for indexing the operating range, which has to be maximized. We show that 2 off-design incidences are sufficient for the Pareto optimization and that the computation of a complete loss polar is not necessary. In addition, this approach answers the question of how to set the incidence values by defining them as design variables of the optimization.

1 Introduction

Real-world application often include multiple and conflicting objectives. A solution to such a problem represents always a compromise between the different objectives. The set of the best compromise solutions is referred as the Pareto set, characterized that starting from a Pareto solution, one objective can only be improved at the expense of at least one other objective.

Evolutionary Algorithms (EAs) are a standard tool for Pareto optimization. EAs perform a population-based search, which allows approximating the Pareto

J.J. Merelo Guervós et al. (Eds.): PPSN VII, LNCS 2439, pp. 122–131, 2002.

front in a single optimization run by evolving the population in a cooperative search towards the Pareto front.

In Pareto optimization, work has been performed in the development of selection operators and especially fitness assignment techniques for Pareto optimization, while the recombination and mutation operators are often copied from simple single-objective algorithms and are non-adaptive over the optimization run.

In single objective optimization, however, the key component is the mutation operator. The operator exploits knowledge from the evolution to adapt a mutation distribution in order to focus on areas of promising solutions. The Covariance Matrix Adaptation (CMA) [5] embeds information about the path of successful mutations (evolution path) in a covariance matrix. This leads to a significant performance gain, compared to non-adaptive EAs.

The evolution path for multi-objective optimization is far more unclear than for single objective optimization, since the population converges in a cooperative search towards the Pareto front and not to a single optimum. Since different solutions in the population converge towards different locations on the Pareto front, the mutation distribution can not be described by one covariance matrix and has to differ between the solutions.

Thus, we introduce a new adaptation method for the mutation step size in Pareto optimization using the self-organizing maps (SOM) of Kohonen [6]. The method is inspired by the work of Milano *et al.* [8], who develop the first SOM for single objective optimization. The SOM is continuously trained on the current best solutions and thus is tracking the evolution path in a learning process. The SOM adapts the mutation step size such that it focuses on areas of promising solutions in order to generate an accelerated convergence.

The aerodynamic design of modern aircraft wings as well as rotor blades from various areas as turbo machinery, helicopters, and wind energy plants relies significantly on the design of 2D cuts (profiles), which are then stacked to a 3D wing or blade. This simplification omits 3D flow effects but is adequate for the design due to the large aspect ratios between span and chord. Real 3D calculations are often performed for the design assessment, after the actual design process. The profiles can be designed individually or can be taken from a profile family.

The aerodynamic performance of a profile is mainly characterized by the thermodynamic losses at the design operating condition and by the operating range. The operating range can be described by the possible variation of the inlet flow angle from the design condition (incidence variation) until separation or stall occurs. These two characteristics are conflicting, thus requiring a set of profile designs for different compromises manifested on a Pareto front.

This paper is organized as follows: First, the principles of EAs for Pareto optimization are introduced, followed by a description of the SOM and the modifications for the learning in an EA. The new algorithm is illustrated on a test problem. Finally, an automated loop for the Pareto optimization of an aerodynamic profile concerning losses and operating range is described. The loop comprises the new optimization algorithm, a profile generation tool and a CFD analysis tool. The properties of the resulting profiles are discussed.

2 Multi-objective Evolutionary Algorithms

The selection operator is often considered as a key operator for multi-objective evolutionary algorithms (MOEAs). It consists of the fitness assignment and the selection mechanism itself. The dominance criterion in combination with niching techniques is most popular for the fitness assignment [11], in order to select on average the less dominated solutions and preserve diversity in the population, respectively. Another key element is elitism [11], a technique of storing always the current best solutions in an archive. For a multi-objective problem, the elite solutions are the current nondominated solutions. The archive is then participating in the selection process.

The Nondominating Sorting Genetic Algorithm (NSGA-II) [2] and the Strength Pareto Evolutionary Algorithm (SPEA2) [12] are two common representatives of MOEAs and implement all of the previously stated techniques.

These algorithms, however, do not describe a mutation or recombination operator. To compare the performance on continuous problems, the authors of SPEA2 and NSGA-II use the polynomial distributed mutation and the simulated binary crossover of Deb *et al.* [1]. Both methods do not implement any learning process, so they do not exploit any knowledge from the evolution path.

2.1 Self-organizing Maps

Self-organizing maps (SOM) [6] define a mapping of a highly dimensional input space $\mathbb{R}^n$ onto a regular lattice of s reference vectors (neurons). The lattice contains a fixed m-dimensional connectivity between the neurons, which is usually of lower dimension than the input space. Figure 1 illustrates a SOM with 25 neurons and a two-dimensional quadrilateral lattice, i.e. $n = 2$, $s = 25$, and $m = 2$. A reference vector $w_i \in \mathbb{R}^n$ is associated to each neuron i.

The response of the network to an input $x_j \in \mathbb{R}^n$ is defined as the best matching neuron c:

$$c(x_j) = \arg\min_i \{||x_j - w_i||\} \tag{1}$$

The SOM can be trained on a set of input data x_j. To each x_j the response c is computed and all SOM neurons are *updated* so as to become closer to the input x_j by the update rule:

$$w_i^{new} = w_i^{old} + h(c,i) \cdot (x_j - w_i), \quad i = 1...s, \tag{2}$$

where $h(c,i)$ is the so-called *neighborhood kernel*, defined so as $h(c,c) = 1$, $h(c,i) \geq 0 \ \forall \ w_i$. We use a kernel, which is known as *bubble kernel*, defined by:

$$h(c,i) = \begin{cases} \alpha \text{ , if } r(c,i) < r_0 \\ 0 \text{ , otherwise} \end{cases}, \tag{3}$$

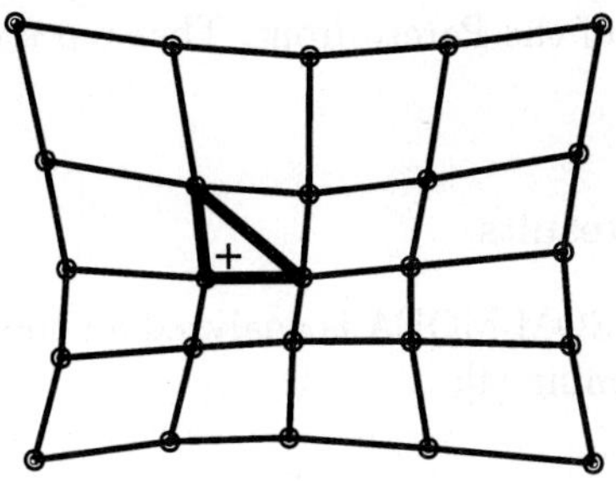

Fig. 1. SOM with 25 neurons [circles] and 2D quadrilateral lattice [thin lines]. A random simplex of adjacent neurons is created [bold line]. Within the simplex a uniformly distributed random point [plus symbol] is generated.

where α is the learning rate, $r(c,i)$ is the Euclidean distance between node c and i in the lattice and r_0 defines the bubble size. The neighborhood function allows approximating a given distribution in an ordered fashion, by preserving neighborhood relationships. One update with all input dates is referred to as one training epoch.

2.2 Self-organizing Maps for Multi-objective Evolutionary Algorithms

Here we use the SOM for the approximation of the Pareto front. To this aim we set connectivity m of the lattice to one dimension less than the objective space, that is the same dimension as the Pareto front; also, since the SOM is defined in design space, the dimension n of its reference vectors is equal to the number of design variables. The SOM is trained on the current parent population of the optimization process in order to approximate the parent population in an ordered fashion. Any *selection operator* like SPEA2 or NSGA-II can select the parent population.

We define a *recombination operator* by using the SOM. The SOM is trained on the design variables of the parent population. Thus, choosing a random point within the area that is covered by the SOM represents an intermediate recombination of the parent population. The recombination procedure chooses randomly a simplex of adjacent neurons in the lattice, and generates a recombined point u from a uniform probability distribution within the simplex (Figure 1).

In addition, a *mutation operator* is defined in order to generate points outside the area covered by the SOM. Normally distributed random numbers are added to the new point u by:

$$u_k \leftarrow u_k + \frac{\sigma}{\sqrt{n}} \mathrm{N}(0,1), \quad k = 1...n, \tag{4}$$

where σ is the step size and is set equal to the Euclidean length of a randomly chosen edge of the simplex. This leads to an adaptive step size over the optimization process, since the SOM is adapting from a random initialization to an ordered approximation of the Pareto front. The step size differs for all possible simplexes of the SOM.

2.3 Experimental Results

The performance of the SOM-MOEA is analyzed for the two-objective test function of Fonseca and Fleming [4]:

$$f_{1/2} = 1 - \exp\left(-\sum_{i=1}^{n}\left(x_k \pm \sqrt{1/n}\right)^2\right) \tag{5}$$

with $x_{1...n} \in [-1, 1]$. The exact Pareto front is obtained for $x_{1...n} = t, \quad -\sqrt{1/n} \leq t \leq \sqrt{1/n}$. The number of design variables is set to $n = 10$. An optimization run is started with the SOM-MOEA and a population size of 60 individuals. A simple selection operator is used, which selects only the current nondominated solutions in an elitistic fashion. In order to keep diversity within the selected set, the clustering algorithm of SPEA2 is used, allowing a maximum number of 30 nondominated solutions.

A one-dimensional SOM is initialized with $s = 20$ neurons, a learning rate $\alpha = 0.05$ and random values for the reference vectors w_i. After each generation, the SOM is trained with 30 training epochs on the selected set. The initial population is randomly generated. Figure 2 shows the initial population and SOM for two dimensions of the design space and for the objective space. Consider that a simplex for this SOM is a straight line.

The optimization run is started computing in total 3.000 solutions. The final population is shown in Figure 3. The figure shows that the SOM is aligned along the analytical Pareto front in design space, and the objective values of the final population are well distributed along the Pareto front. The step size σ of the mutation in Equ. 4 is related to the length of a simplex, i.e. for this one-dimensional network it is equal to the distance between two adjacent neurons. The ratio of the initial and final step size is equal to the distance between adjacent neurons of the SOM in Figure 2 and Figure 3.

3 Automated Design of Aerodynamic Profiles

We consider the automated profile design in the context of a constraint Pareto optimization. An optimization loop is implemented comprising the SOM-MOEA, a profile generation tool and a computational fluid dynamics (CFD) analysis tool.

The profile generator describes the profile by a set of Bezier splines. The spline parameters are coded in a set of engineering parameters, specifying e.g. the profile length, the nose and trailing edge radius, the curvature distribution, etc. Subdividing the profile in several splines is common in profile design [7] [10]. The transformation of the spline parameters to engineering parameters simplifies the comparison and illustration of different profile parameter sets.

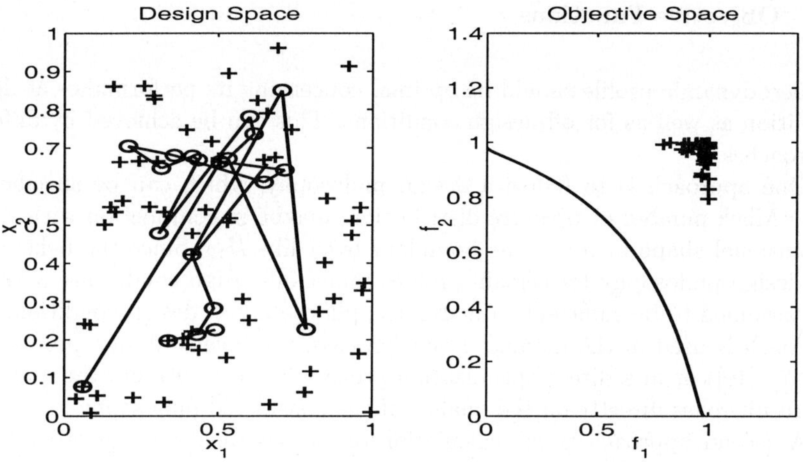

Fig. 2. Initialization of the SOM-MOEA for the two-objective problem of Fonseca and Fleming with 10 design variables: Random population [crosses], random 1-dim. SOM [connected circles] and analytical Pareto front [line] in a 2-dim subspace of the design space and in the objective space.

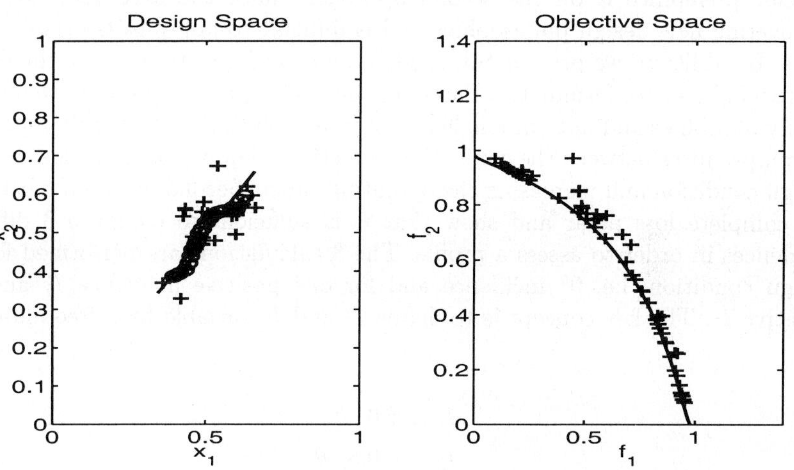

Fig. 3. SOM-MOEA after 50 generations (3000 evaluated solutions). The SOM alignes in the design space along the Pareto front.

The flow analysis is performed with MISES [3], a quasi 3D computational fluid dynamics (Q3D CFD) solver, which solves the Euler equation with an integral, viscous boundary layer formulation. It takes into account the change in the streamline thickness along the profile (quasi 3D).

3.1 Objective Functions

An aerodynamic profile should be optimal concerning its performance at design condition as well as for off-design conditions. This can be achieved by different approaches.

One approach is to follow a design philosophy, which can be specified by e.g. a Mach number or pressure distribution, maybe in conjunction with a non-dimensional shape factor of the boundary layer like H_{12}. Since the behavior of the design philosophy for certain profiles under off-design conditions is known, it is assumed to be sufficient to match the philosophy at design condition. This approach is used in the manual design process, the inverse design process (see e.g. [7], [10]) or in a direct optimization process (see e.g. [9]) and the quality of the result relies directly on the quality of the considered philosophy.

A second approach is the calculation of various incidences in order to approximate the loss polar of the profile as given in Fig. 4, which specifies the behavior of the profile over the complete operating range. A disadvantage is the large number of flow calculations, which are needed to specify the polar as in the optimization of [7]. Furthermore, there is the problem of how many incidences should be computed and for which values.

Our preference is on the second approach, since the first does not allow discovering new design philosophies and is difficult to apply to transonic or 3D flows. In addition, we present two modifications in order to improve the second approach. First, we formulate a Pareto optimization problem in order to obtain a family of profiles and not one single compromise solutions. The family represents all compromises between the conflicting objectives of minimizing the losses at the design condition and increasing the operating range. Second, we do not compute the complete loss polar and show that it is sufficient to compute 3 different incidences in order to assess a profile. The 3 calculations are performed for the design condition, i.e. $0°$ incidence and for one positive incidence I_1 and one negative I_2. The key concept is to define I_1 and I_2 variable by a free multiplier θ:

$$I_1 = 1.0 \cdot \theta \tag{6}$$

$$I_2 = -0.8 \cdot \theta \tag{7}$$

This definition takes into account that the positive incidence I_1 is more critical for stall than I_2. The incidence multiplier θ is an additional design variable.

The profile losses for the 3 incidences are summed to the first objective function f_1. For small values of θ, the losses are computed at small incidences. An optimization for small values of θ leads profiles which have minimal losses in the vicinity of the design condition, while for large values of θ, profiles are optimized for a large incidence range. Thus, θ is not only used as free design variable, but also as second objective function f_2. To both objectives penalties are added for violated constraints and the exact objective functions are given by:

$$\max(f_1), \quad f_1 = \theta - p_1 \tag{8}$$

$$\min(f_2), \quad f_2 = \sum_{i=1}^{3} (l_i) + p_1 + p_2 + p_3 + p_4, \tag{9}$$

where l_i is the profile loss for the incidence i and p_1 to p_4 are 4 penalties, which are non-zero, if the corresponding constraint is violated.

p_1 is a penalty for the convergence of the CFD solver and is equal to the number of failed incidence calculations. Especially for large incidences, the convergence may fail due to flow separation. p_2 is a linear penalty on the deviation of the exit flow angle β to the design exit flow angle β_{design} at design condition, if the deviation exceeds an absolute value of $|\beta - \beta_{design}| > \delta\beta$. p_3 is a linear penalty on the profile area A, if A is smaller than the minimal area A_{min}. The minimal area is defined by the mechanical forces on the profile and the stress limit.

The free design variables are the parameters from the profile generator and the incidence multiplier θ. In total there are 15 design variables.

3.2 Optimization Results

An optimization run is performed for a profile design at an inlet Mach number of 0.67, a desired flow turning of 12° and $\delta\beta = 0.1°$. The SOM-MOEA of Sec. 2.3 is used, except the maximal size of the selected set is increased to 50. In total 10.000 solutions are evaluated. Among all evaluated solutions, 5.461 solutions do not violate any constraints and generate a Pareto front of 283 solutions (Fig. 4). Consider that the incidence multiplier is to be maximized and the losses are to be minimized. The Pareto front underlines the conflict in optimizing the two objectives. For small incidence multipliers, the losses are low, since all 3 incidences are computed almost at the design point. For large incidence multipliers, the loss increases for two reasons. First, the flow is computed at larger incidences leading to higher losses and second, the profile losses are higher at the design condition, since the design has to be more robust for converging at the high incidences. Two Pareto solutions are marked in the figure and their loss polar is given in Fig. 4. The minimal losses are at about 1.4%. The attainable operating range is considered to be bounded by the double of the minimal losses [7]. Solution A contains the smaller incidence multiplier and the loss polar shows lower losses close to the design incidence than solution B, but comprises a smaller operating range. For solutions A and B, the operating range is about 14.4° and 15.5°, respectively. Both polars are characterized by a smooth and continuous increase of losses over the absolute incidence. This indicates a soft stall behavior. Fig. 5 contains the profile shape. Solution A shows the smaller nose radius as well as the smaller maximal thickness.

4 Conclusions

A self-organizing map (SOM) is introduced as mutation and recombination operator for Pareto optimization. The network is actively learning from the current

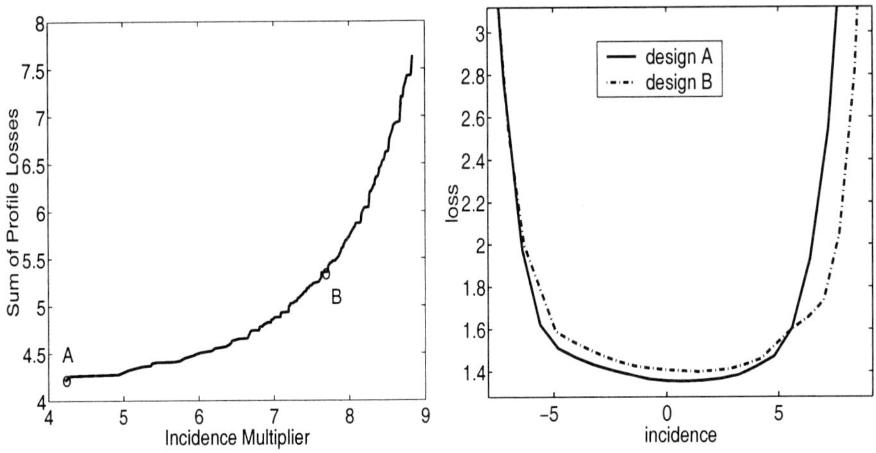

Fig. 4. Pareto front [left] for the profile optimization, and loss polar [right] for two selected Pareto solutions.

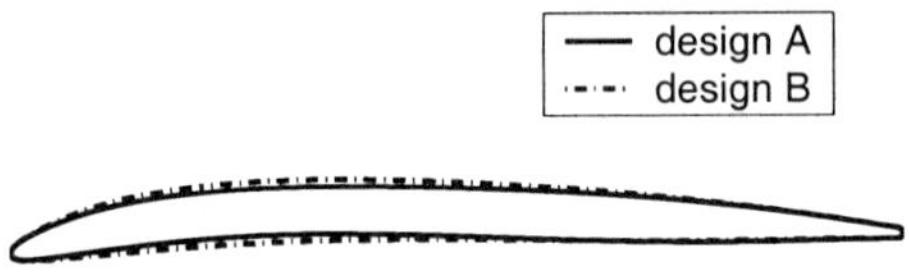

Fig. 5. Profile shape for the two selected Pareto solutions.

nondominated solutions. The SOM comprises the principle of cooperative search by interpolating the current nondominated front, thus it is sharing information about successful design variables values along the Pareto front. The mutation step size is related to the distance of neighboring neurons in the SOM. It varies within the network and is adaptive over the optimization run. The SOM represents a first step in the direction of developing mutation and recombination operators which are especially designed for Pareto optimization and which are able to learn from the evolution in order accelerate the convergence.

The second part describes a formulation for the Pareto optimization of compressor profiles for minimal losses and maximal operating range, which operates with a minimal number of incidence calculations. The key feature is the definition of a multiplier for the incidences, which is used at the same time as design variable and objective function. This optimization introduces the concept of generating profile families in a single cooperative optimization run by using a Pareto optimization algorithm.

Acknowledgments

The first author would like to acknowledge support from the Commission for Technology and Innovation, Switzerland (Project 4817.1) and Alstom (Switzerland) AG.

References

1. Deb, K., Agrawal, R.B.: Simulated binary crossover for continuous search space. Complex Systems, No. 9 (1995) 115–148
2. Deb K., Agrawal, S., Pratap, A., Meyarivan, T.: A fast elitist nondominated sorting genetic algorithm for multi-objective optimization: NSGA-II. Parallel Problem Solving from Nature VI Conference (2000) 849-858.
3. Drela, M., Youngren, H.: A User's Guide to MISES 2.53. MIT (1998)
4. Fonseca, M.C., Fleming, P.J.: Multi-objective genetic algorithms made easy: Selection, sharing and mating restrictions. Proceedings of the 1st International Conference on Genetic Algorithms in Engineering Systems: Innovations and Applications, London, UK, (1995) 45-52
5. Hansen, N., Ostermeier, A.: Completely Derandomized Self-Adaptation in Evolution Strategies. Evolutionary Computation, Vol. 9, No. 2 (2001) 159-195
6. Kohonen T.: Self-organizing maps. Springer series in information sciences, 3rd ed. (2001)
7. Köller U., Mönig, R., Küsters, B., Schreiber, H.-A.: Development of Advanced Compressor Airfoils for Heavy-Duty Gas Turbines, Part I: Design and Optimization. ASME Journal of Turbomachinery, Vol. 122, No. 3 (1999) 397-405
8. Milano, M., Schmidhuber, J., Koumotsakos P.: Active Learning with Adaptive Grids. International Conference on Artificial Neural Networks, Vienna, Austria (2001)
9. Naujoks, B., Willmes, L., Haase, W., Bäck, T., Schütz, M.: Multi-Point Airfoil Optimization Using Evolution Strategies, ECCOMAS 2000, Barcelona, Spain (2000)
10. Trigg, M.A., Tubby, G.R.,Sheard, A.G.: Automatic Genetic Optimization Approach to Two-Dimensional Blade Profile Design for Steam Turbines ASME Journal of Turbomachinery, Vol. 121, No. 1, (1999) 11-17
11. Van Veldhuizen, D.A., Lamont, G.B.: Multiobjective evolutionary algorithms: Analyzing state-of-the-art, Evolutionary Computing, Vol. 8, No. 2., MIT Press, Cambridge, MA (2000) 125-147
12. Zitzler, E., Laumanns, M., Thiele, L.: SPEA2: Improving the Strength Pareto Evolutionary Algorithm for Multiobjective Optimization. EUROGEN 2001, Athens, Greece (2001)

On Fitness Distributions and Expected Fitness Gain of Mutation Rates in Parallel Evolutionary Algorithms

David W. Corne[1], Martin J. Oates[2], and Douglas B. Kell[3,*]

[1] Department of Computer Science, University of Reading, UK
d.w.corne@reading.ac.uk

[2] Evosolve Ltd, Stowmarket, Suffolk, UK
moates@btinternet.com

[3] Institute of Biological Sciences, University of Wales, Aberystwyth, UK
dbk@aber.ac.uk

Abstract. Setting the mutation rate for an evolutionary algorithm (EA) is confounded by many issues. Here we investigate mutation rates mainly in the context of large-population-parallelism. We justify the notion that high rates achieve better results, using underlying theory which notices that parallelization favourably alters the fitness distribution of a mutation operator. We derive an expression which sets out how this is changed in terms of the level of parallelization, and derive further expressions that allow us to adapt the mutation rate in a principled way by exploiting online-sampled landscape information. The adaptation technique (called RAGE - Rate Adaptation with Gain Expectation) shows promising preliminary results. Our motivation is the field of Directed Evolution (DE), which uses large-scale parallel EAs for limited numbers of generations to evolve novel proteins. RAGE is highly suitable for DE, and is applicable to large-scale parallel EAs in general.

1 Introduction

Setting the mutation rate for an evolutionary algorithm (EA) is complicated by the fact that much depends on various details of the EA and the application. Nevertheless, much published work provides generally accepted guidelines. An overall consensus, justified by theory [3,4,9] is that a rate of $1/L$ (where L is chromosome length) is often near-optimal (this can also be said of experimental biology [5]). When they have engaged in comprehensive parametric investigations in specific cases, researchers (e.g. [4]) have sometimes found that higher rates are better. Bäck also notes [4] that the optimal rate seems to increase with λ in a $(1 + \lambda)$ evolution strategy, but that no useful analytical results are known. However, the general suitability of $1/L$ in standard (i.e. serial implementation) settings has been more often confirmed than challenged. Thus, Oates et al [10] find a wide range of optimal mutation rates, with this range tending to *include* $1/L$. Meanwhile, recent theoretical work of note has studied the competent design of parallel EAs [7], but the issue of mutation rate in this context has been little explored.

* Present address: Dept Chemistry, UMIST, PO Box 88, MANCHESTER M60 1QD

J.J. Merelo Guervós et al. (Eds.): PPSN VII, LNCS 2439, pp. 132–141, 2002.

When we consider mutation rate setting in parallel EAs, there is a straightforward intuitive argument for high rates. These generally make higher-fitness mutants available, but with low probabilities; however, the larger the population, the better the chances of a high-rate yielding such mutants. In particular, parallelization means that this benefit is not at the expense of time. If we can evaluate P mutants in parallel, then we can regard the mutation operation as having effectively changed in nature. That is, we can evaluate P mutants in unit time, and can take the 'result' of the parallelized operation to be the fitness of the *best* of them. The mathematics of this follow.

1.1 Notes on Relevance and Applicability

Improved and cheaper hardware, and the wider availability and use of cluster-based computation, now makes the use of parallel implementations of EAs more feasible, and indeed such is now increasingly widespread. This heralds a need for better understanding of parallel EA design. Cantú-Paz [7], among others, is paving the way regarding several aspects of parallel EA design. Here we focus on mutation rate setting.

One relevant parallel EA application (which the authors are working on) is in the protein engineering/biotechnology community, and is called 'Directed Evolution' (DE). This refers to (what amounts to) the application of EAs to the discovery of novel proteins [1,2,12]. Consider, for example, the task of finding a protein which can bind to a particular area of a virus, and remain thermostable at body temperature. To address this in DE, a population of proteins is artificially evolved. They undergo selection, mutation and other operations in the normal way. There are, of course, many problems with this as regards its *in silico* implementation, since we know far too little about protein folding to implement the representation correctly, let alone estimate a novel protein's fitness. The trick, however, is that DE works entirely biologically. The 'representation' of a protein is actually via a gene encoding its overexpression (i.e. many copies are present) within a suitable cell (typically the bacterium *E. coli*). The cell's metabolism naturally produces the protein itself, and fitness comes from direct measurement of the protein's efficacy regarding the target activities of interest. Mutation is typically done by using so-called 'sloppy' or 'error-prone' versions of the Polymerase Chain reaction (PCR) in an *in vitro* step. Many details are of course omitted here, but none which alters the fact that the range of potentially applicable DE strategies essentially matches much of the space of EA designs. In particular, the biotechnology (with 'high-throughput screening') currently allows up to some 1,000,000 mutants to be evaluated per generation, and mutation rate is fully controllable (essentially by varying the conditions of the PCR reaction). DE is a technology with immense potential for novel and highly beneficial protein products, but the 'search spaces' are massive, and much depends on appropriate design and parameterization of DE strategies.

1.2 A Note on Related Work

Seminal work by Fogel and co-authors (e.g. [8] and subsequent papers) relates closely to that described here, but there are subtle differences in approach and applicability which are worth noting. In [8] and related work, Fogel *et al.*, like us, essentially recognize that repeated applications of genetic operators yield valuable information which can be employed to choose and parameterize the operator(s) more wisely. In [8] however, this is done essentially using extensive offline prior sampling (consuming significant numbers of fitness evaluations). Apart from our focus on the way that applicable mutation rates increase with parallelization, the main difference in this work is that we derive an approach which can be applied online, exploiting theory (relying on our restricted attention to standard mutation of *k*-ary chromosomes) which enables us to choose from among many appropriate rates having only sampled (in principle) a single rate previously This work is hence more applicable in cases where time saving in highly-parallelized EAs) is a premium concern.

2 Fitness Distributions and Expected Fitness Gain

Imagine a mutation operator which, with respect to a given parent, has n possible fitness outcomes ($f_1, f_2, \ldots, f_n$). The *fitness distribution* gives, for each f_1, the chance the mutant will have that fitness. E.g. consider a mutation operator which flips a single randomly chosen bit in a binary string of length L, and where the fitness function is MAX-ONES, in which we seek to maximize the number of 1s in the string. If L=100 and the parent has fitness 80, then the fitness distribution can be written as ((79, 0.8),(81, 0.2)); i.e. the chance of the mutant having fitness 79 is 0.8, and the chance of it having fitness 81 is 0.2 (and fitness 82 zero). In contrast, an operator which flips two randomly chosen (but distinct) genes gives approximately: ((78, 0.64), (80, 0.32), (82, 0.038)).

'Expected fitness gain' essentially means what we expect the result of the mutation to yield in terms of a fitness improvement, keeping in mind that mutants less fit or equally fit as the parent will yield zero gain. Throughout, we assume a simple (but powerful) EA model which corresponds to a (1+ λ)-ES where λ is essentially the population size, which (by default) we assume is fully parallelized. Thus raising λ, up to the limit of parallelization, does not increase the elapsed time between generations. Consequently, an operator's fitness distribution is changed (since it now returns the 'best of λ' mutants), and so therefore is the expected fitness gain per generation. We argue that this expected gain increases faster for high-rate operators than for low-rate operators, and consequently there is a point or threshold (in terms of λ, which we will hereon simply call P) at which the higher rate becomes preferable.

2.1 Exploiting Parallelism Leads to Improved Fitness Distributions

From hereon we will work with the *gain* distribution, which only considers non-worse mutants. In the first case above, this is ((80, 0.8), (81, 0.2)), showing, for example, that the result of a mutation will be no change in fitness 80% of the time. For two-gene mutation it is ((80, 0.962), (82, 0.038)). It is natural to ask, what is the expected gain in one application of the operator? This is the sum of outcomes weighted by their probabilities. In the one-gene case this is 80.2, and in the two-gene case it is 80.076.

However, if we can suitably exploit parallelism, the gain distributions of higher rate operators become more favourable, and 'overtake' those with conservative distributions. Some simple intuition for this can follow from our running example. Imagine that we are able to evaluate 20 fitnesses in parallel. A *single* operator application (i.e. in unit time) now yields 20 mutants, from which we take the best. Now, the gain distribution of the single-gene operator is approximately ((80, 0.0115), (81, 0.9885)), in which the chance of achieving 81 is precisely the chance that not all of the 20 mutants were neutral (i.e. $1 - (1 - 0.2)^{20}$). But the gain distribution of the two-gene operator is now ((80, 0.461), (82, 0.539)). The expected fitness achieved after one time unit is therefore 80.9885 in the one-gene case and 81.078 in the two-gene case, so the higher rate operator has the more favourable distribution.

More generally, the 'P-parallelization of μ' is an operator which applies μ to the same parent P times in parallel, and the returned result is the best of these P (or the parent). P-parallelized μ has a gain distribution which we can state as follows, using Blickle and Thiele's analysis of tournament selection [6]. We will assume r distinct fitness outcomes, and give the gain distribution of μ in the form $((f_1,p_1),(f_2,p_2),\ldots,(f_r,p_r))$ where the terms' meanings are obvious from the examples above. An intermediate step is to note the 'cumulative gain distribution' $((f_1,\pi_1),(f_2,\pi_2),\ldots,(f_r,\pi_r))$ where π_i is the probability of the mutant's fitness being either f_i or worse, hence:

$$\pi_i = \sum_{j=1}^{r} f_j \tag{1}$$

and we can also note of course that $\pi_1 = f_1$ and $\pi_r = 1$.

We can now write, following [6], the gain distribution of the P-parallelised operator as $((f_1,q_1),(f_2,q_2),\ldots(f_r,q_r))$ where $q_i = \pi_i^P - \pi_{i-1}^P$, in which it is implicit that $\pi_0 = 0$. What is of particular interest is the expected fitness per application of the operator. P-parallelisation changes this, from:

$$f_E = \sum_{i=1}^{r} f_i p_i \quad \text{to:} \quad f_E = \sum_{i=1}^{r} f_i q_i = \sum_{i=1}^{r} f_i(\pi_i{}^P - \pi_{i-1}{}^P) \tag{2}$$

A key point is that the P-parallelisation of a 'high-rate' operator μ_H will often achieve a better expected gain than the P-parallelisation of its 'low-rate' counterpart μ_L. By simple calculations and approximations (which we shall omit), we can show, for example, that in a case with just three fitness outcomes and distributions as follows:

$$((f_1,L_1),(f_2,L_2),(f_3,L_3)),\ ((f_1,H_1),(f_2,H_2),(f_3,H_3))$$

where $H_3 > L_3$ and $L_3 \approx 0$ then P-parallelized μ_H exceeds P-parallelised μ_L in expected fitness gain when $P > (H_1^P - L_1^P)/H_3$, from which two observations are apparent: first, if $L_1 > H_1$ then any value of P will lead to a better expected gain for the 'high-rate' operator. At first this seems odd, but notice that L_1 and H_1 denote the probabilities in the respective cases that the gain will be zero. Hence, since $L_1 > H_1$ the chance of at least *some* gain is necessarily higher for the high-rate operator. Otherwise, in the more normal situation $H_1 > L_1$, the expression reflects arguably modest needs in population size for parallelized μ_H to outperform parallelized μ_L.

To express the more general case for two operators μ_1 and μ_2, where (without loss of generality) μ_1 has a better expected fitness gain than μ_2 when both are P-parallelised, first, we can rearrange equation (2) to become:

$$f_E = (f_1 - f_2)\pi_1{}^P + (f_2 - f_3)\pi_2{}^P + \ldots + (f_{r-1} - f_r)\pi_{r-1}{}^P + f_r$$

and we can simplify this by considering only cases where fitness levels increase in units (hence $f_k - f_{k+1}$ always equals –1), and obtain:

$$f_E = f_r - \sum_{i=1}^{r-1} \pi_i{}^P$$

Now, given two operators μ_1 and μ_2, we can simply derive:

$$f_E(\mu_1) - f_E(\mu_2) = \sum_{i=1}^{r-1} \pi_i(\mu_2)^P - \sum_{i=1}^{r-1} \pi_i(\mu_1)^P$$

When this exceeds zero, the P-parallelization of μ_1 will have a better expected gain than that of μ_2. One general observation can now be made. In the limit as P becomes very large, the dominant terms are those involving the cumulative probabilities with the highest indices, and we can write:

$$f_E(\mu_1) - f_E(\mu_2) \approx \pi_{r-1}(\mu_2)^P - \pi_{r-1}(\mu_1)^P \approx P(p_r(\mu_1) - p_r(\mu_2))$$

Hence, in the limit, the superiority of μ_1 over μ_2 after P-parallelization is guaranteed as long as μ_1 has a better chance than μ_2 of finding the highest fitness.

Finally, we mention some illustrative calculations in the context of MAX-ONES with L=100. Space issues preclude a fuller display, but we note for example that for a parent with fitness 50, P-parallelized mutation at $9/L$ starts to outperform (in terms of expected gain) a similar parallelization of $1/L$ at $P = 3$. When fitness of parent is 80, the expected gain of P-parallelised $1/L$ is outperformed by that of $10/L$ at $P = 362$.

2.2 Adapting Rates in Parallel EAs Based on Expected Fitness Gain

A wider applicability emerges from recasting entries in the gain distribution in terms of numbers of point mutations and the fitness/distance correlation. That is:

$$p_{1+i}(m,x) = \sum_{j=1}^{L} d_j(m) \cdot c_{j,i}(x) \tag{3}$$

in which we assume the operator under consideration is per-bit flip mutation on binary strings with rate m, and x stands for a specific parent, rather than a fitness. Meanwhile, $p_{1+i}(m,x)$ stands for the chance of a mutant of x having the ith fitness better than that of x, while $d_j(m)$ gives the chance of the operator yielding a mutant j Hamming units distant from x, and $c_{j,i}(x)$ gives the proportion of mutants j Hamming units away from the parent which have the fitness indexed by i. The summation goes up to L, which is the highest Hamming distance attainable. Notice that:

$$d_j(m) = m^j (1-m)^{L-j} \binom{L}{j} \tag{4}$$

In particular, it does not depend on the landscape under consideration, while $c_{j,i}(x)$ expresses the detailed fitness/distance correlation map in the region of x, and does not depend on the mutation rate. Also, we reserve p_1 to stand for the following

$$p_1 = 1 - \sum_{i=2}^{H} p_i \tag{5}$$

Where H is the highest fitness attainable. Now, imagine the requirement to set suitable parameters for a parallel (1+P)-EA. By substituting equations (3) and (5) into (2) (via (1) and (4)), we can find the expected fitness gain per generation for any parent and any mutation rate, and we will suppose that a good rate to set per generation is one which maximizes this expected gain. However we need data for equation (3). The term $d_j(m)$ is analytically accessible, but $c_{j,i}(x)$ will generally be unknown. Data pertinent to it will normally be available, however, and we now propose a method for approximating $c_{j,i}(x)$ from online sampled fitnesses. This leads to a principled technique for adaptively resetting the mutation rate in such EAs after each generation. The dependence on straightforward bit-flip (in general, k-ary) mutation is partly a restriction on applicability, but also the key enabling factor, since this makes equation (4) available, which in conjunction with sampled data allows us to estimate gains for arbitrary rates, even though we may have sampled at only one rate.

We outline the approach first, and then set it out in detail. The essential idea is that mutation rate will be reset between generations based on expected gain. We assume, in the present work, a (1+P)-EA. In generation 0, P mutants of a randomly generated initial solution are generated; by the time we complete generation g, we have generated gP mutants, and have thus obtained P items of data from which to build an approximation of $c_{j,i}(x)$ for each of g parents x. This is used, together with equations (1–5), to find a good rate to use for generation g+1. A rather necessary further approximation stems from the fact that our sample model of $c_{j,i}(x)$ is silent with regard to individuals which are fitter than the current best – but the current best is (in a (1+P)-EA) the parent from which we will be generating mutants. To get around this, we set the mutation rate one generation 'out of phase'. Another difficulty is that we do not yet have a direct analytical route to find the m with maximal expected gain; how we deal with this and other issues is set out in the pseudocode description which follows. Before that, some further notation will serve to clarify the way that we handle approximations to $c_{j,i}(x)$.

Given an arbitrary problem, but assuming a binary encoding, and per-bit flip mutation, let $c_{f,j,g}$ stand for the proportion of individuals which have fitness g among those which are j Hamming distant from an individual with fitness f. Notice that $c_{j,i}(x)$ is generally an approximation to $c_{j,g}(x)$ for any given x with fitness f. In cases such as MAX-ONES (and many others, including some classes of rugged landscapes) the approximation is exact, but in general note that, where $X = \{x \mid f(x) = f\}$:

$$c_{f,j,g} = \frac{1}{|X|} \sum_{x \in X} c_{j,g}(x)$$

i.e. it is an average over all x with the same fitness. Intuitively, we might expect the approximation to improve with j. Next we define: $n_{f,j,g}$ to be the number of points *sampled* by a search algorithm which have fitness g, are mutants of a point with fitness f, and are j Hamming distant from their parent. By also defining $s_{f,j}$ as the total number of samples found so far which are j units distant from a parent with fitness f, we can now note that the operational approximation to $c_{f,j,g}$ is: $n_{f,j,g}/s_{f,j}$. We simplify matters by assuming a modestly-sized integer range.of fitnesses. In some cases in practice, however, it may be pragmatic (at least for the purposes of the calculations for setting the mutation rate), to map the range of fitnesses found so far onto a limited number of 'fitness levels', each standing for a fixed range, e.g. J_3 may capture all fitnesses between 0.7 and 0.8.

Now we can describe our routine for adaptively setting mutation rates in a fully-parallel (1+P)-EA. We assume that the time between generations (fitness evaluation) is significant enough for us to ignore the modest overhead in this routine.

1. *Initialise:* start with a randomly generated initial solution (our 'best-so-far' b), and set an initial rate m . Initialize z values $s_{i,j}$, for i from 1 to z, and reserve space for z^2L values $n_{f,j,g}$, initialized to 0.
2. *Generate:* Produce a set $M = \{m_1, m_2, \ldots, m_p\}$containing P mutants of the best-so-far solution, and evaluate the fitness $f(m_i)$ of each m_i.
3. *Calibrate:* We now have P items of data with which we can improve (or initially construct) an approximation to $c_{f(b),j,g}$. For each individual m_i:
   Where h is the Hamming distance between b and m_i, increment $s_{f(b),h}$ by 1.
   If $f(m_i) > f(b)$, increment $n_{f(b),h,f(mi)}$ by 1.
4. *Adapt:* We now reset the mutation rate as follows, essentially by calculating what the best rate 'should' have been in the current generation, and setting that for the next generation.
   With reference to equation (5), approximate $p_{1+i}(m,b)$, for $i > 1$, by setting $c_{j,g}(b) = n_{f(b),j,g}/s_{f(b),j}$ for all $g > f(b)$ and all j up to and including the most distant mutants of b which have been sampled. Then, assigning $i \in \{1,2,3\ldots\text{H}\}$ for convenience, such that fitness i is the ith in a ranked list of fitnesses in improving order starting with $f(b)$, calculate:

   $$p_{1+i}(m,b) = \sum_{j=1}^{L} d_j(m) \cdot n_{1,j,i} / s_{1,j} \quad \text{for } i > 1 \text{ and then} \quad p_1 = 1 - \sum_{i=2}^{H} p_i$$

   for each of a range of values of m from $1/L$ to $10/L$.
   Using the results, and equations (1–6), we can then calculate $f_E(m)$ for the P-parallelised version of each rate m. Although requiring precision, the calculations are essentially straightforward and speedy, and arbitrarily many rates may be tried (e.g. 100), within an arbitrary range which perhaps goes beyond $10/L$. Finally, set m to be that rate which returned the best value of $f_E(m)$.
5. *Book-keeping:* At this stage we reset the best-so-far solution b to be the fittest individual from the set $M \cup \{b\}$.
6. *Iterate:* If a termination condition is reached, stop. Else, Return to step 2.

We will call this technique RAGE (rate-adaptation with gain-expectation)

3 Experiments

Here we report on preliminary testing of RAGE to establish proof-of-principle. We see its primary niche as being large-scale-parallel EAs, with limited numbers of generations. In these experiments we test the straightforward hypothesis that the theoretical basis of RAGE, and hence the justification behind each renewed rate setting per generation, should improve results over elementary methods. Later work will compare RAGE against other suitable mutation-rate adaptation techniques.

We used RAGE on four test problems: MAX-ONES with $L = 100$, and three simple deceptive problems with block sizes 3, 4, 5, with $L = 90$, 100, 100 respectively. For each, we experiment with two-scenarios: a (1+100)-EA run for 20 iterations, and a (1+1000)-EA run for 10 iterations. For each of these 8 cases, we try 10 versions of RAGE, differing only in the initial mutation rate in the first iteration (after which RAGE 'kicks in'), which ranged from $1/L$ to $10/L$ in steps of $1/L$. Our comparative technique is a straightforward fixed mutation rate throughout, again trialled for each of the 10 rates between $1/L$ and $10/L$. Each experiment was repeated for 50 trials.

Table 1. Comparison between RAGE and fixed-rates on MAX-ONES and Deceptive (block sizes 3, 4 and 5) using (1+100) and (1+1,000)-EAs

Problem	Method (population size)	RAGE vs Fixed	Best RAGE/ Best fixed	Best fixed rate
MAX-ONES	(1+100), 20 gens	8 / 2 / 0	96.98 / 95.88	$2/L$
	(1+1000), 10 gens	7 / 3 / 0	94.74 / 94.48	$4/L$
Deceptive,	(1+100), 20 gens	9 / 1 / 0	105.98/105.82	$2/L$
block size 3	(1+1000), 10 gens	7 / 3 / 0	104.54 / 104.02	$4/L$
Deceptive,	(1+100), 20 gens	6 / 4 / 0	105.14 / 105.06	$2/L$
Block size 4	(1+1000), 10 gens	7 / 3 / 0	106.65 / 105.84	$4/L$
Deceptive,	(1+100), 20 gens	9 / 1 / 0	103.6 / 100.66	$2/L$
Block size 5	(1+1000), 10 gens	7 / 3 / 0	101.52 / 102.26	$4/L$

Table 1 summarises the results, in the following way. Taking for example the row for the deceptive problem, block size 3, using a (1+100)-EA, column 3 summarises the results of 10 pairwise statistical comparisons, one for each mutation rate in the set $\{1/L, 2/L, \ldots, 10/L\}$. In the comparison for $3/L$, for example, a standard statistical test was performed comparing 50 trials of RAGE using $3/L$ as the initial rate, with 50 trials using $3/L$ as the fixed rate in each generation. We score 1 for a 'win' if RAGE was found superior with confidence at least 99%, 1 for a 'loss' if the fixed rate was superior, and 1 for a tie if the comparison was not conclusive. Column 3 adds these scores for each of the 10 rates. In column 4, the best RAGE mean result is shown (best of the 10 RAGE experiments with different initial rates) and is compared with the best fixed-rate mean result (best of the 10 fixed-rate experiments with different fixed rates). Column 5 indicates which rate gave the 'best fixed-rate' result in column 4.

Clearly, RAGE consistently outperforms fixed-rates, whatever the fixed rates are set at. The prospects for RAGE are therefore quite promising, Also, as generally expected, significantly higher rates than $1/L$ work best, increasing with population size, although there is too little data here on that topic to allow any further discussion Finally, though there is evidence that RAGE is a worthwhile technique, insufficient tests have been performed so far to establish it as a generally useful rate adaptation scheme. We discuss this point further below.

4 Concluding Discussion

By considering the concept of the 'gain' distribution of the per-bit flip mutation operator, we have been able to derive expressions which allow us to see how the gain distribution varies with mutation rate m and how it changes when the mutation operation is parallelised in the context of a (1+P)-ES. These investigations are particularly applicable to rate setting in parallel EA implementations, insofar as expected fitness gain is a good measure of the quality of an operator. By using online sample approximations to the exact expressions, we have proposed a routine called RAGE (Rate Adaptation with Gain Expectation), which is suitable for setting mutation rates on a per-generation basis in parallel EAs. Preliminary results show fairly convincingly that RAGE outperforms fixed-rate schemes for a wide range of fixed rates.

The background to the theory included here, and the subsequent proposed RAGE method, is the authors' interest in finding a principled way to control mutation in very-large-scale-parallel evolutionary algorithms. The application of chief interest is Directed Evolution (as discussed in section 1.3), and the RAGE method can be used in that context. The DE application brings with it certain constraints and preferences which affect the choice of rate adaptation technique used (and EA technique in general) One major point is that very large populations are possible in DE, and consequently a considerable amount of appropriate landscape (fitness) data are available at each generation, so it would seem to be sensible to employ a technique which exploited this data as far as possible (rather than use, for example, deterministic fixed rates). A further issue is that using distinct and adaptive rates per individual (such as employed in modern evolution strategies) or distinct and adaptive rates per gene, are both (although ultimately possible) currently infeasible in the context of large-scale DE. The choice of adaptive mutation strategies in DE is thus practically limited to *per-generation* adaptation.

The essential point about the theory in section 2 is the ability to estimate the gain distribution of any mutation rate based on online-sampled landscape information seen to date (section 2.3). Calculating expected fitness gain is one way to exploit this estimated gain distribution, but ongoing work is exploring other methods, since expected gain *per se*, which is essentially an average, is likely to be unduly influenced by high probabilities for modest gains, hence perhaps unwisely favouring lower rates.

Finally, although we focus on a (1+P)-ES, the ideas underlying RAGE are certainly not restricted to this specific selection method. By building a model of landscape structure information as time progresses, RAGE-like adaptation can in theory be used to infer a promising rate with which to mutate any selected parent, via appealing to landscape information pertaining to the fitness of that parent. Developing similar techniques for *recombination* operators is less straightforward, however this is possible and is the topic of ongoing work.

Acknowledgements

We thank the BBSRC (the UK Biotechnology and Biological Sciences Research Council) for financial support, and Evosolve (U.K. registered charity number 1086384) for additional support during this work.

References

1. Arnold, F. M. (ed). Evolutionary protein design. *Advances in Protein Chemistry*, vol. 55. Academic Press, San Diego, 2001.
2. Arnold F. Combinatorial and computational challenges for biocatalyst design. *Nature* 2001;409:253-7.
3. Bäck T, Optimal Mutation Rates in Genetic Search, *Proc. 5th ICGA*, pp 2 – 9, 1993.
4. Bäck T, Evolutionary Algorithms in Theory and Practice, OUP, 1996.
5. Baltz, RH. Mutation in *Streptomyces*. In: Day L, Queener S, editors. The Bacteria, Vol 9, Antibiotic-producing *Streptomyces*. Academic Press, 1986:61-94.
6. Blickle, T., Thiele, L. (1995). A Mathematical Analysis of Tournament Selection, in L.J. Eshelman (ed.) *Proc. 6th International Conference on Genetic Algorithms*, Morgan Kaufmann, pp. 9–16.
7. Cantú-Paz, E. (2000). *Efficient and Accurate Parallel Genetic Algorithms*, Kluwer Academic Publishers.
8. Fogel, D.B. and Ghozeil, A. (1996). Using Fitness Distributions to Design More Efficient Evolutionary Computations, in *Proceedings of the 3rd International Conference on Evolutionary Computation*, IEEE, pp. 11-19.
9. Mühlenbein, H. How genetic algorithms really work: I. Mutation and Hillclimbing, in R.Manner, B. Manderick (eds), Proc. 2nd Int'l Conf. on Parallel Problem Solving from Nature, Elsevier, pp 15-25.
10. Oates, M. and Corne, D. Overcoming Fitness Barriers in Multi-Modal Search Spaces, in Foundations of Genetic Algorithms 6 (2000), Morgan Kaufmann.
11. Rechenberg I, Evolutionsstrategie: Optimierung technischer Systeme nach Prinzipien der biologischen Evolution, Frommann-Holzboog, Stuttgart,1973
12. Voigt CA, Kauffman S & Wang ZG. Rational evolutionary design: The theory of in vitro protein evolution. In: Arnold FM, editor. Advances in Protein Chemistry, Vol 55, 2001:79-160.

Opposites Attract: Complementary Phenotype Selection for Crossover in Genetic Programming

Brad Dolin[1,2], M.G. Arenas[2], and J.J. Merelo[2]

[1] Computer Science Department
Stanford University
Stanford, CA 94305 (USA)
[2] Department of Architecture and Computer Technology
University of Granada
CP 18071 – Granada (Spain)
{brad,maribel,jmerelo}@geneura.ugr.es

Abstract. Standard crossover in genetic programming (GP) selects two parents independently, based on fitness, and swaps randomly chosen portions of genetic material (subtrees). The mechanism by which the crossover operator achieves success in GP, and even whether crossover does in fact exhibit relative success compared to other operators such as mutation, is anything but clear [14]. An intuitive explanation for successful crossover would be that the operator produces fit offspring by combining the "strengths" of each parent. However, standard selection schemes choose each parent independently of the other, and with regard to overall fitness rather than more specific phenotypic traits. We present an algorithm for choosing parents which have complementary performance on a set of fitness cases, with an eye toward enabling the crossover operator to produce offspring which combine the distinct strengths of each parent. We test Complementary Phenotype Selection in three genetic programming domains: Boolean 6-Multiplexer, Intertwined Spirals Classification, and Sunspot Prediction. We demonstrate significant performance gains over the control methods in all of them and present a preliminary analysis of these results.

1 Introduction

Standard crossover in genetic programming (GP) selects two parents independently, based on fitness, and swaps randomly chosen portions of genetic material (subtrees). The mechanism by which the crossover operator achieves success in GP, and even whether crossover does in fact exhibit success over other operators such as mutation, is anything but clear [14]. Intuitively, we reason that this operator allows two individuals to combine their "strengths" to produce even more highly fit offspring.

In bit string genetic algorithms (GA), the idea has firm theoretical grounding. The most basic form of GA crossover consists of selecting a single crossover

J.J. Merelo Guervós et al. (Eds.): PPSN VII, LNCS 2439, pp. 142–152, 2002.

point along the bit string, dividing each parent's bit string at that point, and swapping the resultant fragments to produce two children. Holland [10] defines "schemata" as genotypically similar bit string patterns. Goldberg [7] goes on to show that crossover allows different individuals to combine these successful building blocks in order to produce even fitter offspring. Mutation takes on the rather auxiliary role of ensuring diversity. As such, the functional significance and practical performance gains of crossover in genetic algorithms are widely accepted.

Although some preliminary analogues to GA schema theory have begun to emerge for GP [8, 18, 17, 20], a theory of crossover's evolutionary role in GP is not nearly as complete. As such, the idea that GP crossover works by combining building blocks of possible solutions is appealing, though as yet, speculative. Even still, GP "building block"-esque hypotheses are used in the literature [11, 19, 22].

Indeed, there is even some work which aims to discredit the building block hypothesis for GP crossover. Angeline [2] demonstrates with his "headless chicken" experiment that crossing over one parent with a randomly generated tree performs as well or better than standard crossover. His conclusion is that crossover plays the functional role of "macromutation" rather than "building block engine," as in GAs.

Recent research [14] demonstrates that standard crossover, overall, presents a small advantage over just mutation, though the advantage is highly dependent on parameter settings and problem domain. Koza [12] demonstrates for the Boolean 6-multiplexer problem that when crossover is used, the addition of the mutation operator presents little or no advantage. In another experiment with the same problem, Koza shows that crossover and cloning significantly outperforms mutation and cloning; however, the results are made less interpretable by the fact that a 90% crossover rate is used in the former case, but only a 10% mutation rate in the latter.

Traditional selection methods include fitness proportionate (roulette wheel), rank, and tournament; see Table 1. Fitter individuals have a higher probability of selection, but no effort is made to "match" crossover parents based on genotypic or phenotypic characteristics. We conjecture that crossover could achieve higher performance if parents were chosen to have complementary abilities.

This work presents a selection method, Complementary Phenotype Selection (CPS), which chooses parents whose strengths – as measured over a finite set of fitness cases – are complementary. The goal is to produce offspring (via genetic recombination) which combine the distinct abilities of each parent (phenotypic recombination). We test the new operator in three distinct genetic programming domains – Boolean 6-Multiplexer, Two-Spirals Classification, and Sunspot Prediction – and demonstrate significant performance gains in all of them.

The rest of the paper is organized as follows: Sec. 2 discusses related work; Sec. 3 describes the CPS algorithm; Sec. 4 presents three experiments; Sec. 5 gives some conclusions and a brief analysis; and Sec. 6 discusses ideas for future work.

2 Related Work

Researchers have experimented with several variants to both parent selection and the workings of the crossover operation itself. Ryan maintains one subpopulation of highest-performing individuals, and another subpopulation of adept but smaller individuals; crossover parents are selected one from each population, to reduce premature convergence [21]. Hinterding and Michalewicz [9] augment a highly-modified GA for a constrained parameter optimization problem with a selection procedure that takes the parents' overlap of satisfied constraints into account. However, the selection procedure is only a minor detail of the algorithm and its effects are not analyzed experimentally. Watson and Pollack [24] experiment with variable size individuals to represent subsets of genes. Crossover, then, simply produces the union of the subsets, so that genetic building blocks are directly combined. O'Reilly [15] combines crossover with local search techniques.

An extensive literature review has uncovered only one work in GP dealing with crossover parent selection based on complementary performance on fitness cases. Abrams [1] experiments with a method of parent selection which attempts to maximize total fitness and fitness case coverage using a constant linear weighting scheme. The selection method is tested in 50 independent runs on each of 80 different Boolean functions of arity 3. Population sizes of 1000 and 5 (for diagnostic purposes) are utilized. The control method is tournament selection, but the paper lacks crucial information about the implementation, such as the tournament size. When the population size is 1000, the complementary selection method outperforms the tournament method on nearly all problems; when the population size is 5, the two methods are essentially indistinguishable.

Unfortunately, we cannot readily draw conclusions from these results because of the lack of details regarding the control method. The complementary selection implementation itself suffers from the weakness that it is only defined to work on Boolean "hit-or-miss" fitness cases. An additional shortcoming is the presence of a weight constant used in evaluating possible parent pairs, which seems to be problem- dependent. In this work, we present a flexible complementary selection method which works with fitness cases of arbitrary (and even heterogeneous) data types. Furthermore, our method requires no special weights or constants to be set. Most importantly, we test our method in a variety of distinct GP domains, and against a variety of control selection methods.

3 Complementary Phenotype Selection Algorithm

It should first be noted that the parent selection algorithm we present here is relevant only in GP domains which make use of "fitness cases." This means that each individual in the population is tested over some finite, representative set of problem examples, and each individual's fitness is based on performance over this set. Use of fitness cases is typical in genetic programming; see Sec. 4 for examples. The selection algorithm can be described simply. Assume that there are N total fitness cases, and each individual's performance against each case is recorded. $Better(A[i], B[i])$ means that individual A's score on fitness case i is better than B's score on fitness case i. $Fitness(A)$ gives the fitness which results from having the array of scores A. Then, to select the parents for each crossover operator application:

1. Choose the first parent (the mother) using a typical selection method. Here we use the fitness proportionate selection method. Let M be the array of scores against each fitness case obtained by the mother.
2. Consider each individual in the population as a father candidate. Let F be the array of scores against each fitness case obtained by a given father candidate. We define $B_{F,M}$ as the imaginary "best-case offspring" array of scores against each fitness case. The score for each fitness case, $0 \leq i < N$, is computed as:
$$B_{F,M}[i] = \begin{cases} F[i] & \text{if } Better(F[i], M[i]) \\ M[i] & \text{otherwise} \end{cases} .$$
3. Choose the father F which maximizes $Fitness(B_{F,M})$. If there is a tie, choose the father with the highest overall fitness. If there is still a tie, choose randomly from among these best.

In other words, the algorithm chooses a mother using a standard selection method, then chooses a father whose strengths complement the mother's weaknesses. Note that, as presented above, the algorithm can be used regardless of the data type used to record performance on each fitness case (e.g., Boolean, double). Indeed, different fitness cases may even record performance using different data types. Furthermore, the algorithm makes no restrictions on how total fitness is calculated (if at all) based on fitness cases: a sum or weighted average may be appropriate, though other methods – such as niching (fitness sharing) [6] and/or Pareto multi-objective optimization [7] – may be just as easily utilized.

The computational cost of evaluating a father candidate for selection is minimal: we must determine the N fitness case values for $B_{F,M}$ (normally, taking the maximum or minimum value for each case), and compute $B_{F,M}$'s fitness (normally, simply summing the array). Most notably, no additional tree executions (fitness evaluations) are required. However, father candidate evaluations will grow with the square of population size, since each crossover operation requires that all father candidates be evaluated. Although in practice this computer time is minimal when compared to fitness evaluations, for truly large populations one could – for example – search for father candidates from a fixed-size random sample of the population [1].

Note that the functionality of the crossover operator itself is standard. Following [12], in each parent we randomly select an internal node with 90% probability or a leaf node with 10% probability, and swap the subtrees rooted in these nodes.

4 Experiments

The parameter settings here closely follow Koza [12]. We note here those which are common to all experimental conditions. The population size is 1000. We use

Table 1. Selection schemes. CPS is the new experimental condition; the remaining methods serve as controls

CPS	Complementary Phenotype Selection, as presented above. The mother is chosen using FPS, below
FPS	Fitness Proportionate Selection with "greedy over-selection"[1] [12]. In standard FPS, the probability of selection is directly proportional to fitness
RANK	Linear rank selection [4]. The probability of selecting the ith most fit individual is inversely proportional to i
TOURN-2	Tournament selection [5]. Two individuals are chosen at random, and the highest-fitness individual is selected. A tournament size of 2 is used extensively in the GA literature, and is not very greedy
TOURN-5	Tournament selection with a tournament size of 5
TOURN-7	Tournament selection with a tournament size of 7. A tournament size of 7 is used extensively in the GP literature, and is very greedy

90% crossover and 10% reproduction (cloning). The maximum tree depth is 17. A crossover operation which results in a deeper tree is attempted again after redoing random node selection, for a maximum of 5 tries; if the tree is still too deep, we select new parents and try again. The generative method for the initial random population is "ramped half- and-half," [12] with a maximum initial tree depth of 6. Elitism is used, so that the best individual from each generation is automatically copied into the next.

Results are based on statistics taken over 30 independent runs for each experimental condition. For each experiment, we report results for the 6 different selection schemes given in Table 1.

Java Evolving Objects (JEO) [3], part of the DREAM project [16], was used to implement the experiments. Source code is available at:
`http://dr-ea-m.sourceforge.net`.

4.1 Boolean 6-Multiplexer

This problem is taken from [12]. The goal is to learn an arity-6 Boolean function which functions like an electronic circuit. The first two bits are address bits which together single out one of the remaining four data bits. Each input to the function is a setting for all six bits; each output is the Boolean value of the data bit singled out by the address bits. We test each individual over all possible bit settings, yielding $2^6 = 64$ fitness cases.

The set of terminals is a set of 6 Boolean variables, one for each bit {`A1, A2, D1, D2, D3, D4`}. The set of functions is {`AND, OR, NOT, IF-THEN-ELSE`}. `IF-THEN-ELSE` returns `ARG2` if `ARG1`, else `ARG3`.

[1] When using fitness proportionate selection, we use "greedy over-selection" so that we have an 80% chance of choosing an individual from the top 32% of the population, and a 20% chance of choosing an individual from the bottom 68% of the population. At least with FPS, Koza [12] has demonstrated significant performance gains with this technique.

An individual scores a "hit" on a fitness case if its Boolean output matches the target value for the function. A run is terminated if any individual scores all 64 hits, with a limit of 20 generations. Fitness is computed as

$$\frac{1}{1+(64-NumHits)}.$$

For complementary crossover selection, $Better(A[i], B[i])$ is true whenever $A[i]$ constitutes a hit but $B[i]$ does not.

We plot predicted probability of run success at each generation in Fig. 1. Note that CPS significantly outperforms the other methods, solving the problem with greater than 90% probability by generation 8; by this generation, the next best method (TOURN-7) has roughly a 50% probability of success. We also calculate Koza's "computational effort" [12]: the least number of fitness evaluations one would need to perform in order to have 99% probability of a successful run, using optimal run restarts. This figure is 12,185 for CPS and 27,079 for the next best method, TOURN-7, or more than double the number of fitness evaluations.

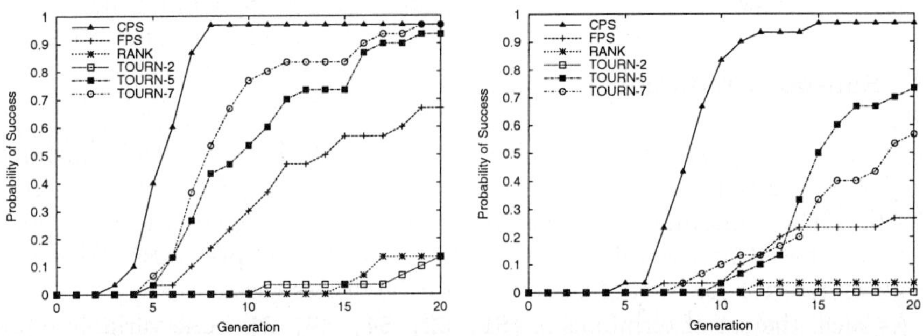

Boolean 6-Multiplexer   Intertwined-Spirals Classification

Fig. 1. Cumulative probability of run success by each generation for the Boolean 6-Multiplexer Problem (left) and the Intertwined Spirals Classification Problem (right). Equivalently, the fraction of runs we would expect to have completely solved the problem by each generation

4.2 Intertwined Spirals Classification

Lang and Whitbrock [13] introduced the rather difficult problem of using an artificial neural network to distinguish between two intertwined spirals. Koza [12] has used GP to solve the problem, using a population size of 10,000. We solve a reduced version of the problem which is more suitable to gathering reliable statistics, using only the innermost 60 points out of the original 194 in the data set. The goal is to evolve a function which takes as input the (x, y) coordinate of a point and returns a negative value if it belongs to one of the intertwined spirals and a non-negative value if it belongs to the other. Each of the 60 data points is a fitness case for this problem.

The set of terminals consists of {X, Y, R}, the variables for the given coordinate, as well as a double-float constant R between, −1 and 1, which takes on a fixed random value each time it is newly created in the initial generation. The set of functions is {+, -, *, %, SIN, COS, IF-LTE}. The modulus (%) function returns 1 when the denominator is 0. The special arity-4 function IF-LTE returns argument ARG3 if (ARG1 ≤ ARG2), and returns ARG4 otherwise.

An individual scores a "hit" on a fitness case if and only if its output is positive on a point from the first spiral, or negative on a point from the second spiral. A run is terminated if any individual scores all 60 hits, with a limit of 20 generations. Fitness is computed as

$$\frac{1}{1 + (60 - NumHits)}.$$

For complementary crossover selection, $Better(A[i], B[i])$ is true whenever $A[i]$ constitutes a hit but $B[i]$ does not.

We plot cumulative probability of run success in Fig. 1. CPS so clearly dominates the other methods that we need not calculate computational effort.

4.3 Sunspot Prediction

The goal of this problem is to evolve a function which models the number of sunspots in a given year, based on an initial part of Wolf Sunspot Data (from 1700 to 1763). Following Angeline [2], we attempt to predict the value for a given year based on the values from 1, 2, 4, and 8 years previous[2]. We record the absolute error for each of these $64 - 8 = 56$ fitness cases.

As such, the set of terminals is {S1, S2, S4, S8, R} – one variable representing the number of sunspots in each of the specified previous years, as well as the double-float constant R described in Sec. 4.2. The set of functions is the arithmetic operators {+, -, *, %, SIN, COS}.

We do not define "hits" for this problem. Fitness is computed as

$$\frac{1}{1 + NMSE},$$

where $NMSE$ is the normalized mean squared error over all fitness cases. A run is terminated after 50 generations. For complementary crossover selection, $Better(A[i], B[i])$ is true whenever $A[i]$ is smaller (less absolute error) than $B[i]$.

We plot the expected $NMSE$ (taken over 30 experiments) of the best individual in the population at each generation in Fig. 2. Of course, lower error is better, and the CPS method reaches lower values, and reaches these values more quickly, than the other methods.

[2] Actually, since, like Angeline, we do not divide the data into a test set and training set, this problem is properly a modeling rather than prediction task. Of course, it suffices for the purpose of comparing selection methods.

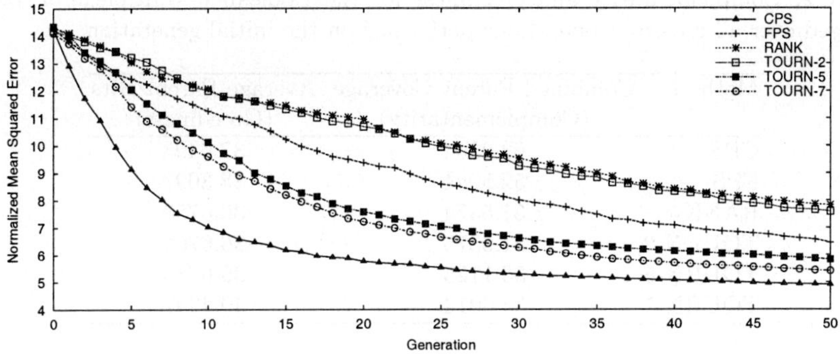

Fig. 2. Expected value for the most fit individual's normalized mean squared error at each generation, for the Sunspot Prediction Problem

5 Analysis and Conclusions

The experiments above demonstrate that on the given set of problems, the Complementary Phenotype Selection (CPS) method has substantial performance gains over the other selection methods. Intuitively, we might reason that by selecting parents which have complementary performance over a set of fitness cases, we are able to produce fitter or more robust offspring. Indeed, that CPS chooses complementary parents can be easily verified empirically. For example, Table 2 gives the number of fitness case hits (out of 64) covered by one or both parents for the Boolean 6-Multiplexer Problem, averaged over all crossover operations done on the first generation[3]. Because this statistic measures the combined strength of the average parent pair, we call it "complementarity." It can be plainly seen that CPS very quickly chooses parents which together cover much of the set, whereas the other methods choose less complementary pairs.

One might raise the objection that the high performance of CPS is not due to complementarity, but merely to a side effect of the selection algorithm – namely that it may be greedier than the others. However, an empirical analysis on the Boolean 6-Multiplexer proves otherwise. CPS's greediness, as measured by the average number of hits of selected parents, is more or less in the middle of the studied control methods, as seen in Table 2; but CPS's performance, as we have demonstrated, is top-notch. Thus, greediness alone cannot account for CPS's high performance.

6 Future Work

The current implementation of CPS uses fitness proportionate selection to select the mother. We will test other selection methods for the mother.

[3] The analysis at generation 0 is uncomplicated by already terminated runs and results of selection pressures from previous generations. An investigation of more long-term run dynamics is forthcoming.

Table 2. Complementarity and greediness for the Boolean 6-Multiplexer Problem, averaged over all crossover operations performed on the initial generation

Method	Combined Parent Coverage (Complementarity)	Average Parent Hits (Greediness)
CPS	62.9364	37.0498
FPS	52.5041	38.3097
RANK	51.5374	36.6758
TOURN-2	51.6375	36.6769
TOURN-5	53.1725	39.6783
TOURN-7	53.6014	40.4397

Since CPS relies on combining distinct parent phenotypes, it may be beneficial to explicitly "cultivate" phenotypically distinct individuals in the population. To this end, we will experiment with combined use of CPS and niching (fitness sharing) [6].

CPS increases the chances of having a total solution "disjunctively distributed" between the two selected parents. Trying many crossover operations between such parents, and giving more tries to more complete complementary pairs, is intuitively appealing. Thus, it may be useful to experiment with "brood selection" methods [23].

The analysis we present above of the functional underpinnings of CPS is far from complete. We know that CPS poses advantages beyond those of greediness, but the question of whether we are actually combining "building blocks" remains unanswered. Future work will attempt to better understand the dynamics of the selection algorithm and possible reasons for its success. One important area for exploration will consider how CPS affects population diversity.

Finally, we have seen that the CPS algorithm is highly domain-independent. It will be interesting to test its effectiveness on more problems, especially more difficult ones than those included for the above analysis. Moreover, CPS need not be restricted to just Genetic Programming. We will see how it fares in Genetic Algorithms applications, as well as other techniques from Evolutionary Algorithms.

Acknowledgments

Brad Dolin thanks the US Fulbright Program and the Spanish Fulbright Commission for supporting this research. The JEO source code [3] is part of the *Distributed Resources Evolutionary Algorithm Machine* (DREAM IST-1999-12679) project [16].

References

[1] Zoe Abrams. Complimentary selection as an alternative method for population reproduction. In John R. Koza, editor, *Genetic Algorithms and Genetic Programming at Stanford 2000*, pages 8–15. Stanford Bookstore, Stanford, California, 94305-3079 USA, June 2000.

[2] Peter J. Angeline. Subtree crossover: Building block engine or macromutation? In John R. Koza, Kalyanmoy Deb, Marco Dorigo, David B. Fogel, Max Garzon, Hitoshi Iba, and Rick L. Riolo, editors, *Genetic Programming 1997: Proceedings of the Second Annual Conference*, pages 9–17, Stanford University, CA, USA, 13-16 July 1997. Morgan Kaufmann.

[3] M.G. Arenas, B. Dolin, J.J. Merelo, P.A. Castillo, I. Fdez de Viana, and M. Schoenauer. JEO: Java evolving objects. In *Proceedings of the Genetic and Evolutionary Computation Conference, GECCO 2002*, 2002.

[4] J.E. Baker. Reducing bias and inefficiency in the selection algorithm. In *Intl. Conf. on Genetic Algorithms and their Applications*, 1985.

[5] Tobias Blickle and Lothar Thiele. A comparison of selection schemes used in genetic algorithms. TIK-Report 11, TIK Institut fur Technische Informatik und Kommunikationsnetze, Computer Engineering and Networks Laboratory, ETH, Swiss Federal Institute of Technology, Gloriastrasse 35, 8092 Zurich, Switzerland, December 1995.

[6] David Goldberg and J. Richardson. Genetic algorithms with sharing for multimodal function optimization. In *Proceedings of the Second International Conference on Genetic Algorithms*, pages 41–49, 1987.

[7] David Goldberg. *Genetic Algorithms in Search, Optimiazation, and Machine Learning.* Addison-Wesley, 1989.

[8] Thomas Haynes. Phenotypical building blocks for genetic programming. In Thomas Back, editor, *Genetic Algorithms: Proceedings of the Seventh International Conference*, pages 26–33, Michigan State University, East Lansing, MI, USA, 19-23 July 1997. Morgan Kaufmann.

[9] R. Hinterding and Z. Michalewicz. Your brains and my beauty: parent matching for constrained optimisation. In *Proceedings of the 5th Int. Conf. on Evolutionary Computation*, 1998.

[10] John H. Holland. *Adaptation in Natural and Artificial Systems.* University of Michigan Press, 1975.

[11] Hitoshi Iba and Hugo de Garis. Extending genetic programming with recombinative guidance. In Peter J. Angeline and K. E. Kinnear, Jr., editors, *Advances in Genetic Programming 2*, chapter 4, pages 69–88. MIT Press, Cambridge, MA, USA, 1996.

[12] John R. Koza. *Genetic Programming: On the Programming of Computers by Means of Natural Selection.* MIT Press, Cambridge, MA, USA, 1992.

[13] K.J. Lang and M.J. Witbrock. Learning to tell two spirals apart. In *Proceedings of the 1988 Connectionist Model Summer School*, pages 52–59. Morgan Kaufmann, 1988.

[14] Sean Luke and Lee Spector. A revised comparison of crossover and mutation in genetic programming. In John R. Koza, Wolfgang Banzhaf, Kumar Chellapilla, Kalyanmoy Deb, Marco Dorigo, David B. Fogel, Max H. Garzon, David E. Goldberg, Hitoshi Iba, and Rick Riolo, editors, *Genetic Programming 1998: Proceedings of the Third Annual Conference*, pages 208–213, University of Wisconsin, Madison, Wisconsin, USA, 22-25 July 1998. Morgan Kaufmann.

[15] Una-May O'Reilly and Franz Oppacher. Hybridized crossover-based search techniques for program discovery. Technical Report 95-02-007, Santa Fe Institute, 1399 Hyde Park Road Santa Fe, New Mexico 87501-8943 USA, 1995.

[16] B. Paechter, T. Baeck, M. Schoenauer, M. Sebag, A.E. Eiben, J. Merelo, and T.C. Fogarty. DREAM: A distributed resource evolutionary algorithm machine. In *Congress on Evolutionary Computation, CEC 2000*, volume 2, pages 951–958, 2000.

[17] Riccardo Poli and W. B. Langdon. A new schema theory for genetic programming with one-point crossover and point mutation. Technical Report CSRP-97-3, School of Computer Science, The University of Birmingham, B15 2TT, UK, January 1997. Presented at GP-97.

[18] Riccardo Poli. General schema theory for genetic programming with subtree-swapping crossover. Technical Report CSRP-00-16, University of Birmingham, School of Computer Science, November 2000.

[19] Justinian P. Rosca and Dana H. Ballard. Discovery of subroutines in genetic programming. In Peter J. Angeline and K. E. Kinnear, Jr., editors, *Advances in Genetic Programming 2*, chapter 9, pages 177–202. MIT Press, Cambridge, MA, USA, 1996.

[20] Justinian P. Rosca. Analysis of complexity drift in genetic programming. In John R. Koza, Kalyanmoy Deb, Marco Dorigo, David B. Fogel, Max Garzon, Hitoshi Iba, and Rick L. Riolo, editors, *Genetic Programming 1997: Proceedings of the Second Annual Conference*, pages 286–294, Stanford University, CA, USA, 13-16 July 1997. Morgan Kaufmann.

[21] Conor Ryan. Pygmies and civil servants. In Kenneth E. Kinnear, Jr., editor, *Advances in Genetic Programming*, chapter 11, pages 243–263. MIT Press, 1994.

[22] Terence Soule, James A. Foster, and John Dickinson. Using genetic programming to approximate maximum clique. In John R. Koza, David E. Goldberg, David B. Fogel, and Rick L. Riolo, editors, *Genetic Programming 1996: Proceedings of the First Annual Conference*, pages 400–405, Stanford University, CA, USA, 28–31 July 1996. MIT Press.

[23] W. A. Tackett and A. Carmi. The unique implications of brood selection for genetic programming. In *Proceedings of the 1994 IEEE World Congress on Computational Intelligence*, Orlando, Florida, USA, 27-29 June 1994. IEEE Press.

[24] R.A. Watson and J.B. Pollack. Symbiotic combination as an alternative to sexual recombination in genetic algorithms. In *Parallel Problem Sovling From Nature, PPSN 2001*, pages 425–434, 2001.

Theoretical Analysis of the Confidence Interval Based Crossover for Real-Coded Genetic Algorithms

C. Hervás-Martínez, D. Ortiz-Boyer, and N. García-Pedrajas

Department of Computing and Numerical Analysis
University of Córdoba, Córdoba 14071, Spain
{chervas,ma1orbod,npedrajas}@uco.es

Abstract. In this paper we study some theoretical aspects of a new crossover operator for real-coded genetic algorithms based on the statistical features of the best individuals of the population. This crossover is based on defining a confidence interval for a localization estimator using the L_2 norm. From this confidence interval we obtain three parents: the localization estimator and the lower and upper limits of the confidence interval. In this paper we analyze the mean and variance of the population when this crossover is applied, studying the behavior of the distribution of the fitness of the individuals in a problem of optimization. We also make a comparison of our crossover with the crossovers BLX-α and UNDX-m, showing the robustness of our operator.

1 Introduction

Binary-coded genetic algorithms (GAs) find the optimization of real valued functions too difficult. The binary representation of the chromosomes evenly discretizes the real domain, so, binary substrings representing each parameter with a desired precision are concatenated to form a chromosome. The main problem is that, if the number of variables is large, the length of the chromosome is too long. A solution to this problem is the real codification of the chromosomes, where each gene is a floating point number that represents a variable of the function to be optimized.

In recent years, several real-coded GAs for non-linear function optimization of continuous variables, using floating-point representation, have been studied [1][2][3][4][5][6][7][8][9]. These GAs have outperformed binary-representations in function optimization. Theoretical studies of real-coded GAs have also been performed [10][11][12][13]. These studies analyze the distribution of the offspring generated by crossover operators with the assumptions that the population size is sufficiently large, so it can be treated by a probability distribution. A recent work [9] analyzes the sampling bias of the crossover operator in optimization problems where the optimum is in a corner of the search space. Multi-parental extension of crossover has been studied in [14][15][16].

In this paper we focus our attention on the dynamic of the distribution of the population subject to the confidence interval based crossover. We make

J.J. Merelo Guervós et al. (Eds.): PPSN VII, LNCS 2439, pp. 153–161, 2002.

two hypothesis: (1) The genes are normally distributed and, (2) the genes of the fittest individuals are also normal and independent from the distribution of the genes of the whole population. We propose a parameter estimator of the genes belonging to the fittest individuals based on minimizing the dispersion function induced by the L_2 norm. Then, using the associated distribution of the statistical estimator we construct $(1-\alpha)$ confidence intervals (CI). Using this estimator and the lower and upper limits of the confidence interval we propose a new multi-parental crossover operator, which is based on learning the statistical features of localization and dispersion of the best individuals of the population. The algorithm is adaptive, as on each generation the parameters of localization and dispersion must be recalculated for obtaining the new confidence intervals of each gene.

This paper is organized as follows, Section 2 describes in detail the proposed crossover operator using L_2 norm; Section 3 analyzes the statistical features of the population once the crossover operator is applied; Section 4 defines an optimization problem in terms of a real-coded genetic algorithm and shows the results of applying our operator to different functions, and a comparison among CIXL2, BLX-α, and UNDX-m crossovers; finally Section 5 states the conclusions of our work.

2 Confidence Interval Based Crossover

In this study we will work with the i-th gene without loss of generality. Let $\boldsymbol{\beta}$ be the set of N individuals that forms the population and $\boldsymbol{\beta}^* \subset \boldsymbol{\beta}$ the subset of the n fittest individuals, and q the number of genes on each chromosome. Let us assume that the genes, β_i, of the chromosomes of the individuals in $\boldsymbol{\beta}^*$ are independent random variables, then we can consider β_i a random variable with a continuous distribution $H(\beta_i)$. With a localization parameter of the form μ_{β_i}, we have the model $\beta_i = \mu_{\beta_i} + e_i$, for each $i = 1, \ldots, q$, being e_1 a random variable.

If we assume that the n fittest individuals form actually a simple random sample $(\beta_{i_1}, \beta_{i_2}, \ldots, \beta_{i_n})$ of the distribution of the fittest individuals of the population β_i^b, the model can be written:

$$\beta_{ij}^b = \mu_b + e_{ij}, \qquad \text{for j= 1, 2, \ldots, n.} \tag{1}$$

Let us consider the L_2 norm, defined as $\|\beta_i^b\|_2 = \sqrt{\sum_{j=1}^n (\beta_{ij}^b)^2}$, then we can define the dispersion function, D_2, induced by the L_2 norm as: $D_2(\mu_b) = \sqrt{\sum_{j=1}^n (\beta_{ij}^b - \mu_b)^2}$. The estimator using this dispersion function of the localization parameter is the mean of the distribution β_i^b, that is, $\hat{\mu_b} = \bar{\beta_i^b}$.

The sample mean estimator is a good linear estimator, so it has the properties of unbiasedness and consistency, and when the distribution of the genes is normal, it follows a normal distribution $N(\mu_b, \sigma_b^2/n)$. Under these assumptions we have a bilateral confidence interval for the localization of the genes using the sample mean as localization parameter. This confidence interval, for a confidence coefficient $1-\alpha$, has the form:

$$I_{1-\alpha}(\mu_b) = \left[\bar{\beta}_i^b - t_{n-1,\alpha/2}\frac{S_b}{\sqrt{n}}; \bar{\beta}_i^b + t_{n-1,\alpha/2}\frac{S_b}{\sqrt{n}}\right], \tag{2}$$

where $S_b = \left[\sum_{j=1}^{n}(\beta_{ij}^b - \hat{\beta}_i^b)^2/(n-1)\right]^{1/2}$ is the standard deviation, and t_{n-1} is a Student t of $n-1$ degrees of freedom.

2.1 Crossover Operator Method

From the confidence interval of 2 we build three individuals that are considered the parents of the proposed crossover. These three parents are formed by: all the lower limit values of the confidence intervals of the genes, individual CILL; all the upper limit values of the confidence intervals of the genes, individual CIUL; and all the means of the confidence intervals of the genes, individual CIM. These individuals divide the domain of the gene values into three subintervals: $I_i^L \equiv [a_i, CILL_i)$, $I_i^M \equiv [CILL_i, CIUL_i]$, and $I_i^R \equiv (CIUL_i, b_i]$.

The interval based crossover operator using L_2 norm, CIXL2, creates an offspring β^s, from an individual of the population, $\beta^f = (\beta_1^f, \beta_2^f, \ldots, \beta_p^f)$, and the three individuals CILL, CIUL, and CIM obtained from the confidence interval. We consider these four individuals and their fitness (being $f(\beta)$ the fitness value of individual β) and distinguish three cases depending on the position of β^f in one of the three subintervals defined above. These three cases are:

Case 1: $\beta_i^f \in I_i^L$. If $f(\beta^f) > f(CILL)$ then $\beta_i^s = r(\beta_i^f - CILL_i) + \beta_i^f$ else $\beta_i^s = r(CILL_i - \beta_i^f) + CILL_i$.
Case 2: $\beta_i^f \in I_i^M$. If $f(\beta^f) > f(CIM)$ then $\beta_i^s = r(\beta_i^f - CIM_i) + \beta_i^f$ else $\beta_i^s = r(CIM_i - \beta_i^f) + CIM_i$.
Case 3: $\beta_i^f \in I_i^R$. If $f(\beta^f) > f(CIUL)$ then $\beta_i^s = r(\beta_i^f - CIUL_i) + \beta_i^f$ else $\beta_i^s = r(CIUL_i - \beta_i^f) + CIUL_i$.

where r is a uniform random number belonging to [0, 1].

3 Analysis of CIXL2 Crossover

Theoretical results so far are developed for genetic algorithms using binary-codified chromosomes. Due to the fact that the domain of real-valued genes is infinite and uncountable, there are serious difficulties in investigating the evolution of this kind of GAs from a theoretical point of view. Qi and Palmieri [10][12] have derived some properties of genetic operators. In these works, the object of study is the changes in the population density function along the genetic evolution considering a population of infinite size.

We focus our attention on the change of the population density function after a CIXL2 crossover has been performed, and on the change of the localization parameters mean and variance.

Let us assume that the gene involved in the crossover, β_i^f, is within the confidence interval, $\beta_i^f \in I_i^M$. We consider that in the j-th generation $\beta_i^f \in N(\mu, \sigma^2)$, and that, provided that the population is sufficiently large, the sample of the fittest individuals of the population is also normally distributed, $\beta_i^b \in N(\mu_b, \sigma_b^2)$; and that the two distributions are independent. With these assumptions the distribution of the mean of the confidence interval, CIM_i, follows a normal distribution, $CIM_i = \bar{\beta}_i^b \in N(\mu_b, \sigma_b^2/n)$.

We have two possible events, $f(\beta_f) > f(CIM)$ with a probability p, and $f(\beta_f) < f(CIM)$ with a probability $1-p$. In the first case the distribution of the offspring is normal, as it is a linear combination of normal distributions, $\beta_i^s = (1+r)\beta_i^f - rCIM_i$:

$$\beta_i^s \in N((1+r)\mu - r\mu_b, (1+r)^2\sigma^2 + r^2\sigma_b^2/n). \tag{3}$$

At the beginning of the evolution $\sigma^2 \gg \sigma_b^2$ and $\mu \neq \mu_b$, due to the fact that the subpopulation of the n fittest individuals is a subset of the population of individuals. Along the evolution, due to the selection process, $\mu \to \mu_b$ and $\sigma^2 \to \sigma_b^2$, yielding:

$$\beta_i^s \to N(\mu_b, \sigma_b^2((1+r)^2 + r^2/n)). \tag{4}$$

In the second case, $f(\beta^f) < f(CIM)$, the distribution of the offspring is normal, as it is a linear combination of normal distributions, $\beta_i^s = (1+r)CIM_i - r\beta_i^f$:

$$\beta_i^s \in N((1+r)\mu_b - r\mu, (1+r)^2\sigma_b^2/n + r^2\sigma^2), \tag{5}$$

and, following the same reasoning above:

$$\beta_i^s \to N\left(\mu_b, \sigma_b^2\left((1+r)^2/n + r^2\right)\right). \tag{6}$$

So, it follows that the distribution of the generated offspring by means of this crossover will be a mixture of normal distributions, and so a normal itself, of mean:

$$E(\beta_i^s) = p\mu_b + (1-p)\mu_b = \mu_b, \tag{7}$$

and variance:

$$\begin{aligned} V(\beta_i^s) &= p^2\sigma_b^2((1+r)^2 + r^2/n) + (1-p)^2\sigma_b^2((1+r)^2/n + r^2) \\ &= \sigma_b^2\left[p^2\left((1+r)^2 + \frac{r^2}{n}\right) + (1-p)^2\left(\frac{(1+r)^2}{n} + r^2\right)\right]. \end{aligned} \tag{8}$$

We can conclude that the variance of the fittest individuals decreases every generation if we choose n in function of the values of p and r. For example, for $p = 0$, n must fulfill, $n > \frac{(1+r)^2}{1-r^2}$, for $p = 1/2$, $n > \frac{1+2r+2r^2}{3-2r-2r^2}$, and for $p = 1$, $n > \frac{r^2}{-2r-r^2}$. So, if we always choose an individual whose fitness value is above

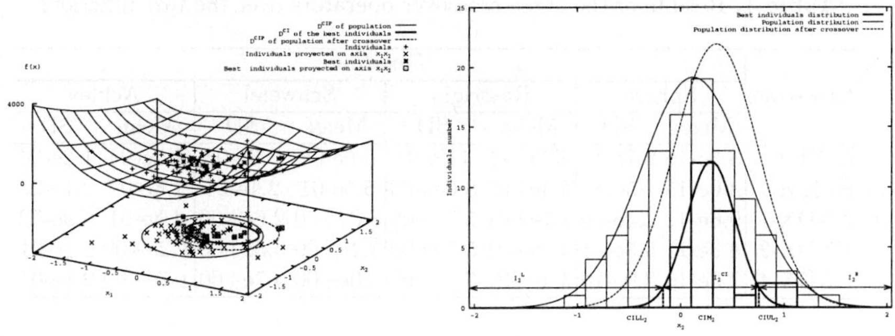

Fig. 1. Effect of the CIXL2 crossover over a population

the average fitness of the population, $p = 1$, the variance of the offspring will decrease in the next generation.

Figure 1 shows the effect of the CIXL2 crossover over the population of individuals. The crossover moves the population to the fittest individuals, that is $\mu \rightarrow \mu_b$, and $\sigma^2 \rightarrow \sigma_b^2$.

4 Optimization of Functions by Means of a Real-Coded GA

We have applied CIXL2 to an optimization problem of real functions, $g : S \rightarrow R$, where $S \subset R^q$. We have individuals of length q and an initial population of $N = 100$ individuals randomly generated in the space S. Each individual is a real-valued vector $\mathbf{x}_k, k \in 1, 2, \ldots, N$.

The optimization problem consists of the minimization of f, so the fitness of each individual, $f(\mathbf{x})$, is computed using an exponential scaling $f(\mathbf{x}) = \frac{1}{g(\mathbf{x})+1}$.

The new population is obtained selecting the best individual of the population, the rest $N - 1$ individuals are selected by means of a tournament selection method with two individuals. Over this population we applied crossover with a probability of 0.6. The offspring always substitutes its parent.

An individual is selected for mutation with a probability of $p_{mi} = 0.1$. Once selected, each gene of the individual is mutated with a probability $p_{mg} = 0.5$. The mutated individual always substitutes its parent. We have used three mutation operators: Non-uniform, continuous and modal discontinuous. The evolution is stopped after 5000 generations.

4.1 Experimental Results

We have compared CIXL2 crossover with two of the most interesting current crossovers UNDX-m [5] and BLX-α [17]. We have considered 4 functions of the set proposed by Eiben and Bäck [18]. These functions are:

Table 1. Results of the three crossover operators over the test functions

Crossover	*Function*							
	Sphere		Rastrigin		Schwefel		Ackley	
	Mean	SD	Mean	SD	Mean	SD	Mean	SD
CIXL2	3.3e-16	9.7e-17	6.2e-02	1.4e-01	6.6e-08	7.3e-08	1.3e-04	1.2e-04
BLX-α	3.0e-15	1.5e-15	3.4e+01	8.0e+00	5.5e-01	2.1e-01	1.8e-07	8.1e-08
UNDX-1	9.6e-11	2.4e-10	2.7e+01	5.7e+00	3.8e+00	2.6e+00	4.8e-01	6.9e-02
UNDX-2	4.8e-11	5.3e-11	3.8e+01	5.5e+00	6.4e+00	3.5e+00	1.2e+00	2.1e-01
UNDX-4	2.3e-10	2.9e-10	4.4e+01	6.7e+00	9.0e+00	4.7e+00	1.7e+00	2.1e-01

Table 2. Statistical Tamhane tests of the three crossover operators. The table shows the difference, $I-J$, between the mean with CIXL2 crossover, I, and the other crossover J, and the significance of this difference using a Tamhane test

Crossover	*Function*							
	Sphere		Rastrigin		Schwefel		Ackley	
	$(I-J)$	Sign.	$(I-J)$	Sign.	$(I-J)$	Sign.	$(I-J)$	Sign.
BLX-α	-2.6e-15	0.003	-3.4e+01	0.000	-5.5e-01	0.000	1.2e-04	0.076
UNDX-1	-9.5e-11	0.930	-2.7e+01	0.000	-3.7e+00	0.013	-4.7e-01	0.000
UNDX-2	-4.8e-11	0.171	-3.8e+01	0.000	-6.3e+00	0.003	-1.1e+00	0.000
UNDX-4	-2.2e-10	0.279	-4.3e+01	0.000	-8.9e+00	0.002	-1.6e+00	0.000

Function	*Expression*	*Range*
Hypersphere	$g_1(\mathbf{x}) = \sum_{i=1}^{q} x_i^2$	$x_i \in [-5.12, 5.12]$
Rastrigin	$g_2(\mathbf{x}) = \sum_{i=1}^{q}(x_i^2 - 10\cos(2\pi x_i) + 10)$	$x_i \in [-5.12, 5.12]$
Schwefel	$g_3(\mathbf{x}) = \sum_{i=1}^{q}\left(\sum_{j=1}^{i} x_j\right)^2$	$x_i \in [-65.536, 65.536]$
Ackley	$g_4(\mathbf{x}) = 20 + e - 20e^{\left(-0.2\sqrt{1/q\sum_{i=1}^{q} x_i^2}\right)} - e^{\left(1/q\sum_{i=1}^{q}\cos(2\pi x_i)\right)}$	$x_i \in [-30, 30]$

For all the functions $q = 30$, the minimum is in $\mathbf{x}_m = (0, 0, \ldots, 0)$ and $g(\mathbf{x}_m) = 0$. g_1 is unimodal and separable, g_2 is multimodal and separable, g_3 is unimodal and nonseparable, and g_4 is multimodal and nonseparable.

In a previous work[19] the optimal values of the confidence coefficient, $1-\alpha$, and the number of fittest individuals, n, were obtained. For all the functions $(1-\alpha) = 0.7$ and $n = 5$, except for the Ackley function where $n = 30$. The evolution was repeated ten times for each test function, and the best individual was selected as the final result of the evolution. The results of applying the operators CIXL2, BLX-0.5 and UNDX-m ($m \in \{1, 2, 4\}$), are shown on Table 1. For each crossover we show the mean and standard deviation of the 10 experiments.

The comparison of the results of the different crossover operators has been made by means of an analysis of variance considering the crossover operator as factor. Table 2 summarizes the results obtained with a Tamhane test. This test was used because a previous Levene test showed that the covariance matrix of the populations were different. From these tests we can conclude that CIXL2

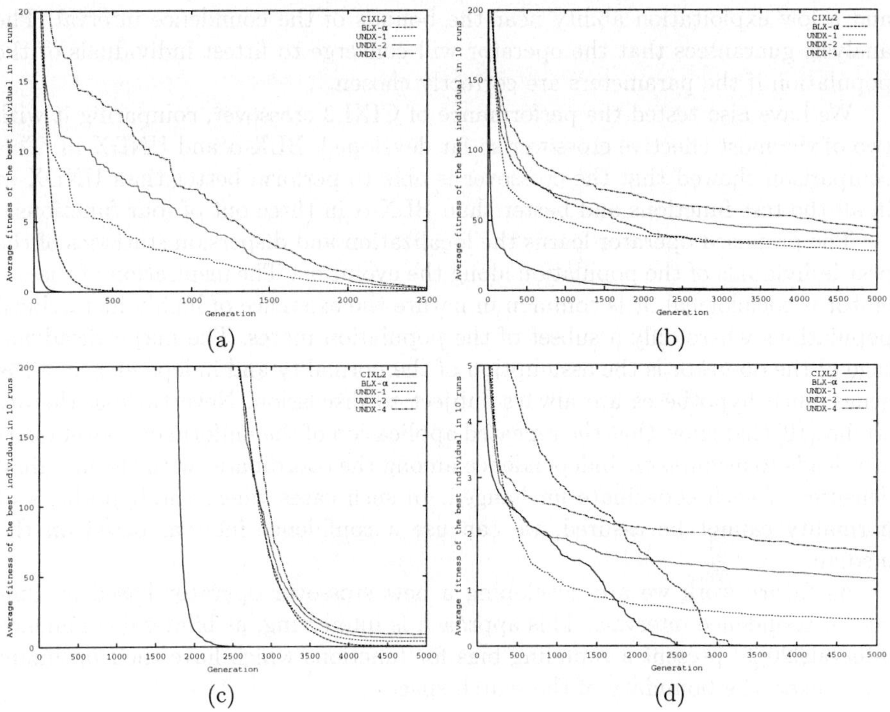

Fig. 2. (a) Averaged fitness of the best individual in 10 runs for Sphere, (b) Rastrigin, (c) Schwefel, and (d) Ackley

crossover achieved the best results for the functions g_1, g_2, and g_3, and BLX-0.5 is better than CIXL2 in the function g_4.

For the Ackley function we carried out an ANOVA III test with three factors: mutation operator, number of fittest individuals (n), and confidence coefficient $(1 - \alpha)$. The best results where achieved with $n = 30$, $1 - \alpha = 0.7$, and a continuous modal mutation. The value of $n = 30$ shows that for multimodal and nonseparable functions we need more information about the best individuals of the population, and not only the information contained in a small subset of the best individuals.

Figures 2abc show that CIXL2 crossover converges faster the other two crossovers. Figure 2d shows how CIXL2 and BLX-0.5 achieved the best results, being their performance better than the results of UNDX-m.

5 Conclusions

We have shown that the CIXL2 crossover has the effect of driving a population to the subset of the fittest individuals. This property makes this crossover a very robust operator with a high exploitation ability within the confidence interval, and a low exploitation ability near the bounds of the confidence interval. The analysis guarantees that the operator will converge to fittest individuals of the population if the parameters are correctly chosen.

We have also tested the performance of CIXL2 crossover, comparing it with two of the most effective crossover so far developed, BLX-α and UNDX-m. This comparison showed that the crossover is able to perform better than UNDX-m in all the test functions and better than BLX-α in three out of four functions.

The proposed operator learns the localization and dispersion statistics of the best individuals of the population along the evolution. The inspiration of the operator is sociological, it is common in nature the existence of highly hierarchical populations where only a subset of the population mates. The major disadvantage of the operator is the assumption of the normality and independence of the genes, such hypotheses are always subject to discussion. Nevertheless, the are studies[10] that show that the repeated application of the uniform crossover operator leads to asymptotic independence among the coordinates with the marginal densities of each coordinate unchanged. In such cases where the hypothesis of normality cannot be assured, we can use a confidence interval based on the median.

As future work we are developing a new crossover operator based on unilateral confidence intervals. This approach is interesting, as bilateral confidence intervals[20][9] present a sampling bias for functions which have their optimum at, or near, the boundary of the search space.

Acknowledgement

This work has been financed in part by the project TIC2001-2577 of the Spanish CICYT and FEDER funds.

References

1. Goldberg, D.E.: Real-coded genetic algorithms, virtual alphabets, and blocking. Complex Systems (1991) 139–167
2. Eshelman, L.J., Schaffer, J.D.: Real-coded genetic algorithms and interval-shemata. In Whitley, L.D., ed.: Foundation of Genetic Algorithms 2, San Mateo, Morgan Kaufmann (1993) 187C3.3.7:1–C3.3.7:8.–202
3. Janikow, C.Z., Michalewicz, Z.: An experimental comparison of binary and floating point representations in genetic algorithms. In: Proc. of the Fourth International Conference on Genetic Algorithms, Morgan Kaufmann Publishers, San Mateo (1991) 31–36
4. Michalewicz, Z.: Genetic Algorithms + Data Structures = Evolution Programs. Springer-Verlag, New York (1992)
5. Ono, I., Kobayashi, S.: A real-coded genetic algorithm for function optimization using unimodal normal distribution crossover. In: 7th International Conference on Genetic Algorithms, Michigan, USA, Michigan State University, Morgan Kaufman (1997) 246–253

6. Ono, I., Kita, H., Kobayashi, S.: A robust real-coded genetic algorithm using unimodal normal distribution crossover augmented by uniform crossover: Effects of self-adaptation of crossover probabilities. In Banzhaf, W., Daida, J., Eiben, A.E., Garzon, M.H., Honavar, V., Jakiela, M., Smith, R.E., eds.: Genetic and Evolutionary Computation Conf. (GECCO'99), San Francisco, CA, Morgan Kaufmann (1999) 496–503
7. Herrera, F., Lozano, M., Verdegay, J.L.: Tackling real-coded genetic algorithms: Operators and tools for behavioural analysis. Artificial Inteligence Review (1998) 265–319 Kluwer Academic Publisherr. Printed in Netherlands.
8. Tsutsui, S., Yamamura, M., Higuchi, T.: Multi-parent recombination with simplex crossover in real coded genetic algorithms. In: PPSN. Volume VI., Springer-Verlag (1999) 657–664
9. Tsutsui, S., Goldberg, D.E.: Search space boundary extension method in real coded genetic algorithms. Information Science **133** (2001) 229–247
10. Qi, X., Palmieri, F.: Theoretical analysis of evolutionary algorithms with an infinite population size on continuous space. part i: Basic properties of selection and mutation. IEEE Trans. Neural Networks **5** (1994) 102–119
11. Qi, X., Palmieri, F.: Theoretical analysis of evolutionary algorithms with an infinite population size on continuous space. part ii: Analysis of the diversification role of crossover. IEEE Trans. Neural Networks **5** (1994) 120–128
12. Nomura, T.: An analysis on crossorvers for real number chromosomes in an infinite population size. In: Fifteenth International Joint Conference on Artificial Intelligence (IJCAI'97), NAGOYA, Aichi, Japan (1997) 936–941
13. Kita, H., Ono, I., Kobayashi, S.: Theoretical analysis of the unimodal normal distribution crossover for real-coded genetic algorithms. In: IEEE International Conference on Evolutionary Computation ICEC'98, Anchorage, Alaska, USA (1998) 529–534
14. Eiben, A., Raué, P.E., Ruttkay, A.: Genetic algorithms with multi-parent recombination. In Davidor, Y., Schwefel, H.P., Männer, R., eds.: The 3rd Conference on Parallel Problem Solving from Nature. Number 866 in Lecture Notes in Computer Science. Springer-Verlag (1994) 78–87
15. Eiben, A., Schippers, C.: Multi-parent's niche: n-ary crossovers on nk-landscapes. In Voigt, H.M., Ebeling, W., Rechenberg, I., Schwefel, H.P., eds.: The 4rd Conference on Parallel Problem Solving from Nature. Number 1141 in Lecture Notes in Computer Science. Springer, Berlin (1994) 319–328
16. Tsutsui, S., Ghosh, A.: A study of the effect of multi-parent recombination in real coded genetic algorithms. In: Proc. of the ICEC. (1998) 828–833
17. Eshelman, L.J.: The CHC adaptive search algorithm: How to safe search when engaging in non-traditional genetic recombination. In: Foundations of Genetic Algorithms. Morgan Kaufman Publisher, San Mateo (1991) 256–283
18. Eiben, A., Bäck, T.: Multi-parent recombination operators in continuous search spaces. Technical Report TR-97-01, Leiden University (1997)
19. Hervás, C., Ortiz, D.: Operadores de cruce basados en estadísticos de localización para algoritmos genéticos con codificación real. In Alba, E., Fernandez, F., Gomez, J.A., Herrera, F., Hidalgo, J.I., Lanchares, J., Merelo, J.J., Sánchez, J.M., eds.: Primer Congreso Español De Algoritmos Evolutivos y Bioinspirados (AEB'02), Mérida, Spain (2002) 1–8
20. Eshelman, L.J., Mathias, K.E., Schaffer, J.D.: Crossover operator biases: Exploiting the population distribution. In: Proceedings of the Seventh International Conference on Genetic Algorithms. (1997) 354–361

Deterministic Multi-step Crossover Fusion: A Handy Crossover Composition for GAs

Kokolo Ikeda[1] and Shigenobu Kobayashi[1]

Interdisciplinary Graduate School of Science and Engineering
Tokyo Institute of Technology
psyche@fe.dis.titech.ac.jp, kobayasi@dis.titech.ac.jp

Abstract. Multi-step crossover fusion (MSXF) is a promising crossover method using only the neighborhood structure and the distance measure, when heuristic crossovers are hardly introduced. However, MSXF works unsteadily according to the temperature parameter, like as Simulated Annealing. In this paper, we introduce deterministic multi-step crossover fusion (dMSXF) to take this parameter away. Instead of the probabilistic acceptance of MSXF, neighbors are restricted to be closer to the goal solution, the best candidate of them is selected definitely as the next step solution. The performance of dMSXF is tested on 1max problem and Traveling Salesman Problem, and its superiority to conventional methods, *e.g.* uniform crossover, is shown.

1 Introduction

Genetic Algorithm (GA) is one of the most effective approximation algorithm for optimization problems, and many applications had been done [Nagata 97]. An important characteristic of GA, comparing with other optimization methods like simulated annealing (SA), is that GA holds multiple search points and their information is exchanged. New solutions, called children, are generated by a **crossover** operator, and they are expected to inherit favorable features of parent solutions.

However, in order to apply GA to a new problem, it is often very difficult to design adequate crossover operator for the problem, especially in discrete or combinatorial domains. Even for the traveling salesman problem (TSP), a benchmark problem which has the simple definition and slight constraints, over ten years and ton of efforts were required to develop better crossovers generating children which satisfy constraints and inherit favorable features of parents [Reinelt 1994]. For most cases, to develop adequate crossover on a problem, the user must become familiar with the problem deeply, not only familiar with GA. Though the first choice for many users *may* be the uniform crossover (UX) or the multi-point crossover, they often generate many lethal children on the problem with severe constraints.

Multi-step crossover (MSX)[Yamada95] is a crossover which is defined in a problem-independent manner using a neighborhood structure and a distance measure. It is relatively easier to provide both of them, as they are very common

J.J. Merelo Guervós et al. (Eds.): PPSN VII, LNCS 2439, pp. 162–171, 2002.

for most problems, than to design a special, heuristic crossover operator. Further, MSX generates children step by step tracking their neighborhood, then new candidate can be easily modified to satisfy constraints. We consider that these characteristics greatly encourage GAs to be used on various fields.

Multi-step Crossover Fusion. (MSXF)[Yamada 96] introduced Metropolis criterion to MSX, regarding the quality of a candidate, and performed very well on scheduling problems. However, for using MSXF the user must fix the temperature parameter T, and it is immediately predicted that this parameter affects intensified impact on the performance of MSXF. As the scaling of fitness value differs depending on the stage of search, it is often difficult, or troublesome, to settle the adequate temperature. We consider this difficulty prevents the possibility of the wider utilization of MSXF.

In this paper, we propose a modified MSXF which works without sensitive parameter T; this will help users to utilize GA easier. Though it is one of sales points and easily accepted that such MSXFs can be simply mounted with two common notions, the neighborhood structure and the distance, we also claim one other point that our proposal method can search efficiently, and performs better than conventional crossover methods.

2 About MSXF, and the Proposal of dMSXF

Given parent solutions p_1 and p_2, MSXF executes a short-term navigated local search from p_1 to p_2; p_1 changes step by step tracking its neighborhood, and p_2 is referred to which direction the search is navigated. The outline of MSXF is described as follows.

1. Set the search point $x_1 = p_1$.
2. */Step k/* prepare the neighborhood of x_k, $N(x_k)$. (*Note*: when the number of the neighborhood is too big to enumerate, $N(x_k)$ should be limited to the size μ randomly sampled neighbors.)
3. Sort $y_i \in N(x_k)$ in ascending order of distance $d(y_i, p_2)$.
4. Select y_i from $N(x_k)$ with a probability inversely proportional to the rank.
5. */Metropolis criterion/* Accept y_i if $f(y_i)$ is superior to $f(x_k)$, otherwise accept with probability $\exp(-\Delta V/T)$, where $\Delta V = f(x_k) - f(y_i)$.
6. Go to the process 4. if not accepted.
7. Let the next step point x_{k+1} be y_i, and go to the process 2. until some termination condition, *e.g.* step k reaches to k_{max}, is satisfied.
8. The best solution among $x_1 ... x_{k_{max}}$ is used for the next generation (following some rules).

If the distance to p_2 *were not* regarded at the process 3., this crossover would behave as SA. On the other hand, if the quality of solutions *were not* regarded in the process 5., this procedure just connect p_1 and p_2, as MSX. MSXF is considered as the algorithm that has favorable characteristics of both SA and crossovers. Figure 1(left) shows the aspect of MSXF search.

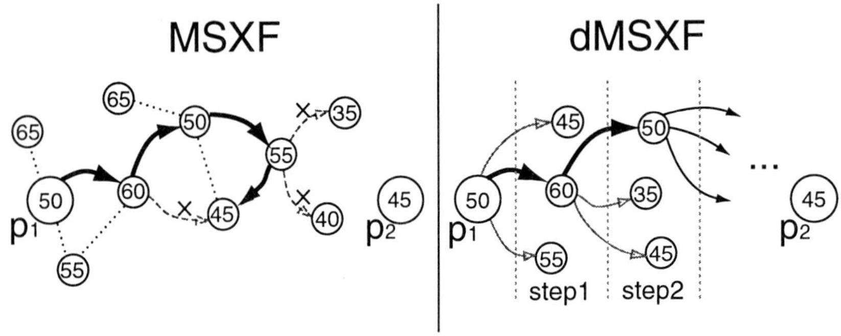

Fig. 1. The aspect of MSXF(left), dMSXF(right): Each circle represents a solution, the number in the circle is the value of the solution to be maximized. Each edge (or arrow) represents a connection as neighbors, bold arrow is a transition of one step. For MSXF, $\times$ means the rejection of a transition

However, the behavior of MSXF greatly depends on the parameter T. *If T is too high*, almost all candidate y_i is accepted independent on its quality, and children just walk from p_1 to p_2, as MSX. On the other hand, *if T is too low*, almost all deterioration of y_i nearer to p_2 is rejected, and children just mill around a sub-optimum, as a local search, because MSXF not necessarily assure every candidate y_i is nearer than x_k to p_2. Adequate temperature varies depending not only on the problem but also the stage of search. Even in the study of SA, the control of temperature is still one of the most important topics to effect its performance.

Consequently, we modify MSXF to work without T; just by selecting the best solution of neighborhood candidates in the deterministic manner. For navigating its search to the direction toward p_2, every member of neighborhood candidates y_i **must be restricted** to satisfy $d(y_i, p_2) < d(x_k, p_2)$. We call this modified version deterministic Multi Step Crossover Fusion, *dMSXF*.

dMSXF searchs regarding both quality of solutions and distance to p_2, without being troubled by temperature parameter T. We note that dMSXF is not the case of MSXF with $T = 0$; though MSXF with $T = 0$ cannot accept any deterioration and should be caught to a suboptimal, dMSXF necessarily move toward p_2 even in the case that all candidates are inferior to the current solution. So, even when p_1 and p_2 are both suboptimal, dMSXF can find better interspace solutions. The procedure of dMSXF is simpler than MSXF, described as follows.

1. Set search point $x_1 = p_1$.
2. (Step k) prepare the μ neighbors of x_k, $N(x_k)$. All of $y_i \in N(x_k)$ must satisfy $d(y_i, p_2) < d(x_k, p_2)$.
3. Select the best solution $y_i \in N(x_k)$.
4. Let the next step point x_{k+1} be y_i, and go to 2. until step k reach to k_{max} or x_k equals to p_2.
5. The best solution among $x_1 ... x_{k_{max}}$ is introduced instead of p_1. Next p_2 will pick up another solution p_3, and p_2 search to the direction toward p_3.

Figure 1(right) shows the aspect of dMSXF search. To claim not only that dMSXF is easily designed but also that the performance of dMSXF can be superior to conventional methods, *e.g.* uniform crossover, we apply our proposal method dMSXF to 1max problem and TSP in the following sections. Though dMSXF *may* lose a favorable aspect of SA which original MSXF holds, as the price for paying for its handy composition by taking the parameter T away, the algorithm will show the sufficient and robust performance on these problems.

3 Analysis on 1max Problem

1max problem is the most primitive benchmark problem on bitstrings. No correlation exists between any bits, the fitness value of a solution is calculated as the number of 1.

3.1 The Case of Conventional Crossovers

For 1max problem, it is predicted that uniform crossover (UX) performs better than multi-point crossovers, because it is no sense to deal adjacent loci in a mass. To confirm this, we apply a GA with these crossovers to 1max problem that the length of string $L_{bit} = 1000$. MGG model [Satoh 1996] is used as the alternation; in each generation,

1. All solutions are coupled randomly, *i.e.* $N_{pop}/2$ couples are made, where N_{pop} is the population size.
2. Each couple, parents, generate C_{cross} children by a crossover.
3. The best and the second-best solutions in the family, *i.e.* parents and children, are introduced instead of parents.

Within one generation, $C_{cross} \cdot N_{pop}/2$ children are created and evaluated. Here we set $N_{pop} = 20$ and $C_{cross} = 200$. To judge clearly, no mutation operator is introduced. We run these GAs 50 times, and the performance is measured as the average of final best values, the standard deviation of them, and the averaged number of generations for one convergence. Here a GA is considered to be converged when no progress of its best value is found within 20 generations. Results are summarized in Table 1, and the superiority of UX is shown.

Table 1. Performance of conventional GA on 1max problem

operater	best value av.	std.	generation
Uniform Crossover	992.16	2.09	47.66
20point Crossover	924.90	7.65	55.14
10point Crossover	858.45	11.07	55.54
5point Crossover	803.85	10.62	56.09

It is of course that the more children are generated in a family, the better solution is found and introduced. Further, this fact directly effects on the final performance of GAs. We apply the GA with UX, changing $C_{cross} = 10,\ 20,\ 50,\ 200,$

and summarize the results in Table 2. The reason why GA with smaller C_{cross} fails is that the amount of progress from parents to the best child is too small.

Table 2. Effect of the number of children C_{cross}

operater	C_{cross}	best value av.	std.	generation
UX	200	992.16	2.09	47.66
	50	983.64	3.79	55.24
	20	970.50	4.52	63.95
	10	946.30	8.54	74.40

Progress Rate (PR) is used to measure the performance of an algorithm, especially of Evolutionary Strategy (ES) [Markon 2001]. PR is the expectation value of the progress per a step, under a condition. Here we define PR of a crossover as the averaged amount of progress, from the better of the parents to the best of the family. For example, the fitness of a parent $p1$ is 70, $p2$ is 80, and three children with the fitness 65, 75, 90 are generated, the amount of progress is $90 - 80 = 10$. It is considered that the bigger PR of a crossover is, the better the crossover is.

Here, we set $L_{bit} = 100$, and $p_1 = \overbrace{000..000}^{1..x}\overbrace{111..111}^{x+1..100}$, $p_2 = \overbrace{111..111}^{1..x}\overbrace{000..000}^{x+1..100}$ complementarily. x is the parameter to decide the bias of parents quality, p_1 and p_2 are comparable in quality when $x = 50$. PR of UX($C_{cross} = 100$) is measured on condition $1 \le x \le 99$ and shown in Figure 2(left, solid line).

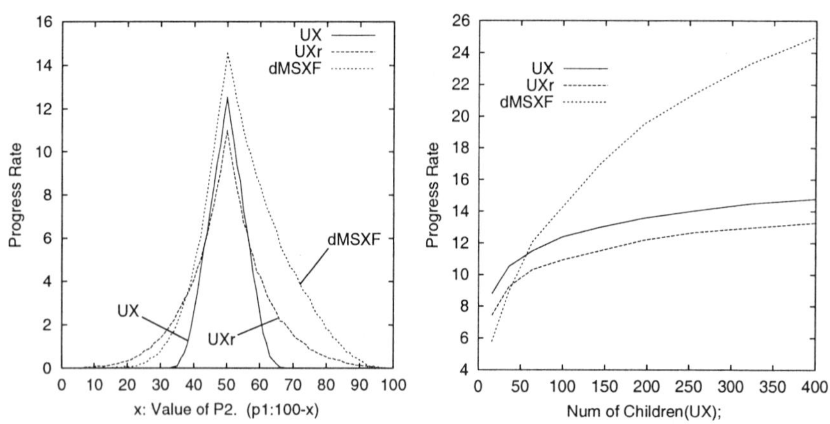

Fig. 2. Progress Rate of UX, UXr, and dMSXF; on the case changing parents quality x (left) and on the case changing the number of children C_{cross} (right)

We can observe that, when one of parents is much superior to another, $x \le 30$ or $70 \le x$, UX can produce no solution better than the parent. The reason why is, the sampling of UX is concentrated to the middle region of parents, and cannot

Table 3. An Example of dMSXF on 1max problem: $k_{max} = 3$, $\mu = 3$. $L_{bit} = 10$, p_1 =0000111111, p_2 =1111110000, here m=8. Bars on bits of x_k means the difference between x_k and p_2, hats on bits means the introduction from p_2

p_1	step 1	step 2	p_2
(base solution x_k)	$\bar{0}\bar{0}\bar{0}\bar{0}11\bar{1}\bar{1}\bar{1}\bar{1}$ (6)	$11\bar{0}\bar{0}11\bar{1}\bar{1}0\bar{1}$ (7)	$111111\bar{1}00\bar{1}$ (8)
0000111111 (6)	$\hat{1}\hat{1}001111\hat{0}1$ (**7**) ↪ $000\hat{1}11\hat{0}111$ (6) $000011\hat{0}\hat{0}\hat{0}1$ (3)	$110011\hat{0}\hat{0}0\hat{0}$ (4) $11\hat{1}011\hat{0}101$ (7) $11\hat{1}\hat{1}111\hat{0}01$ (**8**) ↪	1111110000 (6)

produce children similar to one of them. To fix this difficulty, modified UX (we call UXr temporally) can be introduced; the probability of selecting locus of $p1$ (50% in UX) is randomly selected in every crossovers. PR of UXr($C_{cross} = 100$) is shown in Figure 2(left, broken line). UXr can produce better solution even in $x = 20$ or $x = 80$, but when $x = 50$ PR of UXr is smaller than of UX.

3.2 dMSXF for 1max Problem

To apply our proposal method dMSXF on 1max problem, a neighborhood structure must be introduced. When p_1 differs from p_2 at m loci, $y_j \in N(x_k)$ is created by introducing about m/k_{max} loci of p_2 that differs from x_k. y_j is surely nearer to p_2 than x_k in the sense of Hamming distance. After k_{max} steps, $x_{k_{max}}$ will be almost same to p_2. Table 3 shows an example.

After k_{max} steps, $k_{max}{\cdot}\mu$ children are created, and p_1 is replaced with the best of family. In a generation, every solution fix its navigating solution p_2, create children, and are replaced. So, the total number of children per a generation is $k_{max} \cdot \mu \cdot N_{pop}$. To compare fairly, this number should be equal to of UX, $C_{cross} \cdot N_{pop}/2$. Herewith, we set $k_{max} = 10$, $\mu = 5$. PR of dMSXF is shown in Figure 2(left, dotted line). We can see that dMSXF can produce better solutions than UX in all x.

Next, we investigate the effect of the number of children. We set (k_{max}, μ) to $(4, 2), (6, 3), \ldots, (20, 10)$, and $C_{cross} = 2{\cdot}k_{max}{\cdot}\mu$. Figure 2(right) shows the PR of dMSXF, UX, and UXr where $x = 50$; the more children can be created, the greater superiority of dMSXF to UX or UXr is shown.

We showed PR of dMSXF is greater than UX. Finally, we show GA using dMSXF performs better than of UX. As 6 sets of (k_{max}, μ) are tested, all of algorithms are comparable in the meaning of the calculated amount. L_{bit} is set to 1000, N_{pop} is 20, and 50 trials each are done, results are summarized in Table 4. Original MSXF is also applied, with the same neighborhood as dMSXF, and the temperature parameter $T = 1,\ 2$.

The error of the best dMSXF is 0.16%, this is 1/5 of UX. Further, it is shown that dMSXF is robust to the parameter (k_{max}, μ). In contrast, original MSXF with $T = 2$ performs much worse than with $T = 1$. In conclusion, our proposal method performs robustly better than UX even in the problem in which building-blocks are well separated, like as 1max problem.

Table 4. Performance of UX, UXr, dMSXF, and MSXF using $T = 1, 2$

operater	parameters	children *per* generation	best value av.	std.	generations
UX	$C_{cross} = 200$	2000	992.16	2.09	47.66
UXr	$C_{cross} = 200$	2000	986.50	3.52	51.30
dMSXF	$k_{max} = 5,\ \mu = 20$	2000	996.13	2.27	41.36
	$k_{max} = 7,\ \mu = 14$	1960	997.23	1.58	39.50
	$k_{max} = 10,\ \mu = 10$	2000	997.89	1.22	37.75
	$k_{max} = 14,\ \mu = 7$	1960	**998.40**	1.36	36.66
	$k_{max} = 20,\ \mu = 5$	2000	998.25	1.08	36.10
	$k_{max} = 50,\ \mu = 2$	2000	997.25	1.75	39.55
MSXF	$k_{max}=14,\ \mu=7,\ T=1$	1960	996.39	2.97	48.69
	$k_{max}=14,\ \mu=7,\ T=2$	1960	985.79	4.70	55.60

4 An Application for Traveling Salesman Problem

Traveling Salesman Problem (TSP) is the problem to find a Hamilton cycle that has the shortest tour length on the graph, where the length is the sum of weights of edges on a tour. TSP is a classic NP-hard problem, and a lot of approximation algorithms have been proposed [Reinelt 1994].

The **Edge Assembly Crossover** (EAX) proposed in [Nagata 97] is the state of the art crossover, GA with EAX performs extremely well as against other solvers [Shimodaira 1999]. In this section, we perceive the original EAX as the uniform crossover of elements, *AB-cycles*, and improve its performance by using dMSXF.

4.1 About the Edge Assembly Crossover

When the parent tour-A and tour-B are given, EAX creates children from them as follows. Figure 3 is a sample of it.

1. Divide edges on G' into *AB-cycles*, where G' is the graph obtained by overlapping tour-A and tour-B. A cycle generated by tracing edges of tour-A and tour-B **alternatively** on graph G' is defined as *AB-cycle.* (Fig. 3c)
2. Construct a *E-set* by choosing *AB-cycles* on the basis of some criterion.
3. Generate an intermediate individual by applying the *E-set* to tour-A in the **XOR** manner, *i.e.* by removing edges of tour-A included in the *E-set* from tour-A, and adding edges of tour-B included in the *E-set* to it. (Fig. 3d)
4. Modify the intermediate individual into a valid one by merging its sub-tours. This modification is deterministic. (Fig. 3e)

Since several criterions of choosing *AB-cycles* are conceivable, the random selection and a heuristic selection were introduced in [Nagata 97]. At the random selection, each *AB-cycles* is chosen randomly with probability 0.5, and the EAX that adopts this method is called EAX-rand (Fig. 4a). EAX-rand is taken as the uniform crossover of building blocks, *AB-cycles.* On the other hand, the heuristic selection chooses all *AB-cycles* that satisfy *a* condition regarding its tradeoff between exploitation and exploration.

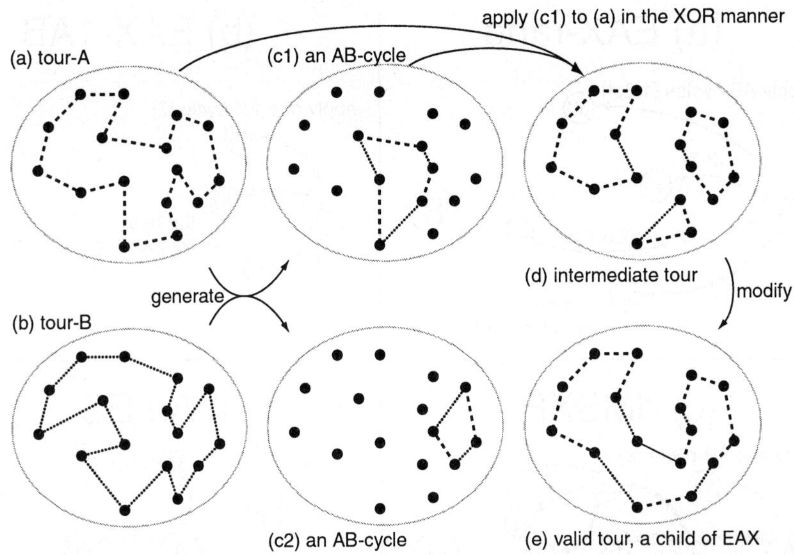

Fig. 3. A sample of EAX: (a) tour-A and (b) tour-B are given, AB-cycles (c1)(c2) are generated, (d) an intermediate tour is created by mixing tour-A and a set of AB-cycles with XOR manner, (e) the valid tour is generated by modifying the intermediate tour

4.2 dMSXF Applied for TSP

Another selection method is introduced in [Nagata 2000], to choose only one from *AB-cycles* that satisfy a condition. By applying an *AB-cycle* on tour-A, the child similar to tour-A is gained; the child walks *a* step from tour-A to tour-B. The EAX that adopts this method is called EAX-1AB (Fig. 4b).

EAX-1AB produces many children around tour-A, and the best one is introduced to the next generation instead of tour-A. Though this manner resembles to ES (Fig. 4d), it is remarkable that all children y_j around tour-A is nearer to tour-B than tour-A, *i.e.* $d(y_j, \text{tour-B}) \leq d(\text{tour-A}, \text{tour-B})$, where the distance is measured as the number of edges mismatch between A and B. Therefore, EAX-1AB should be taken as the case $k_{max} = 1$ of dMSXF, rather than ES.

Herewith we extend EAX-1AB to the multi-step version (Fig. 4c), and compare these algorithms on 7 benchmarks. C_{cross} is set to 10 for EAX-rand and EAX-1AB, $(k_{max}, \mu) = (4, 6) or (5, 8)$ for dMSXF respectively. We run them 30 times on each settings, here the number of finding the optimum, the averaged error(%), and the averaged time(sec) for one run by a Windows PC with 1GHz CPU, are recorded.

Table 5 shows the results. We can see that EAX-1AB and dMSXF perform better than original EAX. Further, dMSXF works efficiently compared with 1-step version, EAX-1AB. Though $C_{cross} = 10$ (recommended by Dr. Nagata) may be too small, experiencially larger C_{cross} for EAX-1AB yields little improvement and much computational cost.

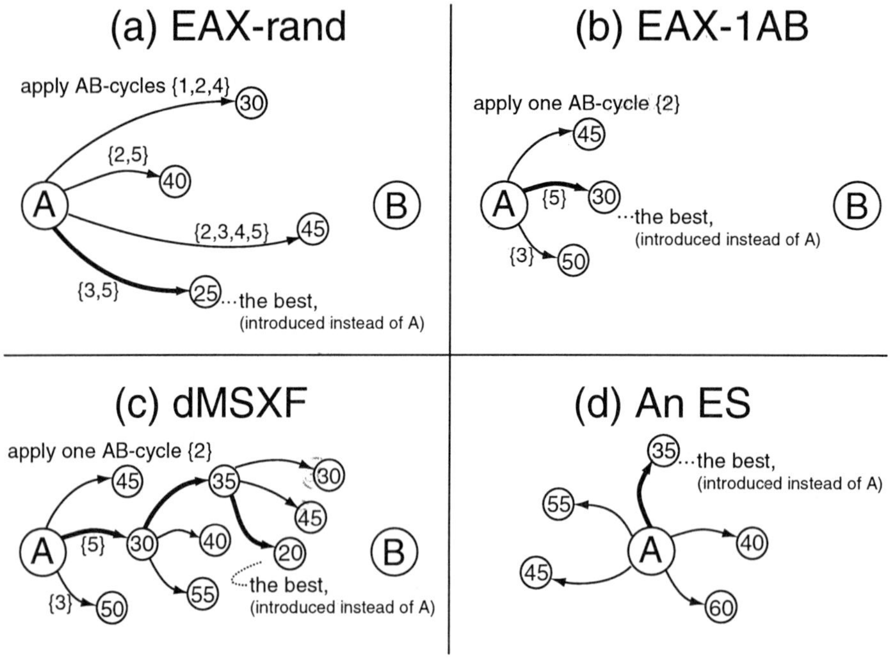

Fig. 4. The aspect of search: (a)EAX-rand, (b)EAX-1AB, (c)dMSXF, and (d)an ES. Now parent A is focused, some children (small circles with their tour length) are created with each manner: (a) apply about a half set of AB-cycles to A, (b) apply one AB-cycle to A, (c) /1st step/ apply one AB-cycle to A, select the best of children, x_2, /2nd step/ apply one AB-cycle to x_2, ... (d) introduce a mutation to A regardless of other solution

Table 5. Performance of EAX-original, EAX-1AB, and dMSXF on TSP

		EAX-rand			EAX-1AB			dMSXF		
instances	N_{pop}	opt.	error	time	opt.	error	time	opt.	error	time
att532	300	10	0.025	29	23	0.006	20	27	0.002	24
rat575	300	3	0.020	47	8	0.012	27	17	0.006	30
rat783	300	20	0.006	72	28	0.001	42	30	0	49
pcb1173	300	2	0.023	172	10	0.017	71	12	0.010	79
pr2392	300	0	0.034	657	18	0.048	365	23	0.026	473
fl3795	300	2	0.028	2716	11	0.009	1768	17	0.006	2301
fnl4461	1000	0	0.031	10876	3	0.002	6952	8	0.001	9784

In this paper we compared only three GAs based on similar operators, they are considered to be extremely superior to other approaches for TSP. For example, branch-and-cut algorithm [Padberg 1991] solved the *att532* problem exactly, but required over than 5 hours on IBM 3090/600 computer, and a GA for TSP Asparagos96 [Schleuter 1997] requires 95 minutes on UltraSparc 170MHz machine.

5 Conclusion and Future Work

Multi-step crossover fusion (MSXF) is a promising method using only the neighborhood structure and the distance measure, when heuristic crossovers are hardly introduced. However, MSXF works unsteadily according to the temperature parameter, like as SA. In this paper, we introduced deterministic multi-step crossover fusion (dMSXF) to take this parameter away. The performance of dMSXF was tested on 1max problem and Traveling Salesman Problem, and its superiority to conventional methods, *e.g.* uniform crossover, was shown. We believe dMSXF is very promising in three senses, the design simplicity, the search efficiency, and the behavior robustness.

For the future work, the selection of the neighborhood structure still remains as a problem to be solved. In common with other neighborhood searches as SA, the performance of dMSXF varies depending on its neighbors to be sampled. So, some guidance for introducing an adequate neighborhood structure to a new problem should be structured.

References

[Markon 2001] Sandor Markon, Dirk V.Arnold, Thomas Bäck, Thomas Beielstein, and Hans-Georg Beyer : Thresholding - a Selection Operator for Noisy ES, *Congress on Evolutionary Computation*, pp. 465-472 (2001)

[Nagata 97] Nagata, Y. and Kobayashi, S. : Edge Assembly Crossover: A High-power Genetic Algorithm fot the Traveling Salesman Problem, *Proceedings of the 7th International Conference on Genetic Algorithms*, pp. 450-457 (1997)

[Nagata 2000] Nagata, Y. : Genetic Algorithm for Traveling Salesman Problem using Edge Assembly Crossover: its Proposal and Analysis, *Doctoral thesis* (2000)

[Padberg 1991] Padberg, M. and G.Rinaldi: A Branch-and-Cut Algorithm for the Resolution of Large-Scale Symmetric Traveling Salesman Problems, *SIAM Review, Vol.33, No.1* pp.60-100 (1991)

[Reinelt 1994] G.Reinelt, The Traveling Salesman: Computational Solutions for TSP Applications. Vol.840 of Lecture Notes in Computer Science, Springer-Verlag (1994)

[Satoh 1996] H.Satoh, M.Yamamura, and S.Kobayashi : Minimal Generation Gap Model for GAs Considering Both Exploration and Exploitation, *Proc. of IIZUKA*, pp.494-497 (1996)

[Schleuter 1997] Gorges-Schleuter, M. : Asparagos96 and the Traveling Salesman Problem, *Proc. of the 1997 IEEE International Conference of Evolutionary Computation,* pp.171-174 (1997)

[Shimodaira 1999] Hisashi Shimodaira: A Diversity Control Oriented Genetic Algorithm (DCGA): Development and Experimental Results, *Proceedings of the Genetic and Evolutionary Computation Conference*, pp. 603-611 (1999)

[Yamada 96] Yamada, T. and Ryohei, N. : Scheduling by Generic Local Search with Multi-Step Crossover, *Proceedings of the 4th conference on 4th PPSN*, pp. 960-969 (1996);

[Yamada95] Yamada, T. and Nakano, R.: A GA with multi-step crossover for job-shop scheduling problems, *Proc. of Int. Conf. on GAs in Engneering Systems: Innovations and Applications (GALESIA) '95*, pp.146-151 (1995)

Operator Learning for a Problem Class in a Distributed Peer-to-Peer Environment*

Márk Jelasity[1,**], Mike Preuß[2], and A.E. Eiben[1]

[1] Free University of Amsterdam, Amsterdam, The Netherlands
jelasity@cs.vu.nl, gusz@cs.vu.nl
[2] University of Dortmund, Dortmund, Germany
mike.preuss@uni-dortmund.de

Abstract. This paper discusses a promising new research direction, the automatic learning of algorithm components for problem classes. We focus on the methodology of this research direction. As an illustration, a mutation operator for a special class of subset sum problem instances is learned. The most important methodological issue is the emphasis on the generalisability of the results. Not only a methodology but also a tool is proposed. This tool is called DRM (distributed resource machine), developed as part of the DREAM project, and is capable of running distributed experiments on the Internet making a huge amount of resources available to the researcher in a robust manner. It is argued that the DRM is ideally suited for algorithm learning.

1 Introduction

This paper discusses a promising new research direction, the automatic learning of algorithm components for problem classes. The main contribution of this paper is threefold. First, we emphasize the importance of an appropriate methodology that allows researchers to produce generalizable knowledge over a problem class rather than a problem instance. Second, we propose a tool that might be ideally suitable for generating such knowledge automatically. Finally, both the methodology and our proposed tool are illustrated via an example: a search operator for a special class of subset sum problem instances is learned.

In the recent years much research effort has been devoted to methods that try to improve heuristic search through some form of learning. To motivate our approach, let us elaborate on its relationship with these methods. The way different approaches generate knowledge can be categorized along at least two dimensions. The first is defined by the distinction between research performed manually and automatic learning. The second is defined by the scope of the

* This work is funded as part of the European Commission Information Society Technologies Programme (Future and Emerging Technologies). The authors have sole responsibility for this work, it does not represent the opinion of the European Community, and the European Community is not responsible for any use that may be made of the data appearing herein.

** Also in RGAI, MTA-SZTE, Szeged, Hungary

J.J. Merelo Guervós et al. (Eds.): PPSN VII, LNCS 2439, pp. 172–183, 2002.

knowledge. This scope can be a single problem instance or a whole class of problems. We will call the later type of knowledge *durable* knowledge referring to its generality and long term relevance. As it should be clear already we will focus on automatically generating durable knowledge here. It is useful however to examine the three other classes generated by the two dimensions described above.

Problem instance specific knowledge generated manually is the unfamous fine tuning process of a given algorithm on a specific problem. Besides of being time consuming, the scientific relevance of such knowledge is questionable due to its restricted scope.

The idea of automatically learning problem instance specific knowledge is very common. It includes all approaches that apply some form of adaptation or learning during the optimization process. The generated knowledge is normally not considered scientifically important and is discarded after the completion of the algorithm. The large field of self-calibrating algorithms (which include the meta-GA approach) belongs to this class [7, 3]. The main idea is that the values of different algorithm parameters are automatically tuned on the fly to achieve optimal performance while solving a problem. Another successful idea is building probabilistic representations of the fitness landscape based on the solutions evaluated during the search and generating new candidate solutions based on this knowledge [14]. If applied as heuristic optimizers, cultural algorithms can be classified into this category as well. They do not fix any knowledge representation offering a more abstract framework for learning based on the performance of the developing population [15].

Durable knowledge is often not generated automatically, partly due to the large computational requirements. One example is [5], where different operators were tested on many random instances of an infinite problem class (NK-landscapes). Using these results it is possible to *predict* the behavior of these operators on *unseen* instances of the problem class. Before optimizing an instance of this class, one can directly chose the best operator.

This paper will argue for the *usefulness* and *feasibility* of *automatically* generating durable knowledge. Usefulness does not need too much explanation: in the case of (practically or theoretically) interesting problem classes this learning can provide us with better algorithms over a whole problem class and can also help us understand this problem class better through the analysis of the collected knowledge. The importance of durable knowledge was emphasized also in [12]. The question of feasibility is not so evident. Producing durable knowledge can be an expensive and slow process. However, distributed systems on wide are networks offer a natural solution to problems that require a huge amount of resources.

With the overall success of the Internet distributed computation is getting more and more attention [6, 17]. Systems exist that can utilize resources available in the form of e.g. the idle time of computers on the Internet [16, 18]. As part of the DREAM project ([13]) such a computational environment was developed, the DRM (distributed resource machine). The DRM—unlike e.g. SETI@home—

is based on cutting-edge peer-to-peer (P2P) and Java technology. This allows it to be scalable, robust and flexible [11].

If we consider that many research and engineering institutions solve instances from the same problem class routinely on a large scale anyway, with an additional layer on top of their network (in the form of a distributed computational environment like the DRM) durable knowledge can be generated even more efficiently.

The outline of the paper is as follows. In Section 2 we elaborate on the methodology that allows us to learn durable knowledge. Section 3 is devoted to the DRM, the tool we will use to learn an operator for our example problem class. We also describe our empirical results on the example learning problem. Section 4 concludes the paper.

2 Methodology

In this section we outline a methodology that supports the generation of durable knowledge. We interpret knowledge as algorithmic knowledge, i.e. we want to learn what type of algorithm is optimal on (or at least well-tailored to) a given problem class. More specifically, we are after a good evolutionary algorithm for a problem class, so the search space of our learning task – which we call the *algorithm space* – consists of all EAs. Here we can make further choices and reduce the task to finding good variation operators, recombination and/or mutation. In case of mutation, the algorithm space would be the space of all possible unary operators acting as part of the evolutionary algorithm solving problems of the given class. The size and complexity of the algorithm space depends on how the possible mutation operators are represented. In general, there are no restrictions on this representation. It is possible to define the space simply by a single parameter, e.g. p_m of a fixed bit-flip operator, but using arbitrary expressions from a suitable syntax, e.g. LISP known from genetic programming, is also possible.

Once the algorithm space is defined the learning algorithm cooperates with the basic EA. In general, there are no restrictions on the learning algorithm. In this paper we use an evolutionary algorithm for this purpose, so our method works roughly as follows. A number of problem instances are provided (e.g., by a random instance generator) and different variants of the basic EA is run on these instances. The variants are defined by the mutation operator it uses and these mutation operators form the points of the learning space. The learning EA performs evolutionary search over these points. Evaluating such a point amounts to evaluating the mutation operator it represents by performing independent runs with the given operator on many problem instances. The value of the operator is obtained by the quality of solutions of the basic EA using it. Clearly, this implies very extensive computations which motivates our usage of DRM, cf. section 3.

2.1 An Example Problem Class

An important precondition of gaining generalizable knowledge is a well defined problem class on which algorithms can be compared [4]. For the purpose of this paper we use the subset sum problem as an example. It is an NP-hard combinatorial optimization problem. Besides, it is known where the hard instances lie [1].

In the subset sum problem we are given a set $W = \{w_1, w_2, \ldots, w_n\}$ of n positive integers and a positive integer M. The task is to find a $V \subseteq W$ such that the sum of the elements in V is closest to, without exceeding, M. For this problem it is possible to define a problem instance generator. Let us assume that a triple (n, k, d) is given where n is the set size as above and $d \in [0, 1]$ is called the *density* of the problem. The set W is created by drawing w_i randomly from the interval $[0, 2^k]$ with a uniform distribution for $i = 1, \ldots, n$ and let M be the sum of a random subset of W of size $\lceil nd \rceil$. We will denote the generated problem class by SubSum(n, k, d). Our running example in the paper will be SubSum$(100, 100, 0.1)$.

It is essential that the problems in a problem class exhibit some common structure in some sense, some similarity that can be exploited and converted to durable knowledge. Note that structure as meant here arises from a combination of an algorithm and a problem class. For a fixed algorithm space the problem class is structured (non-random) if the behavior (performance) of the different algorithms shows some regularities on different instances from the problem class. The lack of such regularities would indicate that the algorithm space in question and the problem class are not related: there is nothing to be learned.

The graph in Figure 1 shows such regularities here with respect to a simple (1+1) EA. The EA uses a binary representation with every bit representing the presence of a given set element in the candidate subset, and a a bit-flip mutation. Different EAs correspond to different mutations, defined by the mutation rate. Figure 1 plots the performance of such mutation rates, defined as the distance from the desired sum (to be minimized) of their "hosting" EA. Each point in the curve was generated using 100 runs on different instances, all runs until 10000 evaluations. The instances for different mutation probability values were different as well.

Figure 1 provides a confirmation that there is a link between mutation operators and algorithm quality: there are some regularities to be exploited, there is something to learn. It is important to note that this is not necessarily all the regularity we can find. Therefore, we are also interested in other algorithms that might exploit other regularities. The very essence of our approach is that we try to explore and discover as much similarity as possible, using other, richer algorithm spaces.

2.2 An Example Algorithm Space

In the case of SubSum$(100, 100, 0.1)$ let us fix the (1+1) EA applied until 10000 evaluations with the binary representation mentioned above. The algorithm space we will use is given by the mutation operator which is defined by two parameters: p_{01} defines the probability of flipping a 0 to 1, and p_{10} defines the probability of flipping a 1 to 0. If $p_{01} = p_{10}$ then we get the traditional bit flip mutation. This choice of representation involves domain knowledge as well [9].

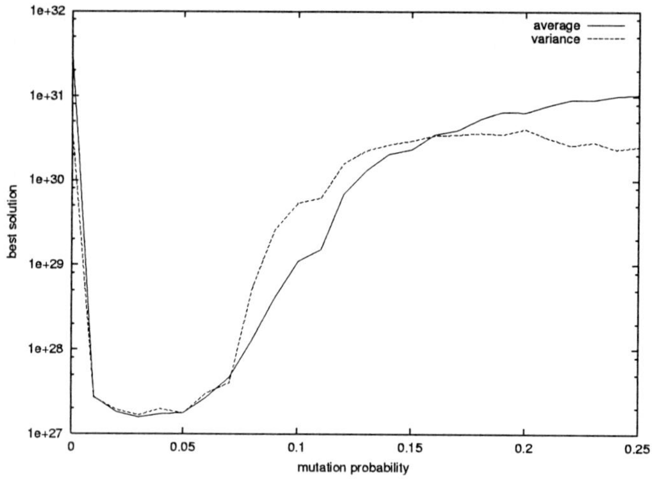

Fig. 1.

Since the density of the problem is relatively low, it might be better if the operator can express a bias towards solutions that contain more 0s than 1s. Our mutation operator can express this bias.

3 A Suitable Tool: DRM

3.1 DRM Structure

The interested reader is kindly asked to refer to [11, 10] for a detailed description of the DRM. A main feature of the DRM is that applications are implemented on it in the conceptual framework of multi-agent systems. In fact, the agents are the threads of the distributed applications. They are mobile, they can communicate with each other and they can make decisions based on the state of the DRM or the application they participate in.

As applications are multi-agent applications, the main task of the DRM is to support these agents. The framework is implemented in Java language which provides mobility and security. Furthermore, to maximize scalability and robustness, the DRM is a pure P2P system, which relies mainly on epidemic protocols [2]. This means that the computers participating in a DRM know only a limited number of other computers from the same DRM. Via exchanging information with these peers only, information spreads as gossip (or epidemic) through the DRM.

3.2 Experiment Structure

To run an application on the DRM one has to think in terms of a multi-agent system. Designing an experiment involves designing the behavior of one or more types of autonomous agents that should perform our experiment. The main concerns in our case are robustness also at the experiment level, not only at the level of the DRM, and the mechanism of distributing information and computation over the DRM.

Our parallelization approach is roughly based on island models. That is, every agent hosts an island where a local population is evolved. The major challenge is however to design a communication mechanism that is robust and scales well in a wide area network environment.

Island Communication. The overall structure of the proposed design mirrors the structure of the DRM itself. We use completely identical islands that communicate using epidemic protocols. Note that it is not an evident design decision; even though the DRM itself is pure P2P, it is possible to design an agent system where one of the agents plays the role of the server of the experiment while the others are the clients.

Our decision is based on our and others' recent findings that indicate the power of our epidemic protocol concerning scalability and robustness [10]. These features are most important in a highly distributed environment like the DRM where network communication goes over a wide-area network and the number of available machines is not known in advance.

At first, the root island is started up by the experimenter who also provides a maximum number of tasks to be computed as a parameter. Each task is an island which is supposed to run the high-level learning EA until a given termination criterion.

Note that the actual number of involved machines depends on their availability and reachability from the root island. The latter is influenced by network separating devices like firewalls between the available machines. Due to the utilized P2P technology, machines do not necessarily need a direct path to each other. Information spreads like an epidemic throughout the DRM in an undirected manner via a chain of islands to finally reach the destination.

Task distribution is done by self-replication of islands according to the load balancing algorithm laid out in [11]. In short, an island that encounters an empty machine sends half of its own tasks there. In contrast to other possible scenarios of parallel experiment startup, we distribute tasks dynamically during experiment run.

Just like the DRM layer which utilizes incomplete databases for node communication, the islands have an incomplete experiment database called the *job-cache.* Every job-cache entry holds the address of another island and its attached value. The address/value table is merged with the one of a remote island by exchanging messages using a randomly selected address and cut to the defined maximum database size. While the expected number of islands doing an experiment may be lower than the number of DRM nodes and thus the job-cache can be smaller, it inherits the advantageous properties of the DRM database, namely information spreading speed and the very low probability of the experiment getting partitioned. Algorithms running within an island may put data into this

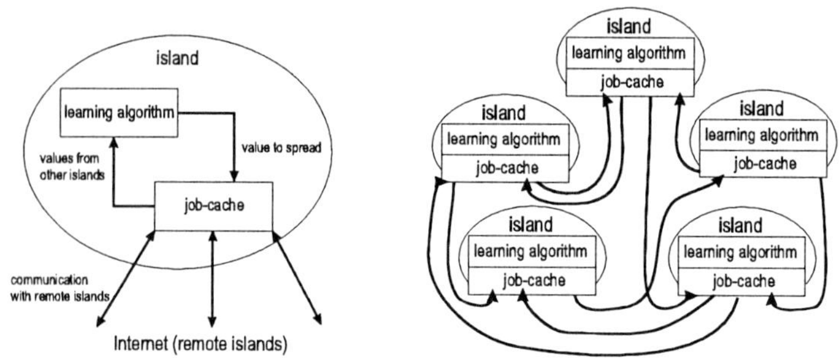

Fig. 2. left: learning algorithm and job-cache communicating within an island, right: example of an incompletely connected set of islands running the same experiment

repository and regular job-cache message interchange will spread these values to other islands in logarithmic time. Figure 2 illustrates messaging between the algorithm and the job-cache (left) and the interconnection of islands via their job-caches (right). For experiments with many islands, these do not necessarily know all other islands because values are not only transferred directly but also indirectly by traversing several job-caches.

Prerequisites of Learning. Learning a good mutation operator for our example, SubSum(100,100,0.1), while varying only two parameters (p_{01} and p_{10}) may not seem very hard, but a first analysis utilizing a grid search over the most interesting area (see Figure 3) reveals at least three difficulties. First, evaluations of one specific operator on different problem instances display a dangerously high variance. The learning algorithm therefore has to average the results of several runs. Second, the approximated fitness landscape as depicted in Figure 3 clearly indicates a wide, almost flat area for $p_{01} > 0.25$. A path-based optimization method depends on a good start point if the neighborhood of the actual solution does not offer any progress. An exploration-based technique can solve this problem although it typically needs more evaluations. Third, the parameter p_{10} has only minor influence on the fitness of the operator for most regions. It increases the search-space but gives little help to finding good solutions if the first parameters value is too high.

To summarize, a successful learning algorithm for this problem class has to be able to cope with a very noisy function and should apply exploration as well as local search.

Learning Algorithm. Based on these prerequisites, we chose a population-based EA as learning algorithm. In a parallel environment there are at least two ways of implementing a population. First, every island may run a pure local search algorithm and integrate search results of the remote islands from time to time. Second, each island may keep a population on its own and add remote results as they become available. Concerning the DRM, the second approach

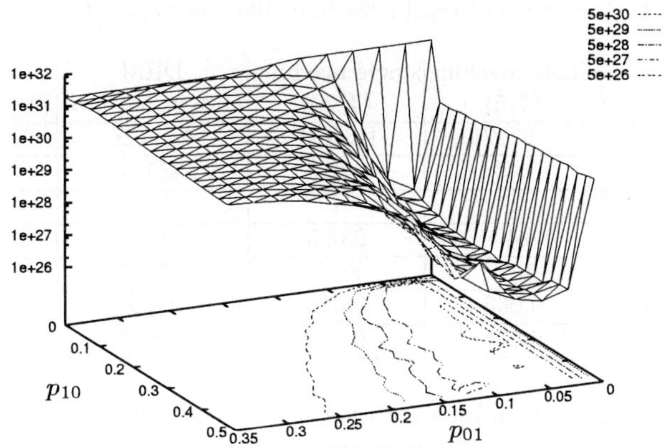

Fig. 3. Performance of the operator described in Section 2.2. Average of 100 runs (each on a new randomly generated problem instance from SubSum(100, 100, 0.1)) per point

is favorable because wide-area network communication is unreliable and rather slow. Nevertheless, both methods have been tested. To prevent the search from getting stuck, the employed selection/replacement operator does not allow for old solutions to survive and thus equals EA comma-selection.

As described above, islands can distribute preliminary results to their neighbors by putting them into the job-cache. Solutions exchanged in our case are the best mutation operators available in the local population.

In the following, we discuss the details of the learning algorithm from the viewpoint of a single island. Every new generation starts with reading the available remotely computed solutions from the job-cache and merging them with the current parent population. Offspring is generated by each time choosing two solutions from the result and applying intermediate recombination, followed by Gaussian mutation. We evaluate the newly created mutation operator by selecting the appropriate (1+1) EA from the defined algorithm space and applying it 5 times to randomly generated instances of the SubSum(100,100,0.1) problem class.

The next generation's parent population is then constructed by selecting the best mutation operators found in the offspring. Note that each solution sent by another island will influence the learning process until replaced by its originator. This rule resembles a *shared memory* concept known from parallel hardware and differs significantly from migration. As a consequence, new islands starting up inherit a set of solutions with acceptable quality after connecting to any other island.

Let m denote the number of remotely computed solutions available. As long as $m = 0$, the learning algorithm equals a (μ, λ) ES, μ and λ meaning the sizes of the local parent population and the offspring, respectively. In the experiments we used $\lambda = 5$ and $\mu = 4$. However, as soon as solutions from remote islands arrive, (μ, λ) changes to a $(\mu + m, \lambda)$ ES. To keep up selection pressure and thus ensure a good local search capability we also increase the number of offspring solutions

Table 1. Accumulated results for four different types of experiments

	single machine $(1,5)$	single machine $(4,20)$	DRM $(1+m,5+4m)$	DRM $(4+m,20+4m)$
best fitness	1.67e26	1.56e26	1.92e26	2.038e26
best evaluation no.	640	682.6	451.5	521.5
concurrent islands	1	1	3.9	4.2
AES	240	232.5	187.5	195.5
success rate	50%	80%	100%	100%
best operator p_{01}	0.066	0.048	0.048	0.066
best operator p_{10}	0.634	0.515	0.545	0.589

created within a generation by a factor λ_r (=4), resulting in a $(\mu+m, \lambda+\lambda_r m)$ ES. Note that the upper bound of m equals the maximum number of entries of the job-cache.

A possible danger is that bad solutions may remain in the job-cache if islands become unavailable due to e.g. network failure. However, for a realistic scenario this is not true on the long run because usually the number of tasks exceeds the size of the job-cache so that old information is gradually removed from the system.

3.3 Experimental Results

Table 1 shows the accumulated results for four types of experiments, namely the $(1, \lambda)$ and (μ, λ) ES on a single machine and the $(1+m, \lambda+\lambda_r m)$ and $(\mu+m, \lambda+\lambda_r m)$ ES running on the DRM. All values are averages over 10 runs per experiment type. The *best fitness* measures the quality of the best mutation operator in terms of the best subset sum solution found by the (1+1) EA using this operator. Let us note that according to the t-test the differences between the values of the best solutions for the four algorithms are not significant (for $\alpha = 0.05$). *best evaluation no.* gives the number of root island evaluations needed to generate the best solution. As mentioned in the previous section 5 tests are done with each operator variant so the number of mutation operators tested is *best evaluation no.*/5. From the grid search depicted in Figure 3 we knew in advance that mutation operators with an average fitness below $1e27$ exist and are already near the optimum. We therefore computed the *success rate* as the fraction of experiments with a best fitness below this upper bound. The *average number of evaluations to solution (AES)* gives the number of root island evaluations needed to reach this level. The last two rows show the found mutation operator parameters p_{01} and p_{10}.

The best results are found sooner on the DRM, this is a side-effect of the fact that many populations work in parallel. The table does not directly reveal the actual speedup of the system. However it was proven in [11] that the speedup is almost linear for any application that is distributed the way our present application was.

The optimal mutation parameters found by the algorithm show a consistent pattern. The ratio p_{01}/p_{10} is close to 0.1 which is as a matter of fact the density parameter of the problem class SubSum(100,100,0.1). It is also notable that the mutation operator corresponding to the optimal parameters outperforms the optimal one-parameter mutation. This can be seen clearly by comparing the value of the best solutions found to the performance of the one parameter mutation shown in Figure 1. This illustrates the possible benefits of learning more complex operators.

Even in the case of our simple example when the operator space was defined by two real parameters it is quite clear that searching this space is much more effective with the automatic approach than using an exhaustive search. To illustrate this, consider that visualizing the whole $[0,1]^2$ space in the quality shown in Figure 3 would require 160000 runs of the underlying (1+1) EA. Even this quality is hardly enough to draw any consistent conclusions over the location of the global optimum due to the high variance. Compared to this our DRM application needed only around 500 runs to learn a very good quality operator. It is of course not surprising that evolutionary search is much better than exhaustive search. This remark is useful only to point out that this quite trivial fact holds in the case of operator learning as well, which is another argument for the automatic acquisition of durable knowledge.

Finally, let us note that all experiments have been run from behind a firewall, using a 56k dial-in network connection. This may increase island distribution time and block several messages directly targeted at the root island but demonstrates the applicability of the DRM technology even under the worst conditions.

4 Conclusions and Future Perspectives

In this paper we have argued for the importance of learning algorithm components for problem classes. The durable knowledge that results from this process can be directly applied to improve algorithms or it can be analyzed further to gain scientific insight into a problem class. A tool was suggested that is suitable for performing this kind of expensive learning by utilizing the idle time of any computers connected to the Internet. The feasibility of the approach was demonstrated on the problem class SubSum(100,100,0.1).

An interesting possibility for future applications is the possibility of developing self-improving software packages. The different copies of the software could communicate through the Internet and exchange information with the help of the DRM or a similar P2P environment that uses epidemic protocols for communication. This process can be completely transparent to the user and due to the P2P approach it could scale very well to millions of copies without any special investment. At this level a learning algorithm could gradually improve the performance of the software via evolving durable knowledge based on the experience on the problems people are trying to solve with the software. Note that this application area is not restricted to learning heuristic operators. An arbitrary component of a software can be improved provided an appropriate evaluation method is available.

Acknowledgments

The authors would like to thank the other members of the DREAM project for fruitful discussions, the early pioneers [13] as well as the rest of the DREAM staff, Maribel García Arenas, Emin Aydin, Pierre Collet and Daniele Denaro.

References

1. M. J. Coster, A. Joux, B. A. LaMacchia, A. M. Odlyzko, C.-P. Schnorr, and J. Stern. An improved low-density subset sum algorithm. *Computational Complexity*, 2:111–128, 1992.
2. A. Demers, D. Greene, C. Hauser, W. Irish, J. Larson, S. Shenker, H. Sturgis, D. Swinehart, and D. Terry. Epidemic algorithms for replicated database management. In *Proceedings of the 6th Annual ACM Symposium on Principles of Distributed Computing (PODC'87)*, pages 1–12, Vancouver, Aug. 1987. ACM.
3. A. E. Eiben, R. Hinterding, and Z. Michalewicz. Parameter control in evolutionary algorithms. *IEEE Transactions on Evolutionary Computation*, 3(2):124–141, July 1999.
4. Á. E. Eiben and M. Jelasity. A critical note on experimental research methodology in EC. In *Proceedings of the 2002 Congress on Evolutionary Computation (CEC 2002)* [8], pages 582–587.
5. A. E. Eiben and C. A. Schippers. Multi-parent's niche: n-ary crossovers on NK-landscapes. In W. Ebeling, I. Rechenberg, H.-P. Schwefel, and H.-M. Voigt, editors, *Parallel Problem Solving from Nature - PPSN IV*, volume 1141 of *Lecture Notes in Computational Science*, pages 319–328. Springer-Verlag, 1996.
6. I. Foster and C. Kesselman, editors. *The Grid: Blueprint for a New Computing Infrastructure.* Morgan Kaufmann Publishers, 1999.
7. J. J. Grefenstette. Optimization of control parameters for genetic algorithms. *IEEE Transactions on Systems, Man, and Cybernetics*, 16(1):122–128, 1986.
8. IEEE. *Proceedings of the 2002 Congress on Evolutionary Computation (CEC 2002).* IEEE Press, 2002.
9. M. Jelasity. A wave analysis of the subset sum problem. In T. Bäck, editor, *Proceedings of the Seventh International Conference on Genetic Algorithms*, pages 89–96, San Francisco, California, 1997. Morgan Kaufmann.
10. M. Jelasity, M. Preuß, M. van Steen, and B. Paechter. Maintaining connectivity in a scalable and robust distributed environment. In H. E. Bal, K.-P. Löhr, and A. Reinefeld, editors, *Proceedings of the 2nd IEEE/ACM International Symposium on Cluster Computing and the Grid (CCGrid2002)*, pages 389–394, Berlin, Germany, 2002.
11. M. Jelasity, M. Preuß, and B. Paechter. A scalable and robust framework for distributed applications. In *Proceedings of the 2002 Congress on Evolutionary Computation (CEC 2002)* [8], pages 1540–1545.
12. M. Jelasity. Towards automatic domain knowledge extraction for evolutionary heuristics. In M. Schoenauer, K. Deb, G. Rudolph, X. Yao, E. Lutton, J. J. Merelo, and H.-P. Schwefel, editors, *Parallel Problem Solving from Nature - PPSN VI*, volume 1917 of *Lecture Notes in Computational Science*, pages 755–764. Springer-Verlag, 2000.

13. B. Paechter, T. Bäck, M. Schoenauer, M. Sebag, A. E. Eiben, J. J. Merelo, and T. C. Fogarty. A distributed resoucre evolutionary algorithm machine (DREAM). In *Proceedings of the 2000 Congress on Evolutionary Computation (CEC 2000)*, pages 951–958. IEEE, IEEE Press, 2000.
14. M. Pelikan, D. E. Goldberg, and F. Lobo. A survey of optimization by building and using probablistic models. Technical Report 99018, Illinois Genetic Algorithms Laboratory, 1999.
15. R. G. Reynolds. Cultural algorithms: Theory and applications. In D. Corne, M. Dorigo, and F. Glover, editors, *New Ideas in Optimization*, Advanced Topics in Computer Science, pages 367–377. McGrow-Hill, 1999.
16. SETI@home. http://setiathome.ssl.berkeley.edu/.
17. A. S. Tanenbaum and M. van Steen. *Distributed Systems: Principles and Paradigms.* Prentice Hall, 2002.
18. United Devices^{tm}. http://ud.com/.

Crossover Operator Effect in Function Optimization with Constraints*

D. Ortiz-Boyer, C. Hervás-Martínez, and N. García-Pedrajas

Departament of Computing and Numerical Analysis, University of Córdoba
C2 Building, Campus of Rabanales s/n, E-14071 Córdoba, Spain
{ma1orbod,chervas,npedrajas}@uco.es

Abstract. Most real-world optimization problems consist of linear cost functions subject to a set of constraints. In genetic algorithms the techniques for coping with such constraints are manifold: penalty functions, keeping the population in the feasible region, etc. Mutation and crossover operators must take into account the specific features of this kind of problems, as they are the responsible of the generation of new individuals.
In this work, we make an analysis of the influence of the selection of the crossover operator in the problem of function optimization with constraints. We focus our work on the crossover operator because this operator is the most characteristic of genetic algorithms. We have used a test set that includes functions with linear and non-linear constraints.
The results confirm the importance of crossover operator, as great differences are observed in the performance of the studied operators. The crossover based on confidence intervals shows the most robust behavior.

1 Introduction

There is no method for determining the global optimum of a general nonlinear programming (NLP) problem. This kind of problems only have a solution when the cost function and the constraints have certain properties, many of such properties are not common in real-world problems. Genetic algorithms (GAS) in general, and most specifically real coded genetic algorithms (RCGAs), are parallel stochastic search algorithms, robust and widely used in optimization of functions without constraints in such cases where classical methods fail in finding an optimum.

Using GAs in the optimization of functions with constraints requires mechanisms to incorporate the constraints into the evolutionary process. A major group of methods maintains all the individuals in the feasible region, and does not allow the existence of individuals that do not fulfill any constraints. The other approach is to allow the existence of individuals outside the feasible region, but penalizing them heavily.

In this context the mechanisms of mutation and crossover have a major impact as the creation of new individuals relies on them. Operators suitable for

* This work has been financed in part by the project TIC2001-2577 of the Spanish CICYT and FEDER funds

J.J. Merelo Guervós et al. (Eds.): PPSN VII, LNCS 2439, pp. 184–193, 2002.

a problem of optimization without constraints may not be suitable for a problem that restricts the search space. The concept of exploration and exploitation must be reformulated in the problems of optimization with constraints. So, it is very interesting a study of the behavior of the most successful crossover operators in an environment where there are constraints in the search space.

In this work, we center our interest in the crossover operator as this operator plays a central role in RCGAs. In fact, it may be considered to be one of the algorithm's defining features, and it is one of the components to be borne in mind to improve the RCGAs behavior [1]. In this context, it is fundamental the ability of crossover operators to solve the balance between the exploration an exploitation of the search space, which is associated to the intervals determined by the extremes of the domain of the genes and by the corresponding alleles of the parents chosen for crossing [1].

This study encompasses different kind of crossovers widely used for optimization problems without constraints, and a new crossover based on confidence intervals whose features make it appropriate for coping with constraints. We have used two set of functions for comparing the operators. The first one is formed by functions whose constraints define a convex search space where it is easy to keep the individuals without introducing additional penalty functions. The second one consists of real-world problems with non-linear constraints where penalty are used.

This paper is organized as follows. Section 2 formulates the problem of nonlinear programming and explains succinctly the most common techniques for treating with constraints. Section 3 is dedicated to a brief description of the tested crossovers with a special attention to the confidence interval based crossover. Section 4 explains the experimental setup and describes the test problems. Section 5 shows the results obtained with the different operators. Finally, Section 6 states the conclusions of our work.

2 Nonlinear Programming Problem

The general nonlinear programing problem is stated as follows:

$$\begin{array}{l} \text{find } \mathbf{x} \text{ which optimizes } f(\mathbf{x}),\ \mathbf{x} = (x_1, ..., x_p) \\ \text{subject to:} \\ \quad g_i(\mathbf{x}) \leq 0,\ j = 1, ..., q \\ \quad h_i(\mathbf{x}) = 0,\ j = q+1, ..., p \end{array} \tag{1}$$

g_i and h_i can be linear or non-linear functions. f and the constraint functions g_i and h_i must fulfill certain properties to ensure that a global optimum could be found with some probability.

Different methods have been developed for the solution of problems of optimization with constraints, these methods are included into two broad categories: direct and indirect methods. Indirect methods try to convert the NLP problem into several linear problems, but that is not always feasible. Direct methods are based on obtaining successive solutions using classical methods modified in a certain way to consider the constraints in order to avoid the generation of non-feasible solutions. Most of these methods are local search methods that depend on the existence of the derivative of the cost function. That means that these methods are not robust in search spaces that are discontinuous, multimodal or noisy.

One of the most interesting tools for solving NLP is the use of real-coded genetic algorithms. For coping with constraints RCGAs use different techniques, such as, preservation of feasibility, penalty functions, searching for feasibility and other hybrids [2] [3]. The first two are the most commonly used and both of them count with a wide variety of algorithms. We will briefly explain these two philosophies.

2.1 Maintaining the Population in the Feasible Region

These methods define mutation and crossover operators that guarantee that the generated individuals fulfill the constraints, this way the population is kept in the feasible region.

One of the most used of this kind of methods is GENOCOP (GEnetic algorithm for Numerical Optimization for COnstrained Problems), developed by Michalewicz [4]. This method takes advantage of the features of convex spaces and can be used for any optimization problem with linear constraints. GENOCOP consists of removing as much variables of the problem as the number of equality equations, simplifying the search space. The remaining inequality equations define a convex search space where intervals could be defined for the variable to take values. These intervals are dynamic and depend on the removed variables.

The main disadvantages of this method are: (i) it needs an initial feasible population and so a method for generating it, and (ii) it can be applied only to problems with linear constraints.

2.2 Penalty Functions

This technique is one of the most used when there are non-linear constraints. It consists of introducing cost functions that penalize the individuals that are outside the feasible region. The fitness of j individual is defined as $\text{fitness}^j(\mathbf{x}) = f^j(\mathbf{x}) \pm Q^j$, where Q^j represents the penalty of a non-feasible individual or the cost to make it feasible. If the individual is in the feasible region $Q^j = 0$.

Most penalty methods use a set of functions, f_i $(1 \leq i \leq p)$, for constructing the penalty function. Function f_i measures the violation of constraint i in the following way:

$$f_i(\mathbf{x}) = \begin{cases} \max\{0, g_i(\mathbf{x})\}, & \text{if } 1 \leq i \leq q \\ |h_i(\mathbf{x})|, & \text{if } q+1 \leq i \leq p \end{cases} \tag{2}$$

There is a wide variety of penalty methods. In this work, as the form of the penalty function is not the object of our study, we will use the most straightforward, $Q^j = \sum_{i=1}^{p} f_i(\mathbf{x})$.

3 Crossover Operators

GAs are search methods of general purpose whose operators establish a balance between exploitation and exploration of the search space. Crossover operator plays the major role, it combines the features of two or more parents to generate one or more offsprings. The underlying idea is that the exchange of information among good parents will produce better offsprings.

Most crossover operators generate individuals within the limits of the parents. In this way, the crossover operator implements a depth search or exploitation, leaving the wide search or exploration in the hands of the mutation operator. This politics, although intuitively natural, makes the population converge to inner points of the search space, producing a fast diminishing of the diversity of the population that usually ends in the premature convergence to a suboptimal solution.

Recent studies on the application of BLX-α [5] and fuzzy connectives based crossovers [6] in optimization of functions without constraints have confirmed the interesting performance of the crossover operators that generate individuals in both exploitation and exploration regions. The exploration introduced by these operators is restricted to the neighborhood of the parents, so it is not an alternative to the wide exploration that is carried out by the mutation operator. If the operator establishes a good balance between exploration (or extrapolation) and exploitation (or interpolation) it is possible to avoid the loss of diversity and the premature convergence.

However, in optimization with constraints, it is not clear whether the use of crossover operators with and exploration component is and advantage, as it could produce too many non-feasible individuals. That is the reason why we consider interesting to carry out a study of the influence of crossovers with and without exploration component in the optimization of functions with constraints.

Let $\beta^{f_1} = \{\beta_1^{f_1}, \beta_2^{f_1}, ..., \beta_i^{f_1}, ..., \beta_p^{f_1}\}$ and $\beta^{f_2} = \{\beta_1^{f_2}, \beta_2^{f_2}, ..., \beta_i^{f_2}, ..., \beta_p^{f_2}\}$ be two parents with p genes. We consider in out study the crossover operators that are described in the following paragraphs.

3.1 Discrete Crossover

An offspring $\beta^s = \{\beta_1^s, \beta_2^s, ..., \beta_i^s, ..., \beta_p^s\}$, is obtained where β_i^s is chosen randomly from the set $\{\beta_i^{f_1}, \beta_i^{f_2}\}$ [7]. It is an exclusively exploiting operator.

3.2 Aritmetic Crossover

Two offsprings $\beta^{s_1} = \{\beta_1^{s_1}, \beta_2^{s_1}, ..., \beta_i^{s_1}, ..., \beta_p^{s_1}\}$ and $\beta^{s_2} = \{\beta_1^{s_2}, \beta_2^{s_2}, ..., \beta_i^{s_2}, ..., \beta_p^{s_2}\}$ are created, where $\beta_i^{s_1} = \lambda\beta_i^{f_1} + (1-\lambda)\beta_i^{f_2}$ and $\beta_i^{s_2} = \lambda\beta_i^{f_2} + (1-\lambda)\beta_i^{f_1}$, where λ is a constant [4]. This crossover tends to generate solutions near the center of the search space. In our experiments, following the bibliography, we have set $\lambda = 0.5$.

3.3 BLX-α Crossover

An offspring $\beta^s = \{\beta_1^s, \beta_2^s, ..., \beta_i^s, ..., \beta_p^s\}$ is generated where β_i^s is chosen randomly in the interval $[\beta_{min} - I \cdot \alpha, \beta_{max} + I \cdot \alpha]$. Being $c_{max} = max(\beta_i^{f_1}, \beta_i^{f_2})$, $c_{min} = min(\beta_i^{f_1}, \beta_i^{f_2})$ and $I = c_{max} - c_{min}$ [5]. For $\alpha = 0.5$, the probability that the genes of the offsprings take values within and without the interval of the values of their parents is the same. In [1] different values of α are tested obtaining a best value of $\alpha = 0.5$.

3.4 Logic Crossover

Four monotone non-decreasing functions are defined: F, S, M and L from $[a_i, b_i] \times [a_i, b_i]$ to $[a_i, b_i]$, where $a_i, b_i \in \Re$ are the bounds of the gene's values. For obtaining F, S, M and L the fuzzy connectives t-norm, t-conorm, averaging functions and a generalized compensation operator $\hat{C}$ are used respectively [8]. Of these four families of fuzzy connectives the best results are obtained using *logic* family [9].

If we consider $Q \in \{F, S, M, L\}$ we can generate $\beta^s = \{\beta_1^s, \beta_2^s, ..., \beta_i^s, ..., \beta_p^s\}$ where $\beta_i^s = Q(\beta_i^{f_1}, \beta_i^{f_2}), i = 1, 2, ..., p$. M function is clearly exploiting while F and S are more exploring, L is relaxedly exploiting. Of the four offsprings generated the best two substitute their parents.

3.5 Extended Fuzzy Crossover

This operator [10] is an extension of the *fuzzy recombination operator* [11]. In this operator, the probability that a gene β_i^s of an offspring takes a value z_i is given by the distribution $p(z_i) \in \{\phi_{\beta_i^{f_1}}, \phi_\mu, \phi_{\beta_i^{f_2}}\}$, where $\phi_{\beta_i^{f_1}}$, ϕ_μ and $\phi_{\beta_i^{f_2}}$ are triangular probability distributions. Three offsprings are generated, each one using one the probability distributions, and the two best are chosen. The probability of generating genes within the exploiting interval $[\beta_i^{f_1}, \beta_i^{f_2}]$ is higher than the probability of generating genes in the exploring intervals $[a_i, \beta_i^{f_1}]$ and $[\beta_i^{f_2}, b_i]$.

3.6 Crossover Based on Confidence Intervals

Let $\boldsymbol{\beta}$ be the set of N individuals that form a population and let $\boldsymbol{\beta}^* \subset \boldsymbol{\beta}$ be the set of the n best ones, according to their fitness value. If we consider that each one of the genes of the chromosomes of $\boldsymbol{\beta}^*$ is normally distributed, we can define three individuals: those formed by the lower bounds (CILL), upper bounds (CIUL) and means (CIM) of the confidence intervals of each gene:

$$CILL_i = \overline{\beta}_i - t_{n-1,\alpha/2}\frac{S_{\beta_i}}{\sqrt{n}};\ CIUL_i = \overline{\beta}_i + t_{n-1,\alpha/2}\frac{S_{\beta_i}}{\sqrt{n}};\ CIM_i = \overline{\beta}_i$$

being $\overline{\beta}_i$, $i = 1, \ldots, n$, the mean of each gene, S_{β_i} the standard deviation of the individuals of the population, $t_{n-1,\alpha/2}$ a value obtained from a student´s t distribution with $n-1$ degrees of freedom and α the probability of a gene of not belonging to its confidence interval.

The individuals CILL and CIUL divide each gene's domain, D_i, into three subintervals I^L, I^{CI} and I^R (see Figure 1a), such that $D_i \equiv I^L \cup I^{CI} \cup I^R$ and

$$I^L \equiv [a_i, CILL_i); \quad I^{CI} \equiv [CILL_i, CIUL_i]; \quad I^R \equiv (CIUL_i, b_i]$$

The interval I^{IC} is a confidence interval built from the best n individuals of the population. The probability of a gene of belonging to the confidence interval (the exploitation interval) is $1-\alpha$.

So, both parameters α and n, set the adequate balance between exploration and exploitation for each kind of problem. In a previous work [12] we have found optimum values for the parameters: $(1-\alpha) = 0.7$ and $n = 5$.

This crossover operator will create, from an individual of the population $\beta^f = (\beta_0^f, \beta_1^f, \ldots, \beta_i^f, \ldots \beta_p^f) \in \boldsymbol{\beta}$, and the individuals CILL, CIUL and CIM, a single offspring β^s in the following way:

- $\beta_i^f \in I^L$: if the fitness of β^f is higher than CILL then $\beta_i^s = r(\beta_i^f - CILL_i) + \beta_i^f$, else $\beta_i^s = r(CILL_i - \beta_i^f) + CILL_i$.
- $\beta_i^f \in I^{CI}$: if the fitness of β^f is higher than CIM then $\beta_i^s = r(\beta_i^f - CIM_i) + \beta_i^f$, else $\beta_i^s = r(CIM_i - \beta_i^f) + CIM_i$.
- $\beta_i^f \in I^R$: if the fitness of β^f is higher than CIUL then $\beta_i^s = r(\beta_i^f - CIUL_i) + \beta_i^f$, else $\beta_i^s = r(CIUL_i - \beta_i^f) + CIUL_i$.

Where r is a random number belonging to [0,1]. From this definition it is clear that the genes of the offspring always take values between the best parent β^f and one of CILL, CIUL or CIM. If β^f is far from the other parent, the offspring will probably suffer an important change, and vice-versa. The first circumstance will appear mainly in the first stages of evolution and the second one in the last stages.

4 Experimental Setup

4.1 Problems with Linear Constraints

We have chosen three functions f_1, f_2, and f_3 [4] that cover the three possible situations: constraints that are only inequalities (f_1), constraints that are only equalities (f_2), and constraints that are a combination of both (f_3). These functions and their optimum values are shown on Table 1.

For the optimization of these functions we will use the GENOCOP method. Mutation and crossover operators must be modified in order to generate only feasible individuals.

4.2 Problems with Non-linear Constraints

We have chosen two problems both for their complexity and their interest in the field of experimental sciences. These problems are the distribution of electrons in a sphere and the shape optimization of a cam [13].

Table 1. Functions with linear constraints.

Function	Optimum
$f_1(\mathbf{x}, y) = -10.5x_1 - 7.5x_2 - 3.5x_3 - 2.5x_4 - 1.5x_5 - 10y - 0.5\sum_{i=1}^{5} x_i^2$ subject to: $6x_1 + 3x_2 + 3x_3 + 2x_4 + x_5 \leq 6.5$ $\quad 10x_1 + 10x_3 + y \leq 20$ $0 \leq xi \leq 1$ $\quad 0 \leq y$	$f_1(\mathbf{x}^*, y^*)$=-213
$f_2(\mathbf{x}) = \sum_{i=1}^{10} x_i \left(c_i + ln \frac{x_i}{\sum_{i=1}^{10} x_i}\right)$ subject to: $x_1 + 2x_2 + 2x_3 + x_6 + x_10 = 2$ $\quad x_3 + x_7 + x_8 + 2x_9 + x_10 = 1$ $x_4 + 2x_5 + x_6 + x_7 = 1$ $\quad x \geq 0.000001, \ (i = 1, ..., 10)$ $c_1 = -6.089, c_2 = -17.164, c_3 = -34.054, c_4 = -5.914, c_5 = -24.721,$ $c_6 = -14.986, c_7 = -24.100, c_8 = -10.708, c_9 = -26.662, c_{10} = -22.179$	$f_2(\mathbf{x}^*)$=-47.760765
$f_3 = (\mathbf{x}) = x_1^{0.6} + x_2^{0.6} - 6x_1 - 4x_3 + 3x_4$ $-3x_1 + x_2 - 3x_3 = 0$ $\quad x_1 + 2x_3 \leq 4$ $x_2 + 2x_4 \leq 4$ $\quad x_1 \leq 3$ $x_4 \leq 1$ $\quad 0 \leq x_i, \ (i = 1, 2, 3, 4)$	$f_3=(\mathbf{x})$=-4.5142

Distribution of the Electrons in an Sphere

This problem, known as the Thomson problem, consists of finding the lowest energy configuration of p point charges on a conducting sphere, originated with Thomson's plum pudding model of the atomic nucleus. This problem is representative of an important class of problems in physics and chemistry that determine a structure with respect to atomic positions.

Potential energy for p points (x_i, y_i, z_i) is defined as

$$\begin{aligned} & f(\mathbf{x}, \mathbf{y}, \mathbf{z}) = \sum_{i=1}^{p-1} \sum_{j=i+1}^{p} \left((x_i - x_j)^2 + (y_i - y_j)^2 + (z_i - z_j)^2\right)^{-\frac{1}{2}} \\ & \text{subject to :} \\ & \qquad x_i^2 + y_i^2 + z_i^2 = 1, \ i = 1, ..., p \\ & \qquad (-1, -1, -1) \leq (x_i, y_i, z_i) \leq (1, 1, 1) \end{aligned} \tag{3}$$

This problem has a lot of local minima, whose number increases exponentially with p. For our experiments $p = 25$[13].

Shape Optimization of a Cam

The problem consists of maximizing the arc of the valve opening for one rotation of a convex cam with constraints on the curvature and on the radius of the cam. The function to optimize is:

$$\begin{aligned} & f(\mathbf{r}) = \pi r_v^2 \left(\frac{1}{p} \sum_{i=1}^{p} r_i\right) \\ & \text{subject to :} \\ & \qquad 2r_{i-1}\dot{r}_{i+1}\dot{c}os(2\pi/5(p+1)) \leq r_i(r_{i-1} + r_{i+1}), \ i = 0, ..., p+1 \\ & \qquad r_{-1} = r_0 = r_{min}, \ r_{p+1} = r_{max}, \ r_{p+2} = r_p \\ & \qquad -\alpha \leq \left(\frac{r_{i+1} - r_i}{2\pi/5(p+1)}\right) \leq \alpha \end{aligned} \tag{4}$$

$r_{min} = 1.0$, $r_{max} = 2.0$, $r_v = 1.0$ and $\alpha = 1.5$[13].

We assume that the shape of the cam is circular over an angle of $6/5\pi$ of its circumference, with radius r_{min}. The design variables r_i, $i = 1, ..., p$, represent the radius of the cam at equally spaced angles distributed over an angle of $2/5\pi$.

Table 2. Results for the test problems. AF: Averaged fitness of the best individual on the 10 experiments; SDF: Standard deviation of AF; AG: Averaged number of generations; SDG: Standard deviation of AG; BF: Best fitness; BG: Generation when the best individual was achieved.

Crossover	NLP with linear constraints																	
	f_1						f_2						f_3					
	AF	SDF	AG	SDG	BF	BG	AF	SDF	AG	SDG	BF	BG	AF	SDF	AG	SDG	BF	BG
Conf. Int.	-210.82	2.17	15.8	0.63	-213.00	18	-47.66	0.00	338	48.51	-47.66	390	-4.48	0.01	5.8	0.63	-4.51	8
BLX-α	-209.13	4.13	60.4	46.18	-212.94	130	-47.65	0.01	321.6	22.91	-47. 66	334	-4.50	0.02	23.2	4.23	-4.51	12
Discrete	-204.05	2.35	54.6	21.97	-207.95	76	-47.65	0.02	378	16.22	-47.66	382	-4.47	0.01	24.6	2.36	-4.50	28
Arithmetic	-185.01	8.73	32.8	14.80	-203.34	20	-47.43	0.14	271.4	28.77	-47.27	226	-4.42	0.04	24.2	1.85	-4.49	24
Ext. Fuz.	-198.01	6.04	48.6	12.31	-205.58	54	-47.65	0.01	316.4	12.18	-47.66	318	-4 .49	0.02	17.4	7.26	-4.51	12
Logical	-183.40	8.59	19	2.56	-198.08	20	-47.60	0.06	235.8	99.07	-47.66	292	-4.48	0.0 2	25	2.56	-4.50	28

Crossover	NLP with non-liner constraints							
	Sphere				Cam			
	AF	SDF	BF	BG	AF	SDF	BF	BG
Conf. Int.	281.69	2.38	279.21	2990	3.97	0.25	4.35	5000
BLX-α	280.50	0.82	279.34	3000	2.94	0.31	3.35	5000
Discrete	310.49	4.30	303.32	2980	3.96	0.21	4.40	5000
Arithmetic	328.73	5.69	322.06	2.930	4.67	0.02	4.69	5000
Ext. Fuz.	288.26	1.36	285.72	3000	3.47	0.99	4.28	5000
Logical	321.56	4.85	314.02	2770	3.73	0.13	3.86	4990

4.3 Setup of the RCGA

We will have a population of 100 individuals, with a crossover probability of $p_c = 0.6$, a mutation probability of $p_g = 0.005$, and a tournament selection method with elitism. As mutation operator we will use uniform mutation as its versatility makes it suitable for a wide range of problems. For every problem we made ten experiments using the library of evolutionary computation *jclec*.

For the problems with linear constraints the stop criterion establishes a threshold of minimum improvement in the averaged fitness during a predefined number of generations. For f_1 we will use a threshold of 0.005 in 5 generations, for f_2 0.001 in 15 generations, and for f_3 0.05 in 15 generations. These values of the threshold reflect the different complexities of the problems.

For the problems with non-linear constraints the stop criterion is a fixed number of generations, 2000 for the problems of electron distribution and 5000 for the shape optimization of a cam.

5 Results

Table 2 shows the results obtained for the test problems. For f_1 function the crossover based on confidence intervals achieves the best results and converges faster than the other crossovers. For f_2 function all crossovers achieve comparative results, with an AF slightly better of the crossover based on confidence intervals. For f_3 the results are also very similar, BLX-α achieves the best value of AF, but the crossover based on confidence intervals converges faster. BLX-α, extended fuzzy and based on confidence intervals crossovers reach the optimum value in one of their experiments.

For the electron distribution problem, BLX-α and confidence interval based crossover achieve similar results, the former has a slightly better AF and the latter a slightly better BG. For the shape optimization of a cam problem BLX-α obtains the worst results, arithmetic crossover obtains the best results, followed by the crossover based on confidence intervals.

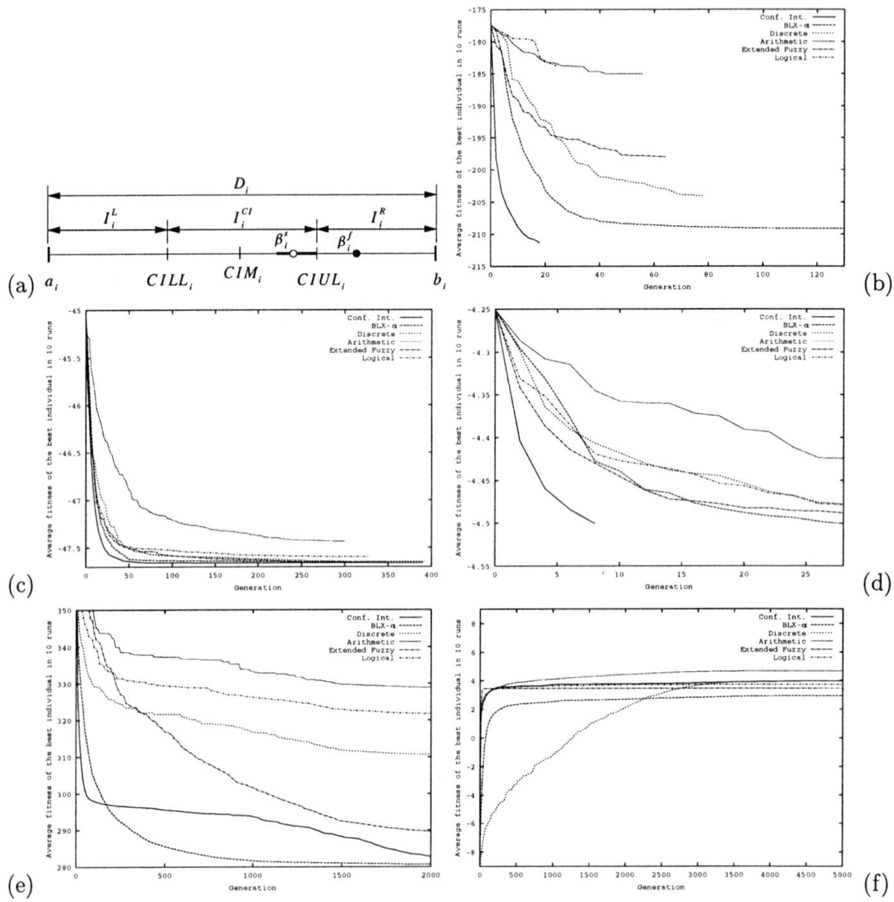

Fig. 1. (a) Graphic representation of the crossover based on confidence intervals; (b) Average fitness of best individuals in 10 runs for f_1; ; (c) Idem for f_2; (d) Idem for f_3; (e) Idem for electron distribution problem; (f) Idem for shape optimization of a cam.

Figures 1bcd show the convergence of the crossovers for the f_1, f_2 and f_3 functions. The figures show clearly that the confidence interval based crossover converges faster than the other crossovers, specially for f_2 and f_3 functions. Figures 1e and 1f show the same effect for the two problems with non-linear constraints. It is interesting to note that of the analyzed crossovers the crossover based on confidence intervals is the most robust, although it could be outperformed in some problems.

6 Conclusions

In this work we have shown the influence that the crossover operator has over a problem of optimization with linear and non-linear constraints. We have proved that the crossover based on confidence intervals is the most robust. That result shows that the dynamic balance between exploration and exploitation of this operator is suitable for this kind of problems.

BLX-α crossover, whose performance in optimization problems without constraints is very good, fails in problems with non-linear constraints. This behavior gives us a hint of its possible vulnerability in this environment.

Our future research work is centered in deepening the study of the crossover operators adding new test problems, and considering a larger number of crossovers. Also, the relation between mutation and crossover operators must be settle as they are closely related.

References

1. Herrera, F., Lozano, M., Verdegay, J.L.: Tackling real-coded genetic algorithms: Operators and tools for behavioural analysis. Artificial Inteligence Review (1998) 265–319 Kluwer Academic Publisherr. Printed in Netherlands.
2. Michalewicz, Z., Schoenauer, M.: Evolutionary algorithms for constrained parameter optimization problems. Evolutionary Computation **4** (1996) 1–32
3. Koziel, S., Michalewicz, Z.: Evolutionary algorithms, homomorphous mappings, and constrained parameter optimization. Evolutionary Computation **7** (1999) 19–44
4. Michalewicz, Z.: Genetic Algorithms + Data Structures = Evolution Programs. Springer-Verlag, New York (1992)
5. Eshelman, L.J., Schaffer, J.D.: Real-coded genetic algorithms and interval-shemata. In Whitley, L.D., ed.: Foundation of Genetic Algorithms 2, San Mateo, Morgan Kaufmann (1993) 187C3.3.7:1–C3.3.7:8.–202
6. Herrera, F., Herrera-Viedma, E., Lozano, E., Verdegay, J.L.: Fuzzy tools to improve genetic algorithms. In: Second European Congress on Intelligent Techniques and Soft Computing. (1994) 1532–1539
7. Mühlebein, H., Schlierkamp-Voosen, D.: Predictive models for breeder genetic algorithm i. continuos parameter optimization. Evolutionary Computation (1993) 25–49
8. Mizumoto, M.: Pictorial representations of fuzzy connectives. part i: Cases of *t*-norms, *t*-conorms and averaging operators. Fuzzy Sets Systems **31** (1989) 217–242
9. Herrera, F., Lozano, M., Verdegay, J.L.: Fuzzy connectives based crossover operators to model genetic algorithms population diversity. Fuzzy Sets Systems **92** (1997) 21–30
10. Herrera, F., Lozano, M.: Gradual distributed real-coded genetic algorithms. IEEE Transactions on Evolutionary Computation **4** (2000) 43–63
11. Voigt, H.M., Mühlenbein, H., Cvetkovic, D.: Fuzzy recombination for the breeder genetic algorithms. In Eshelman, L., ed.: The 6th International Conference Genetic Algorithms, San Mateo, CA, Morgan Kaufmann (1995) 104–111
12. Hervás, C., Ortiz, D.: Operadores de cruce basados en estadísticos de localización para algoritmos genéticos con codificación real. In Alba, E., Fernandez, F., Gomez, J.A., Herrera, F., Hidalgo, J.I., Lanchares, J., Merelo, J.J., Sánchez, J.M., eds.: Primer Congreso Español De Algoritmos Evolutivos y Bioinspirados (AEB'02), Mérida, Spain (2002) 1–8
13. Dolan, E.D., More, J.J.: Benchmarking optimization software with cops. Technical Report ANL/MCS-TM-246, Argonne National Laboratory (2000)

Reducing Random Fluctuations in Mutative Self-adaptation

Thomas Philip Runarsson

Science Institute, University of Iceland
tpr@hi.is

Abstract. A simple method of reducing random fluctuations experienced in step-size control under mutative self-adaptation is discussed. The approach taken does not require collective learning from the population, i.e. no recombination. It also does not require knowledge about the instantiations of the actual random mutation performed on the object variables. The method presented may be interpreted as an *exponential recency-weighted average* of trial strategy parameters sampled by a lineage.

1 Introduction

The paper discusses a method of reducing mean step-size (σ) fluctuations under mutative self-adaptation. Any proposed method capable of reducing these fluctuations will contribute to improved performance [3, p. 325]. One means of achieving this goal is by using collective learning based on the current population, that is by using recombination in the strategy parameter space. The method presented relates to this technique, but by using an *exponential recency-weighted average* of trial strategy parameters sampled via the lineage, *not* by the population. Alternatively, a more sophisticated *derandomized approach to self-adaptation* [4] could be employed. The method presented here is also related to this technique, but *does not* require knowledge about the realized steps (z),

$$\begin{aligned} z_\iota^{(g+1)} &= \sigma_\iota^{(g+1)} N(0,1) \\ x_\iota^{(g+1)} &= x_{r;\lambda}^{(g)} + z_\iota^{(g+1)} \end{aligned} \tag{1}$$

for the given step-size variation $\sigma_\iota^{(g+1)}$. Here x denotes the object variable and index $r;\lambda$ represents the individual assigned the rank r (the rth best out of λ), at any given discrete generation (g).

Reductions in step-size fluctuations are especially important in the case where individual step sizes are used, i.e. each object variable has its own corresponding mean step-size. For simple evolution strategies with small populations the self-adaptation, of individual step sizes, will often fail [5]. Another important domain of application are noisy environments. For example, the fitness function may be noisy [2], or simply the ranking itself may be noisy [7].

J.J. Merelo Guervós et al. (Eds.): PPSN VII, LNCS 2439, pp. 194–203, 2002.

The work presented is motivated by the need for a simple and efficient parallel implementation of evolutionary algorithms using mutative self-adaptation. What is implied by a more efficient implementation is that the communication between individuals in a population should be minimal. The paper is organized as follows. In section 2 a short overview will be given on mutative step-size self-adaptation. This is followed by a section describing the proposed technique for reducing random fluctuations in mutative self-adaptation. An experimental study is then presented in section 4, in order to evaluate the technique empirically. Finally, a discussion and some conclusions are given in section 5.

2 Mutative Step-Size Self-adaptation

In mutative step-size self-adaptation the mutation strength is randomly changed. It is only dependent on the parent's mutation strength, the parent step-size multiplied by a random number. This random number is commonly log-normally distributed. Other distributions are equally plausible [2, 4] and the techniques described in the following section will still be applicable.

The *isotropic* mutative self-adaptation for a (μ, λ) evolution strategy (σES), using the log-normal update rule, is as follows [8],

$$\begin{aligned} \eta_\iota &= \sigma^{(g)}_{r;\lambda} \exp(N(0,\tau_o^2)), \\ \boldsymbol{x}^{(g+1)}_\iota &= \boldsymbol{x}^{(g)}_{r;\lambda} + \boldsymbol{N}(0,\eta_\iota^2), \quad \iota = 1,\ldots,\lambda \end{aligned} \tag{2}$$

where the parent is randomly sampled, $r \in [1, \mu]$, anew for each ι, $\tau_o \simeq c_{(\mu,\lambda)}/\sqrt{n}$ and the step size is updated in the following discrete generation by setting $\sigma^{(g+1)}_{r;\lambda} = \eta_{r;\lambda}$. Similarly, the *non-isotropic* mutative self-adaptation rule is [8],

$$\begin{aligned} \eta_{\iota,j} &= \sigma^{(g)}_{(r;\lambda),j} \exp\left(N(0,\tau'^2) + N_j(0,\tau^2)\right), \\ x^{(g+1)}_{\iota,j} &= x^{(g)}_{(r;\lambda),j} + N_j(0,\eta^2_{\iota,j}), \ \iota = 1,\ldots,\lambda, j = 1,\ldots,n \end{aligned} \tag{3}$$

where $\tau' = \varphi/\sqrt{2n}$ and $\tau = \varphi/\sqrt{2\sqrt{n}}$. The step-size is updated as before by setting $\boldsymbol{\sigma}^{(g+1)}_{r;\lambda} = \boldsymbol{\eta}_{r;\lambda}$.

The primary aim of the step-size control is to tune the search distribution so that maximal progress in maintained. For this some basic conditions for achieving optimal progress must be satisfied. The first lesson in self-adaptation is taken from the 1/5-*success rule* [6, p. 367]. The rule's derivation is based on the probability w_e that the offspring is better than the parent. This probability is calculated for the case where the optimal standard deviation is used $\hat{w}_e$, from which it is then determined that the number of trials must be greater than or equal to $1/\hat{w}_e$ if the parent using the optimal step-size is to be successful. Founded on the sphere and corridor models, this is the origin of the 1/5 value.

In a mutative step-size control, such as the one given by (2), there is no single *optimal* standard deviation being tested, but rather a series of trial step sizes η_ι, $\iota = 1, \ldots, \lambda/\mu$ centered[1] around the parent step size $\sigma_{r;\lambda}$. Consequently, the number of trials may need to be greater than that specified by the 1/5-success rule. If enough trial steps for success are generated near the optimal standard deviation then this trial step-size will be inherited via the corresponding offspring. This offspring will necessarily also be the most likely to achieve the greatest progress and hence be the fittest. The fluctuations on $\sigma_{r;\lambda}$ (the trial standard deviations η_ι) and consequently also on the optimal mutation strength, will degrade the performance of the ES. The theoretical maximal progress rate is impossible to obtain. Any reduction of this fluctuation will therefore improve performance [3, p. 315]. If random fluctuations are not reduced, then a larger number of trials must be used (the number of offspring generated per parent) in order to guarantee successful mutative self-adaptation[2]. This may especially be the case for when the number of free strategy parameters increases, as in the non-isotropic case.

3 Reducing Random Fluctuations

Random fluctuations are reduced for the strategy-parameters, generated by the mutative rules (2) or (3), by letting the following weighted average be inherited to the next generation,

$$\sigma_{\iota,j}^{(g+1)} = \sigma_{(r;\lambda),j}^{(g)} + \chi\big(\eta_{\iota,j}^{(g+1)} - \sigma_{(r;\lambda),j}^{(g)}\big), \quad j = 1, \ldots, n_\sigma \tag{4}$$

where $n_\sigma = 1$ or $n_\sigma = n$ respectively. By considering the $(1, \lambda)$ strategy, one notices that this is an exponential recency-weighted average for a given lineage,

$$\sigma_{(1;\lambda),j}^{(g+1)} = (1-\chi)^{g+1}\sigma_{(1;\lambda),j}^{(0)} + \sum_{i=1}^{g+1} \chi(1-\chi)^{g+1-i}\eta_{(1;\lambda),j}^{(i)} \quad j = 1, \ldots, n_\sigma \tag{5}$$

since $(1-\chi)^{g+1} + \sum_{i=1}^{g+1} \chi(1-\chi)^{g+1-i} = 1$, and when $(1-\chi)$ is less than 1, the weight decreases exponentially according to the exponent of $(1-\chi)$. When $\chi = 1$ the method is equivalent to the canonical approach presented in the previous section. As the average number of trials generated increases (λ/μ) the more likely it will become that the optimal step-size is generated and successful. In this case a value of χ closer to 1 is reasonable. However, if the generated step-size is only an approximation of the optimal one, then a value of χ around 0.2 is more appropriate.

This averaging has the effect of reducing the step-size variations passed on between generations although the variation within a generation remains the same. If one would like to retain the same variation between generations a larger learning rate must be used. That is $E\big[\exp\big(N(0, \tau^2)\big)\big]$ should be the same as

[1] The expected median is $\sigma_{r;\lambda}$.

[2] Some algorithms force a lower bound on the step-size to avoid search stagnation, at this point the mutative self-adaptation is ineffective.

$E\big[1+\chi(\exp\left(N(0,\bar{\tau}^2)\right)-1)\big]$. For the isotropic mutation the corrected learning rate would then be,

$$\bar{\tau}_o^2 = 2\ln\left(\tfrac{1}{\chi}\Big(\exp\left(\tfrac{\tau_o^2}{2}\right)-(1-\chi)\Big)\right) \tag{6}$$

and so if $\chi=1$ then $\bar{\tau}_o=\tau_o$. Similarly, for the non-isotropic mutation,

$$\bar{\varphi}^2 = \tfrac{2}{\upsilon}\ln\left(\tfrac{1}{\chi}\Big(\exp\left(\tfrac{\varphi^2\upsilon}{2}\right)-(1-\chi)\Big)\right) \tag{7}$$

where $\upsilon=\frac{1}{2n}+\frac{1}{2\sqrt{n}}$, and for $\chi=1$ then $\varphi=\bar{\varphi}$.

The new update rules are equivalent to that of (2) and (3), with the corrected learning rates (6) and (7), and the inclusion of (4). For example, the new isotropic mutative self-adaptive rules becomes,

$$\begin{aligned}\eta_\iota &= \sigma_{r;\lambda}^{(g)}\exp\left(N(0,\bar{\tau}_o^2)\right)\\ x_{\iota,j}^{(g+1)} &= x_{(r;\lambda),j}^{(g)}+N_j(0,\eta_\iota^2), \quad j=1,\ldots,n\\ \sigma_\iota^{(g+1)} &= \sigma_{r;\lambda}^{(g)}+\chi\Big(\eta_\iota-\sigma_{r;\lambda}^{(g)}\Big), \quad \iota=1,\ldots,\lambda.\end{aligned} \tag{8}$$

where $0<\chi\le 1$ and $r\in[1,\mu]$.

At this point it may appear more intuitive, given the multiplicative nature of the mutative self-adaptation, to use geometric averaging. In this case the geometrical equivalent of (4) is,

$$\sigma_{\iota,j}^{(g+1)} = \left(\sigma_{(r;\lambda),j}^{(g)}\right)^{(1-\chi)}\left(\eta_{\iota,j}^{(g+1)}\right)^{\chi}, \quad j=1,\ldots,n_\sigma \tag{9}$$

and the learning rates may be simply corrected by setting $\bar{\tau}=\tau/\chi$ or $\bar{\varphi}=\varphi/\chi$. This is again a weighted average over the entire lineage. The same technique is used by *rescaled mutations* [6, p. 197], but with some subtle differences. The aim here is to reduce random fluctuations by averaging. The intention of rescaled mutation is, however, to produce larger σ changes which should result in a more reliable detection of the direction of change [2] for noisy fitness functions. The rescaled mutations use a variable $\kappa_\sigma\equiv 1/\chi$ and scale up the trial mutation strength by a factor $k\approx\sqrt{n}$ during mutation. This will not be done here.

In a noisy environment it may be necessary to reduce random fluctuations for the estimated object variables. This can be done in an equivalent manner to that of the strategy parameters in (4). The inherited object variables are then,

$$x_{\iota,j} = x_{(r;\lambda),j}^{(g)}+N_j(0,\eta_\iota^2) \tag{10}$$

$$x_{\iota,j}^{(g+1)} = x_{(r;\lambda),j}^{(g)}+\chi\Big(x_{\iota,j}-x_{(r;\lambda),j}^{(g)}\Big) \tag{11}$$

where *the fitness is computed based on object vector* $\boldsymbol{x}_\iota$ in (10). Alternatively, one may view this as some form of *genetic repair* [3] based on the lineage, *not* on the population.

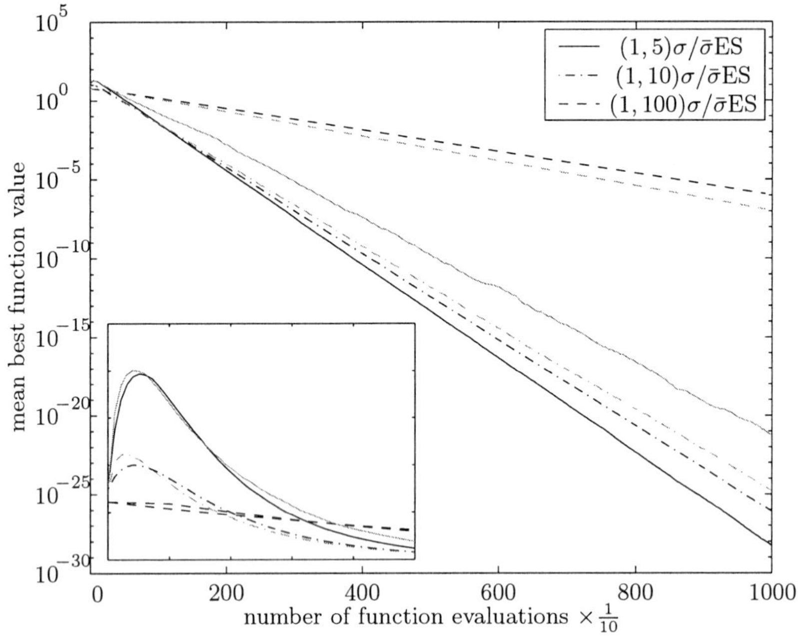

Fig. 1. Noiseless sphere model experiment using the isotropic mutation. The dark lines correspond to the new scheme $\bar{\sigma}$ES ($\chi = 0.2$) and the gray lines to the canonical σES. The sub-figure shows the first few generation in linear scale.

4 Experimental Studies

In this section the behavior of the proposed approach will be examined for the sphere model,

$$\min \quad f(\boldsymbol{x}) = \sum_{k=1}^{n} x_k^2 \tag{12}$$

where $n = 30$, and the noisy sphere model [2],

$$\max \quad g(\boldsymbol{x}) = 1 - \sum_{k=1}^{n} x_k^2 + N(0, \sigma_\epsilon^2) \tag{13}$$

where $n = 10$. Although these models are simple, they do serve the present purpose of verifying the technique.

4.1 Noiseless Sphere Model

Isotropic Mutation. The first experiment aims to examine whether the canonical $(1, \lambda)\sigma$ES using the isotropic mutative self-adaptive rule, (2) or (8) for $\chi = 1$, can be improved by setting χ to a smaller value, say $\chi = 0.2$. The runs using

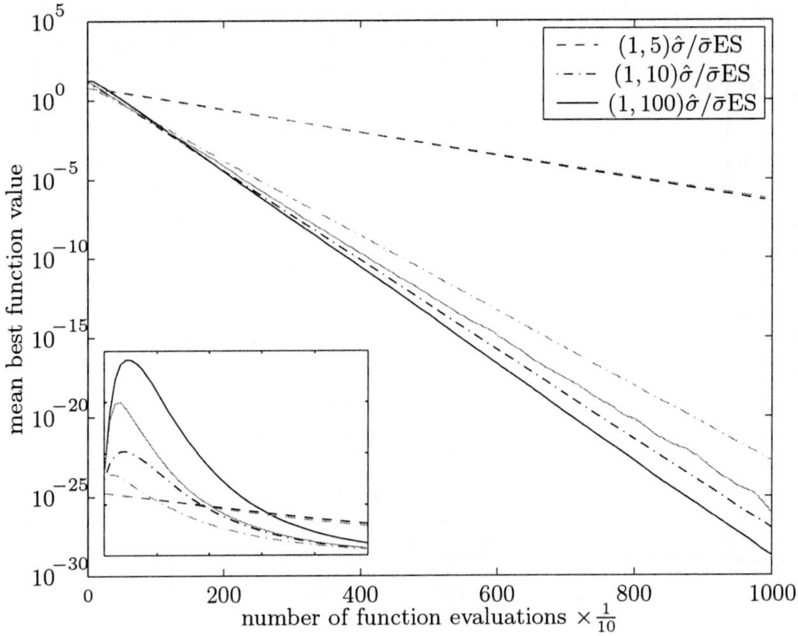

Fig. 2. Noiseless sphere model experiment using the isotropic mutation. The dark lines correspond to the weighted average $\bar{\sigma}$ES ($\chi = 0.3$) and the gray lines to the geometrical average $\hat{\sigma}$ES ($\chi = 0.5$). The sub-figure shows the first few generation in linear scale.

the averaging of σ according to (4) are denoted by $\bar{\sigma}$. The initial step-size $\sigma^{(0)}$ is $2/\sqrt{12}$ and the object variables are randomly initialized in $[-1, 1]$.

A total of 1000 independent runs are performed and the average best function value versus number of function evaluations plotted in figure 1, for $\lambda = 5, 10$, and 100. For $\chi = 0.2$ there is a performance enhancement for the $(1,5)\bar{\sigma}$ES and $(1,10)\bar{\sigma}$ES strategies. However, as the number of trials is increased to 100, the canonical approach $(1,100)\sigma$ES (or $(1,100)\bar{\sigma}$ES with $\chi = 1$) is better. This is as predicted, the greater the number of trials performed the likelier it becomes that the trial mutation created is close the optimal. As a consequence more weight should be put on this estimate, that is χ should be closer to 1.

Notice that the performance of the $(1,10)$ is better than that of the $(1,5)$ strategy[3] in the canonical case, but this is not the case when the random fluctuations have been reduced by setting $\chi = 0.2$.

The same experiment is again repeated and this time the arithmetical style averaging of (4) is compared with the geometrical style averaging of (9). This version is denoted by $(1,\lambda)\hat{\sigma}$ES. In this experiment an attempt is made to optimize χ. It is found that in the case of $\bar{\sigma}$ES the performance is best and

[3] Theoretically, for the sphere model, optimal progress is obtained when $\lambda \approx 5$ and the mutation strength is optimal.

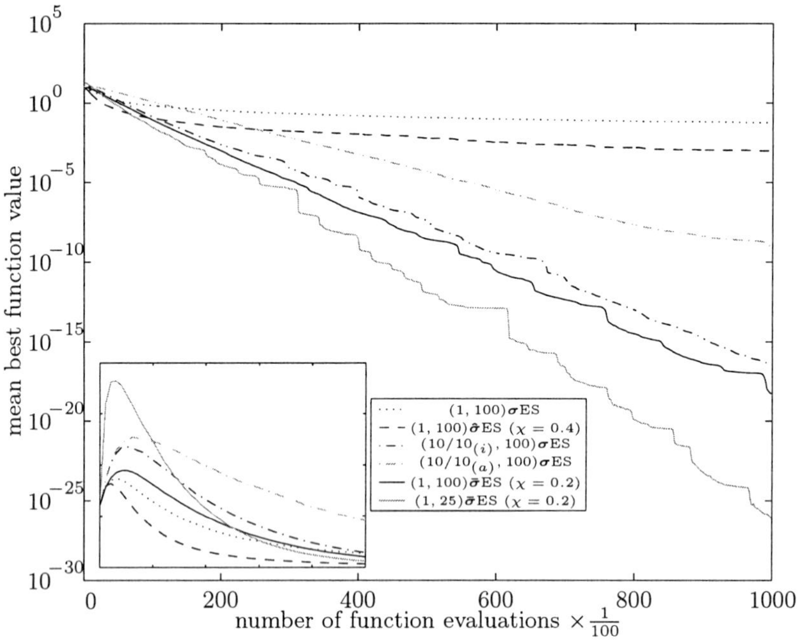

Fig. 3. Noiseless sphere model experiment using the non-isotropic mutation. The subfigure shows the first few generation in linear scale. The legend lists the different strategies in order from worst to best.

similar in the range $0.2 < \chi < 0.4$. For the $\hat{\sigma}$ES the best performance is in the range $0.4 < \chi < 0.6$. The $\bar{\sigma}$ES ($\chi = 0.3$) and $\hat{\sigma}$ES ($\chi = 0.5$) are therefore compared. The average of 1000 independent runs are plotted in figure 2, where one can see that, for the noiseless sphere model, the $\bar{\sigma}$ES is slightly better.

Non-isotropic Mutation. In the second experiment the first experiment is repeated using the non-isotropic mutation with $\varphi = 1$. Again 1000 independent runs are performed and the result plotted in figure 3. This time the mutative self-adaptation fails for the canonical $(1, 100)\sigma$ES – the search stagnates. In order to avoid this problem two new strategies are introduced, both using recombination in the strategy parameter space. The first one uses intermediate recombination,

$$\eta_{\iota,j} = \Big(\frac{1}{\mu}\sum_{k=1}^{\mu}\sigma_{(k;\lambda),j}^{(g)}\Big)\exp\big(N(0,\tau'^2) + N_j(0,\tau^2)\big) \qquad (14)$$

and is denoted by $(\mu/\mu_{(i)}, \lambda)$. The non-isotropic mutation is included on the right-hand-side of (14). The second is an arithmetical recombination, including the mutation,

$$\eta_{\iota,j} = \tfrac{1}{2}\big(\sigma_{(r;\lambda),j}^{(g)} + \sigma_{(k;\lambda),j}^{(g)}\big)\exp\big(N(0,\tau'^2) + N_j(0,\tau^2)\big) \qquad (15)$$

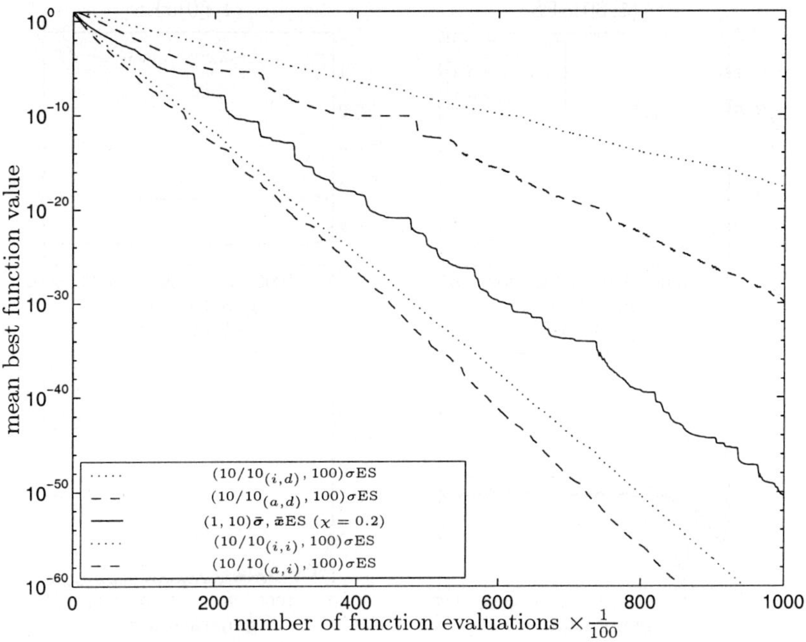

Fig. 4. Noiseless sphere model using non-isotropic mutation and recombination in both strategy and object parameter space. The versions listed in the legend are ordered based on performance, from worst to best.

denoted $(\mu/\mu_{(a)}, \lambda)$, where $k \in [1, \mu]$ is sampled randomly and anew for each j. By using recombination on the strategy parameters, search stagnation is avoided. It is also noticed that arithmetical recombination performs better than intermediate recombination. However, in terms of the number of function evaluation needed, the newly proposed scheme still performs best. For the $(1, 100)$ strategy the search does not stagnate for the case when $\chi = 0.2$. In order to achieve faster progress the number of trial is reduced down to a $(1, 25)$ strategy, but going down to $(1, 10)$ will result in failure[4]. Clearly a greater number of trials will be needed for success as the number of free parameters increases. An attempt to use the geometrical averaging of (9) resulted in faster initial progress and then search stagnation. It seems that arithmetical averaging has a tendency towards larger mutation strengths than geometrical averaging.

The final experiment introduces recombination in the object parameter space. Two versions are tested. The first is again an intermediate recombination and mutation,

$$x_{\iota,j}^{(g+1)} = \frac{1}{\mu} \sum_{k=1}^{\mu} x_{(k;\lambda),j}^{(g)} + N_j(0, \eta_{\iota,j}^2) \tag{16}$$

[4] Unpredictable behavior is observed, where the step-size may keep growing in size.

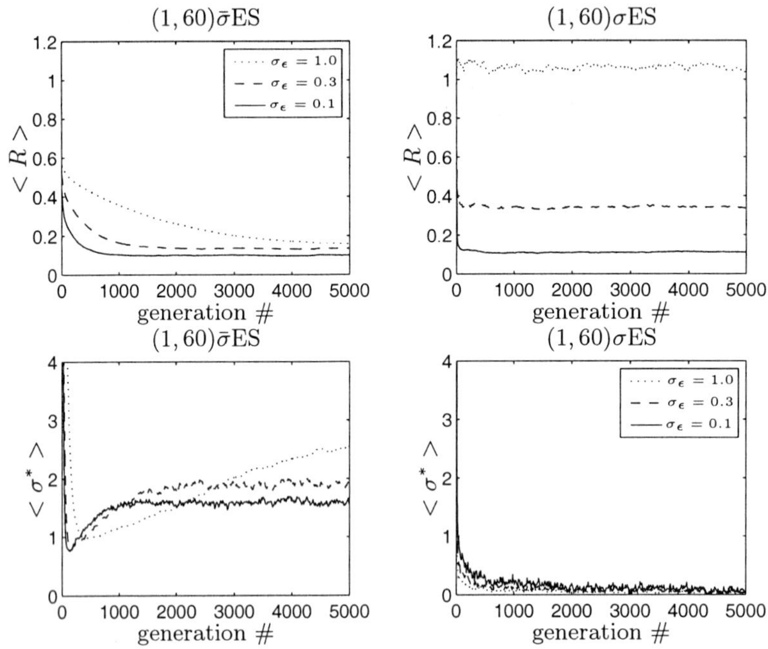

Fig. 5. Noisy sphere model experiments. The plots on the left are from the new technique $\bar{\sigma}$ES and on the right the canonical approach σES.

denoted here by $(\mu/\mu_{(,i)}, \lambda)$ and the second dominant recombination and mutation,

$$x_{\iota,j}^{(g+1)} = x_{(k;\lambda),j}^{(g)} + N_j(0, \eta_{\iota,j}^2) \tag{17}$$

where $k \in [1, \mu]$ is sampled randomly and anew for each j. This version is denoted by $(\mu/\mu_{(,d)}, \lambda)$. The recombinations in object parameter space are used in conjunction with recombination in strategy parameter space resulting in four different strategies: $(\mu/\mu_{(i,d)}, \lambda)$, $(\mu/\mu_{(a,d)}, \lambda)$, $(\mu/\mu_{(i,i)}, \lambda)$, and $(\mu/\mu_{(a,i)}, \lambda)$. In order to make a fair comparison, the new technique is allowed to use the weighted average for the object variables as defined by (11). It is interesting to observe that now it becomes possible to switch to a $(1, 10)$ scheme without any problems. Using geometrical averaging (9) will also result in search stagnation for this experiment and is therefore omitted. The results, averaged over 1000 independent runs are presented in figure 4. The best results are, nevertheless, achieved using intermediate recombination on the object variables.

4.2 Noisy Sphere Model

The final experiment is based on simulations performed in [2]. This involves the noisy sphere model (13), where the object parameters are initialized in $[-\frac{1}{\sqrt{10}}, \frac{1}{\sqrt{10}}]$, and an isotropic mutation with $\sigma^{(0)} = 1$ and $\tau_o = 0.7$ is used.

In this experiment the value of χ is varied as a function of the noise level σ_ϵ in (13). The noise levels σ_ϵ are 0.1, 0.3, and 1.0 and the corresponding values of χ chosen are 0.1, 0.06, and 0.02. For the three different noise levels 300 independent runs are performed using a $(1, 60)$ strategy. The new method uses the averaging (11) for the object variables.

Instead of plotting mean function values, the mean distance to the optimum, $R = \|x\|$, is depicted versus the generation number. Additionally, the mean normalized step-size, $\sigma^* = n\sigma/R$, is plotted against the generation number. The behavior of the new scheme is similar to that of the rescaled mutations technique in [2], as seen in figure 5. The canonical approach on the other hand has difficulties coping with the noisy function.

5 Discussion and Conclusion

The update rule of (4) is not unlike the update rules used in reinforcement learning, where χ is a step-size parameter. When using a constant χ more weight is put on the strategy parameters used recently rather than long-past ones. This makes sense when tracking the nonstationary problem of self-adaptation. Furthermore, the greater the number of trials generated ($\approx \lambda/\mu$) the more reliable the selected strategy parameters become.

For noisy functions smaller χ values were used. At the later stage of search, say after generation g', the problem of self-adaptation becomes essentially stationary. In this case it may be reasonable to set $\chi = 1/(g - g')$, and so all estimates after g' are equally weighted. That is, (4) becomes an incremental implementation of a simple-average.

A technique for reducing of random fluctuations in mutative self-adaptation has been presented and some experimental results given. Self-adaptive rules using individual step sizes commonly fail. The proposed method may be a simple means of alleviating this problem.

References

1. H.-G. Beyer. Toward a theory of evolution strategies: self-adaptation. *Evolutionary Computation*, 3(3):311–347, 1996.
2. H.-G. Beyer. Evolutionary algorithms in noisy environments: theoretical issues and guidelines for practice. *Computer Methods in Applied Mechanics and Engineering*, 186(2–4):239–267, 2000.
3. H.-G. Beyer. *The Theory of Evolution Strategies.* Springer-Verlag, Berlin, 2001.
4. N. Hansen and A. Ostermeier. Completely derandomized self-adaptation in evolution stategies. *Evolutionary Computation*, 2(9):159–195, 2001.
5. A. Ostermeier, A. Gawelczyk, and N. Hansen. A derandomized approach to self-adaptation of evolution strategies. *Evolutionary Computation*, 2(4):369–380, 1995.
6. I. Rechenberg. *Evolutionstrategie '94.* Frommann-Holzboog, Stuttgart, 1994.
7. T. P. Runarsson and X. Yao. Stochastic ranking for constrained evolutionary optimization. *IEEE Transactions on Evolutionary Computation*, 4(3):284–294, September 2000.
8. H.-P. Schwefel. *Evolution and Optimum Seeking.* Wiley, New-York, 1995.

On Weight-Biased Mutation for Graph Problems

Günther R. Raidl[1], Gabriele Kodydek[1], and Bryant A. Julstrom[2]

[1] Vienna University of Technology, Vienna, Austria
[2] St. Cloud State University, St. Cloud, MN, USA
{raidl,kodydek}@ads.tuwien.ac.at, julstrom@eeyore.stcloudstate.edu

Abstract. Many graph problems seek subgraphs of minimum weight that satisfy the problems' constraints. Examples include the degree-constrained minimum spanning tree and traveling salesman problems. Low-weight edges predominate in optimal solutions to these problems, and the performance of evolutionary algorithms for them is often improved by biasing their operators to favor these edges. From the distributions of edges' ranks in optimal solutions to these two problems, we identify probabilities for edges that minimize the average expected time until mutation chooses them for inclusion in a solution. On instances of the degree-constrained minimum spanning tree problem, an evolutionary algorithm performs better with this operator than with alternative mutations. These results are not replicated on instances of the traveling salesman problem, where the inclusion of one edge in a tour requires the inclusion of another dependant edge.

1 Introduction

Given a weighted, undirected graph, many graph problems seek a subset S of the graph's edges that satisfies a set of constraints and has minimum total weight. These include the familiar traveling salesman problem (TSP), in which the edges in S form a Hamiltonian tour; the degree-constrained minimum spanning tree problem (d-MSTP) [6, 7], in which S is a spanning tree with degree no greater than a bound d; the leaf-constrained spanning tree problem [1], in which S is a spanning tree with at least L leaves; the biconnectivity augmentation problem [8], in which S augments a spanning tree so that the resulting network is 2-connected; and many others.

Some of these problems, such as the unconstrained minimum spanning tree problem and the identification of the shortest path between two vertices, can be solved to optimality in polynomial time. Most, including those listed above, are NP-hard. It is not likely that fast algorithms exist to solve these problems exactly, so we turn to heuristics, including evolutionary algorithms (EAs).

It is not surprising—and we verify below—that low-weight edges predominate in optimal solutions to such problems. This suggests that in EAs, crossover and mutation, which build representations of novel solutions from existing representations, should be biased so as to favor edges of lower weight. Several researchers have investigated such schemes [3, 4, 9, 10].

This work is supported by the Austrian Science Fund (FWF), grant P13602–INF.

J.J. Merelo Guervós et al. (Eds.): PPSN VII, LNCS 2439, pp. 204–213, 2002.

Among them, Julstrom and Raidl examined weight-biased crossover operators in EAs for the TSP and the d-MSTP [5]; favoring low-weight edges improved the performance of these algorithms. We extend that inquiry to mutation and derive probabilities for selecting edges to be incorporated into candidate solutions. These probabilities are optimal in the sense that they minimize the expected time to include edges of optimal solutions. For the d-MSTP, we compare a mutation based on this analysis to four others. This theoretically approximately optimal scheme increases the probability of finding optimal solutions and reduces the number of iterations usually used. Applied to the TSP, the advantages of weight-biased approaches are generally smaller because mutation that introduces one edge into a tour necessarily introduces a second as well.

2 Distribution of Edges in Optimal Solutions

It is reasonable that optimally low-weight trees, tours, and other structures in weighted graphs should contain high proportions of low-weight edges. This section confirms and quantifies this observation for the degree-constrained minimum spanning tree and traveling salesman problems on complete graphs $G = (V, E)$ with $n = |V|$ nodes and $m = |E| = n \cdot (n-1)/2$ edges. Let S be the set of edges in a solution, so that $|S| = n - 1$ for the d-MSTP and $|S| = n$ for the TSP.

We examine two kinds of instances of both problems. Euclidean instances consist of distinct points chosen randomly in a square region of the plane; edge weights are the Euclidean distances between each pair of points. Uniform instances consist of edge weights chosen randomly and independently from a specified interval. 1 000 instances of each type were generated with $n = 20$, 50, and 100 nodes. For the d-MSTP, the degree bound d was set to three for the Euclidean instances; note that for such instances there always exists an unconstrained minimum spanning tree whose degree does not exceed five. On the uniform instances, we consider $d = 3$ and $d = 5$.

All these instances were solved to optimality by an algorithm found in the ABACUS branch-and-cut solver [11]. We assign each edge a rank r, $1 \leq r \leq m$, by sorting the edges of an instance according to increasing weights (ties are broken randomly). Figure 1 plots the probabilities $p(r)$ that an edge of rank r appears in an optimal solution. Only the portions of the curves where $p(r)$ is significantly larger than zero are plotted.

Note that the sum of the probabilities $p(r)$ is $|S|$:

$$\sum_{r=1}^{m} p(r) = |S| \,. \tag{1}$$

As predicted, the optimal solutions consist mostly of low-rank—*i.e.*, short—edges. Further, for each kind of problem and each fraction $k \in (0, (n-1)/2]$, the probability $p(\lceil k \cdot n \rceil)$ is approximately constant across all values of $n \gg 1$.

Table 1 illustrates this by listing, for each problem kind and size, the number R of least-cost edges among which $\alpha = 50$, 90, and 99 percent of the edges of an

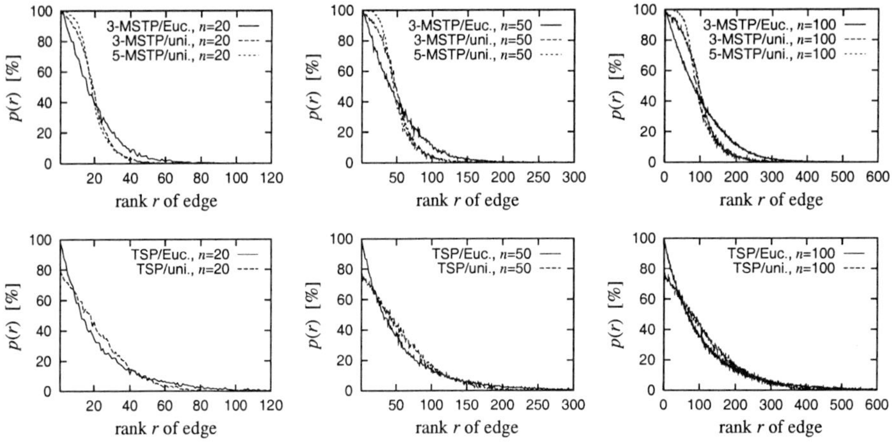

Fig. 1. The probability that an edge appears in an optimal solution as a function of its rank, shown for the 3-MSTP, the 5-MSTP, and the TSP on Euclidean and uniform instances of size $n = 20$, 50, and 100.

Table 1. Numbers R of least-cost edges for each problem class and size $n = 20, 50, 100$ among which $\alpha = 50\%$, 90%, and 99% of optimal solutions' edges are found and corresponding fractions $k = R/n$.

		3-MSTP/Euc.			3-MSTP/uni.			5-MSTP/uni.			TSP/Euc.			TSP/uni.		
$\alpha \setminus n$		20	50	100	20	50	100	20	50	100	20	50	100	20	50	100
50%	R	12	31	63	11	27	54	10	25	51	15	37	72	16	40	80
	k	0.6	0.6	0.6	0.6	0.5	0.5	0.5	0.5	0.5	0.8	0.7	0.7	0.8	0.8	0.8
90%	R	33	89	179	24	63	126	23	60	120	53	134	257	41	107	217
	k	1.7	1.8	1.8	1.2	1.3	1.3	1.2	1.2	1.2	2.7	2.7	2.6	2.1	2.1	2.2
99%	R	63	165	323	41	113	228	40	110	223	107	296	586	67	183	373
	k	3.2	3.3	3.2	2.1	2.3	2.3	2.0	2.2	2.2	5.4	5.9	5.9	3.4	3.7	3.7

optimal solution are located, i.e., the smallest rank R for which the cumulated probability

$$\sum_{r=1}^{R} p(r) \geq \alpha \cdot |S| \,. \tag{2}$$

Table 1 also shows corresponding fractions $k = R/n$, which are quite independent of n for each problem class and each α.

An effective heuristic mutation operator should introduce edges depending on the probabilities with which they appear in optimal solutions. To do this, we identify a closed-form expression $p_A(r)$ that approximates $p(r)$.

Fig. 1 shows that $p(r)$ decreases approximately exponentially with r, particularly in the Euclidean instances of the two problems. Thus we choose

$$p_A(r) = a^r \quad \text{with } 0 < a < 1 \,. \tag{3}$$

The base a should be chosen so that

$$\sum_{r=1}^{m} p_A(r) = \sum_{r=1}^{m} a^r = \frac{a - a^{m+1}}{1-a} = |S| \,. \tag{4}$$

Since a^{m+1} is negligible for problems of even moderate size, we ignore it to obtain

$$\sum_{r=1}^{m} p_A(r) \approx \frac{a}{1-a} \quad \Rightarrow \quad a \approx \frac{|S|}{|S|+1} \,. \tag{5}$$

Fig. 2(a) in Sect. 3.2 plots $p_A(r) = a^r$ with $a = |S|/(|S|+1)$ for the 3-MSTP instances with 100 nodes. It approximates the empirical probabilities $p(r)$ with high accuracy; the mean square error is less than 0.076%. For the 100-node Euclidean instances of the TSP, the mean square error is less than 0.014%. For uniform instances, the error is slightly larger.

3 Optimal Edge-Selection Probabilities

In genetic algorithms, mutation is understood to (re)introduce into the population novel or lost genetic material. In graph problems like the d-MSTP and the TSP, the $m = \binom{n}{2}$ edges of the graph comprise the pool from which this material is drawn.

Purely random mutation chooses each edge to include in a solution according to uniform probabilities; each edge may be chosen with probability $1/m$. We apply the analysis of Section 2 to identify non-uniform probabilities, associated with the edges' ranks, that are optimal in the following sense: Over all edges e^* in an optimal solution S^*, the average expected number of edge selections until e^* is chosen is minimal.

Let $q(r)$ be the probability that an edge-selection scheme chooses the edge e_r whose rank is r. The expected number of selections until e_r is chosen for the first time is

$$EX(e_r) = 1/q(r) \,. \tag{6}$$

Let e^* be an edge in an optimal solution S^*. The probability that e^* has rank r $(1 \le r \le m)$ is $p(r)/|S|$, where $p(r)$ is the probability that e_r appears in an optimal solution. The expected number of edge selections until e^* is chosen for the first time is the weighted sum

$$EX(e^*) = \sum_{r=1}^{m} \frac{p(r)/|S|}{q(r)} = \frac{1}{|S|} \sum_{r=1}^{m} \frac{p(r)}{q(r)} \,. \tag{7}$$

Because $\sum_{r=1}^{m} q(r) = 1$, we can replace $q(m)$ by $1 - \sum_{i=1}^{m-1} q(i)$ in (7) and write

$$EX(e^*) = \frac{1}{|S|} \left(\sum_{r=1}^{m-1} \frac{p(r)}{q(r)} + \frac{p(m)}{1 - \sum_{i=1}^{m-1} q(i)} \right) . \tag{8}$$

To identify selection probabilities $q(r)$ that minimize $EX(e^*)$, we partially differentiate $EX(e^*)$ with respect to each $q(r)$ and set these derivatives equal to zero:

$$\frac{\partial EX(e^*)}{\partial q(1)} = \frac{1}{|S|}\left(-\frac{p(1)}{q(1)^2} + \frac{p(m)}{(1-\sum_{i=1}^{m-1} q(i))^2}\right) = 0$$

$$\frac{\partial EX(e^*)}{\partial q(2)} = \frac{1}{|S|}\left(-\frac{p(2)}{q(2)^2} + \frac{p(m)}{(1-\sum_{i=1}^{m-1} q(i))^2}\right) = 0$$

$$\dots$$

$$\frac{\partial EX(e^*)}{\partial q(m-1)} = \frac{1}{|S|}\left(-\frac{p(m-1)}{q(m-1)^2} + \frac{p(m)}{(1-\sum_{i=1}^{m-1} q(i))^2}\right) = 0 \tag{9}$$

This system of $m-1$ equations can be simplified to

$$\frac{p(1)}{q(1)^2} = \frac{p(2)}{q(2)^2} = \dots = \frac{p(m-1)}{q(m-1)^2} = \frac{p(m)}{(1-\sum_{i=1}^{m-1} q(i))^2} = \frac{p(m)}{q(m)^2}. \tag{10}$$

Let $\varphi = p(r)/q(r)^2$. Then

$$q(r) = \sqrt{\frac{p(r)}{\varphi}} \tag{11}$$

and since

$$\sum_{i=1}^{m} q(i) = 1 = \frac{1}{\sqrt{\varphi}} \sum_{i=1}^{m} \sqrt{p(i)}, \tag{12}$$

we conclude that

$$\varphi = \left(\sum_{i=1}^{m} \sqrt{p(i)}\right)^2 \quad \text{and} \quad q(r) = \frac{\sqrt{p(r)}}{\sum_{i=1}^{m} \sqrt{p(i)}}. \tag{13}$$

3.1 $EX(e^*)$ for Three Edge-Selection Strategies

The optimal edge-selection probabilities $q(r)$ identified in (13), when substituted into equation (7), yield the following average expected number of edge-selections until an edge e^* of an optimal solution is chosen:

$$EX^*(e^*) = \frac{1}{|S|} \sum_{r=1}^{m} \frac{p(r)}{\sqrt{p(r)} \big/ \sum_{i=1}^{m} \sqrt{p(i)}} = \frac{1}{|S|}\left(\sum_{r=1}^{m} \sqrt{p(r)}\right)^2. \tag{14}$$

We replace $p(r)$ by the approximation $p_A(r) = a^r$ to obtain

$$EX^*(e^*) \approx \frac{1}{|S|}\left(\sum_{r=1}^{m} \sqrt{a^r}\right)^2 = \frac{1}{|S|}\left(\frac{\sqrt{a} - a^{(m+1)/2}}{1-\sqrt{a}}\right)^2. \tag{15}$$

Since $a^{(m+1)/2}$ is orders of magnitudes smaller than $\sqrt{a}$ even for moderate problem sizes, we ignore it. Further, replacing a by $|S|/(|S|+1)$ according to (5), we obtain:

$$EX^*(e^*) \approx \frac{a}{|S|\,(1-\sqrt{a})^2} = \frac{1}{(|S|+1)\left(1-\sqrt{\frac{|S|}{|S|+1}}\right)^2} = \tag{16}$$

$$= \left(\sqrt{|S|} + \sqrt{|S|+1}\right)^2 . \tag{17}$$

Thus, $EX^*(e^*) = O(|S|) = O(n)$.

Consider the same expected value when edges are selected according to uniform probabilities: for all $r = 1, \ldots, m$, $q_U(r) = 1/m$. Since $\sum_{r=1}^{m} p(r) = |S|$,

$$EX^U(e^*) = \frac{1}{|S|}\sum_{r=1}^{m}\frac{p(r)}{1/m} = \frac{m}{|S|}\sum_{r=1}^{m} p(r) = m\,. \tag{18}$$

Similarly, let edges' probabilities be proportional to $p(r)$: for all $r = 1, \ldots, m$, $q_P(r) = p(r)/|S|$. Then

$$EX^P(e^*) = \frac{1}{|S|}\sum_{r=1}^{m}\frac{p(r)}{p(r)/|S|} = \frac{|S|}{|S|}\sum_{r=1}^{m} 1 = m\,. \tag{19}$$

For both, uniform and $p(r)$-proportional probabilities, $EX(e^*) = m = O(n^2)$, while for the optimal probabilities, $EX^*(e^*) = O(n)$.

3.2 Approximately Optimal Edge-Selection Probabilities

Replacing $p(r)$ by the approximation $p_A(r) = a^r$ in (13) yields a closed-form expression for the optimal edge-selection probabilities $q_A(r)$:

$$q_A(r) = \frac{\sqrt{p_A(r)}}{\sum_{i=1}^{m}\sqrt{p_A(i)}} = \frac{\sqrt{a^r}}{\sum_{i=1}^{m}\sqrt{a^i}} = \frac{\sqrt{a^r}}{\frac{\sqrt{a}-a^{(m+1)/2}}{1-\sqrt{a}}} = \frac{(1-\sqrt{a})\,a^{r/2}}{\sqrt{a}-a^{(m+1)/2}} \tag{20}$$

Again, $a^{(m+1)/2} \ll \sqrt{a}$, and we ignore it. Again, we replace a with $|S|/(|S|+1)$ according to (5). Thus:

$$q_A(r) \approx \frac{(1-\sqrt{a})\,a^{r/2}}{\sqrt{a}} = a^{\frac{r}{2}}\left(\frac{1}{\sqrt{a}}-1\right) = \left(\frac{|S|}{|S|+1}\right)^{\frac{r}{2}}\left(\sqrt{\frac{|S|+1}{|S|}}-1\right) \tag{21}$$

Fig. 2(b) plots the probabilities $q_A(r)$, $q_U(r)$, and $q_P(r)$ for instances of the 3-MSTP on $n = 100$ nodes.

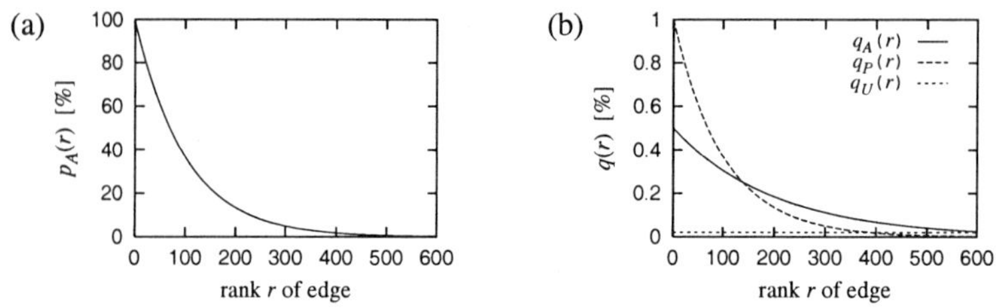

Fig. 2. (a) Approximation of $p(r)$ by $p_A(r) = a^r$ for the 3-MSTP on $n = 100$ nodes, and (b) corresponding edge-selection probabilities $q_A(r)$ (approximately optimal), $q_P(r)$ (proportional to $p_A(r)$), and $q_U(r)$ (uniform).

4 Biasing Mutation for the d-MSTP

We consider an EA for the d-MSTP as described in [9]. Mutation is performed by including a random new edge into a feasible solution and removing another randomly chosen edge from the introduced cycle such that the degree constraint is never violated. If a selected edge already appears in the current solution or the degree constraint cannot be met, the selection is repeated. We apply the following strategies for selecting the edge to be included.

UNIF: Each edge is randomly chosen with probability $q_U(r) = 1/m$.

OPTEX: Edges are selected according to the approximately optimal selection probabilities $q_A(r)$ with respect to $EX(e^*)$.

To perform this edge-selection efficiently, we derive a random edge-rank $\mathcal{R} \in \{1, 2, \ldots, m\}$ from a uniformly distributed random number $\mathcal{U} \in [0, 1)$.

In order to ensure that R has the approximate probability density $q_A(r)$ of (21), we use the inverse of the corresponding cumulative distribution function $F(r)$:

$$F(r) = \sum_{i=1}^{r} q_A(i) \approx \sum_{i=1}^{r} a^{\frac{i}{2}} \left(\frac{1}{\sqrt{a}} - 1 \right) = \frac{\sqrt{a} - a^{(r+1)/2}}{1 - \sqrt{a}} \left(\frac{1}{\sqrt{a}} - 1 \right) =$$

$$= 1 - a^{r/2} = 1 - \left(\frac{|S|}{|S| + 1} \right)^{\frac{r}{2}} . \tag{22}$$

The inverse of $F(r)$ is

$$r = \frac{2 \log(1 - F(r))}{\log |S| - \log(|S| + 1)} . \tag{23}$$

$\mathcal{R}$ can be calculated from $\mathcal{U}$ by setting $F(r) = \mathcal{U}$ in (23) and rounding:

$$\mathcal{R} = \left\lfloor \frac{2 \log(1 - \mathcal{U})}{\log |S| - \log(|S| + 1)} \right\rfloor \bmod m + 1 . \tag{24}$$

Finding the modulus and adding one ensures that $\mathcal{R}$ will be a valid edge rank.

PROPP: Each edge is selected with probability $q_P(r) = p(r)/|S| \approx a^r/|S|$. As with OPTEX, the implementation uses a uniform random number $\mathcal{U}$ transformed by the inverse of the distribution function:

$$F(r) = \sum_{i=1}^{r} \frac{a^i}{|S|} = 1 - \left(\frac{|S|}{|S|+1}\right)^r . \tag{25}$$

This yields

$$\mathcal{R} = \left\lfloor \frac{\log(1-\mathcal{U})}{\log|S| - \log(|S|+1)} \right\rfloor \bmod m + 1 . \tag{26}$$

N$_\beta$: This edge-selection strategy is based on normal distributions as proposed in [9]. The rank of a selected edge is

$$\mathcal{R} = \lfloor |\mathcal{N} \cdot \beta \cdot n| \rfloor \bmod m + 1 , \tag{27}$$

where $\mathcal{N}$ is a normally distributed random number with mean zero and standard deviation one. β controls the biasing towards low-cost edges.

INVW: Each edge $e \in E$ is selected according to probabilities inversely proportional to the edge weights $w(e)$. Such a technique was used in [3] for choosing edges during recombination for the TSP.

5 Experiments on the *d*-MSTP

The five mutation operators were compared in a steady-state EA for the 3-MSTP as described in [9]. The algorithm represents candidate solutions as sets of their edges. Feasible initial solutions are created by a random spanning tree algorithm based on Kruskal's MST algorithm. A new feasible offspring is always derived by performing edge-crossover and mutation. Edge-crossover is based on a random spanning tree algorithm applied to the united edge-sets of two parents. In contrast to [9], no heuristics are used during initialization and recombination. Parents for crossover are selected in binary tournaments with replacement. Each offspring replaces the worst solution in the population except when it duplicates an existing solution.

In the experiments, we considered 50 randomly created Euclidean instances of each size $n = 50$, 100, and 200. The population size was $2\,n$, and the EA terminated if an optimal solution (determined by branch-and-cut) had been reached or the number of evaluations exceeded $5000\,n$.

Runs were performed on each instance with each mutation. For normal-distribution-based edge-selection N$_\beta$, β was set to 0.75, 1, 1.5, 2, and 3. Table 2 shows, for each size n and each operator, the percentage of runs that identified optimal solutions and the average number of evaluations in these runs. The best values are printed in bold.

Table 2. Results of the EA for the 3-MSTP with each mutation on Euclidean instances of size n = 50, 100, and 200: percentages of runs that found optimal solutions (*%-hits*) and average numbers of evaluations of those runs in thousands (*eval/*1000).

n		UNIF	OPTEX	PROPP	$N_{0.75}$	N_1	$N_{1.5}$	N_2	N_3	INVW
50	*%-hits*	98	**100**	**100**	48	68	94	**100**	98	98
	*eval/*1000	54.0	11.7	16.9	40.9	25.0	**10.6**	15.0	11.4	20.4
100	*%-hits*	66	**100**	88	20	46	82	90	94	86
	*eval/*1000	288.2	49.1	67.0	160.8	96.1	93.8	**46.8**	61.7	115.4
200	*%-hits*	8	**96**	78	6	46	64	76	66	46
	*eval/*1000	887.5	198.7	246.1	452.3	382.0	223.9	216.9	**195.4**	618.6

OPTEX performed best on all three sizes; it found optimal solutions on nearly all the instances, and its average numbers of evaluations are among the lowest. Those mutations with lower numbers of evaluations exhibit significantly poorer hit rates. UNIF needed on average the most evaluations, followed by INVW and $N_{0.75}$. Experiments on uniform instances showed similar tendencies.

6 Biased Mutation for the TSP

In comparison to the d-MSTP, incorporating biased edge-selection techniques into mutation of an EA for the TSP is more difficult. A commonly used mutation operator for the TSP acting on a permutation representation is inversion.

This operator can be modified to include a specific new edge selected by one of the above strategies: We invert the substring beginning after the selected edge's first node and ending with the selected edge's second node.

Note, however, that in addition to the selected edge, a second new edge is implicitly included. This second edge depends on the first edge and the current tour; it cannot be chosen according to the edge-selection strategy.

This side-effect strongly influences the idea of biased mutation and is expected to affect performance. Experiments with the TSP similar to those with the d-MSTP did not show significant differences among the edge-selection methods.

7 Conclusions

The rank-based probabilities with which edges appear in optimal solutions of Euclidean and uniform instances of the d-MSTP and the TSP were empirically analyzed and approximated by an exponential function. We then derived probabilities $q_A(r)$ for selecting edges to be incorporated into candidate solutions of an EA during mutation such that the average expected number of edge-selections until finding an edge of an optimal solution is minimized.

Using the degree-constrained minimum spanning tree problem, five different edge-selection strategies for mutation were described and compared. With the scheme using the approximately optimal probabilities $q_A(r)$, the EA identified optimal solutions most often and with comparatively few iterations.

On the traveling salesman problem, however, mutation that introduces one edge always introduces a second as well. While the first may be chosen according to certain probabilities, the second depends on the first and on the current tour. This side-effect overwhelms the differences between the various mutation operators. This analysis nonetheless suggests that mutation that includes edges according to probabilities derived in the proposed way might be effective in EAs for graph problems in which the introduction of one edge does not require the inclusion of others.

References

1. N. Deo and P. Micikevicius. A heuristic for a leaf constrained minimum spanning tree problem. *Congressus Numerantium*, 141:61–72, 1999.
2. C. Fonseca, J.-H. Kim, and A. Smith, editors. *Proceedings of the 2000 IEEE Congress on Evolutionary Computation.* IEEE Press, 2000.
3. J. J. Grefenstette. Incorporating problem specific knowledge into genetic algorithms. In L. Davis, editor, *Genetic Algorithms and Simulated Annealing*, pages 42–60. Morgan Kaufmann, 1987.
4. B. A. Julstrom. Very greedy crossover in a genetic algorithm for the Traveling Salesman Problem. In K. M. George, J. H. Carroll, E. Deaton, D. Oppenheim, and J. Hightower, editors, *Proceedings of the 1995 ACM Symposium on Applied Computing*, pages 324–328. ACM Press, 1995.
5. B. A. Julstrom and G. R. Raidl. Weight-biased edge-crossover in evolutionary algorithms for two graph problems. In G. Lamont, J. Carroll, H. Haddad, D. Morton, G. Papadopoulos, R. Sincovec, and A. Yfantis, editors, *Proceedings of the 16th ACM Symposium on Applied Computing*, pages 321–326. ACM Press, 2001.
6. J. Knowles and D. Corne. A new evolutionary approach to the degree constrained minimum spanning tree problem. *IEEE Transactions on Evolutionary Computation*, 4(2):125–134, 2000.
7. M. Krishnamoorthy, A. T. Ernst, and Y. M. Sharaiha. Comparison of algorithms for the degree constrained minimum spanning tree. Technical report, CSIRO Mathematical and Information Sciences, Clayton, Australia, 1999.
8. I. Ljubic and J. Kratica. A genetic algorithm for the biconnectivity augmentation problem. In Fonseca et al. [2], pages 89–96.
9. G. R. Raidl. An efficient evolutionary algorithm for the degree-constrained minimum spanning tree problem. In Fonseca et al. [2], pages 104–111.
10. G. R. Raidl and C. Drexel. A predecessor coding in an evolutionary algorithm for the capacitated minimum spanning tree problem. In C. Armstrong, editor, *Late Breaking Papers at the 2000 Genetic and Evolutionary Computation Conference*, pages 309–316, Las Vegas, NV, 2000.
11. S. Thienel. *ABACUS – A Branch-And-CUt System.* PhD thesis, University of Cologne, Cologne, Germany, 1995.

Self-adaptive Operator Scheduling Using the Religion-Based EA

René Thomsen and Thiemo Krink

EVALife Group, Dept. of Computer Science, University of Aarhus, Bldg. 540
Ny Munkegade, DK-8000 Aarhus C, Denmark
{thomsen,krink}@daimi.au.dk

Abstract. The optimal choice of the variation operators mutation and crossover and their parameters can be decisive for the performance of evolutionary algorithms (EAs). Usually the type of the operators (such as Gaussian mutation) remains the same during the entire run and the probabilistic frequency of their application is determined by a constant parameter, such as a fixed mutation rate. However, recent studies have shown that the optimal usage of a variation operator changes during the EA run. In this study, we combined the idea of self-adaptive mutation operator scheduling with the Religion-Based EA (RBEA), which is an agent model with spatially structured and variable sized subpopulations (religions). In our new model (OSRBEA), we used a selection of different operators, such that each operator type was applied within one specific subpopulation only. Our results indicate that the optimal choice of operators is problem dependent, varies during the run, and can be handled by our self-adaptive OSRBEA approach. Operator scheduling could clearly improve the performance of the already very powerful RBEA and was superior compared to a classic and other advanced EA approaches.

1 Introduction

Using evolutionary algorithms (EAs) entails the issue of setting various parameters, such as population size, mutation and crossover probabilities, as well as choosing the right variation operators. Previous studies have shown that these settings are dependent on the actual optimisation problem and the optimal settings change during the run of the EA (see e.g. [1]).

Many authors have previously tried to tackle this problem. One possibility is to use self-adaptation in which the EA parameters become a part of the genome and are evolved during the evolutionary process (e.g. [2], [3], and [4]). Another approach related to self-adaptation is to model the EA individuals as mobile agents that can decide, which operator settings they prefer by interpreting their spatial position in a n-dimensional world as operator parameter values [1]. The idea in this approach is that individuals are attracted to good parameter settings and can track and cluster around good settings if the optimal settings change during the run.

The problem of choosing the most suitable variation operators is two-fold. First, the most promising types of operators (e.g. n-point crossover, arithmetic

J.J. Merelo Guervós et al. (Eds.): PPSN VII, LNCS 2439, pp. 214–223, 2002.

crossover, etc.) have to be found that can deal with the problem. Second, the parameters of the operator (e.g. how often they are applied) are usually problem dependent and change over time. For instance, uniform or n-point crossover might be valuable in the beginning of the evolutionary process, but their ability to create useful solutions gradually diminishes. Towards the end of the EA run they often become obsolete, since there is only very little genetic variance left.

Lawrence Davis introduced a scheme that can deal with the problem of utilising the operators [5]. His idea was to adapt the operator probabilities based on their observed performance during the run, i.e. reward operators for being successful. Perhaps, the closest approach to our technique presented in this paper, is the operator scheduling with variable sized subpopulations by Schlierkamp-Voosen and Mühlenbein [6]. Their approach is to structure the population into as many subpopulations as types of operators, such that each subpopulation uses its own operator. The goal is that operators compete for control by shifting subpopulation sizes, i.e. the most successful subpopulation (having the currently best operator) gains individuals (resources) and can utilise its specific operator when needed. The size of the subpopulations is determined globally by a quality and a gain criterion and updated every n-th generation. The subpopulation model corresponds to a classic island or tagging model with no structure inside the subpopulation.

In a previous study, we introduced the so-called Religion-Based EA (RBEA) [7], partly resembling ideas from the diffusion model [8] (a two-dimensional world with interactions restricted to neighboured individuals), patchwork models (individuals as mobile multi-agents) [1], and island models (semi-separation of subpopulations; here: religions) [9, C6.3]. Our motivation for the study presented in this paper was the superior performance of the RBEA compared to other simple and spatial EA techniques and the possibility to extend the approach further by assigning specific mutation operators to the subpopulations (religions). We call our new self-adaptive method the operator scheduling RBEA (OSRBEA).

Compared to Schlierkamp-Voosen and Mühlenbein's approach, the subpopulations in the OSRBEA are spatially structured and self-organised inside and in relation to each other. Furthermore, the subpopulation sizes are a result of local competition and recruitment of individuals from different subpopulations.

2 The Operator Scheduling Religion-Based EA

Apart from the interest in human culture and religion, religious concepts and the history of their impact on human civilisations can serve as a great source of inspiration for population models. The OSRBEA was inspired along these lines regarding the emergent effects of religions on populations.

Religious rules typically include the commitments to reproduce, to believe in no other religions, and to convert non-believers, which provide the means for a propagation of the religion. New members are acquired by reproduction and religious education or by conversion. Not surprisingly, there is strong competition among different religions.

The two main inspirations from real world religions in the OSRBEA are the competition for new believers by conversion and the restriction of matings to individuals that belong to the same religion. The result of these considerations is a hybrid of an evolutionary algorithm and a spatial multi-agent system with competing subpopulations.

The OSRBEA consists of three components: a world, religions, and a population of individuals. The world is represented as a two dimensional grid (see figure 1), in which each cell has eight neighbours and the edges of the grid are wrapped to form a torus. The size of the world is determined by the *world_dimension* parameter (e.g. 25 refers to a 25×25 grid). Furthermore, a cell can contain either one or no individual. The religions define the subpopulations in which the individuals are grouped.

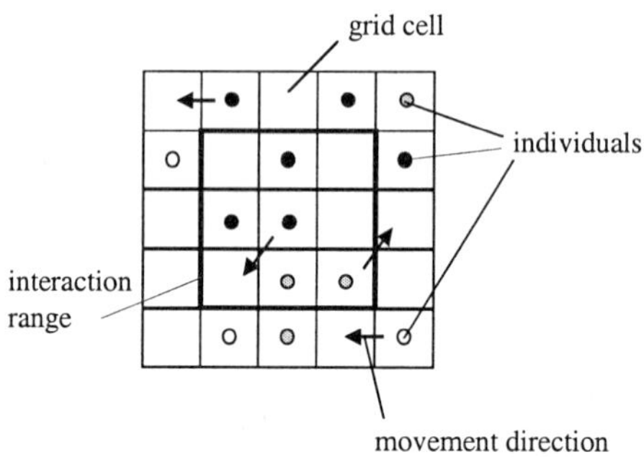

Fig. 1. Example of a 5×5 OSRBEA world. Gray-scales indicate religion memberships. The individual in the centre can only interact with other individuals within its interaction range, i.e. it can mate with either of the two black individuals to its top and left. Furthermore, it can try to convert one of the two gray individuals below and to its right-bottom.

At initialisation, all religions have the same size, which varies during the run within the bounds of the *minimum_believers* and *maximum_believers* thresholds, while the overall *population_size* remains constant. Each individual consists of the attributes *religion*, *location*, and the *genome*. The *religion* attribute defines to which religion the individual belongs and can only be changed if the individual is converted to another religion. The individual's position in the world is determined by its *location* attribute, whereas the *genome* defines its position in the search space. Furthermore, we extended the original RBEA by associating a specific mutation operator to each religion, such that individuals are mutated by the operator of their religion. In the study presented in this paper, the number of religions was three using Gaussian mutation with fixed variance, Gaussian mutation with a variance annealing scheme, and Gaussian mutation with SOC-based variance. The operators are further described in section 3. The pseudo-code for the OSRBEA algorithm is shown in figure 2.

```
procedure OSRBEA
begin
  initialise(p) //p is the EA population
  evaluate(p)
  while (not termination-condition) do
    begin
      random_walk(p)
      convert(p)
      mate(p)
      for each individual i in p do
        mutate(i) with mutation operator
         specified by individual i's religion
      evaluate(p)
    end
end
```

Fig. 2. The pseudo-algorithm of the OSRBEA.

The OSRBEA works as follows: First, all individuals are initialised and evaluated according to the objective function. Afterwards, all individuals consecutively perform the actions *random_walk*, *convert*, *mate*, *mutate*, and *evaluate*.

The *random_walk* action moves an individual to a randomly selected neighbour cell, provided that it is vacant. Further, individuals can change the religion of other believers by using the *convert* action, such that individual a converts a neighbour b of a different religion if its fitness is better. However, conversions are constrained such that an individual cannot be converted if its religion has only *minimum_believers* many members left or the religion size of the converting individual is already as large as *maximum_believers*. The purpose of these constraints is to avoid extinction/super growth of religions and thereby premature loss/dominance of mutation operators. This convert scheme is simpler than the probabilistic scheme of the original RBEA (see [7] for further details) and turned out to be superior in our preliminary experiments.

The execution of the *mate* action creates an offspring with a neighboured individual of the same religion using arithmetic crossover. If the offspring is fitter than one or both of its parents then it will substitute the parent with the worst fitness. However, if an individual belongs to a small religion with a size equal to *minimum_believers* then it can also try to mate with individuals of other religions in its neighbourhood. The *mate* action is executed after the *random_walk* and *convert* action to ensure that the offspring cannot replace its parent before the parent gets the opportunity to convert other individuals.

Finally, some of the individuals are mutated according to the mutation probability p_m using the mutation operator specified by their religion attribute and their fitness scores are re-evaluated.

The entire process is repeated for a fixed number of fitness evaluations. Note that the indices of the individuals are shuffled after each transformation to avoid artefacts from a sequential application of the *convert* and *random_walk* actions.

3 Mutation Operators

The mutation operators used in this study were Gaussian mutation operators using three different variance schemes: (i) a fixed variance scheme with a variance of one, (ii) an annealing scheme using $1/(1 + generation)$, and (iii) a SOC-based variance scheme introduced in [10]. The SOC-based variance is non-fixed and non-decreasing with $\sigma^2 = PL(2)$, where PL is the power law distribution. Power law distributed numbers can be generated by the so-called sandpile model as shown in [10] or simply by using $PL(\alpha) = 1/u^{1/\alpha}$, where $u \sim U(0,1)$ is uniformly distributed, and α is a parameter, which determines the shape of the distribution. In this study we used a fixed value of $\alpha = 2$.

4 Experiments

In order to evaluate the OSRBEA we performed several experiments on commonly used numerical benchmark problems (see section 4.3 for a short description of the problems).

The algorithms used for comparison were a simple EA with SOC-based variance mutation (SOCEA), the original RBEA with SOC-based variance mutation (SOCRBEA), and the new OSRBEA. Each algorithm was tested regarding the 20 and 50 dimensional instances of the benchmark problems presented in section 4.3. The number of fitness evaluations were 500.000 and 1.500.000 for the 20 and 50 dimensions respectively. Each of the experiments was repeated 30 times, and the average fitness of the best individual throughout the evolutionary process was recorded.

For a fair comparison between the algorithms, we tried all combinations of the following variation operators on each of the problems: one-point crossover, uniform crossover, arithmetic crossover, Gaussian mutation with fixed variance, annealed variance, and SOC-based variance. We found that arithmetic crossover with SOC-based variance mutation yielded superior results compared to other operator choices for all benchmark problems. Therefore, we only report the results of the SOCEA and the SOCRBEA for comparison, which used this combination of operators.

4.1 SOCEA Settings

We used the following parameters in the SOCEA: *population_size* = 400, mutation probability (p_m) = 0.75, crossover probability (p_c) = 0.90, and tournament selection with a tournament size of two. As mentioned in section 4 the SOCEA used Gaussian mutation with SOC-based variance and arithmetic crossover as variation operators. Further, we used elitism with an elite size of one to keep the overall best solution found in the population.

4.2 SOCRBEA and OSRBEA Settings

In both the SOCRBEA and the OSRBEA, we used the following parameters: *population_size* = 399, *world_dimension* = 22, *minimum_believers* = 50, *maxi-*

mum_believers = 299, and mutation probability (p_m) = 0.75. The p_c parameter was not used, since the crossover operation was part of the *mate* action. The SOCRBEA used Gaussian mutation with SOC-based variance, whereas the OS-RBEA used all three variants of the Gaussian mutation operator (see section 3 for details). Both algorithms used arithmetic crossover and elitism of size one was applied within each religion.

4.3 Benchmark Problems

In our experiments we used the following six numerical benchmark problems with 20 and 50 dimensions:

Ackley: $f(\overline{x}) = 20 + e - 20 \cdot exp(-0.2 \cdot \sqrt{\frac{1}{n} \cdot \sum_{i=1}^{n} x_i^2}) - exp(\frac{1}{n} \cdot \sum_{i=1}^{n} cos(2\pi x_i)$ where $-30 \leq x_i \leq 30$

Rastrigin: $f(\overline{x}) = \sum_{i=1}^{n}(x_i^2 - 10 \cdot cos(2\pi x_i) + 10)$ where $-5.12 \leq x_i \leq 5.12$

Griewank: $f(\overline{x}) = \frac{1}{4000} \cdot \sum_{i=1}^{n}(x_i - 100)^2 - \prod_{i=1}^{n} cos(\frac{x_i - 100}{\sqrt{i}}) + 1$ where $-600 \leq x_i \leq 600$

Rosenbrock: $f(\overline{x}) = \sum_{i=1}^{n-1}(100 \cdot (x_{i+1} - x_i^2)^2 + (x_i - 1)^2)$ where $-100 \leq x_i \leq 100$

Sphere function: $f(\overline{x}) = \sum_{i=1}^{n} x_i^2$ where $-100 \leq x_i \leq 100$

Step function: $f(\overline{x}) = \sum_{i=1}^{n}(\lfloor x_i + 0.5 \rfloor)^2$ where $-100 \leq x_i \leq 100$

All problems are minimisation tasks and have their global optimum at 0.0.

5 Results

To access the quality of the new operator scheduling extension we compared the OSRBEA with the SOCEA and the SOCRBEA using the six numerical benchmark problems described in section 4.3. Table 1 summarises the results of the various experiments. The mean column shows the mean best fitness value found after 30 runs using 500.000 (20D) and 1.500.000 (50D) fitness evaluations respectively and best column shows the best found solution among the 30 runs. The standard deviations were very small compared to the differences between the mean fitness values for the three algorithms (data not shown).

The OSRBEA found much better solutions for all benchmark problems compared with the SOCEA and the SOCRBEA. However, as expected, the convergence speed of the SOCRBEA and the OSRBEA turned out to be slower in the beginning of the runs compared to the SOCEA, because of the much lower gene flow in the RBEA models caused by restricted mating.

Regarding the Ackley 50D problem, the OSRBEA seems to converge slower than the SOCEA and the SOCRBEA, which is because of a few runs with inferior results indicating occasional problems with performance robustness. However, the final OSRBEA results at the end of the run are better than in all other tested algorithms.

The performance of the SOCEA was always worse compared to both the SOCRBEA and the OSRBEA. Figures 3(a) to 3(h) show the average fitness

Table 1. Results obtained from the experiments with the numerical benchmark problems.

	SOCEA		SOCRBEA		OSRBEA	
Problem	Mean	Best	Mean	Best	Mean	Best
Ackley (20D)	0.22195	0.13659	0.14065	0.09681	0.01860	0.01317
Ackley (50D)	0.38523	0.33497	0.24596	0.19433	0.12607	0.00696
Rastrigin (20D)	1.19592	0.13746	0.56267	0.05364	0.00215	0.00143
Rastrigin (50D)	3.76457	0.86996	3.06669	0.33156	0.58143	0.00086
Griewank (20D)	0.53399	0.34956	0.39838	0.21496	0.01185	0.00745
Griewank (50D)	0.89486	0.79609	0.64339	0.37256	0.00468	0.00159
Rosenbrock (20D)	54.491	45.308	36.933	27.478	19.292	15.458
Rosenbrock (50D)	241.783	189.246	146.518	116.864	48.436	41.943
Sphere (20D)	0.29622	0.18239	0.00427	0.00226	0.00440	0.00271
Sphere (50D)	2.14901	1.45932	0.00205	0.00139	0.00201	0.00136
Step (20D)	0.0	0.0	0.0	0.0	0.0	0.0
Step (50D)	0.3	0.0	0.0	0.0	0.0	0.0

curves of the best individuals in the three algorithms for four of the six benchmark problems, which illustrate these characteristics. In case of the Sphere and Step benchmark problems (not shown in figure 3) the overall performance and convergence speed of the SOCRBEA and OSRBEA were very similar.

During the evolutionary process we observed a shifting balance of power (number of believers in each religion) between the religions, thus changing the frequency of applying specific mutation operators. Typically, the course-grained operators, such as the Gaussian mutation with fixed and SOC-based variance were favoured in the beginning of the run, whereas the more fine-grained mutation with annealed variance took over towards the end, fine-tuning the candidate solutions. Figure 4(a) and 4(b) show examples of this shifting-balance between the operators during the execution of the OSRBEA with the Griewank (50D) and the Rastrigin (50D) problems.

6 Discussion

In this paper we have introduced a new spatial operator scheduling technique called the OSRBEA, which extends our original Religion-Based EA (RBEA). The algorithm is a hybrid between an EA and a spatial multi-agent system with competing subpopulations (religions), where each subpopulation is associated with a specific mutation operator.

The results of our experiments show that this extension can improve the performance significantly. The OSRBEA was capable of obtaining good solutions on all six benchmark problems and its convergence speed towards the end of the run was remarkably fast compared to the SOCEA and the SOCRBEA. Note that the latter two algorithms were already superior compared to a simple EA and the original RBEA regarding all six benchmark problems.

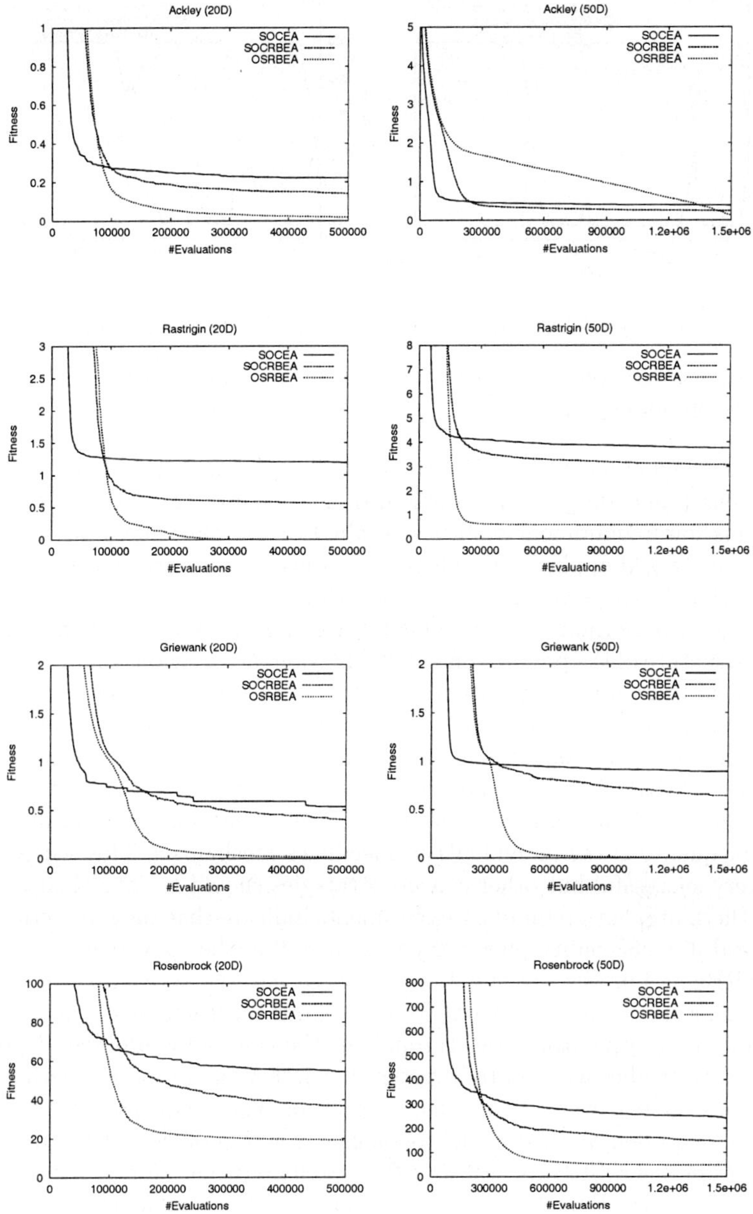

Fig. 3. Performance comparison between the SOCEA, SOCRBEA, and OSRBEA. The graphs show the average fitness of the best individual in 30 runs versus the number of fitness evaluations.

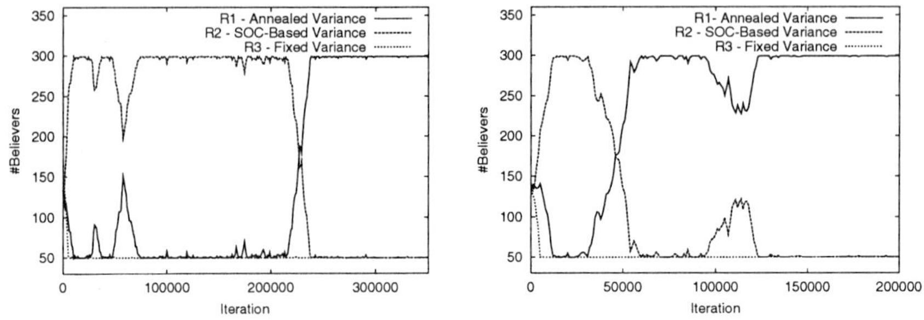

(a) Griewank (50D) (b) Rastrigin (50D)

Fig. 4. Religion size (number of believers) dynamics for each religion R1, R2, and R3 during the optimisation process.

In most experiments the OSRBEA used less than 300.000-600.000 fitness evaluations (depending on the problem dimensionality) to obtain a good solution. This low requirement of fitness evaluations is particulary interesting regarding real world applications where fitness evaluations are typically the most computational expensive part of the algorithm.

The good performance of the OSRBEA can be partially contributed to the spatial structure of the population and its division into religions (subpopulations) with the additional mating restriction reducing the flow of genetic information, thus keeping the genetic variance in the population high (see [11] for more details). Moreover, the assignment of special operators to each of the religions allows the OSRBEA to take advantage of the most suitable operator. In this respect the scheduling of the operators was controlled in a self-adaptive manner using the quality of the individuals as a feedback mechanism. This scheme seems to be very successful but other convert strategies should be investigated.

Furthermore, our preliminary experiments indicate that the interaction range and world dimensionality play a key role regarding the convergence abilities of the OSRBEA. Further investigations of these parameters could lead to additional performance improvements. Moreover, our results indicate that the optimal choice of an operator varies throughout the run and is problem dependent.

In future studies we plan to investigate whether the proposed method is advantageous on real world problems, such as multiple sequence alignment and evolution of hidden Markov model topologies. These problems would most likely benefit from the OSRBEA utilising the specialised operators during the evolutionary process. Finally, it would be interesting to compare the OSRBEA with other operator scheduling techniques, such as the competing subpopulations scheme [6] and the operator reward strategy by Davis [5].

7 Acknowledgements

The authors would like to thank the colleagues at EVALife for valuable comments on early versions of the manuscript. This work was supported by the Danish Natural Science Research Council.

References

1. Krink, T. and Ursem, R.K.: Parameter Control Using the Agent Based Patchwork Model. In: Proceedings of the Second Congress on Evolutionary Computation (CEC-2000), Vol. 1. San Diego, CA, USA (2000) 77–83
2. Bäck, T.: Self-Adaptation in Genetic Algorithms. In: F. J. Varela, P.B. (ed.): Proceedings of 1st European Conference on Artificial Life. MIT Press (1992) 263–271
3. Eiben, A.E., Sprinkhuizen-Kuyper, I.G., and Thijssen, B.A.: Competing Crossovers in an Adaptive GA Framework. In: Proceedings of the 1998 IEEE International Conference on Evolutionary Computation (1998) 787–792
4. Smith, J.E. and Fogarty, T.C.: Self Adaptation of Mutation Rates in a Steady State Genetic Algorithm. In: Proceedings of the 1996 IEEE International Conference on Evolutionary Computing, IEEE Press (1996) 318–323
5. Davis, L.: Adapting Operator Probabilities in Genetic Algorithms. In: Schaffer, J.D. (ed.): Proceedings of the Third International Conference on Genetic Algorithms (ICGA III) (1989) 61–69
6. Schlierkamp-Voosen, D. and Mühlenbein, H.: Strategy Adaptation by Competing Subpopulations. In: Davidor, Y., Schwefel, H.P., and Maenner, R. (eds.): Proceedings of the Third International Conference on Parallel Problem Solving from Nature (PPSN III) (1994) 199–208
7. Thomsen, R., Rickers, P., and Krink, T.: A Religion-Based Spatial Model For Evolutionary Algorithms. In: Schoenauer, M., Deb, K., Rudolph, G., Yao, X., Lutton, E., Merelo, J.J., and Schwefel, H.P. (eds.): Parallel Problem Solving from Nature (PPSN VI) (2000) 817–826
8. Whitley, D.: Cellular Genetic Algorithms. In: Forrest, S. (ed.): Proceedings of the Fifth International Conference on Genetic Algorithms, Morgan Kaufman (1993) 658
9. Bäck, T., Fogel, D.B., Michalewicz, Z., et al. (eds.).: Handbook on Evolutionary Computation. IOP Publishing Ltd and Oxford University Press (1997).
10. Krink, T., Rickers, P., and Thomsen, R.: Applying Self-Organised Criticality to Evolutionary Algorithms. In: Schoenauer, M., Deb, K., Rudolph, G., Yao, X., Lutton, E., Merelo, J.J., and Schwefel, H.P. (eds.): Parallel Problem Solving from Nature (PPSN VI) (2000) 375–384
11. Thomsen, R. and Rickers, P. Introducing Spatial Agent-Based Models and Self-Organised Criticality to Evolutionary Algorithms. Master's thesis, University of Aarhus, Denmark (2000)

Probabilistic Model-Building Genetic Algorithms in Permutation Representation Domain Using Edge Histogram

Shigeyoshi Tsutsui

Department of Management and Information Science, Hannan University
5-4-33 Amamihigashi, Matsubara, Osaka 580-5802 Japan
tsutsui@hannan-u.ac.jp,
http://www.hannan-u.ac.jp/~tsutsui/

Abstract. Recently, there has been a growing interest in developing evolutionary algorithms based on probabilistic modeling. In this scheme, the offspring population is generated according to the estimated probability density model of the parent instead of using recombination and mutation operators. In this paper, we have proposed probabilistic model-building genetic algorithms (PMBGAs) in permutation representation domain using *edge histogram based sampling algorithms (EHBSAs)*. Two types of sampling algorithms, without template (EHBSA/WO) and with template (EHBSA/WT), are presented. The results were tested in the TSP and showed EHBSA/WT worked fairly well with a small population size in the test problems used. It also worked better than well-known traditional two-parent recombination operators.

1 Introduction

Recently, there has been a growing interest in developing evolutionary algorithms based on probabilistic models [Pelikan 99b], [Larranaga 02]. In this scheme, the offspring population is generated according to the estimated probabilistic model of the parent population instead of using traditional recombination and mutation operators. The model is expected to reflect the problem structure, and as a result it is expected that this approach provides more effective mixing capability than recombination operators in traditional GAs. These algorithms are called *probabilistic model-building genetic algorithms (PMBGAs)* or *estimation of distribution algorithms (EDAs)*. In a PMBGA, better individuals are selected from an initially randomly generated population like in standard GAs. Then, the probability distribution of the selected set of individuals is estimated and new individuals are generated according to this estimate, forming candidate solutions for the next generation. The process is repeated until the termination conditions are satisfied.

Many studies on PMBGAs have been performed in discrete (mainly binary) domain and there are several attempts to apply PMBGAs in continuous domain. However, a few studies on PMBGAs in permutation representation domain are found. In this paper, we propose an approach of PMBGAs in permutation representation domain, and compare its performance with traditional recombination operators. In this approach, we develop an *edge histogram matrix* from the current population, where an edge is a link between two nodes in a string. We then sample nodes of a new string according to the edge histogram matrix. We will call this method the *edge histogram based sampling algorithm* (EHBSA). We tested the algorithm in the Traveling Sales-

J.J. Merelo Guervós et al. (Eds.): PPSN VII, LNCS 2439, pp. 224–233, 2002.

man Problem (TSP), a typical, well-known optimization problem in permutation representation domain. The results showed EHBSA worked fairly well on the test problems used. Section 2 of this paper gives a brief overview of PMBGAs. In Section 3, the two proposed EHBSAs are described. The empirical analysis is given in Section 4. Section 5 concludes the paper.

2 A Brief Overview of PMBGAs

According to [Pelikan 99b], PMBGAs in binary string representation can be classified into three classes depending on the complexity of models they use; (1) no interactions, (2) pairwise interactions, and (3) multivariate interactions. In models with no interactions, variables are treated independently. Algorithms in this class work well on problems which have no interactions among variables. These algorithms include the PBIL [Baluja 94], cGA [Harik 98], and UMDA [Mhlenbein 96] algorithms. In pairwise interactions, some pairwise interactions among variables are considered. These algorithms include the MIMIC algorithm [De Bonet 97], the algorithm using dependency trees [Baluja 97]. In models with multivariate interactions, algorithms use models that can cover multivariate interactions. Although the algorithms require increased computational time, they work well on problems which have complex interactions among variables. These algorithms include ECGA [Harik 99] and BOA [Pelikan 99a, 00].

Studies to apply PMBGAs in continuous domains have also been made. These include continuous PBIL with Gaussian distribution [Sebag 98] and a real-coded variant of PBIL with iterative interval updating [Servet 97]. In [Gallagher 99], the PBIL is extended by using a finite adaptive Gaussian mixture model density estimator. The UMDA and MIMIC were introduced in continuous domain. All above algorithms do not cover any interactions among the variables. In EGNA [Larranaga 99], a Gaussian network learns to estimate a multivariate Gaussian distribution of the parent population. In [Bosman 99], two density estimation models, i.e., the normal distribution, and the histogram distribution are discussed. These models are intended to cover multivariate interaction among variables. In [Bosman 00a], it is reported that the normal distribution models have shown good performance. In [Bosman 00b], a normal mixture model combined with a clustering technique is introduced to deal with non-linear interactions. In [Tsutsui 01a, b], an evolutionary algorithm using marginal histogram models in continuous domain was proposed.

A study on PMBGAs in permutation domain is found in [Robles 02]. In it, PMBGAs are applied to solving TSP using two approaches. One is to use discrete PMBGAs and the other is to use continuous PMBGAs. In applying discrete PMBGAs, several different Bayesian network structures are compared. In applying continuous PMBGAs, the correct tour is obtained by sorting the vectors of real numbers. PMBGAs are also applied to solve job shop scheduling problems and graph matching problems [Larranaga 02].

3 Edge Histogram Based Sampling Algorithm (EHBSA)

This section describes how the edge histogram based sampling algorithm (EHBSA) can be used to (1) model promising solutions and (2) generate new solutions by simulating the learned model.

3.1 The Basic Description of the Algorithm

An *edge* is a link or connection between two nodes and has important information about the permutation string. Some crossover operators, such as Edge Recombination (ER) [Whitley 89] and enhanced ER (eER) [Starkweather, 91] which are used in traditional two-parent recombination, use the edge distribution only in the two parents string. The basic idea of the edge histogram based sampling algorithm (EHBSA) is to use the edge histogram of the whole population in generating new strings.

The algorithm starts by generating a random permutation string for each individual population of candidate solutions. Promising solutions are then selected using any popular selection scheme. An *edge histogram matrix (EHM)* for the selected solutions is constructed and new solutions are generated by sampling based on the edge histogram marix. New solutions replace some of the old ones and the process is repeated until the termination criteria are met. This algorithm can be seen as a permutation version of the algorithm which uses marginal histogram models proposed in [Tsutsui 01a, b].

3.2 Developing Edge Histogram Matrix

Let string of kth individual in population $P(t)$ at generation t represent as $s^t_k = (\pi^t_k(0), \pi^t_k(1), ..., \pi^t_k(L\text{-}1))$. $(\pi^t_k(0), \pi^t_k(1), ...,$ and $\pi^t_k(L\text{-}1))$ are the permutation of $(0, 1, ..., L\text{-}1)$, where L is the length of the permutation. Edge histogram matrix $EHM^t\ (e^t_{i,j})\ (i, j = 0,1, .., L\text{-}1)$ of population $P(t)$ is symmetrical and consists of L^2 elements as follows:

$$e^t_{i,j} = \begin{cases} \sum_{k=1}^{N} (\delta_{i,j}(s^t_k) + \delta_{j,i}(s^t_k)) + \varepsilon & \text{if } i \neq j \\ 0 & \text{if } i = j \end{cases} \tag{1}$$

where N is the population size, $\delta_{ij}(s^t_k)$ is a delta function defined as

$$\delta_{i,j}(s^t_k) = \begin{cases} 1 & \text{if } \exists h\, [h \in \{0,1,\cdots L-1\} \wedge \pi^t_k(h) = i \wedge \pi^t_k((h+1) \bmod L) = j] \\ 0 & \text{othersise} \end{cases} \tag{2}$$

and ε (ε>0) is a bias to control *pressure* in sampling nodes just like those used for adjusting the selection pressure in the proportional selection in GAs. The average number of edges of element $e^t_{i,j}$ in EHM^t is $2LN/(L^2\text{-}L) = 2N/(L-1)$. So, ε is determined by a bias ratio B_{ratio} ($B_{ratio} > 0$) of this average number of edges as

$$\varepsilon = \frac{2N}{L-1} B_{\text{ratio}} \tag{3}$$

A smaller value of B_{ratio} reflects the real distribution of edges in sampling of nodes and a bigger value of B_{ratio} will give a kind of perturbation in the sampling (see Section 3.3). An example of EHM^t is shown in Fig. 1.

$$\begin{aligned} s^t_1 &= (0,\ 1,\ 2,\ 3,\ 4) \\ s^t_2 &= (1,\ 3,\ 4,\ 2,\ 0) \\ s^t_3 &= (3,\ 4,\ 2,\ 1,\ 0) \\ s^t_4 &= (4,\ 0,\ 3,\ 1,\ 2) \\ s^t_5 &= (2,\ 1,\ 3,\ 4,\ 0) \end{aligned}$$

(a) $P(t)$

$$\begin{pmatrix} 0 & 3.1 & 2.1 & 2.1 & 3.1 \\ 3.1 & 0 & 4.1 & 3.1 & 0.1 \\ 2.1 & 4.1 & 0 & 1.1 & 3.1 \\ 2.1 & 3.1 & 1.1 & 0 & 4.1 \\ 3.1 & 0.1 & 3.1 & 4.1 & 0 \end{pmatrix}$$

(b) EHM^t

Fig. 1. An example of symmetric edge histogram matrix for $N = 5$, $L = 5$, $B_{ratio} = 0.04$

Although we defined a symmetric EHM^t, i.e., $e_{i,j} = e_{j,i}$, which is applicable to problems such as a symmetric TSP, but here we must note that we need to define an asymmetric EHM^t for problems such as a asymmetric TSP or scheduling problems with permutation representation. An asymmetric EHM^t can be easily defined by a equation similar to Eq. 1.

3.3 Sampling Methods

In this subsection, we describe how to sample a new string from the edge histogram matrix EHM^t. We propose two types of sampling methods; one is an *edge histogram based sampling algorithm without template (EHBSA/WO)*, and the other an *edge histogram based sampling algorithm with template (EHBSA/WT)*.

3.3.1 Edge Histogram Based Sampling Algorithm without Template (EHBSA/WO)

In EHBSA/WO, a new individual permutation $c[]$ is generated straightforwardly as follows:

1. Set the position counter $p \leftarrow 0$.
2. Obtain first node $c[0]$ randomly from $[0, L-1]$.
3. Construct a roulette wheel vector $rw[]$ from EHM^t as $rw[j] \leftarrow e^t_{c[p],j}$ $(j=0, 1, .., L-1)$.
4. Set to 0 previously sampled nodes in $rw[]$ ($rw[\ c[i]\ ] \leftarrow 0$ for $i=0, .., p$).
5. Sample next node $c[p+1]$ with probability $rw[x]/\sum_{j=0}^{L-1} rw[j]$ using roulette wheel $rw[]$.
6. Update the position counter $p \leftarrow p+1$.
7. If $p<L-1$, go to Step 3.
8. Obtain a new individual string $c[]$.

Here, note that the EHBSA/WO is only applicable to problems where the absolute position of each node in a string has no meaning, such as in the TSP. This sampling method is similar in part to the sampling in Ant Colony Optimization [Dorigo 96].

3.3.2 Edge Histogram Based Sampling Algorithm with Template (EHBSA/WT)

EHM^t described in Section 3.2 is in a marginal edge histogram. It has no explicit graphical structure. EHBSA/WT is intended to make up for this disadvantage by using a template in sampling a new string. In generating each new individual, a *template individual* is chosen from $P(t)$ (normally, randomly). The n ($n > 1$) cut points are applied to the template randomly. When n cut points are obtained for the template, the template should be divided into n segments. Then, we choose one segment randomly and sample nodes for the segment. Nodes in other n–1 segments remain unchanged. We denote this sampling method by EHBSA/WT/n. Since average length of one segment is L/n, EHBSA/WT/n generates new strings which are different L/n nodes on average from their templates. Fig. 2 shows an example of EHBSA/WT/3. In this example, nodes of new string from after cut[2] and before cut[1] are the same as the nodes of the template. New nodes are sampled from cut[1] up to, but not including, cut[2] based on the EHM^t.

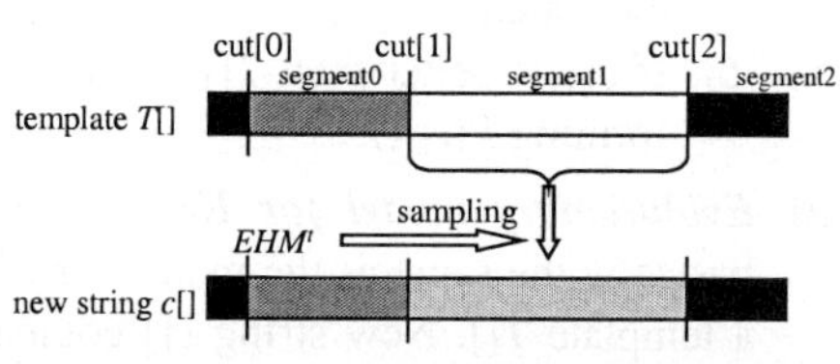

Fig. 2. An example of EHBSA/WT/3

The sampling method for EHBSA/WT/n is basically the same as that of the EHBSA/WO as follows:

1. Choose a template $T[]$ from $P(t)$.
2. Obtain sorted cut point array cut[0], cut[1], .., cut[n–1] randomly.
3. Choose a cut point cut[l] by generating random number $l \in [0, n-1]$.
4. Copy nodes in $T[]$ to $c[]$ from after cut[(l+1) mod n] and before cut[l].
5. Set the position counter $p \leftarrow$ (cut[l] – 1 + L) mod L.
6. Construct a roulette wheel vector $rw[]$ from EHM^t as $rw[j] \leftarrow e^t_{c[p],j}$ (j=0, 1, .., L–1).
7. Set to 0 copied and previously sampled nodes in $rw[]$ ($rw[c[i]] \leftarrow 0$ for i = cut[(l+1) mod n], .., p).
8. Sample next node $c[(p+1) \bmod L]$ with probability $rw[x]/\sum_{j=0}^{L-1} rw[j]$ using roulette wheel $rw[]$.
9. Update the position counter $p \leftarrow (p+1) \bmod L$.
10. If $(p+1) \bmod L \neq$ cut[(l+1) mod n], go to Step 6.
11. Obtain a new individual string $c[]$.

4 Empirical Study

4.1 Experimental Methodology

4.1.1 Evolutionary Models

Here, we describe evolutionary models for EHBSA/WT, EHBSA/WO, and two-parent recombination operators, respectively. All these models are basically the same as *steady state models*.

(1) Evolutionary model for EHBSA/WT:

Let the population size be N, and let it, at time t, be represented by $P(t)$. The population $P(t+1)$ is produced as follows (Fig. 3):

1. Edge histogram matrix EHM^t described in Subsection 3.2 is developed from $P(t)$
2. A template individual $T[]$ is selected from $P(t)$ randomly.
3. EHBSA/WT described in Subsection 3.3.2 is performed using EHM^t and $T[]$, and generate a new individual $c[]$.
4. The new $c[]$ individual is evaluated.
5. If $c[]$ is better than $T[]$, then $T[]$ is replaced with $c[]$, otherwise $T[]$ remains, forming $P(t+1)$.

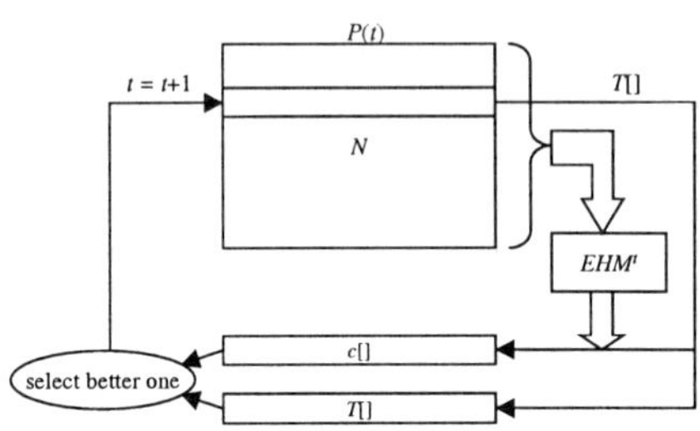

Fig. 3. Evolutionary model for EHBSA/WT

(2) Evolutionary model for EHBSA/WO: Evolutionary model for EHBSA/WO is basically the same as the model for EHBSA/WT except EHBSA/WO does not use a template $T[]$. New string $c[]$ compares randomly selected individual $i[]$ in $P(t)$, and *if* $c[]$ is better than $i[]$, $i[]$ is replaced with $c[]$.

(3) ***Evolutionary model for two-parent recombination operators:*** To compare the performance of proposed methods with the performance of traditional two-parent recombination operators, we designed an evolutionary model for two-parent recombination operators. For fair comparison, we design it as similar as possible to that of the EHBSA. We generate only one child from two parents. Using one child from two parents is already proposed for designing the GENITOR algorithm by Whitley et al. [Whitley 89]. In our generational model, two parents are selected from *P*(*t*) randomly. No bias is used in this selection. Then we apply a recombination operator to produce one child. This child is compared with its parents. If the child is better than the worst parent, then the parent is replaced with the child.

4.1.2 Test Suit and Performance Measures

We tested the algorithm in the Traveling Salesman Problem (TSP), a typical, well-known optimization problem in permutation representation domain. The following well-known data files have been used in this empirical study: 24 cities gr24, 48 cities gr48, and 76 cities pr76. The gr24 and gr48 are used in the study of TSP with EDA in [Robles 02]. We compared EHBSA with popular order based recombination operators, namely, the original order crossover OX [Oliver 87], the enhanced edge recombination operator eER [Starkweather 91], and the partially mapped crossover [Goldberg 89]. We also tried to compare EHBSA with results in [Robles 02] on gr24 and gr48.

Ten runs were performed. Each run continued until the optimum tour was found, the population was converged, or evaluations reached E_{max} Values of E_{max} were 50000, 500000, and 1000000 for gr24, gr48, and pr76, respectively. Population sizes of 60, 120, 240 were used for EHBSA, and 60, 120, 240, 480,960 for other operators, respectively. As to the bias ratio B_{ratio} in Eq. 3, B_{ratio} values of 0.03, 0.015, and 0.005 were used for gr24, gr48, and pr76, respectively.

We evaluated the algorithms by measuring their *#OPT* (number of runs in which the algorithm succeeded in finding the optimum tour), *ANE* (average number of evaluations to find the global optimum in those runs where it did find the optimum), and *Aver* (average length of best solution in each run). Here, a lower value of ANE means a more effective search capability of an algorithm.

4.1.3 Blind Search

In solving TSP using GAs, mutation operators play an important role. Several types of mutation operators are proposed. Also, it is well known that combining GAs with local optimization methods or heuristics greatly improve the performance of the algorithms. Many kinds of heuristics for TSP are proposed [Johnson 02]. For example, in [Ulder 90], Ulder et al. combined GAs with 2-opt heuristics and the algorithm showed greatly improved performance. In [Nagata 97], Nagata et al. proposed a high-power crossover operator for TSP which includes a kind of heuristics in the operator.

In this experiment, we use no mutation and no heuristic to see the pure effect of applying proposed algorithms. Thus, the algorithm is a *blind search.*

4.2 Empirical Analysis of Results

Results in gr24 are shown in Table 1. EHBSA/WO found the optimum tour 7, 9, and 6 times with N = 60, 120, and 240, respectively. On the other hand, EHBSA/WT/*n* found the optimum tour 10 times for all experiments. The ANEs of EHBSA/WT/2 and EHBSA/WT/3 were 9141, and 9523, respectively, showing good performance. Thus, we can see the performance of EHBSA/WT is much better than EHBSA/WO. In the other operators, eER showed good performance. The eER with N = 240, 480, and 960 found the optimum tour 10 times and the ANE for N = 240 was 13394, which is a little larger than EHBSA/WT/n with N = 60. OX showed worse performance than eER although PMX showed the worst performance. Comparing the performance of EHBSA/WT with other operators, EHBSA/WT is slightly better than eER and is much better than OX and PMX. One big difference between EHBSA/WT and eER is that EHBSA/WT requires a smaller population size to work than eER. To compare EHBSA with results in [Robles 02] we see the results with discrete representation (here referred to as *discrete EDA*). In [Robles 02], it is shown that the discrete EDA without local optimization does not find the optimum tour.

Table 1. Results of gr24

Model	Population Size N														
	60			120			240			480			960		
	#OPT	ANE	Aver	#OPT	ANE	Aver	#OPT	ANE	Aver	#OPT	ANE	Aver	#OPT	ANE	Aver
EHBSA/WO	7	16328	1281	9	23637	1273	6	44853	1280						
EHBSA/WT/2	10	**9141**	1272	10	17978	1272	10	35604	1272						
EHBSA/WT/3	10	**9523**	1272	10	18251	1272	10	32956	1272						
EHBSA/WT/4	10	**10677**	1272	10	17652	1272	10	33606	1272						
EHBSA/WT/5	10	**11170**	1272	10	20489	1272	10	36078	1272						
OX	0	-	1345	1	22449	1303	4	34140	1296	1	48674	1301	0	-	1484
eER	1	4738	1299	7	6237	1276	10	**13394**	1272	10	23785	1272	10	42767	1272
PMX	0	-	1492	0	-	1414	2	23191	1341	1	49442	1316	0	-	1572
Other PMBGA*	#OPT=0, ANE is not available, best length = 1328 with MIMIC, best Aver = 1439 with EBNA														

Optimum: 1272

E_{max} = 50000, B_{ratio} = 0.03

* Best data without heuristic using discrete EDA in [Robles 02] . Maximum evaluation is 50000

Results in gr48 are shown in Table 2. EHBSA/WO could not find the optimum tour in gr48. On the other hand, EHBSA/WT/n again found the optimum tour 10 times for all experiments except EHBSA/WT/4 and EHBSA/WT/2 with N = 60. The ANEs for EHBSA/WT/3 and EHBSA/WT/5 were 85387 and 89799, respectively, showing good performance. Thus, we can see again the performance of EHBSA/WT is much better than EHBSA/WO. In the other operators, eER showed weaker performance than EHBSA/WT/n in gr48, but better performance than OX. The best #OPT of eER is 5 with N = 960 and the ANE of this case is 166286, much larger than EHBSA/WT/n. PMX could not find the optimum tour. Comparing the performance of EHBSA/WT with discrete EDA in [Robles 02] is impossible because both termination conditions are different in gr48.

Results in pr76 are shown in Table 3. EHBSA/WO could not find the optimum tour in pr76. On the other hand, EHBSA/WT/n found the optimum tour several times. With N = 60, EHBSA/WT/2, 3, 4, and 5 found the optimum tour 4, 4, 9, and 10 times, respectively. With N = 120, EHBSA/WT/2, 3, 4, and 5 found the optimum tour 9, 9, 9, and 10 times, respectively, showing the best performance. Thus, we can see the performance of EHBSA/WT is much better than the performance of EHBSA/WO in this experiment, too. In the other operators, eER found the optimum tour only 1 time with $N = 480$ and 3 times with $N = 960$, showing worse performance than EHBSA/WT. OX and PMX could not find the optimum tour.

Table 2. Results of gr48

Model	Population Size N														
	60			120			240			480			960		
	#OPT	ANE	Aver	#OPT	ANE	Aver	#OPT	ANE	Aver	#OPT	ANE	Aver	#OPT	ANE	Aver
EHBSA/WO	0	-	5212	0	-	5316	0	-	5773						
EHBSA/WT/2	6	102691	5053	10	174125	5046	10	299391	5046						
EHBSA/WT/3	10	**85387**	5046	10	134597	5046	10	240391	5046						
EHBSA/WT/4	9	82701	5047	10	126444	5046	10	237260	5046						
EHBSA/WT/5	10	**89799**	5046	10	157041	5046	10	257486	5046						
OX	0	-	5527	0	-	5268	0	-	5200	1	162154	5099	2	287852	5082
eER	0	-	5653	0	-	5233	0	-	5098	2	95075	5072	5	166286	5058
PMX	0	-	8285	0	-	7374	0	-	6859	0	-	6116	0	-	5860
Other PMBGA*	#OPT=0, ANE is not available, best length = 6104 with MIMIC, best Aver = 6717 with MIMIC														

Optimum: 5046

$E_{max} = 500000$, $B_{ratio} = 0.015$

* Best data without heuristic using discrete EDA in [Robles 02] . Maximum evaluation was set to 50000

Table 3. Results of pr76

Model	Population Size N														
	60			120			240			480			960		
	#OPT	ANE	Aver	#OPT	ANE	Aver	#OPT	ANE	Aver	#OPT	ANE	Aver	#OPT	ANE	Aver
EHBSA/WO	0	-	119136	0	-	128208	0	-	142206						
EHBSA/WT/2	4	360128	108352	9	457147	108174	7	871319	108201						
EHBSA/WT/3	4	248091	108385	9	472719	108171	8	853801	108201						
EHBSA/WT/4	9	341482	108247	9	607544	108247	0	-	108496						
EHBSA/WT/5	10	**494674**	108159	10	797963	108159	0	-	108807						
OX	0	-	129603	0	-	121642	0	-	116591	0	-	113412	0	-	112259
eER	0	-	142003	0	-	122217	0	-	111839	1		109119	3	394887	108507
PMX	0	-	236827	0	-	213528	0	-	187601	0	-	164883	0	-	158515
Other PMBGA	not available														

Optimum: 108159

$E_{max} = 1000000$, $B_{ratio} = 0.005$

From the results described above, we can see that EHBSA/WT/n worked fairly well in the test problems used. It also worked better than popular traditional two-parent recombination operators. In EHBSAs, the population size appears to be a crucial parameter as with traditional GAs. But one interesting feature of EHBSA/WT/n is that it requires smaller population size than traditional two parent recombination operators. This may be an important property of EHBSA/WT/n. In our experiments, we used a blind search. When we combine EHBSA/WT/n with some heuristics, we may expect that it works well with a smaller population size. As to the number of cut points, a smaller number of n work well with problems with smaller numbers of cities, and a larger number of n work well with problems with larger numbers of cities; i.e., gr24: $n = 2$, gr48: $n = 3$, and pr76: $n = 5$.

5 Conclusions

In this paper, we have proposed probabilistic model-building genetic algorithms (PMBGAs) in permutation representation domain using the Traveling Salesman Problem (TSP), a typical, well-known optimization problem in permutation representation domain and compare its performance with traditional recombination operators. In this approach, we developed an edge histogram model from the current population. Two types of sampling algorithms, EHBSA/WO and EHBSA/WT, were presented. The results showed EHBSA/WT worked fairly well with a smaller size of population on the test problems used. It also worked better than well-known traditional two parent recombination operators.

There are many opportunities for further research related to the proposed algorithms. The effect of parameter values of B_{ratio}, number of cut point of the template n, and size of population N, on the performance of the algorithm must be further investigated. We experimented with EHBSAs using a blind search to test the pure mixing capability of the proposed algorithms. But we must test the algorithms with appropriate heuristics in problems with large numbers of cities. Analyzing the time complexity of the algorithm, and applying EHBSAs to other permutation problems, such as job shop scheduling problems, also remain for future work.

Acknowledgments

The authors gratefully acknowledge Prof. David E. Goldberg and Dr. Martin Pelikan for their valuable comments on PMBGAs during my stay at IlliGAL in 2001.

This research is partially supported by the Ministry of Education, Culture, Sports, Science and Technology of Japan under Grant-in-Aid for Scientific Research number 13680469, and a grant to RCAST at Doshisha University from Ministry of Education, Culture, Sports, Science and Technology of Japan.

References

[Baluja 94] Baluja, S.: Population-based incremental learning: A method for interacting genetic search based function optimization and coemptive learning, *Tech. Rep. No. CMU-CS-94-163*, Carnegie Mellon University (1994).

[Baluja 97] Baluja, S. and Davies: Using optimum dependency-trees for combinatorial optimization: learning the structure of the search space, *Tech. Rep. No. CMU-CS-97-107*, Carnegie Mellon University (1997)

[Bosman 99] Bosman, P. and Thierens, D.: An algorithmic framework for density estimation based evolutionary algorithms, *Tech. Rep. No. UU-CS-1999-46*, Utrecht University (1999).

[Bosman 00a] Bosman, P. and Thierens, D.: Continuous iterated density estimation evolutionary algorithms within the IDEA framework, *Proc. of the Optimization by Building and Using Probabilistic Models OBUPM Workshop at the Genetic and Evolutionary Computation Conference GECCO-2000*, pp.197-200 (2000).

[Bosman 00b] Bosman, P. and Thierens, D.: Mixed IDEAs, *Tech. Rep. No. UU-CS-2000-45*, Utrecht University (2000).

[De Bonet 97] De Bonet, J. S., Isbell, C. L. and Viola, P.: MIMIC: Finding optima by estimating probability densities, In Mozer, M. C., Jordan, M. I., and Petsche, T. (Eds): *Advances in neural information processing systems*, Vol. 9, pp. 424-431. (1997).

[Goldberg 89] Goldberg, D. E.: *Genetic algorithms in search, optimization and machine learning*, Addison-Wesley publishing company (1989).

[Harik 98] Harik, G., Lobo, F. G., and Goldberg, D. E.: The compact genetic algorithm, *Proc. of the Int. Conf. Evolutionary Computation 1998 (ICEC 98)*, pp. 523-528 (1998).

[Harik 99] Harik, G: Linkage learning via probabilistic modeling in the ECGA, *Tecnical Report IlliGALReport 99010*, University of Illinois at Urbana-Champaign, Urbana, Illinois (1999).

[Larranaga 99] Larranaga, P., Etxeberria, R., Lozano, J.A., and Pena, J.M.: Optimization by learning and simulation of Bayesian and gaussian networks, *University of the Basque Country Technical Report EHU-KZAAIK -4/99* (1999).

[Dorigo 96] Dorigo M., Maniezzo, V. and Colorni, A.: The Ant System: Optimization by a Colony of Cooperating Agents, *IEEE Trans. on Systems, Man, and Cybernetics-Part B*, Vol. 26, No. 1, pp. 29-41 (1996).

[Mhlenbein 96] Mhlenbein, H and Paa, G.: From recombination of genes to the estimation of distribution I. Binary parameters, *Proc. of the Parallel Problem Solving from Nature - PPSN IV*, pp. 178-187 (1996).

[Pelikan 99a] Pelikan, M., Goldberg, D. E., and Cantu-Paz, E.: BOA: The Bayesian optimization algorithm, *Proc. of the Genetic and Evolutionary Computation Conference 1999 (GECCO-99)*, Morgan Kaufmann, San Francisco, CA (1999).

[Pelikan 99b] Pelikan, M., Goldberg, D. E., and Lobo, F. G. : A survey of optimization by building and using probabilistic models, *Technical Report IlliGAL Report 99018*, University of Illinois at Urbana-Champaign (1999).

[Pelikan 00] Pelikan, M., Goldberg, D. E., and Cantu-Paz, E.: Linkage problems, distribution estimate, and Bayesian network, *Evolutionary Computation*, Vol. 8, No. 3, pp. 311-340 (2000).

[Sebag 98] Sebag, M. and Ducoulombier, A.: Extending population-based incremental learning to continuous search spaces, *Proc. of the Parallel Problem Solving from Nature - PPSN V*, pp. 418-427 (1998).

[Servet 97] Servet, I. L., Trave-Massuyes, L., and Stern, D.: Telephone network traffic overloading diagnosis and evolutionary computation techniques, *Proc. of the Third European Conference on Artificial Evolution (AE 97)*, pp. 137-144 (1997).

[Larranaga 00] Larranaga, P., Etxeberria, R., Lozano, J. A., and Pena, J. M.: Optimization in continuous domains by learning and simulation of Gaussian networks, *Proc. of the 2000 Genetic and Evolutionary Computation Conference Workshop Program*, pp. 201-204 (2000).

[Robles 02] Robles, V., Miguel, P. D., and Larranaga, P.: Solving the traveling salesman problem with EDAs, *Estimation of Distribution Algorithms*, Larranaga, P. and Lozano, J. A. (eds), Kluwer Academic Publishers, Chapter 10, pp. 211-229 (2002).

[Larranaga 02] Larranaga, P. and Lozano, J. A. (eds): *Estimation of distribution algorithms*, Kluwer Academic Publishers (2002).

[Tsutsui 01] Tsutsui, S., Pelikan, M., and Goldberg, D. E.: Evolutionary Algorithm using Marginal Histogram Models in Continuous Domain, *Proc. of the 2001 Genetic and Evolutionary Computation Conference Workshop Program*, pp. 230-233 (2001).

[Nagata 97] Nagata, Y. and Kobayashi, S.: Edge assembly crossover: A high-power genetic algorithm for the traveling salesman problem, *Proc. of the 7th Int. Conf. on Genetic Algorithms*, Morgan Kaufmann, pp. 450-457 (1997).

[Johnson 02] Johnson, D. S, and McGeoch, L. A.: Experimental analysis of heuristics for the STSP, *The Traveling Salesman Problem and its Variations*, Gutin and Punnen (eds), Kluwer Academic Publishers, Chapter 1 (to appear).

[Oliver 87] Oliver, I., Smith, D., and Holland, J.: A study of permutation crossover operators on the travel salesman problem, *Proc. of the 2nd Int. Conf. on Genetic Algorithms*, pp. 224-230 (1987).

[Starkweather, 91] Starkweather, T., McDaniel, S., Mathias, K, Whitley, D, and Whitley, C.: A comparison of genetic sequence operators, *Proc. of the 4th Int. Conf. on Genetic Algorithms*, Morgan Kaufmann, pp. 69-76 (1991).

[Whitley 89] Whitley, D., Starkweather, T., and Fuquay, D.: Scheduling problems and traveling salesman problem: The genetic edge recombination operator, *Proc. of the 3rd Int. Conf. on Genetic Algorithms*, Morgan Kaufmann (1989).

[Ulder 90] Ulder, N., Pesch, E., van Laarhoven, P., Bandelt, and Aarts, E.: Improving TSP exchange heuristics by population genetics, *Proc. of the Parallel Problem Solving from Nature - PPSN* (1990).

From Syntactical to Semantical Mutation Operators for Structure Optimization

Dirk Wiesmann

FB Informatik, LS 11, Univ. Dortmund, 44221 Dortmund, Germany
wiesmann@LS11.cs.uni-dortmund.de

Abstract. The optimization of structures is important for many industrial applications. But the problem of structure optimization is hardly understood. In the field of evolutionary computation mostly syntactical (pure structure-based) variation operators are used. For this kind of variation operators it is difficult to integrate domain-knowledge and to control the size of a mutation step. To gain insight into the basic problems of structure optimization we analyze mutation operators for evolutionary programming. For a synthetic problem we are able to derive a semantical mutation operator. The semantical mutation operator makes use of domain knowledge and has a well-defined parameter to adjust the step size.

1 Introduction

The problem of structure optimization occurs in many practical applications. As an example take the synthesis of chemical plants, where various processing units, interconnected by material streams, form a complex network, which structure has to be optimized [4]. Another example is the evolution of artificial neural network structures [10]. But the problem of structure optimization is hardly understood. In the field of evolutionary computation mostly syntactical (pure structure-based) variation operators are used. For this kind of variation operators it is difficult to integrate domain-knowledge and to control the size of a mutation step. To gain insight into the basic problems of structure optimization with evolutionary algorithms, we analyze mutation operators for evolutionary programming systems. Within the scope of evolutionary programming (EP) the evolution of Mealy automata is studied. An automaton is represented as a directed graph. The nodes are representing the states of the automaton and the edges are representing the state transitions. Every edge is labeled with an input and an output symbol. In the original work of Fogel et al. [6] the mutation operator is working on the graph structure of the automaton. There are five random mutation operators that affect the graph structure in different ways. The effect of a mutation event on the input/output behavior of the automaton is not obvious. In this paper we propose two variation operators which are not motivated by a random variation of the graph structure, but by the effect of a variation on the input/output behavior of the automaton. For simplification we regard only deterministic finite automata (DFA). In the following we first give

J.J. Merelo Guervós et al. (Eds.): PPSN VII, LNCS 2439, pp. 234–243, 2002.

a short overview on the topic of finite automata and present some well known properties we will refer to later on. After that, we discuss the traditional mutation operators used in EP. Then two alternative approaches are presented and evaluated.

2 Deterministic Automata

Finite automata are a formal representation for the analysis of sequential logic systems. For each point in time a finite automaton is in a state q of a finite nonempty set of states Q. In every step the automaton reads a symbol $w_i \in \Sigma$, writes a symbol $y_i \in \Omega$ and changes its state according the mapping $\delta : Q \times \Sigma \to Q$. Automata of this kind are called deterministic Mealy automata and can be described by the system $(Q, \Sigma, \Omega, q_o, \delta, \gamma)$. Where Q is a finite nonempty set of states, Σ the finite input alphabet, Ω the finite output alphabet, $q_0 \in Q$ the initial state, $\delta : Q \times \Sigma \to Q$ the state transition mapping, and $\gamma : Q \times \Sigma \to \Omega$ the output function. Thus, a Mealy automaton computes a function $f : \Sigma^* \to \Omega^*$, where Σ^* denotes the set of all finite strings of symbols from the alphabet Σ. In the following we will focus on decision problems. An input string $w \in \Sigma^*$ is said to be accepted, iff the automaton is in a final state after reading w. An automaton A of this kind is denoted as a DFA and can be described as a system $A = (Q, \Sigma, q_0, \delta, F)$, where $F \subseteq Q$ is the set of final (accepting) states. The set of all strings accepted by A is denoted as the regular language $L(A)$. The language $L^n \subseteq \Sigma^n$ consists only of strings of a fixed length n. Thus, $L^n(A)$ is the set of all strings of fixed length n accepted by the DFA A.

We will propose a mutation operator which is based on the operations intersection, union, and negation of regular languages. All algorithms can work efficiently on DFAs [2]:

Theorem 1. *Given a DFA A accepting the language $L(A)$ a DFA A' for the complement $\overline{L(A)}$ can be computed in linear time $O(|Q|)$.*

Proof. The set F of the final states needs only to be interchanged with the set $Q \setminus F$. □

Theorem 2. *Given two DFAs A_1 and A_2 accepting the languages $L(A_1)$ and $L(A_2)$ a DFA A accepting the language $L(A_2) \cup L(A_2)$ can be computed in time $O(|Q_1||Q_2||\Sigma|)$.*

Proof. The DFA A is constructed as follows. Let $Q = Q_1 \times Q_2$, and q_0 the pair of the initial states from A_1 and A_2. Then let $F = \{(q_i, q_j) \mid q_i \in F_1 \vee q_j \in F_2\}$ and $\delta((q_i, q_j), a) = (\delta_1(q_i, a), \delta_2(q_j, a))$ with $0 \leq i < |Q_1|, 0 \leq j < |Q_2|$. □

Theorem 3. *Given two DFAs A_1 and A_2 accepting the languages $L(A_1)$ and $L(A_2)$ a DFA A accepting the language $L(A_2) \cap L(A_2)$ can be computed in time $O(|Q_1||Q_2||\Sigma|)$.*

Proof. $L(A_1) \cap L(A_2) = \overline{(\overline{L(A_1)} \cup \overline{L(A_2)})}$. □

In the following we will apply these operations to minimum state DFAs only, i.e. DFAs with the minimum number of states.

Theorem 4. *If at first the unreachable states are eliminated from a DFA A, and then the equivalence class automaton A′ is constructed, then A′ is equivalent to A and has the minimum number of states.*

The proof and further details can be found in [2]. The set of unreachable states of a DFA can be computed by a depth first search starting in the initial state, in time $O(|Q||\Sigma|)$. The non equivalent states can be computed in time $O(|Q|^2|\Sigma|)$.

3 Evolutionary Programming

In the scope of evolutionary programming (EP), the evolution of finite automata has been studied since the 1960s [6, 5]. The starting point of the research was the question whether a simulated evolution on a population of contending algorithms is able to produce some kind of artificial intelligence. Intelligent behavior was considered to be the ability to predict an event in a given environment and to react on this event to meet a given goal. For the purpose of simplification, the environment was modeled as a sequence of symbols taken from a finite alphabet Σ. The algorithms were represented by Mealy automata reading the sequence of symbols. Every symbol the automaton reads, activates a state transition, and produces one output symbol from the finite alphabet Ω. The task of the EP system was to evolve an automaton that correctly predicts, i.e. produces, the next symbol to appear in the environment on the bases of the sequence of symbols it has previously observed. Thus, the number of wrong predictions was minimized.

An EP system works on the graph representation of an automaton [6]. An automaton is represented by a directed graph. The nodes represent the states of the automaton, and the edges correspond to the state transitions. Every edge is labeled by an input and an output symbol. Five different mutations are derived from the graph description: change of an output symbol, change of a state transition, addition of a state, deletion of a state, and change of the initial state. The mutation operator selects with equal probability a certain mode of mutation and applies it to an individual. Depending on the mode of mutation the nodes and edges are selected with equal probability. The number of mutations per offspring is chosen with respect to a probability distribution [5]. A recombination operator was proposed but not implemented.

The graph representation of an automaton and the resulting five modes of mutation have two advantages. Firstly, every single mode of mutation can be performed efficiently. Provided that the graph is stored as an adjacency list, every change of an output symbol and every mutation of a state transition needs only linear time in the number $|Q|$ of nodes. To add or to delete a state needs quadratic time. The change of the initial state can be done in constant time. Since the deletion of a state and the change of the initial state are only allowed when the parent automaton has more than one state, every mutation leads to a feasible representation of an automaton, but the resulting automaton is not necessarily minimal. In particular, there can be nodes and even whole subgraphs that are not reachable from the initial state.

A potential drawback of the mutation may be that every single mode of mutation is based solely on the structure of an automaton. Thus, the standard mutation is a syntactical operator. The size of a mutation, e.g., the length of a mutation step, is directly related to the complexity of the structural modification. A mutation that deletes a state and changes a state transition has greater influence on the structure than a mutation that only changes a symbol of the output alphabet. Thus, the impact on the input/output behavior is not considered here. Even the influence of two mutations of the same mode may vary significantly (see section 5.1). Moreover, it is difficult to find a suitable distance measure (metric), which measures the structural difference of two automata. Especially for the gradual approximation of a solution in a large search space, it is important that mutation will prefer small steps in the search space (regarding a suitable distance measure). By using EP to evolve programs in the form of symbolic expressions it has been observed that preferring mutations with a small effect has some advantages [1]. A formal approach is presented in [3]. By defining two related distance measures within the geno- and the phenotype space, so that neighboring elements have similar fitness, problem-specific knowledge was incorporated into the variation operators. The metric allows one to reason about the distance of individuals and the size of mutation steps in a formal framework. The size of a mutation is correlated with the change in the fitness value and is not directly based on the structural modifications within the representation of an individual. The requirements on the mutation operator are described in section 5.2. By the example of a synthetic problem, where a Boolean function has to be found based on the complete training set, it was shown, that systems which fulfill the requirements have a significant advantage [3].

To simplify our consideration we focus on DFAs in the following. In general this restriction to decision problems is not too strong [7].

4 Fitness Function and Distance of DFAs

Let $S \subset \Sigma^*$ be a finite subset and $T = \{(w, f(w)) | w \in S\}$ a training set. Based on the training set a DFA has to be found which accepts the language $L(A) := f^{-1}(1)$ for a function $f : \Sigma^* \to \{0,1\}$. The EP system has to evolve DFAs which will generalize from the training set to $L(A)$.

For simplicity, we restrict the problem space in two ways. First, we only consider languages with strings of fixed length n. Furthermore, the search will be based on the complete training set T_v. For a function $f : \Sigma^n \to \{0,1\}$ and the training set $T_v = \{(w, f(w)) \mid w \in \Sigma^n\}$ a DFA has to be found which accepts the language $L^n(A) := f^{-1}(1)$. Thus, A must achieve:

$$\forall (w, 1) \in T \text{ is } w \in L^n(A) \text{ and } \forall (w, 0) \in T \text{ is } w \notin L^n(A).$$

The effects of the restrictions will be discussed later. The fitness function $F(A) := |\{(w, f(w)) \in T_v \mid f(w) = 1 \Leftrightarrow w \in L^n(A)\}|$ counts the number of strings on which the DFA A classifies correctly.

Now, how can we measure the similarity of two DFA A and B? The distance d^n of A and B should be the number of strings on which A and B disagree:

$$d^n(A, B) = |L^n(A)| + |L^n(B)| - 2|L^n(A) \cap L^n(B)|$$

The maximum difference in fitness values of two DFA is $d^n(A, B)$. Note that the fitness calculation is based on T_v. The distance measure $d^n : \Sigma^n \times \Sigma^n \to \mathbb{N}$ is a metric. Imagine that all strings from Σ^n are sorted in lexicographical order. A language L^n can then be represented as a bit-string of length $|\Sigma^n|$. The i-th bit equals 1, if the i-th string from Σ^n is in the language L^n. Otherwise the i-th bit equals 0. Thus, d^n equals the hamming distance between the bit-strings belonging to the corresponding languages. Obviously, this distance measure can not distinguish between two structural different DFAs accepting the same language.

5 Proposals for EP Mutation Operators

5.1 Weighted Mutation

The first proposal for a new mutation operator is motivated by the observation that the fitness calculation can provide more information than the pure fitness value. To compute the fitness of a given DFA for every word in the training set a path beginning at the initial state has to be traversed.

In order to assess the influence of a mutation event every node (state) is assigned a weight index with initial value 0. Every time a node is visited during fitness calculation the weight index is incremented by 1. After fitness calculation on the complete training set, the weights give an upper bound for a change in the fitness value caused by mutation.

Let us consider the following example. Let $\Sigma^4 = \{0, 1\}^4$ and let $L^4 = \{0000, 0011, 0101, 0110, 1001, 1010, 1100, 1111\}$ be the language of all strings with an even number of 0's and an even number of 1's of length 4. Figure 1 shows the graph representation (state diagram) of an automaton with the corresponding weights after fitness calculation on the complete training set T_v. In three cases the DFA draws the wrong decision on T_v.

Now, let us discuss the impact of different mutations: The state q_2 has a relative high weight of 15. If the state q_2 is deleted by a mutation event, then the fitness changes by at most 15. In comparison, the deletion of state q_5 can change the fitness by the value of 3 at most. By ranking the states according to their weights, states with a lower weight can be mutated with higher probability than states with a higher weight. State transitions can be mutated likewise. Transitions beginning in a state with a lower weight will be mutated with higher probability than transitions beginning in a state with a higher weight. The insertion of new states will take place with higher probability between states with a lower weight.

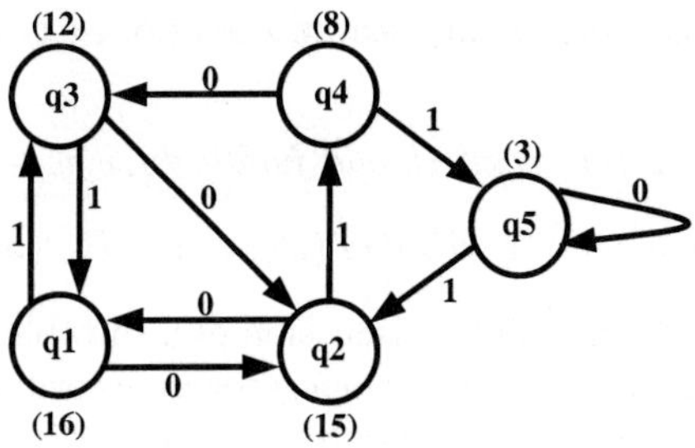

Fig. 1. DFA with weighted states (weights in parenthesis) after fitness evaluation on the complete training set T_4 for the language L^4 of all strings with an even number of 0's and an even number of 1's. State q_1 is initial and final state.

This approach allows the definition of a reasonable probability distribution on mutation events for every single mode of mutation. But it is not obvious how the different modes of mutation should be weighted among each other. For example, should transitions be mutated with higher probability than states? Should we mutate states with a low weight more often than transitions beginning in a state with a high weight? Additionally, even with respect to T_v the upper bound may turn out to be a bad estimate for the real change in fitness. E.g., an improvement and a decline of the fitness may cancel out each other. These observations show once again the problems of variation operators purely based on the structure, even when additional information is available.

5.2 Metric Based Mutation

In order to overcome the deficiencies described above, we first post some formal requirements on the mutation operator. Let $\mathcal{G}$ be the genotype space. Here $\mathcal{G}$ consists of all graph representations of DFAs A accepting a language $L^n(A) \subseteq \Sigma^n$. Since we consider minimum state automata only, $\mathcal{G}$ is finite. Let $d_\mathcal{G} : \mathcal{G} \times \mathcal{G} \to \mathbb{N}$ be a suitable metric on $\mathcal{G}$. Without loss of generality we restrict our discussion to the reduced mutation operator $m' : \mathcal{G} \times \Omega_{m'} \to \mathcal{G}$ with the finite probability space $(\Omega_{m'}, P_{m'})$. With probability $P_{m'}(m'(u) = v) := P_{m'}(\{\omega \in \Omega_{m'} \mid m'(u,\omega) = v\})$ the mutation operator m' changes an element $u \in \mathcal{G}$ to a new element $v \in \mathcal{G}$. The first rule assures that from each point $u \in \mathcal{G}$ any other point $v \in \mathcal{G}$ can be reached in one mutation step.

Guideline M 1 *The mutation m' should fulfill:*

$$\forall u, v \in \mathcal{G} : \; P_{m'}(m'(u) = v) > 0.$$

Moreover small mutations (with respect to $d_\mathcal{G}$) should occur more often than large mutations.

Guideline M 2 *The mutation m' should fulfill:* $\forall u, v, w \in \mathcal{G}$:

$$(d_\mathcal{G}(u,v) < d_\mathcal{G}(u,w)) \Rightarrow (P_{m'}(m'(u) = v) > P_{m'}(m'(u) = w))$$

The mutation should not prefer any search direction, e.g. should not induce a bias by itself.

Guideline M 3 *The mutation m' should fulfill:* $\forall u, v, w \in \mathcal{G}$:

$$(d_{\mathcal{G}}(u,v) = d_{\mathcal{G}}(u,w)) \Rightarrow (P_{m'}(m'(u) = v) = P_{m'}(m'(u) = w)) .$$

A motivation of the guidelines and a discussion of a suitable metric can be found in [3, 9]. We will now design a mutation operator in accordance to the guidelines which uses the metric d^n defined in section 4. The mutation will make use of the efficient synthesis operations for DFAs presented in section 2.

The first step in mutating the DFA A to a DFA b is to randomly choose a step size K with $0 \leq K \leq |\Sigma^n|$. The mutation operator will choose a subset $M^n \subseteq \Sigma^n$ with $|M^n| = K$. Every word $w \in \Sigma^n$ will be selected with the same probability p for the set M^n. Thus we have

$$P(K = k) = p^k \cdot (1-p)^{|\Sigma^n|-k}.$$

To obtain a fast selection of the set M^n we assume that every word can be addressed by a index $i \in \{0, \ldots, |\Sigma^n| - 1\}$. If the word at position i was already selected for the set M^n, then the word with index $(i + l)$ will be chosen next with probability $p \cdot (1-p)^{l-1}$. The random variable L with

$$P(L = l) = p \cdot (1-p)^{l-1}$$

describes the relative position of the next word that will be chosen. Thus L is geometrical distributed with parameter p. If $p = 1/|\Sigma^n|$ then the expected size of the set M^n will be 1.

The set M^n is split in two sets X^n and Y^n with:

$$\forall x \in X^n : x \in L^n(A), \quad \forall y \in Y^n : y \notin L^n(A) \text{ and } X^n \cup Y^n = M^n.$$

No $x \in X^n$ should be accepted by the DFA B. The DFA B should only accept all $y \in Y^n$. On every other input string A and B should agree. Thus, it is $d^n(A, B) = K$. For the partitioning in the sets X^n and Y^n the DFA A has to be tested K times (cost: $K \cdot n$). To obtain B two DFA A_X and A_Y are constructed with:

$$L^n(A_X) = \Sigma^n \setminus X^n \text{ and } L^n(A_Y) = Y^n.$$

With this, we get B as $L^n(B) = (L^n(A) \cap L^n(A_X)) \cup L^n(A_Y)$.

A_X and A_Y are constructed as follows. For every $x^i \in X^n = \{x^1, \ldots, x^{|X^n|}\}$ we construct an automaton A_{x^i} that accepts only the string x^i, thus $L^n(A_{x^i}) = \{x^i\}$. This automaton has $n + 2$ states. Figure 2 shows the structure of a DFA only accepting the string $a = a_1 \ldots a_n$. Thus, we have:

$$L^n(A_X) = \overline{L^n(A_{x^1}) \cup \ldots \cup L^n(A_{x^{|X^n|}})}.$$

For every $y^i \in Y^n = \{y^1, \ldots, y^{|Y^n|}\}$ we construct an automaton A_{y^i} only accepting the string y^i as well. With $L^n(A_{y^i}) = \{y^i\}$ we have:

$$L^n(A_Y) = L^n(A_{y^1}) \cup \ldots \cup L^n(A_{y^{|Y^n|}}).$$

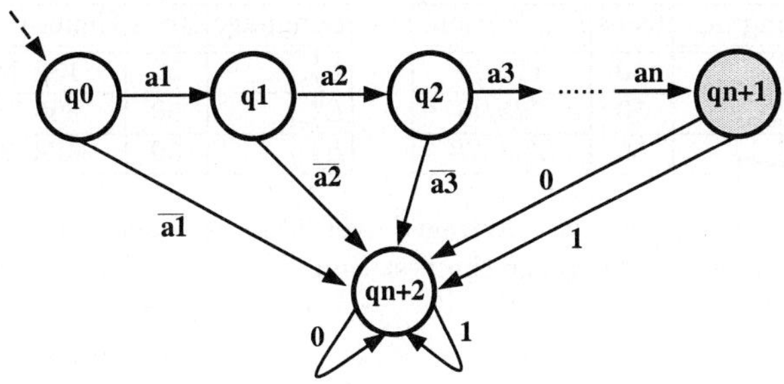

Fig. 2. The structure of a DFA on $\Sigma^n = \{0,1\}^n$. The DFA only accepts the string $a = a_1 \ldots a_n$. The final state is hatched and q_0 is the initial state.

After each synthesis operation the resulting DFA will be minimized. An EP system using this mutation operator is called a MBEP system.

Theorem 5. *The constructed mutation operator fulfills the guidelines M1, M2 and M3.*

Proof. The guideline M 1 is fulfilled because for every step size $k \in \{0, \ldots, |\Sigma^n|\}$ every subset $M^n \subseteq \Sigma^n$ with $|M^n| = k$ has a probability $p^k \cdot (1-p)^{|\Sigma^n|-k} > 0$ to be chosen. According to the design of the operator, every language $L^n \subseteq \Sigma^n$ can be generated. The guideline M 2 is fulfilled because for $k_1 < k_2$ it is guaranteed that $P(K = k_1) > P(K = k_2)$, and all subsets $M^n \subseteq \Sigma^n$ with $|M^n| = K$ have an equal probability of being chosen. This also implicates that guideline M 3 is also fulfilled. □

6 Experiments

For reasons discussed in section 7 a direct comparison between EP and MBEP is not possible. Due to its design the MBEP system searches for languages with strings of fixed length n. An EP system can operate on strings of arbitrary length. A (1+1)-MBEP system was tested on two different languages. The first language L^n_{even} consists of all strings of length n with an even number of 0's and an even number of 1's. The second language L^n_{fel} consists of all strings of length n where the last symbol equals the first. The initial start point was chosen by random selection of an element from the set of all languages with strings of length n with equal probability. We used a constant setting $p = 1/2^n$, but a dynamic adaptation of the parameter p is also possible. The number of generations (mutations) until the language was found the first time were averaged over 50 independent runs (Table 1). One has to keep in mind that the time needed for a mutation depends on the length n of the strings, the step size K, and the size of the DFAs. The mutation operator is efficient in these sizes but more time-consuming than standard EP mutation (see sections 3 and 5.2 and [3]). For 500 mutations the MBEP system needs for $n = 4$, $n = 6$, and $n = 8$, about 1, 4, and 16 seconds, respectively (on a Sparc Ultra 10/300). It is not surprising that the evolution process for both languages need similar amounts of time.

Language	Runs	Generations
L^4_{even}	50	118.09
L^6_{even}	50	725.06
L^8_{even}	50	3956.78

Language	Runs	Generations
L^4_{fel}	50	133.18
L^6_{fel}	50	696.34
L^8_{fel}	50	4056.26

Table 1. Number of generations averaged over 50 independent runs until the $(1+1)$-MBEP system found the language the first time.

To explain this observation, recall the bit-string representation from section 4. Assume a bit-string of length $|\Sigma^n|$ is given for every DFA. At the i-th position the bit-string has a 1 if the DFA draws the right decision for the i-th string. Otherwise this position holds a 0. The fitness function is identical to counting the number of 1's in the bit-string. Thus, the fitness function equals the counting-ones problem [8] on a string of length $|\Sigma|^n$. Since the MBEP mutation operator is based on the metric d^n, all languages L^n have the same difficulty to be found.

7 Problems

The MBEP system is subject to substantial restrictions. The system can only work on regular languages with strings of fixed length n. But this restriction could be weakened. Prior to a mutation step a string length could be chosen with respect to a probability distribution. Then the mutation operator operates only on strings of the chosen length. The restriction that the MBEP system can only work on the complete training set is much stronger. In its current implementation the system has no generalization ability. Due to the construction of the DFAs (see Figure 2), cycles over final states cannot occure. Strings that are too long or too short are kept in a non-accepting state. This problem may be solved by setting transitions starting in state q_{n+1} (see Figure 2) randomly to states in $\{q_0, \ldots, q_{n+1}\}$. Additionally states in $\{q_0, \ldots, q_{n+1}\}$ have to be final states with a certain probability. Unfortunately, first experiments have shown that under this condition the resulting DFAs may become very large. The size of a DFA depend on the size of the incomplete training set. If the training set is too small the DFAs may become too large.

8 Conclusion

In this work we have discussed the mutation operator in evolutionary programming as an example of structure optimization. We proposed two alternative mutation operators for structure optimization. The operators are using additional information to improve the search process. The weighted mutation operator uses information that results from the fitness calculation. The ("semi-syntactical") weighted mutation operator clarifies the problems of structure variations. Even when additional information is available it remains difficult to choose an appropriate mode of mutation. The (semantical) metric-based mutation operator was able to control these problems. A MBEP system has shown its performance on a synthetic problem. For practical (industrial) applications it will be difficult to translate the available domain-knowledge into a metric. Furthermore the existence of synthesis operations for semantical variations is not always ensured. Nevertheless, with certain restrictions it is possible to apply the metric-based design approach [11, 4].

Acknowledgements

This research was supported by the Deutsche Forschungsgemeinschaft as part of the collaborative research center "Computational Intelligence" (531).

References

[1] K. Chellapilla. Evolving computer programs without subtree crossover. *IEEE Transactions on Evolutionary Computation*, 1(3):209–216, 1997.

[2] P. J. Denning, J. B. Dennis, and J. E. Qualitz. *Machines, Languages, and Computation.* Prentice-Hall, Englewood Cliffs, 1979.

[3] S. Droste and D. Wiesmann. Metric based evolutionary algorithms. In R. Poli, W. Banzhaf, W. B. Langdon, J. F. Miller, P. Nordin, and T. C. Fogarty, editors, *Genetic Programming, Proc. of EuroGP'2000, Edinburgh, April 15–16, 2000*, volume 1802 of *LNCS*, pages 29–43, Berlin, 2000. Springer.

[4] M. Emmerich, M. Grötzner, and M. Schütz. Design of graph-based evolutionary algorithms : A case study of chemical process networks. *Evolutionary Computation*, 9(3):329–354, 2001.

[5] D. B. Fogel. *Evolutionary Computation:* Toward a New Philosophy of Machine Intelligence. IEEE Press, New York, 1995.

[6] L. J. Fogel, A. J. Owens, and M. J. Walsh. *Artificial Intelligence through Simulated Evolution.* Wiley, New York, 1966.

[7] M. R. Garey and D. S. Johnson. *Computers and Intractability: A Guide to the Theory of NP-Completeness.* W. H. Freeman, New York, 1979.

[8] G. Rudolph. *Convergence Properties of Evolutionary Algorithms.* Verlag Dr. Kovač, Hamburg, 1997.

[9] G. Rudolph. Finite Markov chain results in evolutionary computation: A tour d'horizon. *Fundamenta Informaticae*, 35(1-4):67–89, 1998.

[10] B. Sendhoff. *Evolution of Structures: Optimization of Artificial Neural Structures for Information Processing.* Shaker, Aachen, 1998.

[11] T. Slawinski, A. Krone, U. Hammel, D. Wiesmann, and P. Krause. A hybrid evolutionary search concept for data-based generation of relevant fuzzy rules in high dimensional spaces. In *Proc. of the Eighth Int'l Conf. on Fuzzy Systems (FUZZ-IEEE'99), Seoul, Korea, Aug. 22–25, 1999*, pages 1431–1437, Piscataway, NJ, 1999. IEEE Press.

Evolutionary Techniques: Coevolution

Parameter Control within a Co-operative Co-evolutionary Genetic Algorithm

Antony Iorio and Xiaodong Li

School of Computer Science and Information Technology
RMIT University, Melbourne VIC 3001, Australia
{iantony,xiaodong}@cs.rmit.edu.au

Abstract. Typically GAs have a number of fixed control parameters which have a significant effect upon the performance of the search. This paper deals with the effects of self-adapting control parameters, and the adaptation of population size within the sub-populations of a coevolutionary model. We address the need to investigate the potential of these adaptive techniques within a co-evolutionary GA, and propose a number of model variants implementing adaptation. These models were tested on some well known function optimisation problems. The experimental results show that one or more of the model variants yield improvements over the baseline co-evolutionary model.

1 Introduction

Recently, ecological models of the co-evolution of species have inspired new approaches for GAs [1–3]. Potter and De Jong have proposed a Co-operative Co-evolutionary Genetic Algorithm (CCGA-1) [1] architecture involving multiple co-evolving sub-populations, each dealing with separate parameters of the problem. Sub-populations are evolved much like a typical GA with selection, recombination and mutation operations. The co-evolutionary aspect of this model results from individuals evaluated in terms of individuals from the other sub-populations. This leads to co-operation between sub-populations to attain a mutual goal. By evolving the sub-populations in a modular fashion the problem is also being broken down, assisting the progress of the search, especially with separable problems [1]. This model has been applied to a variety of problem domains with notable success by Potter and De Jong [1, 4].

Another technique which has been found to improve the performance of GAs is the adaptation of algorithm control parameters. Control parameters of GAs are usually fixed during a run, whereas self-adaptation occurs when control parameters such as mutation or cross-over rate are represented within the chromosome as a parameter. As a result, they undergo the same evolutionary processes as the problem parameters within the chromosome. Like co-evolution, this is another means of providing a GA with more flexibility to conduct it's search effectively and allows the algorithm to adapt itself to the problem during a run [7, 8]. For more information the reader can refer to Eiben *et al.* [5] where a survey of work in the field of parameter control is provided.

J.J. Merelo Guervós et al. (Eds.): PPSN VII, LNCS 2439, pp. 247–256, 2002.

Intuitively, the two combined approaches of co-evolution and parameter control should improve the performance of a GA. In this paper we propose a number of model variants of the CCGA-1 which utilise the GAVaPS adaptive population sizing rule originally proposed by Arabas [6]. Bäck *et al.* [9] has also demonstrated the improvements which result from combining the GAVaPS with self-adaptation of control parameters.

We study the performance of the adaptive CCGA-1 variants on a number of test function optimisation problems. For the purposes of our comparative study, we use the same test functions that Potter and De Jong used in their original paper [1].

The paper is organized as follows: Section 2 describes the CCGA-1 originally proposed by Potter and De Jong [1], along with our extensions to the CCGA-1 using self-adaptive genetic operators (e.g., mutation and crossover) and adaptive sub-population size. Section 3 describes the experimental setup including the performance metrics. Section 4 presents the results, showing that one or more adaptive CCGA-1 variants always performs better than the baseline CCGA-1 upon all of the test functions. The Conclusion summarizes our observations and discusses directions for future research.

2 The Co-operative Co-evolutionary Algorithm with Adaptation

The CCGA-1 model provides a simple model in which it is relatively easy to integrate a variety of evolutionary approaches, while still maintaining the fundamental co-evolutionary behaviour of the algorithm. Consequently, we have chosen the CCGA-1 as a suitable approach to evaluate the performance of self-adaptive crossover and mutation, and the adaptation of sub-population sizes according to the Relative Life-Time and auxiliary population rules [9].

2.1 The CCGA-1 Model

We begin with a description of the fundamental co-evolutionary algorithm employed within our study. The co-operative co-evolutionary genetic algorithm in Figure 1, provides separate populations or species *(s)*, (where the subscript *(s)* represent the sub-population number) for each parameter in the problem domain.

The co-dependent evolution of the parameters is consistent with biological mutualism in that the fitness evaluation of a new individual is done in terms of how much the fittest individuals from the other sub-populations (previous generation) contribute towards it's fitness. In other words the algorithm determines how well an individual co-operates with individuals from other sub-populations; the ultimate outcome being a maximisation in co-operation and overall fitness.

2.2 Self-adaptive Mutation, Crossover, and the Selection Process

We have applied self-adaptation similar to that used by Bäck *et al.* [9]. This requires control parameters of mutation and cross-over to be represented within

```
gen = 0
for each sub-population s do begin
    Pop_s(gen) = randomly initialised population
    evaluate fitness of each individual in Pop_s(gen)
end for
while termination condition = false do begin
    gen = gen + 1
    for each sub-population s do begin
        select Pop_s(gen) from Pop_s(gen - 1) based on fitness
        apply genetic operators to Pop_s(gen)
        evaluate fitness of each individual in Pop_s(gen)
    end for
end while
```

Fig. 1. The Co-operative co-evolutionary genetic algorithm of Potter and De Jong [1].

a chromosome. Each sub-population represents a population of individuals for a particular parameter from the problem domain. The self-adaptive control parameters are encoded in the last two sub-strings of each chromosome, within each sub-population.

The process of self-adaptive mutation we have applied is a two step process; the bits encoding mutation rate at the end of the chromosome are mutated, and the resulting new mutation rate is decoded to be used upon the remaining bits of the chromosome.

Self-adaptive crossover rates are decoded from a sub-string within a chromosome as well. Crossover occurs with a two-point crossover over the entire chromosome after a proportion of the least fit individuals ranked in terms of their fitness, is killed from the sub-population. Two parents are chosen from the remaining sub-population using tournament selection. The probability of crossover for two parents is determined from the average of the two encoded crossover rates. A random number is generated between 0 and 1. If the number is within the range of the average self-adaptive cross-over rate, the parents are mated. The resulting two offspring are mutated with their respective mutation parameters, and reinserted into the existing population. If the crossover test does not succeed the parents are mutated with their respective mutation parameters and reinserted into the sub-population. This process is repeated until the number of individuals which were killed off are replaced. If the number of individuals which are reinserted is odd, one of the replacement individuals is automatically mutated and reinserted into the population.

2.3 Adaptive Sub-population Size

The *GAVaPS*[6] rule applied to the sub-populations of the CCGA-1, gives individuals a relative life-time (*RLT*) at the time of creation, which is in proportion to the average fitness of individuals within the population. Bäck *et al.* [9] reconstituted the *RLT* rules for the problem of function minimisation, which is the variation of the *RLT* rules we will be applying (equations (1), (2), and (3)).

MinLT and *MaxLT* refer to the static minimum and maximum life-times an individual can have. An individual i with a fitness worse than, or equal to the average, can expect a *RLT* closer to *MinLT*. Likewise, an individual which has a fitness better than the average, can expect a *RLT* closer to the maximum value. *AvgFit* is a measure of the average fitness within the sub-population, and *WorstFit* and *BestFit* are measures of fitness for the least and most fit individuals respectively.

$$RLT(i) = MinLT + \eta . \frac{WorstFit - fitness(i)}{WorstFit - AvgFit} \quad iffitness(i) \geq AvgFit \; . \tag{1}$$

$$RLT(i) = \frac{1}{2}(MinLT + MaxLT) + \eta . \frac{AvgFit - fitness(i)}{AvgFit - BestFit} \tag{2}$$
$$iffitness(i) < AvgFit \; .$$

$$\eta = \frac{1}{2}(MaxLT - MinLT) \; . \tag{3}$$

$$SubPopSize(t+1) = SubPopSize(t) + AuxSubPopSize(t) - D(t) \; . \tag{4}$$

$$AuxSubPopSize(t) = SubPopSize(t) \cdot p \; . \tag{5}$$

With every generation the *RLT* value of each individual is decremented by 1. When the *RLT* reaches zero, the individual is removed from the sub-population, unless the individual is the fittest individual. This is equivalent to an elitist strategy.

Within the approach of Bäck *et al.* [9] eventually all individuals die from old age. Bäck conjectured the high selection pressure resulting from population size minimisation contributed to finding good self-adaptive cross-over and mutation rates. However, this approach can potentially lead to premature convergence with an insufficiently large population. This issue was addressed by applying the auxiliary replacement equations (4) and (5) proposed by Arabas *et al.* [6]. Combined with a relatively small maximum life-time for individuals, an appropriate level of selection pressure can be maintained, while introducing new individuals with each generation. At each generation t, an auxiliary population *AuxPopSize(t)* is added, and the least fit individuals *D(t)* are deleted. The fraction of new individuals added to the current sub-population is determined by p, where *SubPopSize(t)* is the sub-population's size at that generation. The auxiliary population is added on using the same selection process applied to the replacement of culled individuals.

Table 1. Descriptions of the algorithm designations.

Algorithm designation	description
CCGA-1	Baseline co-evolutionary model
CCGAM	CCGA-1 with self-adaptive mutation rate
CCGAMX	CCGAM with self-adaptive crossover rate
CCGAAP	CCGA with adaptive sub-population sizing
CCGAMAP	CCGAAP with self-adaptive mutation rate
CCGAMXAP	CCGAAP with self-adaptive crossover rate

3 Experimental Design

The CCGA-1 was evaluated with a number of adaptive variations as shown in Table 1 using the parameters of Table 2. We employed a cap of 500 individuals for the adaptive sub-population size to make sure the sub-populations did not increase in size to an unnecessarily large number. The models we investigated were tested on the same set of functions used by Potter and De Jong [1] (see Table 3).

For each run of a particular algorithm, we terminated after 200,000 evaluations. For each function we conducted 30 runs and acquired the best mean fitness and standard deviation for each of the 200,000 evaluations over those 30 runs. A count of the number of function evaluations is used as a time metric for establishing a comparative baseline performance between the models. This was done for two reasons, firstly the CCGA-1 model has a significantly larger number of evaluations per generation, although its overall performance is typically better than a single population GA. Secondly, the adaptation of population size within a number of the models results in a varying number of evaluations per generation.

4 Results

Table 4 presents the best mean fitness after 200,000 evaluations for each of the models upon each test function. A number of comparative plots of each algorithm's performance are provided for each test function investigated in Figure 2. The performance of each algorithm is ranked in terms of the best results found for a particular algorithm variation and test function within Table 5[1]. The ranked value is between 0 and 5, where a ranking of 5 represents the best performance of a particular algorithm and test function. The ranked values are summed across test functions, for each algorithm. The sum provides an indication of the overall ranking of the algorithms investigated.

[1] The CCGAMAP and CCGAMXAP are so close in their best individual found for the Rosenbrock function, that the ranking is split evenly between them.

Table 2. Algorithm parameters and features.

Parameter	Value
chromosome representation	16 bits for a function variable, and 8 bits for each self-adaptive parameter.
chromosome length	32 bits
selection	rank selection and tournament selection
genetic operators	two-point crossover with bit-flip mutation
static crossover probability	0.6
static mutation probability	1/(chromosome length)
self-adaptive mutation probability range	0.001 to 0.25
self-adaptive crossover probability range	0.6 to 1.0
kill percentage	10%
sub-population size	100 (initial size which is fixed for all CCGA-1 model variants which do not incorporate adaptive sizing)
maximum sub-population size	500 (the maximum size of a sub-population within the CCGA-1 models incorporating adaptive sizing)
sub-population initialisation	random
MinLT	1
MaxLT	4
p	0.7

Table 3. Test functions utilised to evaluate the performance of the algorithms. n is the dimension of the function, and R is the range of the function variables.

Name	Function	n	R
Rastrigin	$f_1(\overrightarrow{x}) = 10.0n + \sum_{i=1}^{n}(x_i^2 - 10.0\cos(2\pi x_i))$	20	[-5.12,5.12]
Schwefel	$f_2(\overrightarrow{x}) = 418.9829n - \sum_{i=1}^{n} x_i \sin(\sqrt{\lvert x_i \rvert})$	10	[-500,500]
Griewangk	$f_3(\overrightarrow{x}) = 1 + \sum_{i=1}^{n} \frac{x_i^2}{4000} - \prod_{i=1}^{n} \cos(\frac{x_i}{\sqrt{i}})$	10	[-600,600]
Ackley	$f_4(\overrightarrow{x}) = 20 + e - 20.\, e^{(-0.2\sqrt{\frac{1}{n}\sum_{i=1}^{n} x_i^2})} - e^{(\frac{1}{n}\sum_{i=1}^{n}(\cos 2\pi x_i))}$	30	[-30,30]
Rosenbrock	$f_5(\overrightarrow{x}) = 100(x_1^2 - x_2)^2 + (1 - x_1)^2$	2	[-2.048,2.048]

We observe from Figure 2 that one or more of the adaptive CCGA-1 models evaluated, performs better than the baseline CCGA-1 upon all of the test functions.

Table 4. Best fitness and standard deviation after 200,000 evaluations.

Algorithm	Rastrigin	Schwefel	Griewangk	Ackley	Rosenbrock
CCGA-1	0.1540 ±0.3687	0.08393 ±0.09581	0.04734 ±0.02994	0.01227 ±.005645	0.02046 ±0.03322
CCGAM	1.921 ±1.317	1.874 ±6.833	0.3494 ±0.1364	1.247 ±0.5214	0.005109 ±0.01043
CCGAMX	2.708 ±1.425	1.045 ±4.594	0.3017 ±0.1431	0.9333 ±0.5245	0.004421 ±0.005681
CCGAP	0.001311 ±0.005653	0.0006013 ±0.0008367	0.03431 ±0.03394	0.005765 ±0.001606	0.03454 ±0.07332
CCGAMAP	2.078 ±4.411	0.02420 ±0.05113	0.06615 ±0.02763	7.646 ±2.579	0.002995 ±0.005983
CCGAMXAP	0.9237 ±1.914	0.01351 ±0.03134	0.05635 ±0.02054	6.359 ±3.228	0.002943 ±0.006692

Table 5. Comparative ranking of the algorithms upon the test functions, in terms of the best individual found after 200,000 evaluations.

Test function	CCGA-1	CCGAM	CCGAMX	CCGAAP	CCGAMAP	CCGAMXAP
Rastrigin	2	4	6	1	5	3
Schwefel	4	6	5	1	3	2
Griewangk	2	6	5	1	4	3
Ackley	2	4	3	1	6	5
Rosenbrock	4	3	2	5	1.5	1.5
Total	14	23	21	9	19.5	14.5
End Ranking	2	5	4	1	3	2

Bäck *et al.* [9] hypothesized that the main source of the improvement in his experimentation of adaptive and self-adaptive GAs was the adaptation of the population size. This is consistent with our observations as well, in that the CCGAAP outperforms the CCGA-1 upon the Schwefel, Rastrigin, Griewangk, and Ackley functions. From Table 5 we see that the CCGAMXAP also performs as well as the CCGA-1 baseline. It is apparent that the CCGAMXAP performs significantly better upon the Rosenbrock and Schwefel functions, where the CCGA-1 performs more poorly.

5 Conclusion

We have established that one or more of the adaptive models performs better than the baseline CCGA-1 on all of the test functions we investigated.

To summarise, we have made the following observations:

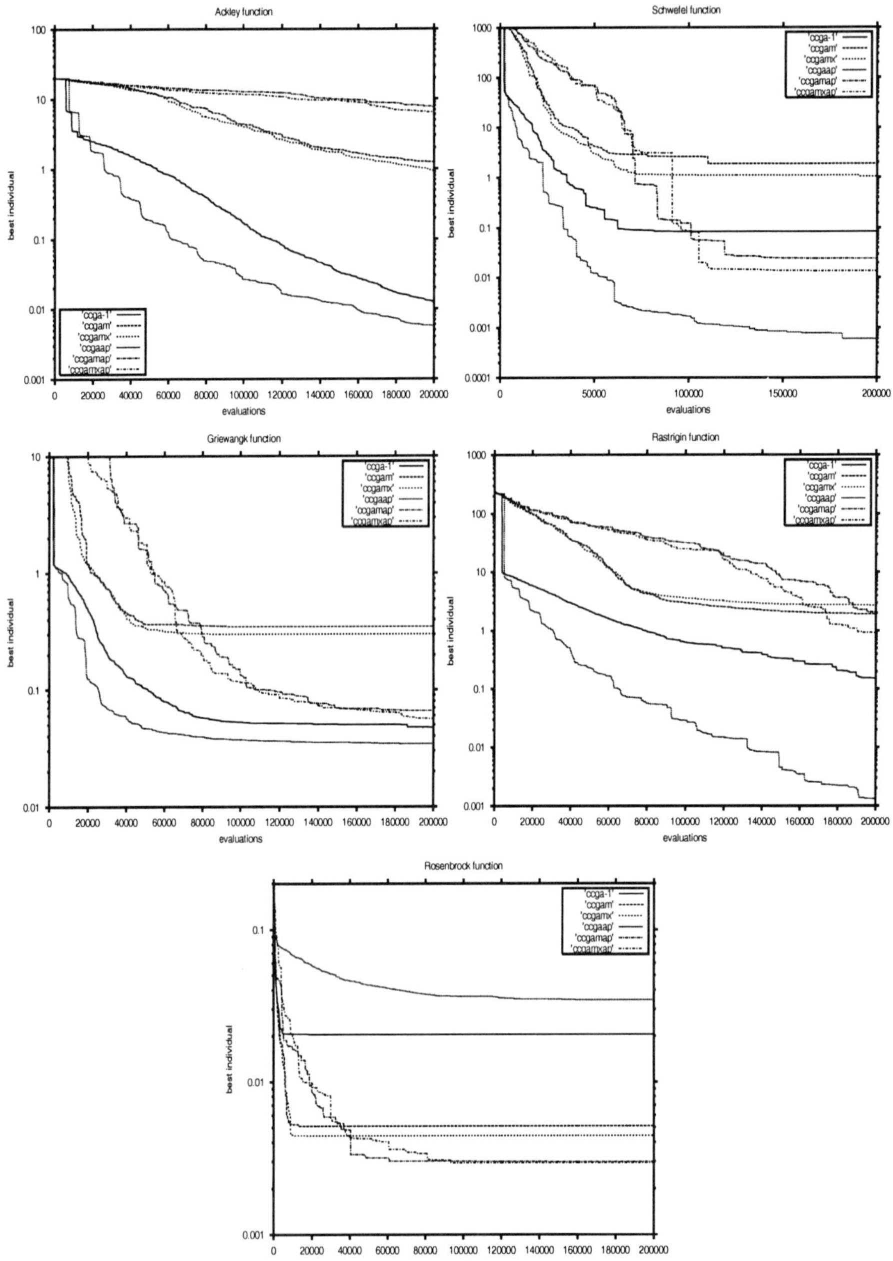

Fig. 2. Best fitness performance of the adaptive CCGA-1 algorithms upon the Ackley, Schwefel, Griewangk, Rastrigin, and Rosenbrock functions.

- The CCGAAP model performs as well or better on all functions except the Rosenbrock function. This suggests that some form of adaptive population sizing can generally yield improvements to the CCGA-1. The application of the *RLT* rule assists in removal of individuals from within sub-populations which are deemed to make a poor contribution towards the fitness evaluation. At the same time, the lifespan of these individuals is sufficient for any beneficial future contributions to be made.
- All the models incorporating self-adaptation obviously involve some degree of learning to establish appropriate mutation and/or crossover rates as the algorithm progresses. This learning is part of the search process, and it is undertaken by each individual in each sub-population. We can see from the experimental results (Figure 2) that self-adaptive CCGA-1 models generally converge quite slowly and also prematurely. We hypothesize that the poor performance of these models is the result of the large number of self-adaptive parameters involved in the search. Within these models we are not just searching for good solutions, but also good control parameters. As a result the search space is significantly larger than it would be otherwise. Within the self-adaptive CCGA-1 models we are searching a much larger number of parameter values, which suggests significant time is taken away from the search for good solutions.
- We observe improvements with both the CCGAMAP and CCGAMXAP over the baseline CCGA-1 upon the Schwefel and Rosenbrock function. Compared with the approach of Bäck *et al* [9] where the population diminishes over successive generations, the sub-populations in our approach do not diminish, but increase in size towards a cap of 500 individuals. It is reasonable to conclude in our case that the selection pressure resulting from the comparatively smaller maximum life-time for individuals helps to remove individuals with unsuitable self-adaptive mutation and crossover values.
- Because of the nature of the CCGA-1 model which separates the object variables into sub-populations, we mutate each of the object variables independently within a self-adaptive scheme. This results in a greater exploration of the search space. It also suggests that good results upon uniform fitness landscapes such as the Rosenbrock function can be expected, where a high degree of exploration is desirable to find a good direction in the search. In contrast, CCGAAP is not as effective upon the Rosenbrock function although it demonstrated effective performance upon the other test functions.
- The RLT rule, with a short maximum lifetime of 4 introduces significant selection pressure to the search, by allowing individuals which have not contributed after 4 generations to die off and be replaced. This is the element of the CCGAMXAP and CCGAMAP model which provides a force for the survival of relatively fitter individuals within sub-populations.

Overall we see that adaptively sizing the sub-populations using the *RLT* rule yields the largest improvements, and further research in the area of adaptively sizing sub-populations would be desirable. The highly adaptive nature of the models we investigated suggests that they may be suitable for optimisation problems involving non-stationary environments [10].

References

1. Potter, M. A., De Jong, K. A.: A Cooperative Co-evolutionary Approach to Function Optimization. In: Davidor, Y., Schwefel, H.-P., Manner R. (eds.): Parallel Problem Solving from Nature PPSN-III., Proceedings of the International Conference on Evolutionary Computation, Lecture Notes in Computer Science Vol. 866. 249–257 (1994)
2. Husbands, P. and Mill, F.: Simulated Co-Evolution as The Mechanism for Emergent Planning and Scheduling. In: Belew, R. and Booker, L. (eds.): Proceedings of The Fourth International Conference on Genetic Algorithms 264–270 (1991)
3. Hillis, W. D.: Co-evolving Parasites Improve Simulated Evolution as an Optimization Procedure. In: Langton, C. G., Taylor, C., Farmer, J. D., and Rasmussen, S. (eds.): Artificial Life II, Santa Fe Institute Studies in the Sciences of Complexity Vol. X 313–324 (1991)
4. Potter, M. A., and De Jong, K. A.: Cooperative Co-evolution: An Architecture for Evolving Coadapted Subcomponents. In: Evolutionary Computation, Vol. 8., No.1, 1–29, (2000)
5. Eiben, A., Hinterding, R. and Michalewicz, Z. Parameter Control in Evolutionary Algorithms. In: IEEE Transactions on Evolutionary Computation Vol. 3 No. 2 124–141 (1999)
6. Arabas, J., Michalewicz Z., and Mulawka, J.: GAVaPS - A Genetic Algorithm with Varying Population Size. In: IEEE World Congress on Computational Intelligence., Proceedings of the First IEEE Conference on Evolutionary Computation Vol. 1. 73–78 (1994)
7. Bäck, T. and Schutz, M.: Intelligent mutation rate control in canonical genetic algorithms. In: Ras, Z. W. and Michalevicz, M. (eds.): Foundations of Intelligent Systems, Ninth International Symposium ISMIS'96, Lecture Notes in Artificial Intelligence, Vol. 1079. 158–167 (1996)
8. Hinterding, R.: Gaussian mutation and self-adaptation for numeric genetic algorithms. In: Proceedings of 1995 IEEE International Conference on Evolutionary Computation 384–389 (1995)
9. Bäck, Th., Eiben, A. E. and van der Vaart, N. A. L.: An empirical study on GAs without parameters. In: Schoenauer, M., Deb, K., Rudolph, G., Yao, X., Lutton, E., Merelo, J. J., and Schwefel, H.-P.(eds.): Parallel Problem Solving from Nature — PPSN V, Lecture Notes in Computer Science Vol. 1917 315–324 (2000)
10. Grefenstette, J.J.: Genetic Algorithms for changing environments. In: Manner, R. and Manderick, B. (eds.): Parallel Problem Solving from Nature — PPSN II. Elsevier Science Publishers B.V., 137-144 (1992)

The Effects of Representational Bias on Collaboration Methods in Cooperative Coevolution

R. Paul Wiegand, William C. Liles, and Kenneth A. De Jong

George Mason University, Fairfax, VA 22030, USA
paul@tesseract.org, {wliles,kdejong}@gmu.edu

Abstract. Cooperative coevolutionary algorithms (CCEAs) have been applied to many optimization problems with varied success. Recent empirical studies have shown that choices surrounding methods of collaboration may have a strong impact on the success of the algorithm. Moreover, certain properties of the problem landscape, such as variable interaction, greatly influence how these choices should be made. A more general view of variable interaction is one that considers epistatic linkages which span population boundaries. Such linkages can be caused by the decomposition of the actual problem, as well as by CCEA representation decisions regarding population structure. We posit that it is the *way* in which represented problem components interact, and not necessarily the existence of cross-population epistatic linkages that impacts these decisions. In order to explore this issue, we identify two different kinds of *representational bias* with respect to the population structure of the algorithm, *decompositional bias* and *linkage bias*. We provide analysis and constructive examples which help illustrate that even when the algorithm's representation is poorly suited for the problem, the choice of how best to select collaborators can be unaffected.

1 Introduction

Coevolutionary Algorithms (CEAs) are interesting extensions to traditional Evolutionary Algorithms (EAs). While fitness in an EA is determined objectively, fitness in a CEA is determined subjectively based on how an individual interacts with other individuals. In cooperative coevolution, individuals that participate in successful interactions (*collaborations*) are rewarded while unsuccessful collaborations are punished.

In this paper, we focus on the particular class of cooperative coevolutionary algorithms (CCEAs) defined by [1, 2]. A standard approach to applying CCEAs to an optimization problem starts by trying to identify some reasonable static decomposition of the problem representation into components represented by each population. So, for example, if given a function of m variables to optimize, one might choose to put each variable in a separate CCEA population. Once a decomposition is established, the fitness of components in one population is estimated by choosing one or more collaborators from the other populations.

J.J. Merelo Guervós et al. (Eds.): PPSN VII, LNCS 2439, pp. 257–268, 2002.

There have been several attempts to understand how collaborators are best chosen in this paradigm. Early work suggested that these choices were tied to the amount of epistatic interaction among the function variables [1, 3]. In a CCEA that uses an N-variable decomposition for such fitness landscapes, this leads to the notion of cross-population epistasis, and to the simple intuition that increasing amounts of cross-population epistasis will require more complex collaboration mechanisms. Unfortunately, the issue is more complicated than this. For example, the simplest collaboration method seems to work best when applying a CCEA to a non-linearly separable, quadratic problem despite the existence of cross-population epistasis [4].

This paper extends and clarifies these issues by focusing on how optimization problems are represented in a CCEA. We identify two kinds of representational bias, *decompositional bias* and *linkage bias*, and show how these biases affect choices of collaboration method. The paper is laid out as follows. In the next section, we will briefly outline some background regarding existing analyses of coevolutionary algorithms, the cooperative coevolutionary architecture on which we will be focusing, and some of the choices surrounding collaboration in this algorithm. The third section discusses what we believe to be the important characteristics of problems, namely decomposability and epistasis. The fourth and fifth sections take the two kinds of representational bias in turn, first discussing the implications of decompositional bias with respect to collaboration, then those of linkage bias. The final section will conclude by discussing our improved understanding how problem characteristics affect choices of collaboration methodology.

2 Cooperative Coevolution

2.1 Existing Analysis of Coevolutionary Algorithms

Much of the analysis of coevolutionary algorithms has focused on their complicated dynamics. For example, considerable effort has been spent trying to understand how one can measure progress in a system where individual fitnesses are subjective in order to help identify some of the pathological behaviors exhibited by these algorithms [5–7]. Additionally, some basic theoretical work uses ideas from simple genetic algorithm theory provided by [8], and applies them to competitive coevolution [9]. That work explores the mechanics of a simple competitive coevolutionary algorithm from an evolutionary game theoretic viewpoint. [10] extended this model to analyze cooperative coevolution for the purposes of investigating their potential as optimizers.

One of the issues of considerable practical significance for coevolutionary algorithms is how to assess the fitness of an individual in one population (species) when that fitness depends in part on the individuals in other populations (species). The theoretical models typically assume "full mixing" in the sense that an individual's fitness is determined by a complete set of "interactions" with all other species. In the simple case that each of the coevolving p populations contains i individuals, the number of "interactions" per fitness evaluation is i^{p-1}.

As a consequence, there is strong practical motivation to estimate fitness by selecting only a small subset of the possible interactions, particularly when $p \gg 2$. This process is usually accomplished by selecting a small number "partners" [3] or "collaborators" [1] from the other species whose interactions are the basis of fitness estimates. Immediate questions arise as to: 1) how many partners, 2) how to select the partners, and, in the case of multiple interactions, 3) how to combine the results of multiple interactions into a single fitness estimate. The answers to these questions are far from simple and depend on a number of factors including 1) whether the coevolutionary model is competitive or cooperative, and 2) the degree and type of cross-population epistasis present [1, 11, 2–4, 12].

2.2 Collaboration in CCEAs

Our focus is on understanding these issues for cooperative coevolutionary models, in particular for the CCEA architecture developed in [1, 2]. In this model there is a single global fitness function for the entire system, but the search space has been decomposed into a number of independent subspaces, each of which is explored in parallel by independent EA populations. In this case, in order to evaluate the fitness of an individual in one population, one or more collaborators must be selected from the other populations in order to assemble a complete object for a global evaluation.

A variety of studies including [1, 3, 4] suggest that, if the degree of cross-population epistasis is not too strong, then selecting the current best individual from each of the other populations and performing a single global fitness evaluation is a surprisingly robust collaboration mechanism. When there is a strong degree of epistasis, a more complex strategy involving more than one collaborator from each of the other populations and a less greedy selection method for those collaborators can improve performance.

Finally, if multiple function evaluations are involved for each subcomponent, how can one combine these results to obtain a single fitness value? Here the literature is fairly consistent in recommending assigning the maximum value obtained as the fitness.

2.3 Representation in CCEAs

However, for the practitioner, there are still a number of questions to be answered including *how* to decompose a complex problem in a way that leads to an effective CCEA-based solution. This is a familiar question for standard EAs, e.g. deciding how best to represent problems (such as Traveling Salesperson problems) to achieve good EA-based solutions. For CCEAs, representations must additionally involve decompositions into separately evolving subcomponents.

The difficulty for the CCEA practitioner is that there is seldom sufficient *a priori* information to select an "optimal" representation or even one with sufficient knowledge to make informed choices about the appropriate collaboration mechanism to use. However, it is clear that any choice of representation introduces a bias in the system that can potentially have strong effects on a particular collaboration strategy.

Intuitively, CCEA representation biases are a result of the granularity of the decomposition (how many subcomponents) and the resulting epistatic interactions among the subcomponents. The focus of this paper will be on understanding these biases and their effects on collaboration mechanisms.

3 Research Methodology

In order to better control for the representational properties that we feel are important to choices of CCEA collaboration mechanisms, this paper focuses on *pseudo-boolean* functions, that is the objective value is assessed by mapping binary strings of length n to real values, $\mathcal{F} : \{0,1\}^n \rightarrow \Re$. The two representational properties of interest are decomposability and epistasis.

3.1 Decomposability

For our purposes, a function is considered decomposable if it can be decomposed into a sum of some number of smaller independent functions. These functions do not necessarily have to be identical. More formally, a function $\mathcal{F}$ is *decomposable* if there exist a set of independent functions, $\{f_1, f_2, \ldots, f_m\}$ such that $\mathcal{F}(\boldsymbol{x}) = \sum_{i=1}^{m} f_i(x_i)$. Some types of problems are rendered more tractable for optimization because of this property since optimization can be done as m independent optimizations [13].

For pseudo-boolean functions, each x_i refer to partitions of the main string, or *building blocks* of the problem. Of particular interest are m-decomposable functions, i.e., those for which an m block decomposition exists, but no finer grained partition exists.

An example of a bit-wise decomposable problem (i.e., $m = string\_length$) is the classic `OnesMax` problem in which fitness is the sum of all the 1 bits in the string. A familiar example of a problem that is not decomposable (i.e., $m = 1$) is the `LeadingOnes` problem. Both of these problems are defined formally in the following sections.

Ideally, for such problems, one would like to map each decomposable unit into a CCEA population. In general, however, the practitioner and the CCEA don't have explicit decomposability information. As a consequence, a CCEA can be mismatched in terms of the number of populations and what gets mapped into each population. We refer to this as the *decompositional bias* of a CCEA.

3.2 Epistasis

This notion of decomposability is intimately tied to the ideas of epistasis. For pseudo-boolean functions we define *epistasis* to mean non-linear interactions between bit positions in the problem [14]. So, for example, a highly epistatic pseudo-boolean function like `LeadingOnes` is 1-decomposable while the non-epistatic `OnesMax` problem is fully decomposable.

Of interest here is the kind of epistasis that a decomposition exhibits (e.g., [12]). Epistatic interactions can be *positive* in the sense that the contribution of a piece of the representation due to its non-linear interaction with other pieces have the same kind of impact on fitness as the contribution of the piece itself; however, the *magnitude* of that impact depends on the value of the other pieces involved in the interaction. Similarly, epistatic interactions can be *negative* in the sense that the impact of the individual pieces involved have the opposite effect on fitness as the contribution of their non-linear combinations, but again the magnitude of this opposing contribution depends on the pieces involved. In addition, one can have epistatic interactions in which *neither the sign nor the magnitude* of the effect can be predicted from the individual components. As we will see, the particular form of epistasis has an important effect on collaboration mechanisms.

3.3 Experimental Framework

In the following sections we will construct several problems that exhibit these properties and we analyze them both formally and empirically. For the empirical studies we used a CCEA with the following properties. A steady state GA was used for evolving each of the populations. Each steady state GA uses a ranked-based selection method, such that a single offspring replaces the worst individual each generation Bit-flip mutation was applied at a rate of $1/r$, where r is the number of bits of individuals in a given population. Parameterized uniform crossover was applied 100% of the time to produce a single offspring in such a way that there was a 0.2 probability of swapping any given bit. As part of the reported experiments, we varied the number of populations, p, but in all cases the number of individuals in each population was 10.

During preliminary sensitivity experiments, various algorithm parameter values were used. These included generational versus steady state models, proportional versus ranked-based selection methods, different population sizes, different variational operators (and rates). Though not reported in this paper, the results were consistent with our reported findings. Our choices for the final algorithm were based on performance results.

4 The Effects of Decompositional Bias on Collaboration

Obviously, different problems have different degrees of decomposability that may or may not be reflected in the particular CCEA representation chosen. Since decomposability information is not generally available for difficult optimization problems, our focus here is on the case where there is some kind of mismatch between the CCEA representation and a problem's "natural" decomposition. The question here is not whether such mismatches make the problem harder to solve. This will clearly happen for some problems [1, 3, 12, 4]. Instead, the question is whether adopting more complex collaboration methods can alleviate such mismatches.

4.1 Controlling Decompositional Bias Experimentally

In order to answer this question, we need to have problems in which we can explicitly control their inherent decomposability. For pseudo-boolean functions this is not difficult to do. One simply defines a function in terms of the sum of m independent subfunctions which are themselves not decomposable. A simple example is obtained by choosing a non-decomposable function of k bits and concatenating m of these k-bit blocks to form a km-bit problem the value of which is just the sum of the individual functional blocks.

More formally, let $\mathcal{F} : \{0,1\}^n \rightarrow \Re$ be some objective function over a binary string of length $n = mk$ and $f : \{0,1\}^k$ be some non-decomposable function over a binary string of length k. Then, given $\boldsymbol{x} \in \{0,1\}^n$,

$$\mathcal{F}(\boldsymbol{x}) = \sum_{i=0}^{m-1} f(m_i)$$

where m_i represents the ith block of k bits in $\boldsymbol{x}$.

From a practitioner's point of view, barring any problem specific knowledge, the simplest way to represent pseudo-boolean functions is to break up bit strings of length n into p blocks and assign each block to a different population. Hence, a decompositional mismatch of the representation may be due to over-decomposition ($p > m$) or under-decomposition ($p < m$). If there are more populations than there are decomposition blocks, there is likely to be strong interaction between populations in the system with respect to the problem, i.e., cross-population epistasis. If $p < m$, the advantage of the parallelism of coevolutionary search is sacrificed.

4.2 Effects of Collaboration Methods

We begin to answer the question of whether problems introduced by decompositional bias can be alleviated by more complex collaboration methods by observing that, if there is no cross-population epistasis, a simple selection method for collaboration is sufficient. If the two populations represent independent pieces of the problem, then optimizing one population independently of the other will result in the complete optimization of the problem. As long as the collaborators chosen for each population member are the same, it doesn't matter how we chose the collaborator. However, it does matter how many collaborators we choose, since picking more than one will incur more unnecessary computational cost in the way of objective function evaluations. Therefore, in the absence of cross-population epistasis, selecting the single best individual [1] from the other populations for collaboration is sufficient. In fact, one could pick this individual randomly, as long as it was the same individual for each member of the population during a given generation. The point isn't that any partnering scheme will

[1] "Best" here means the most fit individual(s) in other population(s) from previous evaluation.

Table 1. Steady state CCEA results on the `LeadingOnes` problem. Each value represents the mean number of evaluations needed to reach the optimum out of 50 trials. From the top left conrner, proceding clockwise, the tables represent data for decompositional biases created using two, four, eight and sixteen populations.

$p = 2$	# Collaborators		
	1	2	3
s-best	8016.7	16015.4	24654.2
s-rand	8989.8	17247.7	25191.8

$p = 4$	# Collaborators		
	1	2	3
s-best	8801.1	17602.1	26198.9
s-rand	10155.5	18757.9	28372.5

$p = 8$	# Collaborators		
	1	2	3
s-best	9821.52	19825.08	29018.32
s-rand	12734.86	22134.34	32052.24

$p = 16$	# Collaborators		
	1	2	3
s-best	11247.20	22350.32	33468.88
s-rand	19233.78	30089.90	41995.38

result in a better collaboration than another, but that since each population can essentially be optimized independently, we only need a *consistent* sample from which to establish a collaboration.

So why would a more complicated collaboration selection method be needed? Recall that how one chooses collaborators is essentially a choice about how one *samples the potential interactions* with the other population. There has to be a reason to believe that more than one sample is needed, or that sampling with a particular bias (say choosing the best) will result in a poor characterization of the relationship between the populations. Either way, some interaction between the populations is certainly needed to justify this. More than simply having such epistasis is at issue, however.

Consider the `LeadingOnes` problem, $f(\boldsymbol{x}) = \sum_{i=1}^{k} \prod_{j=1}^{i} x_j$. This problem is certainly not decomposable. Further, if we aggregate m of them, we can study the effects of running a CCEA when the number of populations p is $>= m$.

In order to study the effects of collaboration on such situations, we constructed the following experiment. Using the CCEA described in the Methodology section, we experimented with a concatenated `LeadingOnes` problem. In this particular case there were 128 bits in the total bit string of the problem, subdivided evenly into $m = 2$ blocks. A total of 6 collaboration selection methods were used. The number of collaborators chosen for a given evaluation was varied $(1, 2, \&3)$ and two selection biases were used: s-best and s-random (without replacement). We varied the number of populations, p, but in all cases the number of individuals in each population was 10. The results for $p = 2$ through $p = 16$ in Table 1 show the average number of evaluations it took the algorithms to reach the optimum (50 trials each). Unless otherwise stated, confidence levels for all tests are 95%.

The Tukey-Means test indicates that in all cases choosing one collaborator is clearly better. This might at first be puzzling since there is clearly cross-population epistasis present when $p > 2$. However, note that a mutation which turns some bit to 1 in an individual in the first population will always result either a neutral or positive change in fitness, regardless of the contents of the other population. The reverse is true, as well. In addition, this decomposition is

Table 2. Steady state CCEA results on the `LeadingOnes-OnesMax` problem. Each value represents the mean number of evaluations needed to reach the optimum out of 50 trials. From the top left conrner, proceding clockwise, the tables represent data for decompositional biases created using two, four, eight and sixteen populations.

$p = 2$	# Collaborators		
	1	2	3
s-best	15592.9	30846.8	48259.6
s-random	16000.7	31320.6	46406.9

$p = 4$	# Collaborators		
	1	2	3
s-best	15549.4	30974.6	45445.7
s-random	16381.6	30319.4	44736.0

$p = 8$	# Collaborators		
	1	2	3
s-best	14014.40	28969.24	44172.40
s-random	17343.58	28795.78	45252.94

$p = 16$	# Collaborators		
	1	2	3
s-best	14247.10	27911.04	41735.98
s-random	21653.86	33975.28	47792.74

also asymmetric for $p = 4$ in that the second and fourth populations will remain relatively unimportant to the first and third populations for some time during the evolution, since each `LeadingOnes` subproblem is solved left-to-right.

One observation that can be made is that by changing the number of populations, the mutation rate is effectively being increased (recall that the mutation is $1/r$, where r is the number of bits per individual in each population). Such issues may be relevant, consequently we ran all population oriented experiments in the paper (including the preceding one) using a constant 1/64 mutation rate. The results (not reported) remain consistent with those reported here.

We can make the problem slightly more interesting by making the right-hand side of the bit string play a more important role. We will do this by scaling the `LeadingOnes` part by k and subtracting `OnesMax` from the total, i.e., $f(\boldsymbol{x}) = k \sum_{i=1}^{k} \prod_{j=1}^{i} x_j - \sum_{i=1}^{k} x_i$. Now not only will the right side matter, but there is some tension between individual bit contributions to fitness and those of their non-linear interactions. Moreover, this tension is "one directional" in a sense. Take the string: "11 00 00 ..." as an example. Flipping the fourth bit to a one will decrease the fitness slightly if the third bit remains 0, while flipping both the third and fourth bits will increase the fitness. However, the same is not true on the other side. Flipping the third bit while the fourth bit remains 0 will also increase fitness. So some of the interactions have this property of sign-dependent epistasis, while others will not. In addition, the linear effects of the bits are very muted compared to the non-linear effects due to the scaling issue.

Using the same experimental setup as before, we studied the effects of collaboration on `LeadingOnes-OnesMax`. The results for $p = 2$ through $p = 16$ in Table 2 show the average number of evaluations it took the algorithms to reach the optimum (50 trials each).

In all cases there was no statistical reason to choose another collaboration method other than the single best individual from the other populations. Not only does this increased decompositional bias not alter the collaboration methodology, it appears as though this problem becomes *easier* for the CCEA to solve, not harder. This turns out to be statistically significant only for the $p = 8$ and $p = 16$ cases (not shown) where there is one or two "best" collaborators chosen.

So far, these experiments confirm what we see in practice, namely that the simple collaboration method involving just the best individuals from each population is quite robust even when there is cross-population epistasis. However, what is still not clear is when it fails. To understand that better, we focus on the the various forms of cross-population epistasis.

5 The Effects of Linkage Bias on Collaboration

The decompositional bias of the previous section focused on potential mismatches between a problem's "natural" decomposition into m components and the number of CCEA populations p used. Even if $p = m$, there is still the question as to whether the CCEA breaks up the string so that each "natural" block is assigned its own population. If not, breaking up tightly linked bits can result in significant cross-population epistasis. In general, the degree to which linked bits in a block are assigned to the same population for the purposes of representation can be thought of as linkage bias.

5.1 Controlling Linkage Bias Experimentally

Again it isn't hard to construct a way of controlling this bias. We define a mask over the entire bit string which specifies to which population a given bit belongs, $\mathcal{M} \in \{1, 2, \ldots, p\}^n$. Note that in the case of these mask definitions, the superscript suggests repetition, and not an exponent. For problems like those in the previous section involving a series of m concatenated non-decomposable r-bit blocks, a mask which corresponds to the most biased linkage (i.e. is more closely aligned with the real problem) is $\mathcal{M}_s = 1^r 2^r \ldots p^r$. Coming up with a mask which is highly pathological is very problem dependent, but a mask which will turn out to be commonly quite bad is $\mathcal{M}_h = (123 \ldots p)^r$. Here every bit in a block is distributed to every population, resulting in the likelihood of a high degree of cross-population epistasis.

As noted earlier, any increase in the amount of cross-population epistasis is likely to make a problem more difficult to solve using a CCEA. The question at hand is whether adopting a more complex collaboration method can alleviate these difficulties. By applying different types of masks, which distribute different pieces of the blocks of the problem to different degrees, we can explore the affect that varying degrees of linkage bias have on collaboration methods.

5.2 Effects of Collaboration Methods

We begin by considering again the `LeadingOnes-OnesMax` problem, assuming $m = p = 2$. Using the $\mathcal{M}_s = 11 \ldots 1 \;\; 22 \ldots 2$ mask presents us the same problem we've already discussed, where there is no cross-population epistasis, while the mask $\mathcal{M}_h = 1212 \ldots 12 \;\; 2121 \ldots 21$ creates a situation with very strong cross-population epistasis. Using the same experimental setup as before, we studied the effects that these two masks had on the choice of collaboration methods. The results are presented in Table 3.

Table 3. Steady state CCEA results on the `LeadingOnes-OnesMax` problem. Each value represents the mean number of evaluations needed to reach the optimum out of 50 trials. The left table represents a linkage bias which uses the $\mathcal{M}_s$ mask, while the right uses the $\mathcal{M}_h$ mask.

$\mathcal{M}_s$	# Collaborators		
	1	2	3
s-best	15592.9	30846.8	48259.6
s-random	16000.7	31320.6	46406.9

$\mathcal{M}_h$	# Collaborators		
	1	2	3
s-best	15305.4	28939.1	45756.5
s-random	17862.5	32802.2	47439.4

Table 4. Steady state CCEA results of the `LeadingPairedOnes-OnesMax` problem. Each value represents the mean value obtained after 100,000 function evaluations out of 50 trials. The left table represents a linkage bias which uses the $\mathcal{M}_s$ mask, while the right uses the $\mathcal{M}_h$ mask.

$\mathcal{M}_s$	# Collaborators		
	1	2	3
s-best	3328.9	2086.5	1590.1
s-random	3366.7	2152.0	1784.1

$\mathcal{M}_h$	# Collaborators		
	1	2	3
s-best	647.6	705.6	761.0
s-random	1247.3	1383.4	1469.1

Differences between the means for s-best and s-random groups for $\mathcal{M}_h$ are significant for one and two collaborators, but not for three. There are no statistically significant differences between these groups (for the same number of collaborators) for the simpler linkage bias.

Once again simply distributing the epistatic linkages across the population boundaries is insufficient to require that a more complicated collaboration method be used. This may seem surprising at first, but note that, for this particular problem, introducing such masks does not change the type of cross-population epistasis, only its degree. Moreover, although not germane to our question, it is interesting to note that in this particular case it seems that increasing the mixing seems to *improve* performance versus the $\mathcal{M}_s$ mask (this is significant for all but the 3-random case). In the case of our generational CCEA experiments, this was true to statistical significance for all groups.

What we have failed to construct so far is a case of the most difficult form of cross-population epistasis: namely, when neither the sign nor the magnitude of the interaction can be reliably predicted. There is a simple change to the current problem that will result in cross-population epistasis of this type, namely by changing the LeadingOnes component to count only paired ones. More formally, we defined a `LeadingPairedOnes-OnesMax` problem given by $f(\boldsymbol{x}) = 2k \sum_{i=1}^{k} \prod_{j=1}^{\frac{i}{2}-1} (x_j x_{j+1}) - \sum_{i=1}^{k} x_i$. Interestingly, this problem is so much more difficult that the optimum was frequently never found. As a consequence Table 4 represents the mean fitness values obtained after a constant 100,000 evaluations (here higher is better).

Now we see exactly the reverse situation. Although it is clearly better to select a single collaborator when there are no cross-population epistatic linkages of this type, as soon as those linkages are spread across populations a more complex collaboration mechanism is required. In the latter case, increasing the number of collaborators does in fact result in statistically improved performance, as does picking the individual randomly rather than greedily. However, in the $\mathcal{M}_s$ case, with the exception of the three collaborator groups, there is no statistical reason to pick randomly over picking the best.

6 Conclusions

Choosing an effective representation of a problem is a critical design decision for any EA-based problem solving. In the case of CCEAs, representation decisions involve additional choices as to how to decompose a problem effectively. Without sufficient *a priori* knowledge about a particular problem (the usual case), particular representations can introduce biases related to the degree to which they match the underlying problem characteristics. This in turn can affect the choice of collaboration mechanism to be used.

Using several well understood pseudo-boolean functions, we explored the effects of two kinds of representational bias (decomposition bias and linkage bias) on collaboration mechanism choices. We were able to show, somewhat surprisingly, that decompositional bias does not appear to dramatically affect the choice of collaboration mechanisms. For the practitioner, this means that the standard collaboration mechanism of choosing the single best individual from each of the other subpopulations is reasonably robust across decompositional biases.

Equally surprising was the fact that this same collaboration mechanism can be robust across different linkages biases as well, but not always. To understand this better, the cross-population epistatic interactions resulting from these linkage biases were studied in more detail. In some cases these interaction are positively or negatively correlated with fitness, in the sense that, though the magnitude of the fitness change depends on the relationship between the linear and non-linear pieces of the problem, the sign does not. The standard collaboration mechanism worked fine for these cases.

However, in the case where both the sign and the magnitude of the fitness contribution are uncorrelated, the standard collaboration mechanism breaks down and a more complex mechanism is required. Intuitively, in such situations additional samples are required to obtain reasonable estimates of fitness.

Clearly, the results presented here are preliminary in nature, and a more thorough examination of these issues is needed. However, we believe these results already provide useful guidance to the CCEA practitioner. An interesting open question for EA design in general is whether this notion of different types of epistasis will also help clarify the effects of gene linkages within a genome.

References

1. M. Potter and K. De Jong. A cooperative coevolutionary approach to function optimization. In Y. Davidor and H.-P. Schwefel, editors, *Proceedings of the Third International Conference on Parallel Problem Solving from Nature (PPSN III)*, pages 249–257. Springer-Verlag, 1994.

2. M. Potter. *The Design and Analysis of a Computational Model of Cooperative CoEvolution.* PhD thesis, George Mason University, Fairfax, Virginia, 1997.
3. L. Bull. Evolutionary computing in multi-agent environments: Partners. In Thomas Baeck, editor, *Proceedings of the Seventh International Conference on Genetic Algorithms (ICGA)*, pages 370–377. Morgan Kaufmann, 1997.
4. R. Paul Wiegand, William Liles, and Kenneth De Jong. An empirical analysis of collaboration methods in cooperative coevolutionary algorithms. In Spector [15], pages 1235–1242.
5. R. Watson and J. Pollack. Coevolutionary dynamics in a minimal substrate. In Spector [15], pages 702–709.
6. S. Ficici and J. Pollack. Challenges in coevolutionary learning: Arms–race dynamics, open–endedness, and mediocre stable states. In Adami et al, editor, *Proceedings of the Sixth International Conference on Artificial Life*, pages 238–247, Cambridge, MA, 1998. MIT Press.
7. D. Cliff and G. F. Miller. Tracking the red queen: Measurements of adaptive progress in co–evolutionary sumulations. In *Proceedings of the Third European Conference on Artificial Life*, pages 200–218. Springer–Verlag, 1995.
8. M. Vose. *The Simple Genetic Algorithm.* MIT Press, 1999.
9. S. Ficici and J. Pollack. A game-theoretic approach to the simple coevolutionary algorithm. In M. Schoenauer, K. Deb, G. Rudolph, X. Yao, E. Lutton, J.J. Merelo, and H.-P. Schwefel, editors, *Proceedings of the Sixth International Conference on Parallel Problem Solving from Nature (PPSN VI)*, pages 467–476. Springer-Verlag, 2000.
10. R. Paul Wiegand, William Liles, and Kenneth De Jong. Analyzing cooperative coevolution with evolutionary game theory. In D. Fogel, editor, *Proceedings of CEC 2002.* IEEE Press, 2002. (To appear).
11. J. Paredis. Coevolutionary computation. *Artificial Life Journal*, 2(3), 1996.
12. L. Bull. On coevolutionary genetic algorithms. *Soft Computing*, 5:201–207, 2001.
13. R. Salomon. Performance degradation of genetic algorithms under coordinate rotation. In L. Fogel, P. Angeline, and T. Bäck, editors, *Proceedings of the Fifth Annual Conference on Evolutionary Programming*, pages 153–161. MIT Press, 1996.
14. Yuval Davidor. Epistasis variance: A viewpoint on ga-hardness. In G. Rawlins, editor, *Foundations of Genetic Algorithms (FOGA)*, pages 23–35. Morgan Kaufmann, 1990.
15. L. Spector, editor. *Proceedings of the Genetic and Evolutionary Computation Conference (GECCO) 2001.* Morgan Kaufmann, 2001.

Multiobjective Optimization

Parallel and Hybrid Models for Multi-objective Optimization: Application to the Vehicle Routing Problem

Nicolas Jozefowiez[1], Frédéric Semet[2], and El-Ghazali Talbi[1]

[1] LIFL, USTL, 59655 Villeneuve d'Ascq CEDEX, France
{jozef,talbi}@lifl.fr
[2] LAMIH, UVHC, 59393 Valenciennes CEDEX, France
frederic.semet@univ-valenciennes.fr

Abstract. Solving a multi-objective problem means to find a set of solutions called the Pareto frontier. Since evolutionary algorithms work on a population of solutions, they are well-adapted to multi-objective problems. When they are designed, two purposes are taken into account: they have to reach the Pareto frontier but they also have to find solutions all along the frontier. It is the intensification task and the diversification task. Mechanisms dealing with these goals exist. But with very hard problems or benchmarks of great size, they may not be effective enough. In this paper, we investigate the utilization of parallel and hybrid models to improve the intensification task and the diversification task. First, a new technique inspired by the elitism is used to improve the diversification task. This new method must be implemented by a parallel model to be useful. Second, in order to amplify the diversification task and the intensification task, the parallel model is extended to a more general island model. To help the intensification task, a hybrid model is also used. In this model, a specially defined parallel tabu search is applied to the Pareto frontier reached by an evolutionary algorithm. Finally, those models are implemented and tested on a bi-objective vehicle routing problem.

1 What is to Solve a Multi-objective Problem?

The solution of a multi-objective problem (MOP) is not a unique optimal solution but a set of solutions called the Pareto frontier. These solutions, called Pareto optimal solutions, are the non-dominated solutions. A solution $\boldsymbol{y} = (y_1, y_2, \ldots, y_n)$ dominates[1] a solution $\boldsymbol{z} = (z_1, z_2, \ldots, z_n)$ if and only if $\forall\ i \in \{1 \ldots n\},\ y_i \leq z_i$ and $\exists\ i \in \{1 \ldots n\},\ y_i < z_i$.

But obtaining the complete set of Pareto optimal solutions for a multi-objective problem may be impossible to attain. That fact tends to discard exact methods. Instead, a *good* approximation to the Pareto set is sought. In this case, while solving a MOP, two purposes must be reached. On one hand, the algorithm

[1] In this paper, we assume that all the objectives must be minimized.

J.J. Merelo Guervós et al. (Eds.): PPSN VII, LNCS 2439, pp. 271–280, 2002.

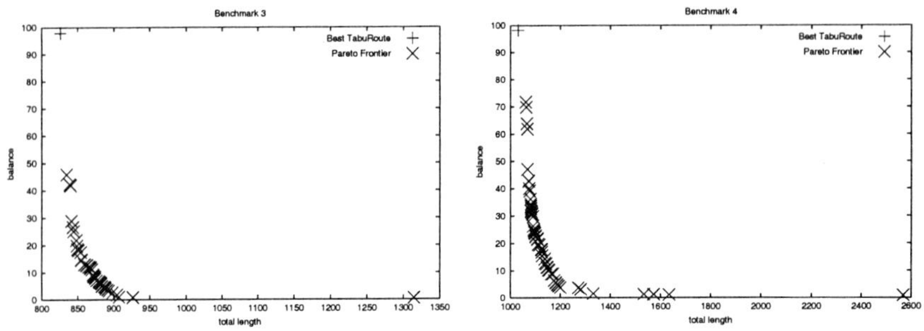

Fig. 1. Comparison between the Pareto frontier and the best solution of TabuRoute.

must converge to the optimal Pareto frontier. It is called the *intensification task*. On the other hand, a good approximation of the optimal Pareto frontier is required. The identified solutions should be well diversified along the frontier. It is called the *diversification task*.

Existing mechanisms are used to take those two goals into account. For example, one way to improve the intensification task is to arrange the solutions according to the Pareto dominance. It is what *ranking* methods like NSGA [16] do. The diversification task can be improved by ecological niche methods like the sharing [6]. However, with very hard instances or large scale benchmarks, these methods may not be sufficient. In this paper, we investigate the utilization of the parallelization and the hybridization to improve the intensification task and the diversification task. Section 2 describes the parallel multi-objective evolutionary algorithm (MOEA) we use as well as a new method to help the diversification task. Section 3 presents the hybrid model and a specially designed multi-start tabu search. Section 4 shows an implementation of those techniques for a bi-objective Vehicle Routing Problem (VRP). Finally, in section 5, the contribution of the different mechanisms is evaluated in order to show their interest.

2 A Parallel MOEA

While developing a MOEA for a bi-objective VRP, the authors observe that the best-known solutions for one of the criteria were bad for the other objective. It seemed that the algorithm tended to converge prematurely to an area of the objective space. Then the used sharing method [6] was not able to fill the gap to the best-known solutions for one of the criteria (figure 1). In this article, we propose a technique, the *Elitist Diversification*, whose purpose is to maintain the population of the MOEA diversified. It is inspired by the elitism. The elitism is a way to speed up and improve the intensification task. It consists in maintaining an archive that will contain the Pareto solutions encountered during the search. Some solutions of this archive are included into the main population of the MOEA at each generation. In the elitist diversification, other archives are added. Those archives contain the solutions that are Pareto optimal when one of the

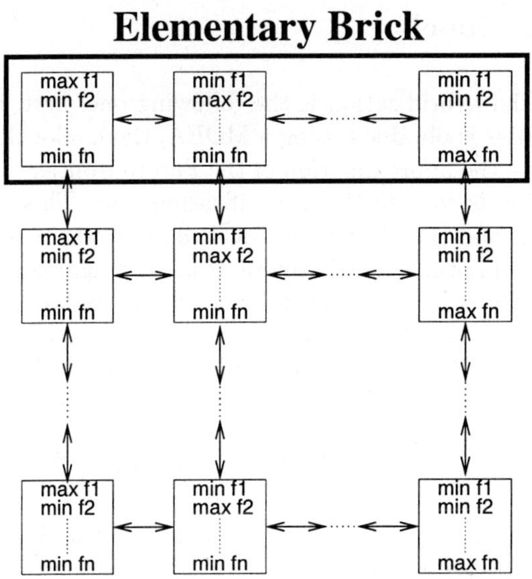

Fig. 2. The parallel model - the toric structure is not shown in order not to obfuscate the figure. The standard elitist archive is present in each island.

criteria is maximized instead of being minimized. For example, in the case of a bi-objective problem where f_1 and f_2 are the two objective functions, there are three archives. One is the standard elitist archive which contains the solutions that are Pareto optimal when both f_1 and f_2 are minimized. A second archive corresponds to the Pareto front when f_1 is minimized and f_2 is maximized. The last archive corresponds to the case where f_1 is maximized and f_2 is minimized. As in the elitism strategy, solutions from these new archives are included into the population of the MOEA at each generation. The role of the solutions of the new archives is to attract the population toward unexplored areas, and then to avoid the algorithm to converge prematurely to an area of the objective space. However, if all the archives are embedded in the same MOEA, the improvement of the diversification task is less important. This leads to the co-operative model that is the elementary brick in figure 2. In this model, an island has only one new archive. The standard elitist archive is present in each island. With a certain period in terms of the number of generations, the islands exchange their standard elitist archives. The communication topology is shown in figure 2.

To speed up the search and help the intensification task and the diversification task, a more general island model is defined. It consists in using more than one elementary brick. The connection between the islands is shown in figure 2. Therefore, an island has four neighbors. Two of them have the same kind of second archive. The communication between the islands is defined as follows: an island sends its standard elitist archive to all its neighbors. But it only sends the elitist diversification archive to the two neighbors that have the same kind of archive. It means the neighbors that are not in the same elementary brick.

3 A Hybrid Model

The principle of the hybridization is the following one: first, an approximation of the Pareto frontier is obtained using a MOEA; then, a local search[2] is applied in order to improve the approximation of the Pareto frontier. Therefore, the role of the hybridization is to help the intensification task. This model has already been studied in [4] and [17]. In [4], the local method was a simple local search, it cannot avoid local optima. Furthermore, it uses an aggregation method which needs to be correctly tuned and which is not able to find all the Pareto optimal solutions. Finally, for each solution of the frontier of the MOEA, only one new solution can be found. The method introduced in [17] has the main drawback to be very costly and it cannot be parallelized to reduce the computational time needed. Moreover, this method is not able to avoid the local optima.

The method introduced now, Π^2-TS (*Parallel Pareto Tabu Search*) deals with those difficulties. It is based on a tabu search approach. The starting points are the solutions of the Pareto set found by the MOEA. Thus, the meta-heuristic can escape from local optima. Furthermore, each tabu search correspond to a parallel process. No aggregation method is used. The selection of the next solution is based on the Pareto dominance. The neighborhood associated with the current solution can be partitioned into three subsets N_1, N_2 and N_3. N_1 contains the neighbors that dominate the current solution. N_2 includes the neighbors that do not dominate the current solution and are not dominated by the current solution either. N_3 is the subset of neighbors that are dominated by the current solution. Then, the next solution is the more dominating individual of N_1. If N_1 is empty, N_2 and N_3 are considered in sequence.

The tabu search algorithm does not provide a unique solution, but a set of solutions that are not dominated. Therefore, a small archive is associated with each tabu search. It contains all the non-dominated solutions found during the search.

To intensify the search, each tabu search focuses on a limited area of the objective space. An example of space restriction in a bi-objective case is shown in figure 5.

4 Application to a Bi-objective VRP

4.1 A Bi-objective VRP

The Vehicle Routing Problem (VRP) is a well-known problem often studied since the end of the 50's. Many practical applications exist for various industrial areas (eg. transport, logistic, workshop problem ...). The VRP has been proved NP-hard [10] and applied solution methods range from exact methods [7] to specific heuristics [8], and meta-heuristics [8][14][13].

[2] Here, the term *local search* is used in a general way. As a matter of fact, it can be a local search, a tabu search, a simulated annealing ...

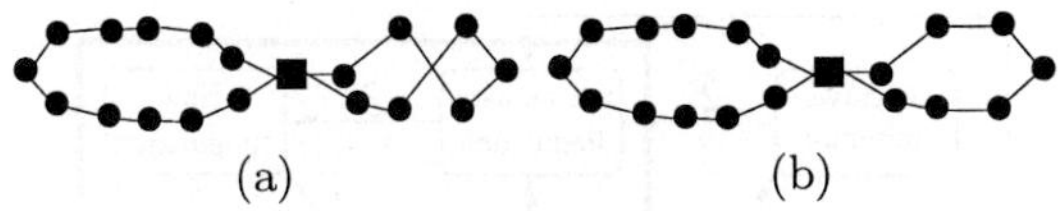

Fig. 3. (a) is better-balanced than (b), but (b) does not artificially improve the balance.

The most elementary version of the vehicle routing problem is the *Capacitated Vehicle Routing Problem* (CVRP). The CVRP is a graph problem that can be described as follows : n customers must be served from a unique depot a quantity q_i of goods ($i = 1, \dots, n$). To deliver those goods, a fleet of vehicles with a capacity Q is available. A solution of the CVRP is a collection of tours where each customer is visited only once and the total tour demand is at most Q.

Existing studies of the VRP are almost all concerned with the minimization of the total distance only. The model studied here introduces a second objective whose purpose is to balance the length of the tours. This new criterion is expressed as the minimization of the difference between the length of the longest tour and the length of the shortest tour. As far as we know, the balancing of the tours as a criterion has been studied in two other cases [9][15]. However, the balance of a solution was not expressed in the same way.

4.2 A Parallel Pareto Genetic Algorithm

The implemented MOEA is based on a generational Genetic Algorithm (GA). The GAs have widely been used to solve multi-objective problems, as they are working on a population of solutions [3]. Two implementations of the GA were implemented. They differ by the crossovers they use. In a first version, the RBX [13] and the split crossover, which is based on the GA defined by C. Prins [14], are used. The split crossover does not work well on benchmarks with clustered customers. With those benchmarks, the split crossover is replaced by the SBX crossover [13]. The mutation operator is the Or-opt. It consists in moving 1 to 3 consecutive customers from a tour to another position in the same tour or to another tour. A 2-opt local search is applied to each tour of each solution. It has three purposes: it allows the solution to be less chaotic, it can improve the total length, and it does not allow the second criterion to be distorted (as shown in figure 3).

In order to favor the intensification task and the diversification task, multi-objective mechanisms are used. The ranking function NSGA [16] is used. To maintain diversity along the Pareto frontier, we use a sharing technique that aims to spread the population along the Pareto frontier by penalizing individuals that are strongly represented in the population. The used elitism is the one defined in section 2.

Preliminary experiments have shown that the second criterion is easier to solve than the first one. Therefore, to save computational resources, only the part of the parallel model corresponding to the minimization of the total length and the maximization of the second criterion was implemented. Moreover, main-

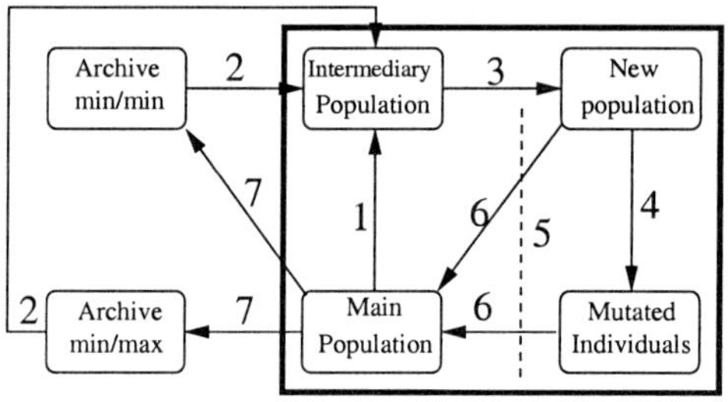

Fig. 4. The genetic algorithm.

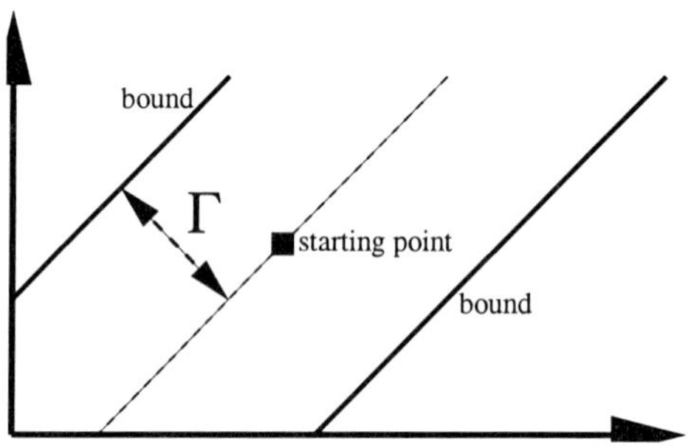

Fig. 5. Restriction of the objective space.

taining the archives is costly. To avoid that and a too strong pressure, we use a small archive, as shown by Zitzler and Thiele [19]. To reduce the size of the archive, a clustering algorithm, *average linking method* [12], is used. It has been proved to work well with data such as the Pareto frontier [19].

The structure of the GA is summarized in figure 4. The different steps are:

1. Selection: SUS.
2. Elitism.
3. Recombination.
4. Mutation.
5. 2-opt local search.
6. Replacement.
7. Updating of the archives.

4.3 A Parallel Pareto Tabu Search for the VRP

The model proposed in section 3 is used. The following implementation of Π^2-TS is made. The starting points are the Pareto frontier found by the GA. The neighborhood operator is the or-opt. The tabu list is defined as follows: when a customer is moved from a tour, it cannot be put back into that tour for N iterations. As suggested in [5], the solutions that violate the capacity constraint can be accepted. The zone from the objective space associated to a tabu search is defined as follows: they are the solutions so that the distance between a solution and the line of slope $\frac{1}{2}$ which goes through the starting point is smaller than a value Γ. It is illustrated in figure 5.

Table 1. Contribution to the diversification task.

Problem	E(W, Wo)	E(Wo, W)
1 (50)	0.84	0.94
2 (75)	0.73	0.67
3 (100)	0.83	0.74
4 (150)	0.85	0.74
5 (199)	0.86	0.79
11 (120)	0.92	0.91
12 (100)	0.93	0.82

5 Evaluation

The evaluation[3] was done on the standard benchmarks of Christofides [2]. More precisely on the benchmarks numbers 1, 2, 3, 4, 5, 11 and 12. They correspond to CVRP instances. The first five benchmarks correspond to maps where the customers are uniformly distributed on the map, while benchmarks 11 and 12 correspond to clustered maps.

To evaluate the contribution of the elementary brick parallel model, we use the *entropy* measure [11][1]. The entropy indicator gives an idea about the diversity of a Pareto front in comparison with another Pareto front. It is defined as follows: Let PO_1 and PO_2 be two sets of solutions. Let PO^* be the set of optimal Pareto solutions of $PO_1 \cup PO_2$. Let N_i be the cardinality of solutions of $PO_1 \cup PO^*$ which are in the niche of the ith solution of $PO_1 \cup PO^*$. Let n_i be the cardinality of solutions PO_1 which are in the niche of the ith solution of $PO_1 \cup PO^*$. Let C be the cardinality of the solutions of $PO_1 \cup PO^*$. Let $\gamma = \sum_{i=1}^{C} \frac{1}{N_i}$ be the sum of the coefficients affected to each solution. The more concentrated a region of the solution space, the lower the coefficient of its solutions. Then the following formula is applied to evaluate the entropy E of PO_1 relatively to the space occupied by PO^*:

$$E(PO_1, PO_2) = \frac{-1}{\lg \gamma} \sum_{i=1}^{C} \left(\frac{1}{N_i} \frac{n_i}{C} \lg \frac{n_i}{C}\right) \quad (1)$$

The results are shown in table 1. In this table, E(W, Wo) is the entropy of the algorithm with the elitist diversification compared to the algorithm without. E(Wo, W) is the reverse. Except for the first benchmark, the new mechanism improves the diversity. For the first benchmark, the GA was able to reach the best solution for the first objective without the new mechanism which is therefore useless. Moreover, the new mechanism has the effect to slow down the convergence for the second objective (figure 6). However, this consequence is compensated by the tabu search of the hybridization. It can also be avoided by using the full parallel model described in section 2. To save computational resources for the convergence to the total distance objective which is more difficult, only half of the model was implemented. Furthermore, it can be dealt by the hybridization.

[3] The results can be found at the url http://www.lifl.fr/~jozef.

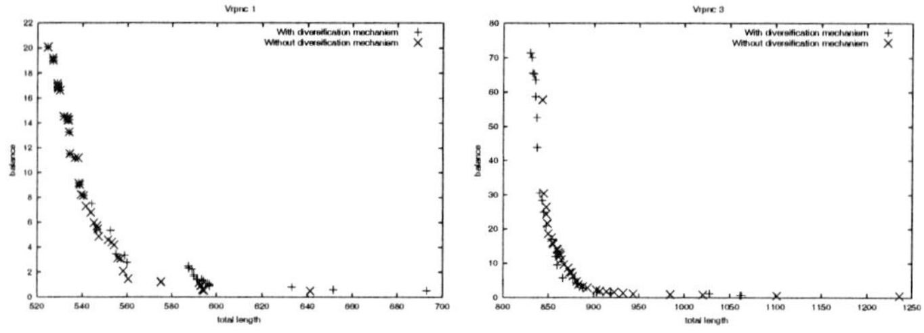

Fig. 6. Two examples of Pareto frontiers for the instances 1 and 3.

Table 2. Contribution of the parallelization.

Problem	E(P, NP)	E(NP, P)	C(P, NP)	C(NP, P)
1 (50)	0.86	0.78	0.75	0.25
2 (75)	0.85	0.69	0.91	0.08
3 (100)	0.90	0.73	1.00	0.00
4 (150)	0.90	0.78	0.85	0.15
5 (199)	0.91	0.71	0.88	0.12
11 (120)	0.94	0.70	0.73	0.27
12 (100)	0.92	0.83	0.70	0.30

In order to show the interest of the general parallel model, we have compared with the results obtained when only one island is used. The measures used are the *entropy* measure E, which has already been used before, and the *contribution* measure C. It quantifies the domination between two sets of non-dominated solutions. The contribution of a set of solutions PO_1 relatively to a set of solutions PO_2 is the ratio of non-dominated solutions produced by PO_1. Let D be the set of solutions in $PO_1 \cap PO_2$. Let W_1 (resp. W_2) be the set of solutions in PO_1 (resp. PO_2) that dominate some solutions of PO_2 (resp. PO_1). Let L_1 (resp. L_2) be the set of solutions in PO_1 (resp. PO_2) that are dominated by some solutions of PO_2 (resp. PO_1). Let N_1 (resp. N_2) be the other solutions of PO_1 (resp. PO_2): $N_i = PO_i \setminus (C \cup W_i \cup L_i)$. Let PO^* be the set of Pareto solutions of $PO_1 \cup PO_2$. So, $\|PO^*\| = \|D\| + \|W_1\| + \|N_1\| + \|W_2\| + \|N_2\|$. The *contribution* of a set PO_1 relatively to PO_2 is stated as follows:

$$C(PO_1, PO_2) = \frac{\frac{\|D\|}{2} + \|W_1\| + \|N_1\|}{\|PO^*\|} \tag{2}$$

The results are given in table 2, where P is for *Parallel algorithm* and NP for *Non-Parallel algorithm*.

To quantify the contribution of the local search, the generational distance [18] is used. Normally it is used between a front and the optimal one. But regarding the GA front, the front given by the local search can be considered as an optimal one. The results are reported in table 3 for five different runs for each problem.

One of the main problems in multi-objective optimization is that optimal Pareto frontiers are not known. We can only compare the value obtained for the

Table 3. Contribution of the local search.

Problem	Run 1	Run 2	Run 3	Run 4	Run 5
1 (50)	0.37	0.48	0.29	0.53	3.39
2 (75)	1.89	0.60	2.33	1.46	1.15
3 (100)	0.41	0.81	1.13	0.62	0.85
4 (150)	24.16	1.73	3.67	3.71	1.82
5 (199)	6.11	20.40	5.13	2.89	4.54
11 (120)	12.44	3.44	1.10	2.18	4.50
12 (100)	1.31	0.99	0.41	2.01	0.62

Table 4. Comparison with the best-known total length.

Problem	Best-known	Best found	%	Avg. found	%
1 (50)	524.61	524.61	0.00	525.89	0.24
2 (75)	835.26	840.00	0.56	846.19	1.38
3 (100)	826.14	829.43	0.39	832.62	0.78
4 (150)	1028.42	1059.09	2.98	1069.32	3.97
5 (199)	1291.45	1353.52	4.80	1365.56	5.73
11 (120)	1042.11	1042.11	0.00	1047.49	0.51
12 (100)	819.56	819.56	0.00	819.56	0.00

total length criterion with the best-known solutions. The results are shown in table 4.

6 Conclusion

In this paper, we have investigated the utilization of parallel and hybrid metaheuristics to improve the intensification task and the diversification task. First, a new mechanism, the elitist diversification, was proposed to favor the diversification. Its utilization leads us to the design of a parallel model that improves the intensification and the diversification. Second, a tabu search, Π^2-TS, was specially designed for the hybridization with a MOEA. These methods were implemented and tested on a bi-objective Vehicle Routing Problem. The measures show that the proposed techniques ensure a better convergence to the Pareto frontier and help the diversification of the found set. Furthermore, the implemented algorithm leads to good results for the VRP studied.

References

1. Basseur, M., Seynhaeve, F., Talbi, E-G.: Design of multi-objective evolutionary algorithms: application to the flow-shop scheduling problem. Proceedings of 2002 Congress on Evolutionary Computation (2002)
2. Christofides, N., Mingozzi, A., Toth, P., Sandi, C.: Combinatorial optimization. Chapter 11. John Wiley (1979)
3. Coello Coello, C. A.: An updated survey of GA-based multiobjective optimization techniques. Technical Report RD-98-08, LANIA, Mexico (1998)

4. Deb, K., Goel, T.: A hybrid multi-objective evolutionary approach to engineering shape design. Proceedings of Evolutionary Multi-criterion Optimization Conference (2001) 385–399
5. Gendreau, M., Hertz, A., Laporte, G.: A tabu search heuristic for the vehicle routing problem. Management Science, Vol. 40 (1994) 1276–1290
6. Goldberg, D. E., Richardson, J.: Genetic algorithms with sharing for multi-modal function optimization. Proceedings of the Sec. Int. Conf. on Genetic Algortihms (1987) 41–49
7. Laporte, G.: Exact algorithms for the traveling salesman problem and the vehicle routing problem. Les Cahiers du GERAD G-98-37 (1998)
8. Laporte, G., Gendreau, M., Potvin, J-Y., Semet, F.: Classical and modern heuristics for the vehicle routing problem. Les Cahiers du GERAD G-99-21 (1999)
9. Lee, T., Ueng, J.: A study of vehicle routing problems with load-balancing. Int. J. Physical Distribution and Logistics Management, Vol. 29 No. 10 (1999) 646–658
10. Lenstra, J. K., Rinnooy Kan, A. H. G.: Complexity of vehicle routing and scheduling problem. Networks, Vol. 11 (1981) 221–227
11. Meunier, H., Talbi, E-G., Reininger, P.: A multiobjective genetic algorithm for radio network optimization. Proceedings of the 2000 Congress on Evolutiorary Computation (2000) 314–324
12. Morse, J. N.: Reducing the size of non-dominated set: Prunning by clustering. Computers and Operations Research, Vol. 7 No. 1-2 (1980) 55–56
13. Potvin, J-Y., Bengio, S.: The vehicle routing problem with time windows part II: genetic search. INFORMS Journal on Computing, Vol. 8 No. 2 (1996)
14. Prins, C.: A simple and effective evolutionary algorithm for the vehicle routing problem. In Proceedings of MIC'2001 (2001)
15. Ribeiro, R., Loureno, H. R.: A multi-objective model for a multi period distribution management problem. In Proceedings of MIC'2001 (2001)
16. Srivinas, N., Deb, K.: Multiobjective optimization using non-dominated sorting in genetic algorithms. Evolutionary Computation, Vol. 2 No. 3, (1995) 279–285
17. Talbi, E-G., Rahoual, M., Mabed, M. H., Dhaennens, C.: A hybrid evolutionary approach for multicriteria optimization problems: Application to the flow shop. Proceedings of Evolutionary Multi-criterion Optimization Conference (2001) 416–428
18. Van Veldhuizen, D. A., Lamont, G. B.: On measuring multiobjective evolutionary algorithm performance. Proceedings of 2000 Congress on Evolutionary Computation (2000)
19. Zitzler, E., Thiele, L.: Multiobjective evolutionary algorithms: a comparative case study and the strength pareto approach. IEEE Transaction on Evolutionary Computation, Vol. 3 No. 4 (1999) 257–271.

Multiobjective Design Optimization of Merging Configuration for an Exhaust Manifold of a Car Engine

Masahiro Kanazaki[1], Masashi Morikaw[2],
Shigeru Obayashi[1], and Kazuhiro Nakahashi[2]

[1] Institute of Fluid Science, Tohoku University, Sendai 980-8577, Japan
kanazaki@mail.cc.tohoku.ac.jp
http://www.reynolds.ifs.tohoku.ac.jp/edge/
[2] Dept. of Aeronautics and Space Engineering, Tohoku University, Sendai 980-8579, Japan

Abstract. Multiobjective design optimization system of exhaust manifold shapes for a car engine has been developed using Divided Range Multiobjective Genetic Algorithm (DRMOGA) to obtain more engine power as well as to achieve less environmental impact. The three-dimensional manifold shapes are evaluated by the unstructured, unsteady Euler code coupled with the empirical engine cycle simulation code. This automated design system using DRMOGA was confirmed to find Pareto solutions for the highly nonlinear problems.

1 Introduction

To improve intake/exhaust system performance of a car engine, many design specifications are required. In addition, car engines today are required not only to have more engine power, but also to be more environmentally friendly. Exhaust gas should be kept at high temperature in the exhaust pipe especially at low rpm conditions because the catalyst located at the end of the exhaust pipe will absorb more pollutant in high temperature conditions. Exhaust gas should also be led from the piston chambers to the exhaust manifold smoothly to maximize the engine power especially at high rpm conditions. Such design usually has to be performed by trial and error through many experiments and analyses. Therefore, an automated design optimization is desired to reduce technical, schedule, and cost risks for new engine developments.

In the previous study, the design system that could account for multiple design objectives has been developed and the exhaust manifold excellent at the emission control was obtained [1]. However, the engine power was not improved very well, because the baseline manifold was for the car engine of a popular car. In this paper, the high power engine of a sports car is considered for multiobjective optimization to increase the engine power as well as to reduce the environmental impact. The baseline manifold is shown in Fig. 1.

To further improve the design optimization system, this paper employs Divided Range Multiobjective Genetic Algorithm (DRMOGA) [2]. DRMOGA have the advantage over the previous MOGA [1], because it can retain diversity of the population better than MOGA.

J.J. Merelo Guervós et al. (Eds.): PPSN VII, LNCS 2439, pp. 281–287, 2002.

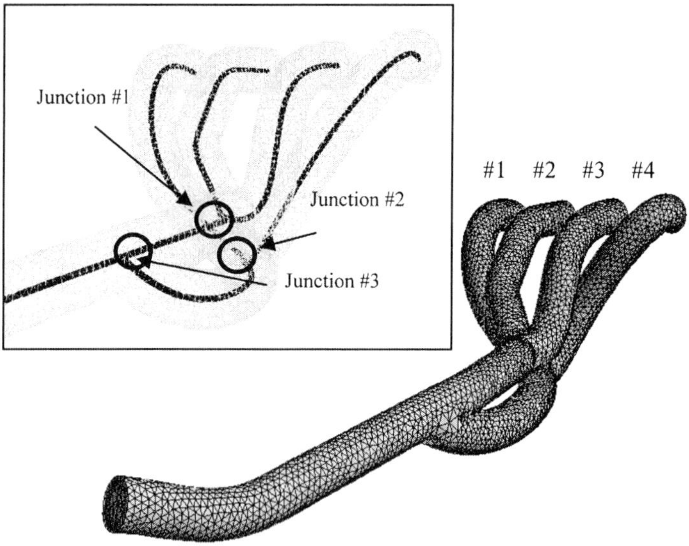

Fig. 1. The initial manifold shape and design variables as junction positions on pipe centerlines

2 Formulation of the Optimization Problem

2.1 Objective Functions

The objective functions considered here are to maximize the gas temperature at the end of the exhaust pipe at 1,500 rpm and to maximize the charging efficiency at 6,000 rpm, where the charging efficiency indicates the engine power. These two objectives are function of a flow over an engine cycle. A flow field of a manifold shape is computed by solving a unsteady three-dimensional inviscid flow code [3]. Unsteady boundary conditions for a flow to and from a manifold are simultaneously computed by using the one-dimensional, empirical engine cycle simulation code [1, 4].

2.2 Divided Range Multiobjective Genetic Algorithm

In this study, the automated design optimization system is developed by using DRMOGA [2]. DRMOGA is characterized by the operation where the individuals are divided into subpopulations.

DRMOGA procedure (Fig. 2) can be explained as follows. First, initial individuals are produced randomly and evaluated. Second, the division of individuals is performed based on the ranking of individuals sorted by values of a focused objective function f_i. Assuming m subpopulations for N individuals, N/m individuals will be allocated to each subpopulation. Then in each subpopulation, the existing MOGA is performed. After MOGA is performed for k generations, all of the individuals are gathered and they are divided again according to another objective function f_j. This focused function will be chosen in turn.

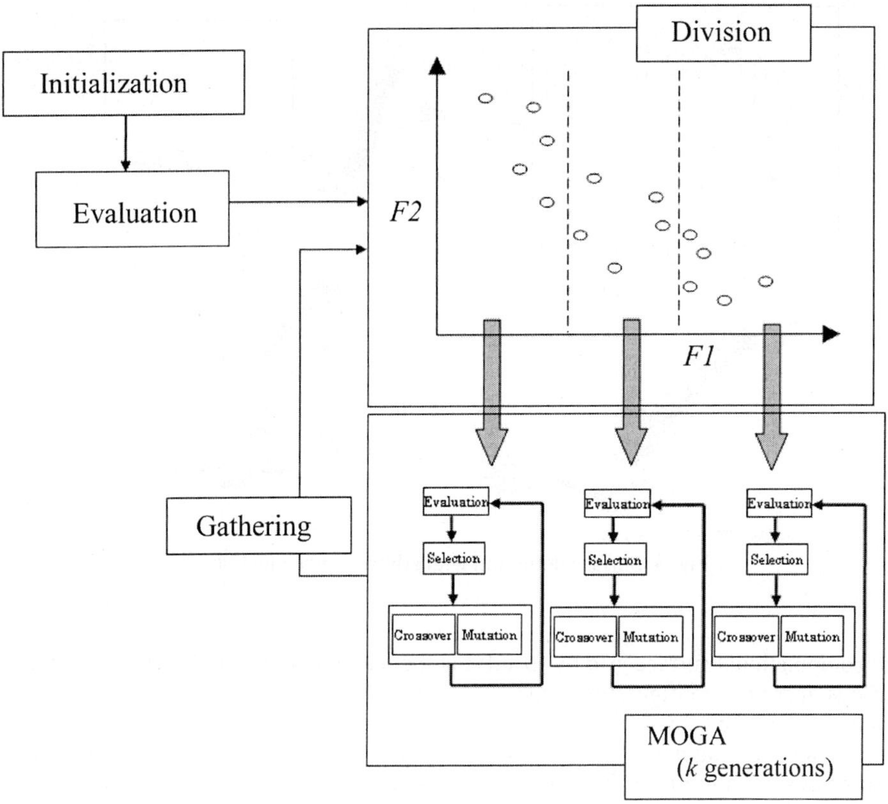

Fig. 2. Procedure of DRMOGA

DRMOGA is known to enhance the population diversity and to produce a better Pareto front [2]. The subdivision of the population based on alternative objective functions prevents the premature convergence to a Pareto front segment and introduces migration of individuals to neighboring Pareto front segments.

In this study, MOGA utilized real-number cording [5], the Pareto ranking method [6], BLX-0.5 [5] and Best-N selection [7] and mutation rate was set to 0.1. Function evaluations in MOGA were parallelized on SGI ORIGIN2000 supercomputer system at the Institute of Fluid Science, Tohoku University. For DRMOGA, k was set to 8 and number of subpopulation was set to 2.

2.3 Geometry Definition

To generate a computational grid according to given design variables, an automated procedure to find a pipe junction from pipe centerlines was developed in the previous study [1] as shown in Fig. 3. In this method, temporary background grids are first generated from the given centerlines. Then the overlap region of the pipes is calculated and removed. The advancing-front method [8] is then applied to generate

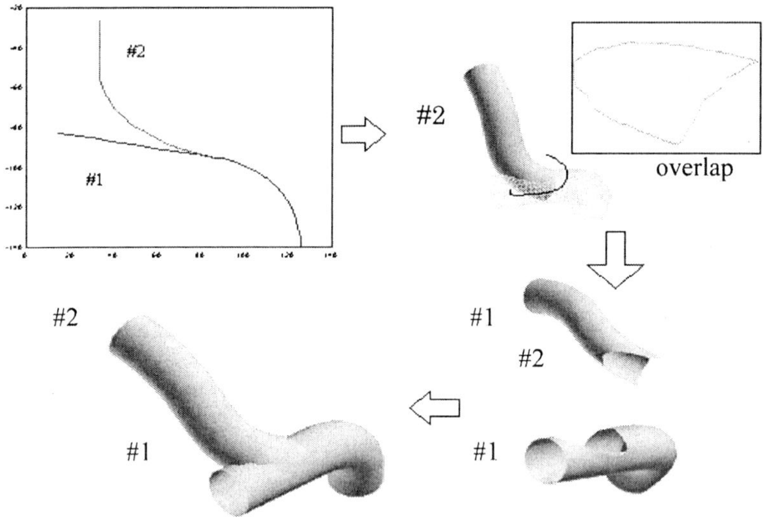

Fig. 3. Surface definition with arbitrary pipe junction

the computational surface grid by specifying the junction as a front. With this method, various merging configurations can be generated only by specifying the merging points on the pipe centerline.

In this study, the initial manifold shape is taken from an existing engine with four pistons as shown in Fig. 1. Topology of the merging configuration is kept unchanged. The pipe shape traveling from the port #2 to the outlet is also fixed. Three merging points on the pipe centerlines, junctions #1-3, are considered as design variables. Pipe centerlines of #1, 3 and 4 are then deformed similarly from the initial shapes to meet the designed merging points. The pipe shapes are finally reconstructed from the given pipe radius. This method allows the automated grid generation for arbitrary merging configuration defined by the pipe centerlines.

This study considered two design cases. The first case assumes a constant pipe radius for all pipes, therefore only three merging points are to be designed. In the second case, the pipe radius of the entire exhaust manifold is considered as a design variable because the pipe radius is known important for the performance of the exhaust manifold from the experiences at the industry. The pipe radius will change from 83% to 122% of the original radius. In the second case, three merging points and the pipe radius are to be designed simultaneously.

3 Design Optimization of an Exhaust Manifold

3.1 Design Problems

In this study, two design problems were considered. First, the design optimization of merging points was performed (Case 1). The population size was set to 32. The evolution was advanced for 25 generations.

Second, the merging points and pipe radius were optimized at the same time (Case 2). In this case, the population size was set to 64. The evolution was advanced for 29 generations.

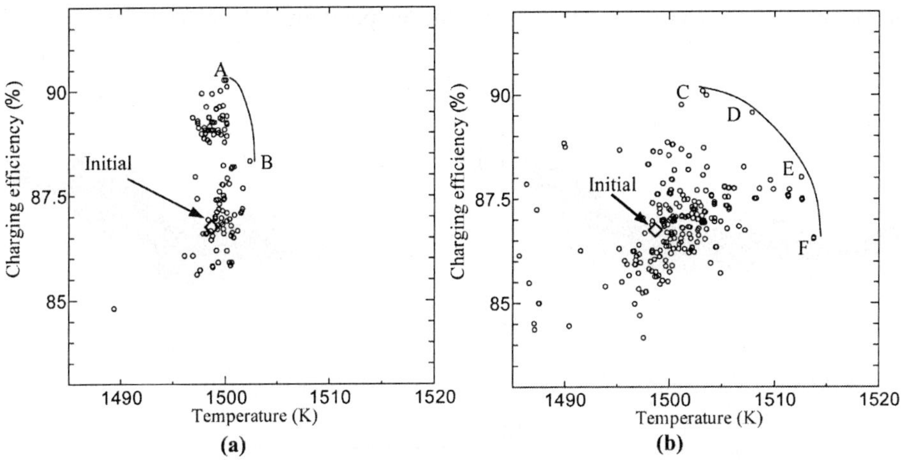

Fig. 4. All solutions produced by DRMOGA plotted in the objective function space; (a) Case 1, merging points optimization, (b) Case 2, merging points and pipe radius optimization

3.2 Comparison of Solution Evolutions

In Case 1, Pareto solutions were found as shown in Fig. 4(a). Many solutions achieve much higher charging efficiency than the initial geometry. These results suggest that the merging points are effective design variables to improve in the charging efficiency that indicates the engine power. However, the improvement in the temperature remained marginal.

In Case 2, Pareto solutions were found as shown in Fig. 4(b). Improvements in both objective functions were achieved. The Pareto front also confirms the tradeoff between the two objectives. This result suggests that the pipe radius is effective to maximize the temperature at the end of the exhaust manifold.

3.3 Comparison of Designed Shapes of Selected Pareto Solutions

Manifold geometries taken from two Pareto solutions in Case 1 are shown in Fig. 5(a). The initial shape is drawn with the mesh in a dark color. The solution A achieved the highest charging efficiency and the solution B achieved the highest temperature. The distance from the merging point #1 to #3 of the solution A became longer than that of the initial manifold. Such a merging shape is expected to reduce the interaction of the exhaust gas led from chambers and thus to lead to a high charging efficiency. On the other hand, the solution B has the tendency such that the distance from one junction to others becomes shorter.

Manifold geometries taken from four Pareto solutions in Case 2 are shown in Fig. 5(b). The solution C in Case 2 shows the same tendency as the solution A in

Case 1. The pipe radii of solutions C and D remained almost unchanged compared with that of the initial manifold. On the other hand, the solutions E and F achieved much higher temperature than the solutions B in Case 1. Moreover, their pipe radii became larger than that of the initial manifold. These comparisons reveal the tradeoff in maximizing the charging efficiency and the temperature of the exhaust gas.

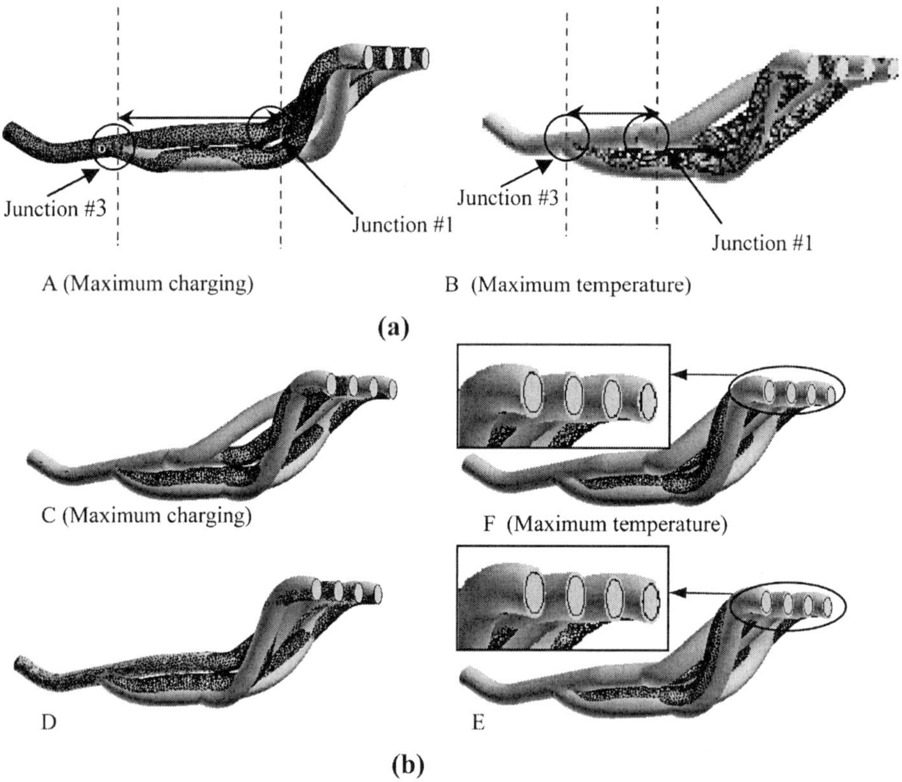

Fig. 5. Manifold shapes of selected from Pareto solutions; (a) Case 1, merging points optimization, (b) Case 2, merging points and pipe radius optimization

4 Concluding Remarks

An improved design optimization system of an exhaust manifold of a car engine has been developed. The design system employs DRMOGA. The three-dimensional manifold shapes are evaluated by the unstructured, unsteady Euler code coupled with the empirical engine cycle simulation code. Computational grids were automatically generated from the designed merging points on pipe centerlines. The initial configuration of the manifold was taken from an existing high power engine with four cylinders.

At first, the manifold shape was optimized by three merging points on the pipe centerlines, assuming the pipe radius constant. The present system found optimal solutions mainly improved in the charging efficiency. This result suggests that the merging configuration is very effective to improve the charging efficiency.

The second case optimized both the pipe radius and merging points. Not only the charging efficiency but also the exhaust gas temperature was improved in this case. This result suggests that the pipe radius is important to improve the exhaust gas temperature. The present system has successfully found solutions that have less environmental impact and more engine power simultaneously than the initial design. The resulting Pareto front also reveals the tradeoff between the two objectives.

Acknowledgements

We would like to thank Powertrain Research Laboratory in Mazda Motor Corporation for providing the one-dimensional empirical engine cycle simulation code and the engine data. The calculations were performed by using the supercomputer, ORIGIN 2000 in the Institute of Fluid Science, Tohoku University.

References

1. M. Kanazaki, S. Obayashi and K. Nakahashi, "The Design Optimization of Intake/Exhaust Performance of a Car Engine Using MOGA," EUROGEN 2001, Athens, Sep. 19-21, 2001, postproceedings in print.
2. T. Hiroyasu, M. Miki and S. Watanabe, "The New Model of Parallel Genetic Algorithm in Multi-Objective Optimization Problems (Divided Range Multi-Objective Genetic Algorithm)," *IEEE Proceedings of the Congress on Evolutionary Computation 2000*, Vol. 1, pp.333-340, 2000.
3. D. Sharov, and K. Nakahashi, "Reordering of 3-D Hybrid Unstructured Grids for Lower-Upper Symmetric Gauss-Seidel Computations," AIAA J., Vol. 36, No. 3, pp. 484-486, 1998.
4. K. Ohnishi, H. Nobumoto, T. Ohsumi and M. Hitomi, "Development of Prediction Technology of Intake and Exhaust System Performance Using Computer Simulation," MAZDA Technical Paper (in Japanese), No. 6, 1988.
5. L. J. Eshelman and J. D.Schaffer, "Real-coded genetic algorithms and interval schemata," *Foundations of Genetic Algorithms2*, Morgan Kaufmann Publishers, Inc., San Mateo, pp. 187-202, 1993.
6. C. M. Fonseca and P. J. Fleming, "Genetic algorithms for multiobjective optimization: formulation, discussion and generalization," 5th International Conference on Genetic Algorithms, Morgan Kaufmann Publishers, San Francisco, pp. 416-423, 1993.
7. K. A. De Jong, "An Analysis of the Behavior of a Class of Genetic Adaptive System," Doctoral Dissertation, University of Michigan, Ann Arbor, 1975.
8. Y. Ito and K. Nakahashi, "Direct Surface Triangulation Using Stereolithography (STL) Data," AIAA Paper 2000-0924, 2000.

Multi-objective Co-operative Co-evolutionary Genetic Algorithm

Nattavut Keerativuttitumrong[2], Nachol Chaiyaratana[1], and Vara Varavithya[2]

[1] Research and Development Center for Intelligent Systems
[2] Department of Electrical Engineering
King Mongkut's Institute of Technology North Bangkok
1518 Piboolsongkram Rd., Bangkok 10800, Thailand
{nchl,vara}@kmitnb.ac.th

Abstract. This paper presents the integration between two types of genetic algorithm: a multi-objective genetic algorithm (MOGA) and a co-operative co-evolutionary genetic algorithm (CCGA). The resulting algorithm is referred to as a multi-objective co-operative co-evolutionary genetic algorithm or MOCCGA. The integration between the two algorithms is carried out in order to improve the performance of the MOGA by adding the co-operative co-evolutionary effect to the search mechanisms employed by the MOGA. The MOCCGA is benchmarked against the MOGA in six different test cases. The test problems cover six different characteristics that can be found within multi-objective optimisation problems: convex Pareto front, non-convex Pareto front, discrete Pareto front, multi-modality, deceptive Pareto front and non-uniformity in the solution distribution. The simulation results indicate that overall the MOCCGA is superior to the MOGA in terms of the variety in solutions generated and the closeness of solutions to the true Pareto-optimal solutions. A simple parallel implementation of MOCCGA is described. With an 8-node cluster, the speed up of 2.69 to 4.8 can be achieved for the test problems.

1 Introduction

A *genetic algorithm* has been established as one of the most widely used technique for multi-objective optimisation. This is because the parallel search nature of genetic algorithm makes the task of approximating Pareto front of optimal solutions in one optimisation run becomes possible. From early developments in the eighties (Schaffer, 1984; Fourman, 1985) to the introduction of a direct relationship between Pareto-optimality and a fitness function (Horn and Nafpliotis, 1993; Fonseca and Fleming, 1993), research interests in this branch of evolutionary computation remain strong. Regardless of the methodology employed, the ultimate aim of multi-objective optimisation using a genetic algorithm remains the same: to identify the solutions that approximate the true Pareto-optimal solutions.

Various types of genetic algorithm are currently available for use in multi-objective optimisation (Hajela and Lin, 1992; Horn and Nafpliotis, 1993; Fonseca

J.J. Merelo Guervós et al. (Eds.): PPSN VII, LNCS 2439, pp. 288–297, 2002.

and Fleming, 1993; Zitzler and Thiele, 1999). Although a number of applications have been benefited from the use of such algorithms, as the search space or problem size increases, the performance of the algorithm always degrades. As a result, the non-dominated solution set identified by the algorithm may highly deviate from the true Pareto front. In addition, the coverage of the Pareto front by the solutions generated may also be affected. This is because the issue regarding the relationship between the problem representation and the size of search space has rarely been discussed in the context of algorithm development for use in *multi-objective optimisation.* Nonetheless, various techniques for single-objective optimisation using a genetic algorithm are available for solving the problem induced by an increase in problem size. One possible technique involves an insertion of a *co-operative co-evolutionary* effect into the search algorithm. A genetic algorithm that exploited such strategy is known as a co-operative co-evolutionary genetic algorithm or CCGA (Potter and De Jong, 1994). In contrast to other co-evolutionary genetic algorithms where the co-evolutionary effect found among sub-populations is the result of a competition for survival by the individuals, the co-evolutionary effect in the CCGA is produced by a co-operation among all species. In brief, a species member in the CCGA represents a part of the decision variable set where all species will co-operatively produce complete solutions to the problem. Each species member will then independently evolve using a standard genetic algorithm mechanism. By partitioning the problem in this manner, the search space that each sub-population has to cover would significantly reduce. Although the CCGA is originally developed for use in single-objective optimisation, the co-operative co-evolutionary effect can also be embedded into a genetic algorithm which is designed for multi-objective optimisation.

In this paper, the co-operative co-evolutionary effect as described by Potter and De Jong (1994) will be integrated into a genetic algorithm called a multi-objective genetic algorithm or MOGA (Fonseca and Fleming, 1993). The MOGA is chosen for this case because of its simplicity and the clear definition regarding the relationship between the Pareto-optimality level of a solution and its corresponding fitness value. The modified MOGA will be referred to as a multi-objective co-operative co-evolutionary genetic algorithm or MOCCGA throughout the paper. Note that the co-operative co-evolutionary effect can also be integrated into other types of genetic algorithm that are designed for use in multi-objective optimisation. In addition to the proposed integration between two genetic algorithms, possible approach for parallel implementation of the developed algorithm will also be discussed in this paper.

The organisation of this paper is as follows. In section 2, the background on the multi-objective genetic algorithm (MOGA) and the co-operative co-evolutionary genetic algorithm (CCGA) will be discussed. The integration between the MOGA and the CCGA will be explained in section 3. In addition, the test problems that will be used to assess the performance of the MOGA will also be given in this section. In section 4, the benchmarking results and discussions are given. Section 5 describes the parallel implementation of MOCCGA and its test results. Finally, conclusions are drawn in section 6.

2 Multi-objective Genetic Algorithm and Co-operative Co-evolutionary Genetic Algorithm

Two types of genetic algorithm that will be integrated together are the multi-objective genetic algorithm (MOGA) and the co-operative co-evolutionary genetic algorithm (CCGA). A brief description of the algorithms follows.

2.1 Multi-objective Genetic Algorithm

The multi-objective genetic algorithm (MOGA) was first introduced by Fonseca and Fleming (1993). The MOGA functions by seeking to optimise the components of a vector-valued objective function. Unlike single-objective optimisation, the solution to a multi-objective optimisation problem is a family of points known as the Pareto-optimal set. Each point in the set is optimal in the sense that no improvement can be achieved in one component of the objective vector that does not lead to degradation in at least one of the remaining components. Given a set of possible solutions, a candidate solution is said to be Pareto-optimal if there are no other solutions in the solution set that can dominate the candidate solution. In other words, the candidate solution would be a non-dominated solution. Assuming, without loss of generality, a minimisation problem, an n-dimensional cost vector $\mathbf{a}$ is said to be dominating another n-dimensional cost vector $\mathbf{b}$ if, and only if, $\mathbf{a}$ is partially less than $\mathbf{b}$ ($\mathbf{a}\ \ p< \mathbf{b}$), i.e.

$$\mathbf{a}\ \ p< \mathbf{b} \leftrightarrow \forall i = 1, \ldots, n : a_i \leq b_i \wedge \exists i = 1, \ldots, n : a_i < b_i \tag{1}$$

By identifying the number of solutions in the solution set that dominate the solution of interest, a rank value can be assigned to the solution. In other words, the rank of a candidate solution is given by the number of solutions in the solution set that dominate the candidate solution. After a rank has been assigned to each solution, a fitness value can then be interpolated onto the solution where a genetic algorithm can subsequently be applied in the optimisation procedure. Note that since the aim of a search by the MOGA is to locate Pareto-optimal solutions, in essence the multi-objective optimisation problem has also been treated as a multi-modal problem. Hence, the use of additional genetic operators including the fitness sharing and mating restriction procedures is also required. However, in addition to the usual application of the fitness sharing and mating restriction procedures in the decision variable space (Fonseca and Fleming, 1995), they can also be carried out in the objective value space (Fonseca and Fleming, 1993). A comprehensive description of the MOGA which covers other advanced topics including goal attainment and priority assignment strategies can be found in Fonseca and Fleming (1998).

2.2 Co-operative Co-evolutionary Genetic Algorithm

The co-operative co-evolutionary genetic algorithm (CCGA) was first introduced by Potter and De Jong (1994). The CCGA functions by introducing an explicit notion of modularity to the optimisation process. This is done in order to provide reasonable opportunity for complex solutions to evolve in the form of interacting co-adapted sub-components. In brief, the CCGA explores the search space by utilising a population which contains a number of species or sub-populations. In contrast to other types of genetic algorithm, each species in the CCGA represents a variable or a part of the problem which is needed to be optimised. A combination of an individual from each species will lead to a complete solution to the problem where the fitness value of the complete solution can be found in the usual way. This value of fitness will be assigned to the individual of interest that participates in the formation of the solution. After the fitness values have been assigned to all individuals, the evolution of each species is then commenced using a standard genetic algorithm. Although the CCGA has been successfully used in various applications, its performance can reduce in the circumstance where there are high interdependencies between the optimisation function variables. In order to solve this problem, Potter and De Jong (1994) have suggested that the fitness of a species member should be obtained after combining it with the current best individuals or the randomly selected individuals from the remaining species depending upon whether which combination yields a higher fitness value. This helps to increase a chance for each individual to achieve a fitness value which is appropriate to its contribution to the solution produced. A comprehensive description of the CCGA and the summary of its applications can be found in Potter and De Jong (2000).

3 MOCCGA and Test Problems

By combining the MOGA and the CCGA together, the resulting algorithm can be referred to as a multi-objective co-operative co-evolutionary genetic algorithm or MOCCGA. Similar to the CCGA, each species in the MOCCGA represents a decision variable or a part of the problem which is needed to be optimised. However, instead of directly assigning a fitness value to the individual of interest which participates in the construction of the complete solution, a rank value will be determined first. Similar to the MOGA, the rank of each individual will be obtained after comparing it with the remaining individuals from the same species. Then a fitness value can be interpolated onto the individual where a standard genetic algorithm can be applied within each sub-population. Note that in this investigation, the fitness sharing strategy utilised in the MOCCGA is similar to the one described in Fonseca and Fleming (1993) where the fitness sharing is carried out in the objective space.

In order to assess the performance of the MOCCGA, the MOCCGA will be benchmarked against the MOGA in six optimisation test cases. These six test problems are developed by Zitzler et al. (2000) for use in multi-objective optimisation benchmarking. The problems are minimisation problems with m decision variables and two objectives. Brief descriptions of the test problems are summarised in Table 1. Each test problem represents different aspects of multi-objective optimisation problems. The benchmarking results will be displayed and discussed in the following section.

Table 1. Descriptions of the test problems.

Test Problem	Characteristic
T_1	Convex Pareto front
T_2	Non-convex Pareto front
T_3	Discrete Pareto front containing several non-contiguous convex parts
T_4	Multi-modality
T_5	Deceptive Pareto front
T_6	Non-uniformity in the solution distribution in the search space

Table 2. Common parameter settings for both the MOGA and the MOCCGA.

Parameter	Value
Selection method	Stochastic universal sampling (Baker, 1989)
Crossover probability	0.7
Mutation probability	0.01

4 Benchmarking Results and Discussions

The MOCCGA will be benchmarked against the MOGA in six test cases described in section 3. The common parameter settings for both the MOGA and the MOCCGA are displayed in Table 2.

The parameter settings which vary from one test problem to the others are the number of generations, the number of species (for the case of the MOCCGA), the number of individuals and the length of binary chromosome. These parameters are set to accommodate the size of search space in each problem. Although the settings are different for each problem, the parameter values are chosen such that the total numbers of objective evaluations are the same for both the MOGA and the MOCCGA. Also recall that the objective values are evaluated twice for each individual in the MOCCGA using the strategy based on the one described in Potter and De Jong (1994).

In this investigation, each algorithm is run five times using different initial populations. Then the search results from all runs are combined where the non-dominated solutions are subsequently extracted from the overall results. The non-dominated solutions located by the MOGA and the MOCCGA from all test cases are displayed in Figures 1-3.

Firstly, the range of variety in solutions found is considered. The MOCCGA can identify the solutions which cover the whole Pareto front in the cases of T_1, T_2 and T_4. In contrast, the MOGA can locate the solutions which indicate the boundary of Pareto front only in the case of T_5. From these observations, the performances of the MOGA and MOCCGA are lowest with the problem that has a discrete Pareto front (T_3) or a non-uniform distribution of solutions in the search space and along the Pareto front (T_6). This means that although the co-operative co-evolutionary effect can help improving the performance of the

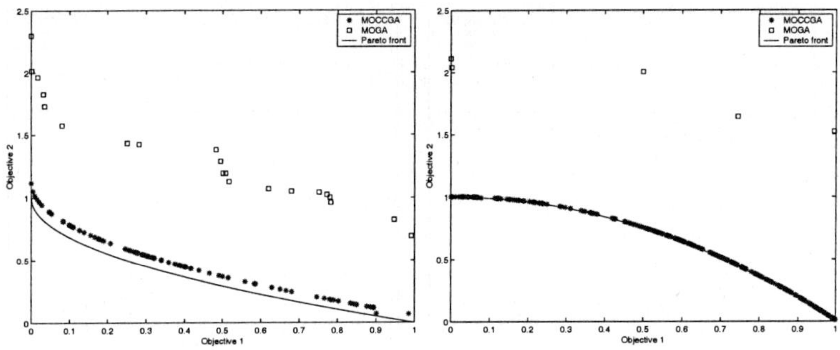

(a) (b)

Fig. 1. (a) Non-dominated solutions and the true Pareto front of the test problem T_1, (b) Non-dominated solutions and the true Pareto front of the test problem T_2.

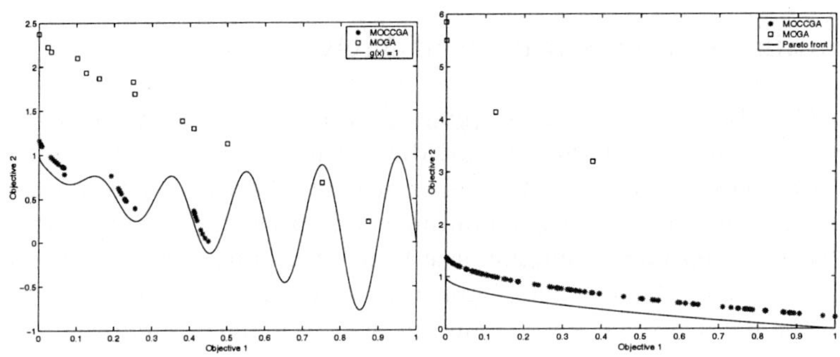

(a) (b)

Fig. 2. (a) Non-dominated solutions and the boundary for $g(\mathbf{x}) = 1$ in the test problem T_3, (b) Non-dominated solutions and the true Pareto front of the test problem T_4.

search algorithm in the context of the Pareto front coverage, this additional effect is insufficient for the algorithm to cope with the discrete and non-uniformity features of the optimisation problems.

Moving onto the consideration on the closeness of non-dominated solutions to the true Pareto-optimal solutions. The MOCCGA has proven to be highly efficient in all test problems. In particular, most of solutions identified by the MOCCGA in the test problems T_1, T_2, T_3, T_4 and T_6 dominate the corresponding solutions identified by the MOGA. The only test problem at which the MOGA can identify non-dominated solutions that are reasonably close to the true Pareto-optimal solutions is the problem T_5. Moreover, the non-dominated solutions identified by the MOCCGA in the cases of T_2 and T_6 are also the true Pareto-optimal solutions. In overall, the introduction of co-operative co-evolutionary effect can improve the genetic algorithm performance in terms of the Pareto front coverage and the number of solutions found which are close to the true Pareto-optimal solutions.

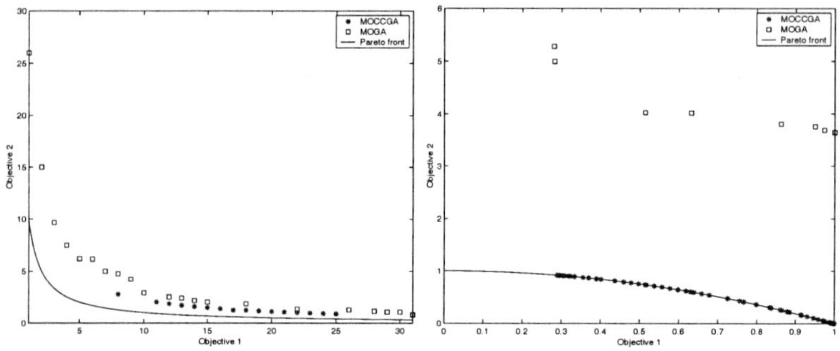

(a) (b)

Fig. 3. (a) Non-dominated solutions and the true Pareto front of the test problem T_5, (b) Non-dominated solutions and the true Pareto front of the test problem T_6.

5 Parallel Implementation of MOCCGA

Genetic algorithms inherit a high degree of parallelism. A large portion of tasks that operated on a population of individuals can be performed without dependency. Examples of parallel tasks in GAs include objective function evaluations and genetic operators. Parallel computing can improve the performance of GAs in two ways, increasing population size to solve complex problem or reducing computation time. Many parallel implementations of GAs have been purposed. Some of the parallel GAs can be found in Spiessens and Manderick (1991) and Gorges-Scheleuter (1990).

In parallel GAs, the entire population is partitioned into sub-populations. The sub-population can consist of complete strings representing the whole decision variable sets or portion of strings. Each sub-population is processed by an assigned processor. At each iteration, sub-populations interact with each other, according to GAs, by exchanging information between processors. In multiprocessor systems, the cost of exchanging information is considered as an overhead. The performance of parallel GAs is therefore limited by the amounts of information that need to be exchanged between sub-populations. For this reason, reducing information exchanges between processors is one approach of improving the performance of parallel GAs. In simple GAs, the algorithm has access to all individuals of the population. However, the information of all individuals is not required to maintain the performance of GAs. The parallel GAs can take advantage of this fact and exchange only a subset of available information. For example, only the best individuals are exchanged between sub-populations. The performance improvement is obtained not only by less communication overhead but also high diversity of chromosome to avoid premature convergent. Hart, et al. (1996) studied the effects of relaxed synchronization on parallel GAs in which an improvement on execution time was shown. The effects of chromosome migration was investigated by Matsumura, et. al. (1997). Each processor executes the genetic operations on a set of chromosome and exchange information to only its neighbours based on different network topologies.

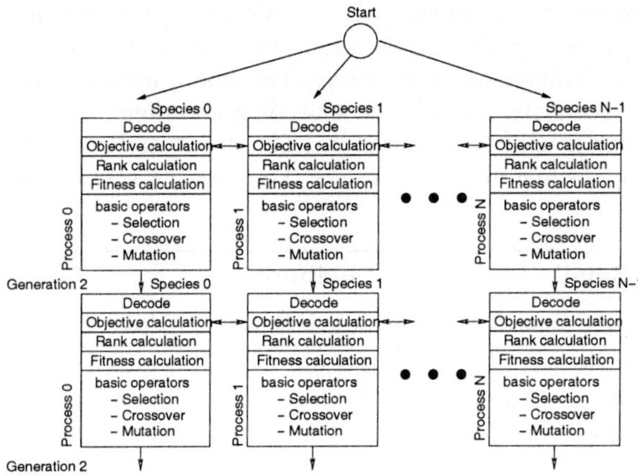

Fig. 4. Implementation of parallel MOCCGA. The communication occurs before the objective function calculation.

As the price to performance ratio of computers has dropped in a rapid rate, high performance computing platforms can be built by interconnecting a group of PCs, so called a cluster. The clusters have been implemented widely and their application domains have been extended into many new areas. Patrick, et. al. (1997) has proposed a distributed genetic algorithm for cluster environment in which a set of library functions for performing parallel genetic operations is provided. However, the parallel implementation of MOCCGA requires different details. In MOCCGA, operations in ranking calculation, selection, and crossover require information accesses within the same species. Information from different species is required only in the objective value calculation. Therefore partitioning the MOCCGA based on species can reduce amounts of information exchange and simplify the program coding. Figure 4 shows the problem partitioning based on Single Program Multiple Processors paradigm. The parallel process starts from a single node and creates parallel process on a set of computers. Each process responsible for one species or a group of species. The information exchange as well as synchronisation is carried out at the beginning of the objective calculation stage.

In original MOCCGA, the fitness calculation strategy prevents communication reduction since fitness calculation for each species require the whole species information in order to have a complete gene pool. The objective values are calculated from two combinations; firstly,the decision variable (after decode the string) of individual in the current species is combined with the best decision variable from other species, and secondly, the decision variable of the current species is combined with the decision variable of individual from other species in random. The objective values of both cases are then compared and the better one is selected. The communication at the beginning of objective calculation involves the broadcast of the best decision variable of individuals and the whole decision variable of the species to all other processes. The communication performance

Table 3. Performance results of the parallel MOCCGA: The execution time is shown in second. The number of individuals in small, medium, and large problems is equal to 100, 500, and 1000 individuals, respectively. The small, medium, and large problems are processed, respectively, for 100, 20, and 10 generations. In Bcast, the standard MPI_bcast() library is used in the program. The divide and conquer broadcast using send() and receive() is used in Log_P results. The speedup results are calculated using Bcast results.

	Small			**Medium**			**Large**		
	Bcast	Log_P	SpeedUp	Bcast	Log_P	SpeedUp	Bcast	Log_P	SpeedUp
1	53.83	59.05	1	299.55	331.95	1	721.48	754.92	1
2	32.16	34.14	1.6	194.53	205.49	1.5	490.72	529.50	1.4
4	19.17	19.15	2.7	129.12	128.31	2.3	355.32	352.28	2.0
8	11.90	10.41	4.8	89.99	86.55	3.3	268.61	260.54	2.69

can be improved using divide and conquer approach in which $\log_2 P$ communication steps are adequate for P processes. The parallel version of MOCCGA was developed using C language and MPI library. The performance results were measured from a 8-node cluster. The number of test species is fixed to 32. An individual is encoded as a binary of 10 bits. The stochastic universal sampling was used in the selection process. The test problem T_1 is selected in the test. We have considered three problem sizes, small, medium, and large problem. The test results are shown in Table 3

The speed up of parallel MOCCGA is quite satisfactory. We considered two approaches of broadcasting objectives value, using MPI_Bcast function library and customize decision value broadcast based on the divide and conquer technique. The results show slightly different in performance. The custom broadcast has better performance results for the 4 and 8 processors. The performance improvement results from the parallelism in communication.

6 Conclusions

In this paper, the integration between two types of genetic algorithms are preseted: an MOGA and a CCGA. The MOCCGA has been benchmarked against the MOGA in six test cases. The search performance of the MOGA can be improved by adding the co-operative co-evolutionary effect to the algorithm. A parallel implemetation of MOCCGA was described.

References

Baker, J. E.: An analysis of the effects of selection in genetic algorithms, Ph.D. Thesis. Computer Science Department, Vanderbilt University, Nashville, TN (1989)

Fonseca, C. M. and Fleming, P. J.: Genetic algorithms for multiobjective optimization: Formulation, discussion and generalization. Genetic Algorithms. Proceedings of the Fifth International Conference, Urbana-Champaign, IL (1993) 416–423

Fonseca, C. M. and Fleming, P. J.: Multiobjective genetic algorithms made easy: Selection, sharing and mating restriction. The Second International Conference on Genetic Algorithms in Engineering Systems: Innovations and Applications (GALESIA'95), Sheffield, UK (1995) 45–52

Fonseca, C. M. and Fleming, P. J.: Multiobjective optimization and multiple constraint handling with evolutionary algorithms - Part 1: A unified formulation. IEEE Transactions on Systems, Man, and Cybernetics 28(1) (1998) 26–37

Fourman, M. P.: Compaction of symbolic layout using genetic algorithms. Proceedings of the First International Conference on Genetic Algorithms and Their Applications (1985) 141–153

Gorges-Scheleuter, M.: Explicit parallelism of genetic algorithms through population structures. Parallel Problem Solving from Nature (1990) 150–159

Hajela, P. and Lin, C. Y.: Genetic search strategies in multicriterion optimal design. Structural Optimization 4 (1992) 99–107

Hart, W. E.,Baden, S., Bekew, R. K. and Kohn, S.: Analysis of the numerical effects of parallelism on a parallel genetic algorithm. Proceeding of the Tenth International Parallel Processing Symposium (1996) 606–612

Horn, J. and Nafpliotis, N.: Multiobjective optimization using the niched pareto genetic algorithm. IlliGAL Report No. 93005, Department of Computer Science, Department of General Engineering, University of Illinois at Urbana-Champaign, Urbana-Champaign, IL (1993)

Matsumura, T., Nakamura, M. and Okech, J.: Effect of chromosome migration on a parallel and distributed genetic algorithm. International Symposium on Parallel Architecture, Algorithm and Networks (ISPAN'97) (1997) 357–612

Patrick D., Green, P. and York, T.: A distributed genetic algorithm environment for unix workstation clusters. Genetic Algorithms in Engineering Systems: Innovations and Applications (1997) 69–74

Potter, M. A. and De Jong, K. A.: A cooperative coevolutionary approach to function optimization. Proceedings of the Third International Conference on Parallel Problem Solving from Nature (PPSNIII), Jerusalem, Israel (1994) 249–257

Potter, M. A. and De Jong, K. A.: Cooperative coevolution: An architecture for evolving coadapted subcomponents. Evolutionary Computation 8(1) (2000) 1–29

Schaffer, J. D.: Some experiments in machine learning using vector evaluated genetic algorithms, Ph.D. Thesis. Vanderbilt University, Nashville, TN (1984)

Spiessens, P., and Manderick, B.: A massively parallel genetic algorithm: Implementation and first analysis. The Fourth International Conference on Genetic Algorithms (1991) 279–285

Zitzler, E. and Thiele, L.: Multiobjective evolutionary algorithms: A comparative case study and the strength Pareto approach. IEEE Transactions on Evolutionary Computation 3(4) (1999) 257–271

Zitzler, E., Deb, K. and Thiele, L.: Comparison of multiobjective evolutionary algorithms: Empirical results. Evolutionary Computation 8(2) (2000) 173–195

Bayesian Optimization Algorithms for Multi-objective Optimization

Marco Laumanns[1] and Jiri Ocenasek[2]

[1] ETH Zürich, Computer Engineering and Networks Laboratory, CH–8092 Zürich
laumanns@tik.ee.ethz.ch
http://www.tik.ee.ethz.ch/aroma

[2] VUT Brno, Faculty of Information Technology, Bozetechova 2, CZ - 612 66 Brno
ocenasek@fit.vutbr.cz
http://www.fit.vutbr.cz/~ocenasek

Abstract. In recent years, several researchers have concentrated on using probabilistic models in evolutionary algorithms. These Estimation Distribution Algorithms (EDA) incorporate methods for automated learning of correlations between variables of the encoded solutions. The process of sampling new individuals from a probabilistic model respects these mutual dependencies such that disruption of important building blocks is avoided, in comparison with classical recombination operators. The goal of this paper is to investigate the usefulness of this concept in multi-objective optimization, where the aim is to approximate the set of Pareto-optimal solutions. We integrate the model building and sampling techniques of a special EDA called Bayesian Optimization Algorithm, based on binary decision trees, into an evolutionary multi-objective optimizer using a special selection scheme. The behavior of the resulting Bayesian Multi-objective Optimization Algorithm (BMOA) is empirically investigated on the multi-objective knapsack problem.

1 Introduction

The Estimation of Distribution Algorithms (EDAs) [5, 8] also called probabilistic model-building evolutionary algorithms have attracted a growing interest during the last few years. Recombination and mutation operators used in standard EAs are replaced by probability estimation and sampling techniques to avoid the necessity to specify certain EA parameters, to avoid the disruption of building blocks and to enable solving of non-linear or even deceptive problems having considerable degree of epistasis.

In multi-objective optimization the usual goal is to find or to approximate the set of Pareto-optimal solutions. Research in the design of multi-objective evolutionary algorithms has mainly focused on the fitness assignment and selection part. In contrast, the variation operators could be used from the single objective case without modification, which gave them only little attention. Some studies indicate, however, that the existence of multiple objectives influences the success probabilities of mutations which in turn has consequences for the choice of

J.J. Merelo Guervós et al. (Eds.): PPSN VII, LNCS 2439, pp. 298–307, 2002.

the mutation strength [7, 3]. For recombination it is unclear whether combining parents that are good in different objectives improve the search as they could create good compromise offspring [2], or whether they contain such incompatible features that a combination does not make sense, thus advocating mating restrictions. The fact that recombination generally is a "contracting" operator might also conflict with the goal to reach a broad distribution of Pareto-optimal solutions.

In this study we investigate the use of EDAs for multi-objective optimization problems to overcome the aforementioned difficulties when creating offspring from a set of diverse parents from different trade-off regions. The next section introduces the Bayesian Optimization Algorithm (BOA) as a special EDA based on the Bayesian network model. It summarizes disadvantages of Bayesian network and introduces BOAs based on decision trees. We derive our own incremental equation for construction of decision trees and demonstrate the algorithm for decision tree construction. In section 3 a new Multi-objective BOA is proposed. We derive design guidelines for a useful selection scheme in connection with the construction of the probabilistic model and develop a new operator based on ϵ-archives [4]. Section 4 demonstrates the applicability of the approach using the multi-objective knapsack problem and compares the results to other multi-objective evolutionary algorithms.

2 Bayesian Optimization Algorithm (BOA)

One of the most general probabilistic models for discrete variables used in EDAs is the Bayesian network (BN). It is able to encode any dependencies of variables that can obtain one out of a finite set of values.

2.1 Structure of Bayesian Network

A Bayesian network for the domain of possible chromosomes $X = (X_0, \ldots, X_{n-1})$ represents a joint probability over X. The BN representation consists of 2 parts, a set of conditional independence assertions and a set of local conditional distributions, that allow us to construct a global joint probability distribution of chromosome from the local distributions of genes.

The first part, the set of conditional independence assertions, is expressed by a dependency graph, where each gene corresponds to one node in the graph. If the probability of the value of a certain gene X_i is affected by value of other gene X_j, then we say that "X_i depends on X_j" or "X_j is a parent variable of X_i". This assertion is expressed by existence of edge (j, i) in the dependency graph. A set of all parent variables of X_i is denoted Π_i, it corresponds to the set of all starting nodes of edges ending in X_i.

In the example of Fig. 1 (left), genes X_0, X_2 are independent and the value of X_1 is affected by X_0 and X_2. Under this assertion we can write the probability of whole chromosome (X_0, X_1, X_2) as the product of local distributions:

$$p(X_0, X_1, X_2) = p(X_0) \cdot p(X_2) \cdot p(X_1|X_0, X_2) \qquad (1)$$

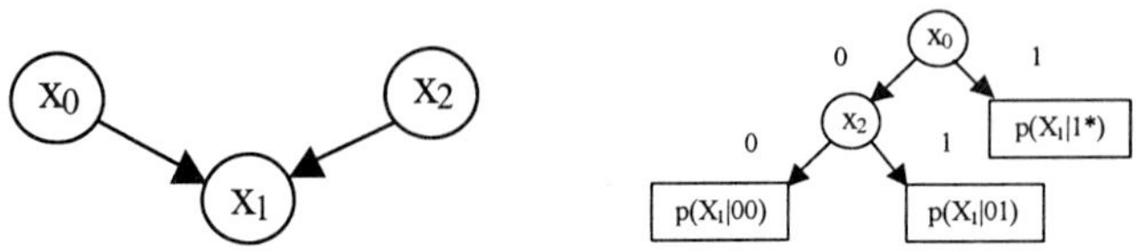

Fig. 1. Example of dependency graph for $n = 3$ (left) and decision tree (right)

Now we will focus on the second part of BN representation – the set of local distributions. For simplicity let's consider binary genes. In the case of gene X_0 resp. X_2 from our example this local distribution is unconditional and can be estimated from the population by simple counting individuals where $X_0 = 1$ resp. $X_2 = 1$ and dividing it by the size of population N. In the case of gene X_1 the gene depends on X_0 and X_2, so the local distribution is conditional and can be expressed by the following table:

$$\begin{array}{r|cccc} X_0, X_2 & 00 & 01 & 10 & 11 \\ \hline p(X_1 = 1|X_0, X_2) & \dots & \dots & \dots & \dots \end{array} \qquad (2)$$

The dots in the second row denote values estimated from the population by

$$p(X_1 = 1|X_0, X_2) = m(X_1 = 1, X_0, X_2)/m(X_0, X_2).$$

With this Bayesian network we are able to determine the probability of each concrete chromosome $X = (x_0, x_1, x_2)$.

The most important and also most time-consuming part of EDA is the algorithm for construction of the probabilistic model from the population. Most methods for automated learning of probabilistic models have been adopted from the area of data mining. When a significant building block is detected among the solutions in the population, the information about dependency of its genes is added to the model.

The original BOA algorithm uses a hillclimbing algorithm to step-by-step improve the Bayesian network. It starts with independent genes (no edges are present between nodes of dependency graph), such that the local probabilities are unconditional. Then in each step the algorithm examines all possible edges and it adds the "best edge" to the network. By the term "best edge" we mean the edge which does not introduce any cycle in the dependency graph and which improves the score most. The quality of each edge is expressed by the Bayes-Dirichlet metric (BDe, see [1]). This equation measures the bias in the probability distribution of combinations of genes in the population. For further details on various types of EDAs see the exhaustive survey [5].

2.2 Binary Decision Trees Based BOA

The problem with the BN approach is that after introducing one more parent variable of X_i, the number of columns of the conditional distribution table of X_i doubles, making the computation time of the BDe metric exponentially increasing with the number of parents. In previous versions of BOA the number of possible parents was limited to k, making the BDe computable in real time.

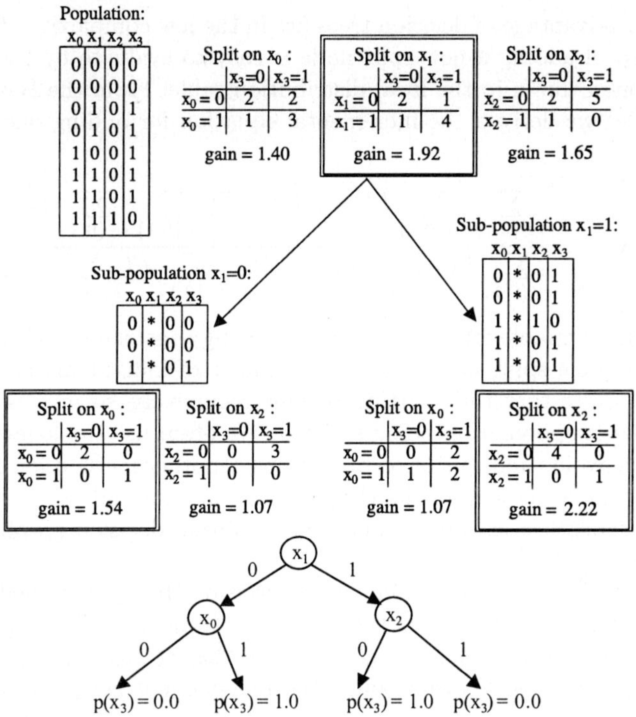

Fig. 2. Example of construction of the decision tree for variable $X_3, n = 4$, and final decision tree for variable X_3

A better divide-and-conquer approach is based on binary decision trees, firstly proposed for EDAs in [6].

The model is composed of the set of trees, one tree is for each variable. The dependence assertions are expressed by existence of splitting nodes and the local probabilities are stated in leaf nodes. A set of all parent variables of X_i denoted Π_i corresponds to the set of all decision nodes of i-th tree.

Decision trees are a more accurate model than Bayesian network. They allow us to describe the dependencies of alleles (gene values). Let us consider our first example from Fig. 1. Assume that if $X_0 = 1$, then the value of X_1 does not depend on X_2, but when $X_0 = 0$, then the value of X_1 depends on both X_0 and X_2. Because $p(X_1|10) = p(X_1|11) = p(X_1|1*)$, this would reduce the number of table columns in (2), as can be seen in the following table:

$$\begin{array}{r|lll} X_0, X_2 & 00 & 01 & 1* \\ \hline p(X_1 = 1|X_0, X_2) & \dots & \dots & \dots \end{array} \quad (3)$$

This situation can be expressed by a decision tree (see Fig. 1, right). Each variable which determines the X_1 value corresponds to one or more split nodes in the tree, each leaf determines $p(X_1)$ among the individuals fulfilling the split conditions on the path from the root.

A further advantage of decision trees lies in the low complexity of their building. The step of adding a new split node is easy to evaluate by the metric – it splits only one column in the local distribution table. From the Bayes-Dirichlet metrics (BDe) we derived the incremental equation for adding one new binary split:

$$Gain(X_i, X_j) = \frac{\sum\limits_{r\in\{0,1\}} \sum\limits_{s\in\{0,1\}} \Gamma(m_{r,s}+1) \cdot \Gamma(\sum\limits_{r\in\{0,1\}} \sum\limits_{s\in\{0,1\}} (m_{r,s}+1))}{\sum\limits_{r\in\{0,1\}} \Gamma(\sum\limits_{s\in\{0,1\}} (m_{r,s}+1)) \cdot \sum\limits_{r\in\{0,1\}} \Gamma(\sum\limits_{s\in\{0,1\}} (m_{r,s}+1))} \tag{4}$$

where X_i is the variable for which the tree is being constructed, X_j is the parent variable – possible split, and $m_{r,s}$ is the number of individuals having $X_i = r$ and $X_j = s$. Note that the splitting is done recursively, so $m_{r,s}$ is determined only from the subpopulation being splitted. We often use the logarithm of this metric, which avoids multiplication operations. Another method for construction of decision trees straight from the BDe metric can be found in [6]. Additionally this paper proposed a leaf-merge operator to obtain decision graphs instead of decision trees.

As result of model construction a set of decision trees is obtained, see Fig. 2. The dependencies are acyclic, so there exists a "topological" ordering of genes $o_0, o_1, \ldots, o_{n-1}$ such that parents Π_i are from the set $\{o_0, o_1, \ldots, o_{i-1}\}$. When generating a new binary chromosome, the independent gene X_{o_0} is generated first, by "flipping the coin" according to its single leaf node. Then, other genes are generated in the $o_1, \ldots, o_{n-1}$ order. The generation of X_i is driven by actual values of parent variables Π_i, the decision tree traversal ends up in one of the leaf nodes which describe the probability $p(X_i = 1)$.

3 Multi-objective BOA

For the design of a multi-objective BOA some important aspects have to be taken into account, some due to the existence of multiple objective, others from the necessity of the probabilistic model building techniques. Preliminary tests with a simple $(\mu + \lambda)$-strategy and fitness assignment based on the dominance grade have shown that a trivial multi-objective extension leads to poor performance. The population is likely to converge to an "easy to find" region of the Pareto set, as already noticed by [9], and duplicate solutions are produced repeatedly. The resulting loss of diversity leads to an insufficient approximation of the Pareto set and is especially harmful for building a useful probabilistic model. Therefore the following design requirements are essential:

1. Elitism (to preclude the problem of gradual worsening and enable convergence to the Pareto set)
2. Diversity maintenance in objective space (to enable a good approximation of the whole Pareto set)
3. Diversity maintenance in decision space (to avoid redundancy and provide enough information to build a useful probabilistic model)

Algorithm 1 *Select*(A, P, μ, ϵ)

Input: old parent set A, candidate set P, minimum size μ, approximation factor ϵ
Output: new parent set A'
for all $x \in P$ **do**
  $B := \{y \in A | \lfloor \frac{\log f_i(y)}{\log \epsilon} \rfloor = \lfloor \frac{\log f_i(x)}{\log \epsilon} \rfloor \quad \forall$ objective functions $i\}$
  if $B = \emptyset$ **then**
    $A := A \cup \{x\}$
  else if $\exists y \in B$ such that $x \succ y$ **then**
    $A := A \setminus B \cup \{x\}$
  end if
end for
$A' := \{y \in A | \nexists z \in A : z \succ y\}$
$D := A \setminus A'$
if $|A'| < \mu$ **then**
  Fill A' with $\mu - |A'|$ individuals $y \in D$ in increasing order of $|\{z \in A' \cup D | z \succ y\}|$
end if
Return: A'

Algorithm 2 $(\mu + \lambda, \epsilon)$-BMOA

$A := \emptyset$
while $|A| < \mu$ **do**
  Randomly create an individual x.
  $A :=$ *Select*$(A, \{x\}, \mu, \epsilon)$
end while
while Termination criteria not fulfilled **do**
  Create Bayesian Model M from A.
  Sample λ new individuals from M.
  Mutate these individuals and put them into B.
  $A :=$ *Select*(A, B, μ, ϵ)
end while

3.1 A New Selection Operator

From the existing selection/archiving operators in evolutionary multi-objective optimization, the ϵ-Archive [4] has been designed to meet the requirements 1 and 2 above. This method maintains a minimal set of solutions that ϵ-dominates all other solutions generated so far. However, as this set can become very small, the scheme has to be modified to provide enough decision space diversity. The new selection operator is described in Alg. 1. The idea is that now also dominated individuals are allowed to survive, depending on the number of individuals they are dominated by.

3.2 The $(\mu + \lambda, \epsilon)$-BMOA

The combination of the selection operator (Alg. 1) and the variation based on the probabilistic model described in Section 2.2 results in a Bayesian Multi-objective Optimization Algorithm described in Alg. 2. In this $(\mu + \lambda, \epsilon)$-BMOA, μ denotes

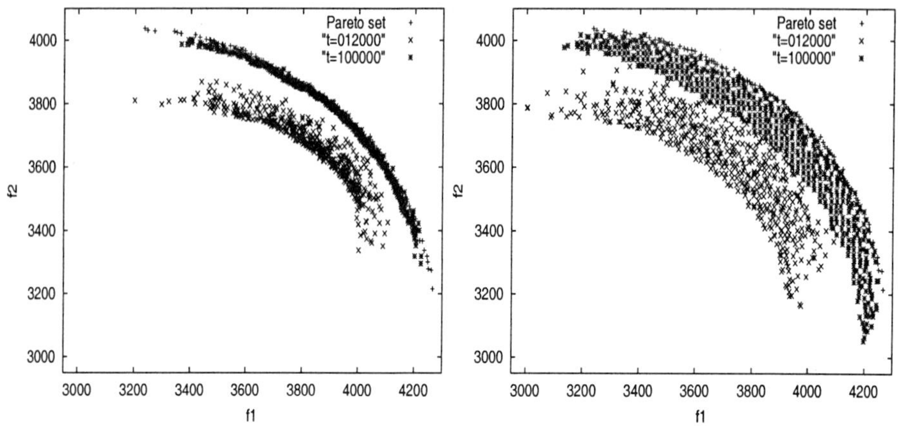

Fig. 3. Development of the population of $(500+500,\epsilon)$-BMOA on the KP-100-2 for $\epsilon = 10^{-6}$ (left) and $\epsilon = 0.005$ (right) after $t = 12000$ and $t = 100000$ function evaluations

the (minimum) number of parents that survive to the next generation being the input to build the model, λ the number of samples from the model in one generation and ϵ the factor that determines the granularity of the approximation. As the Bayesian model M we used the set of decision trees described in section 2.2.

4 Experimental Results

In recent research activities in the field of multi-objective meta-heuristics the 0/1 knapsack problem has become a standard benchmark. Results of several comparative case studies are available in the literature, accompanied by test data through the Internet. The problem can be stated as KP-n-m:

$$\begin{array}{rl} \text{Maximize} & f_j(x) = \sum_{i=1}^{n} x_i \cdot p_{i,j} \\ \text{s.t.} & g_j(x) = \sum_{i=1}^{n} x_i \cdot w_{i,j} \leq W_j \\ & x_i \in \{0,1\}, 1 \leq i \leq n, 1 \leq j \leq m \end{array} \tag{5}$$

where $p_{i,j}$ and $w_{i,j}$ are the elements of the profit and the weight matrices, respectively, and W_j the j-th weight constraint. n denotes the number of binary decision variables and m the number of objectives and constraints. The representation of a solution as a bit string chromosome of length n is straightforward. Infeasible solutions are decoded using a greedy repair mechanism for the calculation of the objective values without changing the genotype of the individuals. The problem is NP-hard, and the exact Pareto set can only be computed for small instances.

4.1 Results of $(\mu + \lambda, \epsilon)$-BMOA

In this section we report results of the BMOA on the knapsack problem (5) with $m = 2$ objectives to demonstrate the applicability of the approach. Each

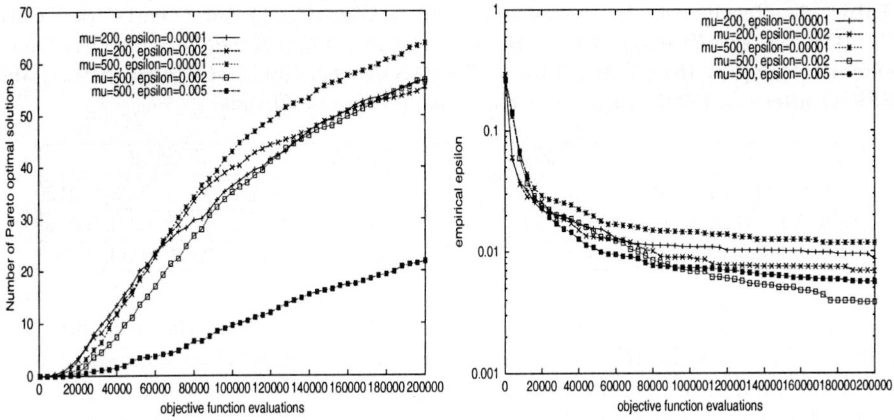

Fig. 4. Average number of Pareto-optimal solution contained in the population (left) and empirical approximation quality ϵ_{min} (right) for KP-100-2 for different μ and ϵ values

individual sampled from the model is additionally mutated by flipping each bit independently with probability $1/n$. Together with the results from [4] this guarantees that in the limit the non-dominated population members converge to a representative subset of the Pareto set of (5) in the sense that each Pareto-optimal point is ϵ-dominated by at least one population member.

Fig. 3 shows the development of the population for different ϵ values for a typical run. With a small ϵ, it can be seen that the population is more concentrated near the middle of the Pareto set compared to the larger ϵ value, where the population is distributed more evenly and broadly. Fig. 4 displays the number of Pareto-optimal solutions found and the empirical approximation quality ϵ_{min} [1] over time. It indicates how the algorithm can be tuned by different settings of the μ and ϵ values. For larger values of both parameters, more dominated individuals will be allowed in the population: Whereas a large ϵ value means that the individuals have to compete for less available "slots", a larger μ simply enlarges the storage. More individuals lead to a more accurate model estimation, but if the fraction of dominated individuals is large, a lot of sampling (and search) effort is wasted on exploring previously visited regions and thereby increasing the running time. The possibility to tune the approximation resolution via the ϵ value is an advantage compared to other existing strategies for diversity maintenance.

4.2 Comparison with Other MOEAs

In order to evaluate BMOA with respect to other MOEAs we use the results of the comparative case study from [10] and focus on the large instances of the knapsack problem with $n = 750$.

[1] $\epsilon_{min} := \text{Min}\{\epsilon \in \mathbb{R}^+ \mid \forall x \in \{0,1\}^n \; \exists y \in \mathbb{R} \,\text{with}\, (1+\epsilon) f(y) \geq f(x)\}$

Table 1. Results of the coverage measures $\mathcal{C}$(BMOA,*) (first entry per cell), $\mathcal{C}$(*,BMOA) (second entry) and of the hypervolume difference $\mathcal{S}$(*)-$\mathcal{S}$(BMOA) (third entry) to compare the $(3000 + 3000, 10^{-6})$-BMOA with NSGA-II, PESA, SPEA, and SPEA2 after $t = 480000$ function evaluations, median of 30 runs

*	NSGA-II	PESA	SPEA	SPEA2
KP-750-2	0.71, 0.00, 0.006	0.71, 0.00, 0.008	0.52, 0.00, 0.009	0.58, 0.00, 0.013
KP-750-3	0.56, 0.00, 0.009	0.64, 0.00, 0.015	0.63, 0.00, 0.014	0.48, 0.00, 0.016
KP-750-4	0.72, 0.00, 0.010	0.97, 0.00, −0.003	0.99, 0.00, −0.003	0.80, 0.00, 0.008

Table 1 compares BMOA to different algorithms using the *Coverage* ($\mathcal{C}$) and the *Hypervolume* ($\mathcal{S}$) metric. $\mathcal{C}(\mathcal{A}, \mathcal{B})$ denotes the relative number of non-dominated individuals contained in the population of algorithm B that are dominated by at least one individual from the population of algorithm A of a given point in time. $\mathcal{S}(\mathcal{A})$ denotes the relative volume of the objective space dominated by the solutions of algorithm A. The $(3000+3000, 10^{-6})$-BMOA is able to dominate more than half of the other algorithms' populations on nearly all instances, with the best results on the four-objective problem. The other algorithms are not able to dominate any of BMOA's non-dominated points, but they generally find a broader distribution as the hypervolume values indicate. Because of its relatively large population size, the BMOA proceeds much slower and it requires more CPU time due to the estimation of the probabilistic model.

5 Conclusion

In this paper we discussed the use of Estimation of Distribution Algorithms for optimization problems involving multiple criteria. A Bayesian Multi-objective Optimization Algorithm $(\mu+\lambda, \epsilon)$-BMOA has been designed using a probabilistic model based on binary decision trees and a special selection scheme based on ϵ-archives. The convergence behavior of the algorithm can be tuned via the values of μ, the minimal population size to estimate the probabilistic model, and ϵ, the approximation factor.

The empirical tests on the 0/1 multi-objective knapsack problem show that the BMOA is able to find a good model of the Pareto set for the smaller instances. In order to find also the outer region of the Pareto set, large μ and ϵ values are required, which slows down the optimization process considerably. Further research could assess MBOA on other multi-objective combinatorial optimization problems with stronger variable interactions and on continuous problems.

From the decision-aid point of view it would be interesting to exploit the Bayesian model also outside the algorithm itself. The compact description of the model could assist a decision maker who can analyze the decision trees to get more insight into the structure of the Pareto set and to learn about correlations in the decision problem at hand.

Acknowledgments

This research has been carried out under the financial support of the Research intention no. CEZ: J22/98: 262200012 - "Research in information and control systems" (Ministry of Education, CZ), the research grant GA 102/02/0503 "Parallel system performance prediction and tuning" (Grant Agency of Czech Republic) and the Swiss National Science Foundation (SNF) under the ArOMA project 2100-057156.99/1.

References

1. D. Heckerman, D. Geiger, and M. Chickering. Learning bayesian networks: The combination of knowledge and statistical data. Technical report, Microsoft Research, Redmont, WA, 1994.
2. J. D. Knowles and D. W. Corne. M-PAES: A memetic algorithm for multiobjective optimization. In *Congress on Evolutionary Computation (CEC 2000)*, volume 1, pages 325–332, Piscataway, NJ, 2000. IEEE Press.
3. M. Laumanns, G. Rudolph, and H.-P. Schwefel. Mutation control and convergence in evolutionary multi-objective optimization. In *MENDEL 2001. 7th Int. Conf. on Soft Computing*, pages 24–29. Brno University of Technology, 2001.
4. M. Laumanns, L. Thiele, K. Deb, and E. Zitzler. Combining convergence and diversity in evolutionary multi-objective optimization. *Evolutionary Computation*, 10(3), 2002.
5. M. Pelikan, D. E. Goldberg, and F. Lobo. A survey of optimization by building and using probabilistic models. IlliGAL Report No. 99018, 1999.
6. M. Pelikan, D. E. Goldberg, and K. Sastry. Bayesian optimization algorithm, decision graphs, and occams razor. IlliGAL Report No. 2000020, 2000.
7. G. Rudolph. On a multi-objective evolutionary algorithm and its convergence to the pareto set. In *IEEE Int'l Conf. on Evolutionary Computation (ICEC'98)*, pages 511–516, Piscataway, 1998. IEEE Press.
8. J. Schwarz and J. Ocenasek. Multiobjective bayesian optimization algorithm for combinatorial problems: Theory and practice. *Neural Network World*, 11(5):423–441, 2001.
9. D. Thierens and P. A. N. Bosman. Multi-objective mixture-based iterated density estimation evolutionary algorithms. In *Proc. of the Genetic and Evolutionary Computation Conference (GECCO-2001)*, pages 663–670. Morgan Kaufmann, 2001.
10. E. Zitzler, M. Laumanns, and L. Thiele. SPEA2: Improving the strength pareto evolutionary algorithm for multiobjective optimization. In K. Giannakoglou et al., editors, *Evolutionary Methods for Design, Optimisation, and Control*, 2002.

An Evolutionary Algorithm for Controlling Chaos: The Use of Multi–objective Fitness Functions

Hendrik Richter

Fraunhofer–Institut für Produktionstechnik und Automatisierung
Nobelstrasse 12, D–70569 Stuttgart, Germany
hir@ipa.fraunhofer.de

Abstract. In this paper, we study an evolutionary algorithm employed to design and optimize a local control of chaos. In particular, we use a multi–objective fitness function, which consists of the objective function to be optimized and an auxiliary quantity applied as an additional driving force for the algorithm. Numerical results are presented illustrating the proposed scheme and showing the influence of employing such a multi–objective fitness function on convergence of the algorithm.

1 Introduction

Recently, it has been shown that evolutionary algorithms are a promising tool for analyzing and controlling nonlinear dynamical systems, see, e.g. [15, 11, 13]. Evolutionary algorithms are remarkably useful if the objective function to be maximized (or minimized) cannot be written as a mathematical expression and has unknown mathematical properties such as continuity and existence of derivatives. This frequently occurs if the objective function can be determined only numerically. In nonlinear dynamics, particularly if chaotic behaviour is involved, we often face the situation that closed–form (algebraic) solutions cannot be found and we thus have to base our experiments on numerical studies. So, the problem of identification [10, 17], modelling [14] and forecasting [15] chaotic system as well as designing and optimizing a control of chaos [16, 6, 13, 7] has been successfully solved using evolutionary algorithms.

In this paper we study some refinements for an evolutionary algorithm recently proposed to design and optimize a local control of chaos [13]. In particular, we use a multi–objective fitness function to improve convergence and speed of optimization. This multi–objective fitness function consists of the objective function to be optimized and an auxiliary quantity used as an additional driving force for the algorithm. Our main result is that employing such a fitness function has a positive effect on convergence speed, particularly if this additional driving force supports optimizing the objective function in terms of the physical meaning.

The paper is structured as follows. In Sec. 2 local control of chaos and controller design is briefly summarized. The evolutionary algorithm and the multi–objective fitness function are discussed in Sec. 3. Our numerical results are presented in Sec. 4, and conclusion are drawn in Sec. 5. The following notation

J.J. Merelo Guervós et al. (Eds.): PPSN VII, LNCS 2439, pp. 308–317, 2002.

is used throughout. $\mathbb{R}^n$ denotes the n–dimensional Euclidean space; $\mathbb{R}^{n\times m}$ the set of all real $n \times m$–matrices; $eig_i(A)$, $i = 1, 2, \ldots, n$ is the i–th eigenvalue of $A \in \mathbb{R}^{n\times n}$; A^T is the transpose of a matrix A and if A is positive–definite, we write $A > 0$.

2 Summary of Local Control of Chaos

We consider a discrete–time dynamical system

$$x(k+1) = f(x(k), u(k)), \tag{1}$$

where $x(k) \in \mathbb{R}^n$ is the state vector, $u(k) \in \mathbb{R}^m$ is the input vector and $k = 0, 1, 2, \ldots$, which displays chaotic behavior for a constant input $u(k) = \bar{u}$ and builds the chaotic attractor A_R. Furthermore, we assume that the system (1) has an unstable equilibrium point $\bar{x}$ (such that $f(\bar{x}, \bar{u}) - \bar{x} = 0$), which we intend to stabilize by means of local control. To that end, we define a feedback matrix $G \in \mathbb{R}^{m\times n}$ and a state–space region of control (SSRC) $W \subset \mathbb{R}^n$, which contains the equilibrium point $\bar{x}$, to employ the local control

$$u(k) = \bar{u} - G(x(k) - \bar{x}), \forall x(k) \in W \tag{2}$$

$$u(k) = \bar{u} = const. \qquad \forall x(k) \notin W. \tag{3}$$

The SSRC W is specified by

$$W = \{x \in \mathbb{R}^n | (x - \bar{x})^T P (x - \bar{x}) < c\}, \tag{4}$$

where $P \in \mathbb{R}^{n\times n}$, with $P = P^T > 0$, and $c > 0$. So, the matrices G and P as well as the constant c have to be determined for local controller design. In the following we outline the design procedure, referring to [12, 13] for detailed discussion. We expand the right hand side of (1) into a Taylor series about $\bar{x}$ and insert the state feedback (2) to obtain

$$x(k+1) - \bar{x} = (A - BG)(x(k) - \bar{x}) + h(x(k)), \tag{5}$$

where $A = \left.\frac{\partial f}{\partial x}\right|_{\substack{x=\bar{x}\\ u=\bar{u}}}$, $B = \left.\frac{\partial f}{\partial u}\right|_{\substack{u=\bar{u}\\ x=\bar{x}}}$ and the map $h(x)$ comprises the terms of second and higher order of the Taylor expansion. So, we can formally write

$$x(k+1) - \bar{x} = H_L \cdot (x(k) - \bar{x}) + H_N(x(k)) \cdot (x(k) - \bar{x}) \tag{6}$$

with $H_L = A - BG$ and $H_N(x(k)) \cdot (x(k) - \bar{x}) = h(x(k))$. For a controllable matrix pair (A, B), a feedback matrix G can be calculated to place the eigenvalues of H_L within the unit disc, which defines the (stabilizing) matrix G, e.g. [5], p.198. For such a stable matrix H_L, the Lyapunov equation

$$H_L^T P H_L - P = -Q \tag{7}$$

is fulfilled for any given $Q = Q^T > 0$. This defines the matrix P. To obtain finally the constant c, we put the matrix

$$Q(x) = Q - H_N(x)^T P H_L - H_L^T P H_N(x) - H_N(x)^T P H_N(x) \tag{8}$$

for the previously defined P and calculate its leading principal minors

$$\Delta_i(x), \quad i = 1, 2, \ldots, n, \tag{9}$$

that is, the determinants of the $(i \times i)$–matrices in the upper left–hand side corner of $Q(x)$. Finally, we get the constant c by solving the constraint optimization problem

$$\max_c \; c > 0, \text{ subject to } \Delta_i(x) > 0, \; i = 1, 2, \ldots, n, \; \forall x : (x - \bar{x})^T P (x - \bar{x}) \leq c, \tag{10}$$

which can be done by Sequential Quadratic Programming (SQP), e.g. [3], p.304.

So, all feasible matrices $Q = Q^T > 0$ (which define the matrix $P = P^T > 0$ via eq. (7)) may be regarded as a bounded subset of a parameter space. An optimal SSRC W can be found by looking for the optimal parameter values within this set. In this context, optimal means to find a SSRC W which allows (in average) fastest control from a given initial state on the chaotic attractor A_R. The evolutionary algorithm considered in the next section aims to reach this goal.

3 The Evolutionary Algorithm

3.1 The Multi–objective Fitness Function

The average time to achieve control can be specified by the relative sojourn; that is, the percentage of the trajectory on the chaotic attractor A_R being within the SSRC W. This relative sojourn can be approximated by the natural measure, [9], p.78,

$$\mu(A_R, W) = \lim_{K \to \infty} \frac{k_W(A_R \cap W)}{K}, \tag{11}$$

where $k_W(A_R \cap W)$ is the quantity of time the trajectory on A_R stays in W during the time interval $0 \leq k \leq K$. Since the goal of optimization is to achieve control as fast as possible, the measure (11) needs to be maximized. So, eq. (11) is the objective function and the optimization problem to be solved reads

$$\max_Q \mu(A_R, W(Q)) \text{ subject to } Q = Q^T > 0. \tag{12}$$

In our previous works [13] we used this measure as fitness function. However, it turned out that there might be cases where it is advantageous to use information about the volume of the SSRC W as well to drive the evolutionary algorithm. In other words, we intend to use an auxiliary quantity (the volume of the SSRC) as a driving force for the evolutionary algorithm. The physical interpretation of this idea is the following. The measure (11) is large if the overlap between the chaotic attractor A_R and the SSRC W is large. Such an overlap can be obtained by either a large SSRC or a (smaller) SSRC which is oriented in such a way that it covers large parts of A_R. In this line of reasoning, orientation is mainly addressed by the driving force measure while size is addressed by volume. The idea is to employ both driving forces in the fitness function.

The volume V of the SSRC W assigned by (4) is

$$V(Q) = \frac{\pi^{\frac{n}{2}}}{\Gamma(\frac{n}{2}+1)} \sqrt{\frac{c^n}{\prod_{i=1}^{n} eig_i(P)}}, \tag{13}$$

where $\Gamma(\frac{n}{2}+1)$ is the Gamma function of $\frac{n}{2}+1$. Here, $V(Q)$ is linked to P by the Lyapunov equation (7) and to c by the solution of (10). So, we can define the multi–objective fitness function

$$F(Q) = \mu(A_R, W(Q)) + \alpha V(Q) \tag{14}$$

with α being a scaling parameter to be specified. To summarize, in order to solve the optimization problem (12) we use an evolutionary algorithm which solves the problem

$$\max_{Q} F(Q) \text{ subject to } Q = Q^T > 0. \tag{15}$$

3.2 Algorithmic Structure

An evolutionary algorithm is a systematic yet probabilistically driven optimization method for finding the largest (or smallest) value of an arbitrarily given function, which bears some features of natural evolution [4, 1, 8]. Following this outline we use the following algorithmic structure to solve the given optimization problem (15).

Algorithm.

1. Generate a set M^1 of symmetric, positive-definite matrices Q which are used to determine the fitness function $F(Q)$. Their entries are initialized randomly and cover the domain of feasible Q uniformly. Define the scaling factor α. Put $k := 1$.
2. Calculate the fitness function $F(Q)$,(14), for every matrix Q of the set M^k.
3. Use tournament selection to select matrices for which the value of the fitness function is above average.
4. Create a new set from the matrices chosen in (3.) by recombination.
5. Do random alteration in the elements of the matrices created in (4) by mutation. Rename the resulting set M^{k+1}.
6. Go back to (2.) and put $k := k + 1$, until a maximal value of the fitness function or another termination criterion is reached.

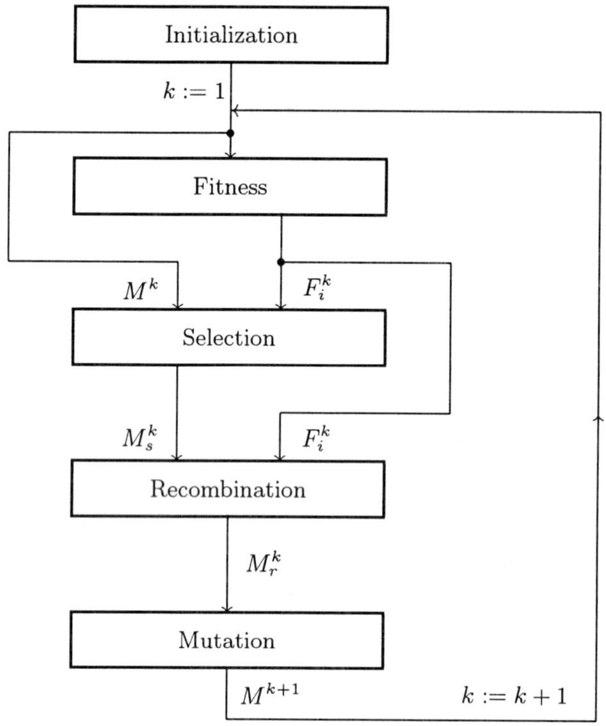

Fig. 1. Structure of the evolutionary algorithm

3.3 Representation and Operators

In this subsection we detail the algorithm given above (see also Fig. 1).

(i) Representation and initialization.
The set of matrices to be optimized over (population of l individuals) consists of l elements at steps $k = 1, 2, \ldots, K$:

$$M^k = \left(Q_1^k \; Q_2^k \; \ldots \; Q_l^k \right),$$

where

$$Q_i^k \in \mathbb{R}^d, \quad Q_i^k = \left(Q_i^k\right)^T > 0, \quad i = 1, 2, \ldots, l, \quad d = \frac{n}{2}(n+1).$$

Here, the matrices Q_i^k are matrices Q at step k. For initialization, symmetric matrices are generated whose entries are realizations of a random variable normally distributed on $[0, \gamma]$. Matrices which are positive–definite are included into M^1. Such a representation results in a search space of $\frac{n}{2}(n+1)$ in which l individuals are supposed to evolve toward the optimum.

(ii) Fitness.
The fitness function

$$F_i^k : \mathbb{R}^d \to \mathbb{R}$$

delivers a quantitative evaluation about every matrix $Q_i^k, i = 1, 2, \ldots, l$ of M^k at every step k by the following procedure: Given Q_i^k, we use (7) to compute P_i^k, (10) to determine c_i^k, (4) to calculate W_i^k, and (14) gives finally F_i^k.

(iii) Selection.
Tournament selection uses the selection operator

$$s : \mathbb{R}^h \times \mathbb{R}^l \to \mathbb{R}^h, \qquad h = ld$$

that produces the set M_s^k from the set M^k and its l fitness values F_i^k. For this purpose, a matrix pair (Q_i^k, Q_j^k) is taken from M^k at random l times where the subscripts i, j, $i \neq j$ are independently chosen as realizations of an integer random variable uniformly distributed on $[1, l]$. If $F_i^k \geq F_j^k$ then Q_i^k is included in M_s^k, otherwise Q_j^k.

(iv) Recombination.
The recombination operator

$$r : \mathbb{R}^h \times \mathbb{R}^l \to \mathbb{R}^h$$

produces a set M_r^k from the elements of M_s^k Again, l random choices of the matrix pairs (Q_i^k, Q_j^k) are made, now with Q_i^k, Q_j^k being element of M_s^k. Each time, a new matrix $Q := \frac{F_i^k}{F_i^k+F_j^k} Q_i^k + \frac{F_j^k}{F_i^k+F_j^k} Q_j^k$ is calculated, and inserted into M_r^k. Such a rule for calculating the new Q is motivated by the assumption that a matrix from M_s^k (and therefore from M^k) with a higher fitness should have more influence on the elements of M_r^k (and hence on M^{k+1}) than a matrix with lower fitness. This reflects the common sense argument that it is more likely to find a matrix with higher fitness near a matrix with high fitness than near a matrix with low fitness. Since $\frac{F_i^k}{F_i^k+F_j^k} > 0$ and $\frac{F_j^k}{F_i^k+F_j^k} > 0$, it is guaranteed that all matrices from M_r^k are again symmetric and positive–definite.

(v) Mutation. The mutation operator

$$m : \mathbb{R}^d \to \mathbb{R}^d$$

is applied to every matrix Q_i^k in M_r^k. It randomly performs one of the following operations with the probabilities p_1, p_2 and p_3, where $p_1 + p_2 + p_3 = 1$.

1 The matrix Q_i^k remains unchanged.
2 A diagonal element of Q_i^k is changed. An entry q_{jj} of Q_i^k is updated as follows: The index j is a realization of an integer random variable uniformly distributed on $[1, n]$. Then, a realization of a random variable normally distributed on $[0, \beta]$ is added to the previous value of q_{jj}.

3 An off diagonal element of Q_i^k is changed. To update an entry q_{jh} of Q_i^k, the indices j, h are first independently chosen as realizations of an integer random variables uniformly distributed on $[1, n]$. Then, a realization of a random variable normally distributed on $[0, \beta]$ is added to the previous value of q_{jh}. To preserve the symmetry of the updated matrix Q_i^k , we set $q_{hj} = q_{jh}$.

After the alteration of entries of Q_i^k (action (2) and (3)) we must check if the changed matrix is still positive–definite. If this is not the case, the action is repeated (most probably the indices of the chosen entry and the value of entry alteration will then be different) until the renewed matrix becomes positive–definite. After the mutation operator is applied to all matrices of M_r^k, the stored best matrix is inserted by replacing a randomly chosen matrix. This set is called M^{k+1}; the procedure starts again with the evaluation of fitness.

4 Numerical Results

As a numerical example we consider the generalized Hénon map [2]

$$x(k+1) = \begin{pmatrix} u(k) - x_3(k)^2 - 0.1x_4(k) \\ x_1(k) \\ x_2(k) \\ x_3(k) \end{pmatrix} . \tag{16}$$

With the nominal input $u(k) = \bar{u} = const.$, the system has the equilibrium points

$$\bar{x}^{(1,2)} = \frac{-1.1 \pm \sqrt{1.21 + 4\bar{u}}}{2} (1, 1, 1, 1)^T .$$

The system (16) exhibits chaotic behavior for $\bar{u} = 1.76$ for which the Lyapunov exponents $\lambda_1 = 0.1538$, $\lambda_2 = 0.1418$, $\lambda_3 = 0.1189$, and $\lambda_4 = -2.7171$ can be determined. In the following we consider the local control of the equilibrium point $\bar{x}^{(1)} = \frac{-1.1+\sqrt{1.21+4\bar{u}}}{2} (1, 1, 1, 1)^T$.

Table 1. Settings of the evolutionary algorithm

Setting	Symbol	Value
Population size	l	16
Maximum number of generations	K	100
Width of initial population	γ	20000
Probability for no mutation	p_1	0.3
Probability for mutation of diagonal element	p_2	0.35
Probability for mutation of off diagonal element	p_3	0.35
Mutation strength	β	1250

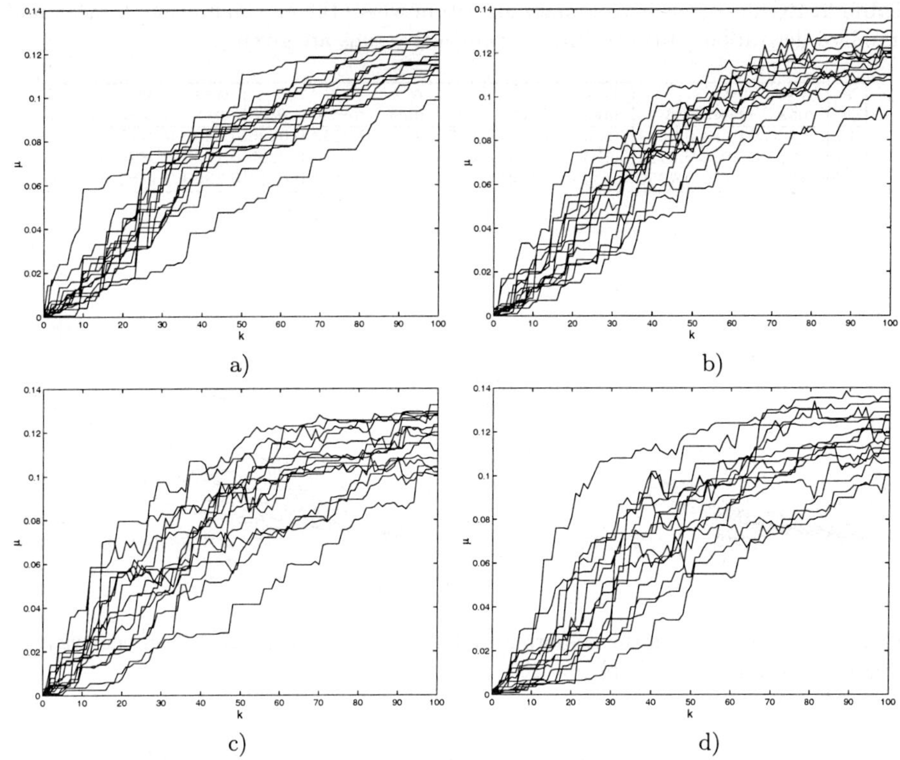

Fig. 2. Time evolution of the objective function (12) in 15 consecutive runs of the algorithm, a) $\alpha = 0$, b) $\alpha = 1$, c) $\alpha = 10$, d) $\alpha = 100$

The feedback matrix $G = (0, 0, -1.7723, -0.1)^T$ was chosen, so that $eig_i(A - BG) = 0, i = 1, \dots, 4$. Now, we optimize the local control using the evolutionary algorithm proposed in the previous section. Fig. 2 shows the results of the time evolution for the objective function (12) for different values of the scaling factor α, where the best individual per generation and run is recorded. Tab. 1 gives a listing of the used parameter values. These values were found in introductory numerical experiments and ensure satisfactory performance for all tested α as well as observability of certain effects discussed in the following. As a comparison, the time evolution for not using a multi–objective function ($\alpha = 0$) is shown in Fig. 2a. For each value of α, 15 consecutive experiments were carried out with the evolutionary algorithm for different randomly chosen initial populations. The evolution starts from low values of the objective function. This is plausible when we take into account that the likelihood for finding a matrix Q with high measure μ at random, scales with the dimension of search space. This also indicates that a pure random search for Q is not likely to be successful option for solving such problems. Furthermore, it can be seen in Fig. 2 that using the additional information about the volume of W has generally a positive effect on the speed with which the measure $\mu(A_R, W)$ moves towards the supposed optimum. There

Table 2. Results of the evolutionary algorithm where the maximum (max), mean and standard derivation (std) over the 15 consecutive runs are given

k	$\alpha =$ 0			$\alpha =$ 1			$\alpha =$ 10			$\alpha =$ 100		
	max	mean	std	max	mean	std	max	mean	std	max	mean	std
10	0.0400	0.0123	0.0091	0.0330	0.0150	0.0075	0.0390	0.0152	0.0096	0.0300	0.0115	0.0082
50	0.0955	0.0805	0.0123	0.1035	0.0824	0.0131	0.1140	0.0859	0.0205	0.1190	0.0825	0.0209
100	0.1300	0.1186	0.0085	0.1345	0.1174	0.0114	0.1325	0.1174	0.0113	0.1360	0.1183	0.0112

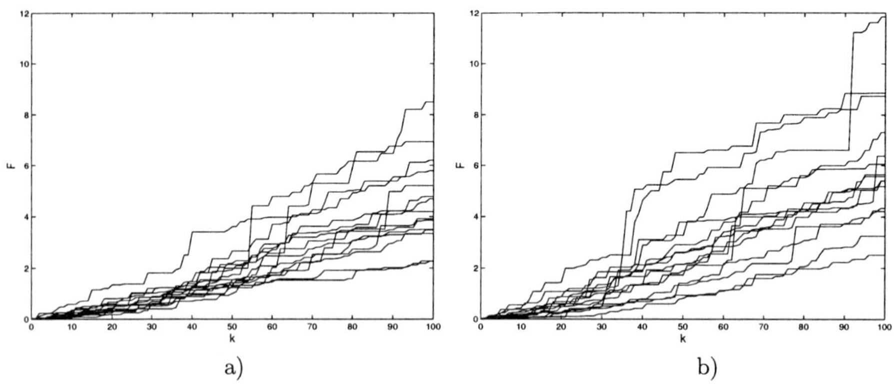

Fig. 3. Time evolution of the re–normalized fitness function (15) in 15 consecutive runs of the algorithm, a) $\alpha = 1$, b) $\alpha = 100$

are at least some evolutions which show faster convergence, while evolutions starting from another initial population tend to evolve slower, see also Tab. 2. An interpretation for these results is that a heavy weighting of the volume as an auxiliary driving force for the algorithm leads to a larger diversity in the individuals and thus to a larger variety if only the measure for these individuals is taken into account. In other words, there might be evolutions that produce matrices Q for which a large volume but a small measure is associated. For small values of α this is regulated via the fitness function and selection. For large α this is not mandatory. As a result, some evolutions tend to speedup in term of convergence of the objective function, while other slowdown. To illustrate this point of view, we consider the re–normalized fitness function $F := \frac{F}{\sqrt{1+\alpha^2}}$, see Fig. 3. Here, we see that for the (re–normalized) fitness the slowdown effect for evolution can not be observed.

5 Conclusion

We have studied an evolutionary algorithm for designing and optimizing a local control of chaotic systems. It determines the optimal SSRC W using a Lyapunov approach to local control. We have taken advantage of a multi–objective fitness function, which applies information about the volume of the SSRC W as well as the values for the objective function, the measure of the chaotic attractor within the SSRC $\mu(A_R, W)$, to drive the algorithm. Numerical results have been presented to demonstrate the given algorithm. In the numerical experiments we noted that the convergence speed can be increased by employing such an auxiliary driving force for the algorithm, particularly since this additional driving force has a sound physical meaning and supports optimizing the objective function.

References

1. Bäck, T.: Evolutionary Algorithms in Theory and Practice: Evolution Strategies, Evolutionary Programming, Genetic Algorithms. Oxford University Press, New York (1996)
2. Baier, G., Klein, M.: Maximum hyperchaos in generalized Hénon maps. Phys. Lett. **A151** (1990) 281-284
3. Fletcher, R.: Practical Methods of Optimization. John Wiley, Chichester (1987)
4. Goldberg, D.E.: Genetic Algorithms in Search, Optimization, and Machine Learning. Addision-Wesley, Reading MA (1989)
5. Kailath, T.: Linear Systems. Prentice-Hall, Englewood Cliffs NJ (1980)
6. Lin, C.T., Jou, C.P.: Controlling chaos by GA-based reinforcement learning neural network. IEEE Trans. Neural Networks **10** (1999) 846-859
7. Marin, J., Solé, R.V.: Controlling chaos in unidimensional maps using macroevolutionary algorithms. Phys. Rev. **E65** (2002) 026207/1-6
8. Michalewicz, Z.: Genetic Algorithms + Data Structures = Evolution Programs. 3rd edn. Springer-Verlag, Berlin Heidelberg New York (1996)
9. Ott, E.: Chaos in Dynamical Systems. Cambridge University Press, Cambridge (1993)
10. Packard, H.N.: A genetic learning algorithm for the analysis of complex data. Complex Systems **4** (1990) 543-572
11. Paterakis, E., Petridis, V., Kehagias, A.: Genetic algorithm in parameter estimation of nonlinear dynamical systems. In: Eiben, A.E., Bäck,T., Schoenauer, M., Schwefel, H.P. (eds.): Parallel Problem Solving from Nature - PPSN V. Springer-Verlag, Berlin Heidelberg New York (1998) 1008-1017
12. Richter, H., Reinschke, K.J.: Local control of chaotic systems: A Lyapunov approach. Int. J. Bifurcation and Chaos **8** (1998) 1565-1573
13. Richter, H., Reinschke, K.J.: Optimization of local control of chaos by an evolutionary algorithms. Physica **D144** (2000) 309-334
14. Rodriguez–Vázquez, K., Fleming, P.J.: Multi-objective genetic programming for dynamic chaotic systems modelling. In: Congress on Evolutionary Computation, CEC'99, Washington, D.C., USA, (1999) 22-28
15. Szpiro, G.G.: Forecasting chaotic time series with genetic algorithms. Phys. Rev. **E55** (1997) 2557-2568
16. Weeks, E.R., Burgess, J.M.: Evolving artificial neural networks to control chaotic systems. Phys. Rev. **E56** (1997) 1531-1540
17. Yadavalli, V.K, Dahulee, R.K., Tambe, S.S., Kulkarni, B.D.: Obtaining functional form of chaotic time series evolution using genetic algorithm. Chaos **9** (1999) 789-794

Evolutionary Algorithms: New Techniques

On Modelling Evolutionary Algorithm Implementations through Co-operating Populations

Panagiotis Adamidis[1] and Vasilios Petridis[2]

[1] Dept of Informatics, Technological Educational Institute of Thessaloniki, Greece 541 01
[2] Dept of Electrical & Computer Eng., Aristotle University of Thessaloniki, Greece 540 06

Abstract. In this paper we present a framework for modelling Simple and Parallel Evolutionary Algorithm implementations as Co-operating Populations. Using this framework, a method called *Co-operating Populations with Different Evolution Behaviours (CoPDEB)*, for generalizing and improving the performance of Parallel Evolutionary Algorithms (PEAs) is also presented. The main idea of CoPDEB is to maintain a number of populations exhibiting different evolution behaviours. CoPDEB was tested on three problems (the optimization of a real function, the TSP problem and the problem of training a Recurrent Artificial Neural Network), and appears to significantly increase the problem-solving capabilities over PEAs with the same evolution behaviour on each population. This paper also studies the effect of the migration rate (*Epoch*) and the population size on the performance of both PEAs and CoPDEB.

1 Introduction

The effectiveness of Evolutionary Algorithms (EAs), is limited by their ability to balance the need for a diverse set of sampling points, with the desire to quickly focus search upon potential solutions. Exploration is achieved through the recombination of data structures (which represent candidate solutions). The data structures constitute the population of the EA, which is evolved towards the optimum, concentrating search in those areas of higher fitness (exploitation).

The objective of this paper is twofold. First, we introduce a framework for modelling Evolutionary Algorithm implementations. A multi-level recursive model at different levels is defined (*Co-operating populations*). This model allows the implementation of any type of EA in any kind of hardware layout. According to the definition, one has a range of representation methods, population structures, selection methods, parameters, operators and mechanisms to choose from.

Although separating individuals into independent populations is a natural way to achieve a better balance of exploration and exploitation for a particular problem, the above problem of choice still remains. The Parallel EA (PEA) paradigms use the same evolution behaviour on each population. One still has to optimize each EA configuration, trying to select the most appropriate operators, parameters, mechanisms and communication methods [1, 2, 4].

The second objective of this paper is to propose a method for overcoming the problem of choice, using a variety of selection mechanisms, operators, communication methods, and parameters. The optimization method is called *Co-operating*

J.J. Merelo Guervós et al. (Eds.): PPSN VII, LNCS 2439, pp. 321–330, 2002.

Populations with Different Evolution Behaviours (CoPDEB), and according to it, each population uses a different evolution behaviour in order to overcome the problem of operator selection and improve the performance of cgPEAs.

Our results show that CoPDEB significantly outperforms cgPEAs where each population evolves using the same configuration.

2 Definition of "Co-operating Populations"

We define a model called *Co-operating Populations (CoP)*, in order to model a type of complex EA-like system, with the following characteristics:

- The system is composed of a finite number of items at each of several levels
- Each item functions as a global random search method emulating natural evolution.
- The structure of the items at each level depends on the structure of the items of all lower levels.
- The structure of the items at the same level is similar.
- Items on the same level can exchange some information, which might be the result of some processing, while maintaining their structure.

CoP can be informally described as follows. Populations of individuals live on islands that are able to communicate with each other. At a second level, populations of islands can form a larger entity with each island regarded as an individual. At a higher level, these larger entities can also form a population with each one regarded as an individual. The procedure can be continued to the required level. Offsprings are created by applying evolutionary operators to a population of individuals. An individual can be a string, an island of individuals or an entity at a higher level. Genetic evolution takes place according to a set of rules regarding selection, recombination and mutation.

Definition: CoP^k is defined recursively over k as a four-tuple which denotes level-k of the model

$$CoP^k = < CoP^{k-1},\ t,\ Reproduction,\ Structure > \tag{1}$$

with initial level defined for k=0 as:

$$CoP^0 = < P^0,\ t,\ Reproduction,\ Structure > \tag{2}$$

where:

$\boldsymbol{CoP^{k-1}}$ is the level $k-1$ of the model
$\boldsymbol{P^0}$ is the initial population of chromosomes
t is the termination criterion
Reproduction is used to advance one generation at level k:

$$Reproduction = < epoch,\ selection,\ operators,\ mechanisms,\ parameters > \tag{3}$$

- *epoch*: the number of completed generations at the lower level k-1, before a new generation at level k can be generated. At initial level 0 (CoP^0), *epoch=1*.
- *selection*: selection methods used (proportionate selection, tournament selection, ranking selection, steady state selection, Boltzmann selection, etc.)

- *operators*: a tuple of evolutionary operators used in the transition from population at time t, to the population at time t+1. Operators include recombination and mutation, along with their variations, inversion, hill-climbing, etc.
- *mechanisms*: general mechanisms such as elitism, replacement method, fitness scaling, fitness ranking, evolution of operators, migration strategy etc.
- *parameters*: selection methods parameters (selection probabilities, tournament size, binary tournament probability, etc.), and operator parameters (mutation rate, recombination probability, size of an individual, migration rate, etc.).

Structure of the CoP can be described as a six-tuple

$$Structure = < popsize, demesize, demestruct, \lambda, topology, popalloc > \quad (4)$$

- *popsize*: the number of individuals of a population
- *demesize*: the size of each neighborhood/deme.
- *demestruct*: the structure of the neighborhood/deme.
- λ: the number of available processing nodes.
- *topology*: the physical interconnection between processing nodes.
- *popalloc*: the specific configuration used for mapping population(s) to node(s)

3 Initial level of Co-operating Populations *(CoP^0)*

One of the most important characteristics of the *CoP* model is its hierarchical and recursive structure. The definition of each level depends on the previous one. Final level is always the initial level 0 *(CoP^0)*. CoP^0 can describe both massively parallel (fine-grained or cellular) EAs and simple (serial) EA implementations. The specific elements of the *Reproduction* and *Structure* tuples will finally determine the type of the EA that will be implemented. According to (1), we have to define the initial population P^0, the *termination* criterion, the *Reproduction* tuple and the *Structure* tuple.

The definition of population P^0, includes the representation (any type) and the initialization of P^0. The use of adaptive encoding usually improves performance.

The initial level determines the *termination* of the whole model.

The *Reproduction* tuple is defined as follows:

- According to the model definition *epoch* is equal to 1
- *selection* denotes the scheme used to choose the parents that will take part in the reproduction of the children. Appropriate selection schemes can be used [3, 5].
- *operators* refer to recombination and mutation with their variations.
- *mechanisms* refer to the general techniques that improve the performance of the algorithm (e.g. fitness scaling, parent replacement methods, etc.). Elitism is also a mechanism. Very frequent is also the use of a mechanism that adapts the operator probabilities [9].
- *parameters* include all selection scheme parameters (selection probabilities, tournament size, number of reproductive trials etc.), operator parameters (mutation rate, recombination probability), length of chromosomes etc.

The *Structure* tuple is defined as follows:

- *popsize* denotes the population size, which may also be adaptive
- In case of fine-grained PEAs where individuals are allowed to mate only within a neighborhood (deme), *demesize* denotes the size of the neighborhood. When *demesize* is equal to *popsize*, there are no neighborhoods and the population is "panmictic".
- *demestruct* denotes the structure (shape) of the neighborhood. Shape of deme may be a cross, square, line etc. Demes overlap by an amount that depends on their shape and size.
- λ denotes the number of available processors. In serial EAs, the whole population is usually assigned to one processor. On the contrary, in the case of fine-grained PEAs, each individual is ideally assigned to one processor.
- *topology* denotes the actual connection topology between the available processors at level 0.
- *popalloc* defines the way that the population is allocated to each processor.

4 Co-operating Populations at Level 1 (CoP^1)

CoP^1, depends on CoP^0. Each individual of CoP^1 is a CoP^0 and corresponds to an island. Each island has its own population, which evolves in parallel with the population of the other islands. The population of an island constitutes a population of possible solutions of the problem at hand.

The main representative and the most well studied paradigm of CoP^1, is the coarse-grained PEAs (cgPEAs) and specifically the island model. Each island of cgPEAs corresponds to a CoP^0 model, relatively isolated from each other. Isolation helps maintain genetic diversity. Migration of individuals from a CoP^0 to another is a possibility. Each individual of CoP^0 corresponds to a gene. We can have demes at CoP^1 level by allowing migration and reproduction between neighboring CoP^0.

In order to define CoP^1, we have to define CoP^0, the termination criterion, the *Reproduction* tuple and the *Structure* tuple. CoP^0 has been defined in section 3. The *termination criterion* of CoP^1 is the termination of the previous level CoP^0.

The *Reproduction* tuple is defined as follows:

- *epoch* equals to the completed number of generations at level 0, before a new generation is created at level 1. We are at the same generation of CoP^1, as long as the population of each CoP^0 evolves independently. In cgPEA, *epoch* is equal to the frequency of migration between islands.
- *selection* denotes the method used to choose each and every CoP^0 that will be used to produce the next generation. The selection methods used in previous level can be used here, after establishing a CoP^0 fitness. In cgPEAs, *selection* denotes the method used to choose the individuals to migrate.
- the two evolutionary *operators* (recombination and mutation) can be implemented in CoP^1 as well. Here, each CoP^0 corresponds to an individual. So, recombination can be implemented by combining the populations of two (or more) CoP^0 parents. Mutation can be implemented in a lot of different ways (exchanging the position of

two genes (i.e. individuals at CoP^0) of CoP^1, or randomizing the value of a gene (re-initialing an individual of a CoP^0, etc).

- *mechanisms* refer to general techniques that improve the performance of the algorithm (e.g. conditional initialization of a CoP^0 population). Migration scheme must also be defined here.
- *parameters* refer to operator parameters, selection method parameters, migration parameters etc. In cgPEAs it denotes the number of individuals that migrate between islands (migration rate).

The *Structure* tuple is defined as follows:

- *popsize* equals the number of available CoP^0.
- *demesize* equals to the number of CoP^0 that interact during the creation of a new generation.
- *demestruct* denotes the structure of the CoP^0 neighborhoods.
- λ equals the number of processors available to level 1.
- *topology* denotes the actual connections between the available processors.

popalloc refers to the allocation of CoP^0 to the available processors. One or more CoP^0 can be assigned to one or more processors. In cgPEA implementations, each processor is usually assigned one CoP^0.

CoPDEB is an instance of CoP^1 like cgPEAs. The difference in evolution behavior comes from the different configuration of each island, at level 0. A different evolution behavior (appropriate operators, parameters, and mechanisms) for each population. These configurations are integrated in the same evolutionary engine.

5 Implementation Issues of Coarse-Grained PEAs and CoPDEB

Both cgPEAs and CoPDEB are completely specified by defining all the components of CoP^1 four-tuple introduced earlier in section 2. The following is a description of the components of the tuple that we adopted in our implementations.

The *Reproduction* tuple for both cgPEAs and CoPDEB is as follows:

- *epoch* values used in our experiments vary between 1 and 500 generations.
- Best individual from each population is *selected* to migrate
- No *operators* are used to process individuals, prior to migration.
- One individual from each population is migrated (*parameters*).
- Migration strategy: Each island sends its best individual to every population, and receives the best individual from all the other islands. Incoming individuals replace random individuals of receiving island (excluding the best one). (*mechanisms*)

The components of the *Structure* tuple are as follows:

- Eight islands were used (*popsize*)
- Since *demesize* is equal to *popsize*, no demes are defined
- Number of available processors (λ) is eight.
- *topology* is static. All available processors are directly or indirectly connected to each other.

- The algorithms can be implemented on any machine or network. Processors are able to communicate with each other. Each island is allocated on a different processor. Any population can be allocated to any processor (*popalloc*).

The implementation of cgPEAs and CoPDEB would be the same if the model at level 0 (CoP^0) was the same. We tried to choose the best configuration for the CoP^0 model in the case of cgPEAs, where each island evolves under the same evolution behaviour. In the case of CoPDEB, each CoP^0 model evolves using a different configuration.

P^0, the initial population of each island is randomly created, after the encoding for each problem is determined. Since CoPDEB focuses on the effect of different evolution behaviours, we use the same encoding on all populations.

Implementation of *Structure* tuple of CoP^0 for both cgPEAs and CoPDEB: Population size (*popsize*) is equal to the number of individuals per population. As the population of each island is "panmictic" *demesize* is equal to *popsize* (no demes). Since the whole population is allocated to one processor (λ=1), *topology* is not defined.

We focus on the effect of the *Reproduction* tuple, which is responsible for the emergence of the different evolution behaviour exhibited by each CoP^0.

5.1 Reproduction Tuples

By definition the *epoch* value of cgPEAs is equal to 1, since every generation, a new population is produced.

Selection has to be balanced with variation from recombination and mutation. We use fitness proportionate selection with "roulette wheel" (in order to maintain a large genotypic diversity of the population), in combination with linear dynamic scaling method (proportionate selection without scaling was found to be less effective in keeping a steady pressure towards convergence).

Initial results, with different types of recombination on the first two problems (see section 6), has lead us to choose adaptive multi-point recombination.

On line adaptation of recombination parameters, and fitness scaling are also used. In order to accelerate the search we used a hill-climbing like operator called "phenotype mutation" [7] applied only to the best individual. Elitism is also used.

Since we do not want to solve the TSP problem, but rather focus on the better optimization and search capabilities of CoPDEB over cgPEAs, we did not alter substantially the behaviour of the algorithm used in the previous two problems. In this context we did not use any specific operators, codings or mechanisms.

Different evolution behaviour is mainly the result of different *Reproduction* tuples The different parameter settings for each population of CoPDEB are not the best, and are not tuned to the specific test problems. Under these restrictions the specific implementation of CoPDEB is **not optimal**. Thus the better search capabilities and the improved performance of CoPDEB are due to the combination of the different evolution behaviour of each population and not due to better population configurations.

In the case of CoPDEB, since *popsize* of CoP^1 is eight, we have to define eight *Reproduction* tuples.

By definition the *epoch* value is equal to 1, and it is the same with cgPEAs.

In order to keep differences from cgPEAs to a minimum, and choose configurations that are not better from the cgPEAs configuration, we have chosen to use on each population: fitness proportionate selection with "roulette wheel" in combination with linear dynamic scaling method, hill-climbing operator "phenotype mutation", elitism, retaining only the best individual at each generation, adaptive recombination probability and mutation rate (except when stated otherwise).

Their different characteristics are described below. For convenience of reference, the populations are numbered from 1 to 8:

Population 1: Its configuration is identical to configuration of cgPEA populations.

Population 2: Different recombination operator. It handles (recombines) genes and not whole individuals. For each gene, it uses one point recombination separating it into two parts. Then, each new gene of the offspring is created taknig the most significant part from the better parent and the least significant part from the worst. This technique produces new chromosomes with small perturbations of their values in the proximity of the chromosome value of the better parent (MSP_LSP operator).

Population 3: Recombination operator that functions like MSP_LSP, but it creates the offspring with the least significant part of the best parent and the most significant part of the worse. This allows for small perturbations of the chromosome values, in the proximity of the chromosome value of the worse parent (LSP_MSP).

Population 4: It uses two point recombination instead of multi-point recombination.

Population 5: Number of cutting points is set equal to half the number of genes.

Population 6: We use uniform crossover, instead of multi-point recombination

Population 7: Adaptive parameter probabilities in favor of mutation. Here we can maintain a higher value of mutation rate, in order to increase exploration of the search space, and mitigate the problem of premature convergence.

Population 8: Configuration of Population 1, without adaptation of recombination probability and mutation rate. Constant smaller recombination probability (0.2) and larger mutation rate. The configuration has the same objective with Population 7.

6 Optimization Problems – Experimental Results

Three problems were used to test the performance of CoPDEB over cgPEAs.

First problem: Training a fully connected Recurrent Artificial Neural Network (RANN) to generate two limit cycles, in the form of two sinusoidal functions of different amplitudes and frequencies [8]. The RANN has 5 neurons (35 trainable weights), two of which are output units. Since every weight is encoded in a 16 bit string, the genotype strings are 35x16=560 bits long, resulting in a search space of $2^{560}=3.77\times10^{168}$ different values. Different input levels (0.2 and 0.8) had to lead the two output units in different oscillations.

The second problem is the optimization function proposed by Rastrigin [10]:

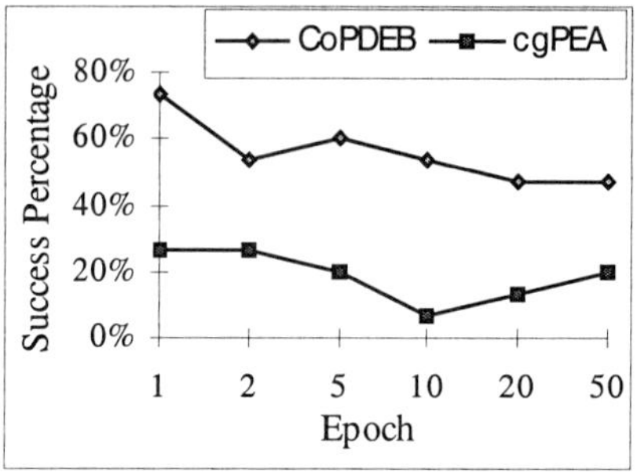

Fig. 1. Successful experiments (RANN)

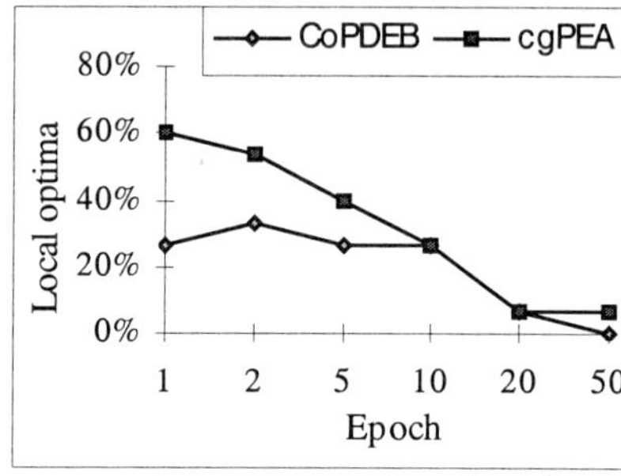

Fig. 2. Experiments stuck at local optima

$$f(x) = nA + \sum_{1}^{n} (x_i^2 - A\cos(\omega x_i)) \tag{3}$$

$n=20; \quad A=10; \quad \omega=2\pi; \quad -5.12 < x_i < 5.12;$

The local minima are located at a rectangular grid with size 1. The global minimum is at $x_i=0$, $i=1...,n$, giving $f(x)=0$.

The third problem is the well-known combinatorial Travelling Salesperson problem. We have taken a symmetric TSP problem from the TSPLIB, a library of travelling salesperson and related problem instances. We have chosen a moderate size problem with 29 cities in Bavaria, in order to be able to conduct a large number of experiments.

Initially, we tested the performance of the eight independent configurations on the first two problems running a set of 100 experiments for each configuration. The results (omitted due to lack of space) show that the configuration of "Population 1" used in cgPEAs has better performance for both problems. Then, we performed two sets of experiments for comparison purposes. The first set concerns CoPDEB, and the second concerns a cgPEA with the evolution characteristics of population 1. Since cgPEAs use the best configuration we can claim that the comparison between cgPEAs and CoPDEB is general, adequate, objective, and not biased to the specific implementations. In the case of RANN, each set comprises 90 experiments, that is 15 experiments for six different epoch values. The epoch values used were: 1, 2, 5, 10, 20, and 50 generations. Population size for each island was set to 200 individuals. The experiments showed an improved performance of CoPDEB over cgPEAs. The success rate of CoPDEB was, in the worst case, doubled (Fig 1).

Using CoPDEB, the percentage of the experiments that were stuck at local optima was decreased, and it kept decreasing as the Epoch value was increasing. CoPDEB usually failed because the generation limit (10000 gen.) was reached. cgPEA experiments failed due to the generation limit even more times (Fig 3), but this is because they have a smaller success percentage than CoPDEB. In the case of cgPEA, most of the failures were due to being stuck at local optima (Fig 2, 3). Thus we can assume that CoPDEB improves the exploration of the search space. CoPDEB requires smaller number of generations to train the RANN as well (Fig 4).

The second problem (less computational expensive than RANN) was also used to test extensively the effect of different epoch values on the performance of CoPDEB

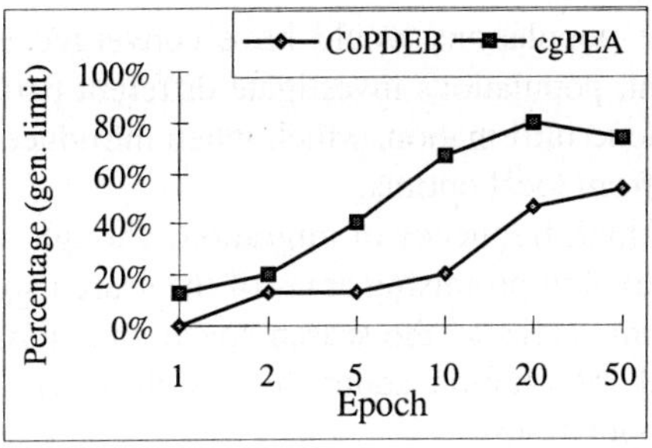

Fig. 3. Stopped at the generation limit (RANN).

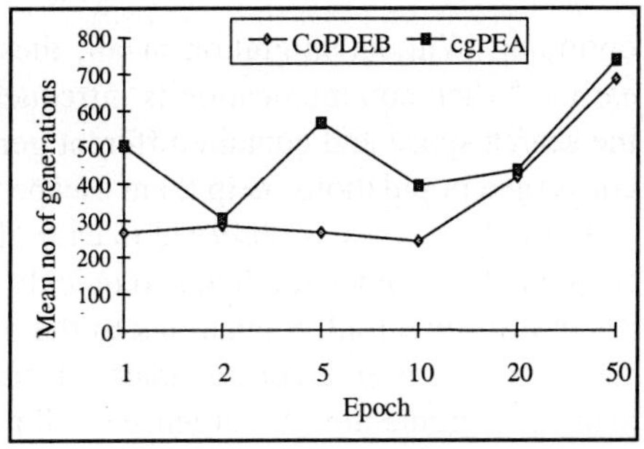

Fig. 4. Successful experiments

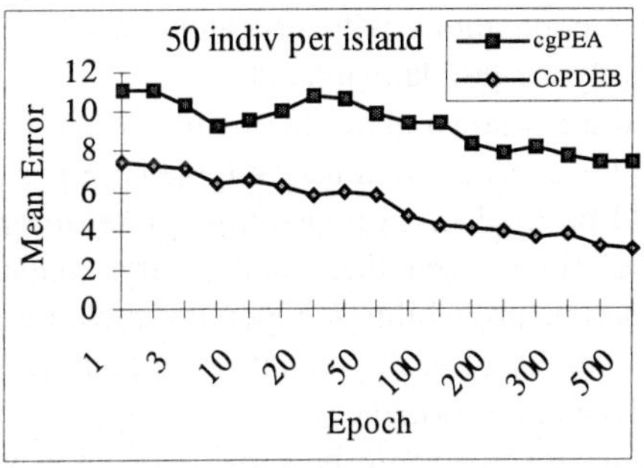

Fig. 5. Performance on Rastrigin funtion

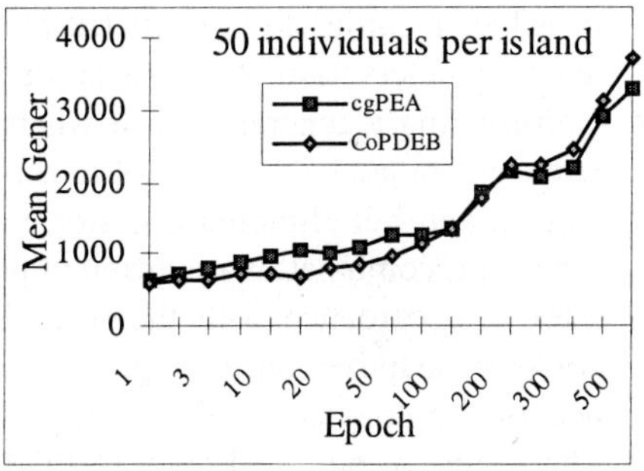

Fig. 6. Average number of generations

and cgPEAs. In this case, each set comprises 4500 experiments. Epoch values ranging from 1 to 500 generations were used. The population values used were: 50, 100 and 200 individuals per island. The results are the mean values from 100 runs.

The results confirm the improved performance of CoPDEB over cgPEAs (Fig 5) which depends on the epoch value. The mean error improves in both paradigms, as the epoch value increases. The results also show a high linear correlation between epoch values and mean errors. The tendency of mean errors to decrease as epoch increases is greater for CoPDEB.

The rate of improvement of the CoPDEB performance with epoch is also increased, as population per island is increased. This is not the case for cgPEAs. As the epoch value increases, the number of generations to converge is also increased. (Fig 6).

The final problem (TSP) was used in order to check the better performance of CoPDEB in combinatorial problems. Epoch values 1 and 10 were tested. The population of each island was 50 individuals. As we have already pointed out, the CoPDEB configuration used is not optimal in solving the TSP problem. Better performance of CoPDEB over cgPEAs is observed in this problem too.

7 Discussion – Conclusions

Our results propose that infrequent migration helps to avoid premature convergence of all populations, though some of them may have converged prematurely at a local optimum. Without migration at all, these populations would have converged even earlier. When communication is infrequent, populations investigate different parts of the search space and contain different genetic information, which when introduced to converged populations, help them escape from local optima.

Loss of diversity is observed in case of high frequency of migration. Though, very frequent communication helps to quickly exploit promising areas of the search space, it can also direct all populations to the same areas of the search space, thus loosing diversity and useful genetic material and information about the search space, and leading to premature convergence of all populations.

Two distinct cases for choosing optimal configurations:

- Solution to a specific problem: One should use as much *a priori* knowledge as possible, in order to construct different populations with operators mechanisms and parameters found to tackle successfully the problem at hand.
- Constructing a general tool, or when little is known about the nature of the problem: Integration of a variety of tools, in the same evolutionary engine is important. Elitism and hill-climbing operator could be used in all populations. Use different types of recombination on different populations. Use either small or large mutation rates in accordance with the recombination probabilities. Operator probabilities could be adaptive, but not against the diversity of each population. Adaptive connection topology can lead to better results and performance.

The results of our work with CoPDEB have shown it to be capable of improving the performance of the cgPEA model, achieving speedup and providing better exploration and exploitation of the search space.

References

1. P. Adamidis and V. Petridis, On the Parallelization of Artificial Neural Networks and Genetic Algorithms, *.Intern. J. Computer Math.* **67** (1998), 105-125.
2. E. Alba, J.M. Troya, A Survey of Parallel Distributed Genetic Algorithms, *Complexity* **4**(4) (1999), 31-52.
3. T. Blickle and L. Thiele, A comparison of selection schemes used in evolutionary algorithms, *Evolutionary Computation* (Winter 96), 4 (4) (1996), 361-394
4. E. Cantú-Paz, A survey of parallel genetic algorithms. Calculateurs Paralleles, Reseaux et Systems Repartis **10**, No. 2 (1998) pp. 141-171
5. U.K. Chakraborty, K. Deb, and M. Chakraborty, Analysis of selection algorithms: A Markov chain approach, *Evolutionary Computation* **4** (2) (Summer 1996), 133-167
6. D.E. Goldberg, H. Kargupta, J. Horn, and E. Cantu-Paz, "Critical Deme Size for Serial and Parallel Genetic Algorithms", IlliGAL Report No. 95002, Illinois Genetic Algorithms Lab., Univ of Illinois at Urbana-Champaign, 1995
7. V. Petridis and S. Kazarlis, Varying quality function in genetic algorithms and the cutting problem, in "Proc First IEEE CEC", vol I, pp. 166-169, IEEE, 1994
8. V. Petridis and A. Papaikonomou, Recurrent Neural Networks as Pattern Generators, in "Proc IEEE International Conference on Neural Networks", pp. 872-875, 1994
9. M. Srinivas and L.M. Patnaik, Adaptive Probabilities of Crossover and Mutation in Genetic Algorithms. *IEEE Transactions on Systems, Man and Cybernetics*, 24 (4).
10. Törn, A., & Zilinskas, A. (1989). Global Optimization. *Lecture Notes in Computer Science*, vol 350, Berlin: Springer-Verlag.

Permutation Optimization by Iterated Estimation of Random Keys Marginal Product Factorizations

Peter A.N. Bosman and Dirk Thierens

Institute of Information and Computing Sciences, Utrecht University,
P.O. Box 80.089, 3508 TB Utrecht, The Netherlands
{Peter.Bosman,Dirk.Thierens}@cs.uu.nl

Abstract. In IDEAs, the probability distribution of a selection of solutions is estimated each generation. From this probability distribution, new solutions are drawn. Through the probability distribution, various relations between problem variables can be exploited to achieve efficient optimization. For permutation optimization, only real valued probability distributions have been applied to a real valued encoding of permutations. In this paper, we present two approaches to estimating marginal product factorized probability distributions in the space of permutations directly. The estimated probability distribution is used to identify crossover positions in a real valued encoding of permutations. The resulting evolutionary algorithm (EA) is capable of more efficient scalable optimization of deceptive permutation problems of a bounded order of difficulty than when real valued probability distributions are used.

1 Introduction

Any reasonable optimization problem has some structure. Using this structure can aid the search for the optimal solution. In black box optimization, this structure is a priori unknown. In order to still be able to exploit problem structure, induction must be performed on previously evaluated solutions.

Holland [9] showed that in the simple GA with one point–crossover, problem structure is only exploited if it is expressed by short substrings of closely located gene positions. If such related genes are not closely located, they are likely to be disrupted by one–point crossover. Thierens and Goldberg [18] showed through the investigation of uniform crossover that the effect of such disruption on problems with non–linear interactions between groups of bits results in an exponential growth of the required population size as the problem length l increases. On the other hand, Harik, Cantú-Paz, Goldberg and Miller [7] showed that for a crossover operator that does not disrupt such building blocks of related genes, the required population size scales with $\sqrt{l}$ instead of exponentially.

In general, to prevent bad scaling behavior of an EA, we need to respect the structure of the optimization problem. Describing the structure of a set of samples can be done by estimating its probability distribution. Subsequently drawing

J.J. Merelo Guervós et al. (Eds.): PPSN VII, LNCS 2439, pp. 331–340, 2002.

more samples from this probability distribution leads to a statistical inductive type of iterated search. Such algorithms have been shown to give promising results for both binary and real valued spaces [2, 8, 12–17].

Our goal in this paper is to show how this technique can be used for permutation problems. The necessity for problem structure exploitation is found in permutation spaces as well, since permutation problems can be constructed that deceive simple permutation GAs [10] in a similar manner as is done for simple binary GAs [5] on which the simple GA is known to scale–up exponentially [18].

The remainder of this paper is organized as follows. In section 2 we present the representation of permutations that we will work with along with two permutation problems that are used in the experiments. A brief overview of IDEAs and its first application to permutation problems is given in section 3. Section 4 describes the estimation of probability distributions over permutations. We test our new IDEAs in section 5. Some conclusions are drawn in section 6.

2 Random Keys and Permutation Optimization Problems

We use the random keys encoding of permutations, which was proposed by Bean [1]. The main advantage of random keys is that no crossover operator can create unfeasible solutions. A random key sequence $\boldsymbol{r}$ is a vector of real values. Each of these real values is usually defined to be in $[0,1]$. The integer permutation $\boldsymbol{p}$ that is encoded by $\boldsymbol{r}$ can be computed in $\mathcal{O}(|\boldsymbol{r}|\log(|\boldsymbol{r}|))$ time by sorting $\boldsymbol{r}$ in ascending order ($\boldsymbol{r}_{\boldsymbol{p}_0} \leq \boldsymbol{r}_{\boldsymbol{p}_1} \leq \ldots \leq \boldsymbol{r}_{\boldsymbol{p}_{|\boldsymbol{r}|-1}}$), which we denote by $\boldsymbol{p} = \textsc{sort}(\boldsymbol{r})$. As an example, we have that $\textsc{sort}(0.61, 0.51, 0.62, 0.31) = (3, 1, 0, 2)$.

We will test our algorithms on two problems introduced by Knjazew [11]. Both problems are a sum of subfunctions f_{sub} that regard random key substrings:

$$f(\boldsymbol{r}) = \sum_{j=0}^{|\boldsymbol{\iota}|-1} f_{sub}(\boldsymbol{r}_{\boldsymbol{\iota}_j}) \tag{1}$$

In eq. 1, $\boldsymbol{\iota}$ is a vector of *index clusters* that defines the indices of the complete random keys sequence that each subfunction will regard. So $\boldsymbol{r}_{\boldsymbol{\iota}_j}$ indicates the random keys found at positions $(\boldsymbol{\iota}_j)_0, (\boldsymbol{\iota}_j)_1, \ldots, (\boldsymbol{\iota}_j)_{|\boldsymbol{\iota}_j|-1}$. Each random key index appears in at least one index cluster. For example, if $\boldsymbol{\iota} = ((0,1),(2),(2,3,4))$, we have $f(\boldsymbol{r}) = f_{sub}(\boldsymbol{r}_0, \boldsymbol{r}_1) + f_{sub}(\boldsymbol{r}_2) + f_{sub}(\boldsymbol{r}_2, \boldsymbol{r}_3, \boldsymbol{r}_4)$.

In the problems that we use in this paper, each index cluster has length κ^ι. The optimum for the subfunction is a random keys sequence that encodes the permutation $(0, 1, \ldots, \kappa^\iota - 1)$. To define the subfunction, a distance measure is used. This distance from any permutation $\boldsymbol{p}$ to the optimum equals $\kappa^\iota - |\textsc{lis}(\boldsymbol{p})|$, where $\textsc{lis}(\boldsymbol{p})$ is the *longest increasing subsequence* in $\boldsymbol{p}$. For example, if $\boldsymbol{p} = (1, 2, 0, 4, 3)$, then $\textsc{lis}(\boldsymbol{p}) \in \{(1,2,4),(1,2,3)\}$. Furthermore, $\kappa^\iota - |\textsc{lis}(\boldsymbol{p})| = 5 - 3 = 2$. Note that the reverse permutation $(\kappa^\iota - 1, \kappa^\iota - 2, \ldots, 0)$ is the only permutation with a distance of $\kappa^\iota - 1$. The subfunction is defined as follows:

$$f_{sub}(\boldsymbol{r}) = \begin{cases} 1 - \frac{|\textsc{lis}(\textsc{sort}(\boldsymbol{r}))|}{\kappa^\iota} & \text{if } |\textsc{lis}(\textsc{sort}(\boldsymbol{r}))| < \kappa^\iota \\ 1 & \text{if } |\textsc{lis}(\textsc{sort}(\boldsymbol{r}))| = \kappa^\iota \end{cases} \tag{2}$$

The fully deceptiveness of the subfunction makes the problem hard for any EA that doesn't identify the boundaries of the index clusters [10, 18] and as a result disrupts the non–linear relations between the genes imposed by f_{sub}.

In the first problem that we shall use to experiment with, the index clusters are *mutually disjoint.* This problem is therefore *additively decomposable.* To avoid the possibility that static crossover operators are biased in optimization because the genes contributing to each subfunction are closely located, the locations for each $\boldsymbol{\iota}_j^{loose}$ are chosen *loosely,* meaning as well spread as possible.

$$\boldsymbol{\iota}_j^{loose} = (j, j + |\boldsymbol{\iota}|, j + 2|\boldsymbol{\iota}|, \ldots, j + (\kappa^{\iota} - 1)|\boldsymbol{\iota}|) \tag{3}$$

In the second problem, the j–th index cluster *shares* its first position with index cluster $j-1$ and its last position with index cluster $j+1$. This overlapping property makes the problem *significantly* more difficult as there are no clear index cluster boundaries. For simplicity, the overlapping index cluster vector $\boldsymbol{\iota}^{overlap}$ is encoded tightly. Optimal solutions are now only given by random key sequences $\boldsymbol{r}$ such that $\text{SORT}(\boldsymbol{r}) = (0, 1, \ldots, l-1)$, where l is the problem length.

$$\boldsymbol{\iota}_j^{overlap} = (j(\kappa^{\iota} - 1), j(\kappa^{\iota} - 1) + 1, j(\kappa^{\iota} - 1) + 2, \ldots, j(\kappa^{\iota} - 1) + \kappa^{\iota} - 1) \tag{4}$$

3 IDEAs and ICE

We assume to have a cost function $C(\boldsymbol{z})$ of l problem variables $z_0, z_1, \ldots, z_{l-1}$ that, without loss of generality, should be minimized. For each z_i, we introduce a stochastic random variable Z_i and let $P^{\theta}(\boldsymbol{\mathcal{Z}})$ be a probability distribution that is uniform over all $\boldsymbol{z}$ with $C(\boldsymbol{z}) \leq \theta$ and 0 otherwise. Sampling from $P^{\theta}(\boldsymbol{\mathcal{Z}})$ gives more samples with an associated cost $\leq \theta$. Moreover, if we have access to $P^{\theta^*}(\boldsymbol{\mathcal{Z}})$ such that $\theta^* = \min_{\boldsymbol{z}}\{C(\boldsymbol{z})\}$, drawing only a single sample results in an optimal solution. This rationale underlies the IDEA (*Iterated Density Estimation Evolutionary Algorithm*) framework [2] and other named variants [8, 12–17].

Problem structure in the form of dependencies between the problem variables, is *induced* from a vector of selected solutions by finding a suitable probabilistic model $\mathcal{M}$. A probabilistic model is used as a computational implementation of a probability distribution $P_{\mathcal{M}}(\boldsymbol{\mathcal{Z}})$. A probabilistic model consists of a structure ς and a vector of parameters $\boldsymbol{\theta}$. The elementary building blocks of the probabilistic model are taken to be probability density functions (pdfs). A structure ς describes what pdfs are used and the parameter vector $\boldsymbol{\theta}$ describes the values for the parameters of these individual pdfs. A *factorization* is an example of a structure ς. A factorization *factors* the probability distribution over $\boldsymbol{\mathcal{Z}}$ into a product of pdfs. In this paper, we focus on *marginal product* factorizations (mpfs). In the binary and real valued case, an mpf is a product of multivariate joint pdfs. This product is represented by a vector of mutually exclusive vectors of random variable indices, which we call the *node vector* $\boldsymbol{\nu}$. An example in the case of $l = 3$ and binary random variables X_i, is given by $P_{\boldsymbol{\nu}}(\boldsymbol{\mathcal{X}}) = P(X_0, X_1)P(X_2)$, meaning $\boldsymbol{\nu} = ((0, 1), (2))$. Once a structure ς is given, the parameters for the multivariate pdfs have to be estimated. The way in which this is done, is predefined on beforehand. Often, this corresponds to a maximum likelihood estimate, such as a frequency count for binary random variables. As a probability distribution can thus be identified using only the structure ς, we denote it by $P_\varsigma(\mathcal{Z})$.

These definitions are used in the IDEA by selecting $\lfloor \tau n \rfloor$ samples ($\tau \geq \frac{1}{n}$) in each iteration t and by letting θ_t be the worst selected sample cost. The probability distribution $\hat{P}_\varsigma^{\theta_t}(\mathcal{Z})$ of the selected samples is then estimated, which is an approximation to the uniform probability distribution $P^{\theta_t}(\mathcal{Z})$. New samples can then be drawn from $\hat{P}^{\theta_t}(\mathcal{Z})$ to replace some of the current samples.

A special instance of the IDEA framework is obtained if selection is done by taking the top $\lfloor \tau n \rfloor$ best samples from the population, $\tau \in [\frac{1}{n}, 1]$, we draw $n - \lfloor \tau n \rfloor$ new samples, and the new samples replace the current worst $n - \lfloor \tau n \rfloor$ samples in the population. This results in the use of *elitism* such that $\theta_0 \geq \theta_1 \geq \ldots \geq \theta_{t_{\text{end}}}$. We call the resulting algorithm a *monotonic* IDEA.

Since the random keys are essentially a real valued domain, real valued IDEAs can directly be applied to permutation problems. Such an approach based upon normal probability distributions was proposed by Bosman and Thierens [3] as well as by Robles, de Miguel and Larrañaga [15]. However, the study by Bosman and Thierens [3] showed that this does not lead to very effective permutation optimization. The main problem with this approach is that solutions are not processed in the permutation space but in the largely redundant real valued space. To overcome this problem, a crossover operator was proposed [3]. This crossover operator reflects the dependency information learned in a factorization. Two parents are first selected at random. In the case of an mpf, the crossover operator then copies the values at the positions indicated by a vector in $\boldsymbol{\nu}$ from one of the two parents. This is repeated until all vectors in $\boldsymbol{\nu}$ have been regarded. Thus, whereas the IDEA is used to find the mpf, crossover is used instead of probabilistic sampling to generate new solutions. The resulting algorithm is called ICE (IDEA *Induced Chromosome Elements Exchanger*). Using ICE instead of a pure real valued IDEA gives significantly better results. The results are comparable with the only other EA that learns permutation structure information, which is the OmeGA by Knjazew [11]. The OmeGA is essentially a fmGA that works with random keys. The dependency information in this normal ICE is however still induced using normal distributions estimated over a largely redundant space, which may introduce false dependency information. To improve induction in ICE, we propose to induce these dependencies in the space of permutations directly by interpreting the random keys as permutations. This is the topic of the next section.

4 Estimating Random Keys Marginal Product Factorized Probability Distributions from Data

For each problem variable r_i we introduce a random variable R_i. Since the random keys encode permutations, the symantics of the R_i differ from those of binary or real valued random variables. We let $\boldsymbol{\mathcal{R}} = (R_0, R_1, \ldots, R_{l-1})$ and write the vector of selected samples as $\boldsymbol{\mathcal{S}} = (\boldsymbol{r}^0, \boldsymbol{r}^1, \ldots, \boldsymbol{r}^{|\mathcal{S}|-1})$, $\boldsymbol{r}^i = (r_0^i, r_1^i, \ldots, r_{l-1}^i)$.

The multivariate joint pdf over a subset of the random keys $\boldsymbol{R_v}$ is defined by the probability at a certain random key subsequence $\boldsymbol{r_v}$. It can be computed by counting the frequency in $\boldsymbol{\mathcal{S}}$ of the permutation $\textsc{sort}(\boldsymbol{r_v})$ represented by $\boldsymbol{r_v}$:

$$\hat{P}(\boldsymbol{R_v})(\boldsymbol{r_v}) = \frac{1}{|\boldsymbol{\mathcal{S}}|}\sum_{i=0}^{|\boldsymbol{\mathcal{S}}|-1}\begin{cases}1 & \text{if } \textsc{sort}((\boldsymbol{r}^i)_{\boldsymbol{v}}) = \textsc{sort}(\boldsymbol{r_v}) \\ 0 & \text{otherwise}\end{cases} \tag{5}$$

To define the random keys marginal product factorized probability distribution, it should be noted that the size of the alphabet for a multivariate joint factor $\hat{P}(\boldsymbol{R}_{\boldsymbol{\nu}_i})$ is $|\boldsymbol{\nu}_i|!$. Since the individual factors are taken to be *independent* of each other, the alphabet size of the whole mpf is $\prod_{i=0}^{|\boldsymbol{\nu}|-1}|\boldsymbol{\nu}_i|!$. However, the total number of possible permutations equals $l!$. Therefore, to construct a probability distribution over all possible permutations of length l, we must normalize the product of the multivariate marginals $\prod_{i=0}^{|\boldsymbol{\nu}|-1}\hat{P}(\boldsymbol{R}_{\boldsymbol{\nu}_i})(\boldsymbol{r}_{\boldsymbol{\nu}_i})$. To illustrate, assume that $r_0 < r_1$ and $r_2 < r_3$. Then there are $\frac{4!}{2!2!} = 6$ permutations of length 4 in which this is so, such as $r_0 < r_1 < r_2 < r_3$ and $r_0 < r_2 < r_3 < r_1$. This implies that the correct factorization of the probability distribution is given by $\hat{P}_{((0,1),(2,3))}(R_0, R_1, R_2, R_3) = \frac{2!2!}{4!}\hat{P}(R_0, R_1)\hat{P}(R_2, R_3)$. Concluding, the marginal product factorized probability distribution over all l variables becomes:

$$\hat{P}_{\boldsymbol{\nu}}(\boldsymbol{\mathcal{R}})(\boldsymbol{r}_{\mathcal{L}}) = \frac{\prod_{i=0}^{|\boldsymbol{\nu}|-1}|\boldsymbol{\nu}_i|!}{l!}\prod_{i=0}^{|\boldsymbol{\nu}|-1}\hat{P}(\boldsymbol{R}_{\boldsymbol{\nu}_i})(\boldsymbol{r}_{\boldsymbol{\nu}_i}) \tag{6}$$

To find a factorization given a sample vector $\boldsymbol{\mathcal{S}}$ of data, we use an incremental greedy algorithm to minimize a metric that represents a trade–off between the likelihood and the complexity of the estimated probability distribution. This is a common approach that has been observed to give good results [2, 8, 12, 14].

The factorization learning algorithm starts from the univariate factorization in which all variables are independent of each other $\boldsymbol{\nu} = ((0), (1), \dots, (l-1))$. Each iteration, an operation that changes $\boldsymbol{\nu}$ is performed such that the value of the penalization metric decreases. This procedure is repeated until no further improvement can be made. For the learning of random keys mpfs, we propose three possible operations. The operation that decreases the penalization metric the most, is actually performed. The first operation is a splice operation that replaces $\boldsymbol{\nu}_i$ and $\boldsymbol{\nu}_j$ $(i \neq j)$ with $\boldsymbol{\nu}_i \sqcup \boldsymbol{\nu}_j$, increasing the complexity of the factorization. The second operation is a swap operation in which two factors $\boldsymbol{\nu}_i$ and $\boldsymbol{\nu}_j$ may exchange an index. This operator allows to correct for lower order decision errors and is therefore *always* preferred over the application of a splice operation. The third operation is a *transfer* operation in which an index is removed from one factor $\boldsymbol{\nu}_i$ and added to another factor $\boldsymbol{\nu}_j$. This last operator is able to correct for some additional special cases of lower order decision errors [4]. A metric that has often proved to be successful, is known as the *Bayesian Information Criterion* (BIC). It scores a model by its negative log–likelihood, but adds a penalty term that increases with the model complexity ($|\boldsymbol{\theta}|$) and the size of the sample vector ($|\boldsymbol{\mathcal{S}}|$) [2]:

$$BIC(\mathcal{M}|\boldsymbol{\mathcal{S}}) = \underbrace{-\sum_{i=0}^{|\boldsymbol{\mathcal{S}}|-1} \ln\left(\hat{P}_{\mathcal{M}}(\boldsymbol{\mathcal{R}})(\boldsymbol{r}^i)\right)}_{\text{Error}(\hat{P}_{\mathcal{M}}(\boldsymbol{\mathcal{R}})|\boldsymbol{\mathcal{S}})} + \underbrace{\frac{1}{2}\ln(|\boldsymbol{\mathcal{S}}|)|\boldsymbol{\theta}|}_{\text{Complexity}(\hat{P}_{\mathcal{M}}(\boldsymbol{\mathcal{R}})|\boldsymbol{\mathcal{S}})} \quad (7)$$

The AIC metric is an alternative to the BIC metric. The AIC metric is similar to the BIC metric, but the complexity term is only given by the number of parameters $|\boldsymbol{\theta}|$. The penalization in the AIC metric is too weak compared to that in the BIC metric to give good results when normal probability distributions are estimated [3]. However, if we use frequency tables to estimate the pdfs in eq. 6, $|\boldsymbol{\theta}|$ grows factorially with an increase of any $\boldsymbol{\nu}_i$. Therefore, the penalization in the AIC will in this case most likely not be too weak. Because of the factorial growth of the number of parameters to be estimated however, we must limit the maximum factor size $\kappa^{\boldsymbol{\nu}}$ in any practical application (we used $\kappa^{\boldsymbol{\nu}} = 7$).

This direct limitation on the maximum order of interaction that can be processed, can be avoided by using *default tables* [6]. In a default table, the probabilities are explicitly specified for a subset of all available entries. For the absent entries, a *default value* is used, which is the average probability of all absent values. One straightforward way to use default tables, is to only specify the average frequency for each entry that occurs in the sample vector. By doing so, no factor can give rise to more parameters than $|\boldsymbol{\mathcal{S}}|$. We may still require $|\boldsymbol{\mathcal{S}}| = \mathcal{O}(\kappa^{\iota}!)$ when there are subproblems with a maximum length of κ^{ι} that need to be exhaustively sampled. However, when we must combine lower order solutions to get solutions of a higher order, the default tables can give us a much more efficient representation of the few good solutions to the subproblems. This latter issue is a important benefit of using local structures in probability distributions. A local structure allows for a more explicit representation of dependencies between values that can be assigned to random variables instead of dependencies between the random variables themselves. As a result, less parameters need to be estimated. Probabilistic models that are capable of expressing more complex dependencies now become eligible for selection when using a penalization metric, whereas otherwise non–local structure models expressing similar dependencies would never have been regarded because of the large number of (redundant) parameters the impose [6]. The use of local structures has been shown by Pelikan and Goldberg [14] to allow for efficient optimization of very difficult hierarchical deceptive optimization problems that exhibit dependencies between combinations of values for large groups of variables.

To use default tables, the random keys for each factor $\boldsymbol{\nu}_i$ are converted into integer permutations. The list of selected permutations is then sorted in $\mathcal{O}(|\boldsymbol{\nu}_i||\boldsymbol{\mathcal{S}}|\log(|\boldsymbol{\mathcal{S}}|))$ time and the frequencies are counted in $\mathcal{O}(|\boldsymbol{\nu}_i||\boldsymbol{\mathcal{S}}|)$ time. Note that we can't map the random keys to integers for faster sorting because the integers would become too large to efficiently represent as the factor size increases.

An additional operation that can be useful when using crossover on random keys, is *random rescaling*. With random rescaling, a block of random keys that is transferred to an offspring in crossover, is scaled to a subinterval of $[0,1]$ with probability p_ϱ. If for instance $(0.1, 0.2, 0.3)$ is scaled to $[0.9, 0.95]$, we get

(0.9, 0.925, 0.95). Note that this doesn't change the permutation that is encoded. Rescaling allows for dependencies *between* the building blocks to be exploited and increases the chance that they are combined properly. To ensure a large enough number of intervals so that the blocks can be ordered, we set this number to l.

5 Experiments

We have tested the new approach to learning a probability distribution over random keys by using it in monotonic 𝕀ℂ𝔼. We applied the algorithms to the additively decomposable deceptive permutation problem with $\kappa^{\iota} = 5$ and the overlapping problem with $\kappa^{\iota} = 4$. We used the rule of thumb by Mühlenbein and Mahnig [12] and set τ to 0.3. All results were averaged over 30 runs.

An mpf over random keys as defined in eq. 6, differs from the traditional definition for binary or real valued domains. This difference results in an additional requirement for using default tables for random keys, which is a cutoff value $\xi \in [0, 1]$ indicating the maximum default table length $\xi|\boldsymbol{S}|$. During the creation of a factorization in the greedy factorization learning algorithm, no operation is allowed to create a factor with a default table longer than $\xi|\boldsymbol{S}|$. Without this restriction, there would be an premature drift towards large factors. In this paper, we use a cutoff value of $\xi = 0.3$. For a thorough explanation, we refer the interested reader to a specialized technical report [4].

To get a good impression of the impact of different model building choices, we tested a varied ensemble of combinations. Figure 1 shows the scale–up behavior of all tested algorithms to obtain the optimal solution in all of the 30 runs on a log–log scale. A linear relationship on this scale indicates a polynomial scale–up behavior. Similar figures are obtained for the minimally required population size and the running time. The actual scaling coefficients are computed with a least squares line fit. The best scaling behavior is obtained when frequency tables are used in combination with the AIC metric. The use of the transfer operation shows a benefit over using only the splice and swap operations.

In figure 2, the results for the overlapping problem are tabulated. The maximum tested population size was $n \leq 10^5$. The results show the minimal requirements on the algorithms to solve the problem optimally or the performance at $n = 10^5$. No algorithm was capable of optimizing the problem without using *random rescaling.* If random rescaling is used, the overlapping deceptive problem can be solved optimally for $n \leq 10^5$ quite efficiently. However, we have now applied random rescaling, knowing that overlapping subproblems exist. Since we normally are not aware of this information, it is of great interest to see the implication of random rescaling on solving the additively decomposable problems. If the structure of additively decomposable problems is correctly found, introducing random rescaling should not matter. However, if the index cluster boundaries are not completely found, random rescaling may introduce additional disruptiveness to the crossover operation. This can indeed be seen in figure 1 as the scaling coefficients worsen as p_{ϱ} increases. The variant of 𝕀ℂ𝔼 that uses normal distributions is not capable of solving the problem at all for $p_{\varrho} = 0.5$. For

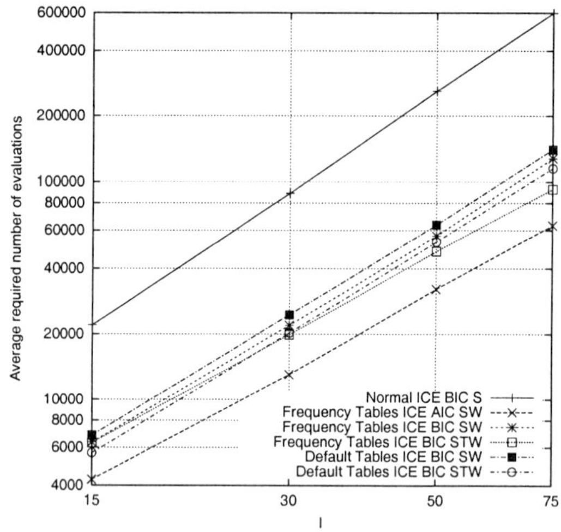

Additively decomposable problem						
Algorithm				$\mathcal{O}(l^{x.xx})$		
ICE	Oper.	Metric	p_ϱ	n	Eval.	Time
N	S	BIC	0.0	1.61	2.06	3.70
N	S	BIC	0.1	1.71	2.79	4.33
N	S	BIC	0.5	—	—	—
P	SW	AIC	0.0	1.23	1.68	4.07
P	SW	AIC	0.1	1.42	2.14	4.58
P	SW	AIC	0.5	1.64	4.14	7.07
P	SW	BIC	0.0	1.34	1.82	4.26
P	SW	BIC	0.1	1.43	1.69	4.51
P	SW	BIC	0.5	2.13	3.87	6.84
P	SWT	BIC	0.0	1.23	1.68	4.08
P	SWT	BIC	0.1	1.33	1.71	4.01
P	SWT	BIC	0.5	1.75	2.77	5.61
PDT	SW	BIC	0.0	1.25	1.87	4.71
PDT	SW	BIC	0.1	1.71	2.21	5.13
PDT	SW	BIC	0.5	3.80	4.36	8.19
PDT	SWT	BIC	0.0	1.32	1.87	4.80
PDT	SWT	BIC	0.1	1.81	2.28	4.88
PDT	SWT	BIC	0.5	3.32	4.45	8.29

Fig. 1. Results on the additively decomposable deceptive problem ($\kappa^\iota = 5$). On the left, the required average number of evaluations as a function of the problem length l on a log–log scale are shown. The straight lines indicate polynomial scale–up behavior. On the right, the polynomial scaling coefficients with respect to the population size n, the average required evaluations and the average actual running time for the tested $\mathbb{ICE}$ variants are shown. The algorithms are indicated by N for the use of normal pdfs, P for the use of permutation pdfs using frequency tables and PDT for the use of permutation pdfs using default tables. The factorization search operations are indicated by S for the splice operation, W for the swap operation and T for the transfer operation.

Additively overlapping problem									
Algorithm				$\|\iota\| = 3$, $\kappa^\iota = 4$			$\|\iota\| = 6$, $\kappa^\iota = 4$		
ICE	Oper.	Metric	p_ϱ	n	subs	Eval.	n	subs	Eval.
N	S	BIC	0.0	100000	2.17	643674	100000	4.07	1579354
N	S	BIC	0.1	100000	2.70	1066013	100000	4.87	2585035
N	S	BIC	0.5	100000	2.87	4111057	100000	5.40	45423314
P	SW	AIC	0.0	21000	3.00	222893	100000	5.20	1654022
P	SW	AIC	0.1	1900	3.00	21022	16000	6.00	328135
P	SW	AIC	0.5	1500	3.00	18982	11000	6.00	442256
P	SW	BIC	0.0	21000	3.00	250336	100000	5.00	1224018
P	SW	BIC	0.1	1600	3.00	19088	22000	6.00	576436
P	SW	BIC	0.5	1500	3.00	22520	18000	6.00	1116807
P	SWT	BIC	0.0	19000	3.00	210534	100000	5.00	1259755
P	SWT	BIC	0.1	1400	3.00	16998	20000	6.00	538037
P	SWT	BIC	0.5	1200	3.00	16759	12000	6.00	798334
PDT	SW	BIC	0.0	70000	3.00	684962	100000	5.00	1374018
PDT	SW	BIC	0.1	4750	3.00	42999	18000	6.00	302783
PDT	SW	BIC	0.5	1500	3.00	13586	80000	6.00	1116019
PDT	SWT	BIC	0.0	50000	3.00	473512	100000	5.00	1416019
PDT	SWT	BIC	0.1	4000	3.00	35371	13000	6.00	202301
PDT	SWT	BIC	0.5	1500	3.00	12641	70000	6.00	1050020

Fig. 2. Results for all algorithms on the overlapping deceptive permutation problem with $\kappa^\iota = 4$; subs stands for the average number of subfunctions solved optimally. The abbreviations are the same as those in figure 1.

$p_\varrho = 0.1$, the scaling behavior is not effected too much, so in general we would suggest the use of a small p_ϱ. However, it would be interesting to see whether the results can be improved by setting p_ϱ adaptively by for instance looking at the rate of success of applying random rescaling and by changing p_ϱ accordingly.

The overhead in the scaling results are smaller for the AIC metric than for the BIC metric. This is a result of the smaller penalization in the AIC metric, which results in larger factors for smaller population sizes. This can lead to overly complex models, for which reason the BIC metric is often preferred. In our case however, it introduces a useful bias for correctly finding the index clusters. The best scaling results are expected to be obtained when the AIC metric is used in combination with the splice, swap and transfer operation.

6 Discussion and Conclusions

We have proposed a new tool for finding and using the structure of permutation problems in evolutionary optimization by estimating mpfs in the space of permutations. By using this probabilistic information to exchange the random keys that encode the permutations in a crossover operator, we obtain the $\mathbb{ICE}$ algorithm. $\mathbb{ICE}$ has been shown to scale up efficiently on deceptive additively decomposable permutation problems of a bounded difficulty and to furthermore give promising results on difficult overlapping deceptive permutation problems.

The use of default tables indicates a slightly larger overall requirement on the computational resources used by $\mathbb{ICE}$. However, the scaling behavior is not effected too much. The advantage of default tables in that they allow more complex models to compete in model selection, is more likely to stand out on hierarchical deceptive problems. Empirical verification of this expectation, combined with the results presented in this paper, would lead us to conclude that the use of default tables with a small probability at using random rescaling is the most effective allround optimization variant of $\mathbb{ICE}$ for permutation problems.

Although good results have been obtained, mpfs are not well suited for problems with overlapping building blocks. To this end, Bayesian factorizations in which gene positions may be dependent on other gene positions are likely to be more appropriate. Using the tools proposed in this paper, Bayesian factorizations may be learned, although it is to be expected that the use of a greedy learning algorithm that introduces one dependency at a time, will also have lower order decision error problems. To overcome this problem, new operators will be required or some effective means of using the distances between random keys.

Finally, although our results are encouraging, we have only used a limited number of test problems. An interesting next effort would be to investigate the performance on real–world problems and provide a comparison with other EAs.

References

1. J. C. Bean. Genetic algorithms and random keys for sequencing and optimization. *ORSA Journal on Computing*, 6:154–160, 1994.

2. P. A. N. Bosman and D. Thierens. Advancing continuous IDEAs with mixture distributions and factorization selection metrics. In M. Pelikan and K. Sastry, editors, *Proceedings of the Optimization by Building and Using Probabilistic Models OBUPM Workshop at the Genetic and Evolutionary Computation Conference GECCO–2001*, pages 208–212. Morgan Kaufmann, 2001.
3. P. A. N. Bosman and D. Thierens. Crossing the road to efficient IDEAs for permutation problems. In L. Spector et al., editor, *Proc. of the Genetic and Evolutionary Computation Conf. – GECCO–2001*, pages 219–226. Morgan Kaufmann, 2001.
4. P. A. N. Bosman and D. Thierens. Random keys on ICE: Marginal product factorized probability distributions in permutation optimization. Utrecht University Technical Report UU–CS–2002–xx., 2002.
5. K. Deb and D. E. Goldberg. Sufficient conditions for deception in arbitrary binary functions. *Annals of Mathematics and Artificial Intelligence*, 10(4):385–408, 1994.
6. N. Friedman and M. Goldszmidt. Learning Bayesian networks with local structure. In E. Horvitz and F. Jensen, editors, *Proc. of the 12th Conference on Uncertainty in Artificial Intelligence (UAI-96)*, pages 252–262. Morgan Kaufmann, 1996.
7. G. Harik, E. Cantú-Paz, D. E. Goldberg, and B. L. Miller. The gambler's ruin problem, genetic algorithms, and the sizing of populations. *Evolutionary Computation*, 7(3):231–253, 1999.
8. G. Harik and D. E. Goldberg. Linkage learning through probabilistic expression. *Comp. methods in applied mechanics and engineering*, 186:295–310, 2000.
9. John H. Holland. *Adaptation in Natural and Artifical Systems.* University of Michigan Press, Ann Arbor, Michigan, 1975.
10. H. Kargupta, K. Deb, and D. E. Goldberg. Ordering genetic algorithms and deception. In R. Männer and B. Manderick, editors, *Parallel Problem Solving from Nature – PPSN II*, pages 47–56. Springer Verlag, 1992.
11. D. Knjazew. Application of the fast messy genetic algorithm to permutation and scheduling problems. IlliGAL Technical Report 2000022, 2000.
12. H. Mühlenbein and T. Mahnig. FDA – a scalable evolutionary algorithm for the optimization of additively decomposed functions. *Evol. Comp.*, 7(4):353–376, 1999.
13. A. Ochoa, H. Mühlenbein, and M. Soto. A factorized distribution algorithm using single connected bayesian networks. In M. Schoenauer, K. Deb, G. Rudolph, X. Yao, E. Lutton, J. J. Merelo, and H.-P. Schwefel, editors, *Parallel Problem Solving from Nature – PPSN VI*, pages 787–796. Springer Verlag, 2000.
14. M. Pelikan and D. E. Goldberg. Escaping hierarchical traps with competent genetic algorithms. In L. Spector, E. D. Goodman, A. Wu, W. B. Langdon, H.-M. Voigt, M. Gen, S. Sen, M. Dorigo, S. Pezeshk, M. H. Garzon, and E. Burke, editors, *Proceedings of the GECCO–2001 Genetic and Evolutionary Computation Conference*, pages 511–518. Morgan Kaufmann, 2001.
15. V. Robles, P. de Miguel, and P. Larrañaga. Solving the traveling salesman problem with EDAs. In P. Larrañaga and J.A. Lozano, editors, *Estimation of Distribution Algorithms. A new tool for Evolutionary Computation.* Kluwer Academic, 2001.
16. R. Santana, A. Ochoa, and M. R. Soto. The mixture of trees factorized distribution algorithm. In L. Spector et al., editor, *Proc. of the GECCO–2001 Genetic and Evolutionary Computation Conference*, pages 543–550. Morgan Kaufmann, 2001.
17. S.-Y. Shin and B.-T. Zhang. Bayesian evolutionary algorithms for continuous function optimization. In *Proceedings of the 2001 Congress on Evolutionary Computation – CEC2001*, pages 508–515. IEEE Press, 2001.
18. D. Thierens and D.E. Goldberg. Mixing in genetic algorithms. In S. Forrest, editor, *Proceedings of the fifth conference on Genetic Algorithms*, pages 38–45. Morgan Kaufmann, 1993.

Advanced Population Diversity Measures in Genetic Programming

Edmund Burke, Steven Gustafson*, Graham Kendall, and Natalio Krasnogor

ASAP Research, School of Computer Science & IT
University of Nottingham, UK
{ekb,smg,gxk,nxk}@cs.nott.ac.uk

Abstract. This paper presents a survey and comparison of significant diversity measures in the genetic programming literature. This study builds on previous work by the authors to gain a deeper understanding of the conditions under which genetic programming evolution is successful. Three benchmark problems (Artificial Ant, Symbolic Regression and Even-5-Parity) are used to illustrate different diversity measures and to analyse their correlation with performance. Results show that measures of population diversity based on edit distances and phenotypic diversity suggest that successful evolution occurs when populations converge to a similar structure but with high fitness diversity.

1 Introduction

Maintaining population diversity in genetic programming is cited as crucial in preventing premature convergence and stagnation in local optima [1][2][3][4][5]. Diversity describes the amount of variety in the population defined by the genetic programming individuals structure or their performance. The number of different fitness values (phenotypes) [6], different structural individuals (genotypes) [7], edit distances between individuals [3] [8], and complex and composite measures [9] [10] [11] are used as measures of diversity.

In this paper, we examine previous uses and meanings of diversity, compare these different measures on three benchmark problems and extend our original study [12] with additional experiments, new analysis and new measures. Population diversity is related to almost every aspect of program evolution and extending the research in [12] will lead to a deeper understanding of evolution in genetic programming. As far as the authors are aware, all the significant diversity measures that occur in the genetic programming literature are reported.

2 Diversity Measures

Measures of diversity attempt to quantify the variety in a population and some methods attempt to control or promote diversity during evolution. The following section surveys both measures that provide a quantification of population

* Corresponding author

J.J. Merelo Guervós et al. (Eds.): PPSN VII, LNCS 2439, pp. 341–350, 2002.

diversity and methods used to actively promote and maintain diversity within genetic programming.

2.1 Population Measures

A common type of diversity measure is that of structural differences between programs. Koza [13] used the term *variety* to indicate the number of different genotypes populations contained. Landgon [7] argues that genotypic diversity is a sufficient upper bound of population diversity as a decrease in unique genotypes must also mean a decrease in unique fitness values.

Keijzer [10] measures program variety as a ratio of the number of unique individuals over population size and subtree variety as the ratio of unique subtrees over total subtrees. Tackett [14] also measures structural diversity using subtress and schemata frequencies. D'haeseleer and Bluming [11] define *behavior* and *frequency* signatures for each individual based on fitness and *gene* frequencies, respectively. The correlation between individuals' respective signatures represents the *phenotypical* and *genotypical* diversity.

When tree representations of genetic programs are considered as graphs, individuals can be compared for isomorphism [5] to obtain a more accurate measure of diversity. Determining graph isomorphism is computationally expensive for an entire population and not straightforward for genetic programs. However, counting the number of nodes, terminals, functions and other properties can be used to determine whether trees are *possible* isomorphs of each other.

McPhee and Hopper [1] investigate diversity at the genetic level by tagging each node created in the initial generation. Root parents, the parents whose tree has a portion of another individual's subtree swapped into it during crossover, are also tracked. McPhee and Hopper found that the number of unique tags dramatically falls after initial generations and, by tracking the root parents, after an average of 16 generations, all further individuals have the same common root ancestor.

Phenotypic measures compare the number of unique fitness values in a population. When the genetic programming search is compared to traversing a fitness landscape, this measure provides an intuitive way to think of how much the population covers that landscape. Other measures could be created by using fitness values of a population, as done by Rosca [5] with entropy and free energy. Entropy here represents the amount of disorder of the population, where an increase in entropy represents an increase in diversity.

2.2 Promoting Diversity

Several measures and methods have been used to promote diversity by measuring the difference between individuals. These methods typically use a non-standard selection, mating, or replacement strategy to bolster diversity. Common methods are neighborhoods, islands, niches, and crowding and sharing from genetic algorithms.

Eschelman and Schaffer [15] use Hamming distances to select individuals for recombination and replacement to improve over hill-climbing-type selection strategies for genetic algorithms. Ryan's [2] "Pygmie" algorithm builds two lists based on fitness and length to facilitate selection for reproduction. The algorithm maintains more diversity, prevents premature convergence and uses simple measures to promote diversity. De Jong et al [8] use multiobjective optimisation to promote diversity and concentrate on non-dominated individuals according to a 3-tuple of <*fitness, size, diversity*>. Diversity is the average square distance to other members of the population, using a specialised measure of edit distance between nodes. This multiobjective method promotes smaller and more diverse trees.

McKay [4] applies the traditional fitness sharing concept from Deb and Goldberg [16] to test its feasibility in genetic programming. Diversity is the number of fitness cases found, and the sharing concept assigns a fitness based on an individual's performance divided by the number of other individuals with the same performance. McKay also studies negative correlation and a *root quartic negative correlation* in [9] to preserve diversity. Ekárt and Németh [3] apply fitness sharing with a novel tree distance definition and suggest that it may be an efficient measure of structural diversity. Bersano-Begey [17] track how many individuals solve which fitness cases and a pressure is added to individuals to promote the discovery of different or less popular solutions.

3 Experiment Design

Our initial study of population diversity measures [12] highlighted that phenotypic measures appeared to better correlate with better fitness. Runs which had better fitness in the last generation also tended to have higher phenotypic diversity measures. This appears to go against conventional wisdom in genetic programming which says that runs must converge to an "exploitation" phase where diversity is lost to focus on better individuals. However, it does agree with the intuitive idea that proper evolution *needs* diversity to be effective.

In this study we extend our original analysis [12] with new experiments and new measures of population diversity which we have adapted from diversity promoting methods. In analysing results, we measure the Spearman correlation [18] between diversity and fitness and examine standard deviations, minimum and maximum values and the diversity of all populations in every run and the best fitness of those populations.

Three common problems are used with common parameter values from previous studies. For all problems, a population size of 500 individuals, a maximum depth of 10 for each individual, a maximum depth of 4 for the tree generation half-n-half algorithm, standard tree crossover and internal node selection probability of 0.9 for crossover is used. Additionally, each run consists of 51 generations, or until the ideal fitness is found.

The Artificial Ant, Symbolic Regression and Even-5-Parity problems are used. All three problems are typical to genetic programming and can be found in many studies, including [13]. The artificial ant problem attempts to find the best strategy for picking up pellets along a trail in a grid. The fitness for this problem is measured as the number of pellets missed. The regression problem attempts to fit a curve for the function $x^4 + x^3 + x^2 + x$. Fitness here is determined by summing the squared difference for each point along the objective function and the function produced by the individual. The even-5-parity problem takes an input of a random string of 0's and 1's and outputs whether there are an even number of 1's. The even-5-parity fitness is the number of wrong guesses for the 2^5 combinations of 5-bit length strings. All problems have an ideal fitness of low values (0=best fitness).

To produce a variety of run performances, where we consider the best fitness in the last generation, we designed three different experiments, carried out 50 times, for each problem. The first experiment, *random*, performs 50 independent runs. The experiment *stepped-recombination* does 50 runs with the same random number seed, where each run uses an increasing probability for reproduction and decreasing probability for crossover. Initially, probability for crossover is 1.0, and this is decreased by 0.02 each time (skipping value 0.98) to allow for exactly 50 runs and ending with reproduction probability of 1.0 and crossover probability 0.0. The last experiment, *stepped-tournament*, is similar but we begin with a tournament size of 1 and increment this by 1 for each run, until we reach a tournament size of 50. In the *random* and *stepped-tournament* experiments, crossover probability is set to 1.0 and the tournament size in *random* and *stepped-recombination* is 7. The *Evolutionary Computation in Java* (ECJ), version 7.0, [19] is used, where each problem is available in the distribution.

The following measures of diversity were introduced previously and are briefly described as they are collected for each generation in every run. **Genotype and phenotype** diversity count the number of unique trees for the genotype measure [7] and the number of unique fitness values in a population represents the phenotype measure [6]. The **entropy** measure is calculated for the population as in [5], where "p_k is the proportion of the population P occupied by population partition k", $-\sum_k p_k \cdot \log p_k$. A partition is assumed to be each possible different fitness value, but could be defined to include a subset of values. **Pseudo-isomorphs** are found by defining a 3-tuple of <terminals,nonterminals,depth> for each individual and the number of unique 3-tuples in each population is the measure. Two identical 3-tuples represent trees which could be isomorphic. **Edit distance 1 and 2** is the edit distance between individuals used by de Jong et al [8] (referred to as "ed 1" in the graphs) and an adapted version of Ekárt and Németh [3] ("ed 2"). Every individual in the population is measured against the best fit individual. This measure is then divided by the population size. The first measure (ed 1) is a standard edit distance measure where two trees are overlapped at the root node. Two different nodes, when overlapping, score a distance of 1 and equal nodes get 0. The edit distance is then the sum of all different nodes and normalised by dividing it by the size of the smaller tree. The second measure (ed 2) is slightly adapted back to its original formulation in [20] where the difference between any two nodes is 1. The difference between two trees is then (defined in [3]):

$$dist(T_1,T_2) = \begin{cases} d(p,q) & \text{if neither } T_1 \text{ nor } T_2 \text{ have any children} \\ d(p,q) + \frac{1}{2} * \sum_{l=1}^{m} dist(s_l,t_l) & \text{otherwise} \end{cases}$$

Where T_1, T_2 are trees with roots p, q and possible children (m total) subtrees s, t. Two trees are brought to the same tree structure by adding "null" nodes to each tree. Note that the differences near the root have more weight, a possibly convenient description for genetic programming as it has been noted that programs converge quickly to a fixed root portion [1].

The Spearman correlation coefficient is computed as [18] $1 - \frac{6\sum_{i=1}^{N} d_i^2}{N^3-N}$. Where N is the number of items (50 runs), and d_i is the distance between each run's rank of performance and rank of diversity in the last generation. A value of -1.0 represents negative correlation, 0.0 is no correlation and 1.0 is positive correlation. For our measures, if we see ideal low fitness values, which will be ranked in ascending order (1=best,...,50=worst) and high diversity, ranked where (1=lowest diversity and 50=highest diversity), then the correlation coefficient should be strongly negative. Alternatively, a positive correlation indicates that either bad fitness accompanies high diversity or good fitness accompanies low diversity.

4 Results

Graphs of the 50 runs for all three experiments and all three problems were examined, along with minimum, maximum and standard deviations of best fitness and population diversity measures. Also, the Spearman correlation coefficient was calculated, correlating the diversity measures with best fitness across each set of 50 runs. This study involved 450 runs of 51 generations each, adding to a previous study [12] of the same size with different random seeds and different measures of diversity. While all three problems showed the same general trends, we focus on the artificial ant and even-5-parity.

Figure 1 shows for the artificial ant problem and *random* experiment, that diversity measures and fitness varied widely. The most dramatic activity occurs early with runs being similar until around generation 10, where they become quite varied. However, from Table 1 we can see several interesting phenomenon. First, by noting the genotype measure and best fitness standard deviations for the artificial ant experiment, we see little variance of best fitness (11.2,15.4,15.9) but large variance of genotype diversity (18.3,120.1,44.2). Also note that the genotype diversity for the random experiment in artificial ant and even-5-parity have very high minimum and maximum values, where the other measures minimum and maximum does not differ across experiments. This information leads us to believe that the genotype diversity measure does not suggest a strong correlation with varying run performance. Note how the other measures have consistent variation, as does fitness.

Using the Spearman correlation coefficient we investigated whether runs that produced good fitness had low/high diversity, where ties in ranks were solved by

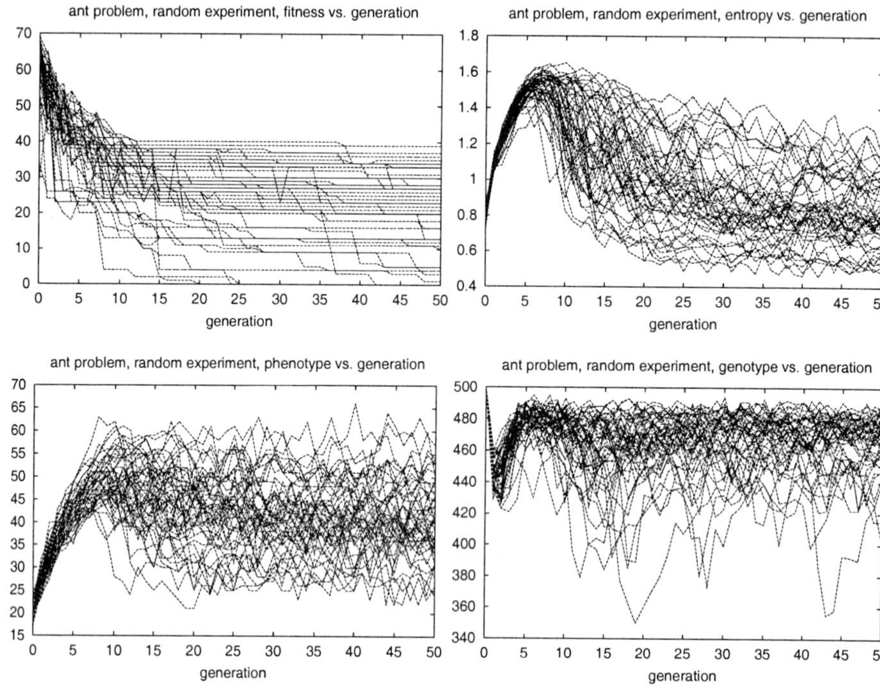

Fig. 1. 50 runs of best fitness per generation for the artificial ant random experiment and a graph for each of the diversity measures of entropy, phenotype and genotype diversity.

splitting the rank among the tying items (add possible ranks and average). Remembering that negative correlation (values close to -1.0) suggest high diversity is correlated with good performance (as we want to minimize fitness). Table 1 shows that high negative correlation is seen most consistently with entropy and phenotype diversity. In fact, only these two measure always produce negative correlation, indicating that a high phenotype variance and entropy values accompany the best fit runs.

Figure 2 shows graphs for the same problem where every populations' best fitness and edit distance diversity measure are plotted. The artificial ant and even-5-parity graphs shown here demonstrate a very interesting phenomenon. Notice that best fitness values (close to 0) also consistently have low edit distance diversity, meaning that for populations containing the best fit individuals, those populations are similar to the best fit individual. The even-5-parity problem indicates that best fitness only occurs in populations that have low edit distance diversity. The artificial ant problem shows that poor fitness tends to occur in populations with higher edit distance diversity and also better populations have low edit distance diversity.

While the phenotypic measures seem to indicate that better performance is accompanied with higher diversity, the edit distance diversity results appear to

Table 1. Problems artificial ant and even-5-parity with experiments *random* (rand), *stepped-tournament* (step-t) and *stepped-recombination* (step-r). Values are from the *final population.* Best fitness ("b.fit") is the best fitness in the final generation. The Spearman coefficient shows perfect correlation with 1.0, negative correlation with -1.0 and no correlation with 0.0. Bold numbers are mentioned in the text and negative correlation indicates that best fitness is correlated with high diversity measure values.

artificial ant problem												
	spearman			min max						standard dev		
	random	step-t	step-r	random		step-t		step-r		random	step-t	step-r
b.fit	-	-	-	0.0	39.0	0.0	50.0	0.0	73.0	**11.221**	**15.378**	**15.944**
gene	0.2673	-0.0533	0.5110	**402.0**	**491.0**	1.0	476.0	271.0	487.0	**18.339**	**120.109**	**44.238**
isom	0.4135	0.0874	0.5816	75.0	339.0	1.0	291.0	67.0	354.0	66.0767	74.3093	63.1815
phene	**-0.2214**	**-0.2029**	**-0.0079**	24.0	57.0	1.0	54.0	13.0	62.0	8.2239	9.4150	7.3103
entro	**-0.358**	**-0.597**	**-0.4506**	0.4829	1.3339	0.0	1.2927	0.6010	1.3498	0.1939	0.2584	0.1958
ed1	-0.0128	-0.4799	-0.1646	0.0876	0.5082	0.0	0.3746	0.0558	0.8245	0.0890	0.0824	0.1110
ed2	0.2874	-0.4196	-0.0606	0.4864	7.0751	0.0	3.5343	0.5201	6.0184	1.3466	0.7675	1.0564
even-5-parity problem												
	spearman			min max						standard dev		
	random	step-t	step-r	random		step-t		step-r		random	step-t	step-r
b.fit	-	-	-	3.0	12.0	4.0	15.0	3.0	15.0	1.8762	2.2670	2.7734
gene	0.1788	-0.3165	0.3295	**412.0**	**482.0**	9.0	470.0	269.0	484.0	**14.0285**	**103.544**	**38.224**
isom	0.2388	0.2221	0.3500	45.0	89.0	1.0	119.0	23.0	123.0	10.0312	19.2771	20.2564
phene	**-0.7326**	**-0.7796**	**-0.8494**	6.0	16.0	1.0	15.0	3.0	15.0	1.8999	2.3756	2.3427
entro	**-0.6978**	**-0.7317**	**-0.763**	0.5431	0.9444	0.0	0.8996	0.0176	0.8829	0.08168	0.1840	0.1439
ed1	0.5628	0.4044	0.5853	0.0737	0.3664	0.0426	0.7840	0.0520	0.8484	0.0644	0.1296	0.1286
ed2	0.3806	0.3344	0.4738	0.3917	2.5040	0.1786	5.277	0.2846	3.4607	0.4327	0.8776	0.7364

contradict that by suggesting that better performance is in populations with low edit distance diversity. In fact, these results indicate something quite interesting, that genetic programming is most successful when populations converge to a similar structure but in a manner which preserves diversity.

Figure 3 demonstrates that when the Spearman correlation is calculated for every population during evolution (150 runs total for each problem) how the different diversity measures correlate with performance *during* evolution. For the different problems some diversity measures correlate better at different times during evolution. Notice the early random behaviour around generations 5-10, the same time of divergence in the graphs in Figure 1 and also the general point of convergence of root ancestors, described in [12] [1].

5 Conclusions

The measures of diversity surveyed and studied here indicate that genotype diversity may not be useful for capturing the dynamics of a population, because of the low correlation. This is also suggested in [2][10]. The fitness based measures of phenotypes and entropy appear to correlate better with run performance. The measures of edit distance diversity, one being a traditional edit distance and the other giving more weight to differences near the root, seem to provide useful information about populations with good/poor performance. Better fit individuals come from populations with low edit distance diversity, meaning that the population is similar to the best fit individual. This information accompanied

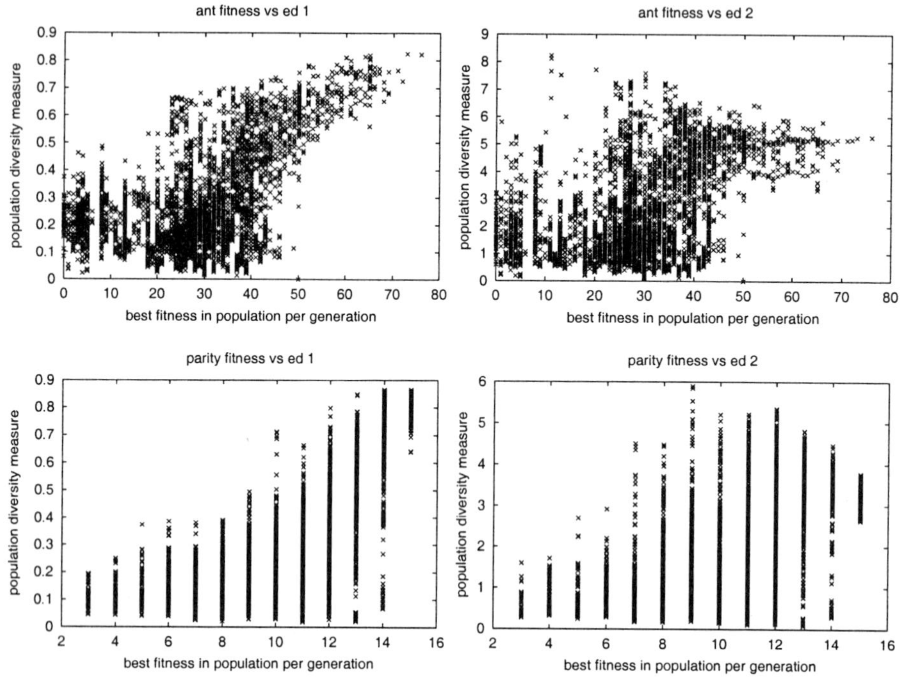

Fig. 2. The best fitness per population is plotted (x axis) against that population's edit distance diversity (ed1, ed2). Here, low fitness is better for both the artificial ant and even-5-parity problems.

with our previous results seem to suggest that populations converge to a similar structure but keep high diversity.

Moreover, if we consider the curves for edit distance diversity (ed1, ed2) *together* with those for phenotype diversity (phenes, entropy) in Figure 3, during most of evolution edit distance diversity correlates positively and phenotype diversity negatively with high fitness. These results taken *together* show that the fitness landscapes defined by the genetic operators chosen and the fitness function used for the problems studied are uncorrelated, i.e., individuals with low edit distance have very different phenotype characteristics (e.g. fitness). This in turn suggests that the search capabilities of the algorithms studied in this paper might be impaired.

While the edit distance measures are expensive, if they prove useful in predicting successful runs we could attempt to find accurate approximations or limit their use to defined generational intervals. Finally, results showed that evolving populations have diversity values which fluctuate between positive and negative correlation with best fitness and this behaviour varied among the studied problems. This paper also indicates the need to carefully define diversity measures and the goal of those measures (high or low values) when using diversity to assess or alter genetic programming evolution.

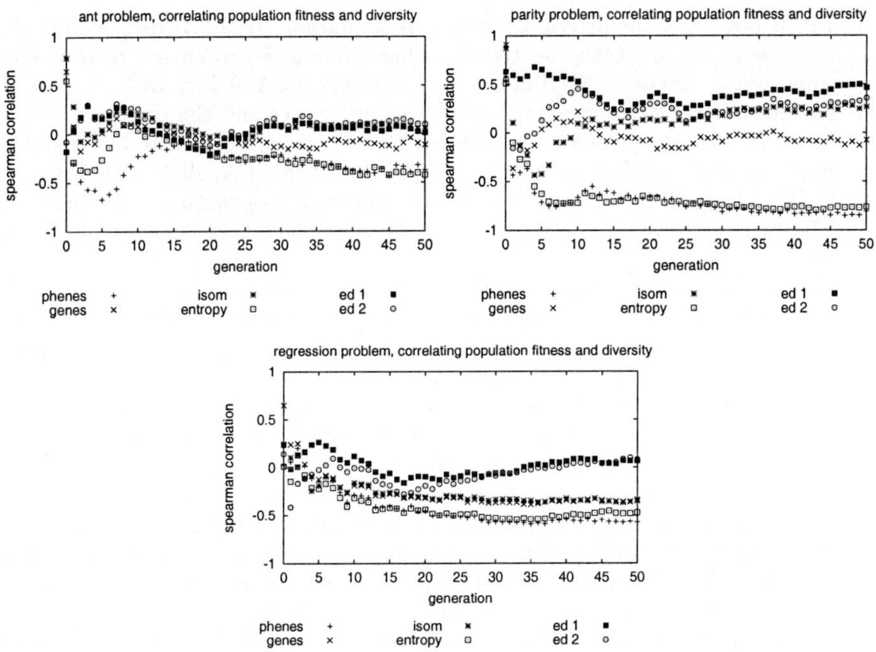

Fig. 3. For each problem, the Spearman correlation between a populations diversity and best fitness is calculated across all runs. Note the fluctuations between negative, no and positive correlation as the populations change during evolution.

6 Future Work

Current research includes studying new problems, tracking root ancestors and other measures during evolution, and applying methods to promote diversity while using different measures to determine their effects. Fitness landscape distance correlation is also being investigated.

References

1. N.F. McPhee and N.J. Hopper. Analysis of genetic diversity through population history. In W. Banzhaf et al., editors, *Proceedings of the Genetic and Evolutionary Computation Conference*, Florida, USA, 1999. Morgan Kaufmann.
2. C. Ryan. Pygmies and civil servants. In K.E. Kinnear, Jr., editor, *Advances in Genetic Programming*, chapter 11, pages 243–263. MIT Press, 1994.
3. A. Ekárt and S. Z. Németh. A metric for genetic programs and fitness sharing. In R. Poli et al., editors, *Proceedings of the European Conference on Genetic Programming*, volume 1802 of *LNCS*, pages 259–270, Edinburgh, 15-16 April 2000. Springer-Verlag.
4. R.I. McKay. Fitness sharing in genetic programming. In D. Whitley et al., editors, *Proceedings of the Genetic and Evolutionary Computation Conference*, pages 435–442, Las Vegas, Nevada, USA, 10-12 July 2000. Morgan Kaufmann.

5. J.P. Rosca. Entropy-driven adaptive representation. In J.P. Rosca, editor, *Proceedings of the Workshop on Genetic Programming: From Theory to Real-World Applications*, pages 23–32, Tahoe City, California, USA, 9 July 1995.
6. J.P. Rosca. Genetic programming exploratory power and the discovery of functions. In J.R. McDonnell et al., editors, *Proceedings of the Fourth Conference on Evolutionary Programming*, pages 719–736, San Diego, CA, 1995. MIT Press.
7. W.B. Langdon. Evolution of genetic programming populations. Research Note RN/96/125, University College London, Gower Street, London WC1E 6BT, UK, 1996.
8. E.D. de Jong, R.A. Watson, and J.B. Pollack. Reducing bloat and promoting diversity using multi-objective methods. In L. Spector et al., editors, *Proceedings of the Genetic and Evolutionary Computation Conference*, San Francisco, CA, 7-11 July 2001. Morgan Kaufmann.
9. R.I. McKay and H.A. Abbass. Anticorrelation measures in genetic programming. In *Australasia-Japan Workshop on Intelligent and Evolutionary Systems*, 2001.
10. M. Keijzer. Efficiently representing populations in genetic programming. In P.J. Angeline and K.E. Kinnear, Jr., editors, *Advances in Genetic Programming 2*, chapter 13, pages 259–278. MIT Press, Cambridge, MA, USA, 1996.
11. P. D'haeseleer. Context preserving crossover in genetic programming. In *Proceedings of the 1994 IEEE World Congress on Computational Intelligence*, volume 1, pages 256–261, Orlando, FL, USA, June 1994. IEEE Press.
12. E. Burke, S. Gustafson, and G. Kendall. Survey and analysis of diversity measures in genetic programming. In *(Accepted as a full paper) Proceedings of the Genetic and Evolutionary Computation Conference*, 2002.
13. J.R. Koza. *Genetic Programming: On the Programming of Computers by Means of Natural Selection.* MIT Press, Cambridge, MA, USA, 1992.
14. W.A. Tackett. *Recombination, Selection, and the Genetic Construction of Computer Programs.* PhD thesis, University of Southern California, Department of Electrical Engineering Systems, USA, 1994.
15. L.J. Eshelman and J.D. Schaffer. Crossover's niche. In S. Forrest, editor, *Proceedings of the Fifth International Conference on Genetic Algorithms*, pages 9–14, San Mateo, CA, 1993. Morgan Kaufman.
16. K. Deb and D.E. Goldberg. An investigation of niche and species formation in genetic function optimization. In J. D. Schaffer, editor, *Proceedings of the Third International Conference on Genetic Algorithms*, Washington DC, 1989.
17. T.F. Bersano-Begey. Controlling exploration, diversity and escaping local optima in GP. In J.R. Koza, editor, *Late Breaking Papers at the Genetic Programming Conference*, Stanford University, CA, July 1997.
18. S. Siegel. *Nonparametric Statistics for the Behavioral Sciences.* McGraw-Hill Book Company, Inc., 1956.
19. S. Luke. ECJ: A java-based evolutionary comp-utation and genetic programming system, 2002. http://www.cs.umd.edu/projects/plus/ecj/.
20. S.-H. Nienhuys-Cheng. Distance between Herbrand interpretations: a measure for approximations to a target concept. In N. Lavraĉ and S. Dẑeroski, editors, *Proceedings of the 7th Internations Workshop on Inductive Logic Programming.* Springer-Verlag, 1997.

Introducing Start Expression Genes to the Linkage Learning Genetic Algorithm

Ying-ping Chen[1] and David E. Goldberg[2]

[1] Department of Computer Science and Department of General Engineering
University of Illinois, Urbana, IL 61801, USA
ypchen@illigal.ge.uiuc.edu

[2] Department of General Engineering
University of Illinois, Urbana, IL 61801, USA
deg@uiuc.edu

Abstract. This paper discusses the use of *start expression genes* and a modified exchange crossover operator in the linkage learning genetic algorithm (LLGA) that enables the genetic algorithm to learn the linkage of building blocks (BBs) through probabilistic expression (PE). The difficulty that the original LLGA encounters is shown with empirical results. Based on the observation, start expression genes and a modified exchange crossover operator are proposed to enhance the ability of the original LLGA to separate BBs and to improve LLGA's performance on uniformly scaled problems. The effect of the modifications is also presented in the paper.

1 Introduction

Regardless of one's point of view, the 1975 publication of Holland's *Adaptation in Natural and Artificial Systems* [1] was an important event in the history of evolutionary computation. That book presented a self-consistent view of genetic algorithm processing. While many remember building blocks, the schema theorem, and the like, most have forgotten or ignored Holland's call for the evolution of tight linkage and its importance in powerful and general adaptation processes. One exception to this widespread set of circumstances is Harik's linkage learning genetic algorithm (LLGA) [2–4]. In that work, the (*gene number*, *allele*) style coding scheme with *introns* [5–12] is used to permit GAs to learn tight linkage of building blocks (BBs) through probabilistic expression (PE). Harik's scheme combines moveable genes, introns, and a special expression mechanism to create an evolvable genotypic structure that makes linkage learning natural and viable for GAs. This combination of data structure and mechanism led to successful linkage learning, particularly on badly scaled problems with badly scaled BBs. Interestingly, the nucleation procedure was less successful on problems with uniformly scaled BBs, and this paper seeks to better understand why this was so and to correct the deficiency.

In particular, this paper introduces the use of *start expression genes* and a modified exchange crossover operator in the LLGA to improve the LLGA's

J.J. Merelo Guervós et al. (Eds.): PPSN VII, LNCS 2439, pp. 351–360, 2002.

performance on uniformly scaled problems. In contrast to the original LLGA, in which every intron (and/or gene) is possibly the point of interpretation of the child, only start expression genes can be the new point of interpretation after crossover in the modified LLGA. Our experiments show that the LLGA with start expression genes and the proposed modified exchange crossover operator is able to overcome the difficulty encountered by the original LLGA.

In this study, we focus on the problems with building-block structure. The problems are assumed decomposable in some way. However, they are not necessarily linearly separable like the test problems used in this paper. For a more detailed definition and discussion of BBs, readers can refer to other materials [13]. Furthermore, by *uniformly scaled BBs*, we mean that all BBs contribute to the fitness equally. By *badly scaled BBs*, we mean that some BBs contribute to the fitness more than the others do, and the contributions might be irregular.

The paper starts with a brief review of the LLGA. Section 2 provides a fundamental background of the previous work on the LLGA. Section 3 investigates the difficulty encountered by the original LLGA when working on uniformly scaled problems. Section 4 proposes a mechanism to overcome the difficulty. It describes start expression genes and the modified exchange crossover operator in detail. Section 5 concludes the paper.

2 Brief Review of the LLGA

In this section, we briefly review the LLGA. Readers who are interested in more detail should refer to other materials [2–4]. The LLGA is generational, and the following topics directly related to this paper are reviewed in this section:

1. chromosome representation;
2. the exchange crossover operator;
3. the mechanisms making the LLGA work.

Each of these topics is discussed in the remainder of this section.

2.1 Chromosome Representation

In the LLGA, chromosome representation is mainly composed of

- moveable genes;
- introns;
- probabilistic expression.

Each of them is described as follows.

The LLGA chromosome consists of moveable genes encoded as (*gene number*, *allele*) pairs and is considered as a circle. The genes in the LLGA are allowed to reside anywhere in any order in the chromosome, while those in traditional GAs are unmoveable and fixed at their own loci.

In order to create a genotypic structure capable of expressing linkage, introns are included in the chromosome. Introns act as non-functional genes, which have no effect on fitness. By using introns, genes no longer have to connect to one another, and linkage can be expressed more precisely. Strictly speaking, the term

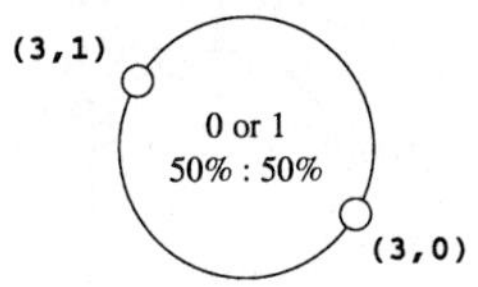

(a) Gene 3 is expressed as 0 with probability 0.5 and 1 with probability 0.5.

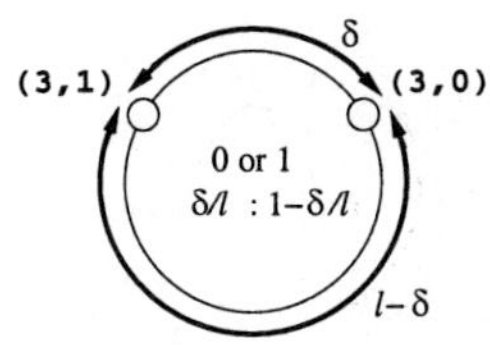

(b) Gene 3 is expressed as 0 with probability δ/l and 1 with probability $1 - \delta/l$.

Fig. 1. Probability distributions of gene 3's alleles represented by PE chromosomes.

intron refers to a noncoding segment of a gene. In our system, genes are discrete entities and either a gene codes or does not, and it is in this sense that we use the term intron as a noncoding gene. In our system, strings of "introns" combine to form *intergenic region*, *extragenic region*, or *pseudogene*, and these terms would perhaps be preferable to the use of the term intron. Nonetheless, our meaning is clear and is accurate in the limited sense indicated.

Furthermore, the method of probabilistic expression (PE) was proposed to preserve diversity at the building-block level. A PE chromosome contains all possible alleles for each gene. For the purpose of evaluation, a chromosome is *interpreted* by selecting a *point of interpretation* (POI) and choosing for each gene the allele occurring first in a clockwise traversal of the circular chromosome. After interpretation, a PE chromosome is expressed as a complete string and evaluated. Besides, the POI is selected on the individual when crossover happens.

As a consequence, a chromosome represents not a single solution but a probability distribution over the range of possible solutions. Figure 1 illustrates the probability distribution over possible alleles of gene 3 of the chromosome. Hence, when different POI are selected, a PE chromosome might be interpreted as different solutions. Figure 2 shows 3 genes of a PE chromosome composed of 6 genes. If point A is the POI, the chromosome will be considered as `((5,1) (4,0) (4,1) (3,0) (3,1) (5,0))` and interpreted as `((5,1) (4,0)` ~~`(4,1)`~~ `(3,0)` ~~`(3,1)`~~ ~~`(5,0)`~~`)` ⇒ ***001, where the struck genes are *shadowed* by their complement genes. And, if point B is the POI, the chromosome will be considered as `((4,0) (4,1) (3,0) (3,1) (5,0) (5,1))` and interpreted as `((4,0)` ~~`(4,1)`~~ `(3,0)` ~~`(3,1)`~~ `(5,0)` ~~`(5,1)`~~`)` ⇒ ***000.

If we consider a PE chromosome as containing exactly one copy of some shadowed gene, we can generalize PE to let a chromosome contain more than one copy of a shadowed gene. Therefore, the extended probabilistic expression (EPE) can be defined with a parameter k encoded with which a chromosome can contain *at most* k copies of a shadowed gene. Figure 3 shows an example of EPE-2 chromosomes.

2.2 The Exchange Crossover Operator

In addition to PE and EPE, the exchange crossover operator is another key feature for the LLGA to learn linkage. The exchange crossover operator is defined

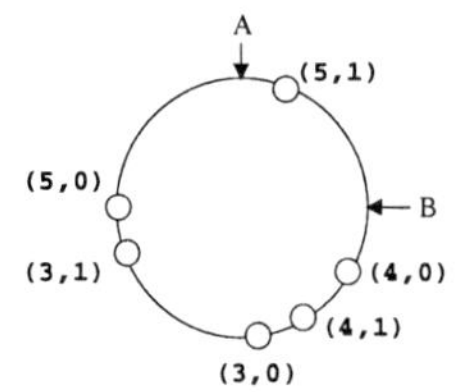

Fig. 2. Different points of interpretation might interpret a PE chromosome as different solutions.

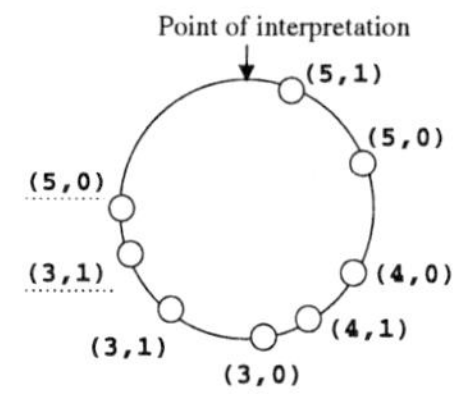

Fig. 3. An example of EPE-2 chromosomes. Each gene can have *at most* 2 complement genes.

on a pair of chromosomes. One of the chromosomes is the *donor*, and the other is the *recipient*. The operator cuts a *random* segment of the donor, selects a grafting point at random on the recipient, and grafts the segment onto the recipient. The grafting point is the POI of the generated offspring. Starting from the POI, redundant genetic material caused by injection is removed right after crossover to ensure the validity of the offspring.

2.3 The Mechanisms Making the LLGA Work

By integrating PE and the exchange crossover operator into traditional GAs, the LLGA is able to solve difficult problems without prior knowledge of good linkage. Traditional GAs have been shown to perform poorly on difficult problems [14] without such knowledge. To better decompose and understand the working of the LLGA, Harik [3] identifies and analyzes two mechanisms of linkage learning: *linkage skew* and *linkage shift*. Linkage skew occurs when an optimal BB is successfully transferred from the donor to the recipient, its linkage gets tighter. Linkage shift occurs when an optimal BB resides in the recipient and survives an injection, its linkage gets tighter. With these two mechanisms, the linkage of BBs can evolve, and tightly linked BBs are formed during the process.

3 A Critique of the Original LLGA

The LLGA was original, provocative, and interesting. It works quite well on badly scaled problems. Unfortunately, it performs poorly on uniformly scaled problems. To better understand that failure, we investigate what the LLGA is supposed to do and observe how and why it fails.

3.1 What Is the LLGA Supposed to Do?

One of the main motivations behind the LLGA was to permit a GA to gather genes belonging to different BBs separately and tightly. Note that there are two things needed here. Not only do individual BBs need to be tight, they also need to be separate to permit effective BB exchange via crossover. Harik's analysis of skew and shift addresses the tight linkage side of the ledger, but it says little or nothing about the evolution of separate BBs.

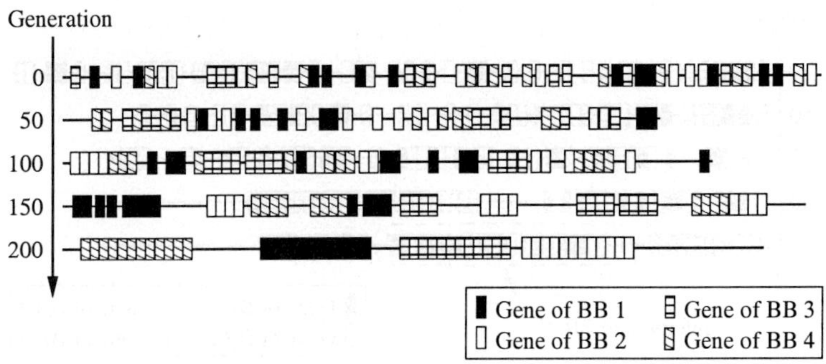

Fig. 4. In a successful run, the original LLGA is able to correctly separate the four BBs and make each of them tightly linked.

In selectionist schemes, sometimes good things happen of their own accord, so we examine a four-BB problem and see if separate and tight linkage can be evolved. In particular, we use the original LLGA with EPE-2 to solve a problem composed of four uniformly scaled order-4 traps [15] with value 3.0 at u=0 ones, value 0.0 at u=3 ones, and value 4.0 at u=4 ones. In this experiment, we use the following parameter settings which are arbitrarily chosen and not fine tuned. The population size is 300, the tournament size is 3, the crossover rate is 1.0, and 800 introns are encoded in the chromosome. Then, we examine typical individuals at generations 0, 50, 100, 150, and 200 shown in Fig. 4.

At generation 0, the genetic materials are randomly distributed in the chromosome. Later, some genes are getting together to form sub-BBs at generation 50. Longer sub-BBs are formed at around generations 100 and 150. At the end of the run, the four BBs are formed in the individual.

As we can see in this case, the LLGA actually constructs four separate and tight BBs as desired. We call the process *nucleation* that correctly separates the BBs and gets them tightly linked. This might seem to put an end to the matter, but we examine a run that did not go this way.

3.2 How Does the LLGA Fail?

The process above yields results along the lines of those hoped for. However, it is not the only result we might get when solving four uniformly scaled BBs. Another possible result is that genes of different BBs get together to form a single, *intertwined building block*.

By examining the results of an unsuccessful run of the same experiment shown in Fig. 5, we can easily observe the process of forming an intertwined building block. We call the process *misnucleation*.

The genetic materials are initially randomly distributed in the chromosome. At generation 50, sub-BBs are being constructed gradually. At generations 100 and 150, we can see that two of the four BBs tend to intertwine each other. Finally, only two BBs are correctly identified and separated. The other two BBs are intertwined as they were a single BB.

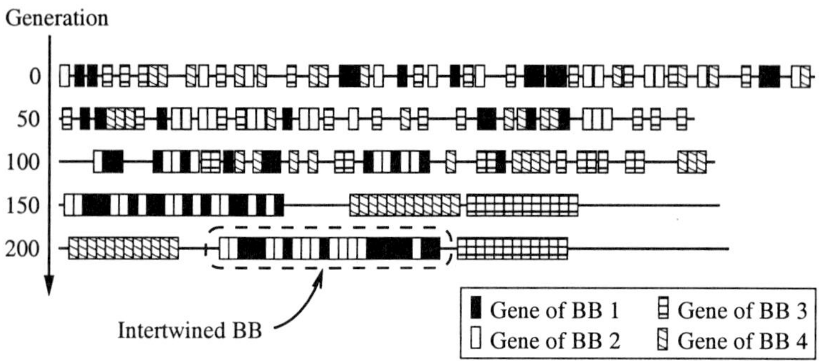

Fig. 5. In an unsuccessful run, the original LLGA cannot correctly separate the four BBs. An intertwined BB of two actual BBs is formed during the process.

In this case, two BBs formed an intertwined BB along the process. The intertwined BB was treated as a single BB in the chromosome, and the LLGA could only make it tight instead of separated. Hence, the real BBs cannot be identified in this situation. The intertwined BB prevented the BBs from mixing well and effectively, and therefore, also prevented the problem from being solved.

3.3 Separation Inadequacy: Key Deficiency of the Original LLGA

Based on the results above and those from a good number of experiments, the original LLGA seems only able to deal with a very small number (around 2 to 5) of uniformly scaled BBs. Misnucleation becomes very likely to happen when the number of BBs gets slightly larger because the nature of uniformly scaled problems requires all BBs to be identified and separated at the same time, but the original LLGA has no mechanism to do that. Therefore, the original LLGA performs poorly on uniformly scaled BBs.

By revisiting the two linkage learning mechanisms, we find that there are only two categories of genetic materials:

1. the genetic materials that affect the solution;
2. the genetic materials that do not affect the solution.

Neither the linkage skew nor the linkage shift separates different BBs and makes each of them tightly linked. What they actually do is make the linkage of those genetic materials that improve the solution quality tighter, no matter the genetic materials belong to the same BB or not. The original LLGA does not have an appropriate mechanism for BB separation.

The same argument is also applicable to the exponentially scaled problems. When the LLGA solves an exponentially scaled problem, it deals with the BBs one by one. Fitness scaling causes the salient BB to be processed first, next most salient, and so on. Therefore, the original LLGA can easily handle exponentially scaled problems because it in fact faces only one BB at a time.

Due to the lack of BB separation mechanism, any power to separate BBs is delivered by the schema theorem [1, 16]. To see this, consider that schemata

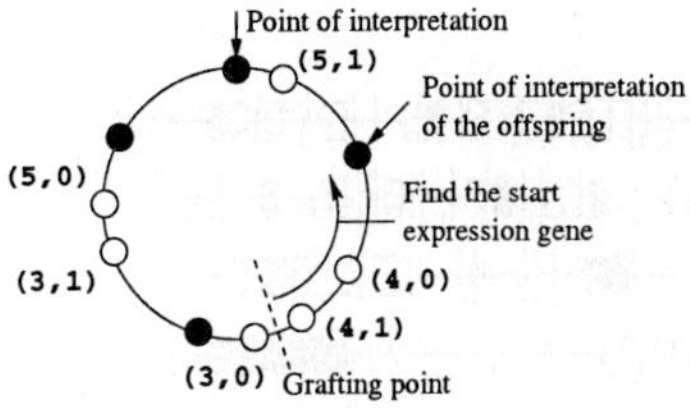

Fig. 6. After selecting the grafting point on the recipient, the nearest start expression gene *before* the grafting point is then the point of interpretation of the offspring.

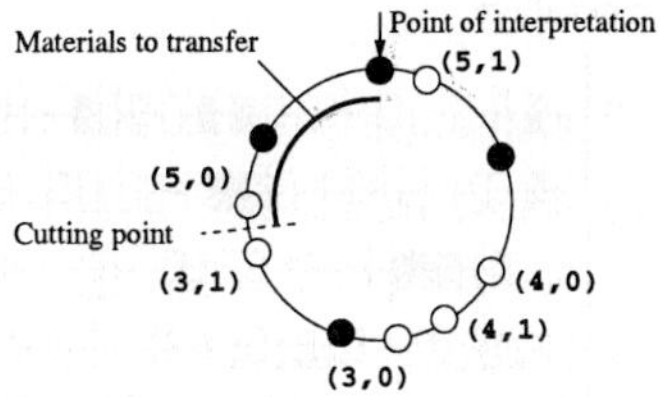

Fig. 7. After selecting the cutting point on the donor, the genetic material *after* the cutting point and before the current point of interpretation is transferred.

with shorter defining lengths are favored and then become more tightly linked so that disruption by crossover can be avoided. Hence, the BBs are sometimes formed under this circumstance. Also because the power to separate BBs from the schema theorem is not enough, the original LLGA usually cannot correctly separate more than five BBs at the same time.

4 Start Expression to Improve Nucleation Potential

Knowing this limitation of the original LLGA, what we should do now is to improve nucleation and separation of BBs. In the original LLGA, any point can be a POI. Such a design provides no advantage for BB separation at the chromosome-representation level because totally random POI create the instability of BB formation. In order to overcome this, we introduce the use of start expression genes, the only possible POI. This should lower the instability of BB formation and improve nucleation potential during the evolutionary process. We also modify the exchange crossover operator so it can work with start expression genes. The modified exchange crossover operator uses the POI of the donor as one of the cutting points to further reduce the randomness of BB formation. Additionally, in the modified LLGA, PE is used instead of EPE-k. In the remainder of this section, we will discuss these modifications in detail. The results show that the modifications indeed make the LLGA able to correctly identify and separate uniformly scaled BBs.

4.1 How Do Start Expression Genes Work?

Start expression genes are special introns. While in the original LLGA, all genes and introns can be the points of interpretation of the child after crossover, only start expression genes can be the points of interpretation in the modified LLGA.

In the original LLGA, when doing the exchange crossover operator, the grafting point selected on the recipient is the new POI of the generated offspring. In the modified LLGA, although the grafting point can still be located at any gene or intron, the POI of the offspring is no longer the grafting point. The new POI is the nearest start expression gene *before* the grafting point. Therefore, after the grafting point is randomly chosen, we find the the first start expression gene just

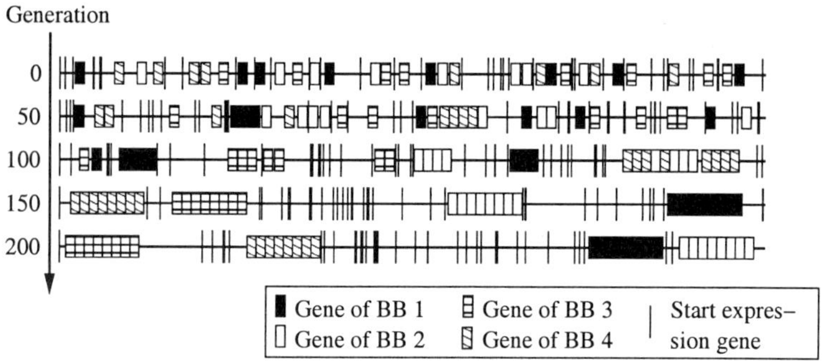

Fig. 8. Using the modified LLGA to solve a four uniformly scaled order-4 traps.

before the grafting point and make it the POI of the child. The genetic material between the start expression gene and the grafting point is first transferred to the child, then the segment from the donor, and finally the rest of the recipient. Figure 6 illustrates how start expression genes work. The black circles are start expression genes of the chromosome.

4.2 The Modified Exchange Crossover Operator

By modifying the exchange crossover operator, we create a further relationship between the genetic operator and the POI. The modified exchange crossover operator can reduce randomness of BB formation and improve nucleation potential.

The original exchange crossover operator cuts a *random* segment from the donor and injects the segment into the recipient. In order to reduce the randomness, the modified exchange crossover operator selects only one cutting point at random. The other cutting point used by the modified exchange crossover operator is always the gene or intron just before the POI of the donor. At the PE-chromosome level, the operation is like the one-point crossover; while at the expressed-string level, the operation is like the uniform crossover. Figure 7 shows the portion of the genetic materials to be transferred.

The modified exchange crossover operator works as follows:

1. select a grafting point at random on the recipient;
2. determine the POI of the child as described in Sect. 4.1;
3. copy the genetic materials between the current POI and the grafting point to the child;
4. select the cutting point at random on the donor;
5. graft the genetic materials between the cutting point and the POI of the donor to the child;
6. copy the rest genetic materials of the recipient to the child;
7. remove the duplicate genes and introns of the child to ensure the validity.

4.3 The Effect of the Modifications

Now we use the modified LLGA to repeat the same experiment in Sect. 3. The modified LLGA has an extra parameter: the number of start expression genes

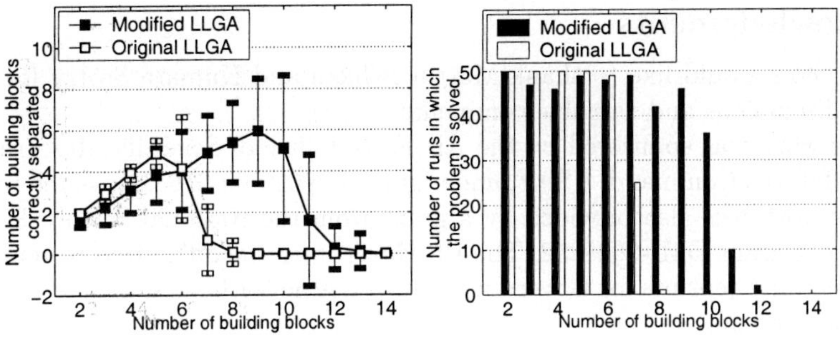

(a) The number of BBs correctly separated at the same time. The results are averaged over 50 runs.

(b) The number of runs in which the problem is solved in the 50 runs.

Fig. 9. Using the modified LLGA to solve problems of different number of BBs.

(n_s). In this experiment, $n_s = 20$. Figure 8 shows that the modified LLGA can correctly separate the four uniformly scaled order-4 traps.

In addition to those experiments on four BBs, we vary the number of BBs to see if the modified LLGA can do better to separate uniformly scaled BBs on larger problems. We still use order-4 traps in this experiment. Except for the number of BBs, all parameters are identical to the previous experiments.

Figure 9 shows that the original LLGA failed if the number of BBs is greater than 6. It cannot separate the BBs and reliably solve the problem. According to the results, equipped with start expression genes and the modified exchange crossover operator, the modified LLGA is capable of separating BBs correctly.

More work along these lines needs to be done to study parameter settings of the LLGA on larger problems. These proof-of-principle results correctly identify a key problem with the original LLGA and demonstrate a promising solution path. We believe that it should lead to scalable linkage learning.

5 Conclusions

Harik took Holland's call for the evolution of tight linkage and developed the LLGA, which employs introns and probabilistic expression to make the linkage learning natural and viable. A drawback of the original LLGA is the need for an exponentially growing population size to solve uniformly scaled problems mainly because the original LLGA has no proper mechanism to separate uniformly scaled BBs. This paper introduces start expression genes to the LLGA for identifying and separating BBs. It also proposes a modified exchange crossover operator to work with start expression genes so that the BB separation capability can be fully exploited. The empirical results show that the modified LLGA indeed has superior ability to correctly identify and separate BBs.

More work needs to be done to understand parameter settings on larger problems. These proof-of-principle results shown in this paper (1) identify a key problem with the original LLGA and (2) show a promising solution path. It should lead to scalable linkage learning.

Acknowledgments

The authors would like to thank Martin Pelikan and Kumara Sastry for many useful discussions and valuable comments.

The work was sponsored by the Air Force Office of Scientific Research, Air Force Materiel Command, USAF, under grant F49620-00-0163. Research funding for this work was also provided by a grant from the National Science Foundation under grant DMI-9908252. The US Government is authorized to reproduce and distribute reprints for Government purposes notwithstanding any copyright notation thereon.

The views and conclusions contained herein are those of the authors and should not be interpreted as necessarily representing the official policies or endorsements, either expressed or implied, of the Air Force Office of Scientific Research, the National Science Foundation, or the U.S. Government.

References

1. Holland, J.H.: Adaptation in Natural and Artificial Systems. University of Michigan Press, Ann Arbor, MI (1975)
2. Harik, G.R., Goldberg, D.E.: Learning linkage. FOGA 4 (1996) 247–262
3. Harik, G.R.: Learning Gene Linkage to Efficiently Solve Problems of Bounded Difficulty Using Genetic Algorithms. Unpublished doctoral dissertation, University of Michigan, Ann Arbor, MI (1997) (Also IlliGAL Report No. 97005).
4. Harik, G.R., Goldberg, D.E.: Learning linkage through probabilistic expression. Computer Methods in Applied Mechanics and Engineering **186** (2000) 295–310
5. Nordin, P., Francone, F., Banzhaf, W.: Explicitly defined introns and destructive crossover in genetic programming. In: Proceedings of the Workshop on GP: From Theory to Real-World Applications. (1995) 6–22
6. Andre, D., Teller, A.: A study in program response and the negative effects of introns in genetic programming. In: Proceedings of GP-96. (1996) 12–20
7. Haynes, T.: Duplication of coding segments in genetic programming. In: Proceedings of NCAI-96. (1996) 344–349
8. Lindsay, R.K., Wu, A.S.: Testing the robustness of the genetic algorithm on the floating building block representation. In: AAAI/IAAI, Vol. 1. (1996) 793–798
9. Wineberg, M., Oppacher, F.: The benefits of computing with introns. In: Proceedings of GP-96. (1996) 410–415
10. Wu, A.S., Lindsay, R.K.: A survey of intron research in genetics. PPSN IV (1996) 101–110
11. Levenick, J.R.: Swappers: Introns promote flexibility, diversity and invention. In: Proceedings of GECCO-99. (1999) 361–368
12. Iba, H., Terao, M.: Controlling effective introns for multi-agent learning by genetic programming. In: Proceedings of GECCO-2000. (2000) 419–426
13. Goldberg, D.E.: The Design of Innovation: Lessons from and for Competent Genetic Algorithms. Kluwer Academic Publishers, Boston, MA (2002)
14. Thierens, D., Goldberg, D.E.: Mixing in genetic algorithms. In: Proceedings of ICGA-93. (1993) 38–45
15. Deb, K., Goldberg, D.E.: Analyzing deception in trap functions. FOGA 2 (1993) 93–108
16. De Jong, K.A.: An Analysis of the Behavior of a Class of Genetic Adaptive Systems. Unpublished doctoral dissertation, University of Michigan, Ann Arbor, MI (1975) (University Microfilms No. 76-9381).

Metamodel–Assisted Evolution Strategies

Michael Emmerich[1], Alexios Giotis[3], Mutlu Özdemir[1], Thomas Bäck[2], and Kyriakos Giannakoglou[3]

[1] Informatik Centrum Dortmund, Center for Applied Systems Analysis, Joseph von Fraunhoferstr. 20, D-44227 Dortmund, Germany
emmerich@icd.de

[2] NuTech Solutions GmbH
Martin Schmeißer Weg 15, D-44227 Dortmund, Germany
baeck@nutechsolutions.de
http://www.nutechsolutions.com

[3] National Technical University of Athens, Dept. of Mechanical Engineering
Athens, Greece
{agiotis,kgianna}@central.ntua.gr

Abstract. This paper presents various Metamodel–Assisted Evolution Strategies which reduce the computational cost of optimisation problems involving time–consuming function evaluations. The metamodel is built using previously evaluated solutions in the search space and utilized to predict the fitness of new candidate solutions. In addition to previous works by the authors, the new metamodel takes also into account the error associated with each prediction, by correlating neighboring points in the search space. A mathematical problem and the problem of designing an optimal airfoil shape under viscous flow considerations have been worked out. Both demonstrate the noticeable gain in computational time one might expect from the use of metamodels in Evolution Strategies.

1 Introduction

Evolution Strategies (ES) are a powerful tool for global optimisation in high-dimensional search spaces. However, their known weakness is that they require a high number of evaluations. Similar problems may be reported for other population–based methods, such as the widely used Genetic Algorithms. In optimisation problems with time–consuming evaluation software (applications in the field of aeronautics are typical examples) this renders the total CPU cost to be prohibitive for industrial use. A way for keeping this cost as low as possible is through the use of a surrogate evaluation tool, i.e. the so–called metamodel; a relevant literature survey can be found in [3], at least from the viewpoint of applications in aeronautics. The Metamodel–Assisted Evolution Strategies (MAES) can be applied for global optimisation with any time–consuming evaluation method, especially in industrial design optimisation. Extending a previous work by the same authors [5], a new enhanced metamodel is employed herein. In the course of evolution, the metamodel's role is to point to the most promising

J.J. Merelo Guervós et al. (Eds.): PPSN VII, LNCS 2439, pp. 361–370, 2002.

individuals that will be re–examined through the time–consuming evaluation software. The use of the metamodel prerequisites the existence of a database and procedure for optimally selecting the database subset to be used for its construction.

Various metamodels can be devised. In the past, Giannakoglou et al [4] utilized radial basis function networks as surrogate models. Other algorithmic variants that use metamodels in evolutionary optimisation can also be found in the literature (for instance, in Jin et al. [6]).

In the present study, we will extend the use of metamodels, as described in [5], by incorporating a local error estimation technique that enables the optimisation method to estimate the reliability of the approximated function values and to exploit further this information. The error estimation is based on the local density and clustering of points and on estimates of local correlations. A search criterion is used, which considers both the local error estimation and the fitness estimation. Furthermore, it will be demonstrated that the step-size self-adaptation is preserved despite the use of the metamodel. Unlike [5], which made use of small population size ES, in the present paper ES with large population size will be employed; to the authors experience, the latter is best suited for global optimisation.

This paper is organised as follows. The metamodel is first introduced and then the ES along with the search criterion based on fitness and error estimations are described. For the assessment of the proposed method, artificial landscapes and an airfoil optimisation problem will be analysed.

2 Kriging and Local Error Estimation

A metamodel approximates a multivariate function using points that have already been evaluated. It is a reasonable assumption to consider that the time required for building the metamodel is negligible compared to the CPU cost for an exact evaluation, at least in real world problems. Thus, a metamodel is to be considered as a fast surrogate model to the exact evaluation software.

Henceforth, $\mathbf{x}_1, \ldots, \mathbf{x}_m \in \mathbb{R}^n$ will denote previously evaluated candidate solutions and $\mathbf{y} = (y_1, \ldots, y_m) := (f(\mathbf{x_1}), \ldots, f(\mathbf{x_m}))$ are results from the exact evaluations associated with each one of the aforementioned solutions. Using any interpolation method, the estimation function $\hat{f}$ returns the exact value y_i, at the data sites $\mathbf{x}_i, i = 1, \ldots, m$. In this paper, the metamodel is based on Kriging techniques, which provide estimates to the fitness values of new candidate solutions. Kriging stands for an isotropic interpolation method which can deal with irregularily distributed points in the search space. We recall that an interpolation method is called isotropic if $\hat{f}$ depends exclusively on distances $\| \mathbf{x} - \mathbf{x}_i \|$ from neighbouring points instead of the absolute value of $\mathbf{x}$ and the direction $\mathbf{x} - \mathbf{x}_i$. Kriging was originated by the mining engineer Krige, who used this method to estimate ore concentrations in gold mines. Later, Kriging was formulated rigorously by Matheron [9]. In recent years it has been used in geostatistics [12] and in metamodelling and optimisation [8, 11, 2]. Kriging assumes that that each measured value of the objective function f is the realisation of an isotropic n-dimensional Gaussian stochastic process with unknown mean $\beta \in \mathbb{R}$ and covariance function of the form $c(\mathbf{s},\mathbf{t}) = \sigma^2 r_\theta(\mathbf{s},\mathbf{t})$. Here $\sigma^2 > 0$ and θ are unknown and

$$r_\theta(\mathbf{s},\mathbf{t}) := \exp(-\theta \parallel \mathbf{s}-\mathbf{t} \parallel^2) \tag{1}$$

Note that this kernel allows to estimate values that are lower than the minimal values of $\vec{y}$. An alternative kernel function would be $r_\theta(\mathbf{s},\mathbf{t}) := \exp(-\theta \parallel \mathbf{s}-\mathbf{t} \parallel)$. For this function the optima of the Kriging approximation are the same as the optima in the set of measured values $\{y_1,\ldots,y_n\}$ (cf. [7]).

In order to construct an approximation by Kriging, unknown parameters (β,σ,θ) have to be estimated by the maximum likelihood method. This is done by solving a minimisation problem with a local search method.

$$\begin{aligned} n\log\hat{\sigma}^2(\hat{\theta}) + \log\det\mathbf{R}(\hat{\theta}) &\to \min \\ \hat{\beta} &= [\mathbf{I}^T\mathbf{R}(\hat{\theta})\mathbf{I}]^{-1}\mathbf{I}^T\mathbf{R}(\hat{\theta})^{-1}\mathbf{y} \\ \hat{\sigma}^2(\hat{\theta}) &= \frac{1}{n}[\mathbf{y}-\mathbf{I}\hat{\beta}]^T\mathbf{R}(\hat{\theta})^{-1}[\mathbf{y}-\mathbf{I}\hat{\beta}] \\ \mathbf{R}(\hat{\theta}) &:= [r_{\hat{\theta}}(\mathbf{x}_i,\mathbf{x}_j)] \end{aligned} \tag{2}$$

Once $\hat{\theta}$ has been obtained, estimations of function values at new points can be computed as follows

$$\hat{f}(\mathbf{x}) := \hat{\beta} + (\mathbf{y}-\mathbf{I}\hat{\beta})^T\mathbf{R}(\hat{\theta})^{-1}\mathbf{r}(\mathbf{x};\hat{\theta}),\ \text{with } \mathbf{r}(\mathbf{x};\hat{\theta}) := [\mathbf{r}_{\hat{\theta}}(\mathbf{x}_i,\mathbf{x})] \tag{3}$$

The mean squared error of this estimation is estimated as follows

$$\hat{\text{MSE}}(\mathbf{x}) = \sigma^2 - \sigma^2[\mathbf{I}; r(\mathbf{x},\hat{\theta})^T]\begin{bmatrix} 0 & \mathbf{I} \\ \mathbf{I} & \mathbf{R}(\hat{\theta})\end{bmatrix}^{-1}\begin{bmatrix}\mathbf{I} \\ \mathbf{r}(\mathbf{x};\hat{\theta})\end{bmatrix} \tag{4}$$

The value of $\hat{\text{MSE}}$ depends on the correlation of the landscape as well as on the local density of points. This has been illustrated in Figure 1. From this figure it comes out that the approximation is precise at the points that have been evaluated and more precise in regions with a high point density. In this region the $\hat{\text{MSE}}$ is low. The approximation is also precise, if a point lies between two data points for which exact measurements exist. This illustrates that Kriging takes into account the clustering of points. The right part in Figure 1 f shows a sinusoidal function with doubled frequency with the same points $\mathbf{x}_i$, $i = 1,\ldots,8$ being evaluated. In this case the correlation between neighbouring points is much weaker, which worsens the quality of the function approximation. The increased difficulty has an effect on the $\hat{\text{MSE}}$ prediction too, which is much more pessimistic in that case.

Although it has been stated that the metamodel's CPU cost could be safely neglected in real word optimisation problems with time–consuming evaluation methods, the real CPU cost of the Kriging method depends mainly on the number of evaluated sites and not so much on the dimension of the search space. In order to estimate θ, repetitive inversions of the covariance matrix $\mathbf{R}(\hat{\theta})$ are needed.

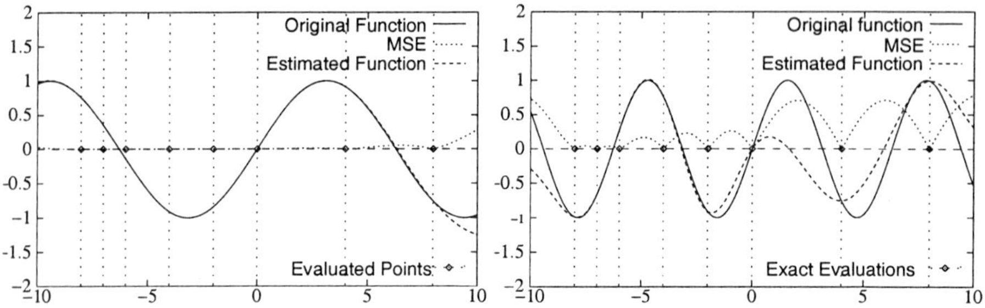

Fig. 1. Interpolation with Kriging: Original function, approximation and error-estimation for the one-dimensional function $\sin(x)$ (left) and $\sin(2x)$ (right).

The CPU cost for this inversion is in $\mathcal{O}(m^3)$ and this determines the asymptotic time complexity of the metamodelling algorithm, which is $\mathcal{O}(N_{opt}m^3 + nm^2)$ (N_{opt} is the number of iterations for the local minimisation of the MSE with $N_{opt} \approx 200$). Note that the time complexity for calculating $M\hat{S}E$ and $\hat{f}$ is in $\mathcal{O}(nm^2)$, once the metamodel has been built.

The Kriging metamodel used in this work is based on the k nearest neighbours at each point, so it will be referred to as *local metamodel.* The value of k has been set to 20. Any further increase in k seems to slightly improve the results but, at the same time, increases the computation time significantly. The Kriging algorithm is the one proposed by Padula et al. [7] with pseudo inversion of the covariance matrix. One building and evaluation of the local metamodel takes about 0.2 seconds on a Pentium III, 1GHz PC. Failed evaluations are treated by penalizing them with the worst feasible value multiplied by 10. It is also recommended to increase k for highly dimensioned search spaces or in case that the exact evaluation tool is time–consuming.

3 Metamodel Assisted Evolution Strategies

It is known that ES are powerful and robust optimisation tools. They are established as standard tools in practical optimisation. In an ES, parameter vectors (search points) $\mathbf{x} \in \mathbb{R}$ together with one or many step-size parameter(s) $\sigma \in \mathbb{R}^+$ form an individual. A set of individuals is termed a population. Modern (μ, κ, λ)-ESs usually work with increased populations. They can be easily adapted to different computing environments and representations. In this study they will be used for continuous parameter optimisation.

The (μ, κ, λ)-ES has first been applied by Schwefel [10]. Within any generation, mutation and recombination operators are applied to generate λ offspring from μ parents. In order to form a new generation, the best from the $\mu + \lambda$ individuals are selected. Individuals that exceed the age of κ generations are eliminated from the selection procedure.

Throughout this study we employed the mutative self adaptation suggested by Ostermeier [10] with a single global step-size. According to [10], a discrete recombination operator has been used for the object variables and an intermediate recombination for the step-size parameter. A recommended strategy variant for complex multimodal problems is the $(15, 5, 100)$-ES. Thanks to the large parent population size, the recombination operator becomes beneficial. Furthermore, the selection pressure $\lambda/\mu \approx 7$ is high enough to enable the self-adaptation of step sizes. The maximum life time of 5 generations makes the strategy robust, even in the presence of discontinuities.

In this work, the metamodel will be used to accelerate the (15, 5, 100)-ES. As in Trosset and Torczon [11], direct optimisation in continuous spaces will be combined with metamodelling techniques. A fitness criterion based on both the estimated value and the estimated local variance of the prediction model is used by setting up a search criterion to estimate the potential outcome of a computer experiment:

$$S_c(\mathbf{x}) := \hat{f}(\mathbf{x}) - w\sqrt{M\hat{S}E(\mathbf{x})}. \tag{5}$$

In the MAES, this criterion S_c is used to pre-select s $(0 < s < \lambda)$ individuals out of λ offspring, in order to evaluate them exactly. Only the individuals with the lowest values for S_c are chosen for this. The behaviour for $w = 1$ and for $w = 0$ will be studied. In the latter case, S_c reduces to $S_{c,w=0}(\mathbf{x}) = \hat{f}(\mathbf{x})$. The difference between both criteria is that with $w = 0$ the most promising candidate solutions are selected whereas with $w = 1$ this concept is extended to still unexplored search areas, by considering the estimation error and additionally selecting the individuals with a potential of good performance. Thus, it is recommended to use S_c in order to make the whole algorithm more robust.

The parameter w can be used to increase the influence of the error term. Its value should be increased in complex problems. On the other hand, the algorithm will converge faster to a local optimum if w is low. The value of $w = 1.0$ is used as default throughout this study.

The number of individuals that are selected for exact evaluations is an important parameter. About 10 exact evaluations per generation is a figure that turned out to perform well. It allows to get enough iterations for making the step-size adaptation work and also to have a sufficient amount of new information for areas of the search space, which are of interest in the forthcoming iteration. Over and above to the individuals which are pre-selected by the search criterion(s), individuals which outperform the so-far best individual are also examined through the exact evaluation tool. By this measure, the algorithm can hardly be trapped in artificial local optima.

With the proposed metamodel assisted selection scheme it is possible for the ES not only to learn from promising individuals but also to memorize and make use of search points with bad performance, by accessing the *long term memory* of the evolution's history.

In order to increase the metamodel's performance each run is started using a randomly initialised population of 100 individuals, which has been exactly evaluated. From this population the 15 best individuals have been selected in order to build the starting population for the ES.

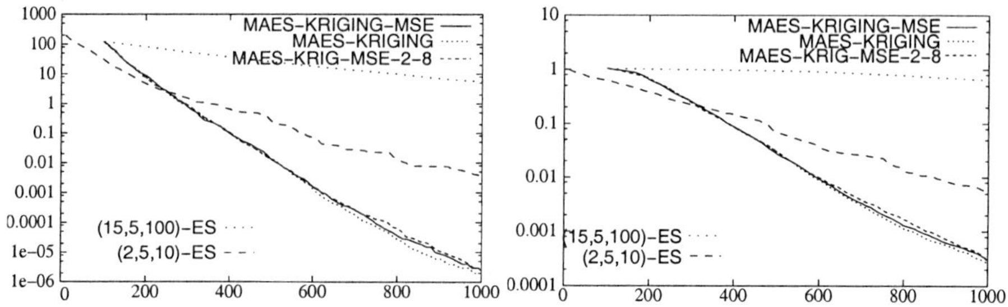

Fig. 2. Average plots of 20 runs on sphere function. Fitness value (right) and the global step-size (left) vs the number of exact evaluations.

Six different strategy variants have been compared in this study:

- (15, 5, 100)-ES: The canonical (15, 5, 100)-ES with random starting population of 100 individuals.
- (2, 5, 10)-ES: The canonical (2, 5, 10)-ES with a random starting population of 10 individuals.
- MAES-KRIGING-MSE: Metamodel–Assisted (15, 5, 100)-ES - 10 individuals are exactly evaluated per generation, selected using criterion $S_{c,w=1}$.
- MAES-KRIG-MSE-2-8: Metamodel–Assisted (15,5,100)-ES - 2 individuals are pre-selected using $S_{c,w=1}$ and 8 individuals by $\hat{f}$, per generation.
- MAES-KRIGING: Metamodel–Assisted (15,5,100)-ES - 10 individuals are pre-selected using the criterion $\hat{f}$.

4 Studies on Artificial Landscapes

In order to prove the general applicability of our approach and learn about their global and local convergence behaviour (speed, reliability), experiments on artificial landscapes have been conducted. The algorithms employed started with a randomly selected population of λ individuals, all of them exactly evaluated. The initial step-size was set to 5% of the variables range. The first test function was a simple sphere function ($\sum_{i=1}^{n} x_i^2, \mathbf{x} \in [-10, 10]^n \subset \mathbb{R}^n$). It has been selected to demonstrate the different local search characteristics of the strategy variants and to investigate their ability to adapt step-sizes.

The result shows that despite the large population size, it is still possible to self-adjust step-sizes by mutative self-adaptation within a comparably low number of exact evaluations. Another conclusion drawn from these computations is that the convergence behaviour is not seriously affected if the search criterion $\hat{f}$ is replaced (partly) by $S_{c,w=1}$. However, it can be seen that the strategies that only employ $\hat{f}$ as pre-selection criterion are slightly better than those also using $S_{c,w=1}$.

The second function is the multimodal Keane's Bump problem 3, which is denoted as follows:

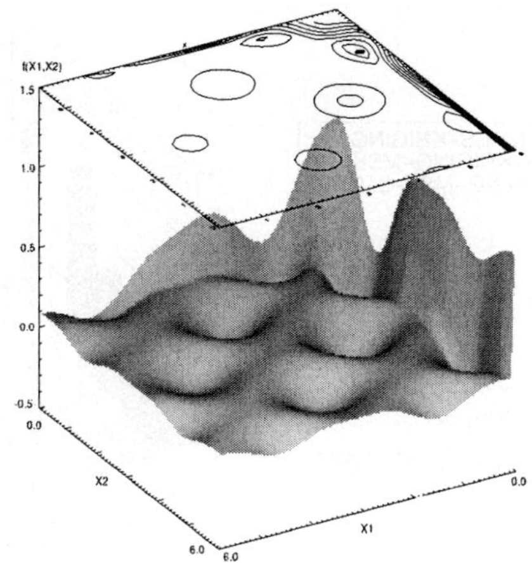

Fig. 3. Keane's function plotted in 2-dimensions for a cut of the search space. The optimum is indicated by the black spot in the upper right corner of the contour plot.

$$\min -\frac{|\sum_{i=1}^{n}(\cos^4 x_i) - 2 * \prod_{i=1}^{n}(\cos^2(x_i))|}{\sqrt{\sum_{i=1}^{n} i * x_i^2}}, \tag{6}$$

$$\prod_{i=1}^{n} x_i > 0.75,\ \sum_{i=1}^{n} x_i < \frac{15n}{2},\ \mathbf{x} \in]0, 10[^n \subset \mathbb{R}^n$$

This problem is characterized by a nonlinear boundary and a high number of local optima. The optimal solution is located near the constraint boundary.

The global convergence behaviour of different strategies was investigated (cf. 4). The $(15, 5, 100)$-ES performs much better than the $(2, 5, 10)$-ES. This indicates that strategies with a small population size are not robust for such problems. In contrast to the previous study, it now matters which search criterion is applied. Runs using $S_{c,w=1}$ perform much better (in average) than those guided only by the function estimation with Kriging $\hat{f}$. Here we get the desired effect that the strategy concentrates not only on the most promising solutions but makes also evaluations in unexplored regions of the search space that have a high potential of containing better solutions.

In the case of complex multimodal functions it is typical that MAES based on Kriging start by yielding a wide margin and that later they are overtaken by the $(15, 5, 100)$-ES. This is contradictory to what often occurs in unimodal functions. The fact that the $(15, 5, 100)$-ES overtakes the Kriging variants in the long term, might be explained by the fact that this strategy adapts the step-size much slower and the high step-size makes it easier to escape from local optima.

Though they work with the same number of iterations the $(2, 5, 10)$-ES leads to very bad results in the Keane function problem. This should be noticed in contrast to what is often believed, viz. that only small populations lead to good results when working with time consuming evaluations. This is certainly applica-

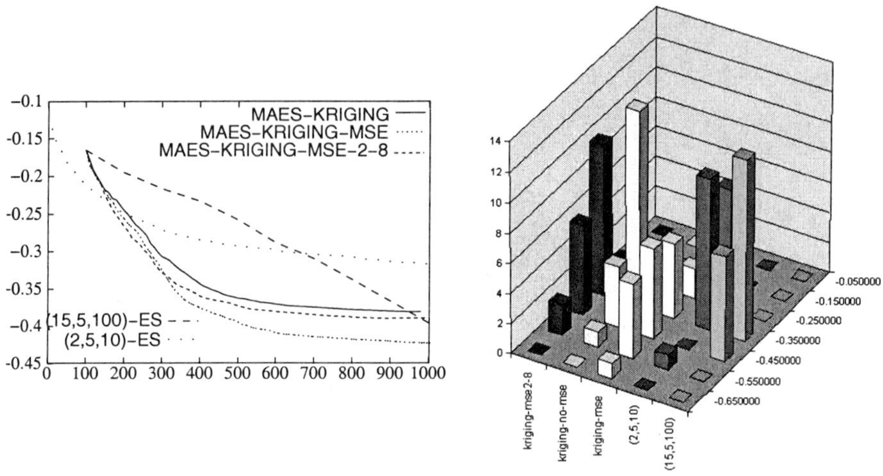

Fig. 4. Results on 20 dimensional Keane Function using the Kriging approximator: Averaged fitness histories for 20 runs (left) and a histogram for the best found values in the 20 runs after 1000 exact evaluations (right) are plotted.

ble for simple convex and unimodal functions. But EAs are not intended to solve problems, where the function topology is simple and for which other (faster) optimization tools can be recommended. ES are used in real-world application with highly nonlinear and multimodal characteristics.

5 Airfoil Shape Optimisation

This problem deals with the redesign of a 2D airfoil (the starting profile is the well known *RAE 2822*) in order to minimize the drag coeficient (C_D) and maximize the lift coefficient (C_L) at certain flow conditions. These are: $Re = 6.2 \times 10^6$, $M_\infty = 0.75$, $\alpha_\infty = 2.734^o$ where Re stands for the freestream Reynolds number based on the chord length, M_∞ for the freestream Mach number and α_∞ for the freestream flow angle. The transition was fixed on both sides at 3% of the chord.

For the parameterization of each shape one circle for the leading edge and two Bezier curves with five control points each have been used. The leading and trailing edge positions were fixed and the total number of design parameters was equal to 22.

The simulation tool was M. Drela's MSES analysis software [1] which at the aforementioned flow conditions yields $C_L = 0.748$, $C_D = 0.0235$. One evaluation takes about $1min$ on a Pentium III 1GHz PC. The cost function was defined as

$$F = C_D + \frac{10}{C_L} \tag{7}$$

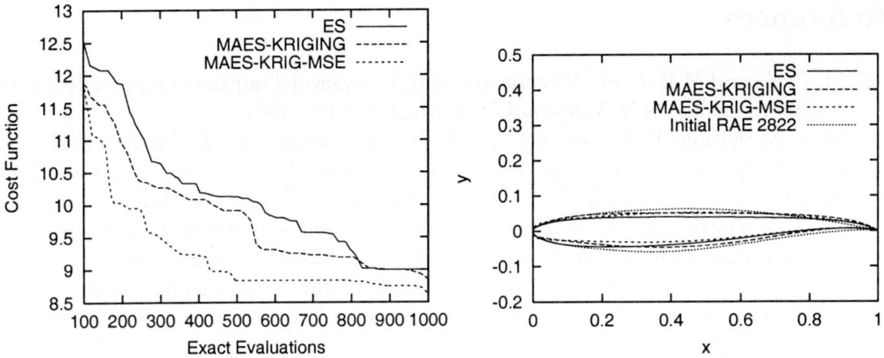

Fig. 5. Left: Average convergence histories of 3 runs of the airfoil shape optimization problem with the ES and MAES using one and both selection criteria. Right: Initial airfoil profile and the optimal ones computed using ES and MAES.

Working with a (15,5,100) strategy, average convergence histories are shown in fig.5 (left) with the conventional ES and MAES using one and both selection criteria. Each curve is the average of three optimization tasks. This plot indicates that the MAES technique is capable of reducing the computing cost compared to the conventional ES. On the right part of figure 5, the initial *RAE 2822* and the new optimal profiles are illustrated.

6 Conclusions

The use of metamodels in the context of ES–based optimisation algorithms was proved to offer economy in computing time. This economy results from the fewer exact evaluations that this method requires. It proved advantageous to use the metamodel not only for predicting the fitness value of new individuals but also for guessing the error associated with these predictions. In particular, the function estimation contributes mostly to reduction of the computational time whereas the error estimation helps to increase the global convergence reliability in complex multimodal problems. It was also proved that the metamodel does not harm the self-adaptivity properties of the method and that populations of increased size with good exploration capabilities can be used with low computing cost.

Acknowledgement

The support from bilateral Personnel Exchange Programme between Greece and Germany (IKYDA 2000) is acknowledged.

References

1. M. Drela and M.B. Giles. Viscous–Inviscid Analysis of Transonic and Low Reynolds Number Airfoils. *AIAA Journal*, 25 (10):1347–1355, 1987.
2. M.A. El-Beltagy, P.B. Nair, and A.J. Keane. Metamodelling Techniques for Evolutionary Optimisation of Computationally Expensive Problems: Promises and Limitations. In A.E. Eiben M.H. Garzon V. Honavar M. Jakiela W. Banzhaf, J. Daida and R.E. Smith, editors, *Proc. of GECCO, Int'l Conf. on Genetic and Evolutionary Computation, Orlando 1999*, pages 196–203. Morgan Kaufman, 1999.
3. K.C. Giannakoglou. Design of optimal aerodynamic shapes using stochastic optimization methods and computational intelligence. *Progress in Aerospace Sciences*, (38(1)):43–76, 2002.
4. K.C. Giannakoglou, A.P. Giotis, and M. Karakasis. Low-cost genetic optimization based on inexact pre-evaluations and the sensitivity analysis of design parameters. *Inverse Problems in Engineering*, (9):389–412, 2001.
5. A. Giotis, M. Emmerich, B. Naujoks, K. Giannakoglou, and Th. Bäck. Low cost stochastic optimisation for engineering applications. In *Proc. Int'l Conf. Industrial Applications of Evolutionary Algorithms, EUROGEN2001, Athens, GR, Sept. 2001*, Barcelona, 2001. CIMNE.
6. Y. Jin, M. Olhofer, and B. Sendhoff. Managing Approximation Models in Evolutionary Aerodynamic Design Optimisation. In *CEC 2001 Int'l Conference on Evolutionary Computation, Las Vegas*, volume 1, pages 592–599, Piscataway NJ, 2001. IEEE Press.
7. A. Padula. Interpolation and pseudorandom function generators. Senior honors thesis, University, Dept. of Computational and Applied Mathematics, Rice University, Houston, TX, 2000.
8. A. Ratle. Accelerating the convergence of evolutionary algorithms by fitness landscape approximations. In A.E. Eiben, Th. Bäck, M. Schönauer, and H.-P. Schwefel, editors, *Parallel Problem Solving by Nature*, volume V of *LNCS*, pages 87–96, Berlin, 1998. Springer-Verlag.
9. J. Sacks, W.J. Welch, W.J. Mitchell, and H.-P. Wynn. Design and analysis of computer experiments. *Statistical Science*, (4):409–435, 2000.
10. H.-P. Schwefel. *Evolution and Optimum Seeking*. Wiley, 1995.
11. M.W. Trosset and V. Torczon. Numerical optimization using computer experiments. Technical report, Institute for Computer Applications in Science and Engineering ICASE TR 9738, NASA Langley Research Center, Hampton Virgina, 1997.
12. H. Wackernagel. *Multivariate Geostatistics*. Springer Verlag, Berlin, 1998.

Limiting the Number of Fitness Cases in Genetic Programming Using Statistics

Mario Giacobini, Marco Tomassini, and Leonardo Vanneschi

Computer Science Institute, University of Lausanne
1015 Lausanne, Switzerland
{Mario.Giacobini,Marco.Tomassini,Leonardo.Vanneschi}@iis.unil.ch

Abstract. Fitness evaluation is often a time consuming activity in genetic programming applications and it is thus of interest to find criteria that can help in reducing the time without compromising the quality of the results. We use well-known results in statistics and information theory to limit the number of fitness cases that are needed for reliable function reconstruction in genetic programming. By using two numerical examples, we show that the results agree with our theoretical predictions. Since our approach is problem-independent, it can be used together with techniques for choosing an efficient set of fitness cases.

1 Introduction

The problem of determining adequate training samples of data is extremely common in the machine learning field and, of course, it is shared by genetic programming (GP). While the issue has received quite a bit of attention in domains such as artificial neural networks [5], it has almost been ignored, to our knowledge, in the case of GP. A noteworthy exception is the book by Banzhaf *et al.* [2] but, although interesting, the discussion is at an introductory level and does not provide quantitative criteria for the practitioner.

Genetic Programming is by its very nature a time consuming program induction method, a drawback that is obviously compensated for in many instances by its flexibility. Now, for most real world applications of GP it is well known that fitness evaluation is by far the most time consuming operation. It would thus be interesting to establish criteria that can help the researcher to limit the time spent in this phase as much as possible without compromising results in terms of quality and, possibly, generalization capability. One is thus confronted with two problems: how to select a sufficient number of fitness cases and how to choose those fitness cases in such a way that they are effective in driving the learning process towards a solution.

Here we approach the former problem i.e., how to significantly bound the number of fitness cases that have to be examined, from a standard statistical and information-theoretical viewpoint. Indeed well-known results of parametric statistics can be brought to bear on the problem directly and allow to reduce the number of fitness cases keeping a fixed significance level with a given, user-determined probability. A second starting point, apparently different from the

J.J. Merelo Guervós et al. (Eds.): PPSN VII, LNCS 2439, pp. 371–380, 2002.

previous one, makes use of the concept of information entropy. By using a couple of simple functions, we show that both approaches allow to fundamentally limit the number of fitness cases and also that they agree in providing the same qualitative bounds.

Several works have approached the second problem i.e., how to choose significant test cases, from an heuristical point of view with the aim of somehow sorting, cutting, or reordering the fitness cases is such a way that they are more useful [3, 4]. Other techniques use co-evolution of the fitness cases [6]. Since our criteria are problem-independent, it is clear that the two approaches are orthogonal; thus, our results can usefully be combined with good heuristics for choosing the current set of fitness cases.

The article is organized as follows. The next section introduces the fundamental statistical concepts that are used and explains their relevance to GP. Section 3 does the same for the quantity of information and entropy as applied to function reconstruction. Section 4 describes the functions to be used for the simulations and the GP parameters. Section 5 presents and discusses the results of the simulations and, finally, section 6 gives our conclusions and hints to future work and possible extensions.

2 Statistics

An important concept of statistics is *parameter estimation* (see e.g. [9]). The fundamental idea is the following: let $\{X_1, \ldots, X_N\}$ be a random sample of observations from a distribution that is specified up to a vector of unknown parameters. Whereas in probability theory it is usual to suppose that all of the parameters of a distribution are known, the opposite is true in statistics, where a central problem is to use the observed data to make inferences about the unknown parameters. Two kinds of parameter estimates are of interest: point estimates and interval estimates. Point estimates provide a single value for a parameter. In the case of interval estimates, rather than stating a certain value for our estimate of a given unknown parameter, we specify an interval in which we estimate that the parameter lies and we attach a degree of *confidence* to the estimate. For example, suppose that $\{X_1, \ldots, X_N\}$ is a sample from a normally distributed population having unknown mean μ and known variance σ^2. It can be shown [9] that $\bar{x} = \frac{1}{N}\sum_{i=1}^{N} x_i$ is the maximum likelihood estimator for μ. However, we don't expect that the sample mean $\bar{x}$ will exactly equal μ, but rather that it will be close. Hence, rather than a point estimate, it is sometimes more valuable to be able to specify an interval for which we have a certain degree of confidence that μ lies within.

2.1 Interval Estimates Applied to GP Fitness Evaluation

Let us consider a classical GP problem where the target function g is defined on N fitness cases, i.e. $g : \{x_1, ..., x_N\} \to \mathbb{R}$. Let also Π be a population of m individuals, $\Pi = \{A_1, ..., A_m\} \subset \Pi^*$, where $A_j : \{x_1, ..., x_N\} \to \mathbb{R}$ for $j \in$

$\{1, ..., m\}$. Let $f : \Pi^* \rightarrow \mathbb{R}$ be the fitness function of the GP, which associates a real value

$$f^{A_j} = \sum_{i=1}^{N} \mid f_i^{A_j} - g_i \mid$$

to each member A_j of a population Π, where $f_i^{A_j}$ is the value for the fitness case x_i of the individual A_j, and g_i is the value of the target function g for the same fitness case. Such a choice of the fitness function, instead of a more classical mean squared error, does not influence the statistical results of this paper. From now on we will write $x_i^{A_j}$ for $\mid f_i^{A_j} - g_i \mid$.

The mean distance of the individual A_j from the target function g is

$$\bar{x}^{A_j} = \frac{1}{N} \sum_{i=1}^{N} x_i^{A_j}$$

so the mean distance of all the individuals $\bar{x}$ is

$$\bar{x} = \frac{1}{m} \sum_{j=1}^{m} \bar{x}^{A_j} = \frac{1}{mN} \sum_{j=1}^{m} \sum_{i=1}^{N} x_i^{A_j} = \frac{1}{N} \sum_{i=1}^{N} \frac{1}{m} \sum_{j=1}^{m} x_i^{A_j}$$

A well-known result in statistics, the Central Limit Theorem [9], tells us that, whatever the original distribution of the fitness cases, since we are summing many independent estimates, the set of the fitness cases will approach a normal distribution. Therefore $\bar{x}$ is normally distributed ($\bar{x} \sim N(\mu, \sigma)$). A standard result for the confidence interval gives [9]

$$P\left(\bar{x} - t_{\alpha/2}\left(\frac{\sigma}{\sqrt{n}}\right) < \mu < \bar{x} + t_{\alpha/2}\left(\frac{\sigma}{\sqrt{n}}\right)\right) \geq 1 - \alpha.$$

Where $1 - \alpha$ $(0 < \alpha < 1)$ is the confidence with which we can expect the mean to be contained in the given interval. That is, $1 - \alpha$ percent of the time the interval will be found to contain the mean. The $t_{\alpha/2}$ is the Student cumulative distribution with $n - 1$ degrees of freedom such that the mean deviates from its true value in the interval $(-t_{\alpha/2}, t_{\alpha/2})$. The t-distribution is tabulated and can be easily computed. The variance σ^2, and thus the standard deviation σ is unknown but can be estimated by the sample variance s.

If we set $K = 2t_{\alpha/2}(s/\sqrt{n})$, the length of the confidence interval, we get a function relating such an interval K and the number of fitness cases n:

$$n = \left(2t_{\alpha/2}\frac{s}{K}\right)^2$$

Thus, n is the number of fitness cases that must be used in order for the mean fitness to be estimated to be in the confidence interval K with a given probability $1 - \alpha$.

3 Entropy

Let Y be a discrete random variable that takes values in the range $\{y_1, ..., y_N\}$, such that $P(Y = y_i) = p_i$, for $i \in \{1, ..., N\}$. The information content is defined as $I(Y = y_i) = -C(\log_a p_i)$, where C and a are two positive constants [1].

Given a discrete random variable Y, we cannot know for sure which of its values $y_1, ..., y_N$ will occur. Consequently, we don't know how much information $I(p_1), ..., I(p_N)$ we will be receiving, so that we may regard the information content of Y itself as a random variable which we denote as $I(Y)$. Clearly it has a range $\{I(p_1), ..., I(p_N)\}$. The expectation value (mean) of $I(Y)$ is called entropy and is denoted by $H(Y)$:

$$H(Y) = E(I(Y)) = -C \sum_{j=1}^{N} p_j \log_a(p_j).$$

In the case where $p_j = 0$, the quantity $p_j \log_a(p_j)$ is not well defined and it is conventionally assumed to be zero.

Such a quantity has two important properties (see [1]):
(i) $H(Y) \geq 0$, and $H(Y) = 0$ iff Y takes one of its values with certainty;
(ii) $H(Y) \leq C \log_a(N)$ with equality iff Y is uniformly distributed.

3.1 Entropy and Function Reconstruction

The concept of entropy of a random variable, described in the previous section, can be easly applied to a discrete target function of a classical GP optimisation problem.

If we take as sample space in a probability space the set of the fitness cases of a GP discrete target function, $S = \{x_1, ..., x_N\}$, we can see the target function $g : \{x_1, ..., x_N\} \rightarrow \{y_1, ..., y_M\} \subset \mathbb{R}$ as a random variable. The function g, in fact, perfectly satifies the definition of a discrete random variable as a mapping from the sample space S of a probability space to a finite subset R of $\mathbb{R}$.

It is therefore possible to calculate the entropy $H(g(x))$ of this function:

$$H(g(x)) = -C \sum_{j=1}^{N} p_j \log_a(p_j),$$

where $p_j = P(g(x) = y_j)$ for $j \in \{1, ..., M\}$. Such a measure will indicate the quantity of information needed to determine the function itself.

The two properties of the entropy of a random variable Y (see section 3) assure that $0 \leq H(g(x)) \leq C \log_a(N)$, where a and C are two positive constants and N is the number of fitness cases. Since we can choose a and C, we can set them to values such that the entropy would always be contained in the interval $[0, 1]$. Taking $a = e$ we have to set $C = 1/\ln(N)$.

For a GP discrete target function it exists a minimum number of sampling points such that the function can be completely reconstructed. The entropy of the target function, i.e. the average amount of information associated to the function, will therefore determine the minimum number of fitness cases to be considered for a reliable reconstruction of the target function itself.

4 Experimental Setting

To test the validity of our assumptions we have decided to set an experimental phase with the evolution, by a classical GP, of two simple functions. Since in this paper we limit ourselves to discrete functions (continuous functions are treated in a similar way), we have decided to evolve a boolean function and a discrete step function.

In the first experiment the target function is the seven variables boolean function g such that $g(x_1, ..., x_7) = ((x_1 \wedge x_2) \vee x_3)$. Such a function has $2^7 = 128$ fitness cases. In the GP evolution we have used as operation set the two boolean connectors *and* ($\wedge$) and *or* ($\vee$), and as terminals the set of all the seven boolean variables $x_1, ..., x_7$.

In the second experiment the target function is the following step function, defined on the interval $[0, ..., 99]$ of the natural numbers, therefore with 100 fitness cases:

$$g(x) = \begin{cases} 40 & \text{if } x \in [0, 40) \\ 70 & \text{if } x \in [40, 70) \\ 95 & \text{if } x \in [70, 95) \\ 100 & \text{if } x \in [95, 100) \end{cases}$$

As terminal set in the GP evolution we have given the variable x and the five constants $0, 40, 70, 95, 100$. For the operation set, four operators $IF1$, $IF2$, $IF3$, $IF4$ are supplied: $IF1(x, \alpha, \beta)$ returns α if $x \in [0, 40)$ and β otherwise, while $IF2$, $IF3$ and $IF4$ do the same when $x \in [40, 70)$, when $x \in [70, 95)$ and when $x \in [95, 100)$ respectively (α and β can either be a terminal symbol or an operator symbol).

For the evolutions we have used a classical GP with a random initial population of 200 individuals with maximum initial depth of 6. In the reproduction phase, a size 10 tournament selection operator is used to select a parent population of 180 individuals to which a standard crossover (with maximum depth of 17) and a standard mutation (with probability 0.1) is applied to produce 180 offsprings. An elitist successor operator is then used to replace the old population with a new one. In each of the 100 generations a set of a given percentage of fitness cases is randomly chosen with uniform probability between all the possible fitness cases (as proposed in [8]), and every individual of each population is evaluated on the same set of fitness cases.

5 Experimental Results

The aim of the two experiments described in the previous section is to show the statistical behavior of the GP evolutions when the number of considered fitness

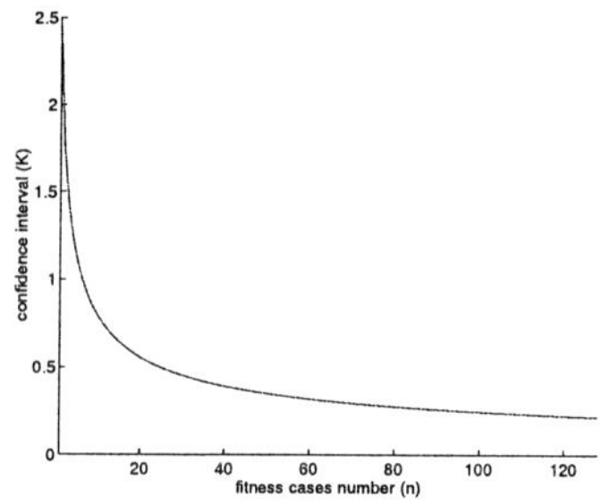

Fig. 1. The curve relating the confidence interval K and the number of fitness cases n for the boolean function $g(x_1, ..., x_7) = ((x_1 \wedge x_2) \vee x_3)$ with $\alpha = 0.01$ and $t_{\alpha/2} = 2.59$.

cases is decreased. For such a purpose the GP tool has been run 50 times for each percentage of fitness cases. Our intent is to show, with the experimental evidence, how the number of times the GP converges (i.e. finds the optimal solutions) decreases when the percentage of the fitness cases used in the evolution decreases.

5.1 Boolean Function

As we have said in section 4, in the first experimental phase the chosen target function is the seven variables boolean function $g(x_1, ..., x_7) = ((x_1 \wedge x_2) \vee x_3)$. For the interval estimation of the function, we need the sample mean $\bar{x}$ and the sample standard deviation s. We have decided to calculate these values over a sample set of 1000 random possible solutions, obtaining $\bar{x} = 0.411492$ and $s = 0.480703$.

On the other hand, this boolean function takes the value 1 on 80 over the 128 fitness cases (and the value 0 on the remaining 48 cases). Thus, the Bernoulli random variable $Y = g(x_1, ..., x_7)$ associated to this function is such that $p_1 = P(Y = 1) = 0.625$ and $p_0 = P(Y = 0) = 0.375$. These information are sufficient to calculate the entropy $H(Y)$ of the function:

$$H(Y) = \frac{1}{\ln(128)}[(p_0 \ln(p_0)) + (p_1 \ln(p_1))] = 0.1363$$

Since the total number of fitness cases is 128, an entropy of the function of 0.1363 tells that the minimal number of fitness cases needed to determine the function itself is the 13.63% of 128, i.e. there exist a set of 17.5 fitness cases by which it is possible to completely determine the function.

A typical behavior of the interval estimate curve relating the confidence interval and the number of fitness cases can be seen in figure 1, for a probability level of $\alpha = 0.01$ (see section 2.1). It is interesting to note that the confidence interval drastically increases in the neighborhood of the entropy of the function.

The experimental results, tabulated in figure 2, confirm our assumptions: when the number of fitness cases is greater than the entropy the behavior of the

fitness cases number / behaviour	128	100	90	80	70	60	50	40	30	20	10	5
convergent	98	84	84	94	86	94	84	90	80	82	48	0
flat	2	16	16	6	14	6	16	6	18	18	8	12
oscillating	0	0	0	0	0	0	0	4	2	0	44	88

Fig. 2. Statistics of the evolutions for the function $g(x_1, ..., x_7) = ((x_1 \wedge x_2) \vee x_3)$: for different numbers of fitness cases the table shows the percentages of the evolutions that converge (i.e. find the optimal solution), have a flat behavior (i.e. no evolution occurs), and have an oscillating behavior (i.e. obtains unreliable results).

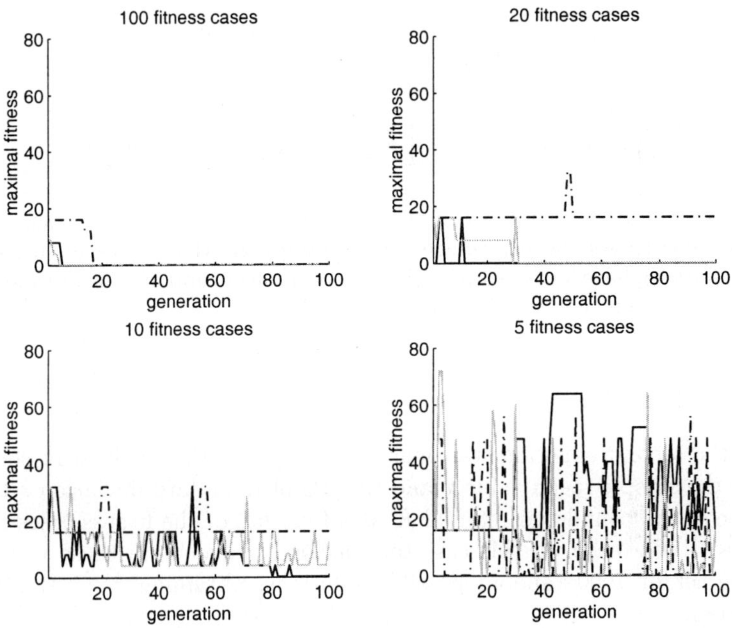

Fig. 3. Three sample GP runs for the boolean function with $n = 100$, $n = 20$, $n = 10$ and $n = 5$ fitness cases respectively. When n is greater than the function entropy (17.5 fitness cases), the curves are mostly convergent. While when n is lower than the function entropy, the curves show an oscillating behavior.

GP runs doesn't significantly change, and therefore the confidence interval stays almost constant. On the contrary, when the number of fitness cases is lower than the entropy of the function, the statistics of the GP runs drastically change. The algorithm doesn't seem to have enough information to build the target function and we start observing an oscillating behavior of the best individual found (see figure 3). Such a result is consistent with the confidence interval which increases drastically below the entropy value (see figure 1).

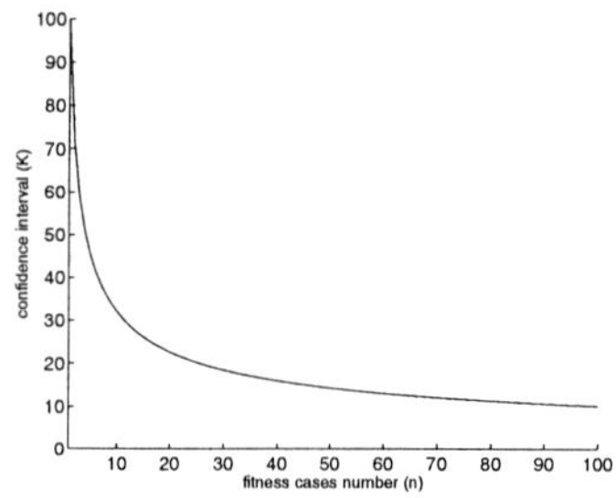

Fig. 4. The curve relating the confidence interval K and the number of fitness cases n for the step function with $\alpha = 0.01$ and therefore $t_{\alpha/2} = 2.59$.

behaviour \ fitness cases number	100	90	80	70	60	50	40	30	20	10	5
convergent	60	74	70	70	80	80	86	78	74	76	60
flat	40	26	30	30	20	20	14	16	16	8	6
oscillating	0	0	0	0	0	0	0	6	10	16	34

Fig. 5. Statistics of the evolutions for the step function: for different numbers of fitness cases the table shows the percentages of the evolutions that converge (i.e. find the optimal solution), have a flat behavior (i.e. no evolution occurs), and have an oscillating behavior (i.e. obtains not unreliable results).

5.2 Step Function

Also in the second experimental phase, the step function defined in section 4, we need the sample mean value $\bar{x}$ and the sample standard deviation s to draw the curve of the confidence interval K as a function of the fitness cases number n (figure 4). We have, in this case too, decided to calculate these values over a sample set of 1000 random possible solutions, obtaining $\bar{x} = 31.1453$ and $s = 19.5185$.

On the other hand, the random variable $Y = g(x)$ associated to this target function takes four different values $\{40, 70, 95, 100\}$ with probability $p_{40} = 0.4$, $p_{70} = 0.3$, $p_{95} = 0.25$ and $p_{100} = 0.05$ respectively. The entropy of the function $H(Y)$ is thus:

$$H(Y) = \frac{1}{\ln(100)} \sum_{i \in \{40,70,95,100\}} p_i \ln(p_i) = 0.2658$$

The minimal number of fitness cases needed to determine the function is therefore 26.58, since the total number of fitness cases is 100.

As can be seen in figure 5, for this test function we have the same statistical behavior as that of the boolean function (see section 5.1). When the number of fitness cases is greater than the entropy of the function we observe a normal convergence behavior. While with a number of fitness cases lower than the entropy

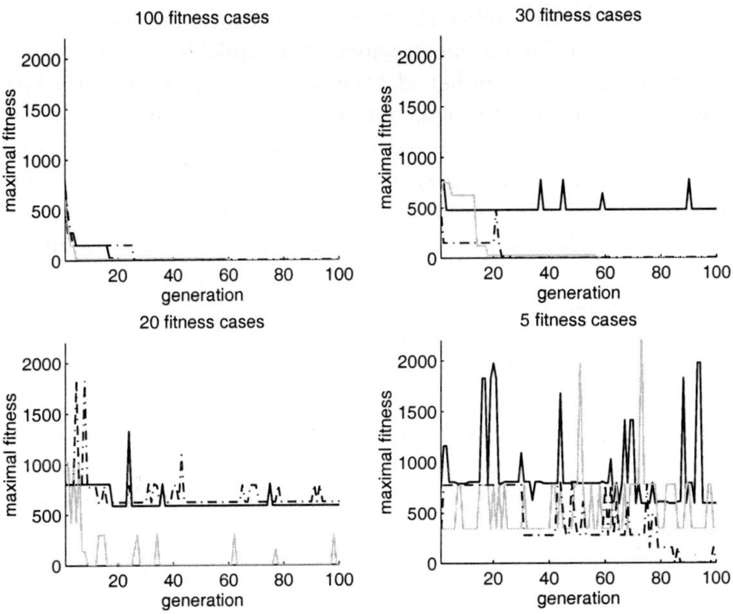

Fig. 6. Three sample GP runs for the boolean function with $n = 100$, $n = 30$, $n = 20$ and $n = 5$ fitness cases respectively. When n is greater than the function entropy (26.5 fitness cases), the curves are mostly convergent. While when n is lower than the function entropy, the curves show an oscillating behavior.

we start observing oscillating curves (see figure 6). Such a result is consistent with the confidence interval which increases drastically below the entropy value (see figure 5).

In light of the results obtained here, it can be said that the approach called *stochastic sampling*, which uses only one fitness case in individual fitness evaluation suggested in [2, 7] cannot, in general, give reliable results. Nevertheless, it can be justified in robotics, the case presented by the authors, given that fitness evaluation is very costly and the environment extremely variable. To asses its validity, one should nevertheless show that time averages can be substituted for ensemble averages, which is true in the limit for ergodic processes only.

6 Conclusions and Future Work

In this paper we have started from the idea that the number of fitness cases that we need to take into account in GP fitness evaluation can fundamentally be limited by statistical and information-theoretic considerations. Some well-established results, such as the Central Limit Theorem and interval estimation of distribution parameters, leads us straightforwardly to formal results which are in agreement with those that are obtained from entropy considerations. These results have been confirmed experimentally for two simple, but typical, function induction problems in GP. Below the lower limit for the number of fitness cases, the confidence interval for the mean widens very quickly and an oscillating behavior sets in, while for a number of fitness cases equal or slightly larger than the minimum, the convergence behavior is reliable and stable.

Our results are of a statistical nature and thus they do not depend on the particular problem. In other words, they tell us how many fitness cases one has to take into account on the average for any problem in order to reach significant results but not how these fitness cases should be chosen for maximum efficacy. Some previous works have tackled the latter problem heuristically, and thus could be used together with our criteria to not only reduce the number of fitness cases, but also to select the most significant ones.

Future work is scheduled to simulate more functions, including the continuous case, in order to further confirm the good results obtained here. Another related problem in machine learning, and thus in GP, is generalization. Generalization, which is the capability of a solution to correctly explain new data not used in the training process, has links to the selection of fitness cases and should thus be tackled next.

References

1. D. Applebaum. *Probability and Information: An Integrated Approach.* Cambridge, Cambridge, UK, 1996.
2. W. Banzhaf, P. Nordin, R. E. Keller, and F. D. Francone. *Genetic Programming, An Introduction.* Morgan Kaufmann, San Francisco CA, 1998.
3. C. Gathercole and P. Ross. Dynamic training subset selection for supervised learning in genetic programming. In Y. Davidor, H.-P. Schwefel, and R. Männer, editors, *Parallel Problem Solving from Nature- PPSN III*, volume 866 of *Lecture Notes in Computer Science*, pages 312–321, Heidelberg, 1994. Springer-Verlag.
4. C. Gathercole and P. Ross. Tackling the boolean even N parity problem with genetic programming and limited-error fitness. In John R. Koza, Kalyanmoy Deb, Marco Dorigo, David B. Fogel, Max Garzon, Hitoshi Iba, and Rick L. Riolo, editors, *Genetic Programming 1997: Proceedings of the Second Annual Conference*, pages 119–127, San Francisco, CA, USA, 1997. Morgan Kaufmann.
5. S. Haykin. *Neural Networks: A Comprehensive Foundation.* Prentice-Hall, London, UK, 1999.
6. W. D. Hillis. Co-evolving parasites improve simulated evolution as an optimization procedure. In C. G. Langton, C. Taylor, J. D. Farmer, and S. Rasmussen, editors, *Artificial Life II*, volume X of *SFI Studies in the Sciences of Complexity*, pages 313–324, Redwood City, CA, 1992. Addison-Wesley.
7. P. Nordin and W. Banzhaf. Genetic programming controlling a miniature robot. In E.V. Siegel and J.R. Koza, editors, *Working Notes for the AAAI Symposium on Genetic Programming*, pages 61–67. MIT Press, Cambridge, MA, 1995.
8. B.J. Ross. The effects of randomly sampled training data on program evolution. In D.Whitley, D.Goldberg, and E. Cantu-Paz, editors, *GECCO 2000 Proceedings of the Genetic and Evolutionary Computation Conference*, pages 443–450. Morgan Kaufmann, 2000.
9. S. M. Ross. *Introduction to Probability and Statistics for Engineers and scientists.* Academic Press, New York, 2000.

Resource-Based Fitness Sharing

Jeffrey Horn

Department of Mathematics and Computer Science
Northern Michigan University
Marquette, Michigan, 49855 USA
jhorn@nmu.edu
http://cs.nmu.edu/~jeffhorn

Abstract. This paper introduces a new algorithm for sharing to induce niching and speciation. *Resource-based fitness sharing* is a compromise between the very natural method of *resource sharing* and the practical technique of *fitness sharing*. Fitness sharing was meant to simulate resource sharing for function optimization problems, in which there are no explicit resources to share. Fitness sharing therefore cannot resolve resource-defined niches as can resource sharing. However, selection operators seem to have great difficulty handling the non-linear interactions among shared fitnesses under "natural resource sharing". To obtain the benefits of both methods, we propose a sharing function that utilizes actual resources but in a form similar to that of fitness sharing, resulting in a set of linear equations for equilibrium, and hence much simpler dynamics under selection. The superiority of this compromise is demonstrated on a resource-coverage problem.

1 Introduction

Resource sharing seems to be nature's way to induce speciation (niching) during evolution. Fitness sharing [1] was inspired by resource sharing, and was meant to simulate it for function optimization, in which explicit resources are not present. Whenever explicit resources, such as reward for classification tasks, can be identified, it has been generally assumed that explicit sharing of the resources was more natural and more appropriate than fitness sharing. But resource sharing appears to induce much more complex dynamics than fitness sharing [2]. It appears that the path to equilibrium can be so difficult for resource sharing that it is likely to lose key species along the way, and so never reach the optimal equilibrium. Fitness sharing, on the other hand, because it does not deal explicitly with the actual resources, cannot in general have the same equilibria as resource sharing, and therefore can only approximate the optimal species distributions. We propose a compromise between the two, *resource-based fitness sharing*, which uses the resources explicitly, so that the equilibrium points are the correct ones, while keeping the equilibrium calculations simple (linear), as in fitness sharing. The result is an algorithm with the natural fit of resource sharing, and the speed and simplicity of fitness sharing.

J.J. Merelo Guervós et al. (Eds.): PPSN VII, LNCS 2439, pp. 381–390, 2002.

2 Background

Sharing methods [3], in which the objective fitness of an individual in the population is reduced because of the need to share limited resources with competitors (including copies of itself), are typically incorporated into artificial evolution in order to promote diversity, to find multiple solutions [3], or to evolve a group of "cooperative" individuals or species [2,4]. In general, members of a single species are similar but may have different genotypes. In this paper however, we limit our analysis to species that consist of a single genotype. Thus each distinct genotype defines a species. A population *P* of size *N* can be partitioned into subsets of identical individuals (e.g., species **A, B, C, …K**) with subset sizes that sum to *N* (e.g., $n_A + n_B + n_C + \ldots + n_K = N$). A *niche* is the set of resources "covered", or "utilized", by a species. The terms *niche* and *species*, and the terms *niching* and *speciation,* are hence used interchangeably in this paper, as they are in much of the literature.

2.1 Fitness Sharing

Fitness sharing was introduced by Goldberg and Richardson in 1987 [1]. Under fitness sharing, the *objective fitness* f_i of an individual *i* is degraded by the presence of nearby individuals:

$$f_{sh,i} = \frac{f_i}{\sum_{j \in P} Sh(i,j)}. \tag{1}$$

Here $f_{sh,i}$ is the "shared fitness" of individual *i*. The sharing function, *Sh(i,j)*, is a decreasing function of the "distance" *d(i,j)* between individuals *i* and *j*, as typically measured in genotypic or phenotypic space. A widely used sharing function is the linearly decreasing *triangular sharing function*:

$$Sh(i,j) = \begin{cases} 1 - \dfrac{d(i,j)}{\sigma_{sh}} & \text{for } d(i,j) < \sigma_{sh} \\ 0 & \text{otherwise.} \end{cases} \tag{2}$$

2.2 Resource Sharing

Resource sharing is applied to problems in which objective fitness is directly proportional to the resources covered. Sharing means that where coverage overlaps, the resources (e.g., credit for examples correctly classified [4]) are divided among the individuals. Thus in the situation in Figure 1, the objective fitness for each copy of species **A** is f_A, while the shared fitness is

$$f_{sh,A} = \frac{f_A - f_{AB} - f_{AC}}{n_A} + \frac{f_{AB}}{n_A + n_B} + \frac{f_{AC}}{n_A + n_C}. \tag{3}$$

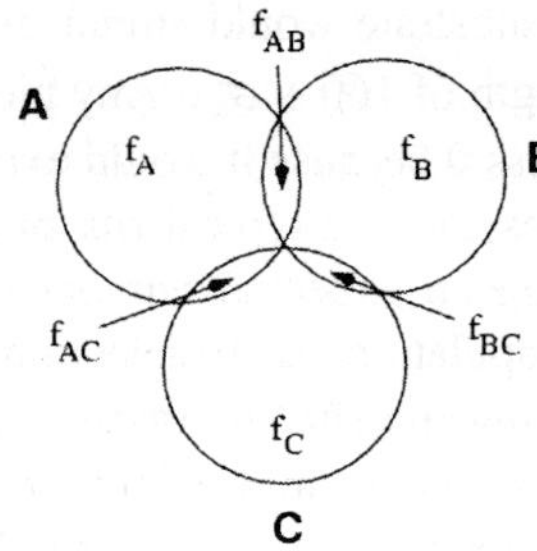

Fig. 1. A resource sharing scenario, with three overlapping species **A, B, C**

2.3 Shape Nesting

The task of placing arbitrary, two dimensional shapes on a two dimensional "substrate", minimizing unused (uncovered) substrate, or maximizing the number of shapes placed, while avoiding overlap of shapes, is a known NP-Hard problem [5,6]. Heuristic algorithms are desired by industries (e.g., paper, steel) in which many small shapes need to be cut from large "blanks" or "cutting stock". In this paper we consider a subclass of the problem, in which the shapes to be nested are identical [6].

3 A Real Test of Niching

David Goldberg originally suggested (personal communication, 1994) that we look at how fitness sharing behaves on a flat fitness landscape, such as a "plateau function". At the time, many new multi-objective genetic algorithms were having great success [7,8] by combining a Pareto selection pressure with fitness sharing. Niching on the Pareto optimal front means competition on a flat fitness function, what Horn, et. al. termed *equivalence class sharing* [7]. Many researchers hoped that niching would promote diversity within this non-dominated set, perhaps "covering" the Pareto front. But what equilibrium population distributions are possible over an equivalence class?

3.1 Fitness Sharing on a Hat Function

We assume a one-dimensional fitness function $f(x) = 1$ for $40 \leq x \leq 200$, and $f(x) = 0$ otherwise, where x is an integer between 0 and 255. Thus f is simply a discrete "hat" function: zero everywhere except on a plateau of high fitness, as in Figure 2 upper left. We assume any encoding of the single decision variable x as a finite length chromosome, such as an eight bit binary coded integer.

In a one-dimensional nesting problem, x can be seen as the coordinate of the center of each piece, with the length of the piece being σ_{sh}. The function can be seen as representing the substrate. The substrate would stretch from location (40 -• σ_{sh} / 2) to (200 + σ_{sh} / 2), with a total length of 160 + σ_{sh}. Any piece with center coordinate $x <$ 40 or 200 $< x$ would have fitness 0 because it would extend beyond the substrate.

In Figure 2 we show the results of a typical run of a GA on the function *f*. The population size is 2560, giving an expected number of 10 copies of each possible species in the initial random population. Only selection is applied (binary tournament selection) each generation, using the shared fitness values, with σ_{sh} set to 20. This means the substrate actually extends from x = 30 to x =210, the total length of each piece is 20, and a maximum of nine of these pieces can be fit onto the substrate.

The four graphs in Figure 2 show the species counts (where each integer value of x is a separate species) over the run. By generation 5 the non-global optima have been eliminated from the population. We also see an "edge effect", where the niches at the very edges of the hat accumulate population share more quickly than other niches. "Internal niches" are in turn affected by the growth of the edge niches. Those niches exactly σ_{sh} units distant from the edges grow almost as quickly as the edge niches. Similarly, niches σ_{sh} units in from these niches grow almost as quickly, and so on. By generation 180, the only niches remaining in the population are those that are exactly a multiple of σ_{sh} distant from both edges. Thus the steady-state population consists of non-overlapping (non-competing) species.

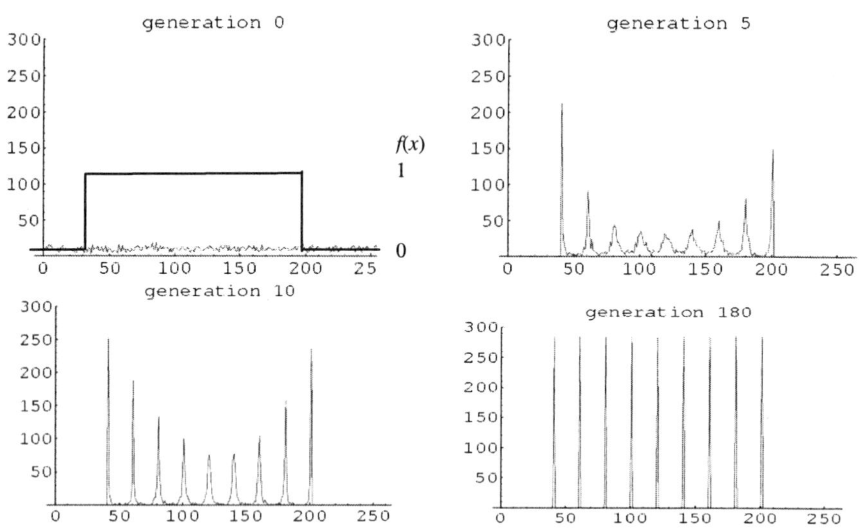

Fig. 2. Performance of fitness sharing (with selection only) on a *hat* function

Fitness sharing appears to have potential to evolve solutions to shape nesting problems. But it is limited by the simplicity of its distance metric (e.g., radial symmetry). It seems appropriate for one-dimensional problems, or multi-dimensional problems with spherical pieces to place (i.e., *sphere packing*). But what about, for example, placing squares in two dimensions?

3.2 Resource Sharing (in One Dimension)

A seemingly more natural approach to this type of problem would be resource sharing, since an explicit resource, the substrate, can be identified. We can simply discretize the substrate via a grid, and for each discrete element of the substrate, share it equally among the pieces "covering" that element. For example, if a particular population distribution resulted in eight pieces overlapping at a particular element of the substrate, then each such piece would receive one eighth of the credit for covering this element, and that one eighth credit would be added to each piece's fitness total.

This approach holds the promise of being more general than fitness sharing (although it could be computationally expensive) since it can accommodate any shape piece, taking into account the exact amount of overlap between pieces, rather than approximating the overlap via a fixed niche radius (i.e., fixed piece diameter, which would likely be that of its bounding sphere).

In Figure 3 we present the results of two typical runs of this algorithm.[1] In both runs we use a small search space: there are only 32 different locations (x value of its center) for each piece, requiring only five bits to encode. Each piece is exactly three units wide. Therefore each piece extends out 1.5 units to the right and left of its x coordinate. Thus two pieces with centers x_0 and x_1 with $| x_1 - x_0 | = 3$ would be up against each other, with no overlap and with no wasted substrate in between. At most 10 pieces (squares) can be lined up on this substrate with no overlap. Population size is 400, binary tournament is the selection method, and no exploration (i.e., no crossover or mutation) is attempted.

At the top of Figure 3, we see the initial random distribution of generation 0. Note that two squares on the left (at positions 0 and 1) and one on the right (at position 31) extend beyond the substrate (the bold rectangle) and have fitness 0. These are soon eliminated, but after 400 generations, the population has not converged to a set of non-overlapping species. We observed this situation as typical of our runs with "regular" resource sharing as described above. Why isn't the performance of resource sharing at least comparable to that of fitness sharing, on this type of problem?

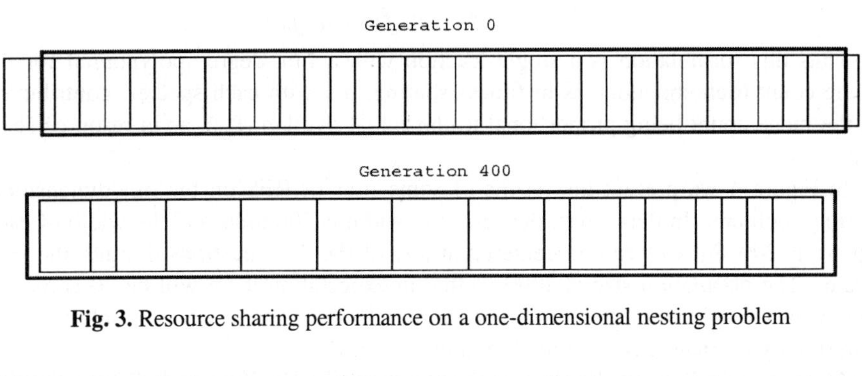

Fig. 3. Resource sharing performance on a one-dimensional nesting problem

[1] We thank Todd Marshall for conducting these experiments as an NMU Directed Study.

Horn [4] finds surprising complexity in the dynamics of resource sharing when more than two overlapping species/niches are involved. The equilibrium equations for fitness sharing and resource sharing, for the general case of k niches, always involve solving a system of k equations in k unknowns. But while the fitness sharing equations are always linear, the degree of the equations for resource sharing grows polynomially in k. Horn noticed more complex convergence even for only three niches, showing overshoots in the expected proportions of species as the population converges to the equilibrium. Perhaps what we have seen here is an indication that nature does not have an easy time with the highly non-linear dynamics of resource sharing among many overlapping niches. Perhaps the computational complexity of equilibrium equations for a sharing method is a direct measure of the complexity of the actual population dynamics of species selection.

3.3 Resource-Based Fitness Sharing (in One Dimension)

What if we could combine the benefits of both types of sharing? What if we could have the resource-based "niche resolution" of resource sharing along with the (apparently) simpler convergence to equilibrium of fitness sharing? In other words, what if we could "linearize" the equilibrium equations of resource sharing, without getting away from the resource-based sharing mechanism entirely? Here is one attempt to do so:

$$f_{Sh,X} = \frac{f_X}{\sum_{\forall \, species \, Y} n_X \, f_{X,Y}} \, . \tag{4}$$

The above is the general equation for the proposed *resource-based fitness sharing* (RFS). A concrete example, for the three overlapping niches **A**, **B**, and **C** in Figure 1, would be:

$$f_{Sh,A} = \frac{f_A}{n_A f_A + n_B f_{AB} + n_C f_{AC}} \, .$$

Thus this formulation is a single fraction with a first degree polynomial for the niche count (denominator), as in fitness sharing, but with each species' contribution to the niche count being proportional to the actual overlap ($f_{X,Y}$), as in resource sharing.

In Figure 4 we present the results of some runs of RFS on the one-dimensional nesting problem. In these runs, the substrate width is 200 units and the width of each square is 20. Squares can be centered at any of the 200 locations through the substrate. The population size is 4000 so that in expectation there will be 20 copies of each possible species in the initial random population. Again, we look only at selection (binary tournament); no mutation or crossover is used.

At generation 0 all of the species are represented. By generation 2 (not shown), the hanging pieces (zero fitness) have been eliminated. Generation 107 shows some evidence of an "edge effect". And by generation 225 the only remaining species are the ten members of the "ideal tiling".[2] Each species is represented by approximately 400 copies, which is the equilibrium distribution. We claim this is a typical run.

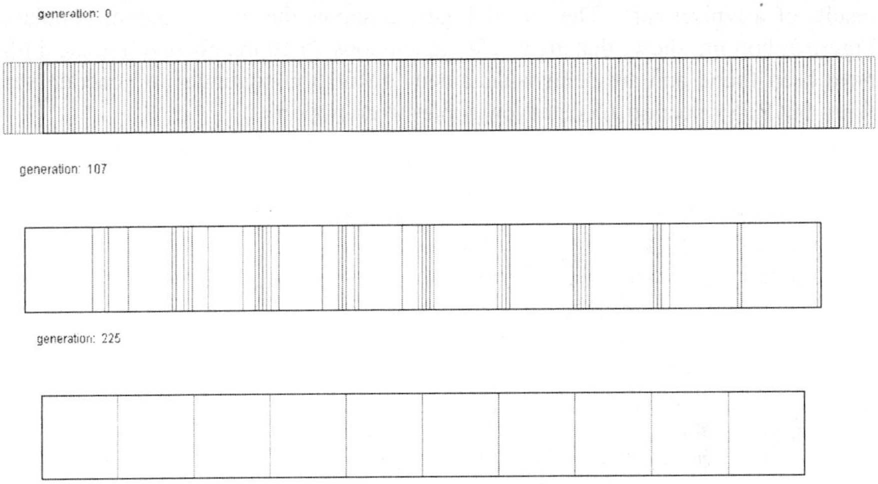

Fig. 4. RFS on a one-dimensional nesting problem

3.4 Resource-Based Fitness Sharing in Two Dimensions

Encouraged by the success of RFS in one dimension, we attempt a two-dimensional problem. Here the substrate is itself a square, of size 40 by 40 units, while the piece size (the smaller squares, to be placed) are of width 10 units. Thus a single ideal tiling exists: a four by four array. There are 40*40 = 1600 different possible locations (species). We conduct two experiments: large population with selection only, and small population with mutation as the discovery operator.

In the first experiment, we look at selection only. A population size of 16000 yields an expected ten copies of each possible species in the initial random population (generation 0 in Figure 5, upper left). In the typical run portrayed in Figure 5, by generation 400 only the sixteen members of the optimal tiling are left, each represented by approximately 1000 copies. In Figure 5 bottom, histograms of the subpopulation sizes for the sixteen species of the ideal tiling show an apparent "corner effect" at generation 25, on the way to a uniform distribution at generation 400.

In our second experiment we test RFS with a discovery operator. In this case we choose Gaussian mutation: each generation, with a probability $p_m = 0.01$ both the x and y coordinates of an individual are individually adjusted[3] by a pseudo-random

---

[2] A *tiling* is an ideal nesting in the sense that no substrate is wasted (unused) between pieces.

[3] A simple chromosome repair mechanism is applied after mutation: if the mutated *x* or *y* coordinate is off of the substrate, it is set equal to the nearest edge of the substrate.

amount, normally distributed around a mean of 0 with a standard deviation of 5. A small population size (N=500) is chosen so as to rely on mutation-based exploration to discover some of the 16 members of the ideal tiling. Figure 5 displays some of the results of a typical run. The top of Figure 5 shows the initial random population. Figure 5, bottom, shows that after 1200 generations all 16 members of the ideal tiling have been found and are being maintained. (Mutation continues to generate variations, but RFS selection allocates few copies to non-optima.)

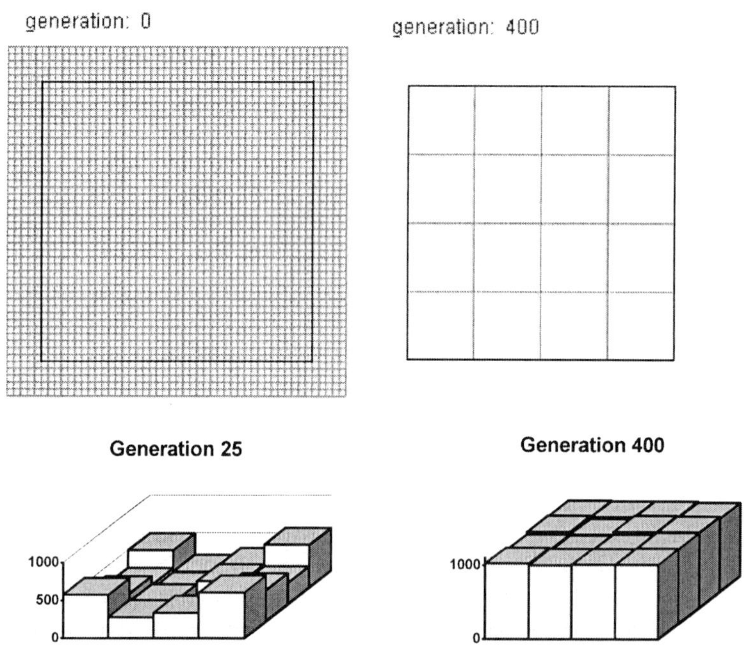

Fig. 5. RFS (with N = 16000 and selection only) on a two-dimensional nesting problem.

4 Discussion

The combination of selection and a sharing method seem to have potential for application to problems in nesting, tiling, layout, packing, and trim minimization [6]. Using "edge effects" and "corner effects", the two operators seem able to select one of many possible subsets of species to cover available resources efficiently. But surprisingly, the more natural-seeming algorithm of explicit resource sharing appears to induce overly complex dynamics, a phenomenon perhaps foretold in the computational complexity of our models of sharing equilibrium. By combining the simplicity of the fitness sharing method with the use of explicit resources as a measure of niche overlap, however, we might be able to obtain the best of both techniques.

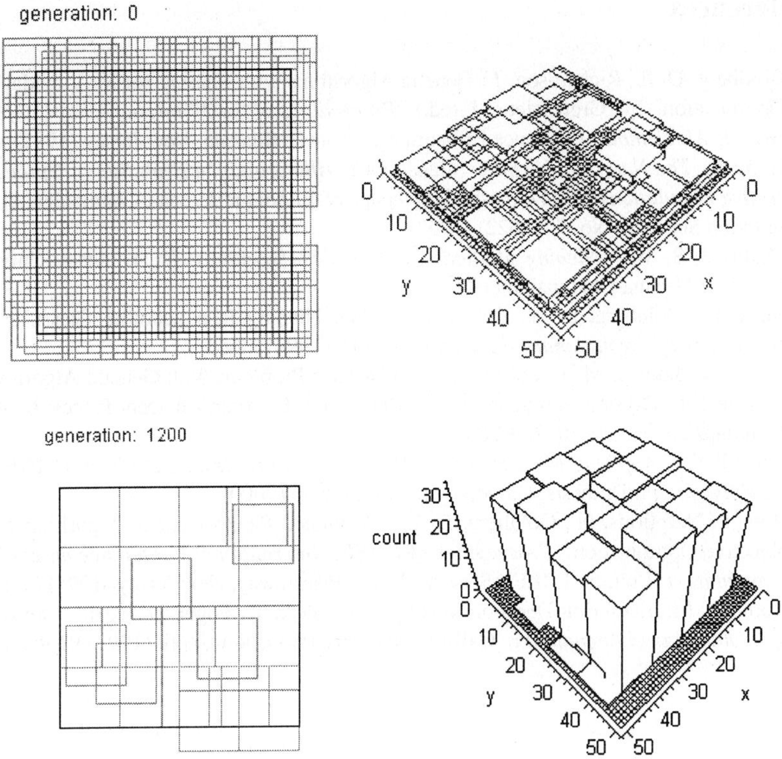

Fig. 6. RFS, with N=500 and using mutation, on the same 2D problem.

As one such synthesis of fitness sharing and resource sharing, RFS needs more investigation, including theoretical analysis as well as empirical work. In the short term, existing theory techniques for niching methods can be applied, including equilibrium analysis, infinite population models, and Markov chains [2,3,4], to verify that RFS is a robust niching method with stable equilibria[4].

Empirical extensions of this work include trying arbitrary and irregular shapes, both for the pieces and the substrate[5]. Long-term directions include 3D layout problems, packing of non-identical pieces, and comparing results with those of alternative algorithms (e.g., clustering) to see if RFS can find a "niche" as a practical algorithm.

---

[4] This should be straightforward to show, as we note that RFS is equivalent to fitness sharing with a distance metric $d(X,Y) = f_X - f_{XY}$ and a niche radius $\sigma_{sh} = f_X$. Substituting these values into Equations 1 and 2 (fitness sharing) yields Equation 4 (RFS) after some algebraic manipulation. Thus we can choose to view RFS as a refinement of fitness sharing for resource-based problems, rather than as an improvement (linearization) of nature's methods.

[5] Such applications would call for rotation as a third decision variable.

References

1. Goldberg, D. E., Richardson, J.: Genetic Algorithms with Sharing for Multimodal Function Optimization. In: Grefenstette, J. (ed.): *Proceedings of the 2nd International Conference on Genetic Algorithms.* Lawrence Erlbaum Associates, Hillsdale, New Jersey (1987) 1-8
2. Horn, J.: *The Nature of Niching: Genetic Algorithms and the Evolution of Optimal, Cooperative Populations.* Ph.D. thesis, University of Illinois at Urbana-Champaign, (UMI Dissertation Services, No. 9812622) (1997)
3. Mahfoud, S. W.: *Niching Methods for Genetic Algorithms.* Ph.D. thesis, University of Illinois at Urbana-Champaign (1995)
4. Horn, J., Goldberg, D. E., Deb. K.: Implicit Niching in a Learning Classifier System: Nature's Way, *Evolutionary Computation, 2*(1) (1994) 37-66
5. Dighe, R., Jakiela, M. J.: Solving Pattern Nesting Problems with Genetic Algorithms: Employing Task Decomposition and Contact Detection Between Adjacent Pieces. *Evolutionary Computation 3*(3) (1996) 239-266
6. Kendall, G.: *Applying Meta-Heuristic Algorithms to the Nesting Problem Utilising the No Fit Polygon.* Ph.D. thesis, University of Nottingham (2000)
7. Horn J., Nafpliotis, N., Goldberg, D. E.: A Niched Pareto Genetic Algorithm for Multiobjective Optimization. *Proceedings of 1st IEEE International Conference on Evolutionary Computation*, Volume 1. IEEE Service Center, Piscataway, New Jersey (1994) 82-87
8. Horn, J.: Multicriterion Decision Making. In: Bäck, T., Fogel, D. (ed.s): *The Handbook of Evolutionary Computation.* Oxford University Press, New York (1997) F1.9:1-15

Evolution Strategy with Neighborhood Attraction Using a Neural Gas Approach

Jutta Huhse[1], Thomas Villmann[2], Peter Merz[1], and Andreas Zell[1]

[1] University of Tübingen, Inst. of Computer Science
D–72076 Tübingen, Sand 1, Germany
{huhse,pmerz,zell}@informatik.uni-tuebingen.de

[2] University of Leipzig, Clinic for Psychotherapy and Psychosomatic Medicine
D–04107 Leipzig, Karl–Tauchnitz–Str.25, Germany
villmann@informatik.uni-leipzig.de

Abstract. In evolution strategies with neighborhood attraction (EN) the concepts of neighborhood cooperativeness and learning rules known from neural maps are transferred onto the individuals of evolution strategies. A previous approach, which utilized a neighborhood relationship adapted from self-organizing maps (SOM), appeared to perform as well as or even better than comparable conventional evolution strategies on a variety of common test functions. In this contribution, an EN with a new neighborhood relationship and learning rule based on the idea of neural gas is introduced. Its performance is compared to the SOM-like approach, using the same test functions. It is shown that the neural gas approach is considerably faster in finding the optimum than the SOM approach, although the latter seems to be more robust for multi-modal problems.

1 Introduction

Evolutionary algorithms (EAs) are stochastic search algorithms inspired by principles of biological evolution working simultaneously on a large number of *potential* problem solutions. Thereby, these solutions are judged by an external cost function, called fitness.

The power of *neural networks* (NNs) arises from complex, dynamic connections between quite simple calculation units (called neurons) working in a massively parallel manner. These features enable neural networks to be both robust and adaptive.

The combination of neural networks and EAs offers new possibilities to increase the power of adaptive approaches. Two main directions can be identified: the first one is to apply EAs to optimize a neural network whereas the second one deals with the transfer of methods from neural networks onto the framework of EAs. For the first type of combining EAs and NNs many examples can be found in the literature. For the second type however, as already stated in [1], there are only a few approaches of incorporating aspects of artificial neural networks into evolutionary algorithms.

J.J. Merelo Guervós et al. (Eds.): PPSN VII, LNCS 2439, pp. 391–400, 2002.

Taking this into account, in this paper we focus on such hybrid systems which combine EAs with *neural maps* as a special type of neural networks. Examples are Kohonen's widely-used *self-organizing map* (SOM) [2] and the *topology representing network* (TRN) [3]. The main paradigm of neural map learning distinguishing it from other network types is the incorporation of neighborhood oriented learning, i. e. neighboring neurons try to adapt in a similar fashion.

Based on previous work, where an *E*volution strategy with *N*eighborhood attraction (EN) on the basis of Kohonen's SOM was introduced and compared to conventional evolution strategies [4], here we present a new neighborhood and attraction scheme for this EN, based on neural gas (NG) [5].

First, an introduction to neural maps is given in section 2. Then, based on the description of the SOM inspired EN (subsection 3.1) an EN with a new, neural gas like neighborhood is explained in subsection 3.2. This new approach is analyzed using several well-known test functions (section 4) and the results are given in section 5.

2 Neural Maps

Neural maps as tools for (unsupervised) topographic vector quantization project data from a (possibly high-dimensional) input space $\mathcal{V} \subseteq I\!R^{D_\mathcal{V}}$ onto a position in some output space (neuron lattice) $\mathcal{A}$, such that a continuous change of a parameter of the input data should lead to a continuous change of the position of a localized excitation in the grid. This property is called *neighborhood preserving* (topographic) mapping. In the most widely used neural map algorithm, the *self-organizing map* (SOM) [2], the N neurons are arranged *a priori* on a fixed grid, i. e. the $\mathbf{r_i} \in \mathcal{A}$ denote the positions of the neurons in the grid $\mathcal{A}$. Usually the grid is chosen as a hypercube, but other structured arrangements are also admissible.

Other algorithms, such as the TRN, based on the neural gas algorithm (NG) [5, 6], do not specify the topological relations among neurons in $\mathcal{A}$ in advance, but dynamically adapt the neighborhood structure according to the data distribution during learning.

Associated with each neuron is a weight vector (or reference vector) $\mathbf{w_r} \in I\!R^{D_\mathcal{V}}$. The mapping is realized by $\Psi_{\mathcal{V}\to\mathcal{A}} : \mathbf{v} \mapsto \mathbf{s}(\mathbf{v}) = \arg\min_{\mathbf{r}\in\mathcal{A}} \parallel \mathbf{v} - \mathbf{w_r} \parallel$, where $\parallel \mathbf{v} - \mathbf{w_r} \parallel$ is the Euclidean distance between a given input signal $\mathbf{v}$ and the weight vector $\mathbf{w_r}$. The neuron $\mathbf{s}$ is referred to as the *winner* or *best–matching unit.* During learning the weight vectors are adapted according to

$$\Delta \mathbf{w_r} = \epsilon \cdot h_\sigma(\mathbf{r}, \mathbf{s}, \mathbf{v})(\mathbf{v} - \mathbf{w_r}) . \tag{1}$$

Thereby $h_\sigma(\mathbf{r}, \mathbf{s}, \mathbf{v})$ is the neighborhood function usually chosen to be of Gaussian shape

$$h_\sigma(\mathbf{r}, \mathbf{s}, \mathbf{v}) = \exp\left(-\frac{(d(\mathbf{r}, \mathbf{s}(\mathbf{v})))^2}{2\sigma^2}\right) \tag{2}$$

where $d(\mathbf{r}, \mathbf{s}(\mathbf{v}))$ is a certain distance measure defined on the set $\mathcal{A}$. For SOM, $d(\mathbf{r}, \mathbf{s}(\mathbf{v}))$ is evaluated in the grid $\mathcal{A}$, whereas for TRN, it is evaluated in the input (problem) space $\mathcal{V}$. With $h_\sigma(\mathbf{r}, \mathbf{s}, \mathbf{v})$, a neighborhood oriented cooperativeness for learning is installed which is triggered by the neighborhood range σ of the neighborhood function. During the learning process usually σ as well as the learning rate ϵ decrease to small values referred to as remaining cooperativeness σ_{final} and learnability ϵ_{final} of the network. In the beginning of the adaptation the neighborhood range and the learning rate are high which corresponds to a *rough but fast learning* whereas in the convergence phase the rough adaptation is replaced by a *fine tuning*. The first phase is indicated by a high cooperativeness between the neurons which realizes a high information transfer. In the fine tuning phase the neurons act more or less separately with an only small (nearest neighbor) but not vanishing communication.

3 Evolution Strategies with Neighborhood Attraction

As mentioned above there are not many approaches that incorporate concepts of neural networks into EA to improve EA-performance [7, 1]. This fact certainly holds also for the special combination of neural maps and ES. Here, we concentrate on evolution strategies with neighborhood attraction (EN) [4]. The idea behind EN is to transfer the neighborhood and the learning rule known from neural maps onto the originally unstructured individuals of an ES to concentrate the individuals around the optimum.

3.1 SOM-Neighborhood Attraction EN

The SOM-Neighborhood attraction EN (SOM-EN) introduced by Huhse and Zell [4, 8] works as follows:

The *neighborhood* between the μ parent individuals is constituted by arranging them on an orthogonal, elastic grid. As known from the SOM, each individual can be identified by its fixed grid position. Two individuals a_i and a_j are grid neighbors if they are directly connected on the grid; i.e. $h_\sigma = 1$ (see equ. 2). If no direct grid connection exists, they are not neighbors ($h_\sigma = 0$).

In contrast to conventional ES and SOM, the initial values of the object variables of the EN individuals are not assigned randomly. Rather, the problem space is divided into equally sized hypercubes, each of them corresponding to one grid position. The object variables of the associated individual are initialized with equally distributed random values within the ranges of its hypercube.

As is customary in ES, the EN individuals are evaluated using a certain fitness function.

For the evolutionary *variation* an EN-specific operator – the neighborhood attraction – is introduced which manipulates the EN individuals according to one learning step in a SOM (cf. 1). Each parent individual a_P is successively selected. Among all its (direct grid) neighbors the best neighbor a_{Nb} is determined. If the fitness value of the neighbor is better than the fitness value of the parent, the EN-specific neighborhood attraction is applied: An offspring individual a_O is produced by attracting the parent individual a_P towards its best neighbor a_{Nb} (see Fig. 1).

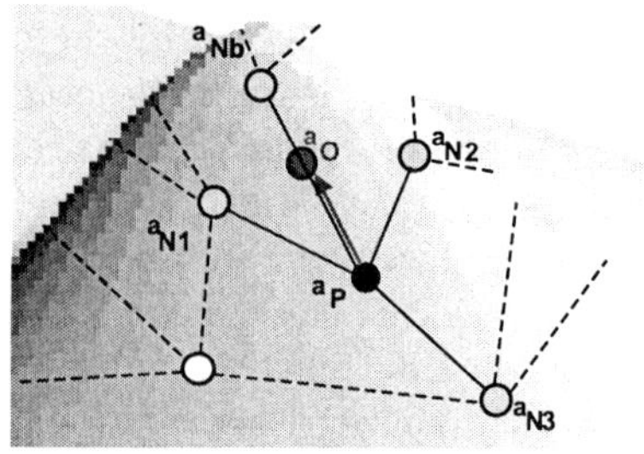

Fig. 1. Neighborhood attraction in SOM-EN

In detail, the variation is realized on the level of the object variables by equation 3, with the neighborhood attraction operator (equ. 4). The neighborhood relations are retained unchanged.

$$\boldsymbol{x}_O = \boldsymbol{x}_P + \Delta\boldsymbol{x}_P \tag{3}$$
$$\Delta\boldsymbol{x}_{\boldsymbol{P}} = \delta \cdot (\boldsymbol{x}_{Nb} - \boldsymbol{x}_P) \tag{4}$$

Thereby, the *neighborhood attraction factor* δ defines the strength of the attraction along the difference vector. One parent individual can produce n_O offspring by varying δ.

If the fitness value of the parent is considered better than those of all its neighbors a_{Nj} a "simple conventional" mutation (referred to as *ES-mutation* here) is performed. n_O offspring are generated according to equ. 5.

$$\begin{aligned} \boldsymbol{z} &= \boldsymbol{\mathcal{N}}(0,1) \\ s_{\mathit{eff}} &= \tfrac{1}{n} d_{min} \\ \Delta\boldsymbol{x}_P &= s_{\mathit{eff}} \cdot \boldsymbol{z} \end{aligned} \tag{5}$$

$\boldsymbol{z}$ is a vector of normally distributed random numbers. The effective step size s_{eff} is determined by $\frac{1}{n}$, the reciprocal number of object variables and by the distance $d_{min} = \min(\|\boldsymbol{x}_P - \boldsymbol{x}_{Nj}\|)$ to the nearest neighbor. Thus, a mutation which jumps over a neighbor and an entanglement of the grid becomes less likely. During the contraction of the grid the effective step sizes decrease due to the influence of d_{min}.

Deviating from the in ES commonly used $(\mu + \lambda)$- or (μ, λ)-selection scheme [9] – where the μ individuals which go to the next generation are selected from all μ parent *plus* all λ offspring individuals or *only* from all λ offspring, resp. – a *parental* selection scheme is used here, which resembles brood selection known from genetic programming [10]. Selection is performed for every parent a_P and its n_O offspring separately. Each parent contributes exactly one individual to the next generation. If the n_O offspring have been produced by neighborhood attraction, the parent is always replaced by its best offspring, even if the offspring's fitness is worse (similar to (μ, λ)-selection). Allowing such temporary fitness decrease, an individual can pass through a local optimum towards better fitness values. If the n_O offspring have been produced by ES-mutation, the best of the parent and its offspring is selected (similar to $(\mu + \lambda)$-selection). Using parental selection, one the one hand, the grid neighborhood is not disrupted by the selection process, and, on the other hand, diversity of the population is better maintained.

3.2 Neural Gas Attraction EN

Now, we propose a new neighborhood and attraction scheme, the neural gas attraction EN (NG-EN). According to the idea in the NG mentioned above, we define the topological neighborhood h_σ for the individuals depending on their distances $d(a_i, a_j)$ in the problem space. The neighborhood radius for each individual a_P is determined by its average (Euclidean) distance $\hat{d_E} = \frac{1}{n}\sum_i (\| \boldsymbol{x_i} - \boldsymbol{x_P} \|)$ to all other individuals. All individuals a_{Nj} inside this radius are considered to be neighbors of a_P.

Like in SOM-EN, *variation* of the individuals is performed on each parent individual a_P separately: All its neighbor individuals a_{Nj} are ranked according to their increasing distances to a_P, as it is known from NG. The neighborhood h_σ decreases with the neighbor's rank and with time (equ. 7). Then a_P places λ offspring in different directions, offspring a_{Oi} in direction to the neighbor a_{Ni} with rank i.

In analogy to the neural gas algorithm [5, 6, 3], the attraction rules from equations (1), (2) and (4) are rewritten as

$$\Delta \boldsymbol{x_P}(i) = \epsilon(t) \cdot h_\sigma(i,t) \cdot (\boldsymbol{x_{Ni}} - \boldsymbol{x_P}) \tag{6}$$

$$h_\sigma(i,t) = \exp\left(\frac{-\text{rank}(i)}{\sigma(t)}\right) \tag{7}$$

$$\sigma(t) = \sigma_i \cdot (\sigma_f/\sigma_i)^{(t/t_{max})}$$

$$\epsilon(t) = \epsilon_i \cdot (\epsilon_f/\epsilon_i)^{(t/t_{max})}$$

with

$$\delta(i,t) = \epsilon(t) \cdot h_\sigma(i,t) \quad \text{and} \quad \delta \in [\delta_{min}, \delta_{max}].$$

Fig. 2 shows the neighborhood attraction δ decreasing with time and rank. Two applicable parameter settings were chosen: While $\sigma_i = 5$, $\sigma_f = 0.05$ and $t_{max} = 30000$ are constant, ϵ_i and ϵ_f were set to either $\epsilon_i = 0.0075$ and $\epsilon_f = 0.00375$ (left) or $\epsilon_i = 0.02$ and $\epsilon_f = 0.01$ (right). The initial value for the resulting δ is in compliance with suggestions made for the δ used in SOM-EN [8].

As in the SOM-EN, this neural gas neighborhood attraction is performed only if a_P is not better than its neighbors. Otherwise the ES-mutation known from (5) is applied.

4 Test Functions

To investigate the new neighborhood attraction scheme we conducted thorough test series on an exhaustive test set of well-known optimization tasks also used

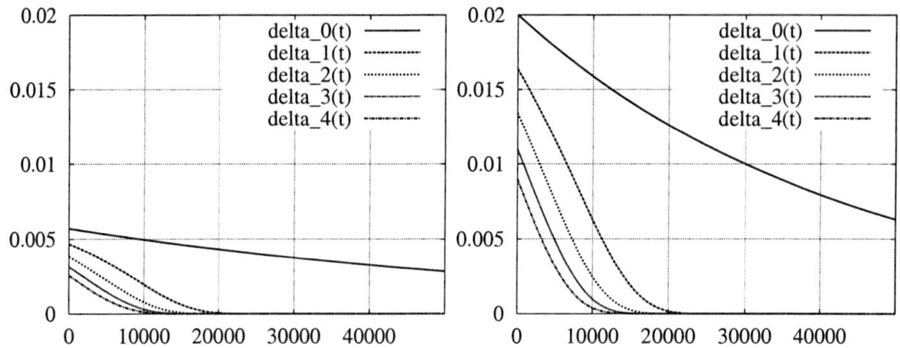

Fig. 2. Neighborhood attraction factor δ_i decreasing with time t and rank i. **Left:** for the neural gas parameter settings ($\sigma_i = 5$, $\sigma_f = 0.05$, $\epsilon_i = 0.0075$, $\epsilon_f = 0.00375$), **right:** ($\sigma_i = 5$, $\sigma_f = 0.05$, $\epsilon_i = 0.02$, $\epsilon_f = 0.01$)

Table 1. Test functions

app.	name	reference	dim.	opt.
f_1	Sphere model	[11]	var.	0
f_2	Generalized Rosenbrock's function	[11]	var.	0
f_3	Step function	[11]	var.	0
f_6	Schwefel's double sum (function 1.2)	[14, 13]	var.	0
f_9	Ackley's function	[15]	var.	0
f_{15}	Weighted sphere model	[16]	var.	0
f_{16}	Fletcher and Powell const.	[17]	5	0
f_{18}	Shekel-5	[18]	4	-10.1532
f_{19}	Shekel-7	[18]	4	-10.4029
f_{20}	Shekel-10	[18]	4	-10.5364
f_{21}	Griewangk-n2	[18]	var.	0
f_{22}	Griewangk-n10	[18]	var.	0
f_{23}	Galar	[19]	var.	2.00686
f_{24}	Kowalik	[20, 13]	4	0.0003075
q_1	Hyperellipsoid, parallel to axes	[21]	var.	0
q_2	Hyperellipsoid, randomly oriented	[21]	var.	0
sdp	Sum of different powers	[22]	var.	0

in e. g. [11–13] and in former investigations on the SOM-EN (see Table 1). The dimensions were set to 20, as long as there were no other ranges predefined. In those cases the highest acceptable dimension was chosen.

5 Test Series

For each function of the test bed, two variants of the NG-EN were compared to the SOM-EN[1]. The parameter settings are given in Table 2.

[1] Comparisons of the SOM-EN to conventional ES concerning convergence velocity and robustness can be found in [4] and [23].

Table 2. Parameter settings

strategy	μ	n_O	$t_{max} = \gamma_{max}$	δ	σ_i	σ_f	δ_{min}	δ_{max}	ϵ_i	ϵ_f
SOM-EN	10	4	30000	0.01	-	-	-	-	-	-
NG-EN	10	4	30000	-	5	0.05	1e-5	0.02	0.02	0.01
NG-EN	10	4	30000	-	5	0.05	1e-5	0.02	0.0075	0.00375

These settings are based on previous parameter studies made for NG-EN [24] and on experience in using SOM-EN [8].

The results were averaged out over 20 runs.

Table 3 summarizes the results of the test series. For each optimization task (column *app*), the table shows how the three tested strategies (grid-neighborhood, neural-gas neighborhood with $\epsilon_i = 0.02$, $\epsilon_f = 0.01$ and neural-gas neighborhood with $\epsilon_i = 0.0075$, $\epsilon_f = 0.00375$) perform according to the number of function evaluations needed to reach the optimum (*# evals*) and according to how often the method converged, i. e. how often the optimum was found (*# conv.*). The best results are indicated by (•).

Due to the preliminary limited number of generations, none of the strategies was able to find the optimum (defined as $f_{\text{opt}}(\boldsymbol{x}) = $ 1e-10) of the applications f_2, q_1, q_2 and *sdp* in time (although for e.g. *sdp* a value $f(\boldsymbol{x}) = $ 1e-8) could be reached). Therefore, no comparisons are shown for those test functions.

Concerning the convergence velocity, it can be seen that the NG-EN with $\epsilon_i = 0.02$ is almost always the fastest, the only exception is function f_3. For most test functions, it is considerably faster than SOM-EN, outperforming SOM-EN by a factor ≈ 2 (f_1, f_{21}, f_{22}, f_{23}), factor ≈ 3 (f_9, f_{16}) or even factor ≈ 5 (f_{19}). The parameter setting $\epsilon_i = 0.0075$, however, results in a noticeably slower convergence. It is almost slower than NG-EN with $\epsilon_i = 0.02$ (again, f_3 is the only exception), and often comparable to or a bit slower than SOM-EN. For the test functions f_9, f_{16}, f_{19} and f_{24}, it is by a factor $\approx 1.5 - 3$ faster than SOM-EN.

Having a look at the robustness of the three strategies, it can be seen that for some test functions all strategies are rather good in finding the optimum. For f_1, f_6 and f_{15} the optimum was always found; for f_{16}, f_{19} and f_{21} all three strategies showed a similar good reliability. But it can also be seen, that in seven out of ten cases, the SOM-EN was more often able to find the optimum. Especially for function f_9 and f_{22} the differing reliability is obvious, while for f_{16} and f_{24} the NG-EN shows better results.

Since the functions where NG-EN shows less reliability are multi-modal ones, these observations have led to the assumption, that the NG-EN gets more often stuck in local optima. Due to the predefined grid-neighborhood in SOM-EN, two neighboring individuals in different local optima can still attract each other and pull each other out of the local optimum. In the NG-EN approach, however, the neighborhood is defined depending on the distance of the individuals. Therefore, it is more likely that only those individuals which are stuck in the same local optimum are considered neighbors. Thus, an attraction that leads out of the local optimum is not likely.

Table 3. Comparing NG neighborhood and SOM neighborhood

app.	**neighb.**	ϵ_i	**# evals**	**# conv.**
f1	grid	-	72010	• 20
f1	n-gas	0.02	• 30010	• 20
f1	n-gas	0.0075	72010	• 20
f3	grid	-	• 503687	• 17
f3	n-gas	0.02	765943	15
f3	n-gas	0.0075	565642	16
f6	grid	-	168010	• 20
f6	n-gas	0.02	• 157210	• 20
f6	n-gas	0.0075	195610	• 20
f9	grid	-	248010	• 15
f9	n-gas	0.02	• 72010	6
f9	n-gas	0.0075	168010	9
f15	grid	-	88810	• 20
f15	n-gas	0.02	• 72816	• 20
f15	n-gas	0.0075	108010	• 20
f16	grid	-	170833	17
f16	n-gas	0.02	• 64210	• 20
f16	n-gas	0.0075	132010	18
f18	grid	-	18676	• 18
f18	n-gas	0.02	• 12010	15
f18	n-gas	0.0075	20010	14
f19	grid	-	54325	• 19
f19	n-gas	0.02	• 11510	16
f19	n-gas	0.0075	18510	16
f20	grid	-	15610	• 20
f20	n-gas	0.02	• 10410	15
f20	n-gas	0.0075	17904	19
f21	grid	-	72010	• 19
f21	n-gas	0.02	• 40010	16
f21	n-gas	0.0075	96010	• 19
f22	grid	-	83086	• 13
f22	n-gas	0.02	• 44454	9
f22	n-gas	0.0075	102010	4
f23	grid	-	23010	• 20
f23	n-gas	0.02	• 12010	13
f23	n-gas	0.0075	29010	18
f24	grid	-	752010	12
f24	n-gas	0.02	• 556010	11
f24	n-gas	0.0075	556245	• 17

6 Conclusions

An evolution strategy with neighborhood attraction (EN) incorporates the neighborhood and learning rule known from neural maps into evolution strategies. Starting from the SOM inspired EN which defines a grid neighborhood between the individuals, an EN with a new neighborhood attraction scheme (NG-EN) was introduced in this paper. The NG-EN dissolves the fixed SOM neighborhood structure and uses a neural gas approach which dynamically adapts the neighborhood structure during learning. The NG-EN approach was compared to the SOM-EN on a basis of exhaustive test series. Although the SOM-EN is still a bit more reliable, especially in solving highly multi-modal optimization tasks with deep and steep local optima, the NG-EN was able to find the optimum of all optimization tasks. It could be shown that for almost all test functions the NG-EN is apparently faster than SOM-EN in finding the optimum.

References

1. Pal, S.K., Mitra, S.: Neuro fuzzy pattern recognition : methods in soft computing. Wiley, New York (1999)
2. Kohonen, T.: Self-Organizing Maps. Springer Verlag, Berlin (1995)
3. Martinetz, T., Schulten, K.: Topology representing networks. Neural Networks **7** (1994)
4. Huhse, J., Zell, A.: Evolution strategy with neighborhood attraction. In Bothe, H., Rojas, R., eds.: Proceedings of the Second ICSC Symposium on Neural Computation – NC 2000, ICSC Academic Press, Canada/Switzerland (2000) 363–369
5. Martinetz, T., Schulten, K.: A "Neural-Gas" network learns topologies. In Kohonen, T., Mäkisara, K., Simula, O., Kangas, J., eds.: Proc. International Conference on Artificial Neural Networks (Espoo, Finland). Volume I., Amsterdam, Netherlands, North-Holland (1991) 397–402
6. Martinetz, T.M., Berkovich, S.G., Schulten, K.J.: 'Neural-gas' network for vector quantization and its application to time-series prediction. IEEE Trans. on Neural Networks **4** (1993) 558–569
7. Villmann, T.: Evolutionary algorithms and neural networks in hybrid systems. In Verleysen, M., ed.: Proceedings of 9th European Symposium on Artificial Neural Networks – ESANN'2001, Evere, Belgium, D-facto (2001)
8. Huhse, J., Zell, A.: Investigating the influence of the neighborhood attraction factor to evolution strategies with neighborhood attraction. In Verleysen, M., ed.: Proceedings of 9th European Symposium on Artificial Neural Networks – ESANN'2001, Evere, Belgium, D-facto (2001) 179–184
9. Schwefel, H.P.: Numerische Optimierung von Computer-Modellen mittels der Evolutionsstrategie. Birkhäuser, Basel, Stuttgart (1977) Volume 26 of Interdisciplinary Systems Research, German.
10. Altenberg, L.: The evolution of evolvability in genetic programming. In Kinnear, K.E., ed.: Advances in Genetic Programming. Complex Adaptive Systems, Cambridge, MIT Press (1994) 47–74
11. deJong, K.: An analysis of the behaviour of a class of genetic adaptive systems. Master's thesis, University of Michigan (1975)

12. Bäck, T.: A user's guide to genesys 1.0. Technical report, University of Dortmund, Department of Computer Science
13. Schwefel, H.P.: Evolution and Optimum Seeking. John Wiley and Sons, New York (1995)
14. Schwefel, H.P.: Numerical Optimization of Computer Models. Wiley, Chichester (1981)
15. Ackley, D.H.: A connectionist machine for genetic hillclimbing. Kluwer Academic Publishers, Boston (1987)
16. Schwefel, H.P.: Evolutionary learning optimum–seeking on parallel computer architectures. In Sydow, A., Tzafestas, S.G., Vichnevetsky, R., eds.: Proceedings of the International Symposium on Systems Analysis and Simulation 1988, I: Theory and Foundations, Akademie der Wissenschaften der DDR, Akademie-Verlag, Berlin (1988) 217–225
17. Fletcher, R., Powell, M.J.D.: A rapidly convergent descent method for minimization. Comp. J. 6 (1963) 163–168
18. Törn, A., Žilinskas, A.: Global Optimization. Volume 350 of Lecture Notes in Computer Science. Springer, Berlin (1989)
19. Galar, R.: Simulation of local evolutionary dynamics of small populations. Biological Cybernetics **65** (1991) 37–45
20. Schwefel, H.P.: Numerische Optimierung von Computer-Modellen mittels der Evolutionsstrategie. Volume 26 of Interdisciplinary systems research. Birkhäuser, Basel (1977)
21. Hansen, N., Ostermeier, A., Gawelczyk, A.: Über die Adaptation von allgemeinen, Koordinatensystem-unabhängigen, normalverteilten Mutationen in der Evolutionsstrategie: Die Erzeugendensystemadaption. Technical report, Technische Universität Berlin (1995)
22. Ostermeier, A., Gawelcyk, A., Hansen, N.: A derandomized approach to self-adaptation of evolution strategies. Evolutionary Computation **2** (1995) 369–380
23. Huhse, J., Zell, A.: Evolution strategy with neighborhood attraction - A robust evolution strategy. In Spector, L., Goodman, E.D., Wu, A., Langdon, W.B., Voigt, H.M., Gen, M., Sen, S., Dorigo, M., Pezeshk, S., Garzon, M.H., Burke, E., eds.: Proceedings of the Genetic and Evolutionary Computation Conference (GECCO-2001), San Francisco, California, USA, Morgan Kaufmann (2001) 1026–1033
24. Huhse, J., Villmann, T., Zell, A.: Investigation of the neighborhood attraction evolutionary algorithm based on neural gas. In: Proceedings of the Sixth International Conference on Neural Networks and Soft Computing (ICNNSC) (to appear). Lecture Notes in Computer Science, Springer (2002)

A New Asynchronous Parallel Evolutionary Algorithm for Function Optimization*

Pu Liu[1], Francis Lau[2], Michael J. Lewis[1], and Cho-li Wang[2]

[1] Department of Computer Science, Binghamton University – SUNY,
Binghamton, N.Y., 13902, USA
{pliu1,mlewis}@binghamton.edu
[2] Department of Computer Science and Information Systems,
The University of Hong Kong, Pokfulam Road, Hong Kong
{fcmlau,clwang}@csis.hku.hk

Abstract. This paper introduces a new asynchronous parallel evolutionary algorithm (APEA) based on the island model for solving function optimization problems. Our fully distributed APEA overlaps the communication and computation efficiently and is inherently fault-tolerant in a large-scale distributed computing environment. For the scalable BUMP problem, our APEA algorithm achieves the best solution for the 50-dimension problem, and is the first algorithm of which we are aware that can solve the 1,000,000-dimension problem. For other benchmark problems, our APEA finds the best solution to G7 in fewer time steps than [16,17], and finds a better solution to G10 than [17].

1 Introduction

Evolutionary Algorithms (EAs) imitate nature's evolutionary process to solve complex problems. Their population-based searching mechanism makes them eminently suitable to be run in parallel on a massive scale. Traditionally, parallel EAs follow one of three models: the *master-slave* (or farming) model, the *island model*, or the *cellular model*. The island model is most popular [4] since it is easy to implement on a local area network with standard workstations. In the island model, the population is divided into subpopulations called *demes*, and each evolves separately on different processors with local memory. With some migration frequency, the demes exchange individuals to prevent premature convergence of subpopulations. This exchange requires communication in implementations of the model.

When implementing the island model, most algorithms adopt a synchronous approach [4]. That is, demes synchronize with one another when migration occurs and individuals are exchanged. The intervals of time between migrations are called *epochs* [4], and all EA processes stop at the end of each epoch to wait for all the others. This barrier synchronization after each epoch results in considerable overhead, and allows

* This work was partially supported by National Science Foundation (NFS) Instrumentation Grant EIA9911099.

J.J. Merelo Guervós et al. (Eds.): PPSN VII, LNCS 2439, pp. 401–410, 2002.

the slowest processor to determine the speed of the algorithm. When loads are not balanced on all processors, especially when the number of processors is large, synchronous algorithms often perform poorly.

Evolution in nature is an asynchronous process. Individuals migrate independently from one population to another without central control or full coordination between populations. This is also the most efficient way to overlap the computation and communication in a parallel algorithm. But most parallel EAs are synchronous [1-7], presumably because of the seemingly complicated exchange of individuals. The few asynchronous algorithms require a central server to manage exchange of individuals [8], which also limits scalability and introduces additional software overhead.

In this paper, we introduce a new *asynchronous parallel evolutionary algorithm* (*APEA*), based on the island model, in which all communication operations are non-blocking. Each processor executes without waiting for network communication between epochs. In contrast to the synchronous algorithms, which scale poorly, our APEA achieves improved performance with additional processors. In addition, APEA is inherently fault-tolerant to failures in underlying computing environments. This is especially important for time-consuming computational problems.

We have implemented our APEA in both PVM [18] and MPI [19]. Using the PVM version of APEA, we solved a scalable optimization BUMP problem [10-16]. The extreme high dimensional BUMP problem allows us to test both the scalability and efficiency of APEA. In the literature, the largest BUMP problem solved is the 50 dimension problem [10]. Our APEA achieves a better solution for the 50 dimension BUMP, and finds a solution for the 1,000,000 dimension BUMP using a 256-node YH-4 MPP supercomputer. With the MPI version of APEA, we tested some widely used benchmark problems, namely G7 [16,17], and G10 [16,17]. Our APEA achieves the best solution for each, and for G10, our APEA finds a solution that is better than the best solution claimed in the literature [17].

The rest of this paper is organized as follows. Section 2 introduces the function optimization problem briefly, then describes our asynchronous parallel evolutionary algorithm and discusses its properties and our approaches to improve convergence characteristics. In the following sections, results of our solution for several benchmark problems are discussed. Concluding remarks are provided in the final section.

2 Asynchronous Parallel Evolutionary Algorithm

2.1 Background

Many real world problems involve complex, non-linear, multi-constraint, mixed format (e.g. integers and floating point numbers) aspects that conventional algorithms cannot solve accurately or in reasonable time. Our goal is to solve numerical optimization problems, which are characterized by what are called the "3 S's", namely (*s*uper non-linear, *s*uper multi-peaked and *s*uper dimensional).

We solve the following general function optimization problem:

$$\max f(X) \qquad X = (x_1, x_2, \ldots, x_n)^T \in R^n$$

subject to

$$l_i \le x_i \le u_i, \quad l_i \in R, \quad u_i \in R, \quad i = 1,2,\ldots,n$$
$$g_i(X):\ 0, \quad i = 1,2,\ldots,q$$

2.2 The New Algorithm

Assuming a subpopulation with N individuals is assigned to each of NPROC processors, each processor executes the same process G, as follows, to steer the asynchronous parallel computation:

```
PROCESS G
1. t=0
2. Initialize P(t)={X_1, X_2, ..., X_n}
3. Evaluate P(t)
4. While (not terminated) do
5.     If  (any  message arrived) then
6.         X_new = Recv_Individual()
7.     else  X_new = Local_Generate()
8.     P'(t)  = P(t)∪{X_new}
9.     P(t)  = select N best individuals from P'(t);
10.    Locate X_best in P(t);
11.    If (t mod T = 0) and (X_best changed)  then
12.        Send X_best to Q other process(es)
13.    t = t+1
14. END WHILE
```

Line 5 checks the receiving buffer to see if a new message has arrived. This operation can be implemented by the "probe" communication primitive, which is provided by both PVM and MPI. Due to the non-blocking communication of Line 5, along with Line 12, the processes never wait for one another. That is, they run completely asynchronously, and the slowest process does not slow down the others. Thus, there is no need for a load-balancing mechanism. Furthermore, APEA is inherently fault-tolerant because it will continue to execute even when some nodes fail. This significant characteristic is especially important for large and time-consuming computational problems.

We also note that if the incoming individual in Line 6 is eliminated through selection in Line 9, then the immigration is useless. Because communication between processors is expensive, we should avoid this situation if possible. Line 12 ensures that only the individual with the best fitness migrates to the other demes. Likewise, Line 9 ensures that the worst individual is eliminated in each iteration. This allows the immigrated individual to have a high probability of surviving in its new deme.

The algorithm is SPMD; all processors run the same process G, and neither central control nor synchronous communication is required. This makes APEA scalable and suitable for solving complex problems via MPP supercomputers.

To be appropriate for parallel systems that are both loosely and tightly coupled, APEA contains two parameters to control the parallel granularity of APEA. Parameter Q indicates the number of processors with which one processor communicates when a new best individual is found within its deme. Parameter T determinates the frequency of individual migration. Together, Q and T determinate the total communication cost during each iteration of APEA. Thus, the parallel granularity and communication costs can be adjusted dynamically by altering the values of these parameters.

Unfortunately, the effects of these parameters on the quality of solutions are not well understood, especially in the asynchronous mode. Cantu-Paz has presented models that predict the effects of the parameters on the population size of demes and on migration rate [4]. However, this investigation considered that migration occurs only after each population converges, which assumes a synchronous algorithm. We discuss the setting of parameters based on the experimental results in Section 3.

Evolutionary programming and genetic algorithms are also applicable to produce a new offspring X'. For efficiency, our Local_Generate () method uses a multi-parent crossover [9]. The offspring X' is generated by parents X_i', i = 1, 2, … , m as follows.

$$X' = \sum_{i=1}^{m} a_i X_i'$$

where the coefficients a_i are chosen randomly and subject to

$$\sum_{i=1}^{m} a_i = 1, \quad -0.5 \le a_i \le 1.5, \quad i = 1,2,\ldots,m$$

To solve constrained optimization problems, Michalewicz presented a comparison study of existing evolutionary algorithms [10]. According to this study, the existing algorithms are grouped as follows: (1) methods based on preserving feasibility of solutions, (2) methods based on penalty functions, (3) methods based on the superiority of feasible solutions, and (4) other hybrid methods. We choose the third method for its simplicity of implementation.

Frequently used termination criteria for EAs include running the algorithm until: (1) a satisfactory solution is found; (2) a fixed time limit is reached; or (3) all the individuals in the population are the same and no further improvement is possible. Our APEA uses the latter criterion. When individuals in a deme are all the same, its process broadcasts a 'termination' message and APEA terminates.

2.3 Related Work

There are many algorithms that explore alternative migration schemes and communication models to try to make parallel genetic algorithms (Ga's) more efficient. Muhlenbein [20] proposed a totally asynchronous cellular PGA, which runs on MIMD parallel computers. In this algorithm, individuals are distributed in a 2-D world. Each individual selects a partner for mating in its neighborhood. The active and intelligent individuals decide when to migrate and control the inter-processor communication. Hart [21] and Alba [22] applied this idea to a coarse-grain PGA whose subpopulations are distributed across *p* processors in a two-dimensional grid. Both report that the asynchronous algorithms outperformed their equivalent synchronous counterparts in real time. In this kind of "active" asynchronous PGA, when migration occurs, at least two messages, a request and a response, are needed. In most cases, communication-fault-handling is also needed to accomplish the migration of one individual. Our APEA uses a larger population of passive individuals. The migration of one single individual only uses one message to send the individual. The extra overhead is the probe, which doesn't incur much overhead.

Grefenstette [23] proposed three variations of the master-slave PGA with the one population maintained by the master, and use synchronous communication. Based on master-slave method, Martin [24] introduced a centralized coarse-grain PGA in which the migration is fully asynchronous. One drawback of this algorithm is its scalability, which is limited by its centralization, which we have eliminated in our APEA.

3 Numeric Experiments

The APEA is implemented by PVM and MPI under different hardware environments. The BUMP problem was solved via a MPP supercomputer YH-4 by PVM, and the other problems were solved on a 16-node PC cluster connected by a 100Mb/s Fast Ethernet switch. Each cluster node consists of a 300MHz Pentium II processor with 128MB of memory, running Linux 2.2.14 with MPICH 1.2.1.

When implemented by MPI, we used functions MPI_Iprobe(), MPI_Recv() and MPI_Isend(). The blocking MPI_Recv() is used only after MPI_Iprobe() returns "true". Both MPI_Iprobe() and MPI_Isend() are non-blocking. For the PVM version, we used similar functions pvm_probe(), pvm_recv() and pvm_mcast().

3.1 The BUMP Problem

The BUMP problem is described in [11] as follows:

$$f_n(X) \equiv \frac{\left|\sum_{i=1}^{n} \cos^4(x_i) - 2\prod_{i=1}^{n} \cos^2(x_i)\right|}{\sqrt{\sum_{i=1}^{n} i x_i^2}} \tag{1}$$

where $$0 < x_i < 10, \quad i = 1, 2, 3, \ldots, n \tag{2}$$

subject to $$\prod_{i=1}^{n} x_i > 0.75 \tag{3}$$

and $$\sum_{i=1}^{n} x_i < 7.5n \tag{4}$$

In 1994, Keane proposed the BUMP problem [11] in optimum structural design, and subsequently published several papers [12, 13, 14] discussing the problem. Because of its scalability, the BUMP problem has been used as a general benchmark problem to test optimization algorithms [16].

There are several potential difficulties in solving this problem. First, the objective function $f_n(X)$ is a nonlinear multimodal function of high dimension. Second, the global optimum is defined by the nonlinear constraint (3) above. When the dimension of the BUMP problem increases, the product in (3) is difficult to calculate (because of overflow). In fact, when *n* is greater than 20, a majority of constraint-handling methods have difficulty returning a high quality solution [9, 15, 16].

El-Beltagy used metamodeling techniques for the BUMP problem and got some numerical results when n=2 and n=5 [15]. By incorporating some problem-specific knowledge, Michalewicz introduced an evolutionary operation called geometrical crossover and reported in [16] that when n=20 the best solution achieved was 0.803553 and when n=50, the best solution was 0.8331937. For n=20, the result Tao Guo reported in [9] is 0.803619. But when n is greater than 50, there is no published result of which we are aware. The expensive computation stunts sequential algorithms. Because evolutionary algorithms are inherently parallel, if we want to solve high dimensional BUMP problem, an obvious promising way is to parallelize the sequential algorithm.

In addition, for a solution vector $X(x_1, x_2,..., x_n)$, the value of the numerator does not change no matter the order of $x_1, x_2,..., x_n$. However, the value of the denominator may change only when the order of $x_1, x_2,..., x_n$ changes. In fact, the value of $f_n(X)$ achieved by a solution vector $X(x_1, x_2,..., x_n)$ is no better than the one achieved by the vector $X^*(x^*_1, x^*_2,..., x^*_n)$ where $x^*_1, x^*_2,..., x^*_n$ is the non-increasing sequence of $x_1, x_2,..., x_n$. So we add a mutation operation sort() to function Local_Generate().

Recalling the nonlinear constraint (3), the sorting process is necessary when *n* is fairly great, i.e. 400, because a product of 400 random numbers ranging from 0 to 10 is liable to overflow or underflow during calculation, even though the final product does not overflow or underflow. For example, if the first 100 random numbers are greater than 2, then the temporary product will be greater than 2^{100} or less than 2^{-100} both of which are far beyond the capability of most architectures. Fortunately, for the descending sequence vector $X^*(x^*_1, x^*_2,..., x^*_n)$, we can use an algorithm to calculate the product and avoid the above situation in many cases. The algorithm multiplies a running product by a small value then the product is in danger of overflowing, and multiplies in a large value when it is in danger of underflowing.

According to the research of Michalewicz [16], the constraint (4) has no effect on the optimal results, so we also ignore constraint (4) in our experiments.

To verify the efficiency and effectiveness of the APEA, we have completed numerical experiments for *n* = 50, 100, 200, 300, 400, 500, 1500.

Table 1. Results for n = 50 to 1500

Dimension	50	100	200	300
Best Result	0.8352620	0.8448539	0.8468442	0.8486441
Dimension	400	500	1500	
Best Result	0.8511074	0.8504975	0.8449622	

From Table 1, when *n* equals 50, we achieved the best results of which we are aware. In order to compare with other algorithms, we display the whole solution here. F(X) is 0.83526201238794 with $\prod_{i=1}^{50} x_i = 0.75000291468818$, where

X = {6.28324314967593, 3.16959135278874, 3.15587932289134, 3.14221834166447, 3.12875141132280, 3.11533720178037, 3.10135016319809, 3.08867068084463, 3.07467526362337, 3.06187302198439, 3.04886636373126, 3.03519777233540, 3.02235650101037, 3.00791087929595, 2.99431484723465, 2.98078207510642, 2.96617535809182, 2.95213454834831, 2.93826263614623, 2.92347051874431, 0.48770508444941, 0.48608797212231, 0.48388281752034, 0.48168154669018, 0.47901882085524, 0.47712972009898, 0.47501098097910, 0.47259800785213, 0.47128211104035, 0.46871206605219, 0.46726830592743, 0.46606878481357, 0.46376534354877, 0.46204633230610, 0.45965521445116, 0.45884822857585, 0.45717243025277, 0.45481532957423, 0.45342381481910, 0.45166163748202, 0.45021480667760, 0.44845064437796, 0.44673980402160, 0.44526519733653, 0.44338085480658, 0.44213030342910, 0.44043946842174, 0.43904935652294, 0.43874135178293, 0.43634175968446}

We set out to determine, via a state-of-the-art MPP supercomputer, the highest dimension BUMP problem we could solve by APEA. We had to speed up the convergence rate without compromising the quality of solution. Parameters T and Q affect both aspects, but unfortunately, it is difficult to predict their exact effects. We performed two groups of experiments to find effective values. The first group observes Q with fixed n and T, while the second group observes T with fixed n and Q. Table 2 contains parts of our experimental results.

Table 2. Results of APEA with NPROC=128, n=1000

T = 200			Q = 2		
Q	Iterations	Results	T	Iterations	Results
2	14653	0.844219	100	11830	0.8410597
2	14115	0.844219	100	8867	0.8410594
4	5187	0.839243	200	14653	0.8442193
4	4998	0.838898	200	14115	0.8442193
6	3130	0.833761	300	12095	0.8448454
6	3996	0.842186	300	12389	0.8448539
8	30169	0.841935	400	11329	0.8410598
8	28877	0.842598	400	11720	0.8421879

We observe that as communication increases, the algorithm converges and the solution degrades. Based on Table 2, we choose Q=10 and T=300 and get the following numerical results in Table 3:

Table 3. Results of APEA with NPROC=256, Q=10, T=300

	N	Terminated t	Time (s)	Solutions
1	10000	20072	313	0.8455810
2	10000	16385	260	0.8456407
3	20000	24236	382	0.8452293
4	20000	22169	344	0.8455882

When n = 100,000, NPROC=128, APEA converged with t = 46252 and solution 0.8448940. The total execution time was 7106 seconds. When n=1,000,000, NPROC=128, we terminated APEA after 16 hours calculation with t=6027 and result 0.8445861. For n=10,000,000, system resources were exhausted.

3.2 Benchmark Problems Solved by APEA Using MPI

1. G7 [16][17]: Minimize

$$f(x) = x_1^2 + x_2^2 + x_1x_2 - 14x_1 - 16x_2 + (x_3 - 10)^2$$
$$+ 4(x_4 - 5)^2 + (x_5 - 3)^2 + 2(x_6 - 1) + 5x_7^2$$
$$+ 7(x_8 - 11)^2 + 2(x_9 - 10)^2 + (x_{10} - 7)^2 + 45$$

subject to:

$$105-4x_1-5x_2+3x_7-9x_8\geq 0$$
$$-3(x_1-2)^2-4(x_2-3)^2-2x_3^2+7x_4+120\geq 0$$
$$-10x_1+8x_2+17x_7-2x_8\geq 0$$
$$-x_1^2-2(x_2-2)+2x_1x_2-14x_5+6x_6\geq 0$$
$$8x_1-2x_2-5x_9+2x_{10}+12\geq 0$$
$$-5x_1^2-8x_2-(x_3-6)^2+2x_4+40\geq 0$$
$$3x_1-6x_2-12(x_9-8)^2+7x_{10}\geq 0$$
$$-0.5(x_1-8)^2-2(x_2-4)^2-3x_5^2+x_6+30\geq 0$$
$$-10\leq x_i\leq 10, i=1,2,...,10$$

The global minimum is known to be f(X*) = 24.3062091 [16][17], where

X* = (2.171996, 2.363683, 8.773926, 5.095984, 0.9906548,
1.430574, 1.321644, 9.828726, 8.280092, 8.375927)

The APEA always achieves the optimal solution with fewer iterations. Table 3 provides the best results under different parameters. The Algorithm CEALM in [17] and other algorithms listed in [17] do not reach the global optimal every time. APEA achieves the optimal value for each of the ten separate runs.

Table 4. The results of G7 with NPROC = 16

NPROC	T	Q	Median time(s)	Result
1	0	0	39	24.3062090690560
2	1	2	18	24.3062090685401
4	30	4	21	24.3062090683975
8	30	8	28	24.3062090683429
16	1	16	43	24.3062090683032

Table 3 shows that increasing communication improves the quality of the solution on average but requires more time to converge.

2. G10 [16][17]: Minimize $f(x)=x_1+x_2+x_3$
subject to

$$1-0.0025(x_4+x_6)\geq 0$$
$$1-0.0025(x_5+x_7-x_4)\geq 0$$
$$1-0.01(x_8-x_5)\geq 0$$
$$x_1x_6-833.33252x_4-100x_1+83333.333\geq 0$$
$$x_2x_7-1250x_5-x_2x_4+1250x_4\geq 0$$
$$x_3x_8-1250000-x_3x_5+2500x_5\geq 0$$
$$100\leq x_1\leq 10000$$
$$1000\leq x_2\leq 10000$$
$$1000\leq x_3\leq 10000$$
$$10\leq x_i\leq 1000, i=4,5,...,8$$

The global minimum is said to be f(X*)=7049.330923 [17], where

X* = (579.3167, 1359.943, 5110.071, 182.0174, 295.5985, 217.9799, 286.4162, 395.5979)

The algorithm CEALM in [17] and other algorithms listed in [16, 17] do not find this value. Our APEA finds a better solution to be f(X) = 7049.24802055468081 where

X = (579.30828537221191, 1359.96826293047116, 5109.97147225199751,
182.01783328980514, 295.60114110998057, 217.98216670925277,
286.41669217921429, 395.60114110996898)

The parameters of APEA are set as follows: NPROC= 4, T=10, Q=2. Each time, APEA converges at the solution above in no more than 26 seconds. In contrast, the sequential simulation of APEA converges at a much worse solution quickly.

4 Conclusion

In this paper, we describe an effective asynchronous parallel evolutionary algorithm and get a series of best solutions for the BUMP problem and other benchmark problems. Still, there are many other issues that should need further research, such as the effect of different Q and T on numerical result, and the topology of neighborhoods.

References

1. H. T. Yang, "A Parallel Genetic Algorithm Approach to Solving the Unit Commitment Problem: Implementation on the Transputer Networks," *IEEE Transactions on Power Systems*, Vol. 12, No.2, pp. 661-668, 1997.
2. A. Wu, K. Y. Wu, R. M. M. Chen, Y. Shen, "Parallel Optimal Statistical Design Method with Response surface modeling using genetic algorithms," *Circuits, Devices and Systems, IEE Proceedings-*, Vol. 145, No.1, 1998.
3. J. D. Lohn, "A Circuit Representation Technique for Automated Circuit Design," *IEEE Transactions on Evolutionary Computation*, Vol. 3, No. 3, pp. 205-219, 1999.
4. E. Cantu-Paz, "Markov Chain Models of Parallel Genetic Algorithms," *IEEE Transactions on Evolutionary Computation*, Vol. 4, No. 3, pp. 216-226, 2000.
5. P. B. Grosso, "Computer Simulations of Genetic Adaptation: Parallel Subcomponent Interaction in a Multi-locus Model," *Ph.D. Dissertation*, University of Michigan, 1985.
6. L. A. Anbarasu et al., "Multiple Sequence Alignment by Parallely Evolvable Genetic Algorithms," in A. S. Wh(ed.) *Proceedings of the 1999 Genetic and Evolutionary Computation Conference Workshop Program*, Orlando, Fla, July 13, 1999, pp. 154-156.
7. H. Lienig, "A Parallel Genetic Algorithm for Performance-Driven VLSI Routing," *IEEE Transactions on Evolutionary Computation*, Vol. 1, No.1, pp. 29-39, 1997.
8. K. Kojima, W. Kawamata, et al, "Network Based Parallel Genetic Algorithm using Client-Server Model", in *Proceedings of the Conference on Evolutionary Computation 2000*, pp. 244-250.
9. T. Guo, L. S. Kang, "A New Evolutionary Algorithm for Function Optimization," *Wuhan University Journal of Natural Sciences*, Vol. 4, No. 4, pp. 409-414, 1999.
10. Z. Michalewicz and M. Schoenauer, "Evolutionary Algorithms for Constrained Parameter Optimization Problems," *Evolutionary Computation*, Vol. 4, No. 1, pp. 1-32, 1996.
11. A. J. Keane, "Experiences with Optimizers in Structural Design," *in Proceedings of the Conference on Adaptive Computing in Engineering Design and Control 94*, ed. (I. C. Parmee, Plymouth, 1994), pp. 14-27.

12. A. J. Keane, "A Brief Comparison of Some Evolutionary Optimization Methods," *Proceedings of Applied Decision Technologies (Modern Heuristic Methods),* 1995.
13. A. J. Keane, "Genetic Algorithm Optimization of Multi-peak Problems: Studies in Convergence and Robustness." *Artificial Intelligence in Engineering*, 1995.
14. A. J. Keane, "Passive Vibration Control Via Unusual Geometric: Application of Genetic Algorithm Optimization to Structural Design," *Journal of Sound and Vibration*, 1995.
15. M. A. El-Beltagy, et al. "Metamodeling Techniques For Evolutionary Optimization of Computationally Expensive Problems: Promises and Limitations." *Genetic Algorithms and Classifier Systems*, 1999.
16. Z. Michalewicz, S. Esguvel et al., "The Spirit of Evolutionary Algorithms." *Journal of Computing and Information Technology*, Vol. 7, pp. 1-18, 1999.
17. M. J. Tahk, B. C. Sun, "Coevolutionary Augmented Lagrangian Methods for Constrained Optimization." *IEEE Transactions on Evolutionary Computation*, Vol. 4, No. 2, 2000.
18. Sunderam, V. S., "PVM: A Framework for Parallel Distributed Computing." *Concurrency: Practice and Experience*, Vol. 2, No. 4, pp. 315-339, December, 1990.
19. "MPI: A Message-Passing Interface Standard." Message Passing Interface Forum, 1994.
20. Heinz Muhlenbein, "Evolution in Time and Space - The Parallel Genetic Algorithm", In Gregory J.E. Rawlins, Editor, *Foudation of Genetic Algorithms 1*, Page 316-337, San Mateo, CA, USA, 1991, Morgan, Kaufmann.
21. William E. Hart, Scott Baden, Richard K. Belew, Scott Kohn , **"**Analysis of the Numerical Effects of Parallelism on a Parallel Genetic Algorithm**"**, *Proc 10th Intl. Parallel Processing Symp.* pp 606-612, 1996.
22. Enrique Alba, Jos M Troya Dpto. de Lenguajes y Ciencias de la Computation, "An Analysis of Synchronous and Asynchronous Parallel Distributed Genetic Algorithms with Structured and Panmictic Islands", *Future Generation Computer Systems*, 17(4):451-465, January 2001.
23. Grefenstette J. J., "Parallel adaptive algorithms for function optimization", Tech. Rep. No. CS-81-19, Vanderbilt University, Computer Science Department, Nashville, TN, 1981.
24. Martin F. J., Trelles-Salazar O., Snadoval F., "Genetic algorithms on LAN-message passing architectures using PVM: Application to the routing problem", In Davidor Y. Schwefel H. –P., Manner R., Eds., *Parallel Problem Solving from Nature, PPSN III*, p.534-543, Springer-Verlag (Berlin), 1994.

Fighting Bloat with Nonparametric Parsimony Pressure

Sean Luke and Liviu Panait

Department of Computer Science, George Mason University
Fairfax, VA 22030 USA
http://www.cs.gmu.edu/~sean/
http://www.cs.gmu.edu/~lpanait/

Abstract. Many forms of parsimony pressure are parametric, that is final fitness is a parametric model of the actual size and raw fitness values. The problem with parametric techniques is that they are hard to tune to prevent size from dominating fitness late in the evolutionary run, or to compensate for problem-dependent nonlinearities in the raw fitness function. In this paper we briefly discuss existing bloat-control techniques, then introduce two new kinds of non-parametric parsimony pressure, Direct and Proportional Tournament. As their names suggest, these techniques are based on simple modifications of tournament selection to consider both size and fitness, but not together as a combined parametric equation. We compare the techniques against, and in combination with, the most popular genetic programming bloat-control technique, Koza-style depth limiting, and show that they are effective in limiting size while still maintaining good best-fitness-of-run results.

1 Introduction

One of the foremost challenges to scaling genetic programming (GP) is *bloat*, the tendency for genetic programming individuals to grow in size as evolution progresses, relatively independent of any justifying improvements in fitness. It is not uncommon for the average size of an individual to grow over three orders of magnitude in just fifty generations. This is a serious stumbling block to genetic programming, as it slows the evolutionary search process, consumes memory, and can hamper effective breeding. Bloat produces a sort of Zeno's paradox, slowing successive generations by so much that it places a cap on GP's useful runtime.

Bloat is not a problem unique to genetic programming. It occurs in a wide variety of arbitrary-length representations, including neural networks, finite state automata, and rule sets. Indeed, the earliest known report of bloating (and of approaches to deal with it) involves evolving Pitt-approach rule systems [1]. However, because GP is the most popular arbitrary-length representation technique, the lion's share of papers on the subject have been in the GP literature.

As discussed in [2], bloating is a hotly debated topic in GP theory; and there is also presently no silver bullet to deal with it. The genetic programming

J.J. Merelo Guervós et al. (Eds.): PPSN VII, LNCS 2439, pp. 411–421, 2002.

literature uses a large number of different techniques to counter bloat, each with its own advantages and disadvantages. By far the most popular such technique is restricting breeding to only produce children less than some maximal tree depth. A distant second is *parsimony pressure*, where the size of an individual is a factor in its probability of being selected.

In this paper we examine two new approaches to parsimony pressure, and report on their success in managing population size while retaining reasonable best-fitness-of-run results as compared to Koza-style tree depth limitation. These techniques are based on modifications of the well-studied *tournament selection* method, which picks N random individuals with replacement, then selects the fittest individual from that pool. We modify tournament selection to consider parsimony as well as fitness, but retain the nonparametric features which make tournament selection popular.

2 Previous Bloat Control Techniques

Modification and Restriction Techniques. The most common approach to bloat control, at least in the GP literature, is *maximal depth restriction* [3]. Here, when a child is created by removing a subtree from a parent and replacing it with another subtree (as is done in subtree crossover or subtree mutation), and the child exceeds a maximal depth limit, then the child is rejected and a copy of the original parent takes its place in the new generation. The standard maximal depth limit for tree-based GP is 17. It is also possible, but much less common, to place size restrictions on the child rather than depth restrictions. *Pseudo-hillclimbing* [4] is a recent restriction approach: when a child is generated, its fitness is immediately assessed. If its fitness is not superior to its parent's fitness, it is rejected and a copy of its parent takes its stead in the new generation. Note that this approach is very similar to depth restriction in mechanism, except that oddly it does not compare size at all, yet is reasonably successful at limiting tree growth. This gives some insight into why depth restriction and pseudo-hillclimbing are successful: they limit growth not only by capping size but by injecting large numbers of parents directly into later generations. As parents are generally smaller than their children (hence the bloat), this has a stunting effect, but at a cost in diversity.

As GP parse trees may generally be viewed as computer functions, one obvious way to counter bloat is to perform *code editing* to optimize those functions. One paper [5] reports strong results with this approach, but there is evidence that editing may lead to premature convergence [6]. Finally, a number of papers (for example [7]) have investigated allowing GP parse trees to adapt, on a component-by-component basis, the probability that a given component will be chosen for crossover. This bloat-control technique is known as *explicitly defined introns*.

Parsimony Pressure. Parsimony pressure is the popular bloat-control technique outside of GP, and is gaining popularity within GP as well. Most such parsimony pressure computes fitness as a linear function of an individual's raw fitness and its size (for example, [8]), though there are some nonlinear examples as well [9]. For a more complete survey of linear and other kinds of parametric parsimony pressure, see [10].

The trouble with parametric parsimony pressure is that it is *parametric*. We mean this in the statistical sense: it considers the actual *values* of size and fitness together in a parametric statistical model for selection: the experimenter must stipulate, in effect, that N units of size are worth M units of raw fitness. Stipulating this function is problematic when fitness is a nonlinear function of actual "worth", as is often the case: fitness functions are typically ad-hoc. It may well be that a difference between 0.9 and 0.91 in fitness is much more significant than a difference between 0.7 and 0.9. Parametric parsimony pressure can thus give size an unwanted advantage over fitness when the difference in fitness is only 0.01 as opposed to 0.2. This is also a problem because the relative significance of exact size and fitness parameters changes during the course of a run. For example, size-parameter dominance may arise late in evolution, when subtle differences in fitness become important. Notice that these issues are similar to those which gave rise to the preference of tournament selection and other nonparametric selection procedures over fitness-proportionate selection.

One approach to fixing this is to adapt the size parameter as the evolutionary run progresses [11], except that such techniques must usually rely on problem-specific analysis. Another recent approach, *pareto parsimony pressure*, eschews parametric techniques and instead treats size as a second objective in a pareto-optimization scheme. Pareto optimization is used when the evolutionary system must optimize for two or more objectives at once, and it is not clear which objective is "more important". An individual A is said to *pareto-dominate* another individual B if A is as good as B in all objectives, and better than B in at least one objective. One possible use of pareto dominance is to stipulate that an individual's fitness is the number of peers it dominates. Unfortunately, the technique has so far had mixed results in the literature. Some papers report smaller trees and the discovery of more ideal solutions [12, 13], but tellingly they omit best-fitness-of-run results. Another reports the mean best-fitness-of-run, but it is *worse* than when not using the technique [14].

3 Two New Parsimony Pressure Techniques

The two new techniques we propose here are modifications of the tournament selection operator. The techniques are *double tournament*, where individuals must pass two layers of tournaments (one by size, one by fitness) to be selected; and *proportional tournament*, where the tournament sometimes picks by size, and sometimes by fitness.

Double Tournament. The double tournament algorithm selects an individual using tournament selection: however the tournament contestants are not chosen at random with replacement from the population. Instead, they were each the winners of *another* tournament selection. For example, imagine if the "final" tournament has a pool size of 7: then seven "qualifier" tournaments are held as normal in tournament selection, and the winners go on to compete in the "final" tournament. Double tournament has been previously used to select for both fitness and diversity [15]. We suggest it may be used for parsimony pressure by having the "final" tournament select based on parsimony while the qualifying tournaments select based on fitness (or vice versa). The algorithm has three parameters: a fitness tournament size S_f, a parsimony tournament size S_p, and a switch (*do-fitness-first*) which indicates whether the qualifiers select on fitness and the final selects on size, or (if false) the other way around.

Our initial experiments revealed that even S_p values as small as 2 put too much pressure on parsimony, and the fitnesses of the resulting individuals were statistically significantly worse than with no parsimony pressure at all. In order to rectify this matter, we permit S_p to hold real values between 1.0 and 2.0. In this value range, two individuals participate to the tournament; with probability $S_p/2$ the smaller individual wins, else the larger individual wins. Ties are broken at random. Thus $S_p = 1$ is random selection, while $S_p = 2$ is the same as a plain parsimony-based tournament selection of size 2.

Proportional Tournament. This technique is even simpler. The proportional tournament algorithm selects an individual using tournament selection as usual, using some fixed tournament size S. However, a proportion of tournaments will select based on parsimony rather than on fitness. A fixed parameter R defines the proportion, where higher values of R imply more of an emphasis towards fitness: $R = 1$ implies that all tournaments will select based on fitness, while $R = 0.5$ implies that tournaments will select on fitness or size with equal probability.

4 Experiments

The bloat-control technique most used in the literature is Koza-style depth limiting, and we, like most of the literature, compare our technique against it. In future work, we will also compare against linear or pareto parsimony pressure. To this end, we performed two experiments. The first experiment compared depth limiting against Double and Proportional tournaments, while the second compared plain depth limiting against depth limiting in combination with the tournaments. We wish to emphasize that although these techniques are being used for GP in this paper, they are general techniques which are representation-independent.

The experiments used population sizes of 1000, with 50-generation runs. The runs did not stop when an ideal individual was found. Runs with plain depth limiting used plain tournament selection with a tournament size of 7. We chose four problem domains: Artificial Ant, 11-bit Boolean Multiplexer, Symbolic Regression, and Even-5 Parity. We followed the Koza-standard parameters specified in these four domains as set forth in [3], as well as its breeding, selection, and tree generation parameters. Artificial Ant used the Santa Fe food trail. Symbolic

Regression used no ephemeral random constants. To compare means for statistical significance, we used ANOVAs with a 95% confidence. The evolutionary computation system used was ECJ 7 [16].

Our results are graphed as follows. For the Double Tournament, we set $S_f = 7$ and let S_p range from 1.0 to 2.0 by increments of 0.1. We experimented with setting *do-fitness-first* to false (leftmost methods in the graphs), and to true (the next set of methods). For the Proportional Tournament, we set $S = 7$, and let R range from 1.0 down to 0.5 by decrements of 0.05. In all graphs, lower fitnesses were better, and the rightmost bar represents plain depth-limiting alone.

4.1 First Experiment

The first experiment compared plain depth limiting against Double and Proportional Tournament, using the four problem domains listed above. We ran 50 runs per technique per problem domain, and plotted the mean best-fitness-of-run and the average tree size per run. The fitness results are shown in Figure 2, and the tree size results in Figure 1.

Results. For most problems, there existed non-extreme settings for both Double and Proportional Tournaments which maintained fitness with significantly smaller tree sizes than plain depth limiting, often by wide margins. In the Symbolic Regression domain, Double and Proportional Tournaments both improved on plain depth limiting in tree size and fitness, but never in a statistically significant manner. As we noted in a previous paper [10], plain depth limiting performs very well in Symbolic Regression. Overall, Double Tournament S_p values in the 1.4–1.6 range did reasonably well. The sweet spot for Proportional Tournament was around $R = 0.7$, which always performed nearly identically to plain depth limiting (other settings could do much better depending on the problem). The particular setting of *do-fitness-first* did not have a significant effect.

4.2 Second Experiment

Reasonable settings of Double and Proportional Tournament either equalled or outperformed plain depth limiting, but not by as wide a margin as we would have hoped. We wondered how well combining each of these two methods with depth limiting would perform against just plain depth limiting alone. In our second experiment we compared the combinations against depth limiting, once again doing 50 runs per technique, then plotted the best fitness per run and the average tree size per run. The fitness results are shown in Figure 4, and the tree size results are shown in Figure 3.

Results. This time judicious settings of S_p or R dramatically outperformed plain depth limiting in all four domains. Overall, Double Tournament S_p values in 1.2–1.6 had equal fitness to plain depth limiting, while significantly outperforming it in tree size, often halving the size. The sweet spot for Proportional Tournament was again around $R = 0.7$, which always halved tree size while maintaining

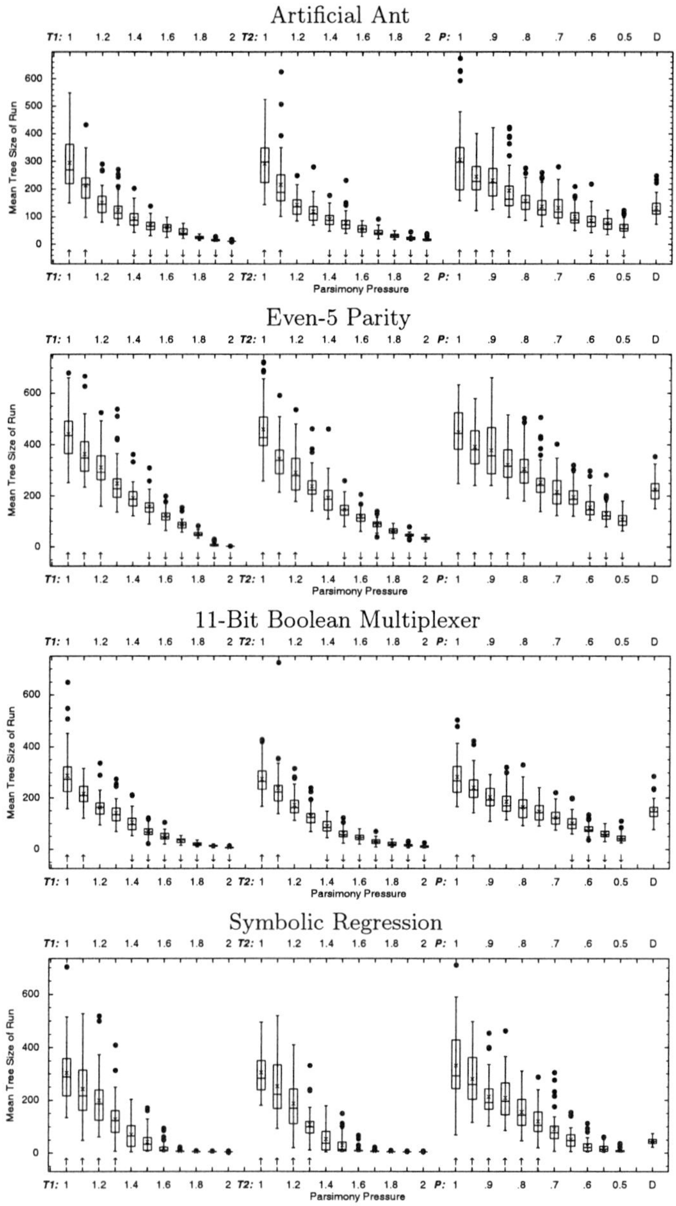

Fig. 1. Mean tree sizes for various parsimony pressure methods, as compared compared to plain depth limiting (labeled *D*). Distributions are plotted with boxplots. Proportional Tournament is labeled *P*, with the given ratio value *R*. Double Tournament is labeled *T1* (*do-fitness-first* false) or *T2* (*do-fitness-first* true), with the given tournament size S_p. The mean of each distribution is indicated with an $\times$. Lower values are better. Techniques statistically superior to plain depth limiting are marked with $\downarrow$; techniques statistically inferior are marked with $\uparrow$

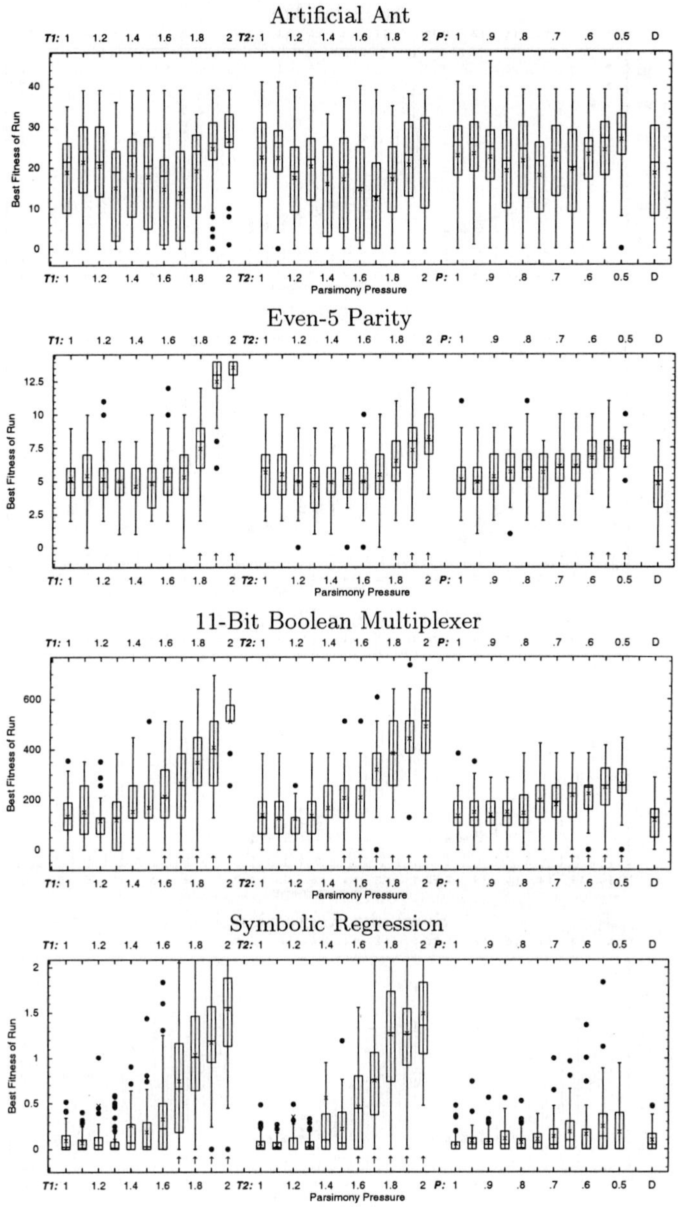

Fig. 2. Best fitness of run for various parsimony pressure methods, as compared compared to plain depth limiting (labeled D). Distributions are plotted with boxplots. Proportional Tournament is labeled P, with the given ratio value R. Double Tournament is labeled *T1* (*do-fitness-first* false) or *T2* (*do-fitness-first* true), with the given tournament size S_p. The mean of each distribution is indicated with an $\times$. Lower values are better. Techniques statistically superior to plain depth limiting are marked with $\downarrow$; techniques statistically inferior are marked with $\uparrow$

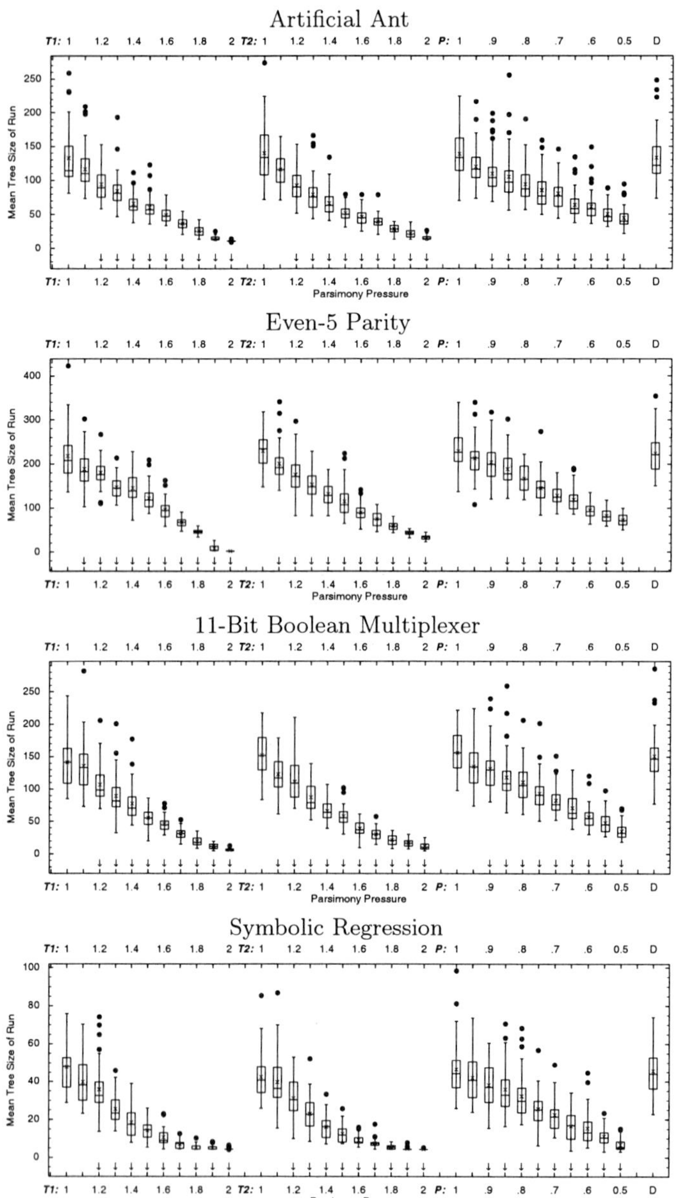

Fig. 3. Mean tree sizes for various parsimony pressure methods in combination with depth limiting, as compared compared to plain depth limiting alone (labeled D). Distributions are plotted with boxplots. Proportional Tournament is labeled P, with the given ratio value R. Double Tournament is labeled *T1* (*do-fitness-first* false) or *T2* (*do-fitness-first* true), with the given tournament size S_p. The mean of each distribution is indicated with an $\times$. Lower values are better. Techniques statistically superior to plain depth limiting are marked with $\downarrow$; techniques statistically inferior are marked with $\uparrow$

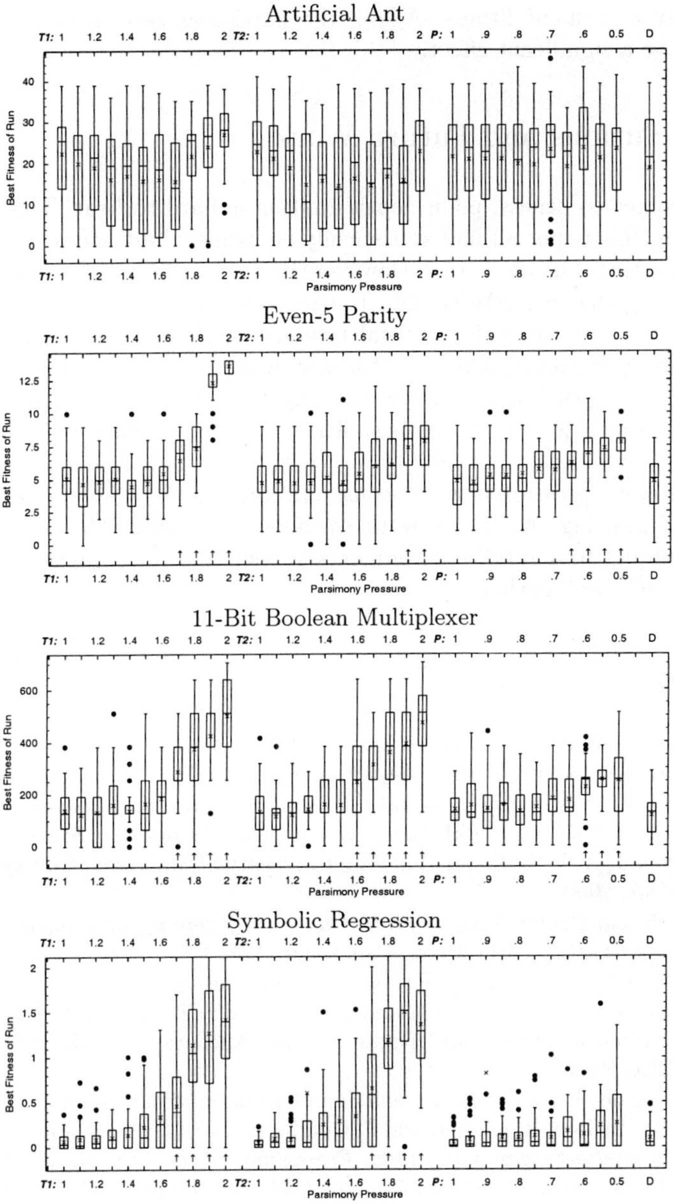

Fig. 4. Best fitness of run for various parsimony pressure methods in combination with depth limiting, as compared compared to plain depth limiting alone (labeled D). Distributions are plotted with boxplots. Proportional Tournament is labeled P, with the given ratio value R. Double Tournament is labeled *T1* (*do-fitness-first* false) or *T2* (*do-fitness-first* true), with the given tournament size S_p. The mean of each distribution is indicated with an $\times$. Lower values are better. Techniques statistically superior to plain depth limiting are marked with $\downarrow$; techniques statistically inferior are marked with $\uparrow$

statistically equivalent fitness. Again, the particular setting of *do-fitness-first* did not have a significant effect.

5 Conclusions and Future Work

When it comes to fitness, plain depth limiting is hard to beat. The techniques discussed in this paper all had statistically equivalent best-fitness-of-run results as depth limiting, but not better. However they were able to achieve these results while lowering the tree size. Double Tournament and Proportional Tournament by themselves could only lower total tree size slightly in comparison to plain depth limiting. However, when *combined* with depth limiting, they significantly outperformed depth limiting alone, yielding tree sizes at half the normal size while maintaining an equivalent best fitness of run. Given their simple implementation and general applicability, we think that nonparametric tournament-based parsimony pressure is worth consideration in a GP system in combination with depth limiting. As future work we hope to examine the applicability of these techniques to non-GP environments as well, and in comparison with other parsimony pressure methods.

References

1. Stephen F. Smith. *A Learning System Based on Genetic Adaptive Algorithms.* PhD thesis, Computer Science Department, University of Pittsburgh, 1980.
2. Sean Luke. *Issues in Scaling Genetic Programming: Breeding Strategies, Tree Generation, and Code Bloat.* PhD thesis, Department of Computer Science, University of Maryland, A. V. Williams Building, University of Maryland, College Park, MD 20742 USA, 2000.
3. John R. Koza. *Genetic Programming: On the Programming of Computers by Means of Natural Selection.* MIT Press, Cambridge, MA, USA, 1992.
4. Terence Soule and James A. Foster. Removal bias: a new cause of code growth in tree based evolutionary programming. In *1998 IEEE International Conference on Evolutionary Computation*, pages 781–186, Anchorage, Alaska, USA, 5-9 May 1998. IEEE Press.
5. Terence Soule, James A. Foster, and John Dickinson. Code growth in genetic programming. In John R. Koza, David E. Goldberg, David B. Fogel, and Rick L. Riolo, editors, *Genetic Programming 1996: Proceedings of the First Annual Conference*, pages 215–223, Stanford University, CA, USA, 28–31 July 1996. MIT Press.
6. Thomas Haynes. Collective adaptation: The exchange of coding segments. *Evolutionary Computation*, 6(4):311–338, Winter 1998.
7. Peter Nordin, Frank Francone, and Wolfgang Banzhaf. Explicitly defined introns and destructive crossover in genetic programming. In Peter J. Angeline and K. E. Kinnear, Jr., editors, *Advances in Genetic Programming 2*, pages 111–134. MIT Press, Cambridge, MA, USA, 1996.
8. Donald S. Burke, Kenneth A. De Jong, John J. Grefenstette, Connie Loggia Ramsey, and Annie S. Wu. Putting more genetics into genetic algorithms. *Evolutionary Computation*, 6(4):387–410, Winter 1998.

9. Jeffrey K. Bassett and Kenneth A. De Jong. Evolving behaviors for cooperating agents. In *International Syposium on Methodologies for Intelligent Systems*, pages 157–165, 2000.
10. Sean Luke and Liviu Panait. Lexicographic parsimony pressure. In W. B. Langdon *et al*, editor, *Proceedings of the Genetic and Evolutionary Computation Conference (GECCO2002)*. Morgan Kaufmann, 2002.
11. Byoung-Tak Zhang and Heinz Mühlenbein. Balancing accuracy and parsimony in genetic programming. *Evolutionary Computation*, 3(1):17–38, 1995.
12. Stefan Bleuler, Martin Brack, Lothar Thiele, and Eckhart Zitzler. Multiobjective genetic programming: Reducing bloat using spea2. In *Proceedings of the 2001 Congress on Evolutionary Computation CEC2001*, pages 536–543, COEX, World Trade Center, 159 Samseong-dong, Gangnam-gu, Seoul, Korea, 27-30 May 2001. IEEE Press.
13. Edwin D. DeJong, Richard A. Watson, and Jordan B. Pollack. Reducing bloat and promoting diversity using multi-objective methods. In Lee Spector, Erik D. Goodman, Annie Wu, W. B. Langdon, Hans-Michael Voigt, Mitsuo Gen, Sandip Sen, Marco Dorigo, Shahram Pezeshk, Max H. Garzon, and Edmund Burke, editors, *Proceedings of the Genetic and Evolutionary Computation Conference (GECCO-2001)*, pages 11–18, San Francisco, California, USA, 7-11 July 2001. Morgan Kaufmann.
14. Aniko Ekart and S. Z. Nemeth. Selection based on the pareto nondomination criterion for controlling code growth in genetic programming. *Genetic Programming and Evolvable Machines*, 2(1):61–73, March 2001.
15. Markus Brameier and Wolfgang Banzhaf. Explicit control of diversity and effective variation distance in linear genetic programming. In James A. Foster, Evelyne Lutton, Julian Miller, Conor Ryan, and Andrea G. B. Tettamanzi, editors, *Genetic Programming, Proceedings of the 5th European Conference, EuroGP 2002*, volume 2278 of *LNCS*, pages 37–49, Kinsale, Ireland, 3-5 April 2002. Springer-Verlag.
16. Sean Luke. ECJ 7: An EC system in Java. http://www.cs.umd.edu/projects /plus/ec/ecj/, 2001.

Increasing the Serial and the Parallel Performance of the CMA-Evolution Strategy with Large Populations

Sibylle D. Müller[1], Nikolaus Hansen[2], and Petros Koumoutsakos[1]

[1] Institute of Computational Science, ETH Zürich
8092 Zürich, Switzerland
{muellers,petros}@inf.ethz.ch
[2] Fachgebiet für Bionik, Technische Universität Berlin
13355 Berlin, Germany
hansen@bionik.tu-berlin.de

Abstract. The derandomized evolution strategy (ES) with covariance matrix adaptation (CMA), is modified with the goal to speed up the algorithm in terms of needed number of generations. The idea of the modification of the algorithm is to adapt the covariance matrix in a faster way than in the original version by using a larger amount of the information contained in large populations. The original version of the CMA was designed to reliably adapt the covariance matrix in small populations and turned out to be highly efficient in this case. The modification scales up the efficiency to population sizes of up to $10n$, where n ist the problem dimension. If enough processors are available, the use of large populations and thus of evaluating a large number of search points per generation is not a problem since the algorithm can be easily parallelized.

1 Introduction

One of the commonly proposed advantages of evolution strategies (ES's) is that they can be easily parallelized, see e.g. Schwefel (1995) or Bäck, Hammel, and Schwefel (1997). ESs with λ children per generation (population size λ) are usually parallelized by distributing the function evaluation for each of the λ children on a different processor. When the number of children is smaller than the number of available processors, the advantage of using ES's in parallel cannot be fully exploited. Consequently, for a large number of processors the algorithm should be able to use a large population efficiently.

In this article, we consider a derandomized ES with covariance matrix adaptation (CMA-ES) for which experimental results (Hansen and Ostermeier, 1997, 2001) show a clear convergence velocity improvement when compared to other ES's. The primary feature of the CMA-ES is its reliability in adapting an arbitrarily oriented scaling of the search space in small populations. The algorithm is in particular independent of any orthogonal transformation of the coordinate system.

J.J. Merelo Guervós et al. (Eds.): PPSN VII, LNCS 2439, pp. 422–431, 2002.

When optimizing considerably complex, e.g. highly nonseparable functions, the adaptation time becomes the limiting factor for the performance of the CMA-ES if the problem dimension n exceeds a certain threshold, usually $n \geq 10$. That is, the number of generations to adapt the covariance matrix of the search distribution to the function topology is the prominent factor for the degraded performance of the algorithm. The reason is that in the CMA-ES $(n^2 + n)/2$ elements of the symmetric covariance matrix $\boldsymbol{C}$ need to be adapted while the search process itself needs only to adjust n variables. Interestingly, for population sizes greater than 20 the adaptation time (i.e., the time to adapt the $(n^2 + n)/2$ elements of the covariance matrix) becomes practically independent of the population size (Hansen 1998) . That means, the performance in number of function evaluations decreases linearly with increasing population size. Alternatively, the implementation of the original CMA-ES on massively parallel computer architectures, e.g. the Beowulf cluster with hundreds of processors, supplies no substantial advantage compared to the use of twenty processors. On remarkably complex functions, the time complexity is of $\mathcal{O}(n^2)$, independent of the population size and the processor number.

How can we use the CMA-ES efficiently on massively parallel architectures with hundreds of processors? How can we increase the efficiency of the CMA-ES, when a large population is preferable to a small one due to other reasons? To increase λ alone does not help shortening the adaptation time as pointed out above. Additionally, a faster adaptation mechanism must be implemented being comparably reliable. The idea that we present in this paper is to increase the adaptation rate of the covariance matrix compared to the original algorithm, without losing its reliablity by exploiting a larger amount of information per generation. This is possible because an increased population should contain more information ready to be exploited in order to obtain a reduced adaptation time. Compared to the original adaptation mechanism the proposed modification would usually require much fewer function evaluations when λ is large but could be slightly less effective within small populations.

The working principles of the original algorithm, the CMA-ES, referred to as Orig-CMA here, are outlined in Section 2 and the modifications are presented in Section 3. In Section 4, the simulation results are discussed and Section 5 provides a conclusion.

2 Algorithm of the CMA-ES

Following Hansen and Ostermeier (2001) , in the (μ_I, λ)-CMA-ES the λ offspring of generation $g + 1$ are computed by

$$\boldsymbol{x}_k^{(g+1)} = \langle \boldsymbol{x} \rangle_\mu^{(g)} + \sigma^{(g)} \boldsymbol{B}^{(g)} \boldsymbol{D}^{(g)} \boldsymbol{z}_k^{(g+1)} \quad , \quad k = 1, \dots, \lambda, \tag{1}$$

where

$$\langle \boldsymbol{x} \rangle_\mu^{(g)} = \frac{1}{\mu} \sum_{i \in I_{sel}^{(g)}} \boldsymbol{x}_i^{(g)} \tag{2}$$

represents the center of mass of the selected individuals of generation g, and $I_{sel}^{(g)}$ is the set of indices of the selected individuals of generation g, with $|I_{sel}^{(g)}| = \mu$. The random vectors $\boldsymbol{z}$ from Equation (1) are $\mathcal{N}(\mathbf{0}, \boldsymbol{I})$ distributed (n-dimensional normally distributed with expectation zero and the identity covariance matrix) and serve to generate offspring for generation $g+1$. They can be used to calculate $\langle \boldsymbol{z} \rangle_{\mu}^{(g+1)}$ analogously to $\langle \boldsymbol{x} \rangle_{\mu}^{(g)}$. The columns of $\boldsymbol{B}^{(g)}$ represent eigenvectors of the covariance matrix $\boldsymbol{C}^{(g)}$. $\boldsymbol{D}^{(g)}$ is a diagonal matrix whose elements are the square roots of the eigenvalues of $\boldsymbol{C}^{(g)}$. Hence, the relation of $\boldsymbol{B}^{(g)}$ and $\boldsymbol{D}^{(g)}$ to $\boldsymbol{C}^{(g)}$ can be expressed by

$$\boldsymbol{C}^{(g)} = \boldsymbol{B}^{(g)} \boldsymbol{D}^{(g)} \left(\boldsymbol{B}^{(g)} \boldsymbol{D}^{(g)} \right)^T \quad \text{and} \quad \boldsymbol{C}^{(g)} \boldsymbol{b}_i^{(g)} = \left(d_{ii}^{(g)} \right)^2 \cdot \boldsymbol{b}_i^{(g)} \tag{3}$$

where $\boldsymbol{b}_i^{(g)}$ represents the i-th column of $\boldsymbol{B}^{(g)}$. Each covariance matrix corresponds to a hyperellipsoid which defines the surface of equal probability to place offspring. Here, the eigenvectors of the covariance matrix define the orientation of the hyperellipsoid and the eigenvalues define the lengths of its axes.

The evolution path $\boldsymbol{p}_c^{(g+1)}$ is calculated by

$$\boldsymbol{p}_c^{(g+1)} = (1 - c_c) \cdot \boldsymbol{p}_c^{(g)} + \sqrt{c_c \cdot (2 - c_c)} \cdot \underbrace{\frac{\sqrt{\mu}}{\sigma^{(g)}} \left(\langle \boldsymbol{x} \rangle_{\mu}^{(g+1)} - \langle \boldsymbol{x} \rangle_{\mu}^{(g)} \right)}_{= \sqrt{\mu}\, \boldsymbol{B}^{(g)} \boldsymbol{D}^{(g)} \langle \boldsymbol{z} \rangle_{\mu}^{(g+1)}} \tag{4}$$

and is used to build the covariance matrix of generation $g + 1$

$$\boldsymbol{C}^{(g+1)} = (1 - c_{\text{cov}}) \cdot \boldsymbol{C}^{(g)} + c_{\text{cov}} \cdot \boldsymbol{p}_c^{(g+1)} \left(\boldsymbol{p}_c^{(g+1)} \right)^T . \tag{5}$$

The update of $\boldsymbol{C}$ is done with a symmetric matrix of rank one (right summand in Equation (5)). The strategy parameters $c_c \in]0, 1]$ and $c_{\text{cov}} \in [0, 1[$ determine the accumulation time for the evolution path $\boldsymbol{p}_c^{(g+1)}$ and the change rate of the covariance matrix, respectively. Note that for $c_c = 1$ in Equation (4) the evolution path reduces to $\sqrt{\mu} \boldsymbol{B}\boldsymbol{D} \langle \boldsymbol{z} \rangle_{\mu}$ which is the mean mutation step of the last generation. Also, the update of $\boldsymbol{C}$ is independent of the adaptation of the global step size.

For the adaptation of the global step size, the evolution path $\boldsymbol{p}_{\sigma}^{(g+1)}$ that is not scaled by $\boldsymbol{D}^{(g)}$ is calculated by

$$\boldsymbol{p}_{\sigma}^{(g+1)} = (1 - c_{\sigma}) \cdot \boldsymbol{p}_{\sigma}^{(g)} + \sqrt{c_{\sigma} \cdot (2 - c_{\sigma})} \cdot \underbrace{\sqrt{\mu}\, \boldsymbol{B}^{(g)} \langle \boldsymbol{z} \rangle_{\mu}^{(g+1)}}_{= \boldsymbol{B}^{(g)} \left(\boldsymbol{D}^{(g)} \right)^{-1} \left(\boldsymbol{B}^{(g)} \right)^{-1} \frac{\sqrt{\mu}}{\sigma^{(g)}} \left(\langle \boldsymbol{x} \rangle_{\mu}^{(g+1)} - \langle \boldsymbol{x} \rangle_{\mu}^{(g)} \right)} \tag{6}$$

and its length is used to compute the step size for generation $g + 1$

$$\sigma^{(g+1)} = \sigma^{(g)} \cdot \exp \left(\frac{1}{d_{\sigma}} \frac{\|\boldsymbol{p}_{\sigma}^{(g+1)}\| - \hat{\chi}_n}{\hat{\chi}_n} \right) , \tag{7}$$

where $\hat{\chi}_n = \mathrm{E}\left[\|\mathcal{N}(\mathbf{0}, \boldsymbol{I})\|\right]$ is the expected length of a $(\mathbf{0}, \boldsymbol{I})$-normally distributed random vector and $\hat{\chi}_n$ is approximated by $\hat{\chi}_n \approx \sqrt{n}\left(1 - \frac{1}{4n} + \frac{1}{21n^2}\right)$. The strategy parameter $c_\sigma \in]0,1]$ determines the accumulation time for the evolution path $\boldsymbol{p}_\sigma^{(g+1)}$, and d_σ is a damping parameter.

The default strategy parameter setting, discussed in Hansen and Ostermeier (2001) in detail, is as follows:

$$c_c = \tfrac{4}{n+4},\; c_{\mathrm{cov}} = \tfrac{2}{(n+\sqrt{2})^2},\; c_\sigma = \tfrac{4}{n+4},\; d_\sigma = c_\sigma^{-1} + 1 \tag{8}$$

Initial values are $\boldsymbol{p}^{(0)} = \mathbf{0}, \boldsymbol{p}_\sigma^{(0)} = \mathbf{0}$ and the initial covariance matrix $\boldsymbol{C}^{(0)}$ is the identity matrix $\boldsymbol{I}$.

3 Modified Algorithm

All modifications solely regard Equation (5) that describes the change of the covariance matrix, i.e., the change of the mutation distribution shape, and c_{cov}. Everything else, in particular the global step size adaptation mechanism remains unchanged in this paper. We add to Equation (5) the following term:

$$\begin{aligned} \boldsymbol{Z}^{(g+1)} &= \frac{1}{\mu} \sum_{i \in I_{sel}^{(g+1)}} \boldsymbol{B}^{(g)} \boldsymbol{D}^{(g)} \boldsymbol{z}_i^{(g+1)} \left(\boldsymbol{B}^{(g)} \boldsymbol{D}^{(g)} \boldsymbol{z}_i^{(g+1)}\right)^T \\ &= \boldsymbol{B}^{(g)} \boldsymbol{D}^{(g)} \left(\frac{1}{\mu} \sum_{i \in I_{sel}^{(g+1)}} \boldsymbol{z}_i^{(g+1)} \left(\boldsymbol{z}_i^{(g+1)}\right)^T\right) \left(\boldsymbol{B}^{(g)} \boldsymbol{D}^{(g)}\right)^T \end{aligned} \tag{9}$$

that is a symmetrical $n \times n$ matrix with rank $\min(\mu, n)$ (with probability one).

The modification of Equation (5) then reads

$$\boldsymbol{C}^{(g+1)} = (1 - c_{\mathrm{cov}}) \cdot \boldsymbol{C}^{(g)} + c_{\mathrm{cov}} \left(\alpha_{\mathrm{cov}} \cdot \boldsymbol{p}_c^{(g+1)} \left(\boldsymbol{p}_c^{(g+1)}\right)^T + (1 - \alpha_{\mathrm{cov}}) \cdot \boldsymbol{Z}^{(g+1)}\right) \tag{10}$$

where $0 \le \alpha_{\mathrm{cov}} \le 1$. Note that for $\alpha_{\mathrm{cov}} = 1$ Equation (10) and Equation (5) are identical and the original CMA-ES is restored. Decreasing α_{cov} changes the parameterized algorithm continuously. Results are presented for $\alpha_{\mathrm{cov}} = 0$ and $\alpha_{\mathrm{cov}} = \frac{1}{\mu}$ and compared with the original CMA algorithm, where $\alpha_{\mathrm{cov}} = 1$, in Section 4.

Since the rank of $\boldsymbol{Z}^{(g+1)}$ is larger than 1, more information is passed to the covariance matrix in each generation. Thus, the adaptation becomes more reliable. Therefore, the adaptation time $1/c_{\mathrm{cov}}$ can be decreased, or in other words, the learning rate c_{cov} can be increased yielding a higher adaptation speed. For $\alpha_{\mathrm{cov}} = 1$, c_{cov} is used as in Equation (8). For $\alpha_{\mathrm{cov}} = 0$, experiments showed that

$$c_{\mathrm{cov}} = \min\left(1, \frac{2\mu - 1}{(n+2)^2 + \mu}\right) . \tag{11}$$

Table 1. Convex quadratic test functions and stopping criteria.

Name	Function
Sphere	$f_{\text{sphere}} = \sum_{i=1}^{n}(x_i)^2$
Ellipsoid	$f_{\text{elli}} = \sum_{i=1}^{n}(1000^{\frac{i-1}{n-1}} x_i)^2$
Cigar	$f_{\text{cigar}} = x_1^2 + \sum_{i=2}^{n}(1000 x_i)^2$

trades of reasonably well the reliability and the adaptation speed of the covariance matrix adaptation, and is thus chosen in the strategies with $0 \le \alpha_{\text{cov}} < 1$. A closed expression for c_{cov} has yet to be found that can be used in both types of strategies represented by $\alpha_{\text{cov}} = 1$ and $0 \le \alpha_{\text{cov}} < 1$.

Equations (9) and (10) are analyzed to motivate the coefficients $\frac{1}{\mu}$ in Equation (9) and α_{cov} in conjunction with $(1 - \alpha_{\text{cov}})$ from Equation (10). Under random selection, the symmetric matrix

$$\sum_{i \in I_{sel}} \boldsymbol{z}_i(\boldsymbol{z}_i)^T = \sum_{i \in I_{sel}} \begin{pmatrix} z_{i1}^2 & z_{i1}z_{i2} & \cdots & z_{i1}z_{in} \\ z_{i2}z_{i1} & z_{i2}^2 & \cdots & z_{i2}z_{in} \\ \vdots & \vdots & \ddots & \vdots \\ z_{in}z_{i1} & z_{in}z_{i2} & \cdots & z_{in}^2 \end{pmatrix} \tag{12}$$

from Equation (9) has diagonal elements that are χ_μ^2 distributed and off-diagonal elements with expectation zero. With $\mathrm{E}\left[\sum_{i \in I_{sel}} z_{ij}^2\right] = \mu$, $j = 1, \ldots, n$, we have that

$$\mathrm{E}\left[\boldsymbol{Z}^{(g+1)}\right] = \frac{1}{\mu}\boldsymbol{B}^{(g)}\boldsymbol{D}^{(g)}\,\mu\boldsymbol{I}\,(\boldsymbol{B}^{(g)}\boldsymbol{D}^{(g)})^T = \boldsymbol{C}^{(g)} \tag{13}$$

under the given selection model. Equation (13) is the reason for choosing the coefficient $\frac{1}{\mu}$ in Equation (9). With $\mathrm{E}\,[\boldsymbol{p}_c^{(g+1)}(\boldsymbol{p}_c^{(g+1)})^T] = \boldsymbol{C}^{(g)}$ (Hansen 1998) and $\mathrm{E}\,[\boldsymbol{Z}^{(g+1)}] = \boldsymbol{C}^{(g)}$, we conclude from Equation (10) that $\mathrm{E}\,[\boldsymbol{C}^{(g+1)}] = \boldsymbol{C}^{(g)}$. This is the reason for choosing $(1 - \alpha_{\text{cov}})$ in conjunction with α_{cov} in Equation (10).

To compare the strategies with different α_{cov}, the functions shown in Table 1 are tested. Tests are carried out in the dimensions $n = [2, 3, 5, 10, 20, 40, 80]$ and for parent numbers $\mu = [2, \lceil n/4 \rceil, \lceil n/2 \rceil, n, 2n, 4n, \lceil n^2/4 \rceil, \lceil n^2/2 \rceil, n^2]$ with a population size of $\lambda = 4\mu$. Initial values are set to $\langle \boldsymbol{x} \rangle_\mu^{(0)} = \boldsymbol{1}$ and $\sigma^{(0)} = 1$. The search is terminated as soon as $f_{\text{stop}} = 10^{-10}$ is reached.

4 Discussion of the Results

Three different strategy variants are presented: New-CMA, where $\alpha_{\text{cov}} = 0$; Orig-CMA, where $\alpha_{\text{cov}} = 1$; and Hybr-CMA, where $\alpha_{\text{cov}} = \frac{1}{\mu}$. Note that for $\mu = 1$, Hybr-CMA is identical to Orig-CMA. The simulation results for the various strategies are analyzed from two different point of views:

Serial performance. We analyze the number of overall function evaluations to reach f_{stop}. This is the appropriate point of view if optimization is performed on a single processor or on a small number of processors that does not exceed the smallest sensible population size λ, usually five to ten. In this case the number of function evaluations is an appropriate measure for the time to reach f_{stop}.

Parallel performance. We analyze the number of generations to reach f_{stop}. When a larger number of processors is available, it becomes interesting to evaluate the number of generations to reach f_{stop}, especially if λ is equal to the number of processors. In this case, the number of generations is an appropriate measure for the time to reach f_{stop} in a single run.

First, we discuss the serial performance on f_{sphere} as shown in Figure 1, upper left. For $\lambda < 10n$ the performance of Orig-CMA and Hybr-CMA are similar as expected. For larger population sizes, Hybr-CMA becomes faster than Orig-CMA by a factor of up to three. This effect cannot be attributed to a faster adaptation of the distribution *shape* because on f_{sphere} the shape is optimal already at the beginning. The reason for the better performance of Hybr-CMA is the faster adaptation of the overall variance of the distribution. Originally, the cumulative path length control (Equations (6) and (7)) facilitates the adaptation of the global step size, i.e. the adaptation of the overall variance. The parameter that tunes the adaptation speed (d_σ in Equation (7)) was (a) chosen conservatively resulting in a somewhat slower but more robust algorithm (Hansen and Ostermeier 2001 , Section 5.1) and (b) chosen w.r.t. small population sizes that realize smaller progress rates than larger populations per generation and therefore demand slower adaptation rates. Consequently, for $\mu > n$ and $\alpha_{\text{cov}} \ll 1$ the distribution adaptation in Equation (10) can successfully contribute to the adapation speed of the overall variance in Hybr-CMA because the change rate $c_{\text{cov}} > \frac{1}{n}$. This is the presumable reason for the observed speed-up on f_{sphere}.

Even though this effect seems to be advantageous at first sight, it may become disadvantageous if the distribution adaptation influences the magnitude of the overall variance significantly. While the path length control is shown to adapt nearly optimal step lengths even for $\mu > 1$ (at least for $\mu < \lambda < n$, Hansen 1998), the distribution adaptation as described in Equation (10) acquires too small step lengths rapidly, if—unlike on f_{sphere}—*the optimal* step length remains constant over time (i.e. in a stationary environment). When the optimal step length decreases significantly fast over time, as on f_{sphere}, this can be an advantage. However, an algorithm that adapts drastically too small variances in a stationary environment is not preferable in general. This suggests an upper limit for reasonable population sizes, arguable at $\lambda \approx 10n$.

Next, the serial performance of the strategy variants on f_{elli} and f_{cigar} is evaluated. The most prominent effect in these results is the dependency on λ. The smallest $\lambda = 8$ (and even a smaller λ for $n = 5$) performs best in all cases. For example, if $n = 20$ the decline of the serial performance between $\lambda = 8$ and $\lambda = 80$ amounts roughly to a factor of 5.5 (on f_{cigar}) and 7.5 (on f_{elli}) for Orig-CMA and of 1.7 (on f_{elli}) and 4.5 (on f_{cigar}) for Hybr-CMA. Considering the serially optimal $\lambda = 8$, the performance difference between Orig-CMA and Hybr-CMA is small. This leads to the conclusion that the introduced modifications in

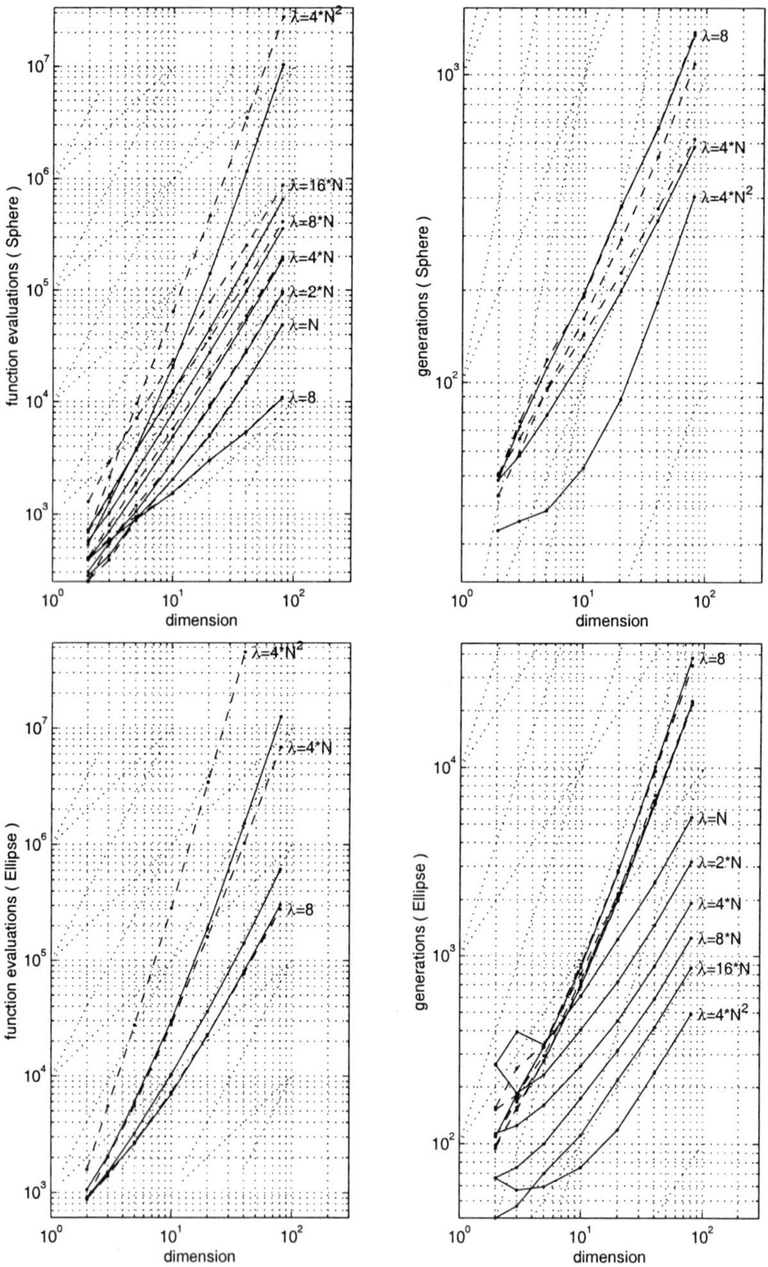

Fig. 1. Number of function evalutations (left) and number of generations (right) over the problem dimension for the sphere function (above) and the ellipse function (below). The Orig-CMA (– – –) and the Hybr-CMA (—) are plotted for $\lambda = 8, n, 2n, 4n, 8n, 16n, 4n^2$ if curves are far apart, or for $\lambda = 8, 4n, 4n^2$.

Hybr-CMA yield at least similar, if not slightly better serial performance even for $\mu = 2$ and $\lambda = 8$ in the tested cases.

Comparing New-CMA with the other strategy variants where $\lambda = 8$ reveals a significant result on f_{cigar} (Figure 2). While Orig-CMA and Hybr-CMA need about $500n$ function evaluations to reach f_{stop}, New-CMA needs about $120n^2$ function evaluations. Using the evolution path $\boldsymbol{p}_c$ in Equation (10) in addition with a cumulation parameter of $c_c \approx \frac{1}{n}$ in Equation (4) yields this impressive speed-up of Orig-CMA and Hybr-CMA. Although detected in earlier investigations (Hansen and Ostermeier 2001), this speed-up is noteworthy in that a completely adaptable covariance matrix with $\frac{n^2+n}{2}$ free parameters can be adapted to certain topologies in $\mathcal{O}(n)$ function evaluations. Since this observation on f_{cigar} is the major difference between Hybr-CMA and New-CMA, the latter algorithm is excluded from the remaining discussion.

While the differences between Orig-CMA and Hybr-CMA are marginal for $\lambda = 8$, the picture changes in favor of Hybr-CMA when the population size is increased. For $\lambda = 8$, the scaling of the needed function evaluations w.r.t. the problem dimension (i.e. the slope of the graphs) is linear on f_{sphere} and f_{cigar}, but it is almost quadratic on f_{elli}. For $\lambda \propto n$, the scaling of Hybr-CMA becomes nearly quadratic, regardless whether the scaling is linear or quadratic for $\lambda = 8$. This is in contrast to Orig-CMA where the scaling *always* deteriorates when λ is increased from a constant value to $\lambda \propto n$. As a result, Hybr-CMA performs never worse and often greatly better than Orig-CMA if $\lambda \propto n$. This clear advantage can of course be expected only if $\mu \propto \lambda$, e.g. $\mu \approx \lambda/4$ as chosen in our investigations. Concluding these observations, Hybr-CMA must be undoubtedly preferred w.r.t. its serial performance if $\mu > 3$ and $n > 5$.

Second, we discuss the parallel performance that comes into play if λ is chosen considerably large. This means that we concentrate the discussion of parallel performance on the cases where $\lambda \propto n$ and $\lambda = 4n^2$.

Certainly, the difference in performance between Orig-CMA and Hybr-CMA discussed above translates to the parallel case. Hybr-CMA outperforms Orig-CMA overall and on any single function, if $\lambda \gg 8$ (assuming $\mu \approx \lambda/4$). Therefore, it is more interesting to interpret the parallel strategy behavior in relation to λ. The parallel performance of Orig-CMA does not change dramatically when λ is increased. The improvement never exceeds a factor of two in dimensions up to 80. In contrast to Orig-CMA, where the parallel performance is less dependent on λ, Hybr-CMA shows a vigorous improvement when λ is increased. For example, increasing λ from n to $8n$, i.e. by a factor of eight, improves the parallel performance by a factor greater than four on f_{elli}.

However, the most impressive result concerns the scaling of Hybr-CMA with respect to the number of generations where $\lambda \propto n$. When $\lambda \leq 10n$, Orig-CMA scales (nearly) quadratically w.r.t. the number of generations on all functions except on f_{sphere}. The same observation holds for Hybr-CMA for $\lambda = 8$. However, when λ is increased to be proportional to n in Hybr-CMA, the number of generations scales *linearly* with the problem dimension on all convex quadratic test functions. On f_{sphere}, the scaling for $\lambda \propto n$ is even slightly sublinear. The

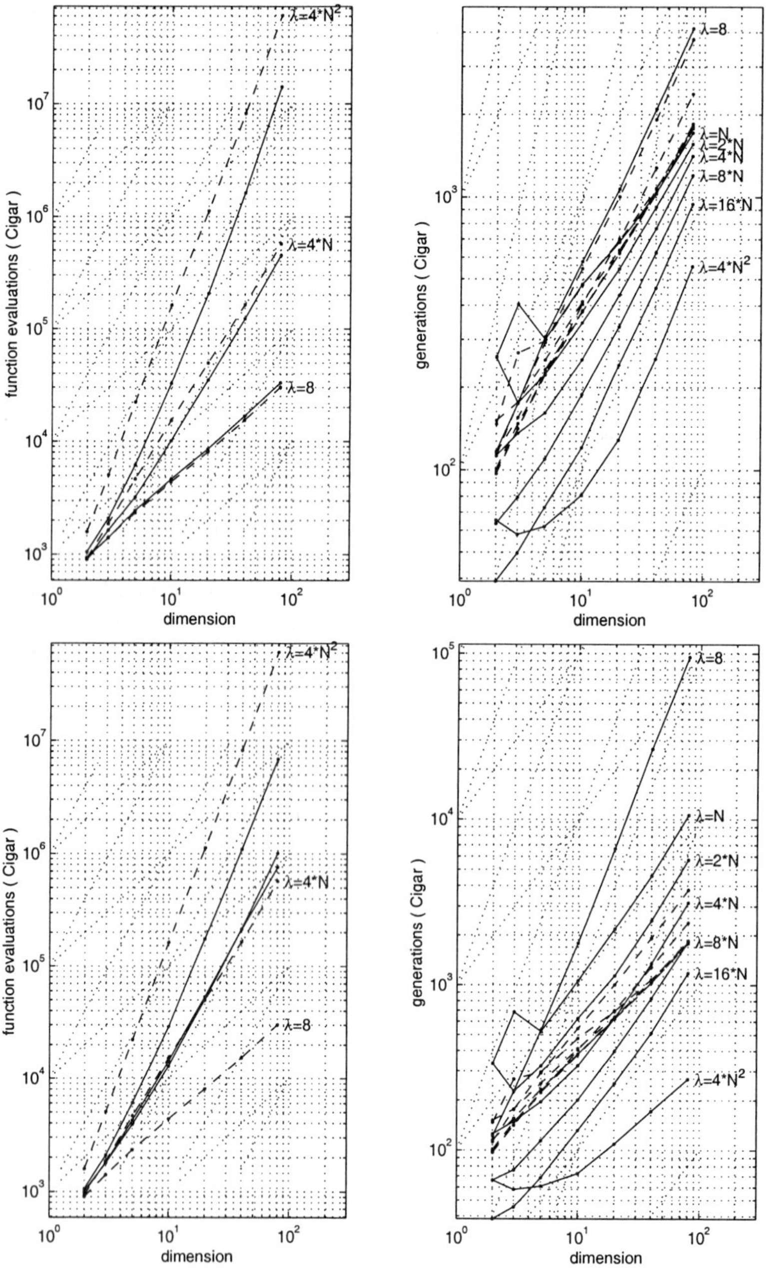

Fig. 2. Number of function evaluations (left) and number of generations (right) over the problem dimension for the cigar comparing Orig-CMA with Hybr-CMA (above) and the cigar comparing Orig-CMA with New-CMA (below). The Orig-CMA (– – –) and the Hybr-CMA or New-CMA (—) are plotted for $\lambda = 8, n, 2n, 4n, 8n, 16n, 4n^2$ if curves are far apart, or for $\lambda = 8, 4n, 4n^2$.

roughly linear scaling for $\lambda \propto n$ is the main improvement of the new Hybr-CMA compared with Orig-CMA.

5 Conclusions

We presented a modified algorithm derived from the derandomized evolution strategy with covariance matrix adapation. Our goal was to devise a technique with which we can optimize in fewer number of generations than with the original strategy, allowing to exploit the often emphasized feature of evolution strategies being easily parallelizable.

This goal is achieved for population sizes up to $\lambda = 10n$. Choosing $\alpha_{\text{cov}} = \frac{1}{\mu}$ and $\mu \approx \lambda/4$, the modified algorithm seems to efficiently exploit the information prevalent in the population and reveals mainly linear time complexity for population sizes proportional to n and up to $10n$, if fully parallelized. This means we were able to reduce the time complexity roughly from $\mathcal{O}(n^2)$ to $\mathcal{O}(n)$.

For $\mu = 1$, the modified algorithm is identical with the original one (assuming identical recombination weights in the latter). With increasing μ, the performance improves remarkably compared with the original algorithm. In our tests, the modified algorithm with $\alpha_{\text{cov}} = \frac{1}{\mu}$ reveals no disadvantage compared with the original one. Only for population sizes larger than $10n$ the adjustment of the overall variance becomes problematic. As a conclusion, the efficiency of the adaptation of the distribution shape in large populations seems to be satisfying for the moment. Future work will address further developments for the adaptation of the global step size.

References

[Bäck et al., 1997] Bäck, T., Hammel, U., and Schwefel, H.-P. (1997). Evolutionary computation: Comments on the history and current state. *IEEE Transactions on Evolutionary Computation*, 1(1):3–17.

[Hansen, 1998] Hansen, N. (1998). *Verallgemeinerte individuelle Schrittweitenregelung in der Evolutionsstrategie. Eine Untersuchung zur entstochastisierten, koordinatensystemunabhängigen Adaptation der Mutationsverteilung.* Mensch und Buch Verlag, Berlin. ISBN 3-933346-29-0.

[Hansen and Ostermeier, 1997] Hansen, N. and Ostermeier, A. (1997). Convergence properties of evolution strategies with the derandomized covariance matrix adaptation: The $(\mu/\mu_I, \lambda)$-CMA-ES. In Zimmermann, H.-J., editor, *EUFIT'97, 5th Europ. Congr. on Intelligent Techniques and Soft Computing, Proceedings*, pages 650–654, Aachen, Germany. Verlag Mainz.

[Hansen and Ostermeier, 2001] Hansen, N. and Ostermeier, A. (2001). Completely derandomized self-adaptation in evolution strategies. *Evolutionary Computation*, 9(2):159–195.

[Schwefel, 1995] Schwefel, H.-P. (1995). *Evolution and Optimum Seeking.* Sixth-Generation Computer Technology Series. John Wiley & Sons Inc., New York.

Adaptive Reservoir Genetic Algorithm with On-Line Decision Making

Cristian Munteanu and Agostinho Rosa

LaSEEB, Instituto de Sistemas e Robotica, Instituto Superior Tecnico
Av. Rovisco Pais 1, Torre Norte, 6.21, 1049-001 Lisboa, Portugal
{cristi,acrosa}@laseeb.ist.utl.pt
http://laseeb.ist.utl.pt

Abstract. It is now common knowledge that blind search algorithms cannot perform with equal efficiency on all possible optimization problems defined on a domain. This knowledge applies also to Genetic Algorithms when viewed as global and blind optimizers. From this point of view it is necessary to design algorithms capable of adapting their search behavior by making use in a direct fashion of the knowledge pertaining to the search landscape. The paper introduces a novel adaptive Genetic Algorithm where the exploration/exploitation is directly controlled during evolution using a Bayesian decision process. Test cases are analyzed as to how parameters affect the search behavior of the algorithm.

1 Introduction

According to the No Free Lunch Theorems [11] there are no algorithms either deterministic or stochastic behaving the same on the complete set of search and optimization problems defined on a finite and discrete domain [11]. A blind approach to the optimization problem, as well as the global optimizer paradigm are therefore out-ruled. We introduce a Genetic Algorithm that adjusts its behavior (exploitation or exploration) based on an on-line decision taken over the fitness landscape on which the search focuses. This algorithm is not anymore blind, in the sense that it uses the current status of the search on the fitness landscape, while modifying its search behavior, according to stagnation or progress encountered during the search process. The algorithm builds on a former variant introduced by the authors in [7] called Adaptive Reservoir Genetic Algorithm (ARGA). The convergence within finite time, with probability 1 of ARGA, was shown to hold in [8], while a real-world application focusing on finding the best Hidden Markov Model classifier for a Brain Computer Interface task, was discussed in [9]. The present article introduces ARGAII, a variant of ARGA, that bears the basic architecture of ARGA, improving however the control mechanism by employing a Bayesian decision process. Based on the current status of the search a decision is taken whether to keep exploiting the search space or switch to exploration. Following a classification of adaptation in Evolutionary Computation (EC) in [4], ARGAII fits within the "dynamic adaptive" class of adaptive Evolutionary Algorithms (EA).

J.J. Merelo Guervós et al. (Eds.): PPSN VII, LNCS 2439, pp. 432–441, 2002.

2 Algorithm Presentation

2.1 ARGA's Basic Architecture

ARGA proposes a novel mechanism for mutating the individuals in a GA population in corroboration with the selection algorithm. In a standard GA [3] each chromosome in the population can be mutated, depending on a given mutation probability. ARGA, however restricts mutations to a subpopulation of chromosomes, called *reservoir* which has its individuals mapped onto a fixed size population. The number of chromosomes in the reservoir is called *diameter* and is adapted during run. If there is no improvement in the best fitness found during a certain number of generations, the diameter of the reservoir grows, in order to obtain a larger diversity in the population and to recast the search in a better niche of the search space. When this event occurs (i.e. an improvement beyond a certain threshold) the diameter of the reservoir is reset to the initial value. The algorithm is given in pseudo-code, as follows:

```
ARGA ()
{-Start with an initial timer counter: t.
 -Initialize a random population: P(t) within specific bounds.
 -Set the initial value of reservoir's diameter Delta to Delta_0.
 -Compute fitness for all individuals.
  while not done do
   -Increase the time counter.
   -Select parents to form the intermediate population by applying
    binary tournament.
   -Perform mutation on reservoir
     {-Select reservoir Rho(t), by choosing in a
       binary tournament the less fitted Delta individuals
       in the intermediate population.
      -Perform mutation on Rho(t) with a random rate in
       (0,P_mutation].
      -Introduce mutants in the intermediate population by
       replacing the former Rho(t) with mutants.}
   -Perform (one-point) crossover on intermediate
    population with a rate P_crossover.
   -Form the population in the next generation by
    applying a k-elitist scheme to intermediate population.
   -Compute the new fitness for all individuals.
   -Adjust the diameter of the reservoir: Delta(t).
  od}
```

The basic structure described in the pseudo-code holds both for ARGA as well as for the improved variant ARGAII, the only difference being the way the diameter is adjusted in ARGAII. More details about ARGA can be found in [7].

2.2 Reservoir Adjustment in ARGAII

In ARGA the adjustment of the reservoir's diameter $\Delta(t)$ is first done by comparing the best individual in the current generation t with the best individual in the previous generation $t-1$. If there is an improvement of the best fitness found beyond a certain threshold ϵ, the diameter is reset to its initial value Δ_0. Otherwise, in ARGA a constant rate $c > 0$ is added to $\Delta(t-1)$ and the integer part of the sum is taken to be the new reservoir's diameter $\Delta(t)$. However, in ARGAII a more complex decision is taken when there is no fitness improvement, and this involves a Bayesian decision process. In both cases, ARGA and ARGAII, if the reservoir $\rho(t)$ becomes bigger than the size of the population again the diameter is reset to its initial value Δ_0. Thus, the reservoir is reset in ARGA on two distinct events: once there is a better than current best individual found in the population, or when the reservoir grows (due to stagnation in finding a better peak) as to fill the whole population. After the reservoir is reset, the algorithm starts exploiting the neighborhood of the already found or current best individual. The reservoir grows while no other better individual is found, the algorithm starts exploring more the search space until the cycle is repeated with the finding of a new better peak, or until the reservoir grows to the size of the population. In ARGAII the following decision is taken: if the algorithm is not finding a better peak, however the region searched by the algorithm has a high fitness landscape ruggedness, the diameter of the reservoir stays the same. In this case, due to the high ruggedness of the search space currently explored, the diameter shouldn't grow, as we estimate that continued exploitation might be productive on such a landscape. If the algorithm is not finding a better peak, but the region where ARGAII explores has a low ruggedness, the diameter grows, as to increase exploration to more promising regions of the search space.

Landscape Ruggedness in ARGAII. In the literature there exist several characteristics of the fitness landscape that differentiate between landscapes that are rugged and multi-peaked, from those that are smooth, or uni-peaked. Such measures are the distribution of the local optima, the modality or the ruggedness of the landscape. In [6] the authors define modality as being the number of the local optima of a fitness landscape and show how this measure is related to the difficulty of finding the global optimum by GAs and hill climbers. In [5] the author following the first fitness landscape analysis by Wright (1932) and a random walk method that extracts useful landscape features, by Weinberger (1990), defines the ruggedness of the fitness landscape by using the auto-correlation function of a time series of the fitness values of points sampled by a random walk on the landscape. A landscape that will have the auto-correlation coefficient close to 1 or -1 will be considered smooth, while if the coefficient is close to zero, the landscape is considered highly rugged. Also based on a random walk on the landscape, several information measures were defined in [10] to better characterize the landscape. For ARGA we propose a more precise measure that is also amenable to a generation-by-generation processing. A random walk on the whole landscape to calculate the correlation coefficient is not feasible computationally, as we would have to apply it in each generation. Also, we are more interested in finding the characteristics of the landscape in the region where the algorithm converges, a correlation coefficient on the landscape where ARGA converges being more difficult and out of hand to define. Thus, a cluster of points Ξ is designated each generation to represents points in the region of convergence of ARGAII. For each point $x \in \Xi$ having fitness value $f(x)$ a local search (LS)[1] is performed as to yield the peak of the basin of attraction in which x lied. This peak has fitness value f^*. We call *drills* the points that suffer a LS process. All *distinct* fitness values f^* found for the drills in Ξ form a set Ψ. Thus, we have the following (generally non-injective) mapping $x \in \Xi \mapsto \Psi$. The measure of ruggedness is the cardinal of the set Ψ, that is $\phi = \mid \Psi \mid$. Most often, drilling yields solutions having better fitness than the current best. In these cases, the best fitted drill replaces the current best individual, on the spot.

Clustering in ARGAII. ARGAII applies a clustering algorithm each generation to determine the sub-population that is currently converging. We define convergence in terms of homogeneity at the genotypic level, thus a convergent sub-population is a sub-population for which the chromosomes are similar at the genotypic levels. As we will use real parameters coded binary, homogeneity will be taken at the coded level of the representation (real valued parameters). The convergent sub-population is considered to be the most populated cluster after performing a cluster-tree algorithm on the whole population. Let this cluster be Θ. The population of drills Ξ is chosen with respect to Θ as follows: A) If the biggest cluster corresponds to the cluster in which the current optimum lies, then choose at random K drills to form the set Ξ. If there are not enough elements in the cluster to choose K drills, then compute the mean and standard deviation of the individuals in the cluster. For each (real-valued) parameter indexed j in the chromosomes pertaining to the cluster, we compute the mean μ_j and standard deviation s_j and we generate the remaining up to K drills, as a string of values taken as samples from the uniform distribution centered on μ_j with deviation s_j, that is $U(\mu_j, s_j)$. B) If the biggest cluster does not contain the current optimum, it is supposed due to takeover, that this optimum will pertain to the biggest cluster in a few next generations. Thus, Ξ is generated uniformly as before, but centered on the current optimum, with the standard deviation computed over the biggest cluster. In both cases the cardinal of the set Ξ is kept to be K. The clustering approach thus yields the subpopulation that converges (i.e. the biggest cluster), and the drills, that should be taken around the space where the algorithm focuses its search. The drills are randomly chosen from individuals in the biggest cluster. After "drilling" the search space we come up with ϕ distinct peaks that where found. ϕ will be an estimate of the ruggedness of the search space in the region where the algorithm converges (i.e. focuses its search).

[1] For this particular implementation we employed a Nelder–Mead simplex direct search method.

The clustering method used is a hierarchical tree clustering with the cutoff value of 0.95, that gives a fairly good consistency of the clustering method [2].

Bayesian Decision in ARGAII. The decision process is taken each generation before adjusting the reservoir size Δ. The hypothesis H_0 is that the optimum is not around in the space searched by ARGAII, while hypothesis H_1 is that the optimum lies somewhere in the space where ARGAII focuses its search. The costs involved in the decision process are C_{10} being the cost of choosing H_1 when H_0 is true, and C_{01} the cost of choosing H_0 when H_1 is true. Thus, we have the following decision:

$$\lambda(\phi) = \frac{p_1(\phi)}{p_0(\phi)} \underset{H_1}{\overset{H_0}{\lessgtr}} \frac{P(H_0)C_{10}}{[1 - P(H_0)]C_{01}}$$

where λ is the likelihood function, ϕ is the cardinal of the set Ψ, the a priori probability of H_0 being $P(H_0)$, and the a priori probability for H_1 being $[1 - P(H_0)]$. However, we don't have any a priori knowledge about the likelihood functions $p_0(\phi)$ and $p_1(\phi)$, only the heuristical argument that as ϕ increases the likelihood that H_0 will be true decreases (i.e. $p_0(\phi)$ decreases), while the likelihood that H_1 will be true, increases (i.e. $p_1(\phi)$ increases). Without any a priori information, we should take the simplest model, that is $p_0(\phi)$ decreases linearly, and $p_1(\phi)$ increases linearly with the same absolute slope. We propose the following likelihood functions to be used:

$$p_0(\phi) = \frac{2}{K+1} \cdot \left(1 - \frac{\phi+1}{K+2}\right); p_1(\phi) = \frac{2}{(K+1)(K+2)} \cdot (\phi+1), \phi = \overline{0 \ldots K}$$

Again, having no a priori information about the probability $P(H_0)$, this will be taken equal to 0.5 (the two hypothesis are a priori equally probable). The parameter of the decision process will be $\gamma = \frac{C_{10}}{C_{01}}$.

Diameter Adjustment in ARGAII. After computing the decision that should be taken for the adjustment of the reservoir's diameter, we proceed with the adjustment itself: we use an intermediate variable δ initialized as $\delta(0) = \Delta(0)$. The initialization is done at the beginning of the run, whenever a better best fitness is discovered, and whenever the size of the reservoir grows to the size of the population N. If no better best fitness value has been found from the last generation to the current and if H_0 was decided to hold true, then:

$$\Delta(t) = \begin{cases} \delta(t), \text{with } \delta(t) = \lceil \delta(t-1) + c \rceil \text{ if } [(\delta(t-1) + c] = \lceil \delta(t-1) + c \rceil \\ \Delta(t-1),\ \delta(t) = \delta(t-1) + c \quad \text{otherwise} \end{cases}$$

If no better best fitness value has been found from the last generation to the current and if H_1 was decided to hold true, then:

$$\Delta(t) = \begin{cases} \delta(t), \text{with } \delta(t) = \lceil \delta(t-1) \rceil \text{ if } [(\delta(t-1)] = \lceil \delta(t-1) \rceil \\ \Delta(t-1),\ \delta(t) = \delta(t-1) \quad \text{otherwise} \end{cases}$$

where $[\cdot]$ denotes the integer part of "·" and $\lceil \cdot \rceil$ is the upper integer towards infinity of "·".

3 Test Problem and Experimental Results

ARGA has been tested on real-valued functions (one defined by the authors, and another classical test function) to analyze the search behavior of ARGAII in comparison to ARGA and to a more standard variant of GA, and also for studying the influence of the γ parameter. For brevity we will call an "optimum" point a local optimum which is the best point found at the end of the run.

3.1 Escaping Trap Local Optima

We construct a multimodal function, called F1 that contains both smooth peaks and wide valleys as well as a highly rugged landscape concentrated in a small region of the search space. The function is ideal for analyzing ARGAII, because decision shifts when passing from smooth peaks, to the highly rugged landscape and therefore, the capacity to escape the local and smooth optima to the more promising region of the search space can be analyzed. The function is given in Fig.1. The initial population is taken within the bounds $x \in [5, 5.5]$ and

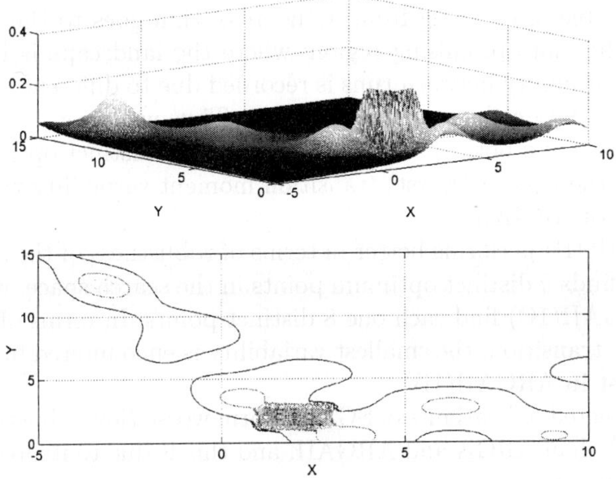

Fig. 1. Function F1

$y \in [14, 14.5]$ as we want to test if the algorithm gets trapped (see, Fig.1). The function is: $\mathrm{F1} = 0.2 \cdot \mathrm{fit1} + \mathrm{fit2}$, where: $a = 1; b = \frac{5.1}{4\pi^2}; c = \frac{5}{pi}; d = 6; e = 10; f = \frac{1}{8\pi}$; $\mathrm{fit1} = \exp(-(x-2)^8 - (y-2)^8) \cdot (\cos(100(X-2)^2 + 50(Y-2)^2))^2$; $\mathrm{fit2} = 1/(a(y - bx^2 + cx - d)^2 + e(1-f)\cos(y) \cdot \cos(x) + \log(x^2 + y^2 + 1) + e)$; We compare ARGAII to ARGA having the same common parameters and to a Standard GA with k-elitist selection on top of a binary tournament selection. The parameters of the strategies are given in Tab.1.

The results are given in Fig.2 which displays statistics over 100 independent runs for each algorithm: ARGAII, ARGA, SGA. For each algorithm we plot

Table 1. Strategies' parameters for F1

Strategy-Parameter	ARGAII (1A,1B,1C)	ARGA	SGA (2A,2B,2C)
Population size: N	30	30	30
Crossover rate: P_c	0.8	0.8	0.8
Mutation rate: P_m	max 0.5	max 0.5	(2A)0.1,(2B)0.01,(2C)0.001
Max. no. generations:	300	300	300
k (elitism)	5	5	5
c	0.4	0.4	not present
Δ_0	5	5	not present
ϵ	10^{-14}	10^{-14}	not present
γ	(1A)0.5,(1B)1,(1C)2	not present	not present
K	5	not present	not present

the best fitness over the run, three curves being generated and representing the maximal, average and minimum value of the best fitness for each generation, over all 100 runs. The optimal points, found at the end of the runs are plotted in a separate graph. From Fig.2 one can note the following: The evolution of the GA comprises two phases: first, the algorithm searches the trap local optima, and when it is capable of escaping from it the algorithm goes to the second phase: searching in the more promising region: where the landscape is highly rugged. The greatest variability between runs is recorded due to different moments when the transition between the two phases occurs. Thus we have:

I) ARGA achieves both the worst robustness (96 distinct optimum points in 100 runs) and the biggest (worst) transition moment variability when compared to all variants of ARGAII.

II) ARGAII(1B) performs better in terms of robustness of the solution found in 100 runs it finds 5 distinct optimum points in the search space, while ARGAII (1A) and ARGAII(1C) find each one 8 distinct points. In terms of variability of the moment of transition, the smallest variability is encountered in ARGAII(1C) and the biggest in ARGAII(1B).

III) As expected, all variants of SGA perform worse (low robustness and quality of solution) than ARGA and ARGAII, and this is due to its low adaptability (fixed mutation rates).

IV) SGA(2A) is able to escape the local optima with relatively low variability but it has also the worst robustness: 5 distinct points. The rest of the instances of SGA most often converge to points outside the most promising region. The lowest variability is achieved by SGA(2B), that is, 3 distinct optimal points.

3.2 Highly Multimodal and Multidimensional Landscapes

Next, the Ackley test function [1] (denominated as F2) was employed for testing ARGAII in comparison to ARGA. The function is highly rugged as well as highly dimensional: a variant with 20 variables (dimensions) being employed. SGA was not used in comparison due to the complexity of F2 as opposed to the lack of potential of SGA. The strategies' parameters are given in Tab.2. The

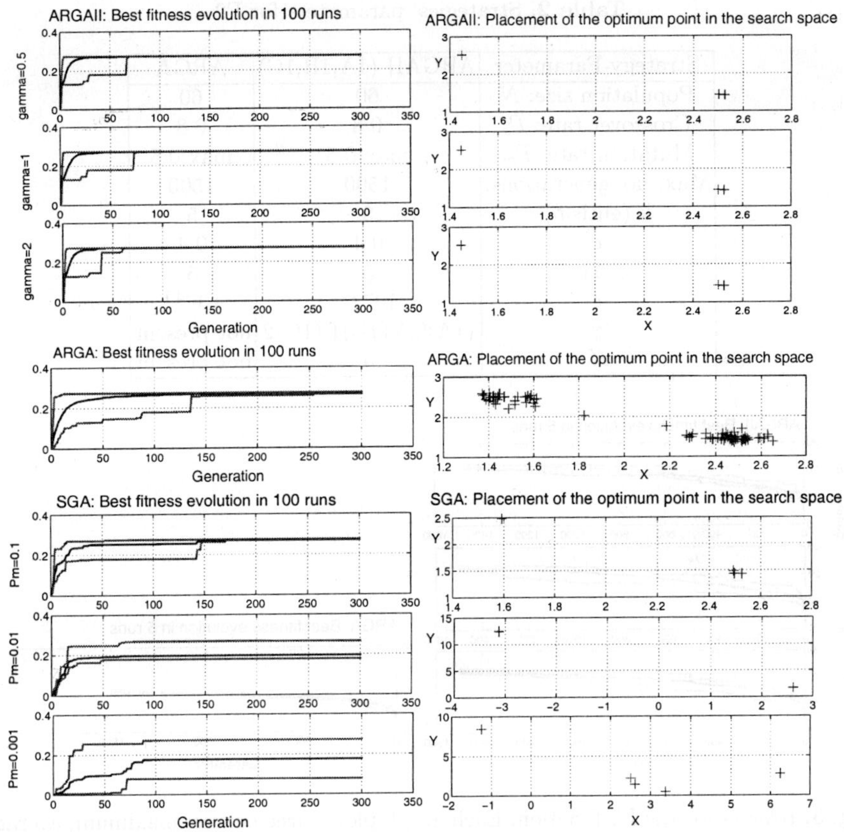

Fig. 2. Results for the F1 function. Left side: Each graph plots three curves: maximum, average, and minimum best fitness, statistics being calculated over 100 runs. Right side: Placement of the optimal solutions encountered in 100 runs

original Ackley function is minimized, however our variant F2 is maximized, and therefore F2 represents an inverted version of Ackley's function scaled by adding the value 25. The optimum is located in the origin, and has fitness value equal to 25.

The results are given in Fig.3 which plots three curves representing the maximal, average and minimum value of the best fitness for each generation, over 5 independent runs. From the plot it follows that: a)ARGA performs poorly compared to ARGAII as one can note that in all cases ARGAII reaches above 20 fitness levels, while ARGA reaches bellow 20, performing the same number of functions evaluations as ARGAII. b) the influence of γ parameter is different from that in F1. Thus, for F2 the influence of the respective parameter is smaller than that for F1, however, the dispersion of the curves plotting the best fitness during evolution, is smallest in the case of ARGAII(1A).

Table 2. Strategies' parameters for F2

Strategy-Parameter	ARGAII (1A,1B,1C)	ARGA
Population size: N	60	60
Crossover rate: P_c	0.8	0.8
Mutation rate: P_m	max 0.5	max 0.5
Max. no. generations:	1500	1500
k (elitism)	5	5
c	0.4	0.4
Δ_0	5	5
ϵ	10^{-14}	10^{-14}
γ	(1A)0.5,(1B)1,(1C)2	not present
K	5	not present

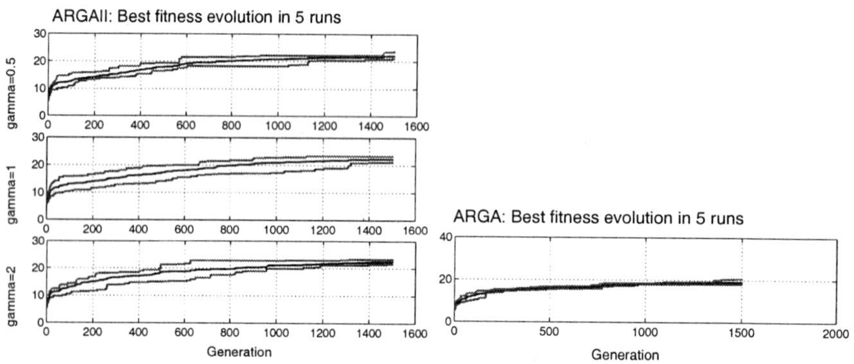

Fig. 3. Results for the F2 function. Each graph plots three curves: maximum, average, and minimum best fitness, statistics being calculated over 5 runs

3.3 Discussion of Results

For F1 the results obtained point out that: employing equal costs in taking the wrong decision (i.e. $\gamma = 1$) was beneficial in terms of robustness of the best solution found at the end of the run (see ARGAII(1B)). When a higher cost was allocated to deciding that the optimum is nearby when actually it was far away (i.e. C_{10}), higher than the cost allocated to deciding that the optimum is far, when it was actually close (i.e. C_{10}), this lead to a more dynamic strategy (ARGAII(1C)). It proved out to be more efficient in terms of escaping the initial local optimum, but worse when it comes to finding a robust solution, as expected; the strategy is more explorative, and might miss the good points from lack of sufficient exploitation. For F2, the influence of the γ parameter is not so dramatic, therefore, we assume that in this case any value of this parameter can be chosen resulting in a similar behavior for ARGAII. The experiment shows that γ has an effect of additional control over the search mechanism. If no information about the search space is known, a unitary value for γ should be adopted: equal costs for the two possible wrong decisions.

4 Conclusions

Employing a direct on–line control to the process of adaptation by including a decision process has proved to be efficient in comparison to a similar algorithm not having incorporated the decision process. The results show that ARGAII is better in terms of discovering good solutions within a small number of generations, for the test functions used. ARGAII proves itself more robust (in the case of F1), and converging to better solutions (in the case of F2), than ARGA. ARGA needs more generations to achieve the same performance as ARGAII[2]. It is important to note that ARGAII makes the assumption that a promising region is a region with high ruggedness. On problems for which this assumption doesn't hold, ARGAII is expected not to work well, as in this case the decision process would be meaningless. The algorithm may be improved at the level where clustering is performed, better and faster variants of clustering should be sought, as this stage gives indirectly the estimate of the ruggedness. Further analysis will be done on a wider test suite, on real world applications, such as the image enhancement problem requiring optimizing complex criteria. ARGAII will be extended to deal with discrete optimization problems.

References

1. Baeck, T.: Evolutionary Algorithms in Theory and Practice. Oxford University Press, New York (1996)
2. Fukunaga, K.: Introduction to Statistical Pattern Recognition. Academic Press (1974)
3. Goldberg, D.: Genetic Algorithms in Search, Optimization, and Machine Learning. Addison–Wesley (1989)
4. Hinterding, R., Michalewicz, Z., Eiben, A.E.: Adaptation in Evolutionary Computation: A Survey. Proceeings of IEEE ICEC97 (1997) 65–69
5. Hordijk, W.: A Measure of Landscapes. Evol. Comput. **4** 4 (1996) 335–360
6. Horn, J., Goldberg, D.: Genetic Algorithm Difficulty and the Modality of Fitness Landscapes. FOGA3, Morgan Kauffman (1995) 243–269
7. Munteanu, C., Lazarescu, V.: Global Search Using a New Evolutionary Framework: The Adaptive Reservoir Genetic Algorithm. Complexity Intnl. **5** (1998)
8. Munteanu, C., Rosa, A.: Adaptive Reservoir Genetic Algorithm: Convergence Analysis. Proceedings of EC'02, WSEAS (2002) 235–238
9. Obermaier, B., Munteanu, C., Rosa, A., Pfurtscheller, G.: Asymmetric Hemisphere Modeling in an Off-line Brain-Computer Interface. IEEE Trans. on Systems, Man, and Cybernetics: Part C. **31** 4 (2001) 536–540
10. Vassilev, V., Fogarty, T., Miller, J.: Information Characteristics and the Structure of Landscapes. Evol. Comput. **8** 1 (2000) 31–60
11. Wolpert, D. H., Macready, W. G.: No Free Lunch Theorems for Optimization. IEEE Trans. on Evol. Comput. **1** 1 (1997) 67–82

[2] The CPU time spent in one ARGAII generation is about 1.7 times bigger than the CPU time spent in one ARGA generation, due mainly to the computational overhead of the LS method

Genetic Algorithm Visualization Using Self-organizing Maps

G. Romero, J.J. Merelo, P.A. Castillo, J.G. Castellano and M.G. Arenas

Department of Architecture and Computer Technology
University of Granada
Campus de Fuentenueva
E. 18071 Granada (Spain)
e-mail: gustavo@geneura.ugr.es URL: http://geneura.ugr.es

Abstract. This paper gives an overview of evolutionary computation visualization and describes the application of visualization to some well known multidimensional problems. Self-Organizing Maps (SOM) are used for multidimensional scaling and projection. We show how different ways of training the SOM make it more or less adequate for the visualization task.

1 Introduction

Evolutionary Algorithms (EA) produce a vast amount of data concerning run performance and progress. Apart from simple convergence towards the solution, the extraction of useful information to get further insight into the state and course of the algorithm is a non-trivial task. Understanding run behavior is difficult due to the fact that EAs adopt a stochastic approach to searching large problem spaces.

EA users often examine how the quality of the solutions found changes over time by using a graph of fitness versus generation number. Although such a graph illustrates the improvements in the quality of the solutions considered during the run, it does not illustrate the structure of the solutions being considered, the regions of the search space being explored, or how to fine-tune the algorithm parameters. Visualization is proposed as a useful method for solving this problem; only by understanding the search behavior of the algorithms can EA users be confident about their individual algorithm components and parameter settings.

The rest of the paper is organized as follows: Section 2 contains a classification and description of visualization techniques. Section 3 describes the state of the art in multidimensional visualization and scaling methods. In sections 4 and 5 we can see the result of applying the SOM based visualization method proposed in this paper to a couple of multidimensional problems, Onemax and Rastrigin, solved through an EA.

J.J. Merelo Guervós et al. (Eds.): PPSN VII, LNCS 2439, pp. 442–451, 2002.

2 Visualization Techniques

Visualization techniques have been used in EA for their ability to represent run data in an intuitive form. As the old saying goes, "a picture is worth a thousand words." These techniques provide a baseline for a better understanding of the evolutionary process.

Visualization techniques can be divided into several categories according to the following criteria:

- *What is seen?* We may graphically represent either the genotype (the representation of a problem solution) or the phenotype (the evaluative interpretation of the representation). In some problems, both of them, genotype and phenotype, can be visualized at once.
- *How many generations?* You can see data produced over many generations, so you can get a picture of the progress of the evolutionary algorithm, or you can see the data produced anew for every generation, so a picture of the current state of the algorithm is obtained.

2.1 Visualizing phenotypes

Visualizing phenotypes is the easiest and oldest visualization technique used in the evolutionary computation field. Normally, an individual's phenotype, is reduced to a single number called fitness. Researchers plot fitness against time or the fitness of other individuals.

Two examples of this are figure 1(c) for the Onemax problem (see section 4) and figure 4(c) for the Rastrigin problem (see section 5). This kind of representation is standard, the same for all problems. Sometimes, when a more complex fitness is used, especially in multi-objective problems, two visualization techniques can be used: plot every objective value separately or plot some combination of all of them.

In the above problems, fitness is the evaluation of the corresponding function substituting x and y for the values the genotype represent. In other problems like the traveling salesman problem (TSP) it can be the length of the path that an integer vector represents. For a genetic programming (GP) problem, it can be the degree of agreement with the objective algorithm or its degree of efficiency. Phenotype visualization lets us see and compare how good several solutions are.

2.2 Visualizing genotypes

The genotype is the representation of a problem solution. For an easy problem like minimizing the Rastrigin function for one dimension, plotted in Figure 4(a), it can be a real number or a binary string representing a real value. In the maximization of the Onemax function for two dimensions, plotted in Figure 1(b), it can be as simple as a pair of real numbers or a bit string representing the values of x and y. For the TSP it can be an integer vector representing a path through the cities. For a Genetic Programming (GP) problem it can be a tree representing an algorithm in any programming language. If possible, genotype visualization allows us to see the structure of the solutions.

This is a less common way of visualizing what is happening with our evolutionary algorithm, and is not usually as easy as showing phenotype progress. For problems involving three or less dimensions, we can plot fitness as a function of genotype. The problem appears when more dimensions are involved. Then a multidimensional scaling method is required to accomplish this task (Section 3).

Unfortunately, this way of visualization depends on the internal representation of genotypes, and as this internal representation is different for different problems, there is no standard way of doing it. So a different way of visualization has to be programmed for every kind of internal representation.

2.3 Visualizing the state of the evolution

Looking at one generation's data will yield a picture of the state of the population at that moment. Many conclusions can be drawn from a look at the state of an evolutionary algorithm. What degree of diversity exists? What areas of the search space are being explored? The sequence of pictures in Figure 3 answer some questions about what solutions look like and how good they are. Similar information can be obtained from any of the pictures in 6.

This gives a lot of information about how the algorithm is working in a single moment, but unfortunately nothing about the whole process. However several of these pictures can be used to extract information about the progress of an algorithm.

2.4 Visualizing the course of the evolution

Sometimes data from many generations can be seen at the same time, showing the course of the evolutionary process. One of the major drawbacks of evolutionary computation is the lack of user interaction during the process. By using some kind of visualization we can follow the process much better and stop it if it is getting lost in the search space or, even better, tune it to improve its work.

The course of evolution can be visualized using genotypes and phenotypes. Figures 1(c) and 4(c) give us a clue about the how is fitness value is evolving. The sequence in Figure 3 shows how genotypes are evolving through the search space. Ideally we would like to see data from many generations and at the same time distinguish which generation these data came from and what their fitnesses are. Unfortunately this is usually difficult to achieve.

3 Multidimensional Visualization

Most techniques for visualization are limited to representing data depending on no more than three variables. This is due to the fact that human vision is limited to three dimensions, but there are two possible extensions to go beyond this limitation: using color for the fourth dimension and time as the fifth dimension.

Neither possibility is very common and requires practice, especially if time is used for visualizing the fifth dimension. However, if the problem incorporates more than five dimensions a method for visualizing an arbitrarily high number of dimensions must be used.

For the visualization of multidimensional data, a method to transform multidimensional data to a lower dimension is needed, preferably to 2 or 3 dimensions. This transformation should provide a lower-dimensional picture where the dissimilarities between the data points of the multidimensional domain correspond to the dissimilarities of the lower-dimensional domain. These transformation methods are referred to as *multidimensional scaling* ([1,2]).

To measure the dissimilarity, the distances between pairs of data points is used. These distances can be genuine distance in the high-dimensional domain such as Euclidean distance.

There are many projection methods with different strengths and weaknesses. In [3] König gives a survey of previous and new unsupervised projection techniques with special attention to their computational complexity and structure preservation. Some of the more well known are principal component analysis (PCA) [4], Sammon's projection [5], curvilinear component analysis (CCA) [6], curvilinear distance analysis (CDA) [7], k-means based ones [8], self-organizing maps (SOM) [9], etc.

Linear methods, like PCA, are computationally light but fail to handle nonlinear structures. Nonlinear methods, such as Sammon's projection, CCA and CDA, can handle nonlinear structures, but they are computationally intensive. They also emphasize local distances over global ones and are therefore excellent for showing the local shape of a data set.

Self-organizing maps (SOM) [9] are a data visualization technique invented by Professor Teuvo Kohonen which reduce the dimensions of data through the use of self-organizing neural networks. The SOM algorithm is based on unsupervised, competitive learning. It provides a topology preserving mapping from the high dimensional space to map units. Map units, or neurons, usually form a two-dimensional lattice and thus the mapping is a mapping from high dimensional space onto a plane. The property of topology preserving means that the mapping preserves the relative distance between the points. Points that are near each other in the input space are mapped to nearby map units in the SOM. The SOM can thus serve as a cluster analyzing tool of high-dimensional data. Also, the SOM has the capability to generalize. Generalization capability means that the network can recognize or characterize inputs it has never encountered before. A new input is assimilated with the map unit it is mapped to. With all the previous methods, once the data points are transformed from high-dimensional space to the lower-dimensional one, no new points can be projected without repeating the whole process. SOM has the advantage of being able to transform new points between spaces very easily once it has been trained.

4 The Onemax Problem

For a binary string of length n this is the problem of maximizing the function

$$f(\vec{x}) = \sum_{i=1}^{n} x_i, x_i \in \{0,1\}^n$$

Although it is a easy problem, it is useful for explaining our visualization method. The problem is a simple unimodal function when the number of dimensions is small. As the number of dimensions increases finding the global unique global maximum gets harder because the number of local optima in flat areas grows exponentially. To take a look of the shape of this function, here are its one and two dimensional plots.

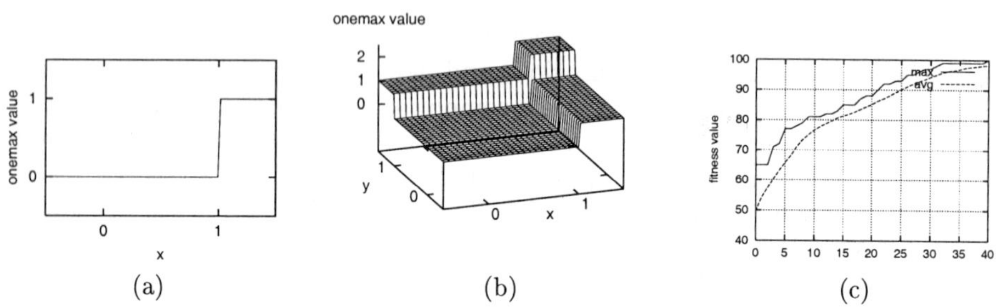

Fig. 1. Plots of the onemax function for one (a) and two (b) dimensions. (c) A fitness versus time plot of the EA run whose projections can be seen in figures 2 and 3.

The search space of this problem is the set of binary strings of fixed length n represented by the binary vector space $\mathbb{Z}_2^n$. The elements of $\mathbb{Z}_2^n$ are the vectors $z = (z_1, z_2, \ldots, z_n)$ with $z_i \in \{0,1\}$. There are a couple of special n-tuples, one consisting entirely of zeros $\mathbf{0}$, and other consisting entirely of ones, $\mathbf{1}$. They are the worst and the best possible solutions to this problem.

When n is greater than 3, the human eye can't see a solution as a vector of $\mathbb{Z}_2^n$ over an n-dimensional function. In this cases a multidimensional scaling method (see section 3) is necessary. As said above, we are going to use SOM for multidimensional scaling and projection. In our experiments we use $n = 100$ for this problem.

The first step is to train the map. As a first approach, the map was trained with random points selected from $\mathbb{Z}_2^{100}$. Then the solutions provided by the evolutionary algorithm (EA) are projected over the SOM. The results when projecting solutions of the problem, with $n = 100$, over the random trained map was not very satisfactory, as can be seen on figure 2.

Obviously, SOM can't do a perfect projection, but can do better than in figure 2. To better the map it has to be trained more carefully. Now, instead

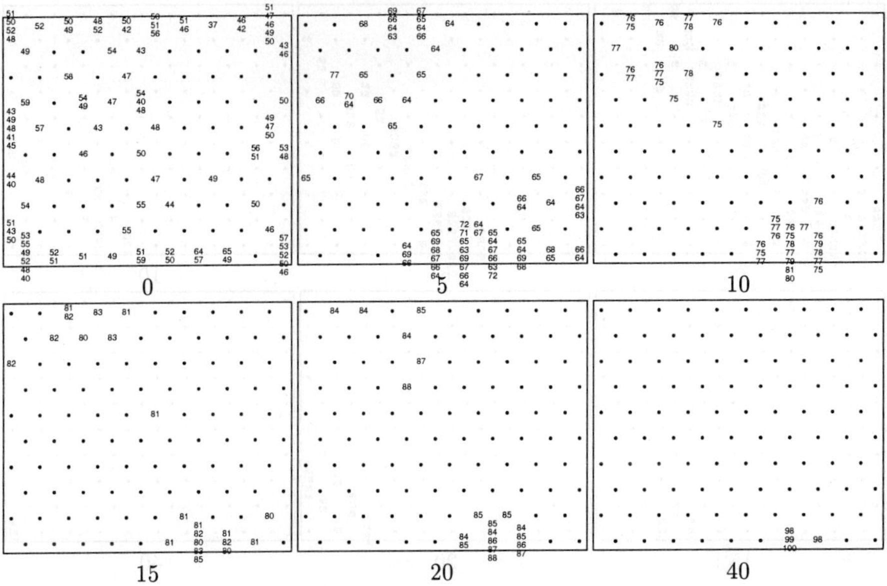

Fig. 2. Projection of onemax solutions over a SOM trained with random points from $\mathbb{Z}_2^{100}$. Numbers inside pictures are fitness value. Numbers below pictures indicate generation number.

of random points from $\mathbb{Z}_2^{100}$, we are going to choose a subset that will improve the projection capabilities of SOM. There are $\binom{100}{x}$ points in $\mathbb{Z}_2^{100}$ with exactly x ones. So the number of different points with equal fitness grows as they move away from **0** or **1**. If we care about this issue when choosing a subset of $\mathbb{Z}_2^{100}$, more points will be selected if they are far from **0** or **1**. Projection the same EA run as as before, the obtained result can be seen in figure 3.

Now we can follow in a better way what the EA is doing. The underlying map has the worse point, **0**, in the top-left corner, and the best, **1**, on the bottom-right one. The first picture of figure 3 shows the initial random population. As individuals are randomly selected with a 50% of posibilities of being 0 of 1 for every position, most of them has a fitness value around 50. Next frames shows how the population evolves to better solutions, until it finds the optimun in generation 40. As better solutions are found, individuals shift to the right-bottom area, where the optimum is located.

Unlike with the previous map, here the convergence process towards optimum is much clearer. With the first map there is not a clear convergence direction, as is with the second. This map also eliminates the problem with the previous attempt, where similar fitness individuals were projected to well-separated areas of the map.

In conclusion, knowledge about the function landscape helps us to improve the SOM results. Now we are going to try the same method with some multidimensional problems of greater difficulty.

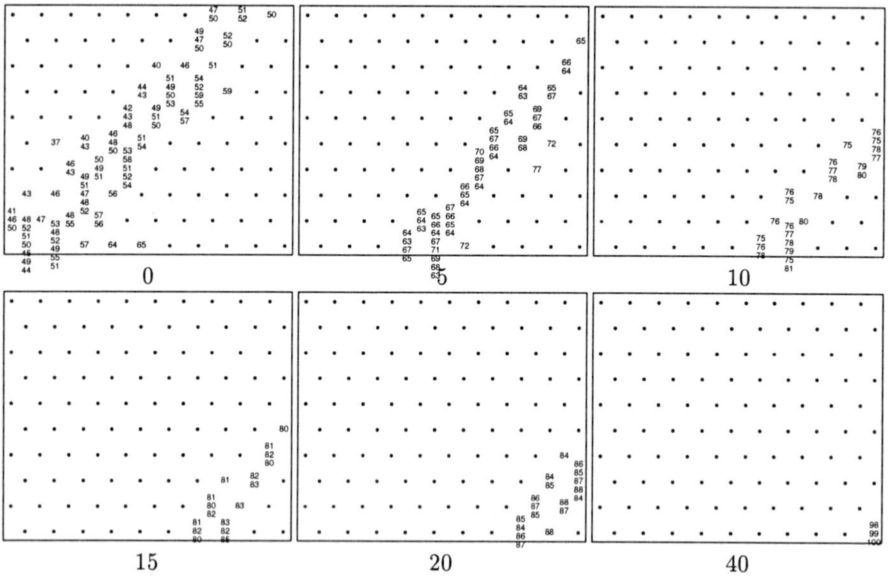

Fig. 3. Projection of onemax solutions over a SOM trained with selected points from $\mathbb{Z}_2^{100}$. Numbers inside pictures are fitness value. Numbers below pictures indicate generation number.

5 The Rastrigin Problem

The Rastrigin function is

$$f(\overrightarrow{x}) = \alpha n + \sum_{i=1}^{n} x_i^2 - \alpha \cos(2\pi x_i), x_i \in [-5.12, 5.12]$$

For the experiments $\alpha = 10$ and $n = 10$. The global minimum of zero is at the point $\mathbf{0} =< 0, 0, \ldots, 0 >$. In figure 4 we can see the plots of the Rastrigin function for one and two dimensions. The main characteristic of this function is the existence of many suboptimal solutions whose values increase as the distance from the global minimum increases.

In this case, the projection over the SOM trained with random points chosen from a hypercube in $\mathbb{R}^{10}$ give us only a small piece of information. From the sequence in figure 5 several conclusions can be extracted:

- The initial random population is adequately selected as it covers all the areas of the search space (see figure 5(0)).
- Diversity decreases with time as the population evolves and converges to the optimum global in $< 0, 0, \cdots, 0 >$.

Like we do with the onemax problem, we look for a better way of training the SOM for our visualization purposes. Now instead of random points from $\mathbb{R}^{10}$

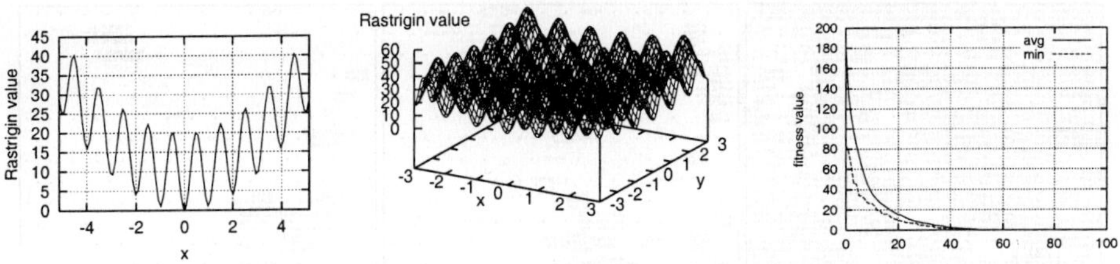

(a) (b) (c)

Fig. 4. Plots of the Rastrigin function for one (a) and two (b) dimensions. (c) A fitness versus time plot of the EA run whose projections can be seen in figures 5 and 6.

we choose more points near the optimum, that we know is in $< 0, 0, \cdots, 0 >$. Figure 6 shows the projection of the same evolutionary process already seen in figure 5. A fitness versus time graph of this evolution can be seen in figure 4(c).

In the second map, figure 6, the center represent $< 0, 0, \cdots, 0 >$ and the values of every cell grows with distance from the center.

The main differences between solutions projections with a random trained map (figure 5) and with a map trained with selected points (figure 6) are:

- The randomly created individuals from the initial population are represented very differently, in figure 5(0) they are spread all over the map, in figure 6(0) they only occupy the borders of the map.
- In the random map is difficult to follow the evolution of the individuals, although there are clusters of similar individuals, as similar individuals may lay in distant areas. With the second map is more easier as it is ordered almost like a star with center 0 and growing values as they get farther from the center.
- Good and bad fitness individuals can be projected over the same area in the first map. This doesn't happen with the non random one.

6 Conclusions and Future Work

SOM is a good for the task of multidimensional scaling and it can be applied to genetic algorithm visualization despite of problems with an arbitrarily high dimensional count.

As is the case with any problem, information about solution improves the visualization task. Initial tests with randomly trained maps over the search space can be helpful to discover some hints about initialization and diversity of the population, but carefully trained maps can do it much better and also give us an idea about the state and the course of the evolutionary process.

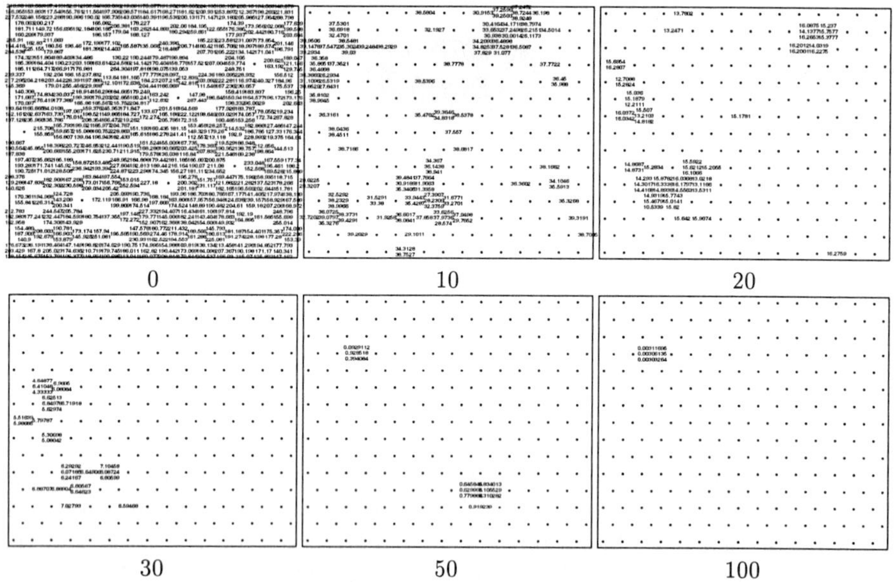

Fig. 5. Projection of Rastrigin solutions over a SOM trained with random points from a hypersphere in $\mathbb{R}^{10}$. Numbers inside pictures are fitness value. Numbers below pictures indicate generation number.

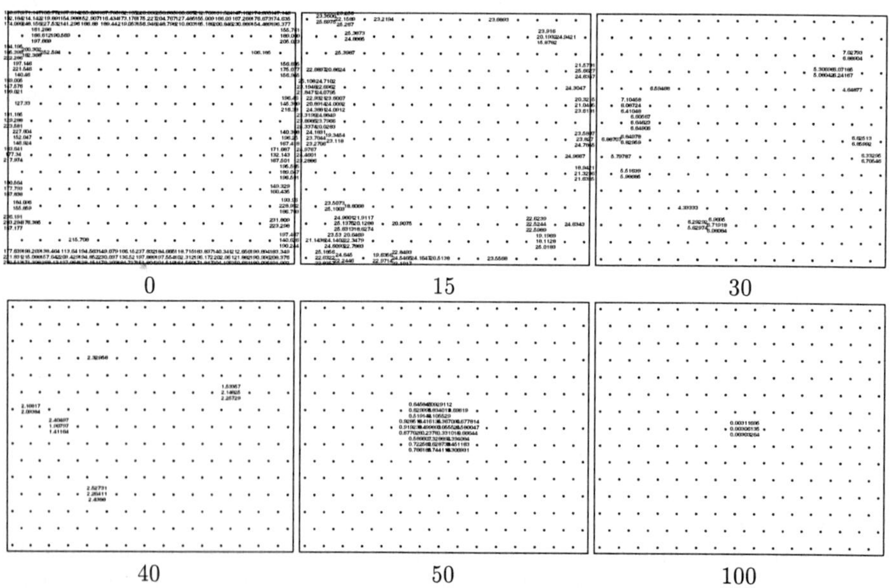

Fig. 6. Projection of Rastrigin solutions over a SOM trained with selected points from $\mathbb{R}^{10}$. Numbers inside pictures are fitness value. Numbers below pictures indicate generation number.

Acknowledgments

This work has been supported in part by the projects CICYT TIC-1999-0550, INTAS-9730950 and IST-1999-12679.

References

1. T.F. Cox and M.A.A. Cox. *Multidimensional Scaling.* London: Chapman & Hall, 1994.
2. B.D. Ripley. *Pattern Recognition and Neural Networks.* Cambridge, GB: Cambridge University Press, 1996.
3. Andreas König. A survey of methods for multivariate data projection, visualisation and interactive analysis. In *Proc. of the 5th International Conference on Soft Computing and Information/Intelligent Systems (IIZUKA'98)*, pages 55–59, October 1998.
4. Christopher M. Bishop. *Neural Networks for Pattern Recognition.* Clarendon Press, Oxford, 1995.
5. J.W. Sammon Jr. A nonlinear mapping for data structure analysis. *IEEE Transactions on Computers*, pages 401–409, 1969.
6. P. Demartines and J. Hérault. Curvilinear component analysis: a self organizing neural network for non lienar mapping of data sets. *IEEE Transactions on Neural Networks*, 8:148–154, 1997.
7. Nicolas Donckers Michael Verleysen John Aldo Lee, Amaury Lendasse. A robust nonlinear projection method. In *European Symposium on Artifial Neural Networks (ESANN2000)*, pages 13–20, April 2000.
8. S. Spangler D.S. Modha and S. Vaithyanathan. Multidimensional cluster visualization using guided tours. Technical Report Reseach Report RJ 10124, IBM Almaden Research Center, San Jose, CA, June 17 1998.
9. T. Kohonen. The Self-Organizing Map. In *Proceedings of the IEEE*, volume 78, pages 1464–1480, 1990.

Generalised Regression GA for Handling Inseparable Function Interaction: Algorithm and Applications

Rajkumar Roy and Ashutosh Tiwari

Department of Enterprise Integration, School of Industrial and Manufacturing Science, Cranfield University, Cranfield, Bedford, MK43 OAL, United Kingdom (UK)
{r.roy,a.tiwari}@cranfield.ac.uk

Abstract. Interaction among decision variables is inherent to a number of real-life engineering design optimisation problems. There are two types of variable interaction: inseparable function interaction and variable dependence. The aim of this paper is to present an Evolutionary Computing (EC) technique for handling complex inseparable function interaction, and to demonstrate its effectiveness using three case studies. The paper begins by devising a definition of inseparable function interaction, identifying the challenges and presenting a review of relevant literature. It then briefly describes Generalised Regression GA (GRGA) for handling complex inseparable function interaction in multi-objective optimisation problems. GRGA is applied to a complex test problem and two real-life engineering design optimisation case studies that exhibit complex inseparable function interaction. It is shown that GRGA exhibits better convergence and distribution of solutions than NSGA-II, which is a high-performing evolutionary-based multi-objective optimisation algorithm. The paper concludes by presenting the future research directions.

1 Introduction

Real-life engineering design optimisation problems, as opposed to the theoretical problems (test cases), are those that are encountered in industry. Some examples of these problems are the design of aerospace structures for minimum weight and the surface design of automobiles for improved aesthetics. Along with multiple objectives, constraints, qualitative issues and lack of prior knowledge, most real-life design optimisation problems also involve interaction among decision variables. In spite of its immense potential for real-life problems, lack of systematic research has plagued the field of interaction for a long time. This can mainly be attributed to the lack of sophisticated techniques, and inadequate hardware and software technologies. However, in the last two decades, some research has been carried out in this area especially in the field of statistical data analysis [1]. This has been further augmented in the recent past with the growth of computational intelligence techniques like Evolutionary Computing (EC), Neural Networks (NNs) and Fuzzy Logic (FL). This paper focuses on the development of an evolutionary-based algorithm for handling complex variable interaction in multi-objective optimisation problems.

J.J. Merelo Guervós et al. (Eds.): PPSN VII, LNCS 2439, pp. 452–461, 2002.

2 Types of Variable Interaction

In an ideal situation, desired results could be obtained by varying the decision variables of a given problem in a random fashion independent of each other. However, due to interaction this is not possible in a number of cases, implying that if the value of a given variable changes, the values of others should be changed in a unique way to get the required results. The two types of variable interaction are discussed below.

Inseparable Function Interaction: The first type of interaction among decision variables, known as inseparable function interaction, is the main focus of this paper. This interaction occurs when the effect that a variable has on the objective function(s) depends on the values of other variables in the function [2] (Fig. 1). This interaction, therefore, manifests itself as cross-product terms in the objective function(s).

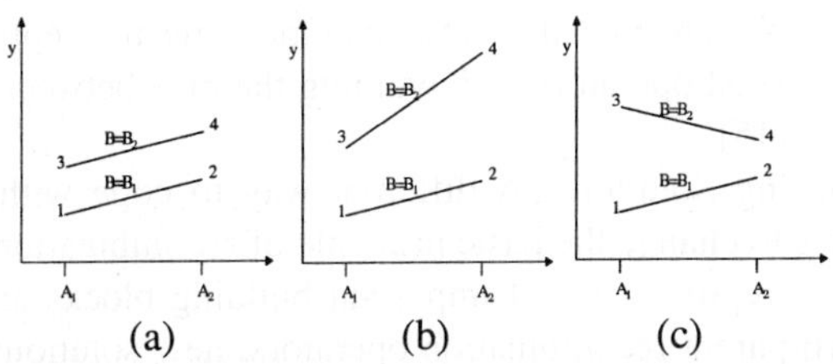

Fig. 1. Examples of inseparable variable interaction (a) No interaction (b) Synergistic interaction (c) Anti-synergistic interaction [3]

Variable Dependence: The second type of interaction among decision variables, known as variable dependence, is discussed in detail by Tiwari and Roy [4]. This interaction occurs when the variables are functions of each other, and hence cannot be varied independently.

3 Challenges Posed by Inseparable Function Interaction

Complex inseparable function interaction poses a number of challenges for multi-objective optimisation algorithms. A GA operates on the building blocks, growing them and mixing them with each other in an attempt to solve the search problem at hand. The inseparable function interaction causes problems for a GA by making it more difficult for it to build these building blocks [5]. Furthermore, in its presence, a multi-objective optimisation problem cannot be decomposed into simpler parts. Hence, a GA requires updating all decision variables in a unique way in order to attain the desired results. With a generic search operator, this becomes a difficult task for the GA. Furthermore, even if a set of Pareto-optimal solutions are obtained, it is difficult to maintain them since any change in one variable must be accompanied by related changes in others in order to remain on the Pareto front. The difficulties that inseparable function interaction may create for a GA are multiple local fronts, deceptive fronts, discontinuities in the Pareto front and inherent bias in the search space [6].

4 Techniques for Handling Inseparable Function Interaction

In GA literature, the inseparable function interaction is termed as epistasis. The GA community defines epistasis as the interaction between different genes in a chromosome [7]. A review of literature reveals that the Evolutionary-based Techniques for handling Inseparable Function Interaction (ETIFIs) can be classified into two broad categories based on the approach used for the prevention of building block disruption. These categories are briefly discussed below.

- Managing Race between Linkage Evolution and Allele Selection: Linkage is defined as the logical grouping of building block components to facilitate their growth and mixing. This strategy handles epistasis on the basis of the observation that the force that causes the evolution of linkage is, in effect, in a race against the force of allele selection. In order to make the GA successful, this strategy proposes three main ways of managing this race: manipulating the representation of solutions, using specialised operators and avoiding the race between linkage evolution and allele selection [5].
- Modelling Promising Solutions: A different way to cope with the disruption of partial solutions is to change the basic principle of recombination. In this approach, instead of implicit reproduction of important building blocks and their mixing by selection and two-parent recombination operators, new solutions are generated by using the information extracted from the entire set of promising solutions. Global information about the set of promising solutions can be used to estimate their distribution, and new solutions can be generated according to this estimate. A general scheme of the algorithms based on this principle is called the Estimation of Distribution Algorithm (EDA) [8].

A number of research questions remain unanswered regarding the theory of inseparable function interaction (epistasis). However, from the practical point of view a number of applications have been developed as discussed above. But these techniques can only deal with single-objective optimisation problems, defined in discrete domains. The few ETIFIs that are available for dealing with real search spaces have limited capability in terms of handling any significant inseparable function interaction. Therefore, this paper addresses the above-identified research gap by presenting an ETIFI 'Generalised Regression GA (GRGA)' for dealing with hybrid-valued (with integer and real variables) multi-objective optimisation problems.

5 Generalised Regression GA (GRGA)

For any continuous portion of the Pareto front, there is a unique relationship involving objective functions. This relationship is difficult to obtain analytically, and even if it is found, it has limited usefulness since mapping from function space to variable space is very complex. However, the existence of a relationship among objective functions of Pareto solutions necessarily implies that corresponding relationship(s) exist among the decision variables of these solutions [9].

To explain the above concept, consider a simple two-objective optimisation problem having f_1 and f_2 as the two objective functions. For any continuous portion of the Pareto front, there exists a Function F involving f_1 and f_2. Suppose the problem has two decision variables x_1 and x_2 that define the functions f_1 and f_2 i.e. f_1 and f_2 can be expressed as $f_1(x_1,x_2)$ and $f_2(x_1,x_2)$, leading to F_1.

$$\begin{aligned} &F(f_1, f_2) = 0, \\ &F(f_1(x_1, x_2), f_2(x_1, x_2)) = 0, \\ &\Rightarrow F_1(x_1, x_2) = 0. \end{aligned} \tag{1}$$

This proves the statement made earlier that there is existence of relationship(s) among the decision variables of the solutions belonging to any continuous portion of the Pareto front. GRGA aims to explore this relationship using non-linear multi-variable regression analysis [1]. It uses the relationship thus obtained [10]:

- To perform periodic and final re-distribution of solutions for aiding their spread over the current front.
- To use history of change of regression coefficients for guiding the search towards the global Pareto front.
- To use rate of change of regression coefficients for determining the termination condition of the algorithm.

As depicted in Fig. 2, GRGA encodes the above-mentioned solution strategy in C++. A high performing evolutionary-based multi-objective optimisation algorithm, NSGA-II [11] has been chosen as the optimisation engine for GRGA. However, since GRGA is completely modular, it can also be used with any other multi-objective optimisation algorithm for enhancing the algorithm performance in handling problems with complex inseparable function interaction. GRGA uses a distribution algorithm for periodically spreading out solutions over their current front [10].

In the following sections, GRGA is applied to a complex test problem and two real-life case studies. The complex inseparable function interaction in these problems makes them particularly difficult for multi-objective optimisation algorithms. These applications compare the performance of GRGA to that of NSGA-II. All the tests reported here correspond to 100 population size, 500 generations, 0.8 crossover probability, 0.05 mutation probability, and simulated binary crossover with 10 crossover distribution index and 50 mutation distribution index. These values are typically used in literature for these parameters. It was observed that the relative results from NSGA-II and GRGA do not change significantly with the change in these values. The results form the typical set obtained from 10 runs with different random number seed values. No major variation was observed in the results with the change in seed values.

6 Test Problem

This test problem is derived using the tuneable (parametric) Reverse Engineered Test Bed (RETB), proposed by Tiwari et al. [12]. The RETB parameters are chosen here with an aim to attain a problem that has two objectives, convex Pareto front, biased search space and multiple local Pareto fronts. The equation of this test problem is given below, and the search space represented by it is visually depicted in Fig. 3.

$$f_1(x_1,x_2) = \frac{1}{\left(1-\exp(-4)\right)}\left[1-\exp\left(-4x_1\right)\right], \forall 0 \le x_1, x_2 \le 1,$$
$$f_2(x_1,x_2) = (2-(f_1/I)^{0.6}) \times (I), \forall 0 \le x_1, x_2 \le 1,$$
$$I(x_1,x_2) = 2-\exp(-2x_2)\cos\left(8\pi x_2\right), \forall 0 \le x_1, x_2 \le 1. \quad (2)$$

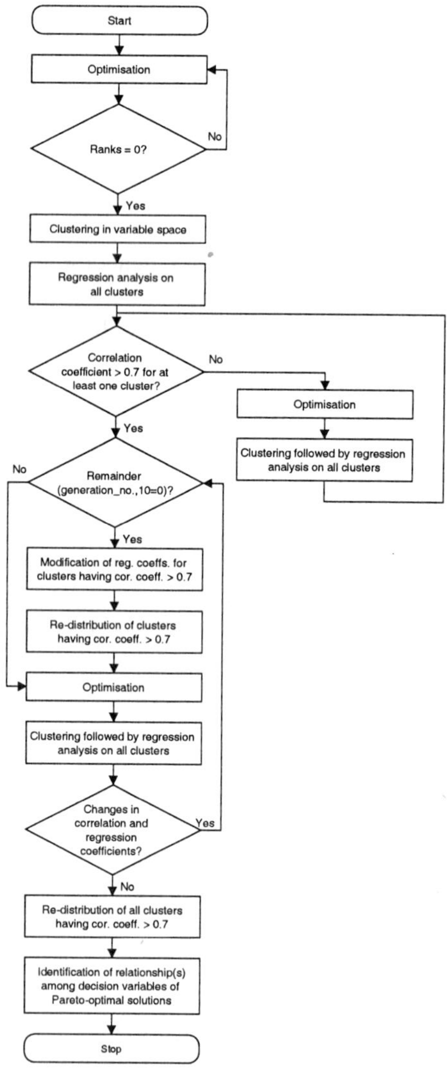

Fig. 2. Generalised Regression GA (GRGA) [10]

Experimental Results: The results obtained by applying GRGA and NSGA-II to this problem are shown in Fig. 3. The γ (convergence metric) and Δ (diversity metric) values corresponding to these results are shown in Table 1 [11].

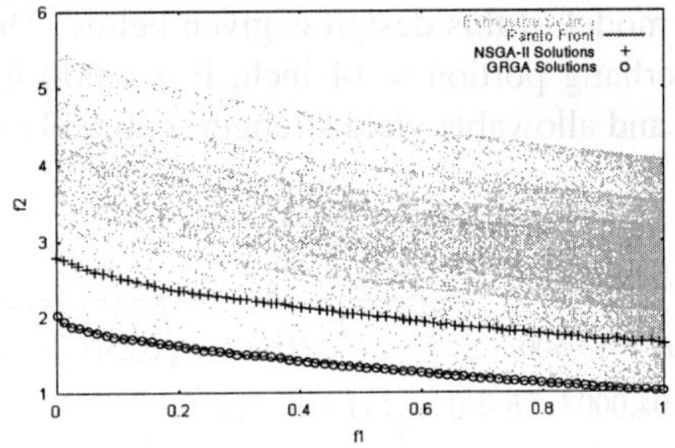

Fig. 3. Experimental results from test problem

Table 1. Performance metrics

Performance metrics		γ	Δ
Optimisation algorithms	NSGA-II	0.856745	0.090002
	GRGA	0.035642	0.080121

Discussion of Results: The following observations can be made from Fig. 3.

- GRGA exhibits better convergence as compared to NSGA-II. This is supported by Table 1 that shows that the γ value of GRGA is an order less than that of NSGA-II. This is because in this problem, the NSGA-II solutions have converged prematurely to one of the local fronts. GRGA addresses this drawback of NSGA-II through periodic modification of regression coefficients using the history of search observed in previous generations. This guides the search towards the global Pareto front by preventing it from getting trapped in local fronts.
- In this problem, both NSGA-II and GRGA exhibit satisfactory diversity.

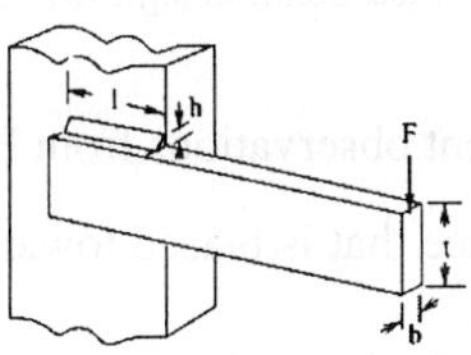

Fig. 4. Welded beam design [13]

7 Design of a Welded Beam

In this problem, a beam needs to be welded on another beam and must carry a certain load (Fig. 4) [13]. The objectives of the design are to minimise the cost of fabrication and the end deflection. Here, the overhang portion of the beam and the applied force (F) are specified, making the cross-sectional dimensions of the beam (b, t) and the weld dimensions (h, l) as the variables. This problem has four constraints, which arise from the allowable strengths of the material and the practicality requirements.

The mathematical model of this design is given below. This equation assumes the following values: overhang portion = 14 inch, F = 6,000 lb force, allowable shear strength = 13,600 psi and allowable yield strength = 30,000 psi.

$$
\begin{aligned}
&\textit{Minimise} \quad Cost = f_1(x) = 1.10471h^2 l + 0.04811tb(14.0 + l), \\
&\textit{Minimise} \quad End\_Deflection = f_2(x) = 2.1952/t^3 b, \\
&\textit{Constraints} \quad g_1(x) = 13{,}600 - \sqrt{\tau'^2 + \tau''^2 + (l\tau'\tau'')/\sqrt{0.25(l^2 + (h+t)^2)}} \geq 0, \\
&g_2(x) = 30{,}000 - 504{,}000/t^2 b \geq 0, \; g_3(x) = b - h \geq 0, \\
&g_4(x) = 64{,}746.022(1 - 0.0282346t)tb^3 - 6{,}000 \geq 0, \\
&\tau' = 6{,}000/\sqrt{2}hl, \; \tau'' = \frac{6{,}000(14 + 0.5l)\sqrt{0.25(l^2 + (h+t)^2)}}{2\left\{0.707hl(l^2/12 + 0.25(h+t)^2)\right\}}.
\end{aligned}
\tag{3}
$$

Experimental Results: The results obtained by applying GRGA and NSGA-II to this problem are shown in Fig. 5.

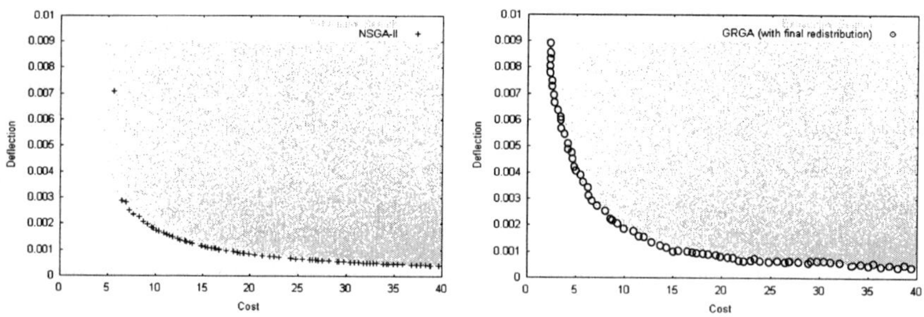

(a) (b)

Fig. 5. Experimental results from welded beam design (a) NSGA-II (b) GRGA (Units: deflection in inches and cost in cost units)

Discussion of Results: The salient observations from Fig. 5 are as follows.

- The problem has a search space that is biased towards high values of cost and low values of deflection.
- In this case, the absence of multiple local fronts makes it possible for both NSGA-II and GRGA to converge to the Pareto front. Due to the inherent bias in the search space, NSGA-II is unable to locate those Pareto-optimal solutions that correspond to low values of cost and high values of deflection. However, since GRGA converges to the Pareto front, it is able to determine the relationships involving decision variables that define the Pareto front. Since GRGA uses these values to redistribute the final solutions, a well-defined Pareto front is attained.
- GRGA reveals that the Pareto front of this problem corresponds to h = 0.422, l = 2.465 and t = 9.990. Therefore, to attain any solution on the Pareto front, the de-

signer needs to fix h, l and t to these values, and choose a value for b based on his/her preferences.

8 Design of a Machine Tool Spindle

This design is shown in Fig. 6 [14]. The four variables in this problem are the dimensions of the spindle (l, d_o, d_a, d_b). Two of these variables are real (l, d_o) and the rest two are discrete (d_a, d_b). This problem involves the minimisation of the volume of the spindle and the static displacement under the force F. This problem has nine constraints that include variable bounds and limits on the maximum radial run-out of the spindle nose.

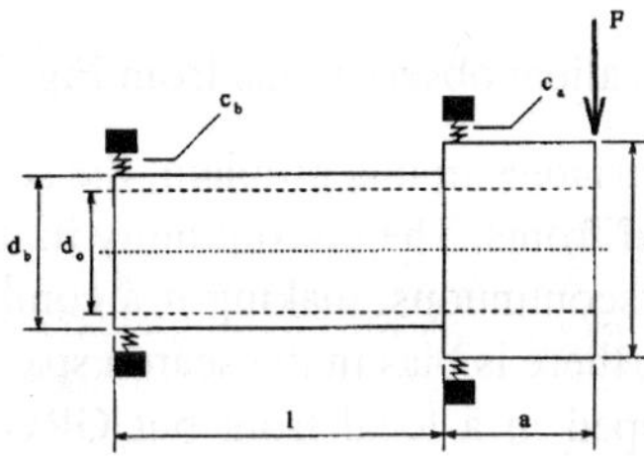

Fig. 6. Design of a machine tool spindle [14]

The mathematical model of this design is given below. This equation assumes the following values: d_{om} = 25 mm, d_{a1} = 80 mm, d_{a2} = 95 mm, d_{b1} = 75 mm, d_{b2} = 90 mm, p_1 = 1.25, l_k = 150 mm, l_g = 200 mm, a = 80 mm, E = 210,000 N/mm^2, F = 10,000 N, $\Delta_a = \Delta_b$ = 0.0054 mm, Δ = 0.01 mm, $d_{ra} = d_{rb}$ = -0.001 mm, d_a must be chosen from {80,85,90,95} and d_b from {75,80,85,90}.

$$Minimise(Volume\_of\_Spindle) \Rightarrow f_1(x) = (\pi/4)[a(d_a^2 - d_o^2) + l(d_b^2 - d_o^2)], \quad (4)$$

$$Minimise(Static\_Displacement) \Rightarrow f_2(x) = \frac{Fa^3}{3EI_a}(1 + \frac{lI_a}{aI_b}) + \frac{F}{c_a}[(1+\frac{a}{l})^2 + \frac{c_a a^2}{c_b l^2}],$$

$$Constraints \Rightarrow g_1(x) = l - l_g \le 0, g_2(x) = l_k - l \le 0, g_3(x) = d_{a1} - d_a \le 0,$$

$$g_4(x) = d_a - d_{a2} \le 0, g_5(x) = d_{b1} - d_b \le 0, g_6(x) = d_b - d_{b2} \le 0,$$

$$g_7(x) = d_{om} - d_o \le 0, g_8(x) = p_1 d_o - d_b \le 0, g_9(x) = \left|\Delta_a + (\Delta_a - \Delta_b)\frac{a}{l}\right| - \Delta \le 0,$$

$$\Delta = Maximum\_Runout\_of\_Spindle\_Nose, \Delta_a = Radial\_Runout\_of\_Front\_Bearing,$$

$$\Delta_b = Radial\_Runout\_of\_Back\_Bearing,$$

$$Moment\_of\_Inertia = I_a = 0.049(d_a^4 - d_o^4), Moment\_of\_Inertia = I_b = 0.049(d_b^4 - d_o^4),$$

$$Bearing\_Stiffness = c_a = 35400|d_{ra}|^{1/9} d_a^{10/9}, d_{ra} = Preload,$$

$$Bearing\_Stiffness = c_b = 35400|d_{rb}|^{1/9} d_b^{10/9}, d_{rb} = Preload.$$

Experimental Results: The results obtained from this problem are shown in Fig. 7.

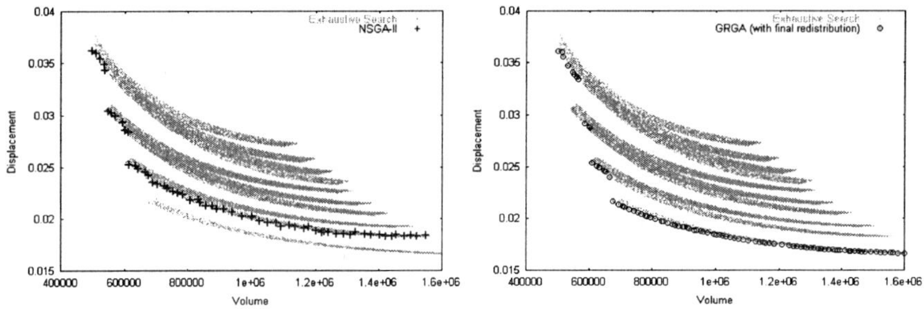

(a) (b)

Fig. 7. Experimental results from design of machine tool spindle (a) NSGA-II (b) GRGA (Units: displacement in mm and volume in mm^3)

Discussion of Results: The salient observations from Fig. 7 are as follows.

- Due to the two discrete variables in this problem, the search space is discontinuous that leads to multiple local fronts. The discontinuity in the search space also makes the global Pareto front discontinuous, making it a combination of parts of several local fronts. Furthermore, there is bias in the search space.
- Here, NSGA-II gets trapped in a local front but GRGA successfully locates the global Pareto front, including all its parts. The reasons for this behaviour are similar to the ones given in Sect. 6. In this problem, both NSGA-II and GRGA exhibit satisfactory diversity. GRGA also reveals to the designer that the Pareto front of this problem corresponds to $l = 187.78$, $d_a = 95$ and $d_b = 90$.

9 Future Research Activities and Conclusions

The current limitations of GRGA and the corresponding future research activities are:

- The performance of GRGA is dependent on how accurately the relationship among decision variables can be represented by RA. Hence, use of more sophisticated non-linear modelling tools have the potential of improving its performance.
- The distribution strategy used in GRGA needs to be further improved to make it more scalable with respect to the number of objectives and dimensions.
- GRGA is not capable of dealing with dependence among decision variables and qualitative issues such as manufacturability and designers' special preferences.

This paper focuses on inseparable function interaction. It presents an evolutionary-based optimisation algorithm, GRGA, which is capable of handling complex inseparable function interaction in multi-objective optimisation problems. GRGA is applied to a test problem and two real-life engineering design optimisation case studies that exhibit complex inseparable function interaction. The performance of GRGA is compared to that of a state-of-the-art multi-objective optimisation algorithm, NSGA-II. It is observed that GRGA exhibits better convergence and distribution of solutions than NSGA-II in handling problems that have complex inseparable function interaction.

Acknowledgements

The authors wish to acknowledge the support of the Engineering and Physical Sciences Research Council (EPSRC) – Grant No. GR/M 71473, Nissan Technical Centre – Europe (NTCE) and Structural Dynamics Research Corporation (SDRC) UK.

References

1. Draper, N.R. and Smith, H.: *Applied regression analysis.* John Wiley and Sons, Inc., New York, USA (1998)
2. Taguchi, G.: *System of experimental design.* Clausing, D. (ed.), UNIPUB/Kraus International Publications, vol. 1 and 2, New York, USA (1987)
3. Phadke, M.S.: *Quality engineering using robust design.* Prentice-Hall International Inc., London, UK (1989)
4. Tiwari, A. and Roy, R.: Variable dependence interaction and multi-objective optimisation. Accepted for publication: *Genetic and Evolutionary Computation Conference (GECCO-2002),* New York (USA), 9-13 July (2002)
5. Harik, G.R.: *Learning gene linkage to efficiently solve problems of bounded difficulty using genetic algorithms.* PhD. thesis, Computer science and engineering, University of Michigan, USA (1997)
6. Deb, K.: 'Multi-objective genetic algorithms: Problem difficulties and construction of test problems'. *Evolutionary computation,* vol. 7, no. 3 (1999) 205-230
7. Beasley, D., Bull, D. and Martin, R.: An overview of genetic algorithms: Part 2, research topics. *University computing,* 15(4) (1993) 170-181
8. Muhlenbein, H. and Paab, G.: From recombination of genes to the estimation of distributions I. Binary parameters. In: *Parallel Problem Solving from Nature IV (PPSN-IV),* Lecture notes in computer science, 46-55, Springer-Verlag, Berlin, Germany (1996)
9. Tiwari, A., Roy, R., Jared, G. and Munaux, O.: Interaction and multi-objective optimisation. In: Spector, L., Goodman, E., Wu, A., Langdon, W.B., Voigt, H.-M., Gen, M., Sen, S., Dorigo, M., Pezeshk, S., Garzon, M. and Burke, E. (eds.). *Proceedings of the Genetic and Evolutionary Computation Conference (GECCO-2001),* 671-678, Morgan Kaufmann Publishers, San Francisco, USA (2001)
10. Tiwari, A.: *Evolutionary computing techniques for handling variable interaction in engineering design optimisation.* PhD Thesis, School of Industrial and Manufacturing Science, Cranfield University, UK (2001)
11. Deb, K., Pratap, A., Agarwal, S. and Meyarivan, T.: *A fast and elitist multi-objective genetic algorithm: NSGA-II.* KanGAL Report No. 200002, Kanpur Genetic Algorithms Laboratory (KanGAL), Indian Institute of Technology (IIT), Kanpur, India (2000)
12. Tiwari, A., Roy, R., Jared, G. and Munaux, O.: Evolutionary-based techniques for real-life optimisation: Development and testing. Accepted for publication: *Applied Soft Computing (ASC) Journal,* Elsevier Science, Netherlands (2002)
13. Deb, K., Pratap, A. and Moitra, S.: Mechanical component design for multi-objective using elitist non-dominated sorting GA. In: *Parallel Problem Solving from Nature V (PPSN-V),* Lecture notes in computer science, 859-868, Springer-Verlag, Germany (2000)
14. Coello, C.A.C.: *An empirical study of evolutionary techniques for multiobjective optimization in engineering design.* PhD Thesis, Department of Computer Science, Tulane University, New Orleans, LA, USA (1997)

Diversity-Guided Evolutionary Algorithms

Rasmus K. Ursem

EVALife, Dept. of Computer Science, University of Aarhus, Bldg. 540
Ny Munkegade, DK-8000 Aarhus C, Denmark
ursem@daimi.au.dk

Abstract. Population diversity is undoubtably a key issue in the performance of evolutionary algorithms. A common hypothesis is that high diversity is important to avoid premature convergence and to escape local optima. Various diversity measures have been used to analyze algorithms, but so far few algorithms have used a measure to *guide the search.*

The diversity-guided evolutionary algorithm (DGEA) uses the well-known distance-to-average-point measure to alternate between phases of exploration (mutation) and phases of exploitation (recombination and selection). The DGEA showed remarkable results on a set of widely used benchmark problems, not only in terms of fitness, but more important: The DGEA saved a substantial amount of fitness evaluations compared to the simple EA, which is a critical factor in many real-world applications.

1 Introduction

A major problem in evolutionary algorithms (EAs) is that simple EAs have a tendency to converge to local optima. This *premature convergence* is caused by several algorithmic features, particularly selection pressure and too high gene flow between population members. First, a high selection pressure will quickly fill the population with clones of the better fit individuals, simply because their survival probability is too high compared to intermediate fit solutions. *Diversity* declines after a short while, and, because the population consists of similar individuals, the algorithm will have difficulties escaping the local optimum represented by the population. However, lowering the selection pressure is rarely an option because this will often lead to an unacceptable slow convergence speed. Second, high gene flow is often determined by the population structure. In simple EAs any individual can mate with any other individual. Consequently, genes spread fast throughout the population and the diversity drops quickly with fitness stagnation as a prevalent outcome.

Several studies have been carried out with the conflicting goals of maintaining a diversity that allows rapid convergence and still avoid premature convergence. Most studies fall in one of the following three categories:

1. Complex population structures to lower gene flow, e.g., the diffusion model [1, C6.3], the island model [1, C6.4], the multinational EA [2], and the religion-based EA [3].

J.J. Merelo Guervós et al. (Eds.): PPSN VII, LNCS 2439, pp. 462–471, 2002.

2. Specialized operators to control and assist the selection procedure, e.g., crowding [4], deterministic crowding [5], and sharing [6].
3. Reintroduction of genetic material, e.g., random immigrants [7], mass extinction models [8], [9], and [10].

Diversity is undoubtably closely related to the performance of evolutionary algorithms, especially when attempts are made to overcome the problems of avoiding premature convergence and escaping local optima. Maintaining high diversity is particulary important for optimization of dynamic and multiobjective problems. For dynamic problems high diversity increases the chances of relocating the peak after a change in the landscape, simply because the population covers a larger part of the search space. Algorithms for multiobjective optimization seek to report many tradeoffs between the conflicting objectives. Hence, higher diversity allows the algorithm to discover a larger part of the so-called Pareto front and thus report multiple tradeoffs between the objectives.

Diversity measures are traditionally used to *analyze* evolutionary algorithms rather than *guide* them. However, diversity measures have been used to control EAs in at least three studies. The Diversity-Control-Oriented Genetic Algorithm [11] use a diversity measure based on Hamming distance to calculate a survival probablility for the individuals. A low Hamming distance between the individual and the current best individual is translated into a low survival probability. Hence, diversity is preserved through the selection procedure. Another approach is the Shifting-Balance Genetic Algorithm [12]. The SBGA calculates a so-called containment factor between two subpopulations, which is based on Hamming distances between all members of the two populations. The distance is calculated between each member of population A and all members of population B. The factor determines the ratio between individuals selected on fitness and individuals selected to increase the distance between the two populations. A third, and more distantly related, approach is the Forking GA, which uses specialized diversity measures to turn a subset of the population into a subpopulation [13]. Two variants of the Forking GA exists. The first variant operates on the genotype, whereas the second type base the division on distances in the search space (on the phenotype).

2 The Diversity-Guided EA

The idea behind the DGEA is simple. Unlike most other EAs the DGEA uses a diversity measure to alternate between exploring and exploiting behavior. To use a measure for this purpose it has to be robust with respect to i) the population size, ii) the dimensionality of the problem, and iii) the search range of each of the variables. An immediate measure for N-dimensional numerical problems is the "distance-to-average-point" measure defined as:

$$diversity(P) = \frac{1}{|L| \cdot |P|} \cdot \sum_{i=1}^{|P|} \sqrt{\sum_{j=1}^{N} (s_{ij} - \overline{s}_j)^2}$$

```
DGEA main
   t = 0
   Initialize population P(0)
   Evaluate population P(0)
   mode = "Exploit"
   while(not(termination condition)) {
     t = t+1
     if(diversity(P(t))< d_low)
          mode = "Explore"
     elseif(diversity(P(t))> d_high)
          mode = "Exploit"

     if(mode == "Exploit")
          Select next generation P(t) from P(t − 1)
          Recombine P(t)
     else
          Mutate P(t)
     Evaluate population P(t)
   }
```

Fig. 1. Pseudocode for the DGEA.

where $|L|$ is the length of the diagonal[1] in the search space $\mathcal{S} \subseteq \mathbb{R}^N$, P is the population, $|P|$ is the population size, N is the dimensionality of the problem, s_{ij} is the j'th value of the i'th individual, and $\overline{s}_j$ is the j'th value of the average point $\overline{s}$. The pseudocode for the DGEA is listed in Fig. 1.

The DGEA applies *diversity-decreasing* operators (selection and recombination) as long as the diversity is above a certain threshold d_{low}. When the diversity drops below d_{low} the DGEA switches to *diversity-increasing* operators (mutation) until a diversity of d_{high} is reached. Hence, phases with exploration and phases with exploitation will occur (see Fig. 2). Theoretically, the DGEA should be able to escape local optima because the operators will force higher diversity regardless of fitness.

If $d_{low} = d_{high}$ the algorithm will maintain a diversity close to the given threshold value, which is particulary useful for dynamic and multiobjective optimization tasks.

An important issue is to apply a mutation operator that rather quickly increases the distance-to-average-point measure. Otherwise, the algorithm will stay in "explore"-mode for a long time. A straightforward idea for a measure-increasing mutation operator is to use the average point of the population to calculate the direction of each individual's mutation. The individual is then mutated with the Gaussian mutation operator, but now with a mean directed away from the average point (see Fig. 3). The purpose of this mutation operator is to *force* the individuals away from the population center. Preliminary results

[1] Assuming that each search variable x_k is in a finite range, i.e., $x_{kmin} \leq x_k \leq x_{kmax}$.

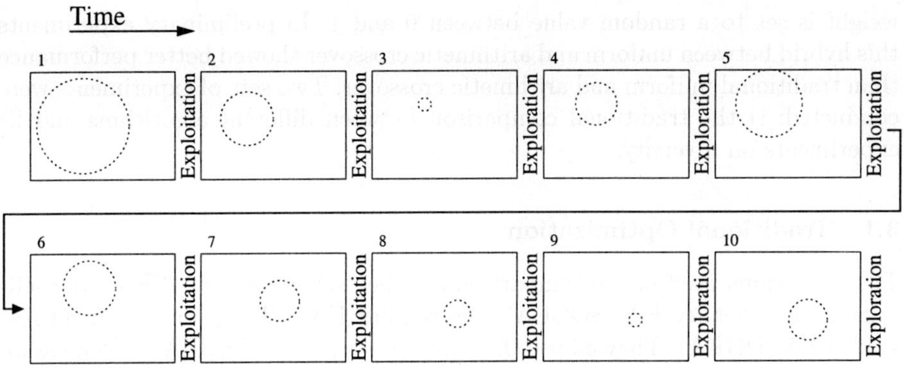

Fig. 2. Phases in the DGEA. The boxes denote the search space, the dotted circles indicate the diversity and position of the population. The mode is shown as the vertical text between the pictures, i.e. exploitation lowers the diversity in picture 1 and transforms it into picture 2.

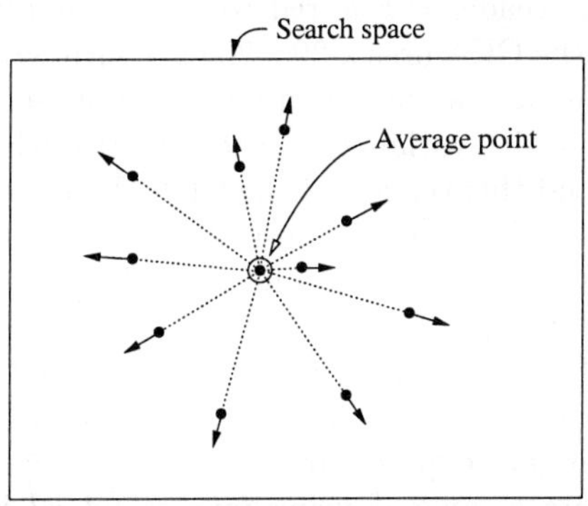

Fig. 3. Directed mutation in the DGEA.

indicated that scaling the normalized direction vector by 0.001 turned out to give the best results.

3 Experiments and Results

The experiments were performed using real-valued encoding, a population size of 400 individuals, binary tournament selection, and an elitism of 1 individual. Probability of mutating an entire genome was $p_m = 0.75$ and probability for crossover was $p_c = 0.9$. The compared algorithms all use variants of the standard Gaussian mutation operator (see below for further details). The mutation operator scales the randomly generated numbers by 20% of the length of the search intervals, which is just to make the operator problem independent. The algorithms use an arithmetic crossover with one weight for each variable. All weights except one are randomly assigned to either 0 or 1. The remaining weight is set to a random value between 0 and 1. In preliminary experiments this hybrid between uniform and arithmetic crossover showed better performance than traditional uniform and arithmetic crossover. Two sets of experiments were conducted: i) the traditional comparison between different algorithms and ii) experiments on diversity.

3.1 Traditional Optimization

The algorithms used in the comparison are the "standard EA" (SEA), the self-organized criticality EA (SOCEA), the cellular EA (CEA), and the diversity-guided EA (DGEA). They all use the above parameters. The SEA uses Gaussian mutation with zero mean and variance $\sigma^2 = 1/\sqrt{t+1}$. The SOCEA is a standard EA with non-fixed and non-decreasing variance $\sigma^2 = POW(10)$, where $POW(\alpha)$ is the power-law distribution[2]. The purpose of the SOC-mutation operator is to introduce many small, some mid-sized, and a few large mutations. The effect of this simple extension is quite outstanding considering the effort to implement it (one line of code). The reader is referred to [10] for additional information on the SOCEA. Further, the CEA uses a 20×20 grid with wrapped edges. The grid size corresponds to the 400 individuals used in the other algorithms. The CEA uses Gaussian mutation with variance $\sigma^2 = POW(10)$, which allows comparison between the SOCEA and this version of the CEA. Mating is performed between the individual at a cell and and a random neighbor from the four-neighborhood. The offspring replaces the center individual if it has a better fitness than the center individual. Finally, the DGEA uses the Gaussian mutation operator with variance $\sigma^2 = POW(1)$ and mean calculated as described in Sect. 2. The diversity boundaries were set to $d_{low} = 5 \cdot 10^{-6}$ and $d_{high} = 0.25$, which proved to be good settings in preliminary experiments.

The algorithms were compared using four standard benchmark problems. Each algorithm was tested on three variants of the problems; a 20 dimensional, a 50 dimensional, and a 100 dimensional. The number of generations was set to 50 times the dimensionality of the test problem, i.e., 20D = 1000 generations, 50D = 2500 generations, and 100D = 5000 generations. The four (minimization) problems are:

$$\text{Ackley F1}(x) = 20 + e - 20\exp\left(-0.2\sqrt{\frac{1}{n}\sum_{i=1}^{n} x_i^2}\right) - \exp\left(\frac{1}{n}\sum_{i=1}^{n} cos(2\pi \cdot x_i)\right) \quad \text{where } -30 \le x_i \le 30$$

[2] Power-law distributed numbers can be generated by $x = 1/u^{1/\alpha}$, where $u \sim U(0,1)$ is uniformly distributed, and α is a parameter determining the shape of the distribution. Another approach used in [10] is to log the avalanche sizes in the so-called sandpile model [14].

$$\text{Griewank F1}(x) = \frac{1}{4000}\sum_{i=1}^{n}(x_i - 100)^2 - \prod_{i=1}^{n}\cos\left(\frac{x_i - 100}{\sqrt{i}}\right) + 1$$
$$\text{where } -600 \le x_i \le 600$$

$$\text{Rastrigin F1}(x) = \sum_{i=1}^{n}(x_i^2 - 10\cos(2\pi x_i) + 10)$$
$$\text{where } -5.12 \le x_i \le 5.12$$

$$\text{Rosenbrock F1}(x) = \sum_{i=1}^{n-1}(100(x_{i+1} - x_i^2)^2 + (x_i - 1)^2)$$
$$\text{where } -100 \le x_i \le 100$$

The results listed in Table 1 clearly show that the DGEA outperforms the other algorithms by several magnitudes. In preliminary tests the DGEA continued to improve the fitness at the end of the fixed number of generations. Hence, better fitness was obtained if the DGEA ran until it stagnated, which is here defined to be 500 generations without fitness improvement. The results for this optimization are listed in the column denoted DGEA*.

From the pseudocode listed in Fig. 1 it seems that the DGEA should be somewhat faster than other algorithms, because the evolutionary operators are used less frequently (although the diversity calculation might be more expensive). Table 2 lists the runtime for the tested algorithms. According to these experiments the DGEA uses less time than even the simple standard EA.

3.2 Investigations on Diversity

Low diversity is often blamed for being the main reason for premature convergence. The principal argument is that a population of clones is likely to get trapped in a local optimum. However, low diversity might increase the chances of producing fitter offspring because the population will be located in the vicinity of an optimum. Table 3 shows the average diversity and mode for each test problem. The data points are recorded at each fitness improvement *after* the first period of fitness stagnation, which is defined to be 20 consecutive generations with no fitness improvement[3].

Two interesting conclusions can be drawn from the results in Table 3. First, the diversity appears to be surprisingly low when improvement occurs[4]. Second, almost no improvement occurs during the exploration phases. This is particulary interesting because it can help save a substantial amount of fitness evaluations during the exploration periods – an important feature for time-consuming evaluation of real-world problems. A variant (DGEA2) without evalution during

[3] This is to eliminate the noise in the beginning of a run.

[4] The range of the diversity measure is 0 to 0.5.

Table 1. Average fitness of the SEA, the SOCEA, the CEA, and the DGEA. 20D problems were optimized for 1000 generations, 50D for 2500, and 100D for 5000 generations. The results in the column DGEA* were obtained by running the DGEA until the best fitness was constant for 500 generations.

Problem	SEA	SOCEA	CEA	DGEA	DGEA*
Ackley 20D	2.49431	0.63380	0.23972	8.05E-4	3.36E-5
Ackley 50D	2.87039	1.52580	0.65169	4.61E-3	2.52E-4
Ackley 100D	2.89336	2.22080	1.14013	0.01329	9.80E-4
Griewank 20D	1.17195	0.93078	0.64280	7.02E-4	7.88E-8
Griewank 50D	1.61642	1.14741	1.03284	4.40E-3	1.19E-3
Griewank 100D	2.25001	1.62948	1.17907	0.01238	3.24E-3
Rastrigin 20D	11.12678	2.87524	1.25016	2.21E-5	3.37E-8
Rastrigin 50D	44.67488	22.46045	14.22400	0.01664	1.97E-6
Rastrigin 100D	106.21298	86.36449	58.38013	0.15665	6.56E-5
Rosenbrock 20D	8292.320	406.490	149.056	96.007	8.127
Rosenbrock 50D	41425.674	4783.246	1160.078	315.395	59.789
Rosenbrock 100D	91250.300	30427.636	6053.870	1161.550	880.324

Table 2. Milliseconds used by the SEA, the SOCEA, the CEA, and the DGEA. Average of 100 runs with 5000 generations on each of the four 100D problems.

Problem	SEA	SOCEA	CEA	DGEA
Ackley 100D	1128405	1528864	2951963	864316
Griewank 100D	1171301	1562931	3656724	969683
Rastrigin 100D	1124925	1513691	2897793	819691
Rosenbrock 100D	1087615	1496164	2183283	883811
Total	4512246	6101650	11689763	3537501
Percentage	100%	135.2%	259.1%	78.4%

exploration was investigated to see if it was possible to save fitness evaluations. The optimization results and the percentage of evaluations used by DGEA2 compared to DGEA and SEA are listed in Table 4. It should be noted that the DGEA2 uses a special kind of elitism during the explorative phases. In this elitism operator the best individual from the previous exploit-phase overwrites a random individual in the population, whereas the worst individual is overwritten during the exploitation phases. This special scheme is used to avoid the evaluation of individuals during the explorative phases.

4 Conclusions

The experiments revealed a number of interesting features of the DGEA in relation to optimization tasks. First, the DGEA outperformed the other algorithms

Table 3. Average diversity and current mode for the DGEA after first stagnation period. Data points are recorded when fitness improvement is detected, i.e., 100% in the Exploit column means that all improvements occured in Exploit mode. Each row is the average of 100 runs.

Problem	Diversity	Exploit	Explore
Ackley 20D	0.000388	100.00%	0.00%
Ackley 50D	0.000764	100.00%	0.00%
Ackley 100D	0.001082	100.00%	0.00%
Griewank 20D	0.000253	100.00%	0.00%
Griewank 50D	0.000662	100.00%	0.00%
Griewank 100D	0.000932	100.00%	0.00%
Rastrigin 20D	0.002056	100.00%	0.00%
Rastrigin 50D	0.002379	100.00%	0.00%
Rastrigin 100D	0.002817	100.00%	0.00%
Rosenbrock 20D	0.000601	99.99%	0.01%
Rosenbrock 50D	0.001134	99.91%	0.09%
Rosenbrock 100D	0.001562	99.91%	0.09%

Table 4. Average fitness of DGEA, average fitness of DGEA2, and number of fitness evaluations in percentage used by DGEA2 compared to DGEA and SEA. Number of generations are: 20D = 1000 generations, 50D = 2500, and 100D = 5000 generations (same as Table 1).

	Fitness		Evaluations in DGEA2	
Problem	DGEA	DGEA2	vs. DGEA	vs. SEA
Ackley 20D	8.05E-4	1.01E-3	70.1%	64.0%
Ackley 50D	4.61E-3	4.36E-3	74.3%	67.9%
Ackley 100D	0.01329	0.01311	77.0%	70.5%
Griewank 20D	7.02E-4	1.11E-3	94.2%	86.7%
Griewank 50D	4.40E-3	3.96E-3	94.8%	87.3%
Griewank 100D	0.01238	9.94E-3	95.4%	87.9%
Rastrigin 20D	2.21E-5	6.88E-4	58.6%	53.2%
Rastrigin 50D	0.01664	0.03699	61.9%	56.3%
Rastrigin 100D	0.15665	0.15613	64.1%	58.4%
Rosenbrock 20D	96.007	86.891	89.9%	82.7%
Rosenbrock 50D	315.395	295.680	90.3%	83.0%
Rosenbrock 100D	1161.550	758.040	90.5%	83.2%
Average			80.1%	73.4%

by several magnitudes on all test problems – it is clearly capable of escaping local optima. Second, the EA part (crossover, selection, and mutation) of the DGEA has lower running time than the standard EA, which again has lower running time than most other EAs. Third, the number of fitness evaluations may be reduced by approximately 25% compared with the standard EA, because the fitness is constant during the explorative phases. Reducing fitness evaluations is highly desireable for real-world applications, because the evaluation is often the time-critical factor in such applications. However, the results showed some variation in the reduction percentages, which indicates that this could be problem dependent.

In a more general context this study show the importance of *both* high and low diversity in optimization. High diversity allows the algorithm to escape local optima whereas low diversity ensures progress when fine-tuning the solutions.

Future work includes testing various variants of the algorithm. For instance, annealing the diversity thresholds d_{low} and d_{high} could lead to improvements, because it may be an advantage to decrease d_{low} near the end of the optimization. Further, investigations on a set of real-world problems are necessary to verify the results in a more realistic context. A number of system identification problems from control engineering are currently being investigated. Preliminary results from these studies are very encouraging.

References

1. Bäck, T., Fogel, D.B., Michalewicz, Z., and others, (eds.): Handbook on Evolutionary Computation. IOP Publishing Ltd and Oxford University Press, (1997)
2. Ursem, R.K.: Multinational Evolutionary Algorithms. In: Proceedings of the Congress of Evolutionary Computation (CEC-99), Angeline, P.J., Michalewicz, Z., Schoenauer, M., Yao, X., and Zalzala, A. (eds.), Vol. 3. 1633–1640 (1999)
3. Thomsen, R., Rickers, P., and Krink, T.: A Religion-Based Spatial Model For Evolutionary Algorithms. In: Parallel Problem Solving from Nature – PPSN VI, Schoenauer, M., Deb, K., Rudolph, G., Yao, X., Lutton, E., Merelo, J.J., and Schwefel, H.P. (eds.), Vol. 1. 817–826 (2000)
4. De Jong, K.A.: An Analysis of the Behavior of a Class of Genetic Adaptive Systems. PhD thesis, University of Michigan, Ann Arbor, MI, (1975). Dissertation Abstracts International 36(10), 5140B, University Microfilms Number 76-9381
5. Mahfoud, S.: Crowding and preselection revisited. Technical Report 92004, Illinois Genetic Algorithms Laboratory (IlliGAL), (1992)
6. Goldberg, D.E. and Richardson, J.: Genetic Algorithms with Sharing for Multimodal Function Optimization. In: Genetic Algorithms and their Applications (ICGA'87), Grefenstette, J.J. (ed.), 41–49. Lawrence Erlbaum Associates, Publishers, (1987)
7. Cobb, H.G. and Grefenstette, J.F.: Genetic Algorithms for Tracking Changing Environments. In: Proceedings of the 5th International Conference on Genetic Algorithms, Forrest, S. (ed.), 523–530 (1993)
8. Thomsen, R. and Rickers, P.: Introducing Spatial Agent-Based Models and Self-Organised Criticality to Evolutionary Algorithms. Master's thesis, University of Aarhus, Denmark, (2000)
9. Greenwood, G.W., Fogel, G.B., and Ciobanu, M.: Emphasizing Extinction in Evolutionary Programming. In: Proceedings of the Congress of Evolutionary Computation, Angeline, P.J., Michalewicz, Z., Schoenauer, M., Yao, X., and Zalzala, A. (eds.), Vol. 1. 666–671 (1999)

10. Krink, T., Thomsen, R., and Rickers, P.: Applying Self-Organised Criticality to Evolutionary Algorithms. In: Parallel Problem Solving from Nature – PPSN VI, Schoenauer, M., Deb, K., Rudolph, G., Yao, X., Lutton, E., Merelo, J.J., and Schwefel, H.P. (eds.), Vol. 1. 375–384 (2000)
11. Shimodaira, H.: A Diversity Control Oriented Genetic Algorithm (DCGA): Development and Experimental Results. In: Proceedings of the Genetic and Evolutionary Computation Conference, Banzhaf, W., Daida, J., Eiben, A.E., Garzon, M.H., Honavar, V., Jakiela, M., and Smith, R.E. (eds.), Vol. 1. 603–611 (1999)
12. Oppacher, F. and Wineberg, M.: The Shifting Balance Genetic Algorithm: Improving the GA in a Dynamic Environment. In: Proceedings of the Genetic and Evolutionary Computation Conference, Banzhaf, W., Daida, J., Eiben, A.E., Garzon, M.H., Honavar, V., Jakiela, M., and Smith, R.E. (eds.), Vol. 1. 504–510 (1999)
13. Tsutsui, S., Fujimoto, Y., and Ghosh, A.: Forking Genetic Algorithms: GAs with Search Space Division Schemes. Evolutionary Computation **5**, 61–80 (1997)
14. Bak, P.: How Nature Works. Copernicus, Springer-Verlag, New York, 1st edition, (1996)

Hybrid Algorithms: Neurogenetic Algorithms, Evolutionary Techniques Applied to Neural Nets

Evolutionary Optimization of Heterogeneous Problems

Lluís A. Belanche Muñoz

Dept. de Llenguatges i Sistemes Informàtics.
Universitat Politècnica de Catalunya.
c/Jordi Girona Salgado 1-3 08034 Barcelona, Spain.
belanche@lsi.upc.es

Abstract. A large number of practical optimization problems involve elements of quite diverse nature, described as mixtures of qualitative and quantitative information, and whose description is possibly incomplete. In this work we present an extension of the *breeder genetic algorithm* that represents and manipulates this heterogeneous information in a natural way. The algorithm is illustrated in a set of optimization tasks involving the training of different kinds of neural networks. An extensive experimental study is presented in order to show the potential of the algorithm.

1 Introduction

Real-world data come from many different sources, described by mixtures of numeric and qualitative variables. These variables include continuous or discrete numerical processes, symbolic information, etc. In particular, qualitative variables might have a different nature. Some are *ordinal* in the usual statistical sense (i.e., with a discrete domain composed by k categories, but totally ordered w.r.t a given relation) or *nominal* (discrete but lacking an order). The data may also come with their own peculiarities (vagueness, uncertainty, incompleteness), and thus may require completely different treatments.

For Evolutionary Algorithms (EA), this translates in the introduction of a coding system (a suitable representation) for all the information not directly representable in the chosen chromosomic language. In practical terms, this involves the use of binary (or small cardinality) alphabets (as in classical genetic algorithms, GA) or real-valued quantities (as in Evolution Strategies, ES). This can be seen as a pre-processing that is not part of the original task, and as such may have deep consequences in the structure of the problem.

On the one hand, the algorithm receives an *indirect* and biased feedback (via the fitness function). One may argue that the decoding (even a complex one) can be seen as part of the genotype-to-phenotype development. While this posture makes sense in living organisms, in artificial systems data representation is a crucial factor for a successful learning process and can have a great impact on performance. The choice of representation should be as faithful as possible, in the sense that the relations between the represented entities should correspond

J.J. Merelo Guervós et al. (Eds.): PPSN VII, LNCS 2439, pp. 475–484, 2002.

to meaningful relations between the original data items. The possibility of injecting knowledge in the genetic operators about the kind of structures they are manipulating is also an appealing idea. Indeed, the choice of a representation scheme should not be made alone, but in conjunction with the genetic operators. A set of genetic operators that allows important partial solutions to propagate and does not permit invalid or unimportant ones to succeed is likely to enhance overall performance. For example, a good tree representation should be able to represent all possible trees (and only trees) and be "continuous" (i.e., similar representations should develop similar trees) [1]. Furthermore, genetic operators should generate only valid trees (or repair invalid trees after generation). In this way, higher-order gene interactions can be worked out in a controlled way.

On the other hand, unwanted gene interactions (epistatic phenomena) are one of the headaches in designing EA solutions. Problems with little or no epistasis are trivial to solve (e.g. by hillclimbing). However, highly epistatic problems may be extremely difficult to solve, since they can prevent the formation of useful building blocks, and are prone to deception. These interactions are very likely to appear as a side effect of the chosen coding scheme. Finally, results involving genes that encode information may be difficult to interpret in isolation, being the case that they do not represent any specific feature.

In this work, we extend the expressive power of the *breeder genetic algorithm* (BGA) [2] allowing it to manipulate heterogeneous information. We then apply it to a set of optimization tasks involving the training of a heterogeneous neural network. The results illustrate how performance and readability are enhanced.

2 The Breeder Genetic Algorithm

The Breeder Genetic Algorithm [2] is in midway between GAs and ESs. While in GA selection is stochastic and loosely inspired in natural selection, the BGA uses *truncation* selection, in which only the best individuals (usually a fixed percentage τ of the population size μ) are to be recombined and mutated. Genetic operators are applied by randomly picking two parents until the number of offspring equals $\mu - q$. Then, the former q best elements are re-inserted into the population, forming a new generation of μ individuals that replaces the previous one. The BGA selection mechanism is then deterministic (there are no probabilities), extinctive (the best elements are guaranteed to be selected and the worst are guaranteed *not* to be selected) and q-elitist (the best q elements always survive from a generation to the next). For the BGA, the typical value is $q = 1$. This is a form of the comma strategy (μ, λ) since the parents are not included in the replacement process, with the exception of the q previous best. Note that, given that q is fixed, only μ needs to be specified, since $\lambda = \mu - q$.

The BGA uses a *direct* representation, that is, a gene is a decision variable (not a way of coding it) and its allele is the value of the variable. An immediate consequence is that, in the absence of other conditionings as constraint handling, the fitness function equals the function to be optimized. In addition, the algorithm does not self-optimize any of its own parameters, as is done in ES and in some meta-GAs [3]. Chromosomes are thus potential solution vectors $\boldsymbol{x}$ of n components, where n is the problem size. This is of crucial importance since:

1. It eliminates the need of choosing a coding function for real numbers
2. It opens the way to the direct manipulation of different kinds of variables, other than real numbers (e.g., fuzzy, discrete) as *single* genes.
3. It permits the design of data-dependent genetic operators.

The BGA is mainly driven by recombination (very much as an ordinary GA), with mutation regarded as an important but background operator intended to reintroduce some of the alleles lost in the population. This view is conceptually right for GAs, because alphabet cardinality is usually very small (two, in most cases). However, for those algorithms that make use of real-valued alleles (like the BGA) mutation has to be seen in the double role of solution fine-tuner (for very small mutations) and as the main discovery force (for moderate ones). In fact, the initial BGA formulation remarked the *synergistic* effect of the combined and iterated application of recombination and mutation to extract the most from an EA [2]. We now briefly describe different possibilities for the genetic operators.

2.1 Recombination

Any operator mixing the genetic material of the parents is called a recombination operator. In a BGA, recombination is applied unconditionally. Let $\boldsymbol{x} = (x_1, \ldots, x_n)$, $\boldsymbol{y} = (y_1, \ldots, y_n)$ be two selected individuals $\boldsymbol{x}, \boldsymbol{y}$ such that $\boldsymbol{x} \neq \boldsymbol{y}$. Let $\boldsymbol{z} = (z_1, \ldots, z_n)$ be the result of recombination and $1 \leq i \leq n$. The following are some of the more common possibilities to obtain an offspring $\boldsymbol{z}$:

Discrete Recombination (DR). Set $z_i \in \{x_i, y_i\}$ (with equal probability).

Line Recombination (LR). Set $z_i = x_i + \alpha(y_i - x_i)$, with a fixed $\alpha \in [0, 1]$.

Extended Intermediate Recombination (EIR). Set $z_i = x_i + \alpha_i(y_i - x_i)$, with $\alpha_i \in [-\delta, 1 + \delta]$ chosen with uniform probability. The $\delta > 0$ parameter expresses the degree to which offspring can be generated out of the parents's scope, an imaginary line that joins them in $\mathbb{R}$. More precisely, $\delta|y_i - x_i|$ is the maximum fraction of the distance between parents where the offspring can be placed, either left to the leftmost parent or right to the rightmost parent. Reasonable values should not exceed $\delta = 0.5$, since the bigger the δ, the more the effect of the parents is diminished in creating offspring.

Fuzzy Recombination (FR). This operator replaces the uniform *pdf* (probability distribution function) by a bimodal one, with the two modes located at x_i, y_i. It thus favours offspring values close to either of the parents, and not in any intermediate point with equal probability, as with previous operators. The label "fuzzy" comes from the fact that the two parts of the *pdf* resemble fuzzy numbers (triangular in the original formulation [4]), fulfilling the conditions:

$$x_i - e|y_i - x_i| \leq t \leq x_i + e|y_i - x_i| \qquad y_i - e|y_i - x_i| \leq t \leq y_i + e|y_i - x_i|$$

stating that offspring t lies in one (or both) of the intervals, being $e > 0$ the fuzzy number's spread, the same for both parts. The favour for offspring values near the parents is thus stronger the closer the parents are. In the simplest case ($e = 0.5$) the two parts meet at the median and this point has zero probability.

2.2 Mutation

A mutation operator is applied to each gene with probability n^{-1} so that, on average, one gene is mutated for each individual. Let $\boldsymbol{z} = (z_1, \ldots, z_n)$ denote the result of mutation of an individual $\boldsymbol{x}$. The elements of $\boldsymbol{z}$ are formed as follows:

Discrete Mutation (DM). Set $z_i = x_i + sign \cdot range_i \cdot \delta$ with $sign \in \{-1, +1\}$ chosen with equal probability, $range_i = \rho(r_i^+ - r_i^-)$, $\rho \in [0.1, 0.5]$ and

$$\delta = \sum_{i=0}^{k-1} \varphi_i 2^{-i}$$

where $\varphi_i \in \{0, 1\}$ taken from a Bernouilli probability distribution such that $\Pr(\varphi_i = 1) = \frac{1}{k}$. In this setting $k \in \mathbb{N}^+$ is a parameter originally related to the *precision* with which the optimum was to be located. In practice, the value of k is related to the *expected* value of mutation steps: the higher k is, the more fine-grained the resultant mutation operator is. The factor ρ sets the *maximum* step that mutation is allowed to produce w.r.t. the range $[r_i^-, r_i^+]$ of variable i.

Continuous Mutation (CM). Same as DM with $\delta = 2^{-k\beta}$, where $\beta \in [0, 1]$ with uniform probability.

3 Extension of the BGA to Heterogeneous Problems

We consider basic data peculiarities most likely to be found in real applications. The BGA representation as well as the workings of the corresponding genetic operators are described. The algorithm manipulates the involved variables as a unique entity at all levels. Obviously, **real-valued** variables are directly treated as such, initialized at random within a pre-declared range, and recombined and mutated with the operators described in (§2.1) and (§2.2).

Ordinal m-valued variables are represented as positive natural numbers in the interval $[1, m]$ an initialized at random within the interval. For recombination, there are three possibilities, which mimic the real-valued operators: DR (generally valid but ignores the order), LR (respects the order), and EIR (idem, needs an δ parameter). Some preliminary investigations lead to the choice of LR ($\alpha = 0.5$), that is, the *median* of the parents. Mutation involves an *increase* (to the immediately following value w.r.t. the linear order) or a *decrease* (idem, but in the opposite sense), and the decision is taken with equal probability.

Nominal m-valued variables are also represented as an interval $[1, m]$, but no order relation is assumed. The clear choice for recombination is DR, being the only one explicitly assuming no underlying order. Mutation is realized by *switching* to a new value in the interval, with equal probability.

Fuzzy quantities. The extension to handle *fuzzy numbers* is given by a tuple of reals (three in the general case, two if the chosen representation is symmetric). *Linguistic variables* are described by their anchor points on the abscissa axis (e.g., four in the case of trapezoidal membership functions).

Recombination of fuzzy numbers is taken as the corresponding extension of the operators for real-valued quantities. In particular, for EIR the mode is obtained following its formula (involving the selection of δ), and the spread is computed using the formula with the *same* α. This makes sense since the spread is usually proportional to the mode. Fig. (1) provides an example.

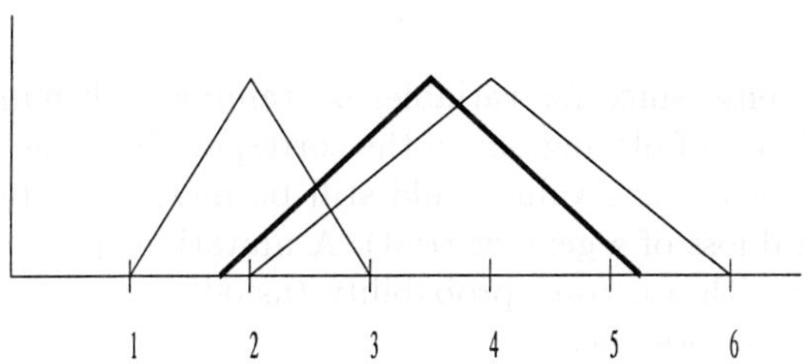

Fig. 1. EIR recombination for fuzzy numbers with $\delta = 0.25$, and $\alpha = 0.75$ uniformly chosen in $[-0.25, 1.25]$. Mode and spread for the two parents are 2.0, 1.0 and 4.0, 2.0. The thicker number is the result of recombination. As for real numbers, the value of α makes offspring resemble its bigger parent more (a factor of $\frac{3}{4}$) than its smaller one. The mode is 3.5 and the spread 1.75.

Mutation of fuzzy numbers is also developed as an extension of the real-valued operators, by taking into account that mode and spread are collectively expressing a single (fuzzy) number. Both continuous and discrete operators can be used, as follows. The change on the mode is determined using the respective formulas. The change on the spread uses the *same* sign and δ (which are the terms depending on probabilities) as used for the mode.

For *linguistic variables*, recombination is also an extension of the operators for real-valued quantities. For EIR, the procedure is analogous as for fuzzy numbers, that is, using the same α for all the involved quantities. In this case, however, the source of uncertainty is different, and there is no need for the offspring spreads to be in a proportion to their modes similar to that of the parents, and other operators could be conceivable. Fig. (2) provides an example. In mutation, a *single* step change is proposed according to the formulas, which affects all the constituting points (modes and spreads) in the same way.

Missing values are dealt with in a specially meaningful way. They are initially generated according to the estimated probability of a missing value in the vari-

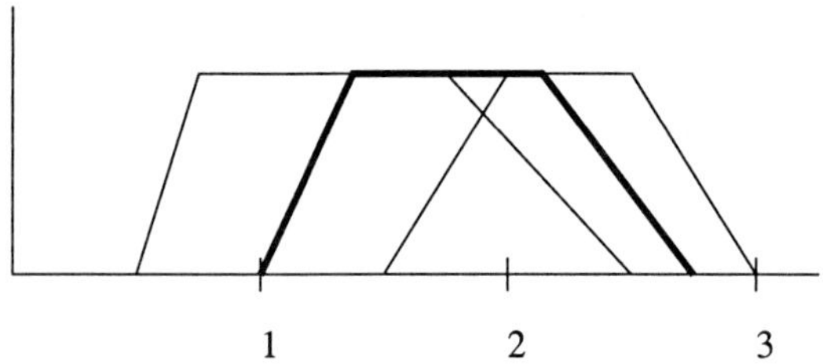

Fig. 2. EIR recombination for trapezoidal linguistic terms, with $\delta = 0.25$ and α uniformly chosen in $[-0.25, 1.25]$, and equal this time to 0.5, for clarity. The thicker set is the result of recombination. In this case, both parents are equally responsible of the obtained offspring ($\alpha = 0.5$).

able. This makes sense since for variables containing high numbers of missing values, the probability of placing one in the corresponding gene increases. If this probability is zero a missing value could still be introduced by mutation (signaling the temporal loss of a gene or trait). A mutation operator sets a missing value in the allele with a certain probability (usually very low). If this change leads to improved performance, it will be retained. A missing value cannot be mutated back to a non-missing one. A definite value can only be recovered by recombination to the (non-missing) gene of another individual.

Recombination is treated as *discrete* (DR) whenever at least one of the parents have a missing trait. This is coherent with the philosophy of EA: recombination stands for the *transmission* of the parents's genetic material to their offspring. If a parent is lacking a gene, this characteristic has to be given the chance to be passed on. Besides, if the trait or gene is lacking for both parents, it will be so for the offspring, since nothing can be "invented from scratch" (this is the role of mutation). In summary, given Ω a recombination operator (possibly heterogeneous), it is extended to a $\Omega_{\mathcal{X}}$ (where $\mathcal{X}$ denotes the missing value) as:

$$\Omega_{\mathcal{X}}(x_i, y_i) = \begin{cases} \Omega(x_i, y_i) & \text{if } x_i \neq \mathcal{X} \wedge y_i \neq \mathcal{X} \\ DR(x_i, y_i) & \text{if } x_i = \mathcal{X} \veebar y_i = \mathcal{X} \\ \mathcal{X} & \text{otherwise} \end{cases}$$

where $\veebar$ denotes exclusive-or. All this manipulation for missing values differs from the one that results by treating it as any other value, for its generation and propagation would be carried out blindly. The proposed treatment has the added appeal of being simple, and natural from the point of view of an EA in the sense that it is taken as a missing *gene* and is independent of the data type.

4 A Case Study in Neural Network Training

4.1 The BGA as a Neural Network Trainer

Evolutionary methods are immediate candidates for neural network optimization, being the case that they may alleviate the problem of local minima. However, when coding an ANN into a GA chromosome, highly complex interactions may develop, due to the influence that a given weight on a hidden unit has on the whole network computation. The usual binary representation of the real-valued weights carries with it extra interactions between non-neighbouring genes, thus inducing strong epistatic effects in GA processing. In these conditions, it is at least doubtful that the *building block hypothesis* can hold. A concise review on the generic use of evolutionary learning algorithms for neural optimization is given in [5, 6]. To the best of our knowledge, the BGA has only been used for some specific neural optimization tasks or application examples [7, 8].

4.2 Heterogeneous Neural Networks

These artificial networks are built out of neuron models defined as a mapping $h : \hat{\mathcal{H}}^n \rightarrow \mathbb{R}$. Here $\hat{\mathcal{H}}^n$ is a cartesian product of an arbitrary number n of *source sets*. These source sets may be extended reals $\hat{\mathcal{R}}_i = \mathbb{R}_i \cup \{\mathcal{X}\}$, extended families of (normalized) fuzzy sets $\hat{\mathcal{F}}_i = \mathcal{F}_i \cup \{\mathcal{X}\}$, and extended finite sets of the form $\hat{\mathcal{O}}_i = \mathcal{O}_i \cup \{\mathcal{X}\}$, $\hat{\mathcal{M}}_i = \mathcal{M}_i \cup \{\mathcal{X}\}$, where each of the $\mathcal{O}_i$ has a full order relation, while the $\mathcal{M}_i$ have not. The special symbol $\mathcal{X}$ extends the source sets and denotes the unknown element (missing information), behaving as an *incomparable* element w.r.t. any ordering relation. According to this definition, neuron inputs are vectors composed of n elements among which there might be reals, fuzzy sets, ordinals, categorical and missing data [11].

An heterogeneous neuron computes a *similarity index*, followed by a classical squashing non-linear function with domain in $[0, 1]$. We use in this work a *Gower-like* similarity index [9] in which the computation for heterogeneous entities is constructed as a weighed combination of partial similarities over single variables. This coefficient has its values in the real interval $[0, 1]$ and for any two objects i, j given by tuples of cardinality n, is given by the expression:

$$s_{ij} = \frac{\sum_{k=1}^{n} g_{ijk}\, \delta_{ijk}}{\sum_{k=1}^{n} \delta_{ijk}}$$

where g_{ijk} is a similarity *score* for objects i, j according to their value for variable k. These scores are in the interval $[0, 1]$ and are computed according to different schemes for numeric and qualitative variables. In particular, for a continuous variable k and any two objects i, j the following similarity score is used:

$$g_{ijk} = \hat{s}\left(\frac{|v_{ik} - v_{jk}|}{\sup_{i,j} |v_{ik} - v_{jk}|} \right)$$

Here, v_{ik} denotes the value of object i for variable k and $\hat{s}(z) = (1 - z)^{\alpha}, \alpha > 0$. The similarity measure used for categorical variables is the *overlap*:

$$g_{ijk} = \begin{cases} 1 \text{ if } v_{ik} = v_{jk} \\ 0 \text{ if } v_{ik} \neq v_{jk} \end{cases}$$

The δ_{ijk} is a binary function expressing whether both objects are *comparable* or not according to their values w.r.t. variable k, as follows:

$$\delta_{ijk} = \begin{cases} 1 \; v_{ik} \neq \mathcal{X} \text{ and } v_{jk} \neq \mathcal{X} \\ 0 \text{ otherwise} \end{cases}$$

For variables representing fuzzy sets, if $\mathcal{F}_i$ is an arbitrary family of fuzzy sets in X, and $\tilde{A}$, $\tilde{B} \in \mathcal{F}_i$, the following similarity relation s is used:

$$s(\tilde{A}, \tilde{B}) = \sup_{u \in X} \{min\ (\mu_{\tilde{A}}(u), \mu_{\tilde{B}}(u))\}$$

Notice that this measure is reflexive in the strong sense and also symmetric. As for the activation function, a modified version of the classical logistic is used [10], which is an automorphism (a monotonic bijection) in $[0, 1]$.

The framework has provision for other types of variables, as ordinal or linguistic, and other kinds of combination for the partial similarities. The resulting heterogeneous neuron is sensitive to the *degree of similarity* between its weight vector and a given input, both composed in general by a mixture of continuous and discrete quantities –possibly with missing data– and can be used for configuring heterogeneous artificial neural networks (HNN).

5 An Experimental Study

Seven learning tasks taken from the Proben repository [12] are studied, altogether representative of the kinds of variables typically found in real problems, while displaying different degrees of missing information (from 0% to 26%). Their main characteristics are displayed in Table 1.

For every data set, the available documentation is analysed in what concerns type and meaning of variables, allowing an assessment on the more appropriate treatment. Missing information is also identified. Specifically, some originally "continuous" variables are treated as ordinal since this makes much more sense than to consider them as continuous. Examples would be *number of times pregnant* or *heart pulse*. There are also variables that, besides being endowed with a total order relation, display a source of vagueness (coming from their *subjective* character) that has to be modeled. This is the case of, for instance, the *temperature of extremities* (`cold, cool, normal, warm`) or the *abdominal distension* (`none, slight, moderate, severe`). These variables are treated as linguistic by respecting the number and order of the (initially crisp) values. In absence of more precise information, the cut points are set at the 0.5 level, as is usually done. Besides, some continuous variables are converted to triangular fuzzy numbers with a low fuzziness (roughly estimated at a 0.5%) reflecting the uncertainty derived from their imprecise measurements.

Table 1. Basic features of the data sets: Def. (default accuracy), Missing (percentage of missing values), Missing-1 (percentage of cases with at least one missing value).

Name	Type	#Cases	Def.	Missing	Missing-1	In→Out	Data
Pima Diabetes	C	768	65.1%	10.6%	48.8%	$8 \to 2$	6R, 0N, 2I
Credit Card	C	690	55.5%	0.65%	5.4%	$15 \to 2$	6R, 9N, 0I
Horse Colic	C	364	61.5%	26.1%	98.1%	$20 \to 3$	5R, 5N,10I
Heart Disease	C	920	55.3%	16.2%	67.5%	$13 \to 2$	3R, 6N, 4I
Solar Flares	R	1066	-	0.0%	0.0%	$9 \to 3$	0R, 5N, 4I
Sinus-Cosinus	R	400	-	0.0%	0.0%	$2 \to 1$	2R, 0N, 0I
SISO-Bench	R	500	-	0.0%	0.0%	$2 \to 1$	2R, 0N, 0I

(C classification R regression) (R real N nominal I ordinal)

Four different architectures are studied, composed of a hidden layer of 4, 8, 12 and 16 neurons plus as many output units as required by the task. The data sets are split in 5 folds, and ten 60%-20%-20% partitions are formed, such that each of the 5 folds appears exactly twice as a validation fold, twice as a test, and 6 times as one of the 3 training folds. For each configuration, ten runs are carried out, varying the initial conditions of the BGA, set to $\mu = 100, \tau = 25$. The task is to minimize the *normalized root square error* (NRSE) on the training part, until 30,000 error evaluations are used in each run. Test errors are computed using the network that produced the lowest error in its validation part.

As a reference, two standard neural networks are also trained in the same conditions, treating all information as real-valued quantities, as is the case in standard neural learning systems. Specifically, the standard scalar-product neuron plus a bias weight, using a logistic as activation function (MLP) and a radial basis function neuron (RBF) based on Euclidean distance, plus a Gaussian with its own variance (to be learned). The BGA is here used in its original formulation. The obtained *generalization results* are summarized in Table 2, where mean test NRSE is displayed, averaged over the four architectures.

Table 2. Results for each data set, averaged over the four architectures. An asterisk (*) means that the result is worse than any of the other two, in the sense that for *all* of the four architectures, Mann-Whitney tests for "greater than" are significant at the 95% level (w.r.t. *both* of the other two models). The average is given as an indication.

Problem	MLP net	RBF net	HNN net
Pima Diabetes	0.692	0.774 (*)	0.700
Horse Colic	0.926(*)	0.714	0.725
Heart Disease	0.661(*)	0.586	0.581
Credit Card	0.677	0.740 (*)	0.533
Solar Flares	1.182(*)	0.855	0.937
Sinus-Cosinus	0.083	0.384 (*)	0.042
SISO-Bench	0.022	0.115 (*)	0.023
Average	0.606	0.595	0.506

Table 3. Heterogeneous weights of a hidden neuron for the *Horse Colic* problem.

#	Name	Type	Value
2	Rectal temperature (celsius)	*fuzzy number*	37.3 ± 2.1
5	Temperature of extremities	*linguistic*	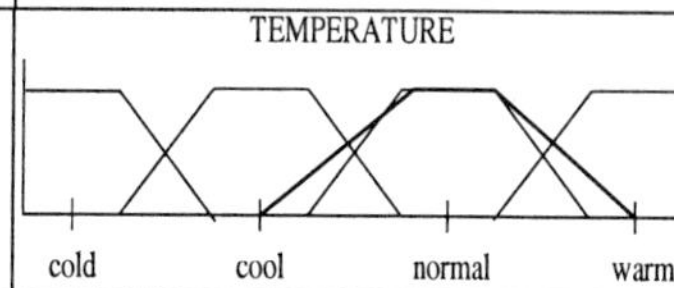
7	Mucous membranes color	*nominal*	“normal pink”
14	Nasogastric reflux PH	*fuzzy number*	2.1 ± 0.3
19	Abdominocentesis appearance	*nominal*	“cloudy”

The *readability* of the obtained solutions is illustrated in Table 3. We show part of the weights of a hidden neuron taken at random from one of the networks delivered by cross-validation for the *Horse Colic* problem. Linguistic terms are shown in graphical form, whereas triangular fuzzy numbers are shown in numerical form (rounded to one decimal) for clarity. Notice the obtained linguistic term is almost symmetric, a characteristic found by the algorithm itself.

Acknowledgements: This work is supported by the Spanish CICYT TAP99-0747.

References

1. Palmer, C.C., Kershenbaum, A. Representing trees in genetic algorithms. In Bäck, Th., Fogel D.B., Michalewicz, Z. (Eds.) *Handbook of Evolutionary Computation.* IOP Publishing & Oxford Univ. Press, 1997.
2. Mühlenbein, H., Schlierkamp-Voosen, D. Predictive Models for the Breeder Genetic Algorithm. *Evolutionary Computation*, 1 (1): 25-49, 1993.
3. Bäck, Th. *Evolutionary Algorithms in Theory and Practice.* Oxford Press, 1996.
4. Voigt, H.M., Mühlenbein, H., Cvetkovic, D. Fuzzy recombination for the continuous Breeder Genetic Algorithm. In Procs. of ICGA'95.
5. Balakrishnan, K., Honavar, V. Evolutionary design of neural architectures - a preliminary taxonomy and guide to literature. Technical report CS-TR-95-01. Dept. of Computer Science. Iowa State Univ., 1995.
6. Yao, X. Evolving Artificial Neural Networks. Procs. of the IEEE, 87(9), 1999.
7. De Falco, I., Iazzetta, A, Natale, P., Tarantino, E. Evolutionary Neural Networks for Nonlinear Dynamics Modeling. In Procs. of PPSN V, Amsterdam, 1998.
8. Zhang, B.T., Mühlenbein, H. Evolving Optimal Neural Networks Using Genetic Algorithms with Occam's Razor. *Complex Systems*, 7(3): 199-220, 1993.
9. Gower, J.C. A General Coefficient of Similarity and some of its Properties. *Biometrics*, 27: 857-871, 1971.
10. Valdés J.J., Belanche, Ll., Alquézar, R. Fuzzy Heterogeneous Neurons for Imprecise Classification Problems. *Intl. Journal of Intelligent Systems*, 15(3): 265-276, 2000.
11. Belanche, Ll. Heterogeneous neural networks: theory and applications. Ph.D. Thesis. Universitat Politècnica de Catalunya, Barcelona, Spain, 2000.
12. Prechelt, L. Proben1: A set of Neural Network Benchmark Problems and Benchmarking Rules. Facultät für Informatik. Univ. Karlsruhe. Tech. Rep. 21/94, 1994.

Automatic Recurrent and Feed-Forward ANN Rule and Expression Extraction with Genetic Programming

Julian Dorado, Juan R. Rabuñal, Antonino Santos, Alejandro Pazos, and Daniel Rivero

Univ. da Coruña, Facultad Informática, Campus Elviña, 15071 A Coruña, Spain
{julian,juanra,nino,ciapazos}@udc.es
danielrc@mail2.udc.es

Abstract. Various rule-extraction techniques using ANN have been used so far, most of them being applied on multi-layer ANN, since they are more easily handled. In many cases, extraction methods focusing on different types of networks and training have been implemented. However, there are virtually no methods that view the extraction of rules from ANN as systems which are independent from their architecture, training and internal distribution of weights, connections and activation functions. This paper proposes a rule-extraction system of ANN regardless of their architecture (multi-layer or recurrent), using Genetic Programming as a rule-exploration technique.

1 Introduction

Artificial Neural Networks (ANN) are systems which are easily implemented and handled. These and other features make them optimal for problem-solving in various areas. However, many developers and researchers avoid their use, since they consider them as "black boxes", that is, they are systems which produce certain response outputs from a series of inputs, while the process through which those outputs are produced remains unknown. For that reason, in fields such as medicine, where their use is highly recommended, people do not consider ANN to be accountable, due to the fact that the reason for their right functioning and the solutions they contribute cannot be explained. An example could be medical diagnosis. A computational system designed for diagnosis should be able to explain the reason for that diagnosis and how it was reached.

Nevertheless, Expert Systems (ES) are able to explain the solution or response achieved, which is their main core and also their guarantee of success. Therefore, this paper tries to develop a system which carries out an automatic extraction of rules from already trained ANN, thus obtaining the knowledge that an ANN obtains from the problem it solves.

Different rule-extraction techniques using ANN have been used so far, always applied to multi-layer ANN due to the fact that they are more easily handled. These networks also have a limited capacity with regard to the knowledge which can be distributed among their connections.

As may be inferred, the extraction of rules and expressions from recurrent ANN is more complicated, due to the fact that past states intervene in neural activation, and

J.J. Merelo Guervós et al. (Eds.): PPSN VII, LNCS 2439, pp. 485–494, 2002.

that their capacity of distributed knowledge is considerably higher than that of multi-layer ANN, since there are no restrictions to neural connectivity. If, besides, recurrent ANN are used in dynamic problems where certain time characteristics such as the prediction of time series intervene, the task of extracting by means of the methods developed so far becomes harder, if not impossible for most of them.
Therefore, the system presented can be applied to every ANN kind. For that reason, the system should comply with a series of requirements [1]:

- ***It should have no ANN architecture requirements***: A rule-extraction mechanism which can work with every type of neural network, including those which are highly interlinked or recurrent.
- ***It should have no ANN training requirements***: Many of the proposed algorithms are based on a given ANN training process for which rule-extraction is designed. Therefore, they are not valid for other training sets.
- ***It should be correct***: It is desirable that the rules describe the ANN as accurately as possible.
- ***It should have a highly expressive level***: Rule language (syntax) characterises the compactness of the extracted symbolic language. Usually, powerful languages are desirable, due to the fact that a very compact and easy to understand rule language can be produced.

2 State of the Art

2.1 Genetic Programming

Some people thinks that Cramer and Fuji, who published on evolving programs in 1985 and 1987 at the very first ICGA conference [2][3], are the pioneers of Genetic Programming (GP). But still others think that Friedberg from 1958 and 1959, who evolved machine language programs [4][5], is really the pioneer.

John Koza created the term which titles the book "Genetic Programming" [6]. This book establishes formally the bases of GP used nowadays. Later, the same author published "Genetic Programming II" [7], and, recently, "Genetic Programming III" [8]. Both explore new possibilities of GP.

Different branches derive form GP. One of the most promising ones with regard to Knowledge Discovery (KD) is that of fuzzy rules [9][10]. This branch derives from the union between fuzzy logic and systems based on rules (SBR). Fuzzy rules can be obtained by means of Evolutionary Computation (EC) with the technique known as Automatically Defined Functions (ADF) [7], which represent an evolution of the concept called "Classical Genetic Programming".

2.2 ANN Rule Extraction

Several authors have studied the relationship between ANN and fuzzy rules [11] [12] [13]. Most results establish that equivalence by means of a process of consecutive approaches. Apart from being purely theoretical solutions, they require a great number of rules in order to approach the ANN functioning [13]. Jang's and Sung's work [11] is rather different, given that they provide an equivalence between radial ANN and fuzzy systems where a finite number of rules or neurons is required, though in this case it is limited to a fixed ANN architecture.

Andrews [14][15] identifies three rule-extraction techniques: "decompositional", "pedagogical" and "eclectic". The first one refers to extraction at the neuron level. The second one treats the ANN as a black box, where, by means of applying inputs to the network, a backward to forward analysis of the neurons in the hidden layers is carried out, extracting the corresponding rules. The third one uses the ANN architecture and the input-output pairs as a complement to a symbolic training algorithm.

Towell and Shavlik [16] apply the first technique using the connections between neurons as rules based on simple transformations. This limits extraction to those networks with a given multi-layer architecture and few process elements [14]. Thrun [17] has developed the main approach by means of using the second technique, titled "Validity Interval Analysis" (VIA). The algorithm uses linear programming (SIMPLEX), applying value intervals to each neuron's activations in each layer. The system extracts "possibly correct" rules through the ANN by means of a backward and forward propagation of those intervals. This algorithm has, in the worst of cases, exponential complexity, due to the fact of using linear programming. Other approaches using the second technique are RULENEG algorithms [18] and DEDEC ones [19], which use an ANN in order to extract rules from another ANN's training set. However, those rule-extraction techniques which focus exclusively on the training data lose the generalization capacity which ANN have. Other rule-extraction techniques are [20] [21] [1], which are based on previously debated approaches.

GAs have recently been used for finding and extracting ANN, due to the advantages offered by evolutionary techniques for searching in complex systems [22]. The second technique (pedagogical) uses a GA where the chromosomes are multi-condition rules based on intervals or value ranks applied to the ANN inputs. These values are obtained from the training parameters. Wong and Leung have used PG for knowledge discovery from databases (KDD) developing a system called LOGENPRO (Logic grammar based GP) [23]. It uses first order logic to represent knowledge. This is the first aproximation that shows the advantages of GP for KDD.

3 Fundamentals

One of the most important aspects of any rule-extraction system is the optimization of the rules obtained from the ANN analysis. It should be kept in mind that the extracted rules may be contained in general rules, and many of the logical expressions obtained may be simplified if they are written in a different way. Therefore, the optimization of rules consists of simplifying and carrying out symbolic operations on the rules. Depending on the extraction method and on the type of rules obtained, various optimization techniques can be applied. They can be classified into two main groups: imbibed optimization methods and a posteriori methods. The latter are usually a syntactic analysis algorithm applied to the rules obtained in order to simplify them. For instance [24] uses Prolog as programming language for a post-processing of the rules obtained. Imbibed optimization techniques are used for rule-extraction algorithms which intrinsically cause the algorithm to produce rules which are more and more optimal. An example may be the technique of depth penalization used in GP. Conceptually, when the adaptation level of a GP individual is evaluated (tree) its capacity is reduced a certain degree according to the number of terminal and non-terminal nodes that the tree has. Thus the existence of simple individuals is dynamically fostered. Therefore, if we are searching for rules (syntactic trees), the appearance of simple (optimization) rules is intrinsically favoured.

Another thing to be taken into account when applying the extraction algorithm is its modus operandi. As previously discussed, extraction techniques can be classified into three main groups: "decompositional", "pedagogical" and "eclectic". EC has been applied in this paper, and specifically GP as building algorithm of a syntactic tree which reflects a set of rules as similar as possible to the functioning of an ANN. A symbolic regression has been applied to the input-output patterns. These patterns are input sequences applied to an ANN and the outputs obtained from it. This type of technique can be termed as "pedagogical", where the ANN is treated as a "black box". This is an advantage, given that it is not necessary to know how an ANN works internally. However, a rule-extraction algorithm which can work with "black box" structures should be able to implement some kind of mechanism which allows a priori incorporation of the knowledge obtained from the "black box", thus reducing considerably the search space of the system rules. These structures are known as "grey box". This is possible thanks to the fact that in GP the number and type of terminal and non-terminal elements which intervene in the search process can be determined. For instance, if we know that an ANN carries out classification tasks, the type of terminal nodes can be determined as Boolean, avoiding floating point levels. This offers a great advantage, given that all the possible combinations of mathematical operations can be eliminated beforehand.

4 Description of the System

The main method proposed for the extraction of ANN rules is the use of EC, due to the fact that it has been proved to be very useful at search tasks, where the solution space increases exponentially with regard to the problem to be solved. The use of GP is proposed since it offers the advantage of having a way of representing and structuring information by means of a semantic tree. This tree diagram is a natural way of representing a rule which can be easily understood by human beings.

The proposed system will be tested contrasting its correct functioning with other existing extraction techniques based on classification tasks. The cases are the breast-cancer diagnosis and the lethal hepatitis one. In these cases, the direct extraction of rules from those data will be proven with the purpose of extracting the rules pertaining to each network by means of training ANN. Finally, the algorithm will be tested with recurrent ANN for time-series prediction tasks. A RANN trained for predicting a laboratory chaotic time series (such as the Mackey-Glass) will be used.

In each case, the work is initially based on obtaining an ANN for solving the problem. Once the ANN are designed and trained, the same test and training values are used for generating a second data pattern which will be used for finding the rules acquired by the ANN in the training process (Fig.1). The rule-extraction algorithm, as discussed before, is based on GP. This search technique allows problem solving by means of the automatic generation of algorithms and expressions. These expressions are codified in the shape of a tree. In order to create this tree, we must specify which nodes will be terminals (leaves) and which will be functions (non terminals). The difference is that some of them will be able to have offspring and the others will not.

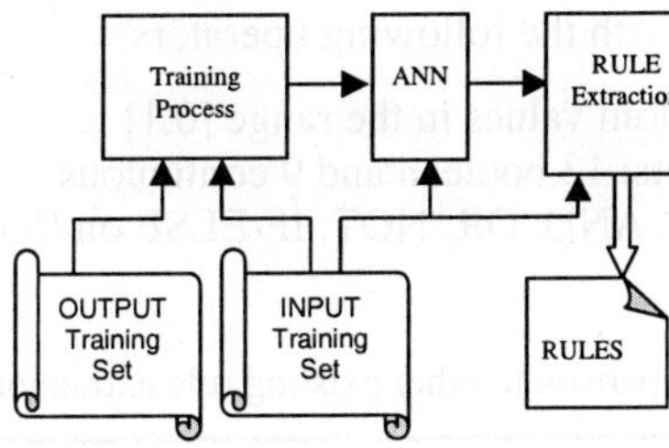

Fig. 1. Rule extraction process

When terminal and functions operators are specified, it is necessary to specify their types: each node will have a type, and the functions will require a specific type to their offspring [25]. This ensures that the trees thus generated satisfy the user's grammar. Besides, both specified operator sets must comply with two requirements: closure and sufficiency, i.e. it must be possible to build correct trees with the specified operators, and the solution to the problem must be able to be expressed by means of those operators. Depending on the problem to be solved, mathematical and trigonometrical functions (sine, cosine, square,...), logic operators (AND, OR, NOT) have been used as operators, together with the typical operations +, -, *, and % which is the protected division (avoid dividing by zero). Thus, logic operators have been basically selected for classification tasks, dispending with mathematical operations. On the opposite side, it is necessary to use them for prediction, due to the nature of the outputs of the ANN which deal with this type of problem.

5 Results

5.1 Classification Problems

ANN have shown their extreme usefulness for tasks in which having an input sequence we must decide whether those values correspond to a certain classification. As it was debated before, various problems of medical diagnosis have been used in order to train ANN so that they produce it. The first data set used was the detection of lethal hepatitis cases. For this purpose we used a database obtained from UCI [26]. It contains 155 examples for two classifications: 32 deaths, and 123 alive cases. There are 19 attributes, 13 of which are Boolean while 6 have discrete values. An multi-layer architecture ANN with 9 neurons in only one hidden layer and with tangent hyperbolic activation functions was trained with these examples, obtaining a fitness of 100% of cases. The extraction of rules has been applied to this ANN and the fitness value obtained is the correct classification of 98.75% of cases.

(IF X_2 THEN 0.8826
ELSE (IF ((IF X_2 THEN 0.8859
    ELSE X_{18})<0.6247)
  THEN (IF (IF ((IF X_6 THEN 0.9800
        ELSE 0.6205)>0.9380)
      THEN (NOT X_{12})
      ELSE X_7)
    THEN X_{18}
    ELSE 0.8826)
  ELSE (IF X_{13} THEN X_{18}
      ELSE 0.2338))) (1)

(X_1) age
(X_2) sex: *men,woman*
(X_3) steroid: *no,yes*
(X_4) antivirals:*no, yes*
(X_5) fatigue: *no, yes*
(X_6) malaise: *no, yes*
(X_7) anorexia: *no, yes*
(X_8) liver big: *no, yes*
(X_9) liver firm: *no, yes*
(X_{10}) spleen palpable: *no, yes*
(X_{11}) spiders: *no, yes*
(X_{12}) ascites: *no, yes*
(X_{13}) varices: *no, yes*
(X_{14}) bilirubin
(X_{15}) alk phosphate
(X_{16}) sgot
(X_{17}) albumin
(X_{18}) protime
(X_{19}) histology: *no, yes*

This results were obtained with the following operators:

- *Constants*: 20 random values in the range [0,1]
- *Variables*: 19 inputs: 13 boolean and 9 continuous
- *Operators*: <, >, =, AND, OR, NOT, IF-ELSE on Boolean and Real values

Table 1. Comparison to other existing rule extraction methods

Method	Accuracy	Ref.	Method	Accuracy	Ref
OUR	98.75 %		ASR	85.0 %	[27]
C-MLP2LN	96.1 %	[24]	FDA	84.5 %	[27]
k-NN, k=18, Manhattan	90.2 %	[24]	LVQ	83.2 %	[27]
FSM + rotations	89.7 %	[24]	CART	82.7 %	[27]
LDA	86.4 %	[27]	MLP with BP	82.1 %	[27]
Naive Bayes	86.3 %	[27]	ASI	82.0 %	[27]
IncNet + rotations	86.0 %	[28]	LFC	81.9 %	[27]
1-NN	85.3 %	[27]	Default	79.4 %	

The next data set corresponds to the detection of breast cancer. We also have a database obtained from UCI [26]. It contains 699 examples for two classifications: 458 cases of benign cancer (65.5%) and 241 cases of malign cancer (34.5%). There are 9 attributes, all of then are discrete.

Various ANN have been trained. However, 100% of classifications was never reached. The rate of success was 98.28% with one hidden layer with 7 neurons and neurons with linear activation function. The same architecture with tangent hyperbolic activation functions has improved the success rate up to 98.68%.

The value of the best fitness obtained for this latter ANN is a right classification in 99.71% of cases (using the outputs produced by the ANN). However, in order to draw a right comparison to other rule extraction techniques, the algorithm is directly applied to the set of initial input-output patterns (dispensing with the ANN outputs). In this case, the global fitness value obtained is the correct classification of 99.28% of cases. The rules obtained are the following (2):

```
(IF (((0.9<X_1) OR  (IF (X_1>0.4)
                  THEN (((0.3>X_7) AND (X_1<>X_4)) AND (X_6>0.2))
                  ELSE FALSE))
                OR (((0.3<X_2) AND (X_5>0.4))
               AND  (IF (X_9<>X4)
                        THEN (X_1<>X_4)
                        ELSE (X_3>0.4))))
THEN   (IF ((X_3>0.4) OR (X_5>0.4))
        THEN (0.0<>X_6)
        ELSE ((0.9<X_1) OR (0.4<X_6)))
ELSE   (IF  (IF (0.4>X_8)
              THEN (0.3>X_3)
             ELSE (X_3=0.6))
        THEN (0.9<X_1)
        ELSE   (IF (X_3>0.4)
                 THEN  (IF (0.3<>X_4)
                          THEN (0.4>X_8)
                          ELSE (0.4<X_6))
                ELSE (X_3>X_5))))                                        (2)
```

- (X_1) clump thickness
- (X_2) uniformity of cell size
- (X_3) uniformity of cell shape
- (X_4) marginal adhesion
- (X_5) single epithelial cell size
- (X_6) bare nuclei
- (X_7) bland chromatin
- (X_8) normal nucleoli
- (X_9) mitoses

Table 2. Comparison to other existing rule extraction methods

Method	Accuracy	Ref.	Method	Accuracy	Ref
OUR	99.28 %		Bayes (pairwise dependent)	96.6	[27]
C-MLP2LN	99.0 %	[24]	Naive Bayes	96.4	[27]
IncNet	97.1	[28]	DB-CART	96.2	[29]
k-NN	97.0	[24]	LDA	96.0	[27]
Fisher LDA	96.8	[27]	LFC, ASI, ASR	94.4-95.6	[27]
MLP with BP	96.7	[27]	CART	93.5	[29]

5.2 Forecast of Time Series

It is necessary to use RANN architectures for the prediction of time series and for modelling this type of problems. The extraction of rules from ANN with recurrent architecture has an additional challenge, since these ANN are characterised by their huge capacity of representation and distributed knowledge among their connections. This can be specifically applied to time and dynamic problems. The problem to be solved will be the prediction of a classical chaotic laboratory time series: the Mackey-Glass series [30]. The following results show that the rules to be obtained from this ANN should incorporate mechanisms for treating time values. Therefore, non-terminal nodes representing mathematical and trigonometrical operations will be used, together with input variables at previous n moments (X_n). Most of these time series cases are structures with a single input and a single output. The input corresponds to a number value at the t moment, while the system's output is the prediction of the number value at t+1. The Mackey-Glass equation is an ordinary differential delay equation (3).

$$\frac{dx}{dt} = \frac{ax(t - \tau)}{1 + x^c(t - \tau)} - bx(t) \tag{3}$$

Choosing $\tau = 30$, the equation becomes chaotic, and only short-term predictions are feasible. Integrating the equation (3) in the rank [t, t + δt] we obtain:

$$x(t + \Delta t) = \frac{2 - b\Delta t}{2 - b\Delta t} x(t) + \frac{\alpha \Delta t}{2 + b\Delta t}\left[\frac{x(t + \Delta t - \tau)}{1 + x^c(t + \Delta t - \tau)} + \frac{x(t - \tau)}{1 + x^c(t - \tau)}\right] \tag{4}$$

The first step is obtaining a RANN which emulates the behaviour of the time series. The RANN that we used has three neurons with tangent hyperbolic activation function with total interconnection. The training files used correspond to the first 200 values of the time series (Fig.3). The RANN resulting from the training process which has yielded the least mean square error (MSE=0.000072) may be seen in Fig.2.

Once we have obtained the RANN, we try to obtain by means of symbolic regression the rules and the expressions which direct its functioning. For this purpose we have used a test file containing the first 1000 values of the time series. These 1000 values are transferred to the RANN, obtaining the corresponding outputs. Using the input-output file, we run the GP algorithm.

Different combinations of terminal and function elements and GP parameters have been tried, and the following have been used:

- *Arithmetic functions*: +, -, *, % (protected division)
- *Constants*: 10 random values in [0,1]
- *Variables*: X_n, X_{n-1}, X_{n-2}, X_{n-3}
- *Selection algorithm*: Tournament
- *Population size*: 1000 individuals
- *Crossover rate*: 95%
- *Mutation rate*: 4%
- *Parsimony level*: 0.0001

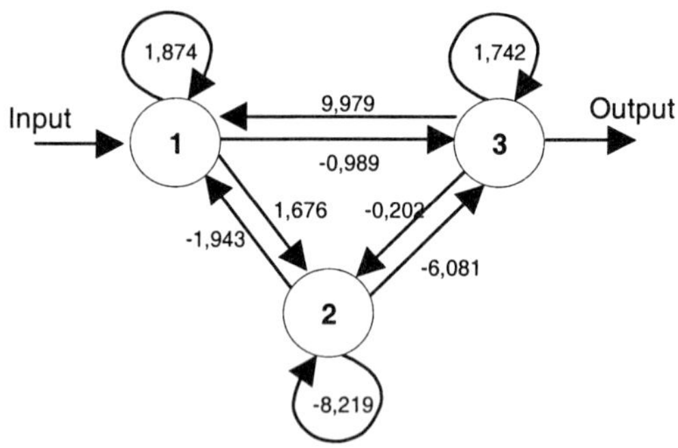

Fig. 2. RANN that emulate the Mackey-Glass function

Besides, the elitist strategy has been used, not allowing the loss of the best individual. The rule expressed as a mathematical function is the following:

$$((X_n * ((((((X_n * (X_n * X_{n-2})) \% X_{n-2}) * (((0.9834 * (((((X_n * (X_{n-2} \% X_{n-3})) * X_{n-3}) * X_{n-3}) \% X_{n-3}) \% (X_n \% X_{n-3}))) \% X_{n-3}) \% ((X_{n-2} * X_{n-2}) \% X_{n-3}))) \% X_{n-3}) * X_{n-2}) \% X_{n-2})) \% ((X_n * X_{n-2}) \% X_{n-3}) * 0.9834)) \quad (4)$$

This function obtains a fitness value (normalised) on the 1000 values produced by the RANN of 0.0029. The next graph compares the values produced by the function forecast (4) and the RANN forecast.

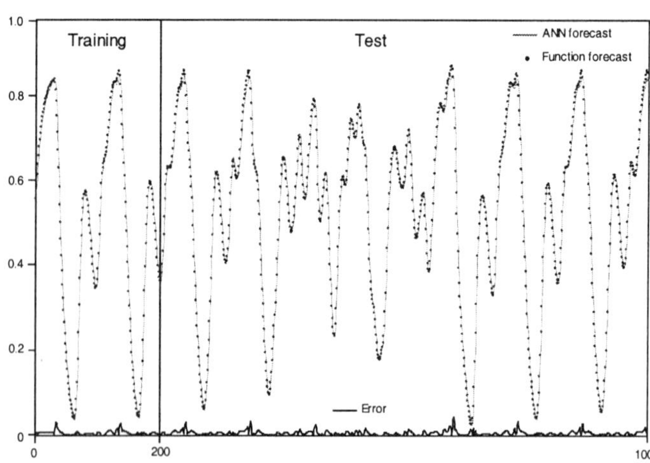

Fig. 3. Comparative between the RANN forecast and the function forecast (4)

6 Conclusions

As inferred from the results presented, the GP-based rule extraction algorithm considerably improves the existing methods, being applicable to any ANN architecture, whether recurrent or feed-forward. Depending on the task for which the ANN has been designed, the types of operators on which GP works might be varied.

As discussed in the introduction, the four features required by a rule-extraction system are faithfully approached in this case. By means of applying EC, ANNs are treated as black boxes, where only the inputs and the outputs they produce are considered. And it does not depend neither on their architecture nor on their training algorithm. The rules extracted from the ANN are so similar to the ANN's functioning as the fitness value produced by the algorithm, and as expressive as the semantic level used by the GP's codified expression trees. Therefore, we may say that GP-based algorithms fit the needs of the rule-extraction systems.

7 Future Works

A possible development is to apply GP not only to obtain rules, but also operability ranks of ANN. It would be like a validation system for the functioning of an ANN for the problem to be solved. Not only the initial training set which carries out the adjustment should be used for this purpose, but also new input data sets should be created simulating all the possible variations of values in which could enter the ANN.

Another future development will be the analysis of the different parameters intervening in the algorithm's correct functioning, according to the type of problem solved by the ANN. It should be treated not as a black box, but as a grey one, where, for instance, the ANN's activation function is known, being incorporated as one of the operators of GP, and analysing which are the rules extracted by this operator.

In order to accelerate the rule-extraction process, it is possible to use a network of computers and so the rule search will be distributed and concurring, exchanging rules.

Acknowledgements

This work has been supported by the Univ. A Coruña project "Extracción de reglas de RNA mediante CE" and CICYT TIC2000-0120-P4-03 of the Spanish government.

References

1. Tickle, A.B.; Andrews, R.; Golea, M.; Diederich, J. "The truth will come to light: directions and challenges in extracting the knowledge embedded within trained artificial neural networks". IEEE Transaction on Neural Networks, vol. 9 nº 6, pp 1057-1068, 1998.
2. Cramer, N.L. "A Representation for the Adaptive Generation of Simple Sequential Programs", Grefenstette: Proc. of 1st International Conference on GA, 1985.
3. Fujiki C. "Using the Genetic Algorithm to Generate Lisp Source Code to Solve the Prisoner's Dilemma", International Conf on GAs, pp. 236-240, 1987.
4. Friedberg R. "A learning machine: Part I", IBM Journal, pp. 2-13, 1958.
5. Friedberg R.M., Dunham B., North J.H. "A learning machine: Part II", IBM Journal of Research and Development, 3(3) 282-287, 1959.

6. Koza J. "Genetic Programming. On the Programming of Computers by means of Natural Selection". The Mit Press, Cambridge, Massachusetts, 1992.
7. Koza J. "Genetic Programming II: Automatic Discovery of Reusable Programs". The Mit Press, Cambridge, Massachusetts, 1994.
8. Koza J., Forrest H., Andre D., Keane M. "Genetic Programming III: Darwinian Invention and Problem Solving". Morgan Kaufmann Publishers, San Francisco, 1999.
9. Fayyad U., Piatetsky-Shapiro G., Smyth P., Uthurusamy R.: "Advances in Knowledge Discovery and Data Mining". AAAI/MIT Press, 1996
10. Bonarini A.: "Evolutionary Learning of Fuzzy Rules: Competition and Cooperation", Fuzzy Modelling: Paradigms and Practice, W. Pedrycz (Ed.), Kluwer Academic Press, 1996
11. Jang J., Sun C. "Functional equivalence between radial basis function networks and fuzzy inference systems". IEEE Transactions on Neural Networks, vol. 4, pp 156-158, 1992.
12. Buckley J.J., Hayashi Y., Czogala E. "On the equivalence of neural nets and fuzzy expert systems", Fuzzy Sets Systems, Nº 53, pp 129-134, 1993.
13. Benítez J. M., Castro J. L., Requena I. " Are artificial neural networks black boxes? " IEEE Transactions on Neural Networks, vol. 8, nº 5, pp 1156-1164, 1997.
14. Andrews R. Diederich J., Tickle A. "A Survey and Critique of Techniques For Extracting Rules From Trained ANN". Knowledge Based Systems 8, pp 373-389, 1995.
15. Andrews R., Cable R., Diederich J., et al: "An evaluation and comparison of techniques for extracting and refining rules from ANN", QUT NRC Technical report, 1996.
16. Towell G., Shavlik J. W. "Knowledge-Based ANN". AI, 70, pp 119-165, 1994.
17. Thrun S. "Extracting rules from networks with distributed representations". NIPS, G. Tesauro, D. Touretzky, T. Leen (eds), MIT Press, 1995.
18. Pop E., Hayward R., Diederich J. "RULENEG: Extracting Rules from a Trained ANN by Stepwise Negation". Queensland University of Technology, Neurocomputing Research Centre. QUT NRC Technical report, 1994.
19. Tickle A. B., Orlowski M., Diedrich J. "DEDEC: A methodology for extracting rules from trained artificial neural networks". Queensland Univ. of Technology, Neurocomputing Research Centre. Technical report, 1996.
20. Chalup S., Hayward R., Diedrich J. "Rule extraction from artificial neural networks trained on elementary number classification task". Queensland University of Technology, Neurocomputing Research Centre. QUT NRC Technical report, 1998.
21. Visser U., Tickle A., Hayward R., Andrews R. "Rule-Extraction from trained neural networks: Different techniques for the determination of herbicides for the plant protection advisory system PRO_PLANT". Proc. of the rule extraction from trained ANN workshop, Brighton, UK, pp 133-139. 1996.
22. Keedwell E., Narayanan A., Savic D. "Creating rules from trained neural networks using genetic algorithms". IJCSS, vol. 1, Nº 1, pp 30-42. 2000.
23. Wong M.L., Leung K.S.: "Data Mining using Grammar Based Genetic Programming and Applications", Kluwer Academic Publishers, 2000.
24. Duch W., Adamczak R., Grabczewski K.: "A new methodology of extraction, optimisation and application of crisp and fuzzy logical rules", IEEE Trans. on N.N., vol. 11, nº 2, 2000.
25. Montana D.J.: "Strongly Typed Genetic Programming", Evolutionary Computation, The MIT Press, pp. 199-200, Cambridge, MA, 1995.
26. Mertz C., Murphy P.: UCI repository of machine learning databases, http://www.ics.uci.edu/pub/machine-learning-data-bases.
27. Ster B., Dobnikar A.: "Neural networks in medical diagnosis: Compararison with other methods", A. Bulsari et al. eds, Proc. Int. Conf. EANN'96, pp. 427-430, 1996.
28. Jankowski N., Kadirkamanathan V.: "Statistical Control of RBF-like Networks for Classification", 7th International Conf. on ANN, pp. 385-390, Lausanne, Switzerland, 1997.
29. Shang N., Breiman L.: "Distribution based trees are more accurate", Int. Conf. On Neural Information Processing, vol. 1, pp. 133-138, Hong Kong, 1996.
30. Mackey M., Glass L.: "Oscillation and chaos in physiological control systems", Science, pp. 197-287, 1977.

Learning and Evolution by Minimization of Mutual Information

Yong Liu[1] and Xin Yao[2]

[1] The University of Aizu
Aizu-Wakamatsu, Fukushima 965-8580, Japan
yliu@u-aizu.ac.jp

[2] School of Computer Science, The University of Birmingham
Edgbaston, Birmingham, UK
X.Yao@cs.bham.ac.uk

Abstract. Based on negative correlation learning [1] and evolutionary learning, evolutionary ensembles with negative correlation learning (EENCL) was proposed for learning and designing of neural network ensembles [2]. The idea of EENCL is to regard the population of neural networks as an ensemble, and the evolutionary process as the design of neural network ensembles. EENCL used a fitness sharing based on the covering set. Such fitness sharing did not make accurate measurement on the similarity in the population. In this paper, a fitness sharing scheme based on mutual information is introduced in EENCL to evolve a diverse and cooperative population. The effectiveness of such evolutionary learning approach was tested on two real-world problems. This paper has also analyzed negative correlation learning in terms of mutual information on a regression task in the different noise conditions.

1 Introduction

Neural network ensembles adopt the divide-and-conquer strategy. Instead of using a single network to solve a task, an neural network ensemble combines a set of neural networks which learn to subdivide the task and thereby solve it more efficiently and elegantly [1]. However, designing neural network ensembles is a very difficult task. It relies heavily on human experts and prior knowledge about the problem. Based on negative correlation learning [1] and evolutionary learning, evolutionary ensembles with negative correlation learning (EENCL) was proposed for learning and designing of neural network ensembles [2]. The idea of EENCL is to regard the population of neural networks as an ensemble, and the evolutionary process as the design of neural network ensembles.

The negative correlation learning and fitness sharing [3] were adopted in EENCL to encourage the formation of species in the population. The idea of negative correlation learning is to encourage different individual networks in the ensemble to learn different parts or aspects of the training data, so that the ensemble can better learn the entire training data. In negative correlation learning, the individual networks are trained simultaneously rather than independently or

J.J. Merelo Guervós et al. (Eds.): PPSN VII, LNCS 2439, pp. 495–504, 2002.

sequentially. This provides an opportunity for the individual networks to interact with each other and to specialize.

Fitness sharing refers to a class of speciation techniques in evolutionary computation [4]. The fitness sharing used in EENCL was based on the idea of the covering set that consists of the same training patterns correctly classified by the shared individuals. This fitness sharing cannot accurately measure the similarity between two individuals. For example, even two individuals have the same covering set, the outputs of two individuals can be quite different. A more accurate similarity measurement between two neural networks in a population can be defined by the explicit mutual information of output variables extracted by two neural networks. The mutual information between two variables, output F_i of network i and output F_j of network j, is given by

$$I(F_i; F_j) = H(F_i) + H(F_j) - H(F_i, F_j) \tag{1}$$

where $H(F_i)$ is the entropy of F_i, $H(F_j)$ is the entropy of F_j, and $H(F_i, F_j)$ is the joint differential entropy of F_i and F_j. The equation shows that joint differential entropy can only have high entropy if the mutual information between two variables is low, while each variable has high individual entropy. That is, the lower mutual information two variables have, the more different they are. By minimizing the mutual information between variables extracted by two neural networks, two neural networks are forced to convey different information about some features of their input.

In this paper, negative correlation learning is firstly analyzed in terms of minimization of mutual information on a regression task. Secondly, a fitness sharing based on mutual information is introduced into EENCL. Through minimization of mutual information, a diverse and cooperative population of neural networks can be evolved by EENCL. The effectiveness of such evolutionary learning approach was tested on two real-world problems.

The rest of this paper is organized as follows: Section 2 explores the connections between the mutual information and the correlation coefficient, and explains how negative correlation learning can be used to minimize mutual information. Section 3 analyzes negative correlation learning via the metrics of mutual information. Section 4 describes EENCL for evolving a population of neural networks, and explores the connections between fitness sharing and mutual information. Section 5 presents experimental results on EENCL by minimizing mutual information. Finally, Section 6 concludes with a summary.

2 Minimizing Mutual Information by Negative Correlation Learning

2.1 Minimization of Mutual Information

Suppose the output F_i of network i and the output F_j of network j are Gaussian random variables. Their variances are σ_i^2 and σ_j^2, respectively. The mutual information between F_i and F_j can be defined by Eq.(1) [5]. The differential entropy $h(F_i)$ and $h(F_j)$ are given by $h(F_i) = \frac{1}{2}[1 + \log(2\pi\sigma_i^2)]$ and $h(F_j) = \frac{1}{2}[1 + \log(2\pi\sigma_j^2)]$. The joint differential entropy $h(F_i, F_j)$ is given by

$$h(F_i, F_j) = 1 + \log(2\pi) + \frac{1}{2}\log[\sigma_i^2\sigma_i^2(1 - \rho_{ij}^2)] \tag{2}$$

where ρ_{ij} is the correlation coefficient of F_i and F_j

$$\rho_{ij} = \frac{E[(F_i - E[F_i])(F_j - E[F_j])]}{\sigma_i^2\sigma_j^2} \tag{3}$$

By substituting F_i, F_j, and Eq. (2) in (1), we get

$$I(F_i; F_j) = -\frac{1}{2}\log(1 - \rho_{ij}^2) \tag{4}$$

From Eq.(4), we may make the following statements:

1. If F_i and F_j are uncorrelated, the correlation coefficient ρ_{ij} is reduced to zero, and the mutual information $I(F_i; F_j)$ becomes very small.
2. If F_i and F_j are highly positively correlated, the correlation coefficient ρ_{ij} is close to 1, and mutual information $I(F_i; F_j)$ becomes very large.

Both theoretical and experimental results [6] have indicated that when individual networks in an ensemble are unbiased, average procedures are most effective in combining them when errors in the individual networks are negatively correlated and moderately effective when the errors are uncorrelated. There is little to be gained from average procedures when the errors are positively correlated. In order to create a population of neural networks that are as uncorrelated as possible, the mutual information between each individual neural network and the rest of population should be minimized. Minimizing the mutual information between each individual neural network and the rest of population is equivalent to minimizing the correlation coefficient between them.

2.2 Negative Correlation Learning

We consider estimating y by forming an neural network ensemble whose output is a simple averaging of outputs F_i of a set of neural networks. Given the training data set $D = \{(\mathbf{x}(1), y(1)), \cdots, (\mathbf{x}(N), y(N))\}$, all the individual networks in the ensemble are trained on the same training data set D

$$F(n) = \frac{1}{M}\Sigma_{i=1}^{M} F_i(n) \tag{5}$$

where $F_i(n)$ is the output of individual network i on the nth training pattern $\mathbf{x}(n)$, $F(n)$ is the output of the neural network ensemble on the nth training pattern, and M is the number of individual networks in the neural network ensemble.

The idea of negative correlation learning is to introduce a correlation penalty term into the error function of each individual network so that the mutual information among the ensemble can be minimized. The error function E_i for individual i on the training data set $D = \{(\mathbf{x}(1), y(1)), \cdots, (\mathbf{x}(N), y(N))\}$ in negative correlation learning is defined by

$$E_i = \frac{1}{N}\Sigma_{n=1}^{N} E_i(n) = \frac{1}{N}\Sigma_{n=1}^{N}\left[\frac{1}{2}(F_i(n) - y(n))^2 + \lambda p_i(n)\right] \tag{6}$$

where N is the number of training patterns, $E_i(n)$ is the value of the error function of network i at presentation of the nth training pattern, and $y(n)$ is the desired output of the nth training pattern. The first term in the right side of Eq.(6) is the mean-squared error of individual network i. The second term p_i is a correlation penalty function. The purpose of minimizing p_i is to negatively correlate each individual's error with errors for the rest of the ensemble. The parameter λ is used to adjust the strength of the penalty.

The penalty function p_i has the form

$$p_i(n) = -\frac{1}{2}(F_i(n) - F(n))^2 \tag{7}$$

The partial derivative of E_i with respect to the output of individual i on the nth training pattern is

$$\begin{aligned}\frac{\partial E_i(n)}{\partial F_i(n)} &= F_i(n) - y(n) - \lambda(F_i(n) - F(n)) \\ &= (1-\lambda)(F_i(n) - y(n)) + \lambda(F(n) - y(n))\end{aligned} \tag{8}$$

where we have made use of the assumption that the output of ensemble $F(n)$ has constant value with respect to $F_i(n)$. The value of parameter λ lies inside the range $0 \leq \lambda \leq 1$ so that both $(1-\lambda)$ and λ have nonnegative values. The standard back-propagation (BP) [7] algorithm has been used for weight adjustments in the mode of pattern-by-pattern updating. That is, weight updating of all the individual networks is performed simultaneously using Eq.(8) after the presentation of each training pattern. One complete presentation of the entire training set during the learning process is called an *epoch*. Negative correlation learning from Eq.(8) is a simple extension to the standard BP algorithm. In fact, the only modification that is needed is to calculate an extra term of the form $\lambda(F_i(n) - F(n))$ for the ith neural network.

From Eq. (8), we may make the following observations. During the training process, all the individual networks interact with each other through their penalty terms in the error functions. Each network F_i minimizes not only the difference between $F_i(n)$ and $y(n)$, but also the difference between $F(n)$ and $y(n)$. That is, negative correlation learning considers errors what all other neural networks have learned while training an neural network.

3 Simulation Results

In order to understand how negative correlation learning minimizes mutual information, this section analyzes it through measuring mutual information on a regression task in three cases: noise free condition, small noise condition, and large noise condition.

3.1 Simulation Setup

The regression function investigated here is

$$f(\mathbf{x}) = \frac{1}{13}\left[10sin(\pi x_1 x_2) + 20\left(x_3 - \frac{1}{2}\right)^2 + 10x_4 + 5x_5\right] - 1 \tag{9}$$

where $\mathbf{x} = [x_1, \ldots, x_5]$ is an input vector whose components lie between zero and one. The value of $f(\mathbf{x})$ lies in the interval $[-1, 1]$. This regression task has been used by Jacobs [8] to estimate the bias of mixture-of-experts architectures and the variance and covariance of experts' weighted outputs.

Twenty-five training sets, $(\mathbf{x}^{(k)}(l), y^{(k)}(l))$, $l = 1, \cdots, L$, $L = 500$, $k = 1, \cdots, K$, $K = 25$, were created at random. Each set consisted of 500 input-output patterns in which the components of the input vectors were independently sampled from a uniform distribution over the interval (0,1). In the noise free condition, the target outputs were not corrupted by noise; in the small noise condition, the target outputs were created by adding noise sampled from a Gaussian distribution with a mean of zero and a variance of $\sigma^2 = 0.1$ to the function $f(\mathbf{x})$; in the large noise condition, the target outputs were created by adding noise sampled from a Gaussian distribution with a mean of zero and a variance of $\sigma^2 = 0.2$ to the function $f(\mathbf{x})$.

A testing set of 1024 input-output patterns, $(\mathbf{t}(n), d(n))$, $n = 1, \cdots, N$, $N = 1024$, was also generated. For this set, the components of the input vectors were independently sampled from a uniform distribution over the interval (0,1), and the target outputs were not corrupted by noise in all three conditions.

Each individual network in the ensemble is a multilayer perceptron with one hidden layer. All the individual networks have five hidden nodes in an ensemble architecture. The hidden node function is defined by the logistic function. The network output is a linear combination of the outputs of the hidden nodes.

For each estimation of mutual information among an ensemble, twenty-five simulations were conducted. In each simulation, the ensemble was trained on a different training set from the same initial weights distributed inside a small range so that different simulations of an ensemble yielded different performances solely due to the use of different training sets. Such simulation setup follows the suggestions from Jacobs [8].

3.2 Measurement of Mutual Information

The average outputs of the ensemble and the individual network i on the nth pattern in the testing set, $(\mathbf{t}(n), d(n))$, $n = 1, \cdots, N$, are denoted respectively by $\overline{F}(\mathbf{t}(n))$ and $\overline{F}_i(\mathbf{t}(n))$, which are given by

$$\overline{F}(\mathbf{t}(n)) = \frac{1}{K}\Sigma_{k=1}^{K} F^{(k)}(\mathbf{t}(n)) \tag{10}$$

and

$$\overline{F}_i(\mathbf{t}(n)) = \frac{1}{K}\Sigma_{k=1}^{K} F_i^{(k)}(\mathbf{t}(n)) \tag{11}$$

where $F^{(k)}(\mathbf{t}(n))$ and $F_i^{(k)}(\mathbf{t}(n))$ are the outputs of the ensemble and the individual network i on the nth pattern in the testing set from the kth simulation, respectively, and $K = 25$ is the number of simulations. The correlation coefficient between network i and network j is given by

$$\rho_{ij} = \frac{\Sigma_{n=1}^{N}\Sigma_{k=1}^{K}\left(F_i^{(k)}(\mathbf{t}(n)) - \overline{F}_i(\mathbf{t}(n))\right)\left(F_j^{(k)}(\mathbf{t}(n)) - \overline{F}_j(\mathbf{t}(n))\right)}{\sqrt{\Sigma_{n=1}^{N}\Sigma_{k=1}^{K}\left(F_i^{(k)}(\mathbf{t}(n)) - \overline{F}_i(\mathbf{t}(n))\right)^2 \Sigma_{n=1}^{N}\Sigma_{k=1}^{K}\left(F_j^{(k)}(\mathbf{t}(n)) - \overline{F}_j(\mathbf{t}(n))\right)^2}} \tag{12}$$

From Eq.(4), the integrated mutual information among the ensembles can be defined by

$$E_{mi} = -\frac{1}{2}\Sigma_{i=1}^{M}\Sigma_{j=1,j\neq i}^{M} log(1 - \rho_{ij}^2) \tag{13}$$

The integrated mean-squared error (MSE) on the testing set can also defined by

$$E_{test\_mse} = \frac{1}{N}\Sigma_{n=1}^{N}\frac{1}{K}\Sigma_{k=1}^{K}\left(F^{(k)}(\mathbf{t}(n)) - d(n)\right)^2 \tag{14}$$

3.3 Experimental Results

The results of negative correlation learning for the different values of λ at epoch 2000 are given in Table 1. For the noise free condition, the results suggest that E_{test_mse} appeared to decrease with increasing value of λ. The mutual information E_{mi} among the ensemble decreased as the value of λ increased when $0 \leq \lambda \leq 0.5$. However, when λ increased further to 0.75 and 1, the mutual information E_{mi} had larger values. The reason of having larger mutual information at $\lambda = 0.75$ and $\lambda = 1$ is that some correlation coefficients had negative values and the mutual information depends on the absolute values of correlation coefficients.

For the small noise (variance $\sigma^2 = 0.1$) and large noise (variance $\sigma^2 = 0.2$) conditions, the results show that there were same trends for E_{mi} and E_{test_mse} in both noise free and noise conditions when $\lambda \leq 0.5$. That is, E_{mi} and E_{test_mse} appeared to decrease with increasing value of λ. However, E_{test_mse} appeared to decrease first and then increase with increasing value of λ. Choosing a proper value of λ is important, and also problem dependent. For the noise conditions used for this regression task and the ensemble architectured used, the performance of the ensemble was optimal for $\lambda = 0.5$ among the tested values of λ in the sense of minimizing the MSE on the testing set.

4 Evolving Neural Network Ensembles

In EENCL [2], an evolutionary algorithm based on evolutionary programming [9] has been used to search for a population of diverse individual neural networks that solve a problem together. Two major issues were addressed in EENCL,

Table 1. The results of negative correlation learning for different λ values at epoch 2000.

	Noise free		Small noise ($\sigma^2 = 0.1$)		Large noise ($\sigma^2 = 0.2$)	
λ	E_{mi}	E_{test_mse}	E_{mi}	E_{test_mse}	E_{mi}	E_{test_mse}
0	0.3706	0.0016	6.5495	0.0137	6.7503	0.0249
0.25	0.1478	0.0013	3.8761	0.0128	3.9652	0.0235
0.5	0.1038	0.0011	1.4547	0.0124	1.6957	0.0228
0.75	0.1704	0.0007	0.3877	0.0126	0.4341	0.0248
1	0.6308	0.0002	0.2431	0.0290	0.2030	0.0633

including exploitation of the interaction between individual neural design and combination, and automatic determination of the number of individual neural networks in an ensemble. The major steps of EENCL are given as follows [2]:

1. Generate an initial population of M neural networks, and set $k = 1$. The number of hidden nodes for each neural network, n_h, is specified by the user. The random initial weights are distributed uniformly inside a small range.
2. Train each neural network in the initial population on the training set for a certain number of epochs using negative correlation learning. The number of epochs, n_e, is specified by the user.
3. Randomly choose a group of n_b neural networks as parents to create n_b offspring neural networks by Gaussian mutation.
4. Add the n_b offspring neural networks to the population and train the offspring neural networks using negative correlation learning while the remaining neural networks' weights are frozen.
5. Calculate the fitness of $M + n_b$ neural networks in the population and prune the population to the M fittest neural networks.
6. Go to the next step if the maximum number of generations has been reached. Otherwise, $k = k + 1$ and go to Step 3.
7. Form species using the k-means algorithm.
8. Combining species to form the ensembles.

There are two levels of adaptation in EENCL: negative correlation learning at the individual level and evolutionary learning based on evolutionary programming (EP) [9] at the population level. Forming species by using the k-means algorithm in EENCL [2] is not considered in this paper.

Fitness sharing used in EENCL is based on the idea of covering the same training patterns by shared individuals. The procedure of calculating shared fitness is carried out pattern-by-pattern over the training set. If one training pattern is learned correctly by p individuals in the population, each of these p individuals receives fitness $1/p$, and the rest of the individuals in the population receive zero fitness. Otherwise, all the individuals in the population receive zero fitness. The fitness is summed over all training patterns.

Rather than using the fitness sharing based on the covering set, a new fitness sharing based on the minimization of mutual information is introduced in

EENCL. In order to create a population of neural networks that are as uncorrelated as possible, the mutual information between each individual neural network and the rest of population should be minimized. The fitness f_i of individual network i in the population can therefore be evaluated by the mutual information:

$$f_i = \frac{1}{\sum_{j \neq i} I(F_i, F_j)} \tag{15}$$

Minimization of mutual information has the similar motivations as fitness sharing. Both of them try to generate individuals that are different from others, though overlaps are allowed.

5 Experimental Studies

This section investigates EENCL with minimization of mutual information on two benchmark problems: the Australian credit card assessment problem and the diabetes problem. Both data sets were obtained from the UCI machine learning benchmark repository. They are available by anonymous ftp at ics.uci.edu (128.195.1.1) in directory /pub/machine-learning-databases.

The Australian credit card assessment problem is to assess applications for credit cards based on a number of attributes. There are 690 patterns in total. The output has two classes. The 14 attributes include 6 numeric values and 8 discrete ones, the latter having from 2 to 14 possible values.

The diabetes data set is a two-class problem that has 500 examples of class 1 and 268 of class 2. There are 8 attributes for each example. The data set is rather difficult to classify. The so-called "class" value is really a binarized form of another attribute that is itself highly indicative of certain types of diabetes but does not have a one-to-one correspondence with the medical condition of being diabetic.

In order to tell the difference between EENCL and EENCL with minimization of mutual information. We name the later approach as EENCLMI. The experimental setup is the same as the previous experimental setup described in [10, 2]. The n-fold cross-validation technique [11] was used to divide the data randomly into n mutually exclusive data groups of equal size. In each train-and-test process, one data group is selected as the testing set, and the other $(n-1)$ groups become the training set. The estimated error rate is the average error rate from these n groups. In this way, the error rate is estimated efficiently and in an unbiased way. The parameter n was set to be 10 for the Australian credit card data set, and 12 for the diabetes data set, respectively.

All parameters used in EENCLMI except for the number of training epochs were set to be the same for both problems: the population size M (25), the number of generations (200), the reproduction block size n_b (2), the strength parameter λ (0.5), the minimum number of cluster sets (3), and the maximum number of cluster sets (25). The number of training epochs n_e was set to 3 for the Australian credit card data set, and 15 for the diabetes data set. The used neural networks in the population are multilayer perceptrons with one

Table 2. Comparison of accuracy rates between EENCLMI and EENCL for the Australian credit card data set. The results are averaged on 10-fold cross-validation. *Mean* and *SD* indicate the mean value and standard deviation, respectively.

	Simple Averaging		Majority Voting		Winner-Takes-All	
Methods	Mean	SD	Mean	SD	Mean	SD
EENCLMI	0.864	0.038	0.870	0.040	0.868	0.039
EENCL	0.855	0.039	0.857	0.039	0.865	0.028

Table 3. Comparison of accuracy rates between EENCLMI and EENCL for the diabetes data set. The results are averaged on 12-fold cross-validation. *Mean* and *SD* indicate the mean value and standard deviation, respectively.

	Simple Averaging		Majority Voting		Winner-Takes-All	
Methods	Mean	SD	Mean	SD	Mean	SD
EENCLMI	0.771	0.049	0.777	0.046	0.773	0.051
EENCL	0.766	0.039	0.764	0.042	0.779	0.045

hidden layer and five hidden nodes. These parameters were selected after some preliminary experiments. They were not meant to be optimal.

5.1 Experimental Results

Tables 2–3 show the results of EENCLMI for the two data sets, where the ensembles were constructed by the whole population in the last generation. Three combination methods for determining the output of the ensemble have been investigated in EENCLMI. The first is simple averaging. The output of the ensemble is formed by a simple averaging of output of individual neural networks in the ensemble. The second is majority voting. The output of the greatest number of individual neural networks will be the output of the ensemble. If there is a tie, the output of the ensemble is rejected. The third is winner-takes-all. For each pattern of the testing set, the output of the ensemble is only decided by the individual neural network whose output has the highest activation. The *accuracy rate* refers to the percentage of correct classifications produced by EENCLMI. In comparison with the accuracy rates obtained by three combination methods, majority voting and winner-takes-all outperformed simple averaging on both problems. Simple averaging is more suitable to the regression type of tasks. Because both problems studied in this paper are classification tasks, majority voting and winner-takes-all are better choices.

Tables 2–3 compare the results produced by EENCLMI and EENCL using three combination methods. Majority voting supports EENCLMI, while winner-takes-all favors EENCL. Since the only difference between EENCLMI and EENCL is the fitness sharing scheme used, the results suggest that combination methods and fitness sharing are closely related to each other. Further studies are needed to probe the relationship of these two.

6 Conclusions

Minimization of mutual information has been introduced as a fitness sharing scheme in EENCL. Compared with the fitness sharing based on the covering set originally used in EENCL [2], mutual information provides more accurate measurement on the similarity. By minimizing mutual information, a diverse population can be evolved.

This paper has also analyzed negative correlation learning in terms of mutual information on a regression task in the different noise conditions. Unlike independent training which creates larger mutual information among the ensemble, negative correlation learning can produce smaller mutual information among the ensemble.

References

1. Y. Liu and X. Yao. Simultaneous training of negatively correlated neural networks in an ensemble. *IEEE Trans. on Systems, Man, and Cybernetics, Part B: Cybernetics*, 29(6):716–725, 1999.
2. Y. Liu, X. Yao, and T. Higuchi. Evolutionary ensembles with negative correlation learning. *IEEE Transactions on Evolutionary Computation*, 4(4):380–387, 2000.
3. Y. Liu and X. Yao. Towards designing neural network ensembles by evolution. In *Parallel Problem Solving from Nature — PPSN V: Proc. of the Fifth International Conference on Parallel Problem Solving from Nature*, volume 1498 of *Lecture Notes in Computer Science*, pages 623–632. Springer-Verlag, Berlin, 1998.
4. D. E. Goldberg. *Genetic Algorithms in Search, Optimization, and Machine Learning.* Addison-Wesley, Reading, MA, 1989.
5. J. C. A. van der Lubbe. *Information Theory.* Prentice-Hall International, Inc., 2nd edition, 1999.
6. R. T. Clemen and R. .L Winkler. Limits for the precision and value of information from dependent sources. *Operations Research*, 33:427–442, 1985.
7. D. E. Rumelhart, G. E. Hinton, and R. J. Williams. Learning internal representations by error propagation. In D. E. Rumelhart and J. L. McClelland, editors, *Parallel Distributed Processing: Explorations in the Microstructures of Cognition, Vol. I*, pages 318–362. MIT Press, Cambridge, MA, 1986.
8. R. A. Jacobs. Bias/variance analyses of mixture-of-experts architectures. *Neural Computation*, 9:369–383, 1997.
9. D. B. Fogel. *Evolutionary Computation: Towards a New Philosophy of Machine Intelligence.* IEEE Press, New York, NY, 1995.
10. D. Michie, D. J. Spiegelhalter, and C. C. Taylor. *Machine Learning, Neural and Statistical Classification.* Ellis Horwood Limited, London, 1994.
11. M. Stone. Cross-validatory choice and assessment of statistical predictions. *Journal of the Royal Statistical Society*, 36:111–147, 1974.

Evolved RBF Networks for Time-Series Forecasting and Function Approximation

V.M. Rivas[1], P.A. Castillo[2], and J.J. Merelo[2]

[1] Dpto. Informática, Univ. de Jaén
E.P.S., Avda. de Madrid, 35, E.23071, Jaén (Spain)
vrivas@ujaen.es
http://wwwdi.ujaen.es/~vrivas

[2] Dpto. de Arquitectura y Tecnología de Computadores, Univ. de Granada
Fac. de Ciencias, Campus Fuentenueva, S/N. E.18071, Granada (Spain)
todos@geneura.ugr.es
http://geneura.ugr.es

Abstract. An evolutionary algorithm with specific operators has been developed to automatically find Radial basis Functions Neural Networks that solve a given problem. The evolutionay algorithm optimizes all the parameters related to the neural network architecture, i.e., number of hidden neurons and their configuration. A set of parameters to run the algorithm is found and tested against a set of different problems about Time-series forecasting and function approximation. Results obtained are compared with those yielded by similar methods.

Keywords: RBF, evolutionary algorithms, EO, functional estimation, time series forecasting.

1 Introduction

Radial Basis Function (RBF) were used by Broomhead and Lowe in [3] to introduce the Radial Basis Function Neural Networks (RBFNN). Traditionally, a RBFNN is thought as a two-layer, fully connected, feed-forward network, in which hidden neuron activation functions are RBF, most times Gaussian functions. The main problem in RBFNN design concerns establishing the number of hidden neurons to use and their parameters: a center point and one or more radii (or widths) per hidden neuron. Once the centers and the radii have been fixed, the weights of the links between hidden and output layers can be calculated analytically using Singular Value Decomposition [17] or any algorithm suitable to solve lineal algebraic equations. So, this paper focuses on the elements of an evolutionary algorithm (EA) designed to find the best components of the RBFNN attempting to solve a given problem. This algorithm is tested against two different kind of problems: time-series forecasting and function approximation.

The rest of the paper is organized as follows: section 2 describes some of the methods used to solve automatic generation of neural nets; section 3 shows our proposed solution, and describes the evolutionary computation paradigm we use

J.J. Merelo Guervós et al. (Eds.): PPSN VII, LNCS 2439, pp. 505–514, 2002.

(EO). Next section (4) briefly describes the procedure followed to tune the EA parameters, as well as the experiments run and the results obtained; and finally, our conclusions and proposals of future work can be read in section 5.

2 State of the Art

There exist several papers that address the problem of automatic neural net design, as for instance the one by Yao [23]. There also some others centering on RBFNN automatic design, and in this sense the paper by Leonardis and Bischof [10] offers a good overview. Generic methods found in bibliography can be divided in three classes:

1. Methods in which **the number of radial basis functions must be given a priori**; after this, computing the values for the centers and the radii is done choosing points randomly from the training set or performing any kind of clustering method with them [13].
2. **Growing methods**. These methods try to build a net by adding neurons to an initially empty hidden layer. An example is Orthogonal Least Squares [5], based on a mechanism called *forward selection*, that adds hidden neurons to an initially empty net until the approximation error is under a prespecified threshold. An improvement over this algorithm is Regularised OLS [6], based on *regularised forward selection*, in which high hidden-to-output weight values are penalized using a penalty term, a process termed regularization [2]. Later, in [14], Orr newly used forward selection as the growing mechanism, and delete-1 (or leave-one-out) and generalized cross-validation as the methods to stop adding neurons. Cross-validation is used to compute the generalization error of many different nets in competition one with each other, halting the algorithm when that error reaches a minimum. Other methods of the same characteristics are Resource Allocation Networks [15], and Growing Cell Structures [7], which try to estimate the centers using hill-climbing techniques, making them liable to fall in local optima.
3. Finally, there are also **pruning methods** which start with an overspecified RBFNN and, iteratively, remove hidden neurons and/or weights. Leonardis and Bischof's algorithm [10] can be classified as a pruning method, in which the Minimum Description Length measure (MDL) is used to achieve a reduction in the complexity of the RBFNN. According to the MDL principle, a set of RBF describing the training data with the shortest possible encoding are selected among a larger set. Pruning methods tend to be very restrictive, and find suboptimal networks.

Special attention must be paid to the use of **evolutionary algorithms** to the construction of neural nets. Some reviews on this topic can be found in [1], and [23].

Initially, evolutionary algorithms usually tried to optimize only one of the parameters of the neural net (number of neurons, topology, or learning rate, amomg others), leaving the rest fixed. Currently, a different methodology can be found in the work by Castillo et al. [18, 19], in which evolutionary algorithms are used to create good Multilayer Perceptrons (MLPs), optimizing all the parameters at the same time, and obtaining very good results. A similar and previous example can be found in Merelo's [11], where a neural network called Learning Vector Quantization (LVQ) is genetically optimized. Nevertheless, both Castillo and Merelo's methods cannot be applied directly to the construction of RBFNN since their genetic operators are geared toward their specific network architecture.

Regarding RBFNN evolution, some good initial works are the papers by Carse and Fogarty in [4], and Whitehead and Choate [22]. Carse determined the RBFNN architecture by using operators adapted from some others previously used in evolution of fuzzy inference. Genes, representing RBFNN, were composed of pairs of centers and widths, and although they were allowed to change in size, a maximum size was fixed when running the algorithm. On the other hand, Whitehead and Choate chose a different way to make RBFNN evolve. In this case only one net was evolved, being the individuals of the evolutionary algorithm the hidden neurons of that net. Individuals were forced to compete and cooperate trying to cover the search space as much as possible while maintaining the minimum number of hidden neurons. The main problem the authors faced was the choice of the fitness function given that a trade-off between cooperation and competition had to be found. The final solution depends on many parameters that have to be properly fixed for the problem being studied. The number of neurons is also an "a priori" given parameter. Section 4 compares results obtained by our algorithm with those obtained by Case and Fogarty, and Whitehead and Choate on a time-series forecasting problem.

Recent papers include Gonzalez et al., as in [9], where expert mutation operators are created to evolve RBFNN in a much more complex way, which also results in higher computational requirement. And it is also very interesting the work by Rivera et al. [20], who offered a new point of view to the problem of building RBFNN. In their work, many hidden neurons cooperate, compete and are modified using *fuzzy evolution* to finally obtain a net that solves the task being studied. The term fuzzy evolution comes from the fact that a table composed of fuzzy rules determines which operator must be applied to a given unit in order to improve its behaviour. Both González and Rivera's works are good examples of how guided search can be carried out to improve the final results. Some of the results showed in section 4 are related to these researchers' work.

3 EvRBF

The method introduced in this work uses an evolutionary algorithm, **EvRBF**, to build RBFNN. EvRBF creates RBFNN with optimum predictive power by finding the number of neurons in the hidden layer, and their centers and radii.

The evolutionary algorithm itself has been programmed using the evolutionary computation paradigm, **EO** (**E**volving **O**bjects), current version is 0.9.2 [21], and can be found at `http://eodev.sourceforge.net`. Its main advantage is that what can be evolved is not necessarily a sequence of genes (bitstring or floating point values), but any data structure implementing any kind of subject to which a fitness or cost function can be assigned. Moreover, in accordance with Michalewicz's ideas published in [12], EO allows evolutionary algorithms to deal directly with the data structures most suited for solving a problem, instead of dealing with representation of that solution (that should then be decoded to obtain the solution itself). As a result, the way operators perform can be defined without regarding the specific way the net is implemented. Thus, in what follows, EvRBF algorithm is defined making no mention to the real implementation of the net.

EvRBF uses a standard evolutionary algorithm with fixed size population, tournament selection, and elitist replacement policy. Two kind of operators, mutation-like and crossover-like, have been specifically created. The algorithm finishes when a previously specified number of generations has been reached.

3.1 Binary Operators

Binary operators are applied to two RBFNN, returning only one. These are the operators currently implemented:

- **Multipoint recombination**. It randomly takes a few hidden neurons from the first net and replaces them by an equal number of randomly chosen neurons from the second net.
- **Average recombination**. This operator chooses hidden neurons from the first net, using a probability of 0.5. For each chosen neuron a random one is chosen in the second net. Then, the values for the center and radii are set to the average of both.

3.2 Unary Operators

Given that in a RBFNN the weights from hidden neurons to output ones can be easily computed, the unary or mutation-like operators affect the hidden neuron components, centers and radii, in their quantities and values, but not to the hidden-to-output weights.

- **Hidden neuron adder**. This operator creates a new neuron with random values for centers and radii (every value in the range allowed for any of the different dimensions of the input space).
- **Hidden neuron remover**. This operator removes a randomly chosen hidden neuron of the net.
- **Center creep**. It changes the values for the centers applying a Gaussian distortion. For each dimension of the center point, the Gaussian function is centered on the current value and it is as wide as the radius applied to that dimension in that neuron.
- **Radius creep**. Changes the values for the radius applying another Gaussian distortion. The Gaussian is centered on the current value and is as wide as the range of each dimension of the input space.
- **Randomize centers**. Changes the values of the centers of the hidden neurons to random values in the range allowed for each dimension of the input space.

- **Randomize radii**. Changes radii values randomly, always with values in the corresponding range of each input space dimension.
- **Closest hidden neuron remover**. Given a hidden neuron, it computes the euclidean distance between all the centers, and deletes one of the closest two that have been found.
- **Half hidden neuron remover**. Removes every hidden neuron of a net applying a probability equal to 0.5.

3.3 The Breeder

Every new generation is created in the following way:

1. Select a few individuals using tournament selection of fixed size.
2. Delete the rest of the population.
3. Generate new individuals applying the operators to individuals selected at step 1.
4. Set the weights of the new individuals using SVD.
5. Remove the useless neurons from the net, i.e., those whose weights to output neurons are very close to 0.
6. Evaluate the net, and set its fitness.

This algorithm always maintains the same number of individuals, thus once tournament selection, performed with 3 individuals, has finished, the operators are applied as many times as needed until the population reaches its fixed size.

As can be seen, a net's fitness is calculated using a three step procedure. Firstly, the net is trained using a set of training samples. Secondly, SVD is used to set the values for the weights and useless neurons are removed. Finally, the fitness value itself is computed as 1 divided by the root mean square error, as eq. 1) shows:

$$fitness = \frac{1}{\sqrt{\frac{\sum_{i=0}^{n-1}(y_i - o(x_i))^2}{n}}} \tag{1}$$

where y_i is the expected output, $o(x_i)$ is the output calculated by the net, and n is the number of input-output pairs composing the validation set of samples. None of these pairs has been used to train the net.

Finally, the full EvRBF algorithm is described as follows:

1. Load a training, a validation and a test sets of samples (if no validation data set is provided, then the training set is split in 75% for the training procedure, and 25% for the validation procedure).
2. Create a first generation of individuals (i.e., RBFNN) with a random number of hidden neurons and with random values for centers and radii.
3. Set the fitness of those individuals belonging to the first generation.
4. Apply the breeder an "a priori" given number of generations.
5. Fully train the nets of the last generation.
6. Compute the generalization error of last generation nets using the test set of samples.

Table 1. Parameters used to run all the experiments described along this section.

Parameter	Value	Parameter	Value
Elitism	0.1	Population	50
Multipoint Recombination	0.5	Average Recombination	1
Neuron Adder	0.5	Neuron Remover	1
Center creep	1	Radius creep	0.25
Random Center	4	Random Radii	2
Closest Remover	2	Half Remover	0.5

4 Experiments and Results

4.1 Determination of the Parameters

A set of guided experiments was carried out to set the values of the different parameters used to run the algorithm. A classification problem was chosen to avoid gearing those values toward the solution of one of the problems discussed later on this section. In order to set the value of a given parameter, the rest of parameters have been fixed to a default value. Then, a set of values for the parameter to determine has been chosen. For each of those values, the algorithm has been run 5 times. When all the values have been used, average values over the 5 runs related to generalisation error and net size have been analysed. Finally, the value that produced the best results is assigned to the parameter. For instance, every operator has been tested using the values 0, 0.25, 0.5, 1, 2, and 4, and leaving the other operators to the value of 1. Table 1 shows the values this method found.

Analysing the values in table 1, it can be concluded that the algorithm is focusing on searching the space, mainly to set the correct values for the centers. At the same time, it is trying to keep the number of centers as short as possible by deleting close neurons. This is an important feature of EvRBF given that the net size is not used to compute the net fitness, but the algorithm avoids growing overspecified nets.

4.2 EvRBF Applied to Time-Series Forecasting and Function Approximation

EvRBF was tested in various forecasting time-series problems and function approximation, using in all the cases the parameters seen above.

B.1. Time-Series Forecasting

As Whitehead and Choate in [22] and Case and Fogarty in [4], the Mackey-Glass time-series has been used as described in [13]. In this case, the value of $t+85$ is being predicted using the values recorded in t, $t-6$, $t-12$, and $t-18$, with a training data set composed of 500 samples.

Table 2. Results obtained by Case & Fogarty, Whitehead & Choate and EvRBF on Mackey-Glass time-series forecasting. EvRBF obtains as good results as Case and Fogarty did, without imposing restrictions to the number of neurons. Whitehead's method behaves better given that it performs a guided search, although only when many hidden neurons are provided the error falls an order of magnitude.

Algorithm	Normalised Error	Number of neurons
Carse & Fogarty	0.25	40
Carse & Fogarty	0.18	60
Carse & Fogarty	0.15	100
Whitehead & Choate	0.18	50
Whitehead & Choate	0.11	75
Whitehead & Choate	0.05	125
EvRBF	$\mathbf{0.177 \pm 0.004}$	$\mathbf{72 \pm 3}$

Table 2 shows the results obtained with the different methods. The error is expressed as the root-mean-squared error over the test set, divided by the standard deviation of the set of correct outputs. The test data set also contains 500 input-output pairs. Values related to EvRBF are averaged over 10 independent runs, and 100 generations were produced on each of such runs.

Results on this problem show that EvRBF behaves as well as the algorithm developed by Case and Fogarty, with the advantage that no restriction have to be imposed to the algorithm regarding the maximum number of neurons allowed. EvRBF can evolve both the number of neurons and the centers and radii for them, and everything done with fast operators and half the number of generations. With respect to results provided by Whitehead and Choate, the error yielded by EvRBF is similar, although not better, when the number of neurons is also comparable. For larger number of neurons the approximation is consequently better. This can only be explained taking into account that EvRBF is currently doing blind search, and it does not include methods for local searching.

Fig. 1 represents the evolution of best, worst and average fitness (figure on the left) as well as size of the best net, maximum, minimum and average sizes (figure on the right) along generations for this problem. As can be seen, size is kept stable while fitness improves during the time EvRBF is run.

B. Function Approximation

The functions we have used have been borrowed from [20], and they are defined as follows:

$$wm2(x) = sin(2\pi x), x \in [-1, 1]$$
$$dick(x) = 3x(x-1)(x-1.9)(x+0.7)(x+1.8), x \in [-2.1, 2.1]$$
$$nie(x) = 3e^{x^2} sin(\pi x), x \in [-1, 1]$$
$$pom(x) = e^{-3x} sin(10\pi x), x \in [-1, 1]$$

Table 3 shows the results obtained with EvRBF compared to those obtained by methods found in [16], [9], and [20]. This results have been obtained running

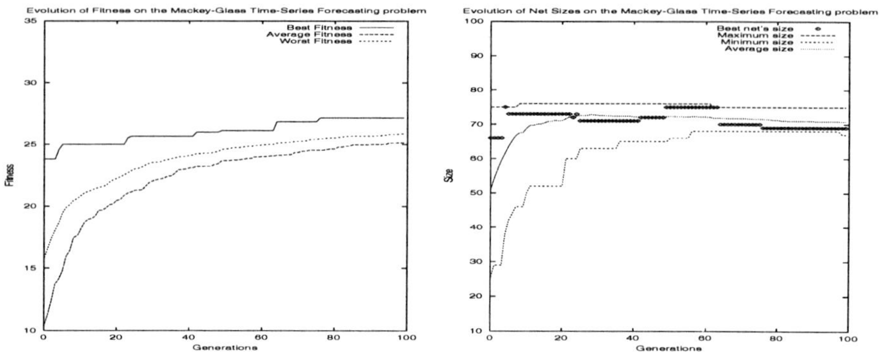

Fig. 1. Figures representing the evolution of fitness (left) and size (right) along generations on the Mackey-Glass Time-Series problem. As it is seen on the right hand side graph, size tends to converge even as there are no check or bounds limit.

Table 3. Comparison of results provided by EvRBF and the methods described by Pomares in [16], González in [8], and Rivera in [20]. In all of the problems, except in the last one, EvRBF obtains better results, although the nets obtained are slightly larger.

Algorithm	*WM2*		*Dick*		*Nie*		*Pom*	
	MSE	Size	MSE	Size	MSE	Size	MSE	Size
Pomares	0.026	10	0.46	7	0.0020	10	0.114	12
González	0.006	6	0.30	5	2.87×10^{-5}	5	**0.066**	8
Rivera	0.08	7	0.22	7	1.34×10^{-4}	6	0.123	8
EvRBF	$\mathbf{1.19 \times 10^{-4}}$	10.5	**0.05**	8.3	$\mathbf{1.03 \times 10^{-5}}$	8.5	0.49	6.10

Table 4. Results obtained by EvRBF on approximation of the functions defined as *wm2*, *dick*, *nie*, and *pom*.

Problem	MSE			Size		
	Min. (Size)	Max. (Size)	Average	Min.	Max.	Average.
wm2	3.40×10^{-5} (12)	3.05×10^{-4} (9)	$1.2 \times 10^{-4} \pm 0.9 \times 10^{-4}$	8	14	11 ± 2
Dick	0.01 (9)	0.14 (8)	0.05 ± 0.04	7	9	8.3 ± 0.8
Nie	6.32×10^{-7} (10)	3.00×10^{-5} (7)	$1.03 \times 10^{-5} \pm 1.07 \times 10^{-5}$	7	10	8.5 ± 1.6
Pom	0.44 (7)	0.52 (4)	0.49 ± 0.03	4	8	6.1 ± 1.5

the algorithm 10 times, always with the same parameters. Only the best net found at the end of each execution has been taken into account. Thus, values in table 3 corresponding to EvRBF shows average values. Table 4 shows detailed information about minimum, maximum, and average error and size, related to the best nets found at the end of the 10 executions.

Table 3 shows how EvRBF obtains better results in all the functions, except the one used by Pomares in [16]. Although EvRBF does not establishes any limit to the number of hidden neurons, this number is quite similar to the rest of methods, but obtains a better approximation to the functions. In this case, despite EvRBF operators are doing blind search, results show that they are covering the search space in a quite good way.

5 Conclusions

An evolutionary algorithm designed to automatically configure RBF Neural Nets has been introduced. The computational framework used to develop this algorithm allows it to handle the solutions to the problems themselves, instead of representations of those solutions. Thus, new operators have been designed without regarding the specific implementation of the individuals, but trying to perform a successful search over the input space.

A set of fixed parameters, found testing the algorithm against a classification problem, has been used in problems of time-series forecasting and function approximation. Results show that this algorithm minimizes differences to expected outputs while maintaining an acceptable number of hidden neurons in every RBFNN. This happens despite the fact that number of hidden neurons is not taking into account when the EA compares two individuals trying to decide which of them is better than the other, a critical point for this kind of methods. Results also show that despite being more parameter-free, the algorithm obtains always reasonable results, and sometimes better results than the other evolving RBF algorithms.

Future work must provide operators that locally tune the solutions found by current operators that make a blind search. New methods for locally training the nets and also well known methods intended to choose centers for the RBFNN, all of them present in literature, are being investigated. This methods will be implemented as operators for the EA, so that more accurate networks be found with the shortest number of hidden neurons.

Acknowledgment

This work has been supported in part by projects CICYT TIC99-0550 and INTAS 97-30950.

References

1. A.V. Adamopoulos, E.F. Georgopoulos, S.D. Likothanassis, and P.A. Anninos. Forecasting the MagnetoEncephaloGram (MEG) of Epilectic Patients Using Genetically Optimized Neural Networks. In *Proceedings of the Genetic and Evolutionary Computation Conference, GECCO'99*, volume 2, pages 1457–1462. Morgan-Kaufmann Publ., July 1999.
2. C.M Bishop. *Neural Networks for Pattern Recognition.* Oxford University Press, 1995. ISBN 0-19-853849-9 (hardback) or 0-19-853864-2 (paperback).
3. D.S. Broomhead and D. Lowe. Multivariable Functional Interpolation and Adaptative Networks. *Complex Systems*, 11:321–355, 1988.
4. B. Carse and T.C. Fogarty. Fast evolutionary learning of minimal radial basis function neural networks using a genetic algorithm. In T.C. Fogarty, editor, *Proceedings of the Evolutionary Computing, AISB Workshop'96*, pages 1–22. Springer-Verlag, 1996.

5. S. Chen et al. Orthogonal Least Squares algorithm for constructing Radial Basis Function Networks. *IEEE Trunsactions on Neural Networks*, 2(2):302–309, 1991.
6. S. Chen et al. Regularised Orthogonal Least Squares Learning for Radial basis function Networks. Submitted to International Journal Control, 1995.
7. B. Fritzke. Supervised learning with growing cell structures. In J.D. Cowan, G. Tesauro, and J. Aspector, editors, *Advances in Neural Information Processing Systems*, volume 6, pages 255–262. Morgan Kaufmann, 1994.
8. J. González, I. Rojas, H. Pomares, and J. Ortega. Rnf neural networks, multiobjective optimization and time series forecasting. *Lecture Notes in Computer Science*, (2084):498–505, 2001.
9. J. González, I. Rojas, H. Pomares, and M. Salmerón. Expert mutation operators for the evolution of radial basis function neural networks. *Lecture Notes in Computer Science*, 2084:538–545, June 2001.
10. A. Leonardis and H. Bischof. And efficient MDL-based construction of RBF networks. *Neural Networks*, 11:963–973, 1998.
11. J.J. Merelo and A. Prieto. G-LVQ a combination of genetic algorithms and LVQ. In D.W. Pearson, N.C. Steele, and R.F. Albrecht, editors, *Artificial Neural Nets and Genetic Algorithms*. Springer-Verlag, 1995.
12. Zbigniew Michalewicz. *Genetic algorithms + data structures = evolution programs*. Springer-Verlag, NewYork USA, 3 edition, 1999.
13. J. E. Moody and C. Darken. Fast learning in networks of locally tuned processing units. *Neural Computation*, 2(1):281–294, 1989.
14. M.J.L. Orr. Regularisation in the Selection of Radial Basis Function Centres. *Neural Computation*, 7(3):606–623, 1995.
15. J. Platt. A resource-allocating network for function interpolation. *Neural Computation*, 3(2):213–225, 1991.
16. H. Pomares, I. Rojas, J. Ortegaa, J. González, and A. Prieto. A systematic approach to self-generating fuzzy rule-table for function approximation. *IEEE Trans. Syst., Man., and Cyber.*, 30:431–447, 2000.
17. W.H. Press, S.A. Teukolsky, W.T. Vetterling, and B.P. Flannery. *Numerical Recipes in C*. Cambridge University Press, 2nd edition, 1992.
18. P.A. Castillo; J. Carpio; J. J. Merelo; V. Rivas; G. Romero; A. Prieto. Evolving multilayer perceptrons. *Neural Processing Letters*, 12:115–127, October 2000.
19. P.A. Castillo; J.J. Merelo; V. Rivas; G. Romero; A. Prieto. G-Prop: Global Optimization of Multilayer Perceptrons using GAs. *Neurocomputing, Vol. 35/1–4, pp.149–163*, 2000.
20. A.J. Rivera, J. Ortega, M.J. del Jesus, and J. Gonzalez. Aproximación de funciones con evolución difusa mediante cooperación y competición de rbfs. In *Actas del I Congreso Español de Algoritmos Evolutivos y Bioinspirados, AEB'02*, pages 507–514, February 2002.
21. J. J. Merelo; M. G. Arenas; J. Carpio; P. Castillo; V. M. Rivas; G. Romero; M. Schoenauer. Evolving objects. pages 1083–1086, 2000. ISBN: 0-9643456-9-2.
22. B. A. Whitehead and T.D. Choate. Cooperative-Competitive Genetic Evolution of Radial Basis Function Centers and Widths for Time Series Prediction. *IEEE Transactions on Neural Network*, 7(4):869–880, July 1996.
23. X. Yao. Evolving artificial neural networks. *Proceedings of the IEEE, 87(9):1423–1447*, 1999.

Hybrid Algorithms: Memetic, Other

Evolutive Identification of Fuzzy Systems for Time-Series Prediction

Jesús González, Ignacio Rojas, and Héctor Pomares

Department of Computer Architecture and Computer Technology
E.T.S. Ingeniería Informática
University of Granada
E. 18071 Granada (Spain)

Abstract. This paper presents a new algorithm for designing fuzzy systems. It automatically identifies the optimum number of rules in the fuzzy knowledge base and adjusts the parameters defining them.
This algorithm hybridizes the robustness and capability of evolutive algorithms with multiobjective optimization techniques which are able to minimize both the prediction error of the fuzzy system and its complexity, i.e. the number of parameters. In order to guide the search and accelerate the algorithm's convergence, new specific genetic operators have been designed, which combine several heuristic and analytical methods. The results obtained show the validity of the proposed algorithm for the identification of fuzzy systems when applied to time-series prediction.

Keywords: Fuzzy systems, evolution, multiobjective optimization.

1 Introduction

A time series is a sequence of values $S = \{s_1, s_2, ..., s_t\}$ that reflect the evolution of a measure over time. Each sample $s_i \in S$ is taken at regular time intervals and is the result of measuring a certain characteristic of a physical or synthetic phenomenon. The goal of time series prediction is to make an accurate forecast of succeeding values in a sequence, based on the current value and on some of the preceding values.

Fuzzy systems have been shown to be efficient tools for this type of problem, due to their local-learning capabilities, their versatility, their simplicity and their interpolation possibilities [20, 12]. However, they require previous knowledge of the number of rules that comprise the knowledge base, the membership functions that define these rules, and their consequents. The determination of the above parameters is not an easy task, due to the non linear dependencies within the system. The bibliography contains references to different algorithms that tackle this problem, with varying degrees of success [20, 16, 14, 10].

Evolutive algorithms [9, 6, 13] are problem-solving techniques based on natural evolution. They work by successively applying genetic operators to a population of potential solutions to a problem, and then selecting the best solutions for the next generation. This process is repeated until the population converges

J.J. Merelo Guervós et al. (Eds.): PPSN VII, LNCS 2439, pp. 517–526, 2002.

to a quasi-optimum solution to the problem. Because of their robustness and easy hybridization, evolutive algorithms are ideal for the identification of fuzzy systems.

This paper proposes the hybridization of an evolutive algorithm using multiobjective optimization techniques and analytical methods to determine the optimum structure of a fuzzy system and to adjust the values of its parameters.

2 Description of the Fuzzy System

Let us consider the sequence of measures S taken from a certain phenomenon $\mathcal{F}$. Let $s_t \in S$ be the last measure in the sequence, henceforth called the current measure, Δ a delay or lag interval and H the prediction interval or horizon. Analytically, the problem of predicting the subsequent values in a time series can be written as:

$$\hat{s}_{t+H} = f(s_t, s_{t-\Delta}, ..., s_{t-(d-1)\Delta}) \tag{1}$$

where f can be taken as a function that models the phenomenon $\mathcal{F}$ and which is capable of estimating the output produced at the instant $t+H$ after the current measure and from the $d-1$ previous measures obtained with a lag of Δ.

After setting the parameters H and Δ, the problem can be approached as one of functional approximation, in which it is necessary to model an unknown function f with d inputs. For a sequence of t measures, we can generate a set of $n = t - H - (d-1)\Delta$ samples of the unknown function f as shown in Table 1.

To simplify the notation, we take $x_k^j = s_{j+(k-1)\Delta}$ and $\boldsymbol{x}^j = (x_1^j, x_2^j, ..., x_d^j) = (s_j, s_{j+\Delta}, ..., s_{j+(d-1)\Delta}) = \boldsymbol{s}^j$. With this notation, the problem can be formulated as the approximation of an unknown function f from a set of samples $X = \{(\boldsymbol{x}^j, y^j) : \boldsymbol{x}^j \in \Re^d, y^j = f(\boldsymbol{x}^j) \in \Re,\ j = 1, 2, ..., n\}$. A fuzzy system to approximate this function can be expressed according to the following rule set:

$$\begin{array}{rl} \textbf{IF} & x_1 \text{ is } X_1^i \textbf{ AND } ... \textbf{ AND } x_d \text{ is } X_d^i \\ \textbf{THEN} & y = R_i \ \forall i = 1, ..., m \end{array} \tag{2}$$

where X_k^i, with $k = 1, ..., d$ is the membership function associated with the input variable x_k in the i-th rule and R_i is the consequent of the rule. Each consequent R_i is a constant or singleton scalar value and represents the contribution to the output by the i-th rule. A fuzzy system with this type of rule is known as a zero-order Takagi-Sugeno fuzzy system [11].

The degree of activation of the i-th rule is calculated by the equation:

$$\alpha_i(\boldsymbol{x}^j) = \prod_{k=1}^{d} \mu_k^i(x_k^j) \tag{3}$$

where $\mu_k^i(x_k^j)$ is the degree of membership of the variable x_k^j within the membership function X_k^i.

Table 1. Generation of samples from a sequence of measures.

j	$\boldsymbol{s}^j$	$y^j = f(\boldsymbol{s}^j)$
1	$(s_1, s_{1+\Delta}, ..., s_{1+(d-1)\Delta})$	$s_{H+1+(d-1)\Delta}$
2	$(s_2, s_{2+\Delta}, ..., s_{2+(d-1)\Delta})$	$s_{H+2+(d-1)\Delta}$
$\vdots$	$\vdots$	$\vdots$
j	$(s_j, s_{j+\Delta}, ..., s_{j+(d-1)\Delta})$	$s_{H+j+(d-1)\Delta}$
$\vdots$	$\vdots$	$\vdots$
n	$(s_{t-H-(d-1)\Delta}, ..., s_{t-H})$	s_t

The system output is obtained by summing the activation of each of the rules weighted by its consequent, and then dividing the result by the total activation of the rules:

$$F(\boldsymbol{x}^j) = \frac{\sum_{i=1}^{m} \alpha_i(\boldsymbol{x}^j) R_i}{\sum_{i=1}^{m} \alpha_i(\boldsymbol{x}^j)} \tag{4}$$

Due to the dispersed nature of the samples within the input space, gaussian membership functions are recommended to resolve this problem. Because they are continuous and derivable, they produce a smoother output and improve the system's interpolation capability. The membership functions used in this study, therefore, are of the following form:

$$\mu_k^i(\boldsymbol{x}^j) = e^{-\left(\frac{\left\| x_k^j - c_k^i \right\|}{r_k^i}\right)^2} \tag{5}$$

It is a complex task to identify the number of fuzzy rules required to model the phenomenon $\mathcal{F}$ in such a way as to achieve a balance between the system's complexity and its prediction error, together with the location of these rules within the input space (values of their centres c_k^i and amplitudes r_k^i). To obtain a fuzzy system with a low prediction error rate, it is normally necessary to possess a large number of fuzzy rules. On the other hand, we generally wish to obtain a simple fuzzy system, even if the prediction error rate is moderately high, because simple systems possess a higher interpolation capacity. Obviously, we must find a compromise between the precision required of the system and its subsequent complexity. To tackle this problem, rather than using classical indices such as Akaike's Information Criterion or the Minimum Description Length parameter [1, 15], or an auxiliary fuzzy system [16, 14], we have opted for a multiobjective evolutive algorithm.

3 Development of a New Evolutive Algorithm

An evolutive algorithm [13] is the result of adapting a genetic algorithm [9] to a particular problem, replacing the bit-chain representation of potential solutions by a data structure that is more natural and closer to the problem. This enables us to obtain more exact solutions and facilitates the design of genetic operators that include specific knowledge and heuristics. For the problem of identifying fuzzy systems, the choice of a flexible representation that provides direct access to any of the system parameters and makes it possible to handle different-structured fuzzy systems in a uniform way is a fundamental necessity for a good solution.

As well as introducing specific knowledge, we have added the capability of carrying out a multiobjective search to minimize both the prediction error of the fuzzy system and the number of parameters it contains.

3.1 Incorporation of Specific Knowledge

The introduction of specific knowledge into an evolutive algorithm helps speed up convergence and also helps with the restrictions that must be met by valid solutions to the problem. This knowledge is introduced by designing specific operators for the problem; these operators study the solutions before they are altered, and attempt to correct or improve them, rather than simply performing a blind modification.

The use of a representation of the solutions based on bit chains greatly hampers this task, as the alteration of a single bit may lead to very different changes in the final solution depending on the position of the bit; moreover, they considerably restrict the precision of the solutions obtained [13]. These disadvantages make it necessary to introduce a representation that provides direct access to any of the parameters of the fuzzy system.

3.1.1 Representation of the Solutions. The most natural representation structure for a fuzzy system F is the set of m fuzzy rules $F = \{\langle \boldsymbol{c}^i, \boldsymbol{r}^i, R_i \rangle : i = 1, ..., m\}$. Each fuzzy rule is determined by its centre $\boldsymbol{c}^i = (c_1^i, ..., c_d^i)$, by its amplitude $\boldsymbol{r}^i = (r_1^i, ..., r_d^i)$ and by its consequent R_i. This structure provides free access to each of the rules that comprise the fuzzy system, and enables us to vary its position, amplitude or consequent, in a simple and efficient way, thus facilitating the design of the genetic operators.

3.1.2 Removing Fuzzy Rules. In order to determine the optimum structure of the fuzzy system, we must incorporate genetic operators to carry out structural alterations to the solutions. The removal of a rule reduces the complexity of the fuzzy system, but also has a greater or lesser effect on the prediction error, depending on which rule is eliminated. The OLS algorithm is applied to the fuzzy system to minimize the effect of this interaction [3, 19] as a prior step to the alteration. The latter algorithm has been mentioned in the literature as a means of calculating the consequents of the fuzzy rules in an optimum way, but it possesses another very important characteristic, too. It calculates an output error reduction coefficient for each rule. The higher the coefficient , the more sensitive is the rule and the greater is the error produced if it is removed.

After calculating the fuzzy system error reduction coefficients, each rule is assigned a probability of being eliminated; this probability is inversely proportional to the error reduction coefficient. A randomly-selected rule is then removed. After this alteration, the Cholesky decomposition is applied to the covariance matrix in order to calculate the consequents of the rules in an optimum way [14].

3.1.3 Division of Fuzzy Rules. This genetic operator is complementary to the previous one. Its purpose is to detect the area within the input space where the greatest prediction error occurs and to add a new fuzzy rule there to minimize the system error.

To do this, it is first necessary to study the contribution of each error to the total system error, by means of the following equation:

$$e^i = \sum_{j=1}^{n} \frac{\alpha_i(\boldsymbol{x}^j)}{\sum_{i_2=1}^{m} \alpha_{i_2}(\boldsymbol{x}^j)} \left| F(\boldsymbol{x}^j) - y^j \right|, i = 1, ..., m \tag{6}$$

where n is the number of training samples and m is the number of rules within the system.

A large value of e^i means that the i-th rule is not sufficient to cover the zone of the input space where it is located, and so a division probability is assigned to each rule. This probability is directly proportional to the rule's contribution to the output error. A rule is then randomly selected to be divided, thus generating two new rules that cover the same input space as did the previous single rule. By this method, the system adds a fuzzy rule where it most contributes to reducing the total system error. After the rule has been divided, the consequents of the system are readjusted by applying the Cholesky decomposition.

3.1.4 Mutation of Fuzzy Systems. This operator is intended to maintain the diversity of the population. When it is applied, it produces random effects on fuzzy systems and helps the algorithm to explore new zones within the search space. The alterations made only affect the location or amplitude of the fuzzy rules, while the structure of the system remains constant.

To choose the fuzzy rule to be altered, the OLS algorithm is used to estimate the sensitivity of the rules and to assign a selection probability, in the same way as was done for the rule-elimination operator. When the rule has been altered, the Cholesky decomposition is used to calculate the optimum consequents for the remaining rules.

3.1.5 Crossover of Fuzzy Systems. This operator receives two fuzzy systems and returns another two, which combine the genetic information stored by their progenitors. The information exchange is carried out at the level of the rules. One of the two progenitors is chosen, and then a random number of rules are selected to be exchanged with the other progenitor. Each rule selected is exchanged for that of the other progenitor that is closest to it in the input space.

In this way the two systems derived from the crossover store information from their two progenitors, thus avoiding the occurrence of systems with overlapping rules, thanks to the locality principle imposed in the exchange process.

3.2 Multiobjective Optimization

Multiobjective optimization consists of finding solutions to a problem in which various objectives must be minimized simultaneously. In principle, no objective is favoured at the expense of any other, and if the objectives are not independent, there is no global minimum for the problem. Such is the case in the identification of fuzzy systems; when the number of system parameters is minimized, the output error normally increases.

In this type of problem, the algorithm must find the set of solutions that achieve an equilibrium between all the objectives so that the final user can choose the best solution according to his/her preferences or necessities.

For an evolutive algorithm to perform a multiobjective search, it is only necessary to change the evaluation stage of the population to take all the objectives into account, while the other stages of the algorithm remain unaltered. In the bibliography, various approaches to carry out such a modification are proposed [17, 5, 18].

3.2.1 Evaluation Function. The evaluation function must assign an aptitude to each individual in the population with respect to each of the objectives to be minimized. As a result of this evaluation, the aptitude of the individuals becomes a vector in which each of the components indicates the aptitude of the individual with respect to a particular objective. For the problem in question, the aptitude of a fuzzy system F takes the form $a(F) = \langle e(F), c(F) \rangle$, in which the first component evaluates its prediction error and the second component, the complexity. The Root Mean Squared Error (RMSE) is used as a measure of the error $e(F)$:

$$e(F) = \sqrt{\frac{\sum_{j=1}^{n} \left(F(\boldsymbol{x}^j) - y^j\right)^2}{n}} \tag{7}$$

where $F(\boldsymbol{x}^j)$ represents the output of the fuzzy system for the input $\boldsymbol{x}^j$ and y^j is the expected output for this input.

The number of fuzzy rules comprising a fuzzy system $c(F)$ is used to measure its complexity.

3.2.2 Assignation of Pseudo-aptitude. The assignation of pseudo-aptitudes is done as in MOGA [5] to distribute the selection pressure between individuals suitably and to enable the evolutive algorithm to continue as if it were carrying out a single-objective search.

3.2.3 Sharing of Pseudo-aptitudes. The means of calculating pseudo-aptitudes described in the previous section favours non-dominated solutions. This form of elitism may lead to a premature convergence to the first non-dominated solution that appears. To avoid this and to maintain the diversity of the population, a pseudo-aptitude sharing mechanism must be introduced into the algorithm [7, 4]. This reduces the aptitude differences between these super-individuals and the rest of the population, so that the former do not reproduce too fast and saturate the population in succeeding generations.

For this purpose, a niche-creation strategy is used within the population. Two individuals are considered members of the same niche if the distance between them is less than a given threshold or minimum distance σ. The following operator was designed to measure the distance between two fuzzy systems:

$$d(F_h, F_l) = \frac{\sum_{i_1=1}^{m_h} min\left\{\left\|c_{i_1}^h - c_{i_2}^l\right\| : 1 \leq i_2 \leq m_l\right\}}{m_h + m_l} + \\ + \frac{\sum_{i_2=1}^{m_l} min\left\{\left\|c_{i_2}^l - c_{i_1}^h\right\| : 1 \leq i_1 \leq m_h\right\}}{m_h + m_l} \quad (8)$$

where m_h and m_l are the number of fuzzy rules in the systems F_h and F_l and where $c_{i_1}^h$ and $c_{i_2}^l$ are their centres.

The aptitude of each individual F_h is multiplied by a factor $\delta_h \geq 1$ that is directly proportional to the number of individuals belonging to its niche. This factor δ_h is calculated by:

$$\delta_h = \sum_{l=1}^{\nu} \mu_{h,l} \quad (9)$$

where ν is the number of systems comprising the population of the evolutive algorithm and $\mu_{h,l}$ measures the degree of membership of the system F_l to the niche defined by F_h. The following equation is used to calculate $\mu_{h,l}$:

$$\mu_{h,l} = \begin{cases} 1 - \left(\frac{d(F_h,F_l)}{\sigma}\right)^2 & \text{si } d(F_h, F_l) < \sigma \\ 0 & \text{otherwise} \end{cases} \quad (10)$$

This mechanism prevents a non-dominated solution from saturating the population in just a few generations, as the more copies that are created, the more its aptitude increases.

4 Results

To test the algorithm presented in the previous section, we chose the Mackey-Glass Chaotic Time Series, generated according to the following equation:

$$s_{t+1} = (1 - b)s_t + \frac{a s_{t-\tau}}{1 + s_{t-\tau}^{10}} \quad (11)$$

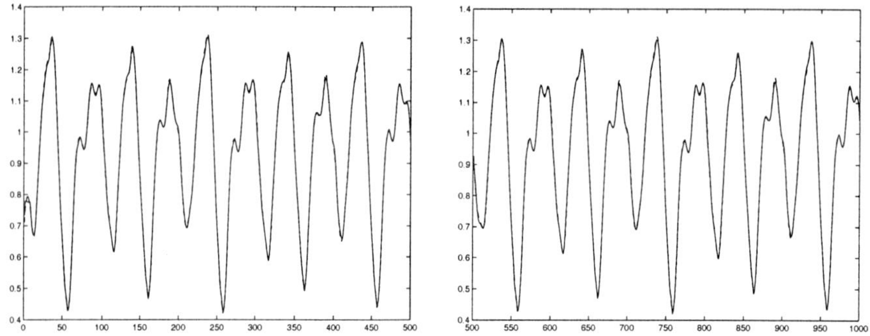

(a) (b)

Fig. 1. Mackey-Glass Time Series (dashed line) and system output with 12 rules (solid line). (a) Training exapmples. (b) Test examples.

Table 2. Different methods and prediction errors.

Method		RMSE
Linear prediction method		0.55
Auto-Regressive Model		0.19
L-X. Wang	T-Norma: Prod.	0.0907
	T-Norma: Mín.	0.0904
NN Cascade Correlation		0.06
6th Order Polynomials		0.04
D. Kim and C. Kim	5 MFs/var.	0.0492
(Genetic algorithm	7 MFs/var.	0.0423
+ Fuzzy system)	9 MFs/var.	0.0379
NN Backpropagation		0.02
ANFIS (NN + Fuzzy Logic)		0.007
proposed algorithm	10 rules	0.009
	12 rules	0.007

When $\tau \geq 17$, the above equation has a chaotic behaviour. Following previous studies [21], we set parameter values at $a = 0.2$, $b = 0.1$ and $\tau = 17$. The lag interval was set at $\Delta = 6$ and the horizon at $H = 6$. In accordance with the procedure described in Section 2 1000 samples were generated, with the form $((s_t, s_{t-6}, s_{t-12}, s_{t-18}), s_{t+6})$. The first 500 were used to train the system and the final 500, to validate it. The evolutive algorithm was run with a population size of 50 for 100 generations and a search was made for solutions with 8-15 rules. Of the final set of solutions, the fuzzy systems with 10-12 rules are outstanding, and the prediction errors of these are presented in Table 2. Figure 1 shows the output of the 12-rule fuzzy system for the training and validation data.

Table 2 shows some of the methods proposed in the bibliography to approach this problem, together with the corresponding prediction error. Evidently, the algorithm presented in the present study obtains much better results than those of the other methodologies, and even equals the excellent results obtained in 1993 by Jang with the ANFIS system [10].

5 Conclusions

This paper proposes an evolutive algorithm that is capable of identifying an appropriate number of rules and parameter values of a fuzzy system, in an automatic fashion. The algorithm proposed can directly evolve a population of fuzzy systems and does not require any intermediate codification, thus facilitating the incorporation of specific knowledge and heuristics into the genetic operators. The algorithm uses analytical methods to calculate the optimum consequents of the rules.

The possibility of carrying out a multiobjective search greatly simplifies the task of identifying the optimum fuzzy system. By reformulating the problem as one of *"finding the fuzzy system that minimizes both the prediction error and the number of fuzzy rules"*, the algorithm automatically searches for solutions that constitute a compromise between the two objectives, and returns a set of non-dominated solutions so that the final user can choose the solution that best suits his/her needs.

The results obtained show that the hybridization of evolutive algorithms with the OLS algorithm and the use of multiobjective optimization techniques can be adapted perfectly to the problem of identifying the optimum structure and the parameters of a fuzzy system.

References

1. H. Akaike. *Information Theory and Extension of the Maximum Likelihood Principle*, pages 267–810. Akademia Kiado, Budapest, 1973.
2. V. Chankong and Y. Y. Haimes. *Multiobjective Decision Making Theory and Methodology.* North-Holland, New York, 1983.
3. S. Chen, C. F. N. Cowan, and P. M. Grant. Orthogonal least squares learning algorithm for radial basis function networks. *IEEE Trans. Neural Networks*, 2:302–309, 1991.
4. K. Deb and D. E. Goldberg. An investigation of niches and species formation in genetic function optimization. In J. D. Schaffer, editor, *Proceedings of the Third International Conference on Genetic Algorithms*, pages 42–50, San Mateo, CA, 1991. Morgan Kaufmann.
5. C. M. Fonseca and P. J. Fleming. Genetic algorithms for multiobjective optimization: Formulation, discussion and generalization. In S. Forrest, editor, *Proceedings of the Fifth International Conference on Genetic Algorithms*, pages 416–423, San Mateo, CA, 1993. Morgan Kaufmann.
6. D. E. Goldberg. *Genetic Algorithms in Search, Optimization and Machine Learning.* Addison-Wesley, 1989.
7. D. E. Goldberg and J. Richardson. Genetic algorithms with sharing for multimodal function optimization. In J. J. Grefenstette, editor, *Genetic Algorithms and Their Applications: Proceedings of the Second International Conference on Genetic Algorithms*, pages 41–49, San Mateo, CA, 1987. Morgan Kaufmann.

8. A. E. Hans. Multicriteria optimization for highly accurate systems. In W. Stadler, editor, *Multicriteria Optimization in Engineering and Sciences, Mathematical Concepts and Methods in Science and Engineering*, volume 19, pages 309–352, New York, 1988. Plenum Press.
9. J. J. Holland. *Adaptation in Natural and Artificial Systems.* University of Michigan Press, 1975.
10. J. S. R. Jang. Anfis: Adaptive network-based fuzzy inference system. *IEEE Trans. Syst., Man, Cybern.*, 23:665–685, May 1993.
11. J. S. R. Jang, C. T. Sun, and E. Mizutani. *Neuro-Fuzzy and Soft Computing.* Prentice–Hall, ISBN 0-13-261066-3, 1997.
12. D. Kim and C. Kim. Forecasting time series with genetic fuzzy predictor ensemble. *IEEE Trans. Fuzzy Systems*, 5(4):523–535, Nov. 1997.
13. Z. Michalewicz. *Genetic Algorithms + Data Structures = Evolution Programs.* Springer-Verlag, 3rd edition, 1996.
14. H. Pomares, I. Rojas, J. Ortega, J. González, and A. Prieto. A systematic approach to a self-generating fuzzy rule-table for function approximation. *IEEE Trans. Syst. Man and Cyber. - Part B*, 30(3):431–447, 2000.
15. J. Rissanen. Modelling by shortest data description. *Automatica*, 14:464–471, 1978.
16. I. Rojas, H. Pomares, J. Ortega, and A. Prieto. Self-organized fuzzy system generation from training examples. *IEEE Trans. Fuzzy Systems*, 8(1), Feb. 2000.
17. J. D. Schaffer. *Some Experiments in Machine Learning using Vector Evaluated Genetic Algorithms.* Ph.d. dissertation, Vanderbilt University, Nashville, TN, 1984. TCGA file No. 00314.
18. N. Srinivas and K. Dev. Multiobjective optimization using nondominated sorting in genetic algorithms. *Evolutionary Computation*, 2(3):221–248, 1995.
19. L. X. Wang and J. M. Mendel. Fuzzy basis functions, universal approximation, and orthogonal least squares learning. *IEEE Trans. Neural Networks*, 3:807–814, 1992.
20. L. X. Wang and J. M. Mendel. Generating fuzzy rules by learning from examples. *IEEE Trans. Syst. Man and Cyber.*, 22(6):1414–1427, November/December 1992.
21. B. A. Whitehead and T. D. Choate. Cooperative-competitive genetic evolution of radial basis function centers and widths for time series prediction. *IEEE Trans. Neural Networks*, 7(4):869–880, July 1996.

HyGLEAM - An Approach to Generally Applicable Hybridization of Evolutionary Algorithms

Wilfried Jakob

Forschungszentrum Karlsruhe, Institute for Applied Computer Science,
P.O. Box 3640, 76021 Karlsruhe, Germany
jakob@iai.fzk.de

Abstract. Most successful applications of Evolutionary Algorithms to real world problems employ some sort of hybridization, thus speeding up the optimization process but turning the general applicable Evolutionary Algorithm into a problem-specific tool. This paper proposes to combine Evolutionary Algorithms and generally applicable local searchers to get the best of both approaches: A fast, but robust tool for global optimization. The approach consists of four different kinds of hybridization and combinations thereof, which are tested and compared using five commonly used benchmark functions and three real world applications. The results show the superiority of two hybridization types, with which reductions in the number evaluations of up to a factor of 100 could be achieved.

1 Motivation

When looking to the papers reporting about real world applications at the major EA conferences in the last ten years, it becomes clear that nearly all of them used some sort of hybridization with problem-specific local searchers or they applied other means of inserting problem-specific knowledge into the evolutionary process like e.g. special genetic operators. The commonly paid price for the achieved speed-up is the lost in generality of the resulting hybrid.

Already Holland suggested in 1975 to use GAs as a kind of preprocessor and finalize the optimization with local searchers [1]. 24 years later Goldberg and Voessner [2] pointed out that is not easy to find an appropriate procedure for distributing the computing time between global and local search and came to the conclusion that this is still an open question. They also stressed that nearly all serious EA applications use some sort of hybridization but that there is still a lack of investigations on suitable types of integration, which are not related to specific applications or limited to certain problem fields.

This paper contributes to this discussion by

- using only general applicable local searchers,
- comparing three well-known types of hybridization, a modified form of one type, and meaningful combinations of them
- introducing a new method of controlling the interaction of the basic algorithms, and
- testing the approach with a wide range of different types of applications to check the generality of the approach.

J.J. Merelo Guervós et al. (Eds.): PPSN VII, LNCS 2439, pp. 527–536, 2002.

2 Methods of Hybridization

For a generally applicable hybridization of an EA, four general alternatives exist:

1. Pre-optimization of the start population
   This can be applied to the entire population or a fraction of it, and it provides the evolution with valid solutions of more or less good quality to start with.
2. Post-optimization of the EA results
   EAs are known to converge slowly. Thus, an improvement can be expected by stopping the evolution after approaching the area of attraction of the global optimum and leaving the rest to the local search.
3. Direct integration
   Optimizing every or the best offspring of one mating only causes the EA to operate over the peaks of the fitness landscape exclusively rather than to treat the valleys and slopes, too. The offspring's genotype can be updated (Lamarckian evolution) or left unchanged (Baldwinian evolution). As both methods which are usually applied to domain-specific local searchers are controversially discussed in literature [3, 4], this was also investigated. Orvosh and Davis recommend updating 5% of the accepted offsprings only [5].
4. Delayed direct integration
   Variant of direct integration, where the evolution works on its own until a criterion similar to the one used for switching from evolutionary to local search for post-optimization is fulfilled.

Pre-optimization can be combined with the other methods, while a fusion of direct integration and post-optimization does not appear to be meaningful.

3 Basic Algorithms Used for Hybridization

To comply with the goal of general applicability, the EA must allow for combinatorial optimization and parameter strings of dynamic length as required by some applications like design optimization [6] or collision-free robot path planning [7]. Especially because of the last requirement, GLEAM (General Learning Evolutionary Algorithm and Method) [7,8], an EA comprising elements of real coded GAs and the Evolution Strategy (ES) was chosen for testing, see also [9]. Among others, GLEAM uses mutation operators influenced by the ES in so far, as small parameter changes are more likely than greater ones. Mutation can also change the gene order and add or delete genes in the case of dynamic chromosomes. GLEAM uses ranking-based selection and elitist offspring acceptance. It is stressed that the introduced hybridization shall work with any other EA, too.

Suitable local search algorithms must be derivation-free and able to handle restrictions in order to preserve the general applicability of the resulting hybrid. Two well-known procedures from the sixties were chosen, since they meet these requirements and are known to be powerful local search procedures: The Rosenbrock algorithm [10] using one start point and the Complex method [11], because it can exploit multiple start points as they are delivered by an EA. The Rosenbrock procedure stops when the rate of changes of the main search direction decreases below a certain value and when the distances covered become too small. This is controlled by an external strategy parameter, here called *precision*. The Complex procedure stops when no improvement is achieved in five consecutive iterations. Schwefel gives a detailed description of both algorithms together with experimental results [12].

As this paper focuses on hybridization and due to the lack of space the basic algorithms have been described here very briefly only and the interested reader is referred to given literature.

4 Controlling the Basic Algorithms

Concerning real world problems neither the structure of the fitness landscape nor the optimum or its area of attraction are known in advance. But as computation of the fitness function frequently is time-consuming, it is possible to perform more sophisticated calculations to estimate when to switch from global to local search.

Fig. 1 shows the typical progress of an EA run. Stagnation phases of the overall quality can be identified easily, e.g. A, B or C. But which one shall be selected for terminating the evolution? This cannot be derived from stagnation only. A better measure is the genotypic diversity within the population. If the population consists of a few genotypically different sub-populations (niches) only, which are of minor difference, then stagnation can be expected, which provides little chance for greater progress. Hence, stagnation phases like in Fig. 1 may be used to trigger a check for niche constitution. Another trigger may be the number of generations without offspring acceptance.

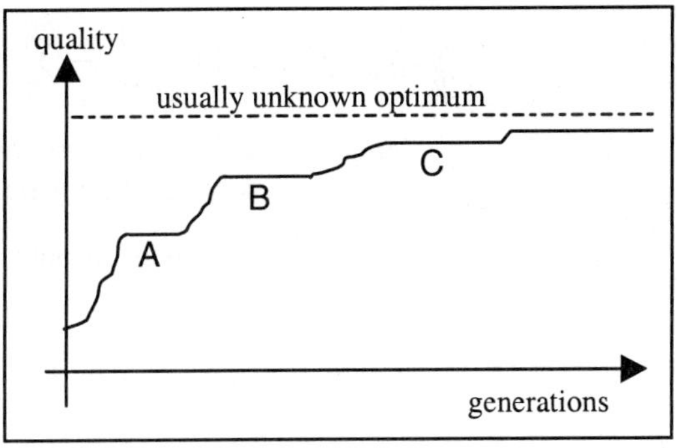

Fig. 1. Typical progress during the run of an EA

4.1 Distance Measures for Chromosomes

To estimate the genotypic diversity, distance measures for chromosomes must be defined, which quantify the parameter distance, the different gene ordering, and the common genes in the case of dynamic chromosomes. Distance functions reported in literature are often too specialized for the problem on hand, rather than to serve as a solution here [9]. Measures should be independent of the application in so far as they must not be influenced by actual parameter ranges or the number of genes and should be within a fixed range. The measures defined here vary between 0 (identity) and 1 (maximum difference). They comply with the four metric axioms, but the proofs and the calculation of $dist_{max}$ are omitted here due to the lack of space. They can be found on the following web page: `http://www.iai.fzk.de/~jakob/hy_gleam/`

Fixed-Length Chromosomes with Irrelevant Gene Order. For this chromosome type, the calculation of the parameter distance Δ_{par} of two chromosomes C_1 and C_2 is sufficient. It is defined as follows:

$$\Delta_{par}(C_1, C_2) = \frac{1}{n}\sum_{i=1}^{n}\frac{|par_{i,1} - par_{i,2}|}{ub_i - lb_i} \tag{1}$$

where n: number of all parameters of all genes
$par_{i,j}$: i-th parameter of chromosome j
lb_i, ub_i: lower and upper limits of the i-th parameter

Fixed-Length Chromosomes with Relevant Gene Order. The positional difference $pd_{1,2}(G_i)$ of one gene G_i of two fixed-length chromosomes C_1 and C_2 is defined by their sequential numbering and the calculation of the absolute difference of their indices. This leads to the positional distance of two chromosomes Δ_{pos}:

$$\Delta_{pos}(C_1, C_2) = \frac{1}{dist_{\max}}\sum_{i=1}^{len} pd_{1,2}(G_i) \tag{2}$$

where len: length of the chromosomes ($len > 1$)
$dist_{\max}$: distance maximum of all genes within one chromosome

For $dist_{\max}$ two cases for odd and even chromosome lengths must be considered:

$$dist_{\max,\, even} = \frac{len^2}{2} \qquad dist_{\max,\, odd} = \frac{len^2 - 1}{2} \tag{3}$$

The overall distance Δ of fixed-length chromosomes with relevant ordering is defined as the mean of Δ_{par} and Δ_{pos}.

Variable-Length Chromosomes. The goal is to determine the precise difference of similar chromosomes, while the exact value of discrepancy of dissimilar ones is of less interest. So the resulting measure may be inexact for more different chromosomes, thus yielding a less complex formula which reduces the computational effort for the fairly frequent distance calculations.

For two chromosomes C_1 and C_2 the set of common genes G_{com} may not be empty. Otherwise, the overall distance is set to 1. As genes in chromosomes of variable length may occur several times, they are treated in the sequence of their indexing and Δ_{par} and Δ_{pos} are defined over G_{com}, where $dist_{\max}$ is taken from the shorter chromosome. This may lead to an oversized value of Δ_{pos}, which may exceed 1 especially for chromosomes with large differences. Thus, Δ_{pos} is limited to 1 and the error is accepted, as it increases with the chromosome difference.

The difference of the presence of genes Δ_{gp} in two non-empty chromosomes is calculated as follows:

$$\Delta_{gp} = 1 - \frac{card(G_{com})}{\max(len(C_1), len(C_2))} \tag{4}$$

The overall distance Δ of variable-length chromosomes is defined as the mean of the three distances, with Δ_{gp} being weighted three times, because Δ_{par} and Δ_{pos} are defined over the common set of genes only.

4.2 Control Criteria for the Basic Algorithms

For estimating the genotypic diversity the individuals are assumed to be in linear order. A niche is formed by adjacent individuals with a fitness greater than half of the global best and with Δ being smaller than the strategy parameter ε. The fitness threshold is introduced to ignore chromosomes of less quality, as they do not contribute to niching of individuals of high fitness. The center individual of a niche is called its representative. Niches, whose representatives differ by less than ε, are merged regardless of their position. The resulting amount of niches *N* and the maximum difference of their representatives $\Delta_{N,max}$ are compared to the strategy parameters ε_{Pop} and N_{max} and the population is considered to be *converged*, if:

$$\Delta_{N,\max} \le \varepsilon_{Pop} \quad \text{and} \quad N \le N_{\max} \tag{5}$$

If this criterion is fulfilled the evolutionary search is stopped and the results are handed over to the local searchers for post-optimization or the local procedures are added to the process of offspring generation in case of delayed direct integration.

5 Experimental Results

In the experiments, five test functions taken from GENEsYs [13] and three real world problems were used. Here, they shall be described very briefly only and the interested reader is referred to the literature.

- **Schwefel's sphere** in the range of $[-10^{10},10^{10}]$ and with a target value of 0.01, a unimodal problem known to be easy for ES, but hard for GA.
- **Shekel's foxholes**, a simple multimodal function, easy for EA, but hard for local searchers (target value: exact minimum)
- Generalized **Rastrigin function** (target value: 0.0001) and
- **Fletcher's function** (target value: 0.00001), both of considerable complexity
- **Fractal function** with an unknown minimum. Target value here: -0.5 (from [13]).
- **Design optimization** of a micro optical communication device considering fabrication tolerances as described in detail in [14]. Despite its only 3 parameters, the task involves some difficulty, because it is of extreme multimodal nature.
- The **resource optimization** is based on the scheduling task of 87 batches in chemical industry, where varying numbers of workers are required during the different batch phases [15]. The maximum number of workers per shift (human resource) and the production time shall be reduced to the largest possible extent. Restrictions like due dates of batches, necessary pre-products from other batches, and the availability

of shared equipment must also be adhered to. Allocation conflicts are solved by the sequence of the batches within a chromosome. But as this can be overwritten by suitable changes of the starting times, the combinatorial aspect is limited.

- The objective of the **robot path planning** task is to move a robot along a line as straightly as possible from a starting to a destination point and avoid collisions with itself and some obstacles by controlling the robot axis motors [7]. As the number of necessary motor commands is not known in advance, the chromosomes must be of dynamic length and the order of the commands is essential. Due to a command which tells the robot control to keep the actual motor settings for a specified amount of control cycles, this task has both integer and real parameters.

Table 1. Properties of the eight test cases and the results for the algorithms applied separately

Experiment	Combi-natorial optim.	Parame-ters	Modality	Implicit restric-tions	Success rate for best run		
					GLEAM	Rosen-brock	Com-plex
Sphere	no	30 real	unimodal	no	0	100	0
Foxholes	no	2 real	multimod.	no	100	3	1
Rastrigin	no	20 real	multimod.	no	100	0	0
Fletcher	no	5 real	multimod.	no	100	10	10
Fractal	no	20 real	multimod.	no	100	0	0
Design	no	3 real	multimod.	no	100	15	12
Resource	(yes)	87 int	multimod.	yes	94	0	0
Robot Path	yes	dynamic mixed	multimod.	yes	100	0	0

These test cases cover a wide range of different application types, as shown in Table 1. The shortcuts and strategy parameters of the algorithms and hybrids, used in the figures below, are as follows:

- G: GLEAM: the population size (p)
- R: Rosenbrock: precision: 10^{-2} (s), 10^{-4} (m), 10^{-6} (l), 10^{-8} (xl), 10^{-9} (xxl)
- C: Complex: none
- Ri: Rosenbrock-initialized start population. Percentage of pre-
- Ci: Complex-initialized start population. optimized individuals
- PR: Rosenbrock post-optimization: see below
- PC1S: Complex post-optimization: each solution is 1 start point (1S)
- PC1C: Complex post-optimization: all solutions form 1 start complex (1C)
- GR, GC: Direct Integration of Rosenbrock or Complex: Lamarck-rate (l), local optimization of all or only the best offspring of one mating (all, best)
- GdR, GdC: Delayed direct integration with Rosenbrock and Complex respectively

The niching is controlled by three parameter settings for ε and ε_{Pop}, P1=(0.005, 0.01), P2=(0.002, 0.005), P3=(0.001, 0.002), and by N_{max} which varies between 2 (p<20) and 5 (p>90) with the population size. The success rate and the average amount of evaluations based on 100 runs (resource and the design task: 50 runs) for each parameterization (job) are taken for comparison. For the resource and the robot problem the local searchers work on the parameters only and leave the combinatorial aspects to the EA. A total of 182,000 jobs consuming 7 cpu-years on 22 sun work-stations (ultra sparc) were needed for the experiments.

5.1 Results of the Test Cases

In Fig. 2 the results of the best jobs with 100% success rates are shown for each test case and hybridization method, and they are compared with the best run of GLEAM. Schwefel's sphere is somewhat exceptional, as GLEAM has no success despite the unimodality of the problem. The success of the Rosenbrock procedure is more or less unrealistic, because no one would start with such an extreme precision, and with less there is no success at all up to high precision. The post-optimization is mentioned in brackets, because even the unsuccessful runs deliver very-high-quality solutions. The results of Shekel's Foxholes show that a fairly easy task for an EA, but hard for local searchers, cannot be improved largely by hybridization. The EA solves the problem so well that the overhead imposed by hybridization is too costly. The Rastrigin function shows an unusual behavior in so far, as GLEAM still works with extremely small population sizes leaving little or no room for improvements by hybridization. In Fig. 3 this case is compared with the typical behavior of an EA using the robot path planning task. With population sizes below a certain value, success rates are expected to drop and the number of evaluations to increase. For the resource and the robot tasks, special precisions of the Rosenbrock algorithm of 0.6 and 0.5 respectively were necessary, as there was no convergence with any standard setting. Due to the extended chromosome types of the last two tasks, further niching control parameterizations were tested, and P0 (0.04, 0.08) led to a remarkable success for resource optimization. Together with Fletcher's function, this task yields the most impressive improvements, while the very difficult fractal function and the design problem show significant success, too.

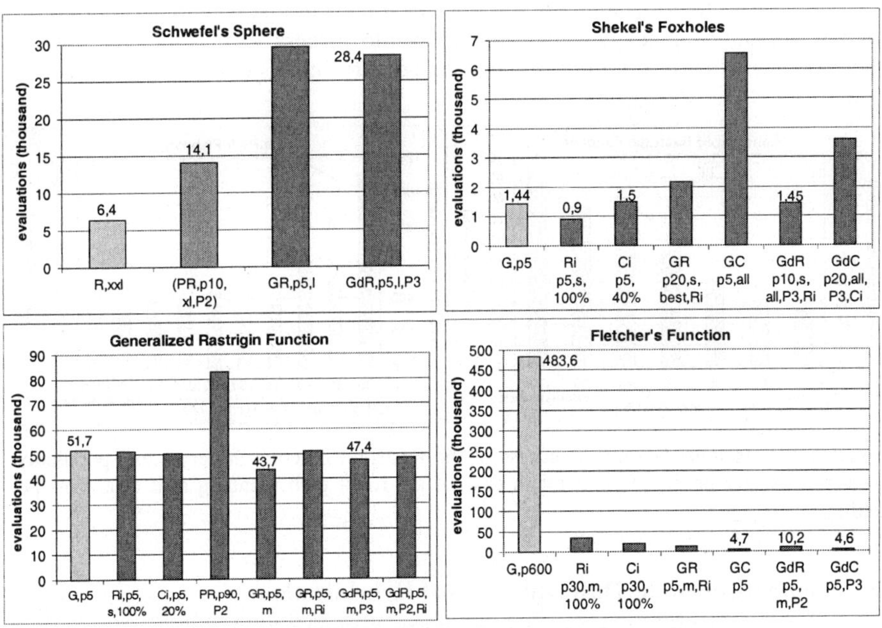

Fig. 2. Results for each test case and hybridization method reaching a 100% success rate. The best jobs per parameterization are shown and compared. Acronyms see above list

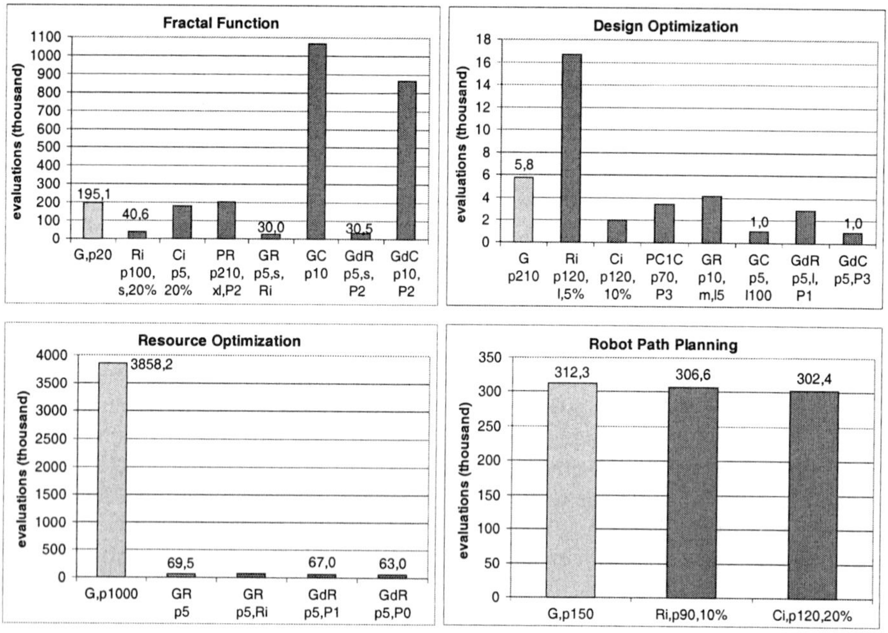

Fig. 2 (cont.). Results for each test case and hybridization method reaching a 100% success rate. The best jobs per parameterization are shown and compared. Acronyms see above list

The examined hybridization approaches reach their limits, when the ordering of a dynamic amount of genes is of vital importance to success, as it is the case with the robot example. No relevant improvement can be stated in this case.

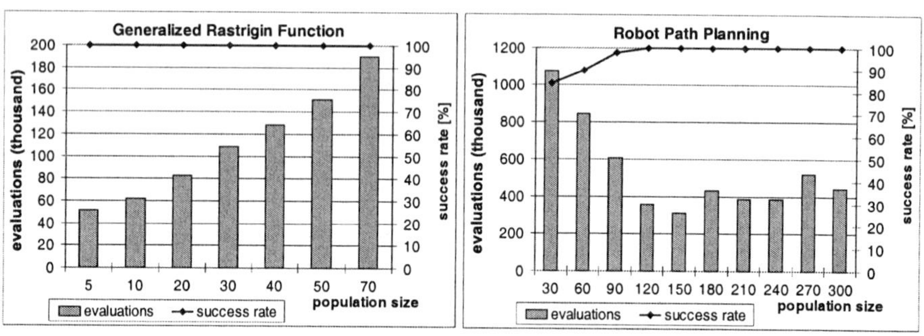

Fig. 3. GLEAM jobs: Untypical success of the Rastrigin function at very small population sizes and usually observed behavior in case of the robot path planning task (100 runs per setting).

5.2 Comparison of the Different Kinds of Hybridization

The different kinds of hybridization are compared on the basis of the achieved improvement of the best job compared to the best GLEAM job. This cannot be done for Schwefel's sphere, because GLEAM was not successful. But it can be stated that post-optimization and (delayed) direct integration yield very good results, see Fig. 2. Furthermore, the robot task will not contribute to the comparison, as there was no improvement worth mentioning. In Fig. 4 the results of the remaining test cases are summarized.

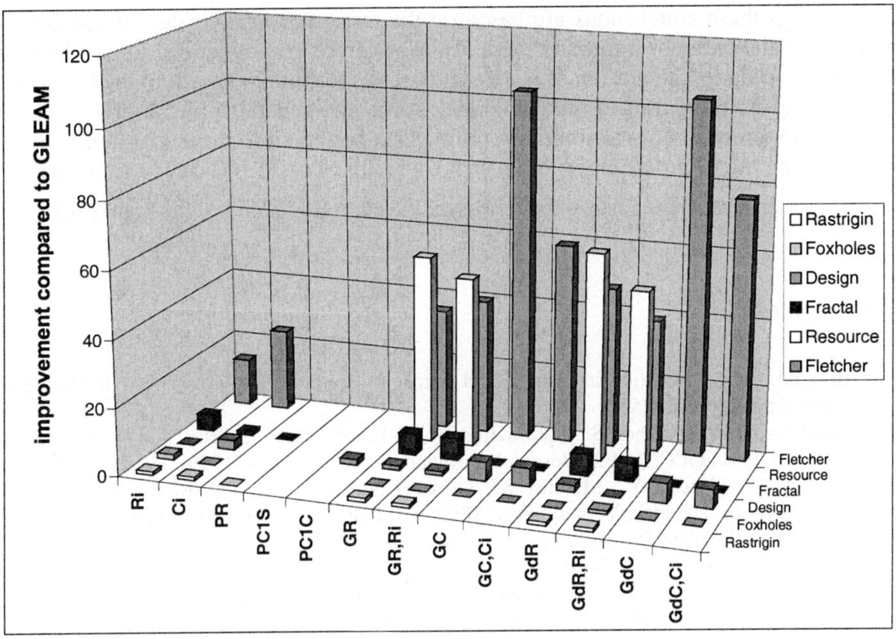

Fig. 4. Comparison of the different kinds of hybridization. Empty fields indicate no sufficient success rate (below 100%), while flat fields stand for fulfilling the optimization task, but with greater effort than GELAM. For PC1S no 100% success rate was reached in any test case.

Conclusions

From the extensive investigation of 13 different kinds of hybridization, based on eight test cases comprising one challenging unimodal problem, simple and complex multimodal tasks, and real world problems involving combinatorial optimization as well as dynamic parameter strings, the following conclusions can be drawn, see also Fig. 4:

1. Though post-optimization improves the results obtained by the EA in most cases, it does not lead to sufficient success, as the introduced control mechanism based on niche detection does not guarantee a stop of evolution only, when the area of the attraction of the global optimum is reached.

2. Direct or delayed direct integration of the Rosenbrock procedure works in all cases and yields very good results as long as the problem is not too simple like in the case of Shekel's foxholes. Pure Lamarckian evolution and local optimization of the best offspring in nearly all cases is superior to Baldwinian evolution or optimization of all offsprings of one mating. The delayed integration based on the niching control algorithm improves the undelayed version by up to 20% less evaluations. Small population sizes between 5 and 20 are sufficient.

3. Pre-optimization helps in most cases but direct integration is better.
4. Hybridization with the Complex algorithm does not always work, but if it does, superior results can be produced. Hybridization with the Rosenbrock procedure is more reliable, but not always as successful as using the Complex.

Although these conclusions are based on the test cases investigated, it can be assumed that they are not limited to them. Thus, it can be recommended to use delayed or undelayed direct integration of the Rosenbrock algorithm to speed up the EA while maintaining the properties of reliable global search and general applicability. With the Rosenbrock procedure, evaluations were found to be reduced by the magnitude of 60 and by using the Complex algorithm instead, a factor of up to 100 can be achieved.

This paper was written using the terms and definitions of VDI/VDE 3550 [16].

References

1. Holland, H.J.: Adaptation in Natural and Artificial Systems. The University of Michigan Press, Ann Arbor (1975)
2. Goldberg, D.E, Voessner, S.: Optimizing Global-Local Search Hybrids. In: W. Banzhaf et al. (eds.): Proc. GECCO'99, Morgan Kaufmann, San Mateo, CA (1999) 220-228
3. Whitley, D., Gordon, V., Mathias, K.: Lamarckian Evolution, The Baldwin Effect and Funct. Opt. In: Davidor, Y. et al.: Proc. PPSN III, LNCS 866, Springer, Berlin (1994) 6-14
4. Gruau, F., Whitley, D.: Adding Learning to the Cellular Development of Neural Networks: Evolution and the Baldwin Effect. Evol. Comp. 1, Vol.3 (1993) 213-233
5. Orvosh, D., Davis, L.: Shall We Repair? Genetic Algorithms, Combinatorial Optimization, and Feasibility Constraints. In: Forrest, S. (ed): 5th ICGA, M. Kaufmann (1993) 650
6. Jakob, W., Quinte, A., et al.: Opt. of a Micro Fluidic Component Using a Parallel EA and Simulation Based on Discrete Element Methods. In: Hernandez, S., et al.: Computer Aided Design of Structures VII, Proc. of OPTI'01, WIT Press, Southampton (2001) 337-346
7. Blume, C.: GLEAM - A System for Intuitive Learning. In: Schwefel, H.P., Männer, R. (eds.): Proc. of PPSN I, LNCS 496, Springer, Berlin (1990) 48-54
8. Blume, C., Jakob, W.: GLEAM – an Evolutionary Algorithm for Planning and Control Based on Evolution Strategy. Conf. Proc. GECCO 2002, Vol. Late Breaking Papers, (2002)
9. Jakob, W.: HyGLEAM: Hybrid GeneraL-purpose Evolutionary Algorithm and Method. In: Callaos, N. et al. (eds.): Proc. SCI'2001, Vol. III, IIIS, Orlando, (2001) 187-192
10. Rosenbrock, H.H.: An Automatic Method for Finding the Greatest or Least Value of a Function. Comp. Journal 3 (1960) 175-184
11. Box, M.J.: A New Method of Constrained Optimization and a Comparison with Other Methods. Comp. Journal 8 (1965) 42-52
12. Schwefel, H.-P.: Evolution and Optimum Seeking. John Wiley & Sons, Chichester (1995)
13. Bäck, T.: GENEsYs 1.0, ftp://lumpi.informatik.uni-dortmund.de/pub/GA
14. Gorges-Schleuter, M., Jakob, W., Sieber, I.: Evolutionary Design Optimization of a Microoptical Collimation System. In: Zimmermann, H.J. (ed.): Proc. Eufit'98, Verlag Mainz, Aachen (1998) 392-396
15. Blume, C., Jakob, W.: Cutting Down Production Costs by a New Optimization Method. In: Proc. of Japan-USA Symposium on Flexible Automation. ASME (1994)
16. Beyer, H.-G., et al.: Evolutionary Algorithms – Terms and Definitions. VDI/VDE-Richtlinie-3550, Blatt 3, Gründruck (Engl. vers. to be published in 2002). VDI, Düsseldorf (2001)

Co-evolving Memetic Algorithms: Initial Investigations

Jim Smith

Intelligent Computer Systems Centre
University of the West of England
Bristol BS16 12QY, UK
james.smith@uwe.ac.uk,

Abstract. This paper presents and examines the behaviour of a system whereby the rules governing local search within a Memetic Algorithm are co-evolved alongside the problem representation. We describe the rationale for such a system, and the implementation of a simple version in which the evolving rules are encoded as (condition:action) patterns applied to the problem representation, and are effectively self-adapted. We investigate the behaviour of the algorithm on a test suite of problems, and show significant performance improvements over a simple Genetic Algorithm, a Memetic Algorithm using a fixed neighbourhood function, and a similar Memetic Algorithm which uses random rules, i.e. with the learning mechanism disabled.
Analysis of these results enables us to draw some conclusions about the way that even the simplified system is able to discover and exploit different forms of structure and regularities within the problems. We suggest that this "meta-learning" of problem features provides a means both of creating highly scaleable algorithms, and of capturing features of the solution space in an understandable form.

1 Introduction

The performance benefits which can be achieved by hybridising Evolutionary Algorithms (EAs) with Local Search(LS) operators, so-called *Memetic Algorithms* (MAs), have now been well documented across a wide range of problem domains such as combinatorial optimisation [1], optimisation of non-stationary functions [2], and multi-objective optimisation [3]. See [4] for a comprehensive bibliography. Commonly in these algorithms, the Local Search improvement step is performed on each of the products of the generating (recombination and mutation) operators, prior to selection for the next population.

There are three principal components which affect the workings of the LS algorithm. The first is the choice of pivot rule, which is usually either *Steepest Ascent* or *Greedy Ascent*. The second component is the "depth" of local search, which can vary from one iteration, to the search continuing to local optimality. Considerable attention has been paid to studying the effect of changing these parameters within MAs e.g. [5].

J.J. Merelo Guervós et al. (Eds.): PPSN VII, LNCS 2439, pp. 537–546, 2002.

The third factor is the choice of neighbourhood function, which can be thought of as defining a set of points that can be reached by the application of some move operator to a point. We can consider the graphs defined by different move operators as "fitness landscapes" [6]. Merz and Freisleben [7] discuss a number of statistical measures which can be used to characterise fitness landscapes, and have been proposed as potential measures of problem difficulty. They show that the choice of move operator can have a dramatic effect on the efficiency and effectiveness of the Local Search, and hence of the resultant MA.

In some cases, domain specific information may be used to guide this choice. However, it has recently been shown that the optimal choice of operators can be not only instance specific within a class of problems [7, pp254–258], but also dependant on the state of the evolutionary search [8]. The idea that periodically changing the move operator may provide a means of escape from local optima by redefining the neighbourhood structure, has been demonstrated in the *Variable Neighbourhood Search* algorithm [9].

The aim of this work is to provide a means whereby the definition of the local search operator used within a MA can be varied over time, and then to examine whether evolutionary processes can be used to control that variation, so that a beneficial adaptation takes place. In order to accomplish this aim, we must address four major issues. The first of these is the representation of LS operators in a form that can be processed by an evolutionary algorithm, and the choice of of initialisation and variation operators. The second is the credit assignment mechanism for assigning fitness to the LS population members. The third is the choice of population structures and sizes, along with selection and replacement methods for managing the LS population. Finally, we require a set of experiments, problems and measurements designed to permit evaluation and analysis of the behaviour of the system. This paper represents the first stage of this analysis.

2 Rule-Based Adaptation of Move Operators

The representation chosen for the LS operators is a tuple *<Search_Depth, Pivot_Rule, Pairing, Move>*. The first two elements are self-explanatory. Values for *Pairing* are taken from the set {*Linked, Random,Fitness_Based*}. When the Local Search phase is applied to a member of the solution population, the value of the *Pairing* in the corresponding member of the LS population is examined. If the value is *Linked* then that LS member is used to act on the solution. If the *Pairing* takes one of the values *Random* or *Fitness_Based* then a selection is made from the available (i.e. unlinked)members of the LS population according. By making this variable part of the rule and subject to evolution, the system can be varied between the extremes of a fully unlinked system,(in which although still interacting the two populations evolve separately), and a fully linked system. In the latter the LS operators can be considered to be extra genetic material which is inherited and varied along with the problem representation, in an exactly analogous way to the inheritance of strategy parameters in Self Adapting

EAs. We note that "Self-Adaptation" can be considered as co-evolution with hereditary symbiosis, i.e. where the two populations share a common updating mechanism.

The representation chosen for the move operators was as *condition:action* pairs, which specify a pattern to be looked for in the problem representation, and a different pattern it should be changed to. Although this representation at first appears very simple, it has the potential to represent highly complex moves via the use of symbols to denote not only wildcard characters but also the specifications of repetitions and iterations. Further, permitting the use of different length patterns in the two parts of the rule gives scope for *cut* and *splice* operators working on variable length solutions.

While the framework that we have describe above is intended to permit a full exploration of several research issues, we shall initially restrict ourselves to considering a simple system, and examining its behaviour on a well understood set of binary encoded test problems. For these initial investigations we therefore restricted the LS operators to a single improvement step, a greedy acceptance mechanism, and full linkage. This restriction to what are effectively self-adaptive systems provides a means of dealing with the credit assignment and population management isssues noted above.

We also initially restrict ourselves to considering only rules where the *condition* and *action* patterns are of equal length and are composed of values taken from the set {0,1,#}.The last of these is a "don't care" symbol which is only allowed to appear in the *condition* part of the rule.

The neighbourhood of a point i is defined by finding the (unordered) set of positions where the substring denoted by *condition* is matched in the representation of i. For each of these a new string is then made by replacing the matching substring with the *action* pattern. To give an example, if we have a solution represented by the binary string 1100111000 and a rule 1#0:111, then this matches the first, second, sixth and seventh positions, and the neighbourhood is the set {1110111000, 11111111000, 1100111100,1100111110}. Note that in this initial work we do not considered the string as toroidal.

3 Related Work

The COMA system can be related to a number of different branches of research, all of which offer different ways of perspectives and means of analysing it's behaviour. Space constraints preclude a full discussion of each of these, so we will briefly outline some of these perspectives.

Although we are not aware of other algorithms in which the LS operators used by an MA are adapted in this fashion, there are other examples of the use of multiple LS operators within evolutionary systems. Krasnogor and Smith [8] describe what they call a "MultiMemetic Algorithm", in they used a simple Self-Adaptive mechanism to evolve the choice of which a fixed set of static LS operators ("memes") should be applied to individual solutions. They report that their systems are able to adapt to use the best meme available for different instances of TSP.

As noted above, if the populations are of the same size, and are considered to be linked, then this instantiation of the COMA framework can be considered as a type of Self Adaptation. The use of the intrinsic evolutionary processes to adapt search strategies, and the conditions necessary, is well documented e.g. for mutation step sizes [10, 11], mutation probabilities [12], recombination operators[13, 14] and general variation operators [15], amongst many others.

If the two populations are not linked, then we have a co-operative coevolutionary system, where the fitness of the members of the LS population is assigned as some function of the relative improvement they cause in the "solution" population. Co-operative co-evolutionary (or "symbiotic") systems have been used with good reported results for function optimisation [16–18] and Bull conducted a series of more general studies on the conditions necessary for co-operative co-evolution to occur [19–21]. These issues will be explored in future work.

If we were to simply apply the rule selected from in the LS population (possibly iteratively) to transform an individual solution, without considering a pivot rule, then we could also view the system as a type of "developmental learning" akin to the studies in the evolution of Genetic Code [22]

COMA differs from the last two paradigms because the LS population can potentially modify the genotypes within the solution population. This phase of improvement by LS can be viewed as a kind of lifetime learning, which leads naturally to the question of whether a Baldwinian approach might be preferable to the Lamarkian Learning currently implemented. However, even if a Baldwinian approach was used, the principal difference between the COMA approach and the co-evolutionary systems above lies in the use of a pivot rule within the LS, such that detrimental changes are rejected.

Finally, and perhaps most importantly, we should consider that if the same rule has an improving effect on different parts of a solution chromosome over as number of generations, then the evolution of rules can be seen as a process of learning generalisations about patterns within the problem representation, and hence the solution space. This point of view is akin to that of Learning Classifier Systems. For the case of unlinked fitness-based selection of LS operators, insight from this field can be used to guide the credit assignment process.

4 The Test Suite and Experimental Set-Up

In order to examine the behaviour of the system it was used with a set of variants of a test function whose properties are well known. This was a sixty four bit problem composed of 16 subproblems of Deb's 4-bit fully deceptive function given in [23]. The fitness of each subproblem i is given by its unitation $u(i)$ (i.e. the number of bits set to 1):

$$f(i) = \begin{cases} 0.6 - 0.2u(i) & : \quad u(i) < 4 \\ 1 & : \quad u(i) = 4 \end{cases} \tag{1}$$

In addition to a "concatenated" version (which we will refer to as 4-Trap), a second "distributed"version (Dist-Trap)was used in which the subproblems were interleaved i.e. sub-problem i was composed of the genes $i, i+16, i+32, i+48$. This separation ensures that even the longest rules allowed in these experiments would be unable to alter more than one element in any of the subfunctions.

A third variant of this problem (Shifted-Trap) was designed to be more difficult than the first for the COMA algorithm to learn a single generalisation, by making patterns which were optimal in one sub-problem, sub-optimal in all others. This was achieved by noting that each sub-problem as defined above is a function of unitation, and therefore can be arbitrarily translated by defining a 4-bit string and using the Hamming distance from this string in place of the unitation. Since we have 16 sub-problems, we simply used the binary coding of the sub-problem's index as basis for its fitness calculation.

We used a generational genetic algorithm, with deterministic binary tournament selection for parents and no elitism. One Point Crossover (with probability 0.7) and bit-flipping mutation (with a bitwise probability of 0.01) were used on the problem representation. These values were taken as "standard" from the literature, bearing in mind the nature of the 4-Trap function. Mutation was applied to the rules with a probability of 0.0625 of selecting a new allele value in each locus (the inverse of the maximum rule length allowed to the adaptive version).

For each problem, 20 runs were made for each population size {100,250,500}. Each run was continued until the global optimum was reached, subject to a maximum of 1 million evaluations. Note that since one iteration of LS may involve several evaluations, this allows more generations to the GA, i.e. we compare algorithms strictly on the basis of the number of calls to the evaluation function.

The algorithms used are: a "vanilla" GA i.e. with no use of Local Search (GA), a simple MA using one iteration of greedy ascent over the neighbourhood at Hamming distance 1 (MA), a version of COMA using a randomly created rule in each application, (i.e. with the learning disabled) (RandComa), variants of COMA using rules of fixed lengths in the ranges $\{1,\ldots,9\}$ (1-Coma,...,9-Coma), and finally an adaptive version of COMA (A-Coma). For A-Coma the rule lengths are randomly initialised in the range [1,16]. During mutation, a value of $+/-1$ is randomly chosen and added with probability 0.0625, subject to staying in range.

5 Comparison Results

Figure 1 shows the results of these experiments as a plot of mean time to optimum for 4-Trap with three different population sizes. When an algorithm failed to reach the optimum in all twenty runs, the mean is taken over the successful runs, and this number is shown. The error bars represent one standard deviation. It should be noted that the scale on the y-axis is logarithmic. We can see that the GA and MA,and 1-Coma algorithms fail to find the optimum as frequently, or when they do as fast, for the smaller population sizes. For all population sizes there is greater variance in the performance of these three algorithms than for the other variants.

Because the variances are unequal, we applied the conservative Tamhane's T2 test to the solution times for the successful runs to detect statistically significant

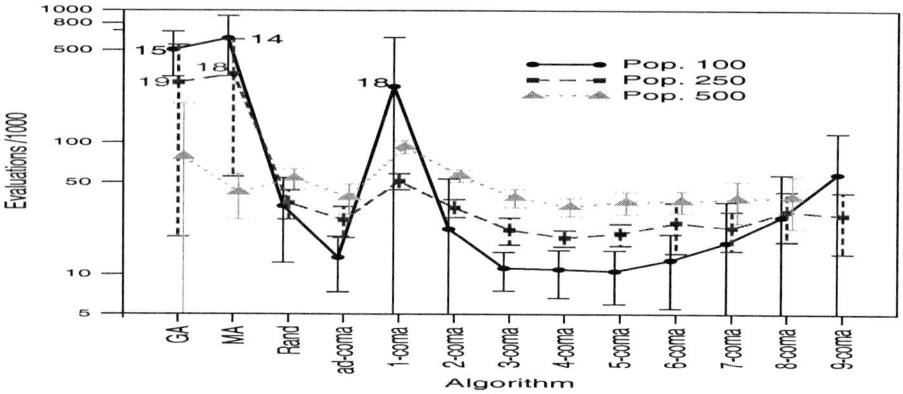

Fig. 1. Times to optimum for the 4-Trap function. Note logarithmic y-axis

differences in performance. The GA, MA and 1-coma algorithms are significantly slower than the rest at the 5% level for a population of 100. For a population size of 250 the GA and MA algorithms are significantly slower. When the population size is increased to 500, the worse performance seen with the GA and MA is no longer significant, according to this conservative test. However, the Rand-Coma is now significantly slower than all but the GA, MA and 2-Coma. 1-Coma is significantly slower than all but the GA, and 2-Coma is slower than all but GA, MA and Rand-Coma.

In short, what we can observe is that for fixed rule lengths of between 3 and 9, and for the adaptive version, the COMA system derives performance benefits from evolving LS rules. Significantly, and unlike the GA and MA, the COMA algorithm does not depend on a certain size population before it is able to solve the problem reliably. This is indicative of a far more scaleable algorithm.

Figure 2 shows the results of the experiments on the variants of the trap functions. For the "Shifted" trap function, the performances of the GA and MA are not significantly different from those on the unshifted version. this refelects the fact that these algorithms solve the sub-problems independently and are "blind" to whether the optimal string for each is different. When we examine the results for the COMA algorithms, we see slower solution times than on the previous problem, resulting from the fact that no one rule can be found which will given good performance in every subproblem. However we see a similar pattern of reliable problem solving for all but 1-Coma and 2-Coma. Analysis reveals that even these last two are statistically significantly better than GA or MA for all but the largest population size. Interestingly, the RandComa algorithm performs well here, probably as a natural consequence of using a random rule every time, so promoting diversity in the rule base.

Considering Dist-Trap, we first note that the GA, MA and Rand-COMA failed to solve the function to optimality in any run, regardless of population size. The poor results of the GA can be attributed to the mismatch between the distributed nature of the representation and the high positional bias of the 1-

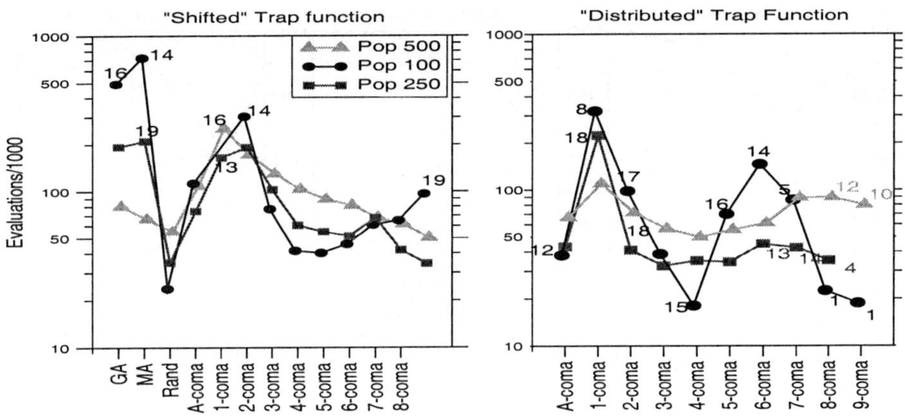

Fig. 2. Times to optimum for the Shifted-Trap and Dist-Trap. Note logarithmic y-axis. Error bars omitted for clarity

point Crossover used. When we consider the action of the bitflipping LS operator on a subproblem, this will lead towards the sub-optimal solution, whenever the unitation is 0,1 or 2, and the greedy search of the neighbourhood will also lead towards the deceptive optimum 75% of the time when the unitation is 3. This observation helps us to understand the poor results of the simple MA, and the 1-Coma algorithm.

When we examine the other COMA results, noting that the success rate is less than for the other problems, we again see the same pattern of better performance for the adaptive version and fixed rulelengths in the range 3-5, tailing off at the extreme lengths. Note that although the mean solution time drops for long rule length, so too does the number of successful runs which we take as our primary criterion. We also note that the failure of the RandComa algorithm indicates that some learning is required here.

6 Discussion and Analysis

The results given above are promising from the point of view of improved optimisation performance, but require some analysis and explanation. The deceptive functions used were specifically chosen because GA theory suggests that they are solved by finding and mixing optimal solutions to sub-problems. When we consider the results, we can see that the performance is best on 4-Trap, with a rule length of 4, which would support the hypothesis that the system is "learning" the structure of the sub-problems. Although not immediately apparent from the logarithmic scales, the solutions times here are less than half those on the other problems.

However we should note that the algorithms are not aware of the sub-problem boundaries. On 4-Trap and Shifted-Trap, for lengths of 4 or less occasionally, and always for lengths greater than 4, the changes made will overlap several sub-problems. This must always happen for Dist-Trap. It is clear from the results

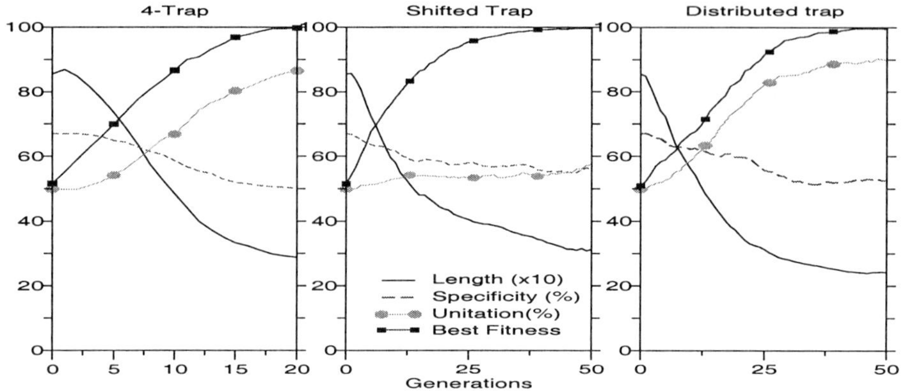

Fig. 3. Analysis of Evolving Rules by Function Type

with different rule lengths, and from the distributed problem, that there is a second form of improvement working on a longer timescale., which does not arise simply from the use of random rules.

In order to examine the behaviour of the algorithm we plotted the population means of the effective rule length (only relevant for A-Coma), the "specificity" (i.e. the proportion of values in the condition not set to #) and the "unitation" (the proportion of bits in the action set to 1), and also the highest fitness in the population (with 100 as the optimum) as a function of the number of elapsed generations. Figure 3 shows the A-Coma results averaged over 20 runs on each of the three problems, with a population of 250. We also manually inspected the evolving rule bases on a large number of runs for each problem.

For the 4-Trap function (left hand graph), the system rapidly evolves medium length $(3-4)$, general (specificity $< 50\%$) rules whose action is to set all the bits to 1 (mean unitation approaches 100%). Note that in the absence of selective pressure (i.e. the pivot rules meant that the solutions were left unchanged), all three of these values would be expected to remain at their initial values, so these changes result from beneficial adaptation. Closer inspection of the evolving rulebase confirms that the optimal subproblem string is being learned and applied.

For the Shifted-Trap function, where the optimal sub-blocks are all different (middle) the rule length decreases more slowly. The specificity also remains higher, and the unitation remains at 50%, indicating that different rules are being maintained. This is borne out by closer examination of the rule sets.

The behaviour on Dist-Trap is similar to that on 4-Trap, albeit over a longer timescale. Rather than learning specific rules about sub-problems, which cannot possibly be happening (since no rule is able to affect more than one locus of any subproblem), the system is apparently learning the general rule of setting all bits to 1.

The rules are generally shorter than for 4-Trap, (although this is slightly obscured by the averaging) which means that the number of potential neighbours is higher for any given rule. Equally, the use of wildcard characters, coupled with the fact that there may be matches in the two parts of the rules, means that length of the rules used defines a maximum radius in Hamming space for the neighbourhood, rather than a fixed distance from the original solution. Both of these observations, when taken in tandem with the longer times to solution, suggest that when the system is unable to find a single rule that matches the problems' structure, a more diverse search using a more complex neighbourhood is used, which slowly adapts itself to the state of the current population of solutions.

7 Conclusions

We have presented a framework in which rules governing LS operators to be used in memetic algorithms can be co-evolved with the population of solutions. A simple version was implemented which used Self-Adaptation of the patterns defining the move operators, and this was shown to give improved performance over both GAs and simple MAs on the test set. We showed that the system was able to learn generalisations about the problem when these were useful. We also noted that the COMA algorithms were far less dependant on a critical population size to locate the global optima, and suggested that this indicates a far more scaleable type of algorithm.

The test set used here was designed to maximise different types of difficulty for COMA, namely deception, inappropriate representation ordering, and multiple different optima. Although space precludes their inclusion, we have repeated these experiments with a wide range of different problem types and found similar performance benefits. Clearly further experimentation and analysis is needed, and there are many issues to be explored, however we believe that this paper represents the first stage in a promising line of research.

References

1. Merz, P., Freisleben, B.: A comparion of Memetic Algorithms, Tabu Search, and ant colonies for the quadratic assignment problem. In: Proceedings of the 1999 Congress on Evolutionary Computation, IEEE Service Center (1999) 2063–2070
2. Vavak, F., Fogarty, T., Jukes, K.: A genetic algorithm with variable range of local search for tracking changing environments. In Voigt, H.M., Ebeling, W., I.Rechenberg, Schwefel, H.P., eds.: Proceedings of the Fourth Conference on Parallel Problem Solving from Nature, Springer Verlag (1996) 376–385
3. Knowles, J., Corne, D.: A comparative assessment of memetic, evolutionary and constructive algorithms for the multi-objective d-msat problem. In: Gecco-2001 Workshop Program. (2001) 162–167
4. Moscato, P.: Memetic algorithms' home page. Technical report, http://www.densis.fee.unicamp.br/~moscato/memetic_home.html (2002)
5. Hart, W.E.: Adaptive Global Optimization with Local Search. PhD thesis, University of California, San Diego (1994)
6. Jones, T.: Evolutionary Algorithms, Fitness Landscapes and Search. PhD thesis, The University of New Mexico, Albuquerque, NM (1995)

7. Merz, P., Freisleben, B.: Fitness landscapes and memetic algorithm design. In Corne, D., Dorigo, M., Glover, F., eds.: New Ideas in Optimization, McGraw Hill (1999) 245–260
8. Krasnogor, N., Smith, J.: Emergence of profitiable search strategies based on a simple inheritance mechanism. In Spector, L., Goodman, E., Wu, A., Langdon, W.B., Voigt, H.M., Gen, M., Sen, S., Dorigo, M., Pezeshk, S., Garzon, M., Burke, E., eds.: Proceedings of the Genetic and Evolutionary Computation Conference, GECCO-2001, Morgan Kaufmann (2001) 432–439
9. Hansen, P., Mladenoviċ, N.: An introduction to variable neighborhood search. In Voß, S., Martello, S., Osman, I.H., Roucairol, C., eds.: Meta-Heuristics: Advances and trends in local search paradigms for optimization. Proceedings of MIC 97 Conference. Kluwer Academic Publishers, Dordrecht, The Netherlands (1998)
10. Schwefel, H.P.: Numerical Optimisation of Computer Models. John Wiley and Sons, New York (1981)
11. Fogel, D.: Evolving Artificial Intelligence. PhD thesis, University of California (1992)
12. Back, T.: Self adaptation in genetic algorithms. In Varela, F., Bourgine, P., eds.: Towards a Practice on Autonomous Systems: Proceedings of the First European Conference on Artificial Life, MIT Press (1992) 263–271
13. Schaffer, J., Morishima, A.: An adaptive crossover distribution mechanism for genetic algorithms. In J.J.Grefenstette, ed.: Proceedings of the Second International Conference on Genetic Algorithms, Lawrence Erlbaum (1987) 36–40
14. Smith, J., Fogarty, T.C.: An adaptive poly-parental recombination strategy. In Fogarty, T.C., ed.: Evolutionary Computing 2, Springer Verlag (1995) 48–61
15. Smith, J., Fogarty, T.: Adaptively parameterised evolutionary systems: Self adaptive recombination and mutation in a genetic algorithm. In Voigt, Ebeling, Rechenberg, Schwefel, eds.: Proceedings of the Fourth Conference on Parallel Problem Solving from Nature, Springer Verlag (1996) 441–450
16. Husbands, P.: Distributed co-evolutionary genetic algorithms for multi-criteria and multi-constraint optimisiation. In Fogarty, T., ed.: Evolutionary Computing: Proceedings of the AISB workshop. LNCS 865, Springer Verlag (1994) 150–165
17. Paredis, J.: The symbiotic evolution of solutions and their representations. In Eshelman, L.J., ed.: Proceedings of the Sixth International Conference on Genetic Algorithms, Morgan Kaufmann (1995) 359–365
18. Potter, M.A., DeJong, K.: A cooperative coevolutionary approach to function optimisation. In Davidor, Y., ed.: Proceedings of the Third Conference on Parallel Problem Solving from Nature, Springer Verlag (1994) 248–257
19. Bull, L.: Artificial Symbiology. PhD thesis, University of the West of England (1995)
20. Bull, L.: Evolutionary computing in multi agent environments: Partners. In Back, T., ed.: Proceedings of the Seventh International Conference on Genetic Algorithms, Morgan Kaufmann (1997) 370–377
21. Bull, L., Holland, O., Blackmore, S.: On meme-gene coevolution. Artificial Life **6** (2000) 227–235
22. Keller, R.E., Banzhaf, W.: The evolution of genetic code in genetic programming. In Banzhaf, W., Daida, J., Eiben, A.E., Garzon, M.H., Honavar, V., Jakiela, M., Smith, R.E., eds.: Proceedings of the Genetic and Evolutionary Computation Conference. Volume 2., Orlando, Florida, USA, Morgan Kaufmann (1999) 1077–1082
23. Back, T., Fogel, D., Michalwicz, Z.: Handbook of Evolutionary Computation. Volume 1. Oxford University Press (1997)

Learning Classifier Systems

Consideration of Multiple Objectives in Neural Learning Classifier Systems

Larry Bull and Matt Studley

Intelligent Autonomous Systems Lab
Faculty of Computing, Engineering & Mathematical Sciences
University of the West of England
Bristol BS16 1QY, U.K.
{larry.bull,matthew.studley}@uwe.ac.uk

Abstract. For effective use in a number of problem domains Learning Classifier Systems must be able to manage multiple objectives. This paper explicitly considers the case of developing the controller for a simulated mobile autonomous robot which must achieve a given task whilst maintaining sufficient battery power. A form of Learning Classifier System in which each rule is represented by an artificial neural network is used. Results are presented which show it is possible to solve both objectives when the energy level is presented as an input along with sensor data. A more realistic, and hence more complex, version of the basic scenario is then investigated.

1 Introduction

If autonomous artificial entities are to be used in complex problem domains they must be able to consider multiple objectives. This paper examines a simple multiobjective scenario using a form of Learning Classifier System (LCS) [16] in which each rule is represented by a neural network (NCS) [6]. The case of a mobile autonomous robot which must find the shortest path to a goal state whilst maintaining appropriate battery power is investigated. To date, there has been very little work considering multiple objectives with LCS or reinforcement learning techniques in general: Dorigo and Colombetti [e.g. 12] presented a hierarchical approach in which a number of pre-trained LCS considered one goal each and a higher-level LCS learned to switch between the controllers appropriately; and Karlsson [17] has presented a similar scheme for multiple Q learners [26]. Others have used reinforcement learning for multiple sequential goals [e.g. 27]. Many artificial life "simulated world" experiments are loosely related to this area since the organisms must often find food to exist or reproduce whilst avoiding predators [e.g. 1]. The use of evolutionary computing techniques for multiobjective optimization [e.g. 10] is also loosely related.

The paper is arranged as follows: the next section describes the LCS architecture used and in Section 3 a simple two-objective maze task is presented with results. In

J.J. Merelo Guervós et al. (Eds.): PPSN VII, LNCS 2439, pp. 549–557, 2002.

Section 4 complexity is added to the scenario, moving it closer to the envisaged robot implementation, with all results and future work discussed in Section 5.

2 A Neural Learning Classifier System

The neural learning classifier system used here is based on Wilson's ZCS [28].

2.1 ZCS

ZCS is a "Zeroth level" Learning Classifier System without internal memory, where the rule-base consists of a population (*P*) of condition/action rules in which the condition is a string of characters from a ternary alphabet {0,1,#} and the action is represented by a binary string. Associated with each rule is a fitness scalar which acts as an indication of the perceived utility of that rule within the system. This fitness of each rule is initialised to a predetermined value termed f_0.

Reinforcement in ZCS consists of redistributing fitness between subsequent "action sets", or the matched rules from the previous time step which asserted the chosen output or "action". A fixed fraction (β) of the fitness of each member of the action set ([A]) at each time-step is placed in a "common bucket". A record is kept of the previous action set $[A]_{-1}$ and if this is not empty then the members of this action set each receive an equal share of the contents of the current bucket, once this has been reduced by a pre-determined discount factor (γ). If a reward is received from the environment then a fixed fraction (β) of this value is distributed evenly amongst the members of [A]. Finally, a tax (τ) is imposed on all matched rules that do not belong to [A] on each time-step in order to encourage exploitation of the fitter classifiers. Wilson [28] notes that this is a change to the traditional LCS bucket brigade algorithm [17] since there is no concept of a rule bid, specificity is not considered explicitly, and the pay-back is reduced by 1-γ on each step.

ZCS employs two discovery mechanisms, a panmictic genetic algorithm (GA) [15] and a covering operator. On each time-step there is a probability *p* of GA invocation. When called, the GA uses roulette wheel selection to determine two parent rules based on fitness. Two offspring are produced via mutation (probability μ) and crossover (single point with probability χ). The parents then donate half of their fitnesses to their offspring who replace existing members of the rule-base. The deleted rules are chosen using roulette wheel selection based on the reciprocal of rule fitness. If on some time- step, no rules match or all matched rules have a combined fitness of less than ϕ times the rule-base average, then a covering operator is invoked.

Bull and Hurst [5] have recently shown that ZCS is capable of optimal performance due to its use of action set fitness sharing (this result contrasts with a number of previous discussions, e.g. [29][8][20][7][21][4]). Here rule fitnesses approach a similar value with rule numerosity indicating utility. However, parameter setting is a crucial aspect of ZCS; to enable the fitness sharing process to occur effectively appropriate settings, particularly for the learning rate, must be found for a given task. Current work is examining just how robust ZCS is to parameters settings.

2.2 A Neural Rule Representation

As LCS are applied to more complex domains the traditional ternary encoding can become limiting (e.g. see [27] for early discussions). Bull and O'Hara [e.g. 6] have examined the use of a neural rule representation scheme for LCS whereby each rule is represented by a multi-layered perceptron (MLP), following from a number of investigations which have made use of other complex rule representations, such as fuzzy logic [e.g. 25] and numerical S-expressions [2]. Since their inception LCS have been compared to neural networks, both conceptually [e.g. 13] and functionally (e.g. [9][24][11]). Bull [3] has suggested that the neural representation scheme is particularly suited to the use of LCS for autonomous robotics as "atoms" of behaviour need not be defined *a priori*, giving the system greater flexibility; actions can be functions of inputs. A new potential benefit from the incorporation of the neural paradigm into LCS is highlighted here.

The representation scheme is related to the use of evolutionary computing techniques to evolve neural networks in general (e.g. see [31] for an overview). In contrast to the majority of that work, an LCS-based approach is coevolutionary, the aim being to develop a number of (small) cooperative neural networks to solve the given task, as opposed to the evolution of one (large) network. That is, a decompositional approach is proposed. Whilst a number of schemes presented to aid the traditional approach could be used within NCS, such as development [e.g. 14], the simple approach appears sufficient here. SANE [22] is most similar to NCS, however SANE coevolves individual neurons within an LCS-based framework rather than small networks of neurons.

The neural scheme works as follows. On receipt of a sensory input (each rule has one input node per input feature), each member of the rule base processes the input in the usual manner for a feedforward MLP using sigmoid transfer functions. All rules have an "extra" output node which signifies whether it should be considered for the current input. Unless this node has the highest activation level on a system cycle, the given rule's condition matches the input and it is tagged as a member of the current match set [M]. Rules propose the action with the highest activation in the output layer node on a cycle, i.e. there is one per action. An action is selected from those advocated by the rules comprising [M] as usual for the chosen underlying LCS architecture.

Rules are represented to the GA as a concatenated string of weights where full connectivity is assumed. Bull and O'Hara [6] discuss the role of the reproduction cost in ZCS, suggesting it may remove rules which never match an input and hence whose fitness remains at f_0 which may disrupt the fitness sharing equilibrium. It can also be seen to reduce the initial "runaway" success of those rules in high payoff niches; the reproductive cost gives the fitness sharing process chance to have an effect within the rule base. It may also be suggested that the reinforcement component, due to the one-step delay in the rule updating of the traditional bucket brigade algorithm, disrupts the fitness sharing process; when the GA fires, the rules in the current [A] have reduced fitnesses, i.e. they will be away from the equilibrium fitness, which makes them susceptible to deletion. However, this can also be viewed as having a beneficial effect on the fitness sharing process as it may aid the suppression of rules in the more frequently visited (updated) niches from gaining a disruptive advantage, particularly early on. Note Wilson's [29] triggered niche GA scheme addresses this problem and the use of such a GA within ZCS and its effects on the fitness sharing process is currently under investigation.

Hence all processing, apart from the matching procedure and the mutation operator (Gaussian on satisfaction of μ per gene), remains the same as that described in Section 2.1 for this paper.

3 A Simple Multiobjective Task

As noted in the introduction, the case of a mobile autonomous robot which must find the shortest path to a goal state whilst maintaining appropriate battery power is considered here. The aim being to move to a real robot platform once expected behaviour is established.

Figure 1 shows a version of the well known Woods 1 [28] maze task which is a two dimensional rectilinear 5x5 toroidal grid. To create two objectives, sixteen cells are blank, seven are blocked (B), one contains a power source (P), and the last is the goal state (G). The NCS is used to develop the controller of a simulated robot which must traverse the maze in search of the goal state. It is positioned randomly in one of the blank cells and can move into any one of the surrounding eight cells on each discrete time step, unless it is blocked. If the robot moves into the goal cell the system receives a reward from the environment, and the task is reset, i.e. food/energy replaced and the robot randomly relocated. Similarly, if the robot moves into the power source cell the system receives a reward from the environment, and the task is reset. The reward received in either case depends upon the robot's internal energy level ε, randomly set from the range [0.0, 1.0] at the start of each trial. If the robot goes to the goal state and $\varepsilon > 0.5$ it receives 1000 otherwise 1. The opposite is true for reaching the power source.

On each time step the robot receives a sensory message which describes the eight surrounding cells. The message is encoded as a 16-bit binary string with two bits representing each cardinal direction. A blank cell is represented by 00, goal by 11, power by 01 and blocked cells by 10. The message is ordered with the cell directly above the robot represented by the first bit-pair, and then proceeding clockwise around it. The robot is also able to sense its internal energy value (single real number).

The trial is repeated 40,000 times and a record is kept of a moving average (over the previous 50 trials, after [28]) of how many steps it takes for the NCS robot to move into a goal cell or a power cell on each trial. Optimum performance is 1.7 steps to the goal under the single objective scenario. The average value of the reward received by the robot per trial is also recorded in this way. For the last 2000 trials the GA is switched off and a deterministic action selection scheme is used whereby the action with largest total fitness in [M] is picked (after [5]). All results presented are the average of ten runs unless otherwise stated.

Figure 2 shows how NCS is able to solve the multiobjective task. Here each rule has seventeen input, six hidden and nine output nodes. The parameters used were: $P = 6000$, $f_0 = 20$, $\beta = 0.9$, $\gamma = 0.3$, $\tau = 0.1$, $\mu = 0.02$, $\chi = 0.5$, $p = 0.25$, $\phi = 0.5$. Figure 2(a) shows a typical run under which, for both goal finding and power source finding, NCS takes around three steps during the first 38,000 trials, before giving roughly optimal performance under the deterministic mode. Note this is not the average of ten runs as the stochastic nature of the energy level assignment causes variance in the number of trials in which either cell is reached and hence averaging obscures underlying behaviour. Figure 2(b) shows how the reward received by the robot rises to about 900 after 10,000 trials and 1000 under deterministic action selection. That is, the NCS controller has learnt when, and the shortest route, to reach either the goal state or power cell depending upon its internal energy level.

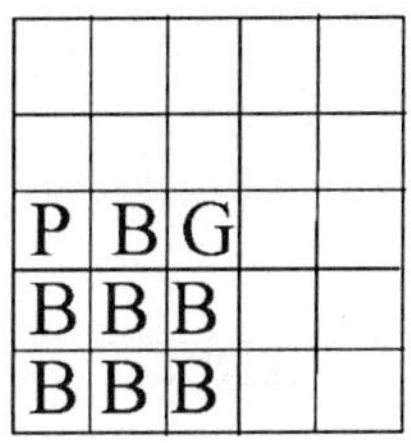

Fig. 1. A simple Multiobjective maze.

4 Towards Robotics: Dynamic Internal State

In the above scenario the (random) level of internal energy remained constant throughout a trial and reward levels were determined simply. In effect two traditional payoff landscapes existed, the robot experiencing either depending upon the assigned ε. More realistically, as the robot moves, the level of internal energy should decrease and the reward received for reaching the power source or goal state should be a *function* of the energy level.

Figure 3 shows the performance of NCS on the simple maze task where the initial random energy level (range [0.5, 1.0]) was decreased by 0.01 per move and the reward received for reaching the power source was $1000(1-\varepsilon)$, whilst that for reaching the goal was 1000ε. All parameters were as before except $\chi=0.1$ and trials were run for longer. Figure 3(a) again shows a typical single run during which NCS learns the shortest path to either objective. Figure 3(b) shows how it gradually learns to improve the amount of reward received. Optimal performance is difficult to ascertain here but with half as many power source trials to goal state trials, each taking the optimal number of steps in either case, an average reward of around 600 is eventually seen.

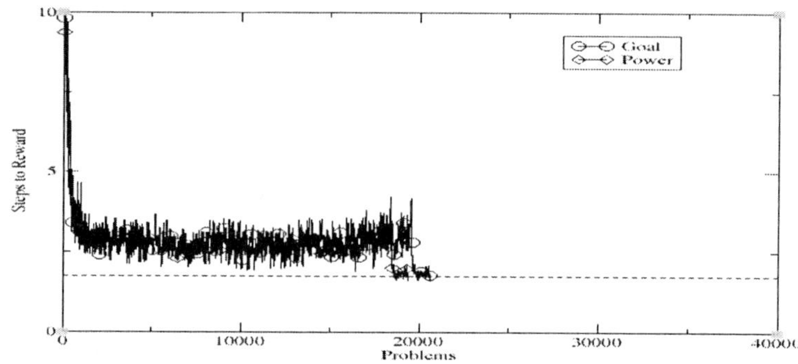

(a)

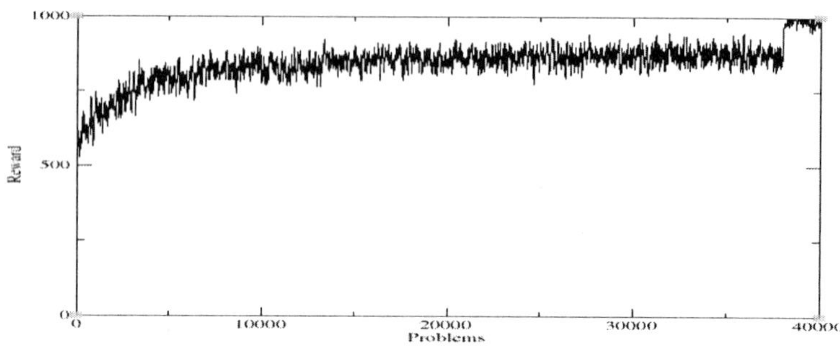

(b)

Fig. 2. NCS performance on the simple multiobjective task.

Therefore a simple approach to the consideration of more than one objective in the learning task of an LCS, where a non-trivial relationship exists between them, is possible. Future work will examine how well this scales to more than two objectives and more complex relationships thereby exploiting the neural representation scheme.

5 Conclusions

For Learning Classifier Systems to be effective in a number of complex problem domains they must be able to consider multiple objectives. This paper has examined developing the controller of a mobile autonomous robot which must achieve a given task whilst maintaining sufficient battery power. Results showed it is possible for NCS to solve both objectives when a random energy level is presented as an input along with sensory data. Here two payoff landscapes are experienced by the learner, one for low internal energy and one for high. A more complex version of the basic scenario was then examined under which the payoff landscape was a function of the internal energy level and hence the position of the robot at any given time was an increasingly significant factor in action choice. NCS was again shown able to perform well under such circumstances. It is important to note just how different this scenario is from single objective tasks: *the payoff landscape is dynamic, changing shape under the influence of the learner's actions.*

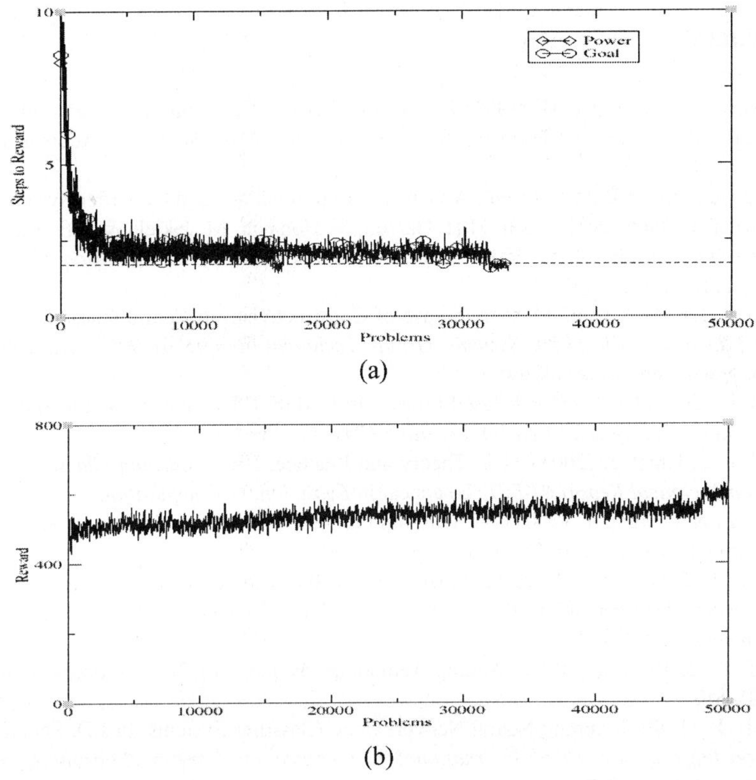

Fig. 3. NCS performance with movement cost and functional reward.

NCS's ability to form generalizations across payoff levels was highlighted in [6]. The dynamic payoff aspect of multiobjective problems makes the use of rule prediction accuracy for fitness [29] problematic, although Wilson's [30] suggestion of using a neural network payoff predictor for each rule represents a way to overcome such difficulties. This is currently under investigation by the authors, as are comparisons with other possible approaches to multiobjective reinforcement learning, such as hierarchies.

The results of this work are now being moved to a real robot platform after [18]. In that work a version of ZCS has been presented which learns to discretize continuous input spaces and considers real-time in the reinforcement process. Their scheme also relies upon the delay in bucket brigade updates to encourage optimal generalizations over the input space. Use of the neural rule structure is being added to the architecture (TCS) to develop controllers for multiobjective tasks.

Acknowledgements

Thanks to the members of the Learning Classifier Systems Group at UWE for many useful discussions during this work.

References

1. Ackley, D. & Littman, M. (1992) Interactions Between Learning and Evolution. In C.G. Langton, C. Taylor, J.D. Farmer & S. Rasmussen (eds) *Artificial Life II*, Addison Wesley, pp487-510.
2. Ahluwalia, M. & Bull, L. (1999) A Genetic Programming-based Classifier System. In W. Banzhaf, J. Daida, A.E. Eiben, M.H. Garzon, V. Honavar, M. Jakiela & R.E. Smith (eds) *GECCO-99: Proceedings of the Genetic and Evolutionary Computation Conference.* Morgan Kaufmann, pp11-18.
3. Bull, L. (2001) A Brief Note on the use of Constructivism in Neural Classifier Systems. *UWE Learning Classifier Systems Group Technical Report 01-006.* Available from http://www.csm.uwe.ac.uk/lcsg.
4. Bull, L. (2002) On Accuracy-Based Fitness. In L. Bull, P-L. Lanzi & W. Stolzmann (eds) *Soft Computing: Special Issue on Learning Classifier Systems* 6 (3).
5. Bull, L. & Hurst, J. (2001) ZCS: Theory and Practice. *UWE Learning Classifier Systems Group Technical Report 01-001.* To appear in *Evolutionary Computation.*
6. Bull, L. & O'Hara, T. (2001) NCS: A Simple Neural Classifier System. *UWE Learning Classifier Systems Group Technical Report 01-005.*
7. Butz, M., Goldberg, D.E. & Stolzmann, W. (2000) The Anticipatory Classifier System and Genetic Generalization. *IlliGAL Report No. 2000032, University of Illinois at Urban-Champaign, USA.*
8. Cliff, D. & Ross, S. (1994) Adding Temporary Memory to ZCS. *Adaptive Behaviour* 3:101-150.
9. Davis, L. (1989) Mapping Neural Networks into Classifier Systems. In J.D. Schaffer (ed) *Proceedings of the Third International Conference on Genetic Algorithms*, Morgan Kaufmann, pp375-378.
10. Deb, K. (2001) *Evolutionary Multiobjective Optimization Algorithms.* Wiley.
11. Dorigo, M. & Bersini, H. (1994) A Comparison of Q-learning and Classifier Systems. In D. Cliff, P. Husbands, J-A. Meyer & S.W. Wilson (eds) *Proceedings of the Third International Conference on Simulation of Adaptive Behaviour: From Animals to Animats 3.* MIT Press, pp248-255.
12. Dorigo, M. & Colombetti, M. (1998) *Robot Shaping.* MIT Press.
13. Farmer, J.D. (1989) A Rosetta Stone for Connectionism. *Physica D* 42:153-187.
14. Gruau, F. & Whitley, D. (1993) Adding Learning to the Cellular Developmental Process: a Comparative Study. *Evolutionary Computation* 1(3):
15. Holland, J.H. (1975) *Adaptation in Natural and Artificial Systems*, University of Michigan Press.
16. Holland, J.H. (1976) Adaptation. In R. Rosen & F.M. Snell (eds) *Progress in Theoretical Biology, 4.* Plenum.

17. Holland, J.H. (1986) Escaping Brittleness. In R.S. Michalski, J.G. Carbonell & T.M. Mitchell (eds) *Machine Learning: An Artificial Intelligence Approach, 2.* Morgan Kauffman, pp48-78.
18. Hurst, J., Bull, L. & Melhuish, C. (2002) ZCS and TCS Learning Classifier System Controllers on Real Robots. *UWE Learning Classifier Systems Group Technical Report 02-002.*
19. Karlsson, J. (1997) *Learning to Solve Multiple Goals.* PhD Dissertation, Rochester.
20. Kovacs, T. (2000) Strength or Accuracy? A Comparison of Two Approaches to Fitness Calculation in Learning Classifier Systems. In P-L. Lanzi, W. Stolzmann & S.W. Wilson (eds) *Learning Classifier Systems: From Foundations to Applications,* Springer, pp194-208.
21. Lanzi, P-L. & Wilson, S.W. (2001) Toward Optimal Classifier System Performance in Non-Markov Environments. *Evolutionary Computation* 8(4):393- 418.
22. Moriarty, D.E & Miikulainen, R. (1997) Forming Neural Networks through Efficient and Adaptive Coevolution. *Evolutionary Computation* 5(2): 373-399.
23. Schuurmans, D. & Schaeffer, J. (1989) Representational Difficulties with Classifier Systems. In J.D. Schaffer (ed) *Proceedings of the Third International Conference on Genetic Algorithms*, Morgan Kaufmann, pp328-333.
24. Smith, R.E. & Cribbs, B. (1994) Is a Learning Classifier System a Type of Neural Network? *Evolutionary Computation* 2(1): 19-36.
25. Valenzuela-Rendon, M. (1991) The Fuzzy Classifier System: a Classifier System for Continuously Varying Variables. In L. Booker & R. Belew (eds) *Proceedings of the Fourth International Conference on Genetic Algorithms.* Morgan Kaufmann, pp346-353.
26. Watkins, C. (1989) *Learning from Delayed Rewards.* PhD Dissertation, Cambridge.
27. Wiering, M. & Schmidhuber, J. (1997) HQ-Learning. *Adaptive Behaviour* 6(2): 219-246
28. Wilson, S.W. (1994) ZCS: A Zeroth-level Classifier System. *Evolutionary Computation* 2(1):1-18.
29. Wilson, S.W. (1995) Classifier Fitness Based on Accuracy. *Evolutionary Computation* 3(2):149-177.
30. Wilson, S.W. (2000) State of XCS Classifier System Research. In P-L. Lanzi, W. Stolzmann & S.W. Wilson (eds) *Learning Classifier Systems: From Foundations to Applications,* Springer, pp63-82.
31. Yao, X. (1999) Evolving Artificial Neural Networks. *Proccedings of the IEEE* 87(9):1423-1447.

On Using Constructivism in Neural Classifier Systems

Larry Bull

Intelligent Autonomous Systems Laboratory
Faculty of Computing, Engineering & Mathematical Sciences
University of the West of England
Bristol BS16 1QY, U.K.
larry.bull@uwe.ac.uk

Abstract. For artificial entities to achieve true autonomy and display complex life-like behaviour they will need to exploit appropriate adaptable learning algorithms. In this sense adaptability implies flexibility guided by the environment at any given time and an open-ended ability to learn novel behaviours. This paper explores the potential of using constructivism within the neural classifier system architecture as an approach to realise such behaviour. The system uses a rule structure in which each is represented by an artificial neural network. Results are presented which suggest it is possible to allow appropriate internal rule complexity to emerge during learning and that the structure indicates underlying features of the task.

1 Introduction

Neural Constructivism (NC) [e.g. 1] proposes a scenario whereby the representational features of the cortex are built through the interactions of the learning entity's development processes and its environment. This paper examines the feasibility of a constructive approach to realize flexible learning within a machine learning architecture which combines neural networks, reinforcement learning and evolutionary computing. Reinforcement learning techniques have been used to control autonomous entities [e.g. 2]. However, in most cases the entities can carry out only pre-specified actions or combinations thereof, where acquired knowledge can sometimes guide the latter. Conversely, biological systems, through sensory input and motor actions, acquire new information and organize it into operational knowledge which then shapes future behaviour. Approaches which generate knowledge not simply by combining existing knowledge but by producing knowledge which brings new functionality to the system are fundamental to the realization of truly autonomous entities. The aim of this paper is to show it may be possible to move toward artificial entities which exhibit such life-like qualities through NC, based around a new version of the Learning Classifier System (LCS) [3] framework - the Neural Classifier System (NCS) [4].

The production of autonomous intelligence requires the consideration of a number of issues including, but not limited to: the *learning architecture*, which must be

J.J. Merelo Guervós et al. (Eds.): PPSN VII, LNCS 2439, pp. 558–567, 2002.

flexible and responsive to environmental change, open ended, with automatic shifts in computational effort; and the *knowledge representation*, needed to provide generalization abilities over the input-action space thereby reducing the size of internal models, the inclusion of dimensions such as temporal context, and be scalable to an increase in sensory input. The field of reinforcement learning is concerned with many of these issues (see [5] for an overview). LCS are a class of reinforcement learner with a number of features which potentially enable them to consider many of the issues within a coherent whole.

It has recently been shown that a very simple LCS based on Holland's ideas can perform optimally, Wilson's ZCS [6], through its use of fitness sharing [7]. Bull and O'Hara [8] have shown how a simple LCS framework can allow aspects of Holland's original framework to be realised effectively through a neural network-based rule representation scheme - NCS. They show NCS is capable of optimal performance, the automatic formation of default hierarchy-like structures and has the potential to maintain multiple lines of internal induction. In this paper the emergence of appropriate rule complexity is considered.

The paper is arranged as follows: the next section briefly describes NCS and Section 3 the simple maze task. In Section 4 the use of neural constructivism within the system is presented. Finally, all results are discussed.

2 NCS: A Simple Neural Learning Classifier System

NCS [8] is based on the ZCS [6] Learning Classifier System. It periodically receives an input from its environment, determines an appropriate response based on this input and performs the indicated action, usually altering the state of the environment. Desired behaviour is rewarded by providing a scalar reinforcement. Internally the system cycles through a sequence of performance, reinforcement and discovery on each discrete time-step.

The NCS rule-base consists of a population of P multi-layered perceptrons (MLPs). Each rule is encoded as a string of connection weights; full connection is assumed. Weights are initialised randomly. Also associated with each rule is a fitness scalar initialised to a predetermined value f_0.

On receipt of a sensory input, each rule has one input node per input feature, all members of the rule-base process the input in the usual manner for an MLP. Each rule has an "extra" output node which signifies whether its output should be considered for the current input. If this node does not have the highest activation level, the given rule's condition "matches" the input and it is tagged as a member of the current match set [M]. An action is selected from those advocated by the rules comprising [M]. Rules propose an action by having their highest activation on a given output layer node. Action selection in NCS is performed by a simple roulette wheel selection policy based on fitness. Once an action has been selected, all rules in [M] that advocate this action are tagged as members of the action set [A] and the system executes the action.

Reinforcement in NCS is done under the implicit bucket brigade [6] which is closely related to Sutton's TD(0) algorithm [9] and consists of redistributing fitness between subsequent [A]. A fixed fraction (β) of the fitness of each member of [A] at each time- step is placed in a "common bucket". A record is kept of the previous action set $[A]_{-1}$ and if this is not empty then the members of this action set each receive an equal share of the contents of the current bucket, once this has been reduced by a pre-determined discount factor (γ). If a reward is received from the environment then a fixed fraction (β) of this value is distributed evenly amongst the members of [A]. Finally, a tax (τ) is imposed on all matched rules that do not belong to [A] on each time-step in order to encourage exploitation of the fitter classifiers. That is, all matching rules not in [A] have their fitnesses reduced by factor τ thereby reducing their chance of being selected on future cycles.

NCS employs two discovery mechanisms, a genetic algorithm (GA)[10] operating over the whole rule-set at each instance (panmictic) and a covering operator. On each time-step there is a probability *p* of GA invocation. When called, the GA uses roulette wheel selection to determine two parent rules based on fitness. Two offspring are produced via mutation and crossover (single point). The parents then donate half of their fitnesses to their offspring who replace existing members of the population. The deleted rules are chosen using roulette wheel selection based on the reciprocal of fitness.

If on some time-step, [M] is empty or has a combined fitness of less than φ times the population average, then a covering operator is invoked. A random rule is created which matches the environmental input. The new rule is given a fitness equal to the population average and inserted into the population over writing a rule selected for deletion as before.

Typical parameters used [8] are: Rule-base (*P*) = 1000, initial rule fitness (f_0) = 20.0, learning rate (β) = 0.8, discount factor (γ) = 0.3, tax (τ) = 0.1, cover trigger (φ) = 0.5, GA rate per time step (*p*) = 0.25, crossover rate = 0.5, and per gene Gaussian mutation rate (μ) = 0.02.

3 A Simple Maze Task

3.1 Woods 1

Figure 1(a) shows the well-known Woods 1 [6] maze task which is a two dimensional rectilinear 5x5 toroidal grid. Sixteen cells are blank, eight contain trees and one contains food. The NCS is used to develop the controller of a simulated robot which must traverse the map in search of food. It is positioned randomly in one of the blank cells and can move into any one of the surrounding eight cells on each discrete time step, unless occupied by a tree. If the robot moves into the food cell the system receives a reward from the environment (1000), and the task is reset, i.e. food is replaced and the robot randomly relocated.

On each time step the robot receives a sensory message which describes the eight surrounding cells. The message is encoded as a 16-bit binary string with two bits representing each cardinal direction. A blank cell is represented by 00, food (F) by 11 and trees (t) by 10 (01 has no meaning). The message is ordered with the cell directly above the robot represented by the first bit-pair, and then proceeding clockwise around it.

The trial is repeated 20,000 times and a record is kept of a moving average (over the previous 50 trials, after [6]) of how many steps it takes for the NCS robot to move into a food cell on each trial. If it moved randomly Wilson calculates performance at 27 steps per trial, whilst the optimum is 1.7 steps. For the last 2000 trials the GA is switched off and a deterministic action selection scheme is used whereby the action with largest total fitness in [M] is picked (after [7]). All results presented are the average of ten runs.

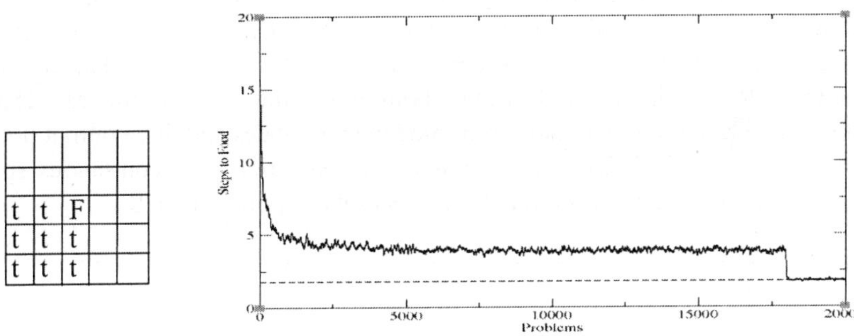

(a) (b)

Fig. 1. Woods 1 (a) and NCS performance in the maze (b).

Figure 1(b) shows how standard NCS is able to solve the maze optimally. Here each rule has sixteen input, four hidden and nine output nodes. The parameters used were as given in Section 2. It can be seen that NCS takes around three steps to food during the first 18,000 trials, before giving optimal performance under the deterministic mode.

3.2 Continuous Inputs and Actions: Less Restricted Behavioural Repertoires

As noted in the introduction, one restriction of many approaches to autonomous entities is the *a priori* fixing of the types of behaviour the system may exhibit. That is, working at a similar level to that of Woods 1, motor control for "go forward" is calculated and fixed, as they are for turns, etc. The learning entity then attempts to combine such "atoms" of behaviour to achieve the set task. However, depending on the level and number of different atoms, potential behaviour is restricted under this scheme. Harvey et al. [e.g. 11] have shown it is possible to evolve neural network controllers where system output is a (complex) function of input; they connect neurons *directly* to motors, with activation converted to power, thereby avoiding the need to prescribe atoms of behaviour.

Very little work exists on the use of LCS to solve multi-step tasks where the input and output spaces are real-valued, those using fuzzy logic being the only known examples (see [12] for an overview). NCS can of course also be used in such cases. To demonstrate this simply, Woods 1 has been slightly modified.

Here the robot receives the same sensory input from its surroundings as before but these are random real values over predefined ranges: a blank cell is represented by the range [$0.0<x<0.1$], food by [$0.9<x<1.0$] and trees by [$0.4<x<0.5$]. Hence each neural rule contains eight input nodes, one for each sensory direction. Actions are given via two output nodes, each imagined as directly controlling one motor of a two-driving-wheeled autonomous robot. To give the eight possible actions of the original task, outputs are discretised into three ranges: low [$-1.0<x< -0.3$], medium [$-0.3<x<0.3$] and high [$0.3<x<1.0$]. Hence with both output neurons giving a high response the robot moves north, when node 1 is high and node 2 is medium it moves northeast, when node 1 is high and node 2 is low it moves east, and so on. The last two combinations (low,medium and low,low) both move the robot northwest. Hence networks consist of eight input nodes, four hidden layer nodes and three output nodes, the latter being linear. Whilst this is a crude approximation of a continuous space model it allows for the demonstration of the capability required of NCS.

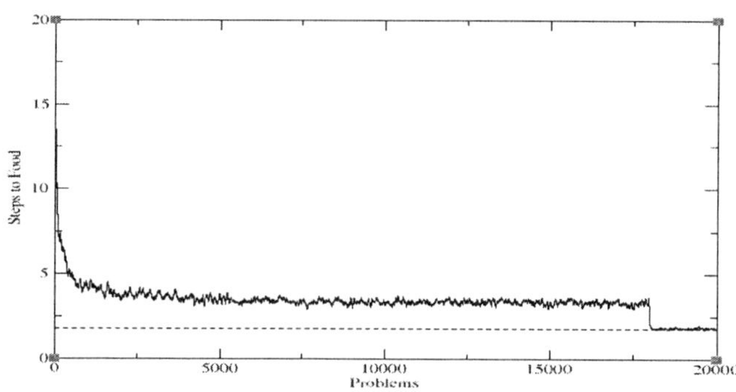

Fig. 2. NCS performance in Woods 1 with continuous valued inputs and actions.

Given the more complex relationship between the inputs and actions the extra matchset membership node mechanism is simplified: when the node gives a positive response to an input the rule does not join [M].

Figure 2 shows NCS gives very similar performance as in the discrete case and gives optimal performance under deterministic action selection. That is, NCS appears able to learn effectively in multi-step tasks where the actions are a function of the inputs implying it may be used without predefined behavioural atoms. Parameters were as before except $\beta = 0.9$.

4 Neural Constructivism in NCS

One neurobiological theory for the emergence of complex reasoning within brains postulates that the dynamic interaction between neural growth mechanisms and the environment drives the learning process, known as neural constructivism [e.g. 1]. This is in contrast to the related evolutionary selectionist idea which emphasises regressive mechanisms whereby initial neural over-connectivity is pruned based on a measure of utility [e.g. 13]. The scenario for constructionist learning is that, rather than start with a large neural network, development starts with a small network. Learning then adds appropriate structure, particularly through growing/pruning dendritic connectivity, until some satisfactory level of utility is reached. In this way the building of a representation of the problem space is both flexible and tailored to the problem by the learner's interaction with it. Therefore suitable, specialized neural structure need not be specified *a priori.*

Redding et al. [14] have used heuristics to add hidden nodes to an MLP during training, showing an ability to develop suitable structure whilst learning a given task. The use of evolutionary computing techniques to allow for the emergence of appropriate complexity in neural networks has been examined by Harvey et al. [e.g. 11]. Here an evolutionary gradualism mechanism is used such that the length of the genotypes can increase to an appropriate size over time; extra nodes can be added to the network through a mutation-like operator during reproduction. Other population-level techniques which allow for an appropriate increase in genotype complexity include duplication [15], where a given genotype has the potential to double in length, and symbiogenesis [e.g. 16], where genotypes from separate coevolving populations can merge.

4.1 A Simple Process of Constructivism

The basic concept of neural constructivism can be used within NCS, termed NCSc, to allow for the emergence of appropriate rule complexity to a given task. Here each rule can have one or more of a fixed number of nodes in its hidden layer fully connected to the input and output layers. With some probability(ψ) such connections for another unconnected node can be added and/or removed from the last connected node with a given probability (ϖ) during reproduction. Hidden nodes are coded on the right-most end of genotypes and the crossover point is chosen using the rule with fewest connected nodes. In this way no other considerations for what is implicitly a variable length representation are needed; the weights for unconnected nodes are simply zero. This aspect of the system is open to future investigation.

4.2 Results

Figure 3(a) shows the performance of NCSc on the simple maze task where rules started with one connected node in the hidden layer. The parameters used were as before but P=2000, ψ=0.01 and ϖ=0.01. It can be seen that optimal performance is obtained in the simple maze. Figure 3(b) shows the average number of connected nodes in the hidden layer of the rules. Here the neural constructivism mechanism causes an increase in the number of hidden nodes connected before achieving optimal performance, although there is an expected cost in terms of time taken whilst average connectivity increases. On average, rules use two or three hidden layer nodes here. Examination of the resulting rules shows a degree of spatial heterogeneity. For example, three of the four locations in the top left corner of Woods 1 (Figure 1a) are typically handled by a rule proposing a move southeast (as noted in [8]). The remaining location is often handled by another rule advocating the necessary move east. Inspection shows that the three-move rule typically contains three hidden nodes, whereas the single move rule contains two hidden nodes. That is, the building of a representation of the problem space has been tailored to the problem by NCSc's interaction with it under the constructivism process.

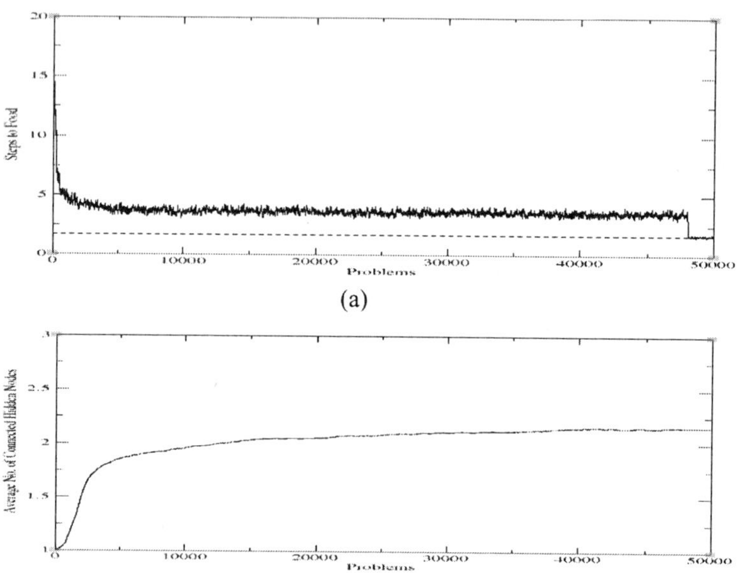

(a)

(b)

Fig. 3. NCSc performance with single initial hidden nodes.

Figure 4 shows the performance of NCSc with the same parameters and started with four hidden nodes connected, i.e. as in Figure 2 but with constructivism sampling rule connectivity. It can be seen that the average number of hidden nodes drops slightly over the trials with the system solving the task. Hence no significant disruption to the learning process appears to occur when NCSc is started with what is known to be an appropriate degree of rule complexity. It can also be noted that there is less pressure to prune average rule connectivity. Future work will examine the sensitivity of the system to ψ and ϖ.

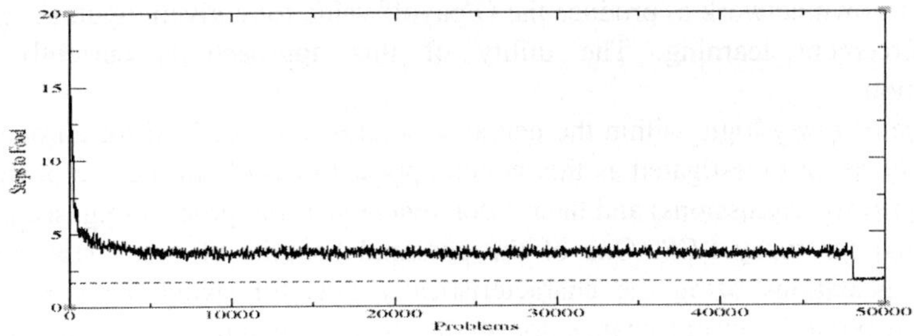

(a)

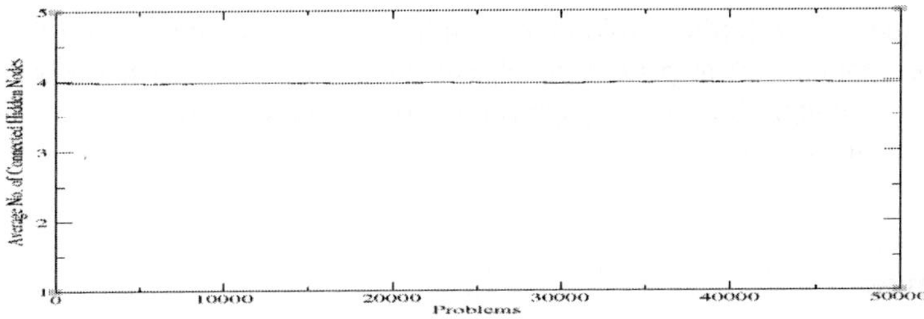

(b)

Fig. 4. NCSc performance with four initial hidden nodes.

5 Conclusions

The LCS approach to artificial learning entities would appear to, potentially at least, encompass many of the key aspects raised in the introduction. This paper has explored the use of constructivism within the neural classifier system architecture as an approach to aid the realization of complex/appropriate autonomous behaviour, exploiting NCS's basis in evolutionary computing. Future work will consider the growth of recurrent connections for non-Markov mazes (after [8]).

NCS uses a neural network to form generalizations over the space of possible inputs and outputs, producing rules for the high payoff actions only. Typically, reinforcement learning approaches attempt to build generalizations over the full input- action space. When the number of such states is large, neural networks have been used to learn the payoff values, one for each possible action [e.g. 5]. However, this cannot easily be used for tasks where the action space is continuous. NCS uses the genetic algorithm to design neural networks to generalize at a level which enables continuous actions. The ability to, and utility of, learning systems to construct full payoff maps in complex problem domains remains open to question; for increasingly complex spaces, sufficient sampling to build accurate, complete models becomes increasingly difficult. However, a version of NCS has been presented which attempts to form complete maps using accuracy-based fitness [4] and future work will explore its behaviour in comparison to the system used here. The payoff-based version has an advantage in that it can form generalizations across payoff levels (see [8] for discussions). Wilson [17] has suggested that each rule in an accuracy-based system maintain its own network to produce the Q/payoff value for a given input, i.e. as done in reinforcement learning. The utility of this approach is currently under investigation.

The use of fuzzy logic within the neural structures (e.g. see [18] for an overview) is currently being investigated as this would appear to ease both the use of memory (e.g. see [19] for discussions) and the reinforcement learning process (e.g. see [12] for discussions) for fuzzy LCS. One of the potential benefits of fuzzy reasoning for autonomous entities being the characteristically graceful switch between actions rather than possibly abrupt change for slight changes in input (see [20] for initial results).

Techniques which further exploit the evolutionary computing basis of LCS are also being investigated, such as parameter self-adaptation [e.g. 21] which has been shown to improve on-line performance by annealing GA disruption and to improve response to environmental change [22].

References

1. Quartz, S.R & Sejinowski, T.J. (1997) The Neural Basis of Cognitive Development: A Constructionist Manifesto. *Behavioural and Brain Sciences* 20(4): 537-596.
2. Mataric, M.J. (1997) Reinforcement Learning in the Multi-Robot Domain. *Autonomous Robotics* 4(1): 73-83.
3. Holland, J.H. (1976) Adaptation. In R. Rosen & F.M. Snell (eds) *Progress in Theoretical Biology, 4.* Plenum.
4. Bull, L. & O'Hara, T. (2001a) A Neural Rule Representation for Learning Classifier Systems. *UWE Learning Classifier Systems Group Technical Report 01-002.* Available from http://www.cems.uwe.ac.uk/lcsg.
5. Sutton, R.S. & Barto, A.G. (1998) *Reinforcement Learning.* MIT Press,
6. Wilson, S.W. (1994) ZCS: A Zeroth-level Classifier System. *Evolutionary Computation* 2(1):1-18.
7. Bull, L. & Hurst, J. (2001) ZCS: Theory and Practice. *UWE Learning Classifier Systems Group Technical Report 01-001.* Available from http:// www.cems.uwe.ac.uk/lcsg. To appear in *Evolutionary Computation.*
8. Bull, L. & O'Hara, T. (2001b) NCS: A Simple Neural Classifier System. *UWE Learning Classifier Systems Group Technical Report 01-005.* Available from http:// www.cems.uwe.ac.uk/lcsg.
9. Sutton, R.S. (1996) Generalization in Reinforcement Learning: Successful Examples using Sparse Coarse Coding. In D.S. Touretzky, M.C. Mozer & M. Hasselmo (eds) *Advances in Neural Information Processing Systems: Proceedings of the 1995 Conference*, MIT Press, pp1038-1044.
10. Holland, J.H. (1975) *Adaptation in Natural and Artificial Systems*, University of Michigan Press.
11. Harvey, I., Husbands, P. & Cliff, D. (1994) Seeing the Light: Artificial Evolution, Real Vision. In D. Cliff, P. Husbands, J-A. Meyer & S.W. Wilson (eds) *From Animals to Animats 3: Proceedings of the Third International Conference on Simulation of Adaptive Behaviour.* MIT Press, pp392-401.

12. Bonarini, A. (2000) An Introduction to Fuzzy Learning Classifier Systems. In P- L. Lanzi, W. Stolzmann & S.W. Wilson (eds) *Learning Classifier Systems: From Foundations to Applications*. Springer, pp83-106.
13. Edelman, G. (1987) *Neural Darwinism: The Theory of Neuronal Group Selection*. Basic Books.
14. Redding, N.J., Kowalcyzk, A. & Downs, T. (1993) Constructive Higher-Order Network Algorithm thet is Polynomial Time. *Neural Networks* 6:997-1010.
15. Lindgren, K. & Nordhal, M.G. (1995) Cooperation and Community Structure in Artificial Ecosystems. *Artificial Life* 1(1) 15-38.
16. Bull, L., Fogarty, T. & Pipe, A. (1995) Artificial Endosymbiosis. In In F. Moran, A. Mereno, J.J. Merelo & P. Chaon (eds) *Advances in Artificial Life - Proceedings of the Third European Conference on Artificial Life*. Springer Verlag, pp273-289.
17. Wilson, S.W. (2000) State of XCS Classifier System Research. In P-L. Lanzi, W. Stolzmann & S.W. Wilson (eds) *Learning Classifier Systems: From Foundations to Applications,* Springer, pp63-82.
18. Tsoukalas, L.H. & Uhrig, R.E. (1997) *Fuzzy and Neural Approaches in Engineering*. Wiley.
19. Furuhashi, K., Nakaoka, K., Morikawa, K. & Uchikawa, Y. (1993) Controlling Excessive Fuzziness in a Fuzzy Classifier System. In S. Forrest (ed) *Proceedings of the Fifth International Conference on Genetic Algorithms.* Morgan Kaufmann, pp635.
20. Bull, L. & O'Hara, T. (2002) Accuracy-based Neuro and Neuro Fuzzy Classifier Systems. *UWE Learning Classifier Systems Group Technical Report 02-001*. Available from http://www.cems.uwe.ac.uk/lcsg.
21. Bull, L. Hurst, J. & Tomlinson, A. (2000) Self-Adaptive Mutation in Classifier System Controllers. In J-A. Meyer, A. Berthoz, D. Floreano, H.Roitblatt & S.W. Wilson (eds) *From Animals to Animats 6 - The Sixth International Conference on the Simulation of Adaptive Behaviour,* MIT Press.
22. Hurst, J. & Bull, L. (2001) A Self-Adaptive Classifier System. In P-L. Lanzi, W. Stolzmann & S.W. Wilson (eds) *Advances in Learning Classifier Systems: Proceedings of the Third International Workshop on Learning Classifier Systems*. Springer, pp70-79.

Initial Modifications to XCS for Use in Interactive Evolutionary Design

Larry Bull, David Wyatt, and Ian Parmee

Faculty of Computing, Engineering & Mathematical Sciences,
University of the West of England, Bristol
{Larry.Bull,David2.Wyatt,Ian.Parmee}@uwe.ac.uk

Abstract. Learning classifier systems represent a technique by which various characteristics of a given problem space may be deduced and presented to the user in a readable format. In this paper we present results from the use of XCS on simple tasks with the general multi-variable features typically found in problems addressed by an Interactive Evolutionary Design process. That is, we examine the behaviour of XCS with versions of a well-known single-step task and consider the speed of learning and the ability to respond to changes. We introduce a simple form of supervised learning for XCS with the aim of improving its performance with respect to these two measures. Results show that improvements can be made under the new learning scheme and that other aspects of XCS can also play a significant role.

1 Introduction

Interactive Evolutionary Design (IED) [14], [12] moves away from the use of evolutionary computing techniques within a rigid optimization environment and instead utilizes them as generators and gatherers of optimal design information. Such information may relate to the characteristics of variables and variable sensitivity, constraints and degree constraint satisfaction and the nature of conflict between multiple design objectives. The hypothesis being that initially ill-defined conceptual design problems can evolve through the close interaction of the designer with machine-based adaptive search processes where solutions generated from such processes provide information that supports a better understanding of the problem domain. The off-line processing of this information promotes related change to the problem space hence supporting the capture of designer experiential knowledge and intuition within subsequent adaptive search. The approach involves the capture of designer experiential knowledge and intuition within adaptive search processes through an iterative designer/machine-based refinement of the design space. Initial machine-based search within a relatively ill-defined design space supports exploration outside of initial constraint, objective and parameter bounds through the designer's consideration of initial results and subsequent re-definition of the space. This last aspect of the process is of interest to us here: we consider a way in which to enhance the presentation of results from a given iteration of the search process through the use of learning classifier systems (LCS) [10]. That is, we are interested in the use of XCS [18] as a data miner in problems where the speed of learning is important as is the

J.J. Merelo Guervós et al. (Eds.): PPSN VII, LNCS 2439, pp. 568–577, 2002.

ability to respond to changes made by the user in the underlying design space, i.e. between iterations of the IED process.

XCS has been shown to perform well on a number of benchmark data mining tasks [7], [17], [11], [8] with the added benefit of producing readable production system rules. In this respect XCS would seem an appropriate technique by which to support designers/decision-makers in the identification of interesting regions of a solution space whilst generating meaningful rules relating to the variables and objectives describing that space. In this paper we consider the use of a simple supervised learning algorithm in XCS to speed learning and thereby replace the incremental Widrow-Hoff updating scheme. That is, data mining need not be treated as a reinforcement learning task. Rather, in this paper, the utility of a given classification is assigned to a newly created rule as indicated by the first training exemplar it experiences. Using versions of the real-numbered multiplexer problem [20] it is shown that the new learning scheme is able to learn near perfect descriptions of the underlying problem space more rapidly than the traditional unsupervised approach; with a reduced amount of training data the supervised learning scheme gives superior performance. It is also shown that, depending on the type of change made, the new scheme is able to respond as well to a change in the task as the standard approach.

The paper is arranged as follows: the next section describes the experimental setup; section 3 presents results from its use on the task; section 4 considers performance in non-stationary versions of the multiplexer task; and finally, all findings are discussed.

2 sXCS

Many real-world problems, such as those in design, contain multiple variables, each of which may be a real number. Wilson [20] has presented a version of XCS for such problems – XCSR. Here conditions consist of interval predicates of the form $\{\{c_1,s_1\}$, ... $\{c_n,s_n\}\}$, where c is the condition's range centre and s is the "spread" from that centre over which the variable is matched by the rule. When a centre with added spread goes outside the defined range it is truncated. All other XCS processing remains as described in [6] except that mutation is done via a random step (range $-0.1 < x < 0.1$), cover produces rules centred on the input value with a range of s_0 and subsumption considers the variable ranges.

Wilson [20] tested XCSR on a real-numbered version of the well-known Boolean multiplexer problem. These single-step functions are traditionally defined for binary strings of length $l = k + 2^k$ under which the first k bits index into the 2^k remaining bits, returning the indexed bit. A correct classification results in a payoff of 1000, an incorrect classification of payoff 0. In the real-numbered version random vectors are formed in the range [0.0, 1.0]. In the first instance, a variable value of 0.5 or less corresponds to a binary 0 in the traditional task. A value greater than 0.5 therefore corresponds to binary 1.

XCS uses the incremental Widrow-Hoff procedure to update expected payoff values. In this paper we introduce a simplified update procedure whereby newly created rules, i.e. those which have never participated in an action set since their creation (via cover or the GA), have their expected payoff value set to that of the first training instance they experience. This value remains constant. All other parameters are initialized and updated as in traditional XCS. With the real-numbered representation scheme this system is here termed sXCSR.

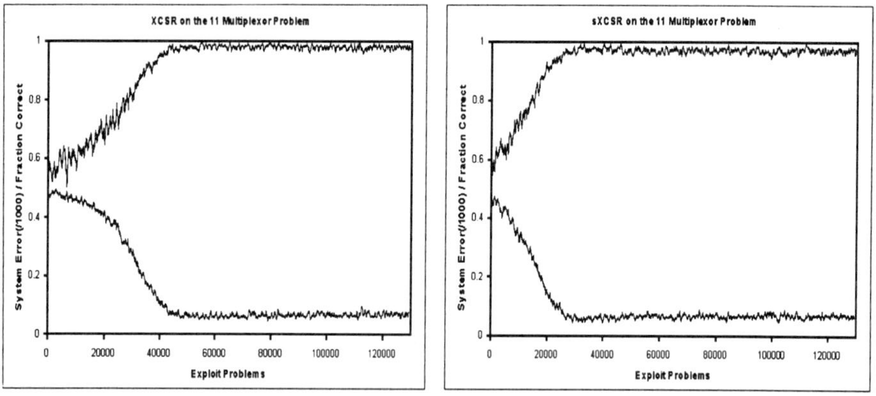

(a) (b)

Fig. 1. XCSR and sXCSR on the 11-variable multiplexer task

We now examine the comparative performance of sXCSR with that of XCSR on versions of the 11-variable multiplexer problem. All parameters are as used in [4] in their investigations of binary versions of the task without payoff landscaping: β=0.2, α=0.1, ε_0=10, ν=5, θ_{GA}=12, χ=0.8, μ=0.04, θ_{del}=20, δ=0.1, θ_{sub}=20, p_I=10, ε_I=0, F_I=0.01, θ_{mna}=2, plus m=0.1 and s_0=1 as defined in [20] and an experimentally determined population size of *N*=5000. Results presented are the average of ten runs.

3 Results

Figure 1(a) shows the performance of XCSR on the 11-variable task. It can be seen that the system achieves approximately 98% correct classification after around 50000 explore trials (running average of exploit trial performance shown in all figures as in [20]). Similarly, the system error drops to around 100 (10%) after 40000 trials. Figure 1(b) shows the performance of sXCSR on the same task. It can be seen that whilst final performance achieved is about the same, around 98%, it reaches this level 20000 trials before XCSR. That is, the initial rate of learning appears quicker under the supervised learning scheme despite XCSR's use of the MAM process [18]. The error drops to 10% around 20000 trials, again, considerably quicker than XCSR. That is, the traditional unsupervised learning approach appears to provide a form of noise to the generalization process; allowing rules to continually adjust their prediction values slows learning.

Under the IED process described above, the ability to form near optimal descriptions of a design space with a reduced amount of training data is beneficial. That is, training data for the classifier may be limited due to the evolutionary algorithm's search/sampling of the given space. Hence sXCSR displays the kind of learning we envisage as particularly suited to IED.

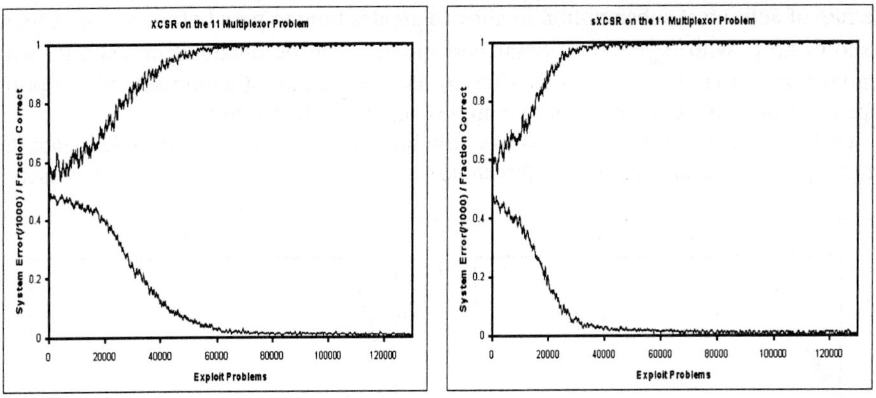

(a) (b)

Fig. 2. XCSR and sXCSR without action-set subsumption

Figure 1 shows how neither system completely solves the 11-variable version of the multiplexer problem. Wilson [20] used a 6-variable version, also finding around 98% performance. He suggests that the system may perform optimally if left to run for longer and/or by use of a more sophisticated mutation operator. However, we have found that optimal performance can be achieved by removing the action-set subsumption process. Figure 2 shows how 100% performance is achieved around 70000 trials for XCSR and 50000 trials for sXCSR. Again, sXCSR shows an increase in learning speed. Therefore, by reducing the "strong" [5] convergence pressure of action-set subsumption, it seems the GA is better able to search the space of real-valued genes.

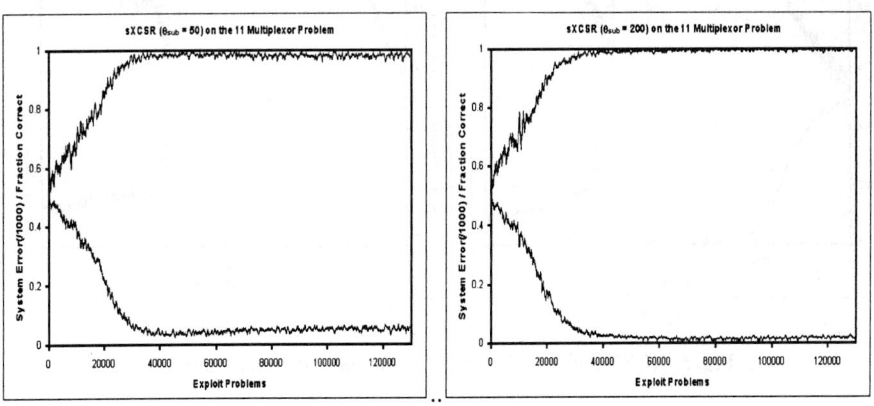

(a) (b)

Fig. 3. The effects of altering the amount of action-set subsumption in sXCSR

Action-set subsumption [19] was introduced primarily to encourage a maximally general solution, i.e. to improve the readability of the resulting rule-base. Since this is of concern to us under the IED application we have examined the effects of varying the rate of action-set subsumption to allow a greater time to pass before its use. Figure 3 shows how, with θ_{sub}=200, optimal performance can be obtained in sXCSR, with similar results found in XCSR (not shown). Hence the rate of action-set subsumption appears critical to XCS with a more complex representation scheme.

Analysis of the resulting rules shows how both forms of XCS learn to delineate the ranges for binary 0 and 1. The 0.5 threshold and generalizations are clear to the user.

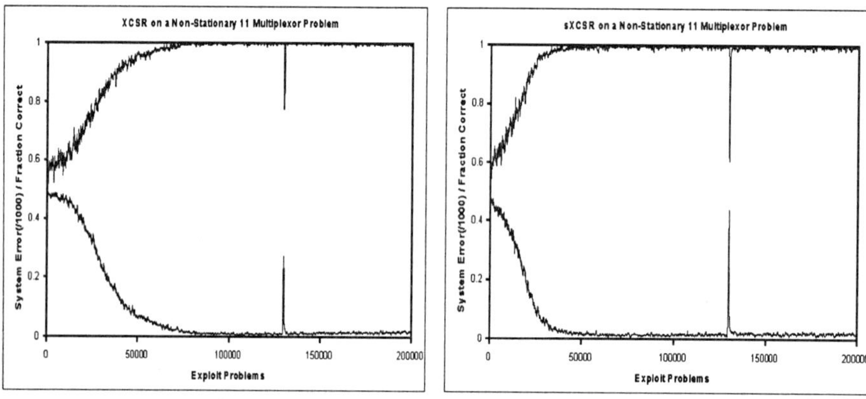

(a) (b)

Fig. 4. XCSR and sXCSR on the first non-stationary 11-variable multiplexer task

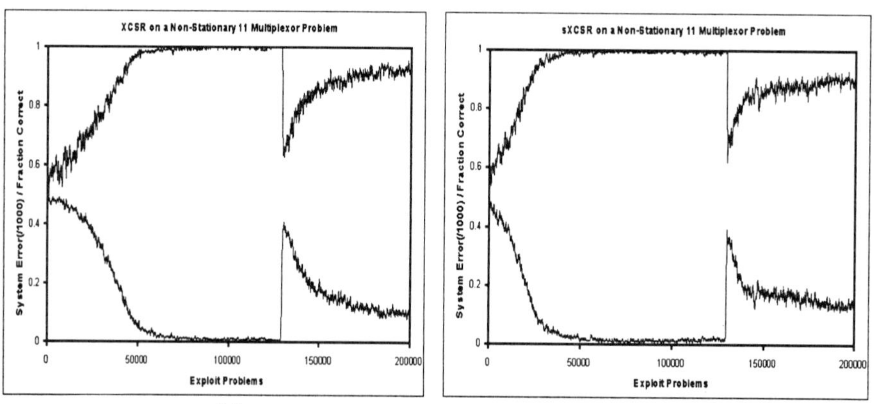

(a) (b)

Fig. 5. XCSR and sXCSR on the harder non-stationary task

4 Results in Non-stationary Tasks

As described above, the IED process is iterative in that the designer may alter aspects of the problem formulation, e.g. relax a constraint, re-define a variable range and / or adjust an objective weighting [13], before re-running the evolutionary search. In this way a non-stationary data mining task is created (we assume that generally some underlying relationships between variables remain after any change in the design space). Hartley [9] has proposed that XCS, through its construction of a full problem mapping, will perform better than a traditional LCS on non-stationary tasks. Bull and Hurst [2] have noted how the benefit of the full mapping depends upon the nature of the change to the task. In the simple case where payoff levels change for the given underlying generalizations, XCS needs only to adjust these values via the reinforcement learning procedure. Conversely, a traditional LCS would probably need to discover new rules or adjust the numerosity of existing rules in response to the change. However, if alterations are made to the underlying generalizations, XCS has no explicit benefit over traditional LCS as it must also discover new rules. We have examined the performance of the supervised learning scheme under both change scenarios using the 11-variable multiplexer problem (θ_{sub}=200).

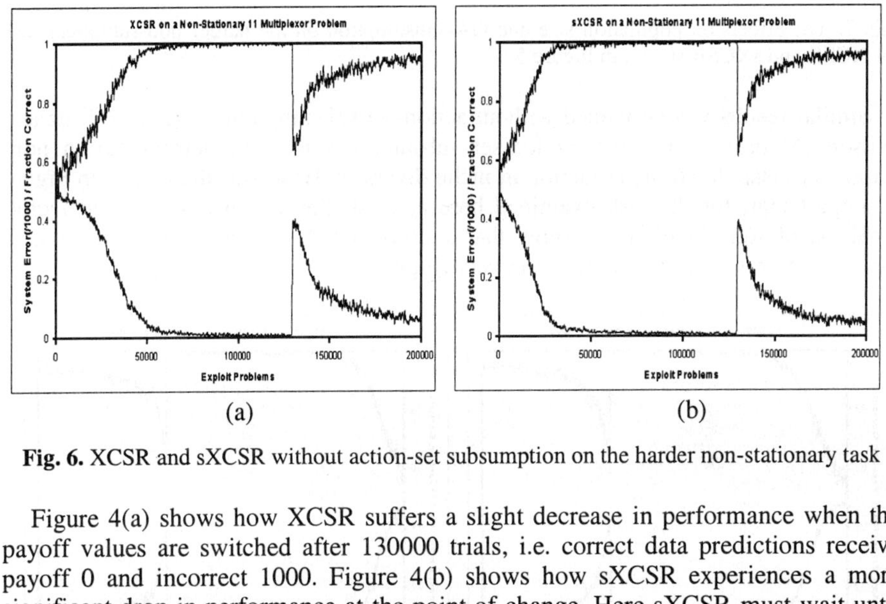

(a) (b)

Fig. 6. XCSR and sXCSR without action-set subsumption on the harder non-stationary task

Figure 4(a) shows how XCSR suffers a slight decrease in performance when the payoff values are switched after 130000 trials, i.e. correct data predictions receive payoff 0 and incorrect 1000. Figure 4(b) shows how sXCSR experiences a more significant drop in performance at the point of change. Here sXCSR must wait until the GA produces new rules within each niche to recover from the alteration which, perhaps not unexpectedly, is a slower process of re-adaptation.

Figure 5 shows the performance of the two systems when the thresholding of the real-numbered multiplexer are altered to the other form presented in [20]. Here every even-numbered locus variable is assigned to binary 0 if it is less than or equal to 0.25 and the odd-numbered loci variables are assigned to binary 0 if they are less than or equal to 0.75. Wilson found this problem harder to solve than the previous version, achieving around 93% performance on a 6-variable version. Figure 5 shows how there is no difference in the rate of recovery of either system and that both suffer far more severely under this form of change; no benefit from the full problem mapping is seen, as suggested in [2]. It can be noted that neither system appears to learn the new task completely over the last 70000 trials indicating, as did Wilson's result [20], the increased difficulty.

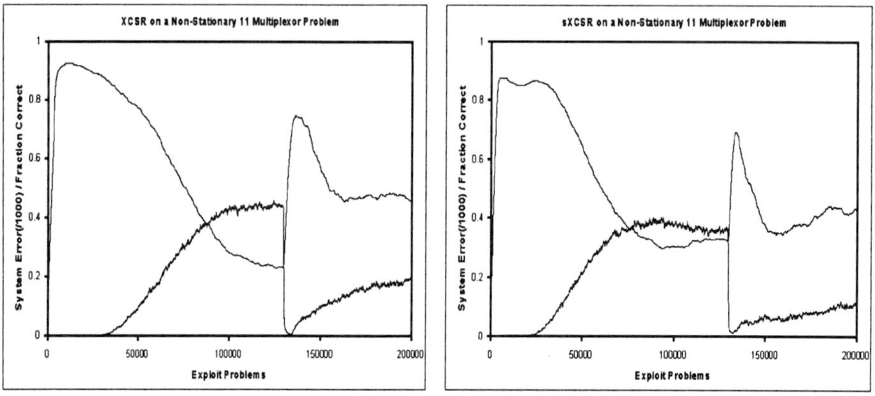

(a) (b)

Fig. 7. The effects on population size and GA-subsumption on the harder non-stationary task for XCSR and sXCSR shown in Figure 5

Similar results were obtained without action-set subsumption (Figure 6). Butz and Wilson [6] briefly note that action-set subsumption may be detrimental in non-stationary tasks due to its reduction in niche diversity. However, the results in Figure 6 suggest that, for the task examined here at least, this is not a significant factor. Analysis of runs in Figure 5 show that a drop in GA subsumption occurs which increases diversity thereby aiding recovery (Figure 7).

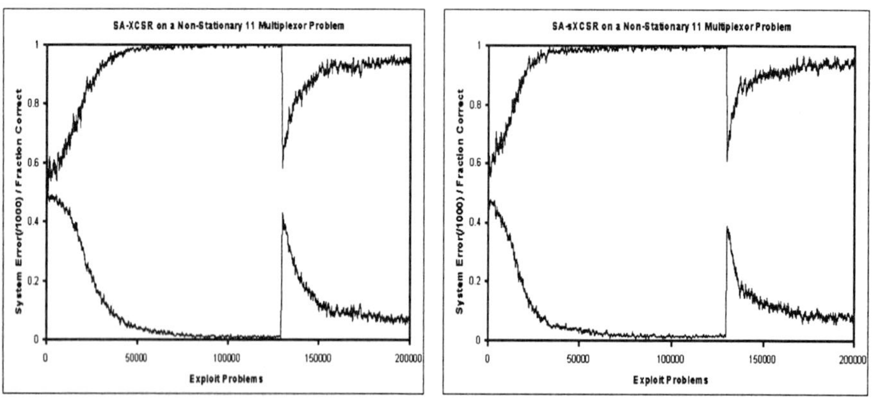

(a) (b)

Fig. 8. The effects of using self-adaptive mutation rates on XCSR and sXCSR on the harder non-stationary task

Bull et al. [3] have used self-adaptive mutation rates within XCS. Self-adaptation has been suggested as beneficial for evolutionary computing techniques in general for non-stationary problem domains [1]. Figure 8 shows the performance of the two systems when a simple form of self-adaptive mutation is added. Here an extra real-coded gene is added to each classifier representing the probability of mutating one of the other genes in the genome using the step-size mutation operator described above. As in [3], the rate itself is mutated before the new rate is used for the remaining genes. Rates are initially assigned randomly from the range [0.0, 1.0]. Results show that self-adaptation provides an increase in learning speed for both XCSR and sXCSR, the latter now recovering slightly quicker than the former after the change. Figure 9 shows how the average mutation rate in the rule base first rises slightly from 0.5 and then decreases as a solution is found, similarly shown in [3] for multi-step tasks, before rising again at the point of change, similarly shown in [22] for multi-step tasks.

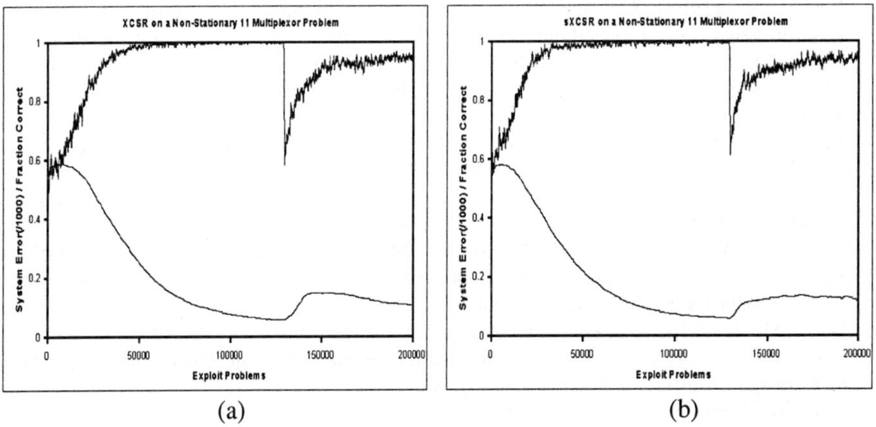

Fig. 9. The average self-adaptive mutation rate in the rule base for both XCSR and sXCSR on the harder non-stationary task

5 Conclusion

In terms of the introduction of a learning technique to the Interactive Evolutionary Design concept the work described can only be considered a preliminary investigation. However, the results strongly indicate a significant potential in the utilisation of learning classifier systems to support designers as part of the IED process. Their preliminary task relates to the identification of interesting regions of a solution space under the IED process through the generation of meaningful rules relating to the variables and objectives describing that space. However, their utility could extend far beyond this initial task in terms of generating rules relating to a wide spectrum of relevant design information. Such rules could be utilised to support a degree of autonomous processing in terms of appropriate succinct data presentation or even to the partial establishment of potential refined problem spaces. This would contribute to one overall objective of the IED research which is to avoid cognitive overload through a machine-based reduction in the amount of information the designer has to deal with ([15],[16]).

Of primary interest is the ability of the presented learning classifier systems to learn fairly accurate models quickly, due to the possibilities of relatively small training sets, and their ability to respond to changes in the underlying problem space as made by the designer. In this paper we have introduced a simple form of supervised learning for XCS with the aim of improving its performance with respect to these two measures. Results show that improvements can be made under the new learning scheme. Future work will examine the scheme's sensitivity to noise in the training data and its application to real-world design data, as well as improving XCS's response to significant changes.

References

1. Baeck, T. (1998) On the Behaviour of Evolutionary Algorithms in Dynamic Environments. In Proceedings of the Fifth IEEE Conference on Evolutionary Computation, pages 446-451. IEEE Press, Piscataway (NJ).
2. Bull, L. & Hurst, J. (2001) ZCS: Theory and Practice. UWE *Learning Classifier Systems Technical Report UWELCSG01-001.*
3. Bull, L., Hurst, J. & Tomlinson, A. (2000) Self-Adaptive Mutation in Classifier System Controllers. In J-A. Meyer, A.Berthoz, D. Floreano, H. Roitblatt & S.W. Wilson (eds) *From Animals to Animats 6 - The Sixth International Conference on the Simulation of Adaptive Behaviour*, MIT Press, Cambridge (MA).
4. Butz, M., Kovacs, T., Lanzi, P & Wilson, S. (2001) How XCS evolves accurate classifiers. In *Proceedings of the Genetic and Evolutionary Computation Conference 2001*, pages 927-934. Morgan Kaufmann, San Francisco (CA).
5. Butz, M. & Pelikan, M. (2001) Analyzing the Evolutionary Pressures in XCS. In *Proceedings of the Genetic and Evolutionary Computation Conference 2001*, pages 927-934. Morgan Kaufmann, San Francisco (CA).
6. Butz, M. V. and Wilson, S. W. (2001) An algorithmic description of XCS. In Lanzi, P. L., Stolzmann, W., and S. W. Wilson (Eds.), *Advances in Learning Classifier Systems. Third International Workshop (IWLCS-2000)*, Lecture Notes in Artificial Intelligence (LNAI-1996). Berlin: Springer-Verlag (2001).
7. Dixon, P.W., Corne, D.W. & Oates, M.J.(2001) A Preliminary Investigation of Modified XCS as a Generic Data Mining Tool. In *Proceedings of the 2001 Genetic and Evolutionary Computation Conference Workshop Program*, pp345-350.
8. Fu, C., Wilson S. & Davis, L. (2001) Studies of the XCSI Classifier System on a Data Mining Problem. In *Proceedings of the Genetic and Evolutionary Computation Conference 2001*, page 985. Morgan Kaufmann, San Francisco (CA).
9. Hartley, A (1999) Accuracy-based fitness allows similar performance to humans in static and dynamic classification environments. In W. Banzhaf, J. Daida, A. E. Eiben, M. H. Garzon, V. Honavar, M. Jakiela, and R. E. Smith, editors, *GECCO-99: Proceedings of the Genetic and Evolutionary Computation Conference*, pages 266-273. Morgan Kaufmann, San Francisco (CA).
10. Holland, J. H. (1986). Escaping brittleness: the possibilities of general-purpose learning algorithms applied to parallel rule-based systems. In Michalski, R. S., Carbonell, J. G., & Mitchell, T. M. (Eds.), *Machine learning, an artificial intelligence approach.* Los Altos, California: Morgan Kaufmann.
11. Lanzi, P. (2001) Mining Interesting Knowledge from Data with the XCS Classifier System. In *Proceedings of the Genetic and Evolutionary Computation Conference 2001*, pages 958-965. Morgan Kaufmann, San Francisco (CA).
12. Parmee I. C., Cvetkovic C., Watson A. H., Bonham C. R. (2000): Multi-objective Satisfaction within an Interactive Evolutionary Design Environment. Evolutionary Computation. **8** (2), pp 197 – 222.
13. Parmee I. C., Cvetkovic C., A. H., Bonham C. R., Packham I. (2001): Introducing Prototype Interactive Evolutionary Systems for Ill-defined Design Environments. Journal of Advances in Engineering Software, **32** (6), Elsevier, pp 429 – 441.

14. Parmee I. C., Bonham C. R. (1999) Towards the Support of Innovative Conceptual Design Through Interactive Designer / Evolutionary Computing Strategies. Artificial Intelligence for Engineering Design, Analysis and Manufacturing Journal; Cambridge University Press, 14, pp 3 – 16.
15. Parmee I. C. (2001a), Poor Definition, Uncertainty and Human Factors – Satisfying Multiple Objectives in Real-world Decision-making Environments. First International Conference on Evolutionary Multi-criterion Optimisation (EMO 2001); Lecture Notes in Computer Science No 1993, Springer; pp 52-66.
16. Parmee I. C., (2001b), Evolutionary and Adaptive Computing in Engineering Design. Springer-Verlag, London.
17. Saxon, S. & Barry, A. (2000) XCS and the Monk's Problem. Lanzi, P. L., Stolzmann, W., and Wilson, S. W., eds. *Learning Classifier Systems. From Foundations to Applications* Lecture Notes in Artificial Intelligence (LNAI-1813) Berlin: Springer-Verlag.
18. Wilson, S.W. (1995) Classifier fitness based on accuracy,. *Evolutionary Computation*, 3(2), 149-175.
19. Wilson, S.W. (1998) Generalization in the XCS classifier system. In *Genetic Programming 1998: Proceedings of the Third Annual Conference* (pp. 665-674), J. Koza et al., eds., San Francisco, CA: Morgan Kaufmann.
20. Wilson, S.W. (2000) Get real! XCS with Continuous-valued inputs. Lanzi, P. L., Stolzmann, W., and Wilson, S. W., eds. *Learning Classifier Systems. From Foundations to Applications* Lecture Notes in Artificial Intelligence (LNAI-1813) Berlin: Springer-Verlag.
21. Wilson, S.W. (2001) Mining Oblique Data with XCS. In Lanzi, P. L., Stolzmann, W., and Wilson, S. W., eds. *Advances in Learning Classifier Systems. Third International Workshop(IWLCS-2000)*, (LNAI-1996) Berlin: Springer-Verlag.
22. Hurst, J. & Bull, L. (2000) Self-Adaptation in Learning Classifier Systems. UWE *Learning Classifier Systems Technical Report UWELCSG00-001*.

First Results from Experiments in Fuzzy Classifier System Architectures for Mobile Robotics

A.G. Pipe and B. Carse

Intelligent Autonomous Systems Laboratory
Faculty of Computing, Engineering and Mathematical Sciences
University of the West of England, Bristol, United Kingdom
Anthony.Pipe@uwe.ac.uk
http://www.ias.uwe.ac.uk

Abstract. We present first results from a comparison between a Fuzzy Classifier System operating at the level of whole rule-bases, and three variants of one that operates at the level of individual rules. The application domain is mobile robotics, and the problem is autonomous acquisition of an "investigative" obstacle avoidance competency. The Fuzzy Classifier Systems operate on the rules of fuzzy controllers with pre-defined fuzzy membership functions. Generally, all of the methods used were capable of producing fuzzy controllers with competencies that exceeded that of a simple hand-coded fuzzy controller that we had devised. The approach operating at the level of whole rule-bases yielded more robust and stable convergence on high performance solutions than any other architecture presented here. It is clear from the results that more work needs to be done to unravel the disappointing convergence dynamics of the algorithms operating at the level of individual rules.

1 Introduction

Evolutionary Computation and Reinforcement Learning are both powerful techniques that can be used to create algorithms capable of autonomously acquiring useful rules about a chosen problem domain. A well-established method that can use both techniques together is the Classifier System [1]. The Classifier Systems approach breaks down into two main techniques, those that operate at the level of individual rules [1] (typified by the Michigan-approach), and those that operate on whole rule-sets as composite entities [2,3] (typified by the Pittsburgh-approach). There are a number of ways to facilitate the use of these systems in solving problems that require real-valued numerical environmental interfaces [4]. Amongst these, those based on fuzzy logic [5,6] are both flexible and powerful. Although Classifier Systems research is in a quite mature state, for *Fuzzy* Classifier Systems there are still deep underlying issues to be settled for a given class of application.

We have chosen to conduct a programme of experimental work in the area of mobile robotics. This application area has characteristics that are complex but easy to visualise, it is widely known, it is a domain with which the authors have previous experience, and the results of the research could have some future use in the real world. This is a real-valued problem domain, and we have chosen fuzzy logic to

J.J. Merelo Guervós et al. (Eds.): PPSN VII, LNCS 2439, pp. 578–587, 2002.

implement local-cued behavioural control of a wheeled robot, the task therefore is to discover good fuzzy rules for implementing a particular competency in an artificial creature, or animat [7].

For the work reported on here, we have imposed some restrictions on its scope, in order to focus the experimental work on some specific topics. First, we allow modification of the fuzzy-rule base only, i.e., the membership function details are presumed already to be set by hand *a priori*, and are *not* the subject of tuning or optimisation. Second, we have looked at "Stimulus-Response" fuzzy systems only, i.e., there is no internal memory. Third, although environmental reinforcement is temporally linked, it is not delayed.

2 Related Work

As early as 1991 Valenzuela-Rendon [5,6] proposed a Fuzzy Classifier System architecture. In 1993 Bonarini had already utilised a Fuzzy Classifier System structure for mobile robot control [8]. Since then Bonarini has gone on to use Fuzzy Classifier systems to investigate a number of important related sub-topics, e.g., hierarchical structures [9], behavior-coordination [10], and reinforcement learning [11]. He has even used his techniques to help build Soccer-playing robot-teams [12]. There are a number of books and introductory texts on the subject of using Evolutionary Computation with fuzzy logic [13,14,15]. In our own work to date [16,17], we have been interested in developing and comparing different Fuzzy Classifier System paradigms in a mobile robot control context. This paper represents the next step in that process.

3 The Application

The work described below concentrates on making investigations into the abilities of an autonomous system to extract useful Stimulus-Response (S-R) behavior from environmental experiences. Such a controller must encapsulate an environmentally reactive competency. We have chosen an "investigative" obstacle avoidance competency for these experiments. Because the behaviors are to be S-R, any linkages between rules are made via the environment itself; there is no need to build internally linked behavioral sequences, and therefore the optimization and/or learning tasks are simplified. The current test harness is based heavily in real robot experimentation carried out in our laboratory. Details of the harness are given briefly below. However, the C source code is freely available on request to the email address above, or directly from our laboratory's web site.

3.1 The Simulated Robot

The following is a general description of the simulated twin-wheeled differential drive robot and its sensorimotor apparatus. The simulated environment assumes that a low-level control system is present, allowing control to be effected by an equivalent steering angle and forward velocity. The robot travels through its environment with a constant forward speed of 0.1 m/s and a maximum continuously variable turning speed of 0.5 rad/s. The robot has an array of five distance sensors. The set of distance measuring sensors form a five element array, set at the following angles from the "straight ahead" position: 0°, 22.5° to the left, 45° to the left, 22.5° to the right, and 45° to the right. The sensors have an 8-metre maximum sensing range.

3.2 The Simulated Environment

The environmental mazes are set on rectangles of any size, although for the experiments reported on in this paper, they are square, being 10-metres on each side. Any number of rectangular obstacles, of any dimension, may be placed in a maze. The start position may also be anywhere inside the maze. All measurements made and movements executed by the robot are continuous real valued, so for this simulation there is no concept of a "grid" or discretised state space.

3.3 Using Fuzzy Logic

The fuzzy controller has five inputs, one from each of the distance sensors and a single output defining steering angle.

The FLS is a "Mamdani"-style system [18]. A conventional distribution of unit-height triangular membership functions was chosen. All functions were identical and equally spaced, with the exception of each function placed at the end of the range of an input or output, as shown in figure 1. For fuzzy AND a product of membership function activations was used for a given rule as opposed to the simpler MIN operator, since it requires little extra processing and is known to produce superior interpolation properties. Defuzzification was performed by conventional centre of gravity calculations. The use of 3 membership functions at each input and 33 at the output was established during a number of preliminary experiments as being appropriate for this type of fuzzy controller in this application and incorporated into this test harness.

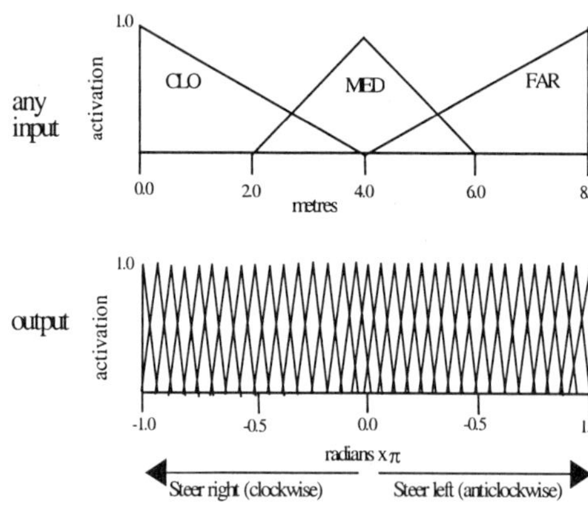

Fig. 1. Membership functions

Table 1. format for a fuzzy rule

0	22.5L	45L	22.5R	45R	OUT

Each fuzzy rule was of the form shown in table 1. Each of the six fields is an integer specifying a fuzzy membership function to use for that angled sensor input or the output in forming a rule. Counting is from left to right on each graph shown in figure 1, i.e., the interval (1-3) for each input and (1-33) for the output.

3.4 The Problem to Be Solved

It is worth stating that these experiments were not designed to test the limits of capability for these algorithms. Rather, they were designed to test the learning and convergence behaviour of the algorithms at only a sufficient level of complexity, so that *comparative* experimental observations have some worth.

The positioning of the obstacles in the maze space was intended to represent a "warehouse"-style environment (see figure 2). There is an area at the top-right that is quite "closed-in", a pair of parallel walls in the centre with two smaller openings, a collection of larger objects at the bottom-left, and some quite large "open" spaces. This environment was chosen for these experiments because it was found that it has some interesting characteristics that make obstacle-avoiding rule acquisition harder than for, say, "corridor"-style mazes. Different categories of rules are required in the "closed-in" areas, compared with the "open" spaces. This means that the architectures must be more efficient in storing knowledge in their rule sets, in order to have adequate behavioral coverage within the fuzzy logic system.

We desired this competency to include a tendency to explore environments, since we have wider aspirations for it as part of a latent learning behavior in the future. In this context, investigation of the environment must be encouraged for such a competency to be useful otherwise a stationary robot could be deemed highly fit for the purpose. Therefore the fitness functions for both architectures included a measure of the maximum "straight-line" distance traveled by the robot from the start location during a trial in the environment. Any environmental trial was terminated under either of two conditions, if a maximum time allocation of 200 simulated seconds was reached, or there was an environmental collision before this time.

4 The Fuzzy Classifier Systems

A major distinction among Classifier Systems is the way that the Evolutionary Algorithm (EA) is applied. With the so-called "Michigan" approach, the individual, as far as the EA is concerned, is a single rule or classifier (It should be noted that, for Michigan-style Classifier Systems the terms "classifier" and "rule", are used interchangeably below). An alternative approach, called the "Pittsburgh" approach, maintains a population of rule-sets: each individual as far as the EA is concerned is a complete assembly of rules encoded on an appropriate genotype. Clearly the role of the EA in the two approaches is different, as are the known difficulties (see [16,17] for a discussion). Indicative works using the Michigan approach include [1,5,6,19] and works using the Pittsburgh approach include [3,20]. In this paper we present results from using each of these techniques, and from using an architecture that is subtly different from both.

4.1 Pittsburgh System

In these experiments, the rule sets are evaluated for fitness by running a trial of the robot five times through a chosen simulated environment for each rule set in the population. When all rule sets have been evaluated in this way, a conventional Genetic Algorithm (GA) applies its operators to produce the next generation. This continues until, either the process is halted by the designer, or the maximum number of generations is reached. The rule set and population sizes were varied, as detailed below, but were initially set to 20 and 40 respectively. Crossover was single-point, with a probability of 0.9, and respecting rule boundaries. Mutation was two-point, one in the input space, and the other in the output space. If input space mutation was to take place, first a rule was selected randomly, then one of the input components with equal probability, and finally its specified membership function was modified with equal probability of selecting any membership function including the "don't care" #-value. Independently, mutation in the output space was evaluated. If this was to occur then one of the output membership functions was selected, with equal probability, to replace the existing one. Quite low values of mutation gave good results, and 0.01 was used throughout the experiments reported, evaluated separately for each point.

The fitness functions used for this architecture was a combination of the maximum "straight-line" distance traveled by the robot from the start location during a trial in the environment, combined with a measure of traveled distance over a route. These two factors were simply combined by multiplying them together. As described above, these values were averaged over five trials with different start locations.

4.2 The Michigan-Style Systems

In order to help alleviate the competition/co-operation problems that this type of system can be prone to, a simple form of population "niching" was used. The rules were divided into sub-populations, where each classifier in each sub-population has the same antecedent (including "don't cares").

In a Michigan-style Classifier System each rule has to have its own fitness value, and therefore an additional factor (to the Pittsburgh approach) related to cumulative activation of the rule during an environmental trial was included. The rule activity was simply accumulated over the trial and then multiplied by the same "distance traveled" factor used as the fitness function for the other architecture.

Although the sizes of the sub-populations were varied, as detailed in following sections, in all experiments reported on here there were 243 sub-populations, with 10 individuals in each sub-population. In these experiments we were interested in investigating the ability of the learning algorithm to derive versatile rule-bases, rather than its ability to tackle very large search spaces. For the problem as it is presently formulated, 243 sub-populations (5 inputs each specifying one of 3 fuzzy membership functions $\rightarrow 3^5$ input states = 243) can cover the entire input search space. However, this still leaves the problems of searching the output space for each rule, the issue of generalisation across the input space, and the subtle problems, mentioned previously, of interplay between rules. All rules in a sub-population begin with identical antecedents, and an output membership function selected randomly from the 33 possibilities. Each rule then has each of its antecedent components potentially modified by randomly changing it to a # "don't care" with 1/5 (i.e. 1/number-of-inputs) probability. The "don't care" policy described above meant that, for these Michigan-style approaches, identical rules could be created in different sub-populations. However, in these experiments this was not monitored. Mutation was carried out in the same way as that described above for the Pittsburgh approach, but with a much higher likelihood. In all experiments reported below the mutation rate was set to 0.1, i.e., ten times that used for the Pittsburgh experiments. Thoughts on why this difference was necessary are given later in the paper.

Of course this problem is greatly simplified by the small number of inputs (5) and the smaller number of input membership functions used in each input (3). It is this that allows for the input space to be completely covered by only 243 population niches. Here are two thoughts on this arrangement. First, we must reiterate that it is not very large problems we are interested in tackling for this work, rather, we are interested in observing convergence dynamics and the quality of solutions found by different approaches. Second, it is clear that, later on, the method could be broadened to larger input spaces by utilising a clustering approach (e.g. k-means or fuzzy clustering) to gather rules into groups. However, this is beyond the scope of the current work.

From this common base, three different approaches were adopted for forming fuzzy controllers from a rule-base of this type. For identification purposes, they are referred to below as the "typeN Michigan System", where N = {1 –3} as itemised below.

1 The first was an off-line method. A rule was selected from each sub-population in a deterministic manner before each environmental trial, thus making a 243-rule fuzzy controller each time. The robot was then run through five trials of the maze environment (from different start locations), acquiring reinforcement. This was repeated up to the size of the sub-populations, until each rule in each sub-population had been teamed together with an individual from another sub-population. The next set of rules were then formed by choosing again from each sub-population and again the robot was run through the maze. This process continued 10 times, i.e., until all rules had been used once as part of a fuzzy controller, and acquired some environmental feedback. The GA was then run within each sub-population, i.e., across these fuzzy-controller rule-sets, and then the whole process was repeated until some stopping condition was reached. This approach is clearly *not* a standard Michigan Classifier System one. However, operation of the GA and reinforcement are both still at the level of individual rules. Rules compete within sub-populations and co-operate across sub-populations.

2 The second method was on-line. A rule was selected from each sub-population on-line, at each time step of an environmental trial. Selection was based on a "bidding" system between members of a sub-population. This approach is more akin to traditional Michigan Classifier System methods. Bid strength was based on a combination of instantaneous rule activation and the cumulative fitness level being acquired by each rule during a trial for evolutionary selection purposes. The relative amounts each of these factors was varied as one of the experimental factors, but they were of roughly equal weight for all the experiments reported on here.

3 The third method was a combination of the first two. For some pre-set number of generations, the system would operate in the first mode, and then switch to operation in the second mode thereafter. This third approach was adopted after observing the convergence dynamics of the first two. The method that was more akin to traditional Michigan Classifier System rule selection, was *very* erratic. The first approach was more stable, perhaps because it deterministically gave *all* rules the opportunity to acquire fitness over a complete maze evaluation. As a "training regime" it was, therefore, quite useful.

5 Experimental Evaluations

There is only space below to describe some selected results from the range of experiments that have been conducted. They each illustrate a noteworthy strength or weakness.

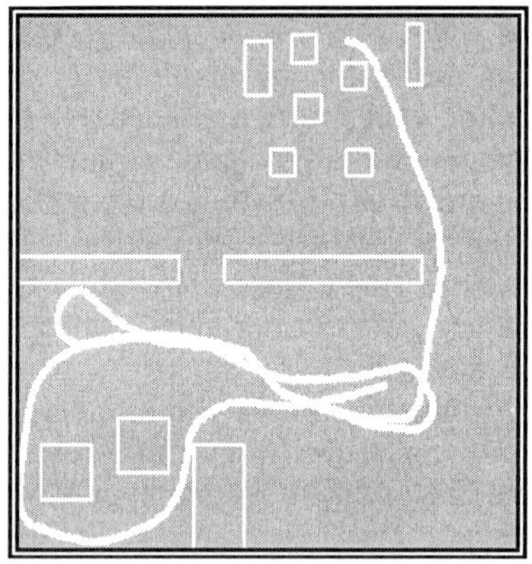

Fig. 2. A typical tuned fuzzy controller

First, it should be pointed out that each of these architectures was capable of generating best fuzzy controllers of similar performance. To save space, we therefore show only a single typical resulting trial of an environment, as shown in figure 2. This figure is actually showing a Type1 Michigan System controller. It was derived by selecting the 8th niche member from each of the 243 sub-populations at generation 16 (a local upward "blip" in the best individual's fitness), i.e., after this group of rules had been through 16 maze trials (each trial consisting of 5 tests at different start locations). The robot trajectory starts at the top right and only stopped when the maximum trial-time was reached. Learning was turned off for this demonstration run, and the start location, whilst similar to one of the five used for fitness evaluation trials, was different from any that had been used during prior learning and evolution. This controller was drawn from the same run illustrated in figure 4 below, which gives an indication of the performance of other controllers; one can see, even in this run, that there were better controllers created in earlier generations.

5.1 Pittsburgh System

Figure 3 shows a typical convergence characteristic for this architecture. Data has been averaged over 5 separate runs of the system, with the population initialised using different random seeds each time. For these runs, the population size was 40 and the trial was run for 20 generations, thus resulting in 800 sets of fitness evaluations for each run (remember that each individual fitness evaluation was in turn averaged over 5 trials from different start locations). In subsequent generations, the population converged fully on a single solution, with only minor disruption from the low mutation rate.

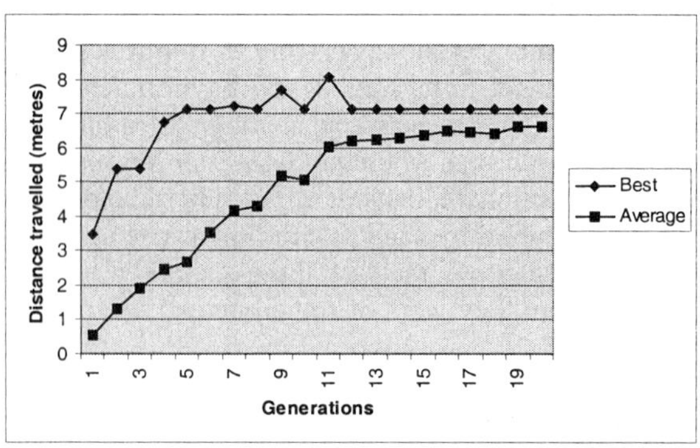

Fig. 3. A typical convergence path for the Pittsburgh System

5.2 Michigan Systems

Although there were minor detailed differences in the resulting best-performing rule-bases for given parameter values across these three architectures, there was no clear advantage to one or another method of choosing a fuzzy controller from the sub-populations. However, there did seem to be some advantage to using the off-line selection technique of the Type1 or Type3 Systems in terms of convergence dynamics (for Type3 we always switched from Type1 to Type2 at generation 15). Although we have not yet fully quantified this observation, it certainly seemed to be the case that these systems were more reliable in finding high performance solutions at some time during a run of the system. This may have been due to the previously mentioned characteristic that *all* rules have a chance at full participation in control for complete environmental trials.

However, the weakness that all these systems displayed, can be seen in the convergence plot shown in figure 4, which was averaged across five runs of the system. It is clear that the population does not have any tendency to converge on a high performance solution. Figure 4 only shows results up to generation 20, but runs 10 times longer than this produced no apparent convergence behavior or significant improvement in average fitness. In fact, in order to generate the occasional occurrence of rule-bases with performance at the same or better level than the Pittsburgh system, it was necessary to have a very high mutation rate of 0.2 or even 0.3. At these levels, mutation was so disruptive that average performance would normally peak somewhere in the early generations, then genetic material would be lost from then on. Figure 4 shows an example where the number of sub-populations was set to 40, i.e., the total number of fitness evaluations sets was 800, as in the similarly illustrated Pittsburgh case. The mutation rate was set to 0.1 for these runs. Neither decreasing the sub-population size, nor increasing it, produced significantly different results.

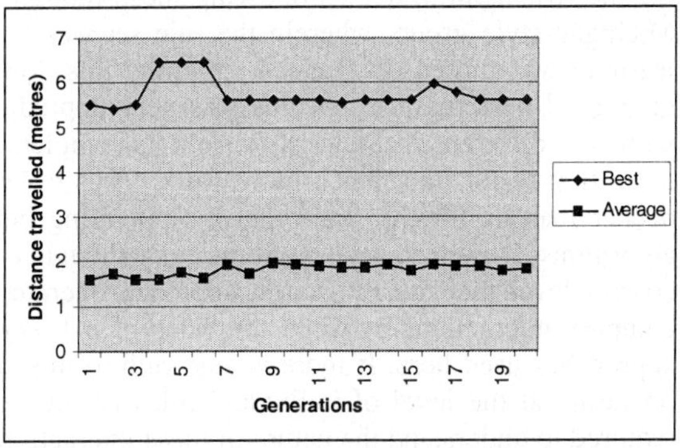

Fig. 4. Typical convergence for Type1 Michigan System

6 Discussion

All four approaches tested were able to produce good robust controllers of comparable performance. Further, they were able to achieve this in comparable numbers of fitness evaluations. However, there were other, more detailed, strengths and weaknesses. Convergence dynamics of the Pittsburgh approach were very stable compared to any of the three Michigan-style approaches, this could augur well for this approach as the problem size is scaled up. Although use of the Type1 algorithm in the Michigan family of approaches by itself, or as part of the "training regime" as in

Type3, improved performance, all three displayed erratic convergence dynamics. In most cases there was a peak of performance, followed by no further improvement, or even gradual loss of genetic material at the high mutation rates required to achieve very high peak performance. Another major difference between the rule bases operated on by the Pittsburgh and Michigan approaches, was the size of rule base used. In the results presented above, the Pittsburgh system was operating on sets of only 20 rules. However, the Michigan-style approaches all used a 243-rule set as the controller. In fact, when we tried a brief experiment on the Pittsburgh system with a rule set size of 240, convergence did seem to be reduced. We tentatively hypothesized that fuzzy controller rule-interaction for a large rule-set like this could be detrimental for the Michigan system. Following these thoughts, we tried an approach with the Michigan-style group, wherein the rule set was "cropped" of its weaker rules after learning was turned off. Indeed, cropping this down to the strongest 20 rules could give good performance, but this was very unpredictable, often yielding a controller with very different characteristics from its complete 243-rule "parent". It is clear from these first results that much more work needs to be done in this area to uncover the deeper reasons for the somewhat disappointing behavior of the Michigan group of algorithms. However, we have some early thoughts. It could be the case that the much coarser level that the Pittsburgh approach operates at is less disruptive to the very complex interactions between the rules of a fuzzy system than the Michigan-style approaches used here. If there is any truth to this statement, this does not mean that operating at the level of individual rules is not worthwhile, just that more work is required to understand the nature of these interactions.

It is worthy of note that, generally, all of these methods were capable of producing fuzzy controllers with competencies that exceeded that of a simple 13-rule hand-coded fuzzy controller that we devised. This controller was unable to tackle the transitions from "closed-in" to "open-spaces" in a robust manner, indicating that these environmental transitions are one of the difficult aspects of the problem posed here.

7 Conclusions and Further Work

Although more work needs to be done in establishing a more capable hand-coded controller, the final comments of the previous section are, we hope, a good sign for the future of these algorithms.

The Pittsburgh approach yielded a more robust and stable convergence on high performance fuzzy rule bases. In the immediate future, it is clear that more work needs to be done to unravel the underlying reasons for some of the observed convergence dynamics, especially for the Michigan-style family of methods.

In the longer term, it is important that we test these learning architectures on their ability to extract more complex competencies, including latent learning tasks. In many cases this will require a solution to the problem of building internally linked behavioral sequences; i.e. a fuzzy rule-base that is able to encapsulate internal state information in some form. For other more complex competencies, there will be a need for a larger number of states for each rule antecedent and possibly a larger number of inputs. This would be a worthwhile test of these learning architectures on larger search spaces.

Finally, we would like to re-focus the application problem as it is presented here, so that it is a more suitable testing platform for a comparison between accuracy-based fitness and strength-based fitness.

References

1. Booker L, Goldberg D & Holland J (1989). Classifier Systems & GAs, AI 40, pp.235-282.
2. Smith S F (1980). A learning system based on genetic adaptive algorithms. PhD thesis, Univ. Pittsburgh.
3. Carse B, Fogarty T C & Munro A (1996). Evolving Fuzzy Rule-based Controllers using Genetic Algorithms. Fuzzy Sets and Systems 80(3), pp.273-293.
4. Bonarini A (1999). Fuzzy and Crisp Representations of Real-valued Input for Learning Classifier Systems. Procs. GECCO 99, Morgan-Kaufmann, pp.52-59.
5. Valenzuela-Rendon M (1991). The Fuzzy Classifier System: Motivations and first results. Parallel Problem Solving from Nature (PPSNII), Springer-Verlag, pp.330-334.
6. Valenzuela-Rendon M (1991). The Fuzzy Classifier System: a Classifier System for Continuously Varying Variables. Procs. 4th Int. Conf. on Genetic Algorithms, pp.346-353.
7. Wilson S W (1987). Classifier Systems & the Animat Problem. Machine Learning 2 (3), pp.199-228.
8. Bonarini A (1993). ELF: Learning incomplete fuzzy rule sets for an autonomous robot. Procs. 1st European Congress on Intelligent Technologies and Soft Computing (EUFIT '93), pp.69-75.
9. Bonarini A (1997). Anytime learning and adaptation of hierarchical fuzzy logic behaviors. Journal of Adaptive Behavior 5(3-4), pp.281-315.
10. Bonarini A & Basso F (1997). Learning to coordinate fuzzy behaviors for autonomous agents. Int. Journal of Approximate Reasoning, F. Herrera (Ed.), 17(4), pp.409-432.
11. Bonarini A, Bonacina C & Matteucci M (2001). An approach to design of reinforcement functions in the real world, agent based applications. IEEE Trans. SMC. In press.
12. Nardi D, Adorni G, Chella A, Clemente G, Pagello E & Piaggio M (2000). ART – Azzuro Robot Team. Robocup99 – Robot Soccer World Cup III, Springer-Verlag.
13. Geyer-Schulz A (1997). Fuzzy Rule-Based Expert Systems and Genetic Machine Learning. Series: Studies in Fuzziness & Soft Comp., vol. 3. Springer-Verlag, ISBN: 3790809640.
14. Pedrycz W (1997). Ed: Fuzzy Evolutionary Computation. Kluwer, ISBN: 0792399420.
15. Bonarini A (2000). An Introduction to Learning Fuzzy Classifier Systems. P.L. Lanzi, W. Stolzmann and S.W. Wilson (Eds.), Learning Classifier Systems- from Foundations to Applications, Lecture Notes in AI, pp.83-104. Springer Verlag Berlin Heidelberg.
16. Pipe A G & Carse B (2000). Acquisition of Fuzzy Rules for Mobile Robot Control: 1st Results from 2 Evolutionary Computation Approaches. GECCO00, pp.849-856.
17. Carse B & Pipe A G, (2001). X-FCS: a fuzzy classifier systems using accuracy based fitness – 1st results. Procs. Int. Conf. Fuzzy Logic and Technology, EUSFLAT, pp.195-198.
18. Mamdani E H & Assilian S (1975). An experiment in linguistic synthesis with a fuzzy logic controller. International Journal of Man-Machine Studies, vol. 7, no. 1, pp.1-13.
19. Parodi A & Bonelli P (1993). A New approach to fuzzy classifier systems. Procs. of 5th International Conference on Genetic Algorithms, pp.223-230.
20. Hwang W & Thompson W (1994). Design of Fuzzy Logic Controllers using Genetic Algorithms. Procs. of the 3rd IEEE Int. Conf. on Fuzzy Systems, pp.1383-1388.

TCS Learning Classifier System Controller on a Real Robot

Jacob Hurst, Larry Bull, and Chris Melhuish

Intelligent Autonomous Systems Laboratory,
University of the West of England,
Bristol BS16 1Q1, UK
jacob.hurst@uwe.ac.uk

Abstract. To date there have been few implementation of Holland's Learning Classifier System (LCS) on real robots. The paper introduces a *Temporal Classifier System* (TCS), an LCS derived from Wilson's ZCS. Traditional LCS have the ability to generalise over the state action-space of a reinforcement learning problem using evolutionary techniques. In TCS this generalisation ability can also be used to determine the state divisions in the state space considered by the LCS. TCS also implements components from *Semi-Mark-Decision Process* (SMDP) theory to weight the influence of time on the reward functions of the LCS. A simple light-seeking task on a real robot platform using TCS is presented which demonstrates desirable adaptive characteristics for the use of LCS on real robots.

1 Introduction

Learning Classifier Systems (LCS) were invented by John Holland (Holland 1976). Traditionally an LCS is a parallel production rule system, which uses a reinforcement learning technique (Sutton & Barto 1998) to change the fitness of the rules contained in the system and a genetic algorithm (GA)(Holland 1975) to search the rule base for new useful and *general* rules. There have been several implementations of LCS on robotic software agents e.g. (Donnart and Myer 1996a) (Stolzmann & Butz 2000). There have been fewer LCS implementations on real robots. Those known to the authors are Dorigo and Colombetti's work on *behavioural shaping* (Dorigio & Colombetti 1997), Donnart and Myer's MonaLysa algorithm (Donnart and Myer 1996b) and Katagami and Yamada's approach with an interactive LCS (Katagami & Yamada 2002).

An LCS can be looked upon as an algorithm that solves reinforcement learning problems and provides generalisations over the state-action space. One of the major problems examined in this paper is that of defining the significant states of the space. This could easily be done by hand but in doing so the programmer is delineating and modelling the world. This is a potential problem for truly autonomous robotic systems. TCS provides a method to bypass this problem by exploiting and extending the evolutionary generalising methodology of LCS to generalise over and *create* state boundaries. TCS also exploits the recent advances in *Semi-Markov-Decision-Process*

J.J. Merelo Guervós et al. (Eds.): PPSN VII, LNCS 2439, pp. 588–597, 2002.

(SMDP) theory to weigh the influence of time on the reinforcement functions of the LCS.

The layout of the paper is as follows: description of the robotic platform, description of the TCS algorithm, description of the robotic task, results from the TCS algorithm. Finally all findings are discussed.

2 The Robot

For this work the robotic platform is a "Linuxbot" developed at the University of the West of England. It consists of a three-wheeled robot, two of which are powered. It is controlled by a 25MHz 386 processor with a 1.8 Mbyte hard drive. Communication between the robot and the outside world is via a radio LAN. The most distinctive characteristic of the platform is that it runs the linux operating system. This platform can run a web server as well as provide access via telnet sessions, while carrying out simple robotic tasks. In the experiments described the sensory input is fairly rudimentary consisting of three light sensitive resistors for the light detection task. This simple sensory setup can easily be extended to include video cameras and other more complex input.

3 The TCS Algorithm

Our approach is very firmly based on the philosophy of elegance and simplicity in ZCS. It differs from ZCS as it contains a different idea of action persistence and the reinforcement-learning algorithm factors in time. Thus, to simplify nomenclature a new name has been introduced - "Temporal Classifier System" (TCS). For a full explanation of the underlying ZCS the reader is referred to (Wilson 1994).

One of the motivations behind basing TCS on ZCS is that ZCS "keeps much of Holland's original framework but simplifies it to increase understandability and performance." (Wilson 1994). Recent work (Bull & Hurst 2002) indicates that ZCS can perform optimally in simple multi-step maze problems when the fitness sharing parameters are set correctly.

Usually an LCS matches an input and then carries out a fixed length action. When the action is finished the reinforcement and performance cycles take place. There are two main problems with cycling the LCS on fixed actions in dynamic robot environments. The first is that while the robot is carrying out the task it is unresponsive to any changes in the environment. The second problem is that after carrying out the fixed length action the robot may not be in a different state, this will result in the algorithm trying to learn in situations when nothing significant has occurred. However, a different approach is possible (Cliff & Ross 1995) (Booker 1990); instead of using a fixed action or a fixed time step the LCS carries out its action until there is a significant change in sensory input between the state when the action is first taken and the current state. For complex sensors there is therefore a distance of significance between the current sensory values and the sensory values representing the next event. Previous approaches using real robots, e.g. (Asada et al 1996), have broken the world up by hand defining a range of input values to represent a certain event type. This approach can work well, but as noted before, it is a limiting factor for truly autonomous learning systems. Further, the programmer's model will have to take into account the idiosyncrasies of the electronic hardware, i.e. sensors often differ in their responses to the same stimuli. The distance of significance may also not be constant in all situations and to a certain extent represents how dynamic the robotic environment is. So, the model when built may be different from robot to robot and situation to situation. One approach to this problem has concentrated on producing mathematical techniques to automate the process of determining the extent of the state-space (Uchibe Asada Hosoda 1997). Hence, in addition to the reinforcement learning technique an *event pre-processor* or filter runs on top of the generalisation system.

TCS works by exploiting and extending the inherent generalization mechanism of LCS from working over a predefined state-action space to defining states in the entire state space. That is by generalizing over the state space the LCS determines what are and what are not critical sensory *events.* The next two sections examine the changes made to ZCS to produce TCS. This extension of the generalization mechanisim has some similarity in approach to use of Hamilton-Jacobi-Bellman equations with function approximators (Doya 2000).

3.1 Time Needs to Be Considered in the Reinforcement Function

When using continuous actions to move from state to state time needs to be considered in the reinforcement function. For example, this happens when more than one action can move the robot from a given state to another state. Consider two hypothetical states, A and B. To move from state A to state B there are two possible actions I and II. Action I takes 5 seconds to get to state B, the other takes 1 hour.

This sort of problem is known as a *Semi-Markov-Decision-Problem.* It is no longer a *Markov* problem where standard reinforcement learning algorithms can be applied. The influence of time needs to be factored into the reinforcement equations. In effect we are trying to model discrete event systems, where significant events occur at discrete intervals, however the amount of time between each events is a real valued number. Explanations of SMDP can be found in (Parr 1998), (Bradtke and Duff 1995) and (Sutton, Percup & Singh 1999).

To tackle such problems Bradtke and Duff (1995) use a variable discount factor. The factor, instead of being a discounted sum of future rewards, becomes a discounted integral of future rewards. Parr(1998) has also presented a continuous time version of standard Q-learning (Watkins 1989) where this integral can be removed from the equation, as essentially the state values are constant when computed (see (Parr 1998) for a full explanation). The equation below is from (Parr 1998), it indicates how Q learning can be modified for SMDP.

"On a transition from state s to s' under action a that has taken time t and received reward r (which is assumed to be appropriately weighted sum of rewards received during t):"

$$Q(s,a)^i \leftarrow Q^{i-1}(s,a)+\alpha^i(s,a)(r+\beta^t V^{i-1}(s')-Q^{i-1}(s,a))$$

β is a discount rate which varies with state and action. Parr(1998) also gives a convergence proof for this equation.

So, a number of changes need to be made to the reinforcement learning equations in LCS. The external reward *r* is discounted by the total time taken t^t to reach the goal:

$$r = e^{-\sigma t^t} r \quad (1)$$

The discount factor γ also factors in time, but here the time transition which occurs *between* events t^i is considered.

$$\gamma^t = e^{-\eta t^i} \quad (2)$$

σ and η are in effect different learning rates to change the emphasis placed on the time taken to obtain the external reward and the time taken between event transitions.

The original update equation for ZCS from (Wilson 1994) is:

$$S_{[A]} \xleftarrow{\beta} r_{imm} + \gamma S_{[A]'} \quad (3)$$

Where $S_{[A]}$ is the fitness of the current action set, r_{imm} is the immediate reward, γ is the discount factor and $S_{[A]'}$ is the fitness of the previous action set. $\xleftarrow{\beta}$ can be considered as the Widrow Hoff Learning procedure (consider the case of two scalars x and y).

$$x \xleftarrow{\beta} y \equiv x \leftarrow x + \beta(y - x)$$

The new update equation for TCS incorporates Equations (1) and (2) into (3):

$$S_{[A]} \xleftarrow{\beta} e^{-\sigma t^t} r_{imm} + e^{-\eta t^i} S_{[A]'} \quad [4]$$

Equation 4 is the update equation for the new learning classifier system TCS. These equations fulfil the requirement of "appropriately weighing the reward" and "discount rate which changes with state and action" (Parr 1998).

3.2 The Action Selection Mechanism of TCS

Initially the action selection mechanism is very much like ZCS, i.e. it uses a roulette wheel to select the action based on the fitness of the rules in the current match set. The difference is how the decision is made to continue with the current action or try to select a new action. First an action is selected and an action set is formed [A]. Input is then sampled continuously from the environment while the robot is carrying out the ordained action.

In the case where none of the current members of [A] match the most recent input the procedure is straight forward: the current members of the action set are removed and either obtain an external reward or are moved to the previous action set where they receive an internal reward. A new action is then selected.

In the case where all the classifiers in [A] match the current input, the effect is simple to continue with the current action.

In the situation where only some of the classifiers in the action set match the current input a decision has to be made, either to continue with the action or stop. The current action set is therefore sub-divided into two sets - those members of the action set still matching the current input are placed in the *continue* set [C] and those which don't are placed in the *drop* set [D].

Roulette wheel selection is then used over the entire action set. If the rule selected is a member of the *continue* set the rules making up the drop set are removed from the current action set and do not receive any reinforcement. The action continues to be taken.

However, if the rule selected is a member of the drop set, all members of the continue set are removed from the action set. The remaining action set is then reinforced with either external reward or moved to the previous action to be reinforced internally. A new action then needs to be selected from the current match set if the robot has not reached the goal state and the cycle continues.

The system creates appropriate generalisations by relying on the existence of a population of varying degrees of generalisation within useful action sets. As the population probabilistically drops out to test more specific rules, the correct less general rules should increase in proportion to the incorrect more general rules. If these more specific rules do not exist the usual covering mechanism compensates.

This system has some similarities to Cobb and Grefenstette's (1991) action persistence scheme, recently implemented in XCS, (Barry 2000). In this setup each action is carried out for a fixed length of time. This length of time can vary between classifiers. A problem with this method is that it is unresponsive to changes in the environment during that time period; this system would be unable to respond to a dynamic environment.

In this paper the task that TCS is being required to do is to move to a goal state. The goal state is the only event predefined by the programmer. Another further complication to the system is that, as the actions are continuous, it could be possible for the robot to discover a general rule forcing it to ignore all stimulus and cycle continuously. As the robot would never drop out of this action cycle to receive reinforcement this would continue forever. To prevent this there is a maximum time that the action is allowed to continue before a drop out is forced. In the current set up this is set to 15 seconds; far longer than the maximal distance the robot is required to travel.

3.3 The Rule Representation for TCS

Representation in LCS traditionally takes the form of ternary strings. This is by no means the only representation open to LCS, a number of other options have been explored; real numbers e.g. (Wilson 2000), S-expressions (Ahluwalia & Bull 1999) (Lanzi 1999), neural networks (Bull & O'Hara 2001), piece-wise linear approximations (Wilson 2001), and fuzzy logic (Valenzuela-Rendon, 1991). The problem that TCS is being asked to solve involves three light sensors each providing real number input to the LCS. The matching part of each classifier therefore consists of six real numbers, each pair of numbers representing a range between which the classifier recognises input for each respective sensor. To cope with this representation the discovery mechanism of TCS has to be altered slightly. When a covering classifier needs to be created ranges are generated randomly to match the environmental conditions. Within the GA the crossover operator is unchanged, and the mutation rate changes to a step-wise creep mutation on the ranges, selected from a random distribution between +/- [0, 0.1]. The action part of the classifier's representation is an integer which can be of three values 0,1,2 , representing the actions move forward continuously, and move right and move left continuously.

The parameter settings used in TCS in this paper are as follows: N= 500, β=0.5, r=10000,χ=0.05, μ=0.05, ϕ=0.5, p=0.25, S_o=25.0, τ=0.1, σ=0.2, η=0.5.

4 The Robotic Task

The task used here to test TCS is a simple one: move from a start state towards the light until the centre sensor reads a specific value. Three light sensitive resistors are placed on top of the robot. Halogen lights are placed at one end of a 2.6m x 1.8m arena. The start state for each experiment is directly in front of the lamps at a distance of 2.4m with random orientation.

The experimental procedure is as follows: the robot is allowed to move about until it reaches the reward state or hits the arena wall. If it hits the arena wall it is returned to the start state. When in the start space, the robot is turned left or right. The extent of this turn is determined randomly between 0 and 80^0. The random turn is limited as much over 80^0 risks starting the robot in a situation where its frontal sensors have no readings. TCS, like ZCS, does not have any mechanism to cope with these sort of non-Markov environments. After the random turn the TCS controller takes over operations.

The reward state occurs when the centre sensor on the robot reads 0.85. This reward state occurs in an area that extends approximately 45 cm around the two-halogen lamps. When it reaches the reward state that run of the experiment is over and the robot is returned to the start state.

Hence there are two principle objectives. The robot must get to the light source as quickly as possible and take as few decisions as possible. Only two actions are needed to solve this problem optimally, a turn to the left or right to correct the random turn at the start (the turn must stop so that the robot is facing straight towards the light) where the next action is to move straight ahead ignoring all changes in sensory input until the robot reaches the reward state. That is, the correcting turn involves recognising a slight change in input as important and the moving ahead action involves ignoring large-scale changes in input.

The time taken on average for the robot to reach the light if the actions are selected optimally should be around 6.5 seconds.

5 Results

The results presented are the averages of five runs. Each run involved recording 150 successful trips from the start position to the reward area close to the light. The data presented is the "raw" values, i.e. unlike most LCS maze experiments a running average is not used. The motivation behind this choice is to clearly indicate the online performance of the system. These runs took around 90 minutes.

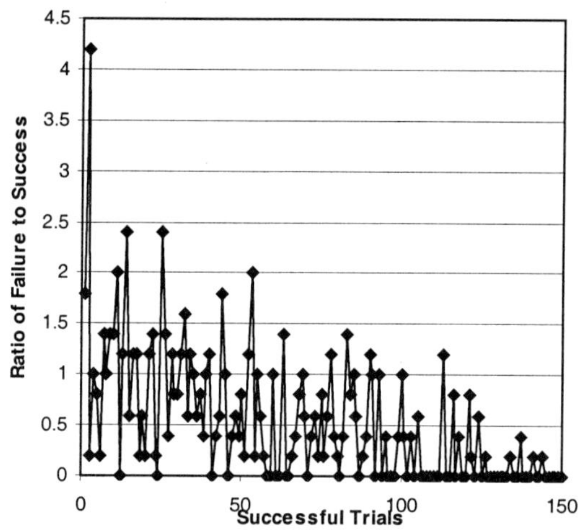

Fig. 1. Showing the ratio of failures to success

Figure 1 displays the ratio of failures to success. At the start of the run there is a large proportion of failure for every success. For each successful trial there are around 4.25 failures where the robot hits the arena wall. As the run proceeds this ratio becomes close to zero. It does not absolutely reach zero as TCS uses a probabilistic action selection mechanism both in the initial selection of an action and in the drop decision as detailed earlier.

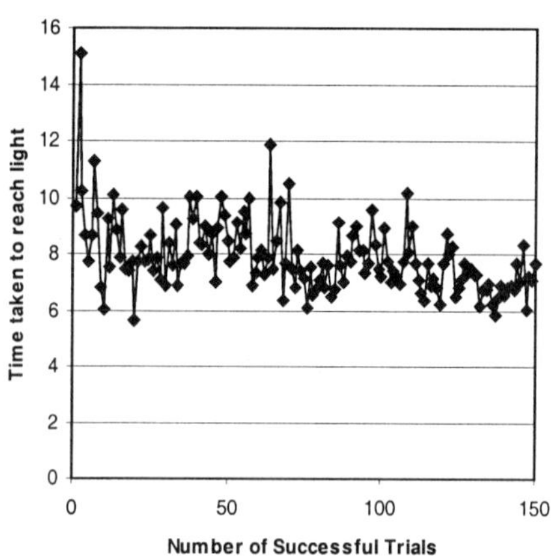

Fig. 2. Time taken to reach reward

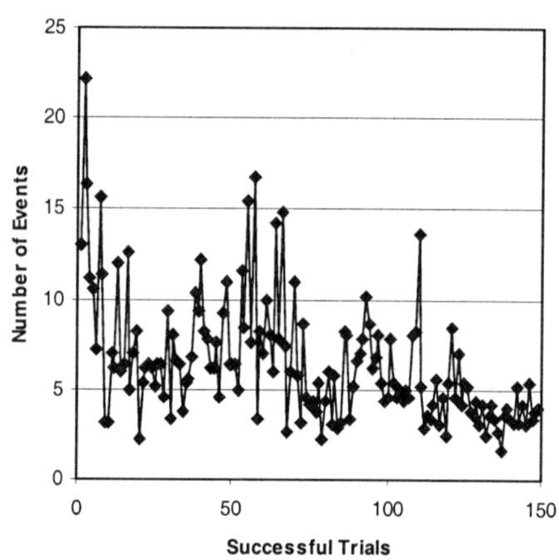

Fig. 3. Number of events taken to reach reward

Figure 2 indicates the time taken for each successful trip to the light. During the course of the run there is a downward trend. At the start of the run there is greater use of more time consuming routes to the reward, towards the end of the run the more efficient routes become more dominant.

Figure 3 shows the number of events needed to move the robot to the reward. At the start of the run there is large amount of variation in the number of events but this variation reduces over time.

The learning in the robot appears to move through three phases. At first the robot learns the corrective actions immediately around the reward area, where the robot first gets to these reward areas by near random action selection. The next step is to learn the rules to reach these areas. This is mainly the move forward rule that is later to be applied further and further from the reward area. When this rule is established the corrective turns at the start can be learnt. At the end of the run, when only a small number of events are occurring, the robot responds to its random turn with a corrective turn to face towards the light. To obtain this precise angle takes at first several corrective turns after which the robot moves directly to the light. Eventually this is reduced to around two events per run, slightly more than optimal since when facing the light the robot actually needs only to move forward. It must be noted that a stochastic action selection policy is used throughout as this is suggested as more appropriate for real autonomous entities (Bull & Hurst 2002). Examination of the resulting rule bases shows optimal knowledge exists.

The normal situation when displaying LCS performance results in simulated mazes is to run the experiment for several thousand trials (e.g. (Wilson (1994)). Hence the results displayed here do not completely converge on balanced solutions (fitness sharing) as described in (Bull & Hurst 2002). The time taken to do these tests makes runs in the thousands an impractical proposition.

There are numerous parameters controlling TCS and all LCS in general. The settings of these parameters are essential to getting ZCS to work optimally (Bull & Hurst 2002). These settings are problem specific and there are no heuristics to guide practitioners. The authors have had some limited success with the use of a self-adaptive mutation rate (Bull *et al* 2000) but this has not occurred with the reinforcement learning parameters (Hurst & Bull 2001). Experimentation with different parameter settings is difficult in a robotic environment and future work will consider the use of simulation in conjunction with the real robots. Two new parameters controlling the emphasis placed on time have been introduced, their precise role and interaction is also a new area for investigation.

Another critical factor is the manner in which over-general rules are coped with by TCS. Consider the situation where an action set contains a number of rules which no longer match the input condition so form the drop set and an over general rule that still matches the current input. If the decision is made to continue with the action the incorrect over-general rule and the correct specific rule receive a similar level of penalty. That is, all members of the action set have paid out β to the bucket. The members of the drop set will receive no further reward, while the over general rule may receive a small level of reward from any subsequent actions. Therefore, the incorrect over general rule may receive a payoff while the correct specific rules only pay a penalty. These issues in TCS are currently under investigation. More complex tasks are also being carried out using TCS and the robotic platform.

6 Conclusion

The introduction and description of the TCS algorithm provides several points of interest. It is the first LCS that has been inspired from current ideas of semi-markov decisions processes that are emerging from current reinforcement learning theory and attempts to cope with time in a systematic way are made. Further it is event driven where this determination of significant events (changes in state) is obtained by utilising the standard evolutionary computing-based generalisation mechanisms of LCS. The results in the robotic environment are promising: the robot first learns to find the light source before it learns to minimise the time taken and the number of states that need to be considered. This implementation is the first robotic implementation using a real numbered classifier representation. It is also the first robotic implementation of an LCS where external reward is presented only at the goal state, so removing the need for a reward function to provide external reward at each "step". The results presented here demonstrate an LCS robot controller operating at a very low level of problem representation. Not only does the input to the robot consist of floating point numbers, but with the exception of problem start and end states no further state definition is carried out by the programmer. Clearly many questions remain open about the use of TCS including the exact manner in which time is weighted, as has been touched upon earlier. Perhaps of more concern is the time taken for TCS to converge to a solution, which is currently long. Another question of more theoretical interest is how TCS fits in with reinforcement learning's *option* theory (Sutton, Percup & Singh 1999). These and other questions, and considerations of more complex tasks, are becoming a focus for our endeavours.

References

Ahluwalia,M & Bull,L (1999) A Genetic Programming Based Classifier System. In W.Banzhaf, J.Daida, A.E.Eihen, M.H.Garzon, V.Honavar, M.Jakiela & R.E.Smith (eds) *Proceedings of the Genetic and Evolution Computation Conference* –GECCO-99. Morgan Kaufmann, pp11-18.

Asada,M, Noda,S, Tawaratsumida,S Hosoda,K (1996) Purposive Behavior Acquisition for a Real Robot by a Vision Based Reinforcement Learning. *Machine Learning Vol23* pp279-303.

Barry, A (2000) Specifying Action Persistence within XCS. In D. Whitley, D.Goldberg, E.Cantu-Paz,L. Spector, I.Parmee & H-G Beyer (eds) *Proceedings of the Genetic and Evolutionary Computation Conference.* Morgan Kaufmann, pp50-57

Booker,L (1990) Instinct as an inductive bias for learning behavioral Sequences In Meyer,J,A and Wilson (eds) *From animals to animats-The first International Conference on Simulation of Adaptive Behavior,* MIT Press

Bull, L., Hurst, J. & Tomlinson, A. (2000) Self-Adaptive Mutation in Classifier System Controllers. In J-A. Meyer, A. Berthoz, D.Floreano, H. Roitblatt & S.W. Wilson (eds) *From Animals to Animats 6 - The Sixth International Conference on the Simulation of Adaptive Behaviour*, MIT Press.

Bull L & Hurst. J (2002) ZCS Redux *EvolutionaryComputation* In Press

Bradtke, S.J & Duff, M.O (1995) Reinforcement Learning Models for continuous-time Markov decision problems, *Advances in Neural information Processing Systems 7* MIT Press pp393-400

Cobb,H & Grefenstette,J (1991)Learning the Persistance of Actions in Reactive Control Rules *Proceedings of the 8th International Machine Learning Workshop*. Morgan Kaufmann pp293-297

Cliff,D & Ross,S (1995) Adding Temporary Memory to ZCS *Adaptive Behavior* 3(2) 101-150

Dorigo, M , Colombetti,M (1997) *Robot Shaping*, MIT Press

Donnart, J.Y. Meyer, J.A. (1996b). Learning Reactive and Planning Rules in a Motivationally Autonomous Animat. *IEEE Transactions on Systems, Man, and Cybernetics - Part B: Cybernetics*. 26(3), 381-395.

Donnart, J.Y et Meyer, J.A. (1996a)Spatial exploration, map learning, and self-positioning with MonaLysa. In Maes, P., Mataric, M., Meyer, J.A., Pollack, J. et Wilson, S. (Eds.). *From animals to animats 4. Proceedings of the Fourth International Conference on Simulation of Adaptive Behavior*. The MIT Press.

Doya,K (2000) Reinforcement learning in continuous time and space *Neural Computation* 12(1) 219-245

Holland,J.H. (1975) *Adaptation in Natural and Artificial Systems*. University of Michigan Press.

Hurst,J ,Bull,L (2001) A Self-Adaptive Classifier System. In P-L Lanzi, W. Stolzmann & S. Wilson (eds) *Proceedings of the Third International Workshop on Learning Classifier Systems*.

Katagami,D, Yamada,S, (2002) Interactive Evolutionary Computation for Real Robot from a view point of observation. In *Proceedings of the The 7th International Conference on Intelligent Autonomous Systems (IAS7)*

Lanzi, P.L (1999) Extending the representation of Classifier Conditions from Messy Coding to S-Expressions. In W.Banzhaf, J.Daida, A.E.Eihen, M.H.Garzon, V.Honavar, M.Jakiela & R.E.Smith (eds) *Proceedings of the Genetic and Evolution Computation Conference –* GECCO-99. Morgan Kaufmann, pp345-352.

Nolfi,S ,Floreano,D (2000) *Evolutionary robotics*, MIT Press

Parr R (1998) *Hierarchical Control and Learning for Markov Decision Processes*, Ph.D. Thesis, University of California, Berkeley.

Santamaria,J Suton,R and Ram,A (1998) Experiments with reinforcement learning in problems with continuous state and action spaces. *Adaptive Behavior* Vol 6 No 2 pp163-217

Stolzmann, Butz.M (2000) Latent learning and action planning in robots with anticipatory classifier systems In P-L. Lanzi, W.Stolzmann & ".W.Wilson (eds) *Learning Classifier Systems: From Foundations to Applications*. Springer.

Sutton, R & Barto,A (1998) *Reinforcement Learning An Introduction*, MIT Press

Sutton,R, Precup,D and Singh,S (1999) Between MDPs and semi-MDPs: A Framework for temporal abstraction in reinforcement learning. *Artificial Intelligence* 112 pp181-211

Uchibi, Eiji, Asada,M and Hosoda,K (1997) Vision Based State Space Construction for Learning Mobile Robots in Multi-Agent Environments. *Proceedings of Sixth European Workshop on Learning Robots* (EWLR-6) pp33-41

Valenzuela-Rendon, M (1991) The Fuzzy Classifier System: a Classifier System for Continuously Varying Variables. Im L.Booker & R.Belew (eds) *Proceedings of the Fourth International Conference on genetic Algorithms*. Morgan Kaufmann pp346-353

Watkins, C (1989). *Learning From Delayed Rewards*. Phd. Cambridge University

Wilson, S.W (1994) ZCS: A Zeroth-level Classifier System. *Evolutionary Computation* 2(1):1-18

Wilson, S.W (2000) Get Real! XCS with Continuous Valued Inputs. In P-L. Lanzi, W.Stolzmann & ".W.Wilson (eds) *Learning Classifier Systems: From Foundations to Applications*. Springer, pp209-222.

Wilson, S.W (2001) Function Approximation with a Classifier System. In L. Spector, M. Gen, S. Sen, M.Dorigo, S,Pezeshk, M.Garzon & E.Burke (eds) *Proceedings of the Genetic and Evolutionary Computation Conference-* GECCO 2001. Morgan Kaufmann, pp974-984.

Comparison of Different Techniques

Comparing Synchronous and Asynchronous Cellular Genetic Algorithms

Enrique Alba[1], Mario Giacobini[2], Marco Tomassini[2], and Sergio Romero

[1] Department of Lenguajes y CC.CC., University of Málaga, Málaga, Spain
eat@lcc.uma.es

[2] Computer Science Institute, University of Lausanne, Lausanne, Switzerland
Mario.Giacobini/Marco.Tomassini@iis.unil.ch

Abstract. This paper presents a comparative study of several asynchronous policies for updating the population in a cellular genetic algorithm (cGA). Cellular GA's are regular GA's with the important exception that individuals are placed in a given geographical distribution (usually a 2-d grid). Operators are applied locally on a set made of each individual and the surrounding neighbors, thus promoting intra-neighborhood exploitation and inter-neighborhood exploration of the search space. Here, we analyze the respective advantages and drawbacks of dealing with this decentralized population in the traditional synchronous manner or in several possible asynchronous update policies. Asynchronous behavior has proven to be better in many domains such as cellular automata and distributed GA's, which, in turn, is also the main conclusion of this work. We will undergo a structured analysis on a set of problems with different features in order to get well grounded conclusions.

1 Introduction

Cellular evolutionary algorithms (cEA) models, also called *diffusion* or *fine-grained* models, are based on a spatially distributed population in which genetic interactions may only take place in a small neighborhood of each individual. Individuals are usually disposed on a regular grid of dimensions $d = 1, 2$ or 3. Cellular evolutionary algorithms were popularized by early work of Gorges-Schleuter [5], and by Manderick and Spiessen [9]. However, we here want to stress the difference between the model and its implementation, and this is why we call them *cellular* and not fine-grained EA's. Cellular EA's are just a new kind of algorithm, and not a parallel implementation on massively parallel machines.

Although fundamental theory is still an open research line for cEA's, they have been empirically reported as being useful in maintaining diversity and promoting slow diffusion of solutions through the grid. Part of their behavior is due to a lower selection pressure compared to that of panmictic EA's. The influence of the neighborhood, grid topology, and grid size/shape on the induced selection pressure has been investigated in detail in [1, 6, 11, 12] (and tested on different applications such as combinatorial and numerical optimization).

J.J. Merelo Guervós et al. (Eds.): PPSN VII, LNCS 2439, pp. 601–610, 2002.

Let us analyze a typical cGA, an important kind of cEA. The cGA iteratively considers groups of individuals belonging to the same neighborhood to work with. In a North-East-West-South (NEWS or Von Newmann) neighborhood type, the central individual plus its 4 neighbors make up a small pool to apply operators on. The cGA iterates through the population in various generations. In each generation, it considers as a central string every individual in the population. Since a string belongs to several neighborhoods, a change in its contents affects its neighbors in a smooth manner, representing a good tradeoff between slow convergence and good exploration of the search space. In a synchronous cGA, we compute the full new generation incrementally onto a temporary population, and them replace the full old population with the new one.

Synchronous Cellular Genetic Algorithm (cGA)

```
proc Reproductive_Cycle (ga):
    for s=1 to MAX_STEPS do
        for x=1 to WIDTH do
            for y=1 to HEIGHT do
                n_list  = Calculate_neigbors  (ga, position (x,y) );
                parent1 = Select (n_list);
                parent2 = Select (n_list);
                Crossover(ga.Pc, n_list[parent1], n_list[parent2], ind_aux.chrom);
                Mutate(ga.Pm, ind_aux.chrom);
                ind_aux.fitness = ga.Evaluate ( Decode ( ind_aux.chrom) ) ;
                Insert_New_Ind(position(x,y),ind_aux,[if better | always], ga, pop_aux);
            end_for;
        end_for;
        ga.pop=pop_aux;
        Collect_Statistics (ga);
    end_for;
end_proc Reproductive_Cycle;
```

Cellular EA's can also be seen as stochastic cellular automata (CA's) [15, 16] where the cardinality of the set of states is equal to the number of points in the search space. CA's, as well as cEA's, usually assume a *synchronous* or parallel update policy, in which *all the cells* are formally updated simultaneously. However, this is not the only option available. Indeed, several works on *asynchronous* CA's have shown that sequential update policies have a marked effect on their dynamics (see e.g. [7, 13, 14]). While asynchronous updating is physically more realistic for CA's due to their finite signal propagation speed, this is not an issue for cEA, unless they are implemented on an actual massively parallel cellular machine, which is seldom the case in practice. However, it would be interesting to investigate asynchronous cEA's and their problem solving capabilities. To our knowledge, the present paper is the first step in that direction. We will thus present a few asynchronous update policies for a cEA, and compare them with the customary synchronous updating on a set of test functions.

The paper is structured as follows. The next section contains some background on asynchronous cEA's and explains the techniques we are considering. Section 3 describes the test problems used, while section 4 gives details on the cEA parameters employed in the simulations, the performance measures, and the statistics used. Section 5 contains a discussion of the experimental results, and section 6 offers our conclusions, as well as some comments on the future work.

2 Asynchronous Cellular Evolutionary Algorithms

There are many ways for sequentially updating the cells of a cEA with a population on a 2-d grid (see an excellent discussion of asynchronous CA's in [13]). The most general one is independent random ordering of updates in time, which consists in randomly choosing the cell to be updated next, with replacement. This corresponds to a binomial distribution for the update probability. The limiting case of such distribution for large n is the Poisson distribution (where n is the number of cells, or individuals, in the grid). This update policy will be called *uniform choice* (UC) in the following.

For comparison purposes we also consider three other update methods: *fixed line sweep*, *fixed random sweep* and *random new sweep* (we employ the same terms as in [13]).

- In *fixed line sweep* (LS), the simplest method, grid cells are updated sequentially $(1, 2 \ldots n)$, line by line of the 2-d grid.
- In the *fixed random sweep* update (FRS), the next cell to be updated is chosen with uniform probability without replacement; this will produce a certain update sequence $(c_1^j, c_2^k, \ldots, c_n^m)$, where c_q^p means that cell number p is updated at time q and $(j, k, \ldots, m)$ is a permutation of the n cells. The same permutation is then used for the following update cycles.
- The *new random sweep* method (NRS) works like FRS, except that a random new cell permutation is chosen anew for each sweep through the array.

A *time step* is defined to be the action of updating n times, which corresponds to updating *all* the n cells in the grid for LS, FRS and NRS, and possibly less than n different cells in the uniform choice method, since some cells might be updated more than once. It should be noted that, with the exception of fixed line sweep, the other asynchronous updating policies are non-deterministic, representing an additional source of non-determinism besides that of the genetic operators. An asynchronous parallel implementation could be easily derived for these algorithms, although we do not explore physical parallelism in this work.

3 Description of the Test Problems

To test the differences between the synchronous and the four asynchronous update models we have decided to use problems representing three large classes of difficulty with interest in evolutionary computation, namely: deception, multimodality, and epistasis. Similarly to the work of Alba and Troya [1], the choice has been driven to the Massive Multimodal Deceptive Problem (MMDP), the Frequency Modulation Sounds Problem (FMS), and the P-PEAKS multimodal generator.

The MMDP has been specifically designed by Goldberg *et al.* [4] to be difficult for an EA. It is made up of k subproblems of 6 bits each (see equation 1). The optimum has a value of k and is attained when the *unitation* of each subproblem, i.e. the number of ones of its 6-bits defining string, is 0 or 6. Thus every subproblem $x_i =< x_{i_1}, ..., x_{i_6} >$ $(i = 1, ..., k)$ contributes to the fitness of a possible solution $\overrightarrow{x} =< x_1 ... x_k >$ according to its *unitation*:

$$f_{MMDP}(\overrightarrow{x}) = \sum_{i=1}^{k} g(unitation(x_i)), \tag{1}$$

where g is such that $g(0) = g(6) = 1$, $g(1) = g(5) = 0$, $g(2) = g(4) = 0.36$, $g(3) = 0.64$. Such function has a quite large number of local optima (22^k), while only 2^k are global solutions, and therefore its degree of multimodality is defined by k. Here we set $k = 40$, obtaining a considerably large degree of multimodality.

Proposed by Tsutsui *et al.* [10], the Frequency Modulation Sounds parameter identification problem (FMS) consists in adjusting a general model $y(t)$ (equation 2) to a basic sound function $y_0(t)$ (equation 3). The problem is to evolve a solution $\overrightarrow{x}$ consisting in 6 real parameters ($\overrightarrow{x} =< a_1, w_1, a_2, w_2, a_3, w_3 >$) each one encoded with 32 bits in the range $[-6.4, 6.35]$, in order $y(t)$ to fit the target function $y_0(t)$.

$$y(t) = a_1 \sin(w_1 t\theta + a_2 \sin(w_2 t\theta + a_3 \sin(w_3 t\theta))), \tag{2}$$

$$y_0(t) = 1.0 \sin(5.0 t\theta + 1.5 \sin(4.8 t\theta + 2.0 \sin(4.9 t\theta))). \tag{3}$$

The goal is, therefore, to minimize the sum of square errors (equation 4):

$$f_{FMS}(\overrightarrow{x}) = \sum_{t=0}^{100} (y(t) - y_0(t))^2. \tag{4}$$

The resulting problem is a complex multimodal function having strong epistasis with minimum value in $\overrightarrow{z}$, where $f(\overrightarrow{z}) = 0$. For our calculations, we consider the algorithm having found an optimum when the error falls below 10^{-2}.

The last optimization task solved in this paper is a problem generator proposed by De Jong *et al.* [8]. This problem generator is an easily parameterizable task which has a tunable degree of epistasis, thus allowing to derive instances with growing difficulty at will. With a problem generator we evaluate our algorithms on a high number of random problem instances, thus increasing the predictive power of the results for the problem class as a whole. Such a characteristic allows a larger fairness when comparing algorithms, since it implicitly removes the opportunity to hand-tune algorithms to a particular problem. In this paper we use the multimodal generator called P-PEAKS (see equation 5).

$$f_{P-PEAKS}(\overrightarrow{x}) = \frac{1}{N} max_{i=1}^{p} \{N - HammingD(\overrightarrow{x}, Peak_i)\} \tag{5}$$

The idea is to generate P strings, each of N random bits, that represent the location of the global optima in the search space. The fitness of a possible solution (a bit string of length N) is the number of bits it has in common with the nearest peak in the Hamming space, divided by the length N of the strings. Problems with a small/large number of peaks are weakly/strongly epistatic. The instance we use has $P = 100$ peaks of $N = 100$ bits each, which represents a medium-high epistasis level.

Table 1. Success rate for the Massive Multimodal Deceptive Problem of the synchronous and the four asynchronous update methods (horizontally) with different grid dimensions (vertically).

MMDP	Synchronous	Line Sweep	Fixed Random Sweep	New Random Sweep	Uniform Choice
20 × 20	0%	0%	0%	0%	0%
32 × 32	74%	32%	38%	44%	56%
40 × 40	98%	86%	92%	84%	94%
50 × 50	100%	98%	100%	100%	100%

Table 2. Success rate for the Frequency Modulation Sounds problem of the synchronous and the four asynchronous update methods (horizontally) with different grid dimensions (vertically).

FMS	Synchronous	Line Sweep	Fixed Random Sweep	New Random Sweep	Uniform Choice
20 × 20	0%	0%	0%	2%	0%
32 × 32	4%	0%	2%	0%	2%
40 × 40	8%	8%	6%	8%	6%
50 × 50	24%	22%	12%	12%	12%

Table 3. Success rate for the P-PEAKS problem of the synchronous and the four asynchronous update methods (horizontally) with different grid dimensions (vertically).

P-PEAKS	Synchronous	Line Sweep	Fixed Random Sweep	New Random Sweep	Uniform Choice
20 × 20	52%	26%	36%	34%	36%
32 × 32	100%	100%	100%	100%	100%
40 × 40	100%	100%	100%	100%	100%
50 × 50	100%	100%	100%	100%	100%

4 Parameters and Statistics Used

As we said in the introduction, many authors have already investigated the influence of the neighborhood, grid topology and grid dimensions on the induced selection pressure [1, 6, 11, 12]. Reduced grid dimensions of 20×20 have been used to design efficient cGA's [1], but, most of the time, larger grids are preferred for the analysis. For the three problems described in section 3 we have investigated four different grid sizes, so as to choose the dimension that, for all the problems, guarantees significant success rate of the five update methods. Each grid size has been tested 50 times for each update method.

All the results are summarized in tables 1, 2 and 3 and were obtained with the same internal parameters. For the parent's selection we have used a roulette wheel operator among the five individuals in the von Neumann neighborhood (the central plus the four neighbors in the North, East, West and South positions). A two point crossover is applied to the two selected parents with probability 1.0, thus producing an offspring individual (the one with the largest proportion of the best parent). Such a new solution is then mutated with different mutation probabilities p_m for the three problems. The obtained offspring

Table 4. Mean number of evaluations for the three problems (vertically) and the five update methods (horizontally).

	Synchronous	Line Sweep	Fixed Random Sweep	New Random Sweep	Uniform Choice
MMDP	277950	201887	216300	217850	238850
FMS	560760	543928	427291	480500	534772
P-PEAKS	243300	189550	191700	201400	203350

will replace the considered individual only if it has a better fitness. We stop the algorithm when a global optimum is found, and we analyze the cost of a successful run of a cEA by measuring the number of evaluations done. In the MMDP, an individual is encoded by a $40 \times 6 = 240$ bit string, in the FMS problem by a $6 \times 32 = 192$ binary chromosome, and in the P-PEAKS problem by a binary vector of length 100. The problems' differences and the different chromosomal lengths determine three mutation probabilities: for the MMDP and the P-PEAKS problem we set p_m to $1/L$, having respectively $p_m = 0.0042$ and $p_m = 0.01$, while for the FMS problem p_m is set to $10/L$, i.e. $p_m = 0.052$.

Capcarrère *et al.* defined a number of statistical measures that are useful for understanding the dynamical behavior of cellular evolutionary algorithms. Two kinds of statistics were used: *genotypic* and *phenotypic*. Genotypic measures embody aspects related to the genotypes of individuals in a population. Phenotypic statistics concern properties of individual performance, essentially fitness (see [3] for the exact definitions). Here, we use the variance and the mean fitness as phenotypic statistics, and the entropy as a statistics pertaining to the genotype (a phenotypic diversity index based on fitness entropy can also be defined but it will not be used here). Differently from the paper of Capcarrère *et al.*, we calculate the entropy in the interval $[0, 1]$, instead of in the interval $[0, \log(N)]$ (where N is the population size). Such a result is obtained setting the multiplicative constant in the entropy formula to $1/\log(N)$ instead of using 1.

5 Experimental Results

In order to have comparable results for all the three problems, we have decided to set the population to the 50×50 grid size (non-square grids analyzed in [1]). The results, for all the runs, are summarized in table 4: for each update method the average number of evaluations needed to solve the three problems is shown.

For the MMDP and the P-PEAKS problem, where success rate is 100%, we can see that the synchronous update method is more expensive than every asynchronous method. We can therefore deduce that for multimodal, deceptive (MMDP) and highly epistatic (P-PEAKS) problems it should be more efficient to implement an asynchronous update method rather than a synchronous one. The behavior of these two problems is slightly different for each asynchronous updating policy. For MMDP and P-PEAKS the Line Sweep method is the fastest policy, and the New Random Sweep is faster than Uniform Choice. The Fixed Random Sweep policy has a convergence speed similar to the Line Sweep for the highly epistatic problem, and to the New Random Sweep for the multimodal

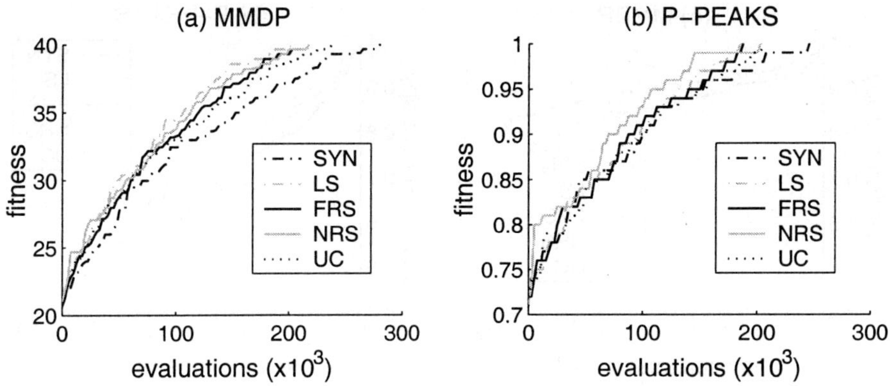

Fig. 1. A sample run of each update method for the MMDP (a), and for the P-PEAKS problem (b).

and deceptive problems. Such a different speed can be seen in figure 1, where a sample curve is drawn for each update method, both for MMDP (a) and for P-PEAKS (b).

This difference in the speed of convergence is followed by a consistent behavior of the entropy curves (see figure 2): the faster the convergence speed is, the lower the entropy is. Such a result confirms the intuitive idea that the genotypic diversity decreases proportionally to the convergence speed, i.e. the faster a cEA is, the bigger the upward thrust of the populations is. The variance and the standard deviation values are consistent with the described behaviors.

For the FMS problem the comparison between the convergence speeds of the different update methods must be coupled with their different success rates. In fact, as it can be seen comparing tables 2 and 4, the synchronous and the Line Sweep methods are slower than the Fixed Random Sweep, the New Random Sweep and the Uniform Choice methods (see also figure 3), but their success rate are twice the percentage of the other three asynchronous update methods. So, Line Sweep seems a good tradeoff between speed and accuracy, at least for problems similar to FMS. FMS showed to be the most difficult problem on which our cEA's have been tested in this study, due to its huge and complex search space. The entropy at the end (not shown here) is always very high, in the interval $[0.9, 0.92]$ for every algorithm.

We have seen that, contrary to the behavior of cellular automata (CA's), the simple Line Sweep policy performs better than the rest in cEA's. Such a result can be explained by the fact that, in cEA's, we don't have a propagation of signals like in CA's, having instead a propagation of information on the solution of the problem. It is true that Line Sweep fixes a preferred direction of propagation in the axes of the grid, but such an order speeds the propagation of information in the population. In fact, if we take our 50×50 toroidal population grids with the chosen von Neumann neighborhood, in synchronous cEA's the information of an individual will take at least 50 time steps (i.e. 125000 evaluations) to reach the farthest individual in the grid, while in a cEA with asynchronous Line

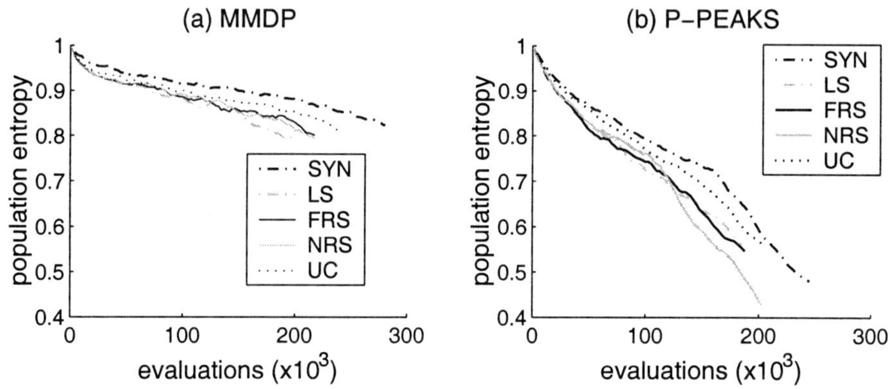

Fig. 2. A sample curve of the entropy of each update method for the MMDP (a) and for the P-PEAKS problem (b).

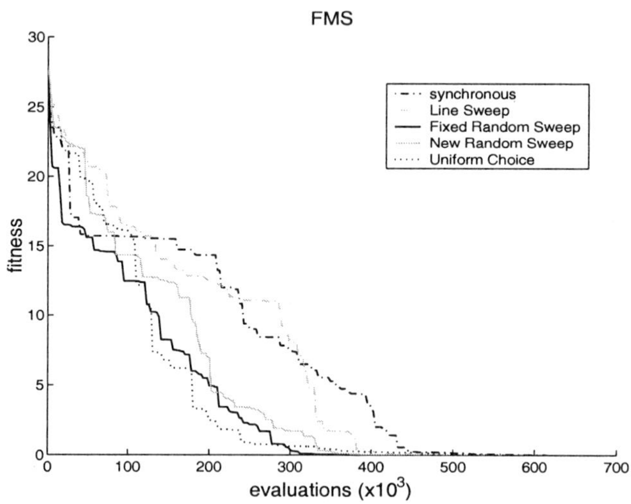

Fig. 3. A sample run of each update method for the FMS problem.

Sweep update method, it can take a small value such as 1 time step (i.e. 2500 evaluations).

6 Conclusions

In this paper we have analyzed the behavior of three alternative policies for asynchronously updating the population of a decentralized cellular GA. We have initiated this research line because we had some preliminary expectations relating asynchronous policies in the field of cellular automata [14] and distributed GA's [2]. We have tackled this study by considering three representative problems: deceptive (MMDP), epistatic (P-PEAKS) and hard (FMS) problems. Our first conclusion is that, for any size of the search grid, the synchronous update policy is the best in terms of percentage of hits, because it always provides an equal or larger success rate with respect to any of the asynchronous policies.

However, if we consider the number of evaluations needed to locate the optimum (efficiency) we got the opposite conclusion: asynchronous methods are faster (sometimes much faster) than the synchronous one. A simple asynchronous policy such that Line Sweep (LS) provides the faster convergence for two of the problems (MMDP and P-PEAKS), while it shows a high and desirable success rate (similar to that of the synchronous update). In the hard FMS problem, Fixed Random Sweep got a considerably faster solution, and can be pointed out as a good approach.

Globally stated, fast convergence means local optima in evolutionary algorithms, but cellular GA's in general, and the Line Sweep policy in particular, offers a good tradeoff solution to this problem without bothering researchers with a large number of parameters to be tuned (maybe only the ratio between the size of the grid and the neighborhood being used [1]).

As a future work, we will enlarge the set of considered problems, include a study of the influence of the shape of the 2-d grid containing the population, and try to better characterize the relationship between the performance measures and the kind of problems.

References

1. E. Alba and J. M. Troya. Cellular evolutionary algorithms: Evaluating the influence of ratio. In M. Schoenauer et al., editor, *Parallel Problem Solving from Nature, PPSN VI*, volume 1917 of *Lecture Notes in Computer Science*, pages 29–38. Springer-Verlag, 2000.
2. E. Alba and J. M. Troya. Analyzing synchronous and asynchronous parallel distributed genetic algorithms. *Future Generation Computer Systems*, 17:451–465, January 2001.
3. M. Capcarrère, M. Tomassini, A. Tettamanzi, and M. Sipper. A statistical study of a class of cellular evolutionary algorithms. *Evolutionary Computation*, 7(3):255–274, 1999.
4. K. Deb D.E. Goldberg and J. Horn. Massively multimodality, deception and genetic algorithms. In R. Männer and B. Manderick, editors, *Proceedings of the PPSN II*, pages 37–46. North-Holland, 1992.
5. M. Gorges-Schleuter. ASPARAGOS an asynchronous parallel genetic optimisation strategy. In J. D. Schaffer, editor, *Proceedings of the Third International Conference on Genetic Algorithms*, pages 422–427. Morgan Kaufmann, 1989.
6. M. Gorges-Schleuter. An analysis of local selection in evolution strategies. In *Genetic and evolutionary conference, GECCO99*, volume 1, pages 847–854. Morgan Kaufmann, San Francisco, CA, 1999.
7. T. E. Ingerson and R. L. Buvel. Structure in asynchronous cellular automata. *Physica D*, 10:59–68, 1984.
8. K.A. De Jong, M.A. Potter, and W.M. Spears. Using problem generators to explore the effects of epistasis. In T. Bäck, editor, *Proceedings of the Seventh ICGA*, pages 338–345. Morgan Kaufmann, 1997.

9. B. Manderick and P. Spiessens. Fine-grained parallel genetic algorithms. In J. D. Schaffer, editor, *Proceedings of the Third International Conference on Genetic Algorithms*, pages 428–433. Morgan Kaufmann, 1989.
10. D. Corne S. Tsutsui, A. Ghosh and Y. Fujimoto. A real coded genetic algorithm with an explorer and an exploiter populations. In T. Bäck, editor, *Proceedings of the Seventh ICGA*, pages 238–245. Morgan Kaufmann, 1997.
11. J. Sarma and K. A. De Jong. An analysis of the effect of the neighborhood size and shape on local selection algorithms. In H. M. Voigt, W. Ebeling, I. Rechenberg, and H. P. Schwefel, editors, *Parallel Problem Solving from Nature (PPSN IV)*, volume 1141 of *Lecture Notes in Computer Science*, pages 236–244. Springer-Verlag, Heidelberg, 1996.
12. J. Sarma and K. A. De Jong. An analysis of local selection algorithms in a spatially structured evolutionary algorithm. In T. Bäck, editor, *Proceedings of the Seventh International Conference on Genetic Algorithms*, pages 181–186. Morgan Kaufmann, 1997.
13. B. Schönfisch and A. de Roos. Synchronous and asynchronous updating in cellular automata. *BioSystems*, 51:123–143, 1999.
14. M. Sipper, M. Tomassini, and M. S. Capcarrere. Evolving asynchronous and scalable non-uniform cellular automata. In G. D. Smith, N. C. Steele, and R. F. Albrecht, editors, *Proceedings of International Conference on Artificial Neural Networks and Genetic Algorithms (ICANNGA97)*, pages 67–71. Springer-Verlag, Vienna, 1997.
15. M. Tomassini. The parallel genetic cellular automata: Application to global function optimization. In R. F. Albrecht, C. R. Reeves, and N. C. Steele, editors, *Proceedings of the International Conference on Artificial Neural Networks and Genetic Algorithms*, pages 385–391. Springer-Verlag, 1993.
16. D. Whitley. Cellular genetic algorithms. In S. Forrest, editor, *Proceedings of the Fifth International Conference on Genetic Algorithms*, page 658. Morgan Kaufmann Publishers, San Mateo, California, 1993.

Satellite Range Scheduling: A Comparison of Genetic, Heuristic and Local Search

L. Barbulescu, A.E. Howe, J.P. Watson, and L.D. Whitley

Computer Science Department
Colorado State University
Fort Collins, CO 80523 USA
{laura,howe,watsonj,whitley}@cs.colostate.edu

Abstract. Three algorithms are tested on the satellite range scheduling problem, using data from the U.S. Air Force Satellite Control Network; a simple heuristic, as well as local search methods, are compared against a genetic algorithm on old benchmark problems as well as problems produced by a generator we recently developed. The simple heuristic works well on the old benchmark, but fails to scale to larger, more complex problems produced by our generator. The genetic algorithm yields the best overall performance on larger, more difficult problems.

1 Problem Description

The U.S. Air Force Satellite Control Network (AFSCN) is responsible for coordinating communications between users and satellites in space. A key mission of the AFSCN is *satellite range scheduling* (SRS), which involves scheduling communications between users on the ground and more than 100 satellites in space. All communications are performed via nine ground stations located around the globe, with an aggregate of sixteen antennas. The AFSCN scheduling center typically receives over 500 user requests for a single day.

Each user request specifies at a minimum an antenna at a particular ground station, a required duration, and a time window within which the duration must be allocated. Requests are classified as either low or high-altitude, corresponding to the orbit of the target satellite. The durations of low-altitude requests are typically equal to the visibility windows, leaving little scheduling flexibility. In contrast, high-altitude satellites are visible to more ground stations for longer periods of time. Consequently, high-altitude requests often specify alternative antennas and/or visibility windows. The objective of the SRS problem is to minimize the number of unsatisfied requests.

The SRS problem is NP-complete[1]: a reduction of the SRS problem to one resource can be shown to be equivalent to a well known NP-complete problem in the scheduling literature, denoted $1|r_j|\sum U_j$ in the three-field notation widely used by the scheduling community. The SRS problem is also over-subscribed in the sense that all requests can rarely be scheduled; to satisfy all user requests,

[1] We are currently working on a paper which presents an NP-completeness proof.

J.J. Merelo Guervós et al. (Eds.): PPSN VII, LNCS 2439, pp. 611–620, 2002.

some form of arbitration process is required. Several algorithms for related over-subscribed scheduling problems have been reported in the literature (e.g., see [6] [11] [13]), but none directly address the peculiarities of the satellite range scheduling problem, including alternative resources and/or time-windows.

Researchers at the Air Force Institute of Technology (AFIT) have developed a number of algorithms for the SRS problem. Gooley and Schalck introduced an algorithm based on a combination of mixed integer programming (MIP) and insertion heuristics [4] [8], which scheduled between 91% and 95% of user requests for small problem instances. Later, Parish used a genetic algorithm called *Genitor* to solve the SRS problem [5]. *Genitor* out-performed the MIP approach, nominally scheduling 96% of user requests.

Both the MIP algorithm and the *Genitor* genetic algorithm were evaluated using the same set of seven real-world problem instances collected in 1992; we refer to this collection of instances as the "AFIT benchmark". In 1992 approximately 300 requests needed to be scheduled for a single day, compared to 500 requests per day in recent years. The need to schedule more requests has a clear impact on problem difficulty. In this paper we investigate whether the problems in the AFIT benchmarks are representative of the kinds of Range Scheduling problems that are solved in the present by AFSCN to determine whether the old results should generalize.

Currently, there is no accepted state-of-the-art algorithm for satellite range scheduling. Because it is an extremely important application, we have been engaged in a study of various algorithms for this problem. In this paper, we replicate the results reported by Parish [5] using *Genitor* to solve the AFIT benchmark problems, and investigate reasons for *Genitor*'s strong relative performance. We identify a simple heuristic that can solve all of the problems in the AFIT benchmark. Finally, we generate new problems by modeling features currently encountered by AFSCN and explore conditions where the heuristic fails. *Genitor* continues to display good results for new problems.

2 Algorithms for Satellite Range Scheduling

In this section, we document the various algorithms considered in this study. We first discuss the method of encoding solutions, and the procedure for decoding solutions into actual schedules. Next, we define the three algorithms used in our analysis: random sampling, local search under a shift neighborhood, and the *Genitor* genetic algorithm. We then conclude by briefly discussing our decision to omit two well-known families of scheduling algorithms in our analysis.

2.1 Solution Representation and Decoding

Each of the algorithms we consider represents solutions as permutations of the integers 1 through N, where N is the total number of requests to be scheduled. A permutation represents the order in which requests are given access to particular resources. A greedy heuristic is then used to generate a schedule from a permutation, by attempting to schedule the requests in the order in which they appear in the permutation. Each request is assigned to the first available resource (from its list of alternatives), and at the earliest possible starting time. If the request cannot be scheduled on any of the alternative resources, it is dropped from the schedule (i.e., bumped). The "fitness" of a schedule is then defined as the total number of requests bumped from the schedule.

2.2 Random Sampling

Random sampling produces schedules by generating random permutations of length N. By randomly sampling a large number of schedules, we can characterize the distribution of solutions in the search space. Further, the performance of random sampling provides a baseline measure of problem difficulty.

2.3 Local Search under the Shift Neighborhood

A key component of any local search algorithm is the move operator. Because problem-specific knowledge for the SRS problem is lacking, we selected the "shift" move operator. The shift operator has been successfully applied to a number of well-known scheduling problems, such as the permutation flow-shop scheduling problem [10]. The neighborhood under the shift operator is defined by considering all $(N-1)^2$ pairs (x, y) of positions in a current solution π, subject to the restriction that $y \neq x-1$. The neighbor π' corresponding to the position pair (x, y) is produced by *shifting* the job at position x into the position y, while leaving all other relative job orders unchanged. If $x < y$, then $\pi' = (\pi(1), ..., \pi(x-1), \pi(x+1), ..., \pi(y), \pi(x), \pi(y+1), ..., \pi(n))$. If $x > y$, then $\pi' = (\pi(1), ..., \pi(y-1), \pi(x), \pi(y), ..., \pi(x-1), \pi(x+1), ..., \pi(n))$.

Given the relatively large neighborhood size, we use the shift operator in conjunction with next-descent search. The neighbors of the current solution are examined in a random order, and the first neighbor with either a lower or equal fitness (i.e., number of bumps) is accepted. Search terminates when a pre-specified number of evaluations is exceeded.

2.4 The *Genitor* Genetic Algorithm

Genitor [12] is a "steady-state" genetic algorithm [2]. Previous studies of the SRS problem at AFIT [5] report good results when using *Genitor* in conjunction with permutation encoding of solutions. In each step of *Genitor*, a pair of solutions is selected and used to generate a single child, which then replaces the worst solution in the current population.

In *Genitor*, the parent solutions are selected based on the rank of their fitness, relative to other solutions in the population. A linear bias is used such that individuals that are above the median fitness have a rank-fitness greater than one and those below the median fitness have a rank-fitness of less than one.

The typical genetic algorithm encodes solutions as bit strings, enabling the use of standard crossover operators such as one-point and two-point crossover [3].

Because we encode solutions as permutations, a special crossover operator is required to ensure that the recombination of two parent permutations results in a child inheriting relevant characteristics of the two parents. We use Syswerda's (relative) order crossover operator [9], which preserves the relative order of the elements in the parents when constructing the child. Syswerda's operator has been successfully applied in a variety of scheduling applications.

2.5 Other Scheduling Algorithms

We also considered straightforward implementations of Tabu search for the SRS problem, but the performance of these algorithms was not competitive. With 500 requests, the number of neighbors under shift or swap-based move operators is roughly 500^2; consequently, Tabu search and other local search algorithms based on steepest descent are simply not practical. We briefly explored methods for reducing the neighborhood size, but in all cases the reduction in neighborhood size severely impacted algorithm performance.

Additionally, we developed constructive search algorithms based on texture-based [1] and slack-based [7] constraint-based scheduling heuristics that select the maximal subset of tasks that can be feasibly scheduled. We found that texture-based heuristics are highly effective when the size of the problem is small (e.g., less than 100 requests) and when alternative or backup requests are not considered. However, on larger problems, the consideration of alternative times makes the straightforward use of constraint-based methods ineffective.

3 The AFIT Benchmark

The AFIT benchmark problems[2] were derived using the ASTRO system, a computer application developed to aid human schedulers. These problems represent the user requests and visibilities for seven days, from October 12 to October 18, 1992. The low-altitude requests in these problems can be scheduled only at one ground station (by assigning it to one of the antennas present at that ground station). The number of requests to be scheduled for the seven problems are 322, 302, 300, 316, 305, 298, and 297 respectively. We note that since 1992, the number of requests received during a typical day has increased substantially.

In our experimental setup we replicated the conditions and the reported results from Parish's study [5]. We ran *Genitor* on each of the seven problems in the benchmark, using the same parameters: population size 200, selective pressure 1.5, order-based crossover, and 8000 evaluations[3] for each run. We also ran random sampling and local search on each AFIT problem, with a limit of 8000 evaluations per run. For each algorithm, we performed a total of 30 independent

[2] We thank Dr. James T. Moore, Associate Professor of Operations Research at the Department of Operational Sciences, Graduate School of Engineering and Management, Air Force Institute of Technology for providing the data.

[3] An increase in the number of evaluations to 50k and of the population size to 400 did not improve the best solutions found for each problem.

Table 1. Performance of *Genitor*, local search, and random sampling on the AFIT benchmark problems, in terms of the best and mean number of bumped requests. All statistics are taken over 30 independent runs. The last column reports the performance of Schalck's Mixed-Integer Programming algorithm [8].

Day	*Genitor*			Local Search			Random Sampling			MIP
	Min	Mean	Stdev	Min	Mean	Stdev	Min	Mean	Stdev	
1	8	8.6	0.49	15	18.16	2.54	21	22.7	0.87	10
2	4	4	0	6	10.96	2.04	11	13.83	1.08	6
3	3	3.03	0.18	11	15.4	2.73	16	17.76	0.77	7
4	2	2.06	0.25	12	17.43	2.76	16	20.20	1.29	7
5	4	4.1	0.3	12	16.16	1.78	15	17.86	1.16	6
6	6	6.03	0.18	15	18.16	2.05	19	20.73	0.94	7
7	6	6	0	10	14.1	2.53	16	16.96	0.66	6

runs on each problem. The results are summarized in Table 1. Included in the table are the results obtained by Schalck using Mixed Integer Programming [8]. As previously reported, *Genitor* yields the best overall performance.

To exploit the differences in scheduling slack and number of alternatives between low and high-altitude requests, we designed a simple greedy heuristic (which we call the "split heuristic") that first schedules all the low-altitude requests (in the order given by the permutation), followed by the high-altitude requests. We show that: (1) for more than 80% of the best known schedules found by *Genitor*, the split heuristic does not increase the number of conflicts in the schedule, and (2) the split heuristic typically produces good (and often best-known) schedules.

We hypothesized that *Genitor* may be learning to schedule the low-altitude requests before the high-altitude requests, leading to the strong overall performance. If true, the evaluation of high-quality schedules should, on average, remain unchanged when the split heuristic is applied. To test this hypothesis, we ran 1000 trials of *Genitor* on each AFIT problem. The results are summarized in Table 2. The second column (labeled "Total Number of Best Known Found") records the number of schedules (out of 1000) with an evaluation equal to the best found by *Genitor* in any run. We then applied the split heuristic to each such schedule. The schedules resulting from the split heuristic fall into three categories. First, the conflicts are identical to those found by *Genitor*; the number of schedules in this category is given in the third column ("Same Evaluation Same Conflicts"). Second, the evaluation is the same but the conflicts are different; the number of schedules in this category is given in column "Same Evaluation Different Conflicts". Third, the evaluation is different; the last column reports the number of schedules in this category. By separating the requests from the schedules produced by *Genitor* into low and high-altitude requests, the evaluation of more than 80% of the schedules remains unchanged. The numbers in the last column of the table also warn that when using the split heuristic only a subspace of the permutations is considered (the permutations that are separated into low and high-altitude requests); this subspace does not contain all

Table 2. The effect of applying the split heuristic when evaluating best known schedules produced by *Genitor*.

Day	Total Number of Best Known Found	Same Evaluation Same Conflicts	Same Evaluation Different Conflicts	Worse Evaluation
1	420	38	373	9
2	1000	726	106	168
3	996	825	115	56
4	937	733	50	154
5	862	800	12	50
6	967	843	56	68
7	1000	588	408	4

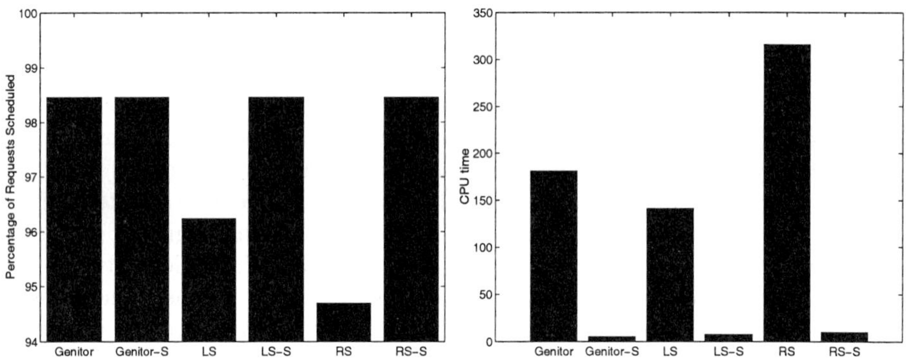

Fig. 1. Algorithm performance for the seven AFIT benchmark problems.

the best-known solutions, and, in fact, for different instances of the problem this subspace could be suboptimal.

Our second hypothesis is that using the split heuristic results in solutions with a small number of conflicts. Figure 1 presents a summary of the results obtained when using the *Genitor*, Local Search and Random Sampling without the split heuristic (30 experiments, 8000 evaluations per experiment), as well as the split versions denoted by *Genitor*-S, Local Search-S and Random Sampling-S. The split versions of the three algorithms were run in 30 experiments with 100 evaluations per experiment. The minimum number of bumps in 30 experiments is recorded for each problem as the percent of requests scheduled. The left half of Figure 1 presents the average percentage of requests scheduled for the seven problems by each algorithm. The corresponding average CPU times (in seconds) appear in the right half of the figure.

For all the problems, Random Sampling-S finds the best known solutions, as illustrated in Table 3. Since the best known solutions were obtained by randomly sampling a small number of permutations, solving the problems in the AFIT benchmark is easy using the split heuristic.

Table 3. Results of running random sampling in 30 experiments, by generating 100 random permutations per experiment. A problem-specific heuristic is used in the evaluation function, where the low-altitude requests are evaluated first.

Day	Best Known	Random Sampling-S		
		Min	Mean	Stdev
1	8	8	8.2	0.41
2	4	4	4	0
3	3	3	3.3	0.46
4	2	2	2.43	0.51
5	4	4	4.66	0.48
6	6	6	6.5	0.51
7	6	6	6	0

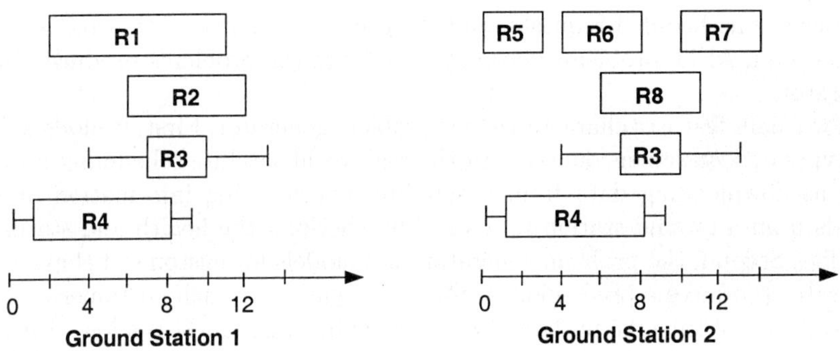

Fig. 2. Problem for which the split heuristic can not result in an optimal solution. Each ground station has two antennas; the only high-altitude requests are $R3$ and $R4$.

However, we can build a simple problem instance for which the optimal solution cannot be found using the split heuristic. Consider the problem represented in Figure 2. There are only two ground stations, and each ground station has two antennas (meaning that at each ground station at most two requests can be scheduled at the same time). There are two high-altitude requests, $R3$ and $R4$, with durations 3 and 7 respectively. $R3$ can be scheduled between start time 4 and end time 13; $R4$ can be scheduled between 0 and 9. Both $R3$ and $R4$ can be scheduled at either of the two ground stations. The rest of the requests are low-altitude requests. $R1$ and $R2$ request the first ground station, while $R5$, $R6$, $R7$, and $R8$ request the second ground station. This problem fits the description of the SRS problems in the AFIT benchmark: the low-altitude requests can be scheduled only at a specific ground station, with a fixed start and end time, while the high-altitude requests have alternative resources and a time window specified. For all the permutation schedules, if the split heuristic is used, $R3$ and $R4$ cannot be scheduled. However, it is possible to find schedules where both $R3$ and $R4$ get scheduled, and only one request ($R1$, $R2$, or $R8$) gets bumped. The subspace containing the permutations with all the low-altitude requests before the high-altitude requests is suboptimal - the global optimum is not necessarily contained in this subspace. The example shows the potential for failure to generate optimal solutions using the split heuristic.

4 Generalizing the AFIT Problems

Does the algorithm performance obtained for the AFIT benchmark transfer to larger sets of similar problems? To explore this question, we built a problem generator which produces problems similar to the AFIT benchmark but also including features encountered in the present-day real-world problems. Then we compare the results of running *Genitor*, local search and random sampling on problems produced by the problem generator to the results reported for the AFIT problems. We show that: (1) *Genitor* consistently results in the smallest number of unscheduled requests, and (2) the performance of the split heuristic on the seven AFIT problems does not transfer to the problems produced by our generator.

Two main features characterize our problem generator. First, it models different types of requests encountered in the real-world satellite scheduling problem, such as downloading data from a satellite, transmitting information or commands from a ground station to a satellite, checking the health and status of a satellite. Second, the problem generator uses models for customer behavior. The generator produces a predefined number of requests for each customer and each request type. With a 0.5 probability we determine if a request is a low-altitude or high-altitude one. For low-altitude requests, we decided to preserve the AFIT definition by assigning the duration equal to the size of the time window. However, we define alternative ground stations for both low and high-altitude requests.

To generate alternatives for a request, we collected data on the Web about the visibilities of various satellites[4] from the locations of the nine ground stations.

We repeat the experiments described for the AFIT problems by running *Genitor*, local search and random sampling for problems produced by our generator. To compare our results to the ones reported for the AFIT problems, but also to generate realistic problems, we ran the experiments for problem sizes 300, 350, 400, 450, and 500. For each size, we generated 30 problem instances.

We again ran *Genitor*, local search and random sampling, with and without the split heuristic, performing 30 runs with 8000 evaluations per run for each problem. An increase in the number of evaluations to 50k and of the population size to 400 did not improve the best solutions found for each problem. We record the number of unscheduled requests for each run. Figure 3 shows that *Genitor* on average outperforms *Genitor*-S and both versions of local search and random sampling. In fact *Genitor* (without the split heuristic) always outperforms all the other algorithms. In Table 4 we first subtract the minimum number of bumped requests for each problem from the minimum number of bumped requests reported by each of the algorithms (with or without the split heuristic) for that

[4] See: http://earthobservatory.nasa.gov/MissionControl/overpass.html for visibilities; thanks to Ester Gubbrud for helping us to compile the databases.

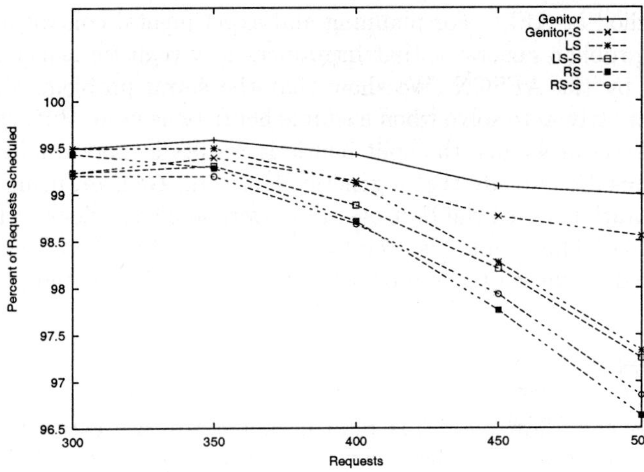

Fig. 3. Average percent of requests scheduled by the no-split and split versions of each of the algorithms.

Table 4. The difference between the minimum number of bumps reported by an algorithm and the minimum number of bumps found by any of the six algorithms (with or without the split heuristic) is averaged over the 30 instances for each problem size.

	Genitor		Local Search		Random Sampling	
Size	Mean	Stdev	Mean	Stdev	Mean	Stdev
300	0.000	0.000	0.000	0.000	0.167	0.213
350	0.000	0.000	0.333	0.368	1.067	1.099
400	0.000	0.000	1.233	1.702	2.833	3.523
450	0.000	0.000	3.667	3.678	5.967	6.240
500	0.000	0.000	8.300	3.941	11.767	7.840
	Genitor-S		Local Search-S		Random Sampling-S	
Size	Mean	Stdev	Mean	Stdev	Mean	Stdev
300	0.767	0.737	0.767	0.737	0.867	0.671
350	0.667	0.851	0.967	1.551	1.367	2.033
400	1.100	1.128	2.167	2.626	2.933	3.168
450	1.467	1.223	3.967	4.309	5.200	6.717
500	2.200	2.097	8.700	8.907	10.667	10.161

problem in 30 runs. Then we average these differences over the 30 instances generated for each size. From both Figure 3 and Table 4, it is clear that the split heuristic always results in an average decrease in performance.

5 Conclusions

Satellite Range Scheduling is an important real world problem that impacts the use of expensive and limited resources. We first considered a version of the problem studied at AFIT. For planning and experimental control purposes, we also built a problem generator that introduces new realistic features, currently encountered by the AFSCN. We show that the seven problems in the AFIT benchmark are trivial to solve when a simple heuristic is used. But, when applied to more realistic problems, the split heuristic results in poor-quality solutions. Finally, our results indicate that a genetic algorithm, *Genitor*, using a permutation representation yields the best overall performance and does so in a modest amount of time. The results also reinforce the notion that benchmarks need to be constructed or chosen to be representative for actual target applications.

References

1. J. C. Beck, A. J. Davenport, E. M. Sitarski, and M. S. Fox.: Texture-based heuristic for scheduling revisited. Proceedings of the Fourteenth National Conference on Artificial Intelligence (AAAI-97), Providence, RI (1997) 241–248
2. Lawrence Davis: Handbook of Genetic Algorithms. Van Nostrand Reinhold, New York (1991)
3. David Goldberg: Genetic Algorithms in Search, Optimization and Machine Learning. Addison-Wesley, Reading, MA (1989)
4. T.D. Gooley: Automating the satellite range scheduling process. Master's thesis, Air Force Institute of Technology (1993)
5. D.A. Parish. A genetic algorithm approach to automating satellite range scheduling. Master's thesis, Air Force Institute of Technology (1994)
6. J.C. Pemberton: Toward scheduling over-constrained remote-sensing satellites. Proceedings of the Second NASA International Workshop on Planning and Scheduling for Space, San Francisco, CA (2000)
7. Steve Smith and C.C. Cheng: Slack-based heuristics for constraint satisfaction problems. Proceedings of the Eleventh National Conference on Artificial Intelligence (AAAI-93), Washington, DC (1993) 139–144
8. S.M. Schalck: Automating satellite range scheduling. Master's thesis, Air Force Institute of Technology (1993)
9. Gilbert Syswerda: Schedule Optimization Using Genetic Algorithms. In Lawrence Davis, editor, Handbook of Genetic Algorithms, chapter 21. Van Nostrand Reinhold, New York (1991)
10. E. Taillard: Some efficient heuristic methods for the flow shop sequencing problem. European Journal of Operations Research **47** (1990) 65–74
11. Gérard Verfaillie, Michel Lemaitre, and Thomas Schiex: Russian doll search for solving constraint optimization problems. Proceedings of the Thirteenth National Conference on Artificial Intelligence (AAAI-96), Portland, OR (1996) 181–187
12. L. Darrell Whitley: The *Genitor* Algorithm and Selective Pressure: Why Rank Based Allocation of Reproductive Trials is Best. J. D. Schaffer, editor, Proc. of the 3rd Int'l. Conf. on GAs, 116–121. Morgan Kaufmann (1989)
13. William J. Wolfe and Stephen E. Sorensen: Three scheduling algorithms applied to the earth observing systems domain. Management Science **46(1)** (2000) 148–166

The LifeCycle Model: Combining Particle Swarm Optimisation, Genetic Algorithms and HillClimbers

Thiemo Krink and Morten Løvbjerg

EVALife Group, Department of Computer Science
Ny Munkegade, Bldg. 540, University of Aarhus
DK-8000 Aarhus C, Denmark
{krink,lovbjerg}@daimi.au.dk
http://www.evalife.dk

Abstract. Adaptive search heuristics are known to be valuable in approximating solutions to hard search problems. However, these techniques are problem dependent. Inspired by the idea of life cycle stages found in nature, we introduce a hybrid approach called the LifeCycle model that simultaneously applies genetic algorithms (GAs), particle swarm optimisation (PSOs), and stochastic hill climbing to create a generally well-performing search heuristics. In the LifeCycle model, we consider candidate solutions and their fitness as individuals, which, based on their recent search progress, can decide to become either a GA individual, a particle of a PSO, or a single stochastic hill climber. First results from a comparison of our new approach with the single search algorithms indicate a generally good performance in numerical optimization.

1 Introduction

In biology, the term life cycle refers to the various phases an individual passes through from birth to maturity and reproduction [1]. This process often leads to drastic transformations of the individual with stage specific adaptations to a particular environment. This phenomenon is particularly amazing considering that the genome remains the same within each cell and life stage, whereas the morphology and behaviour of the phenotype can change drastically in accordance to the requirements of the life stage niche. Some life cycle changes in nature are one-time events such as sexual maturity. Other changes are re-occurring, such as mating seasons. These stages are genetically determined and the individuals have little or no influence on the change of the life cycle stage. The transition between life cycle changes are often triggered by environmental factors. Environmental changes often determine transitions from one life cycle stage to another. Some animals are able to sense and predict these changes and can actively decide to alter their life cycle stage. A particularly interesting animal that has this capability is the microscopically small *Symbion pandora*, which lives as a symbiont on its much larger host - the Norway lobster. The host provides food,

J.J. Merelo Guervós et al. (Eds.): PPSN VII, LNCS 2439, pp. 621–630, 2002.

substrate, and transportation for the symbionts which only inhabit the mouth parts of the host. This rich and diverse environment poses a great challenge to the symbiont because the individual need to evacuate and recolonize the lobster with each lobster moulting. The life stages of *Symbion pandora* include feeding and non-feeding, sedentary and free swimming, as well as sexual and asexual reproduction stages [2].

The ability of an individual to actively decide about its kind of life form in response to its success in its current environment inspired us to the study presented in this paper. The idea behind our LifeCycle model is to create a self-adaptive search heuristic in which each individual (containing the candidate solution) can decide whether it would prefer to belong to a population of a genetic algorithm (GA), a particle swarm optimization (PSO), or become a solitary stochastic hill climber (HC). The decision of the individual depends on its success in searching the fitness landscape. Our motivation for this hybrid approach was that each of these search techniques on its own has its specific problem dependent strengths and weaknesses.

GAs, for instance, are widely applicable, and particularly powerful when domain knowledge can be incorporated in the operator design (see e.g. [3]). However, particle swarm optimisation (PSO) [4] can achieve clearly superior results in many instances of numerical optimization, but there is no general superiority compared to GAs (e.g. [5–9]). Hill climbers, in contrast, are good for local search with a high probability of finding the closest optimum. However, for multimodal functions, their performance is highly dependent on their starting position and hill-climbing techniques often convergence prematurely at local optima. Their main weakness compared to population based approaches, such as GAs and PSOs, is that candidate solutions neither compete nor cooperate [10].

The goal of our LifeCycle model is to make a self-adaptive approach towards a problem invariant search technique that can further take advantage of the changing search requirements during the optimization, such as initial exploration and local fine-tuning towards the end of the run.

2 The LifeCycle Model

The LifeCycle model consists of individuals starting out as PSO particles, which can turn into GA individuals, then hill climbers, then back to particles and so on. The structure of the LifeCycle model is illustrated in fig. 1. In all these heuristics, we use one fitness evaluation per individual per iteration. A LifeCycle individual switches its stage when it has made no fitness improvement for more than 50 iterations.

2.1 The PSO Model

The PSO model used in the LifeCycle model is similar to the traditional PSO model described in [4]. The model consists of a number of particles moving around in the search space, where the position of each particle represents a

```
program LifeCycle_Model
begin
    initialise
    while (not terminate-condition) do
        begin
            for (all individuals)
                evaluate fitness
                switch LifeCycle stage if no recent improvement
            for (PSO particles)
                calculate new velocity vectors
                move
            for (GA individuals)
                select new population
                recombine population
                mutate population
            for (HillClimbers)
                find possible new neighbouring solution
                evaluate fitness for the new solution
                shift to new solution with probability p
        end
end
```

Fig. 1. Structure of the LifeCycle model.

candidate solution to a numerical problem. Each particle has a position vector $\boldsymbol{x}_i$, a velocity vector $\boldsymbol{v}_i$ and the position of the best candidate solution encountered by the particle $\boldsymbol{p}_i$. The PSO also stores the overall best found point $\boldsymbol{p}_g$. The memorized positions are used to attract particles to search space areas with known good solutions.

In each iteration the velocity of each particle is updated in the following way

$$\boldsymbol{v}_i = \chi(w\boldsymbol{v}_i + \varphi_{1i}(\boldsymbol{p}_i - \boldsymbol{x}_i) + \varphi_{2i}(\boldsymbol{p}_g - \boldsymbol{x}_i))$$

where χ is known as the *constriction coefficient* described in [11] and w is the *inertia weight* described in [6]. φ_1 and φ_2 are random values, which are different for each particle and for each dimension. The velocity $\boldsymbol{v}_i$ of each particle is limited by an upper threshold v_{max}. The position of each particle is updated in each iteration by adding the velocity vector to the position vector, such that, $\boldsymbol{x}_i = \boldsymbol{x}_i + \boldsymbol{v}_i$. The particles have no neighbourhood restrictions, meaning that each particle can affect all other particles. This neighbourhood is of type *star* (fully connected network), which has been shown to be a good topology ([5]).

2.2 The GA Model

A classic genetic algorithm consists of a population of individuals refining their candidate solutions through interaction and adaptation. Each individual represents a candidate solution to the given problem. After initialisation the GA enters a loop, in which the population is evaluated, a new population is selected and this new population is altered (p. 151 [10]). The LifeCycle GA model uses binary tournament selection (p. 61 [12]) to generate a new population and elitism to ensure the survival of the individual with the best fitness. The LifeCycle GA model alters the population by crossover and mutation. The crossover operator used in the LifeCycle GA model is the so-called arithmetic crossover. This operator replaces two parent individuals selected for crossover with two child individuals as follows:

$$\begin{aligned} \boldsymbol{x}_{child1} &= w * \boldsymbol{x}_{parent1} + (1-w) * \boldsymbol{x}_{parent2} \\ \boldsymbol{x}_{child2} &= w * \boldsymbol{x}_{parent2} + (1-w) * \boldsymbol{x}_{parent1} \end{aligned}$$

where w is a random value between zero and one. The crossover probability PC determines the probability of an individual to be selected for crossover. For each dimension the probability of mutation PM determines whether or not to mutate. The mutation scheme used in the LifeCycle GA model is the non-uniform mutation described on page 103 in [12]. Entry j of an individual is mutated according to:

$$\Delta x_j = \begin{cases} +(Max - x_j)(1 - r^{(1-t/T)^b}) \\ -(x_j - Min)(1 - r^{(1-t/T)^b}) \end{cases}$$

with a 50% chance each. Max is the search space maximum, Min is the minimum, r is a random number in [0..1], t is the current iteration, T is the total number of iterations and b is a parameter determining the degree of iteration number dependency. Hence, the effect of mutation decreases over the course of the iterations with this scheme.

2.3 The HillClimber

HillClimbers are individuals that refine their candidate solution independently of other individuals by examining the local neighbourhood. The hill-climbing method used in the LifeCycle model is a stochastic hill-climber as described in [10] pp. 118-120. Each HillClimber consists of a solution $\boldsymbol{x}_c$. For each iteration, a new candidate solution $\boldsymbol{x}_n$ is selected within the neighbourhood of $\boldsymbol{x}_c$. $\boldsymbol{x}_c$ is replaced by $\boldsymbol{x}_n$ with probability p given by

$$p = 1/(1 + exp(\frac{eval(\boldsymbol{x}_n) - eval(\boldsymbol{x}_c)}{T})) \qquad (minimisation)$$

where T is a parameter determining the influence of the relative merit (difference in fitness performance). In this paper, we experimented with a constant T=10, which is different from simulated annealing. The better the fitness of the neighbouring point, the higher the chance of replacement.

3 Experimental Settings

In our experiments, we compared the performance of the standard PSO, the standard GA, HillClimbers and the LifeCycle model on five numerical benchmark

Table 1. Test functions.

Sphere	$f_1(x) = \sum_{i=1}^{n} x_i^2$
Rosenbrock	$f_2(x) = \sum_{i=1}^{n-1}(100(x_{i+1} - x_i^2)^2 + (x_i - 1)^2)$
Griewank	$f_3(x) = \frac{1}{4000}\sum_{i=1}^{n}(x_i - 100)^2 - \prod_{i=1}^{n} cos(\frac{x_i - 100}{\sqrt{i}}) + 1$
Rastrigin	$f_4(x) = \sum_{i=1}^{n}(x_i^2 - 10cos(2\pi x_i) + 10)$
Ackley	$f_5(x) = 20 + e - 20e^{-0.2\sqrt{\frac{1}{n}\sum_{i=1}^{n} x_i^2}} - e^{\frac{1}{n}\sum_{i=1}^{n} cos(2\pi x_i)}$

Table 2. Search space and Initialisation ranges.

Function	Search space	Initialisation range
f_1 *Sphere*	$-100 \le x_i \le 100$	$50 \le x_i \le 100$
f_2 *Rosenbrock*	$-100 \le x_i \le 100$	$15 \le x_i \le 30$
f_3 *Griewank*	$-600 \le x_i \le 600$	$300 \le x_i \le 600$
f_4 *Rastrigin*	$-10 \le x_i \le 10$	$2.56 \le x_i \le 5.12$
f_5 *Ackley*	$-32.768 \le x_i \le 32.768$	$16.384 \le x_i \le 32.768$

problems (all minimisation) (see table 1). The first two are unimodal while the latter three are multimodal with many local minima.

The initial population of GAs and PSOs is usually uniformly distributed over the entire search space. According to Angeline [8], this can give false indications of relative performance, especially if the search space is symmetric around the origin where many test functions have their global optimum such as in the classic benchmarks that we used in this paper. In order to prevent this, we used the asymmetric initialisation method by Angeline [8] for all experiments. Search space and initialisation ranges for the experiments are shown in table 2.

In all experiments the population of the LifeCycle model was fixed at 150 individuals. These were all initialised as PSO particles.

3.1 Settings for the PSO

In the PSO model the upper limits for φ_1 and φ_2 were set to 2.0. The inertia weight w was linearly decreased from 0.7 to 0.4 and the constriction coefficient χ was set to 1. The maximum velocity v_{max} of each particle was set to half the length of the search space for each dimension (e.g. $v_{max} = 100$ for f_1 and f_2). Previous research by Shi [6] regarding scalability of the standard PSO showed that the performance of the standard PSO is not sensitive to the population size. This is also our experience from earlier work ([9, 13]). In all PSO experiments, we used a fixed population size of 20 particles.

3.2 Settings for the GA

For the GA, we used the crossover and mutation probabilities shown in table 3. Neither selection nor crossover were performed for population sizes smaller than three. Moreover, we set the mutation loop dependency b to 5 and used a fixed population size of 100 individuals.

Table 3. Crossover and mutation probability used in the GA.

Function	f_1 *Sphere*	f_2 *Rosenbrock*	f_3 *Griewank*	f_4 *Rastrigin*	f_5 *Ackley*
Crossover prob.	0.60	0.50	0.50	0.20	0.50
Mutation prob.	0.30	0.30	0.40	0.02	0.30

Table 4. Experimental results.

Function	PSO	GA	HC	LC
f_1 *Sphere*	$1.7401E-106$	$7.3762E-3$	249.5553	$4.9468E-8$
f_2 *Rosenbrock*	62.4291	99.9857	13908.6493	94.7642
f_3 *Griewank*	$1.0003E-2$	175.1830	269.9571	$5.0235E-3$
f_4 *Rastrigin*	154.4770	0.5387	725.8330	83.4788
f_5 *Ackley*	19.8617	$1.3563E-2$	21.2318	$2.0350E-11$

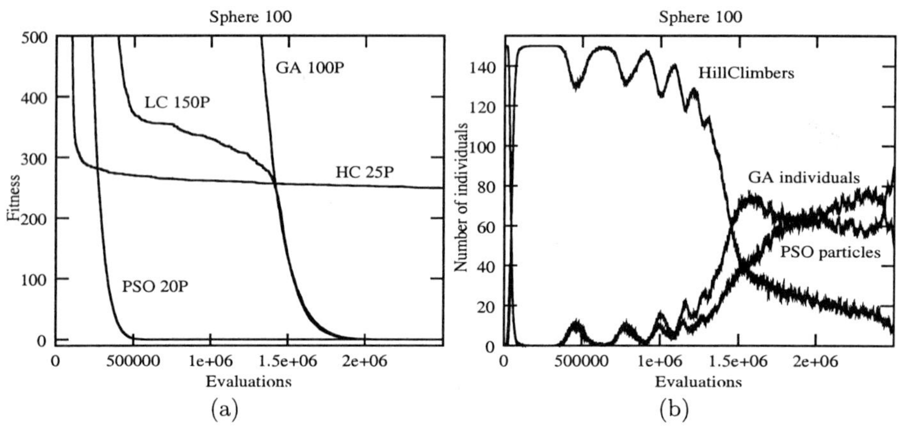

Fig. 2. Sphere function: (a) Performance of the standard GA with population size 100, the standard PSO with population size 20, 25 HillClimbers and the LifeCycle model with population size 150. (b) Composition of LifeCycle individuals.

3.3 Settings for the HillClimbers

In our HillClimber experiments, we used a constant temperature parameter $T = 10$. The size of the search neighbourhood linearly decreased from 1 to 0.001 percent of the search space for each dimension to allow fine-tuning towards the end of the run. In all HillClimbing experiments we used 25 HillClimbers simultaneously and indepedently from each other.

4 Experimental Results

The test functions that we used in our experiments were all 100 dimensional. All experiments were running for 2.500.000 evaluations. Table 4 and fig. 2 to 6 show the fitness vs. the number of evaluations regarding the five benchmark problems. The graphs illustrate the mean values of 50 repetitions for each experiment.

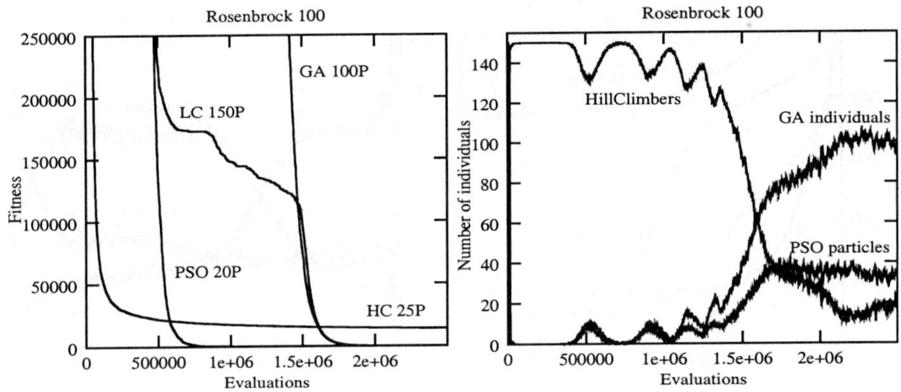

(a) (b)

Fig. 3. Rosenbrock function: (a) Performance. (b) Composition of LifeCycle individuals.

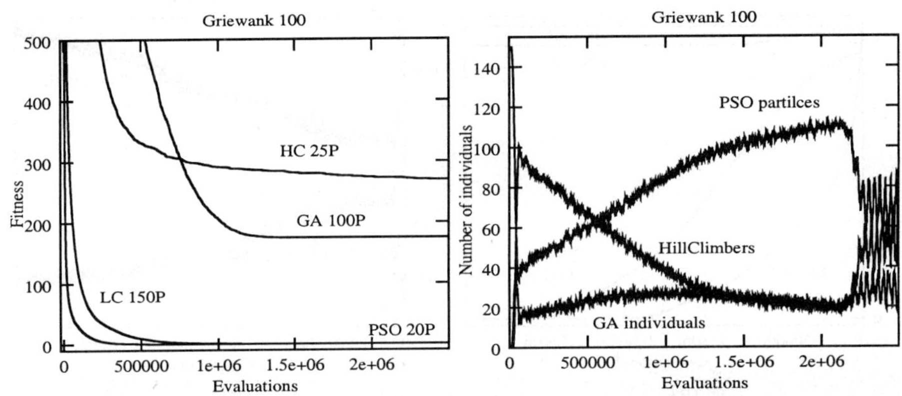

(a) (b)

Fig. 4. Griewank function: (a) Performance. (b) Composition of LifeCycle individuals.

4.1 Performance

Figure 2(a) to 6(a) show the performance of the LifeCycle (LC) model compared to the standard PSO (PSO), the standard GA (GA) and the HillClimbing (HC) algorithm described in section 2.

For the Sphere (fig. 2(a)) and Rosenbrock (fig. 3(a)) function the standard PSO turned out to have the best performance. In both cases HillClimbers yielded the fastest fitness improvements, but failed to find an exact solution. Both the LifeCycle and GA models converged slower than the PSO. The LifeCycle model converged faster and to a better value than the standard GA. Regarding the Griewank function (fig. 4(a)) the standard PSO outperformed the standard GA and the HillClimbing algorithm by far. Here, the search improvements of the LifeCycle model were a bit slower but eventually better compared to the standard PSO. For the Rastrigin function (fig. 5(a)) the standard GA found the best solution, but required more time than the LifeCycle model and the PSO. Here,

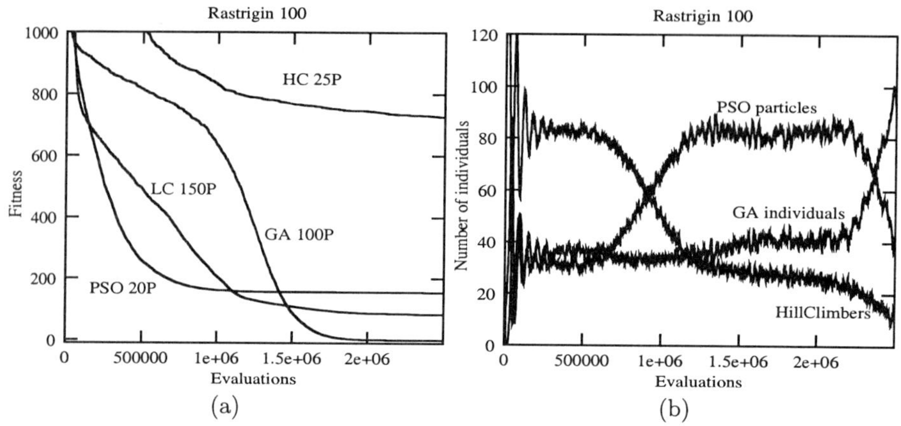

Fig. 5. Rastrigin function: (a) Performance. (b) Composition of LifeCycle individuals.

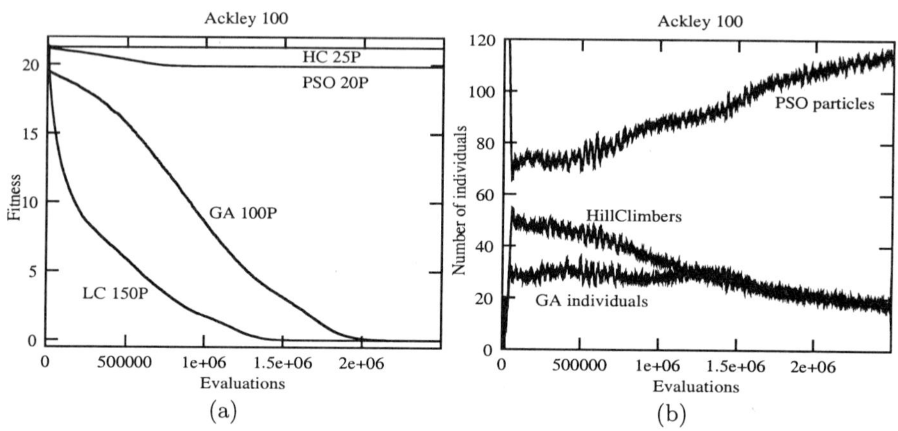

Fig. 6. Ackley function: (a) Performance. (b) Composition of LifeCycle individuals.

the PSO achieved the fastest initial search improvements, but was eventually outperformed by the LifeCycle model towards the end of the run. HillClimbers turned out to be ineffective. Finally, regarding the Ackley function (fig. 6(a)), we found that the PSO and the HillClimbing algorithms were clearly outperformed by the standard GA and the LifeCycle model, where the latter outperformed all other heuristics significantly.

4.2 Composition of the LifeCycle Individuals

Figures 2(b) to 6(b) show the frequency of the life cycle stages over time. All LifeCycle individuals were set to start out as PSO particles. For the Sphere (fig. 2(b)) and Rosenbrock (fig. 3(b)) functions all individuals prefered to become HillClimbers in the beginning of the run. After approximately 50% of the evaluations, a majority of the individuals turned into GA individuals. Moreover, the frequency of the PSO individuals increased towards the end of the run. Regarding the Griewank (fig. 4(b)) and Rastrigin (fig. 5(b)) functions, we found a similar self-adaptation pattern with a high frequency of HillClimbers in the beginning of the run. However, in contrast to the other two test functions, most HillClimbers turned into PSO particles during the run. At the end of the run, we found that GA individuals increasingly took the place of the PSO particles. In case of the Ackley function (fig. 6(b)), we found that the LifeCycle population consisted of a majority of PSO particles, which further increased over time.

5 Discussion and Future Work

Our approach of combining three standard adaptive optimisation algorithms into one self-adaptive hybrid approach turned out to be an improvement over the individual algorithms. Our results show that the LifeCycle heuristic has a generally good performance on all benchmark problems that we used in this study in contrast to the single adaptive algorithms, which have a highly problem dependent performance. For the Sphere (fig. 2) and Rosenbrock (fig. 3) functions, our results suggest that the best strategy is to start with a large number of HillClimbers, which are later turned into PSO particles. On first sight this seems to be counter-intuitive, because of the fine-tuning abilities of the HillClimbers. However, the Stochastic HillClimber has the problem that the acceptance probability of a new candidate solution approaches 0.5 in case of very small fitness differences between neighboured candidate solutions. Furthermore, the neighbourhood step size in continuous optimization has a strong impact on the quality of the results. A too low step size slows down the hill climber unnecessarily whereas a too large step size can result in missing the global optimum. Judging from our experimental results the LifeCycle model helps the standard models to achieve superior results for multimodal testfunctions. This can be seen in figures 4 to 6 for the Griewank, Rastrigin and Ackley functions. For all these three functions the LifeCycle model works best with many PSO particles. However, the pure PSO model is clearly outperformed, especially for the Ackley function (fig. 6(a)). In the present study we focused on the most simple and classic versions of GAs, PSOs, and HillClimbers. An interesting future extension would be to use the life cycle heuristic based on more advanced heuristics, such as spatially extended PSOs [13], SOC EAs [14], and simulated annealing [12]. Another idea would be to introduce tabu search as a fourth search heuristic to the LifeCycle model. Other future work could be to investigate other transition schemes. Furthermore, one could try to exchange one PSO particle with several GA individuals and vice versa, since GAs work better with much larger population sizes than PSOs.

Acknowledgments

The authors would like to thank EVALife colleagues Peter Funch and Matthias Obst of the Dept. of Zoology, Univ. of Aarhus, for the biological inspiration and Rasmus K. Ursem, EVALife, Dept. of Computer Science, Univ. of Aarhus, for reviewing the manuscript. This research was supported by the Danish Natural Science Research Council.

References

1. Lawrence, E. (ed.): Henderson's Dictionary of Biological Terms. Longman (1996).
2. Funch, P., Kristensen, R. M.: Cycliophora is a new phylum with affinities to entoprocta and ectoprocta. In: Nature 378. (1995) 711–714.
3. Filipič, B., Štrancar, J.: Genetic optimization of the EPR spectral parameters: Algorithm implementation and preliminary results. In: Schoenauer, M., Deb, K., Rudolph, G., Yao, X., Lutton, E., Merelo, J. J., Schwefel, H.-P. (eds.): Parallel Problem Solving from Nature – PPSN VI. Springer, Berlin (2000) 693–701.
4. Kennedy, J., Eberhart, R. C.: Particle swarm optimization. In: Proceedings of the 1995 IEEE International Conference on Neural Networks, Vol. 4. (1995) 1942–1948.
5. Kennedy, J.: Small worlds and mega-minds: Effects of neighborhood topology on particle swarm performance. In: Angeline, P. J., Michalewicz, Z., Schoenauer, M., Yao, X., Zalzala, A. (eds.): Proceedings of the Congress of Evolutionary Computation, Vol. 3. IEEE Press (1999) 1931–1938.
6. Shi, Y., Eberhart, R. C.: Parameter selection in particle swarm optimization. In: Porto, V. W., Saravanan, N., Waagen, D., Eiben, A. E. (eds.): Evolutionary Programming VII. Lecture Notes in Computer Science 1447. Springer, Berlin (1998) 591–600.
7. Suganthan, P.: Particle swarm optimiser with neighbourhood operator. In: Angeline, P. J., Michalewicz, Z., Schoenauer, M., Yao, X., Zalzala, A. (eds.): Proceedings of the Congress of Evolutionary Computation, Vol. 3. IEEE Press (1999) 1958–1962.
8. Angeline, P. J.: Evolutionary optimization versus particle swarm optimization: Philosophy and performance differences. In: Porto, V. W., Saravanan, N., Waagen, D., Eiben, A. E. (eds.): Evolutionary Programming VII. Lecture Notes in Computer Science, Vol. 1447. Springer, Berlin (1998) 601–610.
9. Løvbjerg, M., Rasmussen, T. K., Krink, T.: Hybrid particle swarm optimiser with breeding and subpopulations. In: Spector, L., Goodman, E. D., Wu, A., Langdon, W., Voigt, H.-M., Gen, M., Sen, S., Dorigo, M., Pezeshk, S., Garzon, M. H., Burke, E. (eds.): Proceedings of the Genetic and Evolutionary Computation Conference (GECCO-2001). Morgan Kaufmann (2001) 469–476.
10. Michalewicz, Z., Fogel, D. B.: How to Solve It: Modern Heuristics. Springer, Berlin (2000).
11. Clerc, M.: The swarm and the queen: Towards a deterministic and adaptive particle swarm optimization. In: Angeline, P. J., Michalewicz, Z., Schoenauer, M., Yao, X., Zalzala, A. (eds.): Proceedings of the Congress of Evolutionary Computation, Vol. 3. IEEE Press (1999) 1951–1957.
12. Michalewicz, Z.: Genetic Algorithms + Data Structures = Evolution Programs. Springer, Berlin (1992).
13. Krink, T., Vesterstrøm, J. S., Riget, J.: Particle Swarm Optimisation with Spatial Particle Extension. In: Fogel, D. B., Yao, X., Greenwood, G., Iba, H., Marrow, P., Shackleton, M. (eds.): Proceedings of the Fourth Congress on Evolutionary Computation (CEC-2002). (2002).
14. Krink, T., Thomsen, R., Rickers, P.: Applying Self-Organised Criticality to Evolutionary Algorithms. In: Schoenauer, M., Deb, K., Rudolph, G., Yao, X., Lutton, E., Merelo, J. J., Schwefel, H.-P. (eds.): Parallel Problem Solving from Nature – PPSN VI. Springer, Berlin (2000) 375–384.

Metaheuristics for Group Shop Scheduling

Michael Sampels[1], Christian Blum[1],
Monaldo Mastrolilli[2], and Olivia Rossi-Doria[3]

[1] IRIDIA, Université Libre de Bruxelles, CP 194/6
Av. Franklin D. Roosevelt 50, 1050 Bruxelles, Belgium
{msampels,cblum}@ulb.ac.be
[2] IDSIA, Galleria 2, 6928 Manno, Switzerland
monaldo@idsia.ch
[3] School of Computing, Napier University
10 Colinton Road, Edinburgh, EH10 5DT, Scotland
o.rossi-doria@napier.ac.uk

Abstract. The Group Shop Scheduling Problem (GSP) is a generalization of the classical Job Shop and Open Shop Scheduling Problems. In the GSP there are m machines and n jobs. Each job consists of a set of operations, which must be processed on specified machines without preemption. The operations of each job are partitioned into groups on which a total precedence order is given. The problem is to order the operations on the machines and on the groups such that the maximal completion time (makespan) of all operations is minimized. The main goal of this paper is to provide a fair comparison of five metaheuristic approaches (i.e., Ant Colony Optimization, Evolutionary Algorithm, Iterated Local Search, Simulated Annealing, and Tabu Search) to tackle the GSP. We guarantee a fair comparison by a common definition of neighborhood in the search space, by using the same data structure, programming language and compiler, and by running the algorithms on the same hardware.

1 Introduction to the Group Shop Scheduling Problem

A general scheduling problem can be formalized as follows: We consider a finite set of operations O, partitioned into m subsets $\langle M_1, \dots, M_m \rangle =: \mathcal{M}$ $(\dot{\bigcup}_{i=1}^{m} M_i = O)$ and into n subsets $\langle J_1, \dots, J_n \rangle =: \mathcal{J}$ $(\dot{\bigcup}_{k=1}^{n} J_k = O)$, together with a partial order $\preceq \subseteq O \times O$ such that $\preceq \cap J_i \times J_j = \emptyset$ for $i \neq j$, and a function $p : O \to \mathbf{N}$. A feasible solution is a refined partial order $\preceq^* \supseteq \preceq$ for which the restrictions $\preceq^* \cap M_i \times M_i$ and $\preceq^* \cap J_k \times J_k$ are total $\forall\, i, k$. The cost of a feasible solution is defined by

$$C_{\max}(\preceq^*) := \max\{\sum_{o \in C} p(o) \mid C \text{ is a chain in } (O, \preceq^*)\} \ .$$

We aim at a feasible solution which minimizes $C_{\max}$.

M_i is the set of operations that have to be processed on machine i. J_k is the set of operations that belong to job k. Each machine can process at most

J.J. Merelo Guervós et al. (Eds.): PPSN VII, LNCS 2439, pp. 631–640, 2002.

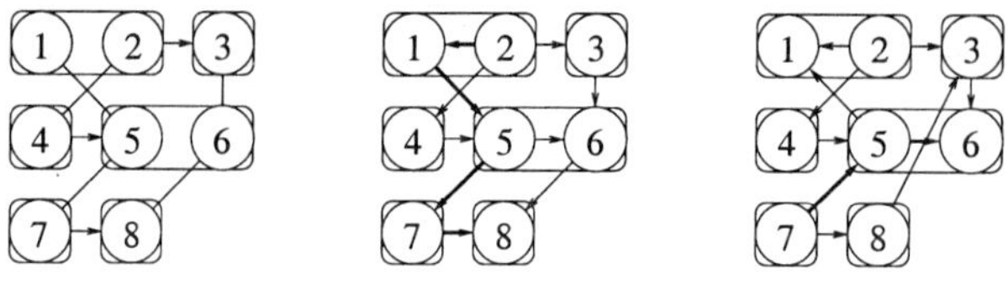

Fig. 1. (a) An example for a GSP instance on 8 operations: $O = \{1, \ldots, 8\}$, $\mathcal{J} = \{J_1 = \{1,2,3\}, J_2 = \{4,5,6\}, J_3 = \{7,8\}\}$, $\mathcal{M} = \{M_1 = \{1,5,7\}, M_2 = \{2,4\}, M_3 = \{3,6,8\}\}$, $\mathcal{G} = \{G_1 = \{1,2\}, G_2 = \{3\}, G_3 = \{4\}, G_4 = \{5,6\}, G_5 = \{7\}, G_6 = \{8\}\}$, $G_1 \prec G_2, G_3 \prec G_4, G_5 \prec G_6$, $p(1) = \cdots = p(4) = 1, p(5) = \cdots = p(8) = 2$; (b) a valid solution with $C_{\max} = 8$ on the chain $2 \preceq^* 1 \preceq^* 5 \preceq^* 7 \preceq^* 8$; (c) an optimal solution with $C_{\max} = 6$ on the chain $7 \preceq^* 5 \preceq^* 6$

one operation at a time. Operations must be processed without preemption. Operations belonging to the same job must be processed sequentially. This is expressed in the constraints for $\preceq^*$. $C_{\max}$ is the makespan of the schedule defined by $\preceq^*$.

This brief problem formulation covers well known scheduling problems: The restriction $\preceq \cap J_i \times J_i$ is total in the Job Shop Scheduling Problem (JSP), trivial ($= \{(o,o) \mid o \in J_i\}$) in the Open Shop Scheduling Problem (OSP), and either total or trivial for each i in the Mixed Shop Problem.

In this paper, we consider a weaker restriction on $\preceq$ which includes the above scheduling problems by looking at a refinement of the partition $\mathcal{J}$ to a partition into *groups* $\langle G_1, \ldots, G_g \rangle =: \mathcal{G}$. We demand that $\preceq \cap G_i \times G_i$ has to be trivial and that for $o, o' \in J$ ($J \in \mathcal{J}$) with $o \in G_i$ and $o' \in G_j$ ($i \neq j$) either $o \preceq o'$ or $o \succeq o'$ holds. We call this problem Group Shop Scheduling Problem (GSP).Note that the coarsest refinement $\mathcal{G} = \mathcal{J}$ (groups sizes are equal to job sizes) is equivalent to the OSP and the finest refinement $\mathcal{G} = \{\{o\} \mid o \in O\}$ (group sizes of 1) is equivalent to the JSP. In the following, for $G \in \mathcal{G}$ we denote $o \in G$ by $g(o) = G$; for $M \in \mathcal{M}$ we denote $o \in M$ by $m(o) = M$. An example for a GSP instance is given in Fig. 1.

2 Common Neighborhood and Local Search

For a feasible solution $\preceq^*$ a chain C is called critical path iff $\sum_{o \in C} p(o) = C_{\max}(\preceq^*)$. $\mathcal{M}$ induces on a critical path $o_1 \preceq^* \cdots \preceq^* o_q$ a subdivision into *machine blocks* of consecutive operations belonging to the same machine, as well as $\mathcal{G}$ induces a subdivision into *group blocks* of consecutive operations belonging to the same group. Brucker et al. [6] proved for the JSP that if there is a feasible solution $\preceq^{*\prime}$ with $C_{\max}(\preceq^{*\prime}) < C_{\max}(\preceq^*)$, then there is a machine block $B^i_M = o^i_1 \preceq^* \cdots \preceq^* o^i_{m_i}$ on a critical path C of $\preceq^*$ such that $\exists\, o \in B^i_M, o \neq o^i_1$ with $o^i_1 \succeq^{*\prime} o$ or $\exists\, o \in B^i_M, o \neq o^i_{m_i}$ with $o^i_{m_i} \preceq^{*\prime} o$. For the GSP, we generalize the above result.

Theorem 1. *Let $\preceq^*$ be a feasible solution to a GSP instance. If there is a solution $\preceq^{*\prime}$ with $C_{\max}(\preceq^{*\prime}) < C_{\max}(\preceq^*)$, then there is a machine or a group block $B^i = o^i_1 \preceq^* \cdots \preceq^* o^i_{n_i}$ in C such that $\exists o \in B^i, o \neq o^i_1$ with $o^i_1 \succeq^{*\prime} o$ or $\exists o \in B, o \neq o^i_{n_i} \in B$ with $o_{n_i} \preceq^{*\prime} o$.*

For the proof we refer to an extended version of this paper [14]. By this theorem it is reasonable to define the neighborhood of a feasible solution $\preceq^*$ as follows: A feasible solution $\preceq^{*\prime}$ is a neighbor of $\preceq^*$ ($\in N(\preceq^*)$) if in a critical path C of $\preceq^*$ for exactly one machine block or exactly one group block $B = o_1 \preceq^* o_2 \preceq^* \cdots \preceq^* o_{n_k-1} \preceq^* o_{n_k}$ on C the order of o_1 and o_2 or the order of o_{n_k-1} and o_{n_k} is swapped in $\preceq^{*\prime}$. This is an extension of the neighborhood which Nowicki and Smutnicki [13] used in their tabu search for the JSP.

A local search procedure can be defined recursively on the neighborhood structure as follows:

$$ls(\preceq^*) = \begin{cases} \preceq^* \text{ if } C_{\max}(\preceq^{*\prime}) \geq C_{\max}(\preceq^*) \;\forall \preceq^{*\prime} \in N(\preceq^*), \\ ls(\preceq^{*\prime\prime}) \text{ with } C_{\max}(\preceq^{*\prime\prime}) \leq C_{\max}(\preceq^{*\prime}) \;\forall \preceq^{*\prime} \in N(\preceq^*) \text{ otherwise.} \end{cases}$$

3 Metaheuristic Approaches

The OSP is an NP-hard problem (Gonzalez and Sahni [11]). The JSP is an NP-hard problem as well, as was first shown by Lenstra et al. [12]. As the GSP contains both problems, it is NP-hard, too. Metaheuristics have shown to be very successful in constructing good solutions to scheduling problems. Błażewicz et al. [2] gave a survey on exact and approximation algorithms for the JSP. Fang et al. presented a genetic algorithm for the OSP, and Taillard [16] published a Tabu Search approach.

In the following, we describe five metaheuristics for the GSP, and we analyze how they compare to each other. We study their behavior on the range of GSP instances from the JSP to the OSP and focus on the influence of the group size on the quality of the algorithms. We aim at a fair comparison of the metaheuristic concepts. Therefore we use a joint implementation of the problem representation and of the neighborhood structure. We explicitly don't aim at the use of sophisticated methods to tune the algorithms.

3.1 Ant Colony Optimization

Ant Colony Optimization (ACO) is a metaheuristic approach proposed by Dorigo et al. in [8]. The basic ingredient of ACO is the use of a probabilistic solution construction mechanism. The best known technique to construct solutions to scheduling problems is the list scheduler algorithm. To construct a solution, this list scheduler algorithm builds a sequence s of all operations from left to right. In every one of the $|O|$ construction steps, the algorithm probabilistically chooses an operation from a set S_t (where $t = 1, ..., |O|$) of admissible operations.

We use the pheromone model called $\mathsf{PH}_{\mathsf{rel}}$ (proposed by Blum and Sampels in [5]) where pheromone values are assigned to pairs of *related* operations. Two operations $o_i, o_j \in O$ are called *related* if they belong to the same group, or if they have to be processed on the same machine. Formally, a pheromone value τ_{o_i,o_j} exists, iff $g(o_i) = g(o_j)$ or $m(o_i) = m(o_j)$. The meaning of a pheromone value τ_{o_i,o_j} is that if τ_{o_i,o_j} is high then operation o_i should be scheduled before operation o_j. The choice of the next operations to schedule is handled as follows. If there is an operation $o_i \in S_t$ with no related and unscheduled operation left, it is chosen. Otherwise we choose among the operations of set S_t with the following probabilities:

$$p(s[t] = o \mid s_{t-1,|O|}, \tau) = \begin{cases} \dfrac{\min_{o_r \in S_o^{rel}} \tau_{o,o_r}}{\sum_{o_k \in S_t} \min_{o_r \in S_{o_k}^{rel}} \tau_{o_k,o_r}} & : \text{ if } \quad o \in S_t \\ 0 & : \text{ otherwise} \end{cases}$$

where $S_o^{rel} = \{o' \in O \mid m(o') = m(o) \vee g(o') = g(o), o' \text{ not scheduled yet}\}$. We also use the earliest starting time of operations with respect to the partial solution $s_{t-1,|O|}$ as heuristic information to bias these probabilities.
We implemented our algorithm in the Hyper-Cube Framework [4]. The Hyper-Cube Framework is characterized by a normalization of the contribution of every solution used for updating the pheromone values. Our algorithm is also implemented as a $\mathcal{MAX}$-$\mathcal{MIN}$ Ant System using an aggressive pheromone update and additional intensification and diversification strategies. To improve the solutions constructed by the ants we apply the local search defined in Sect. 2 and to the iteration best solution we apply a short Tabu Search of length $|O|/2$ based on the same neighborhood. A more detailed description of the this ACO algorithm can be found in [3].

3.2 Evolutionary Algorithm

The evolutionary algorithm (EA) implemented for the GSP is characterized by a steady-state evolution process and a Lamarckian use of local search. We use a best improvement local search on the neighborhood defined in Sect. 2. Tournament selection is used to choose which individuals reproduce at each generation and a "replace if better policy" is used to decide whether or not to accept the offspring for the new population.

The initial population is built using the non-delay version of the list scheduler algorithm introduced in Sect. 3.1. It builds non-delay schedules, i.e. schedules with no unnecessary idle time: No operation can be finished earlier without delaying any other one, and no machine is ever idle when there is an operation that can be started on it. The population size is set to 50. A solution is represented by a list (total order on O), which induces a total order on each $M \in \mathcal{M}$ and each $G \in \mathcal{G}$.

The crossover is a kind of uniform order based crossover respecting group precedence relations. It generates a child from two parents as follows:

1. Produce a partial child list where each position is either filled with the content of the first parent or left free, with equal probability.

2. Insert the missing operations in the partial list in the order in which they appear in the second parent.
3. Put the current operation in the first free position between the last operation of the previous group (first position in the list if the group is the first on the job) and the first of the next ones (last position in the list if the group is the last on the job), if there is any.
4. Otherwise, if there is a free position before the last operation of the previous group, shift backward all operations to fill the first free position and insert the current operation just before the first operation of the next groups.
5. Otherwise put the current operation in the position of the first operation of the next groups shifting forward the following operations until the next free position is filled.

As mutation operator we implemented a variable neighborhood search (VNS) based on the local search described in Sect. 2 for N_k, $k = 1, \ldots, 10$, where $N_k(\preceq^*) = \underbrace{N(N(\cdots N}_{k}(\preceq^*)))$. That means that a random solution in N_1 is chosen first, then the local search is applied, and if no improvement is found, a random solution in N_2 is chosen followed by local search, then a random solution in N_3 and so on until a better solution is found. The mutation rate is set to be 0.5.

3.3 Iterated Local Search

Iterated local search (ILS), in spite of its simplicity, is a powerful metaheuristic that applies a local search algorithm iteratively to modifications of the current solution. A detailed description of ILS algorithms can be found in [15]. It works as follows. First an initial locally optimal solution, with respect to the given local search, has to be built. A good starting point can be important, if high-quality solutions are to be reached quickly. Then, more importantly, a perturbation has to be defined, that is a way to modify the current solution to an intermediate state to which the local search can be applied next. Finally, an acceptance criterion is used to decide from which solution to continue the search process.

The implementation described here for the GSP works with the local search described in Sect. 2. The initial solution is generated using the same non-delay algorithm as in Sect. 3.1. The idea used for the perturbation is to modify slightly the definition of the problem instance data and apply the local search for this modified instance to the current solution regarded as a solution in the new instance; the result is the perturbed solution in the original problem instance. In the GSP the processing times of the operations, unlike group or machine data, can be easily modified so that a solution to one problem instance can be regarded as a solution to the other. For a percentage α of operations the processing time is therefore increased or decreased, with the same probability, by a certain percentage β of its value; then the local search within the modified problem instance is run for the current solution and finally the resulting locally optimal solution to the modified instance, regarded as a solution to the original instance, is the perturbed solution. Note that it is not necessarily a local optimum for the original instance. Now the local search can be applied to the intermediate perturbed solution to reach a locally optimal solution.

Finally the acceptance criterion tells us whether to continue the search from the new local optimum or from our previous solution. Random walk, better, and simulated annealing type acceptance criteria have been tested along with different values for α and β. The random walk acceptance criterion with $\alpha = 40$ and $\beta = 40$ has been selected as it gives the best performance.

3.4 Simulated Annealing

Simulated annealing (SA) is a metaheuristic based on the idea of annealing in physics [1]. This technique can be used to solve combinatorial optimization problems, especially to avoid local minima that cause problems when using simpler local search methods. The algorithm starts out with some initial solution and moves from neighbor to neighbor. If the proposed new solution is equal to or better than the current solution, it is accepted. If the proposed new solution is worse than the current solution, it is even then accepted with some positive probability. For the GSP the latter probability is

$$P_{accept} = \exp(-\frac{\Delta}{T}) = \exp(-\frac{(C_{\max}(\preceq^{*\prime}) - C_{\max}(\preceq^{*}))/C_{\max}(\preceq^{*})}{T}),$$

where $\preceq^*$ denotes the current solution, $\preceq^{*\prime}$ denotes the the proposed next solution, Δ is the percent cost change, and the temperature T is simply a control parameter. Ideally, when local optimization is trapped in a poor local optimum, simulated annealing can "climb" out of the poor local optimum. In the beginning the value of T is relatively large so that many cost-increasing moves are accepted in addition to cost-decreasing moves. During the optimization process the temperature is decreased gradually so that fewer and fewer cost-increasing moves are accepted.

The selection of the temperature is done as follows. We set the initial temperature such that the probability to accept a move with $\Delta = \delta = 0.01$ is $P_{start} = 0.9$. Moreover, at the end of the optimization process, we would like that the probability to accept a move with $\Delta = \delta = 0.01$ is $P_{end} = 0.1$. With this requirements, we constraint the temperature at time x to be $T = r^x \tau_{\max}$, where $\tau_{\max} = -\delta / \ln P_{start}$, $r = \sqrt[t_{\max}]{\delta / (\ln(1/P_{end}) \cdot \tau_{\max})}$, and where $t_{\max}$ denotes the maximum time allowed for computation.

3.5 Tabu Search

Tabu search (TS) is a local search metaheuristic which relies on specialized memory structures to avoid entrapment in local minima and achieve an effective balance of intensification and diversification. TS has proved remarkably powerful in finding high-quality solutions to computationally difficult combinatorial optimization problems drawn from a wide variety of applications [1, 10]. More precisely, TS allows the search to explore solutions that do not decrease the objective function value only in those cases where these solutions are not forbidden. This is usually obtained by keeping track of the last solutions in term of the move used to transform one solution to the next. When a move is performed it is considered *tabu* for the next T iterations, where T is the tabu status length. A solution is forbidden if it is obtained by applying a tabu move to the current solution.

According to the neighborhood defined in Sect. 2, a move for the GSP is defined by the exchange of certain adjacent critical operation pairs. We forbid the reversal of the exchange of a critical operation pair by recording the iteration number on which the exchange was performed and requiring that this number plus the current length T be strictly less than the current iteration number.

The tabu status length T is crucial to the success of the TS procedure, and we propose a self-tuning procedure based on empirical evidence. T is dynamically defined for each solution. It is equal to the number c of operations of the current critical path divided by a suitable constant d (we set $d = 5$). We choose this empirical formula since it summarizes, to some extent, the features of the given problem instance and those of the current solution. For instance, there is a certain relationship between c and the instance size, between c and the quality of the current solution. In order to diversify the search it may be unprofitable to repeat the same move often if the number of candidate moves is "large" or the solution quality is low, in some sense, when c is a "large" number.

With the aim of decreasing the probability of generating cycles, we consider a variable neighborhood set: every non tabu move is a neighbor with probability 0.8. Moreover, in order to explore the search space in a more efficient way, TS is usually augmented with some aspiration criteria. The latter are used to accept a move even if it has been marked tabu. We consider a tabu move as a neighbor with probability 0.3, and perform it only if it improves the best known solution. To summarize, the proposed TS considers a variable set of neighbors and performs the best move that improves the best known solution, otherwise performs the best non tabu move chosen among those belonging to the current variable neighborhood set.

4 Problem Instances

We tested the proposed metaheuristics on the `whizzkids97` instance. This is a GSP instance that was subject to a mathematics competition in The Netherlands in 1997 [18]. It consists of 197 operations on 15 machines and 20 jobs which are subpartitioned into 124 groups. As this is the only established GSP instance, we derived further problem instances from JSP instances.

The most prominent problem instance for the JSP is a problem with 10 machines and 10 jobs which was introduced by Fisher and Thompson [9] in 1963. It had been open for more than twenty years before the optimality of one solution was proved by Carlier and Pinson [7]. Another famous series of 80 problem instances for the JSP and 60 OSP instances was generated by Taillard [16]. We used the Fisher-Thompson-instance `ft10` and Taillard's first JSP instance `tai1` on 15 jobs and 15 machines to generate 10 resp. 15 new benchmark instances for the GSP. For both problems, we refined the job partition into a group partition by subdividing each $J_i = o_1^i \preceq \cdots \preceq o_{j_i}^i$ into b groups of fixed length $g = 1, \ldots, 10$ resp. $= 1, \ldots, 15$ (and possibly one last group of shorter length):

$$\{o_1^i, \ldots, o_g^i\}, \{o_{g+1}^i, \ldots, o_{2g}^i\}, \ldots, \{o_{(b-1)g+1}^i, \ldots, o_{j_i}^i\} \; (b = \lceil j_i/g \rceil) \; .$$

5 Evaluation and Conclusion

We tested the five developed metaheuristics on a PC with an AMD Athlon 1100 Mhz CPU under Linux using the GNU C++ compiler gcc version 2.95.3 [1]. For the `whizzkids97`, we tested each metaheuristic for 30 trials of 18000 seconds each (see Fig. 2). TS, although not finding the best solutions found by SA and ILS, showed the best overall performance, followed by ILS, SA, ACO, and EC. We tested the statistical significance of the differences between the algorithms by a pairwise Wilcoxon rank test, which was adjusted by Holm's method [17] for 5 samples. They are significant at a p-value of less than 0.01.

We further tested our metaheuristics on the 10 GSP instances derived from `ft10` for a time limit of 60 seconds per try (see Fig. 3). TS showed the best result for most group sizes, and ACO was on the second rank. EC performed well for small and large group sizes, but was worse than SA for groups of medium size.

We observed a similar behavior for the 15 instances derived from `tai1`, which we tested for running times of 600 and 1800 seconds per try. ACO is for nearly all group sizes quite close to the performance of TS. However, the TS is the only algorithm that finds (even within 600 seconds) the optimal solution for the original JSP version of `tai1`. EC again performs rather poorly on medium group sizes and performs well on small and big group sizes.

We noticed that the SA in general compared well to the other algorithms. This indicates the power of the neighborhood structure defined in Sect. 2. Although the TS approach yielded the overall best performance, for some group sizes other metaheuristics showed advantages. Our fair comparison showed that depending on the position of the problem instance between the JSP and the OSP different heuristic techniques are helpful. The GSP might best be tackled by a hybrid metaheuristic approach that combines the elements of the algorithms described in this work according to the results of our analysis.

Acknowledgements

We would like to thank Anthony Liekens and Huub ten Eikelder for the implementation of the problem representation and the local search. Our work was supported by the *Metaheuristics Network,* a Research Training Network funded by the Improving Human Potential Programme of the CEC, grant HPRN-CT-1999-00106, and by Swiss National Science Foundation project 20-63733.00/1,

[1] All algorithms, the test instances and a detailed evaluation of the generated results can be found on `http://iridia.ulb.ac.be/~msampels/gsp.data` .

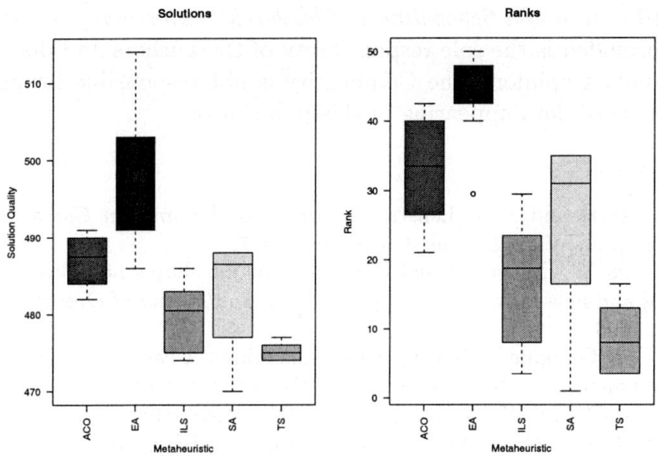

Fig. 2. The absolute values of the solutions to the `whizzkids97` problem instance generated by the five metaheuristics (left) and their relative rank in the comparison among each other (right) are depicted in two boxplots. A box shows the range between the 25 % and the 75 % quantile of the data. The median of the data is indicated by a bar. The whiskers extend to the most extreme data point which is no more than 1.5 times the interquantile range from the box. Extreme points are indicated as circles

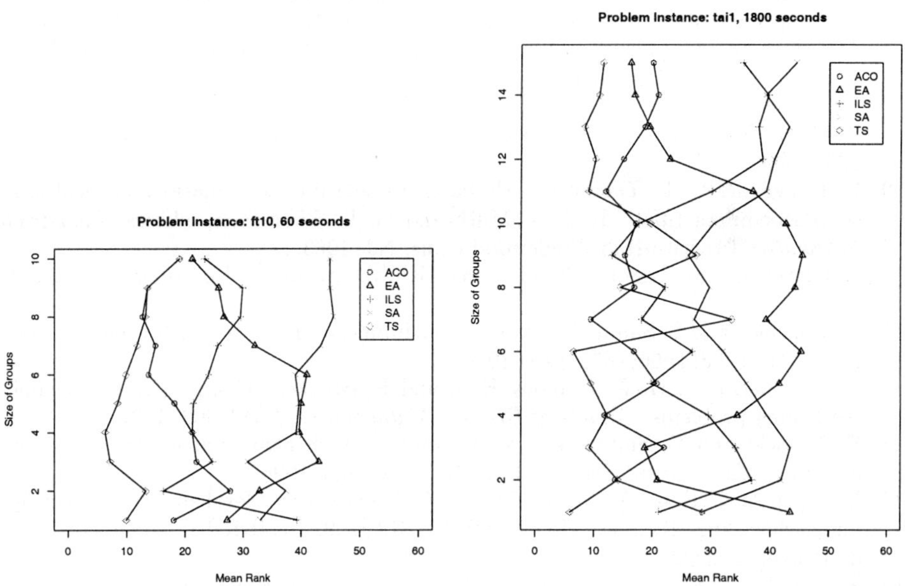

Fig. 3. Mean ranks of the solutions generated by ACO, EA, ILS, SA, and TS to instances derived from `ft10` and `tai1`. For `ft10` (left) the group size was varied from 1 to 10, and for `tai1` (right) it was varied from 1 to 15. For the `ft10` instances we ran the algorithms for 60 seconds, for the `tai1` instances for 1800 seconds

Resource Allocation and Scheduling in Flexible Manufacturing Systems. The information provided is the sole responsibility of the authors and does not reflect the Community's opinion. The Community is not responsible for any use that might be made of data appearing in this publication.

References

1. E. H. L. Aarts and J. K. Lenstra, editors. *Local Search in Combinatorial Optimization.* John Wiley & Sons, Chichester, 1997.
2. J. Błażewicz, W. Domschke, and E. Pesch. The job shop scheduling problem: Conventional and new solution techniques. *European Journal of Operational Research*, 93:1–33, 1996.
3. C. Blum. ACO applied to Group Shop Scheduling: A case study on Intensification and Diversification. In *Proceedings of the 3rd International Workshop on Ant Algorithms (ANTS 2002) (to appear)*, 2002. Also available as technical report TR/IRIDIA/2002-08, IRIDIA, Université Libre de Bruxelles.
4. C. Blum, A. Roli, and M. Dorigo. HC-ACO: The hyper-cube framework for Ant Colony Optimization. In *Proceedings of the 4th Meta-heuristics International Conference (MIC 2001)*, volume 2, pages 399–403, 2001.
5. C. Blum and M. Sampels. Ant colony optimization for FOP shop scheduling: A case study on different pheromone representations. In *Proceedings of the 2002 Congress on Evolutionary Computation (CEC 2002)*, volume 2, pages 1558–1563, 2002.
6. P. Brucker, B. Jurisch, and B. Sievers. A branch and bound algorithm for the job-shop scheduling problem. *Discrete Applied Mathematics*, 49:107–127, 1994.
7. J. Carlier and E. Pinson. An algorithm for solving the job-shop problem. *Management Science*, pages 164–176, 1989.
8. M. Dorigo and G. Di Caro. The Ant Colony Optimization meta-heuristic. In D. Corne, M. Dorigo, and F. Glover, editors, *New Ideas in Optimization.* McGraw-Hill, 1999.
9. H. Fisher and G. L. Thompson. Probabilistic learning combinations of local job-shop scheduling rules. In J. F. Muth and G. L. Thompson, editors, *Industrial Scheduling.* Prentice-Hall, Englewood Cliffs, NJ, 1963.
10. F. Glover and M. Laguna. *Tabu Search.* Kluwer Academic Publishers, Boston et al., 1998.
11. T. Gonzalez and S. Sahni. Open shop scheduling to minimize finish time. *Journal of the ACM*, 23(4):665–679, Oct. 1976.
12. J. K. Lenstra, A. H. G. Rinnooy Kan, and P. Brucker. Complexity of machine scheduling problems. *Annals of Discrete Mathematics*, 1:343–362, 1977.
13. E. Nowicki and C. Smutnicki. A fast taboo search algorithm for the job shop problem. *Management Science*, 42(6):797–813, June 1996.
14. M. Sampels, C. Blum, M. Mastrolilli, and O. Rossi-Doria. Metaheuristics for Group Shop scheduling. Technical Report TR/IRIDIA/2002-07, IRIDIA, Université Libre de Bruxelles, 2002.
15. T. Stützle. *Local Search Algorithms for Combinatorial Problems – Analysis, Improvements, and New Applications.* PhD thesis, TU Darmstadt, Germany, 1998.
16. E. Taillard. Benchmarks for basic scheduling problems. *European Journal of Operational Research*, 64:278–285, 1993.
17. S. P. Wright. Adjusted p-values and simultaneous inference. *Biometrics*, 48:1005–1013, 1992.
18. http://www.win.tue.nl/whizzkids/1997 .

Experimental Investigation of Three Distributed Genetic Programming Models

Marco Tomassini[1], Leonardo Vanneschi[1],
Francisco Fernández[2], and Germán Galeano[2]

[1] Computer Science Institute, University of Lausanne
1015 Lausanne, Switzerland
[2] Computer Science Department, University of Extremadura
C/ Calvario, s/n. 06800 Merida, Spain

Abstract. Three models of distributed Genetic Programming are presented comprising synchronous and asynchronous communication. These three models are compared with each other and with the standard panmictic model on three well known Genetic Programming benchmarks. The measures used are the computational effort, the phenotypic entropy of the populations, and the execution time. We find that all the distributed models are better than the sequential one in terms of effort and time. The differences among the distributed models themselves are rather small in terms of effort but one of the asynchronous models turns out to be significantly faster. The entropy confirms that migration helps in conserving some phenotypic diversity in the populations.

1 Introduction

Parallelism is implicit in nature, where all individuals compete simultaneously for survival. Evolutionary computation has taken inspiration from this observation and several models have been proposed. Some of the first efforts made to investigate the usefulness of parallelism were done in Genetic Algorithms (GAs). Among others, Cantú-Paz [4] modeled the main features of parallel GAs and compared this technique with its sequential counterpart. For reviews of the field see for instance [1,12]. Parallelism in Genetic Programming (GP) is newer and has been less investigated: only a few studies have appeared to date [3,5, 10], and, to our knowledge, no theoretical study has been performed in Parallel and Distributed Genetic Programming (PADGP). Given this situation, we think that an extensive empirical study of several representative test cases is needed in order to attempt to clarify the issues. Although our empirical results are only of a qualitative nature, we believe that such systematic studies are a necessary prerequisite to any efficient use of distributed GP in applications and should pave the way for a better theoretical understanding of co-evolving populations of GP individuals. In this paper we present three original PADGP models and we empirically compare them by means of some experiments of different nature. No investigation of these GP models have to our knowledge already been done. The models have been implemented using the Message Passing Interface standard

J.J. Merelo Guervós et al. (Eds.): PPSN VII, LNCS 2439, pp. 641–650, 2002.

(MPI) [8]. Implementation details can be found in [7]. The article is structured as follows: in section 2 we describe our PADGP models, section 3 gives an account of the benchmarks and the GP parameters used in our experiments, in section 4 we describe the measures used to compare our models and section 5 presents the experimental results. Finally, section 6 offers our conclusions.

2 Description of Distributed Models Used

The first two PADGP models we present in this paper are based on the *Master/Slave* paradigm. In the first one (that we could define *fully synchronized* model), each island runs a standard generational GP and individuals are exchanged from each subpopulation (slave) to the master with synchronous communications at fixed synchronization points between generations. In the second one, asynchronous receive operations are used by the subpopulation processes. In the third one, subpopulations communicate using asynchronous receive operations with no centralized master. In the following paragraphs, the three models are described in more detail.

2.1 Synchronous Communication with Master

In this model, each population executes the following steps:

- Create a random population of programs;
- **While** termination condition not reached **do**
  - Assign a fitness value to each individual;
  - Select a set of individuals for reproduction;
  - Recombine and mutate the new population;
  - **If** communication has to take place at this iteration **then**
    - Select the best n individuals (with $n \geq 0$), pack them and send them to the master (with one MPI_Send operation);
    - Receive a set of n new individuals from the master (with one MPI_Recv operation), unpack them and replace the n worst individuals in the population;

    EndIf

  EndWhile

And the master executes the following steps:

- **For** each iteration in which communication has to take place **do**
  - **For** each population p **do**
    - Receive a set of n individuals from p (MPI_Recv);
    - Send them to another population according to the chosen topology (MPI_Send);

    EndFor

  EndFor

The most used termination criterions are: a satisfactory solution has been found, a maximum computing time is reached or a maximum number of generations have been executed.

2.2 Asynchronous Communication with Master

The behaviour of the populations can be described by the following algorithm:

- Create a random population of programs;
- **While** termination condition not reached **do**
  - Assign a fitness value to each individual;
  - Select a set of individuals for reproduction;
  - Recombine and mutate the new population;
  - **If** communication has to take place at this iteration **then**
    - Select the best n individuals (with $n \geq 0$), pack them and send them to the master (with one MPI_Send operation);
    - Receive a set of n new individuals from the master (with one non-blocking MPI_Irecv operation)

    EndIf
  - Test the completion of all the pending receives;
  - **For** all the terminated receives
    - replace the n worst individuals in the population with the n received individuals

    EndFor

  EndWhile

while the master executes exactly the same steps as in the synchronous model.

2.3 Asynchronous Communication without Master

In this model each population executes the following steps:

- Create a random population of programs;
- **While** termination condition not reached **do**
  - Assign a fitness value to each individual;
  - Select a set of individuals for reproduction;
  - Recombine and mutate the new population;
  - **If** communication has to take place at this iteration **then**
    - Select the best n individuals (with $n \geq 0$), pack them and send them to another process according to the chosen topology (using one MPI_Send operation);
    - Receive a set of n new individuals from another process according to the chosen topology (with one non-blocking MPI_Irecv operation)

    EndIf
  - Test the completion of all the pending receives;
  - **For** all the terminated receives
    - replace the n worst individuals in the population with the n received individuals

    EndFor

  EndWhile

3 Test Functions and GP Parameters

In the absence of theoretical guidance it is difficult to decide which problems should be considered. If only GP theory had the same degree of maturity as GA theory, one could choose a mix of synthetic benchmarks, standard test problems and provably difficult problems. But unfortunately this is not the case yet. Thus, we decided to address a set of problems that have been classically used for testing GP. Obviously, the set is far from complete or even representative of a wide range of typical GP problems. However, these experimental studies are very time consuming and we decided that this would at least represent a useful first step. The following section briefly describes the test problems on which more detailed explanations can be found in [9].

Even Parity 5 Problem. The boolean Even Parity k function of k Boolean arguments returns *true* if an even number of its boolean arguments evaluates to true, otherwise it returns *false*. If $k = 5$, then 32 fitness cases must be checked to evaluate the fitness of an individual. The fitness can be computed as 32 minus the number of hits over the 32 cases. Thus a perfect individual has fitness 0, while a bad individual has fitness 32. the set of functions we employed for GP individuals is the following: $F = \{NAND, NOR\}$. The terminal set in this problem is composed of 5 different boolean variables $T = \{a, b, c, d, e\}$.

Artificial Ant Problem on the Santa Fe Trail. In this problem, an artificial ant is placed on a toroidal grid. Some of the cells from the grid contain food pellets. The goal is to drive the ant on the path to make it eat all the food. We use the same set of functions and terminals as in [9]. As fitness function, we use the total number of food pellets lying on the trail (70) minus the amount of food eaten by the ant during his path.

Symbolic Regression Problem. The problem aims to find a program which matches a given equation. We employ the classic polynomial equation $f(x) = x^4 + x^3 + x^2 + x$, and the input set is composed of the values 1 to 1000. For this problem, the set of functions is the following: F={*,//,+,-}, where // is like / but returns 0 instead of *error* when the divisor is equal to 0, thus allowing syntactic closure. We define the fitness as the sum of the square errors at each test point.

GP Parameters. In all the experiments performed, we used the same set of GP parameters: generational GP, crossover rate 95%, mutation rate 0.1%, tournament selection of size 10, ramped half and half initialization, maximum depth of individuals for the creation phase 6, maximum depth of individuals for crossover 17, no elitism, 1500 individuals (respectively 200 for Symbolic Regression, to avoid the fast convergence typical of this problem) divided in 1, 5 or 10 subpopulations. The subpopulations were connected using a ring topology, and a number of individuals equal to the 10% of the subpopulations size were exchanged between demes at each 10 generations (These last values were found to be suitable in previous work [5, 6]).

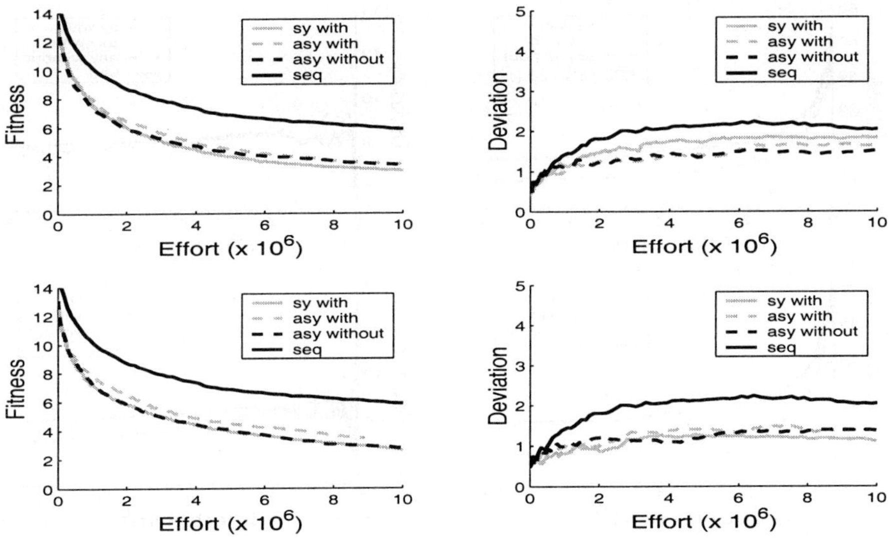

Fig. 1. Fitness and its standard deviation against effort for the Even Parity problem. Upper graphics: 5 populations of 300 individuals, lower graphics: 10 populations of 150 individuals. Sequential version (a panmictic population of 1500 individuals) shown in black.

4 Computational Effort and Population Entropy

In our experiments data are analyzed by means of the *Computational Effort* which has been defined as the total number of nodes evaluated for a given number of generations. To calculate this measure, we must firstly compute the average number of nodes at generation g, taking into account all the populations that are simultaneously working (we will indicate it as avg_length_g) and then compute the partial effort at that generation defined as: $PE_g = i \times p \times avg\_length_g$, where p is the number of populations and i is the number of individuals per population. Finally, we calculate the effort of computation E_g, at generation g, as $E_g = PE_g + PE_{g-1} + PE_{g-2} + ... + PE_0$. This measure is problem-specific but it is useful for comparing different solutions of the same problem.

The measure we used to compute phenotipic diversity in a population P, is the *Population Entropy*, defined as [11]: $H(P) = -\sum_{j=1}^{N} f_j \log(f_j)$, where f_j is the number of individuals in the population P having a certain fitness and N is the total number of individuals in P.

5 Experimental Results

Fitness vs Computational Effort. Due to the stochastic nature of the evolutionary process, all the results presented in this paper were obtained by averaging 60 independent runs of the same experiments. In order to assess statistical significance of the results, graphics of the *Standard Deviation* of fitness between the different runs are also presented. Figures 1, 2 and 3 show the graphics of

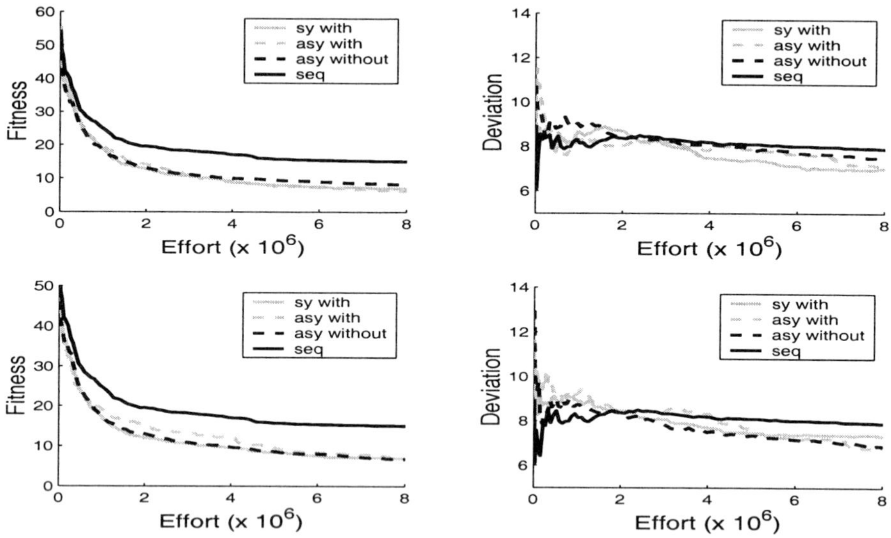

Fig. 2. Fitness and its standard deviation against effort for the Ant problem. Upper graphics: 5 populations of 300 individuals, lower graphics: 10 populations of 150 individuals. Sequential version (a panmictic population of 1500 individuals) shown in black.

fitness and standard deviation against computational effort respectively for the Even Parity 5 problem, the Artificial Ant problem and the Symbolic Regression problem. From these figures, we observe that the distances between the curves of the parallel models and the ones of the sequential versions are always larger than standard deviations, while this is not the case if we compare the curves of parallel models between themselves. This allow us to conclude that, for all the cases presented, the parallel models have a faster convergence than the sequential one, while the speeds of convergence of the parallel models can be considered statistically equivalent.

Population Entropy. Results presented in the following are obtained by averaging 60 independent runs of the same experiments. Curves relative to parallel models are obtained by averaging the entropy of all the subpopulations at each iteration. Figures 4, 5 and 6 show the population entropy against the generation number respectively for the Even Parity 5 problem, the Artificial Ant problem and the Symbolic Regression problem. These figures show that the sequential model has a fast entropy increase in the first few generations and a slow decrease in the rest of the execution. The synchronous with master model shows a sudden decrease of the entropy each time the generation number is a multiple of 10 (messages are synchronously sent each 10 generations), and a sudden increase in the immediately following few generations. This behavior is reasonable since we are sending the best individuals. Thus, it is highly probable that all the individuals sent have similar fitness and, initially, the phenotypic diversity injected in the receiving population is lower. But it is possible that the individu-

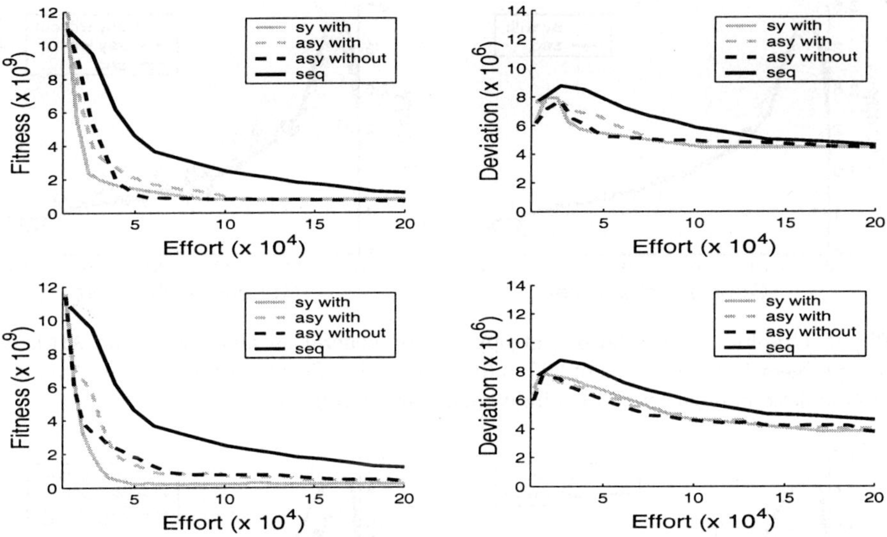

Fig. 3. Fitness and its standard deviation against effort for the Symbolic regression problem. Upper graphics: 5 populations of 40 individuals, lower graphics: 10 populations of 20 individuals. Sequential version (a panmictic population of 200 individuals) shown in black. Note the difference in the scale between fitness curves and standard deviation.

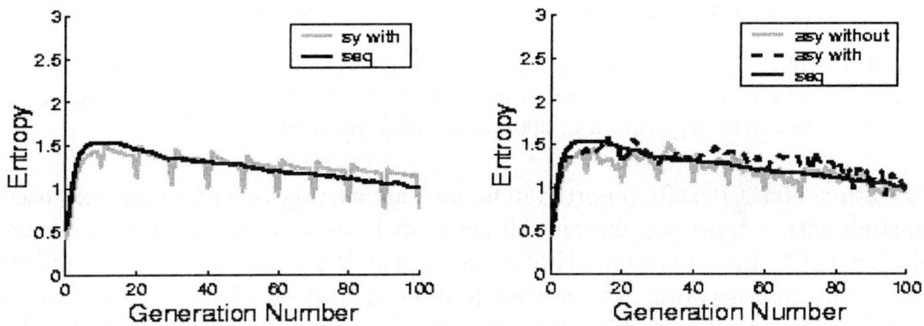

Fig. 4. Graphics of population entropy against generation number for the Even Parity 5 problem (5 populations of 300 individuals each). Curves of the sequential version (a panmictic population of 1500 individuals) are shown too.

als sent have a different syntactic structure from the individuals already present in the receiving population. So, the application of genetic operators in the few successive generations will foster the creation of new individuals with different characteristics (and thus, probably, with different fitness) from the preexisting ones, which leads to a growth of the phenotypic entropy. The asynchronous models show a more irregular behavior, due to the fact that messages are received whenever they arrive, and not at fixed times. In all cases, migrating individuals seems to help in alleviating stagnation in the populations, thus confirming the intuition that is at the basis of distributed evolutionary algorithms.

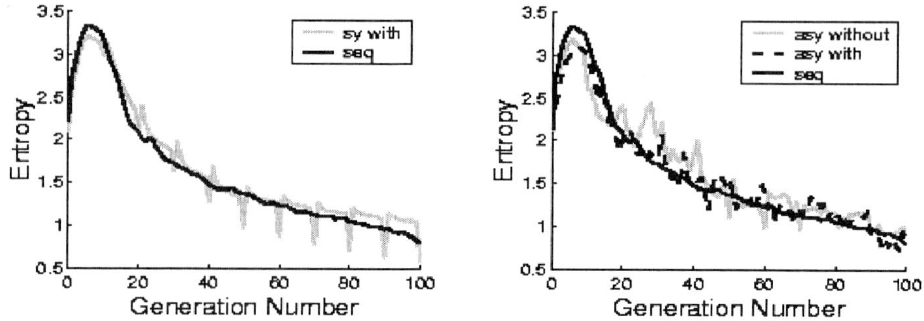

Fig. 5. Graphics of population entropy against generation number for the Artificial Ant problem (5 populations of 300 individuals each). Curves of the sequential version (a panmictic population of 1500 individuals) are shown too.

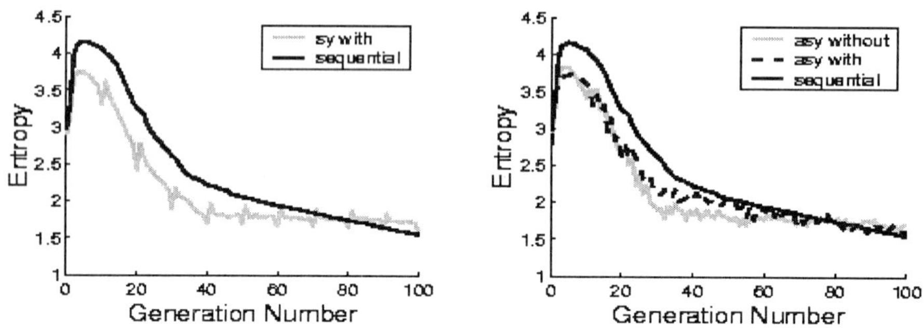

Fig. 6. Graphics of population entropy against generation number for the Symbolic Regression problem (5 populations of 40 individuals each). Curves of the sequential version (a panmictic population of 200 individuals) are shown too.

Execution Time. Results reported in figure 7 are averages over 60 runs, executed starting with a fixed population. All the runs have been executed on a cluster of eleven PCs Intel Pentium III-500 with 128M RAM and an Ethernet 10MB interconnection network. All the runs have been stopped after 100 generations. Since we are not interested in convergence in these experiments, we considered a total number of 1500 individuals for Symbolic Regression too. These tables show that all the parallel models are faster than the sequential version. Moreover, the slower populations for all the parallel models show approximately the same completion times, while there is a larger difference in convergence time among the subpopulations in the asynchronous without master model, which was to be expected due to the characteristics of the model.

6 Conclusions

We have presented four GP models: the standard sequential version and three original parallel and distributed versions (synchronous with master, asynchronous with master and asynchronous without master) and we have compared them from different points of view: computational effort, phenotypic entropy of

Even Parity 5	5 Populations		10 Populations	
	Slower Pop.	Faster Pop.	Slower Pop.	Faster Pop.
Sync. with master	402.75	402.2	191.23	190.72
Async. with mast.	435.21	371.4	207.7	178.47
Asy. without mast.	440.66	101.01	208.66	41.68
Sequential	1688.27			

Artificial Ant	5 Populations		10 Populations	
	Slower Pop.	Faster Pop.	Slower Pop.	Faster Pop.
Sync. with master	325.35	323.59	129.42	124.19
Async. with mast.	341.81	275.63	138.61	113.3
Asy. without mast.	343.42	117.25	140.97	52.75
Sequential	657.02			

Symbolic Reg.	5 Populations		10 Populations	
	Slower Pop.	Faster Pop.	Slower Pop.	Faster Pop.
Sync. with master	446.82	440.13	170.72	157.78
Async. with mast.	455.71	321.46	175.77	146.36
Asy. without mast.	458.16	173.64	178.02	91.05
Sequential	1299.81			

Fig. 7. Completion times (in seconds) for the faster and slower population in the case of 5 and 10 subpopulations (respectively of 300 and 150 individuals each) for Even Parity 5, Artificial Ant and Symbolic Regression problems. Completion times of the sequential version (a panmictic population of 1500 individuals) is shown too.

the populations, and the execution time. No similar studies have to our knowledge already been done. The effort experiments showed a faster convergence of the parallel models compared to the sequential one and a substantial equivalence of the three parallel models among themselves. The entropy experiments confirmed that migration can help in promoting diversity, especially towards the end of the evolution, when it is most useful. The execution time experiments pointed out that the asynchronous models are faster when they find the solution, and they show a greater spread of island completion times with respect to synchronous islands. All the results presented appear to be independent from the benchmark used and they qualitatively agree with results presented in [2] for distributed genetic algorithms. This work is part of a longer term project whose aim is a better understanding of the dynamics of multi-population GP by way of experiment and by theoretical modelling. In the future, we plan to extend the study of synchronous and asynchronous models for GP to other classes of functions, and to extend the capabilities of our systems towards geographically enlarged metacomputing frameworks.

References

1. E. Alba and J. M. Troya. A survey of parallel distributed genetic algorithms. *Complexity*, 4(4):31–52, 1999.
2. E. Alba and J. M. Troya. Analyzing synchronous and asynchronous parallel distributed genetic algorithms. *Future Generation Computer Systems*, 17:451–465, January 2001.
3. D. Andre and J. R. Koza. Parallel genetic programming: A scalable implementation using the transputer network architecture. In P. Angeline and K. Kinnear, editors, *Advances in Genetic Programming 2*, pages 317–337, Cambridge, MA, 1996. The MIT Press.
4. E. Cantú-Paz. *Efficient and Accurate Parallel Genetic Algorithms.* Kluwer Academic Press, 2000.

5. F. Fernández, M. Tomassini, W. F. Punch III, and J. M. Sánchez. Experimental study of multipopulation parallel genetic programming. In Riccardo Poli, Wolfgang Banzhaf, William B. Langdon, Julian F. Miller, Peter Nordin, and Terence C. Fogarty, editors, *Genetic Programming, Proceedings of EuroGP'2000*, volume 1802 of *LNCS*, pages 283–293. Springer-Verlag, 2000.
6. F. Fernández, M. Tomassini, and L. Vanneschi. Studying the influence of communication topology and migration on distributed genetic programming. In J. Miller, M. Tomassini, P. L. Lanzi, C. Ryan, A. Tettamanzi, and W. Langdon, editors, *Genetic Programming, Proceedings of EuroGP'2001*, volume 2038 of *LNCS*, pages 51–63. Springer-Verlag, 2001.
7. F. Fernández, M. Tomassini, L. Vanneschi, and L. Bucher. A distributed computing environment for genetic programming using MPI. In J. Dongarra, P. Kaksuk, and N. Podhorszki, editors, *Recent Advances in Parallel Virtual Machine and Message Passing Interface*, volume 1908 of *Lecture Notes in Computer Science*, pages 322–329. Springer-Verlag, Heidelberg, 2000.
8. Message Passing Interface Forum. MPI: A message-passing interface standard. *International Journal of Supercomputer Applications*, 8(3-4):165–414, 1994.
9. J. R. Koza. *Genetic Programming.* The MIT Press, Cambridge, Massachusetts, 1992.
10. W. Punch. How effective are multiple populations in genetic programming. In J. R. Koza, W. Banzhaf, K. Chellapilla, K. Deb, M. Dorigo, D. B. Fogel, M. Garzon, D. Goldberg, H. Iba, and R. L. Riolo, editors, *Genetic Programming 1998: Proceedings of the Third Annual Conference*, pages 308–313, San Francisco, CA, 1998. Morgan Kaufmann.
11. Justinian P. Rosca. Entropy-driven adaptive representation. In Justinian P. Rosca, editor, *Proceedings of the Workshop on Genetic Programming: From Theory to Real-World Applications*, pages 23–32, Tahoe City, California, USA, 1995.
12. M. Tomassini. Parallel and distributed evolutionary algorithms: a review. In K. Miettinen, M. äkelä, P. Neittanmäki, and J. Périaux, editors, *Evolutionary Algorithms in Engineering and Computer Science*, pages 113–133. J. Wiley, New York, 1999.

Model-Based Search for Combinatorial Optimization: A Comparative Study

Mark Zlochin[1,*] and Marco Dorigo[2]

[1] Dept. of Computer Science, Technion – Israel Institute of Technology, Haifa, Israel
zmark@cs.technion.ac.il
[2] IRIDIA, Université Libre de Bruxelles, Brussels, Belgium
mdorigo@ulb.ac.be

Abstract In this paper we introduce *model-based search* as a unifying framework accommodating some recently proposed heuristics for combinatorial optimization such as ant colony optimization, stochastic gradient ascent, cross-entropy and estimation of distribution methods. We discuss similarities as well as distinctive features of each method, propose some extensions and present a comparative experimental study of these algorithms.

1 Introduction

The necessity to solve $\mathcal{NP}$-hard problems, for which the existence of efficient exact algorithms is highly unlikely, has led to a wide range of heuristic algorithms that implement some sort of search in the solution space. These heuristic algorithms can be classified, similarly to what is done in the machine learning field [15] , as being either *instance-based* or *model-based*. Most of the classical search methods may be considered instance-based, since they generate new candidate solutions using solely the current solution or the current "population" of solutions. Typical representatives of this class are genetic algorithms or local search and its variants, such as, for example, simulated annealing and iterated local search.On the other hand, in the last decade several new methods, which may be classified as *model-based search* (MBS) algorithms, have been proposed. In MBS algorithms, candidate solutions are generated using a parameterized probabilistic model that is updated using the previously seen solutions in such a way that the search will concentrate in the regions containing high quality solutions.

In [20], several MBS approaches, such as *ant colony optimization* (ACO) metaheuristic [6], *stochastic gradient ascent* (SGA) [16, 12], *cross-entropy* (CE) method [18] and *estimation of distribution algorithms* (EDAs) [11], were considered within a common framework, and analysis of their similarities as well as their distinctive features was provided.

* This work was carried out while the author was at IRIDIA, Université Libre de Bruxelles, Belgium.

J.J. Merelo Guervós et al. (Eds.): PPSN VII, LNCS 2439, pp. 651–661, 2002.

In this paper we provide a detailed comparative analysis of these algorithms, in the context of unconstrained binary coded problems. In addition to the analytical comparison, we also present an experimental comparison of the MBS methods discussed in this paper using the MAXSAT problem as a test bed.

2 Model-Based Search

Let us consider a minimization problem$(\mathcal{S}, f)$, where $\mathcal{S}$ is the *set of feasible solutions*, f is the *objective function*, which assigns to each solution $s \in \mathcal{S}$ a cost value $f(s)$. The goal of the minimization problem is to find a feasible solution of minimal cost.

At a very general level, the model-based search approach attempts to solve this minimization problem by repeating the following two steps:

- Candidate solutions are constructed using some parameterized probabilistic model, that is, a parameterized probability distribution over the solution space.
- The candidate solutions are used to modify the model in a way that is deemed to bias future sampling toward high quality solutions.

For any algorithm belonging to this general scheme, two components, corresponding to the two steps above, need to be instantiated:

- A probabilistic model that allows an efficient generation of the candidate solutions.
- A rule for updating the model's parameters.

In this paper we restrict our attention to those problems whose solutions can be coded as unconstrained binary strings, although ACO, SGA and CE can be applied in a more general context as well. A more general discussion about these methods and the relations between them is given in [20]. All of the methods described in this paper employ the following probabilistic model for generating candidate solutions:

- Every bit position, $1 \leq i \leq n$, has an associated parameter τ_i.
- The bits, s_i, are generated independently, with $P(s_i = 1) = F(\tau_i)$[1].

The parameter vector and the resulting probability distribution over the solution space are denoted by $\mathcal{T}$ and $P_{\mathcal{T}}(\cdot)$ correspondingly. The parameters are typically initialized in such a way that the initial distribution is a uniform one.

It should be noted that several algorithms that fall within the MBS framework and that allow for dependencies between the bit positions have been described in the literature (e.g., MIMIC [3], BOA [14] or variants of ACO [17]). Some of these algorithms were reported to yield certain improvement over the simpler algorithms, which generate bits for different positions independently.

[1] In fact, all the algorithms considered in this paper, except for SGA, use simply $F(\tau_i) = \tau_i$.

However, these comparisons were performed on a basis of equal number of iterations, rather than equal computational time. On the other hand, our personal experience, as well as some results in the literature (e.g., [17]), suggest that the considerable computational overhead imposed by using a more complex model renders the "dependencies learning" algorithms uncompetitive. Consequently, these algorithms are not considered in this paper.

3 The Algorithms

In this section we give a brief description of several existing MBS algorithms and discuss the relationships among them. The reader is referred to [20] and references therein for a more detailed discussion. It should be emphasized that the probabilistic model employed by all these algorithms is the same, hence the only difference among them is in the way the parameters are interpreted and modified.

3.1 Ant Colony Optimization

One well-established approach that belongs to the MBS framework is the *ant colony optimization* (ACO) metaheuristic. In ACO algorithms, solutions are generated using stochastic procedures, called *artificial ants*, which construct them by iteratively adding solution components. The components are chosen with a probability which is a function of so called *pheromone* values associated to components. After constructing the solutions, the pheromone values associated with the components belonging to good solutions are increased. This metaheuristic has been successfully applied to the solution of numerous NP-hard problems [6] as well as to time-varying stochastic optimization problems [4]. A particular variant of this metaheuristic, called Hyper-Cube (HC) ACO [2], has been recently proposed in the context of combinatorial problems with binary coded solutions. In HC-ACO the pheromones are bounded between zero and one and the pheromone update rule can be described by the following general scheme:

HC_ACO_UPDATE

$$\tau_i \leftarrow (1-\rho)\tau_i + \rho \frac{\sum_{s\in S_t} Q_f(s)s_i}{\sum_{s\in S_t} Q_f(s)}. \tag{1}$$

where S_t is the sample in the t-th iteration, ρ, $0 \le \rho < 1$, is the learning rate and $Q_f(s|S_1,\ldots,S_t)$ is some "quality function", which is typically required to be non-increasing with respect to f and is defined over the "reference set" $\hat{S}_t$.

In the considered case of unconstrained problems, the pheromones are equal to the marginal probabilities of the corresponding positions, with the bits at different positions being assigned independently[2].

[2] In a preliminary study, we have also considered using heuristic information, similarly to [17]. However, the improvement in performance was negligible, when compared to the improvement obtained with local search. Therefore, it was decided to limit our discussion in this paper to the simpler version of HC-ACO.

Different ACO algorithms may use different quality functions and reference sets. For example, in the very first ACO algorithm — Ant System [5]— the quality function was $1/f(s)$ and the reference set $\hat{S}_t = S_t$. In a more recently proposed scheme, called *iteration best update* [7], the reference set was a singleton containing the best solution within S_t (if there were several iteration-best solutions, one of them was chosen randomly). In the *global-best update* [7, 19], the reference set contained the best among all the iteration-best solutions (and if there was more than one global-best solution, the earliest one was chosen).

In $\mathcal{MAX}$–$\mathcal{MIN}$ Ant System [19], maximum and minimum pheromone trail limits were introduced. With this modification the probability to generate any particular solution is kept above some positive threshold, which helps preventing search stagnation and premature convergence to suboptimal solutions. For HC-ACO, this approach translates into the requirement that the marginal probabilities are kept within the range $[\epsilon, 1-\epsilon]$, where $\epsilon \geq 0$ is the parameter that controls the amount of exploration.

It is worth noting that, as shown in [8], for learning rate $\rho = 1$ and for a particular choice of the quality function, the HC-ACO is equivalent to the *cross-entropy method* [18].

3.2 The Stochastic Gradient Ascent Method

While all the updates described above are of a somewhat heuristic nature, the SGA method allows to derive the parameters update rule in a more principled manner [12].

The SGA method replaces the original optimization problem with the following equivalent continuous *maximization problem*:

$$\mathcal{T}^* = \operatorname*{argmax}_{\mathcal{T}} \mathcal{E}(\mathcal{T}), \tag{2}$$

where $\mathcal{E}(\mathcal{T}) = E_{\mathcal{T}} Q_f(s)$ and $E_{\mathcal{T}}$ denotes expectation with respect to $P_{\mathcal{T}}$. This maximization problem is, in turn, tackled using *stochastic gradient ascent* [16]:

$$\mathcal{T}^{t+1} = \mathcal{T}^t + \alpha_t \sum_{s \in S_t} Q_f(s) \nabla \ln P_{\mathcal{T}^t}(s), \tag{3}$$

where S_t is the sample at iteration t.

In [8] it was demonstrated how the required gradient $\nabla \ln P_{\mathcal{T}^t}(s)$ can be calculated for a general class of probabilistic models. For the model we consider in this paper, namely binary variables without dependencies between different positions, it can be verified that the resulting parameter update rule is:

SGA_UPDATE

$$\tau_i \leftarrow \tau_i + \alpha_t \sum_{s \in S_t} Q_f(s) \frac{F'(\tau_i)}{s_i \cdot F(\tau_i) - (1 - s_i) \cdot (1 - F(\tau_i))} \tag{4}$$

In order to guarantee the stability of the resulting algorithm, it is desirable to have a bounded gradient $\nabla \ln P_{\mathcal{T}}(s)$. For this reason, the use of the "natural" representation $F(\tau_i) = \tau_i$ is inappropriate. Instead, we suggest using the logistic function $F(\tau_i) = \dfrac{1}{1+\exp(-\tau_i)}$. It can be shown that in this case the update rule becomes:

$$\tau_i \leftarrow \tau_i - \alpha_t \sum_{s \in S_t} Q_f(s) F(\tau_i) + \alpha_t \sum_{s \in S_t} Q_f(s) s_i, \tag{5}$$

hence the gradient is indeed bounded.

While the SGA method was originally formulated for iteration-independent quality functions, in [8] it was demonstrated that an alternative derivation of the SGA update through the CE method justifies the use of iteration-dependent quality functions as well. For example, one may use the indicator function $Q_f(s) = I(f(s) < \theta_t)$, where θ_t is a threshold value set to some percentile (say, lower 10%, for minimization problems) of the cost distribution at the last iteration. For this quality function, the expectation $E_{\mathcal{T}} Q_f(s)$ equals the probability of generating solutions, whose cost is below the threshold θ_t, and the threshold is modified adaptively. The update based on this function is henceforth referred to as "top-quality" update.

Similarly to HC-ACO, we may choose to bound the marginal probabilities in order to increase the amount of exploration. In order to keep the marginal probabilities between ϵ and $1-\epsilon$, with the logistic function representation described above, the pheromones should be kept in the range $[\ln(\frac{\epsilon}{1-\epsilon}), \ln(\frac{1-\epsilon}{\epsilon})]$.

3.3 Estimation of Distribution Algorithms

As already mentioned in Section 1, the classical genetic algorithms can be considered an example of the instance-based approach, in which the search is carried out by evolving the population of candidate solutions using selection, crossover and mutation operators. Recently, several new algorithms, which generate new solutions using probabilistic models instead of crossover and mutation, have been proposed within the evolutionary computation community.

In the population-based incremental learning (PBIL) algorithm [1], the population is replaced by a probability vector $\bar{p}$, with all p_i's initially set to 0.5. At every iteration a sample S is generated using the probability vector and then the probability vector is updated as follows:

PBIL_UPDATE
- $S_{top} \leftarrow$ a fixed number of lowest cost solutions from S,
- for every $s \in S_{top}$
  - $p_i \leftarrow (1-\rho)p_i + \rho s_i$,

where ρ is the learning rate.

As it can be easily seen, this update is virtually identical (up to rescaling of the learning rate) to the HC-ACO with top-quality update. In particular, in case only the best solution is used for the update, HC-ACO with iteration-best update is obtained. In [1] two additional updates were suggested. The first was intended to make use of "negative" examples, shifting the probability vector towards the best solution in the positions where it differs from the worst solution:

– if $s_i^{best} \neq s_i^{worst}$ then
$$p_i \leftarrow (1-\rho_{nl})p_i + \rho_{nl} s_i^{best},$$

where s^{best}, s^{worst} are respectively the best and the worst solutions in S and ρ_{nl} is the so-called "negative learning rate". In the second update, the probability vector was randomly perturbed, with an effect similar to that of mutation in standard GA:

– For every i, modify $p_i \leftarrow (1-\rho_{mut})p_i + \rho_{mut} d_i$, with probability p_{mut}.

where ρ_{mut} is the "mutation shift", and, for every i, the mutation direction d_i is 0 or 1, with probability 1/2 each. Both updates were performed in addition to the basic PBIL update described above.

The compact genetic algorithm (cGA) [9] was proposed as a modification of PBIL, intended to represent more faithfully the dynamics of the real GA algorithm. At every iteration, two solutions, a and b, are generated using a probability vector, and then the probability vector is updated as follows (assuming, without loss of generality, that a has lower cost):

cGA_Update

– if $a_i \neq b_i$ then
  if $a_i = 1$ then $p_i \leftarrow p_i + 1/n$,
  else $p_i \leftarrow p_i - 1/n$,
  where n is a parameter, equivalent to the population size in the classical GA.

This basic scheme can be extended to larger samples. Two variants were proposed in [9]. In the first variant, intended to simulate a tournament of size m, a sample S of size m is generated and the basic update above is used for every pair in the set $\{(s^{best}, b) | b \in S, b \neq s^{best}\}$. In the second variant, a "round-robin tournament" is simulated, that is, the basic update is used for every pair of solutions from the sample.

It can be shown that the update for "tournament of size m" cGA can be written as:

$$p_i \leftarrow p_i + \rho \sum_{s \in S} Q(s) s_i - \frac{\rho}{m} \sum_{s \in S} s_i, \tag{6}$$

where $\rho = \frac{m}{n}$ and

$$Q(s) = \begin{cases} 1, & s = s^{best} \\ 0, & \text{otherwise.} \end{cases} \tag{7}$$

For the "round-robin tournament" cGA, it can be shown that the update can also be described by (6), with $\rho = \frac{m(m+1)}{n}$ and $Q(s) = \frac{2 \cdot \text{rank}(s)}{m(m+1)}$ (the highest rank, m, is assigned to s^{best}). It can be easily verified that these two updates are virtually identical to the HC-ACO iteration-best and rank-based updates respectively. The only difference between cGA and HC-ACO is in the form of the evaporation factor. In cGA it is equal to $\frac{\rho}{m} \sum_{s \in S} s_i$, whereas in HC-ACO it is equal to ρp_i, which is simply the expected value of the former.

4 Empirical Comparison

In this section we describe the results of the empirical comparison between the MBS algorithms described above, using MAXSAT as a test bed. MAXSAT is the optimization variant of SAT, the first problem which was shown to be NP-complete. The weighted MAXSAT problem can be formulated as follows. Given k clauses $C_1, \ldots, C_k$ over n binary variables $x_1, \ldots, x_n$, and the weights $w_1, \ldots, w_k$, find an assignment which maximizes the sum of the weights of the satisfied clauses.

4.1 Comparison Setting

The comparison was carried out using randomly generated weighted MAXSAT instances from the SATLIB MAXSAT Benchmark Collection [10]. The benchmark set contains three groups of problems, with 100, 500 and 1000 variables and 500, 5000 and 10000 clauses respectively. Each group contains 10 instances.

The algorithms were evaluated using three different running times. For every problem size, three stopping times, T_1, T_2 and T_3, were chosen as the time that it takes for HC-ACO with population size 10 to perform 50, 200 and 1000 iterations respectively[3].

All of the algorithms described above have one or more parameters, whose choice can clearly affect the performance of the algorithm. Moreover, the optimal parameter setting may depend on the available computational time (e.g., with shorter times a higher learning rate and smaller samples should be more appropriate) and the problem size. Since to-date there are no established methods for the automatic tuning of metaheuristics' parameters, it was decided to use one problem out of every group for tuning the algorithms, and the other 9 for the testing. Specifically, each algorithm was run with a variety of parameter settings (described below), 10 times for each setting, and for every algorithm/problem-size/running-time combination, the configuration with the best average performance was chosen. This automatic tuning procedure insures that the comparison is not biased in favor of one of the algorithms. The separation between the training problem and the test problems guarantees the statistical validity of the performance estimates[4].

For all the algorithms, we considered learning rate $\rho \in \{0.03, 0.1, 0.3, 1\}$, and sample sizes $|S| \in \{10, 50, 200\}$. In HC-ACO and SGA we considered pheromone bounds with $\epsilon \in \{0, 0.01, 0.1\}$. For PBIL, the additional parameters were: "negative learning rate" $\rho_{nl} \in \{0, \rho/10, \rho\}$, mutation probability $p_{mut} \in \{0, 0.01, 0.1\}$ and mutation shift $\rho_{mut} \in \{0.01, 0.1\}$.

---

[3] It should be noted that, due to extensive code reuse in the algorithms' implementation, the running times for all the algorithms (without the local search) were virtually the same.

[4] If, for example, we chose the best performing configuration for *every problem* individually, the resulting average performance would no longer be an unbiased estimate of the actual expected performance.

For HC-ACO, SGA and PBIL we have tested both the iteration-best and the top-quality updates. We have also tested separately the basic CE method, which corresponds to the top-quality HC-ACO algorithm with learning rate 1 and no bounds on probabilities. Finally, both " single tournament" (cGA_{st}) and "round-robin tournament" (cGA_{rr}) versions of the cGA method were tested.

Furthermore, for every algorithm described above, we have also considered a hybrid version, in which the update was based on the population of the elite solutions, that is the highest quality solutions found so far, rather than on the last sample. Finally, all the algorithms (including the hybrid versions) were tested both with and without the use of the local search[5].

4.2 Comparison Results

Every algorithm was run 30 times on every problem with the parameter setting determined using the tuning procedure described in the previous section. Since the optimal solution costs for the benchmark problems used in this study are not known, the cost of the best solution found by *any* of the algorithms in all the test runs were used as estimates. The performance of a single run of the algorithm was evaluated as:

$$\tilde{f} = \frac{f_{best} - f_{opt}}{E\{f\} - f_{opt}}, \tag{8}$$

where f_{best} is the cost of the best solution found during the run, f_{opt} is the estimate of the optimal solution cost and $E\{f\}$ is the expected cost of the solution generated from a uniform distribution[6]. Note that, for a random solution, $E\{\tilde{f}\} = 1$, hence the results coming from different problems are put on a same scale, which allows meaningful averaging over several problems. The results of the comparison are summarized in Tables 1 and 2. Every column corresponds to a comparison with a particular problem size and stopping time. The average score of the empirically best algorithm is printed in bold typeface and the results, which are worse than the best one with 95% confidence[7], are shown in italic. Since the use of the local search leads to a drastic improvement of performance, the results in Table 2 are multiplied by 100 for clarity of presentation. The superscript "h" denotes the hybrid versions of the algorithms, augmented with population, and "n" denotes the problem size, measured by the number of variables.

When local search is not used, in most cases the "round-robin" tournament cGA produces significantly better results. Still, even the performance of cGA

[5] Although using local search is not a common practice in the EDA research field, the results reported next indicate that it certainly should be considered in the future.

[6] Since, for any clause C with d variables, the proportion of non-satisfying assignments is $1/2^d$, it can be verified that, for the 3-MAXSAT problems used in the benchmarks, $E\{f\} = \frac{7}{8}\sum_{j=1}^{k} w_j$.

[7] The statistical analysis was performed using Tukey-Kramer test, which is a modification of the t-test, adapted for the multiple comparisons.

Table 1. Average performance of the algorithm without the local search.

	n=100			n=500			n=1000		
	T_1	T_2	T_3	T_1	T_2	T_3	T_1	T_2	T_3
CE	0.2561	0.1294	0.1033	0.5567	0.4870	0.2026	0.6642	0.5795	0.2585
CE^h	0.2541	0.1497	0.1021	0.5646	0.4854	0.1993	0.6676	0.5784	0.2586
$HC\text{-}ACO_{top}$	0.2039	0.1230	0.0778	0.5567	0.2979	0.1547	0.6439	0.4251	0.2234
$HC\text{-}ACO^h_{top}$	**0.1843**	0.1050	0.0714	0.5138	0.3094	0.1554	0.6301	0.4288	0.2387
$HC\text{-}ACO_{best}$	0.2098	0.1336	0.0762	0.5676	0.3358	0.1670	0.6695	0.4636	0.2503
$HC\text{-}ACO^h_{best}$	0.1980	0.1292	0.1063	0.5558	0.3433	0.1840	0.6719	0.5020	0.3625
SGA_{top}	0.2613	0.1370	0.0879	0.5398	0.3160	0.1588	0.6417	0.4184	0.2153
SGA^h_{top}	0.2349	0.1330	0.0773	0.5397	0.3139	0.1476	0.6475	0.4268	0.2101
SGA_{best}	0.2692	0.1437	0.0808	0.5804	0.3473	0.1625	0.6854	0.4556	0.2242
SGA^h_{best}	0.2709	0.1409	0.1065	0.6005	0.3694	0.1640	0.7049	0.4973	0.2350
$PBIL_{top}$	0.2311	0.1545	**0.0464**	0.5058	0.4015	0.1549	0.6063	0.4816	0.2266
$PBIL^h_{top}$	0.2868	0.1790	0.1038	0.6613	0.3991	0.2133	0.7300	0.4731	0.2491
$PBIL_{best}$	0.2837	0.1206	0.0504	0.5658	0.3072	0.1415	0.6948	0.4380	0.2554
$PBIL^h_{best}$	0.3214	0.1925	0.1094	0.7399	0.4458	0.1576	0.7900	0.5169	0.2635
cGA_{rr}	0.1949	**0.0979**	0.0746	**0.4168**	**0.2606**	**0.1086**	**0.5395**	**0.3947**	**0.1528**
cGA^h_{rr}	0.4625	0.2244	0.1339	0.7500	0.6292	0.4225	0.8163	0.7252	0.5229
cGA_{st}	0.3007	0.1455	0.0814	0.5891	0.3391	0.1535	0.6816	0.4561	0.2498
cGA^h_{st}	0.4627	0.3715	0.3307	0.7687	0.7070	0.6485	0.8351	0.7862	0.7311

is relatively poor (recall that the expected performance score of a randomly generated solution is 1). The use of local search leads to an improvement of almost two orders of magnitude and there seems to be almost no significant differences between the algorithms in this case (note, however, that cGA is often significantly worse than the others). We hypothesize that the differences among the algorithms with the local search could be just an artifact of the particular tuning procedure which we use, rather than an evidence of the advantage of one of the methods. This is the topic of ongoing research.

5 Conclusions

During the last decade a new approach for solving combinatorial optimization problems has been emerging, as already observed in [13]. This approach, which we refer to as model-based search (MBS), tackles the combinatorial optimization problem by sampling the solution space using a probabilistic model, which is adaptively modified as the search proceeds. In this paper we presented a comparative analysis of several existing MBS methods, which construct binary coded solutions by generating every bit independently. Our theoretical analysis revealed considerable structural similarity among these algorithms, and the empirical comparison showed that also the actual performance of the algorithms is quite similar (especially, when the algorithms are hybridized with local search). In the future we hope to extend our analysis to a more general class of MBS algorithms and to compare these algorithms on different types of combinatorial optimization problems.

Table 2. Average performance (multiplied by 100) of the algorithm with the local search.

	n=100			n=500			n=1000		
	T_1	T_2	T_3	T_1	T_2	T_3	T_1	T_2	T_3
CE	0.1816	*0.0300*	**0.0000**	**0.4199**	*0.2416*	*0.1173*	0.8352	*0.4494*	*0.2125*
CE^h	0.2296	*0.0216*	**0.0000**	0.4264	0.2029	*0.1460*	0.7944	*0.4586*	0.1536
$HC\text{-}ACO_{top}$	0.1809	0.0057	0.0016	0.4810	0.2055	0.0817	0.8352	0.3266	0.1203
$HC\text{-}ACO^h_{top}$	0.2331	0.0079	0.0023	0.5066	0.2124	0.0895	0.8440	*0.4586*	**0.0857**
$HC\text{-}ACO_{best}$	0.1303	0.0078	0.0023	0.4846	0.2207	0.0728	*0.9838*	0.3876	*0.1811*
$HC\text{-}ACO^h_{best}$	0.1536	0.0096	*0.0070*	0.4951	0.2160	0.0713	*0.9884*	*0.4991*	0.1324
SGA_{top}	**0.1233**	0.0067	**0.0000**	0.4324	0.2094	0.0823	0.8354	0.3491	0.1247
SGA^h_{top}	0.1934	0.0048	**0.0000**	0.4298	**0.1548**	**0.0485**	0.8230	0.3433	0.1194
SGA_{best}	0.1573	0.0039	0.0054	*0.6292*	*0.2404*	0.0634	0.8929	*0.5087*	0.1416
SGA^h_{best}	0.2104	0.0189	0.0031	*0.6242*	0.2194	*0.1015*	*0.9078*	0.3883	0.1425
$PBIL_{top}$	0.1586	**0.0017**	0.0024	0.4844	0.1899	*0.1107*	0.8353	**0.2925**	0.1113
$PBIL^h_{top}$	0.1729	0.0054	0.0008	0.4472	0.1973	*0.1203*	**0.7821**	0.3026	0.1124
$PBIL_{best}$	0.1606	0.0110	*0.0066*	0.4292	0.1771	0.0673	*0.9078*	*0.4275*	0.1288
$PBIL^h_{best}$	0.1392	0.0192	0.0023	0.4734	*0.2313*	*0.1249*	*0.9074*	*0.4095*	0.1537
cGA_{rr}	0.1737	0.0137	**0.0000**	*0.6502*	*0.2937*	0.0792	*1.3845*	*0.5153*	0.1508
cGA^h_{rr}	0.1580	0.0086	**0.0000**	*0.6792*	0.2269	0.0503	*1.4152*	*0.6443*	*0.1931*
cGA_{st}	0.1600	*0.0257*	0.0047	*0.6212*	*0.2673*	*0.1198*	*0.9264*	*0.4903*	*0.2114*
cGA^h_{st}	0.1568	*0.0248*	0.0031	*0.6713*	*0.2953*	*0.1368*	*0.9054*	*0.6220*	*0.2843*

Acknowledgments

Mark Zlochin is supported through a Training Site fellowship funded by the Improving Human Potential (IHP) programme of the Commission of the European Community (CEC), grant HPRN-CT-2000-00032. Marco Dorigo acknowledges support from the Belgian FNRS, of which he is a Senior Research Associate. More generally, this work was partially supported by the "Metaheuristics Network", a Research Training Network funded by the Improving Human Potential programme of the CEC, grant HPRN-CT-1999-00106. The information provided in this paper is the sole responsibility of the authors and does not reflect the Community's opinion. The Community is not responsible for any use that might be made of data appearing in this publication.

References

[1] S. Baluja and R. Caruana. Removing the genetics from the standard genetic algorithm. In *Proceedings of ICML'95*, pages 38–46. Morgan Kaufmann Publishers, Palo Alto, CA, 1995.

[2] C. Blum, A. Roli, and M. Dorigo. HC–ACO: The hyper-cube framework for Ant Colony Optimization. In *Proceedings of MIC'2001*, volume 2, pages 399–403, Porto, Portugal, 2001.

[3] J. S. de Bonet, C. L. Isbell, and P. Viola. MIMIC: Finding optima by estimating probability densities. In *Proceedings of NIPS'97*, pages 424–431. MIT Press, Cambridge, MA, 1997.

[4] G. Di Caro and M. Dorigo. AntNet: Distributed stigmergetic control for communications networks. *Journal of Artificial Intelligence Research*, 9:317–365, 1998.

[5] M. Dorigo. *Ottimizzazione, Apprendimento Automatico ed Algoritmi Basati su Metafora Naturale.* PhD thesis, Dipartimento di Elettronica, Politecnico di Milano, Milan, Italy, 1992.

[6] M. Dorigo and G. Di Caro. The Ant Colony Optimization meta-heuristic. In D. Corne, M. Dorigo, and F. Glover, editors, *New Ideas in Optimization*, pages 11–32. McGraw Hill, London, UK, 1999.

[7] M. Dorigo and L. M. Gambardella. Ant Colony System: A cooperative learning approach to the traveling salesman problem. *IEEE Trans. on Evol. Comp.*, 1(1):53–66, 1997.

[8] M. Dorigo, M. Zlochin, N. Meuleau, and M. Birattari. Updating ACO pheromones using Stochastic Gradient Ascent and Cross-Entropy methods. In *Proceedings of EvoWorkshops 2002*, pages 21–30. Springer Verlag, Berlin, Germany, 2002.

[9] G. R. Harik, F. G. Lobo, and D. E. Goldberg. The compact genetic algorithm. *IEEE Trans. on Evol. Comp.*, 3(4):287–297, 1999.

[10] H. H. Hoos and T. Stützle. Randomly generated benchmark problems for MAXSAT. Technical Note, Department of Computer Science, University of British Columbia, March 2001.

[11] P. Larrañaga and J.A. Lozano. *Estimation of Distribution Algorithms. A New Tool for Evolutionary Computation.* Kluwer Academic Publishers, 2001.

[12] N. Meuleau and M. Dorigo. Ant colony optimization and stochastic gradient descent. *Artificial Life*, 8(2):103–121, 2002.

[13] N. Monmarché, E. Ramat, G. Dromel, M. Slimane, and G. Venturini. On the similarities between AS, BSC and PBIL: toward the birth of a new meta-heuristic. Technical Report 215, Laboratoire d'Informatique, Université de Tours, 1999.

[14] M. Pelikan, D. E. Goldberg, and E. Cantú-Paz. BOA: The Bayesian optimization algorithm. In *Proceedings of GECCO'99*, volume I, pages 525–532. Morgan Kaufmann Publishers, San Francisco, CA, 1999.

[15] J. Quinlan. Combining instance-based and model-based learning. In *Proceedings of the Twelfth International Conference on Machine Learning (ML-93)*, pages 236–243. Morgan Kaufmann Publishers, San Mateo, CA, 1993.

[16] H. Robbins and S. Monro. A stochastic approximation method. *Annals of Mathematical Statistics*, 22:400–407, 1951.

[17] A. Roli, C. Blum, and M. Dorigo. ACO for maximal constraint satisfaction problems. In *Proceedings of MIC'2001*, volume 1, pages 187–191, Porto – Portugal, 2001.

[18] R. Y. Rubinstein. The cross-entropy method for combinatorial and continuous optimization. *Methodology and Computing in Applied Probability*, 1(2):127–190, 1999.

[19] T. Stützle and H. H. Hoos. $\mathcal{MAX}$–$\mathcal{MIN}$ Ant System. *Future Generation Computer Systems*, 16(8):889–914, 2000.

[20] M. Zlochin, M. Birattari, N. Meuleau, and M. Dorigo. Model-based search for combinatorial optimization. Technical Report TR/IRIDIA/2001-15, IRIDIA, Université Libre de Bruxelles, 2001.

Evolutionary Algorithm Implementations

A Framework for Distributed Evolutionary Algorithms*

M.G. Arenas[3], Pierre Collet[2], A.E. Eiben[1], Márk Jelasity[1], J.J. Merelo[3], Ben Paechter[4], Mike Preuß[5], and Marc Schoenauer[2,**]

[1] Free University of Amsterdam, Amsterdam, The Netherlands
[2] Ecole Polytechnique, Palaiseau, France
[3] Universidad de Granada, Granada, Spain
[4] Napier University, Edinburgh, Scotland
[5] University of Dortmund, Dortmund, Germany

Abstract. This paper describes the recently released DREAM (Distributed Resource Evolutionary Algorithm Machine) framework for the automatic distribution of evolutionary algorithm (EA) processing through a virtual machine built from large numbers of individual machines linked by standard Internet protocols. The framework allows five different user entry points which depend on the knowledge and requirements of the user. At the highest level, users may specify and run distributed EAs simply by manipulating graphical displays. At the lowest level the framework turns becomes a P2P (Peer to Peer) mobile agent system, that may be used for the automatic distribution of a class of processes including, but not limited to, EAs.

1 Introduction

The Distributed Resource Evolutionary Algorithm Machine (DREAM) [1] was first described in [12]. It provides a framework for the production of evolutionary algorithm systems and systems of evolving agents which use the Internet to allow distributed processing in a peer-to-peer scalable fashion. The system also allows the secure sharing of the spare CPU resources of a large number of computers. The scalability of the system will allow new types of problems to be studied which require either very large amount of processing power, or vast numbers of evolving agents competing and co-operating to find a solution to some problem together. The first public release of the software has recently been made, and this paper describes the architecture and functionality of that system.

* This work is funded as part of the European Commission Information Society Technologies Program (Future and Emerging Technologies). The authors have sole responsibility for this work, it does not represent the opinion of the European Community, and the European Community is not responsible for any use that may be made of the data appearing herein.

** The authors are listed in alphabetical order.

J.J. Merelo Guervós et al. (Eds.): PPSN VII, LNCS 2439, pp. 665–675, 2002.

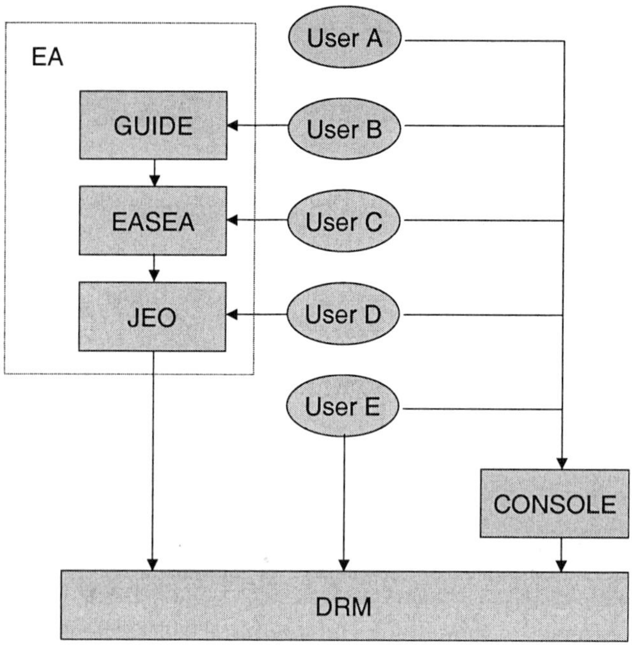

Fig. 1. The DREAM architecture and its user entry points.

2 System Architecture

The system architecture is split into five modules, and each provides a user interface at a different level interaction and abstraction. The architecture is shown in Fig. 1 along with the user entry points. The entry points give a variety of interfaces, with different levels of ease-of-use and power. The upper levels are easier to use, but flexibility is more limited, the lower levels require greater expertise, but give the user greater control over the system. Some of the modules or groups of modules can be used independently of the others, providing additional functionality outside the context of the integrated DREAM system. The five types of user can be categorised as follows:

- *User A* does not wish to use the DREAM for conducting experiments, but simply wishes to donate their spare CPU time or monitor the experiments of others. This type of user, interacts with the DREAM only through the Console, other types of user use the console and (normally) one other user entry point.
- *User B* is either not an experienced programmer or wishes to rapidly prototype a system without the need for textual programming. This type of user interacts with the GUIDE layer which allows distributed evolutionary algorithms to be defined using a fully graphical interface. The GUIDE interfaces with the EASEA layer through the use of the EASEA language. This user should have a knowledge of the workings of evolutionary algorithms.

- *User C* programs the system through the EASEA layer, which provides a high level textual language in which to program distributed evolutionary algorithms. This layer produces Java code through a compiler. The code it produces uses the objects and methods of the JEO (Java Evolutionary Object) library.
- *User D* programs directly in Java and makes use of the JEO library. The library not only provides useful objects and methods for evolutionary computing, but also provides an API (Application Programming Interface) to the DRM (Distributed Resource Machine) core layer. This layer is intended for users with a knowledge of evolutionary algorithms and the Java language.
- *User E* is an expert user, who programs using the DRM API directly. This level of user can use the DRM for many useful distributed processing purposes beyond evolutionary computing, but is not fully protected from the complexities of this type of programming.

The following sections describe each of the architecture layers in turn.

3 GUIDE Layer

The Graphic User Interface for DREAM Experiments is the entry point at the highest possible level of interaction and abstraction. The idea is that even a non-expert programmer should be able to use the DREAM through GUIDE, specifying the algorithm by means of point-and-click in a number of panels referring to different parts of the evolutionary algorithm.

An Evolutionary Algorithm is made of two parts that are almost completely orthogonal: on the one hand are the problem dependent components, including the *genotype structure*, its *initialization*, the *variation operators* (crossover, mutation and the like) that will be applied on the genotypes and of course the *evaluation* (computation of the fitness value). On the other hand, the *evolution engine* implements the artificial Darwinism part of the algorithm and should be able to handle any population of objects that have a fitness, regardless of the actual genotype. Evolution engines are made up of two steps, the *selection* of some parents to become actual *genitors* and generate offspring, and the *replacement* of some individuals by some offspring to build up the next generation.

The structure of GUIDE reflects this point of view, and offers four panels to the user: the Problem Specification panel to define the problem dependent components, the Evolution Engine panel for the Darwinian components, the Distribution Control panel to define the way the different islands will communicate (see section 5) and the Experiment Monitor panel, from where the user can run his experiment and view the temporary and final results.

3.1 Evolution Engine Specification

As far as the ultimate end user is concerned, writing the evolution engine part has nothing to do with the problem being solved – and this is where GUIDE can handle the work completely.

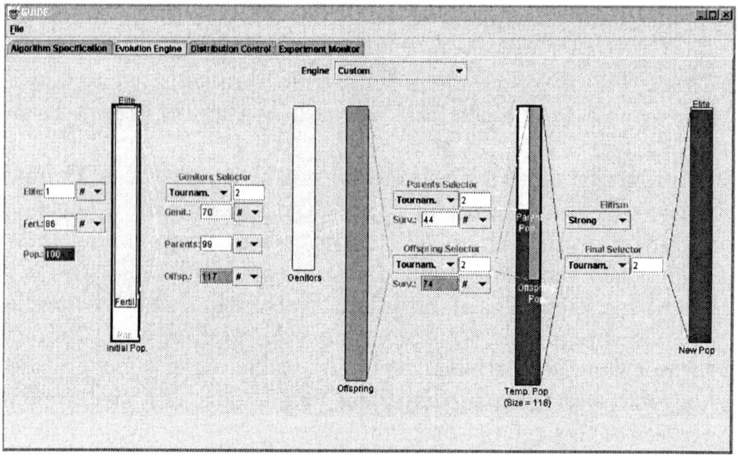

Fig. 2. The Evolution Engine panel of the GUIDE.

Any decent book about EAs describes the main evolutionary engines, namely *Generational* or *Steady-State GA*, *Plus* or *Comma Evolution Strategy* selection, ... It should be possible for a user to say "I want to solve my problem with Generational GA evolution, using 100 generations of 50 infohabitants, initially evenly distributed in 5 different islands, with such and such parameters."

This is part of the functionality offered by the GUIDE. However, restricting the choice to the five mainstream paradigms cited above would have been very restrictive: experience shows that most real-world EAs in fact deviate one way or another from the strict historical paradigms.

The Evolutionary Engine panel of GUIDE (Figure 2) therefore offers a generic set of primitives from which the user can define a huge variety of implementation artificial Darwinism implementations. The already-mentioned "classical" evolution engines are simple instantiations of GUIDE parameters. The pedagogic interest is immense, as one can see the parameters changing when selecting any of the predefined engines. Alternatively, thousands of unexplored engines can be tested, simply by arranging the parameters in an original way.

3.2 Problem-Dependent Components

When using the DREAM, the user refers to a specific problem to be solved. This means that the notion of programming cannot, unfortunately, be completely removed from the description of experiments – at least the computation of the fitness of infohabitants has to be typed in textually. At the moment, this is also true in GUIDE for all problem-dependent parts.

GUIDE therefore provides a panel that allows the user to type in the specificities of his problem in terms of genome structure, genome initialiser, genome recombinator, genome mutator and genome evaluator. This is done thanks to a very high level language with a C/C++/Java like syntax that completely hides the very complex notion of classes/objects that is required to handle populations of genomes. The user can hence concentrate on his problem, not on making the whole thing work in some complex library.

4 EAsy Specification of Evolutionary Algorithms

GUIDE turns mouse driven diagrams representing some complex evolutionary algorithm into runnable code using the genome structure, genetic operators and evaluator specified by the user thanks to a powerful intermediate language associated with a compiler.

The fact that at some point, some representation must be used to, at least, save and reload user experiments, implies that this same representation should be capable of describing virtually any kind of evolutionary algorithm. Rather than designing some obscure internal representation, specific to the DREAM, the decision was taken to create a fully independent human-readable language that could have an existence of its own.

Among other advantages, this meant that EASEA [5, 11] could be developed completely independently from the evolutionary library specific to the DREAM project, simply by using already existing off the shelf evolutionary libraries.

As a result, an evolutionary algorithm specified in the EASEA language (or specified and saved by GUIDE with an EASEA syntax) can be compiled in a C++ source file using the GALib evolutionary library[13], or the EO [2, 10] fully templatised object oriented library; or of course in Java source files using the JEO library of the DREAM.

As a matter of fact, the EASEA language compatible with GALib or EO has been downloaded more than one thousand times in the last 12 months, with the result that even before the DREAM was released, many users around the world have been unknowingly developing potential DREAM applications, that would only need a fully working DREAM system to exploit the distributed resources of the Internet.

The genome structure, as well as its initialiser, recombinator, mutator, and evaluator described in the GUIDE are compiled by the EASEA compiler into a pure Java classes meeting the JEO requirements. The resulting files are ready for compilation by the JAVA compiler. Compiling as well as launching the experiment can be carried out from within the GUIDE Experiment Monitor panel as well as from the command line, depending on the type of user (B or C).

5 JEO Layer

This section explains the Evolutionary Computation layer in DREAM. This layer is called Java Evolutionary Objects, JEO for short.

5.1 JEO from the User's Point of View

JEO is a framework for building evolutionary computation experiments, which sits on top of the DRM layer, but can be used independently from it if no parallel execution is required. It provides a DREAM entry point to *User D* so it is flexible, powerful and extensible enough to allow the users to design, develop and control their experiments in the easiest way. JEO output is an experiment specification that can be run in distributed fashion using the DRM module (see section 6). JEO can also act as the bridge between the EASEA and DRM modules. It hides the physical DRM details, like communication protocols or threads, letting the user concentrate on evolutionary computation concepts.

User D must be familiar with EC concepts. This type of user prefers to work with Java classes to have direct control over the experiment. The procedure to build an experiment is simple. The user implements a Java class to create the objects that will be placed in each Island. The set of Islands that constitutes the experiment will be launched in DRM to distribute the experiment.

For each task to be performed on an island, the user must create one specific object. The requirement on those objects is to meet the corresponding JEO interface, e.g. an *operator* must implement the operators interface, an *individual* the individuals interface... Moreover, JEO provides several classes that actually implement each interface and can be used as examples. However, JEO provides no implementation for the *evaluator* interface, as it is totally problem-dependent.

5.2 Islands and Island Components

JEO provides a general Island class: an island holds one or more *environments.* An environment groups one population and the objects that will be used to evolve that population (i.e. a complete evolutionary algorithm). The different environments possibly interact through the *assessor* objects, that perform the evaluation of all environments simultaneously.

First of all, all populations of the island are initialised using the corresponding initialisers. The following cycle is the run:

1. *Stopping test and statistics calculation* are carried out by *Terminators*: different stopping criteria are available, as well as different on-line or cumulated statistics. These values are sent to the DRM console, passed to GUIDE Experiment Monitor panel, or simply displayed on the screen in the case of JEO being used stand-alone.
2. Each island population then undergoes a single step of evolution through the corresponding *Breeder.* This includes (fitness-based) *selection* and *variation operators.*
3. The *assessor* is then called upon all environments (i.e. all populations). This step can be a simple evaluation in the case of evolutionary optimisation experiment, or can correspond to a few life steps in the case of artificial life simulations. Nevertheless, at the end of this step, all infohabitants are *Rewarded* . This reward mechanism leaves room for forthcoming large-scale simulation of sociological and economical evolution, where every operation (from mating and mutating to being evaluated) will have a cost, and infohabitants will simply fight for survival, without explicit fitness function being defined.

4. *Migration* is then performed by two objects, *Immigrator* and *Emigrator*. *Immigrator* reads from the input immigrant's buffer the recently arrived individuals from some neighbouring island and decides how to include them into the Island populations. *Emigrator* selects the individuals that will emigrate, and selects the neighbour target island.
5. *Clean* is the method that eliminates the "poorest" infohabitants (in term of the rewards mentioned above), the remaining infohabitants becoming the initial population of the next generation.

The experiment can be submitted to the DRM module using the DREAM Console or DRM command line and subsequently the DRM module distributes the experiment through the DREAM machine. The distribution mechanism is completely transparent to the user. The user identifies the experiment using an experiment name and identifies each island using the island name set into the experiment specification.

Parallel evolution is performed asynchronously; every island writes to another island's buffer at any time it is required to; every island reads from its own buffer. There is a thread that receives immigrants and places them in a buffer, from where the thread that runs the evolutionary algorithm can read them.

5.3 JEO as Java Tool

JEO is developed using *jdk 1.3*, like the rest of the DREAM project. JEO classes and interfaces are organized into packages [4, 3]. The main one is the `dream.evolution` which includes the main classes and interfaces as Island class or operators interface. Other important packages are for example `dream.evolution.genomes`, including interfaces and classes to build genomes as linear, tree or graph structures, and `dream.evolution.operators`, that includes variation operators and selectors.

All classes have `javadoc` comments to help the user to understand all variables and methods. Moreover each package includes a `package.html` file where package elements are described.

JEO today includes some basic evolutionary algorithms examples, that solve easy toy problems (e.g. OneMax and Symbolic Regression).

6 DRM Layer

The distributed resource machine (DRM) is composed of (a possibly very large number of) machines all over the Internet forming an environment for distributed applications. At this level of abstraction it is no longer assumed that the application is from the field of evolutionary computation. This section summarizes the basic concepts and the algorithms which implement these concepts. For a more detailed description of the different aspects of the DRM please refer to [8, 9].

6.1 Application Model

In a traditional single-machine environment, an application is composed of one or more threads which are run by an operating system (OS). The OS controls the threads, it assigns resources and takes care of different security aspects. When adapting this approach to very large scale distributed environments, not all aspects can be implemented in exactly the same way due to the relatively high costs of information exchange.

A key feature of our model is that we think of an application as a set of cooperating autonomous agents. For instance, an evolutionary algorithm is implemented in this framework by a set of agents which all host an island. An agent is analogous to a thread in an OS: running an application is done through launching one or more agents that can communicate with each other, can make decisions based on the state of the system as a whole (e.g. available computational resources) or based on the state of each other. The agents can also launch new agents themselves. The DRM controls the agents, the resources they have access to, security, etc.

However these agents have much more freedom: they are also mobile, they can change their physical location while performing computations. On the new location they can continue their task.

Despite the similarities, the applications we have in mind are rather special. Usual things like quickly accessible shared memory are a luxury in the world of large distributed systems. Another problem is the unreliable nature of both the communication channels between the nodes of the environment and the unreliability of the nodes themselves. This restricts the application area to tasks that are robust (not sensitive to loosing a subset of the agents they are composed of) and massively parallel (i.e. only little communication is necessary between the agents). Fortunately evolutionary computation fulfills these requirements and there are additional possible applications as well.

6.2 Implementation

An important implementation decision was that we do not use central servers at all. This is to maximize scalability and robustness. With this restriction even maintaining the connectivity of the network becomes a major challenge. This problem and also the problem of information distribution through the DRM is solved using *epidemic protocols* [6, 7].

In our system this protocol works as follows: each node in the DRM has an *incomplete* database which contains entries on other nodes. These entries contain information over the available resources and the agents there are running on the node. Each node chooses a random node from its database regularly to exchange information. If the size of the database exceeds a given limit, randomly chosen entries are removed to keep the communication costs affordable.

Theoretical and practical results show [7–9] that this protocol is a very effective and scalable way of distributing information over the network. For example if the database of each node contains only 100 random entries then a network of 10^{33} such nodes is partitioned only with a probability of at most 10^{-10}.

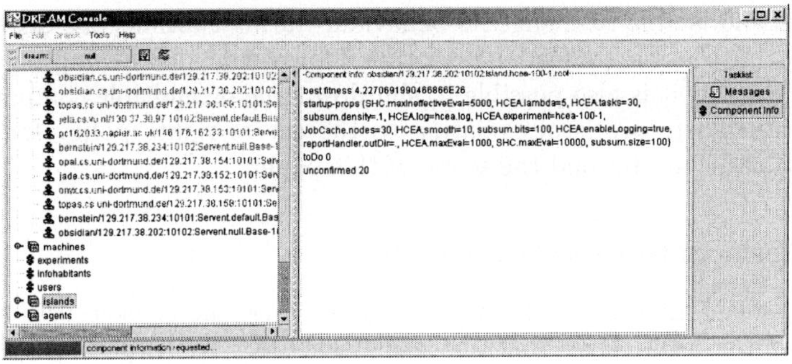

Fig. 3. The Console View.

The question of mobility and security was solved by choosing the Java environment to implement the DRM. This environment offers a natural solution for moving executable code as well as data between hosts and it offers a rich set of security features.

7 DREAM Console

The DREAM console is the primary tool for managing a computer connected to a DRM. It utilises the metaphor of a file manager view onto the DRM to present its known components and the output they produce (see Figure 3) . The list of components is mainly retrieved from the incomplete database kept by the underlying DRM layer (see section 6.2), so that the console usually only listens to the ongoing communication rather than inducing network traffic of its own. Different component types include nodes, experiments, users, islands and agents. Although the console is designed to help the user during execution of an experiment, it does not completely hide DRM layer information. As stated before, agents are analogous to threads in an operating system and islands are always implemented as agents (see section 6.1). However, there are agents performing operating system tasks like shutting down experiments or searching for a piece of information specified by the user. These are visually separated from the island agents.

7.1 Supported User Tasks

User actions supported by the console include, but are not limited to, component inspection and search; experiment startup and control; and visualization and analysis of experiment results. Regardless of the way an experiment has been constructed[1], it can be started using the console. Thereafter, the user may watch creation and evolution of the islands belonging to the new experiment. Actual status or output of a known component can be retrieved by simply clicking on

[1] The procedure of experiment construction is different for users of types B-E, depending on the tools and libraries they chose.

it. If an island is able to change its behaviour during runtime, the experimenter may use the console to send new setup information, for example parameter changes. Finally, it is also possible to shutdown one or multiple components via the console. There are issues relating to the access rights policy for different users, but these are beyond the scope of this paper.

7.2 Interfaces to User/DRM Code

For component information retrieval or messaging, the console uses interfaces defined by the DRM layer that are implemented by the experiment code. Depending on the user entry point, this implementation is done with varying degrees of automation.

8 Conclusion

We have described a system for the easy specification and implementation of highly distributed evolutionary algorithms. This system has already been implemented and the first version released. It includes several independent modules which can be used in an integrated fashion or as stand-alone units. The system allows for the sharing of CPU resources in a secure and scalable fashion. Future work will concentrate on developing algorithms and applications that work particularly well with this architecture. This will not only include applications that require vast amounts of CPU time, but is expected to include solutions to problems that can more easily be partitioned in some way (for example some data mining or scheduling problems). Systems which involve the co-operating and competing of evolving agents, either to solve real world problems, or to model aspects of society, will also be studied.

Acknowledgments

The authors would like to thank the other members of the DREAM project for fruitful discussions, the early pioneers [12] as well as Hans-Paul Schwefel, Emin Aydin and Daniele Denaro.

References

1. http://www.dcs.napier.ac.uk/ benp/dream/dream.htm
2. http://eodev.sourceforge.net.
3. M. Arenas, L. Foucart, J. Merelo, and P. A. Castillo. Jeo: a framework for evolving objects in java. In *Actas Jornadas de Paralelismo*. Universidad Politécnica de Valencia, 2001.
4. M. Arenas, L. Foucart, M. Schoenauer, and J. Merelo. Computacin evolutiva en java: Jeo. pages 46–53. Universidad de Extremadura, Febrero 2002.
5. P. Collet, E. L. M. Schoenauer, and J. Louchet. Take it easea. In *Parallel Problem Solving from Nature VI*, pages 891–901. Springer Verlag, LNCS 1917, 2000.

6. A. Demers, D. Greene, C. Hauser, W. Irish, J. Larson, S. Shenker, H. Sturgis, D. Swinehart, and D. Terry. Epidemic algorithms for replicated database management. In *Proc. 6th Annual ACM Symposium on Principles of Distributed Computing (PODC'87)*, pages 1–12, Vancouver, Aug. 1987. ACM.
7. P. T. Eugster, R. Guerraoui, S. B. Handurukande, A.-M. Kermarrec, and P. Kouznetsov. Lightweight probablistic broadcast. In *Proc. International Conference on Dependable Systems and Networks (DSN 2001)*, Göteborg, Sweden, 2001.
8. M. Jelasity, M. Preuß, and B. Paechter. A scalable and robust framework for distributed applications. In *CEC2002*, pages 1540–1545. IEEE, IEEE Press, 2002.
9. M. Jelasity, M. Preuß, M. van Steen, and B. Paechter. Maintaining connectivity in a scalable and robust distributed environment. In H.E. Bal et al., eds, *Proc. 2nd IEEE Intl Symposium on Cluster Computing and the Grid (CCGrid2002)*, Berlin, Germany, pages 389–394, 2002.
10. M. Keijzer, J. J. Merelo, G. Romero, and M. Schoenauer. Evolving objects: a general purpose evolutionary computation library. In P. Collet et al., eds, *Artificial Evolution'01*. pages 229–241, Springer Verlag, LNCS 2310, 2002.
11. E. Lutton, P. Collet, and J. Louchet. Easea comparisons on test functions : Galib versus eo. In P. Collet et al., eds, *Artificial Evolution'01*. pages 217–228, Springer Verlag, LNCS 2310, 2002.
12. B. Paechter, T. Bäck, M. Schoenauer, M. Sebag, A. E. Eiben, J. J. Merelo, and T. C. Fogarty. A distributed resoucre evolutionary algorithm machine (DREAM). In *CEC2000*, pages 951–958. IEEE, IEEE Press, 2000.
13. M. Wall. http://lancet.mit.edu/ga/.

Optimisation of Multilayer Perceptrons Using a Distributed Evolutionary Algorithm with SOAP

P.A. Castillo[1], M.G. Arenas[1], J.G. Castellano[1],
J.J. Merelo[1], V.M. Rivas[2], and G. Romero[1]

[1] Department of Architecture and Computer Technology
University of Granada. Campus de Fuentenueva. E. 18071 Granada (Spain)
[2] Department of Computer Science
University of Jaén. Avda. Madrid, 35. E. 23071 Jaén (Spain)
pedro@atc.ugr.es

Abstract. SOAP (simple object access protocol) is a protocol that allows the access to remote objects independently of the computer architecture and the language. A client using SOAP can send or receive objects, or access remote object methods. Unlike other remote procedure call methods, like XML-RPC or RMI, SOAP can use many different transport types (for instance, it could be called as a CGI or as sockets). In this paper an approach to evolutionary distributed optimisation of multilayer perceptrons (MLP) using SOAP and language Perl has been done.
Obtained results show that the parallel version of the developed programs obtains similar or better results using much less time than the sequential version, obtaining a good speedup. Also it can be shown that obtained results are better than those obtained by other authors using different methods.

1 Introduction

SOAP is a standard protocol proposed by the W3C ([36], [4]) that extends the remote procedure call, to allow the remote access to objects [9]. SOAP is the evolution of XML-RPC protocol that allows remote procedure call using XML using HTTP as transport protocol. A SOAP client can access remote services using the interface of resident objects in remote servers, using any programming language. At the moment complete implementations of SOAP are available in Perl, Java, Python, C++ and other languages [31]. Opposed to other remote procedure call methods, such as RMI (*remote method invocation*, used by the Java language) and XML-RPC, SOAP has two main advantages: it can be used with any programming language, and it can use any type of transport (HTTP, SHTTP, TCP, SMTP, POP and other protocols).

SOAP sends and receives messages using XML [27, 16, 6], wrapped HTTP-like headings. SOAP services specify the method interfaces that can be accessed using language WSDL (Web Services Description Language) [30, 33]. WSDL is an interface description language that specifies those calls that can be made to the SOAP server and the response it should return.

J.J. Merelo Guervós et al. (Eds.): PPSN VII, LNCS 2439, pp. 676–685, 2002.

Using a WSDL file, it can be specified a service for different languages, so that a Java client can access a Perl server. In some cases, if the client and the server are written using the same language, a WSDL specification is not needed,

SOAP main advantages are:

- it is a lightweight protocol
- it is simple and extensible
- it is used for application communications
- it is designed to use HTTP as transport protocol
- it is not bound to any components technology
- it is not bound to any programming language
- it is based on XML and coordinated by the W3C
- it is the core of the Microsoft's .NET technology

One of the main interests in the knowledge and use of this protocol is its relation with the Microsoft's initiative .NET [5], that bases most of its software supplies in remote services, using the SOAP protocol to carry out the communications (although it is Microsoft's version, incompatible in some cases with the standard). This must probably means that in the near future there will be many clients and servers offering services and using them, sharing SOAP and the .NET specification as common protocol.

The use of SOAP for distributed computation has not been proposed yet, except in some exceptions [20]. Nevertheless, SOAP is a high level protocol, which makes easy the task of distributing objects between different servers, without having to worry about the message formats, nor the explicit call to remote servers.

In this work we propose using SOAP for distributed computation, and we demonstrate how it could be used for evolutionary computation.

A future in which different remote computers offer services to the scientific community can be imagined: by example, all the services available at the moment by means of HTML forms could be implemented easily as SOAP services.

SOAP could be also used for distributed P2P (Peer to Peer) optimisation, where all the computers act as clients and servers to each other, interchanging population elements.

In this paper we intend to explore these abilities, implementing a distributed evolutionary algorithm (EA) using Perl and SOAP, to tune learning parameters and to set the initial weights and hidden layer size of a MLP, based on an EA and Quick Propagation [11] (QP). This paper continues the research on evolutionary optimisation of MLP (*G-Prop* method) presented in [25, 8]. This method leverages the capabilities of two classes of algorithms: the ability of EA to find a solution close to the global optimum, and the ability of the back-propagation algorithm (BP) to tune a solution and reach the nearest local minimum by means of local search from the solution found by the EA. Instead of using a pre-established topology, the population is initialised with different hidden layer sizes, with some specific operators designed to change them (*mutation, multi-point crossover, addition and elimination of hidden units, and QP training* applied as operator).

The EA searches and optimises the architecture (number of hidden units), the initial weight setting for that architecture and the learning rate for that net.

The remainder of this paper is structured as follows: in section 2 it is explained how to implement an EA using SOAP and Perl. Section 3 describes the experiments, and section 4 presents the results obtained, followed by a brief conclusion in Section 5.

2 Distributed Evolutionary Algorithms Implementation Using SOAP and Perl

There are many implementations of distributed genetic algorithms [13], usually using PVM or MPI [17], thus this paper does not intend to innovate in that sense, but in the implementation.

There are many ways to implement a distributed genetic algorithm, one of which is the global paralelization (Farming), in which, as Fogarty and Huang propose [12], Abramson and Abela [2], or Hauser and Männer [26], individual evaluation and/or genetic operator application are parallelized. A master processor supervises the population and select individuals to mate; then slave processors receive the individuals to evaluate them and to apply genetic operators.

Another way to implement paralelization is the migration: the population is divided into small subpopulations of the same size assigned to different processors. From time to time each processor selects the best individuals in its subpopulation and it sends them to his nearer processors, receiving as well copies of the best individuals of his neighbours (migration of individuals). All processors replace the worst individuals of their populations. This kind of algorithms is also known as distributed evolutionary algorithms (Tanese [32], Pettey et al. [14], Cantú-Paz and Goldberg [7]). .

An ideal client-server implementation of a distributed evolutionary algorithm could be a server process with several threads. Each thread would include a population, and would communicate with other threads through the shared code among them. Each thread would use an own tail of individuals to send to other threads. Each thread would evaluate its individuals in different remote computers, carrying out the communication using a SOAP server.

As we cannot use a threaded version of Perl, our implementation is based on a ring topology, with N populations (computers) sending individuals to the next population. The parallel algorithm code has been split in two processes: the evolutionary algorithm and the code that shares individuals (migrator) between populations (islands). The system uses asynchronous communications, that, between EA process and migrator process within an island (as well as between migrators within different islands) are carried out using SOAP on HTTP transport protocol (migrators act as web servers).

Implementation was carried out using the `SOAP::Lite` module [19] for the Perl programming language, for its stability and the familiarity of the authors with this language. In addition, servers are easy to implement using the computer infrastructure that exists in our department.

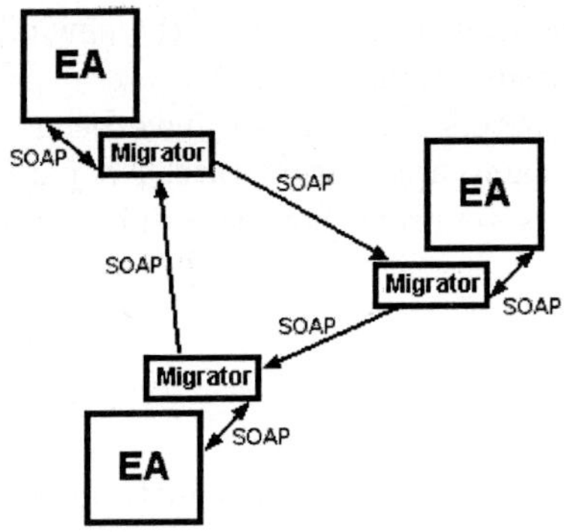

Fig. 1. Architecture: each computer runs two processes, one of them contains the population (represented as a group of people) and the other (Migrator) acts as transporter, being used to write elements of the population; thus, elements of the population are received from other computers; readings and writings are done by the Migrator object, that abstracts the physical architecture of the application.

SOAP was included from the beginning in the OPEAL EC library, and so far, several distributed evolutionary algorithms configurations have been tested on EC benchmark problems such as MaxOnes and tide [23, 18].

2.1 Evolutionary Optimisation of MLP

The evolutionary algorithm has been implemented using the OPEAL library [22], available at http://opeal.sourceforge.net under GPL license.

G-Prop method has been fully described and analysed out in previous papers (see [25, 8]), thus we refer to these papers for a full description. In most cases, evolved MLP should be coded into chromosomes to be handled by the genetic operators, however, *G-Prop uses no binary codification, instead, the initial parameters of the network are evolved* using specific variation operators such as *mutation, multi-point crossover, addition and elimination of hidden units, and QP training* applied as operator to the individuals of the population. The EA optimises the classification ability of the MLP, and at the same time it searches for the number of hidden units (architecture), the initial weight setting and the learning rate for that net.

Although evolved MLP are not coded as bit strings nor other kind of codification (they are made evolve directly) when they are sent from an island to another one, they must be coded as an XML document to be sent using SOAP.

The migration policy is as follows: each n generations the algorithm sends to the migrator e individuals, and takes i individuals. Usually, $e >= i$, in order that always are available individuals so that other islands can take them.

A steady state algorithm was used because it was empirically found to be faster at obtaining solutions than other selection algorithms [35]. For each generation, the best 30% individuals of the population are selected to mate using the genetic operators. After 5 generations, several individuals are taken from the migrator, and they are put together with the new offspring that replace the worst individual in the population.

Only "default" parameters listed above have been used. Genetic operators were applied using the same application rate. No parameter tuning has been done, because our aim is to prove how speedup improves as the number of islands grows.

3 Experiments

The tests used to assess the accuracy of a method must be chosen carefully, because some of them (toy problems) are not suitable for certain capacities of the BP algorithm, such as generalization [10]. Our opinion, along with Prechelt [24], is that, in order to test an algorithm, real world problems should be used.

3.1 Glass

This problem consists of the classification of glass types, and is also taken from [24]. The results of a chemical analysis of glass splinters (percent content of 8 different elements) plus the refractive index are used to classify the sample to be either float processed or non float processed building windows, vehicle windows, containers, tableware, or head lamps. This task is motivated by forensic needs in criminal investigation. This dataset was created based on the glass problem dataset from the UCI repository of machine learning databases. The data set contains 214 instances. Each sample has 9 attributes plus the class attribute: refractive index, sodium, magnesium, aluminium, silicon, potassium, calcium, barium, iron, and the class attribute (type of glass).

The main data set was divided into three disjoint parts, for training, validating and testing. In order to obtain the fitness of an individual, the MLP is trained with the training set and its fitness is established from the classification error with the validating set. Once the EA is finished (when it reaches the limit of generations), the classification error with the testing set is calculated: this is the result shown.

Up to six computers have been used to run the algorithm and to obtain results both in sequential and parallel versions of the program. *Computer speeds range from 400 Mhz to 800 Mhz* and are connected using the *10Mbits Ethernet network* of the University (with a high communication latency). *No experiments using homogeneous computer network* have been done, because our aim is to demonstrate potential of distributed asynchronous EA using web services.

The problem has been studied in the sequential and parallel case using a *ring migration scheme*: sending some individuals to the next island (computer), and getting the best individuals from the previous island.

Each single-computer EA was executed using the parameters shown in Table 1.

Each generation 30 new individual are generated (100 generations), thus 3100 individuals are generated each run. Taking into account that 2 up to 6 computers

Table 1. List of parameters used to execute the EA that runs in a computer (island).

Parameter	Value
number of generations	100
individuals in the population	100
% of the population replaced	30%
hidden units	ranging from 2 to 90
BP epochs to calculate fitness	300

Table 2. Equivalence between number of computers in parallel runs and number of generations in sequential runs (to generate the same number of individuals).

Number of islands	Total number of individuals	Number of generations (sequential version)
1	3100	100
2	6200	200
3	9300	300
4	12400	400
5	15500	500
6	18600	600

have been used, then 3100 to 18600 individual are created and evaluated each run (depending on the number of computers used).

To compare with the sequential version of the method, the sequential EAs have been executed using the number of generations shown in Table 2, so that the number of individuals generated in each run is equivalent to the number of individuals generated using the parallel version:

4 Results

Time was measured using the Unix `time` command. Sequential version of the program was run in the faster machine; and in parallel runs, time spent by the faster machine too was taken and used in results shown. With sequential version, 10 simulations were run, and 5 runs with 2, 3, 4, 5 and 6 machines. Each simulation was carried out adding a new computer from the set of computers used previously.

Results obtained using the sequential version can be shown in Table 3:

Results obtained using the parallel version can be shown in Table 4. As can be seen, better results in time and classification error are obtained dividing the problem between several computers.

Figure 2 shows that speedup does not equals the number of computers used; however, simulation time is improved using several computers. Moreover, up to 6 computers in the ring, the speedup grows, and as can be seen in Figure 2 it could continue growing for a higher number of computers.

Results could be better if a dedicated communication network was used, however, the University Ethernet network is overloaded and that implies a high latency in communications between processes.

Table 3. Results (error % and time) obtained using the sequential version of the method.

Generations	Error (%)	Time (seconds)
100	35 ± 1	20 ± 2
200	35 ± 1	30 ± 5
300	33 ± 2	44 ± 8
400	32 ± 2	70 ± 9
500	32 ± 2	91 ± 4
600	30 ± 3	104± 2

Table 4. Results (error % and time) obtained using the parallel version of the method, and the speedup for each experiment.

Islands	Error (%)	Time (seconds)	Speedup
1	35 ± 1	20 ± 2	1
2	33 ± 2	18 ± 1	1.6
3	33 ± 1	20 ± 1	2.2
4	32 ± 1	20 ± 2	3.5
5	31 ± 2	22 ± 3	4.1
6	30 ± 1	19 ± 2	5.5

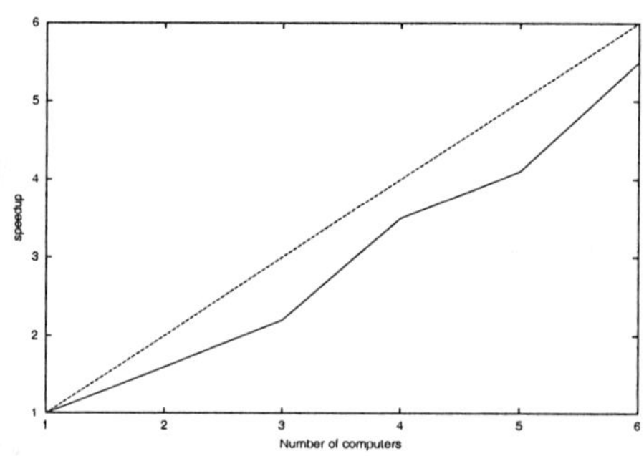

Fig. 2. Plot of the speedup (solid line) and $f(x) = x$ function (dashed line). Although speedup is not lineal, it can be seen that simulation time is improved using several computers. Moreover, up to 6 computers in the ring, the speedup grows, and it could continue growing for a higher number of computers.

Although it is not the aim of this paper to compare the G-Prop method with those of other authors, we do so in order to prove the capacity of both versions of G-Prop (parallel and sequential) to solve pattern classification problems, and how it outperforms other methods. Thus, obtained results (% of error in test) are better than those presented by Prechelt [24] using RPROP [28] (32.08), Grönroos [15] using a hybrid algorithm (32 ± 5), and Castillo et al. [25] using a previous version of the G-Prop method (31 ± 2).

5 Conclusions and Work in Progress

This paper presents a new research line on parallel-distributed computation using SOAP that shows the useful this new protocol can be in the field of the evolutionary computation.

To implement and use communications using SOAP it is not necessary running virtual machines (as in Java programming), nor daemons, just only to install several libraries available for almost any programming language. More over, if a ring topology is used, an arbitrary number of computers (islands) can be added to the network, making the system more efficient.

In these experiments, we have demonstrated that SOAP can be used as communication protocol for distributed evolutionary programming, obtaining a good speedup using a ring topology and adding new computers to the ring. Up to 6 computers in the ring, the speedup grows, and as can be seen in Figure 2 it could continue growing for a higher number of computers. Results could improve using a dedicated communication network instead of the overloaded network of the University.

Although PVM or MPI communications add fewer overheads than SOAP communications since they are done at a lower level, SOAP provides a common interface that can be called from almost any programming language. Thus programs can be written in any language and can share data without the need of worrying about the message formats or communication protocols.

At the same time, as it is a P2P system, it does not overload too much the network. Other distributed systems, such as Jini [34, 21], network traffic is so high that when a high number of computers are used, communication becomes impossible.

As future research, it is very important adding support for SOAP to existing distributed evolutionary algorithm libraries, such as JEO [3], EO [29], and libraries in other languages, in order to allow the implementation of multi-language evolutionary algorithms. It would also be of interest to use data structure description protocols, such as WSDL, to describe data to be evolved, in order to allow using them by any program that understands XML, and be able to send them easily using SOAP.

Another possibility is to test true P2P architectures, where each computer communicates only with one or two computers in the network. It would be very interesting to test asynchronous evolutionary algorithms using random topologies, in such a way that a "servent" (server/client) can enter or leave the network at any moment.

Acknowledgements

This work has been supported in part by projects CICYT TIC-1999-0550, INTAS-9730950 and IST-1999-12679.

References

1. *Actas XII Jornadas de Paralelismo.* Universidad Politécnica de Valencia, 2001.
2. Abramson; Abela J. A. Parallel genetic algorithm for solving the school timetabling problem. *In Proceedings of the Fifteenth Australian Computer Science Conference (ACSC-15), vol. 14, p.1-11*, 1992.
3. M.G. Arenas, L. Foucart, J.J. Merelo, and P. A. Castillo. Jeo: a framework for evolving objects in java. In *Actas Jornadas de Paralelismo* [1].
4. Paco Ávila. SOAP: revoluci=n en la red. *Linux actual*, (19):55–59, 2001.
5. K. Ballinger, J. Hawkins, and P. Kumar. SOAP in the microsoft .NET framework and visual Studio.NET. Available from `http://msdn.microsoft.com/library/-techart/Hawksoap.htm`.
6. D. Box. Inside SOAP. Available from `http://www.xml.com/pub/a/2000/02/09/-feature/index.html`.
7. E. Cantú-Paz and D. E. Goldberg. Modeling idealized bounding cases of parallel genetic algorithms. *In Koza J., Deb K., Dorigo M., Fogel D., GarzOn M., Iba H., Riolo R. Eds. Genetic Programming 1997: Proceedings of the Second Annual Conference, Morgan Kaufmann (San Francisco. CA)*, 1997.
8. P. A. Castillo, J. J. Merelo, V. Rivas, G. Romero, and A. Prieto. G-Prop: Global Optimization of Multilayer Perceptrons using GAs. *Neurocomputing, Vol.35/1-4, pp.149-163*, 2000.
9. DevelopMentor. SOAP Frequently Asked Questions. Available from `http://www.develop.com/soap/soapfaq.htm`.
10. S. Fahlman. An empirical study of learning speed in back-propagation networks. Technical report, Carnegie Mellon University, 1988.
11. S.E. Fahlman. Faster-Learning Variations on Back-Propagation: An Empirical Study. *Proceedings of the 1988 Connectionist Models Summer School, Morgan Kaufmann*, 1988.
12. T. Fogarty and R. Huang. Implementing the genetic algorithm on transputer based parallel processing systems. *Parallel Problem Solving From Nature,p.145-149*, 1991.
13. David E. Goldberg. *Genetic Algorithms in search, optimization and machine learning.* Addison Wesley, 1989.
14. C. B. Pettey; M. R. Leuze; J. J. Grefenstette. A parallel genetic algorithm. *In J. J. Grefenstette (Ed.), Proceedings of the Second International Conference on Genetic Algorithms, Lawrence Erlbaum Associates, pp. 155-162*, 1987.
15. M.A. Grönroos. Evolutionary Design of Neural Networks. *Master of Science Thesis in Computer Science. Dept. of Mathematical Sciences. University of Turku*, 1998.
16. Elliotte Rusty Harold. *XML Bible.* IDG Books worldwide, 1991.
17. J.G.Castellano, M.García-Arenas, P.A.Castillo, J.Carpio, M.Cillero, J.J.Merelo, A.Prieto, V.Rivas, and G.Romero. Objetos evolutivos paralelos. In Universidad de Granada Dept. ATC, editor, *XI Jornadas de Paralelismo*, pages 247–252, 2000.
18. J.J.Merelo, J.G.Castellano, and P.A.Castillo. Algoritmos evolutivos P2P usando SOAP. pages 31–37. Universidad de Extremadura, Febrero 2002.
19. P. Kuchenko. SOAP::Lite. Available from `http://www.soaplite.com`.
20. D. Marcato. Distributed computing with soap. Available from `http://www.devx.com/upload/free/features/vcdj/2000/04apr00/dm0400/-dm0400.asp`.
21. J. Atienza; M. García; J. González; J. J. Merelo. Jenetic: a distributed, fine-grained, asynchronous evolutionary algorithm using jini. pages 1087–1089, 2000. ISBN: 0-9643456-9-2.

22. J. J. Merelo. OPEAL, una librería de algoritmos evolutivos. *Actas del Primer Congreso Español de Algoritmos Evolutivos y Bioinspirados. ISBN:84-607-3913-9. pp.54-59. Mérida, Spain, febrero*, 2002.
23. J. J. Merelo, J.G. Castellano, P.A. Castillo, and G. Romero. Algoritmos gen ticos distribuidos usando soap. In *Actas Jornadas de Paralelismo* [1].
24. Lutz Prechelt. PROBEN1 — A set of benchmarks and benchmarking rules for neural network training algorithms. *Technical Report 21/94, Fakultät für Informatik, Universität Karlsruhe, D-76128 Karlsruhe, Germany. (Also in: http://wwwipd.ira.uka.de/~prechelt/)*, 1994.
25. P.A. Castillo; J. Carpio; J. J. Merelo; V. Rivas; G. Romero; A. Prieto. Evolving multilayer perceptrons. *Neural Processing Letters*, 12:115–127, October 2000.
26. Hauser R.; Männer R. Implementation of standard genetic algorithm on mimd machines. *In Davidor Y., Schwefel H. P., Männer R., Eds., Parallel Problem Solving from Nature, PPSN III, p. 504-513, Springer-Verlag (Berlin)*, 1994.
27. Erik T. Ray. *Learning XML: creating self-describing data.* O´Reilly, January 2001.
28. M. Riedmiller and H. Braun. A Direct Adaptive Method for Faster Backpropagation Learning: The RPROP Algorithm. *In Ruspini, H., (Ed.) Proc. of the ICNN93, San Francisco, pp. 586-591*, 1993.
29. M. Keijzer; J. J. Merelo; G. Romero; and M. Schoenauer. Evolving objects: a general purpose evolutionary computation library. Springer-Verlag, October 2001.
30. A. Ryman. Understanding web services. Available from `http://www7.software.ibm.com/vad.nsf/Data/-Document4362?OpenDocument&p=1&BCT=1&Footer=1`.
31. soaprpc.com. SOAP software. Available from http://www.soaprpc.com/software.
32. R. Tanese. Parallel genetic algorithms for a hypercube. *In J. J. Grefenstette (Ed.), Proceedings of the second International Conference on Genetic Algorithms, Lawrence Erlbaum Associates, pp. 177-184*, 1987.
33. V. Vasudevan. A web services primer. Available from `http://www.xml.com/pub/a/2001/04/04/webservices/index.html`.
34. Bill Venners. Jini FAQ (frequently asked questions). Available from `http://www.artima.com/jini/faq.html`.
35. D. Whitley. *The GENITOR Algorithm and Selection Presure: Why rank-based allocation of reproductive trials is best.* in J.D. Schaffer (Ed.), Proceedings of The Third International Conference on Genetic Algorithms, Morgan Kauffmann, Publishers, 116-121, 1989.
36. D. Box; D. Ehnebuske; G. Kakivaya; A. Layman; N. Mendelsohn; H.F. Nielsen; S. Thatte; D. Winer. Simple Object Access Protocol (SOAP) 1.1, W3C Note 08 May 2000. Available from `http://www.w3.org/TR/SOAP`.

Applications

Off-Line Evolution of Behaviour for Autonomous Agents in Real-Time Computer Games

Eike Falk Anderson

The National Centre for Computer Animation
Bournemouth University
Talbot Campus, Fern Barrow, Poole
Dorset BH12 5BB, UK

Abstract. This paper describes and analyses a series of experiments intended to evolve a player for a variation of the classic arcade game Asteroids™ using steady state genetic programming. The player's behaviour is defined using a LISP like scripting language. While the game interprets scripts in real-time, such scripts are evolved off-line by a second program which simulates the real-time application. This method is used, as on-line evolution of the players would be too time consuming. A successful player needs to satisfy multiple conflicting objectives. This problem is addressed by the use of an automatically defined function (ADF) for each of these objectives in combination with task specific fitness functions. The overall fitness of evolved scripts is evaluated by a conventional fitness function. In addition to that, each of the ADFs is evaluated with a separate fitness function, tailored specifically to the objective that needs to be satisfied by that ADF.

1 Introduction

In recent years the level of realism in computer games has risen dramatically. While the quality of real-time graphics - mainly due to advances in computer graphics hardware - certainly played the major part in this development, one should not forget that artificial intelligence is another important factor for the attainment of realism in games. The playability of computer games is often measured by the quality of the behaviour of intelligent agents in the game environment. If this behaviour appears natural and human-like, the agent seems to be more life-like and real. There is an obvious solution to satisfy the need for the creation of autonomous agents which seem alive. Genetic programming (GP) produces algorithms by using a process that parallels evolution through natural selection, i.e. a simulation of life. GP has so far been applied to a number of different computer game scenarios. Among these are classic videogames like Pac Man® ([Koza 1994]) or Tetris® ([Siegel and Chaffee 1996]). These experiments evolved game playing behaviour in a modified game environment. Most of these game versions are round-based, i.e. the computation of an action in the game is performed while the game is paused. Gameplay resumes only after those computations have finished, and only until the calculated actions

J.J. Merelo Guervós et al. (Eds.): PPSN VII, LNCS 2439, pp. 689–699, 2002.

have been executed. This is in contrast to real-time games in which all actions have to be calculated "on the fly". One of the few attempts to apply GP to a real-time game (RoboCup Soccer) is documented in [Luke et al 1997] and [Luke 1998]. The methods employed for that experiment bear some similarities to the experiments described here. The players are evolved in a modified environment, and not on-line in the game itself. Only the evolved players are used in the real-time application. They also used different algorithm trees for different player objectives, which is similar to the experiments using syntactic constraint to achieve the generation of a separate gene pool for each of the player's objectives as described by [Reynolds 1994]. This approach again resembles the concept of Automatically Defined Functions as described by [Koza 1992]. The experiments described in this paper merge some of these techniques and extend the use of ADFs by calculating a fitness for each of the ADFs in addition to the fitness value calculated for the performance of each individual of the population. The classic arcade game Asteroids is used here as a test case. Asteroids is based on attack and evasion which is a concept that is common to most action oriented video games. The evolution of a successful player could therefore be seen as a proof of concept. It should be possible to transfer a solution of the Asteroids problem to more complex game scenarios by adapting the player's interface with the game to that of another computer game.

In the Asteroids game a "space ship" (the player) has to avoid colliding with a number of "asteroids" to prevent its destruction. At the same time it has to destroy the "asteroids" to win and progress to the next level of the game. For practical reasons a number of modifications to the original game have been made for the version used in the experiments that are described here (Figure 1).

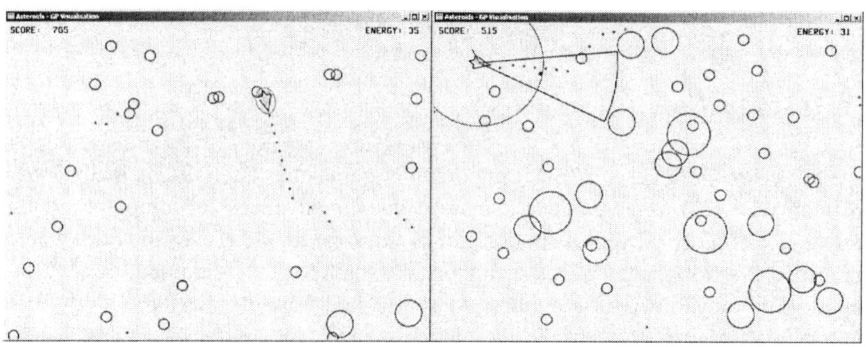

Fig. 1. Screenshots of the Asteroids implementation. The left screenshot shows the player's spacecraft avoiding collision by raising its shields. The right screenshot shows the player's sensor information consisting of asteroid proximity and the player's field of view

In the original arcade version of the Asteroids game the player's spacecraft flies through a two-dimensional field of moving asteroids. The player has a pre-defined number of lives. A collision with an asteroid destroys the spacecraft. The player can shoot at asteroids. If hit, a large asteroid will break up into two medium asteroids which in turn can each be split into two small asteroids. Shooting down an asteroid increases the player's score. The player's only means of defence is to jump into "hyperspace", which removes the player's spacecraft from its current location and randomly repositions it on the screen. The game ends when the player has lost all of his lives. The aim of the game is to stay alive as long as possible and to gain the highest score during that period.

In the implementation used here, the player only has one life to limit the execution time of the game. In addition to that the "hyperspace" escape function has been replaced by a "protective shield", as the results of this feature are unpredictable, i.e. the player's spacecraft could be saved, but just as well be destroyed when using a "hyperspace" jump. The defence "shield" however has a predictable result: it protects the player from destruction during collisions with asteroids by granting invulnerability while it is active. The player is given an initial level of energy. Using shields or firing the gun will drain the player's energy which is replenished over time. The player is therefore encouraged not to waste energy by unnecessary usage of the shields and the gun. Finally a more complex scoring system which gives a more precise reflection of the player's performance is used in this version of the game.

2 Implementation of the Off-Line Game Environment

The environment in which the game is played is a two-dimensional field of 80 x 60 units which continually wraps around. The player starts in the centre. Asteroids are positioned randomly around the player. As a single game of Asteroids can take a few minutes to complete, it is self-evident that trying to evolve the players on-line would take far too long. The obvious solution is to remove the code generation from the real-time application and to evolve the player scripts off-line in a second application which simulates the real-time game. The real-time game itself only interprets and executes the programs that have been evolved by the off-line game simulation. The simulation program contains the GP system that evolves the player scripts. It also contains a copy of the game code that has been stripped of all graphics functions, which is used for evaluating the evolved players' fitness. Without displaying anything on screen, the simulation can run at much greater speed and players evolve much quicker than would be possible in a real-time environment. Whereas it would take several minutes for a single individual of the player population to play a single game in an on-line evolution, using this dual approach, a series of games can be played in a matter of seconds.

To evolve successful players for the game, the code generation program needs to simulate the real-time application as closely as possible. Discrepancies in the frame-rate of the real-time application result in a variable animation step-size for the objects (player, asteroids and bullets). This is simulated in the code generation program by combining the average frame-rate of the real-time game with a random value.

The player interfaces with the game through a LISP like scripting language which implements a number of sensors and controls. In this problem the controls, which are identical to a human player's controls for the space ship, are used as output of the evolved program. The sensors, which reflect the current state of the game, are used as input to the evolved program. The script which controls the autonomous agent is created through evolution, based on the agent's proficiency at playing the game.

3 GP Architecture

The variation of GP used in this project is "strongly typed" GP as introduced by [Montana 1995], which allows for the use of different datatypes. There are two datatypes, one for Boolean values which can be either TRUE or FALSE, and a void datatype which is used for procedures that do not return any data. Three constants (two Boolean: TRUE, FALSE and one void: void) are defined for use in the control structures. The player's sensors and controls make up the terminal set of the GP functions while the control structures are the non-terminal set of functions.

Table 1.

Function	Returns	Description
(targetAhead)	Boolean	TRUE if an asteroid is within the player's field of view, else FALSE
(targetLocked)	Boolean	TRUE if an asteroid is in the player's direct line of fire, else FALSE
(proximityAlert)	Boolean	TRUE if an asteroid is in the player's proximity (within 12 units from the player), else FALSE
(impactAlert)	Boolean	TRUE if the player is about to collide with an asteroid (asteroid is within 3 units from the player), else FALSE
(hasEnergy)	Boolean	TRUE if the player has energy left, else FALSE
(plentyEnergy)	Boolean	TRUE if the player has enough energy for firing more than four shots, else FALSE
(hasShields)	Boolean	TRUE if the player's shields are raised, else FALSE
(lookingAhead)	Boolean	TRUE if the player's direction of movement is identical to the player's heading, else FALSE
(isMoving)	Boolean	TRUE if the player is moving, else FALSE
(accelerating)	Boolean	TRUE if the player has active thrusters, else FALSE
(isTurning)	Boolean	TRUE if the player is turning, else FALSE

3.1 GP Function Set

The sensors of the player's space ship are implemented as a set of Boolean functions. The available sensor information consists of:

- The level of the player's energy.
- The state of the player's movement.
- Approximate positions of targets (asteroids) in relation to the player's position.

The controls for the space ship are implemented as a set of procedures which enable the player to switch its current states. The available instructions are:

- Turning (left, right, not).
- Acceleration (on, off), deceleration (automatically reset for each frame).
- Shields (on, off).
- Firing a single bullet.

A more detailed description of the syntax of these functions and procedures can be seen in Table 1 and Table 2.

Table 2.

Function	Returns	Description
(setThrust)	void	activates the player's thrusters
(noThrust)	void	deactivates the player's thrusters
(decelerate)	void	reduces the player's speed
(setShields)	void	raises the player's shields
(noShields)	void	lowers the player's shields
(rightTurn)/(leftTurn)	void	sets the player to turn clockwise/anti-clockwise
(noTurn)	void	sets the player to stop turning
(fire)	void	fires a single bullet

Player scripts that use this interface are generated using simple Boolean operators (AND, OR, XOR and NOT) which are implemented as non-terminal functions and a small set of control structures which consists of:

- Dyadic selection.
- Comparison.
- Sequence.

See Table 3 for a detailed description of the control structure syntax.

Table 3.

Function	Returns	Description
(if_true b v1 v2)	void	if the Boolean function b returns TRUE the void procedure v1 is executed else if b returns FALSE the void procedure v2 is executed
(if_equal b1 b2 v)	void	if the return values of the Boolean functions b1 and b2 are identical the void procedure v is executed
(sequence v1 v2)	void	executes the two void functions v1 and v2 one after the other

The goal of the game Asteroids is to maximise the player's score. In this implementation of the game the best way to achieve this is to destroy all asteroids as quickly as possible. A precondition for the destruction of all asteroids is the player's survival. This leads to the identification of three distinctive behaviours:

- Aggression - Destroying a target which is in the player's range and line of fire.
- Target Acquisition - Seeking out and finding targets in the shortest possible time.
- Defence - Avoiding collisions with asteroids.

The use of segregated branches of the parse tree for achieving multiple objectives as described in [Reynolds 1994] was the inspiration for the use of ADFs to find a solution that successfully completes the three conflicting objectives of the Asteroids game. This is done by associating each of the objectives with a different ADF. To ensure that each of these three ADFs specialises in satisfying a different objective, the fitness evaluation of individual players is distributed using task specific fitness functions. The GP system uses a separate fitness function for each ADF which evaluates the fitness of that ADF for a specific task. Each of these fitness functions runs a subset of the game simulation which ignores all factors that are not deemed necessary for accomplishing that particular ADF's objective. The error values that are returned by these fitness functions are accumulated and added to the error value of the overall fitness function which evaluates the performance of the player. All ADFs are terminal functions that return void and take no parameters.

The ADFs can contain all of the available functions, procedures and control structures. In the result-producing branch (RPB) which contains the main program however only the control structures (see Table 3) and the three ADFs (Aggression, Defence, Target Acquisition) are available. It was necessary to impose this syntactic constraint, as early experiments showed that otherwise there was a chance of functions and procedures in the RPB cancelling out the results generated by the ADFs.

3.2 Fitness Evaluation

In the main fitness function which evaluates the RPB of each player, the player's fitness is determined using a progressive fitness measure which is similar to the one described by [Siegel and Chaffee 1996]. In that approach successful individuals of a population were re-evaluated in further test cases. Here however the progression is applied from within the fitness function itself. In the fitness function the player's performance is measured in a series of four games of increasing difficulty. Only successful players may proceed to the following game. The games are time limited to prevent infinite loops from occurring. The fitness cases for each game are "Survival", "Speed", "Marksmanship" (*2 levels*), "Aggression" (*2 levels*) and "Score".

After a series of games has been played further fitness cases are tested. These are "Success", "ADF Usage" (*3 cases*), "Execution Speed" (*4 cases*) and "Vitality" (*7 cases*).

The fitness function for the Defence ADF evaluates a special version of the game that runs over a set time and in which destroyed asteroids are constantly regenerated, so that the game never runs out of asteroids. The fitness cases for this ADF are "Energy Conservation" (*2 cases*), "Evasive Manoeuvres" (*2 cases*), "Distance Checks" (*2 cases*) and "Survival".

The fitness function for the Aggression ADF evaluates a special version of the game which runs over a set time. Destroyed asteroids are constantly regenerated, so that the game never runs out of asteroids. The player is automatically moved across the playing area, so it can concentrate on its shooting skills and collisions between the player and asteroids are disabled. The fitness cases for this ADF are "Energy Conservation" (*3 cases*), "Marksmanship" (*4 cases*), "Use of Guns" (*2 cases*) and "Kill Rate".

The fitness function for the Target Acquisition ADF evaluates a special version of the game in which the player automatically fights and defends against asteroids. The fitness cases for this ADF are "Use of Sensors" (*3 cases*), "Steering" (*4 cases*) and "Movement" (*2 cases*).

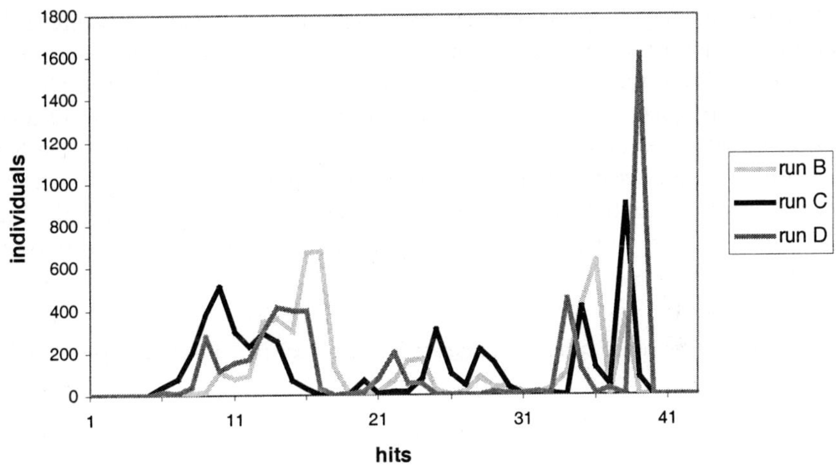

Fig. 2. Conditions that have been satisfied (*hits*) for the individuals of runs B, C and D

4 Experiments and Results

In the experiments performed, a large number of evolution runs were carried out, many of which were only used to verify the effectiveness of refinements of the fitness functions. Six of the runs are of particular interest and are discussed below in some detail. However it has to be said that the experiments are still continuing, so this project is still very much work in progress. One of the earlier experiments evolved a population of 5000 individuals over 2 generations, using higher level primitives instead of ADFs (run A). In four of the experiments described here a population of 5000 individuals was evolved over 10 generations (runs B, C, D and E). A single control run with slightly relaxed fitness conditions was performed with a population of 1500 individuals over 10 generations (run F). All experiments used the same random seed for the generation of the initial population and the possibility of a 6% mutation was introduced to counteract a possible loss of diversity in the player's gene pool.

4.1 Early Experiments without ADFs

The earliest test runs of the code generation program did not use ADFs but were instead set to produce programs that consisted of a single tree. The fitness function used for these experiments was almost identical to the one used for later experiments, except for the fact that it also contained fitness cases that were later moved into the task-specific fitness functions for the ADFs. The player scripts produced by these early experiments were weak and hardly ever survived against more than a single asteroid. To find out if the reason for these poor results was that the function set for the player interface was too small, a set of three higher level primitives (seek, autoprotect, fireAtWill), reflecting the three different objectives of the game and designed to automatically play the game were created as terminal functions. These higher level primitives themselves were just groups of some of the lower level primitives. A hand-coded player script was created by just combining these three higher level primitives with each other using the "sequence" non-terminal function (Table 3). This script proved to be a successful player which never lost a game. If the higher level primitives are used in the code generation program, scripts evolve which are very similar to this initial player script. The following script containing the higher level primitives was generated in run A after two generations (unedited except for addition of comments):

```
(sequence
   seek    ; move until target is in range, then stop
   (sequence
     (sequence
       fireAtWill  ; if a target is in the line of fire, shoot
       autoprotect) ; if collision is immanent raise shields
     autoprotect)) ; if collision is immanent raise shields
```

=

Although possibly not the optimal player, this evolved player employs similar strategies to those used by many human players:

- It keeps turning in one direction until a target is in its line of fire, then stops.
- If a target is in its line of fire, it shoots at the target.
- If no target is in range, it moves forward until a target is in range.

To test if these results could be replicated by just using the low-level primitives, a hand-coded program was created which used three ADFs, each of which emulated one of the higher level primitives.

4.2 Experiments Using ADFs with Separate Fitness Functions

When confronted with the problem of trying to achieve a result that equals the success of the hand-coded program, the question arose how to force the ADFs to each tackle a different objective. The approach that was adopted was to create a separate fitness function for each of the three ADFs. Figure 2 shows the fitness distribution within the populations after each of the runs for this experiment. Although only observed over a series of three runs of ten generations each (runs B, C and D), a convergence of the gene pools of the ADFs, reflecting the make-up of each respective fitness function, becomes apparent. Resulting player scripts show the use of a similar strategy to that of the hand-coded player discussed earlier. It should be noted however that there seems to be a larger loss of diversity in the player's gene pool when this method is used, than can be observed in evolution runs in which the ADFs are not evaluated by separate fitness functions. This might be a serious problem that needs to be addressed in future experiments.

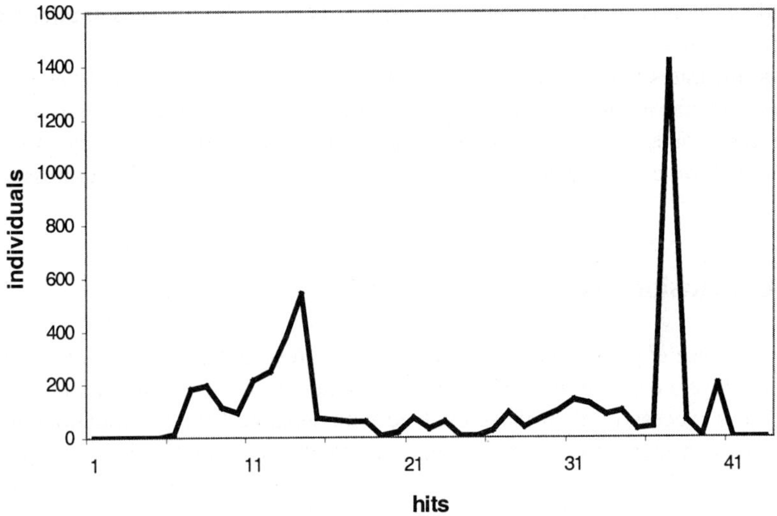

Fig. 3. Conditions that have been satisfied (*hits*) for the individuals of control run E

4.3 Control Runs

A single control run E, using the same parameters and fitness function, however without the additional fitness functions for the ADFs resulted in a player with a similar behaviour and capability to the players evolved using separate fitness functions for the ADFs in runs B, C and D. The overall performance of individuals of run E after ten generations was better than that for players evolved in runs B and C. The fitness distribution in the population of this run is shown in Figure 3. The players in run E needed more generations to evolve to a stage where their performance matched that of the players generated by the earlier runs with separate ADF fitness functions. Although the best individual of this control run displayed a similar behaviour to the best individuals of the earlier runs and completed the same objectives, there is no visible task specialisation in the ADFs.

In another control run F the functions for checking the level of energy, as well as all energy consumption by the players were removed from the program. If energy management is disabled, the resulting players seem to be more successful and are much more likely to survive. A single run over ten generations with a population of 1500 individuals evolved two different kinds of successful player. However the evolved strategies appear a lot less intelligent than those observed in the other runs:

One of the players in run F continuously spun its space ship around while firing its gun. This is an obvious solution, as with unlimited ammunition, there is no pressure regarding accuracy, and if the gun continuously fires in all directions, the chance of any asteroid getting close enough to the player to destroy it without being destroyed itself is minute. The strategy adopted by the other player was less obvious but similarly effective. It raised its shields and flew in a straight line, continuously firing its gun, creating some sort of impenetrable barrier in front of the space ship which destroyed everything in its path. This was only possible with unlimited ammunition and shields.

This illustrates that the pressure created through the player's need for energy management aids the evolution of a more intelligent player. It also shows, that in addition to the three objectives that have been identified, there may be further objectives which need to be completed by a player to succeed, that need to be addressed separately.

5 Conclusion and Future Research

Although these experiments are still work in progress and especially considering the fact that players evolved with ADFs without separate fitness functions do not seem to be any less successful, than those evolved with separate fitness evaluations for each ADF, the latter method looks promising as a possible solution to addressing problems with multiple objectives. The experiments show that this method at least accelerates the evolution of reasonably successful players. However this needs to be verified in further experiments and test runs over more than ten generations each. Furthermore the problem of multiobjective optimisation as described in [Goldberg 1989] needs to be addressed in conjunction with the use of separate fitness functions.

Using GP as a tool for the controlled evolution of autonomous agents for computer games seems an appropriate method for the design of natural agent behaviour. The experiments have shown that GP can be used to evolve the behaviour of an intelligent agent for real-time computer games. The next step will be to improve the agents' intelligence by refining the original problem. Coevolution techniques as used by [Sims 1994], [Reynolds 1994] that use competition between individuals to exert additional evolutionary pressure are a possible solution to this problem, which needs to be explored. Another approach might be to only use GP to evolve the agent's instincts, while other methods could be used to create a higher level intelligence. This might eventually lead to the generation of more complex and realistic agent behaviour than has been achieved so far.

Acknowledgements

The author would like to thank Anargyros Sarafopoulos for inspiration, support and the permission to use his GP system for this work. His ideas and suggestions contributed significantly to this project. Additional thanks go to Prof. Peter Comninos for encouragement and help in the preparation of this paper.

References

[Goldberg 1989] Goldberg, D. E. (1989), *Genetic Algorithms in Search, Optimization, and Machine Learning*, Addison-Wesley, Reading, Massachusetts

[Koza 1992] Koza, J. R. (1992), *Genetic Programming: on the Programming of Computers by Means of Natural Selection*, MIT Press, Cambridge, Massachusetts

[Koza 1994] Koza, J. R. (1994), *Genetic Programming II: Automatic Discovery of Reusable Programs*, MIT Press, Cambridge, Massachusetts

[Luke et al 1997] Luke, S. (1997), Hohn, C., Farris, J., Jackson, G. Hendler, J., *Co-Evolving Soccer Softbot Team Coordination with Genetic Programming*, Proceedings of the RoboCup-97 Workshop at the 15th International Joint Conference on Artificial Intelligence, IJCAI

[Luke 1998] Luke, S. (1998), *Genetic Programming Produced Competitive Soccer Softbot Teams for RoboCup97*, Genetic Programming 1998: Proceedings of the Third Annual Conference, Morgan Kaufmann

[Montana 1995] Montana, D. J., *Strongly Typed Genetic Programming*, Evolutionary Computation, 3(2), 1995, pages 199-230

[Reynolds 1994] Reynolds, C. W. (1994), *Competition, Coevolution and the Game of Tag*, Proceedings of the Fourth International Workshop on the Synthesis and Simulation of Living Systems, MIT Press, Cambridge, Massachusetts

[Siegel and Chaffee 1996] Siegel, E. V. (1996) and Chaffee, A. D., *Genetically Optimizing The Speed of Programs Evolved to Play Tetris*, Advances in Genetic Programming 2, MIT Press, Cambridge, Massachusetts

[Sims 1994] Sims, K. (1994), *Evolving 3D Morphology and Behavior by Competition*, Proceedings of the Fourth International Workshop on the Synthesis and Simulation of Living Systems, MIT Press, Cambridge, Massachusetts

A Parallel Evolutionary Algorithm for Stochastic Natural Language Parsing

Lourdes Araujo

Dpto. Sistemas Informáticos y Programación
Universidad Complutense de Madrid
lurdes@sip.ucm.es

Abstract. This paper presents a parallel evolutionary program for natural language parsing. The implementation follows an island model, in which, after a number of generations, demes exchange some individuals in a round-robin manner. The population is composed of potential parsings for a sentence, and the fitness function evaluates the appropriateness of the parsing according to a given stochastic grammar. Both the fitness function and the genetic operators, which require that the result of their application still corresponds to the words in the input sentence, are expensive enough to make the evolutionary program appropriate for a coarse grain parallel model and its distributed implementation. The system has been implemented in a parallel machine using the PVM (Parallel Virtual Machine) software. The paper describes the study of the parameters in the parallel evolutionary program, such as the number of individuals to be exchanged between demes, and the number of generations between exchanges. Different parameters of the algorithm, such as population size, and crossover and mutation rates, have also been tested.

1 Introduction

Evolutionary algorithms are an efficient method to deal with hard optimization problems [5, 8, 9]. Statistical methods allow posing different aspects of natural language processing as optimization problems of some kind of measurements given by the statistical model. In this way, evolutionary algorithms are a very useful tool to deal efficiently with different problems of natural language processing. Besides, working with statistical models allows an uncertainty margin on the results that matches properly with the nature of the evolutionary algorithms (which do not guarantee the best solution, but one reasonably good).

Natural language parsing is a search problem that requires exploring a tree of possible parsings. The size of this tree increases exponentially with the length of the sentence or text to be parsed. By scoring every parsing in the tree, such a search can be thought of as an optimization problem, since we are looking for the "best" parsing. Stochastic grammars [4, 1] represent an important part of the statistical methods in computational linguistics, which have allowed real progress on a number of issues including disambiguation, error correction, etc. These grammars give us an evaluation of the tree representing a possible parsing

J.J. Merelo Guervós et al. (Eds.): PPSN VII, LNCS 2439, pp. 700–709, 2002.

of a sentence, and we will try to find one of the best parsing according to this measurement.

This work develops a stochastic parallel parser for natural language based on an evolutionary algorithm. The approach adopted herein is based on the evolution programming method, as has been extended by Michalewicz [7] to consider a richer set of data structures for chromosome representation than classic genetic algorithms. In our case individuals are possible parsings, which are evaluated in order to give a measure of how they fit the grammar rules.

Despite the ability of genetic algorithms to find a "good" solution, though perhaps approximate, to optimization problems, when such problems are too hard, they require a very long time to reach the solution. This has led to different efforts to parallelizing these programs in order to accelerate the process. Basically, the approaches to this parallelization can be classified in two groups [2, 6]: *global* parallelization, and *island* of *coarse-grained* parallelization. In the first method there is only one population, as in the sequential model, and it is the evaluation of individuals, and the application of genetic operators what is parallelized. In the island model, the population is divided in subpopulation or *demes*, that usually evolve isolated except for the exchange of some individuals or *migrations* after a number of generations. In this case, we can expect a different behaviour of the parallel model, since this model implies a change in some parameters of the algorithm (such as the population size, which is smaller in each deme), what can result in a faster convergence. Though such a faster convergence may, in principle, reduce the quality of the solutions, results shows [3] that the parallel model with smaller populations but with migration among demes can improve the quality of the sequential solutions, and that there is an optimal number of demes which maximizes the performance. We have adopted the island model, what has allowed us developing a portable implementation valid for both distributed and shared memory platforms.

The rest of the paper proceeds as follows: section 2 describes the main elements in the evolutionary algorithm for parsing; section 3 is devoted to the parallel model; section 4 presents and discusses the experimental results and section 5 draws the main conclusions of this work.

2 Evolutionary Algorithm for a Probabilistic Parsing

Discovering the meaning of a sentence or text for any application (machine translation, database interfaces, etc) requires extracting their grammatical structure, usually represented in the form of a labeled tree, which indicates how words relate to each other and form phrases or clauses. However, many words can play several grammatical roles, what leads to a forest of possible parsing trees for a sentence. Probabilistic Context Free Grammars (PCFG)[4] provide a probabilistic model of the language, as well as a mechanism to select one parsing if there is ambiguity. A PCFG is a context free grammar whose rules are assigned a probability. In our case, the PCFG provides a measure of the fitness of the individuals in the evolutionary algorithm.

Individuals. The chromosomes of our evolutionary algorithm represent potential parsings for an input sentence according to a given probabilistic context free grammar (PCFG). The input sentence is given as a sequence of words. The set of categories for each word is searched in a dictionary (lexicon). A chromosome is represented as a data structure containing the following information:

- Fitness of the chromosome.
- A list of *genes*, each representing the parsing of a different set of words in the sentence.
- The number of genes in the chromosome.
- The depth of the parsing tree.

Each gene represents the parsing of a consecutive set of words in the sentence. If this parsing involves no-terminal symbols, the parsing of the subsequent partitions of the set of words is given in later genes. Accordingly, the information contained in a gene is the following:

- The sequence of words in the sentence to be analyzed by the gene.
- The rule of the grammar used to parse the words in the gene.
- If the right hand side of the rule contains no terminal symbols, the gene also stores the list of references to the genes in which the analysis of these symbols proceeds.
- The depth of the node corresponding to the gene in the parsing tree. It will be used in the evaluation function.

The data structure of Figure 1 represents one possible chromosome for the sentence: "the man sings a song".

Initial Population. In order to create the chromosomes in the initial population, the set of words in the sentence is randomly partitioned, making sure that there is at least one verb in the second part, which corresponds to the main VP. The set of words corresponding to the NP is parsed by randomly generating any of the possible NP rules. The same is done for generating the parsing of the VP with the VP rules. The process is improved by enforcing the application of those rules able to parse the right number of words of the gene. The process continues until there are no terminal symbols left pending to be parsed.

2.1 Individuals Evaluation

The opportunities of an individual to survive depend on the measurements of their adaptation or fitness. In our case, this measurement is given by a couple of values, f_{coher}, which measures the ability of a chromosome to parse the objective sentence, and f_{prob}, which measures the probability of the rules employed in the parsing.

f_{coher} is based on the relative number of *coherent* genes. A gene will be considered *coherent* if

<table>
<tr><td colspan="3">(fitness): X</td></tr>
<tr><td colspan="3">(number of genes): 4</td></tr>
<tr><td colspan="3">(genes):</td></tr>
<tr><td>(gene number)</td><td>(rule)</td><td>(gene decomposition):
(first word, number of words, gene):</td></tr>
<tr><td>(1)</td><td>$S \to NP, VP$</td><td>NP:(1, 2, 2)
VP:(3, 3, 3)</td></tr>
<tr><td>(2)</td><td>$NP \to Det, Noun$</td><td>Det: The
Noun: man</td></tr>
<tr><td>(3)</td><td>$VP \to Verb, NP$</td><td>Verb: sings
NP:(4, 2, 4)</td></tr>
<tr><td>(4)</td><td>$NP \to Det, Noun$</td><td>Det: a
Noun: song</td></tr>
</table>

Fig. 1. Data structure of a chromosome. The first gene tells us that the set of words in the sentence has been partitioned by the third word, i.e., the first two words correspond to the main NP, and the following three words to the main VP. The gene also tells us that the parsing of the NP is given by the gene 2, and the parsing of the VP is given by the gene 3. Since the rule in the gene 2 has only terminal symbols in its right hand side, there is no gene decomposition. On the contrary, the rule for gene 3 presents an NP symbol in its right hand side, whose parsing is done in gene 4.

a) it corresponds to a rule whose right hand side is only composed of terminal symbols, and they correspond to the categories of the words to be parsed by the rule.
b) it corresponds to a rule with non-terminal symbols in its right hand side and each of them is parsed by a coherent gene.

Accordingly, f_{coher} is computed as

$$f_{coher} = \frac{\text{number of coherent genes} - \sum_{i \in \text{incoherent genes}} \frac{\text{penalization}}{\text{depth}(i)}}{\text{total number of genes}}.$$

The formula takes into account the relative relevance of the genes: the higher in the parsing tree is the node corresponding to an incoherent gene, the worse is the parsing. Thus the fitness formula introduces a penalization factor which decreases with the depth of the gene.

f_{prob} is computed as

$$\prod_{i=1}^{n} \text{Prob}(g_i)$$

where $\text{Prob}(g_i)$ is the probability of the grammatical rule of gene g_i in the chromosome.

Fitness is then computed as a linear combination of both:

$$Fitness = w_{\text{coher}} f_{\text{coher}} + w_{\text{prob}} f_{\text{prob}}$$

where w_{coher} and w_{prob} are parameters that allow tuning the computation along the evolution process. In the first generations w_{coher} is higher in order to produce individuals corresponding to possible parsing trees, while later, w_{prob} becomes higher in order to select the most probable individuals.

2.2 Reproduction

Each generation finishes with the creation of new individuals that will substitute other individuals in the population. These new individuals are created by means of the *crossover* and *mutation* operators.

The crossover operator combines two parsings to generate a new one. The part of one parent after a point randomly selected is exchanged with the corresponding part of the other parent to produce two offsprings, under the constraint that the genes exchanged correspond to the same type of parsing symbol (NP, VP, etc) in order to avoid wrong references of previous genes in the chromosome. The operator selects two parent chromosomes. Then, a word is randomly selected from the input sentence, and the inner most gene to which the selected word corresponds in each parent chromosome is identified. If the genes correspond to different sets of words, the next gene in the inner most order is selected. This process continues until the sequences of words whose parsings are to be exchanged are the same, or until the main NP or VP are reached. If the two selected genes parse the same sequence of words, they are exchanged. Otherwise, genes appropriated for the exchange are randomly generated. Mutation is applied to the chromosome resulting of the crossover operation with a probability given by the mutation rate, an input parameter. In this case, a new parsing is generated for a randomly selected gene. At each generation a number of chromosomes equal to the number of offsprings is selected to be replaced. The selection of chromosomes to be replaced in each generation is performed with respect to the relative fitness of the individuals: a chromosome with a worse than average fitness has higher chances to be selected for replacement.

3 Parallel Model and Implementation

We have adopted an island model, which may be implemented in both shared or distributed memory architectures, thus making the model portable. The design of the parallel model is intended to reduce the communications. To this purpose, the following options have been chosen:

- *System components*: The system is composed of a number of processes, called *cooperative parsers*, each of which performs, by evolutionary programming, a parsing for the same input sentence. The initial condition in each deme (random number generation) is different in order to obtain different individuals in different populations. There is a special process, known as *main selector*, which selects the best individual among the best ones of each deme.

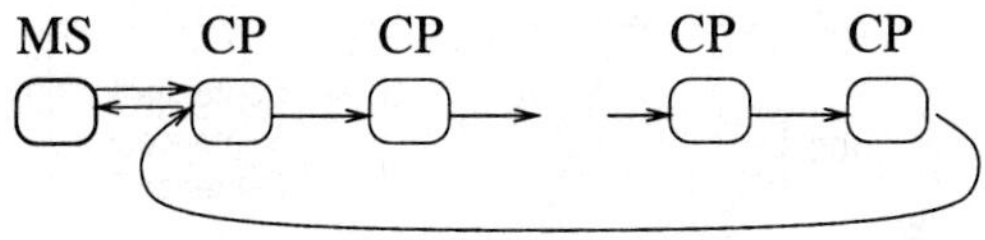

Fig. 2. Migration Policy. CP stands for Cooperative parser, MS for Main Selector.

Table 1. Sentences used in the parsing experiments.

1	Jack(noun) regretted(verb) that(wh) he(pro) ate(verb) the(det) whole(adj) thing(noun)
2	The(det) man(noun) who(wh) gave(verb) Bill(noun) the(det) money(noun) drives(verb) a(det) big(adj) car(noun)
3	The(det) man(noun) who(wh) lives(verb) in(prep) the(det) red(adj) house(noun) saw(verb) the(det) thieves(noun) in(prep) the(det) bank(noun)

- *Migration policy*: To reduce communications, migrations do not take place from one deme to any other, but only to the next one (Figure 2). The N cooperative parsers thus form a ring. Nevertheless, this policy has been compared with an all-to-all policy so as to make sure that this choice does not imply a significant reduction in the quality of the solutions.
- *Synchronism*: We have adopted an asynchronous model, in which a cooperative parser, after a fixed number of generations, sends a fixed number of individuals to the next *cooperative parser* and then continues the evolution, checking in each generation the arrival of the same number of individuals from the previous parser.
- *Convergence policy*: Again with the aim of reducing communications, a *cooperative parser* which reaches convergence sends its best individual to the *main selector*. The main selector takes this solution as the absolute best, giving it to the user and finishing after killing all cooperative parsers. If no parser which reaches the convergence, all of them finish after a number of generations fixed given by the user. Then each one sends its best solution to the selector, which chooses the best among them for the user.
- *Criteria for selection of individuals to migrate*: They are randomly chosen with a probability proportional to their fitness.
- *Criteria for selection of individuals to be replaced by the 'immigrants'*: They are randomly chosen with equal probability.

4 Experiments

The algorithm has been implemented on C++ language with the software PVM on a SGI-Cray ORIGIN 2000. In order to evaluate the performance we have considered the parsing of the sentences appearing in Table 1. The average length of the sentences is around 10 words. However, they present different complexities for the parsing, mainly the length and the number of subordinate phrases.

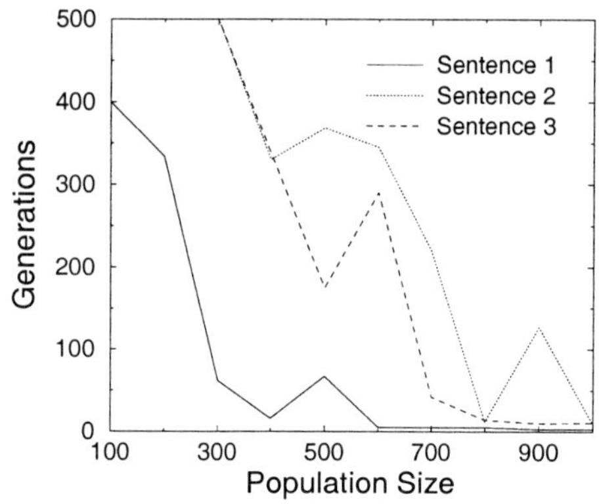

Fig. 3. Number of generations required to reach the correct parsing for different input sentences, when using crossover rate of 50% and mutation rate of 20%.

Table 2. Time in seconds required to reach a correct parsing when processing sequentially and in parallel.

Sentence	Sequential	Parallel				
		2 Proc.	4 Proc.	6 Proc.	8 Proc.	10 Proc.
sentence1	16.55	10.48	3.08	3.09	2.09	2.09
sentence2	50.03	19.12	15.02	10.64	3.48	3.49
sentence3	52.70	25.40	22.71	19.34	14.93	14.79

4.1 Study of the Evolutionary Algorithm Parameters

The parameters of the algorithm determine the influence of two fundamental factors in the success of an evolutionary algorithm: population diversity and selective pressure. The most relevant of them, population size and crossover and mutation rates, have been studied in detail.

Figure 3 shows the number of generations required to reach a correct parsing for each sentence versus the population size, in a sequential execution. In general, the higher the "sentence complexity", the larger the population size required to reach the correct parsing in a reasonable number of steps. The sentence complexity depends on its length and on the number of subordinate phrases it contains. However, large populations lead to replication of chromosomes and slow evolutions, so high percentages of genetic operators are required in order to accelerate the process.

4.2 Evaluating the Parallel Model

Table 2 shows the improvement in performance obtained by increasing the number of processors. This experiment has been carried out with a population size of 200 individuals (the minimum required for the sequential version to reach the correct parsing), a crossover rate of 50%, a mutation rate of 20%, a migrating

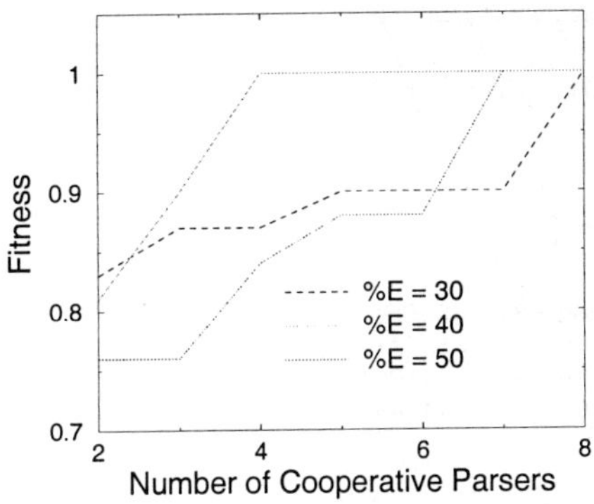

Fig. 4. Performance of the sentence 2 for different sizes of the migrating population, with a population of 50 individual per processor, a crossover rate of 40%, a mutation rate of 20% and an interval between migrations of 10 generations.

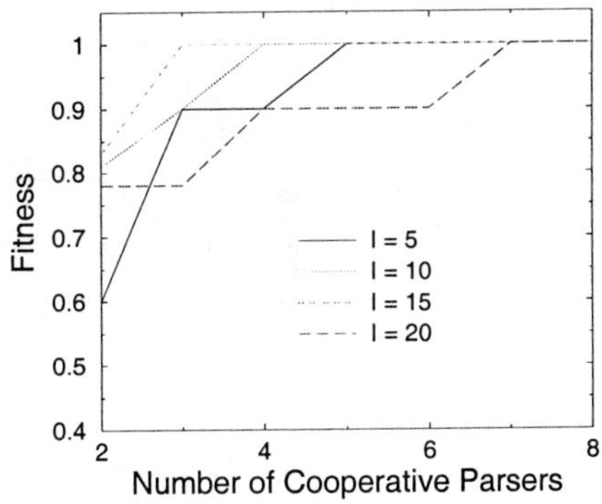

Fig. 5. Performance of the sentence 2 for different intervals of migration, with a population of 50 individual per processor, a crossover rate of 40%, a mutation rate of 20% and a migrating population of 50%.

population of 40 individuals and an interval of migration of 15 generations. We can observe that the parallel execution achieves a significant improvement even with just 2 processors. We can also observe that saturation is reached for some number of processors. However, this number is expected to increase with the complexity of the sentences to be parsed.

4.3 Tuning the Parallel Model

There is a number of options that have to be fixed in order to completely specify the model: rate of the population migrated, number of generations between migrations, parameters of each cooperative parser (size of the deme, crossover and mutation rates, and maximum number of generations), which determine the behaviour of each isolated parser.

Size of the Migrating Population and Migration Interval. Figure 4 shows the results obtained for different sizes of the migrating population for the sentence 2. Results show that a small population, such as that of 50 individuals used in this experiment, requires a large size of the migrating population (50%), in order to reach a high fitness.

The interval of migration is another parameter to fix in the parallel model, closely connected with the size of the migrating population. Experiments have been carried out in order to determine the best values, once the size of the migrating population has been fixed. Figure 5 shows the results obtained for the sentence 2. We can observe that the best results are obtained with an interval of 15 generations. On the contrary, the worst results are obtained for too frequent exchanges (I=5) and also for too spread exchanges (I=20).

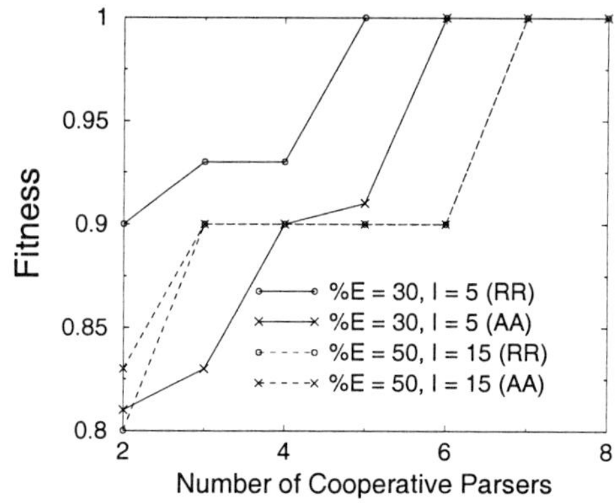

Fig. 6. Comparison of two different communication policies: round-robin (RR) and all-to-all (AA). Results correspond to a population of 50 individual per processor, a limit of 100 generations, a crossover rate 40% and a mutation rate of 20. In the graphics, E stands for Exchange rate and I for migration Interval. Fitness corresponds to the f_coher value.

Round-Robin or All-to-All. We have investigated two different policies migration:

- *Round-Robin* policy: cooperative parsers are organized in a ring sequence, each of them sending the migrating population to the next one in the sequence and receiving it from the preceding one.
- *All-to-all* policy: Each process sends the migrating population to all the other processes and receives population from all of them. In this case, each process sends a migrating population equal to the migrating population size divided by the number of processes in the system.

Figure 6 shows the results obtained with both policies. We can observe that results obtained with both policies are comparable, the round-robin policy slightly outperforming the all-to-all one. Therefore, the round-robin policy is more sensible because with a similar performance considerably reduces communications.

5 Conclusions

The complexity of the parsing process for sentences of natural languages makes it appropriate to use optimization methods such as evolutionary algorithms which provide an approximate solution in a reasonable time. This work presents an evolutionary algorithm which works with a population of potential parsings for a given Probabilistic Context Free Grammar and an input sentence. Probabilistic grammars allow weighting the possible parsings, thus providing a method to evaluate individuals. In this work, the evaluation is given by a combination of the coherence of the parsing, i.e. its ability to match every word and syntactic tag with the grammar rules, and the probability computed from the probabilities of the rules applied.

Results from a number of tests indicate that the evolutionary approach is robust enough to deal with the parsing problem. The tests indicate that the GA parameters need to be suitable for the input sentence complexity. The more complex the sentence (length and subordination degree), the larger the population size required to quickly reach a correct parsing.

The evolutionary algorithm has been parallelized in the form of an island model, in which processors exchange migrating populations asynchronously and in a round-robin sequence. Experiments on the size of the migrating population show that it is necessary to exchange a significant rate of the population. Experiments on the migration intervals show that exchanges should be neither too frequent nor too sparse. Results obtained for these experiments exhibit a clear improvement in the performance, thus showing that the problem has enough granularity for the parallelization, even when applied to artificial sentences. It is expected this improvement to be greater when applied to real sentences as those extracted from a linguistics corpus.

References

1. L. Araujo. Evolutionary parsing for a probabilistic context free grammar. In *Proc. of the Int. Conf. on on Rough Sets and Current Trends in Computing (RSCTC-2000)*, 2000.
2. E. Cantú-Paz. A survey of parallel genetic algorithms. Technical report, Illinois Genetic Algoritms Laboratory, IlliGAL Report No. 97003, 1997.
3. E. Cantú-Paz and D. E. Goldberg. Predicting speedups of idealized bounding cases of parallel genetic algorithms. In T. Back, editor, *Proceedings of the Seventh International Conference on Genetic Algorithms*, pages pp. 113–120. CA: Morgan Kaufmann, 1997.
4. E. Charniak. *Statistical Language Learning.* MIT press, 1993.
5. David E. Goldberg. *Genetic Algorithms in Search, Optimization and Machine Learning.* Addison-Wesly, 1989.
6. R. Poli M. Nowostawski. Review and taxonomy of parallel genetic algorithms. Technical report, School of Computer Science, The University of Birmingham, UK, Technical Report CSRP-99-11, 1999.
7. Z. Michalewicz. *Genetic algorithms + Data Structures = Evolution Programs.* Springer-Verlag, 2nd edition, 1994.
8. A. Ruiz-Andino, L. Araujo, J. Ruz, and F. Sáenz. Parallel evolutionary optimization with constraint propagation. In *Proc. of the Int. Conf. on Parallel Problem Solving from Nature (PPSN)*, volume 1498 of *Lecture Notes in Computer Science*, pages 270–279. Springer-Verlag, 1998.
9. A. Ruiz-Andino, L. Araujo, F. Sáenz, and J. Ruz. A hybrid evolutionary approach for solving constrained optimization problems over finite domains. *IEEE Transactions on Evolutionary Computation*, 4(4):353–372, 2000.

Evolutionary Learning of Boolean Queries by Multiobjective Genetic Programming

Oscar Cordón, Enrique Herrera-Viedma, and María Luque

Dept. of Computer Science and A.I. E.T.S. de Ingeniería Informática
University of Granada. 18071 - Granada (Spain)
{ocordon,viedma}@decsai.ugr.es, mluque@fedro.ugr.es

Abstract. The performance of an information retrieval system is usually measured in terms of two different criteria, precision and recall. This way, the optimization of any of its components is a clear example of a multiobjective problem. However, although evolutionary algorithms have been widely applied in the information retrieval area, in all of these applications both criteria have been combined in a single scalar fitness function by means of a weighting scheme. In this paper, we will tackle with a usual information retrieval problem, the automatic derivation of Boolean queries, by incorporating a well known Pareto-based multiobjective evolutionary approach, MOGA, into a previous proposal of a genetic programming technique for this task.

1 Introduction

Information retrieval (IR) may be defined, in general, as the problem of the selection of documentary information from storage in response to search questions provided by a user [16]. Information retrieval systems (IRSs) are a kind of information systems that deal with data bases composed of information items —documents that may consist of textual, pictorial or vocal information— and process user queries trying to allow the user to access to relevant information in an appropriate time interval. Nowadays, the different world wide web search engines such as Google constitute the main examples of IRSs.

Most of the commercial IRSs are based on the Boolean IR model [18], based on the use of Boolean queries where the query terms are joined by the logical operators AND and OR. This way, the user needs to have a clear knowledge on how to connect the query terms together using the Boolean operators in order to build a query defining his information needs and allowing him to retrieve relevant documents.

The difficulty found by non-expert users to formulate these kinds of queries makes necessary the design of automatic methods for this task. The paradigm of Inductive Query by Example (IQBE) [4], where a query describing the information contents of a set of documents provided by a user is automatically derived, can be useful to assist the user in the query formulation process. Focusing on the Boolean IR model, the only existing approach is that of Smith and Smith [17], which is based on genetic programming (GP) [12]. As usual in the topic, this

J.J. Merelo Guervós et al. (Eds.): PPSN VII, LNCS 2439, pp. 710–719, 2002.

approach is guided by a weighted fitness function combining the two common criteria to measure the performance of an IRS, precision and recall.

In this paper, we will propose a new IQBE algorithm to learn Boolean queries by extending Smith and Smith approach in order to transform it into a multi-objective evolutionary algorithm (EA) not based on a weighted fitness function [9, 5]. This way, we will work as in [15], incorporating MOGA [10] Pareto-based evolutionary multiobjective components into GP. The experimental testbed will be based on one of the most known IR benchmarks, the Cranfield document collection [16, 1].

With this aim, this contribution is structured as follows. Section 2 is devoted to the preliminaries, including the basis of Boolean IRSs, the definition of both precision and recall criteria, and the basis of IQBE techniques. Then, Smith and Smith's proposal is reviewed in Section 3. Section 4 presents the adaptations made to include the Pareto-based multiobjective EA components in the latter algorithm while the experiments developed to test the new proposal are showed in Section 5. Finally, several concluding remarks are pointed out in Section 6.

2 Preliminaries

2.1 Boolean Information Retrieval Systems

An IRS is basically constituted of three main components: *documentary data base*, *query subsystem* and *matching or evaluation mechanism*, whose composition for Boolean IRSs are introduced as follows.

The documentary data base. This component stores the documents and the representation of their information contents. It is associated with the *indexer module*, which automatically generates a representation for each document by extracting the document contents. Textual document representation is typically based on index terms (that can be either single terms or sequences) which are the content identifiers of the documents.

In the Boolean retrieval model, the indexer module performs a binary indexing in the sense that a term in a document representation is either significant (appears at least once in it) or not (it does not appear in it at all). Let D be a set of documents and T be a set of unique and significant terms existing in them. The indexer module of the Boolean IRS defines an indexing function: $F : D \times T \to \{0, 1\}$, where $F(d, t)$ takes value 1 if term t appears in document d and 0 otherwise.

The query subsystem. It allows the users to formulate their queries and presents the relevant documents retrieved by the system to them. To do so, it includes a *query language*, that collects the rules to generate legitimate queries and procedures to select the relevant documents.

Boolean queries are expressed using a query language that is based on query terms and permits combinations of simple user requirements with logical operators AND, OR and NOT [18]. The result obtained from the processing of a query is a set of documents that totally match with it, i.e., only two possibilities are considered for each document: to be or not to be relevant for the user's needs, represented by his query.

The matching mechanism. It evaluates the degree to which the document representations satisfy the requirements expressed in the query, the *retrieval status value* (RSV), and retrieves those documents that are judged to be relevant to it.

As said, the RSV has only two values associated, 0 and 1, in Boolean IRSs. In order to match a query, a document has to fulfill it completely, i.e., it has to include the positive query terms specified in the search expression and not to include those that have been specifically given in that negative way. In order to obtain the set of relevant documents for a query, it is represented as a parse tree and is evaluated from the leaves to the root. Each leaf is associated to the set of documents including (or not including) the corresponding (negative) query term. Then, the retrieved document sets in the inner nodes are computed by applying set arithmetic (with the AND operator being the set intersection and the OR operator standing for the set union). The final set of retrieved documents is that associated to the root when finishing the evaluation of the tree.

2.2 Evaluation of Information Retrieval Systems

There are several ways to measure the quality of an IRS, such as the system eficiency and effectiveness, and several subjective aspects related to the user satisfaction (see, for example, [1], chapter 3). Traditionally, the retrieval effectiveness —usually based on the document relevance with respect to the user's needs— is the most considered. There are different criteria to measure this aspect, with the *precision* and the *recall* being the most used.

Precision is the rate between the relevant documents retrieved by the IRS in response to a query and the total number of documents retrieved, whilst recall is the rate between the relevant documents retrieved and the total number of relevant documents to the query existing in the data base [18]. The mathematical expression of each of them is showed as follows:

$$P = \frac{\sum_d r_d \cdot f_d}{\sum_d f_d} \quad ; \quad R = \frac{\sum_d r_d \cdot f_d}{\sum_d r_d}$$

with $r_d \in \{0,1\}$ being the relevance of document d for the user and $f_d \in \{0,1\}$ being the retrieval of document d in the processing of the current query. Notice that both measures are defined in [0,1], with 1 the optimal value.

Notice that the only way to know all the relevant documents for a query existing in a documentary base (needed to compute the recall measure) is to evaluate them all one by one. Due to this and to the relevence subjectivity, there are several classical documentary bases available, each of them with a set of queries with known relevance judgments, that can be used to test the different new proposals in the field of IR [16,1]. In this contribution, we will deal with the well known Cranfield collection.

As said, up to our knowledge, all the previous applications of machine learning techniques to any of the IRS components trying to optimize both criteria have considered a weighted combination of them. This is why the aim of our contribution is to provide a first step on the application of Pareto-based multi-objective EAs to IR in order to evolve a complete set of Pareto optimal solutions optimizing both criteria simultaneously.

2.3 Inductive Query by Example

IQBE was proposed in [4] as "a process in which searchers provide sample documents (examples) and the algorithms induce (or learn) the key concepts in order to find other relevants documents". This way, IQBE is a process for assisting the users in the query formulation process performed by machine learning methods. It works by taking a set of relevant (and optionally, non relevant documents) provided by a user —that can be obtained from a preliminary query or from a browsing process in the documentary base— and applying an off-line learning process to automatically generate a query describing the user's needs (as represented by the document set provided by him). The obtained query can then be run in other IRSs to obtain more relevant documents. This way, there is no need that the user interacts with the process as in other query refinement techniques such as relevance feedback [16].

There have been proposed IQBE proposals for the different existing IR models. As said, Smith and Smith [17] proposed the GP algorithm to derive Boolean queries that will be considered in this paper. On the other hand, all of the machine learning methods considered in Chen et al.'s paper [4] (regression trees, genetic algorithms and simulated annealing) dealt with the vector space model [16]. Moreover, there are several approaches for the derivation of weighted Boolean queries for fuzzy IRSs [3], such as the GP algorithm of Kraft et al. [13], the niching GA-P method [7] and the simulated annealing-GP hybrid [8]. For descriptions of some of the previous techniques based on EAs refer to [6, 8].

3 The Smith and Smith's Genetic Programming-Based Inductive Query by Example Algorithm for Boolean Information Retrieval Systems

In [17], Smith and Smith proposed an IQBE to derive Boolean queries based on GP. Its components are described next:

Coding Scheme: The Boolean queries are encoded in expression trees, whose terminal nodes are query terms and whose inner nodes are the Boolean operators *AND*, *OR* or *NOT*.

The different expression trees are derived from the following grammar [17]:

$$<QUERY>::=<TERM>|(<QUERY><OPERATOR><QUERY>)$$
$$<OPERATOR>::=AND|OR|NOT$$
$$<TERM>::=t_1|\ldots|t_n$$

Selection Scheme: Each generation is based on selecting two parents, with the best fitted one having a greater chance to be chosen, and generating two offspring from them. Both offspring are added to the current population[1].

Genetic Operators: The usual GP crossover is considered [12], which is based on randomly selecting one edge in each parent and exchanging both subtrees from these edges between the both parents. No mutation operator is considered[2].

Generation of the Initial Population: All the individuals in the first population are randomly generated. A pool is created with all the terms included in the set of relevant documents provided by the user, having those present in more documents a higher probability of being selected.

Fitness function: The following function is maximized:

$$F = \alpha \cdot P + \beta \cdot R$$

where precision P and recall R are computed as showed in Section 2.2, while α and β are the weighting factors. Moreover, when comparing two queries with the same F value, the shorter one is preferred.

4 Incorporating Pareto-Based Multiobjective Components to the Smith and Smith's Algorithm

As said, the Pareto-based multiobjective EA considered to be incorporated to the basic Smith and Smith's GP algorithm in this first work has been Fonseca and Fleming's MOGA [10]. The selection scheme of MOGA is based on dividing the population in several ranked blocks and assigning a higher probability of selection to the blocks with a lower rank, taking into account that individuals in the same block will be equally preferable and thus will receive the same selection probability. The rank of an individual in the population (and consequently of his belonging block) will depend on the number of individuals dominating it.

Therefore, the selection scheme of our multiobjective GP involves the following four steps:

1. Each individual is assigned a rank equal to the number of individuals dominating it plus one (chromosomes encoding non-dominated solutions receive rank 1).
2. The population is increasingly sorted according to that rank.

---

[1] Our implementation differs in this point as we consider a classical generational scheme where the selection probabilities are assigned by the proportional scheme and the reproduction is performed by Baker's stochastic universal sampling [2].

[2] We do use a mutation operator which changes a randomly selected term or operator by a random one, or a randomly selected subtree by a randomly generated one.

3. Each individual is assigned a fitness value which depends on its ranking in the population. In this contribution, we consider the following assignment: $f(C_i) = \frac{1}{rank(C_i)}$.
4. The fitness assignment of each equivalence class (group of individuals with the same rank, i.e., which are non dominated among them) is averaged among them, so that all of them finally receive the same fitness value.

Once the final fitness values have been computed, a usual selection mechanism is applied. In this contribution we consider the proportional assignment and Baker's stochastic universal sampling [2] with an appropriate choice of the parameter values to induce diversity.

It is known that the MOGA selection scheme can cause a large selection pressure that might produce premature convergence. Fonseca and Fleming considered this issue and suggested to use a niching method to appropriately distribute the population in the Pareto [10]. However, as said in [5], one of the main weaknesses of MOGA is that sharing is performed in the objective space, thus making more difficult that two different Pareto solutions with the same objective function values can simultaneously coexist in the population.

For example, this is not a desirable characteristic in our case, as we are interested on obtaining as many queries with the same precision-recall values as possible. Fortunately, as Coello also mentions, there is no specific requirement in the MOGA algorithm to perform sharing in the objective space.

This way, in this paper we apply niching in the parameter (genotypic) space. To do so, we have to keep in mind that we are dealing with chromosomes encoding Boolean queries, and hence we need a metric capable of measuring distances between expression trees. In our case, this is put into effect by the so-called *edit* or *Levenshtein distance* [14], a text metric that computes the distance between two strings as the number of edit (delete, insert or change) steps needed to convert one into the other. In order to compute distances between trees with this metric, we apply it on the strings encoding the preorder representation of the trees.

Let $a = (a_1, \ldots, a_n)$ and $b = (b_1, \ldots, b_m)$ be the two tree preorder strings. The edit distance between them is recursively computed as follows[3]:

$$E((a_1,\ldots,a_n),(b_1,\ldots,b_m)) =
\begin{cases}
n, & \text{if } m = 0\\
m, & \text{if } n = 0\\
min\{E((a_1,\ldots,a_{n-1}),(b_1,\ldots,b_m)) + 1, & \\
E((a_1,\ldots,a_n),(b_1,\ldots,b_{m-1})) + 1, & \text{otherwise}\\
E((a_1,\ldots,a_{n-1}),(b_1,\ldots,b_{m-1})) + d(a_n,b_m)\}, &
\end{cases}$$

with $d(x, y) = 1$ if $x = y$, and 0 otherwise, being the character distance.

Once a valid metric for trees has been defined, it is easy to apply sharing by using the classical Goldberg and Richardson's sharing function [11]:

[3] We should note that in this contribution we have been computed it iteratively by the corresponding Dynamic Programming algorithm.

$$F(C_i) = \frac{f(C_i)}{\sum_{j=1}^{M} Sh(d(C_i, C_j))} \quad ; \qquad Sh(d) = \begin{cases} 1 - (\frac{d}{\sigma_{share}})^\gamma, & \text{if } d < \sigma_{share} \\ 0, & \text{otherwise} \end{cases}$$

with σ_{share} being the niche radius.

5 Experiments Developed and Analysis of Results

As said, the experimental study has been developed using the *Cranfield* collection, composed of 1400 documents about Aeronautics. The 1400 textual documents has been automatically indexed in the usual way by first extracting the non-stop words, thus obtaining a total number of 3857 different indexing terms, and then considering the binary indexing to generate the term weights in the document representations. Among the 225 queries associated to the Cranfield collection, we have selected those presenting 20 or more relevant documents. The resulting seven queries (numbers 1, 2, 23, 73, 157, 220 and 225) have 29, 25, 33, 21, 40, 20 and 25 relevant documents associated, respectively.

Table 1. Results obtained by the basic Smith and Smith's IQBE algorithm

Best					
$\#q$	Run	Sz	P	R	$\#rr/\#rt$
1	7	11	1.0	0.1724	5/5
2	8	17	1.0	0.2	5/5
23	1	17	1.0	0.1515	5/5
73	2	19	1.0	0.2857	6/6
157	1	15	1.0	0.15	6/6
220	5	19	1.0	0.2	4/4
225	2	13	1.0	0.2	5/5

Average						
$\#q$	Sz	σ_{Sz}	P	σ_P	R	σ_R
1	16.8	0.8221	1.0	0.0	0.1379	0.0068
2	17.4	0.9295	1.0	0.0	0.156	0.0068
23	17.6	0.4939	1.0	0.0	0.1212	0.0060
73	18.6	0.2529	1.0	0.0	0.2047	0.0135
157	18.0	0.4242	1.0	0.0	0.1175	0.0086
220	18.8	0.1897	1.0	0.0	0.17	0.0077
225	17.8	0.6449	1.0	0.0	0.164	0.0088

Apart from our Pareto-based Multiobjective proposal, we have also run the basic Smith and Smith's algorithm with a typical setting for the weights in the fitness function ($(\alpha, \beta) = (1.2, 0.8)$). Every algorithm have been run ten times with different initializations during the same fixed number of fitness function evaluations (100000) in an 1GHz Pentium III computer with 256 MB of RAM[4]. The common parameter values considered are a maximum of 20 nodes for the trees, 0.8 and 0.2 for the crossover and mutation probabilities, respectively, and a population size of M =1600 queries. The high value for the latter parameter is because it is well known that GP requires large population sizes to achieve good performance. Finally, the sharing function parameter γ takes value 2 and the niche radius σ_{share} has been experimentally set to 4 (a 20% of the maximum tree size).

[4] The basic algorithm spends more or less 3 minutes whilst our MOGA variant approximately takes 8 minutes.

The results obtained by the basic algorithm are showed in Table 1, while the best results are showed on the left and the average ones on the right. In the left-hand first table, $\#q$ stands for the corresponding query number, *Run* for the number of the run where this result was derived, Sz for the generated query size, P and R for the precision and recall values, respectively, $\#rt$ for the number of documents retrieved by the query, and $\#rr$ for the number of relevant documents retrieved. The columns of the right-hand table stand for the same items showing the averaged values as well as the standard deviations.

Table 2. Results obtained by the proposed multiobjective IQBE algorithm

$\#q$	$\#p$	$\sigma_{\#p}$	$\#dp$	$\sigma_{\#dp}$	M_2	σ_{M_2}	M_3	σ_{M_3}
1	189.7	34.9782	129.1	36.4336	90.4425	17.3184	1.2887	0.0043
2	336.0	60.3004	272.7	59.3562	163.5748	30.2425	1.2835	0.0072
23	203.8	21.6969	139.6	20.5286	97.9832	11.0473	1.3086	0.0032
73	180.1	22.1902	117.4	19.5980	85.0346	11.0898	1.2232	0.0082
157	167.6	18.2176	102.5	16.0163	79.7532	8.9551	1.3202	0.0040
220	162.7	27.0902	96.5	25.1738	75.8301	13.8823	1.2343	0.0060
225	211.9	39.4840	141.3	40.5371	101.1405	19.6761	1.2705	0.0059

	Best Precision						Best Recall					
$\#q$	Sz	σ_{Sz}	P	σ_P	R	σ_R	Sz	σ_{Sz}	P	σ_P	R	σ_R
1	19.0	0.0	1.0	0.0	0.297	0.010	19.0	0.0	0.042	0.003	1.0	0.0
2	19.0	0.0	1.0	0.0	0.32	0.017	19.0	0.0	0.032	0.003	1.0	0.0
23	19.0	0.0	1.0	0.0	0.248	0.009	19.0	0.0	0.039	0.002	1.0	0.0
73	19.0	0.0	1.0	0.0	0.433	0.022	19.0	0.0	0.070	0.007	1.0	0.0
157	19.0	0.0	1.0	0.0	0.212	0.010	19.0	0.0	0.044	0.002	1.0	0.0
220	19.0	0.0	1.0	0.0	0.42	0.014	19.0	0.0	0.056	0.004	1.0	0.0
225	19.0	0.0	1.0	0.0	0.348	0.015	19.0	0.0	0.037	0.003	1.0	0.0

On the other hand, Table 2 shows several statistics corresponding to our multiobjective proposal. The first subtable on the top collects several data about the composition of the ten Pareto sets generated for each query, always showing the averaged value and its standard deviation. From left to right, the columns collect the number of non-dominated solutions obtained ($\#p$), the number of different queries (trees) existing among them ($\#dp$), and the values of two of the usual multiobjective EA metrics M_2 and M_3^* [19]. $M_2 \in [0, \#p]$ measures the distribution of the genotypes of the $\#p$ non-dominated solutions found[5] (i.e., the diversity of the solutions found). M_3^* estimates the range to which the Pareto front spreads out in the objective values. Besides, two queries are selected from each Pareto set, the ones with maximum precision and maximum recall, respectively, and their averaged results are collected in the bottom subtable.

[5] The value of the neighborhood parameter σ considered in this metric has been set to the niche radius value σ_{share}.

In view of these results, the performance of our proposal is very significant. On the one hand, it overcomes the basic Smith and Smith's algorithm in all cases as the results of the latter when considering typical values for the weighted combination are dominated by the solutions in the Pareto front of the former. It seems that the use of a weighting scheme and the lack of a niching scheme make the basic algorithm not to perform appropriately. On the other hand, the main aim of this paper have been clearly fulfilled since the Pareto fronts obtained are very well distributed, as demonstrated by the high number of solutions included in them and the high values in the M_2 and M_3^* metrics.

6 Concluding Remarks

The automatic derivation of Boolean queries has been considered by incorporating the MOGA Pareto-based multiobjective evolutionary approach to an existing GP-based IQBE proposal. The proposed approach has performed appropriately in seven queries of the well known Cranfield collection in terms of absolute retrieval performance and of the quality of the obtained Paretos.

In our opinion, many different future works arise from this preliminary study. On the one hand, more advanced Pareto-based multiobjective EA schemes (such as those elitist ones considering an auxiliary population to better cover the Pareto front [9, 5]) can be incorporated to the basic GP algorithm in order to improve the performance of the multiobjective EA proposed. On the other hand, preference information of the user on the kind of queries to be derived can be included in the Pareto-based selection scheme in the form of a goal vector whose values are adapted during the evolutionary process [10]. Moreover, a training-test validation procedure can be considered to test the real application of the proposed IQBE algorithm. Finally, and more generically, Pareto-based evolutionary multiobjective optimization can be applied either to the automatic derivation of queries for other kinds of IR models (such as the extended Boolean ones tackled in the EAs proposed in [7, 8, 13]) or to other IR problems being solved by EAs [6], thus benefiting from the potential of these techniques in the problem solving.

References

1. Baeza-Yates, R., Ribeiro-Neto, B.: Modern Information Retrieval, Adisson-Wesley (1999).
2. Baker, J.E.: Reducing bias and inefficiency in the selection algorithm, Proc. Second International Conference on Genetic Algorithms (ICGA'87), Hillsdale, NJ, (1987) 14–21.
3. Bordogna, G., Carrara, P., Pasi, G.: Fuzzy approaches to extend Boolean information retrieval, in: P. Bosc, J. Kacprzyk (Eds.), Fuzziness in Database Management Systems (1995) 231–274.
4. Chen, H.: A machine learning approach to inductive query by examples: an experiment using relevance feedback, ID3, genetic algorithms, and simulated annealing, Journal of the American Society for Information Science **49:8** (1998) 693–705.

5. Coello, C.A., Van Veldhuizen, D.A., Lamant, G.B.: Evolutionary Algorithms for Solving Multi-Objective Problems, Kluwer Academic Publishers (2002).
6. Cordón, O., Moya, F., Zarco, C.: A brief study on the application of genetic algorithms to information retrieval (in spanish), Proc. Fourth International Society for Knowledge Organization (ISKO) Conference (EOCONSID'99), Granada, Spain, (April, 1999) 179–186.
7. Cordón, O., Moya, F., Zarco, C.: A GA-P algorithm to automatically formulate extended Boolean queries for a fuzzy information retrieval system, Mathware & Soft Computing **7:2-3** (2000) 309–322.
8. Cordón, O., Moya, F., Zarco, C.: A new evolutionary algorithm combining simulated annealing and genetic programming for relevance feedback in fuzzy information retrieval systems, Soft Computing **6:5** (2002).
9. Deb, K.: Multi-Objective Optimization Using Evolutionary Algorithms, Wiley (2001).
10. Fonseca, C.M., Fleming, P.J.: Genetic algorithms for multiobjective optimization: Formulation, Discussion and Generalization, Proc. Fifth International Conference on Genetic Algorithms (ICGA'93), San Mateo, CA (July, 1993) 416–423.
11. Goldberg, D.E., Richardson, J.: Genetic algorithms with sharing for multimodal function optimization, Proc. Second International Conference on Genetic Algorithms (ICGA'87), Hillsdale, NJ, (1987) 41–49.
12. Koza, J.: Genetic programming. On the programming of computers by means of natural selection, The MIT Press (1992).
13. Kraft, D.H., Petry, F.E., Buckles, B.P., Sadasivan, T.: Genetic algorithms for query optimization in information retrieval: relevance feedback, in: E. Sanchez, T. Shibata, L.A. Zadeh, Genetic Algorithms and Fuzzy Logic Systems, World Scientific (1997) 155–173.
14. Levenshtein, V.I.: Binary codes capable of correcting deletions, insertions and reversals, Sov. Phys. Dokl. **6** (1966) 705-710.
15. Rodríguez-Vazquez, K., Fonseca, C.M., Fleming, P.J.: Multiobjective genetic programming: A nonlinear system identification application, Late Breaking Papers at the Genetic Programming 1997 Conference, Stanford, CA (July, 1997) 207–212.
16. Salton, G., McGill, M.J.: Introduction to modern information retrieval, McGraw-Hill (1989).
17. Smith, M.P., Smith, M.: The use of genetic programming to build Boolean queries for text retrieval through relevance feedback, Journal of Information Science **23:6** (1997) 423–431.
18. van Rijsbergen, C.J.: Information Retrieval (2nd edition), Butterworth (1979).
19. Zitzler, E., Deb, K., Thiele, L.: Comparison of multiobjective evolutionary algorithms: Empirical results, Evolutionary Computations **8:2** (2000) 173–195.

Inferring Phylogenetic Trees Using Evolutionary Algorithms

Carlos Cotta[1] and Pablo Moscato[2]

[1] Dept. Lenguajes y Ciencias de la Computación, ETSI Informática
University of Málaga, Campus de Teatinos, 29071 - Málaga - SPAIN
ccottap@lcc.uma.es
[2] Grupo de Engenharia de Computação em Sistemas Complexos
Dept. de Engenharia de Computação e Automação Industrial
Universidade Estadual de Campinas, C.P. 6101
Campinas, SP, CEP 13083-970, Brazil
moscato@dca.fee.unicamp.br

Abstract. We consider the problem of estimating the evolutionary history of a collection of organisms in terms of a phylogenetic tree. This is a hard combinatorial optimization problem for which different EA approaches are proposed and evaluated. Using two problem instances of different sizes, it is shown that an EA that directly encodes trees and uses *ad-hoc* operators performs better than several decoder-based EAs, but does not scale well with the problem size. A greedy-decoder EA provides the overall best results, achieving near 100%-success at a lower computational cost than the remaining approaches.

1 Introduction

The inference of phylogenetic trees is one of the most important and challenging tasks in Systematic Biology. Such trees are used to represent the evolutionary history of a collection of n organisms (or taxa) from their molecular sequence data, or from other form of dissimilarity information. An accurate estimation of this evolutionary history is a very useful tool in many areas of Biology, such as multiple sequence alignment [4], or molecular epidemiological studies of viruses [11] among others.

The Phylogeny Problem can then be formulated as finding the phylogenetic tree that best –under a certain optimality criterion– represents the evolutionary history of a collection of taxa. Unfortunately, this constitutes a very hard combinatorial optimization problem for most optimality criteria. Exact techniques such as branch-and-bound can be used, but can be computationally unaffordable for even moderate (say, 30-40 taxa) problem instances. Hence, the use of heuristic techniques seems appropriate.

We are concerned in this work about the utilization of evolutionary algorithms (EAs) for tackling the Phylogeny Problem. In this sense, we have initially focused on distance-based measures (Section 2 will provide details about this and other quality measures, as well as about phylogenetic trees in general). From this

J.J. Merelo Guervós et al. (Eds.): PPSN VII, LNCS 2439, pp. 720–729, 2002.

starting point, several EA approaches –based on the use of different representations and/or reproductive operators– have been devised and compared. More precisely, both EAs that directly conduct the search in the space of phylogenetic trees, and EAs that use auxiliary search spaces and decoders have been considered. These EAs are detailed in Section 3. Subsequently, the results of a thorough empirical comparison are reported in Section 4. We conclude by presenting a summary of our conclusions, and outlining future work in Section 5.

2 A Gentle Introduction to Phylogenetic Trees

Assume we are given some molecular-sequence data for a collection S of n taxa. A phylogenetic tree T is a tree with exactly n leaves, each one labeled by a taxon in S. Internal nodes in this tree correspond to hypothetical ancestral organisms, and edges in T represent ancestry-descent relationships. Taxa in S are also termed OTUs (*operational taxonomic units*), while internal nodes are termed HTU (*hypothetical taxonomic units*). Thus, a phylogenetic tree models the evolutionary history of the OTUs, back to their common ancestor (the root of the tree[1]). In the following, we will denote OTUs and HTUs with lowercase letters, and trees with uppercase letters. A LISP-like notation will be used to represent trees, e.g., (hTU) represents the tree rooted at h, with T and U as subtrees[2], and (o) represents a leaf labeled with o.

The goal of the Phylogeny Problem is finding the phylogenetic tree T that best resembles the evolutionary history of OTUs in S. For this purpose, it is clearly necessary to define an optimality criterion. Essentially, such a criterion (and subsequently an inference method) can fall within two major categories, *sequence*-based and *distance*-based. In sequence-based approaches, each node of T is assigned a sequence (known for OTUs, and inferred via pairwise alignments for HTUs). Then, the tree is evaluated using a criterion that –in most situations– is either *maximum likelihood* (ML) or *maximum parsimony* (MP). In the former, a stochastic model of evolution (e.g., the Jukes-Cantor model) is used in order to assess the likelihood that the current tree generated the observed data. On the other hand, an MP criterion specifies that the tree requiring the fewest number of evolutionary changes to explain the data is preferred.

As to distance-based approaches, they are based on transforming the available sequence data into an $n \times n$ matrix M. This matrix is the only information used in the subsequent inference process. More precisely, edges in T are assigned a weight. The basic idea here is that M_{ij} represents the *evolutionary distance* or *dissimilarity* between OTUs i and j. Then, we have an *observed* distance matrix M, and an *inferred* distance matrix $\hat{M}$ (obtained by making $\hat{M}_{ij}$ = distance from i to j in T). The quality of the tree can now be quantified in a variety of ways. Firstly, it is possible to consider some "distance" measure between M and $\hat{M}$;

[1] Biologists often focus on a relaxed model based on *unrooted* trees as well. In this work we will concentrate on rooted trees though. See [6, 8] for heuristic approaches to the inference of unrooted trees.

[2] Non-binary trees are also possible, but they can be easily reduced to binary ones.

usual examples are the $STRESS$ measure (normalized sum of absolute differences), the L_2 metric (least-squares approximation), or the L_∞ metric (minimize maximum absolute difference). Secondly, quality can be directly measured from T. This is typically the case when edge-weighting has been constrained so as to have $\hat{M}_{ij} \geq M_{ij}$; in this situation, minimizing the total weight of T is usually the criterion.

Notice that by taking M_{ij} as the minimum number of evolutionary events needed to transform i in j, this last approach resembles MP. Actually, distance-based methods can be generally considered as an intermediate strategy between ML and MP, exhibiting good performance in practice as well [5]. For these reasons, we have focused in distance-based approaches in this work. To be precise, we have considered the constraint mentioned above, regarding inferred dissimilarities to be greater or equal than observed ones. This is done by forcing $\hat{M}$ to be ultrametric (see [14] for details about ultrametricity); very popular when the molecular-clock hypothesis was in vogue, this condition provides a very good approximation to the optimal solution under more relaxed assumptions (e.g., mere additivity). Furthermore, it has allowed us using exact techniques in order to estimate the absolute quality of the solutions achieved by the different EAs.

3 Evolutionary Approaches to the Phylogeny Problem

In essence, two main approaches can be considered for tackling the Phylogeny Problem with EAs. The first one is the *direct* approach, in which the EA conducts the search in the space $\mathcal{S}_{Ph}$ of all possible phylogenetic trees. The second one is the *indirect* approach, in which an auxiliary $\mathcal{S}_{aux}$ space is used by the EA. In this latter case, a decoder [7] must be utilized in order to perform the $\mathcal{S}_{aux} \longrightarrow \mathcal{S}_{Ph}$ mapping. Either of these approaches requires defining adequate reproductive operators and/or representation schemes. These are described below.

3.1 Direct Search in the Phylogenetic-Tree Space

As mentioned above, a direct approach is characterized by performing the search in the space of all possible phylogenetic trees. Thus, each individual in the EA population directly represents a feasible tree. For this purpose, any of the encoding techniques commonly used in genetic programming (GP) –e.g., LISP-like expressions, preorder traversals, etc.– can be used. Subsequently, appropriate operators must be designed to manipulate this representation.

First of all, consider the recombination operator. This operator must take information pieces from both parents, and combine them to create some offspring. Since individuals directly encode trees in this case, these information pieces naturally emerge in the form of subtrees. Thus, recombination can be expressed in terms of pruning and grafting subtrees, much like it is typically done in GP. However, unlike the most classical GP scenario, subtrees cannot be randomly shuffled, since phylogenetic trees are constrained to have n leaves, each one representing a different OTU. Hence, a slightly different approach must be considered. To be precise, the recombination operator must take care of remov-

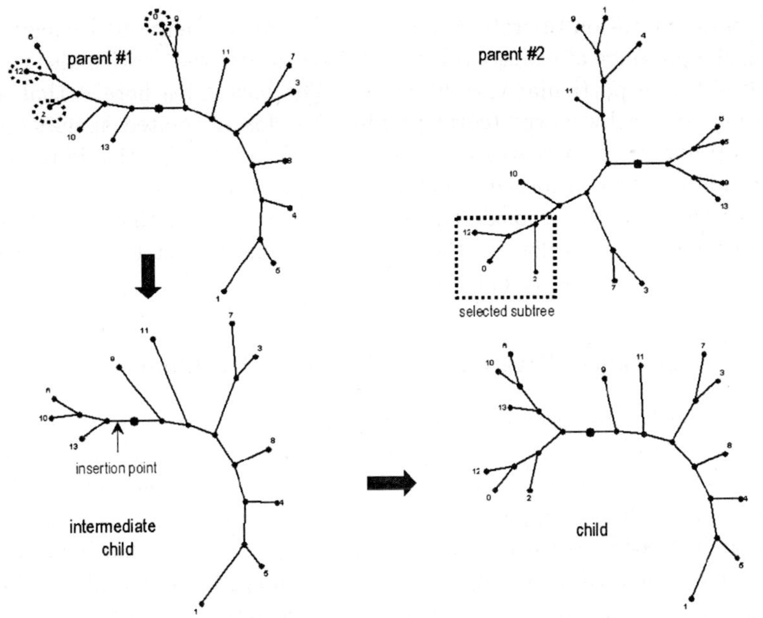

Fig. 1. Functioning of the PDG recombination. A subtree is selected in one of the parents, and inserted at a random point of the other parent, right after having deleted duplicates nodes. The root of the tree is marked with a thick node.

ing duplicates elements before attempting to regraft the selected subtree. Let T_1 and T_2 be the trees being recombined; the whole process would be as follows:

PRUNE-DELETE-GRAFT RECOMBINATION (T_1, T_2)
1. Select a subtree T from T_2.
2. **for each** OTU $o \in T$ **do**
   (a) Find subtree U from T_1 such that $U = (h(o)U')$ or $U = (hU'(o))$.
   (b) Replace U by U' in T_1.
3. Select a random subtree V from T_1.
4. Replace V by $V' = (h'TV)$, where h' is a new HTU.

Figure 1 illustrates the process. This PDG operator has been used by Moilanen [10] in the context of a sequence-based parsimony measure.

As to mutation operators, several options are possible. The following ones have been considered:

– SWAP: two OTUs are selected at random, and their positions are swapped.
– NNI: consider the sequence of leaves $L[T]$ defined as follows:

$$L[(hUV)] = L[U] : L[V]; \quad L[(o)] = \langle o \rangle; \tag{1}$$

  where : is the sequence-concatenation operator. Then, this operator swaps two neighboring leaves in this sequence.
– SCRAMBLE: first, a subtree T' is randomly selected from T. Then, its topology is rearranged at random.

All these mutation operators fulfill the previously-mentioned constraint regarding the presence of exactly n leaves/OTUs in the tree. Notice also that NNI can be seen as a particular case of SWAP. The basic idea here is that a small reordering or nearby leaves (corresponding to closely related species) is more likely to result in an improvement than a larger reordering (this is the case for SWAP, that can be considered somewhat disruptive in this sense). In addition to this, the version of NNI considered in this work checks whether the interchange provides a better –according to the fitness function– tree-structure, and reverts the interchange if this were not the case.

3.2 Decoder-Based EAs for the Phylogeny Problem

As an alternative to directly conducting the search in the solution space, a combination of an auxiliary search space $\mathcal{S}_{aux}$ and a decoder D can be used. This approach is very rich in possibilities, and has several advantages. On one hand, it usually allows utilizing simpler evolutionary operators, due to the fact that $\mathcal{S}_{aux}$ is often an unconstrained search space (unlike $\mathcal{S}_{Ph}$). On the other hand, problem knowledge can be introduced by means of D. Among the potential drawbacks of this approach, one can cite the difficulty in some situations for exploring $\mathcal{S}_{aux}$ in a parsimonious way; the use of a decoder can hinder finding an acceptable notion of locality within $\mathcal{S}_{aux}$.

Basic Setting. The first decoder approach considered resembles in some sense the ordinal (stack-based) representation of the TSP [9]. In this case, OTUs are assumed to be ordered in some sense. Let $S = \langle o_1, \cdots, o_n \rangle$ be the ordered sequence of OTUs[3]. The auxiliary search space $\mathcal{S}_{aux}$ is defined as $\mathcal{S}_{aux} = \prod_{i=1}^{n-2} \mathbb{N}_{2i}$, where $\mathbb{N}_k = \{1, \cdots, k\}$. The decoding process of an individual $s \in \mathcal{S}_{aux}$ is as follows:

Ordinal Decoder(s)

1. Let $T = (h(o_1)(o_2))$
2. **for each** $i \in [1 : n-2]$ **do**
   (a) Let the branches of T be numbered from 1 to $2i$. Let branch s_i join node h and subtree U.
   (b) Replace U by $V = (h'(o_{i+2})U)$ in T, where h' is a new HTU.

As it can be seen, in a phylogenetic tree with k leaves, there exist $2k-1$ nodes, and hence $2k-2$ branches. Each element in s indicates in which of these branches the next OTU in the sequence must be inserted. Thus, the ith element of s ranges from 1 to $2i$. One of the nicest properties of this ordinal representation is its orthogonality [12]. Plainly, this means that any s is feasible as long as each s_i is in its corresponding range. Hence, any positional recombination operator such as single-point crossover (SPX), or uniform crossover (UX) will produce a feasible child when applied to feasible parents.

[3] In this work, a *maxmin* sequence [14] has been considered: the two first elements are those whose distance in M is maximal; each subsequent element o_i ($i > 2$) is the one for which $d(o_i) = \min\{M_{o_i,o_j} \mid j < i\}$ is maximal.

One of the weak points of the ordinal decoder presented above can be found in the use of a fixed sequence: it may be harder to construct good solutions for a certain problem instance using sequence A than using sequence B. This admits two possible solutions. First, a smart method for constructing an appropriate OTU sequence given a certain problem instance could be devised. Alternatively, the EA can make this sequence evolve, along with the ordinal insertion points. In this latter case, the search space is $\mathcal{S}'_{aux} = \mathcal{S}_{aux} \times \mathbb{P}_n$, where $\mathcal{S}_{aux}$ is the space of ordinal sequences described above, and $\mathbb{P}_n$ is the space of n-element permutations. The decoding process would be identical as indicated above, with the sole difference that the OTU sequence would be taken from s as well. The same considerations regarding the use of standard positional operators are applicable in this case too. Additionally, it must be noted that recombination can also be done on the permutation segment of individuals, using any standard permutational operator for this purpose (e.g., OX, UCX [3], etc.)[4].

Using Guidance. The above decoders are essentially blind, i.e., they do not use any phenotypic information in order to guide the construction process; they take all the information they need from the decoder input s. It is reasonable to consider the use of some kind of guidance information though. Two main possibilities are considered here: using this guidance during the decoding stage, or when recombining.

Starting with this latter one, note that recombination is given two individuals s^1 and s^2; rather than simply mixing information from these two individuals in order to create the child, it might be useful to get some assessment on the quality of the partially constructed solution in order to guide the process. A simple way of doing this is following a greedy approach:

GREEDY-INSERTION XOVER(s^1, s^2)

1. Let $T = (h(o_1)(o_2))$
2. **for each** $i \in [1 : n-2]$ **do**
   (a) Let the branches of T be numbered from 1 to $2i$. Let branch s_i^1 (resp. s_i^2) join node h_1 and subtree U_1 (resp. h_2 and U_2).
   (b) Let $T_1, T_2 = T$. Replace U_1 by $V = (h'(o_{i+2})U_1)$ in T_1. Act analogously with U_2 in T_2.
   (c) Let T=best(T_1, T_2).

In the above description, the OTU sequence is not necessarily fixed. Actually, it can be taken from either of the parents, or constructed by recombining (using a standard operator) the parental sequences. Notice also that a symmetric approach can be defined, i.e., assuming a certain insertion sequence, and taking greedy decisions on the structure of the OTU sequence. In this case, the basic units upon which decisions are taken cannot be single OTUs, since the positional representation of permutations is not orthogonal. On the contrary, *blocks*[5] must be considered. Then,

[4] The notation O_1/O_2 will be used to denote that one of $\{O_1,O_2\}$ is applied on an individual, while O_1+O_2 will denote that both operators are sequentially applied.

[5] A block is a compact subsequence of elements such that both parents have the same elements in this segment, although in a possibly-different order (see [3]).

GREEDY-ORDER XOVER(o^1, o^2)

1. Identify block structure in o^1, o^2. Let $B_1^{1|2}, \cdots, B_m^{1|2}$ be the blocks.
2. Let $T_1 = (h(o_1^1)(o_2^1))$. Add remaining OTUs in B_1^1 (do the same with T_2 and B_1^2).
3. Let T=best(T_1, T_2).
4. **for each** $j \in [2:m]$ **do**
   (a) Let $T_1 = T$; **for each** OTU $o_k^1 \in B_j^1$ **do**
      - Insert OTU o_k^1 in branch s_k.

   (b) Act analogously with T_2 and B_j^2.
   (c) Let T=best(T_1, T_2).

This version of the recombination operator has the advantage that requires fewer assessments of partial solutions, since OTUs are inserted in blocks rather than one at a time. Again, the insertion sequence can be fixed, taken from one of the parents, or obtained by recombination.

Finally, consider a similar approach to those above, but focused on the decoder stage rather than on recombination. In this case, the OTU sequence is given, but the insertion sequence is not; the decoder is responsible for finding adequate insertion points for each OTU. The process could be as follows:

PERMUTATIONAL DECODER($\langle o_1, \cdots, o_n \rangle$)

1. Let $T = (h(o_1)(o_2))$
2. **for each** $j \in [1:n-2]$ **do**
   (a) **for each** insertion point $i \in \mathbb{N}_{2i}$ **do**
      - Let $T_i = T$. Insert OTU o_{j+2} in branch i.

   (b) Let T=best($T_1, \cdots, T_{2i}$).

Next section will be devoted to provide empirical evidence regarding the potential usefulness of this and all previous approaches.

4 Empirical Results

The experiments have been done with an elitist generational EA (*popsize* = 100, p_c = .9, p_m = 0.01) using linear ranking selection (η = 2.0). No fine tuning of these standard parameters was attempted. A maximum number of 10^6 evaluations has been enforced. In order to provide a fair comparison, the internal assessments of partial solutions performed by some operators and decoders have been accounted as well.

Two problem instances with 20 and 34 OTUs respectively have been considered [1, 13]. These instances have been obtained by using conditional Kolmogorov complexity to calculate inter-OTU distances[6] [2] from mtDNA sequences. A branch-and-bound algorithm following [14] has been implemented, so as to know the exact optimal solutions. While the 20-OTU instance could be solved rather efficiently (a couple of seconds on a DIGITAL Alpha 400), the 34-OTU instance revealed itself as much harder to solve; it took about six hours and a half on the same machine, and more than half a billion subproblems were evaluated.

Table 1. Results for the 20-OTU instance (averaged for 50 runs).

Algorithm		best	mean ± std.dev.	%success	#evals
PDG	SWAP	8.290368	8.290368 ± 0.000000	100%	246,092
	NNI	8.290368	8.290368 ± 0.000000	100%	85,435
	SCRAMBLE	8.290368	8.290374 ± 0.000042	98%	146,162
Ordinal	SPX	8.337564	8.497876 ± 0.096093	0%	–
	DPX	8.325214	8.466251 ± 0.081407	0%	–
	UX	8.345291	8.471477 ± 0.085760	0%	–
Adaptive ordinal	SPX / OX	8.310436	8.417089 ± 0.066013	0%	–
	SPX / UCX	8.303018	8.400509 ± 0.047586	0%	–
	SPX + OX	8.294011	8.413814 ± 0.067423	0%	–
	SPX + UCX	8.304323	8.409855 ± 0.071146	0%	–
	DPX / OX	8.305980	8.395530 ± 0.064806	0%	–
	DPX / UCX	8.302258	8.415585 ± 0.072114	0%	–
	DPX + OX	8.295929	8.417998 ± 0.064371	0%	–
	DPX + UCX	8.331137	8.416023 ± 0.061231	0%	–
	UX / OX	8.295403	8.409900 ± 0.077820	0%	–
	UX / UCX	8.294327	8.394244 ± 0.059192	0%	–
	UX + OX	8.305803	8.409900 ± 0.077820	0%	–
	UX + UCX	8.290684	8.397843 ± 0.057408	0%	–
Greedy Xover	*insertion*	8.341772	8.481938 ± 0.072921	0%	–
	order	8.299083	8.453090 ± 0.088807	0%	–
	UCX+*insertion*	8.350527	8.460593 ± 0.078182	0%	–
	UX+*order*	8.320109	8.407623 ± 0.052265	0%	–
Permutational Decoder		8.290368	8.290368 ± 0.000000	100%	23,370

The results for the 20-OTU instance are shown in Table 1. The %-success column indicates the number of times the optimal solution is found, and #evals is the mean number of evaluations required in these successful runs. Notice firstly the good results provided by the PDG operator; near 100%-success rates are achieved in combination with any of the mutation operators discussed. While the absolute goodness of these results can be due to the low difficulty of this instance, its relative superiority over most decoder-based approaches is still informative. As it can be seen, the ordinal representation does not manage to find the optimal solution either with SPX, DPX (double-point crossover) or UX. When the OTU sequence is evolved along with the insertion points, performance is clearly improved, although no optimal solution is found. The results for the guided crossover are not satisfactory either. The reason may lie in the high cost of evaluating partial solutions. Actually, the greedy-order crossover yields slightly better results, due to the fact that fewer internal evaluations are needed. Finally, the results for the permutational decoder (using UCX) are the overall best. Unlike the plain decoding of the previous EAs, the greedy decoding is less sensitive to the disruptive effect that genotypic recombination can have on the phenotype. Hence, it manages to find the optimal solution in 100% of the runs,

[6] $M_{ij} = 1 - \frac{K(i)-K(i|j)}{K(ij)}$

Table 2. Results for the 34-OTU instance (averaged for 50 runs).

Algorithm		best	mean ± std.dev.	%success	#evals
PDG	SWAP	14.855763	14.860250 ± 0.002611	2%	933,639
	NNI	14.855763	14.857864 ± 0.001827	18%	427,468
	SCRAMBLE	14.855763	14.858426 ± 0.002104	10%	722,502
Ordinal	SPX	14.959411	15.100436 ± 0.079702	0%	–
	DPX	14.932403	15.141366 ± 0.093397	0%	–
	UX	14.939449	15.112828 ± 0.089895	0%	–
Adaptive ordinal	SPX / OX	14.904175	15.032073 ± 0.060582	0%	–
	SPX / UCX	14.920124	15.034428 ± 0.066467	0%	–
	SPX + OX	14.930042	15.042107 ± 0.074107	0%	–
	SPX + UCX	14.905937	15.031634 ± 0.071966	0%	–
	DPX / OX	14.920697	15.048883 ± 0.080941	0%	–
	DPX / UCX	14.898218	15.042495 ± 0.097559	0%	–
	DPX + OX	14.897718	15.016650 ± 0.073133	0%	–
	DPX + UCX	14.924216	15.043213 ± 0.074294	0%	–
	UX / OX	14.924170	15.050490 ± 0.072886	0%	–
	UX / UCX	14.896434	15.044002 ± 0.071149	0%	–
	UX + OX	14.915946	15.043606 ± 0.077418	0%	–
	UX + UCX	14.923092	15.035905 ± 0.077034	0%	–
Greedy XOver	*insertion*	14.954607	15.127468 ± 0.092056	0%	–
	order	14.951596	15.080942 ± 0.072010	0%	–
	UCX+*insertion*	14.957719	15.107558 ± 0.092509	0%	–
	UX+*order*	14.918159	15.031375 ± 0.067859	0%	–
Permutational Decoder		14.855763	14.855772 ± 0.000049	96%	397,512

using a lower number of evaluations (including internal calculations) than PDG with SWAP, NNI, or SCRAMBLE.

The results for the 34-OTU instance (Table 2) are completely consistent with this analysis. Again, PDG-based EAs perform better than EAs using ordinal decoders or greedy crossover. However, their success rate is notably lower here, due to the higher difficulty of the instance. The performance of the permutational-decoder EA is only marginally affected though. Optimal solutions are found on a regular basis (96%-success) at a lower computational cost than EAs using PDG.

5 Conclusions

A number of EAs for solving the Phylogeny Problem have been developed and compared. An empirical evaluation of these EAs using distance-based measures has shown that directly evolving phylogenetic trees yields better results than indirect approaches using decoders. A notable exception to this rule is provided by the greedy permutational decoder. This approach consistently provides optimal solutions at a lower cost than PDG-based EAs. Moreover, these latter EAs suffer from a scalability problem, exhibiting a clear performance drop when the number of OTUs increases. This does not seem to be the case for the permutational greedy decoder, at least for the instance sizes considered in this work.

Future work will try to confirm these results on larger problem instances. Optimal solutions will not be available in this case, but qualitative assessments will still be possible. Work is in progress here. The use of different evaluation measures is also an interesting line for future developments. In this sense, we plan to test alternative distance-based criteria, as well as maximum-likelihood approaches.

Acknowledgements

The first author is partially supported by Spanish CICYT under grant TIC1999-0754-C03. The second author is supported by Brazilian CNPq under Proj. 52.1100/01-1.

References

1. Y. Cao, N. Okada, and M. Hasegawa. Phylogenetic position of guinea pigs revisited. *Molecular Biology and Evolution*, 14(4):461–464, 1997.
2. X. Chen, S. Kwong, and M. Li. A compression algorithm for DNA sequences and its application in genome comparisons. *Genome Informatics*, 10:51–61, 1999.
3. C. Cotta and J.M. Troya. Genetic forma recombination in permutation flowshop problems. *Evolutionary Computation*, 6(1):25–44, 1998.
4. J. Hein. A new method that simultaneously aligns and reconstructs ancestral sequences for any number of homologous sequences, when the phylogeny is given. *Molecular Biology and Evolution*, 6:649–668, 1989.
5. S. Holmes. Phylogenies: An overview. In Halloran and Geisser, editors, *Statistics and Genetics*, pages 81–119. Springer-Verlag, New York NY, 1999.
6. D. Huson, S. Nettles, and T. Warnow. Disk-covering, a fast converging method for phylogenetic tree reconstruction. *Journal of Computational Biology*, 6(3):369–386, 1999.
7. S Koziel and Z. Michalewicz. A decoder-based evolutionary algorithm for constrained parameter optimization problems. In T. Bäeck, A.E. Eiben, M. Schoenauer, and H.-P. Schwefel, editors, *Parallel Problem Solving from Nature V – LNCS 1498*, pages 231–240. Springer-Verlag, Berlin Heidelberg, 1998.
8. H. Matsuda. Protein phylogenetic inference using maximum likelihood with a genetic algorithm. In *Proceedings of the Pacific Symposium on Biocomputing*, pages 512–523. World Scientific, 1996.
9. Z. Michalewicz. *Genetic Algorithms + Data Structures = Evolution Programs*. Springer-Verlag, Berlin, 1992.
10. A. Moilanen. Searching for the most parsimonious trees with simulated evolution. *Cladistics*, 15:39–50, 1999.
11. C.-K. Ong, S. Nee, A. Rambaut, H.-U. Bernard, and P.H. Harvey. Elucidating the population histories and transmission dynamics of papillomaviruses using phylogenetic trees. *Journal of Molecular Evolution*, 44:199–206, 1997.
12. N.J. Radcliffe. Equivalence class analysis of genetic algorithms. *Complex Systems*, 5:183–205, 1991.
13. A. Reyes, C. Gissi, G. Pesole, F.M. Catzeflis, and C. Saccone. Where do rodents fit? Evidence from the complete genome of *Sciurus vulgaris*. *Molecular Biology and Evolution*, 17(6):979–983, 2000.
14. Bang Ye Wu, Kun-Mao Chao, and Chuan Yi Tang. Approximation and exact algorithms for constructing minimum ultrametric trees from distance matrices. *Journal of Combinatorial Optimization*, 3(2):199–211, 1999.

Towards a More Efficient Evolutionary Induction of Bayesian Networks

Carlos Cotta[1] and Jorge Muruzábal[2]

[1] Dept. Lenguajes y Ciencias de la Computación, ETSI Informática
University of Málaga, Campus de Teatinos, 29071 - Málaga, Spain
ccottap@lcc.uma.es

[2] Grupo de Estadística y Ciencias de la Decisión, ESCET
University Rey Juan Carlos, 28933 - Móstoles, Spain
j.muruzabal@escet.urjc.es

Abstract. Bayesian networks (BNs) constitute a useful tool to model the joint distribution of a set of random variables of interest. This paper is concerned with the network induction problem. We propose a number of hybrid recombination operators for extracting BNs from data. These hybrid operators make use of phenotypic information in order to guide the processing of information during recombination. The performance of these new operators is analyzed with respect to that of their genotypic counterparts. It is shown that these hybrid operators provide notably improved and rather robust results. Some remarks on the future of the area are also laid out.

1 Introduction

A Bayesian Network (BN) is a graphical model postulating a joint distribution for a target set of random variables. One of the main advantageous features of this model is the fact that it provides a neat separation between qualitative and quantitative aspects of this distribution. On one hand, the qualitative aspects are given by the underlying graphical structure, a *Directed Acyclic Graph* (DAG) $\mathbf{G}$. On the other hand, quantitative aspects are provided by the set of probabilities attached to this DAG, say $\boldsymbol{\theta} = \boldsymbol{\theta}(\mathbf{G})$.

Two well-defined problems can be identified within this context: the *network induction* problem (learning an appropriate BN model), and the *inference* problem (determining the predictive conditional distribution at some variable of interest given a BN model and the values taken by certain other variables). While the latter arises when a BN has already been identified and is to be deployed in a given application, the former appears as a previous step. The focus of this work is precisely on this induction problem.

The main issue in the induction problem is learning the structure or DAG $\mathbf{G}$ (a variety of methods can be used to learn the probabilities $\boldsymbol{\theta}$). This turns out to be NP-hard [5], and hence the use of heuristic algorithms is in order [12]. In this sense, evolutionary algorithms [2] (EAs) emerge as interesting candidates for this task. Here we concentrate on the use of EAs for BN induction. More precisely, we explore in detail the role of recombination for this purpose.

J.J. Merelo Guervós et al. (Eds.): PPSN VII, LNCS 2439, pp. 730–739, 2002.

The organization of the paper is as follows. Section 2 presents details of the BN framework and lays out the basic learning problem addressed later. Section 3 introduces the new operators and Section 4 reviews the empirical evidence. Finally, Section 5 closes with some discussion and prospects for future research.

2 Background

This section provides basic ideas and notational details about both BNs and some scoring metrics used for evaluation purposes. A brief overview of EA approaches for evolving DAGs is provided too.

2.1 Bayesian Networks

As mentioned above, a BN is a tuple $(\mathbf{G}, \boldsymbol{\theta})$, where $\mathbf{G}$ is a DAG and $\boldsymbol{\theta} = \boldsymbol{\theta}(\mathbf{G})$ is a set of probability distributions attached to nodes in $\mathbf{G}$. The DAG specifies a number of links or arcs among variables or nodes. If we denote the whole set of variables as $\mathbf{X} = \{X_1, X_2, ..., X_n\}$, each variable X_j has a set of parents denoted by $\Pi_j = \{X_i \in \mathbf{X} \mid \{X_i \to X_j\} \in \mathbf{G}\}$. Then, the DAG $\mathbf{G}$ represents the (skeleton) joint distribution $P(\mathbf{X}) = \prod_{i=1}^{n} P(X_i \mid \Pi_i)$. Note that at least one of the Π_i is empty; we talk of *root* nodes in this case.

A standard BN model arises when data are assumed to follow independent Multinomial distributions, that is, $P(X_i = k \mid \Pi_i = j) = \theta_{ijk}$, where $j = 1, ..., q_i$; $k = 1, ..., r_i$; r_i is the number of distinct values that X_i can assume and q_i is the number of different configurations that Π_i can present. Hence, $\boldsymbol{\theta} = \{\theta_{ijk}\}$ collects all parameters in $\mathbf{G}$ and we have $\sum_k \theta_{ijk} = 1$ for all i and j.

Given a DAG $\mathbf{G}$ and a data matrix $\mathbf{D}$ with n columns and an arbitrary number of exchangeable rows (N), the *likelihood* of the network probabilities $\boldsymbol{\theta}$ is given by the double product of the above Multinomials: $P(\mathbf{D}|\mathbf{G}, \boldsymbol{\theta}) = \prod_{i=1}^{n} \prod_{j=1}^{q_i} \prod_{k=1}^{r_i} \theta_{ijk}^{N_{ijk}}$, where N_{ijk} is the absolute frequency of value k in X_i when its parent configuration Π_i assumes state j. Given maximum likelihood estimators (MLE) $\hat{\boldsymbol{\theta}} = \hat{\boldsymbol{\theta}}(\mathbf{G}, \mathbf{D})$ of $\boldsymbol{\theta}$, $P(\mathbf{D}|\mathbf{G}, \hat{\boldsymbol{\theta}})$ can serve as a rudimentary scoring metric. We nevertheless focus on an alternative *Bayesian* measure. Consider the *marginal* likelihood

$$P(\mathbf{D}|\mathbf{G}) = \int P(\mathbf{D}|\mathbf{G}, \boldsymbol{\theta})\pi(\boldsymbol{\theta}|\mathbf{G})d\boldsymbol{\theta}, \tag{1}$$

where $\pi(\boldsymbol{\theta}|\mathbf{G})$ is a prior distribution on $\boldsymbol{\theta}$. If this measure is combined with a prior distribution on DAG structures $\pi(\mathbf{G})$, the log-posterior

$$F(\mathbf{G}) = \log \pi(\mathbf{G}|\mathbf{D}) = \log \pi(\mathbf{G}) + \log P(\mathbf{D}|\mathbf{G}) \tag{2}$$

is obtained. We will be using this Bayesian measure F for some particular choices of $\pi(\mathbf{G})$ and $\pi(\boldsymbol{\theta}|\mathbf{G})$. To be precise, the former is chosen as $\pi(\mathbf{G}) \propto N^{-g/2}$, where $g = \sum_{i=1}^{n}(r_i - 1)q_i$ is the number of free $\boldsymbol{\theta}$ parameters in the model. This choice penalizes complex (i.e., highly dense) DAGs, and is closely related to the asymptotic *Bayesian Information Criterion* [10] or *BIC* [1]. As to the $\pi(\boldsymbol{\theta}|\mathbf{G})$, it is taken to be the product of independent (conjugate) Dirichlet distributions. In the case of no missing data and noninformative Dirichlet hyperparameters α, it is then possible to perform in closed form the integration leading to $P(\mathbf{D}|\mathbf{G})$ above. This key result adds to the computational tractability of the approach and will be considered here too. The hyperparameters α can be interpreted in terms of *equivalent sample size* [12]. The noninformative choice $\alpha_i = \frac{1}{r_i}$ is usually adopted following theoretical considerations related to *likelihood equivalence* [11].

2.2 Evolutionary Induction of DAGs

Typically, the EA approach for designing BNs evolves DAG structures which –when submitted for fitness calculation– are augmented with $\hat{\boldsymbol{\theta}}$ parameters and fed into a scoring function. The internal representation of DAGs turns out to be a crucial issue here. In essence, choices regarding this aspect can be classified within two main categories: *direct* and *indirect.*

Direct approaches are those in which the search is conducted over the space of all possible DAGs, say $\mathcal{S}_{DAG}$. An obvious potential problem in these approaches is the generation of infeasible solutions (i.e., digraphs with cycles). This can be avoided in two different ways. On one hand, a precedence order among variables can be assumed; then, it suffices to evolve the upper triangular portion of the adjacency matrix of the graph to obtain feasible DAGs (alternatively, closed operators in $\mathcal{S}_{DAG}$ can be defined; we will return to this point below). On the other hand, a repair function can be used to remove cycles before evaluation. See [14] for a comparison of both approaches.

As regards indirect approaches, these use an auxiliary space $\mathcal{S}_{aux}$ to conduct the search. Elements from $\mathcal{S}_{aux}$ are then fed to a suitable (decoder) algorithm to obtain the actual BNs they represent. Consider, for example, the search in the space of $n-$element permutations [13]; the construction heuristic K2 [7] is subsequently used to build the actual BN. This approach has the advantage of filtering out infeasible solutions while it also introduces problem-specific knowledge. There are some drawbacks though. For example, it may be difficult to design parsimonious operators for exploring $\mathcal{S}_{aux}$ in some situations (the $\mathcal{S}_{aux} \longrightarrow \mathcal{S}_{DAG}$ mapping can hinder the design of such operators). Also, some good regions of $\mathcal{S}_{DAG}$ may be unreachable by certain decoders.

Direct approaches do not face these hurdles as long as reproductive operators manipulate *meaningful information units*, that is, the main idea is to depart from the classical (purely syntactic) crossover in order to achieve semantically-sound recombination. This goal can be seen as an upgrading process, and hence the identification of syntactically-correct information units is still required. In this sense, a generic analysis of the syntax of these units has been done in [8]. These authors show that minimal *transmission units*[2] have the following structure:

[1] $BIC = \log P(\mathbf{D}|\mathbf{G}, \hat{\boldsymbol{\theta}}) - \frac{g}{2} \log N$

[2] Transmission units can be seen as minimal –not necessarily elementary– pieces of information that have to transmitted as a whole from parents to offspring so as to ensure feasibility of the latter.

$$\begin{aligned} T(X_i \nrightarrow X_j, \Phi) &= \{X_i \nrightarrow X_j\} \\ T(X_i \rightarrow X_j, \Phi) &= \{X_i \rightarrow X_j\} \cup \{X_r \nrightarrow X_s \mid C^{\oplus}_{sr} = 1\} \end{aligned} \qquad (3)$$

where Φ is the partially-defined descendant DAG at any intermediate step of recombination, $X_i \rightarrow X_j$ (resp. $X_i \nrightarrow X_j$) represents the decision of including in (resp. excluding from) Φ the arc from X_i to X_j, $C^{\oplus} = C^{\infty}_{\Phi}$ XOR $C^{\infty}_{\Phi \cup \{X_i \rightarrow X_j\}}$, and C^{∞}_{Ψ} is the transitive closure of a graph Ψ. The next section shows how the units in Eq. (3) can be endowed with semantic information.

3 Bayesian Network Recombination

Semantically-aware operators are defined in terms of phenotypic information. This implies that recombination no longer takes place at the DAG level, but at the BN level. Two hybrid-operator templates (and phenotypic measures to be used therein) are discussed below.

3.1 Genetic vs. Allelic Recombination

Any DAG $\mathbf{G}$ can be viewed as the composition of a number of basic units η^x_{ij}, where i, j are nodes and $x \in \{1, 0\}$ indicates whether the corresponding directed arc is present in $\mathbf{G}$ or not. Informally speaking, each η_{ij} is a *gene*, whereas each η^x_{ij} is an *allele* for that gene. Two recombination approaches are thus possible.

A first possibility is to focus on individual genes. A recombination operator following this criterion must process all genes (in any suitable –not necessarily fixed– order) to construct a valid solution. For each gene, a decision must be made on whether to use the allele from either the father $\mathbf{G}$ or the mother $\mathbf{H}$ [3]. While a genotypic operator would make these decisions at random, the use of phenotypic information is proposed here. More precisely, a Boolean function β –taking the two parent BNs ($\mathbf{G}$ and $\mathbf{H}$) and the partially-built child (Ξ) as input– determines the value of each gene. The pseudocode of this generic operator (termed PheGT for 'phenotypic gene transmission') is as follows:

1. **for** $i \in \{1..n\}$ **do** $\Pi^{\Xi}_i \leftarrow \emptyset$
2. **for** $i \in \{1..n\}$ **do** $\Upsilon_i \leftarrow (\Pi^{\mathbf{G}}_i \cup \Pi^{\mathbf{H}}_i)$
3. **while** $\exists \Upsilon_j \neq \emptyset$ **do**
   (a) Pick $X_i \in \Upsilon_j$
   (b) $\Upsilon_j \leftarrow \Upsilon_j \setminus \{X_i\}$
   (c) **if** $\beta(\eta_{ij}, \mathbf{G}, \mathbf{H}, \Xi)$ **then**
      i. $\Pi^{\Xi}_j \leftarrow \Pi^{\Xi}_j \cup \{X_i\}$
      ii. **for** $[X_k \nrightarrow X_s] \in T(X_i \rightarrow X_j, \Xi)$ **do** $\Upsilon_s \longleftarrow \Upsilon_s \setminus \{X_k\}$

A different template arises when the emphasis is put on individual alleles. In this case, all η^1_{ij} alleles taken from either parent are put on a common bag. Then, the operator iteratively decides which alleles are extracted and injected (together

[3] Extensions to multiparent recombination are straightforward.

with the corresponding transmission unit) into the child. The operator may decide to terminate transmission at any point along the process; all unspecified genes are given the default value η_{ij}^0 in this case. Phenotypic information can be used here for both deciding the order in which alleles are picked (using a selection function σ), and for determining when to stop (using a Boolean function τ). The corresponding template of this operator (termed PheAT for 'phenotypic allele transmission') is as follows:

1. **for** $i \in \{1..n\}$ **do** $\Pi_i^{\Xi} \leftarrow \emptyset$
2. $\Upsilon \longleftarrow \langle \eta_{ij}^1 \mid X_i \in \Pi_j^{\mathbf{G}} \cup \Pi_j^{\mathbf{H}} \rangle$
3. **while** $\neg\tau(\Upsilon, \Xi)$ **do**
   (a) $\eta_{ij}^1 \leftarrow \sigma(\Upsilon, \Xi)$
   (b) $\Upsilon \leftarrow \Upsilon \setminus \{\eta_{ij}^1\}$
   (c) $\Pi_j^{\Xi} \leftarrow \Pi_j^{\Xi} \cup \{X_i\}$
   (d) **for** $[X_k \nrightarrow X_s] \in T(X_i \rightarrow X_j, \Xi)$ **do** $\Upsilon \leftarrow \Upsilon \setminus \{\eta_{ks}^1\}$

Some possible instantiations of the above templates (β for PheGT; σ and τ for PheAT) are discussed next.

3.2 Phenotypic Measures

The mutual information $MI(X_j, X_i)$ criterion has often been the choice for measuring the *merit* of single alleles η_{ij}^1 [16]. However, this measure is known to have some limitations due to its myopic nature [9]. The *updated* MI measure, namely the *Conditional Mutual Information* measure [9]

$$CMI(X_j, X_i \mid\mid \Pi_j \setminus \{X_i\}) = \sum P(\Pi_j \setminus \{X_i\}) \sum P(X_j, X_i \mid \Pi_j \setminus \{X_i\}) \log \frac{P(X_j, X_i \mid \Pi_j \setminus \{X_i\})}{P(X_j \mid \Pi_j \setminus \{X_i\}) P(X_i \mid \Pi_j \setminus \{X_i\})} \quad (4)$$

reflects the strength of the association between X_j and X_i once the effect of $\Pi_j \setminus \{X_i\}$ is taken into account. While this CMI measure deserves further attention, in this work we have decided to explore a somewhat simpler but nonetheless interesting variant thereof. Specifically, given that $X_i \in \Pi_j$ in either parent, we consider the grand average

$$\mu_{ij} = \frac{r_i}{r_j q_j} \sum Var(X_i, y, w), \quad (5)$$

where the sum ranges across both the $\frac{q_j}{r_i}$ different values w that $\Pi_j \setminus \{X_i\}$ can take and the r_j different values y that X_j can take. The inner term $Var(X_i, y, w)$ is defined as the *variance* of the probabilities $P(X_j = y \mid X_i = z, \Pi_j \setminus \{X_i\} = w)$ across the r_i different values z that X_i can take. These probabilities are of course nothing but $P(X_j = y \mid \Pi_j = (z, w)) = \theta_{j(z,w)y}$ with our earlier notation. As usual, these theoretical μ_{ij} are replaced in practice by their MLE $\hat{\mu}_{ij}$ based on $\hat{\boldsymbol{\theta}}$. If for some (y, w) the estimate of $Var(X_i, y, w)$ is close to 0, we conclude that any $X_i = z$ adds nothing new to what w already tells us about y. It is easy to see that Eq. (4) is also close to 0 in this case. Conversely, if $Var(X_i, y, w)$ is relatively large, then it does matter what X_i has to say in that situation, so we would tend to use both X_i and $\Pi_j \setminus \{X_i\}$ when predicting X_j (in this particular DAG and in general – recall that there is no explicit conservation law for the $\Pi_j's$ from parents to children).

This μ measure can be used within the operator templates presented earlier. We begin with β (central in PheGT). According to Eq. (5), μ_{ij} is always in [0, 0.25]. Thus, a first option is to use $\mu'_{ij} = 4\mu_{ij}$ as our transmission probability: $\beta(\eta_{ij}, \cdot) \equiv URand(0,1) < \mu'_{ij}$. Since this can be a rather demanding criterion for arc transmission, we also consider the more relaxed $\mu''_{ij} = 2\sqrt{\mu_{ij}}$.

As regards PheAT, an allele-selection function σ is required. This admits a direct instantiation since we can always pick the allele with the highest μ_{ij} value (no rescaling required here). A simple (genotypic) criterion has been chosen in turn for the termination function τ. Specifically, a random number is first drawn from a $Binomial(\nu,\phi)$ distribution, where $\phi = 1/2$ and $\nu/2$ approximates the parents' mean number of arcs, and arc transmission is terminated as soon as the child reaches the desired number of arcs (or no transmittable arc remains). With this choice, we can compare PheAT to a pure genotypic version (picking alleles at random).

4 Experimental Results

We have tested a steady-state EA ($popsize = 100$, $maxevals = 15000$, crossover rate $p_X = .9$, mutation rate $p_m = 1/n^2$), using tournament selection (tournament size = 3). No fine tuning of these parameters was attempted. The initial population is obtained by generating DAGs at random[4]. The goal is to minimize the fitness function $-F(\mathbf{G}) = -\log P(\mathbf{D}|\mathbf{G}) + \frac{g}{2}\log N$, see Eq. (2) above.

Two networks have been chosen to benchmark the proposed approach: the ALARM network, a 37-variable network for monitoring patients in the intensive care unit [3], and the INSURANCE network, a 27-variable BN for evaluating car insurance risks [4]. However, due to space constraints, we concentrate here on the former (similar qualitative results have been obtained with the latter). Training sets of $N = 2,000$ examples were simulated from the ALARM network.

We have tried both the phenotypic (PheGT and PheAT) as well as the genotypic (GT and AT) operators. Two variants of each operator have been considered in turn: respectful and non-respectful. The property of *respect* [15] refers in this case to the *initial* transmission of all arcs shared by the parent DAGs. Since acyclicity is enforced at all times, inclusion of (some of) these arcs may be impossible later[5]. Hence, this initial transmission introduces an important qualitative change in behavior. As regards the μ'_{ij} vs. μ''_{ij} choice in PheGT, we consider only the respectful variant and mark the latter option with a subscript.

[4] The size of the sets Π_j is limited by $q_j < 2^{11}$.

[5] For example consider DAGs $\mathbf{G}$ and $\mathbf{H}$ such that $\{X_i \to X_j, X_j \to X_k\} \in \mathbf{G}$, and $\{X_k \to X_i, X_i \to X_j\} \in \mathbf{H}$. If arcs $\{X_j \to X_k, X_k \to X_i\}$ are transmitted to the child, it will be impossible to transmit the common arc $X_i \to X_j$ as well.

Table 1. Results of the different crossover operators on the ALARM network (averaged for 10 runs).

	$-F^*$		$-\log P(\mathbf{D}\|\mathbf{G})$	
Operator	best	mean ± std.dev	best	mean ± std.dev.
GT	28718.18	29888.03 ± 584.18	26931.97	28045.57 ± 631.24
AT	29470.42	29901.69 ± 311.24	27338.98	28085.83 ± 382.90
PheGT	29314.62	30038.57 ± 528.11	27646.22	28614.16 ± 567.53
PheAT	25596.55	26115.35 ± 469.76	24106.94	24726.66 ± 534.44
GT^R	24290.02	24807.84 ± 328.75	22617.82	23111.70 ± 285.17
AT^R	24490.63	24896.07 ± 269.84	22806.67	23139.50 ± 224.46
$PheGT^R$	23929.07	24493.83 ± 317.99	22291.76	22891.72 ± 368.90
$PheAT^R$	23861.68	24430.92 ± 379.92	22216.06	22729.01 ± 299.41
$PheGT_2^R$	23944.63	24245.57 ± 149.02	22422.18	22671.80 ± 170.73
HC	24528.41	24732.48 ± 168.53	22907.42	23000.66 ± 86.04
ALARM	24922.33		22987.90	

Table 2. Structural properties of the networks evolved by the different crossover operators for the ALARM benchmark (averaged for 10 runs).

	#parameters			*BIC*	
Operator	min	mean ± std.dev.	max	best	mean ± std.dev.
GT	415	484.8 ± 65.29	659	28613.96	29779.14 ± 583.73
AT	375	477.8 ± 81.59	629	29372.47	29795.73 ± 308.49
PheGT	266	374.8 ± 62.55	488	29210.28	29946.81 ± 530.89
PheAT	307	365.4 ± 58.89	507	25511.70	26032.53 ± 466.78
GT^R	376	446.3 ± 38.17	505	24194.88	24708.69 ± 324.30
AT^R	425	462.2 ± 30.74	536	24396.27	24796.51 ± 266.89
$PheGT^R$	386	429.2 ± 27.65	483	23842.55	24433.58 ± 329.83
$PheAT^R$	387	451.2 ± 44.28	514	23730.24	24350.72 ± 418.22
$PheGT_2^R$	381	414.1 ± 20.78	441	23859.39	24155.89 ± 147.98
HC	403	455.7 ± 37.12	518	24434.77	24635.45 ± 164.56
ALARM	509			24049.43	

For comparative purposes, a hill climbing (HC) algorithm has also been tested. This HC performs single-arc insertions and deletions and has been run for the same number of evaluations as the EAs (re-start was performed each time stagnation was reached). Table 1 shows the results.

A quick inspection of these results leads to several conclusions of interest. Note first that the introduction of respect yields a substantial improvement in all performance measures. It can also be seen that the phenotypic operators clearly outperform[6] their genotypic counterparts (thus confirming the usefulness of the phenotypic approach), whereas the HC algorithm lies somewhere in between. Additionally, the networks provided by PheAT and PheGT are definitely better (in terms of the selected measures) than the original network. This feature is due

[6] Significantly, using a standard (two-sample) t-test. The same holds when using a test set different from the training set, so as to evaluate overfitting.

to the small size of the training set and indicates that our best operators achieve some refinement in the BNs they produce. Finally, the non-linear mapping leading to μ''_{ij} in PheGT$_2^R$ provides the best overall results. Clearly, this option boosts the ability of PheGT for exploring and finding improved structures.

The structural properties of the evolved BNs are consistent with the analysis above. Table 2 shows the total number of $\boldsymbol{\theta}$ parameters (g) as well as the BIC measure discussed earlier. It can be seen that the networks provided by AT and PheAT are slightly more complex on average than those produced by GT and PheGT, whereas all of them tend to be simpler than the true network. Note also that the phenotypic crossover operators manage to produce networks of similar BIC than the original ALARM network; moreover, in some cases they interestingly provide even lower values.

5 Summary and the Likely Future

We have described and evaluated several new recombination operators for evolving BNs. These operators are based on phenotypic information and thus depart from previously proposed genotypic crossover and phenotypic mutation. It has been shown that our phenotypic variants produce satisfactory results in problems of moderate complexity. In this sense, the observance of the property of respect has revealed itself as a crucial factor for the performance of these operators.

Perhaps the most challenging line of research in the wider BN induction problem refers to the possibility of performing the search over the space of BN *equivalence classes*, say $\mathcal{S}_{Eq}$ (rather than $\mathcal{S}_{DAG}$ as above). Two BNs are (Markov) equivalent if they encode the same statistical model, that is, the same set of independence and conditional independence statements. Let $[\mathbf{G}]$ denote the equivalence class of a DAG $\mathbf{G}$. Given training data generated by $\mathbf{G}$, many DAG-based algorithms use scoring measures that indeed score equally all members of $[\mathbf{G}]$. Hence, all that can be reasonably asked in this case is to reach some DAG in $[\mathbf{G}]$: these algorithms can not be expected to reconstruct $\mathbf{G}$ exactly. Note that the marginal likelihood $P(\mathbf{D}|\mathbf{G})$ in Eq. (1) is one of such metrics, yet our fitness measure $F(\mathbf{G})$ –dependent also on g– is not. It is still interesting to imagine how such an alternative search process could be carried out by an EA similar to the above. For one thing, search strategies that spend most of their time within the same equivalence class would seem rather inefficient (since they inadvertedly keep proposing the same model). We conclude by briefly providing some preliminary insights on this matter.

It turns out that equivalence classes can be compactly represented by (certain class of) *partially* directed acyclic graphs or PDAGs [1, 6]. PDAGs include directed as well as *undirected* arcs. Chickering [6] provides an algorithm that takes a given DAG $\mathbf{G}$ and outputs the PDAG $\bar{\mathbf{G}}$ that uniquely represents its equivalence class $[\mathbf{G}]$. Since $\bar{\mathbf{G}}$ and $\mathbf{G}$ have the same connectivity pattern (ignoring directionality), all undirected arcs in $\bar{\mathbf{G}}$ correspond to *reversible* arcs in $\mathbf{G}$, whereas all directed arcs in $\bar{\mathbf{G}}$ are *compelled*: they show up throughtout $[\mathbf{G}]$. A reasonable assessment of BN quality would require correct directionality for

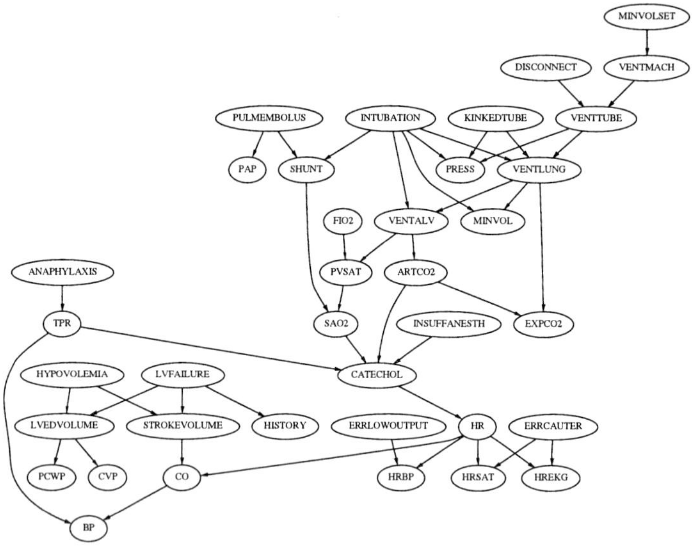

Fig. 1. The ALARM network. The reversible arcs are: MINVOLSET→VENTMACH, PULMEMBOLUS→PAP, ANAPHYLAXIS→TPR and LVFAILURE→HISTORY. We can visualize the key PDAG representing [ALARM] by making these arcs undirected.

compelled arcs but just connectivity for reversible arcs. In the ALARM network, for example, we find 4 reversible and 42 compelled arcs, see Figure 1.

To conclude, consider now potential mutation and crossover operators for some parent PDAGs $\bar{\mathbf{G}}$ and $\bar{\mathbf{H}}$. A first issue refers to PDAG *validity*: not all PDAGs represent equivalence classes. Chickering [6] presents various operators designed to modify a given $\bar{\mathbf{G}}$ so that the resulting PDAG effectively represents a *different* equivalence class. For example, both directed and undirected arcs can be added or deleted (compelled arcs can sometimes be reversed also). The familiar mutation operators found in the evolutionary DAG arena (e.g., [16]) can thus be extended along this way.

As regards crossover, an obvious approach would randomly instantiate $\bar{\mathbf{G}}$ and $\bar{\mathbf{H}}$ so as to obtain DAGs $\mathbf{G}$ and $\mathbf{H}$ from which phenotypic measures could be derived as above. The main challenge remains about how to meaningfully incorporate this DAG-based information when defining the offspring $\bar{\mathbf{K}}$ derived from $\bar{\mathbf{G}}$ and $\bar{\mathbf{H}}$. We are currently exploring some ideas in this direction.

Acknowledgement

The authors are partially supported by grants TIC99-0754-C03-03 and TIC2001-0175-C03-03 from the Spanish CICYT agency. We appreciate the assistance of David Chickering in running his LABEL-EDGES algorithm.

References

1. S.A. Andersson, D. Madigan, and M.D. Perlman. A characterization of markov equivalence classes for acyclic digraphs. *Annals of Statistics*, 25:505–541, 1997.
2. Th. Bäck, D.B. Fogel, and Z. Michalewicz. *Handbook of Evolutionary Computation.* Oxford University Press, New York NY, 1997.
3. I.A. Beinlich, H.J. Suermondt, R.M. Chavez, and G.F. Cooper. The ALARM monitoring system: A case study with two probabilistic inference techniques for belief networks. In J. Hunter, J. Cookson, and J. Wyatt, editors, *Proceedings of the Second European Conference on Artificial Intelligence and Medicine*, pages 247–256, Berlin, 1989. Springer-Verlag.
4. J. Binder, D. Koller, S. Russell, and K. Kanazawa. Adaptive probabilistic networks with hidden variables. *Machine Learning*, 29:213–244, 1997.
5. D.M. Chickering, D. Geiger, and D. Heckermann. Learning bayesian networks is NP-complete. In D. Fisher and H.-J. Lenz, editors, *Learning from data: AI and Statistics V*, pages 121–130, New York NY, 1996. Springer-Verlag.
6. D.M. Chickering. Learning equivalence classes of bayesian-network structures. Submitted manuscript, 2001.
7. G. Cooper and E. Herskovits. A bayesian method for the induction of probabilistic networks from data. *Machine Learning*, 9:309–347, 1992.
8. C. Cotta and J.M. Troya. Analyzing directed acyclic graph recombination. In B. Reusch, editor, *Computational Intelligence: Theory and Applications*, volume 2206 of *Lecture Notes in Computer Science*, pages 739–748. Springer-Verlag, Berlin Heidelberg, 2001.
9. N. Friedman, I. Nachman, and D. Pe'er. Learning bayesian network structures from massive datasets: The sparse candidate algorithm. In H. Dubios and K. Laskey, editors, *Proceedings of the Fifteenth Conference on Uncertainty in Artificial Intelligence*, pages 206–215, San Francisco CA, 1999. Morgan Kaufmann.
10. D. Geiger, D. Heckerman, and C. Meek. Asymptotic model selection for directed networks with hidden variables. In E. Horvitz and F.V. Jensen, editors, *Proceedings of the Twelfth Annual Conference on Uncertainty in Artificial Intelligence*, pages 283–290, San Francisco CA, 1996. Morgan Kaufmann.
11. D. Heckerman, D. Geiger, and D.M. Chickering. Learning bayesian networks: the combination of knowledge and statistical data. *Machine Learning*, 20(3):197–243, 1995.
12. D. Heckerman. A tutorial on learning with bayesian networks. In M.I. Jordan, editor, *Learning in Graphical Models*, pages 301–354. Kluwer, Dordrecht, 1998.
13. P. Larrañaga, C.M.H. Kuijpers, R.H. Murga, and Y. Yurramendi. Learning bayesian network structures by searching for the best ordering with genetic algorithms. *IEEE Transactions on Systems, Man and Cybernetics*, 26(4):487–493, 1996.
14. P. Larrañaga, M. Poza, Y. Yurramendi, R.H. Murga, and C.M. H. Kuijpers. Structure learning of bayesian networks by genetic algorithms: A performance analysis of control parameters. *IEEE Transactions on Pattern Analysis and Machine Intelligence*, 10(9):912–926, 1996.
15. N.J. Radcliffe. Equivalence class analysis of genetic algorithms. *Complex Systems*, 5:183–205, 1991.
16. M.L. Wong, W. Lam, and K.S. Leung. Using evolutionary programming and minimum description length principle for data mining of bayesian networks. *IEEE Transactions on Pattern Analysis and Machine Intelligence*, 21(2):174–178, 1999.

Robust Multiscale Affine 2D-Image Registration through Evolutionary Strategies

Héctor Fernando Gómez García[1], Arturo González Vega[1], Arturo Hernández Aguirre[1], José Luis Marroquín Zaleta[1], and Carlos Coello Coello[2]

[1] Center for Research in Mathematics, Department of Computer Science
Guanajuato, Gto. 36240, Mexico
{hector,gonzart,artha,jlm}@cimat.mx
[2] CINVESTAV-IPN Sección de Computación, México, D.F. 07300, Mexico
ccoello@cs.cinvestav.mx

Abstract. We propose a robust methodology based on multiscale analysis, affine transforms, and evolutionary strategies for solving the image registration problem. The approach is found robust for the affine registration of medical images.

1 Introduction

The goal of image registration is to find the best correspondence between images of the same scene. The intuitive approach to this problem is to find "relevant features" on the images that can be used to bring them into correspondence. As noted in [5], finding relevant features is one of three main components of any image registration problem; the second is similarity metric, and the third is search space and strategy. Our approach to image registration can be described around these components as follows: 1) feature space is not created, nor induced or searched; images are sampled and a few points are used for matching. 2) It uses a metric based on pixel intensity to measure image correspondence, and 3) the search space is kept small by subsampling the images whereas the optimization mechanism is implemented through evolutionary strategies.

2 The Image Registration Problem

For the sake of sufficiency of the article we describe the registration problem (defined elsewhere in the literature). Assume two images $I_1(x, y)$ and $I_2(x, y)$ are available from the same object but the object changes position from one to other. The images are 2-dimensional arrays with some intensity value at every pixel position (x, y). The image registration problem is *to find the mapping between two images* I_1 and I_2 that gives the best correspondence. Equation 1

$$I_2(x, y) = I_1(f(x, y)) \tag{1}$$

is the formal registration model where function $f : I_1(x, y) \rightarrow I_2(x, y)$ performs the mapping between images. Approximations to f can be constructed by some

J.J. Merelo Guervós et al. (Eds.): PPSN VII, LNCS 2439, pp. 740–748, 2002.

transformations: affine and projective amongst several. No transform applies to all problems thus in choosing a suitable transform it is advisable to consider the sources of misregistration. They are generally due to sensor noise, sensor type, and changes in scene conditions [8]. This paper describes an image-to-image registration technique without use of any knowledge about the sensors. Images are taken with the same instrument but (simulated) from different positions, thus the only source of misregistration is related to changes in scene conditions. Note that the image registration problem is clearly a function approximation one. That is, f is unknown but it will be approximated through affine transformations.

2.1 Affine Transforms

An affine transform is a linear transform composed of the following geometric transformations: translation, rotation, scaling, stretching, and shearing [7]. As noted affine transforms are sound basis for our mapping function approximation problem since the source of misregistration can be tackled as follows: distortions due to different sensor orientation are corrected by translation and rotation, changes in altitude are corrected by scaling, and stretching and shearing correct distortions due to changes in the viewing angle [5]. A useful subset of affine transform that combines rotation and translation is called *rigid-body transform.* In Equation 2, (x_2, y_2) is the transformed coordinate (x_1, y_1) after translations (t_x, t_y), rotation by angle θ, and scaling by factor s (an affine transform is only linear when translation is zero [5]).

$$\begin{bmatrix} x_2 \\ y_2 \end{bmatrix} = \begin{bmatrix} t_x \\ t_y \end{bmatrix} + s \begin{bmatrix} cos\ \theta & -sin\ \theta \\ sin\ \theta & cos\ \theta \end{bmatrix} \begin{bmatrix} x_1 \\ y_1 \end{bmatrix} \tag{2}$$

The general $2D$ affine transformation (the basis of our approach) is expressed as shown in Equation 3. The matrix $\begin{bmatrix} a_{1,1} & a_{1,2} \\ a_{2,1} & a_{2,2} \end{bmatrix}$ accounts for rotation, scaling, stretching, and shearing.

$$\begin{bmatrix} x_2 \\ y_2 \end{bmatrix} = \begin{bmatrix} a_{1,3} \\ a_{2,3} \end{bmatrix} + \begin{bmatrix} a_{1,1} & a_{1,2} \\ a_{2,1} & a_{2,2} \end{bmatrix} \begin{bmatrix} x_1 \\ y_1 \end{bmatrix} \tag{3}$$

Hence we should understand the affine registration problem as the problem of finding the parameter set $\{a_{1,1}, a_{1,2}, a_{1,3}, a_{2,1}, a_{2,2}, a_{2,3}\}$ for the affine transform that best mimics the function $f : I_1(x, y) \rightarrow I_2(x, y)$. Provided function approximation is sound we still need to address the question of how well two images match. The similarity measure used affects the matching quality. One popular similarity measures is normalized cross-correlation [14]; since its computation is expensive in this paper we use a simpler and reliable well known approach: the sum of absolute difference of pixel intensities.

3 Related Work

Image registration has been approached from a large variety of techniques, we should only mention in this section those articles closely related to this article. The work of Fitzpatrick and Grefenstette [6] is one of the first works on

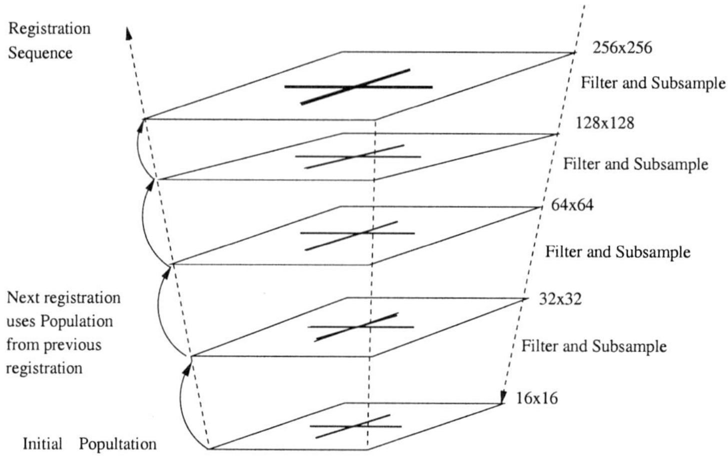

Fig. 1. Multiscale Gaussian pyramid for Image registration

registration of medical images based on Genetic Algorithms. Brown [5] noted that probabilistic methods are more suitable for registration and segmentation of medical images, thus less than 7% of the methods (95) accounted by Maintz [11] use a form of evolutionary algorithm as optimization tool. Ankenbrandt and Buckles [1] use genetic algorithms for scene recognition, Bhanu and Sungkee [4] describe several methods based on evolutionary techniques for image segmentation. Roberts and Howard [12] use genetic programming for orientation detection, Ross et. al. [13] also use genetic programming but for edge detection; Bhandarkar [3] recently compared several techniques for image segmentation using evolutionary computation. Louchet [10] applied evolutionary strategies to stereovision.

4 Multiscale Representation

A multiscale representation, also called Gaussian pyramid, is a set of images generated by the successive application of smoothing and subsampling operations over a source image of dimensionality d. At each step the new image contains only $\frac{1}{2^d}$ pixels of the previous image [9]. A typical multiscale pyramid is shown in Figure 1. The successive application of smoothing and subsampling operations helps to eliminate unnecessary details while keeping important features. These features are very important for the first iteration of our method at the lowest level of resolution (bottom of pyramid). Inherent to multiscaling is the reduction in the size of the image, a property that reduced processing time without altering precision.

5 The Multiscale Affine Image Registration Method

Our approach to image registration is based on a multiscale representation. The sequential application of smoothing and subsampling operations is performed in a top-down fashion. Then registration is performed bottom-up. The whole registration process is as follows and it is shown in Figure 1

- Top-Down step. Apply a smoothing and subsampling procedure K times to both images I_1 and I_2. A set of K subsampled images is computed and stored (K of each image). The smoothing procedure (to prevent alising) is a low-pass filter implemented by image convolution with a Gaussian kernel ($\sigma = 0.5$).
- Bottom-Up step. Register the images at the bottom of the pyramid (lowest resolution). Initial population is seeded with individuals mutated out of the identity transform, that is, individuals are mutations of the identity matrix, and the zero translation vector. Once the $(\mu + \lambda)$-ES algorithm reaches a nominal fitness value or number of iterations, the registration process is repeated but this time using the images at the immediate level above. For seeding the next initial population the object variables of the best individual of the previous registration step are mutated. Control variables are generated anew. At any registration level no more than 256 sampled pixels are used to compute the fitness value. Notice in Figure 1 the image at the bottom has only 256 pixels, but the immediate above must be sampled because it has 1024 pixels (although the image itself was derived by the sampling procedure of the top-down step, it is again sampled to compute the fitness function).

A $(\mu + \lambda)$ Evolutionary Strategy is used for searching and optimization of the six real variables that control the general affine transform. Crossover operation for control and object variables is generalized intermediate, mutation is uncorrelated (no rotation angles), in accordance with that version of the algorithm as described by Bäck [2]. Population size does not change during the process. No knowledge (landmarks) from the image has to be derived or located during the process, our method uses only 256 sample points equally spaced and distributed over the image. Thus, about 0.4 of image pixels are used for registration in our experiments.

Fitness function is based on the similarity measure "absolute difference of intensities", as follows:

$$fitness = \frac{1}{1 + \frac{1}{N}\sum_{\{x,y\}\in\Omega} |I_1(f(x,y)) - I_2(x,y)|} \tag{4}$$

Where Ω is the set of sample points, N is the cardinality of Ω (256), and $f(x,y)$ the required transformation. The fitness function takes values in $[0,1]$ to represent $[nomatch \rightarrow perfectmatch]$. Another strategy to create the set Ω is to take random samples with uniform probability distribution. The set can be generated anew between 3 and 5 times per level, improving the registration ability of the algorithm over noisy images. Since an affine transformation AT maps pixel's integer coordinates of image I_2 into real coordinates on image $I1$, a cubic spline interpolation procedure IP is used to calculate the proper intensity value of each transformed coordinate.

An important issue related to the fitness function is the quality of the registration implied by the fitness value (Equation 4). In Table 1 we show the value of

Table 1. Relation fitness function - summation of intensity error

Fitness Value	Σ term (256 pixels)
1.0	0.0
.99	2.58
.98	5.22
.97	7.91
.96	10.6
.95	13.4
.94	16.3

the summation term for several fitness function values that are used in our experiments. It is clear that a change of .01 in the fitness value implies a change of 2.58 in the summation term. Since the summation term accounts for 256 pixels and each pixel intensity lies in the interval $[0, 255]$, a high value of fitness implies high image matching. Yet, moderate values of fitness also imply good matching.

6 Experiments

After some trials we found that a (100 + 150)-ES finds solutions with average approximation error no greater than 10^{-4} (on each objective variable), in average time of 200 seconds. A (250 + 50)-ES finds solutions with average approximation error no greater than 10^{-2}, in average time of 30 seconds. For all experiments reported in the following sections, the population parameters are: $\mu = 250$, and $\lambda = 50$. Since to each objective variable corresponds one control parameter, an individual is composed by six control variables (Equation 3), and six objective variables. Control parameters (variance) for matrix coefficients are $\sigma_{1,1} = \sigma_{1,2} = \sigma_{2,1} = \sigma_{2,2} \geq 0.1$, and for translation coefficients $\sigma_{1,3} = \sigma_{2,3} \geq 1.0$ Every $\sigma < 5.0$ We used a PC computer with Xeon processor running at 1.7 Ghertz with 1 Gbyte of memory; all algorithms are programmed in C++. In the sequel we describe three experiments.
The goal of the first experiment is to verify the ability of the method to consistently reach the optimum and to measure the error. In the second experiment we contrast overall convergence. Two of the three test images are known to be hard to register: a slice of MRI of the human brain, and an angiogram.

6.1 Robustness and Consistency of the Method

This experiment is designed to measure the approximation error on each of the six parameters. Therefore, a fixed set of affine transformation coefficients (see Equation 3) is randomly generated and used in all 70 runs. The matrix coefficients (that imply rotation, scaling and shearing) are: 0.819684, 0.089627, 0.16434, 0.40437 The coefficients denoting translations over the axes are: 25.532335, and 39.6115 For these experiments we used a MRI image of the brain, as shown in Figure 2.

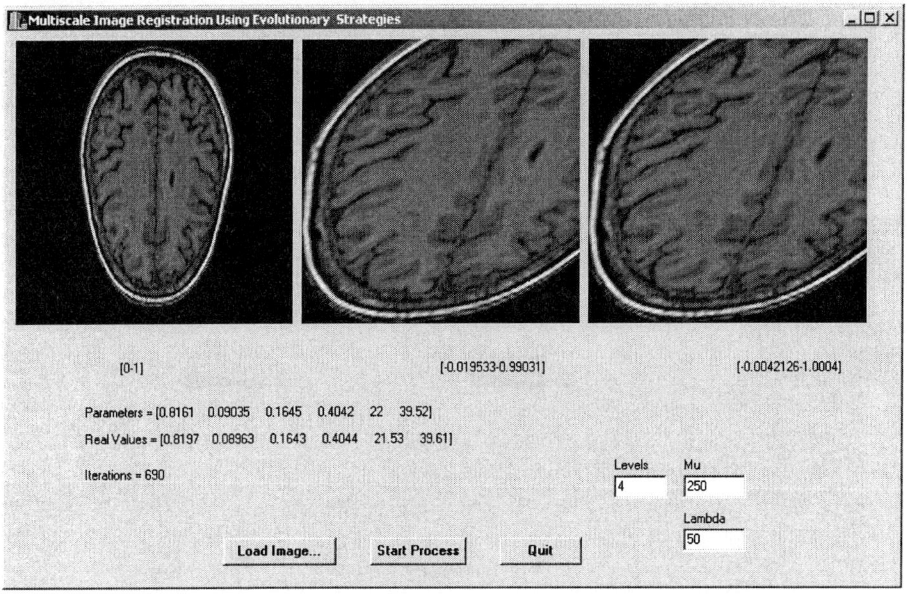

Fig. 2. MRI of brain used in experiment 1

Table 2. Error in transformation parameters by interval

Interval	Measure	**0.8196**	**0.0896**	**0.1643**	**0.4043**	**21.5323**	**39.6115**
0.99-1.0	Avg.	0.8194	0.0904	0.1649	0.4049	21.4508	39.4623
0.99-1.0	Std.Dev.	0.0031	0.0041	0.0025	0.0010	0.9496	0.3939
0.98-1.0	Avg.	0.8190	0.0884	0.1660	0.4050	21.7839	36.6115
0.98-1.0	Std.Dev.	.0031	.0064	.0038	.0010	1.2821	0.5188
0.97-1.0	Avg.	0.8171	0.0830	0.1694	0.4052	22.9012	38.9512
0.97-1.0	Std.Dev.	0.0045	0.0113	0.0072	0.0010	2.3640	0.8696
0.96-1.0	Avg.	0.8164	0.0825	0.1701	0.4050	23.1073	38.8960
0.96-1.0	Std.Dev.	0.0056	0.0126	0.0079	0.0022	2.7110	0.9956
0.95-1.0	Avg.	0.8154	0.0869	0.1703	0.4038	22.4673	39.1022
0.95-1.0	Std.Dev.	0.0063	0.0261	0.0093	0.0057	4.3408	1.5174
0.94-1.0	Avg.	0.7792	0.0803	0.1886	0.4032	27.9592	37.1533
0.94-1.0	Std.Dev.	0.1547	0.0495	0.0491	0.0167	15.1011	4.6558

In Table 2 we resume the robustness and consistency ability of the method to reach the highest fitness value (1.0). Out of 70 runs:

- 24 runs reached $fitness \geq 0.99$
- 29 runs reached $fitness \geq 0.98$
- 44 runs reached $fitness \geq 0.97$
- 47 runs reached $fitness \geq 0.96$
- 52 runs reached $fitness \geq 0.95$
- 63 runs reached $fitness \geq 0.94$

Table 3. Convergence in 20 random experiments for two techniques

Test image	Multiscale Evolutionary	Gauss-Newton
MRI-Brain	19/20	11/20
Angiogram	20/20	6/20
Diana	17/20	1/20

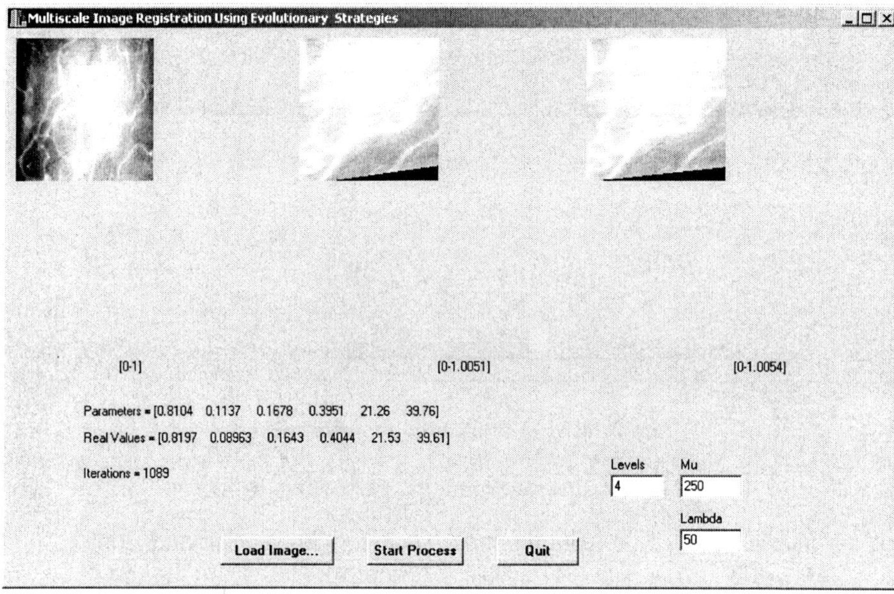

Fig. 3. Evolutionary multiscale registration of an angiogram

The rest of the experiments (7) had $fitness < 0.94$ Information in Table 2 shows that for fitness value of 0.94 the coefficients begin to deteriorate. To the human eye this level of error is not apparent, yet the data for that row indicates that the translation coefficients differ in six pixels (average). Thus, for the experiments of the next section we take a fitness value of up to 0.95 as the minimum required to indicate "good registration".

6.2 Overall Convergence Experiments

In this set of experiments we ran 20 random registrations with each one of the test images: MRI-brain (Figure 2), Diana (Figure 4), and angiogram (Figure 3), and checked for convergence. As explained before, if fitness reaches 0.95 the registration is counted as good. The same problems were also registered using the Gauss-Newton optimization algorithm. In this gradient descent technique we also used the same procedure and fitness function. Table 3 shows the number of successful runs (convergence) of each algorithm for every image.

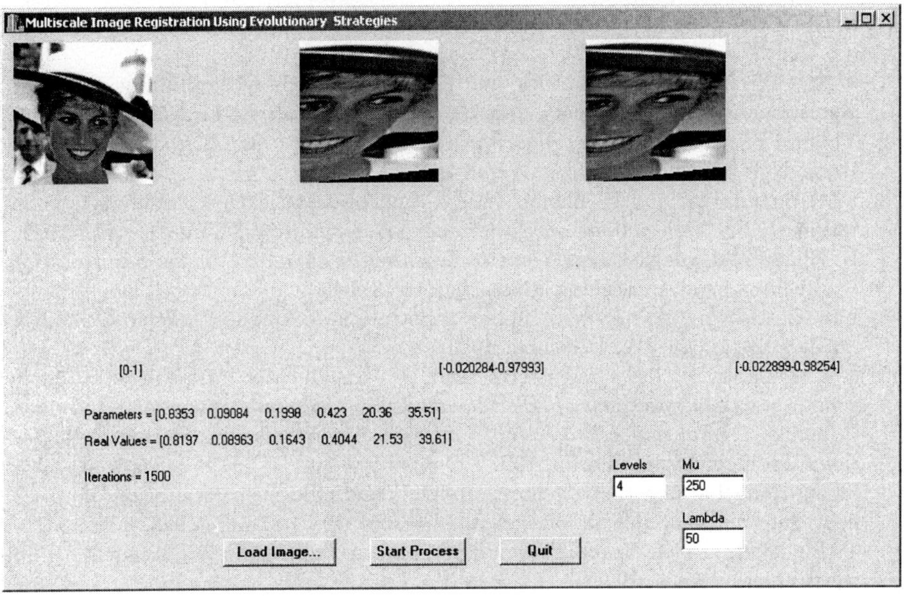

Fig. 4. Evolutionary multiscale registration of "Diana"

7 Discussion and Conclusions

From the first set of experiments we have shown our method consistently finds the solutions with good accuracy. The second set of experiments is a clear proof of the robustness of the method. The Gauss-Newton based method, as any other based on gradient descent, is prone to fall in local minima. The evolutionary approach is a strategy that shares global knowledge among the individuals of the population, thus convergence to the solution occurs with high probability.

In general our approach contradicts several authors [5] who have found weak properties in evolutionary methods for image registration.

Future Work. A fitness function based on normalized cross-correlation (and other approaches) will be studied. Other problem worth of studying is the registration of images with noise. The combination of non-linear transforms with evolutionary techniques is a promising approach to image registration.

Acknowledgments

The first two authors acknowledge support from CONACyT through a scholarship to complete the Master in Science program at CIMAT. The third author acknowledges partial support from CONCyTEG project No. 01-02-202-111 and CONACyT No. I-39324-A. The fourth author acknowledges partial support from CONACyT project No. 34575-A. The last author acknowledges support from CONACyT project No. NSF-CONACyT 32999-A.

References

1. C.A. Ankenbrandt, B.P. Buckles, and F.E. Petry. Scene recognition using genetic algorithms with semantic nets. *Pattern Recognition Letters*, 11:285–293, 1990.
2. Thomas Back. *Evolutionary Algorithms in Theory and Practice.* Oxford University Press, New York, 1996.
3. S.M. Bhandarkar and H. Zhang. Image segmentation using evolutionary computation. *IEEE Transactions on Evolutionary Computation*, 3(1):1–21, April 1999.
4. B. Bhanu and Sungkee Lee. *Genetic Learning for Adaptive Image Segmentation.* Kluwer Academic Publishers, Massachusetts, 1994.
5. Lisa G. Brown. A survey of image registration techniques. *ACM Computing Surveys*, 24(4):325–376, December 1992.
6. J.M. Fitzpatrick, J.J. Grefenstette, and D. Van Gucht. Image registration by genetic search. *Proceedings of IEEE Southeastern Conference*, pages 460–464, 1984.
7. B. Jahne. *Digital Image Processing: Concepts, Algorithms, and Scientific Applications.* Springer Verlag, Berlin, 1997.
8. Shang-Hong Lai and B. C. Vemuri. Reliable and efficient computation of optical flow. *International Journal of Computer Vision*, 29(2):87–105, 1998.
9. T. Lindeberg. *Scale-Space Theory in Computer Vision.* Kluwer Academic Publishers, Netherlands, 1994.
10. J. Louchet. Using an individual evolution strategy for stereovision. *Genetic Programming and Evolvable Machines*, 2(2):101–109, March 2001.
11. J. Maintz and M. Viergever. A survey of medical image registration. *Journal of Medical Image Analysis*, 2(1):1–36, 1998.
12. S.C. Roberts and D. Howard. Genetic programming for image analysis: Orientation detection. *Proceedings of the GECCO Conference 2000*, pages 651–657, 2000.
13. B.J. Ross, F. Fueten, and D.Y. Yashkir. Edge detection of petrographic images using genetic programming. *Proceedings of the GECCO Conference 2000*, pages 658–665, 2000.
14. J. C. Tilton. Comparison of registration techniques for GOES visible imagery data. *Proceedings of IRW, NASA GSFC*, pages 133–136, 1997.

Synthesizing Graphical Models Employing Explaining Away

Ralf Garionis

University of Dortmund, Department of Computer Science XI
44221 Dortmund, Germany
ralf.garionis@uni-dortmund.de

Abstract Graphical models form a successful probabilistic modeling approach: They encode relationships among a set of random variables and provide a representation for the joint probability distribution over these variables. The advantages of the graphical formalism are its origins in probability theory and graph theory, the structural modularity favoring parallel computations, and its visual appeal. In this paper, we discuss a method for constructing a particular instance of graphical models (the Helmholtz machine) by using an evolutionary approach. Particularly, we focus on the explaining away phenomenon difficult to address but potentially improving a graphical model qualitatively. Additionally, we provide initial simulation results for a case study.

1 Introduction

Analysis and understanding of data typically requires the construction or selection of suitable models explaining the data. In both cases, we have to ask how to search for a model, and, whether or not a model found is a good model. When there is a satisfying amount of theory available covering the field the data is rooted in, answering these questions should not be difficult – this is not the case we are interested in. Instead, we target on creating graphical models that provide an explanation of the input data by means of (neural) representations learned.

For doing so, we will use a neural network architecture, the Helmholtz machine [1], gaining its probabilistic interpretation from the two graphical models it is build upon (see section 2). We will address the task of qualitatively specifying the Helmholtz machine's graphical models by using an evolutionary approach tailored to our needs laid out in section 3. Though we are interested in finding architectures of graphical models explaining our input data, we do not target on identifying a certain *layered* graph. Instead, we are especially interested in less symmetric graphical models involving a phenomenon known as *explaining away* [2] (cf. section 1.1). Explaining away has the potential of improving the quality of probabilistic models at the cost of being difficult to access by means of explicit modeling. Our novel approach addresses the automatic integration of explaining away for improving the quality of a graphical model in the context of *unsupervised learning* focusing on representational learning, *ie* transforming

J.J. Merelo Guervós et al. (Eds.): PPSN VII, LNCS 2439, pp. 749–758, 2002.

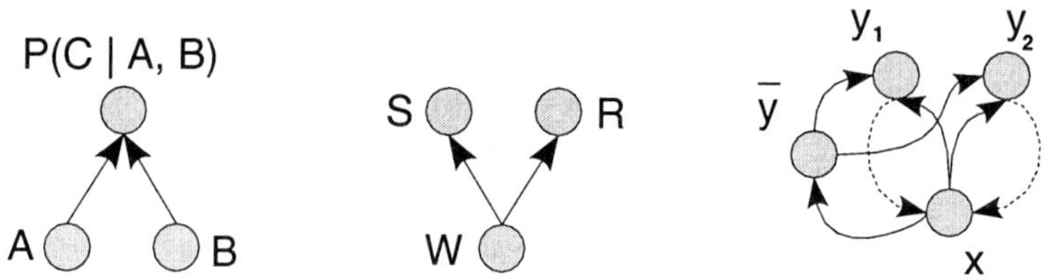

Fig. 1. Connection structures of graphical models. *Left:* Standard structure: A and B are marginally independent, while they are conditionally dependent. *Middle:* Explaining away – S and R are marginally dependent and conditionally independent. *Right:* An implementation of explaining away for the Helmholtz machine by using a dangling unit $\bar{y}$ [3] (recognition connections are shown solid, generative connections dashed): When $\bar{y}$ learns to use the generative weights of y_1 and y_2 (since it does not have its own) it will likely create non-identical connection strengths' (weights) towards y_1 and y_2.

data into neural representations. In section 3.3, we will address how unsupervised learning can find qualitatively and quantitatively good graphical models in terms of representations learned.

1.1 Graphical Models

Graphical models are based on graphs integrating probabilistic variables as nodes with the graph's edges encoding dependencies among these variables [2,4,5,6] (we will focus on *directed acyclic graphs* here). This permits graphical models representing the joint probability distribution over a set of random variables in a way easily accessible to theoretic concepts as well as to visual imagination. Well known synonyms for graphical model are, *eg*, Bayesian networks and belief networks.

By permitting a solidly founded probabilistic interpretation, graphical models provide a framework for understanding and designing learning algorithms, as, for example, the Helmholtz machine's wake-sleep algorithm [7]. While the local wake-sleep update-rules appear desirable since they permit a *parallel* implementation and enjoy as neural learning rules neurobiological plausibility, local semantics are not mandatory for graphical models – though they emphasize the models' character of modularity. They share the aspect of creating complex systems by combining simple units with neural architectures. The goal of graphical models trained by unsupervised learning is to build some *useful* representations of the input data patterns, targeting to finding the hidden causes of the data or to model the data density, *eg*. In case of the Helmholtz machine considered here, we are interested in finding compact representations, *ie* representations that are associated with low coding costs in the sense of the minimum description length (MDL) framework [8]. Then, *useful* may address possible applications like data compression or classification, or studying theories of human perception. Figure 1 explains the essential tool that probabilistic reasoning in the presence of uncertainty comes with: *inference*.

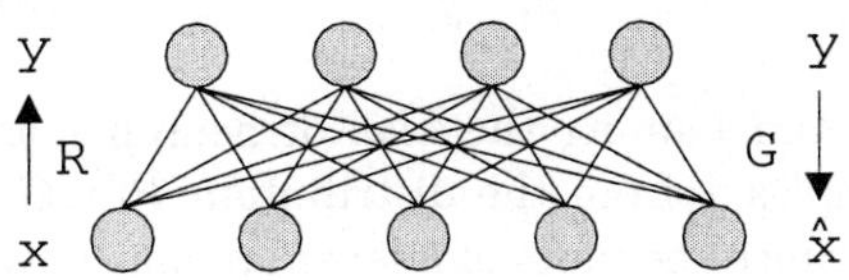

Fig. 2. The Helmholtz machine's two independent graphical models: The bottom-up directed graph is part of the recognition model R, the top-down graph constitutes the generative model G.

1.2 Explaining Away

Explaining away [2] refers to a phenomenon emerging when inferring in graphical models. It allows an explanation of a particular observed event having a low posterior probability if another explanation for the same observation is preferred by the observation – *ie*, the causes compete to explain the observations. For example, when recognizing the event that the grass is wet (W), this may have two possible causes: the sprinkler is on (S) or it is raining (R) [9] (fig. 1). Now, the confirmation of one cause decreases the need to bring up an alternative explanation, *eg*, knowing that it rains reduces the posterior probability that the sprinkler is on. (Note that the cause confirmed *first* "wins".)

In case of the coupled graphical models for the recognition and the generative model of the Helmholtz machine, we can implement this *inhibitory* effect by omitting connections within the generative model among units that remain connected by the recognition model [3] (figure 1, right).

The *inhibition* between two nodes having the same parent node is an important effect for graphical models structured as *directed acyclic graph*, since they do not permit *lateral* connections, that is, connections among child nodes connected to a particular node.

2 The Helmholtz Machine

Designed to form hierarchical representations, the Helmholtz machine [1] is composed of two separate graphical models realizing a forward directed recognition model and a backward directed generative model (though figure 1 shows only one hidden layer y; successive units may be added). While learning, each model is adjusted to improve the likelihood of the observations produced by the other model. Which, for the case of binary stochastic neurons considered here, is performed for both models in alternate succession by the wake-sleep algorithm [7].

For the simple case of a single layered Helmholtz machine, the generative model is parameterized by the weights G while the units inside a specific layer are always assumed to be independent and therefore implement a factorial probability distribution P

$$P(y;G) = \prod_i P(y_i;G) = \prod_i p_i^{y_i}(1-p_i)^{1-y_i} \tag{1}$$

$$P(x|y;G) = \prod_j P(x_j|y;G) = \prod_j p_j^{x_j}(1-p_j)^{1-x_j} \tag{2}$$

with

$$p_i = \sigma(g_i^0) = \frac{1}{1+e^{g_i^0}}\ ; \ p_j = \sigma(g_j^0 + g_j^T \cdot y) \tag{3}$$

using the standard logistic function $\sigma(\cdot)$ and naming the bias of neuron i as g_i^0. The generative model synthesizes samples $\hat{x}$ from the distribution $P(x;G)$ approximating the true distribution $P(x)$. Since the units of a layer are mutually independent and only conditionally depend on the preceding layer's activations (equation 2), the generative model can be run by traversing the layers in top-down order starting with the bias neurons and finishing with the input neurons then fantasizing a sample $\hat{x}$.

Directed bottom-up, the recognition model takes an input pattern x and creates a representational vector y exhibiting possible causes that could be responsible for creating x. The recognition model comes with its own set of parameters R forming another factorial probability distribution Q

$$Q(y|x;R) = \prod_j Q(y_j|x;R) = \prod_j q_j^{y_j}(1-q_j)^{1-y_j} \tag{4}$$

with

$$q_j = \sigma(r_j^0 + r_j^T \cdot x)$$

assuming mutually independent units within a layer. Though independence of the y_i is a valid assumption here, the generative model does not rely on this as it permits the use of *any* vector y (eq. 2). Therefore, Q approximates P.

Focusing on maximum likelihood density estimation, the wake-sleep learning algorithm attempts to maximize the log probability of the data observed under the generative model [3]:

$$\log P(x;G) = \sum_y P(x|y;G)\log P(x|y;G) + H\big(P(x|y;G)\big) \tag{5}$$

$$= \sum_y Q(y|x;R)\log P(y|x;R) + H\big[Q(y|x;R)\big]$$

$$+ D\big[Q(y|x;R)||P(x|y;G)\big] \tag{6}$$

with

$$D(Q||R) = Q\log\frac{Q}{P} \tag{7}$$

naming the Kullback Leibler distance measuring the discrepancy between two distributions and $H(\cdot)$ being the Shannon entropy. According to equation (6), Q approximates P and becomes identical to the original recognition distribution for vanishing $D(Q||P)$.

A maximization of $logP(x;G)$ with respect to G can be achieved by keeping R fixed and running the recognition model starting with an input pattern x producing a sample y. Then changing G proportionally to Δg

$$\Delta g_{ij} \propto \frac{\partial}{\partial G_{ij}} \log P(x|y;G) = (x_j - p_j)y_i \tag{8}$$

achieves stochastic gradient ascent in the log probability. This is called the *wake* phase. During the *sleep* phase the generative model predicts activities of the input units for subsequently adapting R according to

$$\Delta r_{ij} \propto \frac{\partial}{\partial R_{ij}} \log Q(y|x;G) = (y_j - q_j)x_i \qquad (9)$$

Instead of maximizing the Kullback Leibler distance in equation (6), this procedure optimizes $D(P||Q)$ for tractability of computations. This is an approximation due to the lack of the Kullback Leibler distance's symmetry $D(P||Q) \neq D(Q||P)$ for the benefit of resulting in symmetric learning rules for both models.

Now, wake-sleep learning uses the output of the recognition model as target for adapting the generative model and the output of the generative model as reference for the recognition model.

3 Evolving Graphical Models with Explaining Away

The evolution of neural architectures is addressed by a vast amount of literature: The extensive overview of Yao [10] cites more than 80 references in this area. Typically, these approaches deal with feedforward networks or recurrent structures and do not support coupled asymmetric constructions like the Helmholtz machine with explaining away connections. An adaptation of the various methods developed to our constraints is mostly not possible in a straightforward way.

Our design limitation is the derivation of the generative model from the recognition model, *ie*, both connection matrices are symmetric with the generative matrix having particular connections removed being present in the forward-directed recognition network. Therefore it is more helpful to use a genotype encoding scheme permitting a direct mapping of the recognition model's connections into the generative connection matrix. This makes parametric representations finding their origins in the work of Harp *et al* [11] appearing less useful. An evolutionary algorithm manipulating the genotype representation of the Helmholtz machine has to learn the relationship between the Helmholtz machine's two connection matrices, *ie* how the the generative connection matrix is derived from the recognition model. A complex mapping between these matrices weakening causality would take more time for the evolution to discover. Our work presented here is based on the evolution of context-free grammars introduced by Kitano [12]. The advantages of Kitano's suggestions are the compact genotype representation and the flexible semantics of the connectivity matrix evolved.

3.1 Evolving Structural Semantics of Explaining Away

Kitano coupled a matrix rewriting scheme based on context-free grammars with a genetic algorithm [14] (GA): While the connectivity matrix of a graph is developed by the matrix rewriting grammar, the genetic algorithm uses the binary description of the grammar as chromosomes to evolve. Formally, the matrix rewriting scheme is closely related to standard context-free grammars: Let the

$$S = \begin{bmatrix} B & B \\ A & B \end{bmatrix} \xrightarrow{P_N} \begin{bmatrix} B & A & B & A \\ A & B & A & B \\ A & B & B & A \\ A & A & A & B \end{bmatrix} \xrightarrow{P_\Sigma} \begin{bmatrix} 0&0&1&0&0&0&1&0 \\ 1&0&0&1&1&0&0&1 \\ 1&0&0&0&1&0&0&0 \\ 0&1&1&0&0&1&1&0 \\ 1&0&0&0&0&0&1&0 \\ 0&1&1&0&1&0&0&1 \\ 1&0&1&0&1&0&0&0 \\ 0&1&0&1&0&1&1&0 \end{bmatrix}$$

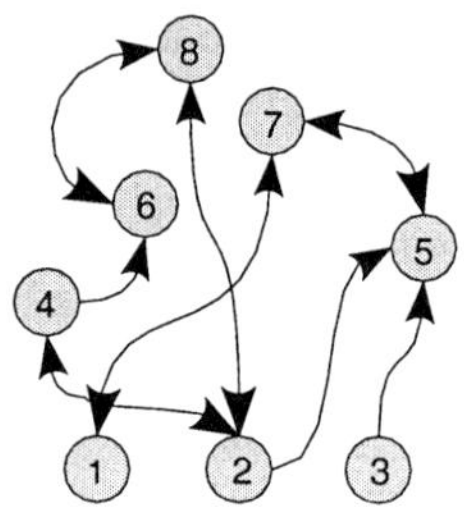

Fig. 3. Rewriting of the matrix grammar defined by equation (10) and (11). *Left:* The start symbol S is rewritten once by P_N and then by P_Σ for termination. *Right:* Helmholtz machine created by the connection matrix shown left – connections with two arrows are present in recognition and generative model (see text for details).

tuple $G = (N, \Sigma, P_N, \Sigma_N, S)$ define a context-free grammar G, where N is a finite set (the *nonterminal symbols*), Σ is the set of *terminal symbols* $\{0,1\}^4$ disjoint from N, P_N is a finite subset of $N \times N^4$ (the *productions*), Σ_N is the set of $N \times \Sigma$, and $S \in N^4$ is the starts symbol.

Obviously, this differs from a classical context-free grammar: For each individual of the GA population, the P_N are used for a constant number of rewriting operations creating a string that is rewritten once by the rules from P_Σ. Additionally, the productions define mappings from symbols to 2×2 matrices, For example, suppose the grammar $G = (N, \Sigma, P_N, \Sigma_N, S)$ is defined by:

$$N = \{A, B\}, \quad \Sigma = \left\{ \begin{bmatrix} 1 & 0 \\ 0 & 1 \end{bmatrix}, \begin{bmatrix} 0 & 0 \\ 1 & 0 \end{bmatrix} \right\}, \quad S = \begin{bmatrix} B & B \\ A & B \end{bmatrix} \tag{10}$$

$$P_N = \left\{ A \rightarrow \begin{bmatrix} A & B \\ A & A \end{bmatrix}, B \rightarrow \begin{bmatrix} B & A \\ A & B \end{bmatrix} \right\} \quad P_\Sigma = \left\{ A \rightarrow \begin{bmatrix} 1 & 0 \\ 0 & 1 \end{bmatrix}, B \rightarrow \begin{bmatrix} 0 & 0 \\ 1 & 0 \end{bmatrix} \right\} \tag{11}$$

Then, one rewriting of S by P_N and a successive rewriting operation performed by P_Σ results in the matrix shown in figure 3. Clearly, we have to choose the number of subsequent applications of P_N in advance to get connection matrices of limited size.

For converting the binary matrix gained by the context-free grammar into the Helmholtz machine's architecture permitting the implementation of explaining away, we have used the following approach: The upper triangular matrix of the connectivity matrix C (created by matrix rewriting) defines the recognition model, *ie* each row assigns connections to a particular unit. The input units correspond to the first m rows of the recognition matrix when considering input data m bits wide. The generative model's connections G are initialized with the transposed recognition matrix. By using elements C_{ji} from the lower triangular matrix of C we remove connections from G if a connection C_{ij} is present in the upper triangular matrix:

$$G_{ij} = \begin{cases} 0 & \text{if } C_{ij} = 1 \text{ and } C_{ji} = 0, \\ 1 & \text{otherwise.} \end{cases} \tag{12}$$

Figure 3 shows an example architecture gained by the matrix on the left side by applying the semantics described above. Our construction scheme does not require the specification of output units since these are defined as the Helmholtz machine's units without outgoing connections.

3.2 Encoding of Chromosomes

We evolve the context-free grammar as binary representation of the set of non-terminal symbols P_N: each right hand side symbol of a production from P_N is encoded by four integer values. Kitano (as quoted by [15]) originally permitted the evolution of the left hand side of the productions as well, at the cost of requiring a repair mechanism when invalid rules were created.

The size of the set P_N may be chosen arbitrary, but for avoiding the GA to generate chromosomes not mappable to phenotypes, we recommend a bijective mapping between genotypes and phenotypes resulting in alphabet sizes being a multiple of the number 2^4 of binary 2×2 matrices. Contrary to the symbols of P_N, the set of productions P_Σ mapping the non-terminals into a maximum of 2^4 terminals is pre-encoded and not changed during evolution.

3.3 Good Models

In section 3 we described how we search for a *good* model of data. Still remains the task of deciding whether or not a model found is good.

The Helmholtz machine's coding costs (this is the difference of Helmholtz free energy of the recognition model and the entropy of the generative model), as minimized by the wake-sleep learning algorithm, is an *indicator* for a proper model. A low Helmholtz energy describes the coding complexity of the data in the sense of the minimum description length framework – given the model structure used. But the recognition and generative models could encode everything but relevant aspects of the input data resulting in a low energy measure. As by definition of the wake-sleep algorithm, a valid recognition model is accompanied by a generative model being able to reproduce patterns from the world of the training data. Therefore, while using the wake-sleep learning algorithm, we measure its success by the closeness of the patterns the generative model has dreamed to the original data (note that the Helmholtz machine's generative model is activated by the bias and not by some test pattern). But the reproduction of patterns similar to those from the world of training patterns is not a sufficient criteria for a valid model: The generative network may reproduce only a single pattern. Therefore we compare the distribution of the dreamt patterns to the distribution of input patterns by calculating the Kullback Leibler distance $D(i||d) = \sum_{x \in X} d(x) \log \frac{i(x)}{d(\hat{x})}$ between the distribution of input patterns $i(x)$ and the distribution of dreamt patterns $d(\hat{x})$. $\hat{x}$ indicates a generated pattern having minimum Hamming distance to pattern x.

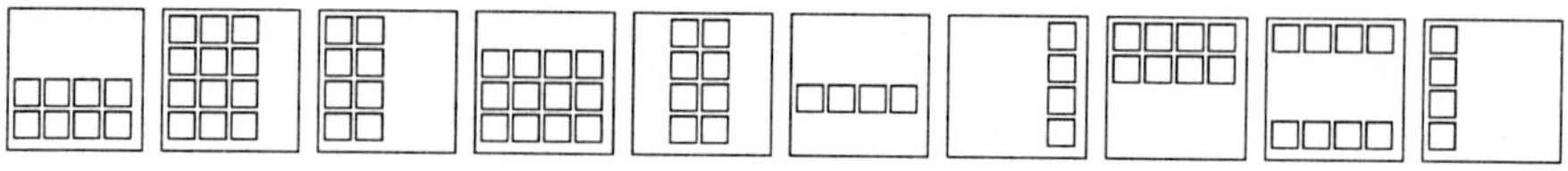

Fig. 4. Examples from the binary bars training set. Matrices are produces by first randomly choosing an orientation (horizontal or vertical) with equal probability. Then, bars of the chosen orientation are randomly created with probability 0.5.

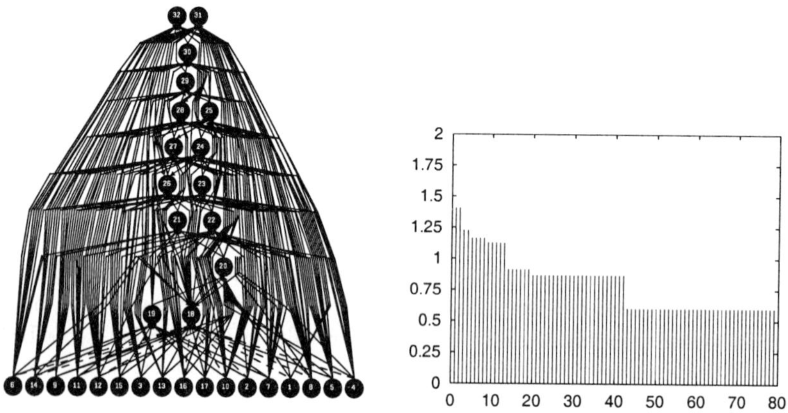

Fig. 5. *Left:* Best evolved individual (Helmholtz machine). Connections that are shared by recognition and generative models are shown solid, explaining away connections are drawn dashed. *Right:* The fitness values used for assessing the evolved Helmholtz machines are in fact products composed of three factors: a) the coding costs, b) fraction of patterns created by the generative model having Hamming distance 0 or 1 to one of the original input patterns, and c) Kullback Leibler distance between input patterns and patterns fantasized by the generative model.

4 Simulations

Along the framework laid out in the preceding sections, our case study aimed at finding good (according to section 3.3) Helmholtz machines for learning compact representations of patterns rooted in a structured world: The training data was composed of binary input vectors of length 16 [7]. Each vector represents a 4×4 matrix where elements were "switched on" (*ie*, set to "1") in groups constituting a bar (see figure 4).

The training was performed by presenting 1000 patterns randomly chosen from a set of 30 unique patterns (created as described above) and updating the recognition and generative weights by the wake-sleep algorithm afterwards (using a learning rate of 0.1 for all units of the network). Learning was completed past ten weight updates (it is possible to improve training results by increasing the number of training cycles). For each Helmholtz machine evolved, we checked the modeling of input data by running the generative model 1000 times activating the input units creating fantasized patterns. When the trained Helmholtz machine has captured (*ie*, successfully learned) the essential structure of the input data, these fantasized patterns are identical with the original training patterns (fig. 4) or very similar to these in their structure. We measured the similarity of pattern sets by the Kullback-Leibler distance (see section 3.3 and figure 5).

The Helmholtz machine's graphical models were evolved by using a genetic algorithm employing a binary representation of the context-free grammar (see section 3.1) having 16 productions P_N and P_Σ. Each symbol N was represented by a binary vector $\{0,1\}^4$ such that a production P_N performing a mapping of symbols into 2×2 matrices of symbols can be encoded by 16 bits resulting in a total of 256 (16×16) bits needed to encode P_N. Therefore, our GA used a "flat" binary vector representation permitting the use of standard crossover and mutation operators.

The GA's population size was 15 individuals creating 100 offsprings. Similar to the (+) selection scheme of a $(\mu + \lambda)$ evolution strategy [16], the sets of parents (μ individuals) and children (λ individuals) were merged, for then placing the μ best individuals (according to their fitness values) of the joint set in the parent population of the next generation. The fitness values for judging about the quality of an evolved Helmholtz machine were products composed of three factors (see figure 5). We used a mutation rate of 0.1 while the results did not depend significantly on the crossover type used (one-point, two-point, uniform).

5 Discussion

In this paper, we have developed a scheme for determining the structure of coupled graphical models by modifying an approach known to support standard feedforward network architectures. In contrast to this, our scheme permits explicit addressing of explaining away connections and asymmetric bidirectional connection structures as used by the Helmholtz machine. While it is simple to evaluate *supervised* trained network structures in terms of their quality, this is much more difficult for networks trained in *unsupervised* fashion. Here, we have provided measures for assessing the quality of graphical models that were subject to *unsupervised* training, resulting in fitness values permitting us to drive an evolutionary optimization process.

First experiments revealed that the evolutionary search for models qualified as "good" according to section 3.3 stabilizes fast (fig. 5, there was no improvement beyond generation 80). The coding costs of 5.1 bits of the models found are close to the optimum of the $\log_2 30 = 4.9$ bits required to encode the 30 training patterns. The best Helmholtz machine as shown in fig. 5 uses 16 connections of the recognition model that are not matched by connections of the generative model. These are connections needed by explaining away. It is difficult to judge about the impact of these connections due to the high connectivity of many network units. We will have to carry out further simulations favoring more sparsely connected structures for recognizing the influence of explaining away. Verification of the importance of particular recognition connections could be performed by matching them by additional generative connections – when the fitness decreases, the explaining away connections are important. However, the results show, that our scheme successfully creates coupled graphical models (Helmholtz machines) with asymmetric connection structure and good coding complexity – further simulations have to show, whether this is the result of the presence of explaining away.

Acknowledgments

We are grateful to René Hoferichter for software support. The author is with the *Collaborative Research Center "Computational Intelligence" (SFB 531)* funded by the Deutsche Forschungsgemeinschaft (DFG).

References

1. Dayan, P., Hinton, G.E., Zemel, R.S.: The Helmholtz machine. Neural Computation **7** (1995) 889–904
2. Pearl, J.: Probabilistic Reasining in Intelligent Systems: Networks of Plausible Inference. Morgan Kaufmann, San Mateo, CA (1988)
3. Dayan, P., Hinton, G.E.: Varieties of Helmholtz machines. Neural Networks **9** (1996) 1385–1403
4. Whittaker, J.: Graphical Models in Applied Multivariate Statistics. Wiley, Chichester (1990)
5. Lauritzen, S.L.: Graphical Models. Clarendon Press, Oxford (1996)
6. Jordan, M.I., ed.: Learning in Graphical Models. MIT Press, Cambridge, MA (1999)
7. Hinton, G.E., Dayan, P., Frey, B.J., Neal, R.M.: The wake-sleep algorithm for unsupervised neural networks. Science (1995) 1158–1161
8. Rissanen, J.: Stochastic Complexity in Statistical Inquiry. Volume 15 of Computer Science. World Scientific, Singapore (1989)
9. Pearl, J.: Embracing causality in default reasoning. Artificial Intelligence **35** (1988) 259–271
10. Yao, X.: Evolving artificial neural networks. Proc. IEEE **87** (1999) 1423–1447
11. Harp, S.A., Samad, T., Guha, A.: Towards the genetic synthesis of neural networks. In Shaffer, J.D., ed.: Proceedings of the Third International Conference on Genetic Algorithms (ICGA), San Mateo, CA, Morgan Kaufmann (1989) 360–369
12. Kitano, H.: Designing neural networks using genetic algorithms with graph generation system. Complex Systems **4** (1990) 461–476
13. Holland, J.H.: Adaptation in Natural and Artificial Systems. The University of Michigan Press, Ann Arbor, MI (1975)
14. Goldberg, D.E.: Genetic Algorithms in Search, Optimization and Machine Learning. Addison Wesley, Reading, MA (1989)
15. Siddiqi, A., Lucas, S.: A comparison of matrix rewriting versus direct encoding for evolving neural networks. In Fogel, D.B., Schwefel, H.P., Bäck, T., Yao, X., eds.: Proc. Second IEEE World Congress on Computational Intelligence (WCCI'98) with Fifth IEEE Conf. Evolutionary Computation (IEEE/ICEC'98), Piscataway, NJ, IEEE Press (1998) 392–397
16. Bäck, T., Schwefel, H.P.: An overview of evolutionary algorithms for parameter optimization. Evolutionary Computation **1** (1993) 1–23

Constructive Geometric Constraint Solving: A New Application of Genetic Algorithms

R. Joan-Arinyo[1], M.V. Luzón[2], and A. Soto[1]

[1] Departament de Llenguatges i Sistemes Informàtics
Universitat Politècnica de Catalunya, Av. Diagonal 647, 8^a, E-08028 Barcelona
{robert,tonis}@lsi.upc.es
[2] Escuela Superior de Ingeniería Informática
Universidad de Vigo, Av. As Lagoas s/n, E-32004 Ourense
{vluzon@ei.uvigo.es}

Abstract. Geometric problems defined by constraints have an exponential number of solution instances in the number of geometric elements involved. Generally, the user is only interested in one instance such that besides fulfilling the geometric constraints, exhibits some addicional properties. Selecting a solution instance amounts to selecting a given root everytime the geometric constraint solver needs to compute the zeros of a multivaluated function. The problem of selecting a given root is known as the *Root Identification Problem.*
In this paper we present a new technique to solve the root identification problem based on an automatic search in the space of solutions performed by a genetic algorithm. The user specifies the solution of interest by defining a set of additional constraints on the geometric elements which drive the search of the genetic algorithm. Some examples illustrate the performance of the method.

1 Introduction

Geometric problems defined by constraints have an exponential number of solution instances in the number of geometric elements involved. Generally, the user is only interested in one instance such that besides fulfilling the geometric constraints, exhibits some additional properties.

Selecting a solution instance amounts to selecting one among a number of different roots of a nonlinear equation or system of equations. The problem of selecting a given root was christianized in [4] as the *Root Identification Problem.*

Several approaches to solve the root identification problem have been reported in the literature. Examples are: selectively moving the geometric elements, conducting a dialogue with the constraint solver that identifies interactively the intended solution, and preserving the topology of the sketch input by the user. For a discussion of these approaches see, for example, references [4, 15] and the references therein.

Adding extra constraints to narrow down the number of possible solutions of geometric problems seems to be a simple approach. However, this approach

J.J. Merelo Guervós et al. (Eds.): PPSN VII, LNCS 2439, pp. 759–768, 2002.

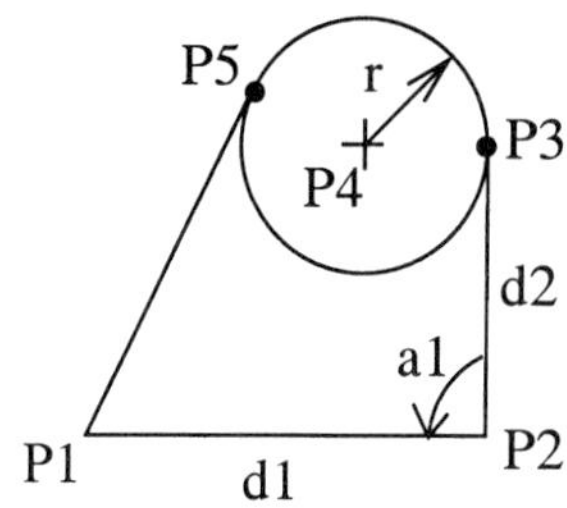

distance(P1, P2) = d1
distance(P2, P3) = d2
angle(segment(P2, P3), segment(P1, P2)) = a1
tangent(segment(P2, P3), circle(P4, r)
on(P3, circle(P4, r))
tangent(segment(P1, P5), circle(P4, r))
on(P5, circle(P4, r))

Fig. 1. Geometric problem defined by constraints.

has been carefully avoided by the field because the resulting over constrained problem is NP hard. Moreover, the set of constraints may be contradictory, [4].

Genetic algorithms have been already applied to a number of global optimization problems. See for example [20] and [6] and references therein. In this paper, for the first time, we apply a genetic algorithm in a new field: Geometric constraint solving. The goal is to solve the root identification problem which can be seen as a constrained optimization problem. The technique is based on an automatic search in the space of solutions performed by a genetic algorithm. The user specifies the intended solution instance by defining a set of additional constraints or predicates on the geometric elements which drive the search of the genetic algorithm. The approach has been implemented and the results are satisfactory, [15].

The outline of the paper is as follows. In Section 2 we briefly review the basic concepts of constructive geometric constraint solving. Section 3 is devoted to the genetic algorithm. As a proof of concept, we present in Section 4 some experimental results. Finally, Section 5 offers a summary and open questions for future work.

2 Constructive Geometric Constraint Solving

In two-dimensional constraint-based geometric design, the designer creates a rough sketch of an object made out of simple geometric elements like points, lines, circles and arcs of circle. Then the intended exact shape is specified by annotating the sketch with constraints like distance between two points, distance from a point to a line, angle between two lines, line-circle tangency and so on. A geometric constraint solver then checks whether the set of geometric constraints coherently defines the object and, if so, determines the position of the geometric elements.

If geometric elements and constraints are like those above, a constraint-based design can be represented by a set of points along with a set of constraints drawn from distance between two points, distance from a point to a line, and angle between two lines, [16]. Figure 1 shows an sketch example of a constraint-based design.

Many techniques have been reported in the literature that provide powerful and efficient methods for solving systems of geometric constraints. For example,

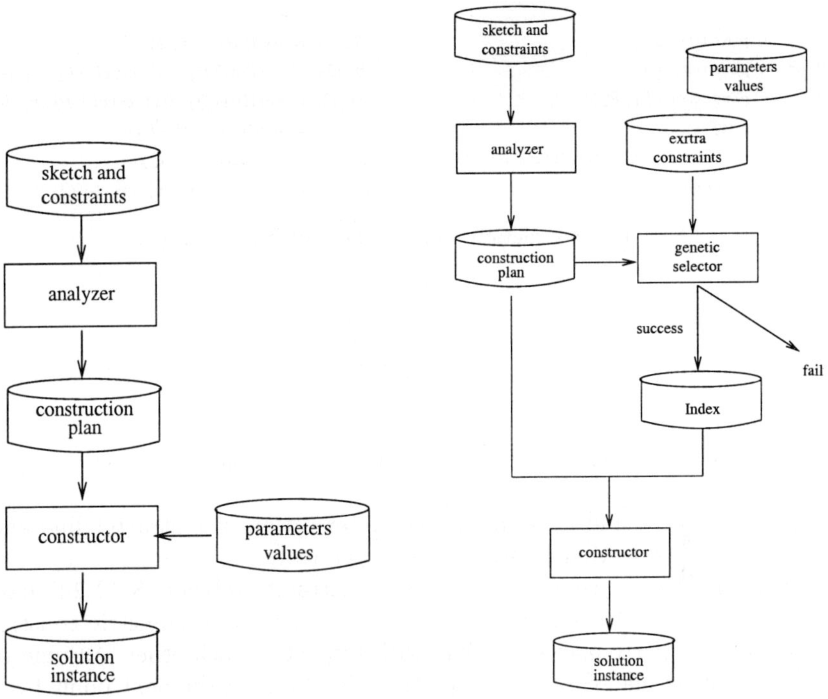

Fig. 2. Left) Basic architecture of constructive geometric constraint solvers. Right) Integrating the genetic algorithm into the solver.

see [5] and references therein for an extensive analysis of work on constraint solving. Among all the geometric constraint solving techniques, our interest focuses on the one known as *constructive.*

Constructive solvers have two major components: the *analyzer* and the *constructor.* The analyzer symbolically determines whether a geometric problem defined by constraints is solvable. If the problem is solvable, the output of the analyzer is a sequence of construction steps, known as the *construction plan*, that places each geometric element in such a way that all constraints are satisfied. After assigning specific values to the parameters, the constructor interprets the construction plan and builds an object instance, provided that no numerical incompatibilities arise. Figure 2 left illustrates the main components in a constructive geometric constraint solver.

The specific construction plan generated by an analyzer depends on the underlying constructive technique and how it is implemented. For example, the ruler-and-compass constructive approach is a well-known technique based on the fact that most useful geometric problems are solvable by ruler, compass and protractor. Figure 3 shows a construction plan for the object of Figure 1, generated by the ruler-and-compass geometric constraint solver reported in [13].

Function names are self explanatory. For example function $adif$ denotes subtracting the second angle from the first one and *asum* denotes the addition of

1. $P_1 = point(0,0)$
2. $P_2 = point(d_1,0)$
3. $\alpha_1 = direction(P_1,P_2)$
4. $\alpha_2 = adif(\alpha_1,a_1)$
5. $P_3 = rc(line(P_2,\alpha_2), circle(P_2,d_2), i_1)$
6. $\alpha_3 = direction(P_2,P_3)$
7. $\alpha_4 = asum(\alpha_3,\pi/2)$
8. $Q_1 = rc(line(P_2,\alpha_4), circle(P_2,r), i_2)$
9. $P_4 = rc(line(Q_1,\alpha_3), circle(P_3,r), i_3)$
10. $Q_2 = midpoint(P_1,P_4)$
11. $r_1 = distance(P_1,Q_2)$
12. $P_5 = cc(circle(P_4,r), circle(Q_2,r_1), i_4)$

Fig. 3. Construction plan for the object in Figure 1.

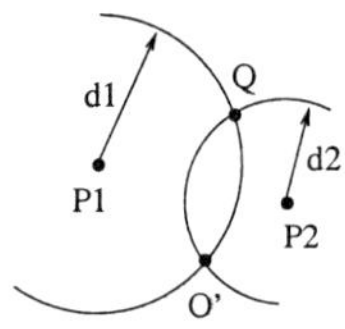

Fig. 4. Possible placements of a point.

two angles while *rc* and *cc* stand for the intersection of a straight line and a circle, and the intersection of two circles respectively.

In general, a well constrained geometric constraint problem, [8, 12, 14], has an exponential number of solutions. For example, consider a geometric constraint problem that properly places n points with respect to each other. Assume that the points can be placed serially, each time determining the next point by two distances from two known points. In general, each point can be placed in two different locations corresponding to the intersection points of two circles. See Figure 4. For n points, therefore, we could have up to 2^{n-2} solutions.

Possible different locations of geometric elements corresponding to different roots of systems of nonlinear algebraic equations can be distinguished by enumerating the roots with an integer index. For a more formal definition see [7, 15].

In what follows, we assume that the set of geometric constraints coherently defines the object under design, that is, the object is generically well constrained and that a ruler-and-compass constructive geometric constraint solver like that reported in [13] is available.

In this solver, intersection operations where circles are involved, *rc* and *cc*, may lead to up to two different intersection points, depending on whether the second degree equation to be solved has no solution, one or two different solutions in the real domain. With each *rc* and *cc* operation, the solver associates an integer parameter, $i_k \in \{-1,0,1\}$, which identifies whether there is no solution, one or two different solutions. For details on how to compute i_k, the reader is referred to [11] and [17].

3 Root Identification

First we define the model used to represent solution instances of the geometric constraint problem. Then we present the genetic algorithm and the genetic selector.

```
Procedure GeneticAlgorithm
  INPUT
    F : Functions in the construction plan.
    C : Values actually assigned to the constraints.
    R : Set of extra constraints.
    ng : Maximum number of generations allowed.
  OUTPUT
    I : Index selected.

  InitializeAtRandom (P)
  Evaluate(P, F, C, R)
  I = SelectCurrentBestFitting (P)
  while not EndCondition (ng, I, R) do
    Selection (P)
    Crossover (P)
    Mutation (P)
    ApplyElitism (P, I)
    Evaluate(P, F, C, R)
    I = SelectCurrentBestFitting (P)
    ng = ng − 1
  endwhile
  return I
EndProcedure
```

Fig. 5. Genetic algorithm.

3.1 The Solution Instance Genetic Model

We are interested only in solution instances that actually are feasible, that is, solution instances where no numerical incompatibilities arise in the constructor. Therefore, we only need to consider integer parameter values i_k in $\{0, 1\}$, where 0 stands for both one solution or two equal solutions, and 1 stands for two different solutions.

Let i_j denote the integer parameter associated by the solver with the j-th intersection operation, either *rc* or *cc*, occurring in the construction plan. We define the *index* associated with the construction plan as the ordered set $I = \{i_1, \ldots, i_j, \ldots, i_n\}$, where n is the total number of *rc* plus *cc* intersection operations in the plan.

A solution instance to the geometric constraint problem is an index I for which the construction plan is feasible. An intended solution instance to the geometric constraint problem is a solution instance for which all the extra constraints hold.

We shall consider populations consisting of a fixed, given number of indexes. Note that the representation of the individuals is binary and that not necessarily each individual is a solution to the geometric constraint problem.

To select the intended solution instance, the type of extra constraint currently available to the user is $PointOnSide(P, line(P_i, P_j), side)$ which specifies that point P must be placed on one of the two open half spaces defined by the straight line through points P_i, P_j, oriented from P_i to P_j. Parameter *side* takes values in {*right, left*}.

3.2 The Genetic Algorithm

The genetic algorithm we have implemented is given in Figure 5. P is the population of indexes at the current generation.

To evaluate the fitness of an index I with respect to the intended solution instance, we have used the function

$$f(I) = \begin{cases} \sum_{i=1}^{|R|} \delta(R_i(I)) & \text{if } I \text{ is a solution instace} \\ MIN & \text{otherwise} \end{cases}$$

where $\delta(R_i(I)) = 1$ if the solution instance associated with index I fulfills the extra constraint R_i and $\delta(R_i(I)) = 0$ otherwise. MIN is the smallest fitness value in the pervious generation. That is, to evaluate an index fitness amounts to counting how many extra constraints its associated solution instance fulfills.

Selection was performed applying linear ranking, [10], and stochastic universal sampling, [2]. Indexes I in the current population including N individuals were sorted according to increasing fitness values, $f(I)$. The rank, $rank(I)$, of the most fit was defined to be 1 and the least fit was defined to be N. The selection probability for an index I was computed by

$$p_s(I) = \frac{1}{N}\left(\mu_{max} - \frac{(\mu_{max} - \mu_{min})(rank(I) - 1)}{N - 1}\right)$$

where $\mu_{min} \in [0, 1]$ is the expected number of offspring to be allocated to the worst index and $\mu_{max} = 2 - \mu_{min}$ is the expected number of offspring to be allocated to the best index in the current generation. This procedure guarantees that the number of copies of any index in the next generation is bounded by the floor and ceiling of the expected number of copies. With the aim of preserving indexes corresponding to good solutions, *elitism* has also been used, [9, 18].

A simple one-point crossover and uniform mutation operations for binary coded populations have been used, [3], Mutation was computed following a simple uniform mutation scheme, [1]. The integer parameter to undergo a mutation, let us say i_j, is selected randomly. Then it mutates into $i'_j = 0$ if $i_j = 1$ and into $i'_j = 1$ otherwise. The algorithm stops when either the current best fitting index corresponds to a solution instance that fulfills all the extra constraints or the number of generations reaches a given maximum number.

3.3 The Genetic Selector

The genetic algorithm is integrated into the constructive solver showed in Figure 2 left through a *genetic selector* as illustrated in Figure 2 right.

As required by the genetic algorithm, the input to the genetic selector includes the construction plan, the set of parameters' values and the set of extra constraints.

The genetic algorithm always returns an index corresponding to the individual in the population showing the best fitness. Three different outputs from the genetic selector need to be distinguished. One output can be an index for which the construction plan is feasible and all the extra constraints hold. In this case an intended solution instance has been found. Notice however that this intended solution is not necessarily unique.

Another output can be an index for which the construction plan is feasible but only a subset of the extra constraints hold. In this case, a message along with the actual solution instance is passed to the user interface interface.

Finally when the index does not correspond to a feasible solution, we allow the selector to fail. This information is passed to the user interface.

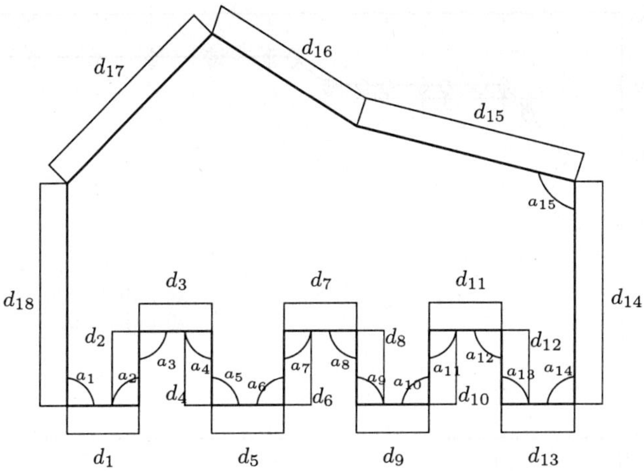

Fig. 6. Geometric problem defined by constraints. Case study A.

4 Experimental Results

To asses the performance of the technique introduced, it has been implemented and thoroughly tested. See [15]. To illustrate this performance, we present a case study and briefly discuss the experimental results.

4.1 Case Study

We consider the geometric constraint problem shown in Figure 6. It consists of 18 straight segments 18 point-point distances and 15 angles. The potential number of solution instances is bounded by $2^{18} = 65{,}536$. The intended solution instance was defined by a set including 27 extra constraints.

An exhaustive evaluation of the set of extra constraints with respect to each solution instance shows that only two solution instances fulfill the set of extra constraints. The construction plan has 16 operations where a root has to be chosen. Therefore each index included 16 binary units. The genetic algorithm was fed with populations consisting of 25, 30, 35 and 40 indexes.

Figure 7 Left shows the number of extra constraints fulfilled by the solution instance selected by the genetic algorithm versus the number of generations. The number of extra constraints fulfilled follows an exponential law and the final value is reached for 30 generations. When the population has 30 individuals or more the selected solution fulfills all the extra constraints.

4.2 Discussion

According to Mühlenbein, [19], at least five parameters are required to describe the initial state and the evolution of an artificial population of a genetic algorithm: Population size, length of the string representing individuals, initial configuration of values in the strings, mutation rate and selection law. Investigating

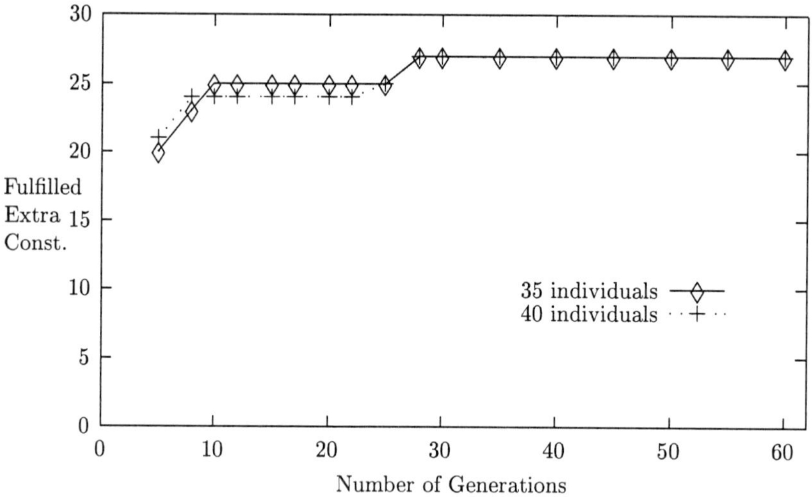

Fig. 7. Extra constraints fulfilled by the selected solution.

n	N	# Extra Constr.	# Solution instances	# Indexes evaluated	%
7	12	6	2^7	61	50
16	40	27	2^{16}	561	2
18	40	39	2^{18}	534	0.3

Fig. 8. Performance of the genetic algorithm.

the behaviour of the genetic algorithm with all five parameters variable would be hard to accomplish. Therefore we have investigated a simpler model where only the population size is variable, where the number of members increases with the number of multiple valuated operations in the geometric constraint problem. The expected number of offspring to be allocated to the worst index was $\mu_{min} = 0.75$. The crossover and mutation probabilities were always 0.6 and 0.2, respectively, [10].

The table in Figure 8 summarizes the results from three different significative experiments reported by Luzón in [15]. The first column shows the number of multi valuated functions in the construction plan. The second column is the number of indexes included in the population. The third column gives the number of extra constraints defined to select the intended solution. The fourth column is the number of indexes in the search space. The fifth column shows the number of indexes actually evaluated. The last column is the ratio between the figures in the fifth and fourth columns.

Results in Figure 8 show that in all cases the number of show that in all cases the number of indexes actually evaluated is a small fraction of the whole search space. Moreover, the performance of the genetic algorithm significantly improves with the problem complexity.

The number of extra constraints fulfilled after six generations was always higher than 66%. As expected in a behaviour that models a natural process, in all cases the number of extra constraints fulfilled by indexes in the current population increased exponentially until reaching an stationary state. The case in the first raw in Figure 8 reached the stationary state after six generations while in the other cases was reached after 30 generations.

Whenever the number of indexes in the population was equal or larger than the number of multi valuated functions in the construction plan, the index selected at the stationary state fulfilled all the extra constraints.

5 Summary and Future Work

In this paper, we have presented a new method to solve the *root identification problem* in two-dimensional constructive geometric constraint solving problems. The technique is based on a genetic algorithm which searches in a potentially exponential large space of solution instances. The user defines the properties of the intended solution by adding a set of extra constraints which are used to drive the genetic algorithm in the search through the space of solution instances. The approach has been implemented on top of an already developed ruler-and-compass geometric constraint solver and has been applied to a number of case studies. The results show that the technique performance is outstanding.

The selection operators used in the algorithm have been linear ranking with universal estochastic samplig. In order to ellucidate the selection behaviour of the algorithm, we will conduct experiments using as selector operators first linear ranking along with stochastic sampling with replacement, then proportional selection with stochastic universal sampling.

Applying genetic algorithms to solve the root identification problem in constructive geometric constraint solving has shown a promising potential. Among others, we plan to explore in the near future the applicability of genetic algorithms to the following issues: Including new types of extra constraints, computing decomposition trees of constraint graphs, fast isolation of minimal overconstrained subgraphs, and automatic addition of extra constraints to underconstrained graphs.

Acknowledgements

This research has been supported by CICYT under the project TIC2001-2099-C03-01. Helpful discussions with S. Vila and J. Vilaplana enriched this work. Comments of anonymous referees improved the paper.

References

1. T. Bäck, D.B. Fogel, and Z. Michalewicz, editors. *Handbook of Evolutionary Computation.* Institute of Physics Publishing Ltd and Oxford University Press, 1997.
2. J. E. Baker. Reducing bias and inefficiency in the selection algorithm. *Proc. Second International Conference on Genetic Algorithms (ICGA'87)*, pages 14–21, 1987.

3. L.B. Booker, D.B. Fogel, D. Whitley, and P.J. Angeline. Recombination. In T. Bäck, D.B. Fogel, and Z. Michalewicz, editors, *Handbook of Evolutionary Computation*, chapter C3.3, pages C3.3:1–C3.3:10. Institute of Physics Publishing Ltd and Oxford University Press, 1997.
4. W. Bouma, I. Fudos, C. Hoffmann, J. Cai, and R. Paige. Geometric constraint solver. *Computer Aided Design*, 27(6):487–501, June 1995.
5. C. Durand. *Symbolic and Numerical Techniques for Constraint Solving.* PhD thesis, Computer Science, Purdue University, December 1998.
6. A. Eiben, P.-E. Raué, and Zs. Ruttkay. GA-easy and GA-hard constraint satisfaction problems. In M. Meyer, editor, *Constraint Processing*, LNCS Series 923, pages 267–284. Springer-Verlag, Heidelberg, 1995.
7. C. Essert-Villard, P. Schreck, and J.-F. Dufourd. Skecth-based pruning of a solution space within a formal geometric constraint solver. *Artificial Intelligence*, 124:139–159, 2000.
8. I. Fudos and C.M. Hoffmann. A graph-constructive approach to solving systems of geometric constraints. *ACM Transactions on Graphics*, 16(2):179–216, April 1997.
9. D.E. Goldberg. *Genetic Algorithms in Search, Optimization Machine Learning.* Addison Wesley, 1989.
10. J.J. Grefenstette. Optimization of control parameters for genetic algorithms. *IEEE Transactions on Systems, Man and Cybernetics*, SMC-16(1):122–128, 1986.
11. R. Joan-Arinyo and N. Mata. A data structure for solving geomegtric cosntraint problems with interval parameters. Technical Report LSI-00-24-R, Department LiSI, Universitat Politècnica de Catalunya, 2000.
12. R. Joan-Arinyo and A. Soto. A correct rule-based geometric constraint solver. *Computer & Graphics*, 21(5):599–609, 1997.
13. R. Joan-Arinyo and A. Soto-Riera. Combining constructive and equational geometric constraint solving techniques. *ACM Transactions on Graphics*, 18(1):35–55, January 1999.
14. G. Laman. On graphs and rigidity of plane skeletal structures. *Journal of Engineering Mathematics*, 4(4):331–340, October 1970.
15. M.V. Luzón. *Resolución de Restricciones Geométricas. Selección de la Solución Deseada.* PhD thesis, Dept. Informática, Universidad de Vigo, December 2001. Written in Spanish.
16. N. Mata. Solving incidence and tangency constraints in 2D. Technical Report LSI-97-3R, Department LiSI, Universitat Politècnica de Catalunya, 1997.
17. N. Mata. *Constructible Geometric Problems with Interval Parameters.* PhD thesis, Dept. LiSI, Universitat Politècnica de Catalunya, 2000.
18. Z. Michalewicz. *Genetic Algorithms + Data Structures = Evolution Programs.* Springer-Verlag, 1996.
19. H. Mühlenbein and M. Georges-Schleuter nad O. Krämer. Evolution algorithm in combinatorial optimization. *Parallel Computing*, 7:65–85, 1988.
20. V. Petridis, S. Kazarlis, and A. Bakirtzis. Varying quality functions in genetic algorithm constrained optimization: The cutting stock and unit commitment problems. *IEEE Transactions on Systems, Man and Cybernetics*, 28, Part B(5), 1998.

Multimeme Algorithms for Protein Structure Prediction

N. Krasnogor, B.P. Blackburne, E.K. Burke, and J.D. Hirst

Automated Scheduling, Optimization and Planning Group
and Computational Biophysics and Chemistry Group
University of Nottingham, Nottingham, United Kingdom
http://dirac.chem.nott.ac.uk/~natk/Public/index.html

Abstract. Despite intensive studies during the last 30 years researchers are yet far from the "holy grail" of blind structure prediction of the three dimensional native state of a protein from its sequence of amino acids. We introduce here a Multimeme Algorithm which is robust across a range of protein structure models and instances. New benchmark sequences for the triangular lattice in the HP model and Functional Model Proteins in two and three dimensions are included here with their known optima. As there is no favourite protein model nor exact energy potentials to describe proteins, robustness accross a range of models must be present in any putative structure prediction algorithm. We demonstrate in this paper that while our algorithm present this feature it remains, in terms of cost, competitive with other techniques.

1 Introduction

A protein's structure determines its biological function. This is the reason why a central component in proteomics is the prediction of a protein's *native* structure from its sequence. This task is called Protein Structure Prediction (PSP). "All-atom" simulations are extremely expensive so researchers often resort to simplified models of the *PSP*, but even the simplified problem still remains computationally intractable in the worst case [2].

The particular simplified models we are concerned with in this paper are the HP model [5] and Functional Model Proteins [10][3] in two and three dimensional lattices. The HP model (and its variants) abstracts the hydrophobic interaction process in protein folding by reducing a protein to a heteropolymer of non-polar or hydrophobic (H) and polar (P) or hydrophilic amino acids. A protein sequence s is represented by a string in a binary alphabet: $s \in \{H, P\}$. Simplified models restrict the space of conformations to self-avoiding paths on a lattice in which vertices are labeled by the amino acids. These lattices may be two-dimensional, e.g. square or triangular, or three dimensional, e.g. diamond. The energy potential in the HP model reflects the fact that hydrophobic amino acids have a propensity to form a hydrophobic core. To capture this feature of protein structures, the HP model adds a value ϵ for every pair of hydrophobes that form a topological contact; a topological *contact* is formed by a pair of amino

J.J. Merelo Guervós et al. (Eds.): PPSN VII, LNCS 2439, pp. 769–778, 2002.

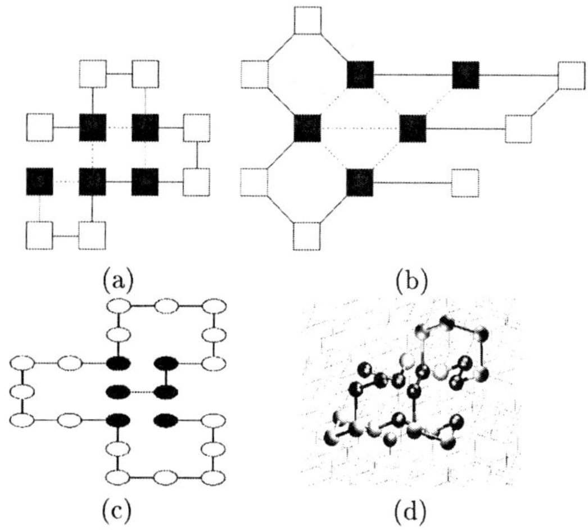

Fig. 1. HP protein embedded in the square lattice (a) and triangular lattice(b). Functional Model protein embedded in the square lattice(c) and diamond (3D) lattice(d). In (c) and (d) native structures are not maximally compact as they must have a "binding pocket".

Table 1. Interaction energy matrix for the standard HP model(a) and Interaction energy matrix for a shifted HP model (b).

(a)	H	P	*(b)*	H	P
H	1	0	H	-2	1
P	0	0	P	1	1

acids that are adjacent on the lattice and not consecutive in the sequence. The value of ϵ is typically taken to be -1. Figure 1 shows sequences embedded in the square and the triangular lattices, with hydrophobic-hydrophobic contacts (HH contacts) highlighted with dotted lines. The conformation in Figure 1 has an energy of -4 in the square lattice embedding and -6 in the triangular lattice embedding. A typical interaction matrix for the HP model is given in table 1(a). The energy interaction in Functional Model Proteins [10],[3] (which introduces repulsive forces) between residues that are in contact is given by table 1(b). Native protein structures in this model are required to have a binding pocket in their native structure (e.g. a hole in their conformation), an energy gap between the minimum energy conformation and the next excited state and to have a unique optimal conformation. Figure 1(c) shows a two dimensional embedding of a Functional Model Protein and 1(d) shows a diamond lattice embbeding.

The construction of effective algorithms for solving structure prediction on simplified models (e.g. the HP model and Functional Model Proteins) is a key stepping-stone towards the structure prediction of real life proteins that cannot be solved by homology or threading methods. Several successful methodologies from the last two Critical Assessment of Structure Prediction [21], CASP3 and

CASP4, employed simplified models for sampling and optimising structures embedded in different lattices [20],[7],[12].

In this paper we will present a novel metaheuristic, called a Multimeme Algorithm, to PSP for four different models: HP model in the square lattice, HP modelin the triangular lattice, Functional Model Proteins in the square lattice and the diamond lattice. To evaluate our algorithm in the first model we will use instances from the public domain that were used by other researchers to test their methods. In the case of the last three models new instances, with their respective optima, will be presented and used as test beds.

2 Evolutionary Algorithms Approaches to Protein Structure Prediction

Several evolutionary algorithms precede the application of Multimeme Algorithms in PSP An early application of Genetic Algorithms (GAs) to PSP due to Unger and Moult [19] is a widely used benchmark. Patton et al. [6] described a standard GA employing, as Unger and Moult did, an internal coordinate representation. They used a penalty method to enforce the self-avoiding constraints. Khimasia and Coveney [11] considered the performance of Goldberg's Simple GA. The objective function was a hybrid between the Random Energy Model [4] and the HP model. Colosimo et al. [18] applied a standard GA to predict the minimum conformational energy of two small real proteins: crambin and ferredoxin. They used the HP model in various 3D cubic grids, where each one increased the spatial resolution. One of us [15][14] explored which kind of encodings, operators, constraint management and energy formulation is more suitable for an evolutionary algorithm designed to tackle minimalist models of PSP and Protein Structure Comparisons. Greenwood et.al. [8] surveyed recent evolutionary approaches to the PSP. More recently Liang and Wong [17] published encouraging results on a hybrid between Monte Carlo optimization and GAs for the square HP model.

3 Multimeme Algorithms for Protein Structure Prediction

Memetic algorithms are evolutionary algorithms that include, as part of the "standard" evolutionary cycle of crossover-mutation-selection, a local search stage. They have been extensively studied and used on a wide range of problems. Multimeme evolutionary algorithms were introduced by Krasnogor and Smith [16] and applied to two bioinformatic problems [14]. The distinction between Memetic and Multimeme Algorithms is that the former uses only one (usually complex) local search while the later employs a *set* of local searchers. Multimeme algorithms self-adaptively select from this set which heuristic to use for different instances, stages of the search or individuals in the population. This kind of algorithm exploits features from Evolutionary Algorithms and Variable Neighborhood Search (by virtue of its multi-operator local search).

In a Multimeme Algorithm an individual is composed of its genetic material and its memetic material. The mechanisms of genetic exchange and variation are the usual crossover and mutation operators but tailored to the specific problem one wants to solve. Memetic transmission is effected using the so called Simple Inheritance Mechanism (SIM) [16]. SIM can be formalized by:

$$L^{t,i} = \begin{cases} L^{t-1,j} & \text{if } \forall k, j \in Parents(i), k \neq j, L^{t-1,j} == L^{t-1,k} \\ L^{t-1,j} & \text{if } F(I_j^{t-1}) > F(I_k^{t-1}) \forall k, j \in Parents(i), k \neq j \\ L^{t-1,k} \text{for any } k \in |Parents(i)| & \text{otherwise} \end{cases} \quad (1)$$

where a meme (local searcher) L, at time $t-1$ that is carried by parent j (or k), will be transmitted to the offspring i if that meme is shared by all the parents. If they have different memes, L is associated to the fittest parent. Otherwise, when fitnesses($F(\cdot)$) are comparable and memes different, a random selection is made. The rationale is to propagate local searchers (i.e. memes) that are associated with fit individuals, as those individuals were probably improved by their respective memes. During mutation, the meme of an individual can also be overridden and a local searcher assigned at random (uniformly from the set of all available local searchers) based on the value of the innovation rate parameter. This is done to introduce novelty in the local search phase of the MMA.

3.1 Tailoring the Multimeme Algorithm for Protein Structure Prediction

The basic evolutionary parameters and settings for the Multimeme Algorithm are now described. Tournament sizes of two and four, a crossover probability of 0.8 and a mutation probability of 0.3 were used. The runs were executed based on a (50,200),(100,400) and (500,1000) replacement strategies. Each generation of the Multimeme Algorithm consisted of a mating stage (two-point crossover with tournament selection), mutation (one and two-point mutation), local search and replacement. Every individual in the population went through an optimization period. The latter was governed by the meme held by the individual. Local search itself was restricted to three iterations in a randomized first improvement fashion, and consequently it was unconverged. For all the experiments reported in this paper the parameters were set according to the criteria described in [14] and [15] and the innovation rate was 0.2.

The memes available to the Multimeme Algorithm can be categorized as follows: pivot (rigid rotation) moves, stretching of a substructure (unfolding), random macro-mutation of a substructure, reflection of a substructure, non-local k-opt and local k-opt. These six local searchers types give rise to several different neighborhoods with which the Multimeme Algorithm will perform its search and were chosen based on previous analysis [14][15]. The Evolutionary Monte Carlo algorithm [17], which represents one of the state of the art systems for two dimensional HP lattice models, employs similar moves as mutation operators (except for the stretch and k-opt operators). Space limitations preclude further description of the local searchers. More details are given elsewhere [14][15]. With the basic ingredients described above, the Multimeme Algorithm performed well on the standard HP model (in two and three dimensional lattices). However, it was not able to reach optimal configurations in the Functional Model Proteins. This was solved by the introduction of a *contact map memory* of current solutions in the mating strategy of the Multimeme Algorithm. With the new mating strategy we were able to solve to optimality instances of both the HP and Functional Model Proteins in two and three dimensional lattices.

3.2 A New Mating Strategy

As mentioned above, a contact map memory was included into the Multimeme Algorithm. During the reproduction phase of the algorithm, each generated offspring was evaluated for compatibility with the contact map memory. An offspring was *compatible* with the memory if at least ϕ_1 of the contacts defined by its structure were themselves compatible. In turn, a contact was *compatible* if not more than ϕ_2 of the individuals already in the population shared that contact. This method involves the determination of the fractions ϕ_1 and ϕ_2 which was done by *ad hoc* experimentation. In this paper $\phi_1 = 25\%, \phi_2 = 66\%$. The inclusion of a memory of the contact maps of already visited solutions has as an advantage (over simply storing fitness evaluations or having an archive of genotypes, i.e. solutions) that the contact maps abstract the geometric details of the structures and keep only the essential topological features of a two dimensional or three dimensional shape. Rotations and symmetries are filtered out and need not be explicitly considered. Given that a contact map can be realized by several different structures, the additional requisite of only accepting offspring that are compatible with the contact map memory pushes the search toward a more exploratory regime, thereby increasing diversity in the population. By holding the information of just a few contact maps in the memory the new mating strategy is actually storing information of a wide area of the whole search space. With this simple strategy we were able to improve on results previously obtained with Multimeme Algorithms [14] on the standard HP model, but more importantly, we were able to solve to optimality instances of the Functional Model Proteins that our previous algorithms were not capable of solving.

4 Results

In this section we will present results obtained with the Multimeme Algorithm using the new mating strategy based on the contact map memory. Functional Model Proteins were introduced in [10]. The optima for the sequences of the Functional Model Protein were obtained by an exhaustive parallel enumeration algorithm. The diamond Functional Model Protein instances and their optima are first published here. Functional Model Proteins are a challenging set of instances, as each one has a unique native state (this is not the case for other well known minimalistic models) which is surrounded by several first excited states.

Moreover, there is an energetic barrier of at least two bonds between the first excited state and the native structure. The Functional Model Proteins presented here are a subset of the available instances with known optima. We computed the native state and first excited states for all of the 2^{23} sequences for the square lattice and the diamond lattice in this model. These can be obtained from [13]. The optima for the triangular lattice instances where obtained by construction in the design process of the sequences. The standard HP lattice sequences were taken from [19],[17],[9] and other references. In all experiments five runs were executed per instance. If the optimum value was not achieved by any of the five runs then we report the best sub-optimum found in bold face. The sequences and results for the *Square Lattice in the Standard HP Model* are shown in table 2. *Two Dimensional Triangular Lattice in the Standard HP Model* instances and results are displayed in table 3 below. The sequences and results for the *Square Lattice in the Functional Model Proteins* can be found in table 4. The number of energy evaluations required to achieve those optimal values is reported. The sequences and results for the *Diamond Lattice in the Functional Model Proteins* can be found in table 4(indexed with letters). To the best of our knowledge, the best algorithm for reduced models is PERM [1]. The results presented here use some of the instances and models tested in [1] and for these cases our results, the positive and the negative ones (e.g. failure to solve instance 4 of the standard, square 2D, HP model), are equivalent. Another point to note is that PERM makes assumptions about "compactness" of the native structure of a protein, which clearly do not apply to the Functional Model Proteins. Indeed, for some instances the optimal structure has one or more binding pockets. Hence, although they do not use the mentioned model, we speculate that their algorithm will not be robust in this domain. Furthermore, it is not possible to compare our algorithm directly with PERM, as our method, like the Evolutionary Monte Carlo method [17], performs a blind search, whereas PERM utilizes information of the sequence being folded. Consequently, we compare our results with other blind methods and assess the robustness of the Multimeme Algorithms across a range of models and instances. Table 5 shows a representative example of the increased robustness of a Multimeme Algorithm when it is compared against a GA and a Memetic Algorithm (that uses only one type of local searcher). In the table, results for instance 15 of table 3 are displayed. The Multimeme approach achieves the optimum solution more frequently than the other approaches and also faster (the GA was given an equivalent number of energy evaluations).

Table 6 shows a comparison of the number of energy evaluations employed by our algorithm and other two well known methods (the GA and Monte Carlo reported in [19]) to solve the square lattice HP instances. Although the Evolutionary Monte Carlo method [17] finds optimum solutions for very challenging instances of the square lattice in the HP model a direct comparison with our algorithm is not possible. Liang and Wong report the number of *feasible* conformations scanned before reaching an optimal structure. However, their algorithm generates thousands of non-feasible structures during the run and this num-

Table 2. Two dimensional square lattice Standard HP instances

#	Sequence	Size	Opt.	MMA
1	$HPHPPHHPHPPHPHHPPHPH$	20	-9	-9
2	$PPPHHPPHHPPPPPHHHHHHHPPHHPPPPHHPPHPP$	36	-14	-14
3	$H^2(PH)^4H^3PHP^3HP^3HP^4HP^3HP^3HPH^4(PH)^4H$	50	-21	-21
4	$H^{12}(PH)^2(P^2H^2)^2(PPH)^2(HP^2H)^2(P^2H^2)^2P^2(HP)^2H^{12}$	**64**	**-42**	**-39**
5	$HPHPPHHPHPPHPHHPPHPH$	20	-9	-9
6	$PPHPPHHPPPPHHPPPPHHPPPPHH$	25	-8	-8
7	$(P^2H)^2HP^2H^2P^5H^{10}P^6(H^2P^2)^2HP^2H^5$	48	-22	-22
8	$PHPPHPHHHPHHPH^5$	18	-9	-9
9	$HPHPHHHPPPHHHHPPHH$	18	-8	-8
10	$HHPPPPPHHPPPHPPPHP$	18	-4	-4
11	$HHHPPHPHPHPPHPHPHPPH$	20	-10	-10

Table 3. Two dimensional triangular lattice Standard HP instances

#	Sequence	Size	Opt.	MMA
1	$HHPPHPHPHPHPHP$	14	-11	-11
2	$HHPPHPPHPHPHPH$	14	-11	-11
3	$HHPHPPHPPHPPHPPH$	16	-11	-11
4	$HHPPHPPHPHPHPPHP$	16	-11	-11
5	$HHPPHPPHPPHPPHPPH$	17	-11	-11
6	$HHPHPHPHPHPHPHPHH$	17	-17	-17
7	$HHPPHPPHPHPHPPHPHPHH$	20	-17	-17
8	$HHPHPHPHPHPPHPPHPPHH$	20	-17	-17
9	$HHPPHPPHPHPPHPHPPHPHH$	21	-17	-17
10	$HHPHPPHPPHPHPHPPHPPHH$	21	-17	-17
11	$HHPPHPHPHPPHPHPPHPPHH$	21	-17	-17
12	$HHPPHPPHPHPHPPHPPHPPHH$	22	-17	-17
13	$HHHPHPHPHPHPHPHPHPHPHHH$	23	-25	-25
14	$HHPPHPPHPPHPPHPPHPPHPPHH$	**24**	**-17**	**-16**
15	$HHHPHPHPPHPHPHPHPHPHPHHH$	24	-25	-25
16	$HHHPHPHPHPPHPHPHPHPHPHHH$	24	-25	-25
17	$HHHPPHPPHPPHPPHPHPPHPHPPHPPHHH$	**30**	**-25**	**-24**
18	$HHHPPHPPHPPHPHPPHPHPPHPPHPPHHH$	**30**	**-25**	**-24**
19	$HHHPPHPPHPPHPHPHPPHPPHPPHPPPPHPHPHHH$	**37**	**-29**	**-26**

ber is not provided in their paper[1]. Nevertheless, their algorithm does solve to optimality all the instances in table 6 and a few longer ones.

Across all the models investigated, our algorithm identifies optimal structures, regardless of considerations of compactness or the size of the protein involved. There are few cases of mis-folding, that is, only a local optimum was found. Unfortunately we were not able to detect any pattern of failure so an improvement cannot be suggested at this time. When comparing the number of energy evaluations of the Monte Carlo and our algorithm we can clearly see the benefits of the Multimeme Algorithm. If we turn to the GA then we find that for one protein (instance 7 in table 6) our approach needed considerably more evaluations.

5 Conclusions and Future Work

The main feature of our algorithm is that it is robust finding optimal structures, across a range of models and difficulty. This is an essential feature needed of

[1] The authors confirmed this with us in a private communication

Table 4. Two dimensional square lattice Functional Model instances (indexed by numbers) and Three dimensional diamond lattice Functional Model instances (indexed by letters) and the number of energy evaluations required by the best run to achieved the optimum or a sub-optimum (in bold face).

#	Sequence	Opt.	MMA	#Evaluations
1	PHPPHPPHHHHPPHPPHPHPPHH	-20	-20	15170
2	PHPPHPPHHHHHPPPPHPPHPPH	-17	-17	61940
3	HPHPHPHHHPPHPPPHPHHPPHH	-16	-16	132898
4	HHHPHHHPPHHPPPHPHPPHHHH	-20	-20	66774
5	PHPPPPPPHPHHPHPHHHHPHPH	-17	-17	53600
6	HHPHPPHPPPPHPPPPHPPPHHH	-13	-13	32619
7	PHPHHPHHHHHHPPHHHPHHHHH	-26	-26	114930
8	HPHPPPHHHHPHPPPPHPHPHHH	-16	-16	28425
9	PHPHHPHHPHHPHPHPHPPPPPH	-15	-15	25545
10	HPHPHPPPPPHHPPPHPHPHPHH	-14	-14	111046
11	PHPPHHHPHPPHPHHPHPPPPPH	-15	-15	52005
A	PHPHPHPHPPPPHPPPPPHPPPH	-11	-11	123979
B	PHPHPHPHPPPPHPPPPPHPPHH	-11	-11	301205
C	PHPHPHPPHHHHHHPHPPHPHPH	**-14**	**-12**	12618
D	PHPHPHPPHHHHHHHPHPPPHPH	-14	-14	1334661
E	PHPHPHPHPPHHPPHPHPHHHHH	-14	-14	482259
F	HPHHPHPHHPPPPPHHPHPHHHH	-15	-15	332842
G	PHPHPHPPHHHHHHPHHHHHHHH	**-16**	**-14**	11132
H	HPPHPPHHPHPHHHPHPHHPHHH	-18	-18	261027
I	HPPHPPHHPHPHHHPHPHHHHPH	-18	-18	550121

Table 5. Number of times and mean first hitting time (in generations) to achieve an optimal solution to instance 15 in table 3. Different algorithms are compared based on 10 independent runs.

Static Memes	Num Optima / Num Runs	Mean First Hitting Time
GA (no memes)	0/10	-
MA with Macro Mutation (r=4)	2/10	27.5
MA with Macro Mutation (r=8)	3/10	53.3
MA with Macro Mutation (r=16)	2/10	43.0
MA with Reflect (r=4)	3/10	20.6
MA with Reflect (r=8)	1/10	79.0
MA with Reflect (r=16)	1/10	45.0
MA with Stretch (r=4)	0/10	-
MA with Stretch (r=8)	0/10	-
MA with Stretch (r=16)	0/10	-
MA with Pivot	5/10	27.0
MultiMeme (all local searchers)	8/10	16.87

Table 6. Energy evaluations used by the Genetic Algorithm and a Monte Carlo approach as quoted from [19] and our MMA for sequences in the HP square lattice.

#	Sequence	GA	MC	MMA
1	$HPHPPHHPHPPHPHHPPHPH$	30492	292443	14621
2	$P(PPHH)^2PPPPPH^7PPHHP^4HHPPHPP$	301339	6557189 (-13)	208233
3	$H^2(PH)^4H^3PHP^3HP^3HP^4HP^3HP^3HPH^4(PH)^4H$	592887	15151203	336763
6	$PPHPPHHPPPPHHPPPPHHPPPPHH$	20400	2694572	18736
7	$(P^2H)^2HP^2H^2P^5H^{10}P^6(H^2P^2)^2HP^2H^5$	126547	9201755 (-20)	1155656

any search method for PSP, as the precise energy formulation that must be optimized is not known. Moreover, the development of energy potentials is a very active area of research and one should expect the frequent publication of new models. A robust search mechanism, such as our Multimeme Algorithm, allows one to change the energy potential without altering too much the algorithmic infrastructure and to investigate folding prediction under the new model. The robustness of our algorithm arises from the evolutionary, population oriented, nature of the search it performs and the amalgamation of several neighborhoods to further improve solutions kept in the population. All of these must be complemented with a suitable "evolutionary memory", that in the work reported here takes the form of a contact map memory.

More experimentation will be undertaken to try to determine the reasons behind the failure to obtain optimum structures in certain sequences. Other models of the PSP will be investigated and the behaviour of our algorithm assessed there.

References

1. U. Bastolla, H. Frauenkron, E. Gerstner, P. Grassberger, and W. Nadler. Testing a new monte carlo algorithm for protein folding. *Proteins: Structure, Function and Genetics*, 32,52, 1998.
2. B. Berger and T. Leight. Protein folding in the hydrophobic-hydrophilic (HP) model is NP-complete. In *Proceedings of The Second Annual International Conference on Computational Molecular Biology, RECOMB 98*, 1998.
3. B.P. Blackburne and J.D. Hirst. Evolution of functional model proteins. *Journal of Chemical Physics*, 115(4):1935–1942, 2001.
4. B. Derrida. Random energy model: Limit of a family of disordered models. *Physical Review Letters*, 45, 1980.
5. Ken A. Dill. Theory for the folding and stability of globular proteins. *Biochemistry*, 24:1501, 1985.
6. A.L. Patton et al. A standard ga approach to native protein conformation prediction. In *Proceedings of the Sixth International Conference on Genetic Algorithms*, pages 574–581. Morgan Kauffman, 1995.
7. M. Feig and C.L. Brooks III. Multiscale modeling protocol for ab initio structure prediction. *In press*, 2000.
8. G.W. Greenwood, B. Lee, J. Shin, and G.B. Fogel. A survey of recent work on evolutionary approaches to the protein folding problem. In *Proceedings of the Congress of Evolutionary Computation (CEC)*, pages 488–495. IEEE, 1999.
9. W.E. Hart. Hp instances. In *http://www.cs.sandia.gov/tech_reports/compbio/tortilla-hp-benchmarks.html.*
10. J.D. Hirst. The evolutionary landscape of functional model proteins. *Protein Engineering*, 12:721–726, 1999.
11. M. Khimasia and P. Coveney. Protein structure prediction as a hard optimization problem: The genetic algorithm approach. In *Molecular Simulation*, volume 19, pages 205–226, 1997.
12. A. Kolinski, M.R. Betancourt, D. Kihara, P. Rotkiewicz, and J. Skolnick. Generalized comparative modeling (genecomp): A combination of sequence comparison, threading, and lattice modeling for protein structure prediction and refinement. *PROTEINS: Structure, Function, and Genetics*, 44:133–149, 2001.
13. N. Krasnogor. Standard hp and functional model proteins instances. *http://dirac.chem.nott.ac.uk/~natk/Public/index.html*, 2001.

14. N. Krasnogor. *Studies on the Theory and Design Space of Memetic Algorithms.* Ph.D. Thesis, Faculty of Computing, Mathematics and Engineering, University of the West of England, Bristol, United Kingdom, 2002.
15. N. Krasnogor, W.E. Hart, J. Smith, and D.A. Pelta. Protein structure prediction with evolutionary algorithms. In W. Banzhaf, J. Daida, A.E. Eiben, M.H. Garzon, V. Honavar, M. Jakaiela, and R.E. Smith, editors, *GECCO-99: Proceedings of the Genetic and Evolutionary Computation Conference.* Morgan Kaufman, 1999.
16. N. Krasnogor and J.E. Smith. Emergence of profitable search strategies based on a simple inheritance mechanism. In *Proceedings of the 2001 Genetic and Evolutionary Computation Conference.* Morgan Kaufmann, 2001.
17. F. Liang and W.H. Wong. Evolutionary monte carlo for protein folding simulations. *Journal of Chemical Physics*, 115(7):3374–3380, 2001.
18. P. Montanari, A. Colosimo, and P.Sirabella. The application of a genetic algorithm to the protein folding problem. In *Proceedings of Engineering of Intelligent Systems (EIS 98)*, 1998.
19. R. Unger and J. Moult. Genetic algorithms for protein folding simulations. *Journal of Molecular Biology*, 231(1):75–81, 1993.
20. Y. Xia, E.S. Huang, M. Levitt, and R. Samudrala. Ab initio construction of protein tertiary structures using a hierarchical approach. *Journal of Molecular Biology*, 300:171–185, 2000.
21. A. Zemla, C. Venclovas, J. Moult, and K. Fidelis. Processing and analysis of casp3 protein structure predictions. *PROTEINS: Structure, Function, and Genetics Suppl*, 3:22–29, 1999.

A Dynamic Traffic Model for Frequency Assignment

Hakim Mabed[1], Alexandre Caminada[1], Jin-Kao Hao[2], and Denis Renaud[1]

[1] FTR&D, 6 Ave des Usines, BP 382, 90007 Belfort, France
{hakim.mabed,alexandre.caminada,denis.renaud}@francetelecom.com
[2] Université d'Angers, 2 Bd Lavoisier, 49045 Angers Cedex, France
Jin-Kao.Hao@univ-angers.fr

Abstract. We are interested in improving the quality of frequency assignment via a more accurate modeling of the traffic over the network. For this purpose, we propose here an original model for FAP, which takes into account both spatial and temporal variation of the traffic. The proposed model is assessed with a hybrid genetic algorithm and compared against a classical model. Experimental results on both artificial and real data show significant improvements of the quality of frequency plan in terms of traffic capacity and robustness of the network.

1 Introduction

In radio mobile networks, the communication is ensured by a radio link. The mobile network operators dispose of a very limited number of frequencies to cover all the network area (limited to 62 frequencies in France). For this reason, the frequency reuse [12] is indispensable to increase the capacity of a network.

A GSM network is composed of a set of sites, each supporting one to three stations [11]. Each station covers an area called cell representing all the points served by this station. According to the quantity of the communications, i.e. the traffic, which may occur in the cell, each station requires a fixed number of frequencies. The frequency assignment problem (FAP) consists in finding an assignment of the available frequency spectrum to the stations of the network, which maximizes the traffic capacity and minimizes interference. Interference is caused by the presence of overlapping areas between cells where several signals of good quality are received. In these areas, traffic satisfaction is highly conditioned by used frequencies. Therefore, traffic modeling constitutes one key aspect of the FAP.

The first works on the FAP are based on a reusing matrix [5, 6, 8, 10] indicating channel separation required between frequencies to completely eliminate the interference. In such a model, interference surface and concerned traffic are ignored. More realistic models were recently proposed, which are based on the quantification of interference risks [3, 4, 9]. This quantification is made on the basis of traffic statistics. More precisely, on each station, traffic intensity recorded at the second busy hour of day (2BH) is considered as traffic reference in interference modeling. We will call this modeling "classical modeling" or "2BH modeling".

In this paper, we discuss about disadvantages of 2BH dimensioning and we propose a dynamic traffic modeling for FAP, which takes into account spatial and

J.J. Merelo Guervós et al. (Eds.): PPSN VII, LNCS 2439, pp. 779–788, 2002.

temporal variation of traffic, in order to improve the traffic capacity modeling and robustness of the network. The dynamic traffic model is tested on both artificial and realistic data, and compared with the classical modeling. To perform those tests a new heuristic based on a hybridization of genetic algorithms and tabu search is elaborated. Experimental results show significant improvements of frequency plan quality both in terms of robustness and traffic capacity.

2 Traffic Modeling in Frequency Assignment Problem

2.1 Traffic Engineering

Traffic evolution can be observed in both spatial and temporal scale [1, 2]. Spatial variation of traffic refers to client mobility and concentration. Time variation of traffic is due to behavioral aspects of clients. Hence, traffic evolution analysis can be carried out either by observing time variation of the traffic over each cell, or by observing the traffic distribution over the network at each time point. Fig 1 shows time variation of traffic over two cells where the traffic (expressed in Erlang) is indicated for each hour. Note that the second busy hour is not the same for the two cells.

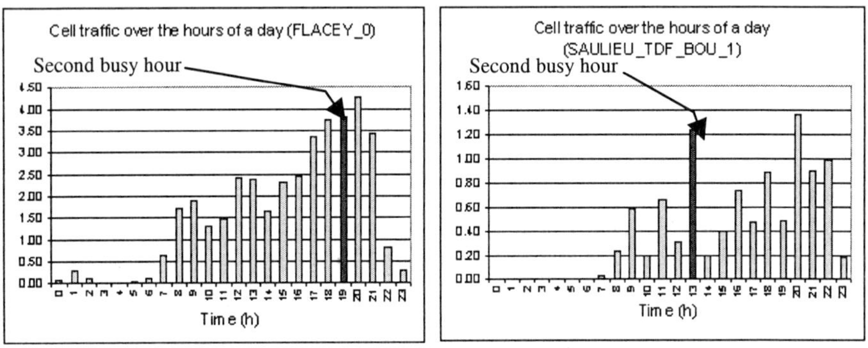

Fig. 1. Time variation of traffic over two cells

2.2 Classical Traffic Modeling for Frequency Assignment

A GSM network is composed of a set of sites, each supporting one to three stations [11, 12]. Each station delimits an area called cell representing all the points served by this station. For each station $S_i \in \{S_1, \ldots, S_N\}$, we know the number of frequencies required, MA_i. Frequency assignment to stations is submitted to constraints of different nature and priority. Those constraints are divided into three classes:

- Co-station constraint (call it C1 hereafter): frequencies assigned to the same station must be spaced by at least 3 channels.
- Co-site constraint (call it C2 hereafter): frequencies assigned to stations located on the same site must be spaced by at least 2 channels.
- Inter-site constraint: frequencies assigned to stations belonging to different sites are spaced according to their mutual interference.

The satisfaction of co-station and co-site constraints is indispensable for a frequency plan to be applicable, while satisfying inter-site constraints is generally impossible. The objective of FAP is then to minimize the potential interference generated by the violation of inter-site constraints.

Inter-site constraints may be modeled by an undirected graph (call it interference graph hereafter) whose nodes correspond to stations and edges represent interference risks. Each edge, connecting two stations S_i and S_j, is weighted by a pair of values ($\beta_{i,j,0}$, $\beta_{i,j,1}$). Where $\beta_{i,j,d}$ measures the importance of interference between S_i and S_j, generated by a pair of frequencies spaced by d channels (interference is considered negligible if d>1).

The impact of traffic on interference is twofold. As jamming station, traffic intensity describes the rate of use of frequencies assigned to the station and hence impacts on the quantity of the generated interference. As interfered station, traffic intensity reflects the importance of the area covered by the station and consequently the interest of interference reduction on this area. Therefore, we can roughly consider $\beta_{i,j,d}$ as a returned value of a function, I, having as arguments the traffic intensity on the two stations and considered inter-channel distance as described by equation (1)

$$\beta_{i,j,d} = I(i, j, t_i, t_j, d) \tag{1}$$

where t_i and t_j correspond to traffic intensity on stations S_i and S_j.

In classical traffic modeling, $\beta_{i,j,d}$ values are calculated on the basis of traffic data recorded at the *second busy hour* (2BH) of the day over each cell. Let t_i^{2BH} be the traffic intensity at the second busy hour over the station S_i. The weights of the interference graph are then calculated using the following expression.

$$\beta_{i,j,d}^{2BH} = I(i, j, t_i^{2BH}, t_j^{2BH}, d) \tag{2}$$

2BH dimensioning suffers from two disadvantages. Firstly, traffic repartition $(t_1^{2BH}, \dots, t_N^{2BH})$ corresponds neither to real traffic cartography nor to a good aggregation of traffic evolution. In others words, 2BH dimensioning causes an alteration in preferential order between interference weights ($\beta_{i,j,d}$) and hence an inaccuracy in traffic capacity measurement. Secondly, 2BH dimensioning ignores time variation of traffic, which is indispensable for elaborating robustness criteria.

3 Dynamic Traffic Modeling for Frequency Assignment Problem

In order to overcome the difficulties encountered with the classical 2BH modeling, we introduce here the notion of dynamic traffic modeling for FAP. To that end, we dispose of data on traffic evolution during np periods. Let t_i^h be the traffic intensity on station S_i at period h and let $\beta_{i,j,d}^h$ be the weights of the interference graph calculated from traffic data at period h. FAP is then defined by np constraint graphs, one per period, such as:

$$\beta^h_{i,j,d} = I(i, j, t^h_i, t^h_j, d) \tag{3}$$

According to those graphs, the quality of a frequency plan will be measured at both a global and local level. The global quality of the frequency plan refers to the sum over times of interference recorded on the network. The local quality measures the performance stability of the frequency plan over the time period where the quality is the lowest. Two criteria are to be retained then: Total interference, and frequency plan robustness. These criteria can be stated more formally as follows.

- Total interference, or global quality of the frequency plan.

$$F_1 = \sum_{h=1}^{np} \sum_{(S_i, S_j)} \sum_{\substack{(f_{i,k}, f_{j,p}) \\ k \in [1..MA_i] \\ p \in [1..MA_j]}} \beta^h_{i,j,|f_{i,k} - f_{j,p}|} \tag{4}$$

where $f_{i,k}$ represents the kth frequency assigned to station S_i.

- Temporal interference distribution or robustness of the frequency plan through a time period. It aims to minimize the worst performance of the frequency plan over the time.

$$F_2 = \underset{h=1}{\overset{np}{MAX}} \sum_{(S_i, S_j)} \sum_{\substack{(f_{i,k}, f_{j,p}) \\ k \in [1..MA_i] \\ p \in [1..MA_j]}} \beta^h_{i,j,|f_{i,k} - f_{j,p}|} \tag{5}$$

According to this dynamic model, the objective of the frequency assignment problem is to find $f_{i,k}$ values which satisfy co-station and co-site constraints and minimize F_1 and F_2.

We turn now to the presentation of a hybrid algorithm for finding frequency plans. This hybrid algorithm combines genetic search and a tabu algorithm and uses the above quality functions (F_1 and F_2) as part of its evaluation function.

4 A Genetic Tabu Search Algorithm for FAP

The following notations will be used in the presentation: *nf* the number of available frequencies, C1 and C2 two binary functions representing the co-station and co-site constraints:

$$C1(i, k, p) = \begin{cases} 1 & \text{if } |f_{i,k} - f_{i,p}| < 3 \\ 0 & \text{else} \end{cases} \tag{6}$$

$$C2(i, j, k, p) = \begin{cases} 1 & \text{if } |f_{i,k} - f_{j,p}| < 2 \\ 0 & \text{else} \end{cases} \tag{7}$$

4.1 Individual Representation and Fitness Evaluation

A frequency plan is coded by a vector $< f_{1,1}, \ldots, f_{1,MA_1}, f_{2,1}, \ldots, f_{2,MA_2}, \ldots, f_{N,1}, \ldots, f_{N,MA_N} >$, representing frequencies assigned to each station. The search space of a problem corresponds therefore to all such configurations where $f_{i,k} \in [1..nf]$.

Constraints C1 and C2 are handled using a penalty-based approach. They, together with the criteria F_1 and F_2, are linearly combined in a single *evaluation (fitness)* function. To stress their importance relative to the quality criteria, co-station and co-site constraints are weighted by a large value ω.

$$\text{minimize } F_{DO} = \omega \left(\sum_{i=1}^{N} \sum_{k=1}^{MA_i-1} \sum_{p=k+1}^{MA_i} C1(i,k,p) + \sum_{i=1}^{N-1} \sum_{\substack{j=i+1 \\ i,j \in \text{ same site}}}^{N} \sum_{k=1}^{MA_i} \sum_{p=1}^{MA_j} C2(i,j,k,p) \right) + F_1 + F_2 \tag{8}$$

4.2 Selection and Replacement Operators

At each iteration, two frequency plans are selected from current population. To favor the selection of good solutions, the population individuals are ordered according to their fitness so that the best solution has the rank 0. Let r_i be the rank of the individual i, the selection probability of i is then calculated following the expression 9.

$$SP_i = \frac{(Pop\_size - r_i) \times 2}{Pop\_size \times (Pop\_size + 1)} \tag{9}$$

After reproduction, new individuals are directly inserted in the population in place of other solutions. Replacement operator favors the elimination of bad frequency plans. Equation 10 represents the replacement probability of the individual i.

$$RP_i = \frac{r_i \times 2}{Pop\_size(Pop\_size - 1)} \tag{10}$$

Let us notice that the best frequency plan is never replaced.

4.3 Crossover and Mutation

The so-called geographic crossover described in [13] is used to generate new frequency plans. This specific crossover operator for the frequency assignment problem works as follow. Given two frequency plans, the first step of the crossover consists in taking randomly a reference station S_i. Let $V(S_i)$ be the set of co-site stations and interfering stations of S_i (S_j interferes with S_i if $\beta_{i,j,d} \neq 0$). Then the frequencies corresponding to stations $S_i \cup V(S_i)$ are exchanged between the two parents generating two new frequency plans (fig 2).

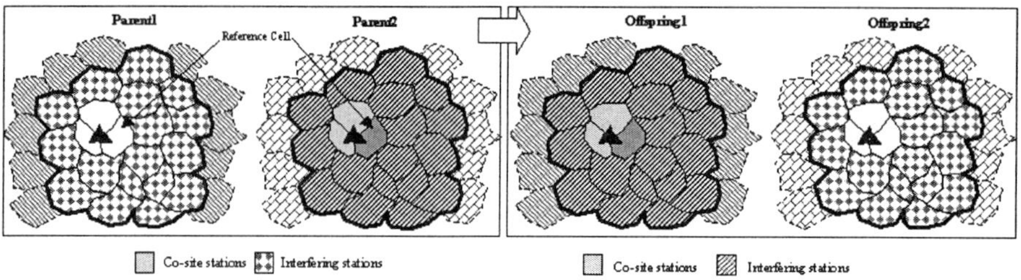

Fig. 2. Crossover operator

For mutation, we use a local search operator based on Tabu search (TS). This operator is basically inspired by the Tabu algorithm described in [7]. The main difference remains at the level of assessing the fitness of frequency assignments.

For a given assignment, a violation score is defined for each frequency of the plan. This score measures the contribution of this frequency to the recorded interference. At each iteration of the TS algorithm, one gene is selected according to its violation score and a new frequency value is affected to it. The pair (gene, old-frequency-value) is then added to the tabu list. Equation 11 and 12 describe respectively the way of calculating the violation scores and gene selection probability of the kth frequency assigned to station S_i.

$$SCORE_{i,k} = \omega \times \left(\sum_{\substack{p=1 \\ p\neq k}}^{MA_j} C1(i,k,p) + \sum_{\substack{j=1 \\ j\neq i \\ i,j\in \text{ same site}}}^{N} \sum_{p=1}^{MA_j} C2(i,j,k,p) \right) + \sum_{h=1}^{np} \sum_{\substack{j=1 \\ j\neq i}}^{N} \sum_{\substack{f_{jp} \\ p\in[1..MA_j]}} \beta^{h}_{i,j|f_{ik}-f_{jp}|} \quad (11)$$

$$GSP_{i,k} = SCORE_{i,k} \Big/ \left(\sum_{j=1}^{N} \sum_{p=1}^{MA_j} SCORE_{j,p} \right) \quad (12)$$

Finally, we give here the main loop of our genetic tabu search algorithm

```
P:=RandomInitPopulation(Pop_size)
For g:=1 to nb_of_generation
          (p1,p2):=SelectParents(P)
          with a Pc probability do
            (f1,f2):=Crossover(p1,p2)
          otherwise f1:=p1; f2:=p2
          f1:=TabuSearch(f1); f2:=TabuSearch(f2);
          (v1,v2):=SelectVictims(P)
          ReplaceBy(v1,f1); ReplaceBy(v2,f2);
```

5 Experimentation and Results

This section is dedicated to the presentation of experimental results of the proposed dynamic model using the hybrid genetic tabu algorithm. Tests are carried out on both fictive and real problems. The results of dynamic traffic model are compared with the classical model based on 2BH dimensioning. The hybrid algorithm uses respectively Equation 8 (dynamic modeling) and Equation 13 (2BH modeling) as its fitness function.

$$F_{H2C} = \omega\left(\sum_{i=1}^{N}\sum_{k=1}^{MA_i-1}\sum_{p=k+1}^{MA_i} C1(i,k,p) + \sum_{i=1}^{N-1}\sum_{\substack{j=i+1 \\ i,j\in \text{ same site}}}^{N}\sum_{k=1}^{MA_i}\sum_{p=1}^{MA_j} C2(i,j,k,p)\right) + \sum_{\substack{(S_i,S_j) \\ i<j\in[1..N]}}\sum_{\substack{(f_{i,k},f_{j,p}) \\ k\in[1..MA_i] \\ p\in[1..MA_j]}} \beta^{2BH}_{i,j,|f_{i,k}-f_{j,p}|} \quad (13)$$

5.1 Fictive FAP Instances

The two fictive FAP instances used in our experimentation represent 63 stations extracted from a real network *B*. The word "fictive" refers only to the data of traffic evolution. The two instances have the following characteristics: 225 frequencies to assign, traffic data over 6 periods and around 1100 inter-site constraints.

Each instance presents a different class of traffic evolution that allows us to study the performance of dynamic traffic modeling on different traffic evolution scenarios. The first network, *B_63_1*, presents synchronous and proportional rises and falls of traffic on the entire network. The second instance, *B_63_2*, stresses the mobility aspect of clients and presents two distinct areas. The rise of traffic on one area is accompanied by a fall of traffic intensity on the other. Table 1 and 2 summarize the traffic evolution for the two instances.

Table 1. Traffic evolution for *B_63_1*

Periods	Traffic situation
Period 0	Traffic load on each station S_i corresponds to 20% of t_i^{2BH}
Period 1	Traffic load on each station S_i corresponds to 50% of t_i^{2BH}
Period 2	Traffic load on each station S_i corresponds to t_i^{2BH}
Period 3	Traffic load on each station S_i corresponds to 110% of t_i^{2BH}
Period 4	Traffic load on each station S_i corresponds to 60% of t_i^{2BH}
Period 5	Traffic load on each station S_i corresponds to 20% of t_i^{2BH}

Table 2. Traffic evolution for *B_63_2*

Periods	Traffic situation
Period 0	In 1st part, traffic load on S_i corresponds to 110% of t_i^{2BH} and to 20% on the 2nd part
Period 1	In 1st part, traffic load on S_i corresponds to 100% of t_i^{2BH} and to 30% on the 2nd part
Period 2	In 1st part, traffic load on S_i corresponds to 70% of t_i^{2BH} and to 50% on the 2nd part
Period 3	In 1st part, traffic load on S_i corresponds to 50% of t_i^{2BH} and to 70% on the 2nd part
Period 4	In 1st part, traffic load on S_i corresponds to 30% of t_i^{2BH} and to 100% on the 2nd part
Period 5	In 1st part, traffic load on S_i corresponds to 20% of t_i^{2BH} and to 110% on the 2nd part

5.2 Real FAP Instance

The proposed traffic model is also tested on a real traffic evolution data (Network D). This network is characterized by: 639 stations, 1411 frequencies to assign, around 30000 inter-site constraints and traffic data over 13 hours (7:00-20:00).

5.3 Performance Criteria

Comparison between dynamic traffic modeling and classical modeling for FAP is made on the basis of *lost traffic*, measured in Erlang. One Erlang corresponds to one hour of communication. We use quality evaluator of PARCELL [1] to measure the lost traffic produced by a given frequency plan. More precisely, given the stations parameters, geographical database, traffic data and a frequency plan, the quality evaluator calculates the lost traffic quantity on each station. Loss in traffic is measured in term of FER (Frame Erasure Rate). The communication is considered bad if this rate exceeds a given threshold. According to required radio quality, we distinguish 3 kinds of thresholds: 2%, 4% and 7%.

5.4 Experimental Results

Tables 3-6 below show experimental results of classical and dynamic traffic modeling on the three FAP instances described above. For each instance, we generate two frequency plans. The first is built on the basis of classical traffic modeling (Equation 13). The second is built on the basis of our dynamic traffic modeling (Equation 8). The performance of each frequency plan, in term of lost traffic, is estimated for each period. We present also at the lower part of the tables, the total lost traffic (global quality criteria), maximal lost traffic on the given time period (robustness criteria) and the gain in Erlang between the two models. Information is given for each of the three quality thresholds (2%, 4% and 7%).

Tests were made using the same parameters of the hybrid algorithm: 100000 iterations for a population of 10 solutions with a crossover rate of 0.3 and tabu search iteration number of 30.

From those three tables, we notice that the dynamic model gives better frequency plans both in terms of global traffic capacity and robustness. Important gains are observed for different traffic evolution scenarios.

Table 6 shows the fitness of the two frequency plans analyzed in table 5. We notice that even if solution generated by classical model is better according to F_{2BH} it is still worse than dynamic model according to F_1 and F_2. This result confirms that classical traffic modeling doesn't allow the production of well-adapted frequency plan for traffic in evolution.

---

[1] Engineering tool for design of mobile radio network, ORANGE society all rights reserved.

Table 3. Results of classical and dynamic traffic modeling for *B_63_1*

		Lost traffic at 2% FER (Erl)		Lost traffic at 4% FER (Erl)		Lost traffic at 7% FER (Erl)	
Periods	**Traffic**	**Classical**	**Dynamic**	**Classical**	**Dynamic**	**Classical**	**Dynamic**
Period 0	116.85	1,43	1,36	0,71	0,67	0,39	0,35
Period 1	292.13	5,14	4,46	2,57	2,32	1,36	1,25
Period 2	584.27	17,63	15,36	9,02	8,27	4,87	4,37
Period 3	642.70	20,87	18,43	11,29	9,98	5,84	5,24
Period 4	350.56	6,96	6,16	3,54	3,11	1,81	1,87
Period 5	116.85	1,43	1,36	0,71	0,67	0,39	0,35
Total		**53,46**	**47,13**	**27,84**	**25,02**	**14,66**	**13,43**
Communication gain		**6.33**		**2.82**		**1.23**	
Maximum		**20,87**	**18,43**	**11,29**	**9,98**	**5,84**	**5,24**

Table 4. Results of classical and dynamic traffic modeling for *B_63_2*

		Lost traffic at 2% FER (Erl)		Lost traffic at 4% FER (Erl)		Lost traffic at 7% FER (Erl)	
Periods	**Traffic**	**Classical**	**Dynamic**	**Classical**	**Dynamic**	**Classical**	**Dynamic**
Period 0	329.29	7,69	7,56	4,16	4,4	2,33	2,32
Period 1	340.51	7,63	7,36	4,04	4,25	2,21	2,26
Period 2	339.27	7,03	6,35	3,47	3,45	1,86	1,84
Period 3	361.64	7,50	6,13	3,50	3,24	1,82	1,80
Period 4	418.86	9,39	7,76	4,65	4,13	2,43	2,35
Period 5	430.06	9,99	8,4	5,34	4,21	2,80	2,40
Total		**49,23**	**43,56**	**25,16**	**23,68**	**13,45**	**12,97**
Communication gain		**5.67**		**1.48**		**0.48**	
Maximum		**9,99**	**7,76**	**5,34**	**4,25**	**2,80**	**2,54**

Table 5. Results of classical and dynamic traffic modeling for *D_639_1*

		Lost traffic at 2% FER (Erl)		Lost traffic at 4% FER (Erl)		Lost traffic at 7% FER (Erl)	
Period	**Traffic**	**Classical**	**Dynamic**	**Classical**	**Dynamic**	**Classical**	**Dynamic**
7:00–8:00	504.11	7.41	6.38	4.42	3.92	2.46	2.21
8:00-9:00	1170.02	21.50	19.62	12.98	11.53	7.26	6.43
9:00-10:00	1747.71	37.90	34.79	22.98	20.63	13.22	11.72
10:00-11:00	2017.26	45.09	41.61	26.94	24.12	15.40	13.29
11:00-12:00	2177.03	50.58	45.82	29.96	26.23	17.08	14.48
12:00-13:00	2104.73	46.58	43.50	27.95	25.17	16.54	13.49
13:00-14:00	1863.42	38.66	37.06	23.60	21.14	13.92	11.57
14:00-15:00	1953.59	43.01	38.42	25.82	22.02	15.25	12.10
15:00-16:00	1984.12	45.44	40.66	27.48	23.37	16.03	12.70
16:00-17:00	2174.47	51.24	47.03	30.90	27.25	18.31	14.99
17:00-18:00	2521.20	60.53	57.10	36.20	33.19	20.97	18.08
18:00-19:00	2792.91	69.73	67.55	42.00	39.17	24.24	21.12
19:00-20:00	2743.83	61.65	61.00	36.88	34.77	21.35	18.88
Total		**579.32**	**540.54**	**348.10**	**312.51**	**202.03**	**171.06**
Communication gain		**38.78**		**35.59**		**30.97**	
Maximum		**69.73**	**67.55**	**42.00**	**39.17**	**24.24**	**21.12**

Table 6. Fitness of the two frequency plans generated for *D_639_1*

Classical solution	Dynamic solution
F_1: 970932,916	F_1: 936139,360
F_2: 115887,751	F_2: 115588,483
F_{H2C}: 128749,617	F_{H2C}: 130377,944

6 Conclusion and Future Works

We have proposed in this paper a finer and more accurate traffic model for the frequency assignment problem of mobile radio networks. Both spatial and temporal aspects of traffic are taken into account, leading to improvements of the traffic capacity and robustness of the frequency plan. We have also presented a hybrid genetic tabu search algorithm for finding frequency plans. Comparisons between the proposed dynamic traffic model and classical 2BH-based traffic model showed significant improvements of the quality of frequency plans both in terms of global traffic capacity and network robustness.

New criteria of robustness are to be studied in the future especially with regard to spatial distribution of interference. It will be also interesting to conceive a multi-objective algorithm to solve the problem. The model might be enriched to give more importance to certain periods (e.g. to favor professional communications).

References

1. Baier, K. Bandelow "Traffic engineering and realistic network capacity in cellular radio networks with inhomogeneous traffic distribution" *IEEE VTC* 46: 780-784, 1997.
2. P. Darwood, I. Oppermann, S. Jakas, W. Linton "Mobile Network Traffic Forecasting" *IEEE VTC 50*: 2000.
3. Eisenblatter, M. Grotschel, A. Koster "Frequency planning and ramifications of coloring" *ZIB-Repport* 00-47: December 2000.
4. M. Fischetti, C. Lepschy, G. Minerva, G. Romanin-Jacur, E. Toto "Frequency assignment in mobile radio systems using branch-and-cut techniques" *European Journal of Operational Research* 123: 241-255, 2000.
5. Gamst "A resource allocation technique for FDMA systems" *Alfa Frequenza* 57(2): 89-96, 1988.
6. W.K. Hale "Frequency assignment: theory and application" *Proceedings of IEEE*, 68(12): 1498-1573, 1980.
7. J.K. Hao, R. Dorne, P. Galinier "Tabu search for frequency assignment in mobile radio networks" *Journal of Heuristics* 4: 47-62, 1998.
8. M. Hellebrabdt, F. Lambrecht, R. Mathar, T. Niessen, R. Starke "Frequency allocation and linear programming" *IEEE VTC* 48: 617-621, 1999.
9. S. Hurley, R.M. Whitaker, D.H. Smith "Channel assignment in Cellular Networks without Channel Separation Constraints" *IEEE VTC* 50: 1714-1718, 2000.
10. F. Jaimes-Romero, D. Munoz-Rodreguez "Channel assignment in cellular systems using genetic algorithms" *IEEE VTC* 45: 741-745, 1996.
11. W. Lee, "Mobile Communications Design Fundamentals" *Wiley Series in Telecommunications*: 1992.
12. V. Mac Donald "Advanced Mobile Phone Service: The Cellular Concept" *The BELL System Technical Journal*: 15-41, January 1979.
13. D. Renaud, A. Caminada "Evolutionary methods and operators for frequency assignment Problem" *SpeedUp Journal* 11(2): 27-32, 1997.

A Parameter-Free Genetic Algorithm for a Fixed Channel Assignment Problem with Limited Bandwidth

Shouichi Matsui, Isamu Watanabe, and Ken-ichi Tokoro

Communication & Information Research Laboratory (CIRL)
Central Research Institute of Electric Power Industry (CRIEPI)
2-11-1 Iwado-kita, Komae-shi, Tokyo 201-8511, Japan
{matsui,isamu,tokoro}@criepi.denken.or.jp

Abstract. Increasing the channel re-usability is necessary for reducing the call-blocking rate in any cellular systems with limited bandwidth and a large number of subscribers. To increase the re-usability, we need an efficient channel assignment algorithm that minimizes the sum of blocking cost and interference cost. We propose a new genetic algorithm for the problem based on the parameter-free GA. The proposed GA finds a good sequence of codes for a virtual machine that produces channel assignment. Results are given which show that our GA, without tedious parameter tuning, produces far better solutions to several practical problems than the existing GAs.

1 Introduction

The channel assignment problem (CAP), or the frequency assignment problem (FAP), is a very important problem today, but is a difficult, NP-hard problem. The radio spectrum is a limited natural resource used in a variety of private and public services, the best-known example would be found in cellular mobile phone systems, or personal communication services (PCS). To facilitate this expansion the radio spectrum allocated to a particular service provider needs to be assigned as efficiently and effectively as possible.

Because the CAP is a very important problem in the real world and an NP-hard problem, a number of heuristic algorithms have been proposed (e.g., [3]), and genetic algorithms (GAs) are applied to minimum span frequency assignment problem (MSFAP) (e.g., [2, 3, 6, 8–10, 13, 14]). To achieve the optimal solution of fixed channel assignment problems, most proposed algorithms try to minimize the amount of necessary channels under satisfying a set of given constraint(e.g., [2, 3, 6, 8–10, 13, 14]).

However, the total number of available channels, or bandwidth of frequencies, are given and fixed in many situations. Minimizing the bandwidth becomes meaningless for such applications. To address the problem, Jin et al. proposed a new cost model [5] in which the available number of channels are given and fixed and the electro-magnetic compatibility constraints and demand constraint

J.J. Merelo Guervós et al. (Eds.): PPSN VII, LNCS 2439, pp. 789–799, 2002.

are relaxed. They also proposed genetic algorithms to solve the problems [4, 5]. But their algorithms use naïve chromosome representation, therefore the search space is too large in scale, thus the results obtained by these algorithms are not good enough.

We propose a new algorithm for the problem based on the parameter-free genetic algorithm (PfGA) proposed by Sawai et al. [7, 12]. The proposed GA is tested using a set of practical benchmark problems, and the performance is far better than the existing GAs. The proposed GA can obtain better solutions than the existing GAs without time-consuming parameter tuning.

2 Fixed Channel Assignment with Limited Bandwidth

We formulate the problem in this section. This section is a brief summary of the paper by Horng et al. [4].

The total amount of blocked calls and interference between cells are the two major considerations of channel assignment. Both blocking and interference brings a certain degree of damage to the system. Thus, an ideal assignment should be a solution that guarantees a low blocking rate and a low degree of interference.

2.1 Notation

Let us consider a cellular system that consists of N cells, and each cell is numbered from 1 to N. A compatibility matrix is a symmetric $N \times N$ matrix $C = (c_{ij})$ with nonnegative integer elements. The value c_{ij} prescribes the minimum frequency separation required between frequencies assigned to cell i and cell j, i.e., if f_i^k and f_j^l are the frequencies assigned to cell i and j respectively, then the following condition $|f_i^k - f_j^l| \geq c_{ij}$ should be satisfied for all i and j.

Radio frequencies are assumed to be evenly spaced, therefore they can be identified with the positive integers. Let $X_{i,j}$ be the variable that takes 1 when the j-th mobile that stays in cell i wishes to own a frequency, and n_i be the number of mobiles that stay in cell i, and let T be the number of channels that each frequency provides under TDMA (Time Division Multiple Access). And let $X_i = \sum_{j=1}^{n_i} X_{i,j}$ be the random variable of required channels in cell i, and let μ_i and σ_i be the expected number and the standard deviation of X_i (no matter the request is failed or success). In general, the load of each cell is maintained by system, therefore μ_i and σ_i could be estimated from the long-term history of base stations' load data.

2.2 Damage of Blocked Calls

Call attempts may fail at a busy station because there are no available channels, and failed calls are called *blocked calls*. The more blocked calls, the more damage caused to the system. Thus, an ideal channel assignment should guarantee that the total amount of blocked calls of all cells be as low as possible.

Let H_i be the number of assigned channels to cell i, then the expected number of blocked calls in cell i is $\sum_{j=H_i+1}^{n_i} P(X_i = j)(j - H_i)$. The objective here is to minimize the cost $\sum_{i=1}^{N} \sum_{j=H_i+1}^{n_i} P(X_i = j)(j - H_i)$. As Horng et al. showed, we can assume the random variable X_i $(1 \le i \le N)$ is close to the normal random variable with parameters μ_i and σ_i [4]. Therefore, as an approximation of

$$\sum_{j=x+1}^{n_i} P(X_i = j)(j - x),$$

we can use

$$\begin{aligned} I_E(x) &= \frac{1}{\sqrt{2\pi}\sigma_i} \int_x^\infty (y - x) \exp\left\{ -\frac{1}{2} \left(\frac{y - \mu_i}{\sigma_i} \right)^2 \right\} dy \\ &= \frac{1}{\sqrt{2\pi}} \sigma_i \exp\left\{ -\frac{1}{2} \left(\frac{x - \mu_i}{\sigma_i} \right)^2 \right\} + \frac{1}{2}(\mu_i - x) \left\{ 1 - \mathrm{erf}\left(\frac{x - \mu_i}{\sqrt{2}\sigma_i} \right) \right\} \\ &= \frac{1}{\sqrt{2\pi}} \sigma_i \exp\left\{ -\frac{1}{2} \left(\frac{x - \mu_i}{\sigma_i} \right)^2 \right\} + \frac{1}{2}(\mu_i - x) \mathrm{erfc}\left(\frac{x - \mu_i}{\sqrt{2}\sigma_i} \right), \end{aligned}$$

where $\mathrm{erf}(x)$ and $\mathrm{erfc}(x)$ are the error and complementary error function defined as

$$\mathrm{erf}(x) = \frac{2}{\sqrt{\pi}} \int_0^x \exp(-t^2) dt, \qquad \mathrm{erfc}(x) = 1 - \mathrm{erf}(x) = \frac{2}{\sqrt{\pi}} \int_x^\infty \exp(-t^2) dt.$$

2.3 Damage from Interference

Interference can occur between a pair of transmitters if the strength of the interfering signal is sufficiently high. Whether a transmitter pair has the potential to interfere depends on many factors, e.g., distance, terrain, power, or antenna design. The higher the potential for interference between a transmitter pair is, the larger the frequency separation required. For example, if two transmitters are sufficiently geographically separated then a frequency can be re-used, i.e., the same frequency can be assigned. At the other extreme if two transmitters are located at the same site then they may require, say, a five-frequency separation.

Violating frequency separation constraint, or EMC (Electro-Magnetic Compatibility) constraint, would bring some degree of disadvantage to the mobiles that experience interference. And the disadvantage should be in proportion to the degree of interference that depends on the frequency distance (i.e., how many Hz between them) and the power it suffered. The degree of damage is defined as follows. Let p be the assigned frequency to cell i, and q be the one to cell j, then the damage caused by interference from this assignment $f(i,j,p,q)$ is defined as follows.

$$f(i,j,p,q) = \begin{cases} 0 & \text{if } |p-q| \geq c_{ij}, \\ f_{i,p} f_{j,q} I_C(c_{ij} - |p-q|) & \text{if } |p-q| < c_{ij} \text{ and } i = j, \\ f_{i,p} f_{j,q} I_A(c_{ij} - |p-q|) & \text{otherwise.} \end{cases}$$

where

$$f_{i,p} = \begin{cases} 1 & \text{if frequency } p \text{ is assigned to cell } i \\ 0 & \text{otherwise,} \end{cases}$$

and I_C and I_A are two strictly increasing functions.

2.4 Objective Function

The objective of the problem is to minimize the total damage, to minimize the sum of the cost of blocked calls and the cost of interference, therefore the problem is defined as follows.

$$\text{Min } O = \sum_{i=1}^{N} \sum_{j=1}^{N} \sum_{p=1}^{Z} \sum_{q=1}^{Z} f(i,j,p,q) + \alpha \sum_{i=1}^{N} I_E(TF_i),$$

subject to $f_{i,p} = 0$ or 1 for $1 \leq i \leq N$ and $1 \leq p \leq Z$ where $F_i = \sum_{p=1}^{Z} f_{i,p}$ for $1 \leq i \leq N$, Z is the allowable number of frequencies, and α is the relative weight of the damage of blocked calls to the damage from interference.

2.5 Related Works

Because CAP (FAP) is an important and very difficult problem to solve exactly, GA based algorithms for the minimum span frequency assignment problem (MS-FAP) have been proposed (e.g., [2, 3, 6, 8–10, 13, 14]).

The performance of the GAs for MSFAP that represent possible assignment directly as a bit-string or a sequence of integers is not good enough, and the permutation based GAs are reported to show good performance [8, 9, 14]. In these GAs, an assignment order of transmitters is represented by a permutation and an assignment is carried out using a sequential algorithm. The scheme has overcome the weakness of the previous two schemes and the performance has improved [8, 9, 14], and a GA with an adaptive mutation rate and a new initialization method was developed and showed very good performance [9]. The performance of the permutation based GAs are high for MSFAP, but they are designed to find an assignment without violating compatibility constraints. Therefore, they cannot be used for the problem shown in this section.

The formulation shown above, which is quite different from MSFAP, is first proposed and a GA for solving the problem was developed by Jin et al. [5], and an improved version was proposed by Horng et al. [4]. However they use naïve representation of $N \times Z$ matrix that is a bad coding and they use simple GA. Rothlauf et al. [11] showed that we should use well-designed GA in the case of bad codings, therefore the performance of previously proposed GAs is not good enough.

3 The Proposed Algorithm

We propose a new genetic algorithm for the FAP formulated in the previous section. The main idea of the proposed GA is to encode a sequence of codes of a virtual machine that performs assignment as a chromosome, and for the GA to search a good sequence of codes that minimizes the total damage cost.

As the genetic algorithm, we use the parameter-free genetic algorithm (PfGA) proposed by Sawai et al. [7, 12] that is outlined in the appendix. The PfGA is not only simple and robust, but also does not need to set almost any genetic parameters in advance that need to be set in other GAs. The performance of PfGA is high for functional optimization problems of 5- or 10-dimensions [7, 12].

3.1 How Many Frequencies Are Necessary for a Cell?

The second term of the objective function decreases as we increase the number of assigned frequencies (F_i), but the first term, interference cost, would increase. Let us consider the cost function

$$C_i(F_i) = \sum_{p=1}^{Z} \sum_{q=1}^{Z} f(i, i, p, q) + \alpha I_E(TF_i),$$

which consists of the blocking cost and the interference cost within a cell i when just considering the co-site compatibility only.

We can find a good candidate of F_i using the $C_i(F_i)$. The frequency separation in a cell decreases as we increase F_i, and the interference cost increases, whereas the $I_E(TF_i)$ decreases. Therefore we can find F_i that minimizes $C_i(F_i)$ by evaluating $C_i(F_i)$ for $F_i = 1, 2, \cdots, \bar{F}_i$, i.e., the optimal F_i^* can be defined as

$$F_i^* = \operatorname*{argmin}_{F_i \in \{1, 2, \cdots, \bar{F}_i\}} C_i(F_i).$$

Because $I_E(x)$ rapidly decreases and becomes close to zero when $x \geq \mu_i + 5\sigma_i$, therefore we can set $\bar{F}_i = \lceil \mu_i + 5\sigma_i \rceil$. The minimal frequency separation S_i^m of each cell i, which minimizes $C_i(F_i)$, is calculated by

$$S_i^m = \min\{\lfloor (Z-1)/(F_i^* - 1) \rfloor, c_{ii}\}.$$

We can also use $LB = \sum_{i=1}^{N} C_i(F_i^*)$ as a (very loose) lower bound of the total cost.

Table 1. Action specification

Action	Aassigning frequency f
0	frequency with separation of S_i^m
1	frequency with separation of $S_i^m + 1$
2	minimum usable frequency

3.2 Virtual Machine

Let us consider a virtual machine that assigns frequencies one by one to cells according to a given sequence of codes. The cost function $C_i(F_i^*)$ is minimized by assigning all frequencies with the separation of S_i^m when $(Z-1)\mathbf{mod}(F_i^*-1) = 0$. When $(Z - 1)\mathbf{mod}(F_i^* - 1) \neq 0$, some frequencies must be assigned with the separation S_i^m+1. Because $C_i(F_i)$ does not consider the compatibility constraint between cell i and j, the assignment with the above rule does not minimize the total cost, therefore some assignments should be with the separation that does not violate the inter-cell constraints.

With these observations, the virtual machine must have at least the three actions/ operations that are shown in Table 1. And the assignment order to cells is also crucial, therefore an instruction of the virtual machine is defined as a pair of integers, namely (*cell_number*, *action*).

The virtual machine assigns a frequency f that is determined by the Table 1 to a cell *cell_number* when $1 \leq f \leq Z$, and does nothing in other cases. After the frequency f is assigned to the cell i, the set of usable frequencies of all cells is updated according to the compatibility matrix C. In this step, we use S_i^m instead of c_{ii}.

3.3 Chromosome Representation

An instruction (p, a) assigns a frequency to a cell p, therefore the length of the sequence becomes $L = \sum_i^N F_i^*$. Thus a valid sequence of codes is expressed as a sequence $S = \{(p_1, a_1), (p_2, a_2), \cdots, (p_L, a_L)\}$. The *cell_number* part in S must be a permutation of cell numbers where a cell number occurs multiple times, therefore we use the random keys representation [1]. The random keys representation encodes a permutation with random numbers. These values are used as the *sort key* to decode a permutation. An example is shown below. Let us consider the sequence of numbers

$$(1,\ 4,\ 0,\ 3,\ 2)$$

From this sequence, by sorting in ascending order, we get the permutation,

$$3 \to 1 \to 5 \to 4 \to 2.$$

Because any ordered set of random keys can be interpreted as a permutation, any crossover and/or any mutation produce valid permutation.

The chromosome of the GA is a sequence of genes that consists of a pair of integers, one for random keys[1] and the other is for action, namely (r, a) where $0 \leq r \leq L-1$, and $a \in \{0, 1, 2\}$.

3.4 Mutation

The original PfGA uses the bit flip mutation. Because the chromosome of our GA uses a pair of integers as a gene, the mutation is done as follows.

The randomly chosen portion of a chromosome is mutated by the following method. Let p and q be the starting and the ending position of genes in a chromosome. The value of random keys part of gene $g_i (p \leq i \leq q)$ is replaced by a randomly chosen integer in the range of $[0, L-1]$. The value of the action part of a randomly chosen gene g_i is replaced by a randomly chosen integer in the range of $[0, 2]$. This mutation is done randomly, i.e., the probability that a gene g_i is mutated or not is 50% for all genes $g_i (p \leq i \leq q)$.

The mutation operator does not generate an invalid chromosome, i.e., a generated offspring is always decoded into a valid sequence of codes for the virtual machine.

3.5 Local Search

After the assignment by the virtual machine, we could improve the assignment by a local search. If a cell has a frequency f_p that violates the interference constraints, and there is a usable frequency f_q, then we can replace f_p by f_q, and the replacement always decreases the total cost. After this local search, the chromosome is modified to reflect the modification.

This modification algorithm changes the order of instructions and the action part according to the result of the local search. The basic ideas are as follows;

- If the frequency f_q is generated by the local search, the corresponding instruction is moved towards the tail to delay the assignment.
- The action that assigns the minimum usable frequency should be kept unchanged.

The modification algorithm is shown in Figure 1. The algorithm returns a modified sequence of codes, therefore the random keys part of a chromosome is modified according to the sequence of codes. The random keys parts are replaced by the orders in the sequence of codes.

4 Experiments and Results

4.1 Benchmark Problems

We tested the algorithm using the problem proposed by Hong et al. [4]. The eight benchmark problems were examined in the experiments. Three compatibility matrices C_3, C_4, C_5 and three communication load tables D_1, D_2, D_3 were

[1] We use integers as random keys instead of real numbers to reduce search space. If we fix the sorting algorithm for decoding, ties of keys cause no problem.

Procedure Modification
BEGIN
Initialize:

(1) Let two sequences of S_1 and S_2 be empty.
(2) Sort assigned frequency in ascending order for each cell, and let the result be $(f_{i,1}, f_{i,2}, \cdots, f_{i,F_i})$. And also let $f_{i,0} \leftarrow 1 - S_i^m$.

Scan genes from head to tail:
LOOP

(1) Let i be the cell number, and a be the action of current gene that corresponds to th k-th assignment to cell i.
(2) Calculate frequency separations $s \leftarrow f_{i,k} - f_{i,(k-1)}$
(3) **If** $(s > S_i^m + 1)$ **or** $(s < 0)$ **then** $t \leftarrow 2$ **else** $t \leftarrow s - S_i^m$
(4) **If** $t = 2$ **then**
  If $f_{i,k}$ is generated by the local search **then** append instruction (i, t) to S_2.
  else append instruction (i, t) to S_1.
 else
  If $a = 2$ **then** append instruction (i, a) to S_1.
  else append instruction (i, t) to S_1.

UNTIL all genes are scanned.
Return the concatenation of S_1 and S_2.
END

Fig. 1. Modification algorithm

combined to make the eight problems shown in Table 2 [4][2]. In Table 2, N denotes the number of cells, Z denotes the number of available frequencies, C denotes the compatibility matrix, and D denotes the communication load table.

The interference cost functions used were $I_C(x) = 5^{x-1}$ and $I_A(x) = 5^{2x-1}$, the weight of blocking cost $\alpha = 1000$, and the value of T (number of channels that each frequency provides under TDMA) was set to 8. The fitness of the individual was defined as the inverse of the objective function, i.e., $1/O$.

4.2 Results

We tested the performance of the GA by running it 100 times for each problem. The results are shown in Table 3. Comparisons with the results by Horng et al. [4] are also given. The column LB denotes the lower bound that is defined in subsection 3.1. For all the experiments, the maximum number of fitness evaluation was set to the same number of the experiment by Horng et al. [4], i.e., $Z \times 1000$.

Table 3 shows that our GA performs very well. It outperforms the previous GA [4] in all cases, and the costs obtained by our GA are very small compared to the ones by Horng et al. [4]. Our GA can find very good assignment that has the same cost of the lower bound for P1 and P2, and the cost is very close to the lower bound for P3 and P4.

[2] According to private communications with an author of the paper [4], there are typos in their paper. The typos are the communication load table of Problem 1, and the weight α. And there is no description of the value of T. We used the correct data that were provided by the author, and they are shown in this paper.

Table 2. Problem specification

P	N	Z	C	D
P1	21	60	C_3	D_1
P2	21	60	C_4	D_1
P3	21	60	C_5	D_1
P4	21	60	C_4	D_2
P5	21	60	C_5	D_2
P6	21	40	C_5	D_1
P7	21	40	C_5	D_2
P8	21	64	C_5	D_3

Table 3. Solutions to the problems

		GA by Horng et al.[4]		Proposed GA	
P	LB	Best	Average	Best	Average
P1	3.7e-4	203.4	302.6	3.7e-04	3.7e-04
P2	4.1	271.4	342.5	4.1	4.1
P3	4.1	1957.4	2864.1	7.2	21.8
P4	231	906.3	1002.4	243.8	247.9
P5	231	4302.3	4585.4	695.9	982.3
P6	190	4835.4	5076.2	820.8	1210.0
P7	2232	20854.3	21968.4	3891.7	5275.3
P8	22518	53151.7	60715.4	34286.5	38605.9

The GA can obtain assignments without any violation of the EMC constraints for the problems P1, P2, and P3. The performance improvement is significant for all problems. And it should be noted that the results were obtained without any parameter tuning. The GA has only one parameter, the maximum number of fitness evaluation, which can be easily determined from the allowable computation time.

5 Conclusions

We have presented here a novel genetic algorithm for a fixed channel assignment problems with limited bandwidth constraint based on the parameter-free genetic algorithm (PfGA). The algorithm uses the GA to find a good sequence of codes for a virtual machine that executes assignment tasks. The proposed GA is tested using a set of benchmark problems, and the performance is superior to the existing GAs without tedious parameter tuning. The proposed GA can obtain very good solutions that were unable to be found using the existing GAs. We can also conclude that the performance of PfGA for the fixed channel assignment problem with limited bandwidth constraint, which is a combinatorial optimization problem, is as good as for functional optimization problems.

Acknowledgments

The authors are grateful to Mr. Ming-Hui Jin of National Central University, Chungli, Taiwan for providing us the data of the experiment.

References

1. Bean, J.C.: Genetics and random keys for sequencing and optimization, *ORSA Journal of Computing*, vol.6, no.2, pp.154–160 (1994).
2. Crompton, W., Hurley, S., and Stephens, N.M.: Applying genetic algorithms to frequency assignment problems, *Proc. SPIE Conf. Neural and Stochastic Methods in Image and Signal Processing*, vol.2304, pp.76–84 (1994).

3. Hurley, S., Smith,D.H., and Thiel, S.U.: FASoft: a system for discrete channel frequency assignment, *Radio Science*, vol.32, no.5, pp.1921–1939 (1997).
4. Horng, J.T., Jin, M.H., and Kao, C.Y.: Solving fixed channel assignment problems by an evolutionary approach, *Proc. of GECCO-2001*, pp.351–358 (2001).
5. Jin, M.H., Wu, H.K., Horng, J.Z., and Tsai, C.H.: An evolutionary approach to fixed channel assignment problems with limited bandwidth constraint, *Proc. of IEEE ICC 2001*, vol.7, pp.2100–2104 (2001).
6. Kim, J.-S., Park, S.H., Dowd, P.W., and Nasrabadi, N.M.: Comparison of two optimization techniques for channel assignment in cellular radio network, *Proc. of IEEE Int. Conf. Commun.*, vol.3, pp.850–1854 (1995).
7. Kizu, S., Sawai, H., and Endo, H.: Parameter-free genetic algorithm: GA without setting genetic parameters, *Proc. of 1997 International Symposium on Nonlinear Theory and its Application*, vol.2 of 2, pp.1273–1276 (1997).
8. Matsui, S. and Tokoro, K.: A new genetic algorithm for minimum span frequency assignment using permutation and clique, *Proc. of GECCO-2000*, pp.682–689 (2000).
9. Matsui, S. and Tokoro, K.: Improving the performance of a genetic algorithm for minimum span frequency assignment problem with an adaptive mutation rate and a new initialization method, *Proc. of GECCO-2001*, pp.1359–1366 (2001).
10. Ngo, C.Y. and Li, V.O.K.: Fixed channel assignment in cellular radio networks using a modified genetic algorithm, *IEEE Trans. Veh. Technol.*, vol.47, no.1, pp.163–172 (1998).
11. Rothlauf, F., Goldberg, D.E., and Heinzl, A.: Bad coding and the utility of well-designed genetic algorithms, *Proc. of GECCO-2000*, pp.355–362 (2000).
12. Sawai, H., Kizu, S.: Parameter-free genetic algorithm inspired by "disparity theory of evolution," *Proc. of PPSN-V*, pp.702–711 (1998).
13. Smith, D.H., Hurley, S., and Thiel, S.U.: Improving heuristics for the frequency assignment problem, *Eur. J. Oper. Res.*, vol.107, no.1, pp.76–86 (1998).
14. Valenzuela, C., Hurley, S., and Smith, D.: A permutation based algorithm for minimum span frequency assignment, *Proc. of PPSN-V*, pp.907–916 (1998).

Appendix: Overview of Parameter-Free Genetic Algorithm (PfGA)

The Parameter-free Genetic Algorithm (PfGA) was proposed by Sawai et al. [7, 12]. The PfGA is a very compact and fast adaptive search algorithm based on the variable-size of population taking a dynamic but delicate balance between exploration, i.e., global search, and exploitation, i.e., local search. The PfGA is not only simple and robust, but also does not need to set almost all genetic parameters in advance that need to be set up in other GAs. The outline of PfGA procedure is shown below [7, 12].

Step 0: Let S be the whole population, and S' be the subpopulation.
Step 1: Select one individual randomly from S, and add it to S'.
Step 2: Select one individual randomly from S, and add it to S'.
Step 3: Select two individuals P_1, P_2 randomly from S' and perform multiple-point crossover to generate two children C_1, C_2. In the crossover n crossover points (n is a random integer $0 < n < L$, where L is the length of chromosome) are randomly selected.

Step 4: For one randomly selected child, perform mutation. In the mutation, a randomly chosen portion of chromosome is inverted, i.e., bit-flipped, at random.

Step 5: Select one to three individuals among P_1, P_2, C_1, C_2 depending on the cases (described later), and feed them back to S'.

Step 6: If the terminating condition is satisfied, then terminate. As the terminating condition, we use the maximum number of fitness evaluation. The algorithm terminates when the number of fitness evaluation reaches the prescribed maximum.

Step 7: If $|S'| > 1$ then go to Step 3, otherwise go to Step 2.

For the selection operation in Step 5, the fitness values f of P_1, P_2, C_1, C_2 are compared, and selection is done according to the following rules.

Case 1: If the fitness values of children $f(C_1)$ and $f(C_2)$ are better than those of the parents, then C_1 and C_2 and $\underset{P_i}{\operatorname{argmax}}(f(P_1), f(P_2)$ are left in S'.

Case 2: If the fitness values of children $f(C_1)$ and $f(C_2)$ are worse than those of parents, then only $\underset{P_i}{\operatorname{argmax}}(f(P_1), f(P_2)$ is left in S'.

Case 3: If the fitness values of either $f(P_1)$ or $f(P_2)$ is better than those of children, then $\underset{P_i}{\operatorname{argmax}}(f(P_1), f(P_2)$ and $\underset{C_i}{\operatorname{argmax}}(f(C_1), f(C_2)$ are left in S'.

Case 4: In all other situations, $\underset{C_i}{\operatorname{argmax}}(f(C_1), f(C_2)$ is preserved and then one individual randomly chosen from S is added to S'.

Real-Coded Parameter-Free Genetic Algorithm for Job-Shop Scheduling Problems

Shouichi Matsui, Isamu Watanabe, and Ken-ichi Tokoro

Communication & Information Research Laboratory (CIRL)
Central Research Institute of Electric Power Industry (CRIEPI)
2-11-1 Iwado-kita, Komae-shi, Tokyo 201-8511, Japan
{matsui,isamu,tokoro}@criepi.denken.or.jp

Abstract. We propose a new genetic algorithm (GA) for job-shop scheduling problems (JSSP) based on the parameter-free GA (PfGA) and parallel distributed PfGA proposed by Sawai et al. The PfGA is not only simple and robust, but also does not need to set almost any genetic parameters in advance that need to be set in other GAs. The performance of PfGA is high for functional optimization problems of 5- or 10-dimensions, but its performance for combinatorial optimization problems, which search space is larger than the functional optimization, has not been investigated. We propose a new algorithm for JSSP based on an extended PfGA, extended to real-coded version. The GA uses random keys for representing permutation of jobs. Simulation results show that the proposed GA can attain high quality solutions for typical benchmark problems without parameter tuning.

1 Introduction

The Genetic Algorithm (GA) is an evolutionary computation paradigm, and has been successfully applied to many practical problems such as functional optimization, combinatorial optimization, and so on [8]. However, the tuning of genetic parameters has to be performed by trial and error, making optimization by GA *ad hoc.*

There have been many research projects for *self-adaptive* GA because such adaptation can tune the parameters while solving a given problem. Nevertheless, it is a very time-consuming task to design an optimal GA in an adaptive way because we have to perform computation many times by trial and error. To address this problem, Sawai et al. have proposed the Parameter-free Genetic Algorithm (PfGA), for which no control parameters for genetic operation need to be set in advance [10, 14, 15]. The performance of the PfGA has been reported to be high compared with other GAs [10, 14, 15]. But as far as the authors know, these papers only show the performance for functional optimization problems, and the performance for combinatorial optimization problems is unknown.

This paper reports the result of empirical evaluation of applicability of PfGA to a combinatorial optimization problem, the job-shop scheduling problem

J.J. Merelo Guervós et al. (Eds.): PPSN VII, LNCS 2439, pp. 800–810, 2002.

(JSSP). The proposed GA keeps the basic framework of PfGA, but the chromosome is extended from bit-string to real numbers. The proposed GA uses random keys to represent the permutation of operations, and uses the hybrid-scheduling method to decode permutation into schedule. Simulation results show that the proposed GA can attain high quality solutions for typical benchmark problems without tedious parameter tuning.

2 Job-Shop Scheduling Problem

In the Job-Shop Scheduling Problem (JSSP), n jobs have to be processed on m different machines. Each job J_i consists of a sequence of tasks $T_{i,1}, \cdots, T_{i,m}$ that have to be completed during an uninterrupted time period of length $p_{i,j} \in \mathbb{N}$ on a given machine ($M(T_{i,j}) \in \{1, \cdots, m\}$). A schedule is an allocation of tasks to time intervals on machines. The goal in the job-shop scheduling problem is to find the sequence of n jobs to be completed on m machines such that the makespan (finish time of the last operation) is minimized.

The job-shop scheduling problem is well known as one of the most difficult NP-hard combinatorial optimization problems. Several approaches have been reported to the JSSP for a few decades. Among them work on GAs for solving the JSSP has a history of only one decade, but they perform well compared with other approaches.

Davis [6] was the first to use Genetic Algorithms to solve JSSP, and many GAs have been proposed and analyzed (e.g., [3, 11–13, 17, 19, 20]). Other heuristics such as Simulated Annealing (SA) and Tabu Search (TS), and exact techniques as branch and bound have also been developed for solving JSSP. Jain and Meeran provide a good description of these techniques [9].

2.1 Chromosome Representation

There are two ways of representing a schedule: indirect and direct. In *indirect* representation, the chromosome contains an encoded schedule. A schedule builder is used to transform the chromosome into a feasible schedule. The indirect representations range from traditional binary representations [12] to domain-specific knowledge representation [1].

In *direct* representation, the chromosome directly represents the production schedule. Bruns [5] shows many ways to deal with direct representations. Direct representation performs efficiently on production scheduling, incorporates domain-specific operations easily, but also requires domain-specific recombination operators.

2.2 Types of Feasible Schedules

There are four types of feasible schedules in JSSP as follows

inadmissible: Inadmissible schedules contain excess idle time, and they can be improved by forward-shifting operations until no excess idle time exists.

semi-active: Semi-active schedules contain no excess idle time, but they can be improved by shifting some operations to the front without delaying others.

active: Active schedules contain no idle time and no operation can be finished earlier without delaying other operations. The optimal schedule is guaranteed to be an active schedule.

non-delay: Non-delay schedules are active schedules, in which operations are placed into the schedule such that the machine idle time is minimized. No machine is kept idle if some operation can be processed.

3 Parameter-Free Genetic Algorithm (PfGA)

3.1 Parameter-Free Genetic Algorithm

The Parameter-free Genetic Algorithm (PfGA) proposed by Sawai et al. [10, 14, 15] is a novel GA inspired by the "disparity theory of evolution." The idea of the theory is based on different mutation rate in double strands of DNA. The PfGA is a very compact and fast adaptive search algorithm based on the variable-size of population taking a dynamic but delicate balance between exploration, i.e., global search, and exploitation, i.e., local search. The PfGA is not only simple and robust, but also does not need to set almost all genetic parameters in advance that need to be set up in other GAs. The performance of PfGA to function optimization problems are reported to be very high [10, 14, 15].

The PfGA procedure is as follows[14].

Step 0: Let S be the whole population, and S' be the subpopulation.

Step 1: Select one individual randomly from S, and add it to S'.

Step 2: Select one individual randomly from S, and add it to S'.

Step 3: Select two individuals P_1, P_2 randomly from S' and perform multiple-point crossover to generate two children C_1, C_2. In the crossover n crossover points (n is a random integer $0 < n < L$, where L is the length of chromosome) are randomly selected.

Step 4: For one randomly selected child, perform mutation. In the mutation, a randomly chosen portion of chromosome is inverted, i.e., bit-flipped, at random.

Step 5: Select one to three individuals among P_1, P_2, C_1, C_2 depending on the cases (described later), and feed them back to S'.

Step 6: If the terminating condition is satisfied, then terminate.

Step 7: If $|S'| > 1$ then go to Step 3, otherwise go to Step 2.

For the selection operation in Step 5, the fitness values f of P_1, P_2, C_1, C_2 are compared, and the selection is done according to the following rules.

Case 1: If the fitness values of children $f(C_1)$ and $f(C_2)$ are better than those of the parents, then C_1 and C_2 and $\operatorname{argmax}_{P_i}(f(P_1), f(P_2)$ are left in S'.

Case 2: If the fitness values of children $f(C_1)$ and $f(C_2)$ are worse than those of parents, then only $\operatorname{argmax}_{P_i}(f(P_1), f(P_2)$ is left in S'.

Case 3: If the fitness values of either $f(P_1)$ or $f(P_2)$ is better than those of children, then $\underset{P_i}{\operatorname{argmax}}(f(P_1), f(P_2)$ and $\underset{C_i}{\operatorname{argmax}}(f(C_1), f(C_2)$ are left in S'.

Case 4: In all other situations, $\underset{C_i}{\operatorname{argmax}}(f(C_1), f(C_2)$ is preserved and then one individual randomly chosen from S is added to S'.

3.2 Distributed Parallel PfGA

Generally speaking, parallel processing in GA aims at reaching better solutions faster than sequential processing by extending the search space. The distributed parallel processing of PfGA is as follows [14].

Let us assume that there are N subpopulations $S_i'(i = 1, 2, \cdots, N)$, and each subpopulation evolves as shown in the previous section. Migration among the subpopulations occurs when a better individual is produced in some subpopulations. One possible migration method is as follows;

If case 1 or 4 happens in some subpopulation, the individual C_i that is the best child is copied to other subpopulations as an emigrant. When other subpopulations receive the immigrant, they add it as an individual and eliminate the worst individual among all individuals in the subpopulation. Sawai et al. proposed many migration methods, master-slave types and uniform-distributed types. In this paper we mainly tested the UD1 where the best child is copied to a randomly chosen other subpopulation, because the UD1 performed best in the preliminary experiments.

4 The Proposed GA

This section describes the proposed GA. The proposed GA uses the random keys for permutation representation, and hybrid scheduling for decoding the permutation.

4.1 Random Keys for Job-Shop Scheduling

The random keys representation [2, 13] encodes a solution with random numbers. These values are used as the *sort keys* to decode the solution. An example of the random keys encoding for the context of job-shop scheduling is shown below. As a simple example, consider the single machine sequencing problem. When a sequence of random numbers is passed to the fitness evaluation routine, sort them and sequence the jobs in ascending order of the sort. For a five job single machine problem, let us consider the chromosome

$$(0.29, 0.96, 0.17, 0.84, 0.48)$$

From this chromosome, we get the sequence,

$$3 \to 1 \to 5 \to 4 \to 2.$$

Because any ordered set of random keys can be interpreted as a job sequence, any crossover and any mutation produce feasible sequence, i.e., all offspring are feasible solutions.

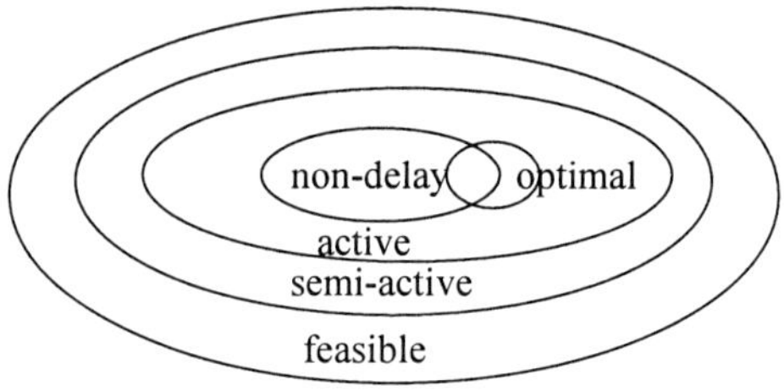

Fig. 1. Relationship of schedule properties [4]

4.2 Real-Coded PfGA (Rc-PfGA)

We extend the PfGA to a real coded version. A gene of the original PfGA takes the value of 0 or 1, i.e., a chromosome is a bit-string. The real coded PfGA is an extension where a chromosome is a string of real numbers. We abbreviate it to Rc-PfGA.

The mutation operator of Rc-PfGA replaces the value of a gene by a random number uniformly distributed in the predefined range.

The random keys representation of permutation can be expressed naturally with the proposed Rc-PfGA.

4.3 Permutation Decoding

Permutations can be decoded into *semi-active*, *active*, or *non-delay* schedules [4]. $\mu_i(k) = j\,(1 \le k \le m_i)$ specifies that M_j is the machine that processes the k-th operation of job J_i, where m_i is the number of operations of J_i. The k-th operation of job J_i processed on machine $M_{\mu_i(k)}$ is denoted as o_{ik}. The required processing time of o_{ik} is denoted by p_{ik}. Let t_{ik} be the starting time of operation o_{ik}, then

$$t_{ik} = \max(t_{i,k-1} + p_{i,k-1}, t_{hl} + p_{hl}) \tag{1}$$

Semi-active Schedule: Schedules are built by successively scheduling the operations by assigning the earliest possible starting time according to Equation (1). The permutation to be decoded serves as a look-up array for the procedure shown below.

An operation o_{ik} found in the permutation is only schedulable if its predecessor operation $o_{i,k-1}$ has been scheduled already. Therefore in a feasible permutation $o_{i,k-1}$ occurs to the left of o_{ik}.

Step 1: Build the set of all beginning operations $A := \{o_{i1} \mid 1 \le i \le n\}$.
Step 2: Select operation o^*_{ik} from A which occurs leftmost in the permutation and delete it from A.
Step 3: Append operation o^*_{ik} to the schedule and calculate its starting time.
Step 4: If o^*_{ik} has successor $o^*_{i,k+1}$, insert it into A.
Step 5: If A is empty, terminate. Otherwise, go to Step 2.

Active Schedule: Active schedules are produced by modification of Step 2 in the above procedure which leads to the well-known algorithm of Giffler and Thompson [7].

Step 2.A1: Find an operation o' from A with the earliest possible completion time.
Step 2.A2: Determine the machine M' of o' and build the set B from all operations in A which are processed on M'.
Step 2.A3: Delete operations in B which do not start before the completion of o'.
Step 2.A4: Select operation o^*_{ik} from B which occurs leftmost in the permutation and delete it from A.

This algorithm produces schedules in which no operation could be started earlier without delaying some other operation or breaking an order constraint. Active schedules are also semi-active schedules.

Non-delay Schedule: Non-delay schedule means that no machine is kept idle when it could start processing some operation. Non-delay schedules are necessarily active and hence also necessarily semi-active.

Schedules are produced similarly as active by using a more rigid criterion for picking an operation from B by the modified step 2.

Step 2.N1: Find an operation o' from A with the earliest possible starting time.
Step 2.N2: Determine the machine M' of o' and build the set B from all operations in A which are processed on M'.
Step 2.N3: Delete operations in B which start later then operation o'.
Step 2.N4: Select operation o^*_{ik} from B which occurs leftmost in the permutation and delete it from A.

Hybrid Schedule [4]: Hybrid schedules can be produced in a flexible way by introducing a parameter $\delta \in [0, 1]$. The setting of δ can be thought of as defining a bound on the length of time a machine is allowed to remain idle.

Step 2.H1: Find an operation o' from A with the earliest completion time $t' + p'$.
Step 2.H2: Determine the machine M' of o' and build the set B from all operations in A which are processed on M'.
Step 2.H3: Find an operation o'' from B with the earliest possible starting time t''.
Step 2.H4: Delete operations in B in accordance with parameter δ such that $B := \{o_{ik} \in B \,|\, t_{ik} < t'' + \delta((t' + p') - t'')\}$.
Step 2.H5: Select operation o^*_{ik} from B which occurs leftmost in the permutation and delete it from A.

Move Search [13]: The move search schedule proposed by Norman and Bean [13] constructs a schedule by using a parameter, *delay factor* DF, that is associated with each operation. The move search schedule is produced as follows.

Step 1: Build the set A of schedulable operation, initially, the first operation of each job is schedulable.
Step 2: Selection.
- **Step 2-D1:** Sorting the random keys of operation in the set A, find the candidate operation o^*_{ik} with the smallest key.
- **Step 2-D2:** Build the set K of operations that requires the same machine as the candidate operation. Order K based on the random key values of the operations.
- **Step 2-D3:** If $|K| = 0$ go to Step 3, otherwise set $a = 1$.
- **Step 2-D4:** Let o_{ij} be the operation associated with element a of K, k_a.
- **Step 2-D5:** If $t_{ij} + DF_{ij} \times p_{ij} < t^*_{ik}$, then o_{ij} becomes the candidate operation. Update t^*_{ik}.
- **Step 2-D6:** If $a < |K|$, let $a = a + 1$ and return to Step 2-D4.

Step 3: Append operation o^*_{ik} to the schedule and calculate its starting time.
Step 4: If o^*_{ik} has successor $o^*_{i,k+1}$, insert it into A.
Step 5: If A is empty, terminate. Otherwise go to Step 2.

4.4 Chromosome

The chromosome of the proposed GA is a sequence of real numbers, and its length is $2 \times n \times m$. The chromosome is divided into two parts, the first part represents the random keys, and the second part represents the δs of the hybrid schedule. Bierwirth and Mattfeld [4] used single δ for their GA, and showed that the performance of GA depends on δ by numerical experiments. They showed that the value of δ to be most suitable is 0.4 for "hard", 0.5 for "moderate", and 0.7 for "easy" problem instances. But it is very hard to classify the given problem instance in advance, therefore we designed our GA to search suitable δs also by the GA.

Bierwirth and Mattfeld used single δ, but we found that the performance of the GA with multiple δ is better than that with single δ by extensible experiments, therefore our GA has $n \times m$ δs, i.e., the hybrid scheduler uses different δ for each operation.

5 Computational Results

The proposed RcPfGA is tested by the benchmark problems from ORLib [18]. As the problem instances, we used the same set that was used by Norman and Bean [13] because their GA and our GA use the same representation, random keys. Our GA uses the hybrid schedule, and the GA by Norman and Bean, RKGA, uses the move search schedule.

Table 1. Simulation results

Prob.	Size	Opt. [9]	RKGA[13] Mean	(%)	Best	(%)	Our GA Mean	(%)	Best	(%)
FT10	10×10	930	945	(1.61)	937	(0.75)	941.85	(1.27)	**930**	(0.00)
FT20	20×5	1165	1176	(0.94)	**1165**	(0.00)	1177.7	(1.09)	1173	(0.69)
LA02	10×5	655	661	(0.92)	**655**	(0.00)	655.46	(0.07)	**655**	(0.00)
LA03	10×5	597	599	(0.34)	**597**	(0.00)	597.84	(0.14)	**597**	(0.00)
LA04	10×5	590	592	(0.34)	**590**	(0.00)	**590.00**	(0.00)	**590**	(0.00)
LA16	10×10	945	958	(1.38)	**945**	(0.00)	945.52	(0.06)	**945**	(0.00)
LA18	10×10	848	850	(0.24)	**848**	(0.00)	**848.00**	(0.00)	**848**	(0.00)
LA19	10×10	842	853	(1.31)	851	(1.07)	852.30	(1.22)	**842**	(0.00)
LA20	10×10	902	908	(0.67)	907	(0.55)	906.80	(0.53)	**902**	(0.00)
LA21	15×10	1046	1062	(1.53)	1055	(0.86)	1071.82	(2.47)	1053	(0.67)
LA22	15×10	927	936	(0.97)	933	(0.65)	943.76	(1.81)	932	(0.53)
LA24	15×10	935	977	(4.49)	966	(3.32)	959.78	(2.65)	938	(0.32)
LA25	15×10	977	995	(1.84)	987	(1.02)	996.80	(2.03)	978	(0.10)
LA26	20×10	1218	**1218**	(0.00)	**1218**	(0.00)	1220.96	(0.24)	**1218**	(0.00)
LA27	20×10	1235	1269	(2.75)	1256	(1.70)	1273.74	(3.14)	1260	(2.02)
LA28	20×10	1216	1241	(2.06)	1241	(2.06)	1244.84	(2.37)	1226	(0.82)
LA29	20×10	1152	1188	(3.13)	1179	(2.34)	1207.64	(4.83)	1184	(2.78)
LA36	15×15	1268	1300	(2.52)	1287	(1.50)	1301.92	(2.68)	1279	(0.87)
LA37	15×15	1397	1432	(2.51)	1418	(1.50)	1440.24	(3.10)	1412	(1.07)
LA38	15×15	1196	1232	(3.01)	1217	(1.76)	1260.92	(5.43)	1208	(1.00)
LA39	15×15	1233	1260	(2.19)	1258	(2.03)	1263.64	(2.48)	1250	(1.38)
LA40	15×15	1222	1256	(2.78)	1234	(0.98)	1258.50	(2.99)	1240	(1.47)

5.1 Results by PfGA

The proposed GA was run using 50 different random seeds where the maximum number of fitness evaluation was set to 1,000,000. The average over 50 and the best solution over 50 seeds are shown in Table 1. The results of Norman and Bean [13] are also shown in the table. Note that their results are from 10 runs, the average are over 10 seeds, and the best is from the first five runs. The numbers in parentheses are the relative error of the makespan, and numbers in bold face means that they are the optimal value.

The GA was tested for the problems FT10, FT20, and LA00–LA40, but the results of problems where average and best makespan are equal to the optimal in both GAs are omitted in the table due to limitations of space. In Table 1, *Prob.* denotes the name of the problem, *Size* denotes the size of the problem, *Opt.* denotes the optimal makespan, *Mean* denotes the average makespan over 50 runs, and *Best* denotes the best makespan over 50 runs.

Table 1 shows the following.

- The performance of the proposed GA is better than that of RKGA for small-sized problems, namely FT10, LA02–LA20, but it is worse than that of RKGA for some larger problems, namely LA27, LA29, LA40.

Table 2. Simulation results of distributed parallel PfGA

	$N=1$			$N=2$			$N=4$			$N=8$		
Prob.	Best	Mean	σ	Best	Mean	σ	Best	Mean	σ	Best	Mean	σ
FT10	**930**	942.2	11.21	**930**	941.1	9.37	**930**	939.2	8.72	**930**	934.0	6.49
FT20	1173	1178.4	1.84	**1165**	1177.6	2.87	**1165**	1177.4	3.90	**1165**	1177.0	3.22
LA02	**655**	655.5	1.99	**655**	**655.0**	**0.00**	**655**	**655.0**	**0.00**	**655**	**655.0**	**0.00**
LA03	**597**	597.8	2.08	**597**	597.1	0.84	**597**	**597.0**	**0.00**	**597**	**597.0**	**0.00**
LA04	**590**	**590.0**	**0.00**	**590**	**590.0**	**0.00**	**590**	**590.0**	**0.00**	**590**	**590.0**	**0.00**
LA16	**945**	945.5	0.50	**945**	945.5	0.50	**945**	945.3	0.44	**945**	945.1	0.32
LA18	**848**	**848.0**	**0.00**	**848**	**848.0**	**0.00**	**848**	**848.0**	**0.00**	**848**	**848.0**	**0.00**
LA19	**842**	852.3	5.86	**842**	852.0	5.04	**842**	847.8	2.77	**842**	846.6	3.57
LA20	**902**	906.8	0.98	**902**	906.5	1.50	**902**	906.8	0.98	**902**	906.1	1.92
LA21	1053	1071.8	11.80	1047	1070.2	11.31	1047	1066.5	9.79	1050	1062.2	7.72
LA22	932	943.8	8.14	**927**	939.5	5.73	**927**	938.3	6.54	**927**	934.5	4.90
LA24	938	959.8	12.57	939	957.1	10.28	938	954.1	10.56	941	951.9	8.44
LA25	978	996.8	10.13	**977**	991.4	7.93	**977**	988.0	7.24	**977**	986.0	6.66
LA26	**1218**	1221.0	4.69	**1218**	1219.5	4.30	**1218**	1218.8	2.24	**1218**	1218.5	2.05
LA27	1260	1273.7	9.53	1255	1267.8	5.45	1250	1265.2	6.88	1250	1263.5	6.32
LA28	1226	1244.8	9.54	1222	1242.8	8.58	1219	1236.2	8.83	**1216**	1232.3	7.88
LA29	1184	1207.6	12.50	1171	1203.5	12.64	1175	1199.8	13.51	1167	1191.0	9.88
LA36	1279	1301.9	6.53	1283	1298.4	6.81	1271	1296.6	6.90	1276	1292.7	5.98
LA37	1412	1440.2	14.39	1407	1432.1	15.51	1407	1427.6	10.64	1400	1422.7	11.63
LA38	1208	1260.9	20.98	1219	1251.3	16.43	1203	1244.1	14.96	1216	1237.7	14.17
LA39	1250	1263.6	11.62	1248	1260.6	11.64	1249	1255.1	7.81	1248	1252.4	4.02
LA40	1240	1258.5	9.21	1228	1254.9	11.27	1228	1249.7	9.03	1231	1245.8	8.03

- The proposed GA can always find the optimal solutions for the problem LA04 and LA18, for which the RKGA sometimes failed to find the optimal. On the other hand, the proposed GA sometimes cannot find the optimal for the problem LA26 for which the RKGA can always find the optimal.
- For the larger size problems, the proposed GA can find better solutions than RKGA, but the average makespans are larger.

5.2 Results by Distributed Parallel PfGA

The distributed parallel version was coded using MPI [16]. It was run using 50 different random seeds where the maximum number of fitness evaluation was set to 1,000,000 for each subpopulation. The number of subpopulations (N) was set to $N = \{2, 4, 8\}$. The average over 50 and the best solution over 50 seeds are shown in Table 2. In Table 2, σ represents the standard deviation of makespan, and other columns have the same meaning as Table 1. The GA was tested for the same problems of serial version and only the results of the problems that are shown in Table 1 are listed due to limitations of space.

Table 2 shows the following.

- As we increase the number of subpopulations, the average makespan decreases, and also the standard deviation of makespan decreases. Therefore, we could easily improve solutions if we could use multiple processors.

– The GA can find the optimal solution as we increase the number of subpopulations. When we set the number of subpopulation to 2, the optimal makespan can be found for FT20, LA22, LA25, which are unable to be found by the single population. The optimal makespan of LA28 can be found when we increase the number of subpopulation to 8.

The number of migration was very small for all problems, it was below 0.01% of the total number of fitness evaluation. Therefore, the overhead of communication by MPI among subpopulations was negligible, and the computation time was almost equal to that of the serial version. The average computation time of the serial version per run, for example, was 42 (sec) for LA02 (10×5), 121 (sec) for FT20 (20×5), 237 (sec) for LA29 (20×10), 225 (sec) for LA40 (15×15) on a PC with an AMD Athlon MP 1800+ (1.53GHz).

6 Conclusion and Future Work

We have proposed a real-coded parameter-free genetic algorithm for JSSP. Numerical experiments showed that the proposed algorithm can attain high quality solutions for typical benchmark problems without tedious parameter tuning. The distributed parallel version can attain better results as we increase the number of subpopulations. The proposed GA is only tested for limited instances of benchmark problems, so we will test it for a wider class of the problems.

The proposed GA searches good permutations, therefore the framework of the proposed GA can be applicable to combinatorial optimization problems that can be formulated as searching good permutations. We will also investigate the applicability to other combinatorial optimization problems.

References

1. Bagchi S., Uckun,S., Miyabe, Y., Kawarnura, K.: Exploring Problem-Specific Recombination Operators for Job Shop Scheduling, *International Conf. Genetic Algorithms (ICGA-91)*, pp.10–17, 1991.
2. Bean, J.C.: Genetics and random keys for sequencing and optimization, *ORSA Journal of Computing*, vol.6, no.2, pp.154–160, 1994.
3. Bierwirth, C., Mattfeld, D., and Kopfer, H.: On permutation representations for scheduling problems, *Proc. of PPSN IV*, pp.310–318, 1996.
4. Bierwirth, C. and Mattfeld, D.C.: Production scheduling and rescheduling with generic algorithms, *Evolutionary Computation*, vol.7, no.1, pp.1–17, 1999.
5. Bruns, R.: Direct chromosome representation and advanced genetic operations for production scheduling, *International Conf. Genetic Algorithms (ICGA-93)*, pp.352–359, 1993.
6. Davis, L.: Job shop scheduling with genetic algorithms, *International Conf. Genetic Algorithms (ICGA-85)*, pp.136–140, 1985.
7. Giffler, B. and Thompson, G., Algorithms for solving production scheduling problems, *Operations Research*, vol.8, pp.487–503, 1960.
8. Goldberg, D.E.: *Genetic Algorithms in Search, Optimization, and Machine Learning*, Addison Wesley, 1989.

9. Jain, A.S., and Meeran, S.: Deterministic job-shop scheduling: past, present and future, *European Journal of Operational Research*, vol.113, pp.390–434, 1999.
10. Kizu, S., Sawai, H., and Endo, H.: Parameter-free genetic algorithm: GA without setting genetic parameters, *Proc. 1997 International Symposium on Nonlinear Theory and its Application*, vol.2 of 2, pp.1273–1276, 1997.
11. Kobayashi, S., Ono, I., and Yamamura, M.: An efficient genetic algorithm for job shop scheduling problems, *Proc. Sixth International Conference on Genetic Algorithms*, pp.506-511, 1995.
12. Nakano, R. and Yamada, T.: Conventional genetic algorithm for job shop scheduling, *Proc. 3rd International Conference on Genetic Algorithms*, pp.474–479, 1991.
13. Norman, B. and Bean, J.C.: Random keys genetic algorithm for job shop scheduling, Engineering Design & Automation, vol.3, no.2, pp.145–156, 1997.
14. Sawai, H., Kizu, S.: Parameter-free genetic algorithm inspired by "disparity theory of evolution", *Proc. of PPSN V*, pp.702–711, 1998.
15. Sawai, H., Kizu., S., and Endo, T.: Parameter-free genetic algorithm (PfGA), *Trans. IEICE, Japan*, vol.J81-D-II, no.2, pp.450–452, 1998 (*in Japanese*).
16. M. Snir, S.W. Otto, S Huss-Lederman, D.W. Walker, J. Dongarra: MPI: The Complete Reference, The MIT Press (1997)
17. Storer, R., Wu, S., and Vaccari, R.: New search spaces for sequencing problems with application to job shop scheduling, *Management Science*, vol.38, pp.1495–1509, 1992.
18. Vaessens, R.J.M.: Operations Research Library of Problems, Management School, Imperial College London, ftp://mscmga.ms.ic.ac.uk/pub/jobshop1.txt, 1996.
19. Vázquez, M. and Whitley, D.: A comparison of genetic algorithms for the static job shop scheduling problem, *Proc. of PPSN VI*, pp.303–312, 2000.
20. Yamada, T. and Nakano, R.: A genetic algorithm applicable to large-scale job-shop problems, *Proc. PPSN II*, pp.281–290, 1992.

Clustering Gene Expression Profiles with Memetic Algorithms

Peter Merz and Andreas Zell

University of Tübingen
Department of Computer Science - WSI-RA
Sand 1, D-72076 Tübingen, Germany
peter.merz@ieee.org

Abstract. Microarrays have become a key technology in experimental molecular biology. They allow a monitoring of gene expression for more than ten thousand genes in parallel producing huge amounts of data. In the exploration of transcriptional regulatory networks, an important task is to cluster gene expression data for identifying groups of genes with similar patterns.
In this paper, memetic algorithms (MAs) – genetic algorithms incorporating local search – are proposed for minimum sum-of-squares clustering. Two new mutation and recombination operators are studied within the memetic framework for clustering gene expression data. The memetic algorithms using a sophisticated recombination operator are shown to converge very quickly to (near-)optimum solutions. Furthermore, the MAs are shown to be superior to multi-start k-means clustering algorithms in both computation time and solution quality.

1 Introduction

In the field of bioinformatics, clustering algorithms have received new attention due to the breakthrough of microarrays. This technology allows monitoring simultaneously the expression patterns of thousands of genes with enormous promise to help genetics to understand the genome [1, 2]. Since a huge amount of data is produced during microarray experiments, clustering techniques are used to extract the fundamental patterns of gene expression inherent in the data. Several clustering algorithms have been proposed for gene expression profile analysis: Eisen *et al.* [3] applied a hierarchical average link clustering algorithm to identify groups of co-regulated genes in the yeast cell cycle. Ben-Dor *et al.* [4] developed the CAST algorithm for this purpose. Tamayo *et al.* [5] used self-organizing maps (SOMs) to identify clusters in yeast and human hematopoietic differentiation data. In [6], the k-means algorithm was used for clustering yeast data and in [7], a comparison of several approaches was made including k-means with average link initialization, which performed well for all tested data sets. In general, these approaches cannot be compared directly, since there is no unique measure of quality. For example, in hierarchical clustering, the clusters are determined by visual inspection using biological knowledge [3], and the number of neurons in the SOM approach [5] is also chosen by visual inspection.

J.J. Merelo Guervós et al. (Eds.): PPSN VII, LNCS 2439, pp. 811–820, 2002.

Clustering can be considered as an optimization problem in which an assignment of data vectors to clusters is desired, such that the sum of squared distances of the vectors to their cluster mean (centroid) is minimal. Let $\mathcal{P}_K$ denote the set of all partitions of X with $X = \{x_1, \ldots, x_N\}$ denoting the data set of vectors with $x_i \in I\!R^m$ and C_i denoting the i-th cluster with mean $\hat{x}_i$. Thus, the objective is

$$\min_{P_K \in \mathcal{P}_K} \sum_{i=1}^{K} \sum_{x_j \in C_i} d^2(x_j, \hat{x}_i), \quad \text{with} \quad \hat{x}_i = \frac{1}{|C_i|} \sum_{x_j \in C_i} x_j \tag{1}$$

where $d(\cdot, \cdot)$ is the Euclidean distance in $I\!R^m$. This problem is called the minimum sum-of-squares clustering (MSSC) problem and is known to be $\mathcal{NP}$-hard [8].

The k-means algorithm is a heuristic which minimizes the sum-of-squares criterion provided an initial assignment/choice of centroids. The number k of clusters is fixed during its run. The k-means heuristic can in fact be regarded as a local search heuristic for this hard combinatorial optimization problem.

Since memetic algorithms are known to be highly effective for several combinatorial optimization problems, including graph bi-partitioning [9], traveling salesman problem [10], or the quadratic assignment problem [11], the application of MAs to the MSSC appears to be promising.

In this paper, memetic algorithms for clustering gene expression profiles using k-means local search are proposed. It is shown that instead of repeatedly starting k-means to find better local optima, the memetic framework is much more effective in finding near optimum solutions quickly.

The paper is organized as follows. The general memetic algorithm framework is described in section 2. In section 3, the new recombination and mutation operator are introduced. In section 4, comparisons of the memetic variation operators are performed on gene expression data and randomly generated data. Furthermore, the MAs are compared with multi-start k-means local search. Section 5 concludes the paper and outlines areas for future research.

2 Memetic Algorithms

Memetic algorithms (MA) [12,13] are population-based heuristic search approaches for combinatorial optimization problems based on cultural evolution. In some respects, they are similar to genetic algorithms which simulate the process of biological evolution. MAs are inspired by Dawkins' notion of a *meme* [14] defined as a unit of information that reproduces itself while people exchange ideas. In contrast to genes, memes are typically adapted by the people who transmit them before they are passed on to the next generation.

According to Moscato and Norman [13], 'memetic evolution' can be mimicked by combining genetic algorithms with local refinement strategies such as local neighborhood search or simulated annealing. Thus, the genetic local search approach proposed in [15] is a special case of a memetic algorithm, which has been shown to be very effective for several combinatorial optimization problems, including the traveling salesman problem (TSP), the graph bi-partitioning problem (GBP), *NK*-Landscapes, and binary quadratic programming [16].

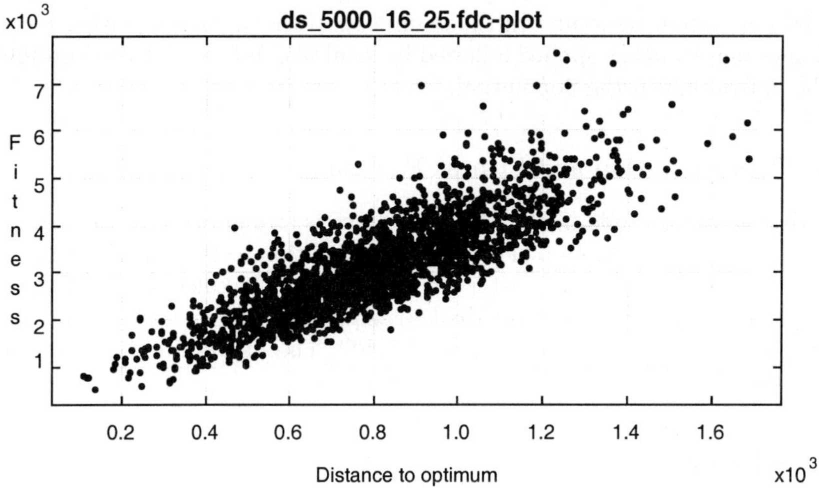

Fig. 1. Fitness-distance scatter plot of the DS-5000 data set. The correlation of fitness and distance to the optimum is shown for 2500 local optima produced with a k-means heuristic for the MSSC.

In contrast to hybrid evolutionary algorithms that use local refinement techniques as additional operators, MAs are designed for searching in the space of locally optimal solutions instead of searching in the space of all candidate solutions. This is achieved by applying local search after each of the evolutionary variation operators. Thus, in every generation, the population of individuals consists solely of local optima.

MAs are known to exploit the correlation structure of the fitness landscape of combinatorial optimization problems [10, 16]: As shown in Fig. 1, many fitness landscapes exhibit a correlation of the fitness of local optima and their distance to the optimum solution. In memetic algorithms, this property can be used in the design of variation operators to achieve a highly effective search approach. However, the balance between disruption and information preservation is important for recombination and mutation: On the one hand, the escape of the local optima must be guaranteed, but on the other hand, disrupting too much may cause the loss of important information gained from previous search.

In a preliminary analysis, we found that the k-means local search heuristic for MSSC produces solutions with a high correlation of fitness and phenotypic distance to the optimum as shown in Fig. 1. Because of the high correlation, the application of memetic algorithms appears to be a promising approach.

3 MAs for Clustering Gene Expression Data

The memetic algorithms described in this contribution are based on a conceptual framework, which is rather simple compared to ASPARAGOS [17] and ASPARAGOS´96 [18], since it uses a single, unstructured population rather than a spatial population structure. In the main loop of the algorithm, either recombination or mutation are applied followed by local search to assure local optimality of the individuals in the population.

3.1 Representation and Fitness Function

The representation used in the MA algorithms is straightforward. The K mean vectors $\hat{x}_i \in \mathbb{R}^m$ in equation (1) constitute a solution to the clustering problem. Given the mean vectors, the cluster memberships can be calculated and therefore have not to be stored in the individuals (genomes). Thus, an individual is simply a k-tuple $a = (a_1, \ldots, a_K)$ of vectors $a_i \in \mathbb{R}^m$. The fitness function used in the MA is the MSSC error provided in equation (1).

3.2 Selection

Selection can be divided into selection for variation and selection for survival. In the recombination and mutation loops, individuals are selected for variation which is done randomly without favoring fitter individuals. Selection for survival is performed on the offspring and parents of the current generation to determine the parents for the next generation. This selection strategy is similar to the selection in the $(\mu + \lambda)$-ES (*Evolution Strategy*): The new population is derived by selecting the best individuals out of the pool of parents and children. Duplicates are eliminated such that a solution is contained no more than once in the population.

3.3 Initialization and Local Search

In all the MAs, local search is performed using the k-means heuristic [19, 20]: Given k centroid vectors a_j, for each input vector $x_i \in X$ the nearest centroid s_j is determined and the vector x_i is associated with the corresponding cluster. Afterwards, all centroid vectors are recalculated by calculating the mean of the vectors in each cluster. This procedure is repeated until the partition does no longer change, and thus a local optimum is reached.

The initial populations of the MAs are produced by applying the k-means algorithm multiple times to randomly generated starting solutions in which the centroids are selected randomly from all input vectors.

3.4 The Mutation Operators

Two mutation operators are used in the MAs. The first is denoted *MM* and simply assigns a randomly chosen input vector x_i $(1 \leq i \leq N)$ to a randomly chosen mean vector a_j $(1 \leq j \leq K)$. The second operator works by randomly selecting two clusters C_i and C_j. Then, the vector $x_{\max} \in C_i$ with the maximum distance to a_i (mean of cluster C_i) is determined and a_j is replaced by $x_{\max}$. This mutation operator is referred to as *FM* in the following.

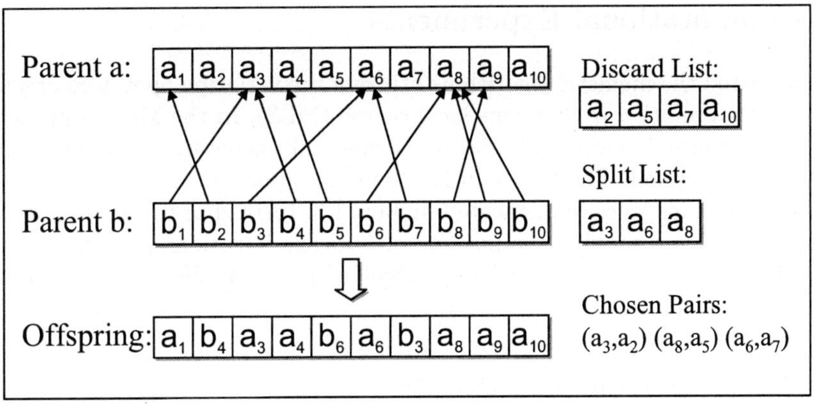

Fig. 2. RX Recombination Operator.

3.5 The Recombination Operators

Two recombination operators are proposed for the MSSC problem. The first one is a uniform crossover and is therefore denoted *UX*. Like in the uniform crossover for binary strings [21], the offspring is generated by copying the mean vectors a_i or b_i from the parents. Each mean vector is copied with probability 0.5 from parent a or parent b.

The second recombination operator is more sophisticated. It replaces the mean vectors in parent a with mean vectors of parent b according to their distribution in the input space X. Consider the example shown in Fig. 2. First, for all mean vectors b_i in parent b the nearest mean vector in parent a is determined. In the example, the nearest mean vector to b_1 is mean vector $a_3 = \text{nearest}(b_1)$ (indicated by the arrow from b_1 to a_3). For each vector a_j, a list of assigned vectors $\mathcal{A}(a_j)$ is maintained. If $a_j = \text{nearest}(b_i)$, b_i will be inserted into $\mathcal{A}(a_j)$. Then, all vectors in parent a with no assignment, i. e. for which the assignment list $\mathcal{A}$ is empty, are stored in a discard list. Next, all vectors of parent a which have more than one assignment are stored in a split list. Afterwards, vectors are chosen in random order from the both split and discard lists. In each step, a pair of split and discard vectors is considered: The discard vector is replaced by a mean vector from parent b by choosing a random element from the assignment list of the split vector. This procedure is repeated until the discard list or the split list is empty. In the first step of the example above, a_3 and a_2 are chosen from the split list and the discard list, respectively. Vector a_2 is then replaced by b_4 which was randomly selected from $\mathcal{A}(a_3) = \{b_1, b_4\}$. In the second step, a_8 and a_5 are chosen and a_5 is replaced by $b_6 \in \mathcal{A}(a_8) = \{b_6, b_9, b_{10}\}$. Analogously, the pair (a_6, a_7) is selected in the third step, and a_7 is replaced by $b_3 \in \mathcal{A}(a_6) = \{b_3, b_7\}$. The resulting offspring produced by the *replacement recombination operator (RX)* is displayed in Fig. 2.

4 Computational Experiments

We compared the memetic algorithms utilizing the variation operators described above with a multi-start k-means local search (MLS). In the MLS, a predefined number of times a start configuration is randomly generated and the k-means algorithm is applied. The best solution found is reported.

All algorithms were implemented in Java 1.3. Instead of comparing absolute running times, we provide the number of k-means iterations performed by the algorithms to have a measure independent of programming language or code optimizations.

4.1 The Gene Expression Data Sets

The first data set denoted as HL-60 is taken from [5] and contains data from macrophage differentiation experiments. The data consists of 7229 genes and expression levels at 4 time points. We applied a variation filter which discarded all genes with an absolute change in expression level less than or equal to 5. The number of genes which passed the filter was 3230. The vectors were normalized afterwards to have mean 0 and variance 1, as described in [6].

The second data set denoted as HD-4CL is also taken from [5] and contains data from hematopoietic differentiation experiments across 4 cell lines. The data consists of 7229 genes, 17 samples each. We applied the same variation filter as above which discarded all genes with an absolute change in expression level less than or equal to 5. The number of genes which passed the filter was 2046. Afterwards, the vectors were normalized to have mean 0 and variance 1.

The third data set is denoted as Cho-Yeast and is described in [22]. It contains the expression levels of 6565 yeast genes measured at 17 time points over two complete cell cycles. As in [6], we discarded the time points at 90 and 100 min, leading to a 15 dimensional space. A variation filter was used which discarded all genes with an absolute change in expression level less than or equal to 50 and an expression level of max/min < 2.0. The resulting number of genes was 2931. Again, the vectors were normalized afterwards to have mean 0 and variance 1.

To study the capability of the MAs to find globally optimum solutions, the fourth and fifth data set were randomly generated with 5000 and 6000 vectors, denoted DS-5000 and DS-6000, respectively. The main vectors were generated by AR(1) autoregressive processes with 16 time points ($x_{i+1} = x_i + \epsilon_i$, where ϵ_i is a normally distributed random number). No variation filter was applied and the vectors were normalized as described above.

For number of clusters for the clustering were taken from [6] for Cho-Yeast, and from [5] for the data sets HL-60, HD-4CL. For the data sets DS-5000 and DS-6000, k was set to the known number of clusters in the experiments.

4.2 Computational Results

In the experiments, all memetic algorithms were run with a population size of $P = 40$. The MAs were terminated before the 100th generation or upon

Table 1. Comparison of MAs and Multi-Start Local Search.

Data Set	Algorithm	Gens	No. LS	Iter LS	Best	Avg. Obj.	Error
HL-60	MLS		2000.0	95896.7	1749.86	1749.90	0.00%
	MA-UX	100	2000.0	64394.6	1749.82	1749.86	0.00%
	MA-RX	19	403.0	2977.1	1749.92	1750.43	0.03%
	MA-FM	100	2000.0	66717.6	1749.86	1749.87	0.00%
	MA-MM	100	2000.0	69126.8	1749.85	1749.86	0.00%
HD-4CL	MLS		2000.0	61435.9	12476.92	12493.51	0.47%
	MA-UX	100	2000.0	38403.0	12436.92	12450.93	0.13%
	MA-RX	83	1684.0	8352.9	12435.85	12444.14	0.07%
	MA-FM	100	2000.0	31033.9	12441.25	12453.87	0.15%
	MA-MM	100	2000.0	32463.3	12435.12	12444.60	0.08%
Cho-Yeast	MLS		2000.0	74632.5	16966.78	16984.32	0.46%
	MA-UX	100	2000.0	51371.7	16908.32	16933.44	0.16%
	MA-RX	98	1979.0	12475.9	16907.06	16916.60	0.06%
	MA-FM	100	2000.0	34405.5	16912.61	16924.54	0.10%
	MA-MM	100	2000.0	35784.3	16909.09	16919.03	0.07%
DS-5000	MLS		2000.0	35954.7	3192.35	3497.14	9.55%
	MA-UX	100	2000.0	26215.0	3192.35	3272.12	2.50%
	MA-RX	52	1064.0	3586.1	3192.35	3192.35	0.00%
	MA-FM	100	2000.0	15492.1	3192.35	3192.35	0.00%
	MA-MM	100	2000.0	24476.5	3192.35	3192.35	0.00%
DS-6000	MLS		2000.0	40929.9	5868.59	6314.74	8.65%
	MA-UX	100	2000.0	27502.6	5811.82	5947.68	2.34%
	MA-RX	55	1110.0	4232.5	5811.82	5833.34	0.37%
	MA-FM	100	2000.0	12886.6	5811.82	5823.11	0.19%
	MA-MM	100	2000.0	25782.8	5811.82	5811.82	0.00%

convergence. The number of mutations or recombinations was set to 20. Since either mutation or recombination is used in the MAs, the number of offspring generated per generation is 20 for all MA runs. Thus, the parameters were chosen to have an identical number of k-means runs in the MLS and the MAs.

In Table 1, the results are displayed for the described data sets. For each algorithm (MLS or MA), the average number of generations (Gens), the average number of k-means local searches (No. LS), the average number of iterations of the k-means local searches (Iter LS), the best objective value found in 20 runs (Best), the average value found in 20 runs (Avg. Obj), and the average percentage excess (Error) are provided.

To give an idea of the computation times required: A run of MA-RX an MLS on the data set Cho-Yeast took 2.6 minutes and 17 minutes, respectively, on a PC with an AMD Athlon CPU (1.2 GHz) compiled with the GNU gcj java compiler (all optimizations enabled).

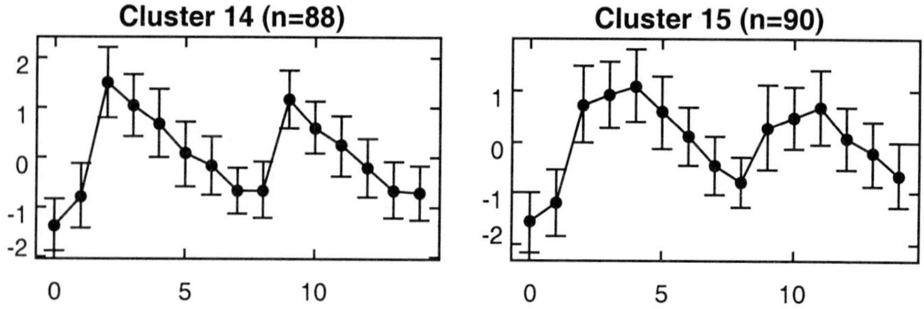

Fig. 3. Two clusters of different partitions of the Cho-Yeast data set. The mean and standard deviation of the expression levels over time are displayed. Left: Cluster 14 of best solution found by MA-RX. Right: cluster 15 of a solution found by the MLS k-means.

All algorithms show similar performance on the first data set (HL-60), all algorithms find solutions with similar fitness, so we believe the best solutions found by the algorithms are optimal. It appears that this data set is easy for k-means based search. Data set HD-4CL and Cho-Yeast show similar properties: the k-means MLS solutions are very close to the best solutions found by the MAs in terms of objective values. However, the MAs are more efficient. Regarding computation times, the MAs MA-RX, MA-FM and MA-MM are much faster. The number of k-means iterations is reduced by a factor of 2 and more compared to MLS. In case of MA-RX, the factors are 7.35 and 5.98 due to the fast convergence of the algorithm (as also indicated by the shorter number of generations compared to the other MAs).

For the larger data sets DS-5000 and DS-6000, the differences of the algorithms become more apparent. MLS finds solutions more than 8.5% above the optimum, while the MAs are capable of finding the optimum solutions. The MA using UX recombination perform consistently worse than the other MAs. MA-RX, MA-FM and MA-MM find the optimum solution in all 20 runs for DS-5000. MA-RX required only 3586.1 iterations of the k-means algorithm which is more than 4.3 times faster than the second best MA. For the largest data set (DS-6000), MA-RX converges still very quickly, providing a good computation time/solution quality trade-off. However, the MAs based on mutation are capable of finding the optimum more often, especially MA-MM: the algorithm finds the optimum in all runs.

From the biologists point of view, it is important how the cluster memberships of the genes change if a MA is used instead of a multi-start k-means. We observed that the clusters produced by the two algorithms differ significantly. Although the solutions have similar fitness values, many genes are assigned to different clusters leading to a surprisingly high phenotypic distance. To illustrate this fact, Two clusters of different solutions of the Cho-Yeast data set are shown in Fig. 3. In each plot, mean and standard deviation (indicated by the error bars) of the expression levels at 15 time points are displayed. On the left, the cluster no. 14 of the best solution found by MA-RX (fitness: 16907) is shown, which consists of 88 genes. In a solution with fitness 16972 produced by multi-start k-means, these genes are distributed over 5 different clusters. For example, 36 of the 88 genes are found in cluster 15, which is displayed on the right side of Fig. 3.

5 Conclusions

In this contribution, memetic algorithms (MAs) for clustering gene expression profiles are proposed. Four problem specific variation operators for minimum sum-of-squares clustering are introduced to be used in a memetic framework using the k-means heuristic as local search. We conducted a comparison study of the MAs and a multi-start k-means clustering algorithm by applying the algorithm to biological data sets as well as to self-generated benchmarking data sets. The results show the superiority of the MA compared to the multi-start k-means. The MA with the newly proposed replacement recombination operator (RX) is shown to produce (near-)optimum solutions and converges very quickly compared to the other MAs. The simple uniform crossover operator frequently used in the evolutionary computation community performed worst. The results indicate that for large data sets ($\geq$ 5000 elements) and a high dimensionality of the input vectors, the MAs perform considerably better than multi-started k-means in terms of objective values. However, for smaller data sets the number of k-means iterations in the MAs is drastically smaller (up to a factor of 7.35 in our experiments). Although the objective values of the solutions lie close together, the cluster memberships change drastically which is very important in the analysis of gene expression data. Hence, the MAs proposed in this paper are shown to be highly efficient and effective for clustering gene expression profiles and thus preferable to (multi-start) k-means clustering.

References

1. Zhang, M.: Large-scale Gene Expression Data Analysis: A New Challenge to Computational Biologists. Genome Research **9** (1999) 681–688
2. Brazma, A., Vilo, J.: Gene Expression Data Analysis. FEBS Letters **480** (2000) 17–24
3. Eisen, M., Spellman, P., Botstein, D., Brown, P.: Cluster Analysis and Display of Genome-wide Expression Patterns. In: Proceedings of the National Academy of Sciences, USA. Volume 95. (1998) 14863–14867
4. Ben-Dor, A., Shamir, R., Yakhini, Z.: Clustering Gene Expression Patterns. Journal of Computational Biology **6** (1999) 281–297
5. Tamayo, P., Slonim, D., Mesirov, J., Zhu, Q., Kitareewan, S., Dmitrovsky, E., Lander, E.S., Golub, T.R.: Interpreting Patterns of Gene Expression with Self-organizing Maps: Methods and Application to Hematopoietic Differentiation. In: Proceedings of the National Academy of Sciences, USA. Volume 96. (1999) 2907–2912
6. Tavazoie, S., Hughes, J.D., Campbell, M.J., Cho, R.J., Church, G.M.: Systematic Determination of Genetic Network Architecture. Nature Genetics **22** (1999) 281–285

7. Yeung, K., Haynor, D., Ruzzo, W.: Validating Clustering for Gene Expression Data. Bioinformatics **17** (2001) 309–318
8. Brucker, P.: On the Complexity of Clustering Problems. Lecture Notes in Economics and Mathematical Systems **157** (1978) 45–54
9. Merz, P., Freisleben, B.: Fitness Landscapes, Memetic Algorithms and Greedy Operators for Graph Bi-Partitioning. Evolutionary Computation **8** (2000) 61–91
10. Merz, P., Freisleben, B.: Fitness Landscapes and Memetic Algorithm Design. In Corne, D., Dorigo, M., Glover, F., eds.: New Ideas in Optimization. McGraw–Hill, London (1999) 245–260
11. Merz, P., Freisleben, B.: Fitness Landscape Analysis and Memetic Algorithms for the Quadratic Assignment Problem. IEEE Transactions on Evolutionary Computation **4** (2000) 337–352
12. Moscato, P.: On Evolution, Search, Optimization, Genetic Algorithms and Martial Arts: Towards Memetic Algorithms. Technical Report C3P Report 826, Caltech Concurrent Computation Program, California Institue of Technology (1989)
13. Moscato, P., Norman, M.G.: A Memetic Approach for the Traveling Salesman Problem Implementation of a Computational Ecology for Combinatorial Optimization on Message-Passing Systems. In Valero, M., Onate, E., Jane, M., Larriba, J.L., Suarez, B., eds.: Parallel Computing and Transputer Applications, Amsterdam, IOS Press (1992) 177–186
14. Dawkins, R.: The Selfish Gene. Oxford University Press (1976)
15. Freisleben, B., Merz, P.: New Genetic Local Search Operators for the Traveling Salesman Problem. In Voigt, H.M., Ebeling, W., Rechenberg, I., Schwefel, H.P., eds.: Proceedings of the 4th International Conference on Parallel Problem Solving from Nature - PPSN IV. Volume 1141 of Lecture Notes in Computer Science., Berlin, Springer (1996) 890–900
16. Merz, P.: Memetic Algorithms for Combinatorial Optimization Problems: Fitness Landscapes and Effective Search Strategies. PhD thesis, Department of Electrical Engineering and Computer Science, University of Siegen, Germany (2000)
17. Gorges-Schleuter, M.: ASPARAGOS: An Asynchronous Parallel Genetic Optimization Strategy. In Schaffer, J.D., ed.: Proceedings of the Third International Conference on Genetic Algorithms, Morgan Kaufmann (1989) 422–427
18. Gorges-Schleuter, M.: Asparagos96 and the Traveling Salesman Problem. In: Proceedings of the 1997 IEEE International Conference on Evolutionary Computation, IEEE Press (1997) 171–174
19. Forgy, E.W.: Cluster Analysis of Multivariate Data: Efficiency vs. Interpretability of Classifications. Biometrics **21** (1965) 768–769
20. MacQueen, J.: Some Methods of Classification and Analysis of Multivariate Observations. In: Proceedings of the Fifth Berkeley Symposium on Mathemtical Statistics and Probability. (1967) 281–297
21. Syswerda, G.: Uniform Crossover in Genetic Algorithms. In Schaffer, J.D., ed.: Proceedings of the 3rd International Conference on Genetic Algorithms, Morgan Kaufmann (1989) 2–9
22. Cho, R.J., Campbell, M.J., Winzeler, E.A., Conway, S., Wodicka, L., Wolfsberg, T.G., Gabrielian, A.E., Landsman, D., Lockhart, D.J., Davis, R.W.: A Genome-wide Transcriptional Analysis of the Mitotic Cell Cycle. Molecular Cell **2** (1998) 65–73

Cellular Automata and Genetic Algorithms for Parallel Problem Solving in Human Genetics

Jason H. Moore and Lance W. Hahn

Program in Human Genetics, Department of Molecular Physiology and Biophysics, 519 Light Hall, Vanderbilt University, Nashville, TN, USA 37232-0700
{Moore,Hahn}@phg.mc.Vanderbilt.edu

Abstract. An important goal of human genetics is to identify variations in genes that are associated with risk of disease. This goal is complicated by the fact that, for common multifactorial diseases such as hypertension, interactions between genetic variations are likely to be more important than the independent effects of any single genetic variation. Attribute interaction is a well-known problem in data mining and is a complicating factor in genetic data analysis. We have previously addressed this problem by developing a parallel approach to problem solving that utilizes one-dimensional cellular automata (CA) for knowledge representation and genetic algorithms (GA) for optimization. In this study, we evaluate the power of this parallel CA approach by simulating gene-gene interactions and adding noise from several common real-world sources. These simulation studies document the strengths of the CA approach and document a weakness that needs to be addressed.

1 Introduction

The identification of DNA sequence variations (i.e. polymorphisms) in genes that increase susceptibility to common complex diseases such as essential hypertension remains a difficult challenge in human genetics [1]. This is because the effects of any given gene on risk of disease is likely to be dependent on interactions with other genetic and environmental factors [2,3]. Thus, for any one genetic variation to be identified, it must be considered in the context of the other factors. Traditional parametric statistical methods such as logistic regression are limited in their ability to deal with interactions among factors because of data sparseness. This is because the data are spread thinly across combinations of factor levels (i.e. the curse of high-dimensionality). The result is a reduced power to identify interactions in relatively small sample sizes. This is a well-known problem in data mining [4] and is becoming increasingly recognized as a problem in human genetics [1].

We have previously addressed this problem by developing a parallel approach to problem solving that utilizes one-dimensional cellular automata (CA) for knowledge representation and genetic algorithms (GA) for optimization [5]. This approach is based on the idea that CA can perform emergent computation as was illustrated in the density estimation problem [6]. It was clear from this example that local interactions

J.J. Merelo Guervós et al. (Eds.): PPSN VII, LNCS 2439, pp. 821–830, 2002.

among spatially and temporally distributed cells can lead to a pattern of cells that contains global information for the problem at hand. We adapted a CA to accept an array of genetic variations as input and produce an output that is predictive of whether that individual is affected with the disease or unaffected. The goal of the present study was to use simulation studies to evaluate the power of this CA approach for identifying gene-gene interactions in the presence of sources of noise that are commonly encountered in genetic studies of human disease.

2 Cellular Automata for Parallel Problem Solving in Human Genetics

Cellular automata (CA) are discrete dynamic systems that consist of an array of cells, each with a finite number of states [7]. The state of each cell changes in time according to the local neighborhood of cells and their states as defined by a rule table. The simplest CA consist of one-dimensional arrays, although some, such as the Game of Life [8], are implemented in two or more dimensions. In this section, we review how CA can be used to perform computations and then how we have exploited this feature to identify complex patterns of genetic variations that are associated with disease.

An intriguing feature of CA is their ability to perform emergent computation [6,9]. That is, the local interactions among spatially distributed cells over time can lead to an output array that contains global information for the problem at hand. For example, Mitchell et al. [6] have used CA to perform density estimation. In that application, a one-dimensional, two-state (1 or 0) CA is given an initial array of states. The goal is to identify a CA rule set such that the CA converges to an array of all 1's if the density of 1's is greater than 0.5 and to an array of all 0's if the density of 1's is less than 0.5. Mitchell et al. found that the CA is able to perform this computation through a series of spatially and temporally extended local computations. This emergent computation feature of CA forms the basis of our method for identifying patterns of genotype variations associated with common complex multifactorial diseases [5].

We have previously developed a CA approach to identifying patterns of genetic variations associated with disease [5]. The general approach is to identify a set of CA operating features that is able to take an array of genotypes (i.e. combinations of DNA sequence variations or alleles inherited from the mother and father) as input and produce an output array that can be utilized to classify and predict whether subjects are affected or unaffected. In this initial study, we fixed the number of cells in the CA to five. Thus, the CA is presented with a maximum of five unique and/or redundant genotypes. We also allowed 'don't care' or wildcard cells to be included. This permits less than five genotypes to be evaluated. Wildcard cells all have the same state and do not contribute any information for discriminating affected from unaffected sib pairs. Assuming each genetic locus has only three possible genotypes, we used a binary encoding with '01' for the first genotype, '10' for the second genotype, '11' for the third genotype, and '00' for the wildcard. Thus, each array presented to the CA consisted of 10 bits with two bits encoding the state of each of the five cells. We used a simple nearest-neighbor rule table that is implemented by looking at the state of the cell in question and the adjacent cell states. With three cells forming a rule and four different states per cell, there are 4^3 or 64 possible rule inputs with four possible output states for each. An important feature of CA is the number of time steps or iterations. This will govern the amount of spatial and temporal information processing or knowledge representation that can be performed. In this study, we allowed a maximum of 128 iterations for each CA. This maximum number of iterations was selected to allow enough time for parallel processing of all the information in an input array without an excessive number of iterations that might complicate interpretation. Thus, there are three essential components to the CA model. First, the correct combination of genetic variations must be selected for initiating the CA cell states. Second, the appropriate rule table that specifies the information processing must be selected. Finally, the number of time steps or iterations for the CA must be selected. We used parallel GAs to optimize selection of these three model components (described below).

How is the output array of the CA used to perform classification and prediction? We first count the number of 1s in the binary encoded output array of the CA run on each set of genotypes for each affected and each unaffected sib in the sample. A classifier is formed by using a frequency histogram of the number of 1s among affected sibs and unaffected sibs. Each histogram bin is labeled affected or unaffected depending on whether the number of 1s represented by that bin were more frequently observed among affected or unaffected individuals. For example, consider the case where 100 affected and 100 unaffected individuals were evaluated. Suppose the number of CA output arrays that contained three 1s was 20 for affecteds and 10 for unaffecteds. This bin would be labeled affected and thus the 10 unaffected individuals would be misclassified. This would contribute 0.05 to the overall misclassification rate. This is performed for each bin and a total classification error is estimated by summing together the individual rates for each bin.

3 Cellular Automata Optimization Using Genetic Algorithms

3.1 Overview of Genetic Algorithms

Genetic algorithms (GAs), neural networks, case-based learning, rule induction, and analytic learning are some of the more popular paradigms in optimization and machine learning [10]. Genetic algorithms perform a beam or parallel search of the solution space that is analogous to the problem solving abilities of biological populations undergoing evolution by natural selection [11,12]. With this procedure, a randomly generated 'population' of solutions to a particular problem are generated and then evaluated for their 'fitness' or ability to solve the problem. The highest fit individuals or models in the population are selected and then undergo exchanges of random model pieces, a process that is also referred to as recombination. Recombination generates variability among the solutions and is the key to the success of the GA search, just as it is a key part of evolution by natural selection. Following recombination, the models are reevaluated and the cycle of selection, recombination, and evaluation continues until an optimal solution is identified.

As with any machine learning approach [10], GAs are not immune to stalling on local optima [13]. Thus, distributed approaches to GAs have been implemented [14]. Here, the GA population is divided into sub-populations or demes. At regular intervals, the best solution obtained by each sub-population is migrated to all other sub-populations. This prevents individual sub-populations from converging on a locally optimum peak because new individuals are periodically arriving to increase the diversity. In biology, it is believed that evolution progresses faster in semi-isolated demes than in a single population of equal size [15]. Indeed, there is some evidence that parallel GAs actually converge to a solution much faster than serial or single-population GAs because of additional selection pressure from choosing migrants based on fitness [14].

3.2 Solution Representation and Fitness Determination

The first step in implementing a GA is to represent the model to be optimized as a one-dimensional binary array. For the CA, we needed to encode five genetic variations and/or wildcards, the CA rule table, and the number of CA iterations. Each of the genetic variations and the number of CA iterations were represented using a total of six 32-bit integers with a modulo operation used to constrain the integer to the desired range (0-9 for the genetic variations and 0-127 for the iterations.) Each CA cell has four possible states and each rule depends on the state of three cells. Encoding this set of 64 rules, with each rule producing one of four two-bit states as output, requires four 32-bit integers. In total, the GA manipulated 10 32-bit integers for a total chromosome length of 320 bits (see Figure 1). Fitness of a particular CA model is defined as the ability of that model to classify subjects as affected or unaffected. Thus, the goal of the GA is to identify a CA model that minimizes the misclassification error. Implementation using cross validation is described below.

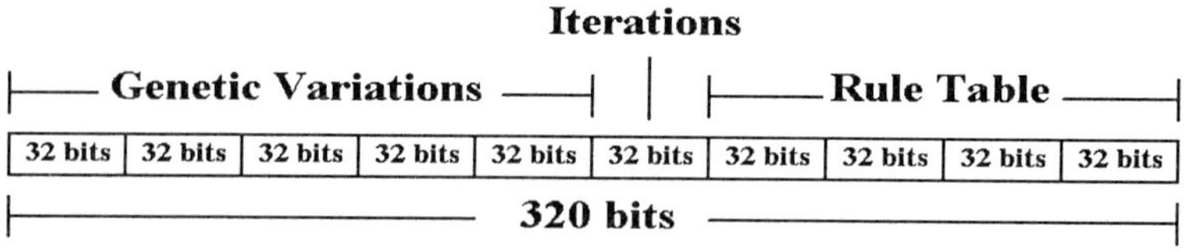

Fig. 1. Encoding of the GA chromosome. The first five 32-bit segments encode genetic variations and/or wild cards while the sixth 32-bit segment encodes the number of iterations. The last four 32-bit segments encode the CA rule table

3.3 Parallel Genetic Algorithm Software and Parameters

The parallel GA used was a modification of the Parallel Virtual Machine (PVM) version of the Genetic ALgorithm Optimized for Portability and Parallelism System (GALLOPS) package for UNIX [16]. This package was implemented in parallel using message passing on a 110-processor Beowulf-style parallel computer cluster running the Linux operating system. Two processors were used for each separate run.

We implemented a parallel GA with two sub-populations or demes undergoing periodic migration. Each GA was run for a total of 200 iterations or generations with migration of the best solutions to each of the other sub-populations every 25 iterations. For recombination, we used a two-point crossover with a DeJong-style crowding factor of three and an incest reduction of three. Selection was carried using stochastic universal sampling. The mutation frequency specifies the probability of a 32-bit field being mutated rather than a single bit. We used a recombination frequency of 0.6 and a mutation frequency 0.02.

3.4 Implementation

The goal of the GA is to minimize the classification error of the CA model. However, from a genetics perspective, the goal is to identify the correct functional genetic variations. That is, from a pool of many candidate genes, the goal is to find those that play a role in influencing risk of disease. We used a 10-fold cross-validation strategy [17] to identify optimal combinations of genetic variations. Cross-validation has been a successful strategy for evaluating genetic models in other studies of common complex diseases [18]. Here, we ran the GA on each 9/10 of the data and retained the CA models that minimized the misclassification rate. Across the 10 retained models, we selected the combination of genetic variations that was observed most frequently. The reasoning is that the functional set of genetic variations should be identified consistently across different subsets of the data [18]. We also estimated the general prediction error of the best set of retained models using each of the independent 1/10 of the data.

4 Data Simulation and Analysis

The goal of the simulation study was to generate a series of datasets where the probability of a subject being affected is dependent solely on interactions among two genetic variations. We then added varying degrees of noise to these datasets from several common real-world sources. Table 1 illustrates the first model (Model 1) that relates two genetic variations, each with two alleles and three genotypes, to the probability of disease. Here, risk of disease is increased by inheriting exactly two high-risk alleles (e.g. *a* and *b* are defined as high risk). Thus, *aa*/*BB*, *Aa*/*Bb*, and *AA*/*bb* are the high-risk genotype combinations. This model was first described by Frankel and Schork [19] and later by Moore et al. [28]. The second model (Model 2) is illustrated in Table 2. This model uses an XOR function and was first described by Li and

Reich [20] and later by Moore et al. [28]. Here, probability of disease is increased if *Aa* or *Bb* is inherited but not both.

Table 1. Probability of disease given different combinations of genetic variations (Model 1)

	Two-genotype probabilities			Single-genotype probabilities
	AA	*Aa*	*aa*	
BB	0	0	.1	.25
Bb	0	.05	0	.25
bb	.1	0	0	.25
Single-genotype probabilities	.25	.25	.25	

Table 2. Probability of disease given different combinations of genetic variations (Model 2)

	Two-genotype probabilities			Single-genotype probabilities
	AA	*Aa*	*aa*	
BB	0	.1	0	.25
Bb	.1	0	.1	.25
bb	0	.1	0	.25
Single-genotype probabilities	.25	.25	.25	

We simulated 100 total datasets using each of these models. Each dataset included the two functional genetic variations and eight non-functional or false-positive genetic variations with 200 affected individuals (cases) and 200 unaffected individuals (controls). A sample size of 200 cases and 200 controls is relatively small in epidemiological studies and is thus a good test of power. Additional sets of data were generated by adding 5% genotyping error (i.e. laboratory measurement error), 5% or 20% phenocopy (i.e. disease risk is due to a random environmental factor and not a genetic factor), or 50% genetic heterogeneity (i.e. subjects generated from two models). These datasets allow the evaluation of the power of the CA approach for identifying gene-gene interactions such as those defined in Table 1 and 2 in the presence of common real-world sources of noise.

We applied the CA approach as described above and by Moore and Hahn [5] to each of the sets of data. The power of the CA approach was estimated as the number of datasets out of 100 in which the correct functional genetic variations were identified.

5 Results

Table 3 summarizes the power of the CA approach for identifying the correct two (A and B) or the correct four (A, B, C, and D) interacting genetic variations in the presence of common sources of noise. In the case of no error, 5% genotyping error, 5% phenocopy, or 20% phenocopy, the CA approach had 100% power to identify each functional genetic variation individually and collectively. However, the power was much lower when 50% genetic heterogeneity was present. In fact, the power of the CA to identify all four genetic variations simultaneously was only 11-12%.

Table 3. Power (%) of the CA approach for identifying each set of interacting genetic variations in the presence of common sources of noise

		Power						
Source of Noise	**Model**	**A**	**B**	**C**	**D**	**AB**	**CD**	**ABCD**
None	1	100	100			100		
None	2	100	100			100		
5% Genotyping	1	100	100			100		
5% Genotyping	2	100	100			100		
5% Phenocopy	1	100	100			100		
5% Phenocopy	2	100	100			100		
20% Phenocopy	1	100	100			100		
20% Phenocopy	2	100	100			100		
50% Genetic Het.	1	55	55	63	65	51	58	11
50% Genetic Het.	2	64	61	60	67	54	56	12

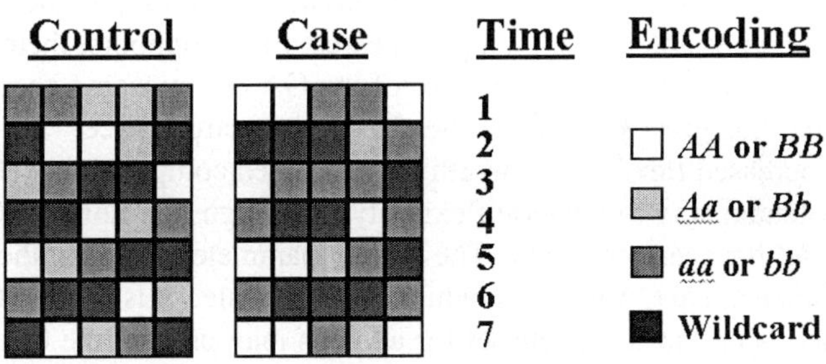

Fig. 2. Example CA models of a control and a case generated by Model 1 with no error. In the first time step, each cell represents a color-encoded genotype for a particular genetic variation. The cells in the subsequent time steps are the result of the computation using the rule table (not shown). The information in cells from the final time step is used as the discriminant score for classification

Figure 2 illustrates example CA models for a control (unaffected subject) and a case (affected subject) generated using Model 1. Since there are only two unique genetic variations in each model, they are repeated to fill the five total inputs. The CA model for the control subject has low-risk input genotypes of *Aa* and *bb* (probability of disease given Aa/bb is 0) while the CA model for the case subject has high-risk input genotypes of *aa* and *BB* (probability of disease given aa/BB is 0.1).

6 Discussion and Conclusions

The identification and characterization of genes that confer an increased susceptibility to common complex diseases is limited by a lack of statistical and computational approaches that are able to model attribute interactions. To address this issue, we have previously developed a parallel problem solving approach that utilizes one-dimensional CA for knowledge representation and GAs for search and optimization [5]. The main conclusion that can be drawn from the present study is that the CA approach has very good power for identifying gene-gene interactions even in the presence of real-world sources of noise such as genotyping error and phenocopy. However, our simulation studies have indicated that this approach seems to be limited in its ability to model a mixture of genetic effects, genetic heterogeneity. Genetic heterogeneity is a well-known complicating factor in genetic studies of common diseases such as type 2 diabetes [21].

An obvious next step in the development of the CA approach will be to improve the power to identify genes in the presence of genetic heterogeneity. One goal of future studies will be to increase the flexibility of the CA approach in an effort to model genetic heterogeneity. There are several ways the flexibility can be increased. First, we will explore the use of more complex rule sets. In this study and our previous studies [5], we used a simple nearest neighbor rule set. That is, the transition of cell state through time is dependent only on that cell state and the states of that cell's immediate neighbors. It is possible that extending the rule set beyond just the nearest neighbors will increase the flexibility. A potential disadvantage of increasing the size of the rule set is increased computational requirements. With just the nearest neighbors, there are 64 possible rules. With the nearest four neighbors, there are 576 possible rules. This greatly increases the size of the GA chromosome needed to encode the rule set and thus greatly increases the size of the search space. We will evaluate whether any increased flexibility is worth the decreased computational efficiency.

Another potential way to increase flexibility is through non-uniform CA that use a different rule set for every cell [22]. The current parameterization of the CA includes a fixed one-dimensional CA lattice in which the same rule set is applied to all cells. It is possible that using the same rule set for all cells may prevent the CA from exploring a richer set of emergent behavior. We will evaluate whether the use of non-uniform CA improves the power to deal with genetic heterogeneity.

A third way to improve flexibility might be to allow more genetic variations to enter the model. In the genetic heterogeneity part of this study, there were two functional genetic variations from each model for a total of four functional variations. We allowed a maximum of five to enter the CA model. It is possible that allowing only five genetic variations into the model was not enough to develop the appropriate knowledge representation for modeling the genetic heterogeneity. Essentially, the CA needed to model a complex OR function (i.e. one set of two genetic variations OR the other set of genetic variations is responsible for disease risk).

A fourth way to improve the power to identify gene-gene interactions in the presence of genetic heterogeneity is to derive more homogeneous subgroups of the data. The use of cluster analysis in case-control studies to identify genetic heterogeneity has been suggested by Schork et al. [29]. Further, Shannon et al. [30] have explored the use of recursive partitioning tree-based methods for identifying homogeneous subgroups. These methods should be explored as a data processing step for the CA.

Human genetics is undergoing an information explosion and a comprehension implosion. In fact, our ability to measure genetic information, and biological information in general, is far outpacing our ability to interpret it. As demonstrated in this study, parallel problem solving strategies such as CA and GAs hold promise for dealing with genetic data that is high-dimensional and complex. The present study is not the first to apply nature-inspired algorithms to a human genetics problem. For example, evolutionary algorithms have been used to optimize data analysis and simulation approaches in genetic epidemiology studies [5, 23-25, 28] and gene expression studies [26, 27]. We anticipate an increase in applications of parallel problem solving algorithms in the field of human genetics as more investigations begin to focus on the challenge of analyzing complex, high-dimensional genetic data.

Acknowledgements

This work was supported by National Institutes of Health grants HL65234, HL65962, GM31304, AG19085, AG20135

References

1. Moore, J.H., Williams, S.M.: New strategies for identifying gene-gene interactions in hypertension. Annals of Medicine 34 (2002) 88-95
2. Templeton, A.R.: Epistasis and complex traits. In: Wade, M., Brodie III, B., Wolf, J. (eds.): Epistasis and Evolutionary Process. Oxford University Press, New York (2000)
3. Schlichting, C.D., Pigliucci, M.: Phenotypic Evolution: A Reaction Norm Perspective. Sinauer Assocuiates, Inc., Sunderland (1998)
4. Freitas, A.A.: Understanding the crucial role of attribute interaction in data mining. Artificial Intelligence Reviews 16 (2001) 177-199
5. Moore, J.H., Hahn, L.W.: A cellular automata approach to detecting interactions among single-nucleotide polymorphisms in complex multifactorial diseases. Pacific Symposium on Biocomputing 2002 (2002) 53-64
6. Mitchell, M., Crutchfield, J.P., Hraber, P.T.: Evolving cellular automata to perform computations: Mechanisms and impediments. Physica D 75 (1994) 361-391
7. Wolfram, S.: Cellular automata as models of complexity. Nature 311 (1984) 419-424
8. Gardner, M.: The fantastic combinations of John Conway's new solitaire game "Life". Scientific American 223 (1970) 120-123
9. Sipper, M.: Evolution of Parallel Cellular Machines. Springer, New York (1997)
10. Langley, P.: Elements of Machine Learning. Morgan Kaufmann Publishers, Inc., San Francisco (1996)
11. Holland, J.H.: Adaptation in Natural and Artificial Systems. University of Michigan Press, Ann Arbor (1975)
12. Goldberg, D.E.: Genetic Algorithms in Search, Optimization, and Machine Learning. Addison-Wesley, Reading (1989)

13. Banzhaf, W., Nordin, P., Keller, R.E., Francone, F.D.: Genetic Programming: An Introduction. Morgan Kaufmann Publishers, San Francisco (1998)
14. Cantu-Paz, E.: Efficient and Accurate Parallel Genetic Algorithms. Kluwer Academic Publishers, Boston (2000)
15. Wright, S.: Isolation by distance. Genetics 28 (1943) 114-138
16. http://garage.cps.msu.edu
17. Ripley, B.D.: Pattern Recognition and Neural Networks. Cambridge University Press, Cambridge (1996)
18. Ritchie, M.D., Hahn, L.W., Roodi, N., Bailey, L.R., Dupont, W.D., Plummer, W.D., Parl, F.F. and Moore, J.H.: Multifactor dimensionality reduction reveals high-order interactions among estrogen metabolism genes in sporadic breast cancer. American Journal of Human Genetics 69 (2001) 138-147
19. Frankel, W.N., Schork, N.J.: Who's afraid of epistasis? Nature Genetics 14 (1996) 371-373
20. Li, W., Reich, J.: A complete enumeration and classification of two-locus disease models. Human Heredity 50 (2000) 334-349
21. Busch, C.P., Hegele, R.A.: Genetic determinants of type 2 diabetes mellitus. Clinical Genetics 60 (2001) 243-54
22. Sipper, M.: Co-evolving non-uniform cellular automata to perform computations. Physica D 92 (1996) 193-208
23. Carlborg, O., Andersson, L., Kinghorn, B.: The use of a genetic algorithm for simultaneous mapping of multiple interacting quantitative trait loci. Genetics 155 (2000) 2003-10
24. Congdon, C.B., Sing, C.F., Reilly, S.L.: Genetic algorithms for identifying combinations of genes and other risk factors associated with coronary artery disease. In: Proceedings of the Workshop on Artificial Intelligence and the Genome. Chambery (1993)
25. Tapadar, P., Ghosh, S., Majumder, P.P.: Haplotyping in pedigrees via a genetic algorithm. Human Heredity 50 (2000) 43-56
26. Moore, J.H., Parker, J.S., Hahn, L.W.: Symbolic discriminant analysis for mining gene expression patterns. In: De Raedt, L., Flach, P. (eds): Lecture Notes in Artificial Intelligence 2167. Springer-Verlag, Berlin (2001).
27. Moore, J.H., Parker, J.S.: Evolutionary computation in microarray data analysis. In: Lin, S. and Johnson, K. (eds): Methods of Microarray Data Analysis. Kluwer Academic Publishers, Boston (2001)
28. Moore, J.H., Hahn, L.W., Ritchie, M.D., Thornton, T.A., White, B.C: Application of genetic algorithms to the discovery of complex genetic models for simulation studies in human genetics. In: W.B.Langdon, E. Cantu-Paz, K. Mathias, R. Roy, D. Davis, R. Poli, K. Balakrishnan, V. Honavar, G. Rudolph, J. Wegener, L. Bull, M.A. Potter, A.C. Schultz, J.F. Miller, E. Burke, and N. Jonoska (eds): Proceedings of the Genetic and Evolutionary Computation Conference, Morgan Kaufmann Publishers, San Francisco (2002)
29. Schork, N.J., Fallin, D., Thiel, B., Xu, X., Broeckel, U., Jacob, H.J., Cohen, D: The future of genetic case-control studies. Advances in Genetics 42 (2001) 191-212
30. Shannon, W.D., Province, M.A., Rao, D.C: Tree-based recursive partitioning methods for subdividing sibpairs into relatively more homogeneous subgroups. Genetic Epidemiology 20 (2001) 293-306

Evolutionary Graph Generation System and Its Application to Bit-Serial Arithmetic Circuit Synthesis

Makoto Motegi, Naofumi Homma, Takafumi Aoki, and Tatsuo Higuchi

Graduate School of Information Sciences, Tohoku University
Aoba-yama 05, Sendai 980-8579, Japan

Abstract. This paper presents an efficient graph-based evolutionary optimization technique called Evolutionary Graph Generation (EGG), and its application to the design of bit-serial arithmetic circuits, which frequently appear in real-time DSP architectures. The potential of the proposed approach is examined through experimental synthesis of bit-serial constant-coefficient multipliers. A new version of the EGG system can generate the optimal bit-serial multipliers of 8-bit coefficients with a 100% success rate in 15 minutes on an average.

1 Introduction

Arithmetic circuits are of major importance in today's computing and signal processing systems. Most of the arithmetic circuits are designed by experienced designers who have specific knowledge of the basic arithmetic algorithms. Even the state-of-the-art logic synthesis tools can provide only limited capability to create structural details of arithmetic circuits.

Addressing this problem, we have proposed an approach to designing arithmetic circuits using a new evolutionary optimization technique called Evolutionary Graph Generation (EGG) [1]–[4]. There are already some approaches to evolutionary design of hardware structures [5]–[8]. The key idea of the proposed EGG system is to employ general graph structures as individuals and introduce new evolutionary operations to manipulate graph structures directly without encoding them into other indirect representations, such as bit strings (used in GA) and trees (used in GP). The potential of EGG has already been investigated through the design of combinational arithmetic circuits, such as high-performance parallel multipliers [1]–[3] and high-radix redundant adders [4].

This paper presents an application of EGG to the design of bit-serial data-parallel arithmetic circuits, which frequently appear in real-time DSP architectures [9]. The potential capability of the proposed approach is examined through experimental synthesis of bit-serial data-parallel constant-coefficient multipliers. A new version of the EGG system can generate the optimal bit-serial multipliers of 8-bit coefficients with a 100% success rate in 15 minutes on an average.

J.J. Merelo Guervós et al. (Eds.): PPSN VII, LNCS 2439, pp. 831–840, 2002.

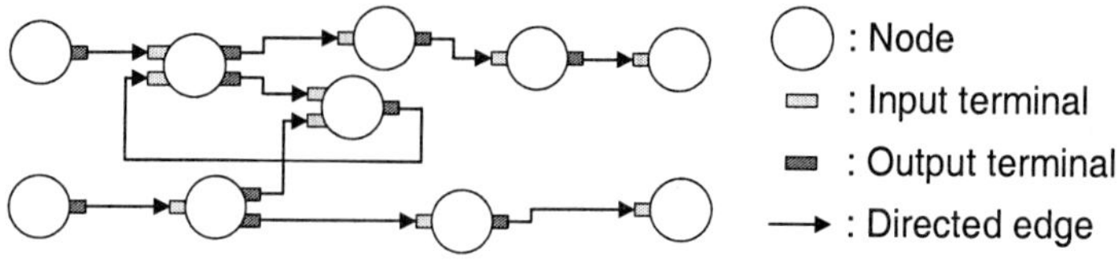

Fig. 1. Example of a circuit graph.

The proposed approach can also be applied to other design specifications including constant-coefficient multiply-adders as demonstrated at the last part of this paper.

2 EGG System

2.1 Basic Model

The Evolutionary Graph Generation (EGG) system employs *circuit graphs* to represent circuit structures. A circuit graph (Fig. 1) consists of nodes and directed edges. Nodes are of two types: functional nodes and input/output nodes. Every node has its own name, the function type and input/output terminals. We assume that every directed edge must connect one output terminal (of a node) and one input terminal (of another node), and that each terminal has one edge connection at most. A circuit graph is said to be *complete* if all the terminals have an edge connection. The EGG system generates only complete circuit graphs in order to avoid needless search for incorrect circuit structures.

More precisely, a circuit graph G is defined as $G = (N^G, T_O^G, T_I^G, \nu_O^G, \nu_I^G, \epsilon^G)$, where

- N^G : the set of nodes,
- T_O^G : the set of output terminals,
- T_I^G : the set of input terminals,
- ν_O^G : mapping from T_O^G to N^G;
  $n = \nu_O^G(u)$ means that the output terminal u $(\in T_O^G)$ belongs to the node n $(\in N^G)$,
- ν_I^G : mapping from T_I^G to N^G;
  $n = \nu_I^G(v)$ means that the input terminal v $(\in T_I^G)$ belongs to the node n $(\in N^G)$,
- ϵ^G : bijection from S_O^G to S_I^G, where $S_O^G \subseteq T_O^G$, $S_I^G \subseteq T_I^G$ and $|S_O^G| = |S_I^G|$;
  $v = \epsilon^G(u)$ means that the output terminal u $(\in S_O^G)$ and the input terminal v $(\in S_I^G)$ have a directed edge connection.

Note here that S_O^G (or S_I^G) is the set of output (or input) terminals having edge connections. The circuit graph G is said to be complete if and only if $S_O^G = T_O^G$ and $S_I^G = T_I^G$.

Fig. 2 shows the overall procedure of the EGG system. All the circuit graphs generated newly in evolution process are evaluated by a symbolic model checking

```
program EGG_System_Flow
 begin
   t := 0;   { t: number of generations }
   initialize(P(t));   { P(t): population }
   evaluate(P(t));
   while t ≤ Max. num. of generations do
     begin
       C(t) := crossover(P(t));   { C(t): offsprings generated by crossover }
       M(t) := mutation(P(t));   { M(t): offsprings generated by mutation}
       evaluate(C(t) ∪ M(t));
       P(t + 1) := select(C(t) ∪ M(t) ∪ P(t));
       t := t + 1
     end
 end.
```

Fig. 2. EGG system flow.

technique [2]. Then, the circuit graphs having higher scores are selected to perform evolutionary operations, *crossover* and *mutation* to produce offsprings for the next generation. Both operations transform the structure of circuit graphs preserving the completeness property, that is, if the parents are complete graphs, the transformed graphs are also complete.

The crossover operation recombines two parent graphs into two new graphs. When a pair of parent graphs G_{p1} and G_{p2} is selected from the population, the crossover operation determines a pair of subgraphs G'_{p1} and G'_{p2} to be exchanged between the parents, and generates offsprings by replacing the subgraph of one parent by that of the other parent as illustrated in Fig. 3(a). In this process, the system selects a subgraph randomly from the mother circuit graph, and selects a *compatible* subgraph from the father circuit graph, where "compatible" means that the cut sets for these subgraphs contain the same number of edges for both negative and positive directions. This ensures the completeness of the generated offsprings. The mutation operation, on the other hand, partially reconstructs the given circuit graph. This operation selects the subgraph randomly and replaces it with a randomly generated subgraph that is compatible with the original subgraph. Fig. 3(b) depicts an example of the mutation operation. This operation also ensures the completeness property of the generated circuit graph.

2.2 EGG System Implementation

We have newly designed a generic object-oriented framework for the EGG system, which is called "EGG framework", and have developed its application to bit-serial arithmetic circuit synthesis. Fig. 4 illustrates the class diagram of the EGG framework, where each rectangular box represents a class template (or a class) and each dashed rectangle shows template arguments. These framework class templates (and a class) contain fundamental components for the EGG system.

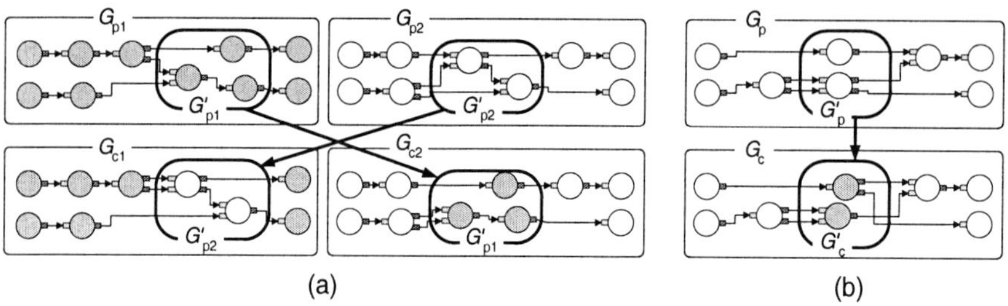

Fig. 3. Examples of evolutionary operations: (a) crossover, (b) mutation.

The overall work-flow is controlled by the `Control` class template, which is associated with `Population`, `Mutation`, `Crossover`, `Selection` and `Evaluation` class templates. The `Population` class template contains the basic individual model defined by the `Graph` class template. The class templates `Mutation`, `Crossover`, `Selection` and `Evaluation` define basic operators for `Population` objects.

The `Graph` class template has `Node`, `SubGraph` and `Fitness`, where `Node` and `SubGraph` inherit from the `AbstractNode` class template. The `AbstractNode` has the `Terminal` class template. The connecting edge at the `Terminal` class template indicates an "association", i.e., one `Terminal` object calls another `Terminal` object to define the network topology. For performing evolutionary operations quickly, the data structure for each `Graph` object is designed to hold a list of data items for a specific number of `SubGraph` objects that are selected from the `Graph` object randomly. In the crossover operation, for example, the system first selects a `SubGraph` object randomly from a list of `SubGraph` objects of the mother `Graph` object, then selects a compatible `SubGraph` object from the father `Graph` object by scanning father's `SubGraph` list.

In order to synthesize bit-serial arithmetic circuits, we have created 6 new classes inheriting from `Control`, `Evaluation`, `Graph`, `Fitness`, `Terminal` and `Node`, respectively. The EGG system can be systematically modified for different design problems by inheriting the framework class templates. We have found that the use of the EGG framework accelerates the development of new applications by providing high-level abstractions of EGG optimization techniques.

3 Synthesis of Bit-Serial Constant-Coefficient Multipliers

We demonstrate the capability of the new EGG system in the synthesis of bit-serial constant-coefficient multipliers. We first summarize some techniques for designing the bit-serial multipliers with constant coefficients. One of the most important techniques in constant-coefficient multiplier design is to encode the target coefficient by the Canonic Signed-Digit (CSD) number representation [10]. The CSD number representation is defined as a special binary Signed-Digit (SD) number representation that contains the least number of non-zero digits. This

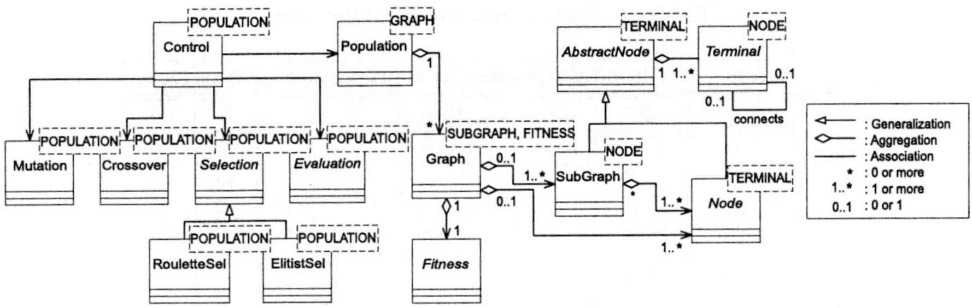

Fig. 4. Class diagram of the EGG framework.

encoding technique makes possible to reduce the number of transistors, which is roughly proportional to the number of non-zero digits. Thus, the known best possible approach to the design of bit-serial constant-coefficient multipliers is to use a one-dimensional systolic array of m processing elements (i.e., full adders), where m is the number of non-zero digits in the given CSD coefficients. In this experiment, the EGG system synthesizes the multiplier structure without using the knowledge of the above techniques.

In our experiments, we have selected a set of nodes shown in Table 1 for synthesizing various bit-serial data-parallel constant-coefficient multipliers and constant-coefficient multiply-adders, which frequently appear in signal processing applications. The circuit graphs generated by the EGG system are evaluated by a combination of two different evaluation metrics, *functionality* and *performance*. The functionality measure F evaluates the validity of the logical function compared with the target function. The performance (i.e., efficiency) measure P, on the other hand, is assumed to be the product of circuit delay D and the number of transistors A.

First, we describe the functionality measure F in detail. In the following discussion, we assume the use of LSB-first bit-serial arithmetic based on two's complement binary number system, where the first bit has the weight 2^0, the second has 2^1, the third has 2^2, and so on. Given a specific circuit graph, the mathematical representation of its function can be derived by symbolic execution using functional definitions given in Table 1.

The function of an n-input 1-output bit-serial arithmetic circuit consisting of the nodes shown in Table 1 can be represented in general as

$$Y = \sum_{i=1}^{n} \hat{K}_i X_i + f(X_1, \cdots, X_n), \tag{1}$$

where X_i $(i = 1, \cdots, n)$ represent bit-serial inputs, Y represents the bit-serial output, $\hat{K}_i$ $(i = 1, \cdots, n)$ are integer coefficients, and $f(X_1, \cdots, X_n)$ is a nonlinear function of input operands. The term f consists of intermediate variables W_j $(j = 1, 2, 3, \cdots)$ which can not be eliminated in the process of the symbolic verification.

Table 1. Nodes used in the experiment.

Name	Symbol	Mathematical representation	Area	Delay
Input	IN$^+$	$Y = X$	0	0
	IN$^-$	$Y = -X$	2	τ
Output	OUT	$Y = X$	33	4τ
Full adder	FA	$2C + S = X_1 + X_2 + X_3$	28	4τ
1-bit register	R	$Y = 2X$	5	—
	2-R	$Y = 4X$	10	
	4-R	$Y = 16X$	15	

We assume that the target function is given by

$$Y = \sum_{i=1}^{n} K_i X_i, \tag{2}$$

where K_i $(i = 1, \cdots, n)$ are integer coefficients. The functionality measure F is calculated by evaluating the similarity between the coefficients $\hat{K}_i$ (in (1)) and the target coefficients K_i. Assume that $\hat{K}_i$ and K_i are written as follows:

$$\hat{K}_i = \hat{k}_{i,0}2^0 + \hat{k}_{i,1}2^1 + \cdots + \hat{k}_{i,j}2^j + \cdots + \hat{k}_{i,\|\hat{K}_i\|-1}2^{\|\hat{K}_i\|-1},$$
$$K_i = k_{i,0}2^0 + k_{i,1}2^1 + \cdots + k_{i,j}2^j + \cdots + k_{i,\|K_i\|-1}2^{\|K_i\|-1},$$

where $\|\hat{K}_i\|$ and $\|K_i\|$ denote the string lengths of the coefficients and $\hat{k}_{i,j}, k_{i,j} \in \{-1, 0, 1\}$. The similarity of these coefficient strings is evaluated by computing their correlation. The correlation $M_{\hat{K}_i,K_i}(s)$ of the two coefficient strings with the shift amount s is defined by

$$M_{\hat{K}_i,K_i}(s) = \begin{cases} \dfrac{1}{\|\hat{K}_i\|} \displaystyle\sum_{l=0}^{\|\hat{K}_i\|-1} \delta\left(\hat{k}_{i,l} - k_{i,l-s}\right) & \text{if } \|\hat{K}_i\| \geq \|K_i\|, \\ \dfrac{1}{\|K_i\|} \displaystyle\sum_{l=0}^{\|K_i\|-1} \delta\left(\hat{k}_{i,l-s} - k_{i,l}\right) & \text{if } \|\hat{K}_i\| < \|K_i\|, \end{cases} \tag{3}$$

where $\delta(x)$ is defined as $\delta(x) = 1$ if $x = 0$ and $\delta(x) = 0$ if $x \neq 0$. In the above calculation, we assume the values of the undefined digit position to be 0 for both coefficient strings. Using this correlation function, the similarity F' between (1) and (2) is defined as

$$F' = \frac{1}{n} \sum_{i=1}^{n} \left[\max_{0 \leq s \leq d} \left\{ 100 M_{\hat{K}_i,K_i}(s) - C_1 s \right\} \right], \tag{4}$$

where $d = \left| \|\hat{K}_i\| - \|K_i\| \right|$ and $C_1 = 10$ in this experiment. The term $C_1 s$ represents the adverse effect due to the shift amount s. Using this similarity, we define the functionality measure F as

Table 2. *DA* product of multipliers: (a) the multipliers generated by the EGG system, (b) the CSD multipliers.

Coefficient K_1	*DA* (a)	*DA* (b)	(a)/(b)	Coefficient K_1	*DA* (a)	*DA* (b)	(a)/(b)	Coefficient K_1	*DA* (a)	*DA* (b)	(a)/(b)
−123	1088	1088	1.00	−35	1024	1024	1.00	55	1197	1197	1.00
−115	1352	1368	0.99	−34	475	475	1.00	58	544	655	0.83
−111	1088	1088	1.00	−29	1008	1152	0.88	66	384	384	1.00
−105	1530	1557	0.98	−26	504	640	0.79	70	516	655	0.79
−101	1530	1557	0.98	−25	1125	1152	0.98	73	1032	1032	1.00
−91	1328	1352	0.98	−23	984	1008	0.98	77	1296	1312	0.99
−87	1328	1352	0.98	−13	968	1107	0.87	84	516	516	1.00
−82	675	675	1.00	−10	425	425	1.00	99	1296	1539	0.84
−79	1064	1064	1.00	−4	180	180	1.00	103	1476	1539	0.96
−78	552	665	0.83	13	912	968	0.94	104	516	680	0.76
−77	1328	1512	0.88	26	476	630	0.76	105	1296	1352	0.96
−58	524	665	0.79	27	1152	1152	1.00	108	592	690	0.86
−54	524	524	1.00	31	837	837	1.00	114	564	680	0.83
−47	1024	1048	0.98	35	992	1134	0.87	120	432	515	0.84
−44	524	524	1.00	43	1256	1512	0.83	123	1242	1242	1.00
−40	400	475	0.84	50	496	655	0.76	127	927	927	1.00
−38	532	640	0.83	52	496	655	0.76				

$$F = F' - C_2 q, \tag{5}$$

where q is the number of delay-free loops in the evolved circuit, and $C_2 = 5$ in this experiment.

On the other hand, the performance measure P is defined as

$$P = \frac{C_3}{DA}, \tag{6}$$

where D is the maximum register-to-register delay estimated with an inverter gate as a unit delay element τ, and A is the total number of transistors [11]. We use $F + P$ as a total fitness function, where the ratio P_{max}/F_{max} is adjusted about 5/100 by tuning the constant C_3.

4 Experimental Results

Table 2 shows the result of a set of evolutionary runs, in which the EGG system generates bit-serial multipliers with 8-bit integer coefficients. We employ the target function of equation (2) with $n = 1$. In this experiment, we assume that the population size is 300, the number of generations is 10000, the maximum number of nodes is 100, the crossover rate is 0.7, and the mutation rate is 0.1. The parameter values are optimized through a set of experiments.

Total 50 coefficients are selected randomly in the range of −128 to 127. We have performed 10 distinct evolutionary runs for each coefficient. As an example,

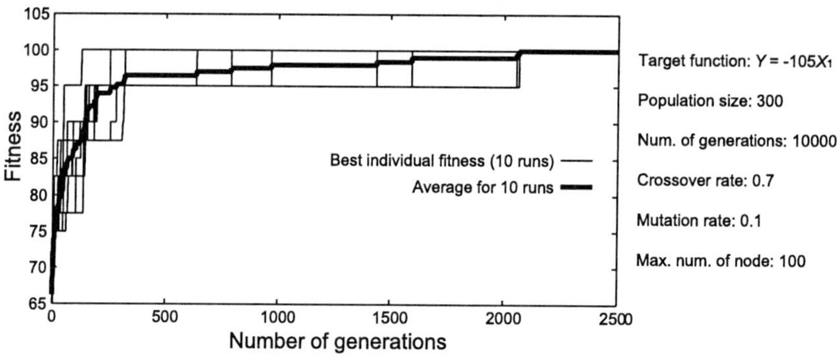

Fig. 5. Transition of the best individual fitness for the case of $Y = -105X_1$.

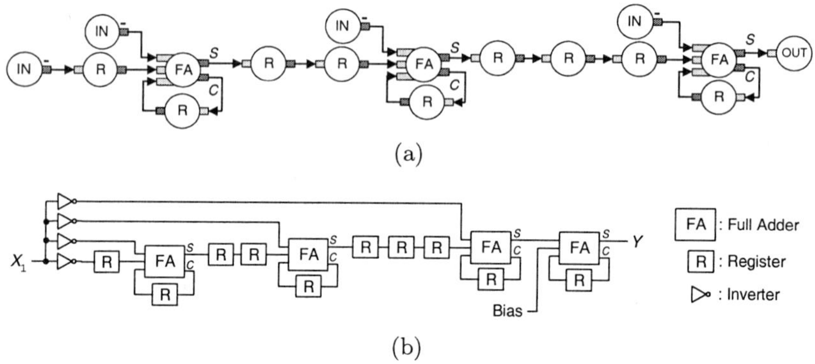

Fig. 6. Example solution for the case of $Y = -105X_1$: (a) the circuit graph generated by the EGG system, (b) the multiplier structure corresponding to graph (a).

Fig. 5 shows the transition of the best individual fitness for the case of $Y = -105X_1$. We can see the staircase improvements of the best individual fitness for every trial. We have obtained the solutions achieving 100% functionality with a 100% success rate in 10 trials. The calculation time for each evolutionary run is about 15 minutes on a Linux PC with 1 GHz Pentium III and 1 GB memory. Percent distribution of the calculation time is approximately as follows: selection-37%, evolutionary operations-30%, evaluation-28%, others-5%.

In Table 2, we show the *DA* product (the maximum register-to-register delay × the total number of transistors) of the multipliers generated by the EGG system (a) and that of the corresponding CSD multipliers (b). From this table, we can confirm that the EGG system can generate efficient multiplier structures whose performances are comparable or sometimes superior to those of systolic CSD multipliers. Fig. 6 shows an example of the solution for the case of $Y = -105X_1$. Note here that the EGG system can synthesize efficient multiplier structure without using the knowledge of arithmetic algorithms. The EGG system can also generate the structural Hardware Description Language (HDL)

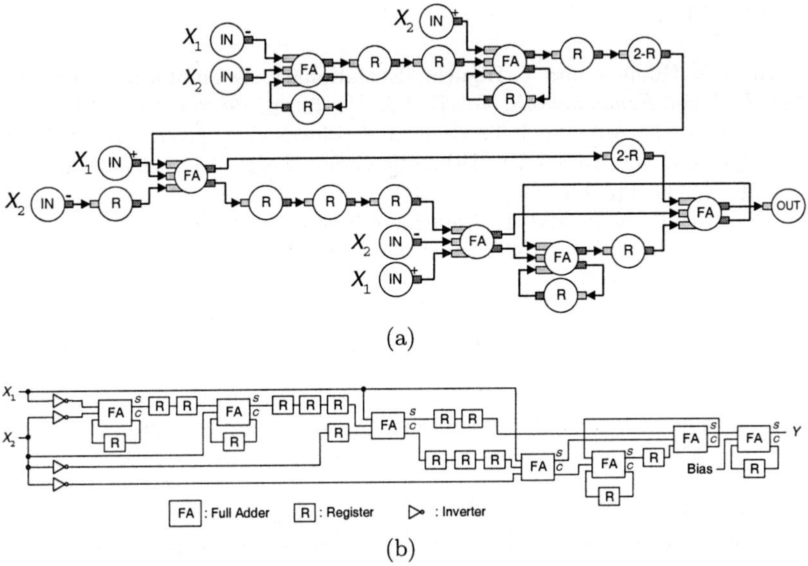

Fig. 7. Example solution for the case of $Y = -123X_1 - 105X_2$: (a) the circuit graph generated by EGG system, (b) the multiply-addition structure corresponding to graph (a).

code for the evolved multipliers, and this may provide a good initial description for state-of-the-art logic synthesis tools in some applications.

The proposed approach can be applied not only to the synthesis of constant-coefficient multipliers but also to other design problems by changing the target function. For example, if we employ the target function (2) with parameters $n = 2$, $K_1 = -123$, and $K_2 = -105$, we can synthesize the constant-coefficient multiply-adder defined by $Y = -123X_1 - 105X_2$. Fig. 7 shows the best solution obtained in the 10000th generation. This clearly demonstrates the potential capability of the EGG system for handling generic synthesis tasks in bit-serial arithmetic.

5 Conclusion

In this paper, we have presented an application of the EGG system to the design of bit-serial constant-coefficient multipliers. The experimental result suggests that the EGG system can solve design problems of serial-pipelined arithmetic circuits without the knowledge of arithmetic algorithms. The research project is now being conducted to apply the proposed EGG system to the hierarchical synthesis of larger-scale arithmetic circuits. The EGG framework used in this paper is open to the public at http://www.higuchi.ecei.tohoku.ac.jp/egg/.

References

1. T. Aoki, N. Homma, and T. Higuchi,"Evolutionary design of arithmetic circuits," *IEICE Trans. Fundamentals*, Vol. E82-A, No. 5, pp. 798–806, May 1999.
2. N. Homma, T. Aoki, and T. Higuchi,"Evolutionary graph generation system with symbolic verification for arithmetic circuit design," *IEE Electronics Letters*, Vol. 36, No. 11, pp. 937–939, May 2000.
3. N. Homma, T. Aoki, and T. Higuchi,"Evolutionary synthesis of fast constant-coefficient multipliers," *IEICE Trans. Fundamentals*, Vol. E83-A, No. 9, pp. 1767–1777, September 2000.
4. M. Natsui, T. Aoki, and T. Higuchi,"Evolutionary graph generation with terminal-color constraint for heterogeneous circuit synthesis," *Electronics Letters*, Vol. 37, No. 13, pp. 808–810, June 2001.
5. F. J. Miller, P. Thomson, and T. Fogarty,"Designing electronic circuits using evolutionary algorithms. arithmetic circuits: A case study," *Genetic Algorithms and Evolution Strategies in Engineering and Computer Science*, pp. 105 – 131, September 1997.
6. R. J. Koza, H. F. III, Bennett, D. Andre, A. M. Keane, and F. Dunlap,"Automated synthesis of analog electrical circuits by means of genetic programming," *IEEE Trans. Evolutionary Computation*, Vol. 1, No. 2, pp. 109 – 128, July 1997.
7. T. Higuchi, M. Iwata, D. Keymeulen, H. Sakanashi, M. Murakawa, I. Kajitani, E. Takahashi, K. Toda, M. Salami, N. Kajihara, and N. Otsu,"Real-world applications of analog and digital evolvable hardware," *IEEE Trans. Evolutionary Computation*, Vol. 3, No. 3, pp. 220 – 235, September 1999.
8. D. J. Lohn and P. S. Colombano,"A circuit representation technique for automated circuit design," *IEEE Trans. Evolutionary Computation*, Vol. 3, No. 3, pp. 205 – 219, September 1999.
9. S. A. White,"Application of distributed arithmetic to digital signal processing: a tutorial review," *IEEE ASSP Magazine*, pp. 4 – 19, July 1989.
10. K. Hwang, *Computer arithmetic: principles, architecture, and design*, John Wiley & Sons, 1979.
11. R. Zimmermann and W. Fichtner,"Low-power logic styles: CMOS versus pass-trasistor logic," *IEEE J. Solid-State Circuits*, Vol. 32, No. 7, pp. 1079 – 1090, July 1997.

Evaluating Multi-criteria Evolutionary Algorithms for Airfoil Optimisation

Boris Naujoks[1], Lars Willmes[2], Thomas Bäck[2,4], and Werner Haase[3]

[1] Chair for Systems Analysis, Computer Science Dept. XI, University of Dortmund
44221 Dortmund, Germany
naujoks@LS11.cs.uni-dortmund.de
[2] NuTech Solutions GmbH
Martin-Schmeißer-Weg 15, 44227 Dortmund, Germany
{willmes,baeck}@nutechsolutions.de
[3] EADS Military Aircraft, Dept. MT63 / Build. 70.N
81663 Munich, Germany
werner.haase@m.eads.net
[4] Leiden Institute of Advanced Computer Science (LIACS), Leiden University
Niels Bohrweg 1, 2333 CA Leiden, The Netherlands

Abstract. A new approach for multi criteria design optimisation is presented in the paper. The problem tackled with this approach is the 2-dimensional design of an aircraft wing. To carry the derandomized step size control also to the multi criteria applications, four different selection schemes are proposed. Furthermore, we present a new method for averaging results of multi objective evolutionary algorithms. This method is then used to compare the results achieved with the proposed algorithms.

1 Introduction

Comparing results from different multi objective optimisation techniques is a difficult task. Different qualities of the results achieved have to be considered as well as the computational effort that was necessary to achieve these results. But these results are needed to compare optimisation techniques and to decide which one is the best for a certain kind of problem.

Using stochastic optimisation techniques like evolutionary algorithms (EAs) makes the task more complicated. Different runs of stochastic optimisation techniques yield different results due to the influence of random numbers. One often used way is to average the results and to compare the averaged results.

Concerning multi objective optimisation, the results are often presented in Pareto fronts. The current paper describes a new method to average Pareto fronts. This new comparison technique is tried on an aviation test case considering different multi objective evolutionary algorithms.

Aviation in general is one of the most important fields in industry and therefore also in science. Due to the many potential savings that are possible in this area many researchers work here on production cost minimization, flight behavior improvement, etc.

J.J. Merelo Guervós et al. (Eds.): PPSN VII, LNCS 2439, pp. 841–850, 2002.

One of the most referenced applications from aviation is aircraft wing design. Due to the development of computational fluid dynamics (CFD) methods, wing design is nowadays mostly done using computers, which provide a scaleable preciseness for different design tasks.

Different methods from optimisation have been carried out on the current design problem, which is presented in detail in section 2. On a simplified problem considering only one flow condition resulting in a single objective optimisation problem, Evolutionary Algorithms (EAs) [1] have been applied very successfully. Especially the derandomized step size control mechanism [2] for evolution strategies (ESs) [3] yielded the best results here. To carry this operator to multi objective tasks and, moreover, use step size adaptation in multi objective applications, four approaches are tackled in section 3.1.

Next to these MOEAs an evolution strategy implementation of NSGA-II (Non-dominated Sorting Genetic algorithm; [4]) is presented for comparison purposes in section 3.1. This new implementation also considers derandomised step size control [2]. The new approach on averaging Pareto-fronts is introduced in section 3.2. First results can be found in section 4, first conclusions are drawn in section 5.

2 Airfoil Design Test Case

In the current investigation a two-objective airfoil design problem for viscous flow is considered. The problem described is one of the test cases from the European research project AEROSHAPE. Here, all modeling issues concerning CFD, e.g. mesh size, mesh generation, used models, pressure calculation, etc. have been fixed and two regimes of flow conditions have been chosen. These regimes vary in the flow parameter settings and a suitable airfoil as a compromise for both conditions is to be designed.

In contrast to the airfoil design problem using only one regime of flow conditions (single point), the current task requires the application of multi criteria decision making methods. Therefore, genetic algorithms as well as evolution strategies have been used in conjunction with Pareto optimisation techniques. In contrast to results already presented on the current test case [5, 6], the parameterization of the airfoil using Bezier points for determining the design has been improved. Here, some x-components of these points have been involved in the optimisation process in addition to the y-components of all points.

The software and technical support for the objective function calculation was provided by the European Aeronautic Defence and Space Company – Military Aircraft (EADS-M), one of the partners in the AEROSHAPE project. The two flow conditions are calculated using different models, namely the Johnson-King model for the subsonic flow (high-lift test case) and the Johnson-Coakley model for the transsonic flow (low-drag test case).

The parameter settings describing the flow conditions in use are given in table 1.

For the objective function calculation two target airfoil designs are given, one for each flow condition. The objective function reads as follows:

Table 1. Summarized design conditions (c=chord length)

Property	Case	high lift	low drag
M_∞	[–]	0.20	0.77
Re_c	[–]	$5 \cdot 10^6$	10^7
$X_{transition}$ (upper / lower)	[c]	3% / 3%	3% / 3%
α	[°]	10.8	1.0

$$F(\alpha_1, \alpha_2, x(s), y(s)) = \sum_{n=1}^{2} \left[W_n \int_0^1 (C_p(s) - C^n_{p,target}(s))^2 ds \right] \quad (1)$$

with s being the airfoil arc-length measured around the airfoil and W_n weighting factors. C_p is the pressure coefficient distribution of the current and $C^n_{p,target}$ the pressure coefficient distribution of the target airfoils, respectively.

Due to the large calculation times for the Navier-Stokes simulations, only the restricted number of 1000 objective function evaluations, being already a lot, is allowed.

3 Multi-objective Evolutionary Algorithms

Evolutionary Algorithms are nowadays a widely spread stochastic optimisation method. They have proven their practicability and efficiency in many optimisation tasks. Nevertheless detailed knowledge about parameterizing these methods accordingly is essential.

Due to their population-based approach, EAs are very promising methods if more than one value to describe the quality of an individual is required. The first overview on multi criteria decision making using EAs was given by Fonseca and Fleming in 1995 [7]. In the following text, their definitions concerning dominance, Pareto-optimality, Pareto-sets, and -fronts are used. The term *Pareto-front* is additionally defined as the set of non-dominated individuals of a given population.

3.1 Multi-objective Derandomized ES (MODES)

Since the single criteria version of the DES approach has been successfully applied to the single point airfoil design problem, a transformation using this approach for multi-criteria design seemed quite natural. Although it is common believe that the main strength of Evolutionary Algorithms in multicriteria optimisation comes from the simultanious optimisation of several parent individuals one parent strategies as e.g. the PAES algorithm [8] proved competitive on a number of problems. So the extensions to the DES are twofold: on the one hand, a multimembered (μ, λ) DES was developed to incorporate the advantages of this approach in MCDM. On the other hand the task of keeping diversity whilst approaching the pareto set had to be added to the $(1, \lambda)$-DES as well.

Selecting one individual from a set of individuals each having more than one objective function value is in the end only a special case of selecting a new population in MOEAs. The first step in the selection scheme is almost clear, if the Pareto concepts are applied. If there is one and only one non-dominated individual, select this one for becoming the parent of the next generation.

Due to hard restrictions concerning the allowed number of objective function evaluations, the elitist (1+10)-DES and (20+20)-DES are used. Combined with the choice of this selection scheme is the hope, that it performs better than the comma strategy, because only improvements with respect to the selection mechanism in use are possible.

In the current investigation, the following selection schemes have been compared:

MODES I: If more than one non-dominated individual is in the population, the number of individuals dominated is taken into account. If there is one individual with a maximum number of dominated individuals, this one is chosen to become the parent of the next generation. If there are more than one with the same maximum number of individuals dominated, the distance to the origin of the objective function space is taken into account. In this special case, the individual with the smallest distance to the origin, which is the global optimum due to objective function formulation, is selected to become the parent of the next generation.

MODES II: This selection scheme is similar to MODES I presented above, but instead of the distance to the origin, the distance to other individuals from the population has been taken into account. More precisely, the individuals with the same number of dominated individuals are compared to each other. Therefore the distance of one individual to the other ones is calculated and added. The one with the greatest sum, thus the one with the greatest sum of distances to the other individuals, is selected and becomes the parent of the next generation.

MODES III: In the third selection scheme investigated here, the second criteria from MODES II, the number of individuals dominated, is omitted. If more than one non-dominated individual is in the population, the one with the greatest distance among all non-dominated solutions is selected to become the parent of the next generation. Therefore the distance of one non-dominated individual to the other non-dominated ones is calculated and added similarly to selection MODES II.

MODES IV: The fourth selection scheme is inspired by the NSGA-II [4] and is used in a multimembered (μ, λ)-DES: First the parent and offspring population are combined to form $R = P \cup Q$. After a nondominated sorting of R the pareto fronts F_i of R are identified. The new parent population P' is filled frontwise with individuals from $F^1, F^2, ...,$ until $|P'| + F^i > \mu$. If the sizes of the first $i - 1$ pareto fronts sum up to μ, the resulting P' contains the correct number of new parent individuals and new offspring can be generated. Otherwise $\mu - |P'|$ individuals must be selected from the i-th pareto front to fill the remaning slots of P'. This task is accomplished by

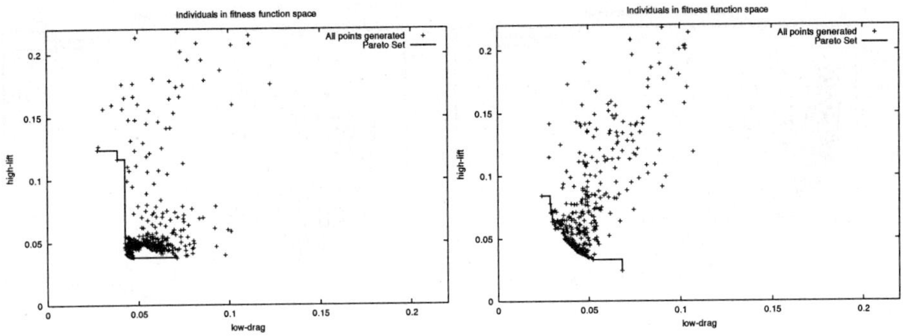

Fig. 1. Pareto-front, MODES I

NSGA-II's crowding sort procedure: Each individual I_k in F^i is assigned a distance value $d_I = 0$. Then for each objective function f_m, $m = 1, \ldots, M$ the Pareto-front F^i is sorted in ascending order so that $f_m(I_k) < f_m(I_{k+1})$ and the distance values for the $k = 1 \ldots |F^i|$ individuals are computed as

$$d_{I_k} = d_{I_k} + \frac{f_m(I_{k+1}) - f_m(I_{k-1})}{f_m^{max} - f_m^{min}}$$

The boundary solutions get a value of $I_1 = I_{|F_i|} = \infty$. The missing slots of P' are filled with the most widley spread individuals of F^i which are found by sorting F^i in ascending order of the distance values and selecting the first $\mu - |P'|$ members of F^i. A new offspring population Q' is generated by discrete recombination of the design variables and the strategy parameters and mutation with the derandomized mutation operator.

For each of the presented selection schemes, different optimisation runs have been performed. Most important for the comparison of results is the obtained Pareto-front showing the most obvious and remarkable results.

A typical Pareto-front obtained by MODES I is presented in figure 1. Here, the search focusses on the path to the origin of the fitness function space.

This behavior changes not until the second selection step, selection considering the number of dominated solutions, is omitted from the selection procedure. Pareto-fronts obtained incorporating MODES II show a compareable behavior like the ones from MODES I and are therfore excluded from further investigations at this point.

Figure 2 presents typical Pareto-front from runs using MODES III. The Pareto-front is covered satisfactorily, showing a wide range of different alternative solutions. This gives the user the possibility to compare many alternative solutions featuring different design aspects.

Furthermore, even the extreme specifications of the fitness function space are explored, where the solutions focus on only one objective. This behavior may be of special interest, if one objective plays an accentuated role among the others.

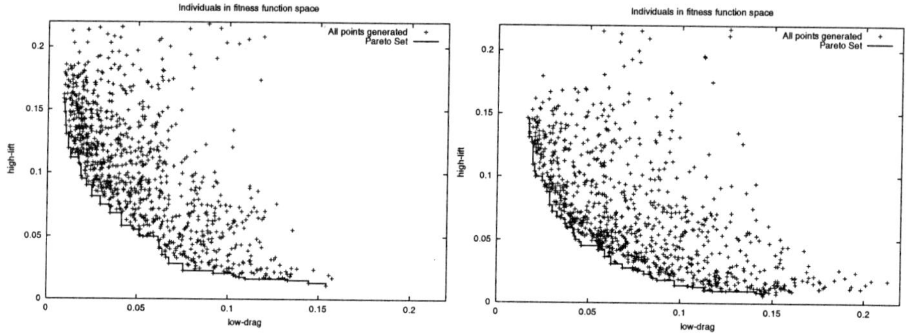

Fig. 2. Pareto-fronts, MODES III

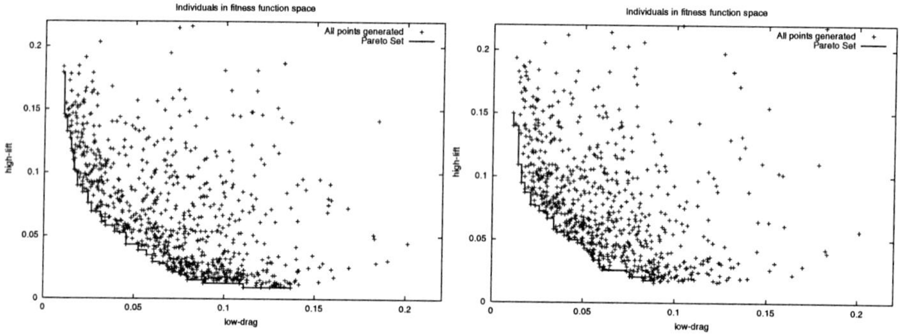

Fig. 3. Pareto-fronts, MODES IV

The same holds for the results obtained using MODES IV presented in figure 3. The results look comparable to the ones received using MODES III and this clarifies the need of a technical instrument to compare the results.

3.2 New Approach Averaging and Comparing MOEA Results

The new approach for averaging Pareto-fronts has to be seen in conjunction with the considered test case. It was developed for this special case and modifications may be necessary for transferring it to other applications. Nevertheless, applied to the test case under investigation it performs very well.

The present method is inspired by the attainment surfaces proposed by Fonseca and Fleming [9] and is intended to supply a method to derive an averaged pareto front from several runs of the same optimisation algorithm. It allows comparing different optimisation algorithms by visually comparing their averaged pareto fronts. In contrast to the Fonseca and Fleming method the present method is based on bisecting lines and parallels to it. First, the interval $[a, b]$ under investigation has to be determined by looking at the objective function values of the Pareto optimal solutions. Afterwards, the number of intersections i through this interval has to be determined by the user.

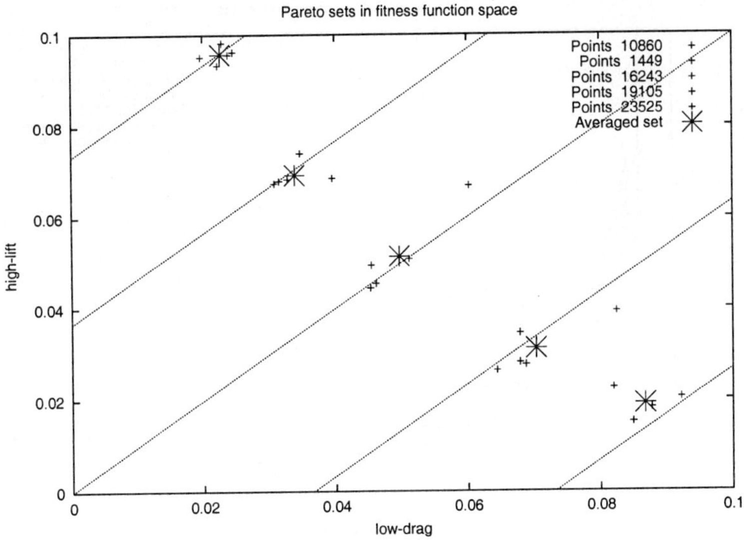

Fig. 4. Point from MODES III considered for the averaging process and the resulting averaged points in a part of the investigated quadrant

In the present case the interval $[0, 0.22]$ which is cut by $i = 5$ intersections is investigated. Furthermore there are corresponding intersections through the interval $[-0.22, 0]$ and the origin. This results in a band of parallel intersections y_s equally distributed around the origin in the quadrant of the search space under investigation:

$$y_s = x + (-b + s \cdot \frac{b-a}{i})$$

with $s = \{-i, \ldots, i\} = \{-5, \ldots, 5\}$ in the current investigation.

Now the Pareto optimal solutions next to these intersections from each optimisation run are determined. This cloud of points according to each intersection is used to calculate an averaged point corresponding to this intersection. This is done by calculating the centroid of this cloud of points. Figure 4 shows the clouds of points corresponding to each intersection from the different optimisation runs in a part of the investigated quadrant.

Figure 5 presents the quadrant under investigation with all defined intersections, the Pareto-fronts obtained using MODES III and the resulting averaged Pareto-front.

4 Results

The averaged fronts of MODES I, MODES III and MODES IV are plotted in figure 6. From this figure it can be seen that the multimembered strategy of MODES IV does not provide an obvious advantage over the single parent approach of MODES III. The overall length of both algorithms' average attainment

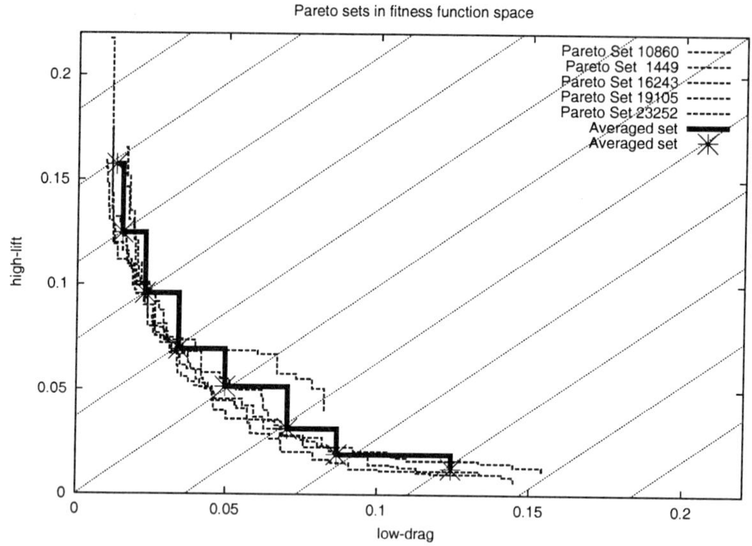

Fig. 5. Pareto-fronts from MODES III and the averaged Pareto-front (*thick line*)

surface is quite comparable while MODES III has advantages on "high lift" and MODES IV has its stronger part on "low drag". Although the (20+20) strategy of MODES IV is only a first shot on the airfoil optimisation problem, a more visible advantage of its multimembered nature might have been expected.

The "closest distance to the origin" selection criteria of MODES I might be of little value in a number of real world applications, but it can be seen from figure 6 that the incorporation of further selective pressure by this restriction increases the algorithm's performance: MODES I gives the best results of all algorithms in the $[0, 0.05] \times [0, 0.05]$ subspace which is closest to the middle intersection. The drawback lies in MODES I's low performance in the bordering regions close to the single criteria optima which results in the shortest average attainment surface of the three algorithms compared.

5 Conclusions

As can be seen in figure 6 the new averaging method provides a reasonable way to compare pareto fronts generated by different algorithms. In contrast to the Fonseca/Flemming approach it does not rely on "virtual" points when the attainment surface is intersected in between two "real" points of the underlieing pareto front but averages "real" points found by the optimisation algorithm. The average attainment surfaces can easily be compared to estimate the quality of different algorithms.

For the airfoil optimisation problem the tested algorithms with the derandomized mutation operator showed a good performance both as one- and multiparent strategies. Especially for the $(\mu + \lambda)$ ES good values for μ and λ are

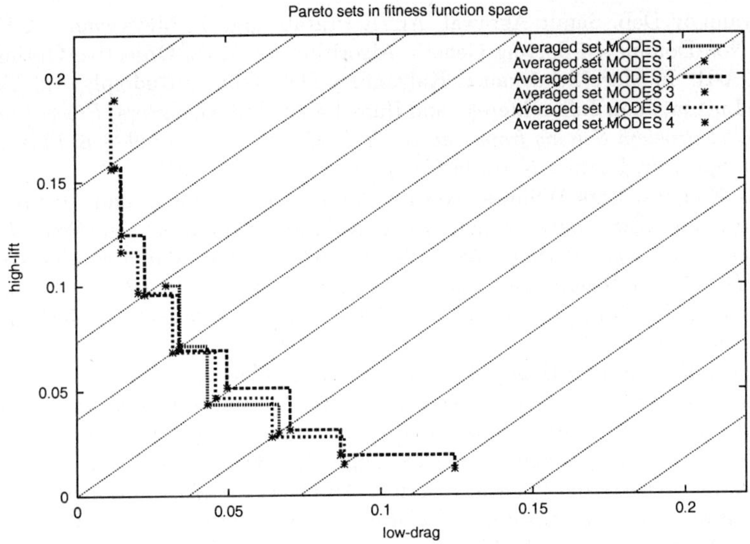

Fig. 6. Averaged Pareto-fronts achieved using MODES I, MODES III, and MODES IV

subject to further research, just as the investigation of selfadaptation in multi-criteria evolutionary optimisation is still at the beginning.

Acknowledgements

This research was supported by the Deutsche Forschungsgemeinschaft as part of the collaborative research center "Computational Intelligence" (531).

The AEROSHAPE project (Multi-Point Aerodynamic Shape Optimisation) is a collaboration between Aerospatiale Matra Airbus, Alenia Aeronautica (Coordinator), Daimler Chrysler Airbus, EADS-M, Dassault Aviation, SAAB, SENER, SYNAPS, CIRA, DERA, DLR, FFA, INRIA, HCSA, NLR, ONERA, and NuTech Solutions. The project is funded by the European Commision, DG Research, under the GROWTH initiative (Project Ref: GRD1-1999-10752).

References

1. Thomas Bäck, David B. Fogel, and Zbigniew Michalewicz, editors. *Handbook of Evolutionary Computation.* Oxford University Press, New York, and Institute of Physics Publishing, Bristol, 1997.
2. Andreas Ostermeier, Andreas Gawelczyk, and Nikolaus Hansen. Step-size adaptation based on non-local use of selection information. In Y. Davidor, H.-P. Schwefel, and R. Männer, editors, *Parallel Problem Solving from Nature — PPSN III International Conference on Evolutionary Computation*, volume 866 of *Lecture Notes in Computer Science*, pages 189–198. Springer, Berlin, 1994.
3. Hans-Paul Schwefel. *Evolution and Optimum Seeking.* Sixth-Generation Computer Technology Series. Wiley, New York, 1995.

4. Kalyanmoy Deb, Samir Agrawal, Amrit Pratab, and T. Meyarivan. A Fast Elitist Non-Dominated Sorting Genetic Algorithm for Multi-Objective Optimization: NSGA-II. In Marc Schoenauer, Kalyanmoy Deb, Günter Rudolph, Xin Yao, Evelyne Lutton, Juan Julian Merelo, and Hans-Paul Schwefel, editors, *Proceedings of the Parallel Problem Solving from Nature VI Conference*, pages 849–858, Paris, France, 2000. Springer. Lecture Notes in Computer Science No. 1917.
5. Boris Naujoks, Lars Willmes, Werner Haase, Thomas Bäck, and Martin Schütz. Multi-point airfoil optimization using evolution strategies. In *Proc. European Congress on Computational Methods in Applied Sciences and Engineering (ECCOMAS'00) (CD-Rom and Book of Abstracts)*, page 948 (Book of Abstracts), Barcelona, Spanien, September 11–14 2000. Center for Numerical Methods in Engineering (CIMNE).
6. Thomas Bäck, Werner Haase, Boris Naujoks, Luca Onesti, and Alessio Turchet. Evolutionary algorithms applied to academic and industrial test cases. In K. Miettinen, M. M. Mäkelä, P. Neittaanmäki, and J. Périaux, editors, *Evolutionary Algorithms in Engineering and Computer Science*, pages 383–397. Wiley, Chichester, 1999.
7. Carlos M. Fonseca and Peter J. Fleming. An Overview of Evolutionary Algorithms in Multiobjective Optimization. *Evolutionary Computation*, 3(1):1–16, Spring 1995.
8. Joshua D. Knowles and David W. Corne. Approximating the Nondominated Front Using the Pareto Archived Evolution Strategy. *Evolutionary Computation*, 8(2):149–172, 2000.
9. Carlos M. Fonseca and Peter J. Fleming. On the Performance Assessment and Comparison of Stochastic Multiobjective Optimizers. In Hans-Michael Voigt, Werner Ebeling, Ingo Rechenberg, and Hans-Paul Schwefel, editors, *Parallel Problem Solving from Nature—PPSN IV*, Lecture Notes in Computer Science, pages 584–593, Berlin, Germany, September 1996. Springer-Verlag.

Hyperheuristics: A Robust Optimisation Method Applied to Nurse Scheduling

Peter Cowling[2], Graham Kendall[1], and Eric Soubeiga[1,*]

[1] ASAP Research Group, School of Computer Science and IT
University of Nottingham, Nottingham NG8 1BB, UK
{gxk,exs}@cs.nott.ac.uk

[2] MOSAIC Research Group, Department of Computing, University of Bradford
Bradford BD7 1DP, UK
Peter.Cowling@scm.brad.ac.uk

Abstract. A *hyperheuristic* is a high-level heuristic which adaptively chooses between several low-level knowledge-poor heuristics so that while using only cheap, easy-to-implement low-level heuristics, we may achieve solution quality approaching that of an expensive knowledge-rich approach, in a reasonable amount of CPU time. For certain classes of problems, this generic method has been shown to yield high-quality practical solutions in a much shorter development time than that of other approaches such as tabu search and genetic algorithms, and using relatively little domain-knowledge. Hyperheuristics have previously been successfully applied by the authors to two real-world problems of personnel scheduling. In this paper, a hyperheuristic approach is used to solve 52 instances of an NP-hard nurse scheduling problem occuring at a major UK hospital. Compared with tabu-search and genetic algorithms, which have previously been used to solve the same problem, the hyperheuristic proves to be as robust as the former and more reliable than the latter in terms of solution feasibility. The hyperheuristic also compares favourably with both methods in terms of ease-of-implementation of both the approach and the low-level heuristics used.

Keywords: Hyperheuristic, Heuristic, Personnel Scheduling, Nurse Scheduling.

1 Introduction

Personnel scheduling deals with the allocation of timeslots and possibly locations and other resources to people. This problem has been extensively addressed in the literature over the past 30 years with a survey in almost every decade [3, 11, 4]. Very often the problem is solved using heuristics. For instance Schaerf [10] tackled a high school timetabling problem formulated as a mathematical programme. He defined two types of neighbourhood moves, the atomic move which swaps two classes of the same lecturer which are scheduled in two different timeslots, and the double move as a pair of atomic moves. The former

* Corresponding author

J.J. Merelo Guervós et al. (Eds.): PPSN VII, LNCS 2439, pp. 851–860, 2002.

is associated with a tabu search and the latter with a randomised nonascendent search (RNA). Both methods are then used alternately. The algorithm produced timetables better than the manual ones for various types of schools. Levine [9] used a hybrid genetic algorithm (GA) to solve an airline crew scheduling problem. The GA was hybridised with a local search heuristic which tries to repair infeasibilities in the solution. Computational experiments compared the hybrid GA with branch-and-cut and branch-and-bound algorithms. Both these latter algorithms produced better solutions than the hybrid GA. Often heuristic methods used to solve personnel scheduling problems make use of sophisticated metaheuristic techniques and problem-specific information to arrive at a good solution. This is the case in [8] and [2] which respectively report the use of tabu search and a genetic algorithm for nurse scheduling.

It is precisely in this context that we proposed a *hyperheuristic* approach [5] as a heuristic that operates at a higher level of abstraction than current metaheuristic approaches. A hyperheuristic controls a set of simple, knowledge-poor, low-level heuristics (for example change, swap, add and drop moves). At each decision point the hyperheuristic must choose which low-level heuristic to apply, without recourse to domain knowledge. Hence we may use hyperheuristics in cases where little domain-knowledge is available (e.g. when dealing with a new, poorly understood or unusual problem) or when a solution must be produced quickly (e.g for prototyping). A hyperheuristic could be regarded as an 'off-the-peg' method as opposed to a 'made-to-measure' bespoke metaheuristic. It is a generic and fast method, which should produce solutions of at least acceptable quality, based on a set of cheap and easy-to-implement low-level heuristics. In order to apply a hyperheuristic to a given problem, we need only a set of low-level heuristics and one or more measures for evaluating solution quality. In [5, 6] and [7] respectively a choice function hyperheuristic of the same type as in section 4 was successfully applied to a real world problem of scheduling business meetings, and to a problem of scheduling undergraduate students' project presentations at a UK academic institution. In this paper, we use our hyperheuristic technique to solve another real-world problem, that of scheduling nurses at a major UK hospital. The problem has been previously solved using tabu search [8] and genetic algorithms [2]. It is our aim to demonstrate that hyperheuristics are not only readily applicable to a wide range of scheduling and other combinatorial optimisation problems, but also can provide very good-quality solutions comparable to those of knowledge-rich sophisticated metaheuristics, while using less development time and simple, easy-to-implement low-level heuristics. In sections 2, 3, 4 and 5 we present respectively the nurse scheduling problem, the solution methods used, experimental results, and conclusions.

2 The Nurse Scheduling Problem

The problem is to create weekly schedules for wards containing up to 30 nurses at a major UK hospital. These schedules must respect working contracts and meet the demands (i.e number of nurses of different grades required) for each day-shift and night-shift of the week, whilst being perceived as fair by the nurses themselves. Nurses work either day-shifts, divided into 'earlies' and 'lates', or night-shifts in a given week. A full week's work typically includes more days than nights[1]. As mentioned in [8], the problem can be decomposed into 3 independent stages. Stage 1 uses a knapsack model to check if there are enough nurses to meet demands. Additional nurses are needed for Stage 2 otherwise. This latter stage is the most difficult and is concerned with the actual allocation of the weekly shift-pattern schedules to each nurse. Then stage 3 uses a network flow model to assign those on day-shifts to 'earlies' and 'lates'. As in [8] and [2] we limit ourselves to the highly constrained problem in stage 2, as stages 1 and 3 can be solved quickly using standard knapsack and network flow algorithms. The stage 2 problem is described as follows. Each possible weekly shift-pattern for a given nurse can be represented as a 0-1 vector of 14 elements (7 day-shifts and 7 night-shifts). A '1'/'0' in the vector represents a day or night 'on'/'off'. For each nurse there is a limited number of shift-patterns corresponding to the number of combinations of the number of days s/he is contracted to work in a week[2]. There are typically between 20 and 30 nurses per ward, 3 grade-bands, and 411 different (F/T and P/T) shift-patterns. Based upon the nurses' preferences, the recent history of patterns worked, and the overall attractiveness of the pattern, a penalty cost is associated to each assignment nurse-shift pattern, values of which were set after agreement with the hospital, ranging from 0 (ideal) to 100 (undesirable) -See [8] for further details. Our decision variables are denoted by x_{ij} assuming 1 if nurse i works shift-pattern j and 0 otherwise. Let parameters g, n, s be the number of grades, nurses and possible shift-patterns respectively. a_{jk} is 1 if shift-pattern j covers shift k, 0 otherwise. b_{ir} is 1 if nurse i is of grade r or higher, 0 otherwise. p_{ij} = penalty cost of nurse i working shift-pattern j; S_{kr} = demand of nurses of grade r or above on day/night (i.e shift) k; and $F(i)$ = set of feasible shift-patterns for nurse i. We may then formulate the problem as follows:

$$Min \quad PC = \sum_{i=1}^{n} \sum_{j=1}^{s} p_{ij} x_{ij} \tag{1}$$

s.t.

$$\sum_{j \in \mathbf{F(i)}} x_{ij} = 1, \quad \forall i \tag{2}$$

$$\sum_{j=1}^{s} \sum_{i=1}^{n} b_{ir} a_{jk} x_{ij} \geq S_{kr}, \forall k, r \tag{3}$$

$$x_{ij} \in \{0, 1\}, \forall i, j \tag{4}$$

[1] e.g. a full-time nurse works 5 days or 4 nights, whereas a part-time nurse works 4 days or 3 nights, 3 days or 3 nights, and 3 days or 2 nights.

[2] For example a F/T nurse contracted to work either 5 days or 4 nights has a total of $C_7^5 = 21$ feasible day shift-patterns and $C_7^4 = 35$ feasible night shift-patterns.

Equations (1), (2) and (3) express respectively the objective of minimising the overall penalty cost associated with the nurses' desirability for the shift-patterns, the constraint that each nurse should work exactly one shift-pattern, and the demand constraints. It should be noted that b_{ir} is defined in such a way that higher-grade nurses can substitute for those at lower grades if needed. The problem is NP-hard [2] and instances typically involve between 1000 and 2000 variables and up to 70 constraints. As noted in [2], the difficulty of a given instance depends upon the shape of the solution space, which in turn depends on the distribution of the penalty cost (p_{ij}) and their relationship with the set of feasible solutions. In this paper we consider 52 data instances, based on three wards and corresponding to each week of the year. These 52 instances, as a whole, feature a wide variety of solution landscapes ranging from easy problems with many low-cost global optima scattered througout the solution space, to very hard ones with few global optima and in some cases with relatively sparse feasible solutions [2]. Optimal solutions are known for each instance as the problem was solved using a standard IP package. However some instances remained unsolved after 15 hours of (Pentium II 200 Mhz PC) run-time. Further experiments with a number of descent methods using different neighbourhoods, and a standard simulated annealing were conducted unsuccessfully, failing to obtain feasibility [2]. The most successful approach which works within the low CPU time available so far is a tabu search which uses chain-moves whose design and implementation were highly problem and instance specific as these moves relied on the way the different factors affecting the quality of a schedule were combined in the p_{ij} as noted in [2]. In [2] a GA which did not make use of chain-moves was also used to solve the problem. Failure to obtain good solutions led to the use of a co-evolutionary strategy which decomposed the main population into several co-operative sub-populations. Problem structure was incorporated in both the way the sub-populations were built, and the way partial solutions were recombined to form complete ones. As a result, the applicability of the co-evolutionary strategy is, likewise, limited to problems with a similar structure.

Here we propose to solve the nurse scheduling problem using a high-level hyperheuristic approach which has been successfully applied to two rather different real-world problems of personnel scheduling. The evaluation function[3] used by the hyperheuristic distinguishes between 'balanced' and 'unbalanced' solutions [8, 2]. Effectively, since nurses work either days or nights it appears that in order for a given solution to be feasible, (i.e enough nurses covering all 14 shifts at each grade) the solution must have sufficient nurses in both days and nights separately. Formally, a solution is balanced in days (or nights) at a given grade r if there are both under-covered and over-covered shifts in the set of days (or nights) at grade r such that the nurse surplus in the over-covered day (or night) shifts suffices to compensate for the nurse shortage of the under-covered day (or night) shifts. In fact, a solution cannot be made feasible until it is balanced [8, 2]. We define $Infeas = \sum_{r=1}^{g}(\rho \times Bal_r + 1)\sum_{k=1}^{14} max\left(\left[S_{kr} - \sum_{i=1}^{n}\sum_{j=1}^{s} b_{ir}a_{jk}x_{ij}\right], 0\right)$,

[3] known as fitness function in the GA literature

where Bal_r is 2 if both day and night are unbalanced at grade r, 1 if either day or night is unbalanced at grade r, and 0 otherwise; ρ is a parameter set to 5, so that a balanced solution with more nurse-shortages is preferred to an unbalanced one with fewer nurse-shortages, as the latter is more difficult to make feasible than the former. Based on this, we define the evaluation function $E = PC + C_{demand} InFeas$ with C_{demand} a weight associated to $InFeas$ as in [2]. The definition of C_{demand} is based on the number, q, of nurse-shortages in the best least-infeasible solution so far, i.e. $q = \sum_{k=1}^{14} \sum_{r=1}^{g} max\left(\left[S_{kr} - \sum_{i=1}^{n} \sum_{j=1}^{s} b_{ir} a_{jk} x_{ij}\right], 0\right)$. Coefficient C_{demand} of $InFeas$ in E is then given by $C_{demand} = \gamma \times q$ if $q > 0$, and $C_{demand} = v$ otherwise; where γ is a preset severity parameter, and v is a suitably small value. The idea is that the weight C_{demand} depends on the degree of infeasibility in the best least-infeasible solution encountered so far, after which it remains at v. We use $\gamma = 8$ and $v = 5$ as given in [2][4]. It is interesting to note that while in [8] unbalanced solutions are repaired, in [2] they are instead avoided through the use of incentives/disincentives to reward/penalise balanced/unbalanced individuals in the population. Here we opt for the former approach and use the same 'balance-restoring' low-level heuristic used in tabu search of [8]. As described in section 4, this low-level heuristic uses a 'change' and a 'swap' type of move. We next describe our hyperheuristic method.

3 A Choice-Function Hyperheuristic Technique

Our hyperheuristic is based upon a *Choice-Function* which adaptively ranks the low-level heuristics. Originally [5], the choice function is determined based on information with regards to individual performance of each low-level heuristic (f_1), joint performance of pairs of heuristics (f_2), and the amount of time elapsed since the low-level heuristic was last called (f_3). Thus we have $f_1(N_j) = \sum_n \alpha^{n-1}(\frac{I_n(N_j)}{T_n(N_j)})$ and $f_2(N_j, N_k) = \sum_n \beta^{n-1}(\frac{I_n(N_j,N_k)}{T_n(N_j,N_k)})$ where $I_n(N_j)/I_n(N_j, N_k)$ (respectively $T_n(N_j)/T_n(N_j, N_k)$) is the change in the objective function (respectively the number of CPU seconds) the nth last time heuristic N_j was called/called immediately after heuristic N_k. Both α and β are parameters between 0 and 1, reflecting the greater importance attached to recent performance. f_1 and f_2 aim at intensifying the search. The idea behind the expressions of f_1 and f_2 is analogous to the exponential smoothing forecast of their performance [12]. f_3 provides an element of diversification, by favouring those low-level heuristics that have not been called recently. Then we have $f_3(N_j) = \tau(N_j)$ where $\tau(N_j)$ is the number of CPU seconds which have elapsed since low-level heuristic N_j was last called. If the low-level heuristic just called is N_j then for any low-level heuristic N_k, the choice function f of N_k is defined as $f(N_k) = \alpha f_1(N_k) + \beta f_2(N_j, N_k) + \delta f_3(N_k)$. In this expression, the choice function attempts to predict the overall performance of each low-level heuristic. In [6], we presented a different choice function which separately predicts the performance of each low-

[4] See [2] for an interesting discussion on the choice of evaluation functions.

level heuristic with respect to each criterion of the evaluation function instead (PC and $Infeas$ in the model above). The choice function f is then decomposed into $f(N_k) = \sum_{l \in \mathbf{L}} f_l(N_k) = \sum_{l \in \mathbf{L}} \left[\alpha_l f_{1l}(N_k) + \beta_l f_{2l}(N_j, N_k) + \frac{\delta}{|L|} f_3(N_k)\right]$ where $\mathbf{L} = \{PC, Infeas\}$ is the set of the evaluation function criteria, and $f_{1l}(N_k)$ (respectively $f_{2l}(N_j, N_k)$) is obtained by replacing $I_n(N_k)$ (respectively $I_n(N_j, N_k)$) with $I_{ln}(N_k)$ (respectively $I_{ln}(N_j, N_k)$) in the expression of $f_1(N_j)$ (respectively $f_2(N_j, N_k)$) above. $I_{ln}(N_k)$ (respectively $I_{ln}(N_j, N_k)$) is the first (respectively second) order improvement with respect to criterion $l \in \mathbf{L}$. Parameter values for α, β and δ are changed adaptively using the procedure in [6]. We will give results for the second approach which works as follows

Do
Choose a search criterion l
- Select the low-level heuristic that maximises f_l and apply it.
- Update choice function f_l's parameters using the adaptive procedure
Until Stopping condition is met.

The probability of choice of criteria PC and $InFeas$ is given by $p_1 = \frac{1}{1+C_{demand}}$ and $p_2 = \frac{C_{demand}}{1+C_{demand}}$ respectively as defined in [6]. We would like to emphasize the fact that the implementation of the hyperheuristic technique was quite fast. In effect the hyperheuristic presented here is a 'standard' approach which was successfully applied to two different real-world problems [5–7]. The hyperheuristic approach only requires a set of low-level heuristics to be added to the hyperheuristic black box, and a formal means of evaluating solution quality. The way the hyperheuristic works is independent of both the nature of the low-level heuristics and the problem at hand. Hence important savings in development time are made possible by the use of the hyperheuristic framework. Development of the framework itself took over eighteen months. For example in [7] high-quality solutions were initially developed in just over two weeks after meeting with the problem owner. Solution development time for the current problem was one-and-a-half months, due to the larger number of instances to be handled and the development of low-level heuristics for this challenging highly-constrained real-world problem. We show in the next section that, despite such a relatively short development time, the hyperheuristic - even when dealing with a very difficult problem - is capable of finding solutions of good quality comparable to those of bespoke metaheuristics within a reasonable amount of time.

4 Experiments

Both our hyperheuristic and its low-level heuristics were coded in Micosoft Visual C++ version 6 and all experiments were run on a PC Pentium III 1000MHz with 128MB RAM running under Microsoft Windows 2000 version 5. In order to compare our results with those of tabu search (TS) and genetic algorithms (GA), our hyperheuristic starts with a solution generated randomly by assigning a random feasible shift-pattern to each nurse as in [8]. All results were averaged over 20 runs. The TS algorithm of [8] used the following 11 low-level heuristics:

[h1] Change the shift-pattern of a random nurse; [h2] Same as [h1] but 1st improving $InFeas$; [h3] Same as [h1] but 1st improving $InFeas$, no worsening of PC; [h4] Same as [h1] but 1st improving PC; [h5] Same as [h1] but 1st improving PC, no worsening of $InFeas$; [h6] Change the shift-pattern type (i.e day/night) of a random nurse, if solution unbalanced; [h7] Same as [h6] but aim is to restore balance[5]; [h8] (shift-chain1): This heuristic considers chains of moves aiming at decreasing both the nurse-shortage in one (under-covered) shift and the nurse-surplus in one (over-covered shift), and leaving the remaining shift unchanged; [h9] (nurse-chain1): Considers chains of moves which move the first nurse in the chain to cover an under-covered shift and move the subsequent nurses to the shift-pattern just vacated by their predecessor in the chain.[6]; [h10] (shift-chain2): Considers a shift-chain of moves aiming at decreasing the penalty cost when the solution is already feasible; [h11] (nurse-chain2): Considers nurse-chains of moves aiming at decreasing the penalty cost when the solution is already feasible[7].

Instead, our hyperheuristic uses 9 low-level heuristics including the first 7 low-level heuristics above and the following: [H8] (Change-and-keep1): This heuristic changes the shift-pattern of a nurse and assigns the removed shift-pattern to another nurse (1st improving PC); [H9] (Change-and-keep2): Same as [H8], but 1st improving PC and no worsening of $InFeas$.

The chain-moves are highly effective moves which were responsible for both feasibility (using shift-chain1 and nurse-chain1) and optimality (using shift-chain2 and nurse-chain2) of the solution in most cases (see [8] for further details). TS can only yield good solutions when equipped with such moves [1, 2]. However, as noted in [1, 2] these moves are highly problem-dependent and, in fact, instance-type dependent. Unlike in TS, the low-level heuristics used by the hyperheuristic are fewer and much simpler than the chain-moves. They are all based around changing, or swapping one or two shift-patterns, thus reflecting what users usually do in practice [5]. In Table 1, we present the results of our hyperheuristic, along with those of both the direct and indirect GA [1, 2] as well as TS [8] and the IP optimal solution [1] for each of the 52 weeks (problem instances) of the year. The stopping condition of the hyperheuristic is 6000 iterations, which corresponds to a CPU time between 44 and 60 seconds on a Pentium II 1000Mhz[8]. We see that for all instances the hyperheuristic is able to find feasible solutions in each of 20 runs. It appears that the hyperheuristic is more reliable than both the direct and the indirect GA in terms of producing practical solutions for the hospital. To confirm the reliability of the hyperheuristic, we ran it on instance 50 (which is a difficult instance for both GA's and appeared to be the most difficult for the hyperheuristic) 100 times and feasibil-

[5] i.e from day to night if night is unbalanced and vice-versa. If both days and nights are unbalanced a swap of shift-pattern type for a pair of nurses, one working days and the other working night is considered. The nurse working day is assigned a night shift-pattern and the nurse working night is assigned a day shift-pattern.

[6] Both [h8] and [h9] chain-moves are defined as paths in a graph. The move is only attempted if the solution is already balanced but not yet feasible.

[7] This time both [h10] and [h11] chains are represented as cycles in a graph.

[8] The TS stopping condition was 1000 moves without overall improvement.

Table 1. Hyperheuristic and metaheuristic performances on the nurse scheduling problem. Results are averaged over 20 runs. Format is proportion of feasible solutions in 20 runs/average penalty cost.

Instances	Hyperheuristic	Direct GA	Indirect GA	Tabu seach	IP cost
Week 1	1/8	1/0	1/0	0	0
Week 2	1/52.8	1/12	1/12	11	11
Week 3	1/50	1/18	1/18	18	18
Week 4	1/17	1/0	1/0	0	0
Week 5	1/11	1/0	1/0	0	0
Week 6	1/ 2	1/1	1/1	1	1
Week 7	1/13.55	0.5/13	1/11	11	11
Week 8	1/14.95	1/11	1/11	11	11
Week 9	1/3.6	0.95/3	1/3	3	3
Week 10	1/5.05	1/1	1/2	1	1
Week 11	1/2	1/1	1/1	1	1
Week 12	1/2	1/0	1/0	0	0
Week 13	1/2	1/1	1/1	1	1
Week 14	1/3.15	1/3	1/3	3	3
Week 15	1/3.05	1/0	1/0	0	0
Week 16	1/40.1	0.95/25	1/25	24	24
Week 17	1/17.6	1/4	1/4	4	4
Week 18	1/20.85	1/7	1/6	7	6
Week 19	1/1.6	1/1	1/1	1	1
Week 20	1/15.45	0.95/5	1/4	4	4
Week 21	1/0	1/0	1/0	0	0
Week 22	1/25.5	1/1	1/1	1	1
Week 23	1/0	0.95/0	1/0	0	0
Week 24	1/1	0.75/1	1/1	1	1
Week 25	1/0.4	1/0	1/0	0	0
Week 26	1/48	0.1/0	1/0	0	0
Week 27	1/3.65	1/2	1/3	2	2
Week 28	1/65.8	1/1	0.95/1	1	1
Week 29	1/15	0.35/3	1/1	2	1
Week 30	1/39.4	1/33	1/33	33	33
Week 31	1/66.9	0.8/66	1/36	33	33
Week 32	1/41.6	1/21	1/21	20	20
Week 33	1/10.6	1/12	1/10	10	10
Week 34	1/42.9	1/17	1/16	15	15
Week 35	1/38.8	1/9	1/11	9	9
Week 36	1/34.85	1/7	1/6	6	6
Week 37	1/8.05	1/3	1/3	3	3
Week 38	1/13.3	1/3	1/0	0	0
Week 39	1/5.1	1/1	1/1	1	1
Week 40	1/9.35	1/5	1/ 4	4	4
Week 41	1/61.3	0.95/27	1/27	27	27
Week 42	1/47.55	1/5	1/8	5	5
Week 43	1/27.35	0.9/8	1/6	6	6
Week 44	1/31.75	0.9/45	1/17	16	16
Week 45	1/5.35	1/0	1/0	0	0
Week 46	1/9.4	0.7/6	1/4	3	3
Week 47	1/3.3	1/3	1/3	3	3
Week 48	1/6.05	1/4	1/4	4	4
Week 49	1/30.4	1/26	0.7/25	24	24
Week 50	1/109.25	0.35/38	0.8/36	35	35
Week 51	1/74.3	0.45/46	1/45	45	45
Week 52	1/62.2	0.75/63	1/46	46	46
Average	1/23.5	0.91/10.8	0.99/9.0	1/8.8	8.7
Run time	< 60 sec	15 sec	10 sec	30 sec	up to hours

ity was again achieved always within 6000 iterations (less than a minute of CPU time). From this point of view, the hyperheuristic appears to be as robust as TS which, too, always found feasible solutions. The hyperheuristic however has the highest average cost of 23.5, though more than 50% of the instances (27 instances) were solved to within 10% of the optimal solution, including 3 instances (weeks 21, 23 and 24) where optimality is reached on each of 20 runs. Also in 9 instances (Weeks 7, 9, 14, 19, 25, 27, 33, 47 and 48) the optimal solution is hit up to 19 times out of 20 runs, corresponding to a probability of optimality of 0.95. This shows that optimal solutions are indeed, within the reach of the hyperheuristic in spite of its simplicity and that of its low-level heuristics, when compared with the problem and instance-specific information used by the TS (chain-moves) and GA (population decomposition and recombination using problem structure) implementations. In terms of cost, we noted that the hyperheuristic performed well for instances with slack demand-constraints and poorly for those with tight constraints (e.g Weeks 26, 28, 42, 50).

Observations of the frequency of call of the low-level heuristics showed that [h2] is called most often (e.g 37% on average for Week 49), followed by [h6] (e.g 10% on Week 49) and all other heuristics are called between 5% and 9%. It appears that each low-level heuristic has a part to play in the search. Observations of $Infeas$ and PC showed that immediately upon finding a feasible solution (i.e $InFeas = 0$) there was a sudden increase in PC. Similar observations mere made in [8]. Regarding choice-function parameters, the hyperheuristic search used a very high δ and a low α and β, thus confirming the need to diversify the search quite frequently, due to the sparse spread of good solutions in the landscape [2]. This was in total agreement with the graph of the variation of $InFeas$ overtime which featured sudden low peaks of $Infeas = 0$, similar to the 'comb' shape graph of the same function in [8]. Typically values of $InFeas = 0$ never lasted more than 41 heuristic calls (compared to a total of 10000 heuristic calls overall) after they were obtained. Values for α_{InFeas} and β_{InFeas} were relatively higher than those of α_{PC} and β_{PC} clearly showing the greater importance attached to feasibility over lowering PC.

5 Conclusions

We have applied a hyperheuristic to an NP-hard highly-constrained problem of scheduling nurses at a major UK hospital. The problem had previously been solved using tabu search and two genetic algorithms. In terms of solution feasibility, our hyperheuristic proved more reliable than both the direct and indirect genetic algorithms and proved to be as robust as tabu search. In terms of cost, over half of the instances were solved within 10% of optimality. In a few instances the hyperheuristic obtained optimal solutions with probability of up to 1, thus proving that optimality is indeed within the reach of the hyperheuristic, in spite of its simplicity and that of its low-level heuristics when compared to the highly problem-specific information used by both TS and the GA's. Because of their problem-specific considerations, both TS and GA implementations for this problem are limited in their applicability to other problems as opposed to the hyperheuristic which has been successfully applied to two other personnel-scheduling problems [5–7]. Moreover, the hyperheuristic does not need any parameter tuning. Hyperheuristics are easy-to-implement and require less domain knowledge than most other heuristic approaches, yet still are able to arrive at good-quality solutions even for very difficult problems within a reasonable amount of CPU and implementation time. It appears that hyperheuristics can be robust and reliable for solving real-world problems of scheduling and optimisation. Ongoing research will investigate other types of hyperheuristics applied to a wider range of real-world problems.

Acknowledgements

We express our gratitude to both Dr Kath Dowsland and Dr Uwe Aickelin for providing us with data and for their valuable support.

References

1. U. Aickelin. Genetic algorithms for multiple-choice optimisation problems. *PhD Thesis, the University of Wales Swansea*, 1999.
2. U. Aickelin and K. A. Dowsland. Exploiting problem structure in a genetic algorithm approach to a nurse rostering problem. *Journal of Scheduling*, 3:139–153, 2000.
3. K. Baker. Workforce allocation in cyclical scheduling problems: A survey. *Operational Research Quarterly*, 27(1):155–167, 1976.
4. D. J. Bradley and J. B. Martin. Continuous personnel scheduling algorithms: a literature review. *Journal Of The Society For Health Systems*, 2(2):8–23, 1990.
5. P. Cowling, G. Kendall, and E. Soubeiga. A hyperheuristic approach to scheduling a sales summit. In E. Burke and W. Erben, editors, *Selected Papers of the Third International Conference on the Practice And Theory of Automated Timetabling PATAT'2000*, Springer Lecture Notes in Computer Science, 176-190, 2001.
6. P. Cowling, G. Kendall, and E. Soubeiga. A parameter-free hyperheuristic for scheduling a sales summit. Proceedings of the 4th Metaheuristic International Conference, MIC 2001, 127-131.
7. P. Cowling, G. Kendall, and E. Soubeiga. Hyperheuristics: A tool for rapid prototyping in scheduling and optimisation. Proceedings of the 2nd European Conference on EVOlutionary COmPutation, EvoCop 2002. To appear.
8. K. A. Dowsland. Nurse scheduling with tabu search and strategic oscillation. *European Journal of Operational Research*, 106:393–407, 1998.
9. D. Levine. Application of a hybrid genetic algorithm to airline crew scheduling. *Computers and operations research*, 23(6):547–558, 1996.
10. A. Schaerf. Local search techniques for large high school timetabling problems. *IEEE Transactions on Systems, Man and Cybernetics Part A:systems and Human*, 29(4):368–377, 1999.
11. J. M. Tien and A. Kamiyama. On manpower scheduling algorithms. *SIAM Review*, 24(3):275–287, July 1982.
12. S. C. Wheelwright and S. Makridakis. *Forecasting methods for management.* John Wiley & Sons Inc, 1973.

Evolving the Topology of Hidden Markov Models Using Evolutionary Algorithms

René Thomsen

EVALife Group, Dept. of Computer Science, University of Aarhus, Bldg. 540
Ny Munkegade, DK-8000 Aarhus C, Denmark
thomsen@daimi.au.dk

Abstract. Hidden Markov models (HMM) are widely used for speech recognition and have recently gained a lot of attention in the bioinformatics community, because of their ability to capture the information buried in biological sequences. Usually, heuristic algorithms such as Baum-Welch are used to estimate the model parameters. However, Baum-Welch has a tendency to stagnate on local optima. Furthermore, designing an optimal HMM topology usually requires *a priori* knowledge from a field expert and is usually found by trial-and-error. In this study, we present an evolutionary algorithm capable of evolving both the topology and the model parameters of HMMs. The applicability of the method is exemplified on a secondary structure prediction problem.

1 Introduction

Hidden Markov models (HMM) are probabilistic models useful for modelling stochastic sequences with an underlying finite state structure. The basic theory of HMMs was developed in the 1960's and has successfully been applied to various speech recognition problems [1]. Recently, HMMs gained significant attention in the bioinformatics community, and have been a preferred choice of method when solving problems, such as protein secondary structure prediction, gene finding, and protein family classification.

Given an initial HMM, the model is trained with known data samples maximising the likelihood of the HMM to accept and generate the samples with high probability. The most popular training method is the Baum-Welch (BW) gradient search, which iteratively estimates new model parameters (transition and emission probabilities) using maximum likelihood estimation. However, as with gradient search in general, BW is sensitive to the initial parameter settings, and often converges to a local optimum.

Standard HMM estimation techniques, such as BW, assume knowledge of the model size and topology. However, for most modelling applications it is not feasible to specify HMMs by hand and it is not always possible to infer the HMM topology from the problem description. Furthermore, as the complexity (many states and transitions) of the HMM increases, designing HMMs manually becomes difficult and often unmanageable. Several authors have suggested to use evolutionary algorithms (EAs) to evolve the model parameters of the HMM

J.J. Merelo Guervós et al. (Eds.): PPSN VII, LNCS 2439, pp. 861–870, 2002.

(e.g. [2] and [3]). EAs and similar stochastic methods can be used to efficiently search large parameter spaces. Furthermore, they are less likely to converge to local optima because they work simultaneously on multiple solutions.

Regarding the problem of designing HMM topologies, a few heuristic methods have been introduced. State splitting [4] and model merging [5]) learn the topologies in either a constructive or a pruning manner, i.e. adding new states to the model, or merging similar states together. However, these techniques are prone to stagnate in structural local optima. Further, the search is usually restricted to subsets of the complete class of HMM topologies.

Recently, EAs have also been applied to evolve HMM topologies. Yada et al. [6] evolved both the topology and the initial model parameters (transition and emission probabilities), which were used as initial guesses in the BW estimation process. The variation operators were insertion and deletion of transitions and states, and two-point crossover to recombine the solutions. Furthermore, the used fitness function was based on the Akaike Information Criterion allowing the EA to balance between prediction accuracy and model complexity. Likewise, Kwong et al. [7] applied an EA to evolve the topologies of left-right HMMs using similar variation operators, although they used the average log likelihood score as fitness function, and thus not penalising or avoiding solutions with high complexity. However, the EA methods introduced so far have either constrained their topologies (e.g. left-right models) or used variation operators that were not tailored to the specific problem. Furthermore, they often use fairly small population sizes (30-50 individuals) and low mutation and crossover probabilities, thereby not taking full advantage of the performance potential of EAs. In this study, we present a HMM EA, which evolves both the topology and the model parameters of HMMs. The HMM EA is used to evolve HMMs for secondary structure prediction as a benchmark to test the applicability of the introduced method.

2 Hidden Markov Models

2.1 General Introduction

A HMM λ is a set of n states, $s_0, s_1, \ldots, s_{n-1}$, connected by m state-transitions. The matrix A contains the transition probabilities, where a_{ij} is the probability for changing from state s_i to s_j. The sum of all transition probabilities leaving a state is 1.0. The first state s_0 is typically used as a *begin* state, with transition probabilities a_{0j} indicating the probability of starting the generation of symbols in state s_j. Further, each state s_i emits a symbol from a finite discrete output alphabet Σ when the state is passed. The matrix E is the emission matrix, where e_{ik} is the probability of emitting a symbol k in state s_i. The sum of all emission probabilities in each state is 1.0. A HMM is often depicted as a directed graph with a node for each state, and an edge between two nodes if a transition exists between the two states. An example of a simple HMM for secondary structure prediction is shown in figure 1. The begin state s_0 and its a_{0i} transition probabilities are not shown. For simplicity, the emission symbols

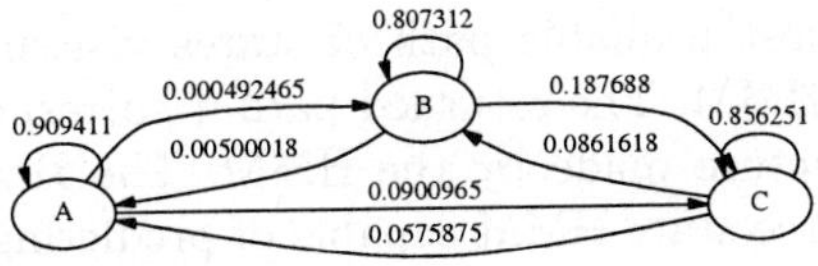

Fig. 1. Example of a simple HMM with three states.

(in this case the amino acid alphabet) and their corresponding probabilities are omitted. A, B, and C represent α-helix, β-sheet, and coil respectively, which are different types of local structures found in proteins (further explained in section 2.2).

HMMs generate observables (strings of symbols) over the output alphabet by non-deterministic walks starting in the begin state and randomly going from state to state according to the transitions between the states. The generation of observables terminates after a specified number of steps (some HMMs have a specific end-state indicating termination). Hence, a single run of the HMM follows a Markovian path $\pi = (\pi_0, \pi_1, ..., \pi_{n-1})$ of visited states and generates a string of symbols, which is the concatenation of the emitted symbols along the path. Only the symbols emitted by the HMM are observable, not the underlying path generating the symbols, hence the term *hidden* Markov model.

In most applications, the objective is to find the path π from a given sequence of symbols. Given an observable o and a HMM λ, efficient algorithms like *forward* and *Viterbi* [1] can be used to determine the probability of generating o given λ and derive the path π of maximal probability generating o. The objective in training HMMs is therefore: Given a set of l observables $o_0, o_1, \ldots, o_{l-1} \in O$, estimate the model parameters (transition a_{ij} and emission probabilities e_{ik}) of the specified HMM λ that maximises the probability $P(O|\lambda)$ of generating O with λ. This task is usually done using the BW technique.

2.2 HMMs for Secondary Structure Prediction

Secondary structure prediction of proteins is an essential problem of Molecular Biology. The goal is to determine the local secondary structures (e.g. α-helix and β-sheet) of a protein given its primary structure (sequence of amino acids).

Asai et al. introduced the idea of using HMMs to predict the secondary structures of protein sequences [8]. A HMM predicting secondary structures contains a number of states representing the different structural categories (e.g. α-helix, β-sheet, and coil). For simplicity, we only modelled α-helix (α), β-sheet (β), and coil. The other known categories such as bend, hydrogen bonded turn, 310 helix, residue in isolated beta bridge and π-helix were set to coil.

Further, the alphabet used in this type of HMM is composed of the letters representing the 20 amino acids, i.e. each state can emit any of the 20 amino acids with a certain probability. After training the model with protein sequences annotated with known secondary structures, new protein sequences with unknown secondary structures can be predicted using the *Viterbi* algorithm [1].

Viterbi derives the most probable path of states visited when generating the new sequence by the HMM. The returned path π corresponds to the prediction of the secondary structure made by the HMM. The HMM shown in figure 1 illustrates a simple three-state HMM capable of predicting secondary structures of protein sequences.

More complex HMMs with several states of each structural type (α, β, coil) would likely improve on capturing the underlying structural properties compared to the simple HMM. The purpose of this study is to use an EA to derive the most suitable HMM topology among the vast number of possible candidates.

3 Methods

3.1 Representation

The HMMs were represented by a $n \times (n-1)$ transition matrix A (no transitions to the begin state are allowed) and an $(n-1) \times k$ emission matrix E (the begin state does not emit symbols), where n is the number of states in the model and k is number of emission symbols (in our study $k = 20$, representing the 20 amino acids). The dimensionality of the matrices is variable, since the number of states can vary. However, a lower limit of three states and the constraint that the HMM should contain at least one α, β, and coil state is imposed to ensure that both α-helix, β-sheet, and coil can be represented and predicted by the HMM.

3.2 Population Initialisation

The population of initial parent HMMs was generated as follows: i) The initial number of states in the HMM was randomly chosen (between three and ten), and at least one state of both α, β, and coil was represented. ii) The entries in the transition matrix A and the emission matrix E were randomly initialised (between 0.0 and 1.0) and each column was normalised so the sum of all entries in a column was 1.0.

3.3 Variation Operators

During the evolutionary process the individuals were exposed to different variation operators in order to alter the candidate topologies and model parameters of the HMMs. Below is a brief description of the various operators used:

The *addState* operator adds a new state to the HMM (either α, β, or coil) and randomly assigns new transition probabilities to and from the state. All columns in the transition matrix are normalised.

The *deleteState* operator randomly chooses a state and deletes it. However, the type of selected state (α, β, coil) has to occur more than once to ensure a valid HMM. All transitions going to and from the deleted state are rewired and all the columns in the transition matrix are normalised.

The *modifyStateType* operator randomly chooses a state and changes its type to either α, β, or coil. Again, the modification is only allowed if the HMM is valid afterwards, i.e. the HMM has to contain both α, β, and coil states.

The *addTransition* operator chooses a random column in the transition matrix and looks for entries with values of 0.0. If any exist, one is chosen randomly and replaced with a new random number (between 0.0 and 1.0). The entire matrix is then normalised, and the solution is accepted if it is valid.

The *deleteTransition* operator randomly chooses a column in the transition matrix and looks for entries with nonzero values. If any exist, one is chosen randomly and set to 0.0. The entire matrix is then normalised, and the solution is accepted if it is valid.

The *swapEntry* exchanges two randomly chosen entries in a column of either the transition or emission matrix.

The *modifyEntry* operator modifies a randomly chosen entry in the transition or emission matrix using standard Gaussian mutation with annealed variance, i.e. variance $= 1/(generation + 1)$. The affected column is then normalised.

The recombination operators used are the standard *uniform crossover*, *arithmetic crossover*, *1-point crossover*, and *2-point crossover*, which take two parent HMMs and create an offspring. In this study the recombination operators were only applied to the emission matrices.

3.4 Fitness Evaluation

The fitness value of a candidate solution is computed with the *forward* algorithm [1]. Given a HMM λ specified by the individual and a set of l observations O the forward algorithm computes the probability $P(O|\lambda)$. The objective is to maximise this probability, i.e. find HMMs with high prediction accuracy while keeping the complexity of the model low. In this study, we used the Bayesian Information Criterion (similar to the Akaike Information Criterion) [9] in the fitness function to balance between these two objectives. The fitness of individual i is thus calculated as:

$$fitness_i = \sum_{j=0}^{l-1} log(P(o_j|\lambda)) + \omega \cdot \frac{log(l+1)}{2} \cdot p_i \ . \tag{1}$$

where ω is a balancing parameter determining the weight of the penalty for model complexity, and p_i is the number of free parameters in the model induced by individual i (here: the total number of entries in the transition and emission matrices).

3.5 The HMM EA

In this study we used a HMM EA as shown in figure 2. The HMM EA works as follows: First, all individuals are initialised and evaluated according to the fitness function (described in section 3.2 and 3.4). Afterwards, the following process is executed as long as the termination condition *current #generations* <

```
procedure HMM EA
begin
  initialise population
  evaluate
  while (not termination-condition) do
    begin
      recombine individuals
      mutate individuals
      evaluate
      selection
    end
  apply Baum-Welch on final solution
end
```

Fig. 2. Pseudo-code of the HMM EA.

max #generations allowed is not fulfilled: Each individual has a probability of being altered, i.e. being exposed to either recombination or mutation (or both). Recombination is applied with probability p_c using one of the four mentioned recombination operators (the choice is made randomly with equal probability for all four operators). Afterwards, mutation is applied with probability p_m using one of the six available mutation operators (the choice is made randomly with equal probability for all six operators). However, an offspring only replaced the parent if it was fitter. Further, individuals that were altered due to recombination or mutation are reevaluated using the fitness function. Finally, tournament selection with a tournament size of two is applied to weed out the least fit individuals. The final solution obtained after max *#generations* is refined using BW (running for five iterations).

4 Experiments

In order to evaluate the quality of our HMM EA method and determine whether it was able to construct useful HMM topologies, the models were trained with 650 protein sequences and validated using 150 protein sequences (see next section for a more detailed description of the data sets used). To evaluate the outcome of these experiments we used Matthew's correlation coefficient (CC) combining standard sensitivity and specificity measures. The CC value is defined in the interval $[-1 : 1]$, the extreme value -1 represents poor prediction accuracy (random) and 1 represents a good prediction accuracy (perfect).

4.1 Sequence Data

The protein sequences used for the training set were taken from the Protein Data Bank (PDB) [10], release 82. This data set, containing 5762 proteins, was further reduced to ensure that all the sequences were not homologous, i.e. entries were excluded if they were too similar. The reduction was performed by the RedHom

Table 1. Results obtained from the experiment.

	Training Set			Validation Set		
	Alpha	Beta	Coil	Alpha	Beta	Coil
True Freq.	0.311561	0.211730	0.476709	0.306837	0.228422	0.464741
Pred. Freq.	0.332781	0.047971	0.619248	0.420303	0.049044	0.530653
CC	0.337795	0.211950	0.275760	0.175357	0.224460	0.175844

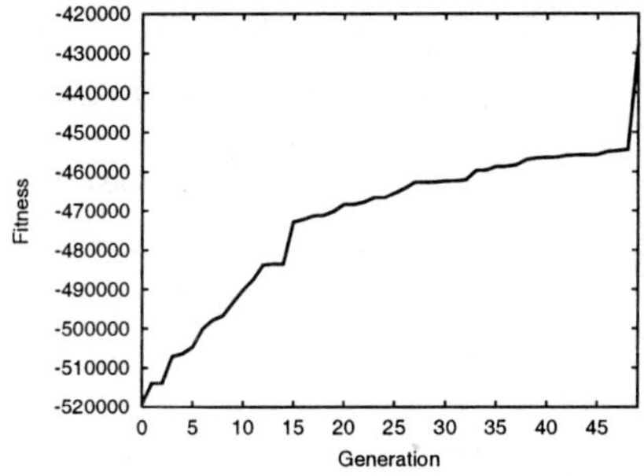

Fig. 3. Mean fitness curve of the experiments with the 650 protein sequences.

program [11], resulting in a final training set containing 650 sequences (see [11] for a more detailed description of the pruning process).

As an independent test, we used protein sequences from the CASP3 and CASP4 targets obtained from the CASP website (`http://predictioncenter.llnl.gov/`) to construct a validation set. These targets do not have sequence similarity with the training set described above. Some of the targets were excluded from the final validation set if they were to similar (using the RedHom program) resulting in a total of 150 sequences.

Both the training and the validation set contained annotated secondary structures for each of the amino acids. Furthermore, since we only focus on predicting α-helix, β-sheet, and coil, the rest of the annotations (bend (S), hydrogen bonded turn (T), 310 helix (G), residue in isolated beta bridge (B) and π-helix (I)) were set to coil.

The PDB entries of the training and validation sets are public available on the world wide web page: `http://www.daimi.au.dk/~thomsen/hmmea/`

4.2 Experimental Setup and Data Sampling

We used the following parameters in the HMM EA: *population size* = 100, mutation probability $p_m = 0.8$, recombination probability $p_c = 0.5$, balancing factor $\omega = 0.9$. The termination criteria of the HMM EA was set to 50 generations. Further, elitism of size one was used to preserve the overall best found HMM topology. The experiments were repeated 25 times, and the average fitness of the best individual throughout the evolutionary process was recorded.

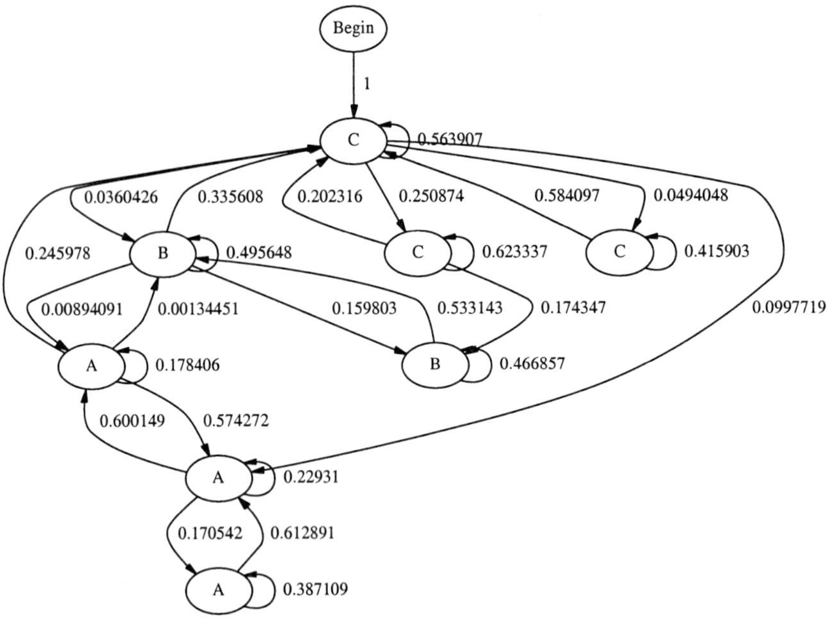

Fig. 4. Best HMM topology found.

5 Results

Table 1 shows the results of the experiments using the training and validation data sets. The *True Freq.* and *Pred. Freq.* rows show the true and the predicted frequency of α-helix, β-sheet, and coil in the data sets respectively. The *CC* row represents the mean of the correlation coefficient value of 25 runs after 50 generations.

In all the runs, we observed similar performance of the HMM EA. Typically, the HMM EA made rapid improvements during the first 15-20 generations. Further improvements occurred throughout the entire optimisation process although the EA slowly settled on a good solution and only improved the topology once in a while. Finally, BW usually fine-tuned the final solution. Figure 3 shows the fitness of the best individual over the number of generations used (averaged over 25 runs), which illustrates these characteristics.

Figure 4 shows an example of the best HMM topology obtained from one of the 25 runs. The begin state (*Begin*) has only one transition to a coil state (C), since all the protein sequences in the training set begin in a coil structure.

6 Discussion

In this paper we have introduced the HMM EA evolving both the topology and the model parameters of HMMs. Further, several new mutation operators specialised for modifying HMM topologies were introduced.

The results of our experiments show that a simple EA is capable of deriving HMMs with good topologies tailored to the task of predicting secondary structures of proteins. The best found HMM obtained a prediction accuracy of 56.8% on the training set and 49% on the validation set using the Q3 measure[1], which is better than the 46% and 40% obtained from the simple three-state HMM shown in figure 1 (optimised using BW). Further, the *True Freq.*, *Pred. Freq.* and *CC* values shown in table 1 indicate that the derived HMM is capable of predicting alpha helices and coil reasonably well. However, the model overestimates coil and thus fails to correctly identify beta sheets.

The performance of the obtained HMM for secondary structure prediction is not competitive to state-of-the-art artificial neural networks (ANNs) having prediction accuracies near 80% [12]. However, these methods usually combine several ANNs and heuristics to weed out wrong predictions substantially improving the overall performance.

The simple HMM approach introduced in this study could be further improved. First, using higher-order HMMs taking previously observed symbols into account would increase the performance even further (Asai et al. obtained accuracies about 58% using a second-order HMM [8]). Second, the variation operators of the EA could be refined to get a better optimisation performance. Finally, the EA parameters used in the experiments introduced in this study were based on a few preliminary runs. Further investigations could lead to additional improvements in performance.

Although one run of the HMM EA typically takes four hours on a 800 MHz Pentium-III PC it is still a viable alternative since the training of the HMMs is usually done offline. Furthermore, exploring just a subset of the possible HMM topologies in a trial-and-error fashion using BW is often infeasible even with a HMM topology of moderate size.

The results confirm that EAs are useful for obtaining improved HMM topologies, which could be advantageous in the context of HMM applications found in bioinformatics.

This study only covers preliminary investigations on the evolution of HMM topologies and model parameters. Future work will include a comparison to other heuristic methods such as model merging [5] and state splitting [4] to access the performance of the proposed method. Finally, it would be interesting to try out other evolutionary algorithms focusing on multi-objective optimisation to balance the trade-off between prediction accuracy and model complexity.

Acknowledgements

The author would like to thank Tejs Scharling for inspiring discussions regarding HMMs and help on the implementation of the HMM framework, and Jan Gorodkin for providing the data sets. This work was supported by the Danish Natural Science Research Council.

---

[1] $Q3 = \text{true positives}/(\text{true positives}+\text{true negatives})$

References

1. Rabiner, L.R.: A Tutorial on Hidden Markov Models and Selected Applications in Speech Recognition. In: Proceedings of the IEEE, Vol. 77, No. 2. (1989) 257–285
2. Kwong, S. and Chau, C.: Analysis of Parallel Genetic Algorithms on HMM Based Speech Recognition System. In: Proceedings of ICASSP (1997) 1229–1233
3. Slimane, M., Venturini, G., de Beauville, J.A., Brouard, T., and Brandeau, A.: Optimizing Hidden Markov Models with a Genetic Algorithm. In: Artificial Evolution (1995) 140–144
4. Tanaka, H., Onizuka, K., and Asai, K.: Classification of Proteins via Successive State Splitting of Hidden Markov Network. In: 13th International Joint Conference on Artificial Intelligence (IJCAI93) (1993) 140–144
5. Stolcke, A. and Omohundro, S.: Hidden Markov Model Induction by Bayesian Model Merging. In: Hanson, S.J., Cowan, J.D., and Giles, C.L. (eds.): Advances in Neural Information Processing Systems, Vol. 5. Morgan Kaufmann, San Mateo, CA (1993) 11–18
6. Yada, T., Ishikawa, M., Tanaka, H., and Asai, K.: Extraction of Hidden Markov Model Representations of Signal Patterns in DNA Sequences. In: Proceedings of the First Pacific Symposium on Biocomputing (1996) 686–696
7. Kwong, S., Chau, C., Man, K., and Tang, K.: Optimisation of HMM Topology and its Model Parameters by Genetic Algorithms. Pattern Recognition, Vol. 34. (2001) 509–522
8. Asai, K., Hayamizu, S., and Handa, K.: Prediction of Protein Secondary Structure by the Hidden Markov Model. Computer Applications in the Biosciences (CABIOS), Vol. 9, No. 2. (1993) 141–146
9. Murphy, K. and Mian, S.: Modelling Gene Expression Data using Dynamic Bayesian Networks. Technical report, Computer Science Division, University of California, Berkeley, CA (1999)
10. Berman, H., Westbrook, J., Feng, Z., Gilliland, G., Bhat, T., Weissig, H., Shindyalov, I.N., and Bourne, P.: The Protein Data Bank. Nucleic Acids Research, Vol. 28, No. 1. (2000) 235–242
11. Lund, O., Frimand, K., Gorodkin, J., Bohr, H., Bohr, J., Hansen, J., and Brunak, S.: Protein Distance Constraints Predicted by Neural Networks and Probability Density Functions. Prot. Eng., Vol. 10. (1997) 1241–1248
12. Petersen, T.N., Lundegaard, C., Nielsen, M., Bohr, H., Bohr, J., and Brunak, S.: Prediction of Protein Secondary Structure at 80% Accuracy. PROTEINS: Structure, Function, and Genetics, Vol. 41. (2000) 17–20

Solving a Real World Routing Problem Using Multiple Evolutionary Agents

Neil Urquhart, Peter Ross, Ben Paechter, and Ken Chisholm

Napier University, School of Computing
10 Colinton Rd, Edinburgh, Scotland
n.urquhart@napier.ac.uk

Abstract. This paper investigates the solving of a real world routing problem using evolutionary algorithms embedded within a Multi-agent system (MAS). An architecture for the MAS is proposed and mechanisms for controlling the interactions of agents are investigated. The control mechanism used in the final solution is based on the concept of agents submitting bids to receive work. The agents are also allowed to alter their bidding strategies as the solution improves. The MAS solves the test problem is solved, which previously could not be solved within the hard constraints.

1 Introduction

1.1 The Problem

Each year in the month of December, the Edinburgh Area of the Scout Association undertakes the collection and delivery of Christmas cards within the City of Edinburgh. The delivery task is then undertaken by Scout Groups across the city. In 2001 a total of 465,000 cards were delivered during a 2 week period. Each group undertakes to deliver everything within a particular geographical area; a group may deliver up to several hundred streets depending on the size of the group. The group must divide up its delivery area into a number of delivery rounds each of which should attempt to satisfy the following constraints:

1. The maximum walking distance for each delivery round should not exceed a set limit, typically 3km.
2. The maximum number of households requiring a delivery per round should not exceed a predefined limit, typically 250.
3. The number of delivery rounds should be minimised.
4. Entire named streets should be allocated to a single round, if possible. Within each round a street may be split if required when constructing the round. Each time a street is split across two or more rounds, this incurs an extra sorting operation.
5. Deliveries that require entry to a building e.g. blocks of flats may not be given to children under 16. It is desirable to group as many of these deliveries into as few rounds as possible.

J.J. Merelo Guervós et al. (Eds.): PPSN VII, LNCS 2439, pp. 871–880, 2002.

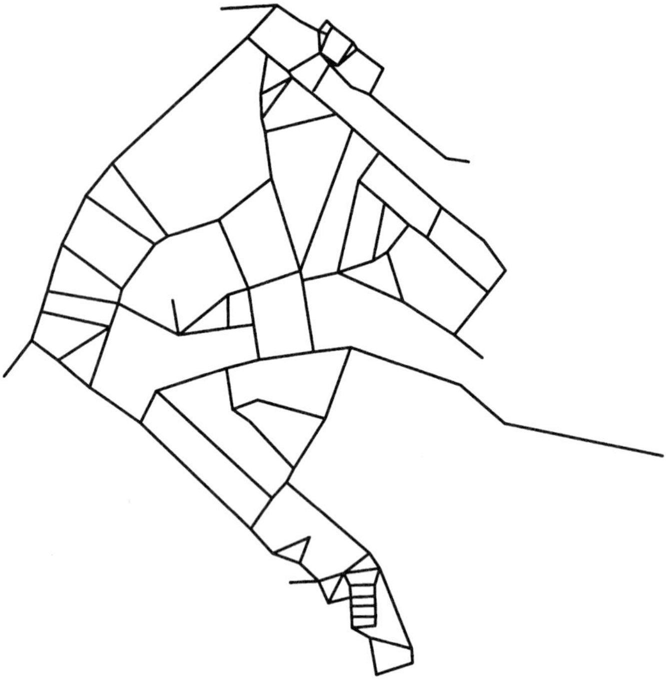

Fig. 1. The Scout Post Test Area

Note that constraints 1 and 2 may be deemed *hard* constraints and the rest *soft* constraints. An area of 1572 deliveries located in the southern suburbs of Edinburgh currently undertaken by the 75th Blackford Scout Group has been identified to use as a trial area. This area is currently problematic as it combines a mixture of tower blocks (with over 100 dwellings per building), semi-detached villas and a street network that has a largely random topology. It has previously been difficult coordinate deliveries to this area and a large number of sorting errors occur (see Fig. 1). The remainder of this paper will examine the use of GAs and Agents in the construction of delivery networks suitable for the solving of the Scoutpost problem.

1.2 Previous Work

Postal delivery based problems have previously been tackled using Street based routing (SBR) in [8] and [7]. SBR employs a representation that is derived from the geographical layout of the houses and streets concerned. Houses are grouped into street sections, each street section being a section of street running between two junctions. The sections may be considered either as single sides, serviced at different times within the round or they mat be both serviced in one single operation. The EA produces a permutation of street sections; to build the final round each street section has a delivery heuristic applied, specifying the order in which the houses contained within the section should be serviced. This method may be contrasted with arc routing based methods as used in [5], [1] and [4]. Some research was undertaken by [2] into the use of a GA for routing garbage collection, this being a related problem, but only covering the construction of single rounds. The arc routing methods use representations based on graphs that do not take account of street and house layouts. The arc routing methods are not flexible enough to take account of situations where it is not necessary to traverse the entire length of an arc. In some instances the optimal delivery pattern will involve starting and finishing at the same end of the street.

The construction of delivery networks consisting of more than one round was discussed in [7]. That paper proposed the use of a method known as Group and Build. Thus consisted of two stages, initially a grouping EA is used to divide up the street sections requiring delivery into adjacent groups (stage 1), and then a round is constructed for each grouping (stage 2) using the SBR EA.

Multi Agent Systems (as discussed in section 3) have been previously applied to flow problems [6]. The system developed by [6] uses agents to represent the transport links, the customers and the goods to be transported. The agents "negotiate" to allocate capacity within the transport links to customers allowing the goods to be transported. Much research on market-based control has been carried out by the authors of [3]. The problems investigated by [3] are mostly based on agents trading a commodity for which the supply and demand characteristics develop dynamically. In the problem being examined in this paper the entire trading stock (i.e. the street sections) is a known quantity from the start.

2 Initial Attempts to Solve the Problem Using Group and Build

The initial attempts to solve the test problem utilised the group and build method [7]. For the purposes of solving the test problem the fitness function of the grouping algorithm was modified to take account of the additional soft constraints of the Scout Association problem (described in section 1.1).

The initial population of the grouping EA is seeded by randomly allocating each street to a group. For instance the street sections HighSt1, HighSt2 and HighSt3 would all be allocated to the same group to satisfy constraint 1. Within each generational cycle 50 child individuals are created by recombining two parents chosen by a tournament of size 2, each child then has a single mutation applied to it. Three different types of mutation are employed, they are selected randomly with a probability p:

$p = 0.25$ A single street section may be selected at random and re-allocated to another delivery round.
$p = 0.25$ A whole street is selected at random and all the sections that make up that street section are re-allocated to a randomly selected delivery round.
$p = 0.49$ Two street sections are selected at random and exchanged between their respective delivery rounds.
$p = 0.01$ A new delivery round is created, initially three complete streets are selected randomly and allocated to it.

Table 1. Summary of Initial Results

Run	Average Round Dist in meters.	No of rounds	Extra Sorts Needed	No of rounds not suitable for children
1	1464.7	8	2	4
2	1421.8	9	3	4
3	1638.1	8	0	4
4	1529.7	8	2	4
5	1469.9	8	1	5
6	1517.5	8	1	5
7	1563.5	10	2	4
8	1552.3	8	4	6
9	1497.0	8	2	4
10	1463.0	8	2	4
11	1513.2	8	2	3
12	1529.3	8	2	4
13	1496.2	8	1	4
14	1400.5	9	2	4
15	1476.7	8	3	4
16	1498.7	8	3	4
17	1540.0	8	4	4
18	1477.0	8	2	4
19	1503.2	8	5	4
20	1440.4	9	2	4
Average	1499.6	8.25	2.25	4.15

The child individuals are placed in the main population, replacing the loser of a tournament between two randomly selected individuals.

The modified fitness function utilises penalties that are allocated for the following reasons:

- Each round that has more than the maximum allowed deliveries.
- The standard deviation of deliveries for each round (to attempt to balance the workloads).
- The number of different rounds converging at a single junction.
- The average distance between streets within a round.
- The number of times a complete street is split between rounds.
- The number of rounds that may not be undertaken by children.
- The number of rounds over and above the minimum required (the minimum being total deliveries / maximum deliveries allowed pen round).

These are each suitably weighted according to their perceived significance. The choice of such weights is a difficult art and this is one reason why this method is not ideal. We do not give the weights in this paper. This group-and-build approach is used for comparison with the main approach described in section 3, and also to provide an initial starting-point solution for the main approach.

The group and build algorithm was applied to the test area, being run 20 times, results may be seen in Table 1. The maximum length value was set at

Table 2. The Distribution of Round Lengths Produced by the Group and Build Algorithm

Length L (in m)	No of rounds	% of total
$L > 3000$	12	7.3
$3000 \geq L > 2500$	35	21.2
$2500 \geq L > 2000$	24	14.5
$L <= 2000$	94	57.0
Total	165	

3km. The maximum number of deliveries was set to 250 households. It may be noted that in each case the average distance was less than 3km. In all cases the number of deliveries was less than 250. Within the 20 runs 165 delivery rounds were constructed giving an average of 8 rounds per solution, although the average length of each round was less than 3km, Table 2 shows the distribution of lengths. Note that 7 % (12 rounds) were breaking the hard 3km constraint. It may be seen from the distribution that if this constraint was lowered the number of rounds breaking the constraint would increase. The Group and Build approach cannot solve the problem beacuse it cannot determine at the grouping stage what the actual distance of the delivery routes will be. Incorporating the SBR algorithm into fitness function so as to determine round length when individual is evaluated was considered impractical by the authors because of the CPU time required to evolve the rounds. The Group and Build Approach can only provide a solution where the density of the housing is such that most groups of houses not exceeding the maximum deliveries constraint are tightly clustered so that the delivery round required to service them is less than the maximum length constraint.

3 The Marketplace Algorithm

3.1 Initial MAS Experiments

The two-stage Group and Build (GAB) approach is unable to exercise any effective control over the length of the delivery rounds produced. Some control over the length may be exercised by altering the maximum deliveries constraint. To overcome this problem the second stage may be re-implemented as a multi-agent system, with each delivery round being constructed by an agent that incorporates a separate copy of the SBR algorithm. Each agent may then evolve its own round separately, to facilitate evolution of the delivery network the agents may exchange work. The exchanging of work is carried out via a coordinator agent to ensure that all street sections are allocated to one agent and one agent only. Each agent also has access to a central data store containing information about the size and layout of the street sections that make up the problem. The street sections that are the basis of the SBR representation are a good unit of exchange when swapping work between agents, rather than exchanging individual delivery points.

Promising results were obtained by allowing the agents who could not evolve a round within the problem constraints to pass a street section to the coordinator, which then selects another agent whose round passes closest to the street section. To allow a more meaningful exchange of work, a flexible control mechanism is required to regulate how the SBR agents and the coordinator communicate.

3.2 Artificial Currency

The use of an artificial "currency" was initially investigated by us as a possible control mechanism. The currency is an abstract entity that is not directly connected with the original problem and is represented by a balance value held by each agent. Street sections may be bought and sold between agents and payments made by currency. Agents may receive a regular payment based on the number of households served and incur operating charges based on the length of round. Agents breaking hard constraints are fined. If an agents balance is negative the agent has to surrender a street section to the coordinator. When the coordinator receives the street section, each agent is invited to submit a "bid", this being a numerical value representing the degree to which the agent wants the work. The agent submitting the highest bid is then allocated the section, and pays the bid by having it deducted from it's balance.

The bids are calculated as follows:

MAXDELS	the maximum deliveries constraint
MAXLEN	maximum length constraint
owner(st)	returns the ID of the agent that st was previously allocated to
balance	returns the agents balance
cLen	the length of the agents current round
cDels	the no of deliveries in the agents current round
me	returns the id of the current agent
avgD(st)	returns the average distance between each street section allocated to the agent and st

```
let st = street section under offer
bid = (MAXLEN - cLen) + avgD(st)
if cDels + dels(st) > MAXDELS then
  bid = bid /2
if owner(st) = me then
  bid = bid /2
if balance < 0 then
  bid = -1
```

In practice, the value of the balance variable has been found to reflect the past performance of the agent rather than its current state. Because agents are identical and are not allowed to modify their bidding strategy or parameters, there is no reason to record this past performance. For instance, if an agent lowers the length of its round, its balance will increase, the agent may then bid for the street section allowing it to exceed the round length constraint. Because of its high balance the agent will not be forced to give up any street sections until the fines imposed on it have decreased the balance to less than 0. As long as its balance is greater then 0 then an agent may continue to bid and acquire street sections whether or not it can undertake the deliveries involved without breaking any constraints. Although unsuccessful, elements of this method have been used in the final algorithm discussed in section 3.3.

3.3 The Market Algorithm

The bidding mechanism used as part of artificial currency mechanism was found to be an effective method of deciding which agent to allocate a surplus street section to. The bidding mechanism may be modified and incorporated in a new control system known as the Market Algorithm. The system is initialised as previously (using the output from the grouping-EA). Each agent may now construct a delivery round based on this initial allocation. Each agent may then evolve a delivery round based on the street sections currently allocated to it. The co-ordinator now requests that the agent with the greatest violation of the hard constraints surrenders a single street section back to the co-ordinator for re-allocation to another agent. Using the bidding process agents may now submit bids for the surplus section. The agent submitting the lowest bid has the section allocated to it.

Assume those functions and variables defined in the currency bidding logic still hold, then bidding logic is as follows:

leastDist(st)	returns the distance to the closest street to st within the current agent
dels(st)	returns the number of deliveries to st
oldR(st)	returns the id of the previous agent that st was allocated to
rdLength(st)	returns the length of street (st)

```
Let st = street to bid for.
theBid = avgD(st);
del = cDels;
newDels = (del + dels(st));
if (newDels > MAXDELS)  then
  theBid = theBid + (newDels*2);
if (oldR(st) = me)  then
  theBid = -1;
if (cLen > MAXLEN)  then
  theBid = -1;
least = (leastDist + rdLength(st))*2;
if ((cLen + least) > MAXLEN)  then thebid = -1;
bid = theBid;
end;
```

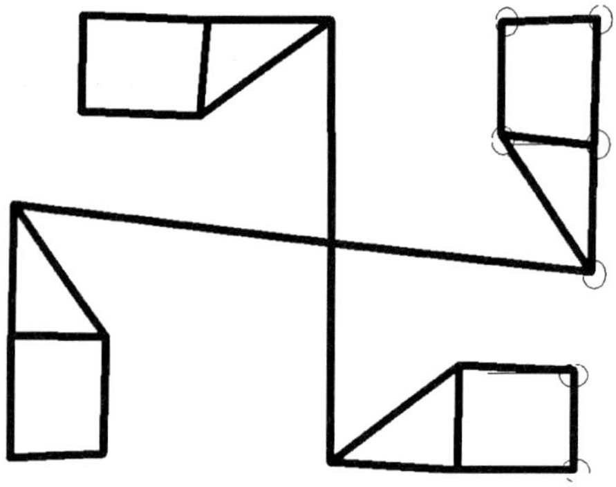

Fig. 2. The Artificial Small Test Data Set

The logic submits a bid of -1 (a -1 bid is never acceptable) if the agent estimates that the addition of the street section will violate any of the hard constraints. When a situation arises where no agent is bidding for a street, a new agent is allowed into the Marketplace. The new agent is formed by taking the agent with the longest round and randomly transferring half its street sections to the new agent.

The method used by an agent to select street sections to be returned to the coordinator was found to have an effect on performance. Several methods were evaluated, the final method chosen removes the street section that has the highest average distance from each of the other street sections allocated to that agent. When street sections have been removed and re-allocated in his manner the two agents affected then evolve updated rounds based on the changes that have taken place. In the results presented here, this cycle of transactions continues for a maximum of 100 transactions.

3.4 2-Stage Bidding

The agent bidding logic described in Section 3.3 may be described as "cautious" as it produces a bid of -1 if the agent estimates that it may break a hard constraint by accepting the street section on offer. In addition to the Scout Post data set, the system was first tested on a small and artificial data set (see Fig. 2). This data set divides into 4 logical areas radiating from a central hub. The total length of streets in each section is 45 units, with 54 households. The maximum deliveries constraint was set to 70, and the Marketplace algorithm was run with maximum length values of 50-90. The results obtained are shown in Fig. 2. Note that the number of delivery rounds required would appear to be excessively high. This data set should be easily split into 4 rounds, especially when the maximum length constraint is as high as 90 units.

An examination of the results shown in Fig. 2 suggests that the marketplace algorithm does not cope well with a constrained problem such as this. If all agents are breaking hard constraints then no exchange of street sections

Table 3. Results Obtained with Market 1 and Market 2, on the Map in Fig. 2

	MAX LEN	90	80	70	60	50
Market1	Avg. length	68.3	60.3	49.0	37.0	28.1
	Avg. % Rounds over	0	0	0	0	0
	Avg. No Rounds	5.7	6.3	7.7	9.0	10.7
Market2	Avg. length	76.6	68.9	56.6	46.3	32.3
	Avg. % Rounds over	0	0	0	0	0.3
	Avg. No Rounds	4.2	4.2	5	6	8.3

Table 4. Results for the Test Area, with Version 1 and Version 2 of the Marketplace Algorithm

	Market 1			Market 2		
Maximum Length Constraint	2K	2.5K	3K	2K	2.5K	3K
Average round Length	1583.2	1980.4	2175.0	1553.6	1955.9	2149.7
Over-length Rounds	0	0	0	0	0	0
Average No of extra sorts	7.1	7.0	4.4	8.4	6.8	4.0
Average No of Supervised Rounds	4.6	4.5	4.4	4.7	4.8	4.3
Average No of Rounds	12.2	9.9	9.1	12.5	10.2	9.1

would be possible without creating a new routing agent, even if the existing agents could cope with the workload simply by exchanging street sections. If the bidding logic is altered to always submit a bid, regardless of whether a hard constraint is broken, then the street sections are divided amongst the initial set of agents regardless of whether the agents are breaking any hard constraints. Using this "relaxed" strategy the system stabilises with adjacent street sections being allocated to the same agent. Finally system reaches an equilibrium with the same street section being exchanged between the same agents. Once this stage has been reached the bidding strategy of agents may then be set back to the original cautious strategy. This allows the rejection of surplus work to create one or more new agents, and reduce each agent's workload to an acceptable level. This change in strategy is facilitated by allowing each agent to keep a list of those street sections last surrendered by the agent. This list is of length 3 and initially empty, but when it contains 3 identical items, the agent realising that it no longer engaged in productive transactions and changes to a cautious strategy. Using the cautious strategy no work should be accepted until some work has been surrendered allowing the agents hard constraints to be fulfilled. The effect of this two stage bidding may be seen in Table 3 under "Market 2".

4 Results and Conclusions

The results for the Scout Post test area (see Fig. 1) may be seen in Table 4, results are shown for the initial Marketplace algorithm (Market1) and the two stage algorithm (Market2). Both versions of the Marketplace Algorithm are capable of solving the ScoutPost problem without breaking any hard constraints.

Although the hard constraints are satisfied, note the increase in average round length and the increased number of extra sorts present in the final solution, the relaxing of the soft constraints being the price that must be paid for the solving of the hard constraints. The results demonstrate that we can solve the Scoutpost problem with the initial 3km limit, and we can also produce solutions for more constrained versions of this problem with 2.5 and 2km limits.

The small contrived dataset (figure 2) shows that for a constrained area, where little surplus capacity exists, relaxing the constraints (by using "relaxed" bidding) allows us to construct an interim solution, which is then modified into a final solution, by the altering of the bidding strategy to "cautious". The success of the Marketplace algorithm is based on the bidding strategy logic and to a lesser extent the logic used by an agent to select items for surrender to the coordinator agent.

Acknowledgements

The authors wish to thank Jonathan Tait and Mark Elliot of the 75th Blackford Scout Troup, Edinburgh for their assistance in this work.

References

1. Jeyakesavan V A Balaji R. A 3/2- approximation algorithm for the mixed postman problem. *SIAM Journal on Discrete Mathematics*, 12(4), 1999.
2. T. Bousonville. Local search and evolutionary computation for arc routing in garbage collection. In et al Spector L, editor, *Proceedings of the Genetic and Evolutionary Computation Conference 2001*. Morgan Kaufman Publishers, 2001.
3. Bruten J Cliff D. Simple bargaining agents for decentralised market-based control. Technical Report HPL-98-17, Hewlet Packard Laboritories., Bristol. United Kingdom., 1998.
4. Han C. Kang M. Solving the rural postman problem using a genetic algorithm with a graph transformation. In *Proceedings of the 1998 ACM Symposium on Applied Computing*. ACM Press, 1998.
5. Ramdane-Cherif W A Lacomme P, Prins C. A genetic algorithm for the capacitated arc routing problem. In Boers E J W et al., editor, *Real World Applications of Evolutionary Computing*. Springer-Verlag, 2001.
6. Wellman M P. A market orientated programming environment and its application to distributed multi-commodity flow problems. *Journal of Artificial Intelligence Research. Morgan Kaufmann Publishers*, 1, 1993.
7. Chisholm K Urquhart N, Paechter B. Street based routing using an evolutionary algorithm. In Boers E J W et al., editor, *Real World Applications of Evolutionary Computing. Proceedings of EvoWorkshops 2001*. Springer-Verlag, 2001.
8. Paechter B Chisholm K Urquhart N, Ross P. Improving street based routing using building block mutations. In *To Appear in: Applications of Evolutionary Computing. Proceedings of EvoWorkshops 2002*. Springer-Verlag, 2002.

Other Bioinspired Algorithms: Cellular Automata, Ant Colony Optimization

An Ant Colony Optimization Approach to the Probabilistic Traveling Salesman Problem

Leonora Bianchi[1], Luca Maria Gambardella[1], and Marco Dorigo[2]

[1] IDSIA, Strada Cantonale Galleria 2, CH-6928 Manno, Switzerland
{leonora,luca}@idsia.ch
http://www.idsia.ch
[2] Université Libre de Bruxelles, IRIDIA
Avenue Franklin Roosevelt 50, CP 194/6, 1050 Brusselles, Belgium
mdorigo@ulb.ac.be
http://iridia.ulb.ac.be/~mdorigo/

Abstract. The Probabilistic Traveling Salesman Problem (PTSP) is a TSP problem where each customer has a given probability of requiring a visit. The goal is to find an a priori tour of minimal expected length over all customers, with the strategy of visiting a random subset of customers in the same order as they appear in the a priori tour.
We address the question of whether and in which context an a priori tour found by a TSP heuristic can also be a good solution for the PTSP. We answer this question by testing the relative performance of two ant colony optimization algorithms, Ant Colony System (ACS) introduced by Dorigo and Gambardella for the TSP, and a variant of it (pACS) which aims to minimize the PTSP objective function.
We show in which probability configuration of customers pACS and ACS are promising algorithms for the PTSP.

1 Introduction

Consider a routing problem through a set V of n customers. On any given instance of the problem each customer i has a known position and a probability p_i of actually requiring a visit, independently of the other customers. Finding a solution for this problem implies having a strategy to determine a tour for each random subset $S \subseteq V$, in such a way as to minimize the expected tour length. The most studied strategy is the a priori one. An a priori strategy has two components: the a priori tour and the updating method. The a priori tour is a tour visiting the complete set V of n customers; the updating method modifies the a priori tour in order to have a particular tour for each subset of customers $S \subseteq V$. A very simple example of updating method is the following: for every subset of customers, visit them *in the same order* as they appear in the a priori tour, skipping the customers that do not belong to the subset. The strategy related to this method is called the 'skipping strategy'. The problem of finding an a priori tour of minimum expected length under the skipping strategy is defined as the Probabilistic Traveling Salesman Problem (PTSP). This is an NP-hard problem [1, 2], and was introduced in Jaillet's PhD thesis [3].

J.J. Merelo Guervós et al. (Eds.): PPSN VII, LNCS 2439, pp. 883–892, 2002.

The PTSP approach models applications in a delivery context where a set of customers has to be visited on a regular (e.g., daily) basis, but all customers do not always require a visit, and where re-optimizing vehicle routes from scratch every day is unfeasible. In this context the delivery man would follow a standard route (i.e., an a priori tour), leaving out customers that on that day do not require a visit. The standard route of least expected length corresponds to the optimal PTSP solution.

In the literature there are a number of algorithmic and heuristic approaches used to find suboptimal solutions for the PTSP. Heuristics using a nearest neighbor criterion or savings criterion were implemented and tested by Jézéquel [4] and by Rossi-Gavioli [5]. Later, Bertsimas-Jaillet-Odoni [1] and Bertsimas-Howell [6] have further investigated some of the properties of the PTSP and have proposed some heuristics for the PTSP. These include tour construction heuristics (space filling curves and radial sort), and tour improvement heuristics (probabilistic 2-opt edge interchange local search and probabilistic 1-shift local search). Most of the heuristics proposed are an adaptation of a TSP heuristic to the PTSP, or even the TSP heuristic itself, which in some cases gives good PTSP solutions. More recently, Laporte-Louveaux-Mercure [7] have applied an integer L-shaped method to the PTSP and have solved to optimality instances involving up to 50 vertices.

No application of nature-inspired algorithms such as ant colony optimization (ACO) [8] or genetic algorithms has been done yet. This paper investigates the potentialities of ACO algorithms for the PTSP. In the remainder of the paper we first introduce the PTSP objective function and notations (section 2), then we describe the ACO algorithms which we tested (section 3), and finally we show the experimental results obtained (section 4).

2 The PTSP Objective Function

Let us consider an instance of the PTSP. We have a completely connected graph whose nodes form a set $V = \{i = 1, 2, ..., n\}$ of customers. Each customer has a probability p_i of requiring a visit, independently from the others. A solution for this instance is a tour λ over all nodes in V (an 'a priori tour'), to which is associated the expected length objective function

$$E[L_\lambda] = \sum_{S \subseteq V} p(S) L_\lambda(S) \ . \tag{1}$$

In the above expression, S is a subset of the set of nodes V, $L_\lambda(S)$ is the distance required to visit the subset of customers S (in the same order as they appear in the a priori tour), and $p(S)$ is the probability for the subset of customers S to require a visit:

$$p(S) = \prod_{i \in S} p_i \prod_{i \in V-S} (1 - p_i) \ . \tag{2}$$

Jaillet [3] showed that the evaluation of the PTSP objective function (eq.(1)) can be done in $O(n^2)$. In fact, let us consider (without loss of generality) an a priori tour $\lambda = (1, 2, \ldots, n)$; then its expected length is

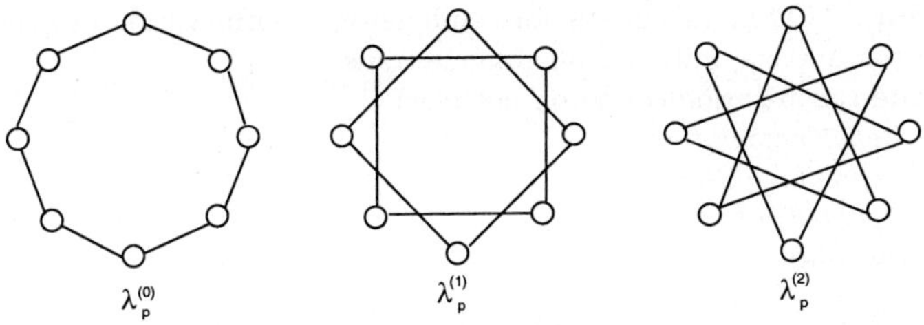

Fig. 1. The lengths of these sets of (sub)tours, $\lambda_p^{(0)}$, $\lambda_p^{(1)}$ and $\lambda_p^{(2)}$, constitute the first three terms of the expected length for the homogeneous PTSP. From left to right, the total length of each set of (sub)tours gives the terms $L_\lambda^{(0)}$, $L_\lambda^{(1)}$ and $L_\lambda^{(2)}$ of equation (4).

$$E[L_\lambda] = \sum_{i=1}^{n} \sum_{j=i+1}^{n} d_{ij} p_i p_j \prod_{k=i+1}^{j-1} (1 - p_k) + \\ \sum_{i=1}^{n} \sum_{j=1}^{i-1} d_{ij} p_i p_j \prod_{k=i+1}^{n} (1 - p_k) \prod_{l=1}^{j-1} (1 - p_l) \,. \quad (3)$$

This expression is derived by looking at the probability for each arc of the complete graph to be used, that is, when the a priori tour is adapted by skipping a set of customers which do not require a visit. For instance, an arc (i, j) is actually used only when customers i and j do require a visit , while customers $i+1, i+2, ..., j$ do not require a visit. This event occurs with probability $p_i p_j \prod_{k=i+1}^{j-1} (1 - p_k)$ (when $j \leq n$). In the special class of PTSP instances where $p_i = p$ for all customers $i \in V$ (the homogeneous PTSP), equation (3) becomes

$$E[L_\lambda] = p^2 \sum_{r=0}^{n-2} (1-p)^r L_\lambda^{(r)} \quad (4)$$

where $L_\lambda^{(r)} \equiv \sum_{j=1}^{n} d(j, (j + 1 + r) \bmod n)$. The $L_\lambda^{(r)}$'s have the combinatorial interpretation of being the lengths of a collection of gcd(n,r + 1) sub-tours[1] $\lambda_p^{(r)}$, obtained from tour λ by visiting one customer and skipping the next r customers. As an example, Fig. 1 shows $\lambda_p^{(0)}$ (i.e., the a priori tour), $\lambda_p^{(1)}$ and $\lambda_p^{(2)}$ for a PTSP with 8 customers.

3 Ant Colony Optimization

In ACO algorithms a colony of artificial ants iteratively constructs solutions for the problem under consideration using artificial pheromone trails and heuristic information. The pheromone trails are modified by ants during the algorithm execution in order to store information about 'good' solutions. Most ACO algorithms follow the algorithmic scheme given in Fig. 2.

[1] The term 'gcd' stays for 'greatest common divisor'.

```
procedure ACO metaheuristic for combinatorial optimization problems
    Set parameters, initialize pheromone trails
    while (termination condition not met)
        ConstructSolutions
        (ApplyLocalSearch)
        UpdateTrails
    end while
```

Fig. 2. High level pseudocode for the ACO metaheuristic.

ACO are solution construction algorithms, which, in contrast to local search algorithms, may not find a locally optimal solution. Many of the best performing ACO algorithms improve their solutions by applying a local search algorithm after the solution construction phase. Our primary goal in this work is to analyze the PTSP tour construction capabilities of ACO, hence in this first investigation we do not use local search.

We apply to the PTSP Ant Colony System (ACS) [9, 10], a particular ACO algorithm which was successfully applied to the TSP. We also consider a modification of ACS which explicitly takes into account the PTSP objective function (we call this algorithm probabilistic ACS, that is, pACS). In the following, we describe how ACS and pACS build a solution and how they update pheromone trails.

3.1 Solution Construction in ACS and pACS

A feasible solution for an n-city PTSP is an a priori tour which visits all customers. Initially m ants are positioned on their starting cities chosen according to some initialization rule (e.g., randomly). Then, the solution construction phase starts (procedure *ConstructSolutions* in Fig. 2). Each ant progressively builds a tour by choosing the next customer to move to on the basis of two types of information, the pheromone τ and the heuristic information η. To each arc joining two customers i, j it is associated a varying quantity of pheromone τ_{ij}, and the heuristic value $\eta_{ij} = 1/d_{ij}$, which is the inverse of the distance between i and j. When an ant k is on city i, the next city is chosen as follows.

- With probability q_0, a city j that maximizes $\tau_{ij} \cdot \eta_{ij}^{\beta}$ is chosen in the set $J_k(i)$ of the cities not yet visited by ant k. Here, β is a parameter which determines the relative influence of the heuristic information.
- With probability $1 - q_0$, a city j is chosen randomly with a probability given by

$$p_k(i,j) = \begin{cases} \frac{\tau_{ij} \cdot \eta_{ij}^{\beta}}{\sum_{r \in J_k(i)} \tau_{ir} \cdot \eta_{ir}^{\beta}}, & \text{if } j \in J_k(i) \\ 0, & \text{otherwise.} \end{cases} \tag{5}$$

Hence, with probability q_0 the ant chooses the best city according to the pheromone trail and to the distance between cities, while with probability $1 - q_0$ it explores the search space in a biased way.

3.2 Pheromone Trails Update in ACS and pACS

Pheromone trails are updated in two stages. In the first stage, each ant, after it has chosen the next city to move to, applies the following local update rule:

$$\tau_{ij} \leftarrow (1-\rho) \cdot \tau_{ij} + \rho \cdot \tau_0, \tag{6}$$

where ρ, $0 < \rho \leq 1$, and τ_0, are two parameters. The effect of the local updating rule is to make less desirable an arc which has already been chosen by an ant, so that the exploration of different tours is favored during one iteration of the algorithm.

The second stage of pheromone update occurs when all ants have terminated their tour. Pheromone is modified on those arcs belonging to the best tour since the beginning of the trial (best-so-far tour), by the following global updating rule

$$\tau_{ij} \leftarrow (1-\alpha) \cdot \tau_{ij} + \alpha \cdot \Delta\tau_{ij}, \tag{7}$$

where

$$\Delta\tau_{ij} = ObjectiveFunc_{best}^{-1} \tag{8}$$

with $0 < \alpha \leq 1$ being the pheromone decay parameter, and $ObjectiveFunc_{best}$ is the value of the objective function of the best-so-far tour. In ACS the objective function is the a priori tour length, while in pACS the objective function is the PTSP expected length of the a priori tour. In the next section we discuss in more detail the differences between ACS and pACS.

3.3 Discussion of Differences between ACS and pACS

Differences between ACS and pACS are due to the fact that the two algorithms exploit different objective functions in the pheromone updating phase. The global updating rule of equations (7) and (8) implies two differences in the way ACS and pACS explore the search space. The first and most important difference is the set of arcs on which pheromone is globally increased, which is in general different in ACS and pACS. In fact, the 'best tour' in eq. (8) is relative to the objective function. Therefore in ACS the search will be biased toward the shortest tour, while in pACS it will be biased toward the tour of minimum expected length. The second difference between ACS and pACS is in the quantity $\Delta\tau_{ij}$ by which pheromone is increased on the selected arcs. This aspect is less important than the first, because ACO in general is more sensitive to the difference of pheromone among arcs than to its absolute value.

The length of an a priori tour (ACS objective function) may be considered as an $O(n)$ approximation to the $O(n^2)$ expected length (pACS objective function). In general, the worse the approximation, the worse will be the solution quality of ACS versus pACS. The quality of the approximation depends on the set of customer probabilities p_i. In the homogeneous PTSP, where customer probability is p for all customers, it is easy to see the relation between the two objective functions. For a given a priori tour L_λ we have

$$\Delta = L_\lambda - E[L_\lambda] = (1-p^2)L_\lambda - \sum_{r=1}^{n-2}(1-p)^r L_\lambda^{(r)}, \tag{9}$$

which implies

$$\Delta \sim O(q \cdot L_\lambda) \tag{10}$$

for $(1-p) = q \to 0$. Therefore the higher the probability, the better is the a priori tour length L_λ as an approximation for the expected tour length $E[L_\lambda]$.

In the heterogeneous PTSP, it is not easy to see the relation between the two objective functions, since each arc of the a priori tour L_λ is multiplied by a different probabilistic weight (see eq.(3)), and a term with L_λ cannot be isolated in the expression of $E[L_\lambda]$, as in the homogeneous case.

ACS and pACS also differ in time complexity. In both algorithms one iteration (i.e., one cycle through the *while* condition of Fig. 2) is $O(n^2)$ [11], but the constant of proportionality is bigger in pACS than in ACS. To see this one should consider the procedure *UpdateTrail* of Fig. 2, where the best-so-far tour must be evaluated in order to choose the arcs on which pheromone is to be updated. The evaluation of the best-so-far tour requires $O(n)$ time in ACS and $O(n^2)$ time in pACS. ACS is thus faster and always performs more iterations than pACS for a fixed CPU time.

4 Experimental Tests

4.1 Homogeneous PTSP Instances

Homogeneous PTSP instances were generated starting from TSP instances and assigning to each customer a probability p of requiring a visit. TSP instances were taken from two benchmarks. The first is the TSPLIB at http://www.iwr.uni-heidelberg.de/groups/comopt/software/TSPLIB95/tsp/. From this benchmark we considered instances with a number of city between 50 and 200. The second benchmark is a group of instances where customers are randomly distributed on the square $[0, 10^6]$. Both uniform and clustered distributions where considered in this case, and the number of cities varied between 50 and 350. For generating random instances we used the Instance Generator Code of the 8th DIMACS Implementation Challenge at http://research.att.com/dsj/chtsp/download.html.

4.2 Computational Environment and ACS Parameters

Experiments were run on a Pentium Xeon, 1GB of RAM, 1.7 GHz processor. In order to asses the relative performance of ACS versus pACS independently from the details of the settings, the two algorithms were run with the same parameters. We chose the same settings which yielded good performance in earlier studies with ACS on the TSP [10]: $m = 10$, $\beta = 2$, $q_0 = 0.98$, $\alpha = \rho = 0.1$ and $\tau_0 = 1/(n \cdot Obj)$, where n is the number of customers and Obj is the value of the objective function evaluated with the nearest neighbor heuristic [10].

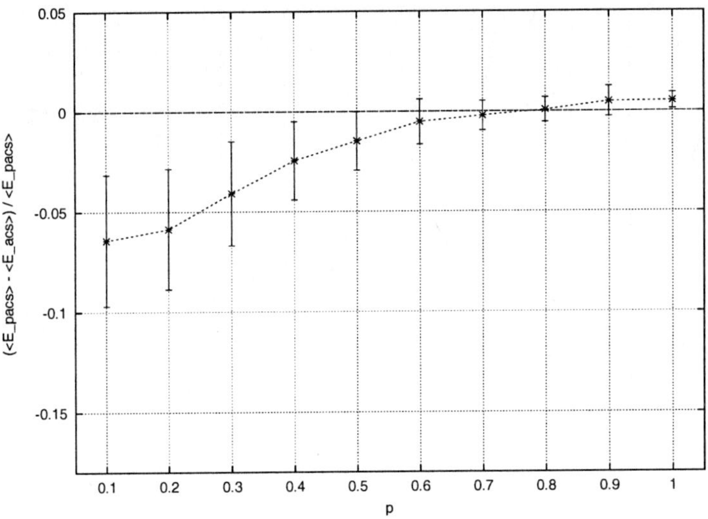

Fig. 3. Relative performance of pACS versus ACS for the homogeneous PTSP. The vertical axis represents $(E[L_\lambda(pACS)] - E[L_\lambda(ACS)])/E[L_\lambda(pACS)]$. On the horizontal axis there is the customer probability p. Each point of the graph is an average over 21 symmetric homogeneous PTSP instances. Error bars represent average deviation, defined as $\sum_{i=1}^{n} |x_i - <x> |/n$, with $n = 21$. Note that for $p = 1$ ACS outperforms pACS, since for a fixed CPU stopping time ACS makes more iterations.

The stopping criterion used in both algorithms is CPU time in seconds, chosen according to the relation $stoptime = k \cdot n^2$, with $k = 0.01$. This value of k lets ACS perform at least $17 \cdot 10^3$ iterations on problems with up to 100 customers, and at least $15 \cdot 10^3$ iterations on problems with more than 100 customers. For each instance of the PTSP, we ran 5 independent runs of the algorithms.

4.3 Results

For each TSP instance tested, nine experiments were done varying the value of the customer probability p from 0.1 to 0.9 with a 0.1 interval. Fig. 3 summarizes results obtained on 21 symmetric PTSP instances, one third of the instances were taken from the TSPLIB, the others were random instances (half of them uniform and half clustered). The figure shows the relative performance of pACS versus ACS, averaged over the tested instances. A typical result for a single instance is reported in Fig. 4.

As it was reasonable to expect, for small enough probabilities pACS outperforms ACS. In all problems we tested though, there is a range of probabilities $[p_0, 1]$ for which ACS outperforms pACS. The critical probability p_0 at which this happens depends on the problem.

The reason why pACS does not always perform better than ACS is clear if we consider two aspects: the complexity (speed) of ACS versus pACS, and

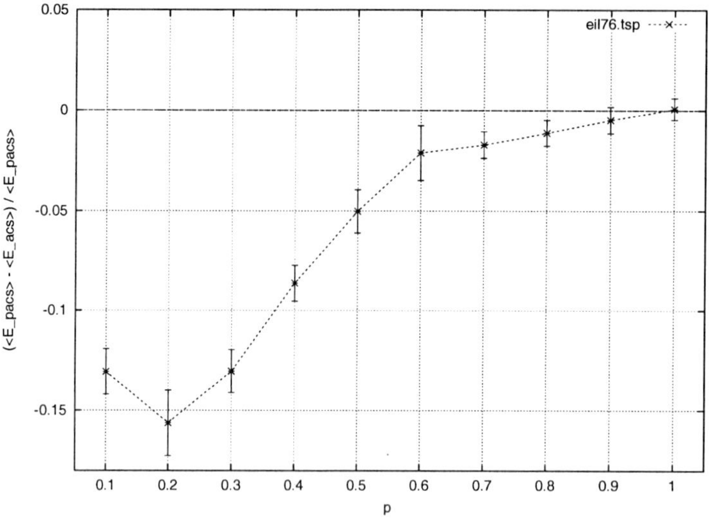

Fig. 4. Relative performance of pACS versus ACS for the eil76.tsp instance of the TSPLIB. Error bars represent the average deviation of the result over 5 independent runs of the algorithms.

the goodness of the PTSP objective function approximation as p approaches 1. When p is near to 1, a good solution to the TSP is also a good solution to the PTSP; therefore, ACS, which performs more iterations than pACS, has a better chance to find a good solution.

4.4 Absolute Performance

For the PTSP instances we tested, the optimal solution is not known. Therefore, an absolute performance evaluation of the pACS heuristic can only be done against a theoretical lower bound, when this is available and tight enough. A lower bound to the optimal solution would give us an upper bound to the error performed by the pACS heuristic. In fact, if LB is the lower bound and $E[L_{\lambda^*}]$ is the optimal solution, then by definition we have

$$E[L_{\lambda^*}] \geq LB \ . \tag{11}$$

If the solution value of pACS is $E[L_\lambda]$, then the following inequality holds for the relative error

$$\frac{E[L_\lambda] - E[L_{\lambda^*}]}{E[L_{\lambda^*}]} \leq \frac{E[L_\lambda] - LB}{LB} \ . \tag{12}$$

For the homogeneous PTSP and for instances where the optimal length L_{TSP} of the corresponding TSP is known, it is possible to use the following lower bound to the optimal expected length [6]

$$LB = pL_{TSP}(1 - (1-p)^{n-1}) \ . \tag{13}$$

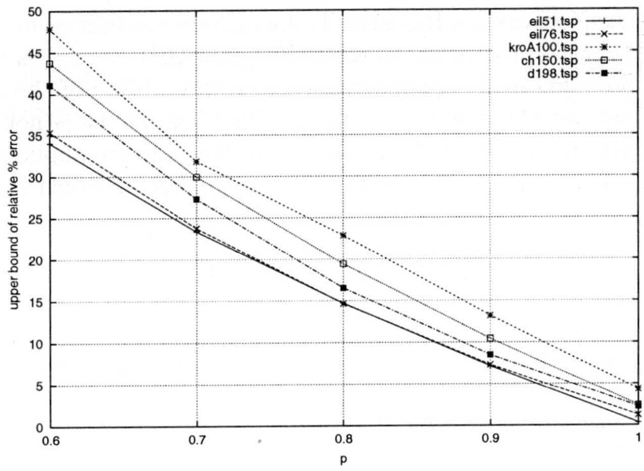

Fig. 5. Upper bound of relative percent error of pACS for 5 TSPLIB instances. The horizontal axis represents the customer probability.

If we put this lower bound into the right side of equation (12), we obtain an upper bound of the relative error of pACS. Fig. 5 shows the absolute performance of pACS, evaluated with this method, for a few TSPLIB instances. From the figure we see that, for instance, pACS finds a solution within 15% of the optimum for a homogeneous PTSP with customers probability 0.9.

5 Conclusions and Future Work

In this paper we investigated the potentialities of ACO algorithms for the PTSP. In particular, we have shown that the pACS algorithm is a promising heuristic for homogeneous TSP instances. Moreover, for customers probabilities close to 1, the ACS heuristic is a better alternative than pACS.

At present we are investigating the heterogeneous PTSP, for different probability configurations of customers. This is an interesting direction of research, since it is closer to a real-world problem than the homogeneous PTSP. We are also trying to improve pACS performance by inserting in the ants' tour construction criterion information about the customers probabilities. Work which will follow this paper also comprehends a comparison of pACS with respect to other PTSP algorithms. Moreover, pACS should be improved by adding to the tour construction phase a local search algorithm. The best choice and design of such a local search is also an interesting issue for the PTSP.

Acknowledgments

This research has been partially supported by the Swiss National Science Foundation project titled "On-line fleet management", grant 16R10FM, and by the

"Metaheuristics Network", a Research Training Network funded by the Improving Human Potential programme of the CEC, grant HPRN-CT-1999-00106. The information provided in this paper is the sole responsibility of the authors and does not reflect the Community's opinion. The Community is not responsible for any use that might be made of data appearing in this publication. Marco Dorigo acknowledges support from the Belgian FNRS, of which he is a Senior Research Associate.

References

1. D. J. Bertsimas, P. Jaillet, and A. Odoni. A priori optimization. *Operations Research*, 38:1019–1033, 1990.
2. D. J. Bertsimas. *Probabilistic Combinatorial Optimization Problems.* PhD thesis, MIT, Cambridge, MA, 1988.
3. P. Jaillet. *Probabilistic Traveling Salesman Problems.* PhD thesis, MIT, Cambridge, MA, 1985.
4. A. Jézéquel. *Probabilistic Vehicle Routing Problems.* Master's thesis, MIT, Cambridge, MA, 1985.
5. F. A. Rossi and I. Gavioli. *Aspects of Heuristic Methods in the Probabilistic Traveling Salesman Problem*, pages 214–227. World Scientific, Singapore, 1987.
6. D. J. Bertsimas and L. Howell. Further results on the probabilistic traveling salesman problem. *European Journal of Operational Research*, 65:68–95, 1993.
7. G. Laporte, F. Louveaux, and H. Mercure. An exact solution for the a priori optimization of the probabilistic traveling salesman problem. *Operations Research*, 42:543–549, 1994.
8. M. Dorigo, G. Di Caro, and L. M. Gambardella. Ant algorithms for discrete optimization. *Artificial Life*, 5(2):137–172, 1999.
9. L. M. Gambardella and M. Dorigo. Solving symmetric and asymmetric TSPs by ant colonies. In *Proceedings of the 1996 IEEE International Conference on Evolutionary Computation (ICEC'96)*, pages 622–627. IEEE Press, Piscataway, NJ, 1996.
10. M. Dorigo and L. M. Gambardella. Ant Colony System: A cooperative learning approach to the traveling salesman problem. *IEEE Transactions on Evolutionary Computation*, 1(1):53–66, 1997.
11. M. Dorigo, V. Maniezzo, and A. Colorni. The Ant System: Optimization by a colony of cooperating agents. *IEEE Transactions on Systems, Man, and Cybernetics – Part B*, 26(1):29–41, 1996.

When Model Bias Is Stronger than Selection Pressure

Christian Blum and Michael Sampels

IRIDIA, Université Libre de Bruxelles, CP 194/6
Av. Franklin D. Roosevelt 50, 1050 Bruxelles, Belgium
{cblum,msampels}@ulb.ac.be

Abstract. We investigate the influence of model bias in model-based search. As an example we choose Ant Colony Optimization as a well-known model-based search algorithm. We present the effect of two different pheromone models for an Ant Colony Optimization algorithm to tackle a general scheduling problem. The results show that a pheromone model can introduce a strong bias toward certain regions of the search space, stronger than the selection pressure introduced by the updating rule for the model. This potentially leads to an algorithm where over time the probability to produce good quality solutions decreases.

1 Introduction

Model-based search (MBS) [7] algorithms are increasingly popular methods for solving combinatorial optimization problems. In MBS algorithms such as Ant Colony Optimization (ACO) [5] or Estimation of Distribution Algorithms (EDAs) [9, 10], candidate solutions are generated using a parametrized probabilistic model that is updated depending on previously seen solutions. The update aims to concentrate the search in regions of the search space containing high quality solutions. In particular, the reinforcement of solution components depending on the solution quality is an important factor in the development of heuristics to tackle hard combinatorial optimization problems. It is assumed that good solutions don't occur sporadically, but consist of good solution components. To learn which components contribute to good solutions can help to assemble them to better solutions. In general, a model-based search approach attempts to solve an optimization problem by repeating the following two steps: 1) Candidate solutions are constructed using some parametrized probabilistic model, that is, a parametrized probability distribution over the solution space. 2) The candidate solutions are used to modify the model parameters in a way that is deemed to bias future sampling toward high quality solutions.
Ant Colony Optimization (ACO) is a metaheuristic approach proposed by Dorigo et al. [5] for solving combinatorial optimization problems. The inspiring source of ACO is the foraging behavior of real ants. In ACO, the probabilistic model is called *pheromone model* and the pheromone values are the model parameters. Often it is implicitly assumed that the average performance of model-based

J.J. Merelo Guervós et al. (Eds.): PPSN VII, LNCS 2439, pp. 893–902, 2002.

algorithms such as ACO is increasing over time. However, during empirical investigations of an ACO algorithm for the Group Shop Scheduling problem[1] in [2], we observed that for some of the pheromone models chosen the performance of the system was decreasing over time. This triggered us to find out the reasons for this phenomenon.
The outline of the paper is as follows. In Section 2 we present the Group Shop Scheduling problem. In Section 3 we briefly outline the ACO algorithm to tackle the Group Shop scheduling problem together with two different pheromone models. In Section 4 we analyse one of the pheromone models on a small example problem instance and we show the existence of a strong model bias. In Section 5 we investigate the interactions between model bias and selection pressure. We show that even increasing the selection pressure might be not enough to eradicate model bias. In Section 6 we finally give some conclusions and an outlook to the future.

2 A General Scheduling Problem

A general scheduling problem can be formalized as follows: We consider a finite set of operations O which is partitioned into subsets $\mathcal{M} = \{M_1, \ldots, M_m\}$ (machines) and into subsets $\mathcal{J} = \{J_1, \ldots, J_n\}$ (jobs), together with a partial order $\preceq \subseteq O \times O$ such that $\preceq \cap J_i \times J_j = \emptyset$ for $i \neq j$, and a function $p : O \to \mathbf{N}$ assigning processing times to operations. A feasible solution is a refined partial order $\preceq^* \supseteq \preceq$ for which the restrictions $\preceq^* \cap J_i \times J_i$ and $\preceq^* \cap M_k \times M_k$ are total $\forall\, i, k$. The cost of a feasible solution is defined by

$$C_{\max}(\preceq^*) = \max\{\sum_{o \in C} p(o) \mid C \text{ is a chain in } (O, \preceq^*)\}$$

where $C_{\max}$ is called the makespan of a solution. We aim at a feasible solution which minimizes $C_{\max}$.
M_k is the set of operations which have to be processed on machine k. Further, J_i is the set of operations which belong to job i. Each machine can process at most one operation at a time. Operations must be processed without preemption. Operations belonging to the same job must be processed sequentially. This brief problem formulation covers well known scheduling problems: The restriction $\preceq \cap J_i \times J_i$ is total in the Job Shop scheduling problem (JSP), trivial ($= \{(o, o) \mid o \in J_i\}$) in the Open Shop scheduling problem (OSP), and either total or trivial for each i in the Mixed Shop scheduling problem (MSP) [3, 11].
For the Group Shop scheduling problem (GSP), we consider a weaker restriction on $\preceq$ which includes the above scheduling problems by looking at a refinement of the partition $\mathcal{J}$ to a partition into *groups* $\mathcal{G} = \{G_1, \ldots, G_g\}$. We demand that $\preceq \cap G_i \times G_i$ has to be trivial and that for $o, o' \in J$ ($J \in \mathcal{J}$) with $o \in G_i$ and $o' \in G_j$ ($i \neq j$) either $o \preceq o'$ or $o \succeq o'$ holds. Note that the coarsest refinement

[1] For historical reasons, this problem was called FOP Shop scheduling in [2]. However, the name Group Shop Scheduling fits better to the structure of the problem.

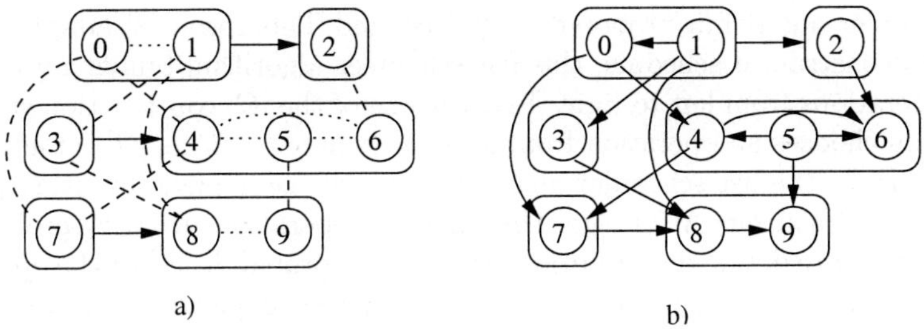

Fig. 1. a) The disjunctive graph representation [8] of a simple instance of the GSP consisting of 10 operations partitioned into 3 jobs, 6 groups and 4 machines (processing times are omitted in this example). In order to obtain a solution, the dashed and dotted links have to be directed without creating any cycles. The directed arcs between operations of successive groups are simplified as inter-group connections. b) A solution to the problem (the arcs undirected in a) are directed such that the resulting graph does not contain any cycles).

$\mathcal{G} = \mathcal{J}$ (groups sizes are equal to job sizes) is equivalent to the OSP and the finest refinement $\mathcal{G} = \{\{o\} \mid o \in O\}$ (group sizes of 1) is equivalent to the JSP. In the following $j(o)$ will denote the job, $g(o)$ the group and $m(o)$ the machine of an operation $o \in O$. See Figure 1 for an example of the GSP.

3 An ACO Algorithm for the GSP

We consider the ACO algorithm already outlined in [2]. The framework of this algorithm is shown in Algorithm 1. The most important features of this algorithm are explained in the following. In Algorithm 1, $\tau = \{\tau_1, ..., \tau_l\}$ is a set of pheromone values, k is the number of ants, and s_j is a solution to the problem (sequences containing all operations), constructed by ant j, where $j = 1, ..., k$.

Algorithm 1 ACO for the GSP

```
InitializePheromoneValues(τ)
while termination conditions not met do
  for j = 1 to k do
    s_j ← ConstructSolution(τ)
  end for
  ApplyOnlineDelayedPheromoneUpdate(τ,s_1,...,s_k)
end while
```

InitializePheromoneValues(τ): In every version of our algorithm we initialize all the pheromone values to the same positive constant value.

ConstructSolution(τ): An important part of an ACO algorithm is the constructive mechanism used to probabilistically construct solutions. We use the well-known list scheduler algorithm proposed by Giffler and Thompson [8] adapted to the GSP. To construct a schedule, this list scheduler algorithm builds a sequence s of all operations from left to right. In every one of the $|O|$ construction steps, the algorithm chooses an operation from a set S_t (where $t = 1, \ldots, |O|$) of admissible operations. In the list scheduler algorithm this choice is made by using priority rules. In contrast, our ACO algorithm picks an operations from S_t probabilistically (for a more detailed description see [2]). The probabilities for the operations in S_t (called transition probabilities) to be chosen, depend on the pheromone model. In the following we present two different pheromone models, the first one from the literature, the second one is a new development.

Learning of a predecessor relation in s: Pheromone model $\mathsf{PH}_{\mathsf{suc}}$ was introduced by Colorni et al. [4] for an Ant System to tackle the JSP. In the following a simple extension of this model to the GSP is outlined. In this model we have a pheromone value τ_{o_i,o_j} on every ordered pair of operations $o_i, o_j \in O$. Additionally we have pheromone values τ_{o_i} $\forall o_i \in O$. The probability to choose an operation $o \in S_t$ in the t-th construction step is the following one:

$$p(s[t] = o \mid s_{t-1,|O|}, \tau) = \begin{cases} \frac{\tau_o}{\sum_{o_k \in S_t} \tau_{o_k}} & : \text{ if } o \in S_t \text{ and } t = 1, \\ \frac{\tau_{o_i,o}}{\sum_{o_k \in S_t} \tau_{o_i,o_k}} & : \text{ if } o \in S_t, t > 1, s[t-1] = o_i, \\ 0 & : \text{ otherwise,} \end{cases}$$

where $s[t]$ denotes the position t in sequence s, and $s_{t-1,|O|}$ denotes a partial sequence of current length $t-1$ and final length $|O|$. In this pheromone model, the choice of the next operation to be scheduled is dependent on the operation scheduled in the previous step.

Learning of relations among operations: In this new pheromone model – which we called $\mathsf{PH}_{\mathsf{rel}}$ – we assign pheromone values to pairs of *related* operations. We call two operations $o_i, o_j \in O$ *related* if they belong to the same group, or if they have to be processed on the same machine. Formally, a pheromone value τ_{o_i,o_j} exists, iff $g(o_i) = g(o_j)$ or $m(o_i) = m(o_j)$. The meaning of a pheromone value τ_{o_i,o_j} is that if τ_{o_i,o_j} is high then operation o_i should be scheduled before operation o_j. The choice of the next operations to be scheduled is handled as follows. If there is an operation $o_i \in S_t$ with no related and unscheduled operation left, it is chosen. Otherwise we choose among the operations of set S_t with the following probabilities:

$$p(s[t] = o \mid s_{t-1,|O|}, \tau) = \begin{cases} \frac{\min_{o_r \in S_o^{rel}} \tau_{o,o_r}}{\sum_{o_k \in S_t} \min_{o_r \in S_{o_k}^{rel}} \tau_{o_k,o_r}} & : \text{ if } \quad o \in S_t \\ 0 & : \text{ otherwise} \end{cases}$$

where $S_o^{rel} = \{o' \in O \mid m(o') = m(o) \vee g(o') = g(o),\ o' \text{ not scheduled yet}\}$. The meaning of this rule to compute the transition probabilities is the following: If at least one of the pheromone values between an operation $o_i \in S_t$ and a related operation o_r (not scheduled yet) is low, then the operation o_i probably should not be scheduled now. By using this pheromone model the algorithm tries to learn relations between operations. The absolute position of an operation in the sequence s is not important anymore. Importance is given to the relative position of an operation with respect to the related operations.

ApplyOnlineDelayedPheromoneUpdate(τ,s_1,...,s_k): We implemented our algorithm in the Hyper-Cube Framework [1], one of the ways of implementing an ACO algorithm. The Hyper-Cube Framework is characterized by a normalization of the contribution of every solution used for updating the pheromone values. This leads to a scaling of the objective function values and the pheromone values are implicitly limited to the interval $[0, 1]$ (see [1] for a more detailed description). For $\mathsf{PH}_{\mathsf{rel}}$, the Ant System [6] updating rule in the Hyper-Cube Framework (henceforth HC-AS) is the following.

$$\tau_{o_i,o_j} \leftarrow (1-\rho) \cdot \tau_{o_i,o_j} + \rho \cdot \sum_{l=1}^{k} \Delta^{s_l} \tau_{o_i,o_j} \tag{1}$$

where

$$\Delta^{s_l} \tau_{o_i,o_j} = \begin{cases} \frac{f(s_l)}{\sum_{r=1}^{k} f(s_r)} & \text{if } o_i \text{ before } o_j \text{ in } s_l, \\ 0 & \text{otherwise,} \end{cases} \tag{2}$$

where $f(s) = 1/C_{\max}(s)$ for all possible solutions s. For $\mathsf{PH}_{\mathsf{suc}}$ the pheromone updating rule is the same, just that for the update of pheromone values τ_{o_i} the notation is different. So, a pheromone value τ_{o_i,o_j} receives update (as shown in equation (2)), if there exists an $t \in \{1, ..., |O| - 1\}$ such that $s_l[t] = o_i$ and $s_l[t+1] = o_j$ (if o_i and o_j are immediate successors in sequence s_l). Accordingly, a pheromone value τ_{o_i} receives update, if $s_l[1] = o_i$.

4 Investigating $\mathsf{PH}_{\mathsf{suc}}$ on a Small Example Instance

In [2] we observed that the performance[2] of the algorithm described in this paper using $\mathsf{PH}_{\mathsf{suc}}$ was strongly decreasing over time for all problem instances tested. At first sight, this seems odd, as the pheromone update rule described in the previous section rewards solution components (successor relationships) that occur in good quality solutions more than it rewards solution components found in worse solutions. To explain the decreasing performance, we consider the small problem instance shown in Figure 2. For the three solutions to the problem (shown in Figure 2b) it holds: $C_{\max}(s_1) = 60$, $C_{\max}(s_2) = 40$ and $C_{\max}(s_3) = 60$. This means that s_2 is the optimal solution to the problem instance. In the following we examine the average update received by any pheromone value τ_{o_i,o_j} of pheromone model $\mathsf{PH}_{\mathsf{suc}}$. The decisive pheromone values are $\tau_{1,2}$ and $\tau_{3,4}$. The algorithm starts with equal pheromone values. $\tau_{1,2}$ receives update by the three sequences $1-2-3-4$ (with value 60), $3-4-1-2$ (with value 60) and $3-1-2-4$ (with value 40). In contrast, pheromone value $\tau_{1,3}$ receives update by the two

[2] We measured the performance of the system by the average quality of the solutions found by the ants in every iteration. This way of measuring the performance of an algorithm is widely accepted in the field of Evolutionary Computation.

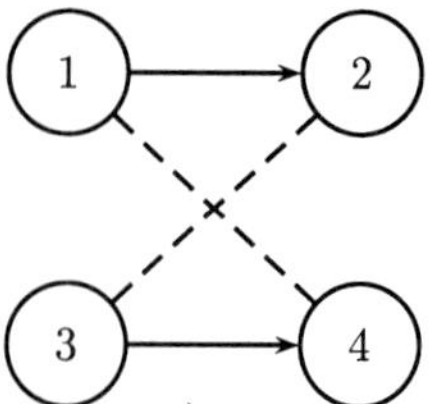

a)

s_1: $1 \preceq^* 4,\ 2 \preceq^* 3$

s_2: $1 \preceq^* 4,\ 3 \preceq^* 2$

s_3: $4 \preceq^* 1,\ 3 \preceq^* 2$

b)

$1 - 2 - 3 - 4 \mapsto s_1$
$1 - 3 - 2 - 4 \mapsto s_2$
$1 - 3 - 4 - 2 \mapsto s_2$
$3 - 4 - 1 - 2 \mapsto s_3$
$3 - 1 - 4 - 2 \mapsto s_2$
$3 - 1 - 2 - 4 \mapsto s_2$

c)

Fig. 2. a) A small example problem instance: $O = \{1,2,3,4\}$, $\mathcal{J} = \{J_1 = \{1,2\}, J_2 = \{3,4\}\}$, $\mathcal{G} = \{G_1 = \{1\}, G_2 = \{2\}, G_3 = \{3\}, G_4 = \{4\}\}$, $\mathcal{M} = \{M_1 = \{1,4\}, M_2 = \{2,3\}\}$, $G_1 \prec G_2$, $G_3 \prec G_4$, $p(1) = 10$, $p(2) = 20$, $p(3) = 20$, $p(4) = 10$. b) The three solutions to this problem instance characterized by their machine orders. c) 6 possible sequences that might be generated by the ACO algorithm and the correspondence to the three solutions shown in b).

sequences $1 - 3 - 2 - 4$ (with value 40) and $1 - 3 - 4 - 2$ (with value 40). This means, that according to the update rule in equation (2), the average update for $\tau_{1,2}$ is higher than the average update for $\tau_{1,3}$ although the partial sequence $1-3$ is only to be found in sequences corresponding to the optimal solution and $1-2$ is mostly to be found in sequences corresponding to sub-optimal solutions. For symmetry reasons, the same holds for pheromone value $\tau_{3,4}$ (corresponding to $\tau_{1,2}$) and $\tau_{3,1}$ (corresponding to $\tau_{1,3}$). The probability for the algorithm to start a sequence with operation 1 or operation 2 remains for symmetry reasons on average the same during the run-time of the algorithm. Therefore, after starting a sequence with operation 1, the probability to schedule 2 immediately afterward (4 immediately after 3, respectively) grows during the course of the algorithm. This leads to a decrease in system performance while the probability to construct the two sub-optimal solutions s_1 and s_3 increases from iteration to iteration. We conclude that the problem constraints (e.g., the partial sequence $1 - 2 - ...$ can only be completed in one possible way, $3 - 4 - ...$, respectively) together with the pheromone model introduce a strong bias, which is stronger than the "drive" toward good solutions introduced by the pheromone updating rule. The first force we henceforth call *model bias* and for the second force we borrow an expression from Evolutionary Computation and call it *selection pressure*. In the following section we study the interactions of model bias and selection pressure for problem instances in the space between Open Shop and Job Shop Scheduling.

5 Model Bias and Selection Pressure

In order to examine the phenomenon of model bias, we generated a problem instance 3x3 with 9 operations, 3 jobs and 3 machines having the same principal structure as the problem instance in Figure 2a. We generated all possible GSP instances in the range between JSP and OSP. Then we applied Algorithm 1 to all instances using pheromone models $\mathsf{PH}_{\mathsf{suc}}$ and $\mathsf{PH}_{\mathsf{rel}}$ with $k = 100$ ants per iteration and evaporation rate $\rho = 0.1$. The results indicate that the model bias

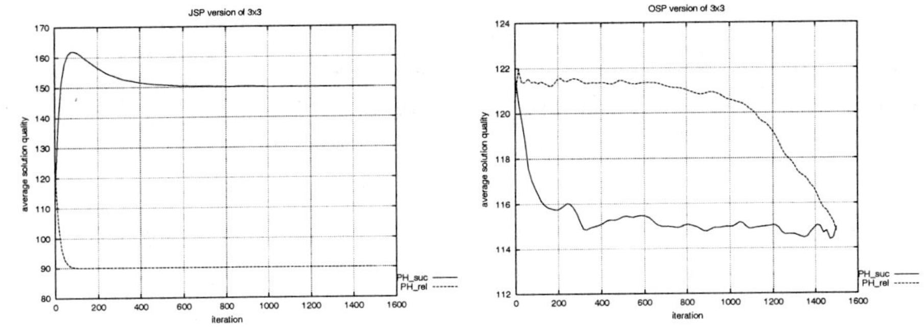

Fig. 3. The graphs show results of the ACO algorithm using different pheromone models on two versions of the problem instance 3x3. The data for this problem instance is as follows (omitting group data): $O = \{1, ..., 9\}$, $\mathcal{J} = \{J_1 = \{1,2,3\}, J_2 = \{4,5,6\}, J_3 = \{7,8,9\}\}$, $\mathcal{M} = \{M_1 = \{1,5,9\}, M_2 = \{2,6,7\}, M_3 = \{3,4,8\}\}$, $p(1) = p(5) = p(9) = 10$, $p(2) = p(6) = p(7) = 20$, $p(3) = p(4) = p(8) = 30$. The graph on the left shows the results for the JSP version and the graph on the right for the OSP version. Every line shows the average performance of the algorithm averaged over 100 runs ($k = 100$ ants per iteration, evaporation rate $\rho = 0.1$, 1500 iterations).

decreases from having a big influence for JSP to having practically no influence for OSP (in Figure 3 we show the results for the JSP and the OSP version of 3x3). This confirms that the problem constraints and the chosen pheromone model constitute **together** the strength of a possible model bias.

In order to show that the analysis performed in the last section not only holds for small example instances, we ran experiments on 3 further problem instances. The whizzkids97 problem instance is a difficult GSP instance with 197 operations which was subject of a competition held in 1997 organized by the TU Eindhoven, The Netherlands. We conducted experiments on the OSP version (henceforth whizzkids97_osp), on the original version (whizzkids_gsp) and on the JSP version (whizzkids_jsp) of this problem. In addition to the pheromone update rule HC-AS, we applied the pheromone updating rule that is only using the best solution found in an iteration to update the pheromone values (henceforth HC-IB). This update rule puts considerably more selection pressure than HC-AS on the system. The results are shown in the three graphs on the left hand side of Figure 4. For $\mathsf{PH_{suc}}$ in conjunction with HC-AS we notice a strong decrease in performance in the first couple of hundred iterations. Then the system takes a sharp turn and the performance is slowly increasing. This effect is weakening from the JSP version whizzkids97_jsp, where the effect is strongest, over the original version whizzkids97_gsp, to the OSP version whizzkids97_osp, where this effect doesn't occur at all. It is very interesting to see that when using pheromone update rule HC-IB instead of HC-AS the overall performance of the system can be considerably improved. Still there is a slowed-down decrease in performance for approximately the first 700 iterations. This means that although a stronger selection pressure can improve the algorithm, it can't eradicate the effect of a

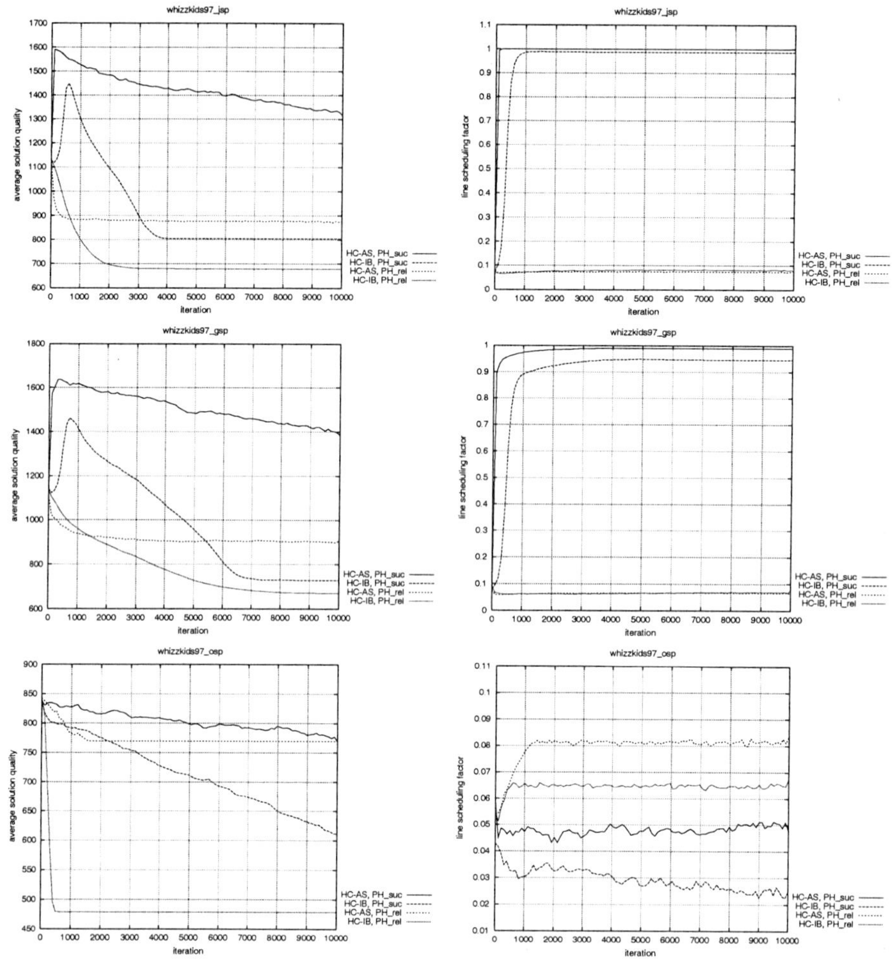

Fig. 4. The three graphs on the left hand side show average solution quality per iteration (averaged over 10 runs) on three different instances (top down: whizzkids97_jsp, whizzkids97_gsp and whizzkids97_osp) for two different pheromone models $\mathsf{PH}_{\mathsf{suc}}$ and $\mathsf{PH}_{\mathsf{rel}}$ and two different pheromone update rules HC-AS and HC-IB. For all experiments we used $k = 10$ ants per iteration and the best evaporation rate from a range of tested evaporation rates. The three graphs on the right hand side show the mean values of the "line scheduling factor" for the experiments on the left hand side.

strong model bias. The performance curves for $\mathsf{PH}_{\mathsf{rel}}$ show for all three problem instances a clear advantage over the performance curves for $\mathsf{PH}_{\mathsf{suc}}$.

Next we explore the reasons for the sharp turn from strongly decreasing performance to slowly increasing performance for $\mathsf{PH}_{\mathsf{suc}}$. The analysis in the previous section indicates that the problem constraints together with the pheromone model introduce a strong bias toward producing sequences (solutions) where the operations of the jobs are scheduled in line (operations of the same jobs are clustered in the sequences produced). To measure the extend of this clustering in a sequence s, we defined the following measure, which we called *line scheduling factor*:

$$f_{ls} : s \mapsto \frac{\sum_{t=1}^{|O|-1} \delta(s,t)}{\sum_{J \in \mathcal{J}} |J| - 1}, \tag{3}$$

where $\delta(s,t) = 1$ if $j(s[t]) = j(s[t+1])$ and $\delta(s,t) = 0$ otherwise. We measured the value of $f_{ls}(s)$ for the experiments on the three whizzkids97 instances. In the three graphs on the right hand side of Figure 4 the mean values of $f_{ls}(s^{ib})$ for the iteration best solutions s^{ib} are shown. It is interesting to see that for pheromone model $\mathsf{PH}_{\mathsf{rel}}$, these values are staying nearly constant, whereas for pheromone model $\mathsf{PH}_{\mathsf{suc}}$ these values are strongly increasing. For the whizzkids97_jsp – when using pheromone update rule HC-AS – it is even reaching its maximum of 1.0 and keeping this value until the algorithm stops. On top of that, the point when the line scheduling factor reaches a stable point coincides with the point when the system takes a sharp turn from decreasing performance to increasing performance. This indicates that there exists a kind of "stable area" in the search space where the strong model bias is leading the system to at the beginning of the search process. Once the system has reached this stable area – which is characterized by a clustering of the operations belonging to the same job in the sequences produced by the algorithm – the model bias is diminishing and the selection pressure – now being the strongest force – is leading the system to find good solutions in the stable area.

6 Conclusions and Outlook

We have shown the possible existence of strong model bias in model-based search algorithms such as Ant Colony Optimization. Model bias is generated by the model in conjunction with the constraints of the problem. We have shown that the model bias introduced by pheromone model $\mathsf{PH}_{\mathsf{suc}}$ used by an ACO algorithm to tackle the Group Shop scheduling problem (GSP) is strongest for JSP and diminishes for OSP (both JSP and OSP are extreme cases of GSP). Furthermore we have shown that it is possible to reduce the effect of model bias by using a model parameter update rule characterized by a strong selection pressure. Nevertheless, even with a strong selection pressure the model bias leads the system initially to a stable area in the search space where the model bias is low and where the selection pressure takes over to guide the system. In this stable area the selection pressure then causes the system to find good solutions with respect to the quality of the solutions in this area. This means that when using a model-based algorithm, the choice of the model is crucial and has a strong influence on the success of the algorithm. In the future we plan a search space analysis for the GSP to further understand the interactions between model bias and selection pressure.

Acknowledgments

This work was supported by the "Metaheuristics Network", a Research Training Network funded by the Improving Human Potential program of the CEC, grant HPRN-CT-1999-00106. The information provided is the sole responsibility of the authors and does not reflect the Community's opinion. The Community is not responsible for any use that might be made of data appearing in this publication.

References

1. C. Blum, A. Roli, and M. Dorigo. HC-ACO: The hyper-cube framework for Ant Colony Optimization. In *Proceedings of MIC'2001 – Meta-heuristics International Conference*, volume 2, pages 399–403, Porto, Portugal, 2001.
2. C. Blum and M. Sampels. Ant Colony Optimization for FOP Shop scheduling: A case study on different pheromone representations. In *Proceedings of the 2002 Congress on Evolutionary Computation, CEC'02*, volume 2, pages 1558–1563, 2002.
3. P. Brucker. *Scheduling algorithms.* Springer, Berlin, 1998.
4. A. Colorni, M. Dorigo, V. Maniezzo, and M. Trubian. Ant system for Job-shop scheduling. *Belgian Journal of Operations Research, Statistics and Computer Science*, 34(1):39–54, 1993.
5. M. Dorigo and G. Di Caro. The Ant Colony Optimization meta-heuristic. In D. Corne, M. Dorigo, and F. Glover, editors, *New Ideas in Optimization*, pages 11–32. McGraw-Hill, 1999.
6. M. Dorigo, V. Maniezzo, and A. Colorni. Ant System: Optimization by a colony of cooperating agents. *IEEE Transactions on Systems, Man and Cybernetics–Part B*, 26(1):29–41, 1996.
7. M. Dorigo, M. Zlochin, N. Meuleau, and M. Birattari. Updating ACO pheromones using Stochastic Gradient Ascent and Cross-Entropy methods. In *Proceedings of the EvoWorkshops 2002*, LNCS 2279, pages 21–30. Springer, 2002.
8. B. Giffler and G.L. Thompson. Algorithms for solving production scheduling problems. *Operations Research*, 8:487–503, 1960.
9. H. Mühlenbein and G. Paaß. From recombination of genes to the estimation of distributions. In H.-M. Voigt et al., editor, *Proceedings of the 4th Conference on Parallel Problem Solving from Nature, PPSN IV*, volume 1411 of *LNCS*, pages 178–187, Berlin, 1996. Springer.
10. M. Pelikan, D.E. Goldberg, and F. Lobo. A survey of optimization by building and using probabilistic models. Technical Report No. 99018, IlliGAL, University of Illinois, 1999.
11. M. Pinedo. *Scheduling: Theory, Algorithms, and Systems.* Prentice-Hall, Englewood Cliffs, 1995.

Evolution of Asynchronous Cellular Automata

Mathieu S. Capcarrere

Logic Systems Laboratory
School of Computer anc Communication Sciences
Swiss Federal Institute of Technology, Lausanne
CH-1015 Lausanne, Switzerland
mathieu.capcarrere@epfl.ch

Abstract. One of the prominent features of the Cellular Automata (CA) model is its synchronous mode of operation, meaning that all cells are updated simultaneously. But this feature is far from being realistic from a biological point of view as well as from a computational point of view. Past research has mainly concentrated on studying Asynchronous CAs in themselves, trying to determine what behaviors were an "artifact" of the global clock. In this paper, I propose to evolve Asynchronous CAs that compute successfully one of the well-studied task for regular CAs: The synchronization task. As I will show evolved solutions are both unexpected and best for certain criteria.

1 Introduction

Though life was an awe-inspiring model for the first Cellular Automata designers, simplification was a prime constraint in all these studies, not only for obvious practical reasons, but also as a guide to extract the quintessential ideas behind self-replication. Nevertheless, I believe that an ever present quality in natural systems was omitted more for the former reasons rather than the latter choice: *robustness*. Robustness to faults and robustness to asynchronous dynamics. This paper will concentrate on the second question, that of asynchronous cellular automata.

One of the prominent features of the Cellular Automata (CA) model is its synchronous mode of operation, meaning that all cells are updated simultaneously. But this feature is far from being realistic from a biological point of view as well as from a computational point of view. As for the former, it is quite evident that there is no accurate global synchronization in nature. As for the latter, it turns out to be impossible to maintain a large network of automata globally synchronous in practice.

In this paper, I propose to co-evolve asynchronous cellular automata. While in the past the research on asynchronous CA concentrated either on studying the phenomena in itself, i.e., see how asynchrony affected the behavior of synchronous CAs or on the theoretical aspects, e.g., what kind of traces do such CA accept, in this study we propose to evolve asynchronous CAs that solve computational problem. More precisely, I will concentrate on a well studied task for

J.J. Merelo Guervós et al. (Eds.): PPSN VII, LNCS 2439, pp. 903–912, 2002.

the synchronous case: *The synchronization task.* This may seem paradoxical, but as we will see with this illustration, it turns out that evolution gives excellent results given the use of spurious states.

In a first section, we will quickly overview the question of asynchronous cellular automata in the past literature and present the model of asynchrony studied hereafter. I will then ,in section 3, expose shortly the evolutionary algorithm used and the task for which the solution were sought. In section 4 we will give the result one may obtain with a straightforward approach, using two-state CAs. The lack of result in that case demonstrates the need for the different approach. I will thus argue in section 5 the necessity for redundant CAs. This will lead us naturally in section 6 to evolve redundant CAs, which give rise to strikingly efficient and unexpected solutions. Finally we conclude in section 7.

2 Asynchronous Cellular Automata

A preliminary study of asynchronous CAs, where one cell is updated at each time step, was carried out by Ingerson and Buwel, [11]. In that paper, the different dynamical behaviors of synchronous and asynchronous CAs were compared and they concluded that some of the apparent self-organization of CAs is, in fact, an artifact of the synchronization of the clocks. Wolfram, [20], also noted that asynchronous updating makes it more difficult for information to propagate through the CA and that, furthermore, such CAs may be harder to analyze. Asynchronous CAs have then been discussed from a problem-solving and/or Artificial Life standpoint in many papers, among them [1,9,12,14,16,19]. All these works devoted to asynchronous cellular automata only concentrated on the study of the effects but not on correcting asynchrony or dealing with it. From a theoretical computer science perspective, Zielonka [21, 22] introduced the concept of asynchronous cellular automata. Though the question attracted quite some interest [5, 13, 15], the essential idea behind them was to prove that they were "equivalent" to synchronous CA in the sense that, for instance, they recognize the same trace languages or could decide emptiness. From these two fields, we thus know that asynchronous CA are potentially as powerful as synchronous CA, computationally speaking, and, nevertheless, that most of the effects observed in the synchronous case are "artifacts" of the global clock. In this paper, we thus propose to develop asynchronous CA exhibiting the same computational behavior as synchronous CA through evolution. As Gacs [7,8] reminded us, asynchrony may be considered as a special case of fault-tolerance. However, if this consideration is nice in its generalization (i.e. a fault-tolerant CA is also asynchronous), it eschews a lot of potential optimization. There is an evident cost in terms of the complexity of the automata or its reliability[1].

There are many ways to model asynchrony. Sipper *et al* [18] have already considered the evolution of asynchronous CAs to solve the density task[2] and

[1] There exists perfect asynchronous CAs while it is impossible to make perfectly fault-tolerant CAs.

[2] In its fixed point form.

the synchronization task. However, they limited themselves to a partial form of asynchrony where full blocks of cell were guaranteed to be synchronized. If the hypothesis was not completely unreasonable from a practical point of view, its main motivation lied surely in the difficulty to evolve binary asynchronous CA. However, in this paper, I am considering a more general model, described below and show that given a certain number of considerations it is possible evolve good solutions for that model.

The model: I use a most general fully asynchronous model. Each cell has the same probability p_f of not updating its state at each step. In this case the cell state remains unchanged. Otherwise the uniform or non uniform CA is perfectly classic. Hence, the probability of a CA of size N of working synchronously for t time steps is $(1 - p_f)^{Nt}$. Obviously for low values of p_f, the standpoint is that of a synchronous CA where faults may occur, while high values of p_f model a fully asynchronous CA.

3 The Co-evolutionary Framework of Non-uniform Cellular Automata

In this paper I investigate the evolution of *non-uniform cellular automata*. Such automata function in the same way as uniform ones, the only difference being in the cellular rules that need not be identical for all cells. Our focus here is on the *evolution* of non-uniform CAs to perform computational tasks using *the cellular programming approach*. In this section, I explain the cellular programming algorithm which is used herein and is detailed by Sipper in [17].

I study q-state, non-uniform CAs, in which each cell may contain a different rule. A cell's rule table is encoded as a string of the alphabet q (the "genome"), containing the next-state (output) for all possible neighborhood configurations. Rather than employ a population of evolving, uniform CAs, as with standard genetic algorithm approaches, the algorithm involves a single, non-uniform CA of size n, where the population of cell rules is initialized at random. Initial configurations are then generated at random, in accordance with the task at hand, and for each one the CA is run for M time steps. Each cell's fitness is accumulated over $C = 300$ initial configurations. After every C configurations evolution of rules occurs by applying crossover and mutation. This evolutionary process is performed in a completely *local* manner, where genetic operators are applied only between directly connected cells. It is driven by $nf_i(c)$, the number of fitter neighbors of cell i after c configurations. If $nf_i(c) = 0$ then the rule of cell i is left unchanged. If it is equal to 1 then it is replaced by the rule of the cell that is fitter. In all other cases, the rule of cell i is then set to the result of the cross-over between two of the fitter cells. Crossover between two rules is performed by selecting at random (with uniform probability) a single crossover point and creating a new rule by combining the first rule's string before the crossover point with the second rule's string from this point onward. Mutation is applied to one position of the string of a rule with probability 0.001 per bit.

Let's now describe the synchronization task for which solutions are evolved for.

The task: The one-dimensional synchronization task was introduced by Das *et al.* [6] and studied among others by Hordijk [10]. Sipper [17] proposed using non-uniform CAs. In this task the CA, given any initial configuration, must reach a final configuration, within M time steps, that oscillates between all 0s and all 1s on successive time steps. As with the density task[3], synchronization also comprises a non-trivial computation for a small-radius CA. In the case of uniform synchronous CA, there is a perfect $r = 3$ CA, in the non-uniform synchronous case, many perfect $r = 1$ CAs do exist.

The fitness used for this task will be described in the forthcoming section when appropriate.

Obviously, in the synchronous non-uniform case there is an immediate solution consisting of a unique 'master' rule, alternating between '0' and '1', whatever the neighborhood, and all other rules being its 'slave' and alternating according to its right neighbor state only. However, as I have shown elsewhere, studying numerous evolution of synchronous non-uniform CA for the synchronization task, it appeared that this "basic" solution was never found by evolution. In fact, the "master" or "blind" rule 10101010, rule 170, was never part of the evolved solutions. This is simply due to the fact that this rule has to be unique for the solution to be perfect. There cannot be two of these rules in a perfect solution. This latter observation is contradictory to the natural tendency of the evolutionary algorithm used to create blocks of rule. See [2], chapter five for details.

4 Co-evolution of Two-State Cellular Automata

If the evolution of asynchronous binary CA to do synchronization was tried out in [18], there were two major differences in the model used then that lead us to test again this kind of evolution, the results of which I am going to present now. The first difference was that, as said before, the asynchrony was considered by block: blocks of adjacent cells were synchronized internally but the blocks were working asynchronously one with another. The second difference was that blocks were almost forcedly working asynchronously, i.e., the possibility of two blocks updating synchronously was very very low.

The evolution of binary asynchronous CA presented now was done using the model of asynchrony presented at the end of section 2 and the evolutionary algorithm presented above. We concentrated on $r = 1$, two-state non-uniform CA, thus each cell rule is encode as an 8 bit strings. The fitness used is the classic one for this type of evolution. After N time steps, the state in which the majority of cells are is chosen as the goal (e.g., 1), and one point of fitness to every cell in this state. Then the CA is updated once, and each cell in the opposite state (e.g., 0) gets an additional point of fitness. This last step is repeated three times, thus each cell can get a maximum of four. This is repeated on C different initial configurations. During the evolution the value of p_f was fixed. Values of 0.001, 0.005, and 0.01 were tried out for evolution. Of about 100 runs two kind of strategies were found that are illustrated in figure 1:

[3] See [4] and [3] for a rather complete computational overview of the density task.

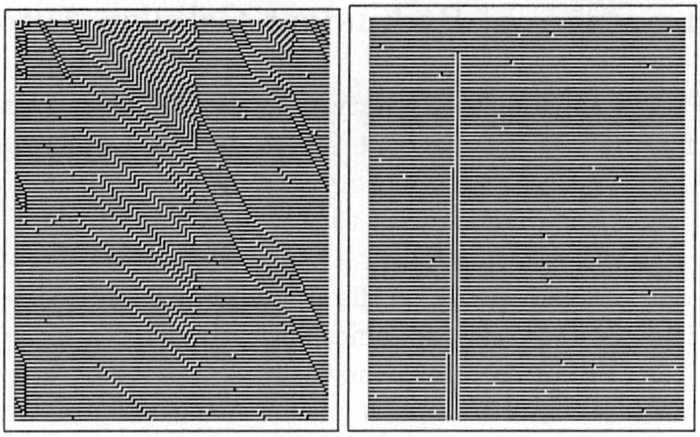

(a) (b)

Fig. 1. Examples of two typical strands of evolution of asynchronous two-state CAs for the synchronization task. The CAs sizes are 159. Time is flowing down. p_f is $\frac{2}{159}$. While in (a) t=0 is at the top, in (b) t=312 is at the top. It demonstrates in (a) the first type of evolutionary run encountered where the errors are corrected almost like the starting configuration, through time, while in (b) one may see the second kind. There the errors are corrected instantly but this *always* lead to the configuration ...11011... and ...00100... leading to a fix point and degrading though time.

The two kind of strategies found by the evolutionary runs are imperfect. Either it dealt with error through a "slow" or "instant" recovery. In the "slow" kind errors are shifted until they are dealt with by an appropriate block. In some way it deals with errors exactly as with any configuration converging eventually to the perfect synchronizing cycle. However that method presents the disadvantage that when p_f gets a little bit higher than about $\frac{5}{159}$ then error cumulates themselves, thereby either creating complete chaos or to the least a huge proportion of unsynchronized cell. This type is illustrated in figure 1.(a). The second strand of evolutionary run, the "instant" kind acts totally differently. After having reached the alternating cycle all 1s, all 0s, it corrects error instantly or in two or three time steps at most. This behavior, that appears to be the perfect behavior sought, seems to be necessitating a fatal misbehavior. If a combination of two or three faults occur in succession in surrounding cells (the precise fatal error depends on the evolutionary runs) it creates local fixed points ...1111010111... or ...000101000.... which never resolves. Unfortunately this leads eventually to complete destruction of the synchronization, as any fault occurring on the neighboring cells of this zone are then never corrected. It always lead, after enough time steps, to the global fixed point 010101010101... This actually shows that the CAs of this kind found through evolution, though synchronizing globally almost on all input configurations perfectly when working in the synchronous mode, are *not* perfect synchronizing CAs.

It thus appears that asynchronous synchronization seems impossible to obtain with binary CAs.

5 The Advantages of Redundant Cellular Automata

The relatively weak capability of binary CAs to cope with even limited asynchrony called for the use of redundant CA to deal with full asynchrony. I call redundant CA, a CA which uses more states in the asynchronous mode than is necessary in the synchronous mode to solve the same task. The idea behind redundancy is that the information in a CA configuration is not only the current state and the topology, but also the timing. For instance, in a synchronous uniform CA with rule 184[4], that classifies perfectly density, a task that requires absolutely no loss of information as shown in [3], if a block of two or more 1's is present at position i after t steps, it means that there are no block of white cells between the positions $i, i+t$, or more exactly that the density has been calculated for the cells $i-t...i+t$. Hence the information that this block carries is not just 111 but rather a 3-tuple $(111, i, t)$. Therefore, to deal with asynchrony, I add redundancy to store, partially, that timing information explicitly in the state rather than implicitly in the global synchronization as is the case in synchronous CA. As I have shown in [2], it is possible to design, for any q-state synchronous CA, a $3{*}q^2$-state asynchronous CA that conserves completely this information. Thus there exists a perfect solution to asynchrony. However, this turns out to be costly and, though it conserves perfectly the information, it does not maintain the visual appearance of the CA.

However the task considered in this paper, the synchronization task, seems to be the perfect example of a lossy task, i.e., there is no need to maintain absolutely the full information present in the synchronous case to solve the problem in the asynchronous case. The idea in evolving asynchronous redundant CA for synchronization is thus to find a good compromise between an acceptable loss of information and the resolution of the task, i.e, between q and $3 * q^2$ states.

6 Co-evolution of Redundant Synchronizing Cellular Automata

As we saw in section 4, the evolution of binary CAs does not produce very good results on real asynchrony. Hence the question to find the good compromise on the number of states needed, i.e., between the 2 states necessary in the synchronous case, and the $3 * 2^2 = 12$ states we know to be sufficient to simulate the synchronous CA in the asynchronous mode. I thus propose here to try to evolve 4-state CAs.

The evolutionary algorithm used is the cellular programming algorithm presented earlier. The crossover used is the standard one-point crossover, thereby the number of states makes no difference in that respect. The 4-state non-uniform CA was evolved in the faulty environment. Asynchrony is then just one of the constraints like the radius or the number of states. Each string of genomes is tested on 100 configurations for 3 probability-of-fault values, $p_f = 0.001, 0.002,$

[4] in Wolfram's notation

and $\frac{1}{159}$. The fitness is the "same" as the one for the binary case. After $1.5 * N$ time steps, for four steps, each cell gets a point of fitness if it alternates correctly between 1_{mod2} and 0_{mod2}. Thus I do not constrain the cycle to be all 0s and all 1s it may also be between all 2s and all 3s or any combinationon. The first state to be in is determined by which state the majority of cells is in.

Globally the evolutionary runs are very successful, and if we consider a fitness of 0.98 as equivalent to a fitness of 1.0 in the synchronous case[5], the success rate is equivalent to the evolution of binary CAs in the synchronous case. Figure 2 presents two successfully evolved CAs.

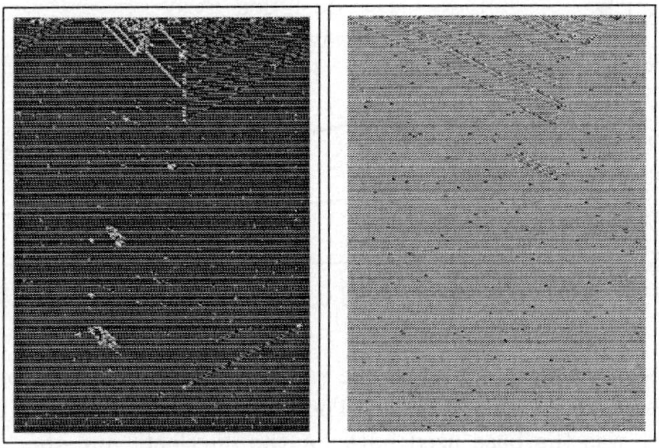

(a) (b)

Fig. 2. Two examples of 4 state non-uniform redundant CAs found through evolution that successfully cope with an asynchronous environment. The probability of synchrony faults here is $p_f = 0.002$ in the two figures. The size of the CAs is 159 and there are 400 time steps shown here. The different strategies in (a), (b) are discussed in the text. In (a) we see a CA that settles in the cycle 0-1 while (b) shows a CA that settles in the cycle 2-3.

It is interesting to look at the strategies found by evolution to cope with asynchrony. As shown in Figure 2, it appears that either the CA settles into the cycle 0,1 or the cycle 2,3 but none of the CA evolved exploited the possibility of falling into the cycle 0,1,2,3 or anything more complex. Thus for most of the working the CA is not really using the extra states. Nevertheless, in the two cases above, the two unused states in the synchronized cycle are used to go from the initial random configuration to the desired unified state. But, more importantly from our viewpoint, each fault of synchrony, produces one or two unused states in the next time step. This, somehow, allows the CA to "detect" the anomaly and correct it very briefly, in the next 1,2 or 3 time steps. In the worst cases, either a "tumor" develops briefly but is resorbed relatively quickly as in figure 2.a or the error drifts until it reaches a rule, or a group of rules able to absorb it, (figure 2.b). It is important to note that neither the cell, nor the neighbor

[5] The faults introduce necessarily some cells in the wrong state.

knows it suffered from a fault. Hence, this detection is only done through the wrong state of the cell concerned. Tested in the synchronous mode these CAs perfectly synchronized on "all" configurations (10^6 configurations tested).

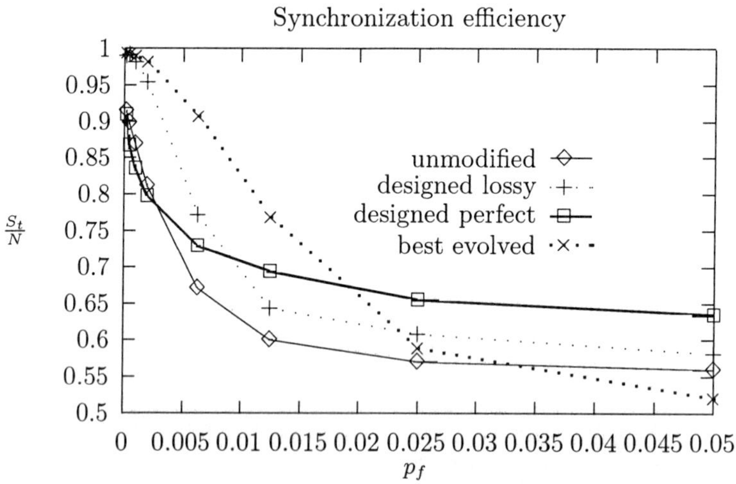

Fig. 3. The experiments were run on the same 10 random initial configurations for 10000 time steps. The CA size N is 159. The "unmodified" version is the binary non-uniform CA that was the base used both for the "perfect" and the "lossy" method. The "best evolved" is the one shown in Figure 2.b.

Actually if we test the best solution found by evolution (2.b), it turns out better than the designed solutions, both the perfect and an imperfect one[6], in terms of the percentage of cells synchronized for low values of p_f's. The test realized was to measure the proportion of cells synchronized in the same state, in an alternating cycle, for different values of p_f's. This test was realized for two reasons. First, this percentage of cell synchronized is what we are really trying to optimize for this task. Besides an asynchronous CA with high percentages will vissually appear similar to synchronous CAs. Interestingly, all evolved solutions fared better for a probability of faults less than or equal to $1/N$ than the deigned solution. Howver for higher p_fs the perfect solution fared the best. The fact that it does not lose information explains this fact. We can note also that the designed lossy method, not presented here, achieves its goal of being very good for low values of p_f, however in the middle range it is worse than the evolved solution but maintains performance better in the long run. Finally, as a control, the solution evolved in a non faulty environment were tested. Obviously when the number of faults is very low, by its nature, it correctly resynchronizes. Nevertheless it falls quickly to low values as p_f increases to settle down close to 0.55. We should not

[6] A simple 4-state asynchronous solution to the synchronization task was designed in [2]. It is not presented here but to put simply it was imperfect but corrected the errors instantly, taking advantage of the knowledge about the alternating cycle sought.

be fooled by the lower score of the evolved 4-state solution for high p_f's . The unmodified version can take only 2 states and thus would fare 0.5 as the lowest possible score, while the evolved solution can take 4 states and thus could settle to percentages as low as 0.25. Figure 3 illustrates these results.

7 Concluding Remarks

CA asynchrony was often studied in itself in the past literature and it was often concluded that the global behavior from a CA, the emergent behavior, was an artifact of the global clock. This conclusion was not wrong in itself but rather reflected a wrong standpoint on a reality. Time *is part* of the visual information contained in a CA. Now if we tackle the asynchrony problem with this idea of restoring all the information, then as said, we can design a totally asynchronous CA that simulates exactly, with no loss of information, any synchronous CA. However this presents two main problems. The first one is that the required number of states is quite higher than the original number of states. The second problem is that it is visually different from the original CA. The visual efficiency of the original CA, the number of cell synchronised in this case, is lost.

Evolution may thus be used to limit both this problem. As presented, the cellular programming algorithm was very successful at finding 4-state solutions that was both economic and still leaved an acceptable visual impression. Actually, it's all a question of the possible compromise between the information loss and the necessity to maintain that information. The synchronization task is a good example of a task not requiring a perfect integrity.

References

1. Hugues Bersini and Vincent Detours. Asynchrony induces stability in cellular automata based models. In R.A. Brooks and P. Maes, editors, *Proceedings of the Artificial Life IV conference*, pages 382–387, Cambridge, MA, 1994. MIT Press.
2. Mathieu S. Capcarrere. *Cellular Automata and Other Cellular Systems: Design & Evolution.* Phd Thesis No 2541, Swiss Federal Institute of Technology, Lausanne, 2002.
3. Mathieu S. Capcarrere and Moshe Sipper. Necessary conditions for density classification by cellular automata. *Physical Review E*, 64(3):036113/1–4, December 2001.
4. Mathieu S. Capcarrere, Moshe Sipper, and Marco Tomassini. Two-state, r=1 cellular automaton that classifies density. *Physical Review Letters*, 77(24):4969–4971, December 1996.
5. Robert Cori, Yves Métivier, and Wieslaw Zielonka. Asynchronous mappings and asynchronous cellular automata. *Information and Computation*, 106:159–202, 1993.
6. Rajarshi Das, James P. Crutchfield, Melanie Mitchell, and James E. Hanson. Evolving globally synchronized cellular automata. In L. J. Eshelman, editor, *Proceedings of the Sixth International Conference on Genetic Algorithms*, pages 336–343, San Francisco, CA, 1995. Morgan Kaufmann.

7. Peter Gács. Self-correcting two-dimensionnal arrays. In Silvio Micali, editor, *Randomness in computation*, volume 5 of *Advances in Computing Research*, pages 223–326, Greenwich, Conn, 1989. JAI Press.
8. Peter Gács. Reliable cellular automata with self-organization. In *Proceedings of the 38th IEEE Symposium on the Foundation of Computer Science*, pages 90–99, 1997.
9. H. Hartman and Gérard Y. Vichniac. Inhomogeneous cellular automata (inca). In E. Bienenstock et al., editor, *Disordered Systems and Biological Organization*, volume F 20, pages 53–57. Springer-Verlag, Berlin, 1986.
10. Wim Hordijk. The structure of the synchonizing-ca landscape. Technical Report 96-10-078, Santa Fe Institute, Santa Fe, NM (USA), 1996.
11. T. E. Ingerson and R. L. Buvel. Structures in asynchronous cellular automata. *Physica D*, 10:59–68, 1984.
12. Yasusi Kanada. Asynchronous 1d cellular automata and the effects of fluctuation and randomness. In R.A. Brooks and P. Maes, editors, *A-Life IV: Proceedings of the Fourth Conference on Artificial Life*, page Poster, Cambridge, MA, 1994. MIT Press.
13. Dietrich Kuske. Emptiness is decidable for asynchronous cellular machines. In C. Palamidessi, editor, *CONCUR 2000*, Lecture Notes in Computer Science, LNCS 1877, pages 536–551, Berlin, 2000.
14. Martin A. Nowak, Sebastian Bonhoeffer, and Robert M. May. Spatial games and the maintenance of cooperation. *Proceedings of the National Academic of Sciences USA*, 91:4877–4881, May 1994.
15. Giovanni Pighizzini. Asynchronous automata versus asynchronous cellular automata. *Theoretical Computer Science*, 132:179–207, 1994.
16. Birgitt Schönfisch and André de Roos. Synchronous and asynchronous updating in cellular automata. *BioSystems*, 51:123–143, 1999.
17. Moshe Sipper. *Evolution of Parallel Cellular Machines: The Cellular Programming Approach.* Springer-Verlag, Heidelberg, 1997.
18. Moshe Sipper, Marco Tomassini, and Mathieu S. Capcarrere. Evolving asynchronous and scalable non-uniform cellular automata. In *Icannga 1997: Proceedings of the Third Bi-Annual Conference*, pages 66–70, Wien, Austria, 1998. Springer-Verlag.
19. W. Richard Stark. Dynamics for fundamental problem of biological information processing. *International Journal of Artificial Intelligence Tools*, 4(4):471–488, 1995.
20. Stephen Wolfram. Approaches to complexity engineering. *Physica D*, 22:385–399, 1986.
21. Wieslaw Zielonka. Notes on finite asynchronous automata. *Informatique théorique et Applications/Theoretical Infomatics and Applications*, 21(2):99–135, 1987.
22. Wieslaw Zielonka. Safe executions of recognizable trace languages by asynchronous automata. In Albert R. Meyer and Michael A. Taitslin, editors, *Logic at Botik'89*, Lecture Notes in Computer Science, LNCS 363, pages 278–289, Berlin, 1989. Springer-Verlag.

Improved Ant-Based Clustering and Sorting in a Document Retrieval Interface

Julia Handl[1] and Bernd Meyer[2]

[1] FB Informatik, Universität Erlangen-Nürnberg
Julia.Handl@gmx.de
[2] School of Computer Science, Monash University, Australia
[3] bernd.meyer@acm.org

Abstract. Sorting and clustering methods inspired by the behavior of real ants are among the earliest methods in ant-based meta-heuristics. We revisit these methods in the context of a concrete application and introduce some modifications that yield significant improvements in terms of both quality and efficiency. Firstly, we re-examine their capability to simultaneously perform a combination of clustering and multi-dimensional scaling. In contrast to the assumptions made in earlier literature, our results suggest that these algorithms perform scaling only to a very limited degree. We show how to improve on this by some modifications of the algorithm and a hybridization with a simple pre-processing phase. Secondly, we discuss how the time-complexity of these algorithms can be improved. The improved algorithms are used as the core mechanism in a visual document retrieval system for world-wide web searches.

1 Introduction

Ant-based sorting and clustering algorithms, introduced in a seminal paper by Deneubourg [Den90] and later extended by Lumer and Faieta [LF94], were among the first meta-heuristics to be inspired by the behavior of ants. Our interest in ant-based clustering is motivated from a concrete application perspective: we employ these methods as the core of a visual document retrieval system for world-wide web searches to classify online documents by contents-similarity.

The capability to perform a combination of clustering and multi-dimensional scaling [CC94], i.e. to generate a distance-preserving embedding, that has been ascribed to ant-based algorithms appears to make them particularly well-suited for our application. However, our experiments do not support the suggestion in the literature that "...a certain distance preserving embedding is guaranteed..." [KS99] (Here the goal was a distance preserving embedding of graphs into a metric space). Instead the results show that the original algorithms perform effective clustering, but have only limited capability for multi-dimensional scaling. We demonstrate how this can be improved upon with some modifications of the algorithms and a simple pre-processing stage. We also demonstrate modifications of the algorithms that significantly improve their run-time, making them acceptable for interactive applications, such as online searches.

J.J. Merelo Guervós et al. (Eds.): PPSN VII, LNCS 2439, pp. 913–923, 2002.

1.1 Contents-Based Document Clustering: Topic Maps

Insufficiently specific world-wide web searches often return thousands of documents. To help the user orient in such large document collections, it has proven useful to classify documents according to contents-similarity and to visualize the classified document collection in the form of a *topic map* [Fab00]. On a topic map semantically similar documents appear in spatial proximity, whereas unrelated documents are clearly separated. Documents are clustered around topics represented by mountains in the landscape. "Height" of a document corresponds to its relevance for the topic and topic labels serve as landmarks.

Our hybrid ant-based clustering method is the core mechanism of a fully implemented search-interface that operates as a front-end to the popular search engines Google and HotBot. Users specify a full text query which is passed on to the back-end search engine. Matching documents returned by the search engine are classified based on either a full text analysis or only on the contents of the snippets returned by the search engine and a topic map is generated to visualize the entire collection of documents (Fig. 1). Starting from these maps, the user can conveniently explore the query results by browsing the documents, which are accessed through a mouse click on their map location. Due to space limitations this paper focuses on the clustering algorithm and will not discuss details of the interface or the pre-processing (for details of these see [Han]).

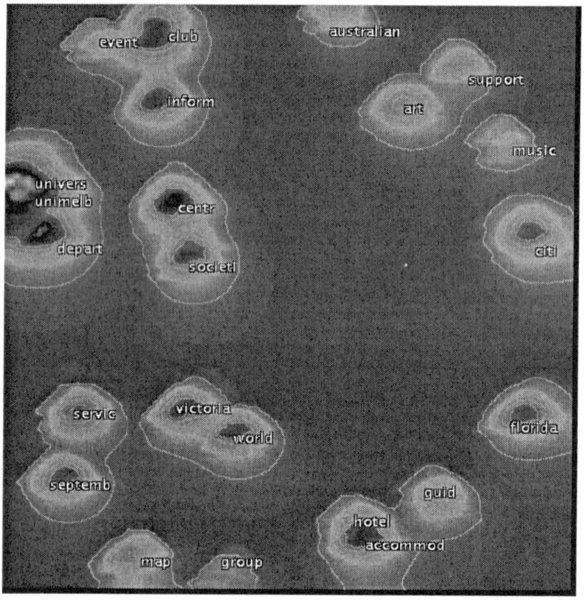

Fig. 1. Ant-generated Topic Map (1000 documents, Query "Melbourne", Search Engine "Google")

Essentially there are two central steps involved in generating topic maps: (1) The documents must be given positions in a high-dimensional conceptual document space that reflects their contents adequately; (2) The document space needs to be embedded into the 2-dimensional visualization space. Ideally, this embedding should combine multi-dimensional scaling with clustering. *Clustering* of the document data helps the user to identify the main structure and to focus on the main topics in the collection. *Multi-dimensional scaling* is necessary, as it provides meaningful inter-cluster relations. On an ideal topic map, clusters with similar contents should be in close proximity, while differences in subject should be reflected by spatial distance.

The idea of a map metaphor for document retrieval as such is not new and quite a few interfaces that use this visualization have been proposed (see Section 4). However, these systems have only dealt with comparatively small sets of documents or with precompiled (offline) data. Systems that have performed fast document classification online "on the fly", such as [CD00,ZE99], had to resort to simpler forms of document classification. To our knowledge, our system is the first to generate large topic maps dynamically for online-queries.

2 Clustering versus Multi-dimensional Scaling

Let us first make precise the embedding task that the application poses. A set of n relevant keyword terms is selected to classify the documents. Each document is positioned in the resulting n-dimensional document space according to its use of keywords. The k-th component of the characteristic vector for document i is computed using normalized IDF weighting [Sal88]:

$$d_{ik} = \frac{tf_{ik} \times log(N_D/n_k)}{\sqrt{\sum_{j=1}^{N}(tf_{ij} \times log(N_D/n_j))^2}}$$

where N_D is the size of the entire document collection; n_i is the number of documents containing the specific term i; and the term frequency tf_{ik} is $tf_{ik} = log(f_{ik})+1$ or $tf_{ik} = 0$ if $f_{ik} = 0$. Here f_{ik} is the frequency of word k in document i. To reduce the dimensionality of the document space we apply *Latent Semantic Indexing* [DDL+90] as a post-process. In the resulting document space, Euclidean distance is an adequate measure of the contents-similarity of two documents.

The second step in generating a topic map is to find a distance-preserving clustering of the high-dimensional document space into the 2-dimensional display space. Ant-based sorting-and-clustering seems a prime candidate to compute this kind of embedding. Let us briefly recall the original algorithm. The ants act on a two-dimensional torus grid on which documents are initially assigned random positions. Each grid cell can at any time only be occupied by a single document. Ants perform two probabilistic actions. (a) *Picking*: if an unloaded ant steps on a field occupied by a document, it may pick up this element; (b) *Dropping*: an ant carrying a document can, at any stage, drop this element on a free cell. At each iteration of the algorithm an ant performs a picking or dropping action or neither according to the probabilistic decision and subsequently steps on a randomly chosen immediately neighboring grid cell. The probability of picking or dropping is influenced by the ant's local perception of its environment: $p_{drop}(i) = \left(\frac{f(i)}{k_d+f(i)}\right)^2$; $p_{pick}(i) = \left(\frac{k_p}{k_p+f(i)}\right)^2$ where k_d and k_p are adjustable parameters ($k_d = k_p \in [0.01, 0.1]$ in our experiments) and

$$f(i) = max\left(0, \frac{1}{|S|}\sum_{\{c_{ij}\in S|d_{ij}\neq nil\}}\left(1 - \frac{d(k, d_{ij})}{\alpha\,\mu}\right)\right)$$

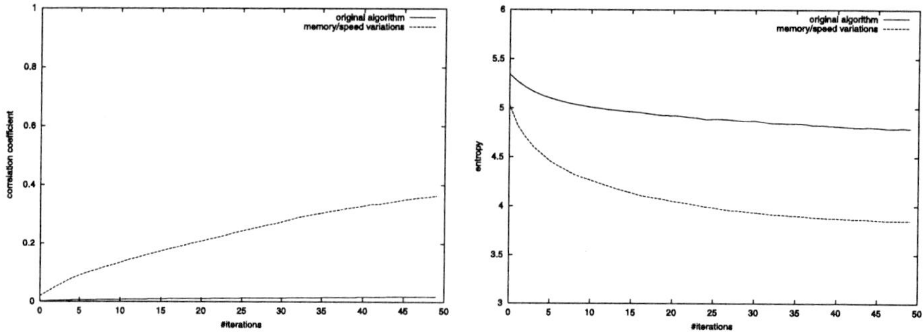

Fig. 2. Inter-cluster correlations (left) and entropy (right) (mean values for 50 runs).

where d_{ij} is the index of the document in cell c_{ij} and S is the neighborhood around the ant's current position (5×5 in our experiments). $d(k, d_{ij})$ is the similarity (distance) between the document with index d_{ij} and the document k currently carried or considered by the ant. N is the number of documents in the collection and the scaling factor is $\mu = \frac{2}{N(N-1)} \sum_{k=1}^{N} \sum_{l=1}^{k-1} (d(k,l))$. The parameter $\alpha \in [0, 1]$ permits further adjustment of the resulting values. It will later play a crucial role in our modifications of the algorithms.

As discussed above, the embedding to generate a proper topic map requires (a) adequate clustering and (b) that the distances in the visualization space strongly correlate with the distances in the document space. *Do ants really generate a distance preserving embedding?* In [LF94] the original algorithm was mainly evaluated using visual observation and by measuring spatial entropy. The quality of sorting was measured by the global fit and a dissimilarity measure. As these do not reflect the global correlation well enough, [KSL98] used the overall Pearson Correlation (the degree of linear relationship between two variables) as an additional performance measure. Here, correlations of up to 0.65 were reported but not analyzed further.

Importantly, it must be noted that the method is very sensitive to the choice of α and correlations like 0.65 are only achieved with the proper choice of α. This problem was already noted in [KSL98], but no method was suggested how the proper α value could be found without resorting to manual experimentation. As we will discuss later, the proper choice of α depends on the structure of the data and the solution is to introduce an adaptive strategy.

To further analyze the performance, we measure Pearson correlation on several levels: (1) Overall correlation of all elements; (2) Inter-cluster correlations, where the weighted average of all cluster elements is used as cluster center; (3) Intra-cluster correlations, which are observed by computing the correlations for the clusters individually. In our experiments we have additionally measured Spearman rank correlation, which is better suited to track non-linear relationships. However, as the experimental results generally confirm the observations made for the Pearson correlation, we do not report them in this paper.

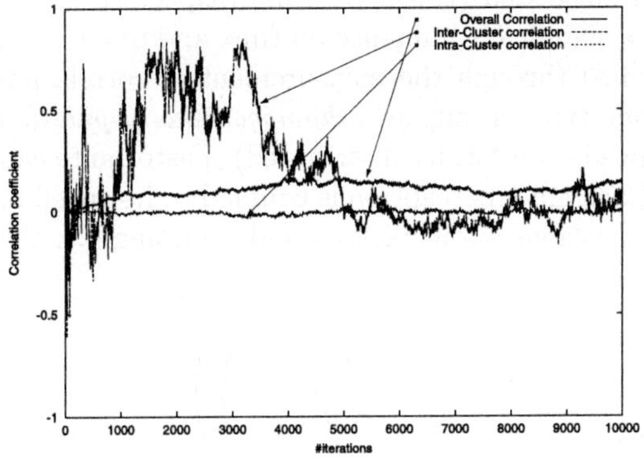

Fig. 3. Overall correlation and inter/intra cluster correlation (original algorithm, single run, 800 element set).

In the original studies in [LF94] good results were reported for an artificial test data set based on visual inspection. Initially, these findings were confirmed by the correlations measured in our experiments. However, more extensive experiments reveal that these results are mostly due to the simplicity of the test data. The test data is composed of only four normal distributed clusters (100 elements each) around the centers (0,0); (0,8); (8,0); (8,8) with an identical standard deviation of 2 and sorted on a grid of size of 52×52. Due to this limited grid size, the limited number of elements and the relatively good spatial separation of the clusters, the algorithm can recover the cluster structure too easily.

In further experiments with larger grid sizes and more complex test data the entropy is still reduced, but the data leaves very serious doubts about the quality of the obtained embedding. The modified test data consists of four clusters of 200 elements each on a grid of size 100×100. Cluster centers are now irregularly distributed at (0,0); (0,80); (8,0) and (30,30).

Figure 2 shows the development of overall correlation and entropy on the modified test data set for the original algorithm and the algorithm with speed and memory improvements (see Section 3) averaged over 50 runs. It is clearly visible that the correlation remains very low. When analyzing the different correlation types further we can see that intra-cluster correlation and overall correlation both stay low. Even more interestingly, in individual runs the inter-cluster correlation is not stabilizing but oscillates (Fig. 3). This effect is also visible in a standard deviation of about 0.4 for the inter-cluster correlation data.

We conclude that the demands of multi-dimensional scaling for generating topic maps cannot sufficiently be met by the original algorithms, as the distances between individual clusters are not represented correctly. We also have to face the problem that a suitable fixed α-value cannot be determined in advance.

3 Algorithm Modifications

We now discuss our modifications to these algorithms that improve both performance and run-time. Lumer and Faieta already suggested two modifications to the algorithm to improve the quality of the results: (1) Each ant keeps a short-term memory of its most recent drop locations. This allows to bias the ant's movement after a pick-up in the direction of the "closest match" among the last

n drop locations. In [LF94] it was suggested (through visual inspection of the results) that this modification has a significant influence on time and quality. As expected, this suggestion is confirmed through the measurement of correlation values. (2) Instead of using only one type of ant, an *inhomogeneous population* is used which moves at different speeds (uniformly distributed). Faster ants can skip several grid cells in one step. Each ant's perception is coupled to its speed v, so that slower ants make "fussier" decisions about picking and dropping, i.e. the perception function becomes:

$$f(i) = max\left(0, \frac{1}{|S|}\sum_{\{c_{ij}\in S|d_{ij}\neq nil\}}\left(1-\frac{d(k,d_{ij})}{\alpha\,\mu\,\frac{v-1}{V_{max}}}\right)\right)$$

When introducing inhomogeneous populations in isolation, our experiments do not reconfirm the performance gain reported in [LF94]. It appears that the improvements were due to introducing larger stepsizes simultaneously with inhomogeneous populations. We adopt the idea of short-term memory and use inhomogeneous populations with the minor but crucial modification of "jumps". We also introduce an adaptive scaling strategy and some further modifications to achieve reliable results and to improve the efficiency.

Adaptive scaling: The artificial test data used in the literature provide idealized conditions. Due to the choice of regularly distributed cluster centers and identical spread, inter-document distances are limited to a small range and smoothly distributed. When conducting experiments on more difficult test sets with irregular inter-cluster distances and standard deviations, more clusters and, in particular, of higher dimensionality, we found that an appropriate α value cannot be determined without a-priori knowledge of the data's structure. In consequence, we introduce an adaptive strategy to determine the α value for a given data set. The sorting process starts with $\alpha = 0.1$ and increases α by 0.01 up to a maximum value of 1.0 after each sequence of 250 steps in which only few dropping or picking actions occur.

This means that our methods starts with a very fine distinction between data elements and reduces it only if necessary. In our experiments, this method reliably determined suitable alpha values. Also note that we scale similarity values in the interval $[1 - \frac{max_{i,j} d(i,j)}{\alpha\,\mu}, 1]$ instead of $[0, 1]$. Thus, *negative influence* of large dissimilarities is permitted which leads to improved cluster separation. Visual inspection of the sorting result for a three cluster data set with irregular inter-cluster distances in Fig. 4 shows that the adaptive strategy manages to deal well with it, whereas the non-adaptive strategy can only separate a single cluster.

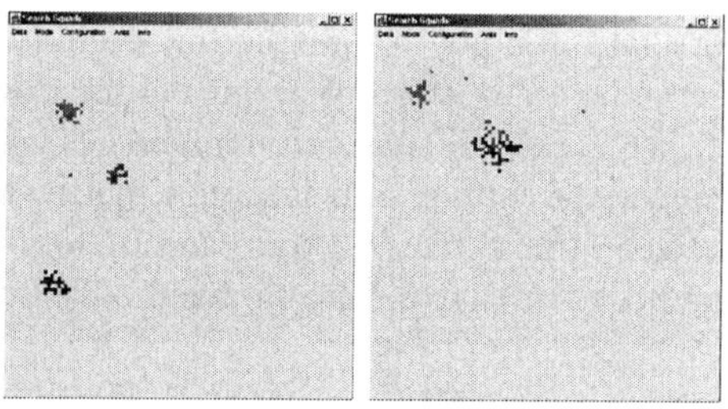

Fig. 4. Separation of Three Clusters (left: adaptive $\alpha \rightarrow 0.32$, right: non-adaptive $\alpha = 1.0$)

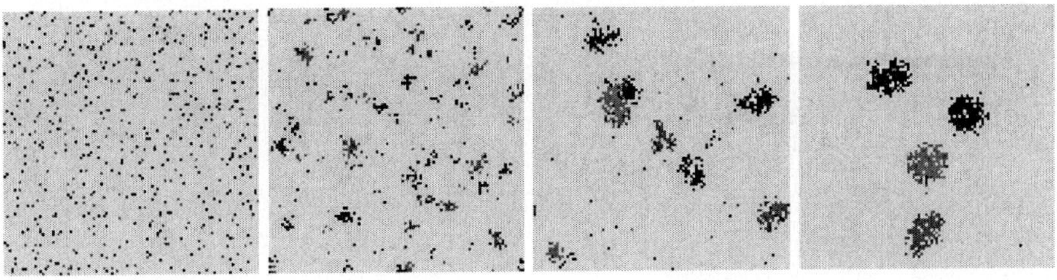

(a) Initial scattering $t = 0$ (b) Original Alg. t=30, corr=0.01 (c) Original Alg. t=1000, corr=0.24 (d) Improved Alg. t=30, corr=0.6

Fig. 5. Results on a four cluster test set (b,c: $\alpha = 1.0$, d: adaptive $\alpha \to 1.0$).

Jumps: Experiments show that the use of inhomogeneous populations only leads to significant improvements in runtime and sorting quality if used with very large speed values (up to 50% of the grid size), as the ants' large steps favor the dissolution of preliminary small clusters. The interesting aspect here is that this introduces "jumping" ants and the smooth moving of an agent through a continuous space is transformed into a form of more global sampling with directional bias.

Stagnation control: With complex data, early stagnation of the whole clustering process can be a problem. This is caused by outliers in the data sets. Due to their high dissimilarity to all other data elements, agents do not manage to dispose of these items once they had been picked. This results in *blocked* ants performing random walks on the grid without contributing to the sorting process. Similar to [MSV99], we therefore use a *failure counter* for each ant. After 100 unsuccessful dropping attempts an ant drops its load regardless of the neighborhood's similarity.

Eager ants: In the original algorithm, ants often spend large amounts of time searching for new documents to pick up. To prevent this time-consuming search we couple each ant with a new document immediately after it drops its load. As soon as this happens, the ant randomly chooses a document from the index of all non-carried documents, moves to its position and tries to pick it up. Failure results in the random choice of another document.

3.1 Evaluation

A comparison of the improved algorithm with the original version shows a significant improvement of both sorting quality and time. As shown in Figure 5, both algorithms start at $t = 0$ with a random disorder of data elements on the grid. The slow clustering progress for the original algorithm can be observed in Figure 5b and Figure 5c. Only little clustering has been obtained after 30 iterations (one iteration consists of 10000 individual actions) and even for $t = 1000$ the four clusters within the data have not been correctly identified. The modified version, in contrast, reliably determines the four main clusters within the first 30 iterations. The corresponding spatial ordering of the test data can be seen in Figure 5d.

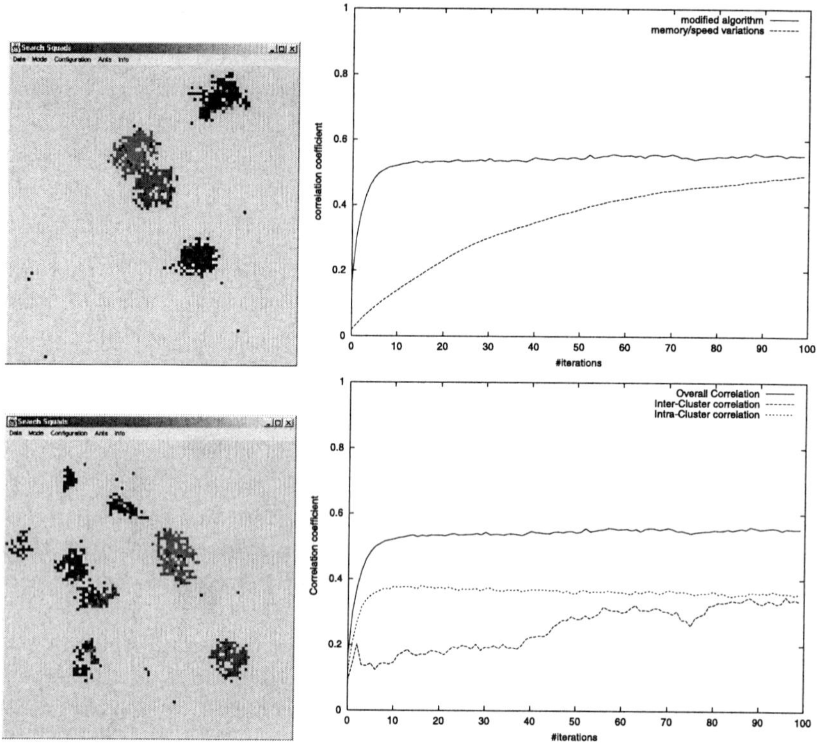

Fig. 6. Performance gain for all improvements (top left) and using only memory and inhomogeneous populations (bottom left). Overall correlation for both algorithms (top right) and individual correlations types for improved algorithm (bottom right). Mean values for 50 runs.

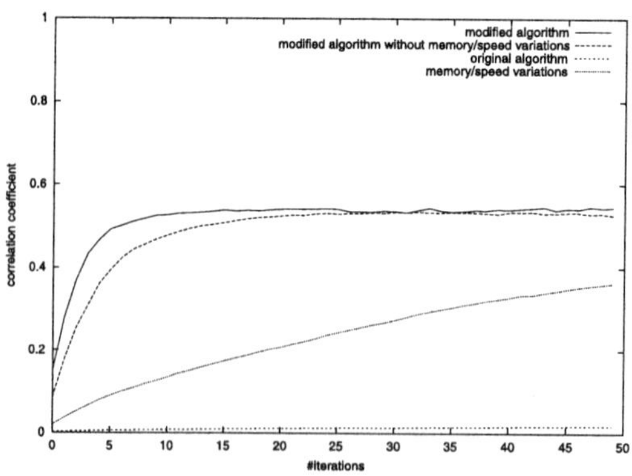

Fig. 7. Overall correlation for four versions of algorithms on 4×200 test data (mean of 50 runs).

These differences in performance are reflected by the analytical performance measurements. While the modified ant-algorithm improves the overall correlation significantly and is soon close to convergence, the plots for the original algorithm show only little improvement in the same runtime (Fig. 7).

A comparison of different versions of the algorithm further isolates the effect of our newly introduced modifications from that of memory and inhomogeneous populations alone (Fig. 6). Overall correlation and cluster identification are clearly improved for the full version. Unfortunately, it has to be noted that the achievable inter-cluster correlations are not superior.

The quality of sorting if there are no distinct clusters in the data deserves individual assessment. For better visual observation we use a test set merely consisting of one cluster with uniformly distributed elements (Fig. 8). Nine different regions within the distribution are colored (left) so that the reconstruction of the order within the cluster can be observed (right). A visual inspection indicates that the algorithm is relatively successful in reproducing the approximate color ordering within the data. Exact measurement shows that the correlation for this test data rarely exceeds 0.7, but given that the goal of our algorithms is to generate embeddings for visual inspection only, it can be argued that this correlation is sufficient. This experiment suggests that it could be possible to achieve better intra-cluster correlations by adding a *second stage* of sorting in which ants with small step sizes are confined to move *within* each cluster.

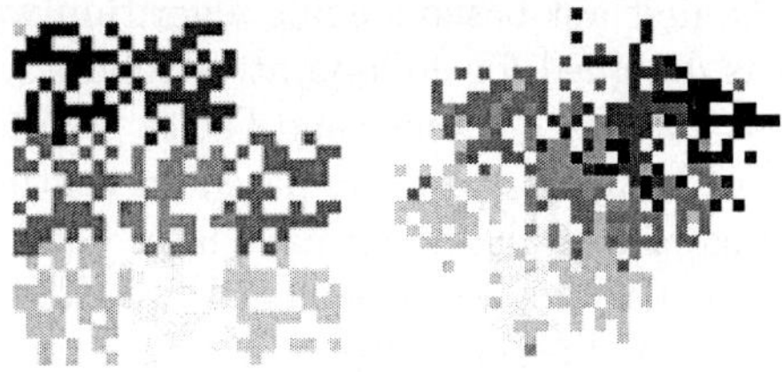

Fig. 8. Dense data: correlation=0.60 ($\alpha \to 0.7, t = 100$)

3.2 Pre-processing

From above we see that the modified algorithm achieves reasonable overall correlations, but that the inter-cluster correlations are not sufficient for the purpose of topic-map generation. To obtain proper inter-cluster distances, we can exploit the fact that document positions are comparatively stable in regard to their initial positions. This allows us to initialize the clustering algorithm with an approximate (unclustered) embedding. While computing an optimal multi-dimensional scaling is a computationally hard problem, it is relatively easy to compute an approximate scaling that is sufficient for visual inspection. We do this with a straight-forward algorithm adopted from [NL01] which iteratively minimizes the sum-squared error between the distances d_{ij} in document space and the distances d'_{ij} in the two-dimensional map space updating the positions p_j via the gradient descent rule: $p_{jk} \leftarrow p_{jk} - \frac{lr(d_{mj}-d'_{mj})}{d'_{mj}} abs(p_{jk} - p_{mk}) sign(p_{jk} - p_{mk})$ for each dimension ($k = 1, 2$) independently. lr is the learning rate (typically 0.05). The minimization stops when the increase of the Pearson correlation between d_{ij} and d'_{ij} falls under a pre-defined threshold. Clustering starts from the approximate scaling, which results in significantly improved correlation values of the final embedding. For the four-cluster test set an average inter-cluster correlation of almost 0.9 is obtained, improving the overall correlation to about 0.7.

4 Related Work

The first ant-based sorting algorithm was presented by Deneuborg [Den90] and was designed for collaborative robotics. In this work, only binary classification was addressed. Lumer and Faeita [LF94] later extended these ideas to clustering more complex data sets based on continuous similarity functions. Modified versions of this algorithm have later been applied to graph clustering, focusing in particular on applications in VLSI [KS94,KLS97,KS99,KSL98]. For an extensive survey of different ant-based meta-heuristics which are based on pheromone-trail communication, the interested reader is invited to refer to [CDG99].

The idea of using a map metaphor for visualizing contents similarity has some tradition and a full survey is beyond the scope of this paper. The idea was originally introduced by Chalmers [Cha93] and later picked up in other projects, among them the commercial systems Cartia (`www.aurigin.com`), Spire (`showcase.pnl.gov`) and Kartoo (`www.kartoo.com`), Lighthouse [LA00] and others. A recent HCI study clearly shows that a landscape metaphor can help to significantly enhance user performance in search tasks [Fab00]. A significant proportion of systems that generate such visualization has relied on self-organizing maps. Since generating topic maps with self-organizing maps requires computation times that are prohibitive in an interactive setting [Lag00], we decided to investigate ant-based sorting/clustering as an alternative. To the best of our knowledge, this is the first system to generate large topic maps for several thousand documents in an interactive setting. Ant-based clustering for a map of 4200 documents, for example, takes 29 seconds on a 997MHz Pentium-III running JDK 1.4.1 and would be considerably faster in an optimized C implementation.

5 Conclusions

We have re-examined ant-based sorting and clustering methods in the context of a real-life application and demonstrated modifications of the algorithms that yield significant improvements in terms of quality and speed.

It is obvious that we have sacrificed most of the original methods' biological plausibility for the performance improvements. However, this should not concern us, as we are not trying to analyze the behavior of real insects, rather we use their behavior as inspiration for the design of an efficient meta-heuristics. In fact, lifting the restrictions imposed by real physical systems (and therefore sacrificing the biological plausibility) leads to some interesting insights: for example, an intuitive explanation of why ant-based sorting works well may assume that the coherence of the space in which the ants operate is crucial to the function of the algorithms. As it turns out with the introduction of jumps, this is not the case.

The use of a stochastic algorithm in a query system poses challenging questions concerning query refinement (incrementallity) and query repetition (stability) which we are planning to investigate. Our evaluation of the algorithm was mainly based on measurements for artificial test data and on visual observation for real query data. Clearly, a thorough evaluation for real test data as well as user studies are important question to be addressed in the future.

References

[CC94] T.F. Cox and M.A.A. Cox. *Multidimensional Scaling.* Chapman & Hall, 1994.

[CD00] Hao Chen and Susan Dumais. Bringing order to the web. In *ACM CHI*, The Hague, April 2000.

[CDG99] D. Corne, M. Dorigo, and F. Glover, editors. *New Ideas in Optimization*, chapter 2: The Ant Colony Optimization Meta-Heuristic, pages 379–387. McGraw-Hill International (UK) Limited, 1999.

[Cha93] M. Chalmers. Using a landscape metaphor to represent a corpus of documents. In A. Frank and I. Campari, editors, *Spatial Information Theory: A Theoretical Basis for GIS*, pages 377–390. Springer-Verlag, September 1993.

[DDL+90] S. C. Deerwester, S. T. Dumais, T. K. Landauer, G. W. Furnas, and R. A. Harshman. Indexing by latent semantic analysis. *Journal of the American Society of Information Science*, 41(6):391–407, 1990.

[Den90] J. L. Deneuborg. The dynamics of collective sorting. robot-like ants and ant-like robots. In *1st International Conference on Simulation of Adaptive Behaviour: From animals to animats 1*, pages 356–363. MIT Press, Mai 1990.

[Fab00] S. I. Fabrikant. *Spatial Metaphors for Browsing Large Data Archives.* PhD thesis, Department of Geography, University of Colorado, 2000.

[Han] J. Handl. Visualising internet-queries using ant-based heuristics. Honours Thesis. Dept. of Computer Science, Monash University, Australia. 2001.

[KLS97] P. Kuntz, P. Layzell, and D. Snyers. A colony of ant-like agents for partitioning in VLSI technology. In *4th European Conference on Artificial Life.* MIT Press, July 1997.

[KS94] P Kuntz and D. Snyers. Emergent colonization and graph partitioning. In *3rd International Conference on Simulation of Adaptive Behaviour: From Animals to Animats 3.* MIT Press, April 1994.

[KS99] P. Kuntz and D. Snyers. New results on an ant-based heuristic for highlighting the organization of large graphs. In *99 Congress on Evolutionary Computation*, pages 1451–1458. IEEE Press, July 1999.

[KSL98] P. Kuntz, D. Snyers, and P. Layzell. A stochastic heuristic for visualising graph clusters in a bi-dimensional space prior to partitioning. *Journal of Heuristics*, 1998.

[LA00] A. Leuski and J. Allan. Lighthouse: Showing the way to relevant information. In *IEEE Information Vizualization*, Salt Lake City, October 2000.

[Lag00] K. Lagus. *Text Mining with the WEBSOM.* PhD thesis, Department of Computer Science and Engineering, Helsinki University of Technology, 2000.

[LF94] E. Lumer and B. Faieta. Diversity and adaption in populations of clustering ants. In *3rd International Conference on Simulation of Adaptive Behaviour: From Animals to Animats 3.* MIT Press, July 1994.

[MSV99] N. Monmarche, M. Slimane, and G. Venturini. On improving clustering in numerical databases with artificial ants. In *Advances in Artificial Life (ECAL'99), LNAI 1674.* Springer-Verlag, 1999.

[NL01] D. J. Navarro and M. D. Lee. Spatial visualisation of document similarity. In *Defence Human Factors Special Interest Group Meeting*, August 2001.

[Sal88] G. Salton. *Automatic Text Processing.* Addison-Wesley, New York, 1988.

[ZE99] O. Zamir and O. Etzioni. Grouper: A dynamic clustering interface to web search results. In *8th World Wide Web Conference*, Toronto, May 1999.

An Adaptive Flocking Algorithm for Spatial Clustering

Gianluigi Folino and Giandomenico Spezzano

ICAR-CNR
Via Pietro Bucci cubo 41C
c/o DEIS, Università della Calabria, 87036 Rende (CS), Italy
{folino,spezzano}@isi.cs.cnr.it

Abstract. This paper presents a parallel spatial clustering algorithm based on the use of new Swarm Intelligence (SI) techniques. SI is an emerging new area of research into Artificial Life, where a problem can be solved using a set of biologically inspired (*unintelligent*) agents exhibiting a collective intelligent behaviour. The algorithm, called SPARROW, combines a smart exploratory strategy based on a flock of birds with a density-based cluster algorithm to discover clusters of arbitrary shape and size in spatial data. Agents use modified rules of the standard flock algorithm to transform an agent into a hunter foraging for clusters in spatial data. We have applied this algorithm to two synthetic data sets and we have measured, through computer simulation, the impact of the flocking search strategy on performance. Moreover, we have evaluated the accuracy of SPARROW compared to the DBSCAN algorithm.

1 Introduction

Clustering spatial data is the process of grouping similar objects according to their distance, connectivity, or their relative density in space [1]. Spatial clustering has been an active area of research into data mining, with many effective and scalable clustering methods developed. These methods can be classified into partitioning methods [2], hierarchical methods [3,4], density-based methods [5], and grid-based methods [6]. Han, Kamber, and Tung's paper [7] is a good introduction to this subject.

Recently, other algorithms based on biological models have been proposed to solve the clustering problem. These algorithms are characterized by the interaction of a large number of simple agents sensing and changing their environment locally. They exhibit complex, emergent behaviour that is robust compared to the failure of individual agents. Ants colonies, flocks of birds, termites, swarms of bees etc. are agent-based insect models that exhibit a collective intelligent behaviour (swarm intelligence) [8] and may be used to define new algorithms of clustering.

In one of the first studies related to this domain, due to Deneubourg et al. [9], a population of ant-like agents randomly moving onto a 2D grid are allowed to move basic objects so as to classify them. This method was further developed by Lumer and Faietta [10] with simple objects that represent records in a numerical data set, and by Kuntz and Snyers [11] who analyzed a real clustering problem in order to efficiently resolve an optimization problem. Monmarchè et al. [12] exploit this existing work from the knowledge discovery point of view with the aim of solving real world

J.J. Merelo Guervós et al. (Eds.): PPSN VII, LNCS 2439, pp. 924–933, 2002.

problems. They introduce a more robust heuristics based on stochastic principles of an ant colony in conjunction with the deterministic principles of the Kmeans algorithm. A flocking algorithm has been proposed by Macgill and S. Openshaw [13,14] as a form of effective search strategy to perform an exploratory geographical analysis. The method takes advantage of the parallel search mechanism a flock implies, by which if a member of a flock finds an area of interest the mechanics of the flock will drive other members to scan that area in more detail.

In this paper, we present a parallel spatial clustering algorithm SPARROW (***SPA**tial Cluste**R**ing Algo**R**ithm thr**O**ugh S**W**arm Intelligence*), which is based on an adaptive flocking algorithm combined with a density-based cluster algorithm, to discover clusters of arbitrary shape and size in spatial data. SPARROW uses the stochastic and exploratory principles of a flock of birds for detecting clusters in parallel according to the density-based principles of the DBSCAN algorithm, and a parallel iterative procedure to merge the clusters discovered.

SPARROW is a multi-agent algorithm where agents use modified rules of Reynolds' standard flock algorithm [15] to transform an agent into a hunter foraging for clusters in spatial data. Each agent searches the clusters in parallel and, by changing colour, signals the presence or the lack of significant patterns in the data to other flock members. The entire flock then moves towards the agents (*attractors*) that have discovered interesting regions, in order to help them, avoiding the uninteresting areas that are instead marked as obstacles. Moreover, each agent has a variable speed, though sharing a common minimum and maximum with the others. An agent will speed up in order to leave an empty or uninteresting region, whereas it will slow down in order to investigate an interesting region more carefully. The variable speed introduces an adaptive behaviour in the algorithm. In fact, the agents adapt their movement by changing their behaviour (speed) according to their previous experience represented by the agents which have stopped to signal an interesting region or an empty one.

We have built a Starlogo [16] simulation of SPARROW to investigate the interaction of the parameters that characterize the algorithm. The first experiments showed encouraging results and a better performance of SPARROW in comparison with the standard flock search and the linear randomised search.

The remainder of this paper is organized as follows: section 2 briefly presents the heuristics of the DBSCAN algorithm used for discovering clusters in spatial data, section 3 introduces the classical flocking algorithm and presents the SPARROW algorithm; section 4 discusses the obtained results while section 5 draws some conclusions and refers to future work.

2 The DBSCAN Algorithm

One of the most popular spatial clustering algorithms is DBSCAN, which is a density-based spatial clustering algorithm. A complete description of the algorithm and its theoretical basis is presented in the paper by Ester et al. [17]. In the following we briefly present the main principles of DBSCAN. The algorithm is based on the idea that all points of a data set can be regrouped into two classes: *clusters* and *noise*. Clusters are defined as a set of dense connected regions with a given radius (*Eps*) and containing at least a minimum number (*MinPts*) of points. Data are regarded as noise if the number of points contained in a region falls below a specified threshold. The two parameters, Eps and MinPts, must be specified by the user and allow to control the density of the cluster that must be retrieved. The algorithm defines two different kinds of points in a clustering: *core points* and *non-core points*. A core point is a point with at least *MinPts* number of points in an Eps-neighborhood of the point. The non-core points in turn are either *border points* if are not core points but are density-reachable from another core point or *noise points* if they are not core points and are not density-reachable from other points. To find the clusters in a data set, DBSCAN starts from an arbitrary point and retrieves all points with the same density reachable from that point using Eps and MinPts as controlling parameters. A point p is density reachable from a point q if the two points are connected by a chain of points such that each point has a minimal number of data points, including the next point in the chain, within a fixed radius. If the point is a core point, then the procedure yields a cluster. If the point is on the border, then DBSCAN goes on to the next point in the database and the point is assigned to the noise. DBSCAN builds clusters in sequence (that is, one at a time), in the order in which they are encountered during space traversal. The retrieval of the density of a cluster is performed by successive spatial queries. Such queries are supported efficiently by spatial access methods such as R*-trees.

3 A Multi-agent Spatial Clustering Algorithm

In this section, we will present the SPARROW algorithm which combines the stochastic search of an adaptive flocking with the DBSCAN heuristics for discovering clusters in parallel. SPARROW replaces the DBSCAN serial procedure for clusters identification with a multi-agent stochastic search that has the advantage of being easily implementable on parallel computers and is robust compared to the failure of individual agents.

We will first introduce Reynolds' flock of birds model to describe the movement rules of the agents from which SPARROW takes inspiration. Then we will illustrate the details of the behavioral rules of the agents that move through the spatial data looking for clusters and communicating their findings to each other.

3.1 The Flock Algorithm

The flock algorithm was originally devised as a method for mimicking the flocking behavior of birds on a computer both for animation and as a way to study emergent behavior. Flocking is an example of emergent collective behavior: there is no leader, i.e., no global control. Flocking behavior emerges from the local interactions. In the flock algorithm each agent has direct access to the geometric description of the whole scene, but reacts only to flock mates within a certain small radius. The basic flocking model consists of three simple steering behaviours:

Separation. gives an agent the ability to maintain a certain distance from others nearby. This prevents agents from crowding too closely together, allowing them to scan a wider area.

Cohesion. gives an agent the ability to cohere (approach and form a group) with other nearby agents. Steering for cohesion can be computed by finding all agents in the local neighbourhood and computing the *average position* of the nearby agents. The steering force is then applied in the direction of that *average position*.

Alignment. gives an agent the ability to align with other nearby characters. Steering for alignment can be computed by finding all agents in the local neighbourhood and averaging together the 'heading' vectors of the nearby agents.

3.2 SPARROW: A Flocking Algorithm for Spatial Clustering

SPARROW is a multi-agent adaptive algorithm able to discover clusters in parallel. It uses a modified version of standard flocking algorithm that incorporates the capacity for learning that can find in many social insects. In our algorithm, the agents are transformed into hunters with a foraging behavior that allow them to explore the spatial data while searching for clusters.

SPARROW starts with a fixed number of agents that occupy a randomly generated position. Each agent moves around the spatial data testing the neighborhood of each location in order to verify if the point can be identified as a *core point*. In case it can, all points of the neighborhood of a core point are given a temporary label. These labels are updated as multiple clusters take shape concurrently. Contiguous points belonging to the same cluster take the label corresponding to the smallest label in the group of contiguous points.

Each agent follows the rules of movement described in Reynolds' model. In addition, our model considers four different kinds of agents, classified on the basis of the density of data in their neighborhood. These different kinds are characterized by a different color: *red*, revealing a high density of interesting patterns in the data, *green*, a medium one, *yellow*, a low one, and *white*, indicating a total absence of patterns. The main idea behind our approach is to take advantage of the colored agent in order to explore more accurately the most interesting regions (signaled by the red agents) and avoid the ones without clusters (signaled by the white agents). Red and white agents stop moving in order to signal this type of regions to the others, while green and yellow ones fly to find more dense clusters. Indeed, each flying agent computes its heading by taking the weighted average of alignment, separation and cohesion.

The following are the main features which make our model different from Reynolds':

- *Alignment* and *cohesion* do not consider yellow boids, since they move in a not very attractive zone.
- *Cohesion* is the resultant of the heading towards the average position of the green flockmates (centroid), of the attraction towards reds, and of the repulsion from whites, as illustrated in figure 1.
- A *separation* distance is maintained from all the boids, apart from their color.

In the following we use the Starlogo language to describe our algorithm and to perform the simulations. SPARROW consists of a setup phase and a running phase shown in Figure 2. During the setup phase agents are created, data are loaded, some general settings are made and the turtles choose their color. In the running phase four distinct procedures are repeated by each turtle for a fixed number of times (*MaxNumberOfGenerations*). In fact, *ask-turtles* is a StarLogo instruction that makes all the turtles execute a procedure in parallel and waits for the completion of the operation before continuing.

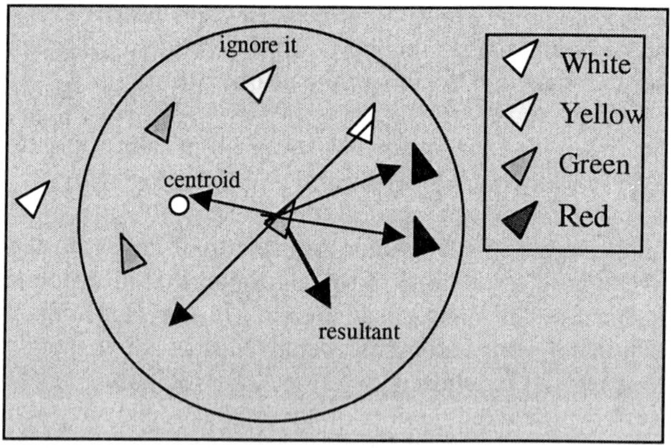

Fig. 1. Cohesion.

The *choiceColor* procedure chooses the color and the speed of the boid with regard to the local density of the clusters in the data. It is based on the same parameters used in the DBSCAN algorithm: MinPts, the minimum number of points to form a cluster and Eps, the maximum distance that the agents can look at. In practice, the agent computes the density (*localdensity*) in a circular neighborhood (with a radius determined by its limited sight) and then executes the following instructions:

if localdensity > MinPts [**set** color red **set** speed 0]
if MinPts/4 < localdensity < MinPts [**set** color green **set** speed 1]
if 0 < localdensity < MinPts/4 [**set** color yellow **set** speed 2]
if localdensity = 0 [**set** color white **set** speed 0]

Thus, red and white boids will stop indicating interesting and desert regions to the others, while greens will move more slowly than yellows since they will explore denser zones of clusters. In the running phase, the yellow and green agents will compute their heading, according to the rules previously described, and will move following this direction and with the speed corresponding to their color. Afterwards, they will compute their new color, deriving from the movement. According to whether they have become red or white, a new boid will be generated in order to maintain a constant number of turtles exploring the data. In case the turtle falls in the same position of an older it will die.

At this point red boids will run the *mergeColor* procedure, which will merge the neighboring clusters. The merging phase considers two different cases: when we have never visited points in the circular neighborhood and when we have points belonging to different clusters. In the first case, the points will be labeled and will constitute a new cluster; in the second case, all the points will be merged into the same cluster, i.e. they will get the label of the cluster discovered first.

```
To setup
   import-data;
   load  the data and the clusters;
   create-turtles number ;
   create turtles in random positions
   . . . . . .
   ask-turtles [choiceColor]
   . . . . . . . .
end

To run
   repeat MaxNumberofGenerations [
 ask-turtles[if color = green or color = yellow [computeDir]

 ask-turtles[if color = green or color = yellow
      [move choiceColor
      if color = red or color = white
           [generateNewBoid
     if count-turtles-here > 1 [die]]]
           ask-turtles [if color = red [mergeCluster]]

 ask-turtles[if color = green or color = yellow
      [set age age + 1
    if age > maxLife [ generateNewBoid die ]]]

   ] ;end repeat

end ; run procedure
```

Fig. 2. Starlogo code of the setup and run procedure of Sparrow.

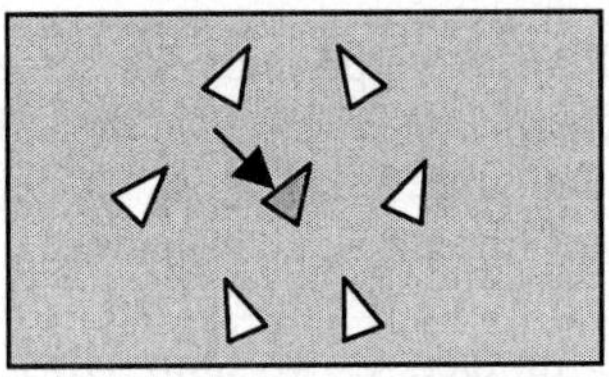

Fig. 3. The cage effect.

The last part of code invoked by ask-turtles was added to the original algorithm to avoid a '*cage effect*' (see figure 3), which occurred during the first simulations; in fact, some boids could remain trapped inside regions surrounded by red or white boids and would have no way to go out, wasting useful resources for the exploration. So, a limit was imposed on their life; hence, when their age exceeded a determined value (*maxLife*) they were made to die and were regenerated in a new randomly chosen position of the space.

4 Experimental Results

We evaluated the accuracy of the solution supplied by SPARROW in comparison with the one of DBSCAN and the performance of the search strategy of SPARROW in comparison with the standard flocking search strategy and with the linear randomized search. Furthermore, we evaluated the impact of the number of agents on foraging for clusters performance. To this purpose, we implemented the three different search strategies in Starlogo and compared their performance with a publicly available version of DBSCAN. For the experiments we used two synthetic data sets. The structure of these data sets is shown in figure 4(a) and 4(b). The first data set, called GEORGE, consists of 5463 points. The second data set, called DS4, contains 8843 points. Each point of the two data sets has two attributes that define the x and y coordinates. Furthermore, both data sets have a considerable quantity of noise.

(a) GEORGE: 5463 points (b) DS4: 8000 points

Fig. 4. The two data sets used in our experiments.

Although DBSCAN and SPARROW produce the same results if we examine all points of the data set, our experiments show that SPARROW can obtain, with an accuracy greater than 95%, the same number of clusters with a slightly smaller number of points for each cluster using a smaller number of spatial queries. The same results cannot be obtained by DBSCAN because of the different strategy of attribution of the points to the clusters. In fact, if we stop DBSCAN before which it has performed the spatial queries on all the points, we should obtain a correct number of points for the clusters already individuated and probably a smaller number of points for the cluster that we were building but of course we will not discover all the clusters. Table 1 and table 2 show, for the two data sets, the number of clusters and the number of points for each cluster found by DBSCAN and SPARROW and the relative error associated with each cluster. In particular, for the GEORGE data set each cluster found in SPARROW has a number of points that is about 2 percent lower than that discovered by DBSCAN and for the DS4 data set about the 3 percent.

The spatial queries performed by SPARROW, with 200 time steps, are for the GEORGE data set about the 27 percent of those performed by DBSCAN and for the DS4 dataset about the 45 percent.

Table 1. Number of clusters and number of points for clusters for GEORGE data set.

Number of clusters	Number of points for cluster (SPARROW)	Number of points for cluster (DBSCAN)	Relative error (percent)
1	832	848	-1.89%
2	690	706	-2.27%
3	778	800	-2.75%
4	782	815	-4.05%
5	814	818	-049%
6	712	718	-0.84%

Table 2. Number of clusters and number of points for clusters for DS4 data set.

Number of clusters	Number of points for cluster (SPARROW)	Number of points for cluster (DBSCAN)	Relative error (percent)
1	844	876	-3.65%
2	920	928	-0.86%
3	216	220	-1.82%
4	1866	1924	-3.01%
5	522	534	-2.25%
6	491	502	-2.19%
7	278	291	-4.47%
8	2308	2406	-4.07%
9	272	280	-2.86%

To verify the effectiveness of the search strategy we have compared SPARROW with the random-walk search (RWS) strategy of the standard flock algorithm and with the linear randomized search (LRS) strategy. Figure 5 gives the number of clusters found through the three different strategies in 250 time steps for the DS4 data set. Figure 5 reveals that the number of clusters discovered at time step 65 from RWS and LRS strategy is slightly higher than that of SPARROW. From time step 66 to 110 the behavior of SPARROW is better than that of RWS but worse than LRS. SPARROW presents a superior behavior on both the search strategies after the 110 time step because of the adaptive behavior of the algorithm that allows agents to learn on their previous experience. A similar behaviour is also present in the GEORGE data set. Finally, we present the impact of the number of agents on the foraging for clusters performance. Figure 6 gives, for the DS4 data set, the number of clusters found in 250 time steps for 25, 50 and 100 agents. A comparative analysis reveals that a 100-agents population discovers a larger number of clusters than the other two populations with a smaller number of agents.

This scalable behaviour of the algorithm determines a faster completion time because a smaller number of iterations are necessary to produce the solution.

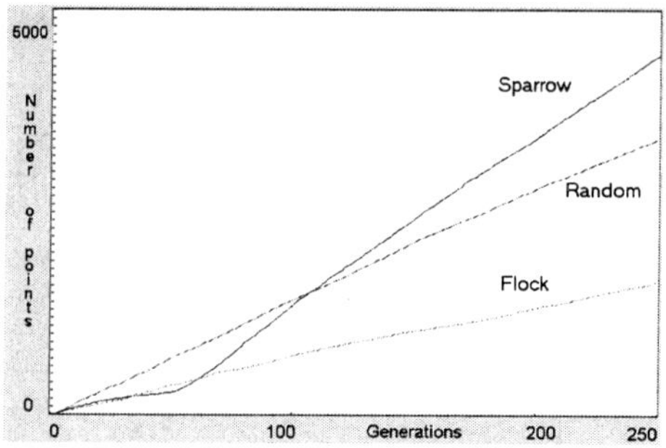

Fig. 5. Number of clusters found for the DS4 dataset.

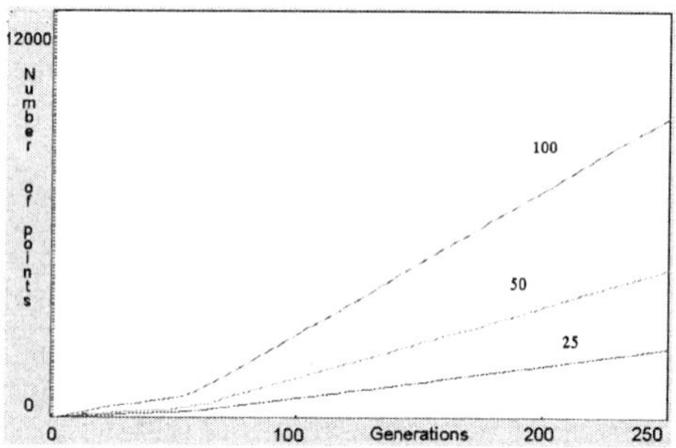

Fig. 6. The impact of the number of agents on foraging for clusters strategy.

5 Conclusions

In this paper, we have described the parallel clustering algorithm SPARROW, which is based on the use of swarm intelligence techniques. The algorithm combines a smart exploratory strategy based on a flock of birds with a density-based cluster algorithm to discover clusters of arbitrary shape and size in spatial data. The algorithm has been implemented in STARLOGO and compared with DBSCAN using two synthetic data sets. Measures of accuracy of the results show that SPARROW exhibits the same behaviour of DBSCAN although it needs a smaller number of spatial queries. Moreover, the adaptive search strategy of SPARROW is more efficient than those of the random-walk search (RWS) strategy of the standard flock algorithm and of the linear randomized search (LRS).

References

1. Han J., Kamber M., *Data Mining: Concepts and Techniques*, Morgan Kaufmann 2000.
2. Kaufman L., Rousseeuw P. J., *Finding Groups in Data: An Introduction to Cluster Analysis*, John Wiley & Sons, 1990.
3. Karypis G., Han E., Kumar V.,: CHAMELEON: A Hierarchical Clustering Algorithm Using Dynamic Modeling, *IEEE Computer*, vol. 32, pp.68-75, 1999.

4. Zhang T., Ramakrishnan R., Livny M.: *Birch: A New Data Clustering Algorithm and its Applications*, in: Data Mining and Knowledge Discovery, vol. 1, n.2, pp. 141-182, 1997.
5. Sander J., Ester M., Kriegel H.-P., Xu X.: *Density-Based Clustering in Spatial Databases: The Algorithm GDBSCAN and its Applications*, in: Data Mining and Knowledge Discovery, vol. 2, n. 2, pp. 169-194, 1998.
6. Wang W., Yang J., Muntz R., STING: A Statistical Information Grid Approach to Spatial Data Mining, *Proc. of Int. Conf. Very Large Data Bases (VLDB'97)*, pp. 186-195, 1997.
7. Han J., Kamber M., Tung A.K.H., *Spatial Clustering Methods in Data Mining: A Survey*, H. Miller and J. Han (eds.), Geographic Data Mining and Knowledge Discovery, Taylor and Francis, 2001.
8. Bonabeau E., Dorigo M., Theraulaz G., *Swarm Intelligence: From Natural to Artificial Systems,* Oxford University Press, 1999.
9. Deneubourg J. L., Goss S., Franks, N., Sendova-Franks A., Detrain C., and Chretien L., The Dynamic of Collective Sorting Robot-like Ants and Ant-like Robots, *Proc. of the first Conf. on Simulation of Adaptive Behavior*, J.A. Meyer et S.W. Wilson (Eds), MIT Press/Bradford Books, pp. 356-363, 1990.
10. Lumer E. D., Faieta B., Diversity and Adaptation in Populations of Clustering Ants, *Proc. of the third Int. Conf. on Simulation of Adaptive Behavior: From Animals to Animats (SAB94)*, D. Cliff, P. Husbands, J.A. Meyer, S.W. Wilson (Eds), MIT-Press, pp. 501-508, 1994.
11. Kuntz P. Snyers D., Emergent Colonization and Graph Partitioning, *Proc. of the third Int. Conf. on Simulation of Adaptive Behavior: From Animals to Animats (SAB94)*, D. Cliff, P. Husbands, J.A. Meyer, S.W. Wilson (Eds), MIT-Press, pp. 494-500, 1994.
12. N. Monmarché, M. Slimane, and G. Venturini, "On improving clustering in numerical databases with artificial ants", in *Advances in Artificial Life: 5th European Conference*, ECAL 99, LNCS 1674, Springer, Berlin, pp. 626-635, 1999.
13. Macgill, J., Openshaw, S., The use of Flocks to drive a Geographic Analysis Machine, in *Proc. of the 3rd Inter. Conf. on GeoComputation*, University of Bristol, UK, 1998.
14. James Macgill, Using Flocks to Drive a Geographical Analysis Engine, *Artificial Life VII: Proceedings of the Seventh International Conference on Artificial Life*, MIT Press, Reed College, Portland, Oregon, pp. 1-6, 2000.
15. Reynolds C. W., *Flocks, Herds, and Schools: A Distributed Behavioral Model*, Computer Graphics vol. 21, n. 4, , (SIGGRAPH 87), pp. 25-34, 1987.
16. V. S. Colella, E. Klopfer, M. Resnick, *Adventures in Modeling: Exploring Complex, Dynamic Systems with StarLogo,* Teachers College Press, 2001.
17. Ester M., Kriegel H.-P., Sander J., Xu X., A Density-Based Algorithm for Discovering Clusters in Large Spatial Databases with Noise, *Proc. 2nd Int. Conf. on Knowledge Discovery and Data Mining (KDD-96)*, Portland, OR, 1996, pp. 226-231, 1996.

Evolution of Asynchronous Cellular Automata for the Density Task

Marco Tomassini and Mattias Venzi

Computer Science Institute, University of Lausanne
1015 Lausanne, Switzerland

Abstract. The evolution of asynchronous automata for the density task is presented and compared with the evolution of synchronous automata. We describe the influence of various asynchronous update policies on the computational strategy. We also investigate how synchronous and asynchronous cellular automata behave when the update policy is gradually changed, showing that asynchronous cellular automata are more robust.

1 Introduction

Cellular automata(CAs) are discrete dynamical systems that have been studied theoretically for years due to their architectural simplicity and wide spectrum of behaviors [12]. In this work we concentrate on the commonly assumed hypothesis of simultaneous updating of the CA cells that is, the *synchronicity* of the CA. This update mode is interesting because of its conceptual simplicity and because it is easier to deal with in mathematical terms. However, perfect synchronicity is only an abstraction: if CAs are to model physical or biological situations or are to be considered physically embodied computing machines then the synchronicity assumption is untenable. For the latter case, in any spatially extended system signals cannot travel faster than light. Hence, it is impossible for a signal emitted by a global clock to reach any two computing elements at exactly the same time. In this study we relax the synchronicity constraint and work with various kinds of *asynchronous* CA updating modes on a well-known problem: density classification by a CA. The few existing studies on asynchronous CAs have shown that asynchronous update often gives rise to completely different time evolutions for the CA. For instance, cyclic attractors are no longer possible and generally there is a loss of the rich structures commonly found in synchronous CAs (see e.g. [2, 6, 4]).

The paper is organized as follows. The following section 2 summarizes definitions and facts about standard CAs and their asynchronous counterparts. Section 3 deals with the artificial evolution of asynchronous CAs for the density task and compares their behavior and solution strategies with those of known synchronous CAs. Section 4 presents the results of studying the behavior of synchronous CAs when the environment becomes gradually asynchronous and, respectively, of asynchronous CAs that become progressively more synchronous. Section 5 presents our conclusions and hints to further work and to open questions.

J.J. Merelo Guervós et al. (Eds.): PPSN VII, LNCS 2439, pp. 934–943, 2002.

2 Synchronous and Asynchronous Cellular Automata

Cellular automata are dynamical systems in which space and time are discrete. A standard cellular automaton consists of an array of cells, each of which can be in one of a finite number of possible states, updated synchronously in discrete time steps, according to a local, identical interaction rule. Here we will only consider Boolean automata for which the cellular state $s \in \{0,1\}$. The state of a cell at the next time step is determined by the current states of a surrounding neighborhood of cells. The regular cellular array (grid) is d-dimensional, where $d = 1, 2, 3$ is used in practice; in this paper we shall concentrate on $d = 1$ i.e., one-dimensional grids. The identical rule contained in each cell is usually specified in the form of a rule table with an entry for every possible neighborhood configuration of states. For one-dimensional CA, a cell is connected to r local neighbors (cells) on either side; each cell thus has $2r + 1$ neighbors. When considering a finite-sized grid, spatially periodic boundary conditions are frequently applied, resulting in a circular grid for the one-dimensional case. The term *configuration* refers to an assignment of ones and zeros at a given time step

There are many ways for sequentially updating the cells of a CA (for an excellent discussion, see [10]). The most general one is independent random ordering of updates in time, which consists in randomly choosing the cell to be updated next, with replacement. This corresponds to a binomial distribution for the update probability, the limiting case of which for large n is the Poisson distribution (n is the number of cells in the grid).

For comparison purposes we also employ two other update methods: fixed random sweep and random new sweep (we employ the same terms as in [10]). In the fixed random sweep update, each cell to be updated next is chosen with uniform probability without replacement; this will produce a certain update sequence $(c_1^j, c_2^k, \ldots, c_n^m)$, where c_q^p means that cell number p is updated at time q and $(j, k, \ldots, m)$ is a permutation of the n cells. The same permutation is then used for the following update cycles. The random new sweep method is the same except that each new sweep through the array is done by picking a different random permutation of the cells. A sweep means updating n times, which corresponds to updating all the n cells in the grid for fixed random sweep and random new sweep, and possibly less than n cells in the binomial method, since some cells might be updated more than once.

It should be noted that our chosen asynchronous updating being non-deterministic, the same CA may reach a different configuration after n time steps on the same initial distribution of states, which is not the case for synchronous CAs, since there is a single possible sequence of configurations for a synchronous CA for a given initial configuration of states.

3 Evolving 1-D Asynchronous CAs for the Density Task

In this section we define the density task and we describe how asynchronous CAs for performing this task can be evolved by genetic algorithms (GA). We also deal with the features of the evolutionary process and compare the evolved CA strategies with those observed in the synchronous case.

3.1 The Density Task

The density task is a prototypical computational task for CAs that has been much studied due to its simplicity and richness of behavior. For one-dimensional finite CA of size n (with n odd for convenience) it is defined as follows: the CA must relax to a fixed-point pattern of all 1s if the initial configuration of states contains more 1s than 0s and, conversely, it must relax to a fixed-point pattern of all 0s otherwise, after a number of time steps of the order of the grid size. This computation is trivial for a computer having a central control. However, the density task is non-trivial for a small radius 1-D CA since such a CA can only transfer information at finite speed relying on local information exclusively, while density is a global property of states configuration [9]. It has been shown that the density task cannot be solved perfectly by a uniform, two-state CA with finite radius [7], although a slightly modified version of the task can be shown to enjoy perfect solution by such an automaton [3]. In general, given a desired global behavior for a CA, it is extremely difficult to infer the local CA rule that will give rise to the emergence of the computation sought because of the possible non-linearities and large-scale collective effects that cannot be predicted from the sole local CA updating rule. Since exhaustive evaluation of all possible rules is out of the question except for elementary ($d = 1, r = 1$) automata, one possible solution of the problem consists in using evolutionary algorithms, as proposed by Mitchell *et al.* [9, 8] for uniform and synchronous CAs and by Sipper for non-uniform (the rules need not be all the same) ones [11].

3.2 Artificial Evolution of Cellular Automata

In this paper we use a genetic algorithm similar to the one described in [9] for synchronous CAs, with the aim of evolving asynchronous CAs for the density task. Each individual in the population represents a candidate rule and is represented simply by the output bits of the rule table in lexicographic order of the neighborhood (see section 2). Here $r = 3$ has been used, which gives a chromosome length of $2^{2r+1} = 128$ and a search space of size 2^{128}, far too large to be searched exhaustively. The population size is 100 individuals, each represented by a 128-bit string, initially randomly generated from a uniform density distribution over the interval $[0, 1]$. The fitness of a rule in the population has been calculated by randomly choosing 100 out of the 2^n possible initial configurations (IC) with uniform density in the same manner as for the initial population and then iterating the rule on each IC for $M = 2n$ time steps, where $n = 149$ is the grid size. The rule's fitness is the fraction of ICs for which the rule produced the correct fixed point, given the known IC density. At each generation a different set of ICs is generated for each rule. After ranking the rules in the current population according to their fitness, the 20% top rules are copied in the next population without change. The remaining 80 rules are generated by crossover and mutation. Crossover is single-point and is performed between two individuals randomly chosen from the top 20 rules with replacement and is followed by single-bit mutation of the two offspring. The best 80 rules after the application of the genetic operators enter the new population.

The performance of the best rules found at the end of the evolution is evaluated on a larger sample of ICs and it is defined as the fraction of correct classifications over 10^4 randomly chosen initial configurations. Moreover, the ICs are sampled according to a binomial distribution (i.e., each bit is independently drawn with probability 1/2 of being 0). Clearly, this distribution is strongly peaked around $\rho_0 = 1/2$ and thus it makes a much more difficult case for the CA (ρ_0 is the density of 0s in the initial configuration).

Due to the high computational cost, we have performed 15 runs, each lasting for 100 generations, for each of the asynchronous update policies. This is not enough to reach very good results, but it is sufficient for studying the emergence of well-defined computational strategies, which has been our main objective here.

3.3 Evolutionary Dynamics and Results: Synchronous CAs

Mitchell and co-workers performed a number of studies on the emergence of synchronous CA strategies for the density task during evolution (see e.g. [8, 9], where more details can be found). In summary, these findings can be subdivided into those pertaining to the evolutionary history and those that are part of "final" evolved automata. For the former, they essentially observed that, in successful evolution experiments, the fitness of the best rules increase in time according to rapid jumps, giving rise to what they call "epochs" in the evolutionary process. Each epoch corresponds roughly to a new, increasingly sophisticated solution strategy. Concerning the final CA produced by evolution, it was noted that, in most runs, the GA found non-sophisticated strategies that consisted in expanding sufficiently large blocks of adjacent 1s or 0s. This "block-expanding" strategy is unsophisticated in that it mainly uses local information to reach a conclusion. As a consequence, only those IC that have low or high density are classified correctly since they are more likely to have extended blocks of 1s or 0s. In fact, these CAs have a performance around 0.6. However, some of the runs gave solutions that presented novel, more sophisticated features that yielded better performance (around 0.77) on a wide distribution of IC. These new strategies rely on travelling signals that transfer spatial and temporal information about the density in local regions through the lattice. An example of such a strategy is given in figure 1, where the behavior of the so-called GKL rule is depicted [9]. The GKL rule is a hand-coded one but its behavior is similar to that of the best solutions found by evolution. In spite of the relative success of the genetic algorithm, there exist hand-coded CAs that have better performance, besides the GKL rule itself. On the other hand, Andre *et al.* [1] have been able to artificially evolve a CA that is as good as the best manually designed CA by using genetic programming. Although they used a great deal more computational resources than Mitchell and coworkers, this nevertheless shows that artificial evolution is a viable solution in this case.

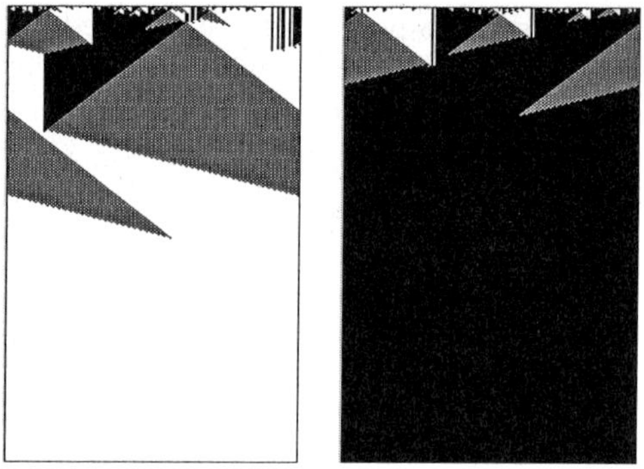

(a) (b)

Fig. 1. Space-time diagram for the GKL rule. CA cells are depicted horizontally, while time goes downward. The 0 state is depicted in white; 1 in black. The density of zeros ρ_0 is 0.476 in a) and $\rho_0 = 0.536$ in b).

Table 1. Performance of the best evolved asynchronous rules calculated over 10^4 binomially distributed initial configurations. Rule numbers are in hexadecimal.

Update Mode	Rule	Performance
IndependentRandom	00024501006115AF5FFFBFFDE9EFF95F	67.2
FixedRandomSweep	114004060202414150577E771F55FFFF	67.7
RandomNewSweep	00520140006013264B7DFCDF4F6DC7DF	65.5

Crutchfield and co-workers have developed sophisticated methodologies for studying the transfer of long-range signals and the emergence of computation in evolved CAs. This framework is known as "computational mechanics" and it describes the intrinsic CA computation in terms of regular domains, particles and particle interactions. Details can be found, e.g., in [5].

3.4 Evolutionary Dynamics and Results: Asynchronous CAs

For the evolution of asynchronous CAs we have used GA parameters as described in section 3.2. As expected, the evolved asynchronous CAs find it more difficult to solve the density task due to their stochastic nature. In fact, a given CA could classify the same initial configuration in a different way depending on the update sequence, and indeed, although synchronous CAs are delocalized systems, because of the presence of a global clock, a kind of central control is still present, which is not the case for asynchronous CAs. Nevertheless, for all the asynchronous update methods CAs with fair capabilities were evolved. In the following table we list the best rules found by the GA for the three update modes. We note that the performance of the solutions are lower than the corresponding figures for synchronous CAs.

(a) (b)

Fig. 2. Space-time diagrams for an epoch 2 rule. a) $\rho_0 = 0.194$, b) $\rho_0 = 0.879$. The rule only classifies low or high densities.

The behavior of CAs evolved with the independent random ordering i.e., binomial cell update mode and random new sweep are very similar, while fixed random sweep gave poorer results. We thus describe the independent random ordering mode here. During most evolutionary runs we observed the presence of periods in the evolution in which the fitness of the best rules increase in rapid jumps. These "epochs" were observed in the synchronous case too (see section 3.3) and correspond to distinct computational innovations i.e., to mayor changes in the strategies that the CA uses for solving the task.

In epoch 1 the evolution only discovers local naive strategies that only work on "extreme" densities (i.e., low or high) but most often not on both at the same time. Fitness is only slightly over 0.5. In the following epoch 2, rules specialize on low or high densities as well and use unsophisticated strategies, but now they give correct results on both low and high densities. This can be seen, for instance, in figure 2.

In epoch 3, with fitness values comprised between 0.80 and 0.90, one sees the emergence of block-expanding strategies, as in the synchronous case, but more noisy. Moreover, narrow vertical strips make their appearance (see figure 3).

The following, and last, epoch 4 sees the refinement of the vertical strips strategy with fitness above 0.9 and steadily increasing. The propagating patterns become less noisy and the strategy is little affected by the intrinsic stochasticity of the update rule. Figure 4 illustrates the best solution found by evolution at the end of epoch 4. The "zebra-like" moving patterns, which represent the most efficient strategies for evolved asynchronous automata, are different from those found in the synchronous case. In fact, the asynchronous updating modes have the effect of destroying or delaying the propagation of the long-range transversal signals that carry information in the synchronous case (see figure 1). Thus, the CA expands 1* and 0* blocks which enter in collision and annihilate and small-block propagate in time, which gives the characteristic zebra-like patterns. These strips are stable and propagate further to the right or to the left.

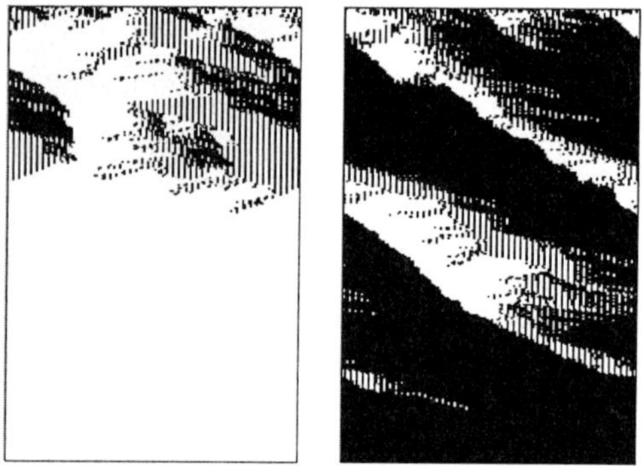

(a) (b)

Fig. 3. Space-time diagrams for an epoch 3 rule. a) $\rho_0 = 0.489$, b) $\rho_0 = 0.510$. Block-expanding and vertical strips make their appearance.

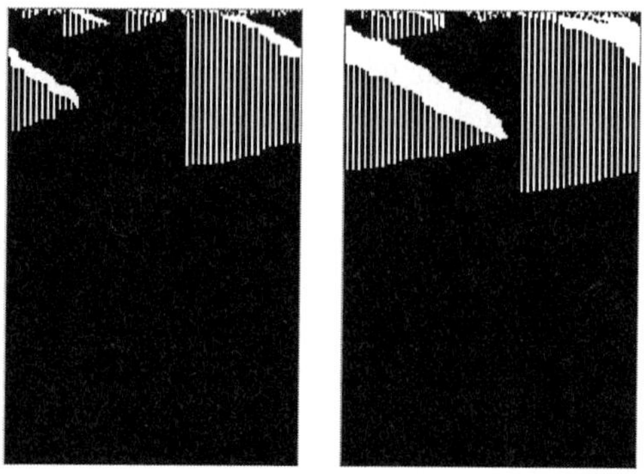

(a) (b)

Fig. 4. Space-time diagrams for the best asynchronous rule found. The density $\rho_0 = 0.55$ in both a) and b) and the initial state configuration is the same. The different time evolution points out the stochasticity of the updating policy.

4 Merging the Synchronous and Asynchronous Worlds

We have seen that evolved synchronous CAs for the density task have a rather better performance than asynchronous ones, as it was expected if one takes into account their deterministic nature. Now, although parallel update is unfeasible, one could obtain a more realistic approximation by subdividing the whole grid into blocks of $c \leq n$ cells that are updated synchronously within the block, while the blocks themselves are updated asynchronously. Thus, if the number of blocks varies from 1 to n the system will go from complete synchrony to complete asynchrony (n=180 here). Let us start with synchronous CA rules becoming progressively asynchronous (random new sweep). Both the best evolved rule as well as the GKL rule gave very poor results. They are extremely sensitive to pertur-

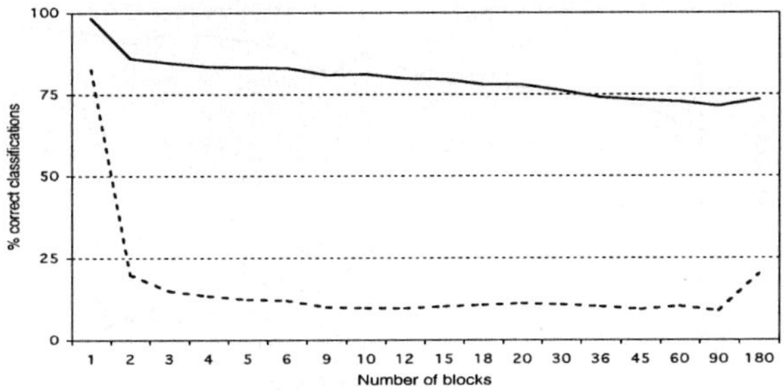

Fig. 5. Percent of correct classifications as a function of the number of blocks in the grid for two choices of initial configurations for the GKL rule.

(a) (b)

Fig. 6. Asynchronous behavior of the GKL rule with a) two and b) 10 blocks respectively.

bations since even a small amount of noise destroys the strict synchronization carried by the propagating transversal signals (see section 3.3). This can be seen in figure 5, where the performance of the GKL rule is shown against number of blocks for two distributions of initial configurations. Performance remains acceptable for ICs chosen uniformly between 0 and 1 but it totally degrades for a binomial distribution, the more difficult and interesting case.

Figure 6 depicts the space-time diagram of the GKL rule with 2 and 10 asynchronous blocks respectively. One sees clearly that, with two blocks already, signals are prevented from travelling and from combining at the block boundaries, as it would be the case if all the cells were updated in parallel (see figure 1 for comparison), which explains why the CA is incapable to perform the task.

Starting now from the other end of the spectrum, we consider the best asynchronous CA rule going progressively more synchronous (i.e., from right to left in figure 7). In this case we see that performances are progressively lower but the loss is gradual and only for the extreme cases of a few large blocks does per-

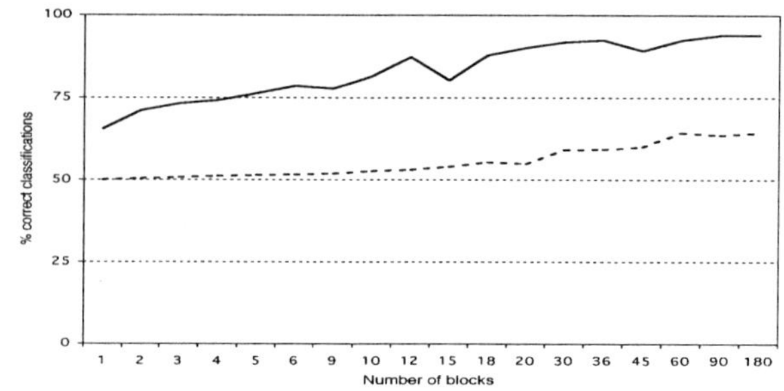

Fig. 7. Percent of correct classifications as a function of the number of blocks in the grid for two choices of initial configurations for the best asynchronous rule.

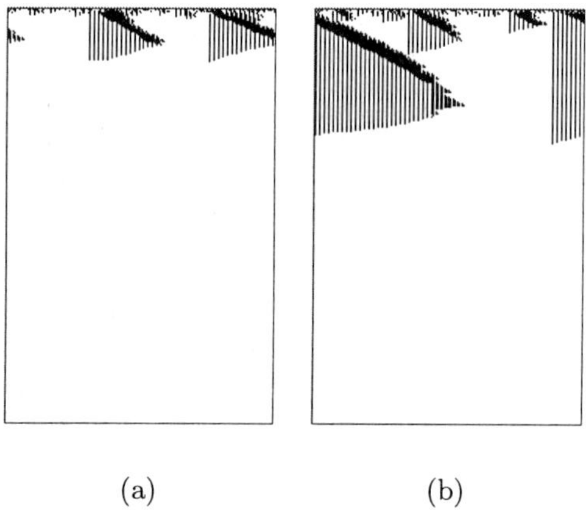

Fig. 8. Synchronous behavior of the best asynchronous rule with a) 90 and b) 60 blocks respectively. The density $\rho_0 = 0.444$ in both a) and b).

formance approach 0.5. This was to be expected, since the rule has been evolved in a completely asynchronous environment.

Figure 8 depicts the case of 90 and 60 blocks and shows that the strategy of solution is not perturbed in these cases.

We can thus conclude that, although synchronous and asynchronous rules have been evolved, respectively, in synchronous and asynchronous environments, asynchronous rules adapt better to changes, i.e., they are more robust.

5 Conclusions

In this work we have shown that physically more realistic asynchronous CAs of various kinds can be effectively evolved for the density task using genetic algorithms, although their performance is lower than that obtained by evolved synchronous CAs. We have also shown that the computational strategies discovered by the GA in the asynchronous case are different from those of synchronous CAs due to the presence of a stochastic component in the update. This very reason makes them more resistant to changes in the environment and thus potentially more interesting as computational devices in the presence of noise. Other important aspects that we are studying, but are not included here, are the scalability properties of evolved CAs and further investigations into their fault-tolerance aspects.

References

1. D. Andre, F. H Bennett III, and J. R. Koza. Discovery by genetic programming of a cellular automata rule that is better than any known rule for the majority classification problem. In J. R. Koza, D. E. Goldberg, D. B. Fogel, and R. L. Riolo, editors, *Genetic Programming 1996: Proceedings of the First Annual Conference*, pages 3–11, Cambridge, MA, 1996. The MIT Press.
2. H. Bersini and V. Detour. Asynchrony induces stability in cellular automata based models. In R. A. Brooks and P. Maes, editors, *Artificial Life IV*, pages 382–387, Cambridge, Massachusetts, 1994. The MIT Press.
3. M. S. Capcarrere, M. Sipper, and M. Tomassini. Two-state, r=1 cellular automaton that classifies density. *Physical Review Letters*, 77(24):4969–4971, December 1996.
4. I. Harvey and T. Bossomaier. Time out of joint: attractors in asynchronous random boolean networks. In P. Husbands and I. Harvey, editors, *Proceedings of the Fourth European Conference on Artificial Life*, pages 67–75, Cambridge, MA, 1997. The MIT Press.
5. W. Hordijk, J. P. Crutchfield, and M. Mitchell. Mechanisms of emergent computation in cellular automata. In A. Eiben, T. Bäck, M. Schoenauer, and H.-P. Schwefel, editors, *Parallel Problem Solving from Nature- PPSN V*, volume 1498 of *Lecture Notes in Computer Science*, pages 613–622, Heidelberg, 1998. Springer-Verlag.
6. T. E. Ingerson and R. L. Buvel. Structure in asynchronous cellular automata. *Physica D*, 10:59–68, 1984.
7. M. Land and R. K. Belew. No perfect two-state cellular automata for density classification exists. *Physical Review Letters*, 74(25):5148–5150, June 1995.
8. M. Mitchell, J. P. Crutchfield, and P. T. Hraber. Evolving cellular automata to perform computations: Mechanisms and impediments. *Physica D*, 75:361–391, 1994.
9. M. Mitchell, P. T. Hraber, and J. P. Crutchfield. Revisiting the edge of chaos: Evolving cellular automata to perform computations. *Complex Systems*, 7:89–130, 1993.
10. B. Schönfisch and A. de Roos. Synchronous and asynchronous updating in cellular automata. *BioSystems*, 51:123–143, 1999.
11. M. Sipper. *Evolution of Parallel Cellular Machines: The Cellular Programming Approach.* Springer-Verlag, Heidelberg, 1997.
12. S. Wolfram. *Cellular Automata and Complexity.* Addison-Wesley, Reading, MA, 1994.

Author Index

Lecture Notes in Computer Science

For information about Vols. 1–2371
please contact your bookseller or Springer-Verlag

Vol. 2372: A. Pettorossi (Ed.), Logic Based Program Synthesis and Transformation. Proceedings, 2001. VIII, 267 pages. 2002.

Vol. 2373: A. Apostolico, M. Takeda (Eds.), Combinatorial Pattern Matching. Proceedings, 2002. VIII, 289 pages. 2002.

Vol. 2374: B. Magnusson (Ed.), ECOOP 2002 – Object-Oriented Programming. XI, 637 pages. 2002.

Vol. 2375: J. Kivinen, R.H. Sloan (Eds.), Computational Learning Theory. Proceedings, 2002. XI, 397 pages. 2002. (Subseries LNAI).

Vol. 2377: A. Birk, S. Coradeschi, T. Satoshi (Eds.), RoboCup 2001: Robot Soccer World Cup V. XIX, 763 pages. 2002. (Subseries LNAI).

Vol. 2378: S. Tison (Ed.), Rewriting Techniques and Applications. Proceedings, 2002. XI, 387 pages. 2002.

Vol. 2379: G.J. Chastek (Ed.), Software Product Lines. Proceedings, 2002. X, 399 pages. 2002.

Vol. 2380: P. Widmayer, F. Triguero, R. Morales, M. Hennessy, S. Eidenbenz, R. Conejo (Eds.), Automata, Languages and Programming. Proceedings, 2002. XXI, 1069 pages. 2002.

Vol. 2381: U. Egly, C.G. Fermüller (Eds.), Automated Reasoning with Analytic Tableaux and Related Methods. Proceedings, 2002. X, 341 pages. 2002 .(Subseries LNAI).

Vol. 2382: A. Halevy, A. Gal (Eds.), Next Generation Information Technologies and Systems. Proceedings, 2002. VIII, 169 pages. 2002.

Vol. 2383: M.S. Lew, N. Sebe, J.P. Eakins (Eds.), Image and Video Retrieval. Proceedings, 2002. XII, 388 pages. 2002.

Vol. 2384: L. Batten, J. Seberry (Eds.), Information Security and Privacy. Proceedings, 2002. XII, 514 pages. 2002.

Vol. 2385: J. Calmet, B. Benhamou, O. Caprotti, L. Henocque, V. Sorge (Eds.), Artificial Intelligence, Automated Reasoning, and Symbolic Computation. Proceedings, 2002. XI, 343 pages. 2002. (Subseries LNAI).

Vol. 2386: E.A. Boiten, B. Möller (Eds.), Mathematics of Program Construction. Proceedings, 2002. X, 263 pages. 2002.

Vol. 2387: O.H. Ibarra, L. Zhang (Eds.), Computing and Combinatorics. Proceedings, 2002. XIII, 606 pages. 2002.

Vol. 2388: S.-W. Lee, A. Verri (Eds.), Pattern Recognition with Support Vector Machines. Proceedings, 2002. XI, 420 pages. 2002.

Vol. 2389: E. Ranchhod, N.J. Mamede (Eds.), Advances in Natural Language Processing. Proceedings, 2002. XII, 275 pages. 2002. (Subseries LNAI).

Vol. 2390: D. Blostein, Y.-B. Kwon (Eds.), Graphics Recognition. Proceedings, 2001. XI, 367 pages. 2002.

Vol. 2391: L.-H. Eriksson, P.A. Lindsay (Eds.), FME 2002: Formal Methods – Getting IT Right. Proceedings, 2002. XI, 625 pages. 2002.

Vol. 2392: A. Voronkov (Ed.), Automated Deduction – CADE-18. Proceedings, 2002. XII, 534 pages. 2002. (Subseries LNAI).

Vol. 2393: U. Priss, D. Corbett, G. Angelova (Eds.), Conceptual Structures: Integration and Interfaces. Proceedings, 2002. XI, 397 pages. 2002. (Subseries LNAI).

Vol. 2394: P. Perner (Ed.), Advances in Data Mining. VII, 109 pages. 2002. (Subseries LNAI).

Vol. 2395: G. Barthe, P. Dybjer, L. Pinto, J. Saraiva (Eds.), Applied Semantics. IX, 537 pages. 2002.

Vol. 2396: T. Caelli, A. Amin, R.P.W. Duin, M. Kamel, D. de Ridder (Eds.), Structural, Syntactic, and Statistical Pattern Recognition. Proceedings, 2002. XVI, 863 pages. 2002.

Vol. 2397: G. Grahne, G. Ghelli (Eds.), Database Programming Languages. Proceedings, 2001. X, 343 pages. 2002.

Vol. 2398: K. Miesenberger, J. Klaus, W. Zagler (Eds.), Computers Helping People with Special Needs. Proceedings, 2002. XXII, 794 pages. 2002.

Vol. 2399: H. Hermanns, R. Segala (Eds.), Process Algebra and Probabilistic Methods. Proceedings, 2002. X, 215 pages. 2002.

Vol. 2400: B. Monien, R. Feldmann (Eds.), Euro-Par 2002 – Parallel Processing. Proceedings, 2002. XXIX, 993 pages. 2002.

Vol. 2401: P.J. Stuckey (Ed.), Logic Programming. Proceedings, 2002. XI, 486 pages. 2002.

Vol. 2402: W. Chang (Ed.), Advanced Internet Services and Applications. Proceedings, 2002. XI, 307 pages. 2002.

Vol. 2403: Mark d'Inverno, M. Luck, M. Fisher, C. Preist (Eds.), Foundations and Applications of Multi-Agent Systems. Proceedings, 1996-2000. X, 261 pages. 2002. (Subseries LNAI).

Vol. 2404: E. Brinksma, K.G. Larsen (Eds.), Computer Aided Verification. Proceedings, 2002. XIII, 626 pages. 2002.

Vol. 2405: B. Eaglestone, S. North, A. Poulovassilis (Eds.), Advances in Databases. Proceedings, 2002. XII, 199 pages. 2002.

Vol. 2406: C. Peters, M. Braschler, J. Gonzalo, M. Kluck (Eds.), Evaluation of Cross-Language Information Retrieval Systems. Proceedings, 2001. X, 601 pages. 2002.

Vol. 2407: A.C. Kakas, F. Sadri (Eds.), Computational Logic: Logic Programming and Beyond. Part I. XII, 678 pages. 2002. (Subseries LNAI).

Vol. 2408: A.C. Kakas, F. Sadri (Eds.), Computational Logic: Logic Programming and Beyond. Part II. XII, 628 pages. 2002. (Subseries LNAI).

Vol. 2409: D.M. Mount, C. Stein (Eds.), Algorithm Engineering and Experiments. Proceedings, 2002. VIII, 207 pages. 2002.

Vol. 2410: V.A. Carreño, C.A. Muñoz, S. Tahar (Eds.), Theorem Proving in Higher Order Logics. Proceedings, 2002. X, 349 pages. 2002.

Vol. 2412: H. Yin, N. Allinson, R. Freeman, J. Keane, S. Hubbard (Eds.), Intelligent Data Engineering and Automated Learning – IDEAL 2002. Proceedings, 2002. XV, 597 pages. 2002.

Vol. 2413: K. Kuwabara, J. Lee (Eds.), Intelligent Agents and Multi-Agent Systems. Proceedings, 2002. X, 221 pages. 2002. (Subseries LNAI).

Vol. 2414: F. Mattern, M. Naghshineh (Eds.), Pervasive Computing. Proceedings, 2002. XI, 298 pages. 2002.

Vol. 2415: J.R. Dorronsoro (Ed.), Artificial Neural Networks – ICANN 2002. Proceedings, 2002. XXVIII, 1382 pages. 2002.

Vol. 2416: S. Craw, A. Preece (Eds.), Advances in Case-Based Reasoning. Proceedings, 2002. XII, 656 pages. 2002. (Subseries LNAI).

Vol. 2417: M. Ishizuka, A. Sattar (Eds.), PRICAI 2002: Trends in Artificial Intelligence. Proceedings, 2002. XX, 623 pages. 2002. (Subseries LNAI).

Vol. 2418: D. Wells, L. Williams (Eds.), Extreme Programming and Agile Methods – XP/Agile Universe 2002. Proceedings, 2002. XII, 292 pages. 2002.

Vol. 2419: X. Meng, J. Su, Y. Wang (Eds.), Advances in Web-Age Information Management. Proceedings, 2002. XV, 446 pages. 2002.

Vol. 2420: K. Diks, W. Rytter (Eds.), Mathematical Foundations of Computer Science 2002. Proceedings, 2002. XII, 652 pages. 2002.

Vol. 2421: L. Brim, P. Jančar, M. Křetínský, A. Kučera (Eds.), CONCUR 2002 – Concurrency Theory. Proceedings, 2002. XII, 611 pages. 2002.

Vol. 2422: H. Kirchner, Ch. Ringeissen (Eds.), Algebraic Methodology and Software Technology. Proceedings, 2002. XI, 503 pages. 2002.

Vol. 2423: D. Lopresti, J. Hu, R. Kashi (Eds.), Document Analysis Systems V. Proceedings, 2002. XIII, 570 pages. 2002.

Vol. 2425: Z. Bellahsène, D. Patel, C. Rolland (Eds.), Object-Oriented Information Systems. Proceedings, 2002. XIII, 550 pages. 2002.

Vol. 2426: J.-M. Bruel, Z. Bellahsène (Eds.), Advances in Object-Oriented Information Systems.Procedings, 2002. IX, 314 pages. 2002.

Vol. 2430: T. Elomaa, H. Mannila, H. Toivonen (Eds.), Machine Learning: ECML 2002. Proceedings, 2002. XIII, 532 pages. 2002. (Subseries LNAI).

Vol. 2431: T. Elomaa, H. Mannila, H. Toivonen (Eds.), Principles of Data Mining and Knowledge Discovery. Proceedings, 2002. XIV, 514 pages. 2002. (Subseries LNAI).

Vol. 2435: Y. Manolopoulos, P. Návrat (Eds.), Advances in Databases and Information Systems. Proceedings, 2002. XIII, 415 pages. 2002.

Vol. 2436: J. Fong, C.T. Cheung, H.V. Leong, Q. Li (Eds.), Advances in Web-Based Learning. Proceedings, 2002. XIII, 434 pages. 2002.

Vol. 2438: M. Glesner, P. Zipf, M. Renovell (Eds.), Field-Programmable Logic and Applications. Proceedings, 2002. XXII, 1187 pages. 2002.

Vol. 2439: J.J. Merelo Guervós, P. Adamidis, H.-G. Beyer, J.-L. Fernández-Villacañas, H.-P. Schwefel (Eds.), Parallel Problem Solving from Nature – PPSN VII. Proceedings, 2002. XXII, 947 pages. 2002.

Vol. 2440: J.M. Haake, J.A. Pino (Eds.), Groupware: Design, Implementation and Use. Proceedings, 2002. XII, 285 pages. 2002.

Vol. 2442: M. Yung (Ed.), Advances in Cryptology – CRYPTO 2002. Proceedings, 2002. XIV, 627 pages. 2002.

Vol. 2443: D. Scott (Ed.), Artificial Intelligence: Methodology, Systems, and Applications. Proceedings, 2002. X, 279 pages. 2002. (Subseries LNAI).

Vol. 2444: A. Buchmann, F. Casati, L. Fiege, M.-C. Hsu, M.-C. Shan (Eds.), Technologies for E-Services. Proceedings, 2002. X, 171 pages. 2002.

Vol. 2445: C. Anagnostopoulou, M. Ferrand, A. Smaill (Eds.), Music and Artificial Intelligence. Proceedings, 2002. VIII, 207 pages. 2002. (Subseries LNAI).

Vol. 2446: M. Klusch, S. Ossowski, O. Shehory (Eds.), Cooperative Information Agents VI. Proceedings, 2002. XI, 321 pages. 2002. (Subseries LNAI).

Vol. 2447: D.J. Hand, N.M. Adams, R.J. Bolton (Eds.), Pattern Detection and Discovery. Proceedings, 2002. XII, 227 pages. 2002. (Subseries LNAI).

Vol. 2448: P. Sojka, I. Kopeček, K. Pala (Eds.), Text, Speech and Dialogue. Proceedings, 2002. XII, 481 pages. 2002. (Subseries LNAI).

Vol. 2451: B. Hochet, A.J. Acosta, M.J. Bellido (Eds.), Integrated Circuit Design. Proceedings, 2002. XVI, 496 pages. 2002.

Vol. 2453: A. Hameurlain, R. Cicchetti, R. Traunmüller (Eds.), Database and Expert Systems Applications. Proceedings, 2002. XVIII, 951 pages. 2002.

Vol. 2454: Y. Kambayashi, W. Winiwarter, M. Arikawa (Eds.), Data Warehousing and Knowledge Discovery. Proceedings, 2002. XIII, 339 pages. 2002.

Vol. 2455: K. Bauknecht, A M. Tjoa, G. Quirchmayr (Eds.), E-Commerce and Web Technologies. Proceedings, 2002. XIV, 414 pages. 2002.

Vol. 2456: R. Traunmüller, K. Lenk (Eds.), Electronic Government. Proceedings, 2002. XIII, 486 pages. 2002.

Vol. 2469: W. Damm, E.-R. Olderog (Eds.), Formal Techniques in Real-Time and Fault-Tolerant Systems. Proceedings, 2002. X, 455 pages. 2002.

Vol. 2470: P. Van Hentenryck (Ed.), Principles and Practice of Constraint Programming – CP 2002. Proceedings, 2002. XVI, 794 pages. 2002.

Vol. 2483: J.D.P. Rolim, S. Vadhan (Eds.), Randomization and Approximation Techniques in Computer Science. Proceedings, 2002. VIII, 275 pages. 2002.